18

2	4.00
—	−272
x	−269
	0.179
He	
helium	

Group 1

Group	13 IIIA	14 IVA	15 VA	16 VIA	17 VIIA	18 VIIIA
2	5 / 10.81 / 2.0 / 2300 / x / 2550 / 2.34 **B** boron	6 / 12.01 / 2.5 / 3550 / x / 4827 / 2.26 **C** carbon	7 / 14.01 / 3.0 / −210 / −196 / 1.25 **N** nitrogen	8 / 16.00 / 3.5 / −218 / −183 / 1.43 **O** oxygen	9 / 19.00 / 4.0 / −220 / −188 / 1.70 **F** fluorine	10 / 20.18 / — / x / −249 / −246 / 0.900 **Ne** neon
3	13 / 26.98 / 1.5 / 660 / 2467 / 2.70 **Al** aluminum	14 / 28.09 / 1.8 / 1410 / x / 2355 / 2.33 **Si** silicon	15 / 30.97 / 2.1 / 44.1 / 280 / 1.82 **P** phosphorus	16 / 32.06 / 2.5 / 113 / 445 / 2.07 **S** sulfur	17 / 35.45 / 3.0 / −101 / −34.6 / 3.21 **Cl** chlorine	18 / 39.95 / — / x / −189 / −186 / 1.78 **Ar** argon

Group	10	11 IB	12 IIB	13 IIIA	14 IVA	15 VA	16 VIA	17 VIIA	18 VIIIA
4	28 / 58.69 / 1.8 / 1455 / 2+ 2730 / 3+ 8.90 **Ni** nickel	29 / 63.55 / 1.9 / 1083 / 2+ 2567 / 1+ 8.92 **Cu** copper	30 / 65.38 / 1.6 / 420 / 2+ 907 / 7.14 **Zn** zinc	31 / 69.72 / 1.6 / 29.8 / 3+ 2403 / 5.90 **Ga** gallium	32 / 72.61 / 1.8 / 937 / 4+ 2830 / 5.35 **Ge** germanium	33 / 74.92 / 2.0 / 817 / 613 / 5.73 **As** arsenic	34 / 78.96 / 2.4 / 217 / 684 / 4.81 **Se** selenium	35 / 79.90 / 2.8 / −7.2 / 58.8 / 3.12 **Br** bromine	36 / 83.80 / — / x / −157 / −152 / 3.74 **Kr** krypton
5	46 / 106.42 / 2.2 / 1554 / 2+ 2970 / 4+ 12.0 **Pd** palladium	47 / 107.87 / 1.9 / 962 / 1+ 2212 / 10.5 **Ag** silver	48 / 112.41 / 1.7 / 321 / 2+ 765 / 8.64 **Cd** cadmium	49 / 114.82 / 1.7 / 157 / 3+ 2080 / 7.30 **In** indium	50 / 118.69 / 1.8 / 232 / 4+ 2270 / 2+ 7.31 **Sn** tin	51 / 121.75 / 1.9 / 631 / 3+ 1750 / 5+ 6.68 **Sb** antimony	52 / 127.60 / 2.1 / 450 / 990 / 6.2 **Te** tellurium	53 / 126.90 / 2.5 / 114 / 184 / 4.93 **I** iodine	54 / 131.29 / — / x / −112 / −107 / 5.89 **Xe** xenon
6	78 / 195.08 / 2.2 / 1772 / 4+ 3827 / 2+ 21.5 **Pt** platinum	79 / 196.97 / 2.4 / 1064 / 3+ 2808 / 1+ 19.3 **Au** gold	80 / 200.59 / 1.9 / −39.0 / 2+ 357 / 1+ 13.5 **Hg** mercury	81 / 204.38 / 1.8 / 304 / 1+ 1457 / 3+ 11.85 **Tl** thallium	82 / 207.20 / 1.8 / 328 / 2+ 1740 / 4+ 11.3 **Pb** lead	83 / 208.98 / 1.9 / 271 / 3+ 1560 / 5+ 9.80 **Bi** bismuth	84 / (209) / 2.0 / 254 / 2+ 962 / 4+ 9.40 **Po** polonium	85 / (210) / 2.2 / 302 / 337 / — **At** astatine	86 / (222) / — / x / −71 / −61.8 / 9.73 **Rn** radon
7	110 / (269) **Uun** ununnilium	111 / (272) **Uuu** unununium	112 / (277) **Uub** ununbiium	113	114	115	116	117	118

Lanthanides (period 6):

63 / 151.97 / — / 822 / 3+ 1527 / 2+ 5.24 **Eu** europium	64 / 157.25 / 1.1 / 1313 / 3+ 3273 / 7.90 **Gd** gadolinium	65 / 158.93 / 1.2 / 1356 / 3+ 3230 / 8.23 **Tb** terbium	66 / 162.50 / — / 1412 / 3+ 2567 / 8.55 **Dy** dysprosium	67 / 164.93 / 1.2 / 1474 / 3+ 2700 / 8.80 **Ho** holmium	68 / 167.26 / 1.2 / 1529 / 3+ 2868 / 9.07 **Er** erbium	69 / 168.94 / 1.2 / 1545 / 3+ 1950 / 9.32 **Tm** thulium	70 / 173.04 / 1.1 / 819 / 3+ 1196 / 2+ 6.97 **Yb** ytterbium	71 / 174.97 / 1.2 / 1663 / 3+ 3402 / 9.84 **Lu** lutetium

Actinides (period 7):

95 / (243) / 1.3 / 994 / 3+ 2607 / 4+ 13.7 **Am** americium	96 / (247) / — / 1340 / 3+ 13.5 **Cm** curium	97 / (247) / — / 3+ / 4+ 14 **Bk** berkelium	98 / (251) / — / 3+ **Cf** californium	99 / (252) / — / 3+ **Es** einsteinium	100 / (257) / — / 3+ **Fm** fermium	101 / (258) / — / 1021 / 2+ 3074 / 3+ **Md** mendelevium	102 / (259) / — / 2+ / 3+ **No** nobelium	103 / (260) / — / 3+ **Lr** lawrencium

Nelson

CHEMISTRY

Frank Jenkins
Science Department Head
Ross Sheppard High School

Hans van Kessel
Math/Science Learning Coordinator
Bellerose Composite High School

Dick Tompkins
Science/Technology Coordinator
Old Scona Academic High School

Oliver Lantz
Science Department Head
Harry Ainlay High School

Contributing Authors

Michael V. Falk
Assistant Principal
Harry Ainlay High School

Michael Dzwiniel
Chemistry Teacher
Harry Ainlay High School

George H. Klimiuk
Chemistry Teacher (Retired)
McNally Composite High School

British Columbia Edition

Nelson Canada

I(T)P An International Thomson Publishing Company

Toronto • Albany • Bonn • Boston • Cincinnati • Detroit • London • Madrid • Melbourne
Mexico City • New York • Pacific Grove • Paris • San Francisco • Singapore • Tokyo • Washington

I(T)P™ International Thomson Publishing
The ITP logo is a trademark under licence

© Nelson Canada,
A Division of Thomson Canada Limited, 1996

Published in 1996 by
Nelson Canada,
A Division of Thomson Canada Limited
1120 Birchmount Road
Scarborough, Ontario M1K 5G4

All student investigations in this textbook have been designed to be as safe as possible, and have been reviewed by professionals specifically for that purpose. As well, appropriate warnings concerning potential safety hazards are included where applicable to particular investigations. However, responsibility for safety remains with the student, the classroom teacher, the school principal, and the school board.

Care has been taken to trace ownership of copyright material contained in this publication. The publisher will gladly receive any information that will rectify any reference or credit line in subsequent editions.

Canadian Cataloguing in Publication Data
Main entry under title:
Nelson chemistry
British Columbia ed.

Includes index.

ISBN 0-17-604985-1

1. Chemistry.   I. Jenkins, Frank, 1944 –

QD33.J45 1996   540   C96–930448–X

Printed and Bound in Canada
3 4 5 6 7 8 9 ML 5 4 3 2 1 0 9 8 7

This book is printed on acid-free paper, approved under Environment Canada's "Environmental Choice Program." The choice of paper reflects Nelson Canada's goal of using, within the publishing process, the available resources, technology, and suppliers that are as environment friendly as possible.

COVER:
D. Hallinan/Masterfile
The photograph shows a water-filled volumetric flask.
© Kenneth Edward/BioGrafx - Science Source
The model shows the arrangement of atoms in the crystalline state of water — ice. Each oxygen atom, in addition to the two covalent bonds formed with hydrogen atoms, forms a weak hydrogen bond with a neighboring hydrogen atom of another water molecule.

CREDITS
BRITISH COLUMBIA EDITION PROJECT TEAM

Kate Baltais
Janis Barr
Marnie Benedict
Laurel Bishop
Cecilia Chan
Vicki Gould
Liz Harasymczuk

Ann Ludbrook
Renate McCloy
Georgina Montgomery
Anne Norman
Winnie Siu
David Steele
Rosemary Tanner

COMPOSITION: First Image

REVIEWERS

BRITISH COLUMBIA

Willy Chow
Teacher
College Heights Secondary School
Prince George, British Columbia

Dexter Horton
Teacher
Rick Hansen Secondary School
Abbotsford, British Columbia

Leslie Johnstone
Chemistry Teacher
Point Grey Secondary School
Vancouver, British Columbia

Axel R. Kellner
Chemistry/Physics Teacher
Point Grey Secondary School
Vancouver, British Columbia

Susan Kovach
Chemistry Teacher
Walnut Grove Secondary School
Fort Langley, British Columbia

Don Lacy
Chemistry Teacher
Stelly's Secondary School
Brentwood Bay, British Columbia

Margaret Redway
Consultant
Fraser Scientific & Business
Services Inc.
Delta, British Columbia

Dennis Secret
Vice-Principal
George Pearkes Secondary School
Port Coquitlam, British Columbia

Cheri Smith
Department Head, Science
Yale Secondary School
Abbotsford, British Columbia

Charles Stewart
Department Head, Science
Pleasant Valley Secondary School
Armstrong, British Columbia

ALBERTA

Dr. Margaret-Ann Armour
Assistant Chair
Department of Chemistry
University of Alberta
Edmonton, Alberta

Ted Doram
Teacher, Science Department
Sir Winston Churchill High
School
Calgary, Alberta

Virginia Grinevitch
Chemistry Teacher
Crowsnest Consolidated High
School
Crowsnest Pass, Alberta

T. M. Hensby
Chemistry Teacher
Lindsay Thurber Comprehensive
High School
Red Deer, Alberta

Charles J. MacKenzie
Chemistry Instructor
Medicine Hat High School
Medicine Hat, Alberta

Deborah Miller
Teacher
Queen Elizabeth Junior and
Senior High School
Calgary, Alberta

Rod Wensley
Department Head, Math-Science
Camrose Composite High School
Camrose, Alberta

John D. Wilkes
Department Head, Science
Calgary Board of Education
Calgary, Alberta

NEW BRUNSWICK

Terri Ann Gibson
Teacher
Woodstock High School
Woodstock, New Brunswick

Rhoda Wilson
Chemistry Teacher
Sussex Regional High School
Sussex, New Brunswick

NOVA SCOTIA

Glenn Josephson
Teacher – International
Baccalaureate Program
Park View Education Centre
Bridgewater, Nova Scotia

Alton MacLeod
Science Department Head
Riverview High School
Cape Breton, Nova Scotia

NEWFOUNDLAND

K. D. Bradley Clarke
Chemistry Teacher
Ascension Collegiate
Bay Roberts, C. B. N.,
Newfoundland

Dana Griffiths
Coordinator of Science and
Technology Education
Avalon Consolidated School
Board
St. John's, Newfoundland

Carol B. Hoskins
Science Teacher
Discovery Collegiate
Bonavista, Newfoundland

John Ivany
Science Teacher
Gander Collegiate
Gander, Newfoundland

Shelley Malcolm
Chemistry Teacher
Labrador City Collegiate
Labrador City, Newfoundland

Table of Contents

UNIT I   MATTER AND ITS DIVERSITY

Demonstrations

UNIT II   CHEMICAL COMBINATION

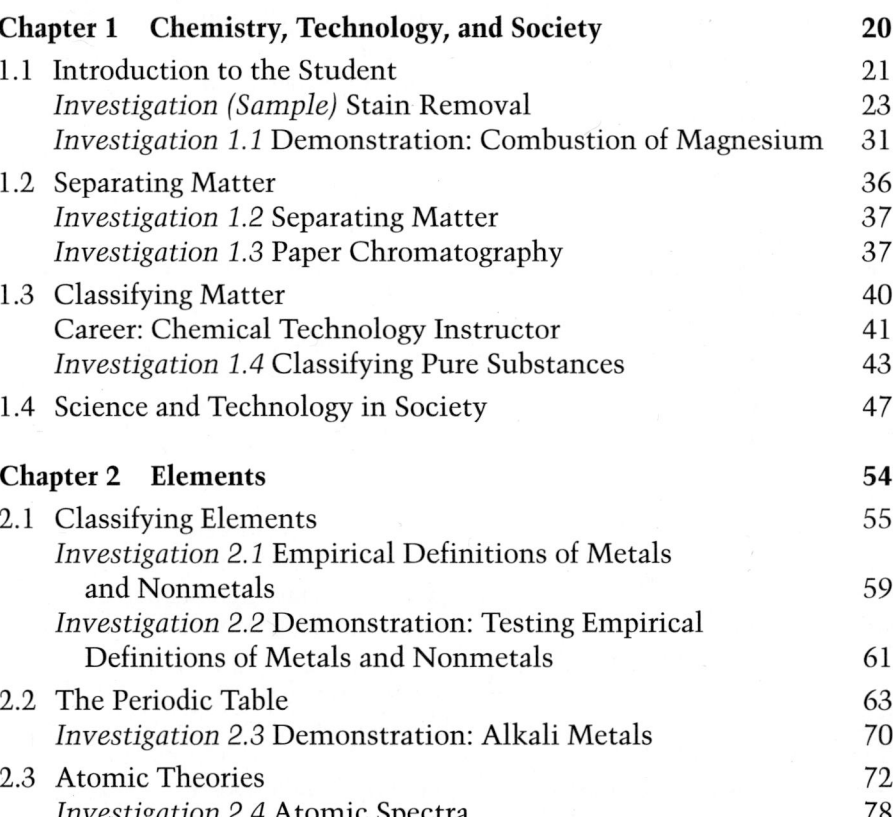

Consumer products

4

Dramatic illustrations

UNIT III   THE MEASURE OF CHEMISTRY

Biographies

Margin features

Careers

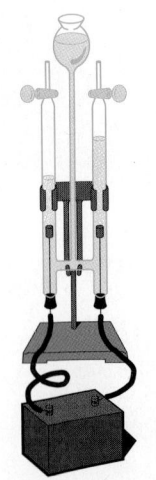

Explaining matter

UNIT IV   CHEMICAL BONDING IN MOLECULAR SYSTEMS

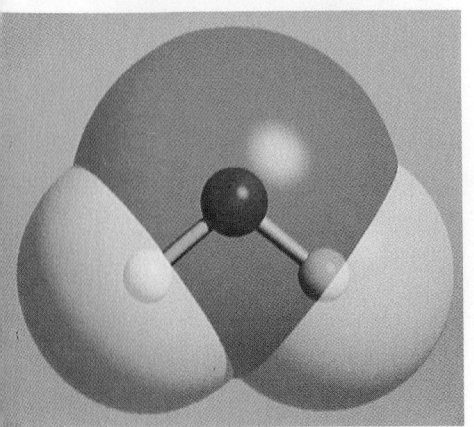

Computer-generated models

UNIT V   ENERGY CHANGES IN CHEMICAL SYSTEMS

Natural products

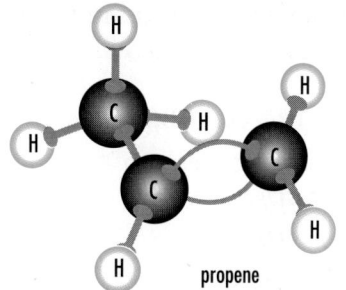

propene

Communication

Technological problem solving

Nobel Prize winners

UNIT VI   REACTION KINETICS AND EQUILIBRIUM

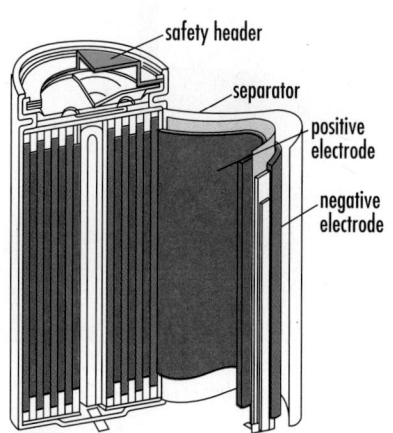

safety header

separator

positive electrode

negative electrode

Technological applications

Empirical foundations

UNIT VII ELECTROCHEMICAL SYSTEMS

Chemicals and society

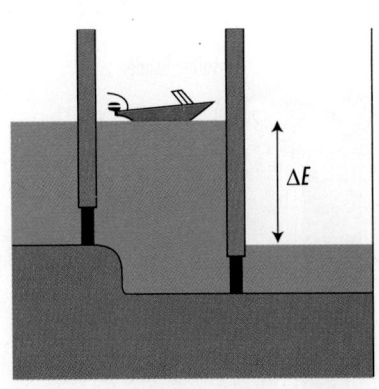

ΔE

Analogies

APPENDICES

Laboratory skills

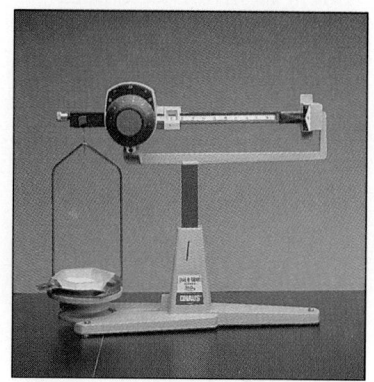

Laboratory equipment

INTEREST FEATURES

FEATURES OF NELSON CHEMISTRY

- Chapters 1 to 4 provide the basic empirical and theoretical concepts and communication conventions upon which almost all of chemistry depends. In all chapters, the chemistry content and skills are developed gradually and integrated with many **examples, exercises, and investigations.**

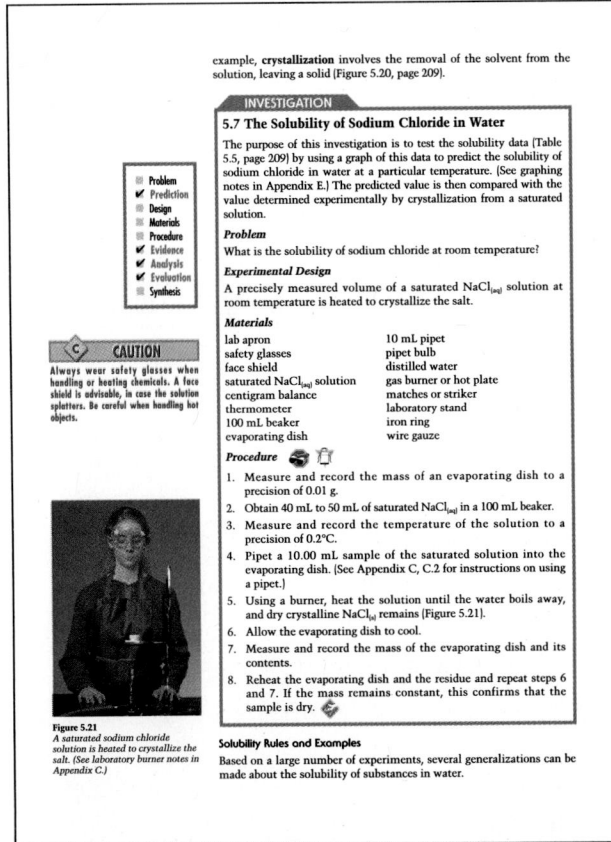

example, **crystallization** involves the removal of the solvent from the solution, leaving a solid (Figure 5.20, page 209).

INVESTIGATION

5.7 The Solubility of Sodium Chloride in Water

The purpose of this investigation is to test the solubility data (Table 5.5, page 209) by using a graph of this data to predict the solubility of sodium chloride in water at a particular temperature. (See graphing notes in Appendix E.) The predicted value is then compared with the value determined experimentally by crystallization from a saturated solution.

Problem

What is the solubility of sodium chloride at room temperature?

Experimental Design

A precisely measured volume of a saturated $NaCl_{(aq)}$ solution at room temperature is heated to crystallize the salt.

Materials

lab apron	10 mL pipet
safety glasses	pipet bulb
face shield	distilled water
saturated $NaCl_{(aq)}$ solution	gas burner or hot plate
centigram balance	matches or striker
thermometer	laboratory stand
100 mL beaker	iron ring
evaporating dish	wire gauze

Procedure

1. Measure and record the mass of an evaporating dish to a precision of 0.01 g.
2. Obtain 40 mL to 50 mL of saturated $NaCl_{(aq)}$ in a 100 mL beaker.
3. Measure and record the temperature of the solution to a precision of 0.2°C.
4. Pipet a 10.00 mL sample of the saturated solution into the evaporating dish. (See Appendix C, C.2 for instructions on using a pipet.)
5. Using a burner, heat the solution until the water boils away, and dry crystalline $NaCl_{(s)}$ remains (Figure 5.21).
6. Allow the evaporating dish to cool.
7. Measure and record the mass of the evaporating dish and its contents.
8. Reheat the evaporating dish and the residue and repeat steps 6 and 7. If the mass remains constant, this confirms that the sample is dry.

Solubility Rules and Examples

Based on a large number of experiments, several generalizations can be made about the solubility of substances in water.

☑ Problem
☑ Prediction
☐ Design
☐ Materials
☐ Procedure
☑ Evidence
☑ Analysis
☑ Evaluation
☐ Synthesis

CAUTION

Always wear safety glasses when handling or heating chemicals. A face shield is advisable, in case the solution splatters. Be careful when handling hot objects.

Figure 5.21
A saturated sodium chloride solution is heated to crystallize the salt. (See laboratory burner notes in Appendix C.)

- **Safety**, an integral part of the text, is emphasized in all of the investigations. Symbols indicate cautionary notes and disposal tips. Safety is highlighted in Chapter 1 (page 26), and safety rules are presented in Appendix D, D.1.

- **Scientific skills** are placed in a broad framework, beginning with an explicit concept of the nature of science, which in turn generates a problem-solving model composed of processes and their component skills.

- **Problem-solving skills** are developed through numerous examples, exercises, problems, and investigations. Investigations are an integral part of this text. Through example and practice, students learn all aspects of scientific problem solving: devising a problem, making a prediction, selecting or creating an experimental design, listing materials, writing a procedure, collecting evidence, performing an analysis, and conducting an evaluation of the scientific concept or other authority being tested.

- The problem-solving approach used in this textbook is unique in its ability to promote **the participation of students in constructing their own knowledge**, and in developing critical and creative thinking skills.

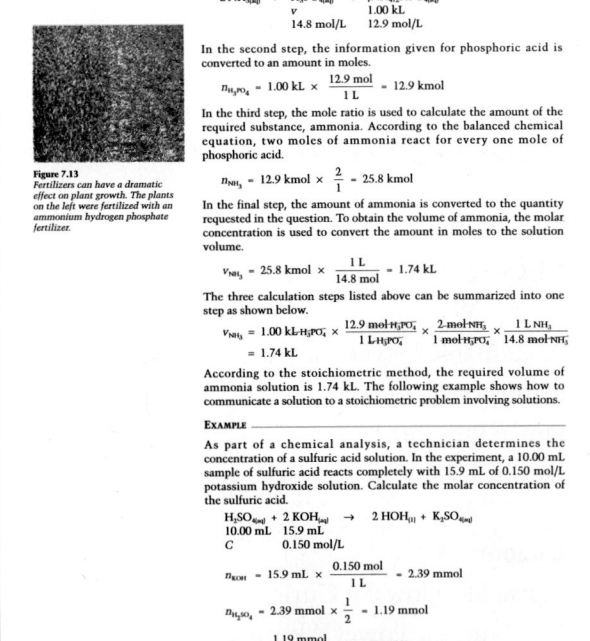

$$2\,NH_{3(aq)} + H_3PO_{4(aq)} \rightarrow (NH_4)_2HPO_{4(aq)}$$

$$\phantom{2\,NH_{3(aq)} + H_3PO_{4(aq)} \rightarrow} v \quad\quad 1.00\ kL$$

$$14.8\ mol/L \quad 12.9\ mol/L$$

In the second step, the information given for phosphoric acid is converted to an amount in moles.

$$n_{H_3PO_4} = 1.00\ kL \times \frac{12.9\ mol}{1\ L} = 12.9\ kmol$$

In the third step, the mole ratio is used to calculate the amount of the required substance, ammonia. According to the balanced chemical equation, two moles of ammonia react for every one mole of phosphoric acid.

$$n_{NH_3} = 12.9\ kmol \times \frac{2}{1} = 25.8\ kmol$$

In the final step, the amount of ammonia is converted to the quantity requested in the question. To obtain the volume of ammonia, the molar concentration is used to convert the amount in moles to the solution volume.

$$v_{NH_3} = 25.8\ kmol \times \frac{1\ L}{14.8\ mol} = 1.74\ kL$$

The three calculation steps listed above can be summarized into one step as shown below.

$$v_{NH_3} = 1.00\ kL\,H_3PO_4 \times \frac{12.9\ mol\,H_3PO_4}{1\ L\,H_3PO_4} \times \frac{2\ mol\,NH_3}{1\ mol\,H_3PO_4} \times \frac{1\ L\,NH_3}{14.8\ mol\,NH_3}$$

$$= 1.74\ kL$$

According to the stoichiometric method, the required volume of ammonia solution is 1.74 kL. The following example shows how to communicate a solution to a stoichiometric problem involving solutions.

Figure 7.13
Fertilizers can have a dramatic effect on plant growth. The plants on the left were fertilized with an ammonium hydrogen phosphate fertilizer.

EXAMPLE

As part of a chemical analysis, a technician determines the concentration of a sulfuric acid solution. In the experiment, a 10.00 mL sample of sulfuric acid reacts completely with 15.9 mL of 0.150 mol/L potassium hydroxide solution. Calculate the molar concentration of the sulfuric acid.

$$H_2SO_{4(aq)} + 2\,KOH_{(aq)} \rightarrow 2\,HOH_{(l)} + K_2SO_{4(aq)}$$

$$10.00\ mL \quad 15.9\ mL$$

$$C \quad\quad 0.150\ mol/L$$

$$n_{KOH} = 15.9\ mL \times \frac{0.150\ mol}{1\ L} = 2.39\ mmol$$

$$n_{H_2SO_4} = 2.39\ mmol \times \frac{1}{2} = 1.19\ mmol$$

$$C_{H_2SO_4} = \frac{1.19\ mmol}{10.00\ mL} = 0.119\ mol/L$$

$$\text{or}\ C_{H_2SO_4} = 1.59\ mL\,KOH \times \frac{0.150\ mol\,KOH}{1\ L\,KOH} \times \frac{1\ mol\,H_2SO_4}{2\ mol\,KOH} \times \frac{1}{10.00\ mL}$$

$$= 0.119\ mol/L$$

The following is the left page sample (first inset):

The Aurora Borealis
The aurora borealis, also called the northern lights, is a natural light display that shimmers from red through green. The farther north you live, the more often you can observe this beautiful display at night. The aurora borealis originates with solar flares, which expel charged particles known as the "solar wind." Some of these particles become trapped in the Earth's magnetic field. Near the poles, these particles spiral downward into the atmosphere and strike, energize, and ionize molecules in the air. In both northern lights and flame tests, high energy ions lose energy in a form that is seen as colored light.

Figure 5.9
Copper(II) ions usually impart a green color to a flame. The green color of the flame and the blue color of solutions of copper(II) ions can be used as diagnostic tests for copper(II) ions.

- Problem
- Prediction
- Design
- Materials
- Procedure
- ✓ Evidence
- ✓ Analysis
- Evaluation
- Synthesis

Qualitative Analysis by Color

Some ions impart a specific color to a solution, a flame, or a gas discharge tube. For example, copper(II) ions produce a blue aqueous solution and usually a green flame. A *flame test*, a test for the presence of metal ions such as copper(II), is conducted by dipping a clean platinum or nichrome wire into a solution and then into a flame (Figure 5.9). The initial flame must be nearly colorless and the wire, when dipped in water, must not produce a color in the flame. Usually, the wire is cleaned by dipping it alternately into hydrochloric acid and then into the flame, until very little color is produced. The colors of some common ions in aqueous solutions and in flames are listed on the inside back cover of this book.

Exercise

11. List two household substances that are purchased as solids but are then made into solutions before use.
12. What color are the following ions in an aqueous solution?
    (a) iron(III)   (c) $Cu^{2+}_{(aq)}$
    (b) sodium   (d) $Ni^{2+}_{(aq)}$
13. What color are the following ions in a flame test?
    (a) calcium   (c) Na^+
    (b) copper(II)   (d) K^+
14. Design a diagnostic test for carbonate ions using a reactant that would not precipitate sulfide ions in the sample. Include a net ionic equation.
15. Design an experiment to determine if acetate and/or carbonate ions are present in a solution. Include net ionic equations.
16. Design an experiment to determine if potassium and/or strontium ions are present in a solution. Include net ionic equations.
17. (Enrichment) Design an experiment to analyze a single sample of a solution for any or all of the $Tl^+_{(aq)}$, $Ba^{2+}_{(aq)}$, and $Ca^{2+}_{(aq)}$ ions.

INVESTIGATION

5.4 Qualitative Analysis by Color

The purpose of this investigation is to do qualitative chemical analyses using solution and flame colors.

Problem

Which of the solutions labelled 1, 2, 3, and 4 contains one of the metal ions listed in the ion colors chart (see inside back cover of this book)?

Experimental Design

The solution and flame color of several known solution samples are observed, and then the unknown solutions are identified.

- **Exercises** provide an immediate opportunity to check understanding and to develop concepts. The questions use a language that is consistent with a modern view of the nature of science.

The following is the right page sample (second inset):

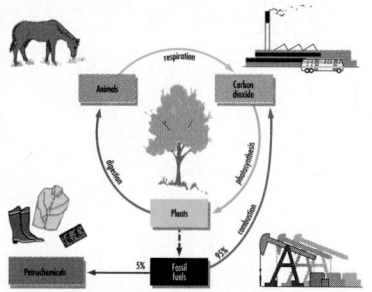

of the amino acids glycine and alanine illustrates the formation of a peptide bond.

Among **carbohydrates** — compounds with the general formula $C_x(H_2O)_y$ — polymerization occurs as well. Simple sugar molecules are the monomers; they undergo a condensation polymerization reaction,

Figure 9.33
Cellulose occurs in wood and many other plant materials. Although humans cannot digest cellulose, many microorganisms can break it down. As the breakdown occurs, wood "rots."

Figure 9.34
The carbon cycle is a unique illustration of the interrelationship of all living things with the environment — a key connection is the bonding of the carbon atom.

Exercise

50. What is the monomer from which polypropylene is made?
51. What other product results when nylon is made?
52. Suggest a reason why the particular type of nylon shown on page 381 is called "Nylon 6-6."
53. Teflon, made from tetrafluoroethene monomer units, is a polymer that provides a non-stick surface on cooking utensils. Write a structural diagram equation to represent the formation of polytetrafluoroethene.
54. Polyvinyl chloride, or PVC plastic, has numerous applications. Write a structural equation to represent the polymerization of chloroethene (vinyl chloride).
55. Alkyd resins used in paints are polyesters. Using structural diagrams, write a chemical equation to represent the first step in the reaction of 1,2,3-propanetriol with 1,2-benzenedioic acid. Note that many possible structures can form as a result of a three-dimensional growth of the polymer.
56. (Discussion) As with most consumer products, the use of polyethylene has benefits and problems. What are some beneficial uses of polyethylene and what problems result from these uses? Suggest alternative substances for each application.

- **Problems**, in the same format as the investigations, are used to reinforce the development of scientific problem solving. They serve as a link between investigations and exercises. This unique approach provides valuable practice and insight into scientific problem solving and introduces students to investigations that may be too dangerous, too difficult, too expensive, or too time-consuming to perform in the laboratory.

- **Marginal notes** provide interesting and informative asides.

- **Definitions and concepts** are introduced at an appropriate pace. Empirical definitions and concepts are always presented before the theoretical ones. Each new term, highlighted in boldface type, is listed in a comprehensive Glossary at the end of the textbook.

- **Features** include biographical profiles of scientists and a wide variety of topics related to science, technology, and STS issues.

- **Careers** include up-to-date interviews with women and men working in science-related fields.

CAREER

ENVIRONMENTAL CHEMIST

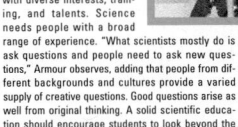

In Scotland, where she grew up, Dr. Margaret-Ann Armour trained as a physical organic chemist. After working for several years as a research chemist in the paper-making industry, where she became interested in the effects of chemicals on the environment, Dr. Armour diversified her work.

According to Armour, good researchers must have special attributes in addition to their fund of knowledge. Perseverance and optimism are essential qualities. Results may be slow to appear, and without optimism it is difficult to persevere. Above all, a researcher must be a good communicator. "A scientist isn't a lone individual working in isolation in the laboratory," she says. Because science has become increasingly specialized, "it's essential to be able to communicate with people in other disciplines." Research groups are often composed of people with diverse interests, training, and talents. Science needs people with a broad range of experience. "What scientists mostly do is ask questions and people need to ask new questions," Armour observes, adding that people from different backgrounds and cultures provide a varied supply of creative questions. Good questions arise as well from original thinking. A solid scientific education should encourage students to look beyond the facts to the underlying ideas.

As the Assistant Chair of the Department of Chemistry at the University of Alberta, Armour has to juggle a heavy load of administration, research, and interaction with students and the public. Her research group develops and tests methods for recycling and disposing of waste and surplus chemicals. As research into hazardous wastes has commanded the attention of government and business, Armour has learned to communicate effectively with non-scientists as well as with other scientists. She has devised strategies to explain complex ideas so that a lay audience can understand them.

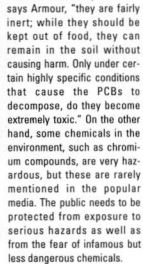

Armour is troubled by the public's poor grasp of some pollution issues. Information imparted by the media is often simplistic or misleading, and it can be distorted by political or business influences. For example, reports on PCBs (polychlorinated biphenyls) have often appeared in the news. These compounds were used as transformer oils in the electrical industry and they're usually described in the media as deadly. "In fact," says Armour, "they are fairly inert; while they should be kept out of food, they can remain in the soil without causing harm. Only under certain highly specific conditions that cause the PCBs to decompose, do they become extremely toxic." On the other hand, some chemicals in the environment, such as chromium compounds, are very hazardous, but these are rarely mentioned in the popular media. The public needs to be protected from exposure to serious hazards as well as from the fear of infamous but less dangerous chemicals.

As an advocate of women's participation in science, Armour helped establish a committee called WISEST (Women in Scholarship, Engineering, Science, and Technology) at the University of Alberta. WISEST encourages young women to aspire to careers in science, partly by means of a summer program that gives students a chance to work with scientists and technologists.

According to Armour, there is a looming shortage of chemistry teachers in universities. Chemistry enrolments in undergraduate programs are down, so the future supply of teachers will be limited. Because professors will be retiring in large numbers, the demand for instructors will increase. Small supply and large demand spell opportunities for young people looking forward to a career in chemistry.

- *According to Dr. Armour, what are some attributes of good researchers?*
- *Why is it important to be sceptical about chemical information in the media?*

- Each chapter ends with an **Overview**, including a Summary, Key Words, and Review, Application, and Extension questions. Each Overview provides an opportunity for students to recall important information and apply new knowledge. Lab-based problems are included in the Overview so students can practice their scientific problem solving.

GLOBAL WARMING

The term "greenhouse effect" refers to the heating of the Earth in a process similar to the heating of a greenhouse when the sun shines on it. Greenhouse gases play the role of the glass panels of the greenhouse. The greenhouse effect has made the Earth habitable. Without this effect, the Earth would be much cooler, probably without life as we know it. However, human activities now appear to be increasing the greenhouse effect.

The major greenhouse gases are carbon dioxide, chlorofluorocarbons (CFCs), and methane. Scientists estimate that 50% to 55% of the greenhouse effect is caused by carbon dioxide, 20% to 25% by CFCs, and 20% to 25% by methane. Smaller contributions are made by dinitrogen oxide and other gases. A large proportion of the human-produced carbon dioxide in the atmosphere is the result of combustion processes. All of the CFCs are thought to be a result of human activities. However, an estimated 40% of atmospheric methane comes from natural sources such as swamps, marshes, lakes, and oceans. Sources of methane from human activities include livestock (15%), rice paddies (20%), coal mining (6%), and oil and natural gas production (6%).

Opinions vary as to the rate and probable extent of the warming. Will the warming occur so slowly that people and natural systems can adapt? Or might conditions shift suddenly, wreaking unimaginable havoc? Based on current emissions of greenhouse gases, computer models suggest an increase of 1.5°C in the average temperature of the Earth by the year 2050, and an increase of more than 3.0°C by the year 2100. Temperature increases of this magnitude are predicted to cause a rise in sea levels of up to one metre over current levels, an extension of frost-free seasons by up to two months at high latitudes, and increased probability of prairie droughts. If the predicted droughts and changes in sea levels occur, much of the best agricultural land in the world will become unproductive and a vast area of populated land will become uninhabitable.

Because the certainty of predictions is low, many people, including some scientists, believe that the threat of the greenhouse effect is minimal. The Earth's temperature has fluctuated in the past, for example, during the ice ages. It is possible that, independent of human interference, the temperature of the Earth is increasing naturally. It is also possi-ble that the Earth is in the midst of a cooling trend and that the human-generated greenhouse effect is preventing another ice age. Models of the atmosphere are complex but inadequate for making precise predictions. The capacity of the oceans to absorb higher levels of carbon dioxide is not known, nor is the effect of an increased concentration of atmospheric carbon dioxide on plant growth understood.

Although the rate and extent of global warming are difficult to predict, it seems reasonable to reduce the production of greenhouse gases to avoid upsetting the delicate balance of the biosphere. We can create technologies to switch from high carbon fuels to low carbon fuels and use conventional fuels more efficiently. We can also practice energy conservation and exploit energy sources that do not produce carbon dioxide, such as solar energy, wind power, fuel cells, and photovoltaic cells.

- *How is the Earth like a greenhouse?*
- *Many people question the predicted temperature increases and have little confidence in the computer models (the authority upon which the predictions are made). Should we just wait and see what happens in the future?*

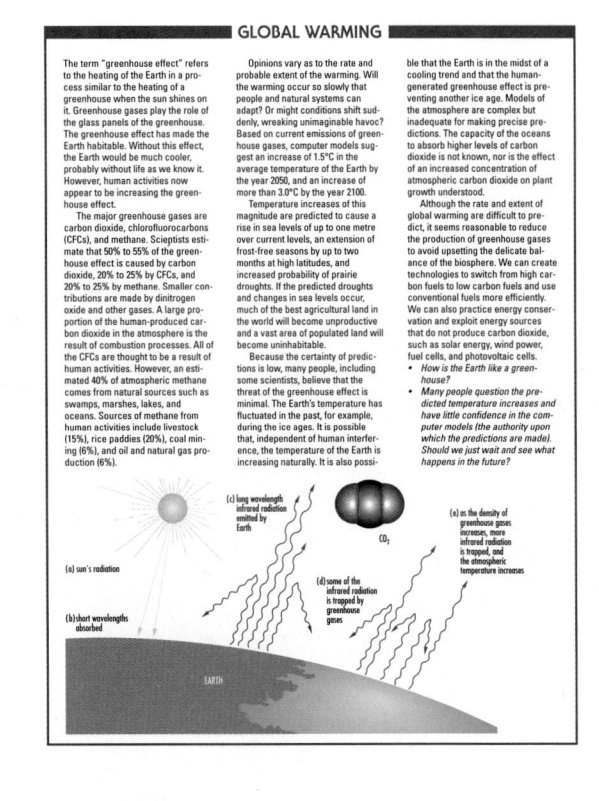

(a) sun's radiation
(b) short wavelengths absorbed
(c) long wavelength infrared radiation emitted by Earth
(d) some of the infrared radiation is trapped by greenhouse gases
(e) as the density of greenhouse gases increases, more infrared radiation is trapped, and the atmospheric temperature increases
CO_2
EARTH

- **Appendix A Answers to Overview Questions** provides answers to Overview questions.

- **Appendix B Scientific Problem Solving** emphasizes the nature of science, scientific problem solving, processes, and skills. A problem-solving model is presented, along with a model for investigation reports. The investigation report is consistent with a modern view of the nature of science and reflects a realistic approach to laboratory work.

- **Appendix C Technological Problem Solving** presents technological equipment, processes, and skills. Laboratory equipment is described, along with the technological skills required to use it. Common procedures, such as diagnostic tests, filtration, and titration, are described in detail for reference.

- **Appendix D Safety and Society** provides waste disposal and laboratory safety information, which supplements the cautions found in each student investigation. Also included in this section is a decision-making model for STS issues. The model promotes a multi-perspective approach to decision making.

- **Appendix F Data Tables** provides essential references. Additional data tables are included on the inside back cover of this book.
- **Appendix G Common Chemicals** lists names and formulas for easy reference.
- A comprehensive **Glossary** provides a useful dictionary of chemical terms and concepts and the **Index** enhances the efficiency of learning from the textbook.
- The **periodic table** on the inside front cover is an additional reference, containing both empirical and theoretical data. For easy reference, a separate student handout with the periodic table on one side and data tables on the other is available.
- There are **parallel streams of empirical and theoretical content** throughout the textbook. Chemistry is an experimental science; the appeal to experiment as the ultimate authority and source of knowledge in science is evident throughout this book.

APPENDIX D

Safety and Society

Science is a human endeavor, technology has a social purpose, and both have always been part of society. Science, together with technology, affects society in a myriad of ways. Society also affects science and technology, by placing controls on them and expecting solutions to societal problems. Within our society, safety for people and the environment is of paramount importance whether in a chemistry laboratory, chemical industry, or home.

D.1 Laboratory Safety Rules

Safety is always important in a laboratory or in other settings that feature chemicals or technological devices. It is your responsibility to be aware of possible hazards, to know the rules — including ones specific to your classroom — and to behave appropriately. Always alert the teacher in case of any accident.

Glass Safety and Cuts

- Never use glassware that is cracked or chipped. Give such glassware to your teacher or dispose of it as directed. Do not put the item back into circulation.
- Never pick up broken glassware with your fingers. Use a broom and dustpan.
- Do not put broken glassware into garbage containers. Dispose of glass fragments in special containers marked "broken glass."
- If you cut yourself, inform your teacher immediately. Imbedded glass or continued bleeding requires medical attention.

Burns

- In a laboratory where burners or hot plates are being used, never pick up a glass object without first checking the temperature by lightly and quickly touching the item. Glass items that have been heated stay hot for a long time but do not appear to be hot. Metal items such as ring stands and hot plates can also cause burns; take care when touching them.
- Do not use a laboratory burner near wooden shelves, flammable liquids, or any other item that is combustible.
- Before using a laboratory burner, make sure that long hair is always tied back. Do not wear loose clothing (wide long sleeves should be tied back or rolled up).

- **Appendix E Communication Skills** includes a primer on scientific language, a review of SI units and prefix symbols, rules for SI use, the rule of a thousand, rules for communicating the precision and certainty of measured and calculated values, and a review of graphs and tables. Communication is also stressed in every example, problem, investigation report, table, and graph presented throughout the text.

"Nature to be commanded, must be obeyed."

Francis Bacon, English philosopher

(1561–1626)

UNIT III

THE MEASURE OF CHEMISTRY

Scientists select an aspect of the world around them and define systems that interest them. A system can be as simple as a solution in a beaker, a gas sealed in its container, or a chemical reaction in a flask. Generally, qualitative observations come first, but ultimately scientists seek to obtain and explain quantitative measures of the system. In both scientific and technological work, quantitative descriptions based on the best available measurements are the most highly valued.

Other systems comprise more than chemicals and their reactions. Social, political, economic, and ecological systems are closely related to science and technology; for informed decisions on STS issues, interdisciplinary knowledge is essential. New manufacturing processes, for example, are extremely complex; we must consider not only the underlying science and technology, but also how these processes affect individuals, politics, the economy, and the environment.

ACKNOWLEDGEMENTS

Nelson Chemistry has evolved from the STSC (Science, Technology, Society, and Communication) Chemistry Project, a seven-year classroom-based curriculum project. Our years of writing, classroom testing, and revisions would have been impossible without the support of many people. First and foremost are our families, whose unselfish support and encouragement have kept us going.

The many teachers with whom each of the authors has taught have all made contributions to this textbook. This is especially true of teachers with whom we did some previous textbook writing — Dean Hunt, Dale Jackson, Eugene Kuzub, Tom Mowat, and Myron Baziuk. Not to be forgotten are the professors of chemistry and science education at the Universities of Alberta and Calgary.

Our students have been enthusiastic and patient and have provided more valuable feedback than they realize. They have been our partners and, as always, are the most important part of the educational process. We thank the many pilot teachers and administrators throughout Alberta, British Columbia, and the Atlantic provinces who have believed in us and supported our project. Their acceptance of an STS (Science-Technology-Society) textbook into their classrooms helped to show others that students can learn STS content simultaneously with rigorous chemistry content.

We are grateful for the timely financial support from the Secretary of State of the Government of Canada (Canadian Studies Directorate and Public Awareness Program for Science and Technology), and from the Alberta Foundation for the Literary Arts. This financial support provided the necessary resources to start the lengthy process that has resulted in this book and to involve teachers and students in more than 50 schools in the initial piloting.

Nelson Canada, and Bill Allan, Martin Goldberg, and David Steele in particular, are to be commended for their willingness to break the mold of traditional chemistry textbooks and enthusiastically support a new generation — the academic STS science textbook. Lynn Fisher, Nelson's Team Leader and Publisher, did an outstanding job in orchestrating all the pieces of the puzzle to meet internal and external deadlines. We are especially grateful for the thorough job and good nature of the editors, Janis Barr and Anne Norman. We also acknowledge the contributions made by West Coast Editorial Associates, Rosemary Tanner, Colin Bisset, Winnie Siu, Cecilia Chan, Ann Ludbrook, the reviewers of this book, and the rest of the team at Nelson Canada. They made many valuable suggestions and corrections. We hope that teachers and students find this book a useful resource to guide them through the challenging network of ideas and activities that constitute a chemistry program of studies.

Frank Jenkins, Ph.D.; Hans van Kessel, M.Sc.; Dick Tompkins, B.Sc.;
Oliver Lantz, Ph.D.; Michael Falk, Ph.D.;
Michael Dzwiniel, M.Sc.; George Klimiuk, M.A.

INTRODUCTION TO NELSON CHEMISTRY

Nelson Chemistry exemplifies the current national and international trends in Science-Technology-Society (STS) education. In this book, scientific and communication skills, the nature of science and technology, and STS issues in our society are integrated with the academic science content, rather than being added as an afterthought.

Each unit opens with a few paragraphs highlighting key science concepts, emphasizing their relationship to chemistry and the world around you. Throughout the book, the major emphasis is on the knowledge, skills, and attitudes that are essential to an understanding of chemistry. In each chapter, the introduction, text, exercise questions, investigations, and features serve to blend chemistry content with the STS components in an interesting and logical way. Knowledge, skills, and attitudes relating to technology and society are secondary emphases, integrated with appropriate chemistry topics.

The concepts presented in this book are tied to a base of empirical knowledge. This empirical background includes everyday experiences, laboratory work, and lab-based problems, and other experimental results that are discussed in the text. *Nelson Chemistry* strives to develop a system of communication that portrays science as an endeavor in which concepts are constructed in the mind in order to explain and predict observations made by the senses. Care has been taken to ensure that the language of presentation in this textbook portrays this view of science.

You are encouraged to use a familiar concept or experience to predict the results of an experiment. After selecting or creating an experimental design, you collect and analyze the evidence. By comparing the predicted result and the experimental result, you evaluate the concept used to make the prediction. In this book, the anomalies — predictions that are falsified — are accentuated rather than ignored, leading to a synthesis of new concepts to replace the unacceptable ones.

A new concept is applied to solve a variety of problems so that you show, by using the new concept correctly, that you have mastered it. Then the process is repeated — the concept that has been mastered is tested for its ability to explain and predict new observations. If the concept passes these tests, it is retained; if not, it is revised or replaced. In this way, experience in the science classroom reflects the experience of chemists as they construct and test scientific knowledge. If you understand how scientific knowledge is constructed and how science relates to technological and social issues, you will be better equipped to make a valuable contribution to life in the 21st century.

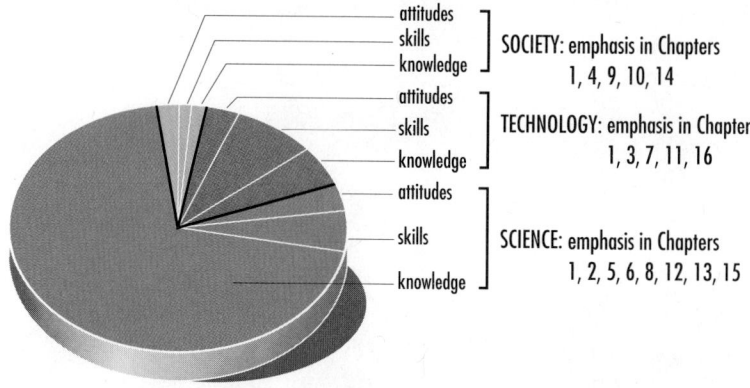

attitudes
skills
knowledge] SOCIETY: emphasis in Chapters 1, 4, 9, 10, 14

attitudes
skills
knowledge] TECHNOLOGY: emphasis in Chapters 1, 3, 7, 11, 16

attitudes
skills
knowledge] SCIENCE: emphasis in Chapters 1, 2, 5, 6, 8, 12, 13, 15

UNIT I

MATTER AND ITS DIVERSITY

When we focus on the natural world at the juncture of Earth and sky, the diversity is staggering. Waterways, mountains, rocks, animals, plants, and their myriad variations can overload our senses. Everywhere we look, there are new sights, new sounds, and new experiences. Scientists such as anthropologists, sociologists, biologists, geologists, and astronomers study the diversity within their own realms of interest.

Chemists focus on matter — on everything from chemicals in outer space, to everyday materials, to the structure of atoms and atomic nuclei. The diversity of matter is both a problem and a blessing: how can we comprehend the extent of this diversity? Organizing knowledge is the key, and the concept of elements and their classification in the periodic table is fundamental to the understanding of chemistry.

Chemistry, Technology, and Society

Like literature, philosophy, art, and music, science provides a framework for understanding the world around us. Each of these branches of knowledge has its own ways of exploring and understanding the world. One of the ways in which science differs from other endeavors, however, is in the methods it uses to test and evaluate ideas. A key characteristic of science is that all concepts must be testable. All scientific concepts can be shown to be acceptable or unacceptable, based on past or future experimental evidence.

Many human activities involve claims of scientific thinking; for example, marketing health and beauty products, making decisions in hospitals and courts of law, and debating issues such as acid rain and global warming. It is sometimes hard to assess claims made in the name of science. Is a brand of shampoo better because it has a lower pH value? What does a manufacturer of headache tablets *mean* by "scientifically tested and proven"?

Information is sometimes misrepresented as being scientific. For example, it has been claimed that people can walk on a bed of hot coals only if they have had training in positive thinking. In order to evaluate both scientific claims and pseudoscientific claims, we need a foundation of scientific knowledge. More importantly, we must know how to analyze and evaluate evidence presented to support the claims for new technologies or scientific advances. In this book you are presented with many opportunities to practice the skills of analysis and evaluation.

Nelson Chemistry was written with you and your education as the focus of the book. You, as a citizen, will have the opportunity to influence society in the 21st century. A solid foundation of scientific knowledge will increase your understanding, not only of science, but also of related political, ecological, economic, social, and technological issues. In this textbook, chemistry is presented in the context of the modern world.

Although the book's scope may seem broad, the content is developed slowly and carefully, with many examples, exercises, and investigations. Your participation — by listening to your teacher, asking and answering questions, performing experiments, and reading this textbook — is needed to maximize your learning. To understand the concepts and the specialized language of chemistry, you need practice. We have provided you with exercises in each chapter and with questions in the Overview that concludes each chapter. The study of chemistry includes concepts and communication skills, such as using the language of chemistry (symbols, formulas, and chemical equations), the language of measurement (quantity symbols and the *Système International*, SI, unit symbols), and the language of mathematics (calculations). Remember that participation, practice, communication skills, and learning the language of science will help ensure your success in chemistry and other subjects.

Ways of Knowing

Students often ask, "How am I supposed to know that?" There are many ways of knowing. One possibility is *memorization* — the key terms in boldface type and the information in the summaries in this textbook may need to be memorized. Some information can be obtained from *references* such as the periodic table (inside front cover) and data tables (Appendices and inside back cover), which contain detailed information that can be referenced, even on exams. Sometimes you are *given* the information in a question, in an exercise, or on a test. Another way of knowing is *empirical* — by experience and experimentation. Performing laboratory investigations, observing teacher demonstrations, and using your personal, out-of-school experience are empirical ways of learning. Finally, a *theoretical* way of knowing involves applying your understanding of scientific ideas to answer questions. You will understand chemistry well when you have memorized the key terms and ideas, when you know where to reference the information that you have not memorized, when you can use given information to answer questions, when you have a wide range of experience with chemical systems, and when you have an adequate grasp of theory. Although your learning style may favor one method over the others, a balanced knowledge of science requires using all five ways of knowing — memorization, reliance on references, given information, empirical work, and theoretical understanding.

Exercise

1. What are some learning processes and skills that you can use in class to help you learn the most from this chemistry course?

2. List five ways of knowing frequently used to answer the question, "How am I supposed to know that?"

3. Answer the following questions about this textbook. Use the Table of Contents, the Index, the Glossary, and the Appendices to help you find the answers in the most efficient way. Use a referenced way of knowing only. Include with your answer the page number where you found the answer.

    (a) What are the names of Chapters 2 to 4 and on what pages do they start?

    (b) How many investigations are there in Chapters 1 to 4?

    (c) What is the answer to question 22 in the Overview for Chapter 5?

    (d) What is the atomic number and melting point for the element, copper?

    (e) From Appendix G, what is the chemical formula and recommended name for malachite?

    (f) Draw a sketch of an Erlenmeyer flask.

    (g) Which variable should be listed on the vertical (y-axis) of a graph?

    (h) Write the definition for "science" as found in two places in the textbook.

    (i) In what chapter and on what page do you find the definition for binary molecular compounds?

    (j) From the reference sheet on the inside back cover, what is the specific heat capacity of iron?

    (k) What is the first step in the procedure for lighting a laboratory burner?

    (l) From Appendix B, what are the headings for writing a laboratory report?

    (m) What is the WHMIS symbol for a flammable chemical?

What Is Science?

Besides the processes of describing, predicting, and explaining, science also includes specialized methods, scientific attitudes, and values.

Science involves *describing, predicting,* and *explaining* nature and its changes in the simplest way possible. Everyone carries out these processes in an informal way, although they may not consciously categorize their thinking in this manner. According to Albert Einstein (1879 – 1955), "The whole of science is nothing more than a refinement of everyday thinking." In our daily conversations, we frequently describe objects and events. We read and hear descriptions in the news media, in literature, in sports, and in music. Scientists refine the descriptions of the natural world so that these descriptions are as precise and complete as possible. In science, reliable and accurate descriptions of phenomena become scientific laws.

We commonly predict such things as the weather or the winners in sports events. Scientists make predictions that can be tested by performing experiments. Experiments that verify predictions lend support to the concepts on which the predictions are based. We try to explain events in order to understand them. Young children ask many "why" questions: "Why is a fire hot? Why is water wet?" Scientists, like young children, try to understand and explain the world by constructing concepts. Scientific explanations are refined to be as logical, consistent, and simple as possible.

In scientific problem solving, descriptions, predictions, and explanations are developed and tested through experimentation. In the normal progress of science, scientists ask questions, make predictions based on scientific concepts, and design and conduct experiments to obtain experimental answers. As shown in Figure 1.1, scientists evaluate this process by comparing the results they predicted with their experimental results.

Every investigation has a purpose — a reason why the experimental work is done. Suppose, for example, a soft drink occasionally gets spilled on your living room rug. Sometimes you can remove the stain with no trouble, but sometimes you find the clean-up much more difficult. Why is this, you wonder? From reference books, you learn that some stains can be removed by neutralization; that is, to prevent stains from becoming permanent, you need to treat spills of acidic substances with a base and spills of basic substances with an acid.

Laboratory Reports

Scientific research is often very complex and involves, by analogy, many roadblocks, road repairs, and detours along the way. Scientists record all of their work, but the formal report submitted for publication often does not reflect the difficulties and circling back that is part of the process. The order of the report headings does not reflect a specific scientific method. The following sample investigation provides a model for scientific problem solving and headings for writing scientific reports.

> "In the field of observation, chance favors only the prepared mind." — Louis Pasteur (1822 – 1895)
>
> Not all science progresses systematically. Occasionally, a scientist has a sudden insight or makes an unexpected — even accidental — discovery. Although these insights and discoveries cannot be planned, they tend to occur most frequently to those scientists who are most immersed in their work.

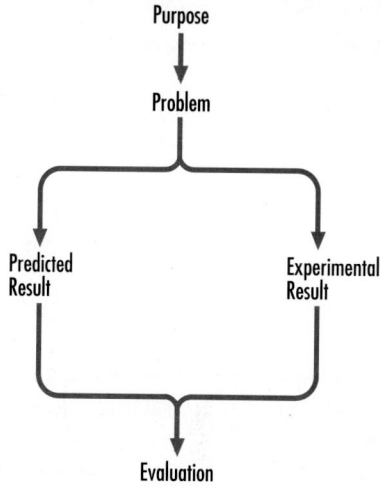

Figure 1.1
Many investigations involve comparing a predicted result with an experimental result.

INVESTIGATION

Stain Removal

To test the concept that neutralization of acids or bases is effective in stain removal, pose a queston, make a prediction, and create an experimental design to test your prediction. The following is an example of how you can approach this investigation scientifically; this is the way you will carry out investigations in this book. Your purpose is to test the concept of neutralization as it applies to stain removal for a soft drink spilled on a carpet.

Problem

Using your purpose as a guide, state the problem as a specific question that you expect to answer.

"What effect does the type of cleaner (acid or base) have on the removal of the stain?"

Prediction

Based on an authority (for example, a scientific concept or a reference book), make a prediction, and provide the reasoning behind the prediction.

In any investigation where the procedure appears in the textbook, symbols remind you to wear safety glasses and lab aprons, occasionally to wear rubber gloves, and to wash your hands before leaving the laboratory. These safety precautions should become a routine part of all laboratory procedures.

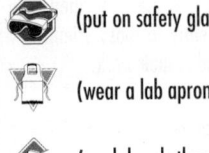 (put on safety glasses)

(wear a lab apron)

(wash hands thoroughly)

(wear rubber gloves)

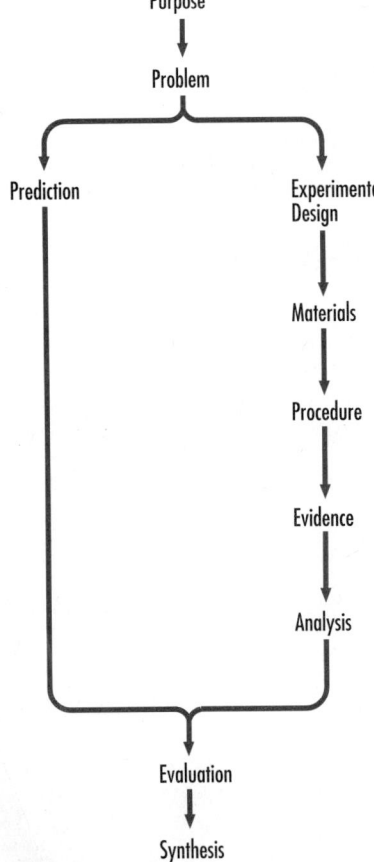

Figure 1.2
A model for scientific problem solving.

"According to the concept of acid-base neutralization, a base should remove the soft drink stain because most foods are acidic and bases are known to neutralize acids."

Experimental Design

You need a specific, carefully designed plan to answer the question experimentally; for example: "A vinegar solution (acid), a baking soda solution (base), and water are used on identical stains. The type of cleaning solution is the manipulated variable and the degree of removal of the stain is the responding variable. Controlled variables include the volume of solution placed on the stain, the length of time the solution is left on the stain, the amount of scrubbing, and the type of soft drink. Water is used as a control because there is water in both the acidic and the basic cleaning agents."

Materials

List everything you need, including quantities and sizes. Also list safety equipment.

3 samples of soft drink (50 mL each)
100 mL vinegar solution
100 mL baking soda solution
water
white rug
etc.

Procedure

List and number the steps in the procedure, including safety precautions and waste disposal methods. Be precise.

" 1. Pour the 3 samples onto three different areas of the white rug.

2. Wait 5 min (or 5 s or 24 h).

3. etc."

Evidence

Include all observations needed to answer the problem statement. In many chemistry experiments, you may have both qualitative and quantitative evidence. Sometimes tables of evidence are used. (In this investigation, you will simply observe the stains.)

Analysis

In investigations with numerical quantitative data, manipulation of the evidence may be required as part of the analysis — including new tables, graphs, and calculations. At the end of this section, answer the question posed as the "Problem."

(In this example you would indicate any differences in the effects of the acid or base cleaners, compared with water.)

Evaluation

In this part of the investigation report, evaluate the experiment and the concept being tested. Were the Experimental Design, the Procedure, and your technological skills adequate to answer the question? You might suggest modifications to the Design or Procedure that would lead to another experiment. In judging this experiment, you might suggest that you should have allowed more time for the solutions to act.

Compare the predicted answer with the experimental answer, and evaluate the authority used to make the prediction. If the authority is judged to be unacceptable, you have to decide what to do next. (In this investigation the concept of acid-base neutralization may be judged acceptable or unacceptable for predicting the type of solution that is successful for stain removal.)

Synthesis

Occasionally, you may need to create a new concept or a new experimental design. Your critical thinking, in the previous Evaluation section, is followed here by some creative thinking.

☑ **Problem**
☑ **Prediction**
☑ **Design**
☑ **Materials**
☑ **Procedure**
☑ **Evidence**
☑ **Analysis**
☑ **Evaluation**
☑ **Synthesis**

In many investigations some of the steps, such as the Problem, Materials, and Procedure, will be provided for you; the parts you are to provide will be indicated with checkmarks in a checklist like the one shown here. These steps are based on the problem-solving model shown in Figure 1.2. By following this model, you will not only learn about chemistry, but also improve your problem-solving ability, a skill that is useful in most aspects of life.

Exercise

4. What is science?

5. List the laboratory report headings in order of their presentation.

6. When actually doing their research, do scientists always follow the headings of their report in a linear (straight-ahead) order?

7. Why might Evidence be a better heading than Observations or Data in a laboratory report?

8. In what two sections of the laboratory report do we find an answer to the question asked in the Problem section of the report?

Laboratory Work

Performing laboratory experiments is a good way to learn about chemistry. Through experimentation, chemists are able to produce scientific generalizations and laws, and to test predictions arising from theories, generalizations, and laws.

Laboratory work should involve minimal hazards to yourself, your fellow students or co-workers, and the environment. Such work should also be cost-efficient, time-efficient, neat, and orderly. Appendix C describes some common laboratory procedures and skills. Correct procedures will help you obtain accurate laboratory evidence with a minimum of time and effort. Adequate preparation is necessary for efficiency in the laboratory.

You will communicate the methods and results of experiments by writing laboratory reports. The Sample Investigation on page 23 presents the format and content of a laboratory report. A more complete summary appears in Appendix B.

Laboratory Safety

To understand chemistry it is necessary to do chemistry. That is why investigations are integrated within each chapter in this book. All chemicals, no matter how common, and all pieces of equipment, no matter how simple, may be potentially hazardous to you, to your classmates, and to society. You are responsible for knowing about all aspects of laboratory safety, including the hazards associated with specific chemicals, and for carrying out all investigations safely. Safety is stressed continuously in this textbook. Be safety-conscious and always

- prepare for each investigation by reading the instructions in the textbook and by following your teacher's instructions
- use common sense to govern your behavior in the laboratory
- know the location of the safety equipment you might need in an emergency
- wear your lab apron and safety glasses while carrying out all investigations in the laboratory
- protect the environment by cleaning up your laboratory area and disposing of wastes as directed by your teacher. (Guidelines are provided in Appendix D.)

Read, understand, and follow

- the following safety notes,
- the questions and answers in the Laboratory Tour Exercise,
- the specific safety cautions in each Investigation, and
- the comprehensive list of laboratory safety rules in Appendix D.

A very important aspect of laboratory safety is your attitude toward laboratory work. Behavior that is acceptable in the classroom might be dangerous in the laboratory. For example, a friendly pat on the back could have serious consequences for someone holding a beaker of acid. Working in the laboratory can be fun, interesting, and productive, but carelessness can lead to serious injury. The most important features of a safe laboratory are the knowledge, skills, and attitudes of the people working in it. Follow these guidelines:

- ***Recognize hazards.***

    The materials and equipment used in laboratories often look harmless, but hazards nevertheless exist. The Workplace Hazardous Materials Information System (**WHMIS**) label on a chemical bottle alerts the user to the potential hazards of the chemical (Figure 1.3). To determine in greater detail the safety of chemicals, a Material

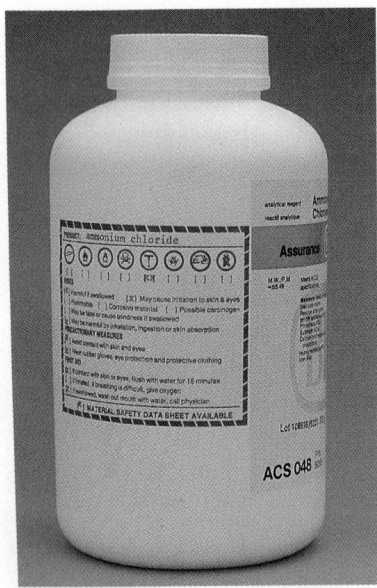

Figure 1.3
WHMIS labels describe the potential hazards of chemicals. Material Safety Data Sheets describe risks, precautions, and first aid.

Safety Data Sheet (**MSDS**) is available. These sheets list the potential hazards of chemicals, both individually and in combination with other chemicals.

- ***Use safe procedures and techniques.***
  Laboratory safety involves using the correct equipment and knowing appropriate handling techniques. For example, lighting and operating a laboratory burner present potential safety hazards. You can minimize hazards by following accepted lab procedures.

- ***Respond to emergencies sensibly.***
  Everyone should know how and when to operate a fire extinguisher, how to react if clothing catches fire, and what to do if a chemical is spilled or splashed on someone's skin or eyes.

WHMIS

The Workplace Hazardous Materials Information System (WHMIS) provides workers and students with complete and accurate information regarding hazardous products. To comply with WHMIS, suppliers must inform consumers of the properties and procedures for safe use of all hazardous materials and workers must learn and apply this information, provided on product labels and Material Safety Data Sheets. This Canada-wide system aims to reduce injuries and illness caused by exposure to hazardous substances. The regulations cover all worksites where chemicals are used, including dry cleaners, photography labs, garages, and schools. The best time to learn WHMIS regulations is before using a potentially hazardous substance.

The key to the WHMIS system is clear and standardized labelling. The supplier is responsible for providing a label when the product is sold, and the workplace is responsible for ensuring that the labels remain intact or are replaced, if necessary. When materials are transferred from stock bottles to smaller containers for student use, the smaller containers are also labelled, listing the hazards and indicating where more information can be found.

 **Class A:** Compressed gas

 **Class B:** Flammable and combustible material

 **Class C:** Oxidizing material

Class D: Poisonous and Infectious Materials

 Materials causing immediate and serious toxic effect

 Materials causing other toxic effects

 Biohazardous infectious material

 **Class E:** Corrosive material

 **Class F:** Dangerously reactive material

Exercise (Laboratory Tour)

Before beginning this exercise, read the Laboratory Safety Rules in Appendix D, D.1 and learn your school's specific safety procedures. As you tour the laboratory, answer the following questions related to safety, efficiency, and attitude. Your teacher will provide information and demonstrations.

9. Where do you put your books, purses, jackets, bags, etc., when you enter the laboratory?

10. When should you wear protective clothing, such as laboratory aprons and safety glasses? Where and how are aprons and safety glasses stored?

11. What items are stored at the safety station?

12. Where is the eye-wash equipment? How and when do you use it?

13. What type of fire extinguisher is in the laboratory and how do you operate it?

14. What should you do if your clothing catches fire? What should you do if someone else's clothing catches fire?

15. Where is the laboratory's fire exit? What is the evacuation procedure in case of an emergency?

16. What should you do immediately if any chemical comes into contact with your skin? Does the laboratory have a safety shower or hoses attached to water taps?

17. Where should chairs be placed while you do experiments?

18. Where is the distilled or purified water stored and how is it distributed?

19. How and where are the equipment and chemicals provided for each investigation?

20. Where are the MSD sheets for the laboratory chemicals kept?

21. Who determines the quantity of chemicals that you will use in each investigation? What should you do if you take too much?

22. What should you do with excess chemicals when the laboratory period ends?

23. What disposal methods are available for toxic and non-toxic wastes and for broken glass?

24. What is your school's policy concerning clean-up, assigned stations or partners, inappropriate behavior in the laboratory, and the sharing of work between partners?

25. Using the information in Appendix C, C.1, list the correct procedure for lighting a burner.

The Nature of Science and Technology

Science and technology are two different but parallel and intertwined human activities. **Science** is the study of the natural world with the goal of describing, explaining, and predicting substances and changes. The goal of science is to *develop or test a scientific concept*. **Technology** is the skills, processes, and equipment required to manufacture useful products or to perform useful tasks. The goal of technology is to get the process or equipment to *work*.

Technology is an activity that runs parallel to science (Figure 1.4). Technology often leads science as in the development of processes for creating fire, cooking, farming, refining metal, and the invention of the battery. The use of fire, cooking, farming, and refining led the scientific understanding of these processes by thousands of years. The invention of the battery in 1800 was not understood scientifically until the early 1900s. Seldom does technology develop out of scientific research, although as the growth of scientific knowledge increases, the number of instances of technology as applied science is increasing.

The invention of fire and the battery provided science with sources of energy with which to conduct experimental designs that would not otherwise be possible. Science would not progress very far without the increasingly advanced technologies available to scientists. Often scientific advances have to wait on the development of technologies for research to be done; for example, glassware, the battery, the laser, and the computer.

Often science is blamed for the effects of technology. Often people say, "Science did this, " or "Science did that." But most often it was a technological development that was responsible. Technologies and scientific concepts are created by people and used by people. We have to learn how to intelligently control and evaluate technological developments and scientific research, but we can't unless we are scientifically and technologically literate.

Science is essentially research (an intellectual pursuit). Concepts start as hypotheses and end as discarded or accepted theories or laws. Science meets its goal of concept development by continually testing concepts to death. Old concepts are revised or replaced when they do not pass the testing process. Scientific knowledge may be the most trusted kind of knowledge that we have, but scientists view it at best

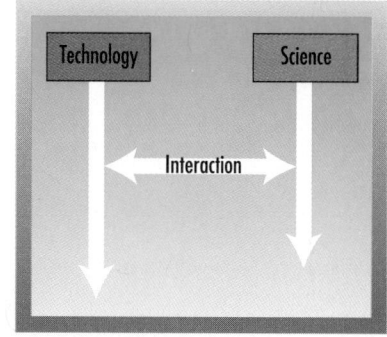

Figure 1.4
Science and technology are parallel streams of activity, where historically the development of a technology has most often led the scientific explanation of the technology. Increasingly, technologies are being developed from an application of scientific knowledge. Certainly, science and technology feed off of one another.

TECHNOLOGY LEADING SCIENCE

The extraction and refining of metals is an evolving technological process, called metallurgy, that has been around for thousands of years. This technology involved people with a high degree of skill and a knowledge of procedures. The technological procedures had been developed over the years from a systematic trial-and-error approach.

It was not until this century that we came to a scientific understanding of the natural processes involved in the technological processes of metallurgy. Laboratory research pro-

duced the concepts of the electron, the ion, and electron transfer that we now use to explain the metallurgical processes.

as having varying degrees of uncertainty. With this sceptical attitude toward the knowledge that they themselves create, no wonder scientists are given a special place of trust in our society.

What Is Chemistry?

Chemistry is the physical science that deals with the composition, properties, and changes in matter. Chemistry is everywhere around you, because you and your surroundings are composed of chemicals with a variety of properties. However, chemistry involves more than the study of chemicals. It also includes studying chemical reactions, chemical technologies, and their effects on the environment. Chemists solve problems in basic research and in technological processes using specific skills such as filtration, crystallization, and titration, as well as analysis, evaluation, and synthesis of scientific knowledge. Safe handling of chemicals at home and in the laboratory, and responsible disposal of non-toxic and toxic wastes are important aspects of chemistry. Finally, chemistry requires attitudes such as open-mindedness, a respect for evidence, and a tolerance of reasonable uncertainty.

Chemistry is primarily the study of changes in matter. For example, coals burning, fireworks exploding, and iron rusting are all changes studied in chemistry. A **chemical change** or **chemical reaction** is a change in which one or more new substances with different properties are formed. Chemistry also includes the study of **physical changes**, such as water freezing to form ice crystals and boiling to form water vapor, during which no new substances are formed. Although the study of unchanging matter can yield new information, most researchers are more interested in observing and interpreting changes in matter. In this book you learn about chemistry through examples that involve chemical reactions. Chemists study how and why one chemical is changed into a different chemical. They may not understand the change completely, but they strive to perfect their descriptions and their explanations.

Branches of Chemistry

Traditionally, there are five subdivisions used to classify the various fields of study within chemistry. This classification helps to organize our knowledge about chemicals and chemical change and to organize our study of chemistry. The following lists the fields of study of these subdivisions.

- *organic chemistry* — hydrocarbons (compounds of hydrogen and carbon) and their derivatives
- *inorganic chemistry* — chemicals other than organic
- *biochemistry* — chemicals and chemical change in living systems
- *physical chemistry* — changes in the structure and energy of chemicals and chemical systems
- *analytical chemistry* — separation, identification, and the quantification of matter

Careers with Chemistry
There are numerous careers and life experiences that require some knowledge of chemical composition and change. Some of these are listed below in no particular order.

pharmacist	cook
forester	dental assistant
farmer	medical technologist
nurse	petroleum engineer
geologist	chemical technologist
forensic chemist	custodian
medical doctor	environmental lawyer
chemical engineer	chemistry teacher
nutritionist	metallurgist
veterinarian	chemical technician
science teacher	cosmetologist
agrologist	homemaker
dentist	science writer or reporter
radiologist	housewares sales
food chemist	environmental technologist
mechanic	food and drug inspector
environmentalist	chemical sales
truck driver	chemistry professor
miner	greenhouse technician
beautician	battery salesperson
astronomer	hardware sales
painter	analytical chemist
carpenter	car owner or driver
fireman	research chemist
trainer	chemical manufacturer
florist	plastics technologist
fisherman	pulp-mill worker
brewer	welder
driller	toxicologist
gemologist	fabric manufacturer
baker	safety inspector
dry cleaner	fertilizer sales
wine producer	biochemist
manufacturer	citizen or voter
consumer	chemistry student

There are also other interdisciplinary chemistry subdivisions that combine chemistry with biology, geology, physics, and astronomy. Chemistry is considered a basic science because it helps one to understand most other science disciplines. The following lists the fields of study of these interdisciplinary subdivisions.

- *environmental chemistry* — the impact of various chemicals on the ecosphere
- *natural-products chemistry* — the chemicals and changes in the natural environment
- *drug chemistry* — the chemical composition and changes of drugs
- *food chemistry* — the chemical composition and changes of foodstuffs
- *geochemistry* — the chemical composition and changes in the Earth's crust
- *petroleum chemistry* — the chemicals and changes in petroleum and petroleum products
- *nuclear chemistry* — the chemicals involved in and produced by changes in the nucleus of an atom
- *radiochemistry* — the effects of nuclear radiation on chemical composition and changes
- *astrochemistry* — the origin and interactions of chemicals in the universe, especially interstellar matter

Exercise

26. Define science, technology, and chemistry.
27. List five branches of chemistry.
28. List nine interdisciplinary divisions of chemistry.
29. List your top five choices of careers that involve chemistry.
30. List a book or electronic reference that provides information on one of the careers in chemistry.

INVESTIGATION

1.1 Demonstration: Combustion of Magnesium

Classification systems are useful for organizing information. In Investigation 1.1, you will observe the combustion of magnesium and classify what you learn from this activity. While observing magnesium burn, record your observations in a table of evidence. Classify your observations according to what you observed before, during, and after the demonstration, and then classify what you have learned as observation or interpretation, and as qualitative or quantitative. (See the Sample Investigation, page 23, and Appendix B, B.3, for an outline of the parts of a laboratory report.)

- Problem
- Prediction
- Design
- Materials
- Procedure
- ✔ Evidence
- ✔ Analysis
- Evaluation
- Synthesis

CAUTION

Figure 1.5
The investigation of magnesium burning in air yields a number of observations. Organizing the observations makes them easier to understand.

This investigation, the first in your study of chemistry, provides you with an opportunity to follow strict safety procedures. The burning of magnesium as demonstrated by your teacher is, like many investigations in chemistry, potentially dangerous. The bright flame present during the burning of magnesium emits ultraviolet radiation that can permanently damage one's eyes. Thus, it is imperative that no one look directly at the burning magnesium. Note that your teacher has taken precautions to ensure that as the magnesium is lighted, eyes are protected from the radiation. The glass beaker reduces the ultraviolet light transmitted to a level that is safe to observe. **Only observe the burning magnesium when it is within the glass beaker**.

Problem

What changes occur when magnesium burns?

Experimental Design

Magnesium is observed before, during, and after being burned in air. All observations are recorded and then classified.

Materials

lab apron
safety glasses
rubber gloves
magnesium ribbon (approx. 5 cm)
steel wool
laboratory burner and striker
crucible tongs
large glass beaker

Procedure

1. Take safety precautions, then light the laboratory burner. (Instructions for lighting a laboratory burner are in Appendix C, C.2.)

2. Use rubber gloves when handling the steel wool.

3. Use tongs to hold the magnesium ribbon.

4. Clean the magnesium ribbon with steel wool; record any observations.

5. Light the magnesium ribbon in the burner flame (Figure 1.5) and hold the burning magnesium inside the glass beaker to observe.

6. Record any observations while the magnesium burns.

7. Record any observations after the magnesium has burned.

8. Classify the observations you have made in as many ways as you can.

Classifying Knowledge

The Evidence section of an investigation report includes all observations related to a problem under investigation. An **observation** is a direct form of knowledge obtained by means of one of your five senses — seeing, smelling, tasting, hearing, or feeling. An observation might also be obtained with the aid of an instrument, such as a balance, a microscope, or a stopwatch. In Investigation 1.1 you may have classified the observations in several ways. Perhaps you classified the knowledge in terms of time, recording observations as the investigation progressed, noting what you saw before the experiment, while the magnesium burned, and after the experiment.

Observations may also be classified as qualitative or quantitative. A **qualitative observation** describes qualities of matter or changes in matter; for example, a substance's color, odor, or physical state. A **quantitative observation** involves the quantity of matter or the degree of change in matter; for example, a measurement of the length or mass of magnesium ribbon. All quantitative observations include a number; qualitative observations do not.

In the investigation, you may also have distinguished between observations and interpretations. An **interpretation**, which is included in the Analysis section of an investigation report, is an indirect form of knowledge that builds on a concept or an experience to further describe or explain an observation. For example, observing the light and the heat from burning magnesium might suggest, based on your experience, that a chemical reaction is taking place. A chemist's interpretation might be more detailed: The oxygen molecules collide with the magnesium atoms and remove electrons to form magnesium and oxide ions. Clearly, this statement is not an observation. The chemist did not observe the exchange of electrons.

Table 1.1

CLASSIFICATION OF KNOWLEDGE	
Type of Knowledge	**Example**
empirical	• observation of the color and size of the flame when magnesium burns
theoretical	• the idea that "magnesium atoms lose electrons to form magnesium ions, while oxygen atoms gain electrons to form oxide ions"

Observable knowledge is called **empirical knowledge**. Observations are always empirical. **Theoretical knowledge**, on the other hand, explains and describes scientific observations in terms of ideas; theoretical knowledge is *not observable*. Interpretations may be either empirical or theoretical, and depend to a large extent on your previous experience of the subject. Table 1.1 gives examples of both kinds of knowledge.

Both graphite, the "lead" in pencils, and diamonds are common forms of the element carbon.

Exercise

31. Classify the following statements about carbon as observations or interpretations.
    (a) Carbon burns with a yellow flame.
    (b) Carbon burns faster if you blow on it.
    (c) Carbon atoms react with oxygen molecules to produce carbon dioxide molecules.
    (d) Global warming is caused by carbon dioxide.

32. Classify the following statements about carbon as qualitative or quantitative.
    (a) The flame from the burning carbon was 4 cm high.
    (b) Coal is a primary source of carbon.
    (c) Coal has a higher carbon-to-hydrogen ratio compared with other fuels.
    (d) Carbon is a black solid at standard conditions.

33. Classify the following statements about carbon as empirical or theoretical.
    (a) Carbon atoms are composed of six electrons and six protons.
    (b) Carbon is found in several forms in nature: for example, charcoal, graphite, and diamond.
    (c) Graphite conducts electricity, but diamond does not.
    (d) Graphite contains some loosely held electrons, whereas the electrons in diamond are all tightly bound in the atoms.

34. Scientific knowledge can be classified as empirical or theoretical.
    (a) What is the key distinction between these two types of knowledge?
    (b) Is the evidence collected in an experiment empirical or theoretical?
    (c) How would you classify the knowledge in the Analysis section of an investigation report?

Experimental Design

An **experimental design** is a plan for obtaining the answer to a specific question. An experimental design also outlines the methods used in an experiment. Often this design is written in terms of **manipulated**, **responding**, and **controlled variables** (Table 1.2). In some experiments a control is necessary to ensure the validity of the experimental analysis.

Table 1.2

VARIABLES	
Type of Variables	**Definition**
manipulated variable (also called the *independent variable*)	the property that is systematically changed during an experiment
responding variable (also called the *dependent variable*)	the property that is measured as changes are made to the manipulated variable
controlled variable	a property that is kept constant throughout an experiment

Write an experimental design for the following investigation.

Problem

How does the mass of magnesium burned relate to the mass of product formed?

Experimental Design

Different masses of magnesium are burned inside a container filled with air. The mass of product is measured for each mass of magnesium used. The mass of magnesium is the manipulated variable and the mass of product is the responding variable. All other variables, such as the composition of the air inside the container, the amount of air available, and the initial temperature, are kept constant.

Exercise

35. List the manipulated and responding variables and one controlled variable of this experimental problem: "How does altitude affect the boiling point of pure water?"

36. Write a brief plan to describe an experiment to answer the problem in question 35.

FIRE WALKING

Training in fire walking is an interesting example of scientific misrepresentation. Fire-walk leaders suggest that training in positive thinking can enable people to walk on a bed of hot coals. Some people pay hundreds of dollars to take such "mind-over-matter" seminars and to demonstrate the success of their training — by taking a fire walk.

The misrepresentation of science is in the method; that is, in the experimental design. A scientific analysis of fire walking might begin with the question, "Is training necessary?" and might continue with the design of an experiment.

To test the hypothesis that mental training is necessary to teach people how to walk unharmed on a bed of hot coals, a scientific experimental design is required. This involves a randomly selected control group of people who receive no training, and another randomly selected experimental group of people who receive several hours of specific instruction. Both groups attempt the fire walk under identical conditions. The results are tabulated and compared. Several trials, with different groups of people, are carried out, in order to ensure that the results are reproducible — an important feature of scientific inquiry. The results of free public fire walks have shown that anyone can walk safely across a bed of hot coals prepared by a specialist.

- *Do you need psychic powers to walk on a bed of hot coals?*
- *What is a control group?*

1.2 SEPARATING MATTER

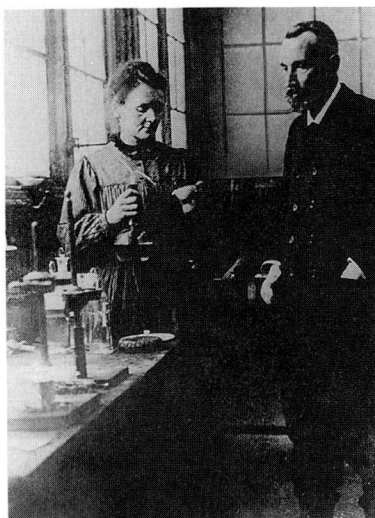

Figure 1.6
Marie Curie and her husband Pierre used a series of thousands of crystallizations over a period of four years to separate a gram of radium from eight tonnes of waste ore. Their perseverance earned Marie her doctorate and won them both the Nobel Prize in physics in 1903.

Most naturally occurring substances are mixtures; very few are even close to being pure substances. The separation of mixtures into pure substances or groups of similar substances is a common technique used in the laboratory, industry, and home. Separation methods range from the very simple to the very complex. As is the case in most of the history of science, separation techniques evolved as the scientific knowledge and technology advanced. For example, around 1900, Marie and Pierre Curie worked for four years to separate one gram of radium from eight tonnes of pitchblende ore (Figure 1.6).

Today, natural products chemists use sophisticated equipment to separate dozens of substances from the leaf of a single plant (Figure 1.13, page 46). The technologies have improved but many of the basic principles of separation have remained more or less the same.

On an industrial scale, thousands of tonnes of metal ore are pulverized and processed until pure metal is eventually obtained. Pure metal is used directly in products such as copper pipes, or in combination with other substances to produce alloys such as stainless steel (iron, chromium, and nickel). Drinking water is purified, sewage is separated and treated, and air is liquefied and separated into its components. Tar sand is separated into sand, water, and tar (bitumen). The bitumen, like crude oil, is distilled — separated into a wide variety of similar components such as gasoline, diesel fuel, and asphalt.

As consumers, we usually buy the purified substances or mixtures of these substances produced for us by industries. We use salt, vinegar, baking soda, and water in cooking. We use methanol (windshield washer antifreeze) and ethylene glycol (radiator antifreeze) along with gasoline and oil in our cars. We also do some separating in our homes — we separate leftover food, dishes, and utensils from the waste. Food scraps in the waste go to the compost pile. Then we separate the glass, metal, plastics, and paper waste that are recyclable. This process is continued at the recycling plants where different metal and paper products are separated from one another.

As you can see, the process of separating and mixing continues. Chemistry in the laboratory, industry, and home involves many instances of separating and then remixing in a combination that serves a useful purpose. In this section, you will be reintroduced to some separation techniques with which you are familiar and introduced to some new techniques useful in chemistry. The more techniques you know, the more flexible and adept you become at problem solving in the laboratory. Some of these techniques are summarized and described in Table 1.3, page 38.

Exercise

37. State one separation technique used in each of the following technologies.
    (a) vacuum cleaner
    (b) separating recycled steel cans from aluminum cans
    (c) separating silt, water, and oil in a large pond

(d) purifying water
(e) making a cup of tea using a tea bag
(f) separating gasoline from crude oil
(g) solution mining of rock (table) salt from an underground deposit
(h) refining sugar from a hot, concentrated syrup
(i) concentrating metal ores

38. Write an experimental design (i.e., a general plan) to separate each of the following mixtures into their components.
(a) granite, salt, and iron
(b) limestone, sugar, nickel, and water
(c) alcohol, water, and salt

Appendix G is a list of common chemicals, their chemical names, formulas, and uses or sources.

INVESTIGATION

1.2 Separating Matter

In research and industrial laboratories, chemical analysis often begins with the separation of a mixture, followed by the identification of one or more of the components. The purpose of Investigation 1.2 is to design and carry out an experiment to separate a mixture into all of its components. At this stage you are not asked to identify the components.

Your teacher will give your group a mixture. Examine it carefully before planning your experimental design and procedure. You may have to use several techniques in order to completely separate all the components. After you have completed your planning, obtain your teacher's approval before proceeding.

Problem

How many components are present in the unknown mixture?

- Problem
- Prediction
- ✔ Design
- ✔ Materials
- ✔ Procedure
- ✔ Evidence
- ✔ Analysis
- Evaluation
- Synthesis

CAUTION

Some of the materials are toxic and irritant. Avoid contact with skin and eyes.

INVESTIGATION

1.3 Paper Chromatography

Chromatography is a versatile and precise method of separating liquid or gas mixtures. It can be used for either qualitative or quantitative analysis. In this experiment you will separate the components of different colored inks.

Problem

How many components are present in each ink sample?

Experimental Design

The method of paper chromatography is used to separate the components of several colored ink samples. Filter paper is used as the porous medium.

- Problem
- Prediction
- Design
- Materials
- Procedure
- ✔ Evidence
- ✔ Analysis
- Evaluation
- Synthesis

Table 1.3

TECHNIQUES USED TO SEPARATE MATTER

Technique	Description	
mechanical	One or more components are picked out of the mixture either manually or by use of a magnet for magnetic substances (iron, cobalt, nickel). 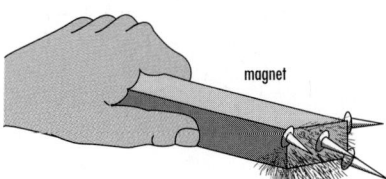	 *The drum behind these recycling workers mechanically separates aluminum cans from plastic. The worker on the left separates the various kinds of plastic by hand, and makes sure no aluminum is in the mixture.*
settling	Some heterogeneous mixtures can be separated by letting one of the components settle to the bottom. Spinning the mixture at high speed (centrifuging) may be used to accelerate this process.	 *A water treatment settling tank.*
flotation	Oil, detergents, or other chemicals are added to the heterogeneous mixture and air is blown through. The froth containing the desired component floats, and is skimmed off the surface. This technique is used to concentrate ores of zinc, copper, nickel, and lead, and to separate bitumen (tar) from sand. 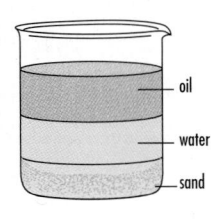	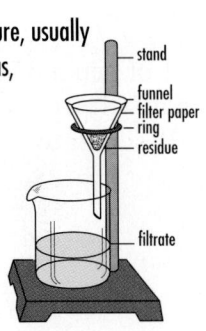 *A flotation tank in a metal refinery.*
filtration	A heterogeneous mixture, usually a solid in a liquid or gas, is passed through a screen or filter. The solid is trapped and separated from the liquid or gas.	 *An ordinary furnace filter separates dust and other particulates from the air.*

Table 1.3 (continued)

TECHNIQUES USED TO SEPARATE MATTER

Technique	Description

extraction — The mixture is mixed with a solvent that dissolves one or more, but not all, components. For example, table salt and sand can be separated by using water to dissolve (extract) the salt.

solvent containing dissolved component

undissolved component

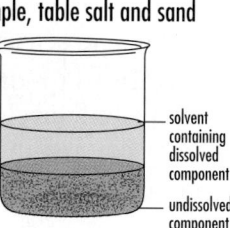

A coffee maker uses hot water to extract some of the components from ground coffee beans.

fractional distillation — A liquid mixture is boiled and one or more components are separated as they vaporize from the mixture at different temperatures.

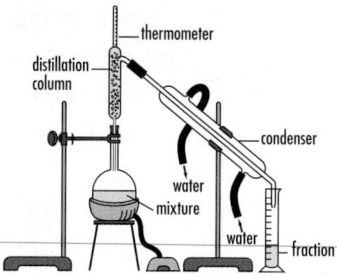

thermometer
distillation column
condenser
water
mixture
water
fraction

An oil refinery has fractional distillation towers to separate the various components of crude oil. (See also Figure 9.8.)

crystallization — A dissolved solid is separated from a solution by cooling or concentrating the solution to crystallize the solid.

crystals
evaporating dish

Crystallization occurs naturally as salt water evaporates.

chromatography — A mixture is carried by a solvent through a stationary, porous medium such as a column of solids or a filter paper. Separation occurs because components of the mixture move at different rates in the porous medium.

ink (mixture)
filter paper (porous medium)
solvent

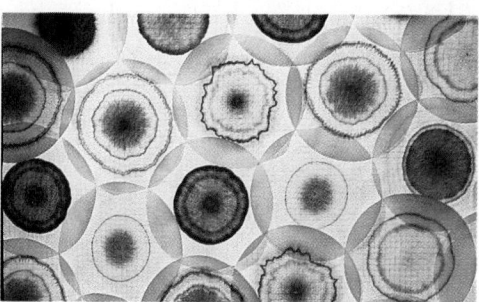

Paper chromatography reveals the components of a variety of industrial dyes.

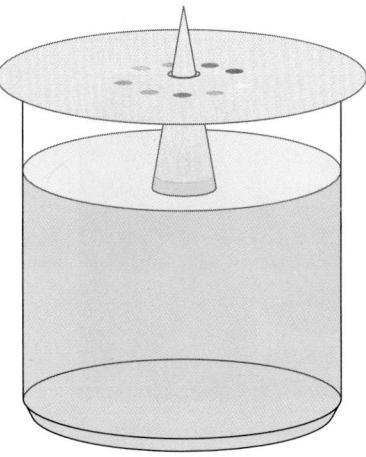

Figure 1.7
The center hole of the filter paper fits over the tip of the cone. The base of the cone takes up water from the beaker.

Materials

lab apron	beaker, plastic cup, or petri dish
safety glasses	2 pieces of filter paper (10 to 12 cm)
distilled water	various water-soluble felt-tip pens
scissors	

Procedure

1. Poke a small hole in the center of a piece of filter paper.
2. Equally space a number of small dots of ink around the circumference of a small circle about 6 to 8 mm from the center hole. (Use different colored inks or alternating colors.)
3. Record the pen brand name and color of each dot.
4. Using the second piece of filter paper, cut out a small triangle with a base of about 2 cm and a height of about 3 cm. Fold or roll it into a cone.
5. Place the pointed end of the cone into the center hole of the spotted piece of filter paper.
6. Fill a container with water so that the base of the cone will dip into the water when placed as shown in Figure 1.7.
7. Dry the rim of the container and then place the spotted piece of filter paper across the rim of the container.
8. Allow the pattern to develop until the water spreads to about 1 cm from the edge of the paper.
9. Remove the filter paper and allow it to dry.
10. Identify the number of components in the ink of each colored pen used.

1.3 CLASSIFYING MATTER

Matter is anything that has mass and occupies space. Anything that does not have mass or that does not occupy space — energy, happiness, and philosophy are examples — is not matter. To organize their knowledge of substances, scientists classify matter (Figure 1.9, page 42). A common classification differentiates matter as **pure substances**, whose composition is constant and uniform, and **mixtures** (impure substances), whose composition is variable and may or may not be uniform throughout the sample. Empirically, **heterogeneous mixtures** are non-uniform and may consist of more than one phase (e.g., the sample in Investigation 1.2). Your bedroom, for example, is a heterogeneous mixture because it consists of solids such as furniture, gases such as air, and perhaps liquids such as soft drinks. **Homogeneous mixtures** are uniform and consist of only one phase. Examples are some inks, tap water, aqueous solutions, and air.

You can classify many substances as heterogeneous or homogeneous by making simple observations. However, some substances that appear homogeneous may, on closer inspection, prove to be heterogeneous (Figure 1.8). Introductory chemistry focuses on pure substances and homogeneous mixtures, commonly known as **solutions**.

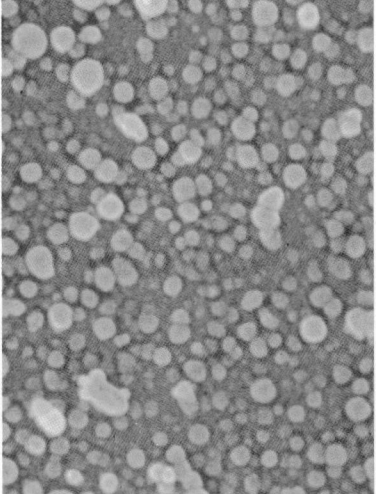

Figure 1.8
Although milk is called "homogenized," close examination through a microscope reveals solid and liquid phases. Milk is a heterogeneous mixture.

CAREER

CHEMICAL TECHNOLOGY INSTRUCTOR

Nyron Jaleel graduated from high school in 1974. He had been especially interested in biology and chemistry in high school, and he decided to pursue a Bachelor of Science degree. After graduation, Nyron did three years of research work investigating plant biochemistry. It became apparent that he was enjoying the parts of his job that involved interaction with other people as much as the scientific procedures and processes he performed, so he returned to university, graduating with a teacher's certificate.

"One of the most interesting experiences of my life was my student teaching assignment," Nyron recalls. "I had no problems relating to the students or with the curriculum, and I really enjoyed the practice teaching sessions in several schools. Then I was assigned to teach one of the rounds in my old high school! As a former student, you're never quite sure that you are really allowed to go in the staff room without knocking, and you have to retrain your brain to realize that your former teachers are colleagues now. The teachers got me through that, though, and many of them are still good friends."

After teaching adult education classes for a while, Nyron applied for a position as an instructor at a technical college. Since 1984 he has taught in the college's chemical technology department. He claims that there are few careers anywhere that can match the one he has chosen.

"There is always the dialogue and discussion with students," he observes, "and my students are generally very keen to learn — because their careers depend on it. So they are interested, informed, and challenging, which keeps me on my toes. As well, because the college is a very large facility that works closely with the community and with industry, I constantly meet interesting and influential people from all

areas who are involved in science education. And, finally, the institute has excellent facilities for every sport or recreational activity you can think of. I don't know too many people who have access to squash courts, gymnasiums, and pools right at their place of employment." Nyron stresses a need to keep active — because cooking is taught at the school, the food available for staff members must be considered another bonus in his job!

Food chemistry is one of the courses Nyron teaches. Other courses in which he is involved include polymer chemistry, biochemistry, and pre-technology courses; the latter are designed to upgrade students' high school backgrounds. As an instructor, Nyron is typically assigned lecture or lab classes for 18 to 20 hours a week, and he spends as much time again in course development, preparation, and marking. The instructional hours vary from day to day, but "I like that structure," he says, "because I can schedule much of what I need to do myself, at times that are convenient for me."

Instructing in a technological college means keeping up with the latest developments in science and industry. Nyron has just completed his Master of Education degree, which he earned by taking summer and night courses. He will probably be seconded soon, which means he will be assigned to work in research or industry for a term, to freshen and upgrade scientific skills pertaining to his job. In Nyron's career, he is always learning.

- *What does Nyron Jaleel like best about his chosen career?*
- *What university degrees has he earned? Why does he continue to study and learn new things?*

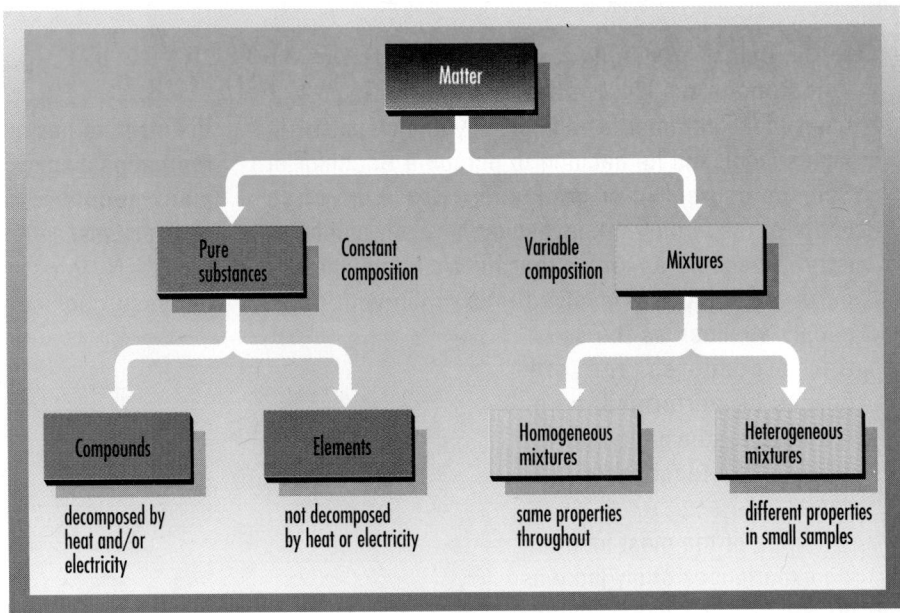

Figure 1.9
A classification of matter.

This empirical (i.e., observable) classification system is based on the methods used to separate matter. The parts of both heterogeneous mixtures and solutions can be separated by physical means, such as filtration; distillation; chromatography; mechanically extracting one component from the mixture; allowing one component to settle; or using a magnet to separate certain metals (see Investigations 1.2 and 1.3). A pure substance cannot be separated by physical methods. A compound can be separated into more than one substance only by means of a chemical change involving heat or electricity. Separating a compound into its elements is called **chemical decomposition**. **Elements** cannot be broken down into simpler chemical substances by any physical or chemical means.

Although the classification of matter is based on experimental work, theory lends support to this system. According to theory, elements are composed entirely of only one kind of atom. An **atom**, according to theory, is the smallest particle of an element that is still characteristic of that element. According to this same theory, **compounds** contain atoms of more than one element combined in a definite fixed proportion. Both elements and compounds may consist of **molecules**, distinct particles composed of two or more atoms.

Table 1.4

DEFINITIONS OF ELEMENTS AND COMPOUNDS			
Substance	**Empirical Definition**	**Theoretical Definition**	**Examples**
element	substance that cannot be broken down chemically into simpler units by heat or electricity	substance composed of only one kind of atom	Mg (magnesium), O_2 (oxygen), C (carbon)
compound	substance that can be decomposed chemically by heat or electricity	substance composed of two or more kinds of atoms	H_2O (water), NaCl (table salt), $C_{12}H_{22}O_{11}$ (sugar)

Solutions, unlike elements and compounds, contain particles of more than one substance, uniformly distributed throughout them.

A pure substance can be represented by a **chemical formula**, which consists of symbols representing the atoms present in the substance. You can use chemical formulas to distinguish between elements, which are represented by a single symbol, and compounds, which are represented by a formula containing two or more different symbols. Examples of formulas, along with empirical and theoretical definitions, are summarized in Table 1.4.

INVESTIGATION

1.4 Classifying Pure Substances

Before 1800, scientists distinguished elements from compounds by heating them to find out if they decomposed. If the products they obtained after cooling had different properties from the starting materials, then decomposition had occurred. This experimental design was the only one known at that time. The purpose of Investigation 1.4 is to evaluate this experimental design.

Problem

Are water, bluestone, malachite, table salt, and sugar empirically classified as elements or compounds?

Prediction

According to the theoretical definitions of element and compound and the given chemical formulas, these substances are all compounds. The reasoning behind this prediction is that the chemical formulas for water, bluestone, malachite, table salt, and sugar include more than one kind of atom.

water	$H_2O_{(l)}$
bluestone	$CuSO_4 \cdot 5\,H_2O_{(s)}$
malachite	$Cu(OH)_2 \cdot CuCO_{3(s)}$
table salt	$NaCl_{(s)}$
sugar	$C_{12}H_{22}O_{11(s)}$

Experimental Design

A sample of each substance is heated using a laboratory burner and any evidence of chemical decomposition is recorded.

Materials

lab apron
safety glasses
face shield
distilled water
bluestone
malachite
table salt
sugar
cobalt chloride paper
250 mL Erlenmeyer flask

laboratory scoop
laboratory burner and striker
ring stand and wire gauze
crucible
clay triangle
hot plate
large test tube (18 × 150 mm)
utility clamp and stirring rod
medicine dropper
piece of aluminum foil

CAUTION

Some of the materials are toxic and irritant. Avoid contact with skin and eyes.

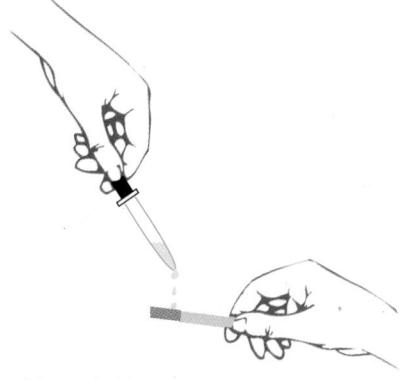

Figure 1.10
A strip of cobalt chloride paper is blue when dry, but turns a pale pink when wet with water.

Problem
Prediction
Design
Materials
Procedure
✔ Evidence
✔ Analysis
✔ Evaluation
Synthesis

Procedure

1. (a) Test some distilled water with cobalt chloride paper and notice the change in color (Figure 1.10).
   (b) Pour distilled water into an Erlenmeyer flask until the water is about 1 cm deep. Set up the apparatus as shown in Figure 1.11 (a).
   (c) Dry the inside of the top of the Erlenmeyer flask. Place a piece of cobalt chloride paper across the mouth of the flask.
   (d) Boil the water. Record any evidence of decomposition of the water.

2. (a) Place some bluestone to a depth of about 0.5 cm in a clean, dry test tube. Set up the apparatus as shown in Figure 1.11 (b).
   (b) Record any evidence of decomposition of the sample.

3. (a) Set a crucible in the clay triangle on the iron ring as shown in Figure 1.11 (c). Add only enough malachite to cover the bottom of the dish with a thin layer.
   (b) Heat the sample slowly at first, with a uniform, almost invisible flame; then heat it strongly with a two-part flame (Appendix C, C.2).
   (c) Record any evidence of decomposition of the sample.

4. (a) Place a few grains of table salt and a few grains of sugar in two separate locations on a piece of aluminum foil. Place the foil on a hot plate.
   (b) Set the hot plate to maximum heat and record any evidence of decomposition.

Figure 1.11
Heating substances. (a) An Erlenmeyer flask is used to funnel vapors. (b) A test tube is used when heating small quantities of a chemical. (c) A crucible is required when a substance must be heated strongly.

(a)
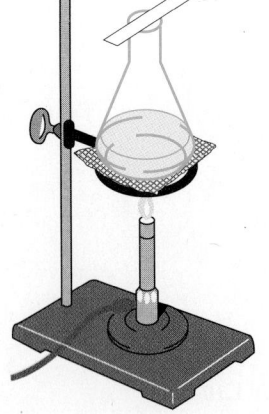

(b)
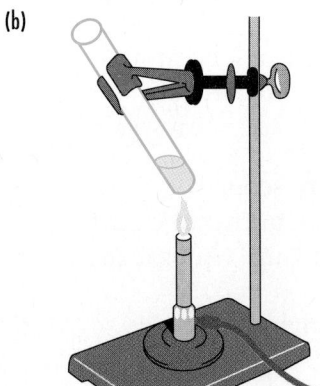

(c)
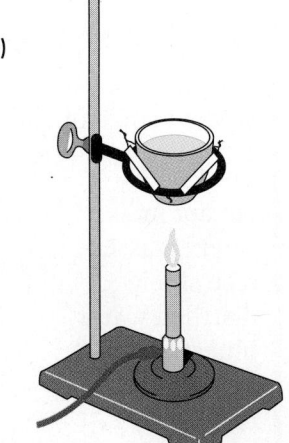

Technological Developments

As you saw in Investigation 1.4, heating is not an adequate experimental design for classifying elements and compounds empirically because it does not work for all compounds. Before 1800, for example, water was considered to be an element. The first conclusive evidence that water was a compound occurred shortly after the invention of the battery by Alessandro Volta in 1800 (Figure 1.12). Water was decomposed into hydrogen gas and oxygen gas by passing an electric current through water. This chemical change can be summarized by the following word equation. Read the word equation as, "Water decomposes to produce hydrogen and oxygen."

$$\text{water} \rightarrow \text{hydrogen} + \text{oxygen}$$

The invention of the battery led to many more successful decompositions of compounds in their pure or solution form. For example, in 1807 a young English chemist, Humphry Davy, decomposed sodium and potassium compounds as well as other substances once thought to be elements. Many new elements were discovered that occur naturally only as compounds. Not only did Davy successfully apply the new experimental design of decomposition by electricity, he also invented a variation of this design: molten-state electrical decomposition. In this variation, a compound is heated until it melts and forms a liquid. If an electrical current is then passed through the liquid, the compound decomposes. Davy is reported to have been an excitable man. Imagine his excitement at having produced potassium metal, then seeing the shiny globules of this metal react vigorously with water and burst into lavender flames!

The battery is an example of a technology that led to scientific breakthroughs. Without the inventive genius of Volta, Davy would have had no starting point for the experimental design for molten-state decomposition. As parallel activities, science and technology work together to advance human knowledge.

While Davy's discovery and isolation of sodium and potassium were scientific breakthroughs, he himself said on his deathbed that: "Michael Faraday (see Chapter 16 feature) was my greatest discovery."

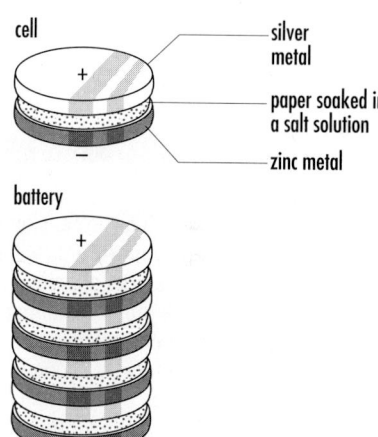

Figure 1.12
Alessandro Volta's "sandwich" of two different metals separated by a conducting solution makes an electric cell. Electric current from batteries like this led to new discoveries in chemistry.

Exercise

39. Describe how to distinguish experimentally between each of the following pairs of substances.

   (a) heterogeneous and homogeneous mixtures

   (b) solutions and pure substances

   (c) compounds and elements

40. According to modern theory, how do elements and compounds differ?

41. Pure iodine contains only iodine molecules. An iodine solution also contains only molecules. Are both iodine and an iodine solution pure substances? Justify your answer.

42. Classify each of the following chemicals as either an element or a compound.

   (a) B   (b) Ba   (c) BaO

   (d) OCl_2 (e) Cl_2 (f) NaCl

43. Using an example, describe the importance of technology in science.

44. (Discussion) Technology is sometimes defined as applied science. Evaluate this definition.

Figure 1.13
Chemists and technologists have identified over 10 million pure substances and the number continues to grow. Our forests, not unlike the tropical rainforests, are a rich source of new chemicals. Natural-products chemists make a career out of separating and identifying natural chemicals.

A CHEMICAL LIFE

No matter what job or profession you choose, you and your friends will pursue lifelong careers as chemical consumers. Here are the comments of a student who has picked up some chemical vocabulary by taking some chemistry courses.

"Hey, my life's just a matter of going from one chemical reaction to another all day long," Dave told some high school friends, as he munched French fries drenched in

acetic acid
and sodium chloride. "Let me expand upon this startling observation.

"This morning, I tumbled out of bed and put calcium carbonate flavored with peppermint on my teeth.

Then I hopped into the shower with a bar of glyceryl tristearate and a bottle of sodium lauryl sulfate. Afterwards, before putting on my polyesters and cottons, I rubbed on some aluminum chloride and esters to keep that just-showered scent all day long.

"For breakfast I spooned some sucrose into my bowl of carbohydrates and essential vitamins and poured in an aqueous solution of fats and proteins. I downed the mixture and my glass of citric acid and rushed into the garage, where I found my front tire in dire need of a gaseous solution of nitrogen and oxygen. I knew that I'd be turfed out of math if I were late for class, so I persuaded a friend to give me a lift — on condition that I buy some octane for the tank.

"I got to math just as the buzzer went and immediately the class was asked to take out their cellulose sheets and graphite sticks for a spot quiz. I got so nervous I took a calcium carbonate and magnesium hydroxide tablet to rid myself of excess hydrochloric acid. I passed that quiz and before I knew it, class was over. At break, Jim slipped me a piece of gum, sweetened with aspartame and flavored with methyl salicylate. Then it was off to computer class, where I

sent messages scurrying through silicon wafers by hitting polyvinyl chloride keys with my protein-laden fingers.

"On my way to biology, I listened to some pounding tunes by running my chromium

oxide medium over a playback head. In biology, Lori and I used stainless steel instruments to probe a formalin-preserved sheep's heart. We put the heart into polyethylene and scrubbed the counters with sodium hypochlorite.

"That was this morning. I wonder what chemistry I'll meet up with this afternoon.... Later — now I have to make it to chem class!"

Problem 1A Decomposition Using Electricity

This problem extends Investigation 1.4, using the substances that were not decomposed by heating. The purpose of the investigation in this problem is to test the experimental design of decomposition by electricity to determine whether a pure substance is an element or a compound. Complete the Analysis and Evaluation of the investigation report.

Problem

Are water and table salt classified as elements or compounds?

Prediction

According to current theoretical definitions of element and compound, as well as the given chemical formulas, water and table salt are classified as compounds. The chemical formulas indicate that water, $H_2O_{(l)}$, and table salt, $NaCl_{(s)}$, are composed of more than one kind of atom.

Experimental Design

Electricity is passed through water and through molten salt. Any apparent evidence of decomposition is noted.

Evidence

PASSING ELECTRICITY THROUGH SAMPLES		
Sample	**Description**	**Observations after Passing Electricity through Sample**
water	colorless liquid	two colorless gases produced
table salt	white solid	silvery solid and pale yellow-green gas formed (Figure 1.14)

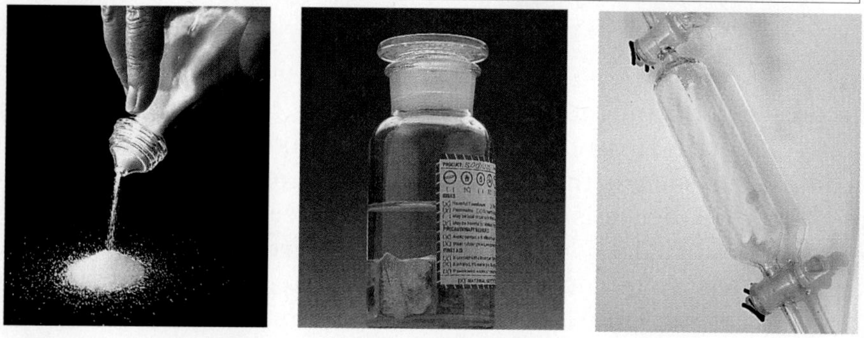

Figure 1.14
Left to right, sodium chloride (table salt), sodium (dangerously reactive metal), chlorine (poisonous, reactive gas).

1.4 SCIENCE AND TECHNOLOGY IN SOCIETY

The science of chemistry goes hand-in-hand with the technology of chemistry: the skills, processes, and equipment required to make useful products, such as plastics, or to perform useful tasks, such as water purification. Chemists make use of many technologies, from test tubes to computers. A particular chemical technology may or may not be understood thoroughly by chemists. For example, the technologies of glass-making and soap-making existed long before scientists could

Many familiar materials, such as plastics, metals, ceramics, and soap, are products of both science and technology.

explain these processes. We now use thousands of metals, plastics, ceramics, and composite materials developed by chemical engineers and technologists. However, chemists do not have a complete understanding of superconductors, ceramics, chrome-plating, and some metallurgical processes. Sometimes technology leads science — as in glass-making and soap-making — and sometimes science leads technology. Overall, science and technology complement one another.

Science, technology, and society (**STS**) are interrelated in complex ways (Figure 1.15). In this chapter and throughout this book, the nature of science, technology, and STS interrelationships will be introduced gradually, so that you can prepare for decision making about STS issues both now and in the 21st century.

You can acquire specialized knowledge for understanding STS issues by studying science. For example, a discussion of global warming becomes an informed debate when you have specific scientific knowledge about the topic; scientific skills to acquire and test new knowledge; and scientific attitudes and values to guide your thinking and your actions. You also need an understanding of the nature of science and of scientific knowledge.

Scientists have indicated that, at present, both the observations and the interpretations of global warming are inadequate for understanding

Figure 1.15
Many issues in society involve science and technology. Both the problems and the solutions involve complex interrelationships among these three categories. An example of an STS issue is the problem of acid rain.

the present phenomenon and for predicting how the situation will change in the future. However, scientists will always state qualifications such as these, even 100 years from now, no matter how much more evidence is available. In a science course you learn that scientific knowledge is never completely certain or absolute. When scientists testify in courts of law, present reports to parliamentary committees, or publish scientific papers, they tend to avoid authoritarian, exact statements. Instead, they state their results with some degree of uncertainty. In studying science, you learn to look for evidence, to evaluate experiments, and to attach a degree of certainty to scientific statements. You learn to expect and to accept uncertainty, but to search for increasingly greater certainty. This is the nature of scientific inquiry.

Chemicals and chemical processes represent both a benefit and a risk for our planet and its inhabitants. Food, water, and air are beneficial chemicals for life, but both human population growth and lifestyles that demand high energy and resource use are placing great stress on the Earth's resources. Chemistry has enabled people to produce more food, to dwell more comfortably in homes insulated with fibreglass and polystyrene, and to live longer, thanks to clean water supplies, more varied diet, and modern drugs. While enjoying these benefits, we also consciously and unconsciously assume certain risks. For example, when chemical wastes are dumped or oil spills into the environment, the effects can be disastrous. Assessing benefits and risks is a part of evaluating advances in science and technology.

The Western world is increasingly dependent on science and technology. Our society's affluence has led to countless technological applications of metal, paper, plastic, glass, wood, and other materials. Thousands of new scientific discoveries and technological advances are made each year. As our society embraces more and more sophisticated technology, we tend to seek technological "fixes" for problems, such as chemotherapy in treating cancer, and the use of fertilizers in agriculture. However, a strictly technological approach to problem solving overlooks the multi-dimensional nature of the problems confronting us.

Deciding how to use science and technology to benefit life on Earth is extremely complex. Most science-technology-society (STS) issues can be discussed from many different points of view, or **perspectives**. Even pure scientific research is complicated by economic and social perspectives. For example, should governments increase funding for scientific research when money is needed for social-assistance programs? Environmental problems such as discharge from pulp mills and air pollution are controversial issues involving many perspectives. For rational discussion and acceptable action on STS issues, a variety of perspectives must be taken into account. Of many possible STS perspectives on air pollution, five are listed below.

- A *scientific perspective* leads to researching and explaining natural phenomena. Research into sources of air pollution and its effects involves a scientific perspective.

- A *technological perspective* is concerned with the development and use of machines, instruments, and processes that have a social purpose. The use of instruments to measure air pollution and the development of processes to prevent air pollution reflect a technological approach to the issue.

- An *ecological perspective* considers relationships between living organisms and the environment. Concern about the effect of a smelter's sulfur dioxide emissions on plants and animals, including humans, reflects an ecological perspective.

- An *economic perspective* focuses on the production, distribution, and consumption of wealth. The financial costs of preventing air pollution and the cost of repairing damage caused by pollution reflect an economic perspective.

- A *political perspective* involves government actions and measures. Proposed legislation to control air pollution involves a political perspective.

A Lack of Control
In the fall of 1994, people living on the east side of Saint John, New Brunswick, experienced several days of the worst air pollution they'd ever known. Homes and cars were coated with black soot. The air stank. Some people suffered breathing problems. Why? A local oil refinery had shut down its pollution control equipment for routine maintenance. The company had been granted permission by the government to continue operating without this equipment. The result? Worse than anyone expected. Environment officials promise it will never happen again. Local residents say it should never have happened in the first place.

"We especially need imagination in science. It is not all mathematics, nor all logic, but it is somewhat beauty and poetry." — Maria Mitchell, American astronomer (1818 – 1889)

A mnemonic that may be used to recall these five STS perspectives is STEEP.

Exercise

45. Identify four or more current STS issues.

46. Classify each of the following statements about aluminum as representing a scientific, technological, ecological, economic, or political perspective.
    (a) Recycled aluminum costs less than one-tenth as much as aluminum produced from ore (Figure 1.16).
    (b) Aluminum ore mines in South America have destroyed the natural habitat of plants and animals.
    (c) Aluminum is refined in Canada using electricity from hydro-electric dams.
    (d) In Quebec, aluminum is refined using hydro-electric power that some politicians in Newfoundland have claimed belongs to their constituents.
    (e) In 1886, American chemist Charles Hall discovered through research that aluminum can be produced by using electricity to decompose aluminum oxide dissolved in molten cryolite.

47. Instead of changing their lifestyles, many people look to technology to solve problems that are often caused by the use of technology! Suggest one technological fix and one lifestyle change that would help to solve each of the following problems.
    (a) Aluminum ore from South America used to produce aluminum metal for beverage cans will be in short supply soon.
    (b) Pure aluminum cans thrown into the garbage are not magnetic and are therefore difficult to separate from the rest of the garbage.
    (c) People throw garbage into bins for recyclable aluminum cans.

48. (Discussion) Aluminum is used extensively for making beverage cans. List some benefits and risks of this practice. Are there any alternatives that might have equal or better benefits and fewer risks? Be prepared to argue your case.

Figure 1.16
Recycling the aluminum in cans benefits the economy and the environment — a win-win situation.

OVERVIEW

Chemistry, Technology, and Society

Summary

- Chemistry involves specialized knowledge, skills, and attitudes.

- Scientific knowledge can be classified as observations or interpretations. It can also be classified as qualitative or quantitative, and as empirical or theoretical.

- An experimental design is a general plan for solving a scientific problem. This design often includes controlled, manipulated, and responding variables.

- Safe and efficient laboratory work is ensured by appropriate attitudes and behavior, knowledge and preparation, international safety symbols, the WHMIS program, and MSD sheets.

- Matter can be classified empirically as heterogeneous or homogeneous; as pure substances or mixtures; and as elements or compounds.

- Chemical compounds can be decomposed by heat or electricity, whereas elements cannot.

- Mixtures can be separated by a variety of techniques including mechanical, settling, flotation, filtration, extraction, fractional distillation, crystallization, and chromatography.

- Science, technology, and society (STS) are interrelated and interdependent.

- Perspectives on STS issues are often classified as scientific, technological, ecological, economic, and political.

Key Words

atom
chemical change (chemical reaction)
chemical decomposition
chemical formula
chemistry
compound
controlled variable
element
empirical knowledge
experimental design
heterogeneous mixture
homogeneous mixture
interpretation
manipulated variable
matter
mixture
molecule
MSDS
observation
perspective
physical change
pure substance
qualitative observation
quantitative observation
responding variable
science
solution
STS
technology
theoretical knowledge
WHMIS

Review

1. In the study of chemistry, what attitudes are useful?

2. When you read a scientific statement, how do you know if the statement is empirical or theoretical?

3. List the steps for lighting a laboratory burner.

4. The eye acts like a sponge and can absorb chemicals that might be splashed in it. What should you do if any chemical is splashed in your eye?

5. If your clothes catch fire in or away from the laboratory, what action should you take?

6. What questions should you ask before disposing of a chemical?

7. Name two examples of each of the following:
   (a) pure substance

(b) homogeneous mixture

(c) heterogeneous mixture

8. Write an empirical and a theoretical definition of (a) element and (b) compound.

9. What technological invention allowed a better experimental design to classify elements and compounds? Describe briefly how this invention is used.

10. Describe the relationship between science and technology.

Applications

11. An important part of chemistry is direct experience. What observations about the burning of magnesium cannot be made from the photograph in Figure 1.5 on page 32?

12. Plastics seem to be everywhere. Classify the following statements about plastics in three ways: observation or interpretation; qualitative or quantitative; and empirical or theoretical.

    (a) Plastic containers are often less rigid than metal containers.

    (b) High density polyethylene is rigid and melts at approximately 135°C.

    (c) Plastics are formed from very long chains of molecules bonded together.

    (d) Plastic soft drink bottles can be recycled to produce more soft drink bottles.

    (e) Plastics are not biodegradable into smaller molecules.

    (f) Only about 5% of fossil fuels are used to produce petrochemicals such as plastics; 95% of fossil fuel use involves burning for energy production.

13. (Discussion) Classification is not restricted to science. To make the world easier to understand, we classify music, food, vehicles, and even people. Give an example of a useful classification system that you have encountered in your life. Describe another example in which you think the effect of the classification is negative. Why do you think it is negative?

14. Classify the following statements as scientific, technological, ecological, economic, or political. Some statements can be classified in more than one way.

    (a) Plastics are generally inexpensive materials compared with other materials.

    (b) Plastics are formed from long chains of molecules called polymers.

    (c) Some provincial governments have actively promoted the establishment of a plastics industry in their province.

    (d) CFCs (chlorofluorocarbons), once used for making plastic foam, are thought to be one cause of the destruction of the ozone layer around the Earth.

    (e) The idea for producing some plastics came from copying natural long-chain molecules such as proteins and cellulose.

    (f) Large factories are constructed to meet consumer, commercial, and industrial demand for plastics.

    (g) Plastic liners may be used in garbage dumps to stop the leaching of toxic materials into the environment.

15. Write an experimental design to determine which plastics are thermoplastics that can be reshaped by heating. Write a brief paragraph describing the general plan, and then list the manipulated, responding, and controlled variables.

16. Write an experimental design to determine whether a substance is an element or a compound. Make the design extensive enough to provide a high degree of certainty in the answer.

17. How many of the separation techniques listed in Table 1.3 (page 38) are used in your home? State an example for each.

18. Federal government regulations allow only about a dozen artificial food colors. Food manufacturers may use any combination of the permitted synthetic colors. Write an experimental design to determine the number of artificial coloring compounds that are used in a package of grape-flavored drink crystals.

Extensions

19. Write a technological design to separate a mixture of aluminum cans, steel cans, and glass bottles for a recycling industry. Assume a large-scale operation without any initial

concern for cost. Include plastic bottles for an even greater challenge.

20. (a) Prepare a table that lists the advantages and disadvantages of aluminum, steel, glass, and plastic for beverage bottles. Include advantages and disadvantages using scientific, technological, ecological, economic, and political perspectives.

    (b) Based upon your decision-making process, which beverage container do you think is best? Which advantage do you value most?

21. Imagine that a vacuum cleaner salesperson comes to your home to demonstrate a new model. The salesperson cleans a part of your carpet with your vacuum cleaner, and then cleans the same area again using the new model. A special attachment on the new model lets you see the additional dirt that the new model picked up. Analysis seems to indicate that the new model does a better job. Evaluate the experimental design and explain your reasoning.

Internet

22. Find a source of MSDS on the internet and print the sheet for concentrated hydrochloric acid.

23. Search the internet for a source of information on careers in chemistry. Print the information on the career that most interests you.

24. The internet has sites that specialize in providing information on chemical separation techniques. Find one of these sites, lists, or news groups and print some information.

25. Find what you can on the internet about beverage container recycling. For example, what are the advantages or alternatives to using aluminum cans?

2 Elements

Long before recorded history, humans used elements for many purposes. Copper, silver, and gold were shaped into beautiful jewellery and other objects, both artistic and practical. Ancient peoples discovered that another element, tin, could be combined with copper to make a much harder material, from which they made stronger cutting tools, more effective weapons, and mirrors. At the dawn of recorded history, about 3000 B.C., the Egyptians learned to extract iron from iron ore. They used cobalt to make blue glass and another element, antimony, in cosmetics. In the first few centuries A.D., the Romans discovered how to use lead to make water pipes and eating utensils.

During the Middle Ages in Europe, alchemy flourished. Alchemists sought a method for transforming metals into gold. Magic, observation, and experimentation all played important roles in alchemy. Secrecy was paramount, as alchemists dreamed of the power and wealth that exclusive knowledge of the transformation process would bring them. Although they failed in this quest, they developed many experimental procedures and discovered new elements and compounds.

Modern scientists also study elements, although in greater detail than the alchemists did, and for different reasons. It is now possible not only to use elements in many ways and to explain their properties in theoretical terms, but also to apply new technologies to create images of the atoms that make up elements.

CLASSIFYING ELEMENTS

Since ancient times, people have known of seven metallic elements. And long before the invention of the telescope, they were also aware of seven celestial bodies. In ancient writings, the same symbols used to represent the elements were used to represent the sun, the moon, and the five "wandering stars" that we now know as planets (Figure 2.1). By the early 1800s, alchemists had discovered new elements, and the complexity of their symbols led to problems in communication. This prompted an English chemist and former schoolteacher, John Dalton (1766 – 1844), to devise simpler symbols for each element (Figure 2.2).

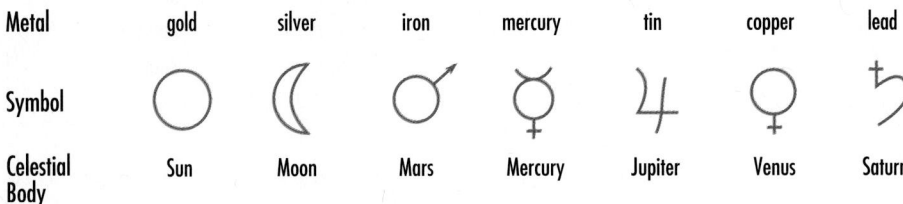

Metal	gold	silver	iron	mercury	tin	copper	lead
Symbol							
Celestial Body	Sun	Moon	Mars	Mercury	Jupiter	Venus	Saturn

Figure 2.1
From ancient times, metals were associated with particular celestial bodies.

In 1814, Swedish chemist Jöns Jacob Berzelius (1779 – 1848) suggested using just letters as symbols for elements. In this system, which is still used today, the symbol for each element consists of either a single uppercase letter or an uppercase letter followed by a lowercase letter. Because Latin was the common language of communication among educated Europeans in Berzelius's day, many of the symbols were derived from the Latin names for the elements (Table 2.1, page 56). Today, although the names of elements are different in different languages, the same symbols are used in all languages. Scientific communication throughout the world depends on this language of symbols, which is characteristically *international*, *precise*, *logical*, and *simple*.

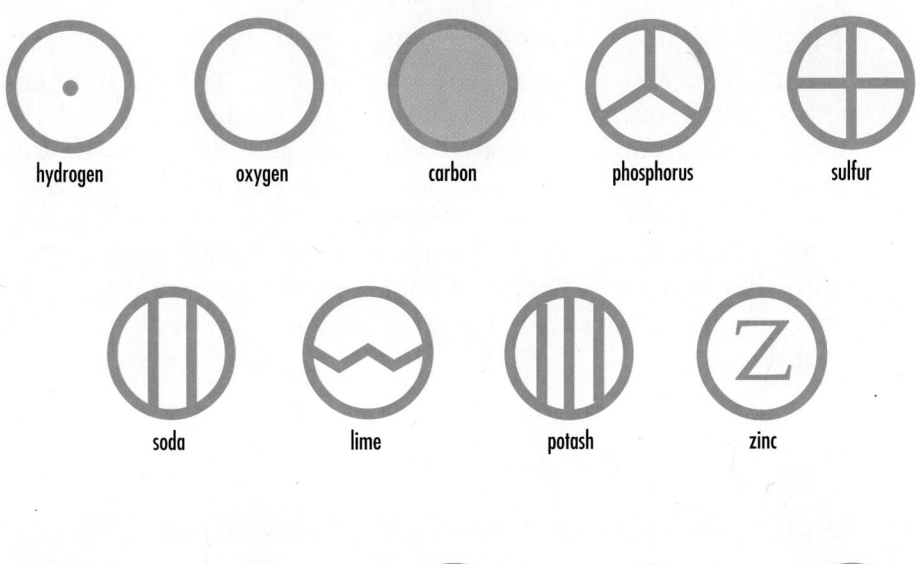

Figure 2.2
Dalton's symbols included both drawings and letters. Compounds such as lime, soda, and potash were, at that time, classified as elements since they could not be decomposed by heating.

Figure 2.3
Besides hundreds of data tables, this common reference book contains summaries of IUPAC rules governing chemical names and symbols.

The Chinese character for hydrogen is unique to that language, but the symbol is recognizable worldwide.

name   symbol

Table 2.2

IUPAC ELEMENT ROOTS	
Number	**Root**
0	nil
1	un
2	bi
3	tri
4	quad
5	pent
6	hex
7	sept
8	oct
9	enn

What is the temporary IUPAC name and symbol for element 111?
Element 111 is temporarily named Unununium, and its symbol is Uuu.

Table 2.1

SELECTED SYMBOLS AND NAMES OF ELEMENTS				
Symbol	**Latin**	**English**	**French**	**German**
Ag	argentum	silver	argent	Silber
Au	aurum	gold	or	Gold
Cu	cuprum	copper	cuivre	Kupfer
Fe	ferrum	iron	fer	Eisen
Hg	hydrargyrum	mercury	mercure	Quecksilber
K	kalium	potassium	potassium	Kalium
Na	natrium	sodium	sodium	Natrium
Pb	plumbum	lead	plomb	Blei
Sb	stibium	antimony	antimoine	Antimon
Sn	stannum	tin	étain	Zinn

Scientists have organized a governing body for scientific communication: the International Union of Pure and Applied Chemistry (IUPAC) specifies rules for chemical names and symbols. The IUPAC rules, which are summarized in many scientific references (Figure 2.3), are used all over the world. In Appendix F, you will find a list of all the English names of the elements in alphabetical order, along with their respective symbols.

SUMMARY: IUPAC RULES FOR ELEMENT SYMBOLS AND NAMES

- Element names should differ as little as possible among different languages. However, only the symbols are truly international.

- The first letter (only) of the symbol is always an uppercase letter (e.g., the symbol for cobalt is Co, not CO, co, or cO).

- The name of any new metallic element should end in -ium.

- Artificial elements beyond atomic number 103 have IUPAC temporary names and symbols derived by combining (in order) the "element roots" for the atomic number and adding the suffix, "-ium." The symbols of these elements consist of three letters, each being the first letter of the three element root names corresponding to the atomic number. (See Table 2.2 and margin note that follows.) The permanent names and symbols are determined by a vote of IUPAC representatives from each country.

Exercise

1. List the four characteristics of scientific communication.
2. What does the acronym IUPAC stand for?
3. The use of two letters from the element name greatly expands the number of distinct possibilities. Using the element list in

Appendix F, write all English element names and international symbols for elements with "s" as the first letter of their names.

4. When element number 114 is discovered, what would be its temporary name and symbol according to the IUPAC rules?

5. Table 2.3 lists modern element symbols and ancient technological applications. Write the English IUPAC name for each element symbol.

Table 2.3

ANCIENT TECHNOLOGICAL APPLICATIONS OF ELEMENTS	
International Symbol	**Technological Application**
Sn	part of bronze (Cu and Sn) cutting tools, weapons, and mirrors
Cu	main part of all bronze and brass (Cu and Zn) alloys
Pb	used by Romans to make water pipes
Hg	liquid metal used as a laxative by the Romans
Fe	produced by Egyptian iron smelters in 3000 B.C.
S	burned for fumigation by Greeks in 1000 B.C.
Ag	gold-silver alloys made by Greeks in 800 B.C.
Sb	ground ore used in early Egyptian cosmetics
Co	used in Egyptian blue-stained glass in 1500 B.C.
Al	part of alum used as a fire retardant in 500 B.C.
Zn	part of brass mentioned by Aristotle in 350 B.C.

6. Which ancient uses of elements listed in Table 2.3 are now considered unsafe and not recommended? Justify your answer.

7. John Dalton erroneously classified lime and several other compounds as elements. What was his evidence for this classification? What does this indicate about the certainty of scientific knowledge?

8. (Discussion) What are some advantages and disadvantages of the IUPAC system of temporary names and symbols for elements beyond number 103?

9. (Internet) Search the internet for the latest information on the discovery and naming of element 104 and beyond.

Communicating Empirical Knowledge in Science

Communication is an important aspect of science. Scientists use several means of communicating knowledge in their reports or presentations. Some ways of communicating empirical knowledge are presented below.

- *Simple descriptions* communicate a single item of empirical knowledge, that is, an observation. In Investigation 1.1, Chapter 1, you might communicate the simple description that magnesium burns in air to form a white, powdery solid.

- *Tables of evidence* report a number of observations. The manipulated (independent) variable is usually listed in the first column, and the responding (dependent) variable is entered in the final column. Table 2.4 shows results from a quantitative experiment similar to Investigation 1.1.

Table 2.4

MASS OF MAGNESIUM BURNED AND MASS OF ASH PRODUCED		
Trial	Mass of Magnesium (g)	Mass of Product (g)
1	3.6	6.0
2	6.0	9.9
3	9.1	15.1

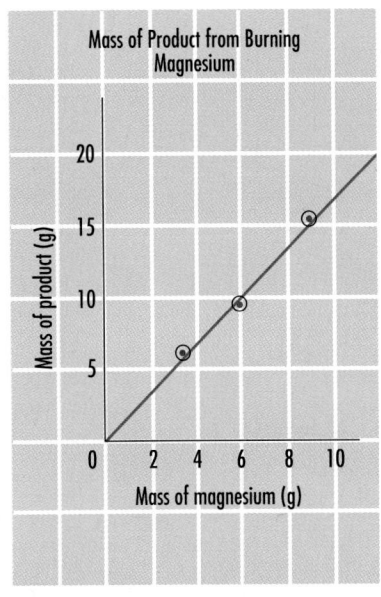

Figure 2.4
The relationship between the mass of magnesium that reacted and the mass of product obtained.

- *Graphs* are visual presentations of observations. According to convention, the manipulated variable is labelled on the *x*-axis, and the responding variable is labelled on the *y*-axis (Appendix E, E.4). For example, the evidence reported in Table 2.4 is shown as a graph in Figure 2.4.

- **Empirical hypotheses** are preliminary generalizations that require further testing. Based on Figure 2.4, for example, you might tentatively suggest that the mass of the product of a reaction will always vary directly with the mass of a reacting substance.

- **Empirical definitions** are statements that define an object or a process in terms of observable properties. For example, a metal is a shiny, flexible solid.

- **Generalizations** are statements that summarize a limited number of empirical results. Generalizations are usually broader in scope than empirical definitions and often deal with a minor or sub-concept. For example, many metals slowly react with oxygen from the air in a process known as corrosion.

- **Scientific laws** are statements of major concepts based on a large body of empirical knowledge. Laws are more important and summarize more empirical knowledge than generalizations. For example, the burning of magnesium, when studied in greater detail (Table 2.5), illustrates the law of conservation of mass.

Table 2.5

MASSES OF MAGNESIUM, OXYGEN, AND PRODUCT OF REACTION			
Trial	Mass of Magnesium (g)	Mass of Oxygen (g)	Mass of Product (g)
1	3.2	2.1	5.3
2	5.8	3.8	9.6
3	8.5	5.6	14.1

According to the evidence in Table 2.5, the total mass of magnesium and oxygen is generally equal to the mass of the product. Similar studies of many different reactions reflect the **law of conservation of mass**: In any physical or chemical change, the total initial mass is equal to the total final mass of material.

For a statement to become accepted as a scientific law, evidence must first be collected from many examples and replicated by many scientists. Even after the scientific community recognizes a new law, that law is subjected to continuous experimental tests based on expected characteristics of all laws. Laws must accurately *describe* current observations and *predict* future events in a *simple* manner. Like other methods of communicating empirical knowledge, which do not provide explanations, scientific laws do not explain *why* the knowledge exists. A law, such as the law of conservation of mass, is a statement of accepted fact. A theory or idea is required to explain why the fact exists.

Exercise

10. List seven ways of communicating empirical knowledge.

11. Using the observations in Table 2.4 and Appendix E, E.4, draw and label a graph showing the mass of oxygen used in the burning of different masses of magnesium.

12. How does a generalization differ from a scientific law?

INVESTIGATION

2.1 Empirical Definitions of Metals and Nonmetals

The purpose of this investigation is to test your previous knowledge of two classes of elements — metals and nonmetals. A common method of organizing information in science is to classify it into categories or classes based on one or more empirical properties common to all members of a class. Classification is the process of gathering evidence, looking for similarities in properties, and identifying useful properties. An empirical definition is written using the unique or defining properties of the class.

Problem
What are the properties of the selected elements?

Prediction
According to my empirical definition of metals and nonmetals, the properties of the metals are ..., and the properties of the nonmetals are ....

Experimental Design
The predicted properties are observed for each of the selected elements.

- ☐ Problem
- ✔ Prediction
- ☐ Design
- ☐ Materials
- ✔ Procedure
- ✔ Evidence
- ✔ Analysis
- ✔ Evaluation
- ☐ Synthesis

 **CAUTION**

(T) (☠) Iodine is very toxic if ingested, and the vapor irritates mucous membranes if inhaled. Do not open the sealed container.

(🔥) Sulfur is combustible and toxic. When handling sulfur, wear rubber gloves.

Metals and Nonmetals

When chemists investigate the properties of materials, they must specify the conditions under which the investigations were carried out. For example, water is a liquid under normal conditions indoors, but it would probably become a solid outdoors in winter. Ordinarily, tin is a silvery-white metal, but at temperatures below 13°C, it gradually turns grey and crumbles easily. For the sake of accuracy and consistency, the IUPAC has defined a set of standard conditions. Unless other conditions are specified, descriptions of materials are assumed to be at **standard ambient temperature and pressure**. Under these conditions, known as **SATP**, the materials and their surroundings are at a temperature of 25°C and the air pressure is 100 kPa.

From many observations of the properties of elements, scientists have found that **metals** are shiny, bendable, and good conductors of heat and electricity (Figure 2.5). The majority of the known elements are metals, and all metals except mercury are solids at SATP. The remaining known elements are mostly nonmetals. **Nonmetals** are not shiny, not bendable, and not good conductors of heat and electricity in their solid form. At SATP, most nonmetals are gases and a few are solids (Figure 2.6). Solid nonmetals are brittle and lack the lustre of metals.

In the definitions for metals and nonmetals, *bendable, ductile,* and *malleable* are often used interchangeably, although they are different properties (Figure 2.5). **Ductile** means that the metal can be drawn (stretched) into a wire or a tube, and **malleable** means that the metal can be hammered into a thin sheet. The word *lustrous* is often used in these definitions in place of *shiny*.

In addition to SATP, scientists also use STP, standard temperature and pressure, conditions. STP refers to exactly 0°C and 101.325 kPa (1 atm).

Malleability of Gold
Gold is one of the most malleable metals. It is used when super thin foil (~0.1 µm thick) is required in experiments and technologies. Gold foil is used for decorative gilding (e.g., on cakes and books) and for coating artificial satellites to reflect infrared radiation.

Figure 2.5
Gold (shown on the left) is easily bendable. Steel's malleability is shown in the center photograph. Steel, which is made from iron, is rolled into flat sheets that are used in cars and home appliances. Copper (shown on the right) is an excellent conductor of electricity. Because it is also ductile, it is used to make electrical wiring.

Problem 2A Testing Empirical Definitions of Metals and Nonmetals

The purpose of this problem is to test the empirical definitions of metals and nonmetals. Complete the Prediction, Analysis, and Evaluation of the investigation report.

Problem

What are the properties of the selected elements?

Experimental Design

Each element is observed at SATP, and the malleability and electrical conductivity are determined for the solid form.
The metals tested are cadmium, chromium, nickel, and platinum.
The nonmetals tested are bromine, chlorine, oxygen, and phosphorus.

Evidence

Figure 2.6
Sulfur is a common nonmetal, obtained from the hydrogen sulfide found in natural gas.

OBSERVATIONS OF SELECTED METALS AND NONMETALS			
Element	**Appearance**	**Malleability of Solid**	**Electrical Conductivity of Solid**
bromine	red-brown liquid	no	no
cadmium	shiny solid	yes	yes
chlorine	yellow-green gas	no	no
chromium	shiny solid	yes	yes
nickel	shiny solid	yes	yes
oxygen	colorless gas	no	no
platinum	shiny solid	yes	yes
phosphorus	white solid	no	no

INVESTIGATION

2.2 Demonstration: Testing Empirical Definitions of Metals and Nonmetals

The purpose of this demonstration is to test the empirical definition of metals and nonmetals. Scientific knowledge is continually being tested. This process usually proceeds by posing a problem, predicting the answer according to known scientific knowledge, designing and conducting an experiment, and finally evaluating the outcome (see page 23). Contrary to what you might expect, scientists purposely look for examples that will disagree with existing knowledge. Significant advances in scientific knowledge are made when predictions are falsified.

Problem

What are the properties of the selected elements?

Experimental Design

Each element is observed at room temperature. The malleability and electrical conductivity are determined for the solids. The metals tested are mercury and sodium. Carbon and silicon, which do not appear to be metals, are also tested.

- ☐ Problem
- ☑ Prediction
- ☐ Design
- ☐ Materials
- ☐ Procedure
- ☑ Evidence
- ☑ Analysis
- ☑ Evaluation
- ☐ Synthesis

 **CAUTION**

Mercury is toxic and must be handled in a sealed container.

Sodium is dangerously reactive. Safety glasses or a face shield along with a safety shield should be used.

13	14	15	16	17
5 B	6 C	7 N	8 O	9 F
13 Al	14 Si	15 P	16 S	17 Cl
30 Zn	31 Ga	32 Ge	33 As	34 Se	35 Br
48 Cd	49 In	50 Sn	51 Sb	52 Te	53 I
80 Hg	81 Tl	82 Pb	83 Bi	84 Po	85 At

Figure 2.7
The properties of some elements have led to the creation of a category of elements called metalloids.

A New Class of Elements

Your previous experience, combined with the results of Investigation 2.1 and Problem 2A, produced empirical definitions of metals and nonmetals. However, Investigation 2.2 showed that at least two elements, carbon and silicon, clearly do not fit the definition for either metals or nonmetals. Scientists have found a few other elements that behave similarly to silicon. These elements are members of a small class known as metalloids (discussed in Chapter 8, page 305). The elements are found near the blue "staircase line" in the periodic table inside the front cover of this book (see Figure 2.7).

METALS IN THE HUMAN BODY

Twelve elements make up more than 99% of the human body (see table below). Of these, five are metals. Calcium is found in bones and teeth; potassium, sodium, and magnesium exist in fluids throughout the body; and iron is found mainly in red blood cells and muscle tissue. These metals are essential for life. Some metals, however, are extremely toxic to humans.

Evidence of lead poisoning has been found in the bodies of English sailors who died in 1847, when the Franklin expedition to find a North West passage was blocked by ice west of King William Island. Samples of bone tissue from the remains of a crew member, preserved in the ice for more than 100 years, were analyzed by scientists at the University of Alberta. The elevated concentrations of lead in the bones may have been a result of eating food from cans sealed with lead solder.

Until the 1960s, lead was used in household water pipes, in lead solder, in glazes on dishes, and in paints. Even more recently, lead was used in gasoline. Unfortunately, young children sometimes ate flakes of lead-based paint by chewing on cribs, toys, or windowsills. People of all ages were harmed by inhaling lead-containing dust from automobile exhausts, as well as eating food from cans soldered with lead alloys.

Symptoms of chronic lead poisoning — headaches, loss of appetite, fatigue — appear when lead concentration in the blood reaches 25 µg/100 mL, but the nervous system can be damaged at much lower levels. Babies and young children are especially susceptible to lead poisoning, and may suffer permanent brain damage following ingestion or inhalation over time.

Mercury is another metal that can damage living things. From the 17th century to the mid-19th century, when felt hats made from beaver fur were stylish, mercury was used in producing felt. Workers in hat factories often developed the nervous system disorders that accompany mercury poisoning, including loss of memory and, eventually, insanity. The expression "mad as a hatter" came from this common malady afflicting felt-hat makers.

In the 20th century, the discharge of waste materials from pulp mills and industrial plants into waterways has caused an increased accumulation of mercury in fish. Humans in some polluted areas are advised not to eat fish, at the risk of developing mercury poisoning.

Since the discovery of a relationship between high blood levels of lead or mercury and nervous system disorders, many countries have drawn up regulations to control the consumer, commercial, and industrial uses of these potentially dangerous metals.

- *What are the six major elements and five major metals found in the human body?*
- *Are the elements of the human body really found as elements?*

COMPOSITION OF THE HUMAN BODY

Element	Percent by Mass (%)
oxygen	65.0
carbon	18.0
hydrogen	10.0
nitrogen	3.0
calcium	1.5
phosphorus	1.0
potassium	0.35
sulfur	0.25
sodium	0.15
chlorine	0.15
magnesium	0.05
iron	0.004

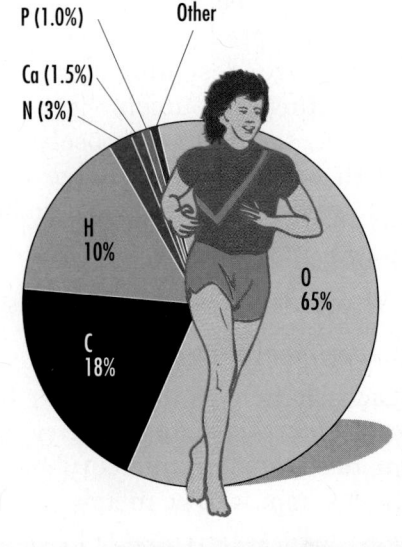

2.2 THE PERIODIC TABLE

With the discovery of more and more elements in the early 1800s, scientists searched for a systematic way to organize their knowledge by classifying the elements. As increasingly accurate instruments were invented, scientists began to make careful measurements of mass, volume, and pressure in the course of their investigations, thus building up a store of empirical knowledge. By studying the reactions of various elements with hydrogen and using the quantitative relationships that emerged, they determined the relative atomic mass of each element. For example, atoms of carbon were found to have a mass 12 times the mass of hydrogen atoms; oxygen atoms have a mass 16 times the mass of hydrogen atoms. Atoms of hydrogen appeared to be the lightest, so a scale was devised in which hydrogen has an atomic mass of 1 unit. The relative atomic masses of some common elements are shown in Table 2.6.

In 1864, the English chemist John Alexander Newlands (1837–1898) arranged all of the elements known then in order of increasing atomic mass. When he did this, he noticed that certain physical and chemical properties recurred in a regular pattern. For example, sodium, potassium, lithium, rubidium, and cesium are all soft, silvery-white metals. They are highly reactive elements, and they form similar compounds with chlorine. There is a strong "family" resemblance among them. The elements that follow these five in Newlands's arrangement — beryllium, magnesium, calcium, strontium, and barium — also exhibit a strong family resemblance. Newlands noticed that various physical and chemical properties of these and other families were repeated periodically in the sequence of elements. He stated this observation as a **periodic law**: *When elements are arranged in order of increasing atomic mass, chemical and physical properties form patterns that repeat at regular intervals.*

Unfortunately, Newlands initially called his law "the law of octaves," an analogy to the octave scale in music. Since there is no logical connection between music and chemistry, many scientists at the time ridiculed this law, and it was not generally accepted by the scientific community. In the same year that Newlands announced his findings, Lothar Meyer also arranged the elements in order of atomic mass and thought he found a repeating pattern in their atomic volumes (Problem 2B, page 64). However, there were gaps in his pattern and Meyer later admitted that he did not have the courage to propose new elements to fill the gaps.

Table 2.6

RELATIVE ATOMIC MASSES OF SELECTED ELEMENTS	
Element	**Relative Atomic Mass**
hydrogen	1
carbon	12
oxygen	16
sodium	23
sulfur	32
chlorine	35.5
copper	63.5
silver	108
lead	207

Today, the order of the elements, rather than being based on atomic mass, is based on theories of atomic structure. There are now many properties of elements that have been shown to have periodicity (Chapter 8).

Problem 2B Testing the Periodic Law

One of the properties of elements that was investigated in connection with the periodic law was atomic volume. This was calculated in the 1800s from the relative atomic mass and the density of each element in the solid state.

$$\text{relative atomic volume} = \frac{\text{relative atomic mass}}{\text{density}}$$

The purpose of this problem is to test the periodic law to see if the periodicity applies to atomic volume. Complete the Analysis of the investigation report, in graph form. (See Appendix E, E.4 for rules of graphing. Use a computer graphing program or spreadsheet if your teacher agrees.)

Problem

How is atomic volume related to the atomic mass of the elements?

Prediction

According to the periodic law, a graph of atomic volume versus atomic mass should have a pattern that repeats at regular intervals.

Experimental Design

The prediction and the periodic law are tested by graphing the relationship between relative atomic mass and relative atomic volume. Consider the relative atomic mass to be the manipulated variable and the relative atomic volume to be the responding variable.

Evidence

RELATIVE ATOMIC MASSES AND VOLUMES OF SELECTED ELEMENTS		
Element	Relative Atomic Mass	Relative Atomic Volume
Li	6.9	13.0
Be	9.0	4.9
B	10.8	4.6
C	12.0	5.3
N	14.0	13.7
O	16.0	11.2
F	19.0	15.0
Na	23.0	23.6
Mg	24.3	14.0
Al	27.0	10.0
Si	28.1	12.1
P	31.0	17.0
S	32.1	15.5
Cl	35.5	19.0
K	39.1	45.4
Ca	40.1	25.9
Sc	45.0	15.0
Co	58.9	6.6
Zn	65.4	9.2
As	74.9	13.0
Se	79.0	16.5
Br	79.9	25.6
Rb	85.5	55.9
Sr	87.6	34.5

Mendeleyev's Periodic Table

In 1869, Russian chemist Dmitri Mendeleyev (Figure 2.8) reported periodic properties in a list of elements similar to the interpretations of Newlands and Meyer. In Mendeleyev's original table of elements, elements were listed in order of atomic mass in vertical columns until the chemical properties of the elements started to repeat themselves and a new column was started. This table did not last very long for several reasons. New empirical evidence showed that some elements were misplaced and the organization of the groups was not simple enough to be useful.

Mendeleyev revised his original table and published a new periodic table of the elements in 1872. In this table, he listed all the elements known at that time, in order of atomic mass. The table is organized in such a way that elements with similar properties appear in the same column or "family" (Figure 2.9).

Mendeleyev's table contains some blank spaces where the known elements did not appear to fit. However, he had such confidence in his table that, where no element existed for a particular set of predicted properties, he assumed that the element had not yet been discovered. For example, in the periodic table in Figure 2.9 there is a blank between silicon, Si (28), and tin, Sn (118). Mendeleyev predicted that an element, which he called "eka-silicon" (after silicon), would eventually be discovered and that this element would have properties related to those of silicon and tin. He made detailed predictions of the properties of this new element. Sixteen years later, a new element named germanium was discovered in Germany. Its properties are listed in Table 2.7, page 66, beside the properties that Mendeleyev had predicted for eka-silicon. The boldness of Mendeleyev's quantitative predictions and their eventual success made him and his periodic table famous.

No one in the scientific community at the time could explain why Mendeleyev's predictions were correct — no acceptable theory of periodicity was proposed until the early 1900s. This mystery must have made the accuracy of his predictions even more astounding.

Figure 2.8
Dmitri Ivanovich Mendeleyev (1834 – 1907) was born in Siberia, the youngest of 17 children. After becoming a chemistry professor, he explored a wide range of interests, including natural resources such as coal and oil, meteorology, and hot air balloons. His work demanded tremendous patience and an extremely methodical approach. Imagine collecting all available information on all the elements, and then searching for patterns that no one else had noticed.

Over the years since Mendeleyev's original periodic table, more than one hundred versions of periodic tables have been suggested.

GROUP	I	II	III	IV	V	VI	VII	VIII
Formula of Compounds	R_2O	RO	R_2O_3	RO_2 H_4R	R_2O_5 H_3R	RO_3 H_2R	R_2O_7 HR	RO_4
1	H(1)							
2	Li(7)	Be(9.4)	B(11)	C(12)	N(14)	O(16)	F(19)	
3	Na(23)	Mg(24)	Al(27.3)	Si(28)	P(31)	S(32)	Cl(35.5)	
4	K(39)	Ca(40)	–(44)	Ti(48)	V(51)	Cr(52)	Mn(55)	Fe(56), Co(59) Ni(59), Cu(63)
5	[Cu(63)]	Zn(65)	–(68)	–(72)	As(75)	Se(78)	Br(80)	
6	Rb(85)	Sr(87)	?Yt(88)	Zr(90)	Nb(94)	Mo(96)	–(100)	Ru(104), Rh(104) Pd(105), Ag(108)
7	[Ag(108)]	Cd(112)	In(113)	Sn(118)	Sb(122)	Te(125)	I(127)	
8	Cs(133)	Ba(137)	?Di(138)	?Ce(140)	——	——	——	
9	——	——	——	——	——	——	——	
10	——	——	?Er(178)	?La(180)	Ta(182)	W(184)	——	Os(195), Ir(197) Pt(198), Au(199)
11	[Au(199)]	Hg(200)	Tl(204)	Pb(207)	Bi(208)	——	——	
12	——	——	——	Th(231)	——	U(240)		

Figure 2.9
Mendeleyev's periodic table, 1872. Later, scientists rearranged the purple boxes to form the middle section of the modern periodic table. (For the formulas shown, "R" is used as the symbol of any atom in that family of elements.)

Table 2.7

GERMANIUM FULFILLS THE PREDICTIONS FOR EKA-SILICON		
Property	**Predicted for Eka-silicon (1871)**	**Observed for Germanium (1887)**
atomic mass	72 (average of Si and Sn)	72.5
specific gravity	5.5 (average of Si and Sn)	5.35
reaction with water	none (based on none for Si and Sn)	none
reaction with acids	slight (based on Si — none; Sn — rapid)	none
oxide formula	XO_2 (based on SiO_2 and SnO_2)	GeO_2
oxide specific gravity	4.6 (average of SiO_2 and SnO_2)	4.1
chloride formula	XCl_4 (based on $SiCl_4$ and $SnCl_4$)	$GeCl_4$
chloride boiling point	86°C (average of $SiCl_4$ and $SnCl_4$)	83°C

Exercise

17. State the periodic law.

18. Does the periodic law explain the relationship between atomic mass and chemical properties? What is needed to provide an explanation?

19. Sulfur is a yellow solid and oxygen is colorless gas, yet both elements are placed in the same column or family. What properties might have led Mendeleyev to place them in the same family?

20. In the 1890s, an entirely new family of elements was discovered. This family consisted entirely of unreactive gases called noble gases. Did this discovery support Mendeleyev's periodic table? Describe briefly.

21. In 1817 a German chemist, Johann Doebereiner, identified several groups of three elements which had similar chemical formulas. In these groups, called triads, the atomic mass of the middle element was equal to the average mass of the other two elements. For example, the mass of sodium (23) is the average of the atomic masses of lithium (7) and potassium (39). Based on Doebereiner's concept, fluorine, chlorine, and bromine should be a triad.

    (a) The atomic mass of fluorine is 19 and of bromine is 80. According to Doebereiner's concept, predict the atomic mass of chlorine.

    (b) Empirically, the atomic mass of chlorine was found to be 35.5. Evaluate the prediction from part (a) and evaluate the concept of triads.

22. (Discussion) Science is considered by many people to be completely objective. However, the history of science shows that this is not always the case. Illustrate the social and personal dimensions of science using Newlands and Meyer as examples.

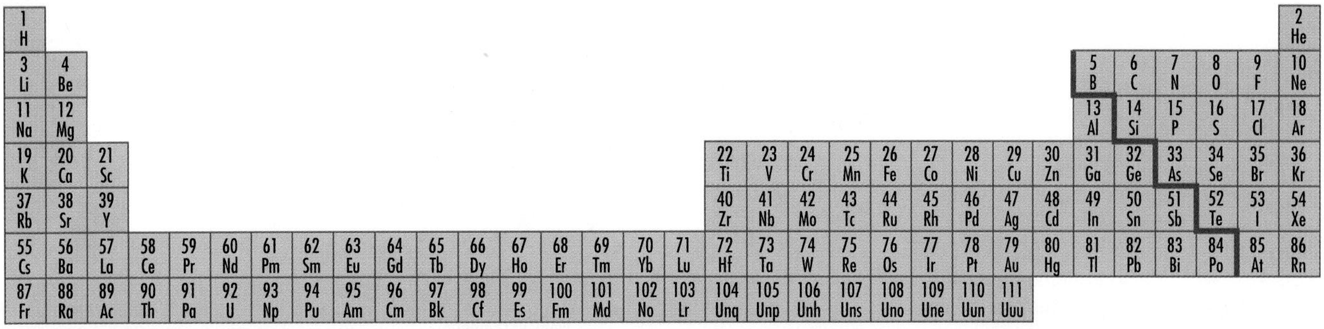

The Modern Periodic Table

Figure 2.10 shows the modern periodic table. In this table, every element is in sequence, but the shape of the table makes it difficult to print on a page and still include useful descriptions of each element. The periodic table is usually printed in the form shown in Figure 2.11, with two separate rows at the bottom. Note the important features of this table.

- A **family** or **group** of elements has similar chemical properties and includes the elements in a vertical column in the main part of the table.

- A **period** is a horizontal row of elements whose properties gradually change from metallic to nonmetallic from left to right along the row.

- Metals are located to the left of the "staircase line" in the periodic table, and nonmetals to the right.

Compare the periodic table in Figure 2.11 to the one on the inside front cover of this book. Periodic tables usually include each element's symbol, atomic number, and atomic mass, along with other

Figure 2.10
Because of its inconvenient shape, this extended form of the periodic table is rarely used.

Figure 2.11
In the modern form of the periodic table, the eight groups of elements in Mendeleyev's table have been split into groups A and B. Because this grouping system is not used consistently throughout the world, the IUPAC has recommended replacing it with an international numbering system. Since 1984, the groups have been numbered from 1 to 18. Periods (horizontal rows) are numbered from 1 to 7.

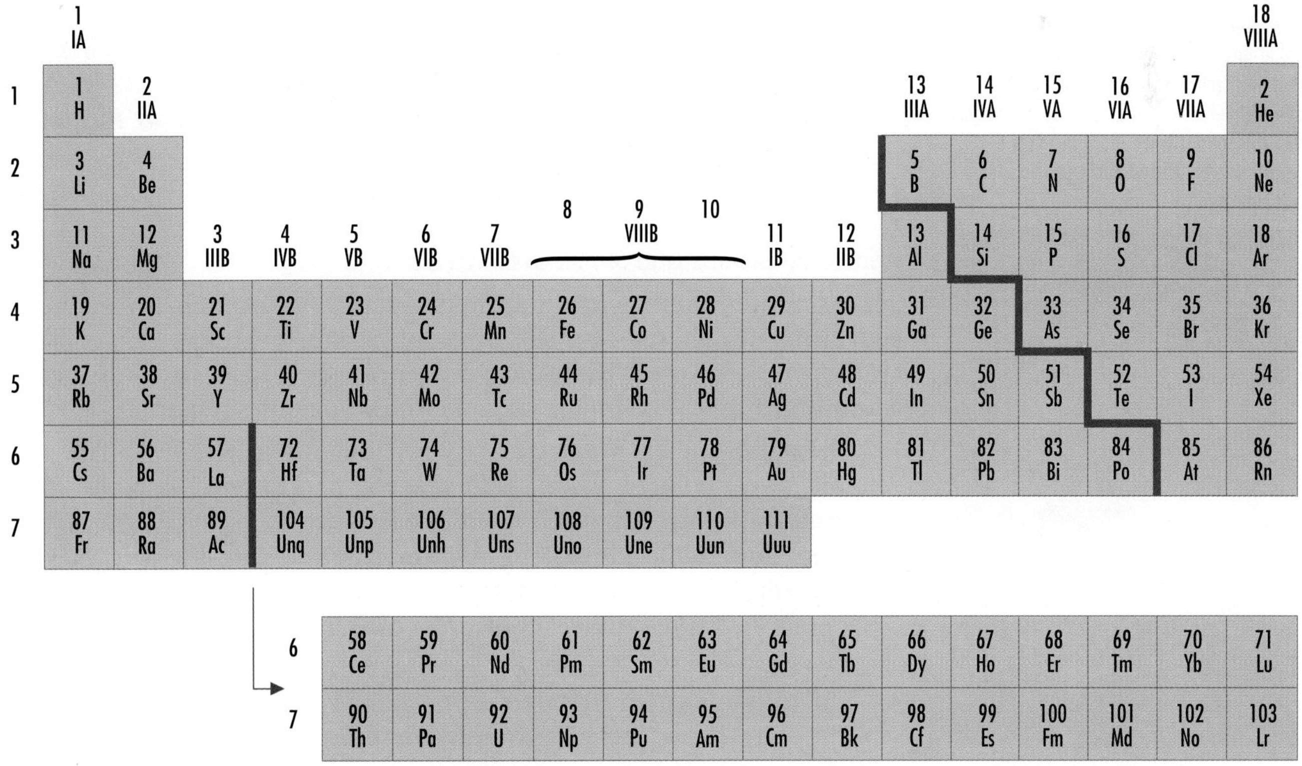

Key

(theoretical)			(empirical)
atomic number	**26**	55.85	atomic molar mass (g/mol)
electronegativity	1.8	1535	melting point (°C)
common ion charge	3+	2750	boiling point (°C)
other ion charge	2+	7.87	density of solids, liquids (g/cm³)
	Fe		density of gases (g/L)
	iron		gases in red
			liquids in green
			synthetic in blue

element symbol
element name

Figure 2.12
This key, which also appears on this book's inside front cover, helps you to determine the meaning of the numbers in the periodic table.

Atomic number is defined empirically (for now) as the number of the place occupied by the element in the periodic table.

Noble Gas Compounds

In 1962, Canadian chemist Neil Bartlett synthesized the first noble gas compound while working at the University of British Columbia. The yellow crystalline compound with the formula $XePtF_6$ shattered the assumption that noble gases could not react with other elements. Before this discovery, the noble gases were called "inert gases" because they were thought to be non-reactive. Bartlett's compound generated enormous interest in noble gas reactions, once scientists realized that such reactions were possible. One year after Bartlett's synthesis, so many other noble gas compounds were produced that a 400-page book on known noble gas compounds was published.

information that varies from table to table. The periodic table on the inside front cover features a box of data for each of the elements, and a key explaining the information in each box. The key is also shown in Figure 2.12. Note that theoretical data are listed in the column on the left, and empirically determined data are listed on the right.

Practice locating the following information in the table.

• English name for each element

• international symbol for each element

• atomic number

• atomic mass (**Atomic mass** is currently defined relative to the mass of a carbon atom, which is assigned a value of 12 atomic mass units. An **atomic mass unit** is defined as 1/12 of the mass of a carbon atom.)

• physical state (solid, liquid, or gas) of each element at SATP (Elements that are gases at SATP have red symbols. Mercury and bromine, the only elements that are liquids at SATP, have green symbols. All other elements are solids at SATP.)

• group number appearing at the top of each column (Two numbering systems appear in your periodic table: the IUPAC numbers go from 1 to 18, and the American group numbers are Roman numerals followed by the letters "A" or "B.")

• period number appearing beside each horizontal row

Names of Groups and Series of Elements

Some families of elements and the two series of elements (those in the two horizontal rows at the bottom of the periodic table) have traditional names that are commonly used in scientific communication. It is important to learn these names (Figure 2.13).

• The **alkali metals** are the family of elements in Group 1. They are soft, silver-colored metals that react violently with water to form basic solutions. All alkali metals react with halogens (Group 17) to form compounds similar to sodium chloride, or table salt, $NaCl_{(s)}$. The most reactive alkali metals are cesium and francium. All alkali metals are stored under oil or in a vacuum to prevent reaction with air.

• The **alkaline-earth metals** are the family of elements in Group 2. They are light, reactive metals that form oxide coatings when

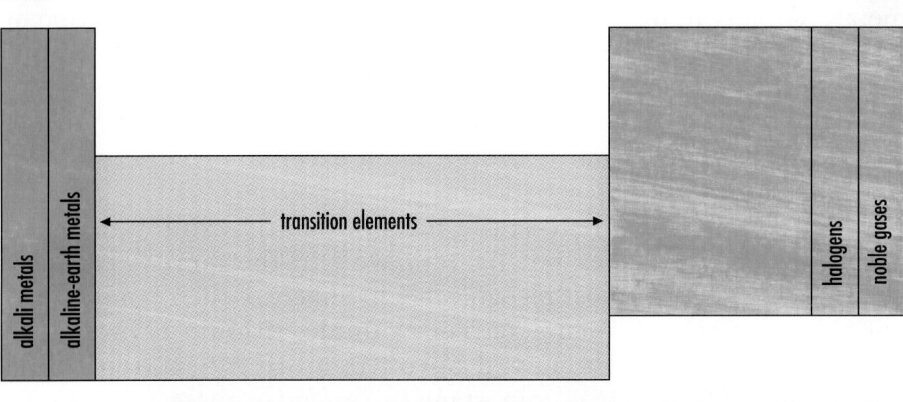

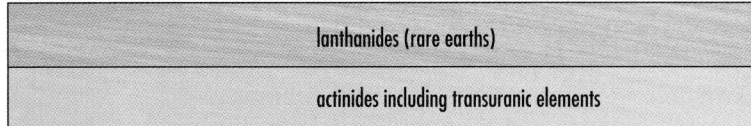

exposed to air. All alkaline-earth metals react with oxygen to form oxides with the general chemical formula, $MO_{(s)}$, such as $MgO_{(s)}$ and $CaO_{(s)}$.

- The **halogens** are the elements in Group 17. They are all extremely reactive, with fluorine being the most reactive. All halogens form hydrogen halides, $HX_{(g)}$, such as $HI_{(g)}$ and $HCl_{(g)}$. Water solutions of hydrogen halides are all acidic.

- The **noble gases** are the elements in Group 18. They are special because of their extremely low chemical reactivity. The noble gases are of special empirical and theoretical interest to chemists.

- The **representative elements** are the elements in Groups 1, 2, and 13 to 18. Of all the elements, the representative elements best follow the periodic law. For the sake of simplicity, the laws and theories presented in introductory chemistry courses are often restricted to these elements.

- The **transition elements** are the elements in Groups 3 to 12 (originally labelled the "B" groups). These elements exhibit a wide range of chemical and physical properties.

In addition to the common classes of elements described above, the bottom two rows in the periodic table also have common names. The *lanthanides* (rare-earth elements) are the elements with atomic numbers 58 to 71. The *actinides* are the elements with atomic numbers 90 to 103. The synthetic (not naturally occurring) elements that have atomic numbers of 93 or greater are referred to as *transuranic elements* (beyond uranium).

Limitations of the Periodic Table

The periodic table was a significant advance and is still an essential part of the study of chemistry. However, there are limitations to the periodic table. Many properties such as melting points, boiling points, and atomic size do not vary smoothly within a family or regularly

Ida Noddack, Discoverer of Rhenium

Mendeleyev's periodic table (Figure 2.9, page 65) shows several gaps in Group VII. Ida Noddack and her husband, Walter Karl Friedrich, made detailed predictions about the properties of undiscovered elements in this group and then searched for these elements. In 1925, they isolated the first gram of rhenium from 650 kg of ore, and conducted extensive studies on the chemistry of this new metal, which is platinum-white, very hard, and stable in air below 600°C.

Less well-known among Ida Noddack's scientific achievements is the initial concept of nuclear fission. She predicted that atoms could be broken apart, but her idea conflicted with theories of atomic structure at that time, so it was ignored for several years.

across the periods of the table. The representative metals (Groups 1 and 2) are more reactive near the bottom of the family, and some transition metals (Groups 10 to 12) are more reactive near the top. The position of hydrogen, the lightest element, is a problem. Hydrogen sometimes behaves like a member of Group 1 (alkali metals), sometimes like Group 17 (halogens), or in its own unique way (acids). It is probably more accurate to say that *hydrogen is an exception to almost every rule in chemistry.*

Generally, the element at the top of a family exhibits unusual behavior and often does not fit very well in the family. Period 2 elements, usually the most recognizable members of their families, are not the most typical members. For example, lithium is more like magnesium than sodium, and boron is more like silicon than aluminum. Although it is important to recognize these limitations, scientists generally agree that the evidence supporting the general principles of the periodic table and its usefulness in chemistry far outweigh the limitations described above.

Problem
✔ Prediction
Design
Materials
Procedure
✔ Evidence
✔ Analysis
✔ Evaluation
Synthesis

INVESTIGATION

2.3 Demonstration: Alkali Metals

The purpose of this demonstration is to test the concept of a chemical family.

Problem

What are the physical and chemical properties of lithium, sodium, and potassium?

Experimental Design

Samples of lithium, sodium, and potassium are observed and tested for mechanical properties and electrical conductivity. A small piece of each element is placed in water and the reactivity observed.

Materials

lab apron	conductivity apparatus
safety glasses or face shield	large beaker
safety shield	distilled water
tongs	wire gauze square
tweezers	lithium
mineral oil	sodium
sharp edge (knife or scoop)	potassium
petri dish	

Procedure

1. Transfer a small piece of lithium into a petri dish containing mineral oil.
2. Using a knife, remove the surface layer from one side and observe the fresh surface.
3. Test the metal for electrical conductivity.
4. Fill the beaker about one-half full with water.
5. Add a small piece (about 2 mm size) of the element into the beaker and cover immediately with the wire gauze.
6. Record observations of any chemical reaction.
7. Repeat steps 1 to 6 using sodium and potassium.

Exercise

23. Among elements, how does the number of metals compare with the number of nonmetals?

24. Describe the positions of the representative elements in the periodic table.

25. List two physical and two chemical properties of the alkali metals.

26. How does the reactivity vary within the alkali metal family compared with the halogen family?

27. Nitrogen and hydrogen form a well-known compound, $NH_{3(g)}$, ammonia. According to the position of phosphorus in the periodic table, predict the most likely chemical formula for a compound of phosphorus and hydrogen.

28. List three limitations of the periodic table.

29. Canada is rich in mineral deposits containing a variety of elements. Table 2.8 lists a few examples from across Canada. In your notebook, complete the element information for each example.

30. (Discussion) List some positive and negative aspects of resource development in Canada. Identify the STS perspective for each statement.

Table 2.8

ELEMENTS AND MINERAL RESOURCES						
Mineral Resource or Use	Element Name	Atomic Number	Element Symbol	Group Number	Period Number	SATP State
(a) high quality ores at Great Bear Lake, NT	radium					
(b) rich ore deposits at Bernic Lake, MB				1	6	
(c) potash deposits in Saskatchewan		19				
(d) large deposits in New Brunswick	antimony					
(e) extracted from Alberta sour natural gas			S			
(f) radiation source for cancer treatment				9	4	
(g) large ore deposits in Nova Scotia	barium					
(h) world scale production in Sudbury, ON		28				
(i) fuel in CANDU nuclear reactors			U			
(j) fluorspar deposits in Newfoundland				17	2	
(k) large smelter in Trail, BC		30				

Elements

Elements have many uses.

hydrogen – fuel
helium – balloons
lithium – batteries
boron – some steel
carbon – pencil lead
nitrogen – air
oxygen – respiration
neon – signs
sodium – photoelectric cells
magnesium – wheels
aluminum – bats, hockey sticks
silicon – semiconductors
phosphorus – fluorescence
sulfur – matches
chlorine – disinfectant
argon – fluorescent tubes
potassium – heat transfer medium
calcium – alloyed in flints
scandium – (none listed)
titanium – some steel
vanadium – some steel
chromium – shiny plating
manganese – several alloys
iron – all steel
cobalt – radiation treatment
nickel – coins
copper – wire and pipe
zinc – galvanizing
gallium – (none listed)
germanium – semiconductors
arsenic – rat poison
selenium – pigment in glass
bromine – for gold extraction
krypton – Superman deterrent
silver – jewellery
tin – plating cans
iodine – disinfectant
tungsten – lightbulb filament
gold – jewellery
mercury – fillings
lead – solder
uranium – reactor fuel
plutonium – bomb

Figure 2.14
"No amount of experimentation can ever prove me right; a single experiment can prove me wrong." This statement illustrates Albert Einstein's view of the nature of science.

Empirical knowledge, the sum of all observations, is the foundation for ideas in science. Usually, experimentation comes first and theoretical understanding follows. For example, the properties of some elements were known for thousands of years before a theoretical explanation was available. Chemical formulas were determined about 100 years before they could be explained. Mendeleyev's periodic law, which was based solely on observations, was about 40 years in advance of any theoretical explanation. This is a common occurrence; scientific laws are usually stated before a theory is developed to explain observations.

So far in this chapter, you have encountered only empirical knowledge of elements, based on what has been observed. Why do the properties of elements vary across the periodic table? Why are families of elements similar in their physical and chemical properties? How do we explain the chemical formulas of compounds formed from elements? To answer these and other questions about elements requires some ideas about what makes up elements.

Curiosity leads scientists to try to explain nature in terms of what cannot be observed. This step — formulating ideas to explain observations — is the essence of theoretical knowledge in science. Albert Einstein (Figure 2.14) referred to theoretical knowledge as "free creations of the human mind."

Scientists communicate theoretical knowledge in several ways:

- *Theoretical descriptions* are specific descriptive statements based on theories or models. For example, "a molecule of water is composed of two hydrogen atoms and one oxygen atom."

- **Theoretical hypotheses** are ideas that are untested or extremely tentative. For example, "protons are composed of quarks that may themselves be composed of smaller particles."

- **Theoretical definitions** are general statements that characterize the nature of a substance or a process in terms of a non-observable idea. For example, a solid is theoretically defined as "a closely packed arrangement of atoms, each atom vibrating about a fixed location in the substance."

- **Theories** are comprehensive sets of ideas based on general principles that explain a large number of observations. For example, the idea that materials are composed of atoms is one of the principles of atomic theory; atomic theory explains many of the properties of materials. Theories are dynamic; they continually undergo refinement and change.

- *Analogies* are comparisons that communicate an idea in more familiar or recognizable terms. For example, an atom may be conceived as behaving like a billiard ball. All analogies "break down" at some level; that is, they have limited usefulness.

- *Models* are diagrams or apparatuses used to simplify the description of an abstract idea. For example, marbles in a vibrating box could be used to study and explain the three states of matter. Like analogies, models are always limited in their application.

"I do not find that anyone has doubted that there are four elements. The highest of these is supposed to be fire, and hence proceed the eyes of so many glittering stars. The next is that spirit, which both the Greeks and ourselves call by the same name, air. It is by the force of this vital principle, pervading all things and mingling with all, that the earth, together with the fourth element, water, is balanced in the middle of space."
— Pliny the Elder (Gaius Plinius Secundus), naturalist and historian (A.D. 23 – 79)

Theories that are acceptable to the scientific community must *describe* observations in terms of non-observable ideas, *explain* observations by means of ideas, *predict* results in future experiments that have not yet been tried, and be as *simple* as possible in concept and application.

Early Greek Theories of Matter

Greek philosophers first proposed an atomic theory of matter in the 5th century B.C. They believed that all substances were composed of small, indivisible particles called *atoms* (from the Greek word for "uncuttable"). Atoms were conceived to be of different sizes, to have regular geometric shapes, and to be in constant motion. Empty space was thought to exist between atoms. Aristotle (384 – 322 B.C.) severely criticized this theory, arguing that atoms in continuous motion in a void is an illogical idea. At that time, belief in a "void" was considered to be atheistic; this probably contributed to his opinion.

Aristotle developed a theory of matter based on the idea that all matter is made up of four basic substances — earth, air, fire, and water. He believed that each basic substance had different combinations of four specific qualities — dry, hot, cold, and moist (Figure 2.15). Aristotle's theory of the structure of matter was the prevailing model for almost 2000 years, including the period of alchemy in the Middle Ages. The demise of Aristotle's model followed the scientific revolution in physics and the new emphasis, in the 18th century, on quantitative measurements. Too many of the predictions and explanations using Aristotle's theory were shown to be false. This led to the revival of the atom concept.

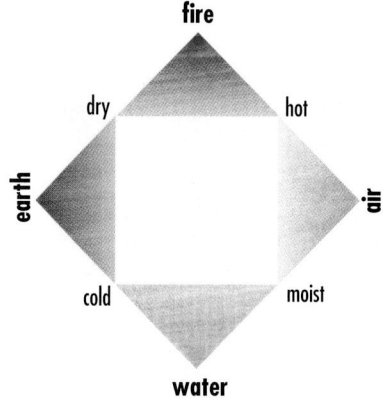

Figure 2.15
In Aristotle's model of matter, each basic substance, or element, possesses two of four essential qualities. For example, earth was dry and cold, but fire was dry and hot. This model was based on logical thinking, but not on experimentation.

Dalton's Atomic Theory

By the beginning of the 19th century, after decades of experimentation, quantitative relationships among substances had been discovered in the laboratory. These relationships appeared to hold true for all chemical reactions and became scientific laws.

- The *law of conservation of mass* states that in any physical or chemical change, the total mass remains constant. For example, 24 g of magnesium reacts with 16 g of oxygen to form 40 g of product.

- The *law of definite composition* states that elements combine in definite proportions by mass. For example, hydrogen and oxygen always react in the same proportion to produce water — 2 g of hydrogen to 16 g of oxygen, or, in other words, in a ratio of 1:8.

- The *law of multiple proportions* states that when a fixed mass of one element combines with a second element to form two different compounds, the masses of the second element will form a simple whole number ratio. For example, 16 g of oxygen may react with either 6 g or 12 g of carbon to form two very different compounds. The two masses of carbon correspond to a 1:2 ratio.

John Dalton, the English scientist who devised a complex system of symbols for the elements, worked out many chemical formulas empirically by conducting experiments on the properties of gases. To

Antoine Lavoisier
Antoine Lavoisier, a French chemist, developed the law of conservation of mass by reacting chemicals in a closed, glass vessel. He was known to use concentrated solar energy as a heat source. He measured the mass of the vessel plus contents before and after the reaction. Lavoisier wrote the first modern chemistry textbook. Because of his association with his tax-collecting father-in-law, Lavoisier was guillotined during the French Revolution.

Figure 2.16
In Dalton's model, an atom is a solid sphere, similar to a billiard ball. This simple model is still used today to represent the arrangement of atoms in molecules.

explain his experimental results, as well as the three laws stated above, he introduced an atomic theory of matter in 1803 (Figure 2.16). This theory came to replace Aristotle's model of matter. Dalton's theory states:

- All matter is composed of tiny, indivisible particles called atoms.
- Atoms of an element have identical properties.
- Atoms of different elements have different properties.
- Atoms of two or more elements can combine in constant ratios to form new substances.

Dalton's theory was very successful in explaining the laws of conservation of mass, definite composition, and multiple proportions. Since atoms are indivisible, and are rearranged only when compounds are formed, you must end up with the same number and kinds of atoms after a chemical reaction. This explains why mass is conserved and is the basis for the balancing of chemical reaction equations discussed in Chapter 4. Since the atoms of an element have identical properties, such as mass, and combine in constant ratios, a definite composition is required. Futhermore, a reacting mass ratio could be interpreted as a relative atomic mass ratio (Table 2.6, page 63) which was required for the development of the periodic law.

Development of Atomic Theory from 1803 to the 1920s

By the late 1800s, several experimental results conflicted with Dalton's atomic theory. As one example, English physicist J. J. Thomson passed electricity through gases in vacuum tubes and found evidence for the existence of negatively charged particles that could be removed from atoms. In 1897, he postulated the existence of **electrons**, subatomic particles possessing a negative charge. With this new idea, Thomson developed a model of the atom that has electrons evenly distributed inside the spherical positive part of the atom (Figure 2.17(a)). In 1904, Japanese scientist H. Nagaoka represented the atom as a large, positively charged sphere surrounded by a ring of negative electrons. This model is shown in Figure 2.17(b). Until 1911, there was no evidence to contradict either of these models.

Electron Charge
Robert Millikan attended a conference in Winnipeg, MB in August, 1909, where he presented a paper to, among others, J. J. Thomson and Ernest Rutherford. The scientific paper described his attempts at determining the charge on the electron by determining the charge on water droplets. On his way home by train to Chicago, Millikan created the experimental design using oil droplets for which he won the Nobel Prize in physics in 1923.

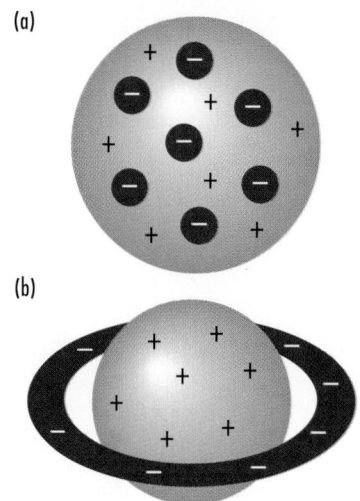

Figure 2.17
(a) In Thomson's model, the atom is a positive sphere with embedded electrons. This can be compared to a raisin bun in which the raisins represent the negative electrons and the bun represents the region of positive charge.
(b) In Nagaoka's model, the atom is compared to the planet Saturn, where the planet represents the positively charged part of the atom, and the rings represent the negatively charged electrons.

From 1898 to 1907, New Zealand-born physicist Ernest Rutherford worked at McGill University in Montreal. Designed to test the current atomic models, his experiments involved shooting alpha particles (small positively charged particles produced by radioactive decay) through very thin pieces of gold foil. Based on J. J. Thomson's model of the atom, Rutherford predicted that all the alpha particles would travel through the foil largely unaffected by the atoms of gold. Although most of the alpha particles did pass easily through the foil, a small percentage of particles was deflected through large angles, as shown in Figure 2.18. Rutherford deduced that an atom must contain a tiny, positively charged core, the **nucleus**, which is surrounded by a mostly empty space containing negative electrons (Figure 2.19, page 76). The nucleus is relatively massive compared with the electrons. From the percentage of alpha particles that was deflected and the deflection angles, Rutherford calculated that the nucleus is only about one ten-thousandth of the total size of the atom (Figure 2.21, page 76).

In 1914, Rutherford coined the word "proton" for the smallest unit of positive charge in the nucleus. **Protons** are subatomic particles with a positive charge. Empirical support for the existence of protons came from one of Rutherford's students, H. G. J. Moseley. His X-ray experiments showed that the positive charge in the nucleus of atoms increases by one unit in progressing from each element to the next in Mendeleyev's periodic table. This discovery led Moseley to the concept of **atomic number**, defined theoretically as the number of protons contained in the nucleus of an atom. Moseley was the first to recognize this relationship between atomic number and nuclear charge. This provided a new insight into the periodic table. Originally, elements were listed and numbered in order of increasing atomic mass. This sequence now had an explanation — a list of elements in order of the number of protons in the nucleus.

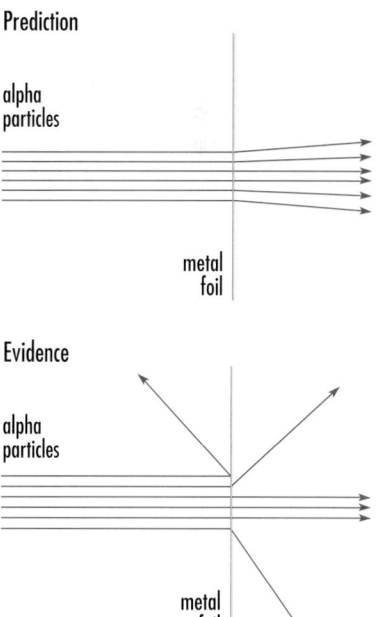

Prediction

alpha particles

metal foil

Evidence

alpha particles

metal foil

Figure 2.18
Rutherford's experimental observations were dramatically different from what he had expected.

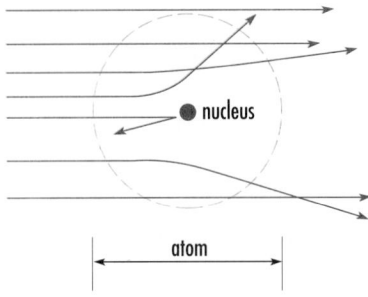

Figure 2.19
To explain his results, Rutherford suggested that an atom consisted mostly of empty space, and that most of the alpha particles passed nearly straight through the gold foil because these particles did not pass close to a nucleus.

Table 2.9

MASSES AND CHARGES OF SUBATOMIC PARTICLES		
Particle	Relative Mass	Relative Charge
electron	1	1−
proton	1836.12	1+
neutron	1838.65	0

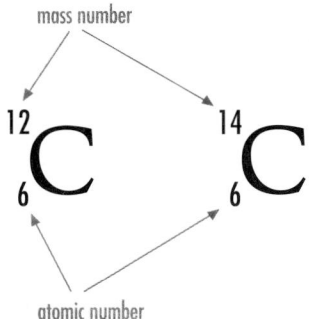

Figure 2.20
Two isotopes of carbon. Carbon-12 is stable, but carbon-14 is radioactive. Carbon-14 is used for carbon dating of old artifacts.

Figure 2.21
This analogy illustrates Rutherford's calculations. If the nucleus of an atom were the size of an ant, the atom would be the size of a football field.

According to Rutherford's model, most of the mass of the atom consisted of positively charged protons in the nucleus. However, the estimated mass of a proton and the number of protons (atomic number) could not account for all of the atomic mass. To explain this discrepancy, Rutherford predicted the existence of a neutral particle similar in mass to a proton. In 1932, James Chadwick demonstrated that atomic nuclei must contain heavy neutral particles as well as positive particles, in order to account for the atom's entire mass. These neutral subatomic particles were called **neutrons**. Most chemical and physical properties of elements can be explained in terms of these three subatomic particles — electrons, protons, and neutrons (Table 2.9). An **atom** is composed of a nucleus containing protons and neutrons and a number of electrons equal to the number of protons; an atom is electrically neutral.

Frederick Soddy, a colleague of Rutherford's at McGill University, was the first to propose that the number of neutrons can vary from atom to atom, even in atoms of the same element. An **isotope** is a form of an element in which the atoms have the same number of protons as all other forms of that element, but a different number of neutrons. For example, carbon atoms (atomic number six) all have six protons in the nucleus. The most common form of carbon, carbon-12, also has six neutrons. Carbon-14 is an isotope of carbon because it has six protons and eight neutrons in the nucleus (Figure 2.20). Different isotopes of the same element have the same chemical properties, but different masses. All elements exist naturally as a mixture of isotopes. The term **mass number**, theoretically defined as the sum of the number of protons and neutrons in an atom, can be used to describe an isotope. For example, the mass number of the most common isotope of carbon, which contains six protons and six neutrons, is 12. This isotope is therefore referred to as carbon-12. This is the basis for the definition of atomic mass unit (page 68).

Rutherford's model of the atom raised some thorny questions. For example, scientists could not understand why the nucleus did not break apart because of the mutual repulsion of the positive protons. Also, they could not explain why atoms did not collapse because of the attraction of negative electrons and positive protons. In response to the first question, Rutherford suggested the idea of a nuclear force — an

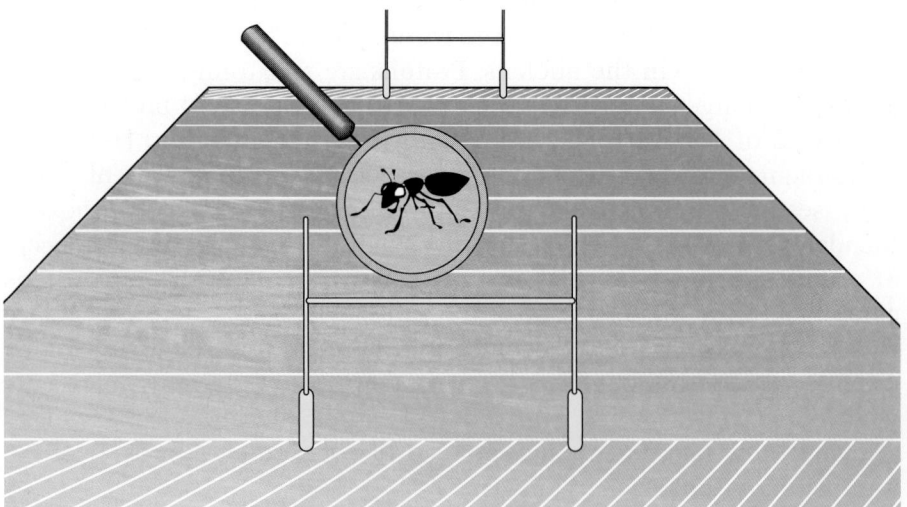

attractive force within the nucleus that was much larger than any electrostatic force of repulsion. The answer to the second question — concerning why the electrons did not fall in to be "captured" by the nucleus — required a bold and creative approach pioneered by a young Danish physicist named Niels Bohr.

RADIOACTIVITY AND RADIATION

The accidental discovery in 1896 of some unknown rays that could penetrate thick layers of paper had a major influence on the progress of science, medicine, and technology. This newly discovered phenomenon, called radioactivity, is the spontaneous emission of radiation by certain substances. At McGill University, Montreal, Ernest Rutherford and his associates determined that radioactivity was the disintegration or decay of certain unstable elements, which transformed them into other elements and released radiation.

> unstable element →
> more stable element + radiation

Rutherford classified this nuclear radiation into three classes: alpha (α), beta (β), and gamma (γ). He received the Nobel Prize for this Canadian work in 1908.

According to modern atomic theory, alpha particles are helium nuclei, beta particles are electrons, and gamma rays are similar in nature to light. A cloud chamber saturated with alcohol vapor (see photo in the next column) shows the actual paths of individual nuclei and nuclear radiation.

The subsequent discovery of neutrons and isotopes led to the realization that only some isotopes of certain elements are unstable. The term **radioisotope** means a radioactive isotope of an element. Nuclear radiation from radioisotopes has many varied applications. Only a few of these are listed in the next column.

- Iridium-192, which emits gamma radiation, is extensively used for

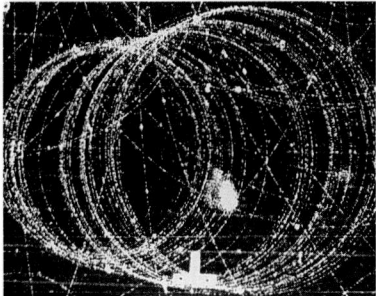

the examination of welds in pipelines and vessels.

- Radiation from certain radioisotopes, such as hydrogen-3, is used with phosphors (phosphorus compounds) to produce luminous dials on watches, gauges, and switches. The radiation given off by the radioisotope causes the phosphors to glow.
- Radiation is routinely used to preserve foods and prevent sprouting of certain vegetables like potatoes. It also kills bacteria that could cause spoilage. This irradiation is very effective and does not make the food radioactive.
- A variety of radioisotope tracers are used in medical diagnosis. A common example is technetium-99 which is injected and used to detect suspected tumors in organs such as the liver, spleen, kidneys, and lungs.
- Radioisotopes are also widely used in the treatment of cancerous tumors. Canadian scientists and engineers pioneered the use of gamma radiation from cobalt-60 as a radiation therapy for cancer (see photo in the next column).

The ability of radiation to destroy living cells explains its functions in food preservation and in cancer

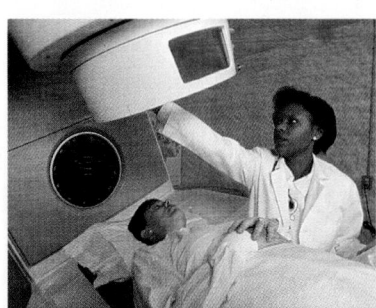

treatment. However, this same ability is responsible for the dangers of radiation. People who work with radiation sources or receive radiation need as much protection as is practical. This includes shielding the body with materials such as lead, minimizing time spent exposed to radiation, and maximizing the distance from any radiation sources. However, no one on Earth can escape low levels of radiation exposure because radiation occurs naturally in our environment. The major source is the cosmic radiation that reaches Earth from outer space. Other natural sources include the soil, food, and building supplies. It is not surprising that the human body is also radioactive, corresponding to about five hundred thousand nuclear decays per minute in an adult. We add to our natural radiation exposure by medical X-rays, cigarette smoke, and airplane trips. Radiation is unavoidable. It adds immensely to our standard of living but also has many risks. We must maintain the delicate balance between the benefits and risks of radiation.

- *List two radioisotopes from the article and calculate the number of neutrons in the nucleus of each.*
- *Describe one benefit and one risk of radiation.*

CHARACTERISTICS OF NUCLEAR RADIATIONS			
Radiation	Approximate Speed	Penetration in Air	Effective Barrier
alpha (α, He^{2+})	variable but relatively slow	a few centimetres	a sheet of paper
beta (β, e$^-$)	variable but relatively fast	a few metres	1–2 mm of metal
gamma (γ)	very fast (speed of light)	unlimited	1 m of lead or concrete

Harriet Brooks (1876–1933)
Ontario-born Harriet Brooks was Rutherford's first graduate student at McGill University, Montreal. She was also the first woman at McGill to receive a master's degree in physics. She discovered radon as a radioactive by-product of radium, and her evidence led to the theory of transmutation proposed by Rutherford and Soddy. She also gathered evidence that Rutherford was first to interpret theoretically as a series of transformations.

Brooks may have been the only scientist to have worked in the laboratories of Ernest Rutherford, J. J. Thomson, and Marie Curie. Rutherford described Brooks as a brilliant empirical scientist. Rutherford often provided the theoretical interpretations of Brooks's work. (Read about Harriet Brooks in the biography bearing her name.)

- Problem
- Prediction
- Design
- Materials
- Procedure
- ✔ Evidence
- ✔ Analysis
- Evaluation
- Synthesis

CAUTION

The light voltage power supply should be operated by your teacher only. Stand away from the gas tubes in case dangerous radiation is produced.

INVESTIGATION

2.4 Atomic Spectra

The purpose of this investigation is to study spectra of light produced by different types of atoms. Visible light originates from many sources, mostly associated with changes in matter. Everyone is familiar with sunlight and has seen the separation of this light into the colors of a rainbow. This is an example of a continuous spectrum; that is, a continuous arrangement of colors. Many light sources, such as an ordinary tungsten light bulb, produce light similar to sunlight. However, gaseous atoms produce light with different characteristics. The light produced by exciting atoms using electricity provided unexpected and important evidence which led to the development of modern atomic theory.

Problem

What effect does the element have on the spectrum of light produced?

Experimental Design

The vapors of several elements are excited by passing electricity through each gas. In each case the spectrum is observed with a spectroscope and compared with a white light spectrum. The element used is the manipulated variable and the appearance of the spectrum is the responding variable. The voltage and precision of the spectroscope are controlled, and the light from an ordinary light bulb is used as a control.

Materials

straight filament light bulb (at least 40 W) and receptacle
spectroscope
gas discharge power supply
hydrogen, helium, and neon gas tubes

Procedure

1. Turn on the light bulb and turn off all other lights in the classroom.

2. Using the spectroscope, observe the spectrum of white light. Record your observations as a labelled diagram.

3. Connect the hydrogen tube to the power supply and turn off all other lights.

4. Observe the spectrum (Figure 2.22), and record your observations as a labelled diagram.

5. Repeat steps 3 and 4 using the other gas tubes.

Hydrogen

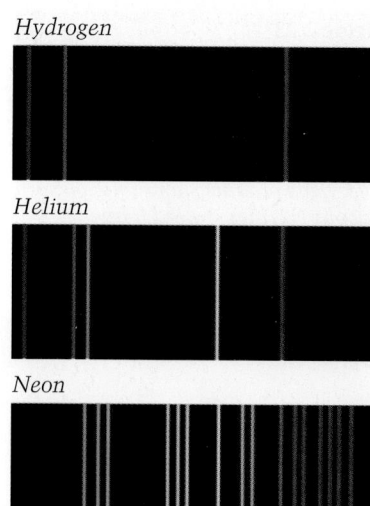

Helium

Neon

Figure 2.22
The line spectra for hydrogen, helium, and neon obtained from a spectroscope. Niels Bohr explained the visible line spectrum of hydrogen by assuming that the energy of electrons is quantized.

Exercise

31. What is the key difference between empirical and theoretical knowledge?

32. List six methods used to communicate theoretical knowledge.

33. What four characteristics of a theory are required in order for the scientific community to accept the theory?

34. What is the difference between a theory and a law?

35. Write a theoretical definition of each of the following terms: atom, electron, nucleus, proton, neutron, isotope, atomic number, and mass number.

36. Distinguish between an isotope and a radioisotope.

37. Cobalt-60 is widely used as a source of gamma rays for industrial and medical applications. Predict the number of protons and neutrons in the nucleus of an atom of cobalt-60.

38. Uranium-235 is the nuclear fuel used in CANDU (Canadian deuterium uranium) nuclear reactors. Predict the number of protons and neutrons in the nucleus of an atom of uranium-235.

39. Calculate the number of neutrons in the radioactive isotopes listed in the margin.

40. (Discussion) The terms "radiation" and "radioactivity" often elicit a strong, negative emotional response from people. Why is it important to have some knowledge about these phenomena?

The following radioisotopes (radioactive isotopes) are produced artificially, for example, within the core of CANDU nuclear reactors. Most of the radioisotopes are used for medical diagnosis or therapy or for industrial or research work.

(a) Ir-192 — analysis of welds
(b) Co-60 — cancer treatment
(c) Tc-99 — monitoring blood flow
(d) I-131 — hyperthyroid treatment
(e) C-14 — archeological dating
(f) Rb-87 — archeological dating
(g) Hg-203 — dialysis monitoring
(h) P-32 — reduces white cells
(i) Cr-51 — tumor treatment
(j) Y-90 — treating Cushing disease
(k) Au-198 — treating Cushing disease
(l) Co-57 — monitoring vitamin B_{12}
(m) Tl-201 — monitoring blood flow
(n) Sr-85 — bone scanning
(o) In-111 — brain tumor scanning
(p) As-74 — brain tumor scanning
(q) Ga-68 — brain tumor scanning
(r) Cu-64 — brain tumor scanning
(s) Se-75 — pancreas tumor scanning

Bohr's Atomic Theory

The genius of Niels Bohr lay in his ability to combine aspects of several theories and atomic models. He created a theory that, for the first time, could explain the periodic law. Bohr saw a relationship between the sudden end of a period in the periodic table and the quantum theory of energy proposed by German physicist Max Planck in 1900 and utilized by Albert Einstein in 1905.

Planck suggested that, just as matter consists of multiples of small units called atoms, energy also consists of multiples of small units. The energy of one packet of energy was later called one *quantum* of energy. Bohr also refined British scientist John Nicholson's idea that electrons in atoms can have only certain specific energies; when electrons orbit around the nucleus, only certain orbits are possible. Experiments with electricity and gases produced spectral evidence that quanta of energy were somehow related to the structure of atoms of

Figure 2.23
When electricity is passed through a gaseous element at low pressure, the gas emits light of only certain wavelengths, which can be seen if the light is passed through a prism. Every gas produces a unique pattern of colored lines, called a line spectrum.

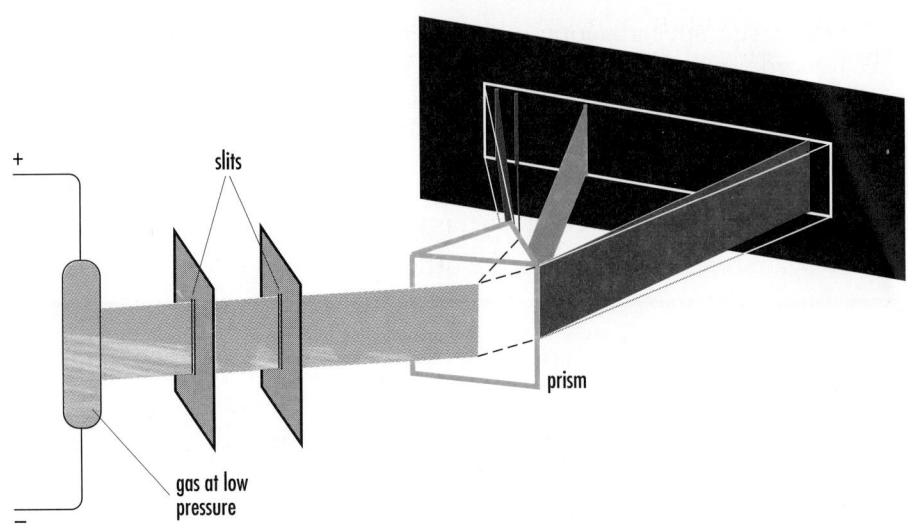

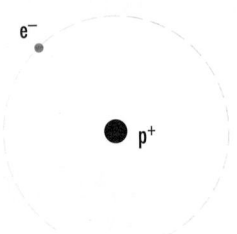

Figure 2.24
The Bohr model of a hydrogen atom in its lowest energy state includes the nucleus (one proton) and a single electron in the first orbit.

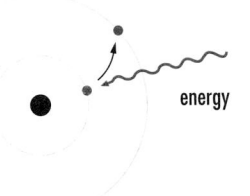

(a) An electron gains a quantum of energy.

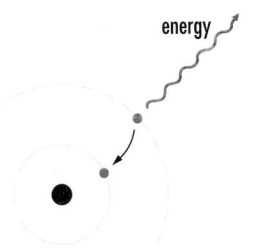

(b) An electron loses a quantum of energy.

Figure 2.25
(a) Energy is absorbed as electrons rise to a higher energy orbit.
(b) Energy is released as electrons fall to a lower energy orbit.

Figure 2.26
The maximum number of electrons in each energy level can be predicted from the number of elements in each period of the periodic table.

different elements (Figure 2.23). Using all of these ideas as well as the periodic law and experimental evidence, Bohr suggested a new theory of atomic structure; here are its basic ideas:

- Each electron has a fixed quantity of energy related to the circular orbit in which the electron is found (Figure 2.24).

- Electrons cannot exist between orbits, but they can move to unfilled orbits if a quantum of energy is absorbed or released (Figure 2.25).

- The higher the energy level of an electron, the further it is from the nucleus.

- The maximum number of electrons in the first three energy levels is 2, 8, and 8 (Figures 2.26 and 2.27).

- An atom with a maximum number of electrons in its outermost energy level is stable; that is, it is unreactive.

Bohr developed his theory mathematically to explain the visible spectrum of hydrogen gas, which is shown in Figure 2.23. He also used the theory to predict the existence of other lines in the ultraviolet and infrared regions of the spectrum, lines which had not yet been observed. Later his predictions were verified. However, even this successful theory would require some changes in order to explain and predict the spectra of larger atoms.

One of the major triumphs of Bohr's theory was its explanation of the periodic law. Bohr suggested that the properties of the elements can be explained by the arrangement of electrons in orbits around the

nucleus. As indicated by the 2, 8, 8 arrangement of elements in the first three periods of the periodic table, orbits may contain only certain numbers of electrons (Figure 2.27). The unreactive nature of the noble gases is explained by the full outer orbits of the atoms. According to Bohr's atomic theory, the reactivity of the halogens is due to the halogen atoms having one electron less than a full outer orbit; the reactivity of the alkali metals is due to these atoms having only one electron in their outer orbits. Similarly, members of other families resemble each other in the arrangement of electrons in their outer orbits. These ideas will be useful in explaining many aspects of chemistry in later chapters.

Bohr was not able to explain why 2, 8, and 8 were "magic numbers" of electrons in the electron orbits. Line spectra dictated the energy levels in his theory, and the periodic law dictated the number of electrons in each energy level. In hindsight, Bohr's atomic theory may seem to be an obvious consequence of the evidence and concepts available to Bohr at the time, but only a well-prepared, creative mind could have put it all together as Bohr did.

At the time, the Bohr model was a tremendous step forward especially in the understanding of the periodic table. The model very successfully explained each colored line in the spectrum of hydrogen, but could not explain all of the lines in the spectra of other elements. The theory was also too simple to account for the relative brightness of spectral lines and some other experimental results. Many scientists, including Bohr, realized that this theory was only an intermediate step. Further refinements would be required to make the theory consistent with the evolving ideas and experiments in the field of quantum physics. As it turned out, Niels Bohr played a significant role in the development of the new and complex theory of quantum mechanics.

Exercise

41. According to the Bohr theory, what is the significance of full, outer electron orbits of atoms? Which chemical family has this unique property?

42. The alkali metals have similar physical and chemical properties. According to the Bohr theory, what theoretical similarity of alkali metal atoms helps to explain their empirical properties?

43. According to the Bohr theory, what happens to an electron in an atom as it absorbs energy and as it releases energy?

44. Based on the periodic table, predict the maximum number of electrons in the fourth through the seventh energy levels.

Period 1		Period 2		Period 3
H 1e⁻		Li	$2e^- + 1e^-$	Na $2e^- + 8e^- + 1e^-$
		Be	$2e^- + 2e^-$	Mg $2e^- + 8e^- + 2e^-$
		B	$2e^- + 3e^-$	
		C	$2e^- + 4e^-$	
		N	$2e^- + 5e^-$	
		O	$2e^- + 6e^-$	
		F	$2e^- + 7e^-$	
He 2e⁻		Ne	$2e^- + 8e^-$	Ar $2e^- + 8e^- + 8e^-$

Figure 2.27
In 1921, Bohr explained the periodicity of chemical properties of elements using the key idea of filled orbits. Filled orbits are shown here in color. This idea remains a key feature of the current atomic theory of quantum mechanics.

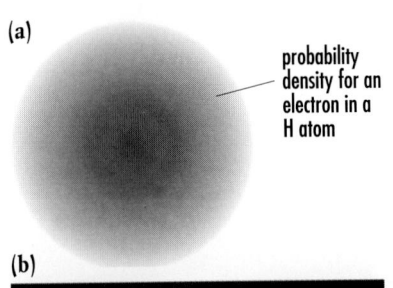

(a)

probability density for an electron in a H atom

(b)

Figure 2.28
(a) According to quantum mechanics, there is a region around the nucleus in which there is a high probability of finding an electron. The motion of the electron is not known.
(b) Similarly, as the blades of a fan rotate rapidly, the position and motion of an individual blade at any instant are unknown.

```
            6e⁻
            8e⁻
            2e⁻
           16p⁺
        sulfur atom
            S
```

Figure 2.29
A diagram of the electron energy levels for a sulfur atom summarizes important theoretical information about this atom.

▌**Examples**
Draw diagrams of the electron energy-level models for hydrogen and silicon atoms.

```
                         4e⁻
                         8e⁻
            1e⁻          2e⁻
            1p⁺         14p⁺
        hydrogen atom  silicon atom
            H             Si
```

Quantum Mechanics

The currently accepted theory of atomic structure, developed in the 1920s, is the theory of **quantum mechanics**. It retains two key ideas from previous atomic theories — the nucleus (Rutherford) and discrete energy levels (Bohr). A major revision in quantum mechanics is the description of the electron and the idea of the existence of numerous sub-levels for the electrons to occupy. According to quantum mechanics, an electron with a specific energy occupies a region of space, called an *orbital* (Figure 2.28). This theory does not provide any information about the path travelled by an electron, and provides only a general description of the region where an electron is likely to be found. In its complete form, quantum mechanics is highly mathematical and has a major drawback — it does not provide an easily communicated model of the atom. However, despite its abstractness, the theory has proven to be the most powerful theory scientists have ever created to explain and predict the properties and reactions of chemicals.

Chemists believe that chemical reactions of elements and many compounds occur by the rearrangement of electrons. In order to understand the chemical properties of an element, one must understand the arrangement of electrons in its atoms before and after a chemical reaction. By restricting the quantum mechanics theory to representative elements and simplifying some of the concepts, it is possible to expand our understanding of atomic structure sufficiently to be able to explain and predict some simple chemical reactions. The main features of this restricted version of quantum mechanics are listed below.

- The number of protons in the nucleus of an atom is equal to the atomic number of the element.

- The number of electrons around the nucleus of an atom is equal to the number of protons.

- An **energy level** represents a specific value of energy of an electron and corresponds to a general location. The number of occupied energy levels in any atom is normally the same as the period number in which the atom appears.

- For the first three energy levels, the maximum number of electrons that can be present are 2, 8, and 8 in order of increasing energy.

- A lower energy level is filled with electrons to its maximum before the next level is started.

- The electrons in the highest energy level that contains any electrons are called **valence electrons**. For representative elements, the number of valence electrons is the same as the last digit of the group number of the atom.

The statements listed above provide the rules for constructing diagrams of electron energy-level models. For example, sulfur has an atomic number of 16. According to current theory, this means that the sulfur atom has 16 protons in the nucleus and 16 electrons outside the nucleus. Sulfur appears in period number 3 on the periodic table. This means that the 16 electrons are distributed among 3 different energy levels (Figure 2.29). Since lower energy levels are filled first, the first level is full (2 electrons), the second level is also full (8 electrons), but the third level is not full (6 electrons). Therefore, a sulfur atom has 6 valence electrons and is represented by the diagram shown in Figure 2.29.

Evaluation of Scientific Theories

"Physical concepts are free creations of the human mind, and are not, however it may seem, uniquely determined by the external world. In our endeavor to understand reality we are somewhat like a man trying to understand the mechanism of a closed watch. He sees the face and the moving hands, even hears its ticking, but he has no way of opening the case. If he is ingenious he may form some picture of a mechanism which could be responsible for all the things he observes, but he may never be quite sure his picture is the only one which could explain his observations. He will never be able to compare his picture with the real mechanism and he cannot even imagine the possibility of the meaning of such a comparison." (Albert Einstein and Leopold Infeld, *The Evolution of Physics*, New York: Simon and Schuster, 1938, page 31.)

It is never possible to *prove* theories in science. A theory is accepted if it logically describes, explains, and predicts observations. A major endeavor of science is to make predictions based on theories and then to test the predictions. Once the evidence is collected, a prediction may be

- *verified* if the evidence agrees within reasonable experimental error with the prediction. If this evidence can be replicated, the scientific theory used to make the prediction is judged to be acceptable, and the evidence adds further support and certainty to the theory;
- *falsified* if the evidence obviously contradicts the prediction. If this evidence can be replicated, then the scientific theory used to make the prediction is judged to be unacceptable.

"When it comes to atoms, language can be used only as in poetry. The poet, too, is not nearly so concerned with describing facts as with creating images." — Niels Henrik David Bohr, Danish physicist (1885 – 1962)

NIELS HENRIK DAVID BOHR

Niels Bohr (1885 – 1962) obtained his doctorate in physics from the University of Copenhagen in Denmark. He then decided to further his education by studying with J. J. Thomson at Cambridge University in England. However, Thomson was difficult to work with, so Bohr transferred to Ernest Rutherford's laboratory at the University of Manchester. In 1913 Bohr published his theory of the atom, for which he received the 1922 Nobel Prize in physics. Probably the most famous debates in science took place between Bohr and Einstein, debates prompted by Einstein's scepticism regarding the new theory of quantum mechanics.

When Hitler rose to power in Germany, Bohr played a significant role in spiriting Jewish physicists out of Germany. In 1943, three years after the German occupation of Denmark,

Bohr escaped imprisonment by fleeing to Sweden. Before leaving Denmark, Bohr took two gold Nobel medals, given to him for safekeeping, and dissolved them in acid to prevent them from falling into German hands. He had already donated his own gold medal to the Finnish war relief. While in Sweden, Bohr helped organize the evacuation of many Jewish Danes from Denmark.

Bohr escaped from Sweden in a tiny plane in which he nearly died from lack of oxygen before landing safely in England. Eventually he reached the United States, where he worked at Los Alamos, New Mexico, on the atomic bomb project. His concern about the consequences of the bomb did not make him popular with some people.

After World War II, Bohr returned to Copenhagen, where he precipitated the dissolved gold from the acid and

recast the Nobel medals. This act symbolized the triumph of freedom and democracy. However, the new threat of atomic weapons was becoming more and more apparent. Bohr worked tirelessly for the development of peaceful uses of atomic energy and organized the first Atoms for Peace Conference in Geneva in 1955. In 1957 he received the first Atoms for Peace Award.

"Seeing" Atoms

In the 1980s two kinds of microscopes were developed that can produce images of atoms and molecules. These microscopes can distinguish objects that are only 10^{-12} m to 10^{-11} m apart and they have magnifications up to 24 million times. The scanning tunnelling microscope (STM) produces images indirectly by using feedback to draw electrons from a material's surface at a constant rate; the atomic force microscope (AFM) probes the surface of a specimen with a diamond stylus whose tip may be only one atom wide. These microscopes were invented in the course of technological research unrelated to atomic theory, and they have led to new scientific discoveries and technological applications.

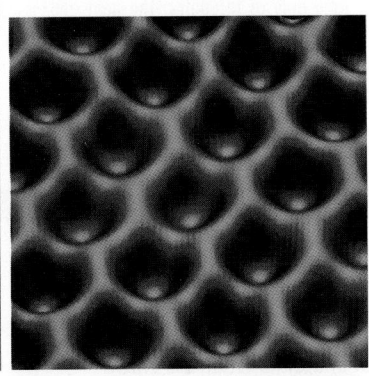

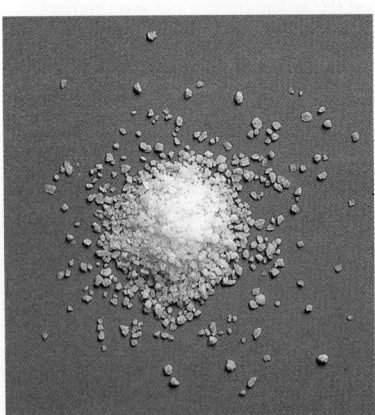

Figure 2.30
A very reactive metal (sodium) reacts with a poisonous, reactive nonmetal (chlorine) to produce a relatively inert compound (sodium chloride).

An unacceptable theory requires further action; there are three possible strategies.

- *Restrict* the theory. Treat the conflicting evidence as an exception and use the existing theory within a restricted range of situations. (This strategy is used in this book with regard to the theory of quantum mechanics, and it is also used frequently in other introductory science courses.)

- *Revise* the theory. This is the most common option. The new evidence becomes part of an improved theory. The development of atomic theories, from Dalton to quantum mechanics, is an example of this process.

- *Replace* the existing theory with a totally new concept. This is the most drastic and least frequently used option. One example is the replacement of Aristotle's theory by Dalton's atomic theory.

Exercise

45. Which two key ideas from previous atomic theories are retained by quantum mechanics?

46. The restricted version of quantum mechanics can only explain or predict electron energy levels for the atoms of some elements. What are these elements?

47. Use the periodic table and theoretical rules to predict the number of occupied energy levels and the number of valence electrons for each of the following atoms: beryllium, chlorine, krypton, iodine, lead, arsenic, and cesium.

48. Draw diagrams of the electron energy-level models like those in Figure 2.29, page 82, for the first 20 elements. Arrange the diagrams in eight columns and four rows, corresponding to the families and periods of the elements in the periodic table.

49. What are the attributes of an acceptable theory?

50. When is a prediction verified, and what effect does this have on the concept used to make the prediction?

51. When is a prediction falsified, and what effect does this have on the concept used to make the prediction?

52. What are three possible strategies when a scientific concept is judged to be unacceptable?

Formation of Monatomic Ions

In the laboratory, sodium metal and chlorine gas can react violently to produce a white solid, sodium chloride, commonly known as table salt (Figure 2.30). Sodium chloride is very stable and unreactive compared with the elements sodium and chlorine. Bohr originally suggested (page 80) that the stable, unreactive behavior of the noble gases was explained by their full outer electron orbits. Logically, any explanation of the difference in the behaviors of sodium and chlorine compared with sodium chloride should be consistent with the explanation for the stable, unreactive character of the noble gases. With this in mind, you could imagine that when a sodium atom collides with a chlorine atom, an electron is transferred from the sodium atom to the chlorine atom.

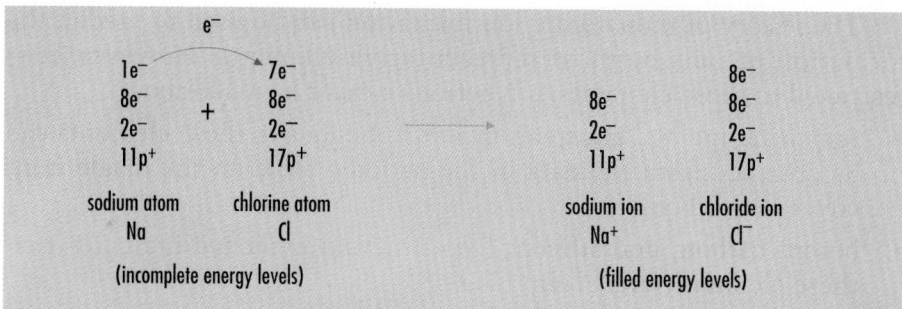

$1e^-$	$7e^-$			$8e^-$
$8e^-$	$8e^-$		$8e^-$	$8e^-$
$2e^-$	$2e^-$		$2e^-$	$2e^-$
$11p^+$	$17p^+$		$11p^+$	$17p^+$

sodium atom   chlorine atom   sodium ion   chloride ion
Na   Cl Na^+ Cl^-

(incomplete energy levels)   (filled energy levels)

Figure 2.31
Energy-level models for the reaction of sodium and chlorine. The models are used to explain how two very reactive elements could react to form a relatively inert compound.

The electron structure of both of these entities is now the same as the structure of the nearest noble gas. The diagrams in Figure 2.31 summarize this reaction. You can logically verify the charge on all particles in the diagrams by comparing the total number of electrons and protons in each model.

According to this theory, when the neutral atoms collide, an electron is transferred from one atom to the other, and both atoms become particles called **ions** which have an electrical charge. Sodium ions and chloride ions are **monatomic ions** — single atoms that have gained or lost electrons. The high reactivity of sodium and chlorine is explained by their incomplete outer energy level. The low reactivity of sodium chloride is explained by the filled outer energy levels for the sodium and chloride ions, as shown in Figure 2.31. This interpretation leads to an important theoretical rule for predicting the number of electrons that atoms will lose or gain in an electron transfer reaction. *Atoms of the representative elements form monatomic ions when losing or gaining electrons to form the same stable electronic structure as atoms of the nearest noble gas.*

Note that sodium (atomic number 11) has one more electron than the nearest noble gas, neon (atomic number 10). Chlorine (atomic number 17) has one fewer electron than its nearest noble gas, argon (atomic number 18). A transfer of one electron from a sodium atom to a chlorine atom will result in both particles having the filled electron energy levels of the noble gas nearest to them in atomic number.

QUARKS

By the 1960s, 24 subatomic particles had been discovered. In much the same way that Mendeleyev had grouped the elements into the periodic table, American physicist Murray Gell-Mann grouped the subatomic particles and used his organization of evidence to predict the discovery of yet more particles. He predicted the existence of fundamental particles that he called quarks. According to his theory, only a few kinds of quarks existed, but when grouped in different combinations, quarks could account for a large number of the 24 known subatomic particles.

Empirical evidence for the existence of quarks followed, and in 1990 the Nobel Prize in physics was awarded to three physicists, Jerome I. Friedman, Henry W. Kendall, and Richard E. Taylor, a Canadian who conducted his research at the Stanford Linear Accelerator Center (SLAC) in California. The work done by Taylor and his colleagues that led to the 1990 Nobel Prize is remarkably similar to the work of

Rutherford that led to Rutherford's 1908 Nobel Prize in chemistry. Rutherford's observations of the deflection of alpha particles striking gold foil provided empirical evidence for the existence of the atomic nucleus. At SLAC, the observations of the deflection of high-energy electrons striking the nuclei of atoms provided empirical evidence

for the existence of quarks within the protons and neutrons. The achievements of Taylor and his colleagues indicate that electrons (e⁻), up-quarks (u), and down-quarks (d) are fundamental building blocks of matter. Protons and neutrons are each made up of three quarks. A proton is "uud" and a neutron is "udd."

Richard Taylor was born and raised in Medicine Hat, Alberta, and received his B.Sc. and M.Sc. from the University of Alberta before earning his Ph.D. at Stanford University in California. Dr. Taylor feels strongly that Canada must increase its funding for scientific research and development. Referring to Canada's economic status, he says the country isn't "going to be able to just dig it up and cut it down for too much longer."

- *Is Richard Taylor primarily an empirical or theoretical scientist?*
- *According to modern theory, what three fundamental entities make up all matter?*

The theory of monatomic ion formation can be used to predict the formation of ions by most representative elements. However, it is restricted to these elements; predictions cannot be made about

- transition metals. Information about the ions of these elements can be obtained from the data in the periodic table on the inside front cover of this book.
- boron, carbon, and silicon. Experimental evidence indicates that these elements rarely form ions.
- hydrogen. Hydrogen atoms usually form positive ions by losing an electron. Although unusual, a negative hydrogen ion can be formed.

Positively charged ions are called **cations**. All of the monatomic cations are formed from the metallic elements when they lose electrons in an electron transfer reaction. The metals in the representative groups commonly form cations with the same number of electrons as atoms of the nearest noble gas. Names for monatomic cations use the full English name of the element followed by the word "ion"; for example, sodium ion.

Negatively charged ions are called **anions**. All of the monatomic anions come from the nonmetallic elements. According to theory, nonmetals tend to gain electrons in an electron transfer reaction, forming anions with the same number of electrons as atoms of the nearest noble gas. Names for monatomic anions use the stem of the English name of the element with the suffix "-ide" and the word "ion" (Table 2.10).

Table 2.10

NAMES AND SYMBOLS OF MONATOMIC ANIONS		
Group 15	**Group 16**	**Group 17**
nitride ion, N^{3-}	oxide ion, O^{2-}	fluoride ion, F^-
phosphide ion, P^{3-}	sulfide ion, S^{2-}	chloride ion, Cl^-
arsenide ion, As^{3-}	selenide ion, Se^{2-}	bromide ion, Br^-
	telluride ion, Te^{2-}	iodide ion, I^-

The number of electrons lost or gained is equal to the difference between the number of electrons in atoms of the representative element and the number of electrons in atoms of the nearest noble gas (Figure 2.32).

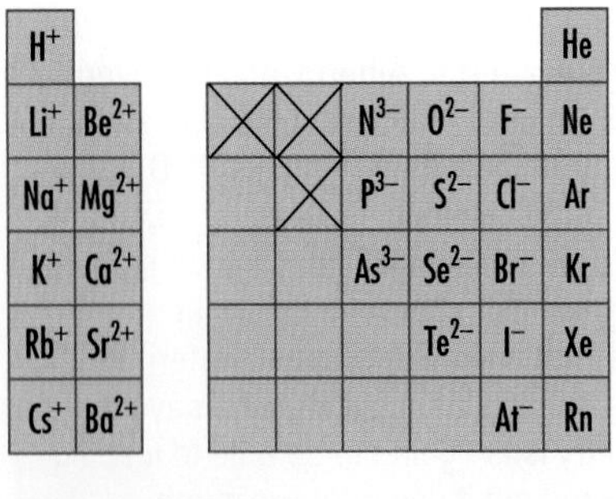

The symbols for monatomic ions include the element symbol with a superscript indicating the net charge. The symbols "+" and "−" represent the words "positive" and "negative." For example, the charge on a sodium ion Na^+ is "positive one," an aluminum ion Al^{3+} is "positive three," and an oxide ion O^{2-} is "negative two."

Figure 2.32
The periodic table can be used to predict the charge on representative element cations and anions.

SUMMARY: THEORETICAL DESCRIPTIONS OF ATOMS AND IONS			
	Atoms	**Positive Ions Formed by Metals**	**Negative Ions Formed by Nonmetals**
Name	element name	element name	element root + -ide
Nucleus	#p$^+$ = atomic number	#p$^+$ = atomic number	#p$^+$ = atomic number
Electrons	#e$^-$ = #p$^+$	#e$^-$ < #p$^+$	#e$^-$ > #p$^+$

Exercise

53. If a scientific theory or other scientific knowledge is found to be unacceptable as a result of falsified predictions, what three options are used by scientists?

54. Write a theoretical definition of cation and anion.

55. Draw diagrams of the electron energy-level models for the ions of the first 20 elements, except boron, carbon, silicon, and the noble gases. Arrange the diagrams in columns and rows corresponding to the families and periods in the periodic table.

56. The alkali metals all react violently with halogens to produce stable white solids. Draw diagrams of the electron energy-level models (like Figure 2.31, page 85) for each of the following reactions.

   (a) lithium + chlorine → lithium chloride

   (b) potassium + fluorine → potassium fluoride

57. Rubidium, cesium, and francium are also members of Group 1 and react similarly to lithium, sodium, and potassium (Figure 2.33). Write the names and symbols for the monatomic ions formed by rubidium, cesium, and francium.

58. The alkaline-earth metals form Group 2 in the periodic table. Draw diagrams of the electron energy-level models for the following reactions.

   (a) magnesium with oxygen

   (b) calcium with oxygen

59. List the ion charges of the monatomic ions for the following families: alkali metals, alkaline-earth metals, Group 13, Group 15, Group 16, halogens.

60. All the electron energy-level diagrams drawn in the previous questions have complete or filled outer energy levels. What is the experimental evidence for these filled outer energy levels?

61. Draw diagrams of the electron energy-level model for a hydrogen ion, H^+, and a hydride ion, H^-. How do these illustrate the unique nature of hydrogen?

62. Write the symbols for the following atoms and ions; e.g., the sodium atom is Na, while the chloride ion is Cl^-.

   (a) sulfur atom

   (b) oxide ion

   (c) lithium ion

   (d) phosphide ion

   (e) aluminum atom

   (f) gallium ion

   (g) rubidium ion

   (h) iodide ion

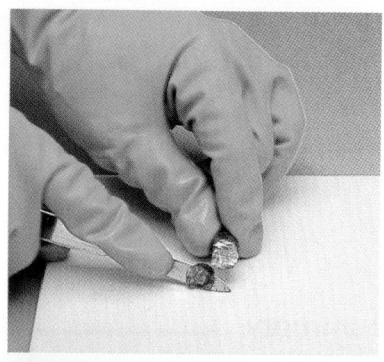
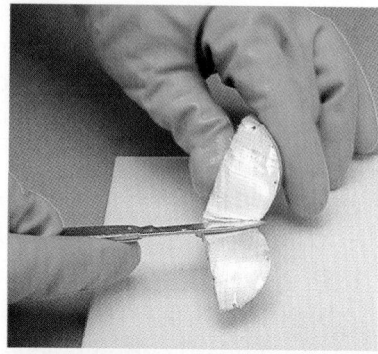
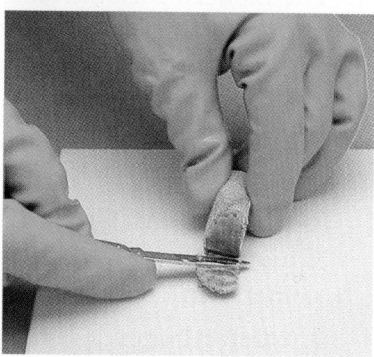

Figure 2.33
The alkali metal family lithium (top), sodium (middle), and potassium (bottom) have similar physical and chemical properties that can be explained by modern atomic theories.

OVERVIEW

Elements

Summary

- Elements have IUPAC symbols and may be classified as metals or nonmetals. Metals are shiny, bendable, and good conductors of heat and electricity. Nonmetals have none of these properties.

- Empirical knowledge can be communicated as simple descriptions, tables of evidence, graphs, empirical definitions, empirical hypotheses, generalizations, or scientific laws.

- The modern periodic table was developed from evidence of periodicity in chemical and physical properties. Elements in the same group have similar properties.

- Alkali metals (Group 1), alkaline-earth metals (Group 2), transition metals (Groups 3 to 12), halogens (Group 17), and noble gases (Group 18) are well-known families of elements in the modern periodic table.

- Theoretical knowledge in science is based on unobservable ideas and is communicated using descriptions, definitions, theories, analogies, and models. Acceptable theories describe, explain, predict, and are as simple as possible.

- In Dalton's theory, matter is composed of tiny, indivisible atoms, identical for one element, but different between elements. Atoms combine in constant ratios to form compounds.

- Rutherford's theory includes the key idea of a nucleus containing positively charged protons. The number of protons is the same as the atomic number and equals the number of negatively charged electrons in the space outside the nucleus.

- The Bohr theory added the new idea of electrons in specific orbits corresponding to certain energy values. This theory produced the first explanation of the periodic law and table.

- The restricted quantum mechanics theory includes three major subatomic particles: protons and neutrons in the nucleus, and electrons in energy levels outside the nucleus.

- Atoms of the representative elements are thought to form monatomic ions by losing or gaining electrons to obtain the same number of electrons as the nearest noble gas.

- Science involves two parallel types of activities — empirical and theoretical.

Key Words

alkali metals
alkaline-earth metals
anions
atom
atomic mass
atomic mass unit
atomic number
cations
ductile
electrons
energy level
empirical definition
empirical hypothesis
energy level
family (group)
generalizations
halogens
ions
isotope
law of conservation of mass
malleable
mass number
metals
monatomic ions
neutrons
noble gases
nonmetals
nucleus
period
periodic law

CHAPTER 2

protons
quantum mechanics
radioisotope
representative elements
SATP
scientific laws
theoretical definition
theoretical hypothesis
theory
transition elements
valence electrons

Review

1. Write the English names for the elements represented by each of the following IUPAC international symbols.
   (a) Fe
   (b) I
   (c) P
   (d) Na
   (e) Cl
   (f) Hg
   (g) Cu
   (h) S

2. For the elements in question 1,
   (a) which element symbols come from old Latin names?
   (b) which elements are metals and which are nonmetals?

3. Is the symbol "CA" an acceptable international symbol for calcium? Justify your answer.

4. List three characteristics of a scientific law acceptable to the scientific community.

5. What is the most important distinction between metals and nonmetals?

6. What are the three most abundant elements by mass in the human body?

7. (a) Which scientist is credited with the classification of elements into the first accepted version of the periodic table?
   (b) How did the periodic law come to be accepted, even though it could not initially be explained?

8. Use the periodic table on the inside front cover of this book to answer the following questions.
   (a) At SATP conditions, which elements are liquids and which are gases?

(b) What is the purpose of the "staircase line" that divides the periodic table into two parts?
(c) Identify by name and symbol the following three elements: period 3, Group 2; period 6, Group 14; period 2, Group 17.
(d) What are the atomic numbers of hydrogen, oxygen, aluminum, silicon, chlorine, and copper?

9. Sketch an outline of the periodic table and label the following: alkali metals, alkaline-earth metals, transition metals, staircase line, halogens, noble gases, metals, and nonmetals.

10. What is the most reactive metal? nonmetal?

11. State two limitations of the periodic table.

12. What is unusual about hydrogen as an element compared with other elements?

13. Distinguish between empirical and theoretical knowledge.

14. List six methods of communicating theoretical knowledge in science.

15. List the four characteristics of theories acceptable to the scientific community.

16. For each of the following atomic theories, sketch a diagram of the model and state a possible analogy.
    (a) Dalton
    (b) Thomson
    (c) Nagaoka
    (d) Rutherford
    (e) Bohr

17. List the three main subatomic particles, including their location in the atom, their relative mass, and their charge.

18. How did Niels Bohr explain the periodic law?

19. If a scientific theory is found to be unacceptable, what three options are available?

20. How is each of the following theoretical descriptions of atoms obtained from the position of the element in the periodic table?
    (a) number of protons
    (b) number of electrons
    (c) number of valence electrons for representative elements
    (d) number of occupied energy levels for representative elements

21. (a) How do you know what monatomic ions form in a reaction between a metal and a nonmetal?
(b) List the charges on the common ions formed by atoms in Groups 1, 2, 13, 15, 16, and 17.

22. Use specific examples to describe how Mendeleyev's development of the periodic table from the periodic law illustrates three characteristics of an acceptable scientific law.

23. Explain, according to Rutherford's model of the atom, why a small percentage of alpha particles was deflected through large angles when fired at a thin sheet of gold foil.

24. Why has there been a series of atomic theories?

Applications

25. According to IUPAC rules, what is the temporary name and symbol for element number 112?

26. Predict the chemical and physical properties of the undiscovered element, ununoctium.

27. State a common use for each of the following elements.
(a) chlorine
(b) aluminum
(c) oxygen
(d) silicon
(e) carbon
(f) neon

28. List the number of protons, electrons, and valence electrons in each of the following atoms.
(a) magnesium
(b) aluminum
(c) iodine

29. List the number of each of the three subatomic particles in atoms of each of the following isotopes.
(a) calcium-42
(b) strontium-90

30. Radioactive iodine is a common radioisotope for medical use. If the iodine is in the form of monatomic iodide-123 ions (as part of sodium iodide), state the number of protons, neutrons, and electrons.

31. Using the restricted theory of quantum mechanics, draw diagrams of electron energy-level models for the atom and ion of each of the following elements.
(a) potassium
(b) oxygen
(c) chlorine

32. What empirical and theoretical characteristics of the noble gas family have made this family especially interesting to chemists?

33. (a) For what elements can the restricted quantum mechanics theory be used to predict the formation of monatomic ions?
(b) For what elements can the monatomic ions not be predicted?

34. Write the chemical name and symbol corresponding to each of the following theoretical descriptions.
(a) 11 protons and 10 electrons
(b) 18 electrons and a net charge of 3⁻
(c) 16 protons and 2 extra electrons

35. Draw diagrams of electron energy-level models to represent the reactant atoms and product ions in the following reaction equation.
magnesium atom + oxygen atom →
magnesium ion + oxide ion

36. Match the element to the empirical or theoretical description or technological or natural use as an atom or ion. Seek the best match overall.

H	(a)	graphite pencil lead
He	(b)	forms 3+ ion
C	(c)	home water pipes
N	(d)	yellow solid
O	(e)	yellow gas
Ne	(f)	the lightest gas
Al	(g)	80% of air
P	(h)	light bulb filament
S	(i)	in potash fertilizer
Cl	(j)	supports respiration
Ca	(k)	two protons
K	(l)	five valence electrons
Fe	(m)	liquid metal
Ni	(n)	noble nonmetal
Cu	(o)	in nuclear reactors

Ag	(p)	in jewellery
I	(q)	soft, dense metal
W	(r)	in all steel alloys
Au	(s)	79 e⁻ in an atom
Hg	(t)	alkaline earth
Pb	(u)	ion has 54 e⁻
U	(v)	in jewellery

Extensions

37. Gallium was not in Mendeleyev's periodic table, as it had not yet been discovered. Use a variety of techniques to predict the density of gallium from the densities of its neighboring elements. For each technique, calculate the percent difference between the accepted value (5.90 g/cm^3) and your predicted value. Judge which technique seems to work best.

38. Write a short paragraph on the nature of science, including the evolution of scientific concepts in the quest for scientific knowledge.

39. Choose an element and write a report on it, including information from both scientific and technological perspectives. Take an issue-oriented approach, and discuss the issue from an ecological, economic, or political perspective. In your opening paragraph or title, communicate directly or indirectly the approach that you are taking.

40. On page 83, Albert Einstein's quote indicates that we can never open the watch to find out what is inside. What is meant by this analogy? If you ever did a "black box" experiment to illustrate the nature of scientific knowledge in school, were you allowed to open the "box"? According to Einstein, should you be allowed to open the box? Discuss this question in terms of the definition of theoretical knowledge presented in this chapter.

41. Isaac Newton indicated that what he had accomplished had been done "by standing on the shoulders of giants" who had come before him in physics. Albert Einstein is quoted similarly below. Write a paragraph in your own words to interpret what these two geniuses are trying to communicate.

"Creating a new theory is not like destroying an old barn and building a skyscraper in its place. It is rather like climbing a mountain, gaining new and wider views, discovering unexpected connections between the starting point and its rich environment. But the point from which we started out still exists and can be seen, although it appears smaller and forms a tiny part of our broad view gained by the mastery of the obstacles on our adventurous way up."

Problem 2C Testing the Theory of Ions

The purpose of this problem is to test the theory of ions presented in this chapter. Complete the Prediction and Evaluation of the investigation report. Evaluate the prediction and the concept only.

Problem

What is the chemical formula of the compound formed by the reaction of aluminum and fluorine?

Prediction

According to the restricted quantum mechanics theory of atoms and ions, the chemical formula of the compound formed by the reaction of aluminum and fluorine is [your answer]. The reasoning behind this prediction, including electron energy-level diagrams, is [your reasoning].

Experimental Design

Aluminum and fluorine are reacted in a closed vessel, and the chemical formula is calculated from the masses of reactants and products.

Analysis

According to the evidence gathered in the laboratory, the chemical formula of the compound formed by the reaction of aluminum and fluorine is AlF_3. The evidence could be interpreted as indicating that one aluminum atom combines with three fluorine atoms to produce a compound with a formula of one aluminum ion and three fluoride ions; i.e., $Al^{3+}F_3$.

UNIT II

CHEMICAL COMBINATION

Although you can learn some things about chemicals by observing their physical properties, laboratory work with chemical reactions reveals a great deal more about the compounds. You can make inferences about chemicals based on the changes that occur in chemical reactions. By studying chemical reactions, you can construct generalizations and laws and, eventually, infer the theoretical structure of the compounds involved.

Initially, theories of the structure of matter attempt to explain the known chemical properties of a substance. The validity of a theory is determined by its ability to both explain and predict changes in matter. How and why do chemicals react? What compounds will form as a result of a reaction? How do we explain the different properties of compounds? Chemical combination represents not only the compounds we know but also the processes or chemical reactions by which we know them.

3 Compounds

Although fewer than 100 elements occur naturally on Earth, millions of compounds are made from them. Chemical compounds by the thousands line the shelves of drugstores, food markets, greenhouses, and hardware and hobby stores. Some of these substances that are similar in appearance and in atomic structure may have completely different applications. For example, water and hydrogen peroxide both contain the same types of atoms. The first of these compounds is essential for all life on Earth, while the second is a bleaching agent that, in sufficiently high concentration, can cause explosions. During debates about fluoridation, the public's confusion of fluoride with fluorine led to controversy about the health effects of adding fluoride to water supplies. Aware that fluorine is a highly corrosive chemical, some people feared the potential hazards of fluorides added to drinking water to reduce tooth decay in children. The debate about the effects of fluoride has subsided now that the success of fluoridation has been documented. But given the tremendous number of chemical compounds that we rely on every day, as well as the ones such as dioxins and cholesterol that we read about in the popular press, confusion and misconceptions are bound to arise. A society that depends on chemicals as much as ours needs as much understanding as possible about the benefits, risks, and responsibilities associated with the use of chemicals.

3.1 CLASSIFYING COMPOUNDS

Before chemists could understand compounds, they had to devise ways to distinguish them from elements. Once they achieved this, they could begin to organize their knowledge by classifying compounds. In this chapter, classification of compounds is approached in three different ways: by convention, empirically, and theoretically.

Classification of Compounds by Convention

Elements are commonly classified as metals or nonmetals (Investigation 2.1, page 59). Given that compounds contain atoms of more than one kind of element, what combinations can result? Three classes of compounds are possible: *metal-nonmetal*, *nonmetal-nonmetal*, and *metal-metal* combinations (Figure 3.1). Theoretical structures related to these classes of compounds are discussed later in the chapter.

- Metal-nonmetal combinations are called **ionic compounds**. An example is sodium chloride (NaCl).

- Nonmetal-nonmetal combinations are called **molecular compounds**. An example is sulfur dioxide (SO_2).

- Metal-metal combinations are called alloys. Alloys range from common metal-metal solutions (for example, silver-gold alloys in coins) to rare inter-metallic compounds (for example, CuZn, Cu_5Zn_8, and $CuZn_3$ in brass at certain temperatures).

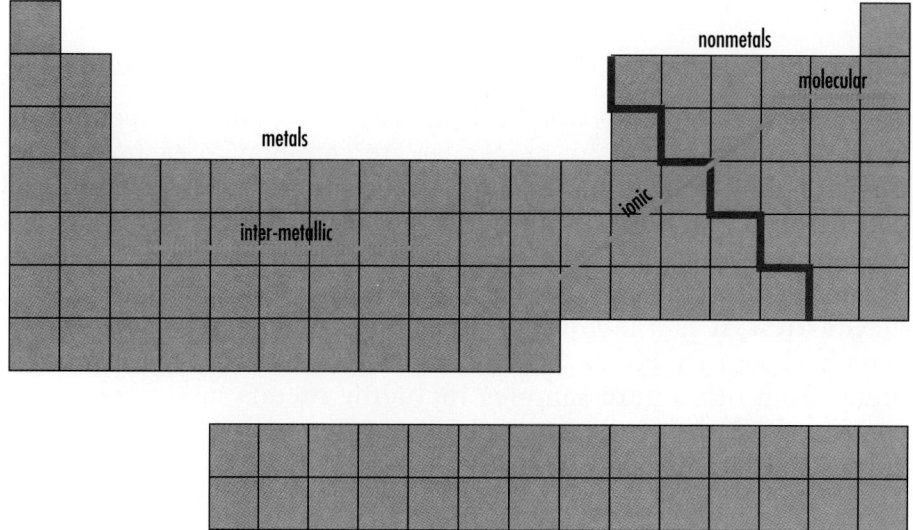

> ■ **Alternate Terms**
> - Molecular compounds are sometimes called *covalent compounds*, but in this textbook the preferred term is *molecular compounds*.
> - *Alloys*, such as brass and amalgams, are solutions of a metal with other metals and, sometimes, nonmetals. Alloys do not combine in definite proportions — inter-metallic compounds do.

Figure 3.1
From two classes of elements there can be three classes of compounds. These classes of compounds are called ionic, molecular, and inter-metallic.

Exercise

1. Distinguish, empirically and theoretically, between an element and a compound.

2. Distinguish, empirically and theoretically, between a metal and a nonmetal.

3. Classify each of the following chemicals as elements or compounds.
   (a) $C_{12}H_{22}O_{11(s)}$ (sugar)
   (b) $Fe_{(s)}$ (in steel)
   (c) $CO_{(g)}$ (poisonous)
   (d) sodium chloride (salt)
   (e) oxygen (20% of air)
   (f) calcium (reactive)

4. Classify each of the following elements as metals or nonmetals.
   (a) lead (poisonous)
   (b) phosphorus (reactive)
   (c) chlorine (poisonous)
   (d) $U_{(s)}$ (nuclear reactors)
   (e) $Hg_{(l)}$ (liquid)
   (f) $Br_{2(l)}$ (liquid)

5. Classify each of the following compounds as ionic or molecular.
   (a) $C_6H_{12}O_{6(s)}$ (glucose)
   (b) $Fe_2O_3 \cdot 3H_2O_{(s)}$ (rust)
   (c) $H_2O_{(l)}$ (water)
   (d) nitrogen dioxide (pollutant)
   (e) potassium chloride (fertilizer)
   (f) zinc sulfide (zinc ore)

6. Draw a flowchart showing the classification of pure substances, elements, metals, nonmetals, compounds, ionic, and molecular.

7. Why do chemists create classification systems for chemicals?

Empirical Classification of Ionic and Molecular Compounds

The properties of compounds can be used to classify compounds as ionic or molecular. Many properties are common to each of these classes, but by focusing on the more important properties, an empirical definition of each class can be found. By further restricting the properties of each empirical definition to those easiest to identify, diagnostic tests for ionic and molecular compounds can be designed. A **diagnostic test** is a laboratory procedure conducted to identify or classify chemicals. For example, electrical conductivity can identify a metal from other pure samples including metals, nonmetals, and compounds. Some of the common diagnostic tests used in chemistry are described in Appendix C, C.3.

■ Problem
■ Prediction
■ Design
■ Materials
■ Procedure
✔ Evidence
✔ Analysis
✔ Evaluation
■ Synthesis

INVESTIGATION

3.1 Empirical Definitions of Compounds

The purpose of this investigation is to test hypothetical empirical definitions of ionic and molecular compounds. By noting the properties of eight substances, you will test an empirical definition of each class of compound. Assume that each substance is typical of its class and has no characteristics that will interfere with the investigation.

Problem

What properties can be used to form the empirical definitions of ionic and molecular compounds?

Prediction

According to previous experience, the following properties could be useful in forming empirical definitions of ionic and molecular compounds: state of matter at SATP, solubility of the compound in water, color of the aqueous (water) solution, and electrical conductivity of the aqueous solution.

Experimental Design

The compounds provided are classified as ionic or molecular, based on the classification of compounds by convention and on the chemical name and formula for each compound. Using these samples of ionic and molecular compounds, evidence is collected for each of the predicted defining properties. The conductivity test includes a control test of the conductivity of distilled water.

Materials

lab apron	sodium bicarbonate, $NaHCO_3$
safety glasses	calcium sulfate, $CaSO_4$
sodium chromate, Na_2CrO_4	50 mL beaker or spot plate
methane, CH_4 (or other hydrocarbon)	distilled water
sucrose, $C_{12}H_{22}O_{11}$	laboratory scoop
sodium chloride, $NaCl$	medicine dropper
copper(II) sulfate, $CuSO_4$	test tube
ethanol, C_2H_5OH	stirring rod
	conductivity tester

Procedure

1. Observe the compound and record its state of matter at the ambient temperature.

2. Add some distilled water to the reaction vessel.

3. Test and note the electrical conductivity of the water.

4. Add a small quantity of the chemical to the water as shown in Figure 3.2, page 98.

5. Stir the mixture, if possible, and record whether the chemical dissolves.

6. Record the color of any solution formed.

7. Test the electrical conductivity of the mixture (record as "yes" or "no," relative to the distilled water reading).

8. Dispose of the mixture as directed and rinse the reaction vessel and conductivity probes.

9. Move to the next station and repeat steps 1 to 8 to test the other chemicals.

▌Development of Diagnostic Tests

general properties
↓
defining properties
↓
empirical definitions
↓
diagnostic tests

▌States of Matter in Chemical Formulas

Chemical formulas include information about the numbers and kinds of atoms or ions in a compound. It is also common practice in a formula to specify the state of matter as a subscript. Four subscripts commonly used are: (s) to indicate "solid"; (l) to indicate "liquid"; (g) to indicate "gas"; and (aq) to indicate "aqueous," which refers to solutions in water. **Aqueous solutions** are readily formed by substances that have high solubility in water.

$NaCl_{(s)}$	pure table salt
$CH_3OH_{(l)}$	pure antifreeze
$O_{2(g)}$	pure oxygen
$C_{12}H_{22}O_{11(aq)}$	aqueous sugar solution

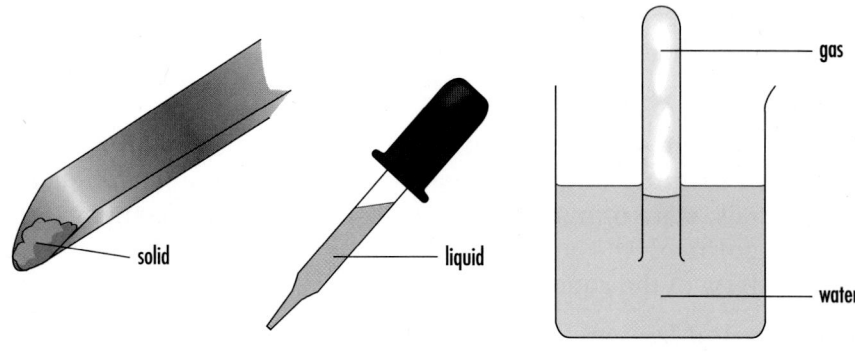

Figure 3.2
For a solid chemical, use a small quantity in a scoop as shown. If the chemical is a liquid, use a dropper full of liquid. If the chemical is a gas, observe the static display or try to dissolve the gas as shown or as instructed by your teacher.

Figure 3.3
Conductivity is used to distinguish between aqueous solutions of soluble ionic and molecular compounds. Solutions of ionic compounds conduct electricity but solutions of molecular compounds do not.

Empirical Definitions of Compounds

In a series of replicated problem similar to Investigation 3.1, scientists have found that **ionic compounds** are all solids at SATP. When dissolved in water, these compounds form solutions that conduct electricity. Scientists have also discovered that **molecular compounds** at SATP are solids, liquids, or gases which, when dissolved in water, form solutions that do *not* conduct electricity. These *empirical definitions* — a list of empirical properties that define a class of chemicals — will prove helpful throughout your study of chemistry. For example, electrical conductivity of a solution is an efficient and effective diagnostic test that determines whether a compound is ionic or molecular (Investigation 3.1, and Figure 3.3).

Problem 3A Testing Empirical Definitions of Compounds

The purpose of this problem is to test the empirical definitions of ionic and molecular compounds (as created from Investigation 3.1). Testing scientific concepts is one of the most important roles of a scientist. The more tests that a concept passes, the greater is the confidence you have in the concept. Complete the Analysis and Evaluation of the investigation report.

Problem

What are the properties of the following compounds?

calcium carbonate, $CaCO_3$ nickel(II) chloride, $NiCl_2$
glucose, $C_6H_{12}O_6$ methanol, CH_3OH
cobalt(II) nitrate, $Co(NO_3)_2$ sodium carbonate, Na_2CO_3
hexane, C_6H_{14} butane, C_4H_{10}

Prediction

According to the empirical definitions of ionic and molecular compounds, the properties of the selected compounds are as shown in the following table.

PREDICTED PROPERTIES OF SELECTED COMPOUNDS

Name and Formula	Class (i/m)	State at SATP (s/l/g)	Conductivity of Aqueous Solution (yes/no)
calcium carbonate, $CaCO_3$	i	s	yes
glucose, $C_6H_{12}O_6$	m	s, l, or g	no
cobalt(II) nitrate, $Co(NO_3)_2$	i	s	yes
hexane, C_6H_{14}	m	s, l, or g	no
nickel(II) chloride, $NiCl_2$	i	s	yes
methanol, CH_3OH	m	s, l, or g	no
sodium carbonate, Na_2CO_3	i	s	yes
butane, C_4H_{10}	m	s, l, or g	no

Experimental Design

The physical state of the pure compound and the electrical conductivity of the aqueous solution are observed at SATP (Figure 3.4).

manipulated variable:	selected compound
responding variables:	pure state at SATP
	electrical conductivity of aqueous solution
controlled variables:	temperature
	quantity of compound
	volume of water

Evidence

OBSERVED PROPERTIES OF SELECTED COMPOUNDS

Name and Formula	Class (i/m)	State at SATP (s/l/g)	Conductivity of Aqueous Solution (yes/no)
calcium carbonate, $CaCO_3$	i	s	n/a
glucose, $C_6H_{12}O_6$	m	s	no
cobalt(II) nitrate, $Co(NO_3)_2$	i	s	yes
hexane, C_6H_{14}	m	l	n/a
nickel(II) chloride, $NiCl_2$	i	s	yes
methanol, CH_3OH	m	l	no
sodium carbonate, Na_2CO_3	i	s	yes
butane, C_4H_{10}	m	g	n/a

"n/a" indicates that the test on a solution is *not applicable* because the compound is not soluble in water (Figure 3.5, page 100).

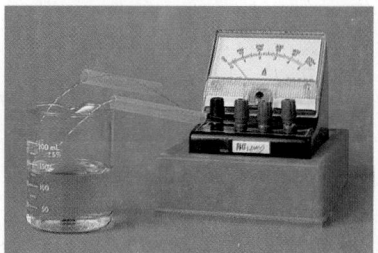

Figure 3.4
An aqueous solution of nickel(II) chloride conducts electricity (top), whereas an aqueous solution of methanol does not conduct electricity (bottom).

Figure 3.5
Hexane does not dissolve in water, as evidenced by the two layers in the test tube. The less dense hexane floats to the top. In the second test tube, butane gas does not dissolve in water. If the butane dissolved, the water level would rise in the tube.

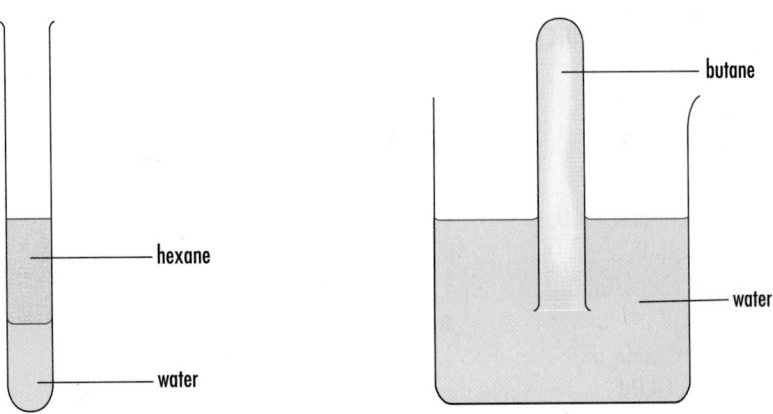

- Problem
- ✔ Prediction
- Design
- Materials
- Procedure
- ✔ Evidence
- ✔ Analysis
- ✔ Evaluation
- Synthesis

INVESTIGATION

3.2 Demonstration: Empirical Definitions of Compounds

The purpose of this demonstration is to test the empirical definitions for ionic and molecular compounds. First, classify the compounds as ionic or molecular and predict the properties of each compound. Then use the evidence collected to answer the Problem statement and to evaluate the prediction and the empirical definitions.

Problem

What are the properties of the following compounds?

potassium chloride, KCl
methanol, CH_3OH
sodium hydroxide, $NaOH$
hydrogen citrate, $C_3H_4OH(COOH)_3$

Experimental Design

The physical state of the pure compound and the electrical conductivity of the aqueous solution are observed at SATP.

manipulated variable: class of compound

responding variables: pure state at SATP
                        electrical conductivity of aqueous solution

controlled variables: temperature
                        quantity of compound
                        volume of water

Materials

lab apron
safety glasses
potassium chloride, KCl
methanol, CH_3OH
sodium hydroxide, $NaOH$
hydrogen citrate, $C_3H_4OH(COOH)_3$

50 mL beaker or spot plate
distilled water
laboratory scoop
medicine dropper
stirring rod
conductivity tester

Procedure

1. Observe and record the pure state of matter at the ambient temperature.
2. Add a small quantity of distilled water to the 50 mL beaker or spot plate.
3. Test and note the electrical conductivity of the water.
4. Add a small quantity of the chemical to the water as shown in Figure 3.2, page 98.
5. Stir the mixture to dissolve the compound.
6. Test and record the electrical conductivity of the mixture (as "yes" or "no," relative to the distilled water reading).
7. Dispose of the mixture as directed and rinse the container and conductivity probes.
8. Repeat steps 1 to 7 to test the other chemicals.

Empirical Definition of Acids

In science, it is not uncommon for new evidence to conflict with widely known and accepted theories, laws, and generalizations. Rather than viewing this as a problem, it is best regarded as an opportunity to improve our understanding of nature. As a result of the new evidence, the scientific concept is either restricted, revised, or replaced. (These three choices are sometimes referred to as the "3 Rs.")

In Investigation 3.2, some of the evidence does not fit with the classification of all compounds as either ionic or molecular. For example, aqueous hydrogen citrate (citric acid) — whose chemical formula is $C_3H_4OH(COOH)_3$ — is a compound composed of nonmetals. You might predict that this compound is molecular. However, a citric acid solution conducts electricity, which might lead you to predict that the compound is ionic (Figure 3.6). This conflicting evidence necessitates a revision of the classification system. A third class of compounds, called acids, has been identified, and the three classes together provide a more complete description of the chemical world. In many ways, acids are a special category distinct from both ionic and molecular compounds. As pure substances they most often resemble molecular compounds, but in solution their conductivity suggests a separate class or subclass of compounds.

The unusual and distinguishing properties of acids are evident when these compounds form aqueous solutions. Although more encompassing definitions of acids appear in Chapters 5, 13, and 14, this simplified definition, restricted to aqueous solutions, is sufficient for classification. **Acids** are solids, liquids, or gases as pure compounds at SATP that form conducting aqueous solutions that make blue litmus paper turn red. Acids exhibit their special properties only when dissolved in water. As pure substances, all acids, at this point in your chemistry education, have the properties of molecular compounds.

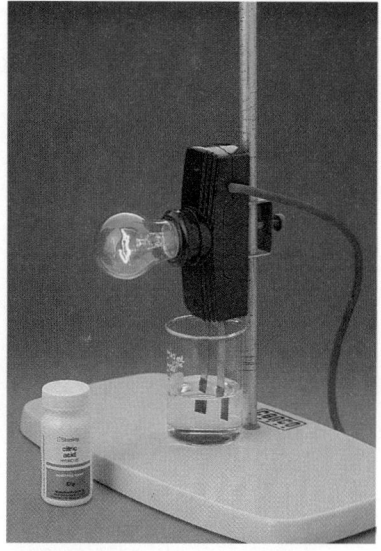

Figure 3.6
Aqueous hydrogen citrate (citric acid) — an acid in the juice of citrus fruits such as oranges and grapefruits — might be predicted to be molecular, but it forms a conducting solution. It is therefore classified as an acid.

Vitamin C and Health
Some acids are essential to human health. Ascorbic acid, also known as vitamin C, is necessary for the formation of bone, teeth, and cartilage, and it also prevents scurvy. Nobel Prize winner Linus Pauling suggested that large doses of vitamin C may help to prevent many illnesses, including the common cold. Because of the scarcity of supporting evidence from experiments, however, Pauling's ideas are not widely accepted by the scientific community.

Experimental work has also shown that some substances make red litmus paper turn blue. This evidence has led to another class of substances: **bases** are empirically defined as compounds whose aqueous solutions make red litmus paper turn blue. Compounds whose aqueous solutions do not affect litmus paper are said to be **neutral**. These empirical definitions will be expanded in Chapters 5, 13, and 14.

The properties of ionic compounds, molecular compounds, acids, and bases are summarized in Tables 3.1 and 3.2.

Table 3.1

PROPERTIES OF IONIC AND MOLECULAR COMPOUNDS AND ACIDS

	Ionic	Molecular	Acids
State at SATP	(s) only	(s), (l), or (g)	(s), (l), or (g)
Conductivity of aqueous solution	high, low	none	high, low

Table 3.2

PROPERTIES OF ACIDS, BASES, AND NEUTRAL COMPOUNDS

	Acid	Base	Neutral Compound
Effect on blue litmus paper	turns red	none	none
Effect on red litmus paper	none	turns blue	none

Litmus
Litmus, a dye obtained from lichens, has been used for over 200 years as a diagnostic test to identify acids. Litmus paper consists of absorbent paper permeated with litmus.

Exercise

8. Write empirical definitions of ionic and molecular compounds.
9. Write empirical definitions of acids and bases.
10. Use the evidence provided to classify each of the following compounds as ionic, molecular, acid, or base.

Compound	SATP State	Solution Conductivity	Litmus Test
(a) A	s	yes	red to blue
(b) B	l	no	no change
(c) C	g	yes	blue to red
(d) D	s	yes	no change
(e) E	s	no	no change

11. Which responding variable could be omitted from the experimental design in the previous question to make the design more efficient (i.e., take less time to complete) and yet still be valid?
12. Use the evidence provided to classify each of the following chemicals as metal, nonmetal, ionic, molecular, acid, or base.

Chemical	SATP State	Pure SATP Conductivity	Solution Conductivity	Litmus Test
(a) A	s	no	yes	red to blue
(b) B	l	no	no	no change
(c) C	l	yes	n/a	n/a
(d) D	g	no	yes	blue to red
(e) E	s	no	yes	no change
(f) F	s	no	n/a	n/a

n/a indicates that the test was not applicable (possible) because the chemical did not dissolve in water.

13. In which two sections of the laboratory report do answers to the Problem statement appear?

14. What is the last thing you write in the Evaluation section of a laboratory report when the prediction is falsified?

15. Once empirical definitions of compounds are established, what kind of knowledge about compounds is likely to follow?

3.2 EXPLAINING IONIC COMPOUNDS

Using well-designed experiments and the technology available to them, chemists empirically determined chemical formulas long before an adequate theory was available to explain or predict them. **Empirical formulas** were determined from measurements made when compounds were formed from elements or decomposed into elements.

The reactive elements sodium and chlorine combine to form the compound sodium chloride, which is not highly reactive. The sodium and chlorine atoms form ions that have the same electron arrangement as the nearest noble gases. The same theory is used to explain the character of unreactive noble gases and ionic compounds. The empirically determined formulas of other ionic compounds can also be explained using the theory of ion formation.

Consider first the simplest class of ionic compounds: the binary ionic compounds of the representative elements. **Binary ionic compounds** are composed of two kinds of monatomic ions. For example, table salt is composed of the Na^+ cation (positively charged ion) and the Cl^- anion (negatively charged ion). Charges on the ions of the representative elements can be predicted and explained using the restricted quantum mechanics theory of atomic structure (Chapter 2). From the predicted ion charges, you can *explain* the formation of unreactive ionic compounds such as sodium chloride, calcium chloride, and aluminum chloride. To explain empirical formulas of compounds such as these, a simple extension of the ion formation theory can be used. This extended theory includes the idea that *the net electrical charge in a theoretical chemical formula is zero* (Table 3.3, page 104). The sum of the charges on the positive ions is equal to the sum of the charges on the negative ions.

$NaCl_{(s)}$ is explained as $\overset{1^+\ 1^-}{Na^+Cl^-_{(s)}}$. $AlCl_{3(s)}$ is explained as $\overset{3^+\ 3(1^-)}{Al^{3+}Cl^-_{3(s)}}$.

Purity of Lab Chemicals
Even "pure" substances are not completely pure. Chemicals can be purchased in grades of purity — technical, lab, and reagent grades — for different purposes. For example, technical grade chemicals are used for less precise work, whereas reagent grade chemicals are required for more exacting work in the laboratory.

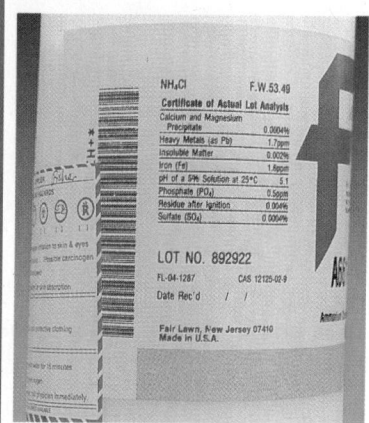

Logical Consistency
Explanations are judged on their ability to explain an observation, generalization, or law in a logical, consistent, and simple fashion. For example, the inert character of noble gases and the inert chemical properties of ionic compounds are explained logically and consistently by using the *same* atomic theory. This atomic theory suggests that there is a magic number of electrons to fill each energy level. When this maximum number is reached, the entity is inert — a nice, simple concept.

Table 3.3

Ion Symbol

The ion symbols in the chemical formula follow an IUPAC convention. Clockwise from the upper right, the element symbol (below) is surrounded by a superscripted ion charge, the subscripted number of "atoms," the subscripted atomic number, and the superscripted mass number. It is customary to omit the mass number and the atomic number in a chemical formula. Also, the ion charge is usually visualized but not written, although it is important to realize that in an ionic compound, the symbols represent ions, not atoms. It is not wrong to write the ion charge, but it is not conventional and you should move toward not including the ion charges in a chemical formula — but do not forget, ionic compounds are composed of ions.

$$^{202}_{80}Hg_2^{2+}$$

Ternary Compounds

Compounds composed of two elements are binary, whereas compounds from three elements are called **ternary**.

EXPLANATIONS OF IONIC FORMULAS			
Chemical Name	Empirical Formula	Ions Involved	Theoretical Formula
sodium chloride	NaCl	$Na^+ \ Cl^-$	Na^+Cl^- or NaCl
calcium chloride	$CaCl_2$	$Ca^{2+} \ Cl^- \ Cl^-$	$Ca^{2+}Cl_2^-$ or $CaCl_2$
aluminum chloride	$AlCl_3$	$Al^{3+} \ Cl^- \ Cl^- \ Cl^-$	$Al^{3+}Cl_3^-$ or $AlCl_3$

To explain an empirically determined ionic formula for a compound, two steps are involved.

- First, predict the charges of the individual ions based on atomic theory and add them to the chemical formula.

- Second, multiply the formula subscripts by the individual ion charges to show that the sum of the positive and negative charges is zero. A **formula unit** of an ionic compound is the smallest amount of the compound that has the composition given by the chemical formula, such as one Al^{3+} ion and three Cl^- ions for aluminum chloride.

A formula unit of aluminum chloride and a crystal of aluminum chloride are both said to have a ratio of aluminum ions to chloride ions of 1:3. A crystal of aluminum chloride that is large enough to see would have an incredibly large number of aluminum and chloride ions, but the overall ratio would still be 1:3. Any explanation of an ionic solid must show the overall ratio of cations to anions such that the net charge is zero (Figure 3.7). There is no such thing as a molecule of an ionic compound. Only molecular substances are composed of molecules.

 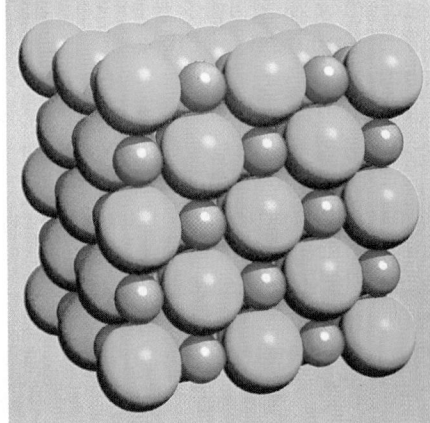

Figure 3.7
To agree with the explanation of the empirical formula for sodium chloride, the model of a sodium chloride crystal must represent both a 1:1 ratio of ions and the shape of the salt crystal.

"The mere formulation of a problem is far more essential than its solution, which may be merely a matter of mathematical or experimental skill. To raise new questions, new possibilities, to regard old problems from a new angle requires creative imagination and marks real advances in science." — Albert Einstein (1879 – 1955)

Explaining Polyatomic Ions

As chemists worked with ionic compounds, they determined some empirical formulas that they had not expected. For example, laboratory work yields formulas such as $NaNO_3$, K_2CO_3, $CaSO_4$, and $Mg(ClO_3)_2$. When scientists tried to explain these formulas in terms of charges on monatomic ions, the results were not logically consistent with previous work.

The unexpected results were based on the assumption that all of the ions are monatomic (composed of only one kind of element/"atom"). When a theory cannot explain empirical results, the theory can be restricted, revised, or replaced. In this case, chemists decided to retain the idea of balanced charges but to revise the theory of ion formation, by establishing a new category of ions. A **polyatomic ion** is a cation or an anion composed of a group of atoms with a net positive or negative charge. The NO_3^-, CO_3^{2-}, SO_4^{2-}, and ClO_3^- ions are illustrated in Figure 3.8. Experimental work indicates that a large number of polyatomic ions exist, most of which have been found to have a negative charge. The concept of polyatomic ions has proven successful in explaining ionic compounds(Figure 3.9). A list of the names and formulas of some common polyatomic ions is provided on the inside back cover of this book. The empirical formulas can now be explained using the ion charge for the simple cations and for the polyatomic anions.

Monatomic Ions	Polyatomic Ions
Na^+	NH_4^+
N^{3-}	NO_3^-
P^{3-}	PO_4^{3-}
S^{2-}	SO_4^{2-}
Cl^-	ClO_3^-

The charge on the polyatomic ion is an overall (net) charge for the group of atoms.

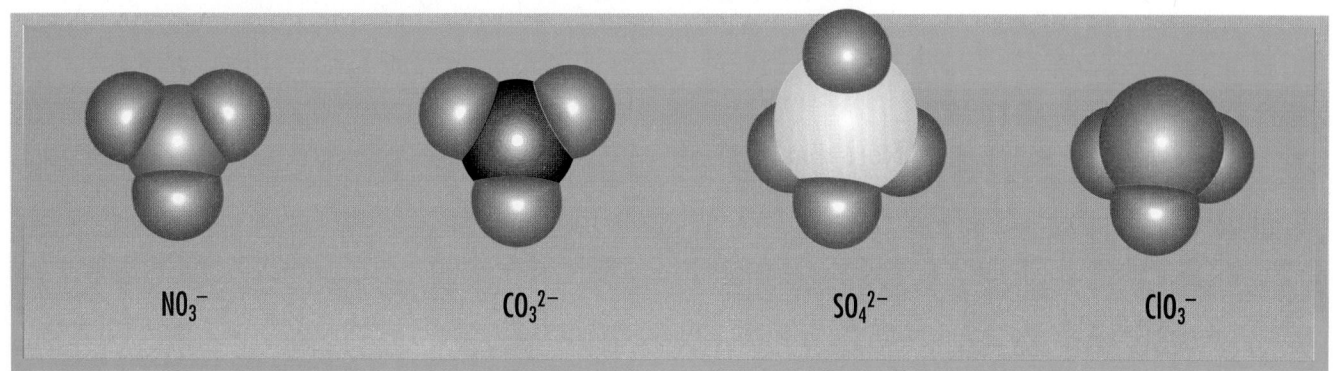

$$NO_3^- \qquad CO_3^{2-} \qquad SO_4^{2-} \qquad ClO_3^-$$

$NaNO_{3(s)}$ is explained as $\overset{1^+}{Na^+}\,\overset{1^-}{NO_3^-}_{(s)}$.

$K_2CO_{3(s)}$ is explained as $\overset{2(1^+)}{K^+_2}\,\overset{2^-}{CO_3^{2-}}_{(s)}$.

$CaSO_{4(s)}$ is explained as $\overset{2^+}{Ca^{2+}}\,\overset{2^-}{SO_4^{2-}}_{(s)}$.

$Mg(ClO_3)_{2(s)}$ is explained as $\overset{2^+}{Mg^{2+}}\,\overset{2(1^-)}{(ClO_3^-)}_{2(s)}$.

Figure 3.8
The concept of polyatomic ions was developed to explain the occurrence of groups of atoms in the empirical formulas of some ionic compounds. A theoretical explanation of polyatomic ions is presented in Chapter 8.

Ions of Multi-Valent Metals

When chemists were determining empirical formulas in the early 1800s, they encountered another result that they could not explain. They discovered that some metals combined with nonmetals in multiple proportions. For example, they found two compounds of iron and oxygen, FeO and Fe_2O_3, and two compounds of copper and oxygen, CuO and Cu_2O. More than 100 years later, an explanation emerged. Chemists now accept that some metals are **multi-valent**; that is, they can form more than one ion, each with its own particular charge. Iron, for example, is said to form the ions Fe^{2+} and Fe^{3+}. This provides an explanation of the two empirical formulas for iron oxide.

$FeO_{(s)}$ is explained as $\overset{2^+}{Fe^{2+}}\overset{2^-}{O^{2-}}_{(s)}$. $\qquad$ $Fe_2O_{3(s)}$ is explained as $\overset{2(3^+)}{Fe^{3+}_2}\overset{3(2^-)}{O^{2-}_3}_{(s)}$.

This explanation is consistent with explanations of the chemical formulas of ionic compounds of representative elements and ionic compounds containing polyatomic ions. It fulfills all of the criteria required for an acceptable scientific explanation — it is *logical*, *consistent*, and *simple*.

Figure 3.9
Not unlike monatomic ions, polyatomic ions, such as the carbonate ions in calcium carbonate, occupy points in space in the ionic crystal lattice.

Multi-valent ions often can be distinguished by their different colored aqueous$_{(aq)}$ solutions.

$Cu^{2+}_{(aq)}$	blue
$Cu^+_{(aq)}$	green
$Fe^{3+}_{(aq)}$	yellow-brown
$Fe^{2+}_{(aq)}$	pale green

See "Ion Colors" in the table on the inside back cover of this textbook.

Hydrates

Hydrates are compounds that decompose at relatively low temperatures to produce water and an associated compound (usually an ionic compound). This evidence indicates that the water, called water of hydration, is loosely held to the ionic compound. In Investigation 1.4, page 44, you decomposed bluestone to produce water and a white substance. A quantitative study of this reaction indicates that five water molecules are included with each formula unit of the compound. When the water of hydration is removed from the hydrate, the product is referred to as **anhydrous** (Figure 3.10). The water molecules are assumed to be electrically neutral.

$$\overset{2^+ \quad 2^- \qquad 5(0)}{Cu^{2+}SO_4^{2-} \cdot 5H_2O_{(s)}} + \text{heat} \rightarrow Cu^{2+}SO_4^{2-}{}_{(s)} + 5H_2O_{(g)}$$

blue   white   colorless

Note that to indicate the presence of water in a hydrate, the formula of the compound is written first, followed by a raised dot and the number of water molecules.

CuSO₄·5H₂O₍s₎

CuSO₄₍s₎

Figure 3.10
When bluestone crystals (CuSO₄•5H₂O₍s₎) are heated, a white powder (CuSO₄₍s₎) is produced. When water is added to the white powder, bluestone is produced.

SUMMARY: IONIC COMPOUNDS

Laboratory investigations indicate that there are several types of ionic compounds:
- binary ionic compounds such as NaCl, MgBr₂, and Al₂S₃
- ternary and other polyatomic ionic compounds such as Li₂CO₃ and (NH₄)₂SO₄
- compounds of multi-valent metals such as CoCl₂ and CoCl₃
- hydrated compounds such as Na₂CO₃•10H₂O and MgSO₄•7H₂O

The empirical formulas of these types of compounds can be explained in a logically consistent way, using two concepts:
- Ionic compounds are composed of two kinds of ions, a cation and an anion.
- The sum of the charges on all the ions is zero.

The value of the theory of ion formation is demonstrated by its success in explaining empirically determined formulas for ionic compounds. The next step is to test the ability of this theory to *predict* the chemical formula of ionic compounds.

Exercise

16. How many different ions are usually found in the chemical formula for an ionic compound?

17. What are the general names given to a positive ion and a negative ion?

18. Write the symbol, complete with charge, for each of the following ions.
    (a) sodium
    (b) chloride
    (c) chlorate
    (d) nitride
    (e) iron(III)
    (f) phosphate
    (g) ammonium
    (h) sulfate
    (i) hydroxide
    (j) sulfite

19. Rewrite the following chemical formulas, indicating the ion charges of the two ions in each.
    (a) $MgCl_{2(s)}$ (found in sea water)
    (b) $Na_2S_{(s)}$ (foul-smelling substance)
    (c) $Al_2O_{3(s)}$ (bauxite)*
    (d) $Li_2CO_{3(s)}$ (drug used to treat some mental disorders)
    (e) $(NH_4)_2SO_{4(s)}$ (fertilizer)
    (f) $Fe_2O_{3(s)}$ (an iron ore)
    (g) $CuSO_4 \cdot 5H_2O_{(s)}$ (bluestone)*
    (h) $SO_{3(g)}$ (pollutant)

Ions
An ionic formula communicates a positive (cation) followed by a negative (anion). You cannot have only positive or only negative ions in a chemical. In the formula, the ratio of ions results in a net charge of zero.

An asterisk in an Exercise question means that you should memorize the formula and description.

3.3 CHEMICAL NAMES AND FORMULAS

Communication systems in chemistry are governed by the International Union of Pure and Applied Chemistry (IUPAC). This organization establishes rules of communication to facilitate the international exchange of knowledge. However, even when a system of communication is international, logical, precise, and simple, it may not be generally accepted if people prefer to stick with old names and are reluctant to change. There are many examples of chemicals that have both traditional names and IUPAC names. Chemical **nomenclature** is the system of names for chemicals. Although names of chemicals are language-specific, the rules for each language are governed by the IUPAC.

The previous section explains the chemical formulas and names of ionic compounds. Besides *explaining* empirical formulas, an acceptable theory must also be able to *predict* future empirical formulas correctly. Experimental evidence provides the test for a prediction made from a theory. A major purpose of scientific work is to test concepts by making predictions.

Nomenclature
Nomenclature refers to a system of names. The word comes from the Latin *nomenclator*, meaning "name-caller." Name-callers in ancient Rome were slaves who accompanied their masters to announce the names of people they met.

PHARMACIST

Jan Thomson works as a pharmacist. Although she set out to become a dental hygienist, she enjoyed the pharmacy courses she took at university and decided to make pharmacy her career goal.

As a pharmacist, Thomson plays a dual role. She is a member of a health-care team who interprets doctors' prescriptions and who provides advice about different medications. She is also a retailer who enjoys the constant stream of customers who come to buy everything from personal care products to candy bars.

The major component of Thomson's work is filling prescriptions. A doctor's prescription contains the patient's name and address, the name and quantity of the prescribed drug, and instructions about how to take the drug. For most people, prescriptions are impossible to read because the instructions are in Latin! For example, "t.i.d." is shorthand for a Latin phrase meaning "three times a day." Similarly, "b.i.d." means "twice a day" and "q.i.d." means "four times a day." A doctor writes "sig" on a prescription, followed by one of these abbreviations, to indicate how to take the medication.

In pharmacy, knowledge of chemistry laboratory techniques is required in order to compound some prescriptions and prepare special creams and ointments. Drugstores are equipped with apparatus familiar to high school science students, such as graduated cylinders, conical flasks, and balances.

Modern pharmacists use computers to store and retrieve information about how drugs interact with each other. However, not all decisions are left to the computer. According to Thomson, "It depends on the computer software system that you are using. Some programs flag the interactions, some don't. You need to understand what interacts with what and the severity of the interaction in order to decide to fill the prescription, or to consult with the physician." All pharmacies keep drug profiles of customers so that drug interactions can be identified. That's why it's a good idea to patronize one pharmacy for all your prescriptions.

To become a pharmacist, a student must enjoy the sciences. University pharmacy programs require students to take courses in biology, organic and inorganic chemistry, biochemistry, anatomy, physiology, and pharmacology. Communication courses train the student to interact effectively with the public, and provide practice in preparing clear, concise instructions about how to take medications. A pharmacist relies on both the label and verbal instructions to ensure that the customer understands exactly when and how to take the medication. Most universities stipulate student enrolment in a general science program before entering the pharmacy program.

Thomson is enthusiastic about her career choice: "It's a good job, a well-paying job, a rewarding job. You assist a lot of people, both in providing more information as to what the doctor is prescribing and in helping to select a non-prescription product to relieve the symptoms of a cold, allergy, diaper rash, or whatever. To give good advice, you need to know what alternatives there are and what advantages and drawbacks they have. If you are willing to help people and to do the hard work to get into pharmacy, it's a really rewarding career."

Another major benefit that pharmacists enjoy is the ability to negotiate their working hours. In this profession, it is possible to work part-time or to choose a shift that allows a pharmacist to do other things, such as pursue further studies or raise a family. Also, pharmacists find it relatively easy to find work almost anywhere, making this an attractive career for people who like to travel or whose spouses are often transferred.

- *What university courses does a pharmacist take?*
- *What equipment in drugstores is similar to equipment in high school laboratories?*

Predicting and Naming Ionic Compounds

Binary Compounds

Charges on monatomic ions of the representative elements are predicted from atomic theory. Charges on monatomic ions of other elements are located in a reference, such as the periodic table on the inside front cover of this book. These charges, along with the idea of balanced charges, are used to make predictions of ionic formulas. To predict an ionic formula from the name of a compound, write the chemical symbol, with its charge, for each of the two ions in the name of the compound. Then predict the simplest whole number ratio of ions to obtain a net charge of zero. For example, for the compound aluminum chloride, the ions are Al^{3+} and Cl^-. For a net charge of 0, the ratio of aluminum ions to chloride ions must be 1:3. The formula for aluminum chloride is therefore $AlCl_3$. This prediction agrees with the chemical formula determined empirically in the laboratory.

A complete chemical formula should also include the state of matter at SATP. Recall the generalization that all ionic compounds are solids at SATP. The complete formula is therefore $AlCl_{3(s)}$.

$$Al_{(s)} + Cl_{2(g)} \rightarrow Al^{3+}Cl^-_{3(s)} \text{ or } AlCl_{3(s)}$$
$$\text{aluminum} + \text{chlorine} \rightarrow \text{aluminum chloride}$$

The name of a binary ionic compound is the name of the cation followed by the name of the anion. The name of the metal ion is stated in full and the name of the nonmetal ion has an *-ide* suffix, for example, magnesium oxide, sodium fluoride, and aluminum sulfide. Remember, name the two ions.

> Note that the focus here is on writing the correct chemical formulas for reactants and products, not on balancing the equation. The chemical equation is shown here along with the word equation.

Exercise

20. What is the rule for naming binary ionic compounds?

21. Use the IUPAC rules to name the following binary ionic compounds.
    (a) table salt, $NaCl_{(s)}$*
    (b) lime, $CaO_{(s)}$*
    (c) road salt, $CaCl_{2(s)}$*
    (d) magnesia, $MgO_{(s)}$
    (e) potash, $KCl_{(s)}$*
    (f) bauxite, $Al_2O_{3(s)}$*
    (g) zinc ore, $ZnS_{(s)}$
    (h) a hydride, $CaH_{2(s)}$

22. What is the rule for writing chemical formulas for binary ionic compounds?

23. Write the chemical formulas and IUPAC names for the binary ionic products of the following chemical reactions. Do not get distracted by the formulas for the nonmetals or try to balance the equations.
    E.g., $Li_{(s)} + Br_{2(l)} \rightarrow Li^+Br^-_{(s)}$ or $LiBr_{(s)}$ (lithium bromide)
    (a) $Sr_{(s)} + O_{2(g)} \rightarrow$
    (b) $Ag_{(s)} + S_{8(s)} \rightarrow$

> **Aluminum Oxide**
> Rubies and sapphires are composed primarily of aluminum oxide. Impurities impart the color to these gems. Aluminum oxide crystals are very hard and are even used in sandpaper. Aluminum oxide is also found in bauxite, the ore that is refined into aluminum at Kitimat, British Columbia. Potter's clay also contains aluminum oxide.

> **Formulas for Elements**
> The chemical formulas for elements have been determined empirically and are listed in Table 3.5 on page 116. For now you will be given the formulas for elements. Ignore the element formulas as in question 23 since they do not affect the formulas of compounds formed.

(c) $Zn_{(s)} + I_{2(s)} \rightarrow$
(d) $Al_{(s)} + Cl_{2(g)} \rightarrow$
(e) $Ca_{(s)} + S_{8(s)} \rightarrow$
(f) $Na_{(s)} + Cl_{2(g)} \rightarrow$
(g) $Ga_{(s)} + O_{2(g)} \rightarrow$

Multi-Valent Metals

Most transition metals and some representative metals can form more than one kind of ion, i.e., they are multi-valent. For example, iron can form an Fe^{3+} ion or an Fe^{2+} ion, although Fe^{3+} is more common. In the reaction between iron and oxygen, two products are possible. The chemical formulas for the possible ionic compounds formed by the reaction are predicted in the standard way, by examining ion charges and balancing charges.

Writing Ionic Formulas
Predict ionic formulas by using a ratio of ions such that the net charge is zero.

$$\overset{2(3^+)\ 3(2^-)}{Fe^{3+}_2O^{2-}_3} \qquad\qquad \overset{2^+\ \ 2^-}{Fe^{2+}O^{2-}}$$

$$Fe_2O_{3(s)} \qquad\qquad FeO_{(s)}$$

In the periodic table on this book's inside front cover, possible ion charges are shown, with the more common charge listed first

▮ PRODUCT LABELS ▮

Consumer products are generally sold under a trade name that often gives few clues to the chemical contents of the product. In Canada, manufacturers are required by law to list any substances added to the initial natural product. Chemical information usually appears in fine print on the container's label.

Reading product labels is important for several reasons, including chemical education, cost comparisons, and product evaluation. Most people do not realize the extent to which we consume chemicals such as aluminum silicate, calcium phosphate, sodium glutamate (MSG), phosphoric acid, iron(II) sulfate, butylated hydroxytoluene (BHT), and others. These chemicals are common additives in foods. The most widely used drug in the world is acetylsalicylic acid, which is sold under its generic name (ASA) or under various brand names such as Aspirin.

Reading and using information printed on product labels enables cost comparisons. Since chemical communication is precise and international, a specific chemical name must always refer to the same chemical substance. If apparently different products contain the same amounts of all the same chemicals, they may be effectively the same, in spite of what advertisements may claim.

You will find that IUPAC rules are often broken when the names of chemicals are listed on product labels. For example, sodium glutamate (a preservative and flavor enhancer) is called monosodium glutamate (MSG); sodium phosphate (a cleaning agent) is called trisodium phosphate (TSP); and calcium hydrogen phosphate in baking powder is called monocalcium phosphate. For safety reasons, this lack of standardization is unfortunate.

Logic and consistency sometimes lose out to tradition in chemical communication. Some people prefer to use common names for chemicals.

You may be familiar with these common names given in Appendix G: brine, Glauber's salt, gypsum, lime, limestone, lye, muriatic acid, potash, sand (silica), slaked lime, and soda ash.

- *What are the corresponding chemical names and formulas?*
- *What regulations, if any, would you recommend for listing and naming chemicals on consumer products?*

(Figure 3.11). If the ion of a multi-valent metal is not specified in a description or an exercise question, you can assume the charge on the ion is the most common one.

(theoretical)

atomic number	**26**	55.85	atomic molar mass (g/mol)
electronegativity	1.8	1535	melting point (°C)
common ion charge	3+	2750	boiling point (°C)
other ion charge	2+	7.87	density of solids, liquids (g/cm³)

(empirical)

density of gases (g/L)
gases in red
liquids in green
synthetic in blue

Fe
iron

element symbol
element name

To name the compounds, name the two ions. In the IUPAC system, the name of the multi-valent metal includes the ion charge. The ion charge is given in Roman numerals in brackets; for example, iron(III) is the name of the Fe^{3+} ion and iron(II) is the name of the Fe^{2+} ion. The Roman numerals indicate the charge on the ion, not the number of ions in the formula. The names of the previously mentioned compounds are:

$$Fe_{(s)} + O_{2(g)} \rightarrow Fe_2O_{3(s)} \quad \text{iron(III) oxide}$$
$$Fe_{(s)} + O_{2(g)} \rightarrow FeO_{(s)} \quad \text{iron(II) oxide}$$

To determine the chemical name from a given chemical formula containing an ion of a multi-valent metal, determine the necessary charge on that ion to yield a net charge of zero. For example, suppose the empirical formula of a compound is found to be MnO_2. The charge on the Mn ion must balance the charge on the two oxide ions, which are known to have a charge of 2^-. Let x represent the charge on the Mn ion.

x	$2(2^-)$	$x + 2(2^-) = 0$
MnO_2		$x + (4^-) = 0$
manganese(IV) oxide		$x - 4 = 0$
		$x = 4^+$

The manganese ion must have a charge of 4^+. As before, the name of the compound is made up of the names of the two ions, so MnO_2 is called manganese(IV) oxide. With practice, you can calculate the ion charge simply by inspecting the formula.

Classical System
An older (classical) system for naming ions of multi-valent metals uses the Latin name for the element with an *-ic* suffix for the larger charge and an *-ous* suffix for the smaller charge. This system is used only for multi-valent metals that were known when Latin was a common language among scientists and for multi-valent metals with no more than two possible ion charges. In this system, iron(III) oxide is "ferric oxide" and iron(II) oxide is "ferrous oxide."

Fe^{3+}	ferric
Fe^{2+}	ferrous
Cu^{2+}	cupric
Cu^+	cuprous
Sn^{4+}	stannic
Sn^{2+}	stannous
Pb^{4+}	plumbic
Pb^{2+}	plumbous
Sb^{5+}	stibnic
Sb^{3+}	stibnous

IUPAC System
You will use the IUPAC system of naming ionic compounds rather than the classical system. Name the compound by naming the two ions. If the metal is multi-valent, the name of the metal ion includes the charge on the ion; for example, Cu^{2+} is called copper(II).

Exercise

24. What is the rule for writing chemical formulas for multi-valent, binary ionic compounds?

25. What is the rule for naming multi-valent, binary ionic compounds?

26. Write the chemical formulas for the following ionic compounds, complete with the SATP pure state of matter.
    (a) mercury(II) sulfide (cinnabar ore)
    (b) molybdenum(IV) sulfide (molybdenite ore)

(c) manganese(IV) oxide (pyrolusite ore)
(d) nickel(II) bromide (forms green solution)
(e) copper(II) chloride (forms blue solution)
(f) iron(III) iodide (forms yellow-brown solution)

27. Write the chemical formula and IUPAC name of the most common ionic product for each of the following chemical reactions. Do not get distracted by the formulas for the nonmetals or try to balance the equations.

E.g., $Bi_{(s)} + O_{2(g)} \rightarrow Bi_2O_{3(s)}$ (bismuth(III) oxide)

(a) $Cu_{(s)} + Br_{2(l)} \rightarrow$
(b) $Ni_{(s)} + O_{2(g)} \rightarrow$
(c) $Pb_{(s)} + S_{8(s)} \rightarrow$
(d) $Sn_{(s)} + I_{2(s)} \rightarrow$
(e) $Hg_{(l)} + Cl_{2(g)} \rightarrow$
(f) $Cr_{(s)} + S_{8(s)} \rightarrow$
(g) $Co_{(s)} + Cl_{2(g)} \rightarrow$
(h) $Fe_{(s)} + O_{2(g)} \rightarrow$

28. Use the IUPAC rules to name the following binary ionic compounds found in metal ores.

(a) cassiterite, $SnO_{2(s)}$
(b) chalcocite, $Cu_2S_{(s)}$
(c) galena, $PbS_{2(s)}$
(d) hematite, $Fe_2O_{3(s)}$
(e) molybdite, $MoO_{3(s)}$
(f) bauxite, $Al_2O_{3(s)}$
(g) argentite, $Ag_2S_{(s)}$
(h) zincite, $ZnO_{(s)}$

29. (Internet) Search for information on metal ores mined by industries in your region of the country.

Compounds with Polyatomic Ions

Charges on polyatomic ions can be found in a table of polyatomic ions, such as the one on the inside back cover of this book. Predicting the formula of ionic compounds involving polyatomic ions is done in the same way as for binary ionic compounds. Write the ion charges and then use a ratio of ions that yields a net charge of zero. For example, to predict the formula of a compound containing copper ions and nitrate ions, write the following:

$$\underset{}{Cu^{2+}(NO_3^-)_2} \qquad\qquad Cu(NO_3)_{2(s)} \qquad \text{copper(II) nitrate}$$

Two nitrate ions are required to balance the charge on one copper(II) ion (Figure 3.12). Note that parentheses are used in the formula to indicate the presence of more than one polyatomic ion. Do not use parentheses with one polyatomic ion or with simple ions. Do not write: $Ag_2(SO_4)_{(s)}$ or $(Ag)_2SO_{4(s)}$.

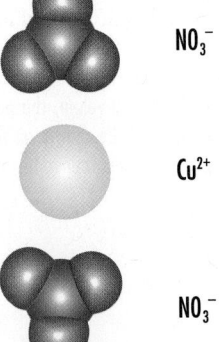

NO_3^-

Cu^{2+}

NO_3^-

Figure 3.12
According to theory, two nitrate groups are required to balance the charge on one copper(II) ion. This theory agrees with observations.

Exercise

30. According to the theory that explains and predicts the chemical formulas for ionic compounds, how many kinds of ions are present in an ionic compound?

31. Sketch diagrams of the sulfate and carbonate polyatomic ions.

32. Write the IUPAC name for each of the following ionic compounds containing polyatomic ions.
    (a) in tobacco, $NaNO_{3(s)}$
    (b) meat preservative, $NaNO_{2(s)}$
    (c) fertilizer, $(NH_4)_2HPO_{4(s)}$
    (d) fertilizer, $NH_4H_2PO_{4(s)}$
    (e) forms blue solution, $Cu(NO_3)_{2(s)}$
    (f) forms green solution, $CuNO_{3(s)}$
    (g) fireproofing, $Na_2SiO_{3(s)}$
    (h) coffee whitener, $Al_2(SiO_3)_{3(s)}$ (see margin note)

33. Write the chemical formula complete with the SATP pure state of matter for each of the following ionic compounds containing polyatomic ions.
    (a) sodium phosphate (rust remover)
    (b) sodium bicarbonate (baking soda) (see margin note)
    (c) sodium carbonate (washing soda)
    (d) calcium hydroxide (slaked lime)
    (e) ammonium nitrate (fertilizer)
    (f) ammonium phosphate (fertilizer)
    (g) copper(II) sulfate (fungicide)
    (h) magnesium hydroxide (milk of magnesia)

34. According to the reported research (see margin note), is aluminum a cause or an effect of Alzheimer's disease?

35. Write empirical and theoretical definitions of an ionic compound.

Ionic Hydrates

There are two common systems of naming ionic hydrates that are acceptable to the IUPAC. The number of water molecules associated with each formula unit is indicated by either a number or a prefix (Table 3.4). For example, bluestone, $CuSO_4 \cdot 5H_2O_{(s)}$, is named:
   copper(II) sulfate pentahydrate or
   copper(II) sulfate-5-water
Washing soda, $Na_2CO_3 \cdot 10H_2O_{(s)}$, is named:
   sodium carbonate decahydrate or
   sodium carbonate-10-water

Hydrates can be made anhydrous by heating. Anhydrous compounds become hydrates by association with water. For example, a simple antiperspirant works as follows:

$AlCl_{3(s)}$ + 6 $H_2O_{(l)}$ → $AlCl_3 \cdot 6H_2O_{(s)}$
(anhydrous)   aluminum chloride hexahydrate

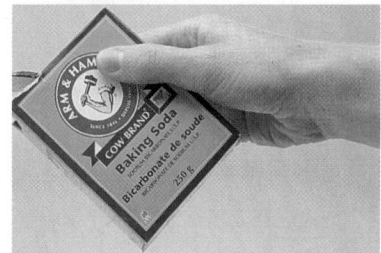

Table 3.4

PREFIXES USED IN CHEMICAL NAMES	
mono	1
di	2
tri	3
tetra	4
penta	5
hexa	6
hepta	7
octa	8
ennea	9
deca	10

Anhydrous $CuSO_{4(s)}$ is a white powder. The hydrate, $CuSO_4 \cdot 5H_2O_{(s)}$ is a blue crystalline solid (see Figure 3.10, page 106).

Classes of Ionic Compounds

Regardless of the class of ionic compounds, they are all named by the same rules, and their formulas are all determined in the same way.

1. Mono-Valent Binary

   $Na_2S_{(s)}$, sodium sulfide

   $AlF_{3(s)}$, aluminum fluoride

2. Multi-Valent Binary

   $PbS_{(s)}$, lead(II) sulfide

   $PbO_{2(s)}$, lead(IV) oxide

3. Ternary (Polyatomic Ions)

   $(NH_4)_3PO_{4(s)}$, ammonium phosphate

   $Cu(NO_3)_{2(s)}$, copper(II) nitrate

4. Hydrates

   $Co(NO_3)_2 \cdot 6H_2O_{(s)}$, cobalt(II) nitrate hexahydrate

   $FeSO_4 \cdot 7H_2O_{(s)}$, iron(II) sulfate-7-water

Exercise

36. Write IUPAC names for the following anhydrous and hydrated ionic compounds.
    (a) white powder, $CuSO_{4(s)}$ (see margin note)
    (b) bluestone, $CuSO_4 \cdot 5H_2O_{(s)}$
    (c) anhydrous $Na_2SO_{4(s)}$
    (d) Glauber's salt, $Na_2SO_4 \cdot 10H_2O_{(s)}$
    (e) anhydrous $MgSO_{4(s)}$
    (f) Epsom salt, $MgSO_4 \cdot 7H_2O_{(s)}$

37. Write the chemical formulas for the following ionic hydrates.
    (a) iron(III) oxide trihydrate (rust)
    (b) aluminum chloride-6-water (antiperspirant product)
    (c) sodium thiosulfate pentahydrate (photographic "hypo")
    (d) cadmium(II) nitrate-4-water (photographic emulsion)
    (e) lithium chloride tetrahydrate (fireworks)
    (f) calcium chloride-2-water (de-icer)

38. How do you convert a hydrate to an anhydrous compound?

39. How do you convert an anhydrous compound to a hydrate?

40. Write a design for an experiment to identify two unlabelled samples as $Na_2SO_4 \cdot 10H_2O_{(s)}$ and $Na_2SO_4 \cdot 7H_2O_{(s)}$.

41. (Enrichment) Use a reference to determine the temperature required to convert bluestone from a solid blue hydrate to a white anhydrous solid.

SUMMARY: IONIC COMPOUNDS

- To name an ionic compound, name the two ions.
- To write an ionic formula, determine the ratio of ions that yields a net charge of zero.

Exercise

42. Provide the chemical formula or IUPAC name, as required, for each of the following ionic compounds. Include the SATP pure state of matter for each compound.
    (a) $Na_2O_{(s)}$ (f) $CaSO_4 \cdot 2H_2O_{(s)}$
    (b) calcium sulfide   (g) lead(IV) oxide
    (c) $KNO_{3(s)}$ (h) sodium sulfate decahydrate
    (d) iron(III) chloride   (i) aluminum oxide
    (e) $HgO_{(s)}$ (j) calcium phosphate

43. For the IUPAC chemical names in each of the following word equations, write the corresponding chemical formulas (including the SATP pure states of matter) to form a chemical equation. (It is not necessary to balance the chemical equation.)

(a) Sodium hypochlorite is a common disinfectant and bleaching agent. This compound is produced by the reaction of chlorine, $Cl_{2(g)}$, with lye.

chlorine$_{(g)}$ + sodium hydroxide$_{(aq)}$ →
      sodium chloride$_{(aq)}$ + water$_{(l)}$ + sodium hypochlorite$_{(aq)}$

(b) Sodium hypochlorite solutions are unstable when heated and slowly decompose.

sodium hypochlorite$_{(aq)}$ →
      sodium chloride$_{(aq)}$ + sodium chlorate$_{(aq)}$

(c) The calcium oxalate produced in the following reaction is used in a further reaction to produce oxalic acid, a common rust remover.

sodium oxalate$_{(aq)}$ + calcium hydroxide$_{(s)}$ →
      calcium oxalate$_{(s)}$ + sodium hydroxide$_{(aq)}$

(d) Blue cobalt chloride paper turns pink when exposed to water.

cobalt(II) chloride$_{(s)}$ + water$_{(l)}$ → cobalt(II) chloride hexahydrate$_{(s)}$

44. Write the international chemical formulas with SATP pure states of matter and the IUPAC names for the compounds formed from the following elements. Unless otherwise indicated, assume that the most common metal ion is formed. (Write the full chemical equations, but it is not necessary to balance the equations.)
  (a) $Mg_{(s)} + O_{2(g)} →$
  (b) $Ba_{(s)} + S_{8(s)} →$
  (c) $Sc_{(s)} + F_{2(g)} →$
  (d) $Fe_{(s)} + O_{2(g)} →$
  (e) $Hg_{(l)} + Cl_{2(g)} →$
  (f) $Pb_{(s)} + Br_{2(l)} →$
  (g) $Co_{(s)} + I_{2(s)} →$

45. For the chemical formulas in each of the following equations, write the corresponding IUPAC names to form a word equation.
  (a) The main product of the following reaction (besides table salt) is used as a food preservative.

$NH_4Cl_{(aq)} + NaC_6H_5COO_{(aq)} → NH_4C_6H_5COO_{(aq)} + NaCl_{(aq)}$

  (b) Aluminum compounds, such as the one produced in the following reaction, are important constituents of cement.

$Al(NO_3)_{3(aq)} + Na_2SiO_{3(aq)} → Al_2(SiO_3)_{3(s)} + NaNO_{3(aq)}$

  (c) Sulfides are foul-smelling compounds that can react with water to produce basic solutions.

$Na_2S_{(s)} + H_2O_{(l)} → NaHS_{(aq)} + NaOH_{(aq)}$

  (d) Nickel(II) fluoride may be prepared by the reaction of nickel ore with hydrofluoric acid.

$NiO_{(s)} + HF_{(aq)} → NiF_{2(aq)} + H_2O_{(l)}$

46. (Extension) Potassium aluminum sulfate, $KAl(SO_4)_2$, is commonly known as alum and is used as a water clarifier. Use the theory of ion formation to explain the formula of this compound, called a double salt.

47. (Discussion) What types of knowledge are used to determine the chemical formulas of polyatomic ions? What major type of scientific knowledge do chemists like to work towards?

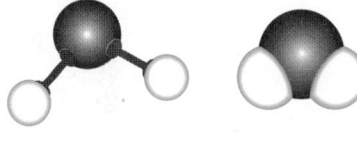

water (H₂O)

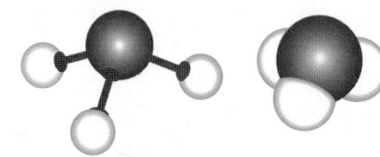

ammonia (NH₃)

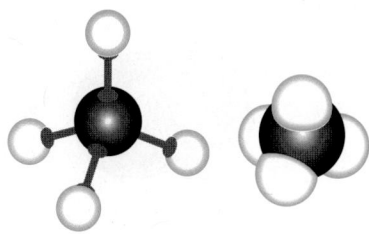

methane (CH₄)

Figure 3.13
Molecular models of H₂O, NH₃, and CH₄ help in understanding the theoretical explanation for the empirical formulas of water, ammonia, and methane. Space-filling models are generally preferred by scientists to ball-and-stick models, although both have advantages and limitations.

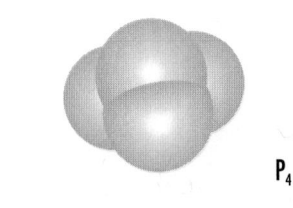

P_4

S_8

H_2

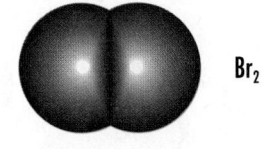

Br_2

Figure 3.14
Models representing the molecular elements P_4, S_8, H_2, and Br_2.

Molecular Formulas

Many empirical molecular formulas, such as H_2O, NH_3, and CH_4, had been determined in the laboratory by the early 1800s, but chemists could not explain or predict molecular formulas using the same theory as for ionic compounds. The theory that was accepted for these compounds was the idea that nonmetal atoms share electrons and that the sharing holds the atoms together in a group called a **molecule**. The chemical formula of a molecular substance — called a **molecular formula** — indicates the number of atoms of each kind in a molecule (Figure 3.13, page 115).

The common feature that this molecular theory shares with ion theory was that the atoms achieved a stable, low-energy state similar to the noble gases. According to molecular theory, nonmetal atoms share electrons in a covalent bond to attain a maximum number of valence electrons rather than gaining electrons from metal atoms. A more complete description of this molecular theory and covalent bonds is presented in Chapter 8. Until you reach Chapter 8, you will have no theoretical way to explain and predict molecular formulas. In the meantime, you will either be given formulas, look them up in references, or memorize them. In the exercises in this section, you will be given the names or the empirically determined formulas for molecular substances. You will start with molecular elements and then move to molecular compounds.

Molecular Elements

As you have seen from the given chemical formulas for elements in the preceding examples and exercises, the chemical formula of all metals is shown as a single atom, whereas nonmetals frequently form diatomic molecules (that is, molecules containing two atoms). Some useful rules are provided in Table 3.5 (the examples in this table should be memorized). An explanation of these rules is in Chapter 8. Some students find they are able to memorize the diatomic elements by grouping the elements ending in *-gen*; for example, hydro*gen*, nitro*gen*, oxy*gen*, and the halo*gens*. $O_{3(g)}$ is a special unstable form of oxygen called ozone. S_8 is called cyclooctasulfur, octasulfur, or usually just sulfur. Figure 3.14 illustrates models of some of these molecules.

Table 3.5

THE CHEMICAL FORMULAS OF METALLIC AND MOLECULAR ELEMENTS		
Class of Elements	**Chemical Formula**	**Examples**
metallic elements	all are monatomic	$Na_{(s)}$, $Hg_{(1)}$, $Zn_{(s)}$, $Pb_{(s)}$
molecular elements (nonmetallic)	some are diatomic	$H_{2(g)}$, $N_{2(g)}$, $O_{2(g)}$, $F_{2(g)}$, $Cl_{2(g)}$, $Br_{2(1)}$, $I_{2(s)}$
	some have molecules containing more than two atoms	$O_{3(g)}$, $P_{4(s)}$, $S_{8(s)}$
	all noble gases are monatomic	$He_{(g)}$, $Ne_{(g)}$, $Ar_{(g)}$
other elements	the rest of the elements can be assumed to be monatomic	$C_{(s)}$, $Si_{(s)}$

3.3 Demonstration: Determining an Empirical Formula

The purpose of this demonstration is to illustrate how the technological process of electrolysis may be used to determine empirically the chemical formula of water. The design of the experiment is to decompose water using an electric current, identify the products (see Diagnostic Tests in Appendix C, C.3), and establish an empirical formula for water. In this experiment, it is assumed that the only products of the decomposition of water are the two gases produced by the electrolysis.

- ▢ Problem
- ▢ Prediction
- ▢ Design
- ▢ Materials
- ▢ Procedure
- ✔ Evidence
- ✔ Analysis
- ✔ Evaluation
- ▢ Synthesis

Problem

What is the chemical formula for water?

Prediction

According to previous memorization, the chemical formula for water is H_2O.

Experimental Design

Water is decomposed in a Hoffman apparatus (Figure 3.15). The volumes of gases produced and the evidence from the diagnostic tests for the presence of hydrogen and oxygen gases are collected. Assume that both hydrogen and oxygen gases are diatomic.

Diagnostic test for hydrogen: If a lit match is held to a gas, and the gas explodes (pops), then the gas is likely hydrogen.

Diagnostic test for oxygen: If a glowing splint is held to a gas, and the glowing splint glows more brightly or relights, then the gas is likely oxygen.

Materials

lab apron	sodium sulfate solution (catalyst)
safety glasses	two 18 × 150 mm test tubes
Hoffman apparatus	wood splints
two connecting wires	matches
6 to 12 V power supply	

Procedure

1. Fill the Hoffman apparatus with the sodium sulfate solution.

2. Connect the apparatus to the power supply and allow it to operate.

3. Measure and record the volume of each gas produced.

4. Collect a sample of each gas in a test tube.

5. Perform the diagnostic test for hydrogen on the larger volume of gas produced. Only a small quantity of the gas is needed for testing.

6. Perform the diagnostic test for oxygen on the smaller volume of gas produced. Only a small quantity of the gas is needed for testing.

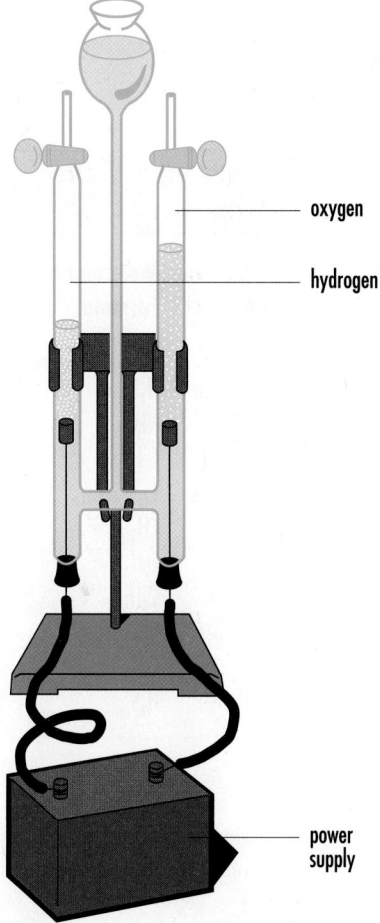

oxygen

hydrogen

power supply

Figure 3.15
A Hoffman apparatus was used in the early 1800s to decompose water into its elements by electricity. The apparatus is designed to collect and measure the volume of gases.

Molecular compounds are sometimes called *covalent compounds* because the molecule contains covalent bonds. The preferred term in this textbook is *molecular* because the substance is composed of molecules. In summary, ionic compounds are composed of ions, whereas molecular compounds are composed of molecules.

Molecular Compounds

The names of some compounds communicate the number of atoms in a molecule. The IUPAC has assigned Greek numerical prefixes to the names of binary molecular compounds. The prefixes are the same as those used in naming hydrates (Table 3.4, page 113). Other naming systems are used when a molecule has more than two kinds of atoms.

EXAMPLE

The following are examples of names of binary molecular compounds. Recall that *binary* refers to compounds composed of only two kinds of atoms and that *molecular* refers to compounds composed only of nonmetals.

Reactants		Product	Name
$C_{(s)}$ + $S_{8(s)}$	→	$CS_{2(l)}$	carbon disulfide
$N_{2(g)}$ + $I_{2(s)}$	→	$NI_{3(s)}$	nitrogen triiodide
$N_{2(g)}$ + $O_{2(g)}$	→	$N_2O_{(g)}$	dinitrogen oxide
$P_{4(s)}$ + $O_{2(g)}$	→	$P_4O_{10(s)}$	tetraphosphorus decaoxide

Table 3.6

COMMON MOLECULAR COMPOUNDS

IUPAC Name	Molecular Formula
water	$H_2O_{(l)}$ or $HOH_{(l)}$
hydrogen peroxide	$H_2O_{2(l)}$
ammonia	$NH_{3(g)}$
glucose	$C_6H_{12}O_{6(s)}$
sucrose	$C_{12}H_{22}O_{11(s)}$
methane	$CH_{4(g)}$
propane	$C_3H_{8(g)}$
octane	$C_8H_{18(l)}$
methanol	$CH_3OH_{(l)}$
ethanol	$C_2H_5OH_{(l)}$
hydrogen sulfide	$H_2S_{(g)}$

Naming Molecular Compounds

According to IUPAC rules, the prefix system is used only for naming **binary molecular compounds** — molecular compounds composed of two kinds of atoms. This rule is similar to that used for ionic compounds; that is, name the two ions. Other systems of communication have been established for naming some common molecular compounds. For now, memorize the names of a few of these common molecular compounds (Table 3.6). Eventually you will supplement memorization with theoretical knowledge of chemical formulas.

For hydrogen compounds such as hydrogen sulfide, $H_2S_{(g)}$, the common practice is *not* to use the prefix system. In other words, we do not call this compound dihydrogen sulfide. Don't be surprised that hydrogen compounds are an exception to the rule — hydrogen is almost always an exception. Some scientists refer to hydrogen as the black sheep of the element family. So far you have seen that, unlike other elements, hydrogen can form both a cation and an anion, some hydrogen compounds can be acids, and hydrogen compounds require a special naming system. Other exceptions for hydrogen will be discussed in Chapter 8.

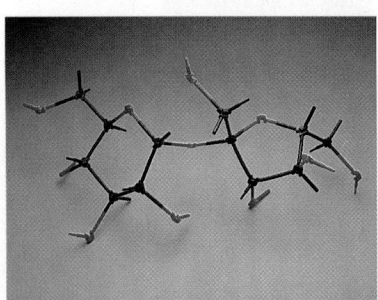

Figure 3.16
This photograph shows a model of sucrose, first synthesized by Raymond Lemieux, a Canadian who was born, grew up, and worked in Alberta.

Exercise

48. Write an empirical and a theoretical definition of molecular compounds.

49. Until a theoretical way of knowing molecular formulas is available, you must be given the formula or name and then rely on memory or use the prefix system to provide the name or formula as required. Provide the names or formulas (complete with the SATP states of matter) for the following substances.

(a) chlorine (toxic)
(b) phosphorus (reacts with air)
(c) helium (inert)
(d) carbon (black)
(e) $NH_{3(g)}$ (smelling salts)
(f) sucrose (sugar) (see Figure 3.16)
(g) $C_2H_5OH_{(l)}$ (alcohol)
(h) methane (fuel — natural gas)

50. Write the chemical formulas for the following molecular substances emitted as gases from the exhaust system of an automobile. Most of these substances may produce acid rain.
(a) nitrogen
(b) carbon dioxide
(c) carbon monoxide
(d) octane
(e) nitrogen dioxide
(f) nitrogen monoxide
(g) dinitrogen oxide
(h) dinitrogen tetraoxide
(i) sulfur dioxide
(j) water

51. Write chemical equations to accompany the given statements or word equations, including the SATP states of matter.

E.g., nitrogen + oxygen → nitrogen dioxide
$$N_{2(g)} + O_{2(g)} \rightarrow NO_{2(g)}$$

(a) Solid silicon reacts with gaseous fluorine to produce gaseous silicon tetrafluoride.
(b) Solid boron reacts with gaseous hydrogen to produce gaseous diboron tetrahydride.
(c) Aqueous sucrose and water react to produce aqueous ethanol and carbon dioxide gas.
(d) Methane gas reacts with oxygen gas to produce liquid methanol.
(e) nitrogen + oxygen → dinitrogen oxide
(f) nitrogen oxide + oxygen → nitrogen dioxide
(g) octane + oxygen → carbon dioxide + water vapor
(h) octane + oxygen →
    carbon dioxide + carbon monoxide + carbon + water vapor

52. (Enrichment) If automobiles are burning gasoline (assume $C_8H_{18(l)}$), why are nitrogen oxide pollutants being produced? There is no nitrogen in gasoline!

Nitrogen Oxides
Nitrogen oxides (NO_x) and carbon oxides (CO_x) are air pollutants emitted from car exhausts. Carbon dioxide is produced from burning automobile fuel: gasoline, propane, natural gas, or alcohol. Carbon dioxide traps radiated heat from Earth, which causes global warming. Carbon monoxide is an odorless, toxic gas that results from incomplete combustion of automobile fuel.

Nitrogen oxides include nitrogen dioxide, nitrogen monoxide (nitric oxide), and dinitrogen oxide (nitrous oxide). These gases are produced as a result of the high temperature and pressure in a car's combustion cylinder. Nitrogen dioxide is the brown gas responsible for the color of smog.

Gaseous Oxides
All carbon, nitrogen, and sulfur oxides are produced as gases in most chemical reactions; e.g., $CO_{2(g)}$, $NO_{2(g)}$, and $SO_{2(g)}$.

SUMMARY: ELEMENTS AND MOLECULAR COMPOUNDS

• Empirically, molecular compounds as pure substances are solids, liquids, or gases at SATP. If they dissolve in water, their aqueous solutions do not conduct electricity.

Elements
1. **Metals**
   - all monatomic; e.g., $Mg_{(s)}$, $Au_{(s)}$, and $Hg_{(l)}$
2. **Nonmetals**
   - memorize formulas; e.g., $H_{2(g)}$, $Cl_{2(g)}$, $P_{4(s)}$, and $S_{8(s)}$

Molecular Compounds
1. **Binary Molecular (two elements)**
   - prefix system; e.g., $P_4O_{10(s)}$, tetraphosphorus decaoxide
   - memorized; e.g., $NH_{3(g)}$, ammonia
2. **Ternary Molecular (three elements)**
   - memorized; e.g., $C_6H_{12}O_{6(s)}$, glucose

- Theoretically, molecular elements and compounds are formed by nonmetal atoms bonding covalently to share electrons in an attempt to obtain the same number of electrons as the nearest noble gas.
- The chemical formulas for all metallic elements are monatomic; e.g., aluminum is $Al_{(s)}$ and iron is $Fe_{(s)}$.
- The chemical formulas for nonmetallic elements are memorized at this time. Memorize the formulas in Table 3.5 on page 116.
- The chemical formulas and/or the names of molecular compounds are given. You will not predict these formulas until you learn Chapters 8 and 9.
- Memorize the chemical formulas, names, and states of matter for the selection of common binary and ternary molecular compounds provided in Table 3.6 on page 118.
- Memorize the prefixes provided in Table 3.4 on page 113.
- The chemical formulas for most binary molecular compounds are obtained from the prefixes in the given names; e.g., dinitrogen tetraoxide gas is $N_2O_{4(g)}$.
- The chemical names for most binary molecular compounds use prefixes to communicate the formula subscripts; e.g., $N_2S_{5(l)}$ is dinitrogen pentasulfide.
- The SATP states of matter of metallic and nonmetallic elements are memorized or referenced from the periodic table inside the front cover of this book.
- The SATP states of matter of selected molecular compounds are memorized from Table 3.6 on page 118. For other molecular compounds referred to in questions, you are given the states of matter.

- ▪ Problem
- ▪ Prediction
- ▪ Design
- ▪ Materials
- ▪ Procedure
- ✔ **Evidence**
- ✔ **Analysis**
- ▪ Evaluation
- ▪ Synthesis

INVESTIGATION

3.4 Molecular Models

The purpose of this investigation is to use molecular models to produce images of molecules. Assume that the models provide acceptable three-dimensional structures. Where more than one structure is possible for a molecule, the most symmetrical structure is accepted.

Problem

What are the three-dimensional structures of the molecular substances water, hydrogen peroxide, hydrogen sulfide, methane, methanol, ethanol, propane, ammonia, chlorine, and sulfur (cyclooctasulfur)?

Experimental Design

Models of different molecules are constructed using molecular model kits. The three-dimensional arrangement of atoms in each molecule is determined by observing the model.

Materials
molecular model kit

Procedure

1. (Prelab) From the names of the substances listed in the Problem statement, write the accepted molecular formulas in a table of evidence.

2. Assemble symmetrical models to represent the molecular formula of each substance. (When assembling or disassembling models that have springs, always twist clockwise to avoid unravelling the spring.)

3. Draw a structural diagram of the model in your evidence table. (See Figure 3.17.)

4. Do not disassemble the models until your teacher has checked them.

5. When you are finished, organize the components in the box provided and have your teacher check the box.

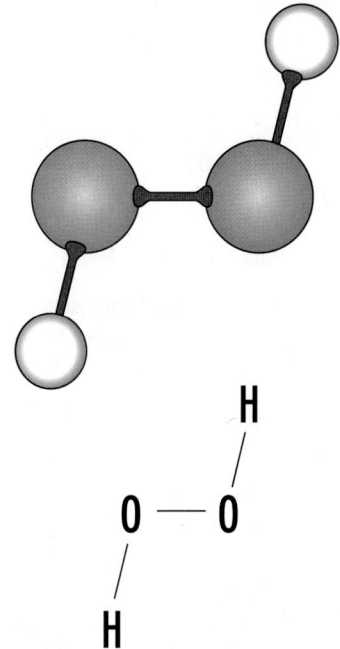

Figure 3.17
The most symmetrical, theoretical structure for hydrogen peroxide can be communicated by a structural diagram using lines and atomic symbols.

Naming and Writing Formulas for Acids and Bases

In this chapter, acids and bases are given very restricted empirical and theoretical definitions. Aqueous hydrogen compounds that make blue litmus paper turn red are classified as acids and are written with the hydrogen appearing first in the formula. For example, $HCl_{(aq)}$ and $H_2SO_{4(aq)}$ are acids. $CH_{4(g)}$ and $NH_{3(g)}$ are not acids, so hydrogen is written last in the formula. In some cases, hydrogen is written last if it is part of a group such as the COOH group; for example, $CH_3COOH_{(aq)}$. These –COOH acids are organic acids and are described in more detail in Chapter 9.

Empirically, acids as pure substances are molecular compounds, as evident from their solid, liquid, and gas states of matter. Theoretically, they are composed of covalently bonded nonmetals. However, the formulas of acids can be explained and predicted by assuming that they are ionic compounds. (Recall that in Section 3.1 it was pointed out that solutions of acids exhibit electrical conductivity, suggesting that ions are present in solution, even if ions are not present in the pure state molecules.) For example, the chemical formulas for the acids $HCl_{(aq)}$, $H_2SO_{4(aq)}$, and $CH_3COOH_{(aq)}$ can be *explained* as follows.

$$H^+Cl^-_{(aq)}, \quad H^+_2SO_4^{2-}_{(aq)}, \quad CH_3COO^-H^+_{(aq)}$$

The chemical formulas for acids can also be *predicted* by assuming that these aqueous molecular compounds of hydrogen are ionic:

aqueous hydrogen sulfide is $H^+_2S^{2-}_{(aq)}$, or $H_2S_{(aq)}$
aqueous hydrogen sulfate is $H^+_2SO_4^{2-}_{(aq)}$, or $H_2SO_{4(aq)}$
aqueous hydrogen sulfite is $H^+_2SO_3^{2-}_{(aq)}$, or $H_2SO_{3(aq)}$

Table 3.7

THE CHLORINE ANION NOMENCLATURE SERIES	
Formula	**IUPAC Name**
ClO_4^-	perchlorate ion
ClO_3^-	chlorate ion
ClO_2^-	chlorite ion
ClO^-	hypochlorite ion
Cl^-	chloride ion

▌Common Acids and Bases

Many electrolytes commonly found around our homes are acids or bases. For example, sulfuric acid is the liquid in car batteries. Hydrochloric acid, phosphoric acid, and acetic acid are found in cleaning agents that remove rust, stains, and scale. Kitchen acids include acetic acid (in vinegar), citric acid (in lemon juice), and lactic acid (in sour milk). Your medicine cabinet might contain acetylsalicylic acid (ASA) and ascorbic acid (vitamin C).

Very few bases are found in foods or are used in food preparation. However, many cleaners contain bases. Sodium hydroxide is a base that is the main ingredient in common drain and oven cleaners. Many of these compounds are very reactive and potentially hazardous, so pay close attention to directions on labels.

Acids are often named according to more than one system because they have been known for so long that the use of traditional names persists. The IUPAC suggests that names of acids should be derived from the IUPAC name for the compound. In this system, sulfuric acid would be named aqueous hydrogen sulfate. However, the classical system of nomenclature is well entrenched, so it is necessary to know two or more names for many acids, especially the common ones.

The common names for acids are derived from the names of the negative ions. The system for naming anions uses the ending "-ide" for monatomic anions, and also for a few polyatomic anions, like the cyanide ion, CN^-, the hydroxide ion, OH^-, and the hydrogen sulfide ion, HS^-. The ending "-ate" is used for the most common polyatomic anions, most of which have one other type of atom bonded to oxygen. The most common sulfur-oxygen anion, for example, is SO_4^{2-}, which is called sulfate. If an anion has one fewer oxygen atom than the most common form, the name ends in "-ite," as in the sulfite ion, SO_3^{2-}. If the anion has one more oxygen atom than the most common form, the prefix "per" is added to the name, as in the perchlorate ion, ClO_4^-. If the anion has two fewer oxygen atoms than the most common form, the prefix "hypo" is added, as in the hypochlorite ion, ClO^-. These naming rules for ions formed from chlorine and oxygen are illustrated in Table 3.7.

The classical names for acids are based on anion names, according to three simple rules.

- If the anion name ends in "-ide," the corresponding acid is named as a "hydro ——— ic" acid. Examples are hydrochloric acid, $HCl_{(aq)}$, hydrosulfuric acid, $H_2S_{(aq)}$, and hydrocyanic acid, $HCN_{(aq)}$.

- If the anion name ends in "-ate," the acid is named as a " ——— ic" acid. Examples are nitric acid, $HNO_{3(aq)}$, sulfuric acid, $H_2SO_{4(aq)}$, and phosphoric acid, $H_3PO_{4(aq)}$.

- If the anion name ends in "-ite," the acid is named as a " ——— ous" acid. Sulfurous acid, $H_2SO_{3(aq)}$, nitrous acid, $HNO_{2(aq)}$, and chlorous acid, $HClO_{2(aq)}$, are examples.

The classical system of acid nomenclature is part of a system for naming a series of related compounds. Table 3.8 lists the acids formed from the chlorine-based anions in Table 3.7 to illustrate this naming system.

Table 3.8

CLASSICAL ACID NOMENCLATURE SYSTEM		
Classical Name	**Systematic IUPAC Name**	**Formula**
perchloric acid	aqueous hydrogen perchlorate	$HClO_{4(aq)}$
chloric acid	aqueous hydrogen chlorate	$HClO_{3(aq)}$
chlorous acid	aqueous hydrogen chlorite	$HClO_{2(aq)}$
hypochlorous acid	aqueous hydrogen hypochlorite	$HClO_{(aq)}$
hydrochloric acid	aqueous hydrogen chloride	$HCl_{(aq)}$

Chemists have discovered that all aqueous solutions of ionic hydroxides make red litmus paper turn blue; that is, these compounds are *bases*. Other solutions have been classified as bases, but for the time being, restrict your definition of bases to aqueous ionic hydroxides such as $NaOH_{(aq)}$ and $Ba(OH)_{2(aq)}$. The name of the base is the name of the ionic hydroxide; for example, aqueous sodium hydroxide and aqueous barium hydroxide.

SUMMARY: ACIDS AND BASES

- Empirically, acids are aqueous molecular compounds of hydrogen that form electrically conductive solutions and turn blue litmus red.
- By convention, the formula for an empirically identified acid is written as H____$_{(aq)}$ or ____$COOH_{(aq)}$.
- As pure substances, acids are molecular compounds, and thus can be solids, liquids, or gases; $HCl_{(g)}$, $H_2SO_{4(l)}$, and $C_3H_4OH(COOH)_{3(s)}$.
- The chemical formulas and electrical conductivity of aqueous solutions of acids can be explained and predicted by assuming that these molecular compounds are ionic; e.g., $H^+_2SO_4^{2-}{}_{(aq)}$ or $H_2SO_{4(aq)}$.
- The classical names for acids follow this pattern: hydrogen ____ide becomes a "hydro____ic" acid; hydrogen ____ate is a "____ic" acid; hydrogen ____ite is a "____ous" acid; and hydrogen hypo____ite is a "hypo____ous" acid.
- The IUPAC name for an acid is aqueous hydrogen_____; e.g., aqueous hydrogen sulfate for $H_2SO_{4(aq)}$.
- Empirically, bases are aqueous ionic hydroxides that form electrically conductive solutions and turn red litmus blue.
- There is no special nomenclature system for bases. They are named as ionic hydroxides; e.g., $KOH_{(aq)}$ is potassium hydroxide.

▌Acids and Bases
1. Acids
   - turn blue litmus red
   - H_____$_{(aq)}$ or _____$COOH_{(aq)}$
   - systematic name; e.g., $HCl_{(aq)}$, aqueous hydrogen chloride
   - classical name; e.g., $HCl_{(aq)}$, hydrochloric acid

2. Bases
   - turn red litmus blue
   - ionic hydroxide, <u>cation</u>$(OH)_{(aq)}$
   - systematic name; e.g., $Ba(OH)_{2(aq)}$, barium hydroxide

---

Chemicals that form neither acidic nor basic solutions are said to be *neutral* (see page 102). Molecular compounds that are not acidic are called *neutral molecular* or simply *molecular*, and ionic compounds that are not basic are called *neutral ionic* or simply *ionic*. Empirically, acids, bases, and neutral compounds can be distinguished by their effect on litmus paper. By a convention of communication established by the scientific community, the character of a chemical is displayed by writing the chemical formula in a special way, as described above.

Exercise

53. Write empirical definitions of an acid and a base.

54. Write empirical definitions of (neutral) ionic and (neutral) molecular compounds.

55. How do you recognize from its chemical formula whether a chemical is an acid or a base?

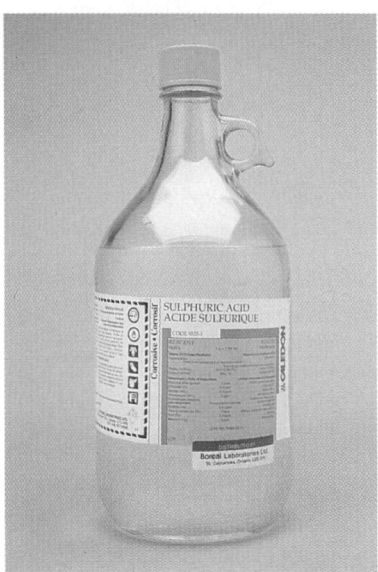

Sulfuric acid is always one of the top ten chemicals manufactured in industrialized countries, where it is used extensively. The quantity of sulfuric acid manufactured in a country is said to indicate the state of development of the country's economy. If countries are ranked on the basis of sulfuric acid production, the order is similar to the ranking of the countries in terms of GNP (gross national product).

Sulfuric acid is a component of acid rain produced from sulfur dioxide emissions in the atmosphere reacting with oxygen and then with rain water.

56. Classify the following as acidic, basic, neutral ionic, or neutral molecular.
    (a) $KCl_{(aq)}$ (fertilizer)
    (b) $HCl_{(aq)}$ (in stomach)*
    (c) $CH_3OH_{(aq)}$ (windshield washer antifreeze)*
    (d) $Ba(OH)_{2(aq)}$ (corrosive)
    (e) $C_{17}H_{35}COOH_{(aq)}$ (fatty acid)
    (f) aqueous hydrogen sulfate (in car battery)*
    (g) sodium hydroxide (oven/drain cleaner)*
    (h) ethanol (beverage alcohol)*

57. Write the chemical formulas for the following acids.
    (a) aqueous hydrogen chloride (from a gas)
    (b) hydrochloric acid (stomach acid)*
    (c) aqueous hydrogen sulfate (from a liquid)
    (d) sulfuric acid (car battery)*
    (e) aqueous hydrogen acetate (from a liquid)
    (f) acetic acid (vinegar)*
    (g) aqueous hydrogen nitrite (from a gas)
    (h) nitric acid (for making fertilizers)

58. Write accepted names for the following acids.
    (a) $H_2SO_{3(aq)}$ (acid rain)
    (b) $H_3PO_{4(aq)}$ (rust remover)
    (c) $HCN_{(aq)}$ (rat killer)
    (d) $HF_{(aq)}$ (etches glass)
    (e) $H_2CO_{3(aq)}$ (carbonated beverages)*
    (f) $H_2S_{(aq)}$ (rotten egg odor)
    (g) $H_3BO_{3(aq)}$ (insecticide)
    (h) $C_6H_5COOH_{(aq)}$ (preservative)

59. List two common consumer acids and two common consumer bases and their uses.

60. Write chemical equations, including the SATP states of matter, for the following reactions involved in the manufacture and use of sulfuric acid.
    (a) Sulfur is reacted with oxygen from the air to produce sulfur dioxide gas.
    (b) Sulfur dioxide reacts with oxygen to produce sulfur trioxide gas.
    (c) Sulfur trioxide gas is reacted with water to produce sulfuric acid.
    (d) Sulfuric acid reacts with ammonia gas to produce aqueous ammonium sulfate (a fertilizer).
    (e) Sulfuric acid reacts with rock phosphorus $(Ca_3(PO_4)_{2(s)})$ to produce phosphoric acid and solid calcium sulfate (gypsum).

61. Write chemical equations, including SATP pure states of matter, for the following reactions involved in the destructive reactions of acid rain.
    (a) Sulfuric acid rain may react with limestone (see Appendix G) to deteriorate buildings, statues, and gravestones and produce aqueous hydrogen carbonate (carbonic acid) and solid calcium sulfate.

(b) Sulfuric acid from rain reacts with solid aluminum silicate in the bottom of a lake to release aqueous hydrogen silicate (silicic acid) and toxic aqueous aluminum sulfate.

62. Write chemical equations, including SATP pure states of matter, for each of the following reactions involved in the control of acid rain.
(a) Sulfur dioxide emissions can be reduced in the exhaust stack of a metal refinery by reacting the sulfur dioxide gas with lime (see Appendix G) and oxygen to produce solid calcium sulfate (gypsum).
(b) Sulfuric acid in an acid lake can be neutralized by adding slaked lime (solid calcium hydroxide) to produce water and solid calcium sulfate.

63. (Discussion) What are some of the benefits and risks of the technological or natural production of sulfuric acid? Try to answer from a variety of perspectives — scientific, technological, ecological, economic, and/or political.

The sulfuric acid content of acid rain or acid lakes can be analyzed and modified by environmental chemists.

INVESTIGATION

3.5 Qualitative Analysis of Solutions

The purpose of this investigation is to identify the solutions provided. Use the empirical definitions of neutral ionic, neutral molecular, acids, and bases to determine the identity of the solutions.

Problem

Which of the solutions provided is hydrochloric acid, sodium hydroxide, sodium chloride, and sucrose?

Experimental Design

Each solution is tested for electrical conductivity and with litmus paper. All observations are recorded in a table of evidence.

manipulated variable: unknown solutions

responding variables: electrical conductivity
litmus color

controlled variables: solution volume
solution concentration
time
temperature
conductivity tester

Materials

apron
safety glasses
solutions labeled A, B, C, and D
four 50 mL beakers or one spot plate

conductivity tester
distilled water
red litmus paper
blue litmus paper

	Problem
	Prediction
	Design
	Materials
	Procedure
✔	Evidence
✔	Analysis
	Evaluation
	Synthesis

CAUTION

Both hydrochloric acid and sodium hydroxide are corrosive. Avoid contact of any unknown solution with skin and eyes.

If the conductivity tester is connected to a main electrical outlet, ensure that the electrical equipment is switched off when handling it (e.g., when rinsing or changing the reaction vessel) to avoid an electrical shock.

Procedure

1. Obtain about 10 mL of each solution in beakers or several drops of each solution in a spot plate.

2. Test each solution for electrical conductivity and rinse the leads.

3. Test each solution with red litmus paper.

4. Test each solution with blue litmus paper.

5. Dispose of the chemicals down the drain with running water.

Exercise

64. Draw a flow chart to classify all pure substances studied.

65. Classify the following chemicals found around the home (Figure 3.18) as ionic, molecular, acid, or base.
    (a) $C_6H_4COOCH_3COOH_{(s)}$ (ASA, Aspirin)
    (b) $(C_6H_4Cl)_2CHCCl_{3(s)}$ (DDT, insecticide)
    (c) $C_7H_5NO_3S_{(s)}$ (saccharin)
    (d) $(CH_3)_2CO_{(l)}$ (acetone)
    (e) $CCl_2F_{2(l)}$ (freon, refrigerant)
    (f) $C_3H_7OH_{(l)}$ (rubbing alcohol)
    (g) $C_5H_7O_4COOH_{(s)}$ (vitamin C)
    (h) $CaSO_4 \cdot \frac{1}{2}H_2O_{(s)}$ (plaster of Paris)
    (i) $NaHCO_{3(s)}$ (in Eno)
    (j) $C_3H_4OH(COOH)_{3(s)}$ (in Eno)
    (k) $C_2H_2(OH)_2(COOH)_{2(s)}$ (in Eno)

66. The following chemicals are addictive and may have dangerous consequences for human life. Based upon the chemical formula, classify the compounds as ionic, molecular, acid, or base.
    (a) $C_{18}H_{21}NO_{3(s)}$ (codeine)
    (b) $C_2H_5OH_{(l)}$ (beverage alcohol)
    (c) $C_8H_{10}N_4O_{2(s)}$ (caffeine)
    (d) $C_{10}H_{14}N_{2(s)}$ (nicotine)
    (e) $C_{21}H_{30}O_{2(s)}$ (THC in marijuana)
    (f) $C_{17}H_{21}NO_{4(s)}$ (cocaine)

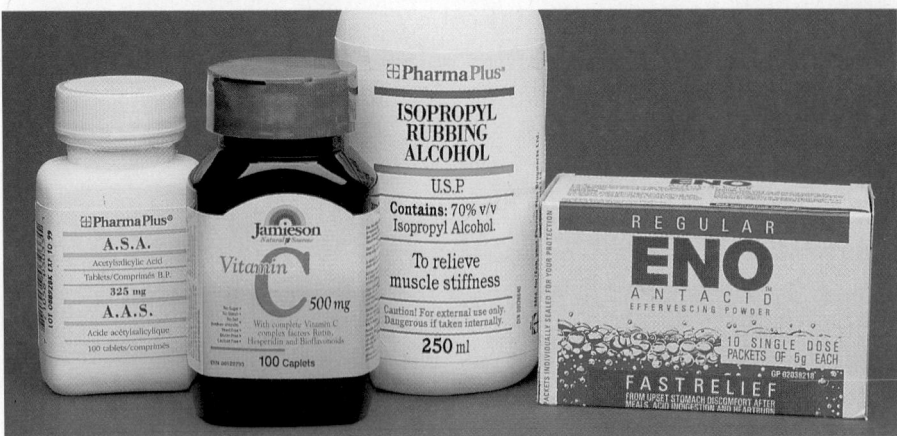

Figure 3.18
Household chemicals can be classified as ionic, molecular, acid, or base.

67. Everything in our world is composed of chemicals. The following chemicals are necessary for human life. Write the names or chemical formulas with SATP pure states of matter for the following chemicals of life. The list is restricted to chemicals for which you can name and write formulas. Organic and biochemistry units in your chemistry and biology courses cover more chemicals of life.
   (a) oxygen (respiration reactant)*
   (b) carbon dioxide (respiration product)*
   (c) water (blood)*
   (d) hydrochloric acid (digestion)*
   (e) sodium hydrogen carbonate (blood buffer)
   (f) $NaH_2PO_{4(s)}$ (blood buffer)
   (g) glucose (carbohydrate)*
   (h) sodium chloride (perspiration)
   (i) $H_2S_{(g)}$ (colon gas)
   (j) ammonia (colon gas)
   (k) $CH_{4(g)}$ (colon gas)
   (l) hydrogen (colon gas)

68. Naturally occurring and technological products are common in our homes. Classify and then write the chemical formulas or names and SATP pure states of matter for the following chemicals commonly found in homes (Figure 3.19).
   (a) silicon dioxide (glass)
   (b) oxygen (air)*
   (c) tungsten (light bulb filament)
   (d) argon (fluorescent lights)
   (e) $CH_{4(g)}$ (natural gas)*
   (f) calcium sulfate-2-water (gypsum)
   (g) aluminum (cans)
   (h) $CaCl_{2(s)}$ (road salt)
   (i) Zn (dry cell)
   (j) NH_4Cl (dry cell)
   (k) manganese(IV) oxide (dry cell)
   (l) sodium hydroxide (alkaline cell)

Figure 3.19
Household chemicals include elements and compounds.

Figure 3.20
Everything around us is made of chemicals. The components of an automobile are composed of a wide variety of chemicals.

69. The automobile industry is a large chemical consumer (Figure 3.20). Write the chemical formulas and SATP pure states of matter or names of the following chemicals.
    (a) $ZnO_{(s)}$ (galvanizing)
    (b) zinc (galvanizing)*
    (c) methane (alternate fuel)*
    (d) hydrogen (alternate fuel)
    (e) $CH_3OH_{(l)}$ (antifreeze)*
    (f) sulfuric acid (battery)*
    (g) lead(II) sulfate (battery)
    (h) hydrogen (battery)
    (i) $PbO_{2(s)}$ (battery)
    (j) lead (battery)
    (k) chromium (decoration)
    (l) $H_2O_{(l)}$ (coolant)
    (m) $C_8H_{18(l)}$ (gasoline)*

70. Chemicals are also used in cooking, which, after all, is just "kitchen chemistry". Write the chemical formulas and SATP pure states of matter or names of the following chemicals.
    (a) water (solvent)*
    (b) table salt (seasoning)*
    (c) $CaHPO_{4(s)}$ (in baking powder)
    (d) $NaHCO_{3(s)}$ (in baking soda and baking powder)*
    (e) aluminum silicate (coffee whitener)
    (f) ice (cooling drinks)*
    (g) ethanol (cooking alcohol)*
    (h) $C_{12}H_{22}O_{11(s)}$ (table sugar)*
    (i) acetic acid (vinegar)
    (j) $NaC_5H_8NO_{4(s)}$ (MSG-flavoring)
    (k) $C_{17}H_{35}COOH_{(aq)}$ (fatty acid)

71. Washing clothes, dishes, and bodies also requires chemicals. Write the chemical formulas with SATP pure states of matter or names of the following.
    (a) $H_2O_{(l)}$ (solvent)
    (b) $Na_5P_3O_{10(s)}$ (detergent)
    (c) sodium carbonate-10-water (water softener)
    (d) $CaCO_{3(s)}$ (product of softening)
    (e) calcium bicarbonate (hard water)
    (f) sodium chloride (water softener)
    (g) sodium hypochlorite (bleach)
    (h) $NaC_{17}H_{35}COO$ (soap)

72. Some chemicals in home medicine cabinets and bathrooms are listed below (Figure 3.21). Write their chemical formulas with SATP pure states of matter or names.
    (a) magnesium hydroxide (antacid)
    (b) $Al(OH)_{3(s)}$ (antacid)
    (c) $NaHCO_{3(s)}$ (antacid)
    (d) $Na_2SO_4 \cdot 10H_2O_{(s)}$ (Glauber's salt)
    (e) $FeSO_{4(s)}$ (iron supplement)

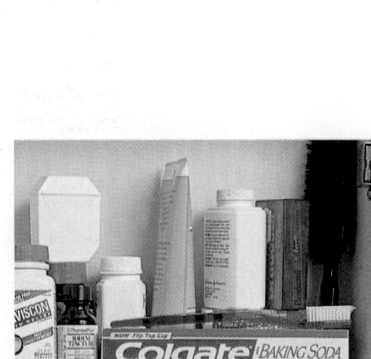

Figure 3.21
Bathroom chemicals help us live a long, pleasant, and healthy life.

(f) $CaCO_{3(s)}$ (scouring agent)

(g) sodium hydrogen sulfate (toilet bowl cleaner)

(h) sodium hydroxide (drain cleaner)*

(i) aluminum chloride (antiperspirant)

(j) hydrogen peroxide (disinfectant)

(k) iodine (disinfectant)

(l) $NaF_{(s)}$ (in toothpaste)

(m) ammonia (cleaner)*

73. Many chemicals are found in the garden and yard (Figure 3.22). Write the chemical formulas (with SATP pure states of matter) or names of those listed below.

(a) copper(II) sulfate pentahydrate (fungicide)

(b) $KCl_{(s)}$ (potash fertilizer)

(c) ammonium sulfate (fertilizer)

(d) ammonium hydrogen phosphate (fertilizer)

(e) $C_3H_{8(g)}$ (barbecue)*

(f) carbon (briquettes)

(g) $Ca(OCl)_{2(s)}$ (swimming pool)

(h) $SiO_{2(s)}$ (sand)

74. The classical names for acids are still used extensively. Provide the classical name or formula for the following:

(a) hydrofluoric acid (etches glass)

(b) acetic acid (vinegar)*

(c) $H_2SO_{3(aq)}$ (acid rain)

(d) hypochlorous acid (bleach)

75. (Discussion) Are chemicals beneficial or are they a risk to human survival? Try to answer from a variety of perspectives — scientific, technological, ecological, economic, and/or political.

Figure 3.22
Propane, $C_3H_{8(g)}$, is a common fuel for barbecues.

OVERVIEW

Compounds

Summary

- Classification in chemistry involves observing properties, identifying defining properties, forming empirical definitions, and designing diagnostic tests.

- The compounds formed from metals and nonmetals are classified as ionic, molecular, or inter-metallic.

- Ionic compounds are solids at SATP and their aqueous solutions conduct electricity. According to theory, they are composed of cations and anions. The chemical formulas for ionic compounds are predicted by making the net charge on a formula unit equal zero. The names for ionic compounds are derived by naming the two ions involved.

- Molecular compounds are solids, liquids, or gases at SATP, and their aqueous solutions do not conduct electricity. According to theory, they are composed of molecules formed from nonmetal atoms.

- Acids are defined in this chapter as aqueous molecular compounds of hydrogen that make blue litmus paper turn red. By convention, the formula for an acid is written to begin with H or end in COOH. Acid nomenclature is referenced, classical, or systematic. Bases are defined as aqueous ionic hydroxides that make red litmus paper turn blue.

Key Words

acid
anhydrous
aqueous solution
base
binary ionic compound
binary molecular compound
diagnostic test
empirical formula
formula unit
hydrate
ionic compound
molecular compound
molecular formula
molecule
multi-valent
neutral (compound)
nomenclature
polyatomic ion
ternary

Review

1. (a) Prepare a table to summarize the properties of ionic and molecular compounds.
    (b) Prepare a similar table for acids and bases.
    (c) Use a key to indicate which of these properties are defining properties that could be included in empirical definitions. Use another key to identify properties suitable for diagnostic tests.

2. What two theoretical ideas are used to explain and predict ionic formulas?

3. (a) List four characteristics of an effective system of scientific communication.
    (b) What international organization establishes the rules for chemical names and formulas?
    (c) In the scientific community, why is a chemical formula a more acceptable way of communicating than a chemical name?

4. What three items of information should be communicated by a chemical formula?

5. Are chemical formulas empirical, theoretical, or both? Explain your answer.

Applications

6. Write the chemical formula for the following substances. Include the pure state of matter at SATP. An everyday name or use is given in

parentheses and some products containing these chemicals are shown below.

(a) sodium hydrogen sulfate (toilet bowl cleaner)
(b) sodium hydroxide (lye, drain cleaner)
(c) carbon dioxide (dry ice, soda pop)
(d) acetic acid (vinegar)
(e) sodium thiosulfate-5-water (photographic "hypo")
(f) sodium hypochlorite (laundry bleach)
(g) octasulfur (vulcanizing rubber)
(h) potassium nitrate (saltpeter, meat preservatives)
(i) phosphoric acid (rust remover)
(j) iodine (disinfectant)
(k) aluminum oxide (alumina, aluminum ore)
(l) potassium hydroxide (caustic potash)
(m) ozone (absorbs ultraviolet radiation)
(n) methanol (gas line and windshield washer fluid)
(o) aqueous hydrogen carbonate (carbonated beverages)
(p) propane (fuel)

7. Write IUPAC names for the following substances. An everyday name, use, or result is given in parentheses.
(a) $CaCO_{3(s)}$ (marble, limestone, chalk)

(b) $P_2O_{5(s)}$ (fertilizer labelling)
(c) $MgSO_4 \cdot 7H_2O_{(s)}$ (Epsom salts)
(d) $N_2O_{(g)}$ (laughing gas, anaesthetic)
(e) $Na_2SiO_{3(s)}$ (water glass)
(f) $Ca(HCO_3)_{2(s)}$ (hard water chemical)
(g) $HCl_{(aq)}$ (muriatic acid, gastric fluid)
(h) $CuSO_4 \cdot 5H_2O_{(s)}$ (copper plating, bluestone)
(i) $H_2SO_{4(aq)}$ (acid in car battery)
(j) $Ca(OH)_{2(s)}$ (slaked lime)
(k) $SO_{3(g)}$ (source of acid rain)
(l) $NaF_{(s)}$ (toothpaste additive)

8. Write the IUPAC name for each reactant and product in a word equation for the following (unbalanced) chemical equations.
(a) $KOH_{(aq)} + H_2CO_{3(aq)} \rightarrow HOH_{(l)} + K_2CO_{3(aq)}$
(b) $Pb(NO_3)_{2(aq)} + (NH_4)_2SO_{4(aq)} \rightarrow PbSO_{4(s)} + NH_4NO_{3(aq)}$
(c) $Al_{(s)} + FeSO_{4(aq)} \rightarrow Fe_{(s)} + Al_2(SO_4)_{3(aq)}$
(d) $NO_{2(g)} + H_2O_{(l)} \rightarrow HNO_{3(aq)} + NO_{(g)}$

9. Write chemical equations for the following reactions. Include the physical state at SATP as part of each formula.
(a) nitrogen + oxygen $\rightarrow$ nitrogen dioxide gas
(b) iron(III) acetate solution + sodium oxalate solution $\rightarrow$ solid iron(III) oxalate + sodium acetate solution
(c) cyclooctasulfur + chlorine $\rightarrow$ liquid disulfur dichloride
(d) copper + silver nitrate solution $\rightarrow$ silver + copper(II) nitrate solution

10. Write the formula and name for the product of each of the following reactions. Assume SATP and the most common ion charges for the ions in the compound produced.
(a) $K_{(s)} + Br_{2(l)} \rightarrow$
(b) $Ag_{(s)} + I_{2(s)} \rightarrow$
(c) $Pb_{(s)} + O_{2(g)} \rightarrow$
(d) $Zn_{(s)} + S_{8(s)} \rightarrow$
(e) $Cu_{(s)} + O_{2(g)} \rightarrow$
(f) $Li_{(s)} + N_{2(g)} \rightarrow$

11. Cement is a complex mixture of chemicals. Write their chemical formulas with SATP pure states of matter or names.
(a) calcium oxide (lime, 62%)
(b) silicon dioxide (silica, 22%)
(c) aluminum oxide (alumina, 7.5%)
(d) $MgO_{(s)}$ (magnesia, 2.5%)
(e) $Fe_2O_{3(s)}$ (an iron oxide, 2.5%)
(f) $SO_{3(g)}$ (a sulfur oxide, 1.5%)

12. To make concrete, you mix cement with sand and gravel in varying ratios — 1:3:6 (lean), 1:2:4 (stronger), or 1:1:2 (under salt water). Water is added as a reactant in a quantity of about 16% by mass. Shown below are just a few of the many complex reactions between the lime in the cement and the sand, gravel, and water mixture that occur to produce concrete. Assuming pure states of matter, write a chemical equation for each reaction.
    (a) calcium oxide reacts with water to produce calcium hydroxide
    (b) calcium oxide reacts with silicon dioxide to produce calcium silicate
    (c) calcium oxide reacts with sulfur trioxide to produce calcium sulfate
    (d) calcium oxide reacts with carbon dioxide to produce calcium carbonate

Extensions

13. Select a compound from this chapter that you find interesting and write a report on it. What is it used for? How much is made in Canada annually? Use your library and other resources to research the compound and tell how it is obtained. Outline its unique properties. Concentrate on its applications to current technology as well as its effects on the environment and on society. If you wish, organize your information according to different perspectives.

14. Write a report on the benefits of universal systems of communication in the sciences. Include a discussion of criteria used to judge systems of communication. Cite examples from this chapter to support your arguments. Speculate on why SI units of measurement and IUPAC nomenclature are not used universally.

15. The following excerpt from a newspaper article is an example of ongoing scientific inquiry and the probability of more exciting discoveries in the future. Assuming that the effects of nitric oxide are somewhat uncertain, rewrite the second paragraph to reflect a higher degree of uncertainty.

    "A simple and familiar chemical, nitric oxide, that is best known as a major precursor of acid rain and smog, is emerging in a surprising new role, as one of the most powerful known substances in controlling bodily functions.... Nitric oxide, the new findings show, is a messenger molecule involved in a wide variety of activities.

    It mediates the control of blood pressure. It helps the immune system kill invading parasites that sneak into cells. It stops cancer cells from dividing. It transmits signals between brain cells....

    Scientists are amazed to discover that nitric oxide is crucial to so many biological systems. As word of nitric oxide's significance spread, researchers first reacted with disbelief and then asked themselves how they could have missed the signs of its presence for so long...." — Gina Kolata, *New York Times*, reprinted in *The Edmonton Journal*, July 28, 1991.

Problem 3B A Chemical Analysis

The purpose of this problem is to use chemical analysis to classify unknown solutions. The experimental design includes diagnostic tests for analysis. Previous empirical definitions of ionic and molecular compounds are assumed to be valid. Complete the Analysis of the investigation report.

Problem

Which of the solutions labelled 1, 2, 3, and 4 is $KCl_{(aq)}$, $C_2H_5OH_{(aq)}$, $HCl_{(aq)}$, and $Ba(OH)_{2(aq)}$?

Experimental Design

Each solution is tested with a conductivity apparatus and with litmus paper to determine its identity. A sample of the water used for preparing the solutions is tested for conductivity as a control. Taste tests are ruled out because they are unsafe.

Evidence

Solution	Conductivity	Litmus Paper
water	none	no change
1	high	no change
2	high	blue to red
3	none	no change
4	high	red to blue

Problem 3C Extending Empirical Definitions of Compounds

The purpose of this problem is to further extend the previously determined empirical definitions of ionic and molecular compounds to include the electrical conductivity of the solid, liquid, and aqueous states of matter. To achieve this purpose, the conductivity of several substances in the three states are measured. Complete the Analysis and Evaluation sections of the investigation report.

Problem

What are the empirical definitions of ionic and molecular compounds?

Prediction

According to the current definitions of ionic and molecular compounds, ionic compounds are all solids at SATP that form electrically conductive solutions, while molecular compounds are solids, liquids, or gases at SATP that form non-conductive solutions.

Experimental Design

Pure samples of water (H_2O), calcium chloride $(CaCl_2)$, sucrose $(C_{12}H_{22}O_{11})$, methanol (CH_3OH), sodium hydroxide $(NaOH)$, and potassium iodide (KI) are tested for electrical conductivity in the pure state at SATP, in the pure molten state, and in aqueous solution.

  manipulated variable: compound tested

  responding variable: electrical conductivity

  controlled variables: temperature
  quantity of chemical
  quantity of water
  conductivity apparatus
  conductivity of the water

Evidence

ELECTRICAL CONDUCTIVITY OF COMPOUNDS IN DIFFERENT STATES				
Chemical	Pure	Conductivity		
Formula	State	Pure	Molten	Aqueous
H_2O	liquid	none	none	n/a*
$CaCl_2$	solid	none	high	high
$C_{12}H_{22}O_{11}$	solid	none	none	none
CH_3OH	liquid	none	none	none
$NaOH$	solid	none	high	high
KI	solid	none	high	high

* n/a - not applicable

4 Chemical Reactions

Chemical reactions are responsible for the smog that exists to some extent in many major cities. Chemical reactions have also provided a partial solution to this problem. The photograph shows Los Angeles on a smoggy day. What causes this thick, brown haze? Even with strict pollution controls now in place, the large number of vehicles driven makes air pollution from car exhausts a serious problem for Los Angeles and many other large cities.

Car engines burn gasoline using air as a source of oxygen. This combustion reaction is rarely complete and carbon monoxide, a major contributor to smog, is formed. The primary component of air is nitrogen, so the pollutant nitrogen monoxide is also formed when air encounters the high temperatures inside car engines. Sunlight provides the energy that promotes a series of secondary reactions producing nitrogen dioxide and ozone.

Chemists and chemical engineers have applied their knowledge of reactions to develop catalytic converters for automobiles. These devices convert some of the carbon monoxide to carbon dioxide, and nitrogen monoxide to nitrogen. Although resolving the air pollution issue involves more than science and technology, by learning about chemical reactions you will better understand the scientific concepts behind such issues.

The explanation of natural events is one of the aims in science. Careful observation, leading to the formation of a concept or theory, and followed by testing and evaluating the ideas involved, defines the basic process scientists use to increase understanding of the changes going on in the world around us. A useful way to begin is to classify the types of changes that occur in matter. Changes in matter can be explained at three levels according to size. Modern scientists study and discuss matter at a *macroscopic* (naked eye observable) level, or at a *microscopic* (too small to see without a microscope) level, or at a *molecular* (smallest particles of a substance) level. As outlined in Chapter 2, chemists are particularly concerned with the molecular level, and to understand their observations, they usually start by basing their explanations on the atomic theory proposed by John Dalton in 1803.

Types of Changes in Matter

Chemists often describe changes in matter as a **physical change**, **chemical change**, or **nuclear change** (Figure 4.1), depending on whether or not they believe that a change has occurred in the molecules, electrons, or nuclei of the substance being changed. The quantity of energy associated with every change in matter can also help classify the type of change.

Physical changes are any changes where the fundamental particles remain unchanged at a molecular level, like the phase changes of evaporation and melting. There is no change in the written formula of the substance involved. Dissolving a chemical is usually classified as a physical change. Other examples include changes in physical structure that change appearance, like grinding a shiny piece of copper into a fine powder, which is black in color but is still copper metal. Physical changes in matter usually involve relatively small amounts of energy change.

A chemical change involves some kind of change in the chemical bonds within the fundamental particles (i.e., atoms and/or ions) of a substance, and is represented by a change in the written formula. At least one new substance is formed, with physical and chemical properties different from those of the original matter. Normally, chemical changes in matter involve larger energy changes than physical changes.

Nuclear changes create entirely new atomic particles. These are represented by formulas that show new atomic symbols, different from those of the original matter. Nuclear changes involve extremely large changes in energy, which allow them to be identified. In 1896, Henri Becquerel noticed the continuous production of energy from a piece of rock that showed no other changes at all. His observation led to the discovery of radioactivity.

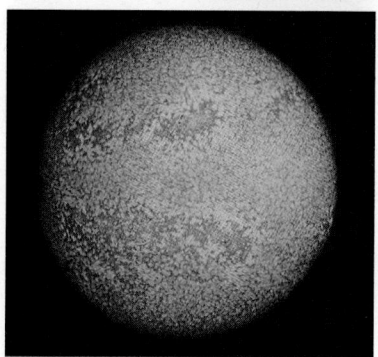

Figure 4.1
Hydrogen may undergo a physical (top), chemical (middle), or nuclear change (bottom). In the first photo hydrogen is boiling at −253°C (at only 20°C above absolute zero). Second, hydrogen is burning — as it does on the space shuttle and in hydrogen-fueled automobiles. Third, hydrogen is undergoing nuclear fusion on the sun and is being converted into helium.
$H_{2(l)} \rightarrow H_{2(g)}$ *a physical change*
$2\,H_{2(g)} + O_{2(g)} \rightarrow 2\,H_2O_{(g)}$ *a chemical change*
$H_{(g)} + H_{(g)} \rightarrow He_{(g)}$ *a nuclear change*

Table 4.1

PHYSICAL, CHEMICAL, AND NUCLEAR CHANGE		
Change	Empirical Description	Theoretical Description
Physical	• state or energy change • solid $\leftrightarrow$ liquid $\leftrightarrow$ gas • no new substance • small energy change	• $H_2O_{(l)} + energy \rightarrow H_2O_{(g)}$ • $H_2O_{(s)} \leftrightarrow H_2O_{(l)} \leftrightarrow H_2O_{(g)}$ • no new molecules • intermolecular forces broken and made
Chemical	• color, odor, state and/or energy change • new substance formed • new permanent properties • medium energy change	• $2 H_2O_{(l)} + energy \rightarrow 2 H_{2(g)} + O_{2(g)}$ • old $\rightarrow$ new molecules • atoms/ions rearranged • chemical bonds broken and made
Nuclear	• often radiation emitted • new elements formed • enormous energy change	• $^2_1H + ^1_1H \rightarrow ^3_2He$ • new atoms formed • nuclear bonds broken and made

The Kinetic Molecular Theory

How could you explain why a drop of food coloring added to a glass of cold water slowly spreads out, or *diffuses*, throughout the water? Or how could you explain why the amount of water in an open container slowly decreases as some of the water evaporates? Scientists would say that the molecules of food coloring and the molecules of water are moving and colliding with each other, and this causes them to mix. Similarly, because of molecular motion, some of the water molecules in the open container obtain sufficient energy from collisions to escape from the liquid. The idea of molecular motion that is used to explain these observations has led to the **kinetic molecular theory**, which has become a cornerstone of modern science.

The central idea of the kinetic molecular theory is that the smallest particles of a substance are in continuous motion. These particles may be atoms, ions, or molecules. As they move about, the particles collide with each other and with objects in their path. Very tiny objects, such as pollen grains or specks of smoke, are buffeted by these particles, and move erratically, as shown in Figure 4.2.

Why Do Chemical Reactions Occur?

In order to explain chemical reactions, the kinetic molecular theory can be expanded to create a theory of chemical reactions. According to the kinetic molecular theory, the particles of a substance are in continuous, random motion. This motion inevitably results in collisions among the particles. If different substances are present, all the different particles will collide randomly with each other. If the collision has a certain orientation and sufficient energy, the components of the particles will rearrange to form new particles. The rearrangement of particles that occurs is the chemical reaction. This

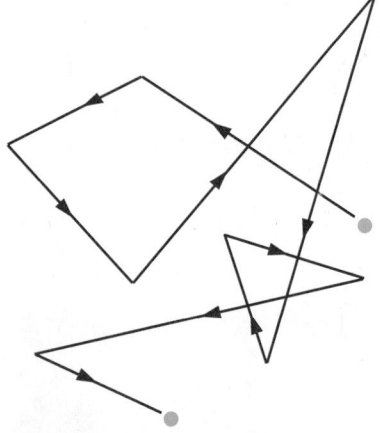

Figure 4.2
Observation of microscopic particles such as pollen grains or specks of smoke shows a continuous, random motion known as Brownian motion, named for Scottish scientist Robert Brown, who first described it. Scientists' interpretations of this evidence led to the formation of the kinetic molecular theory.

general view of a chemical reaction is known as the *collision-reaction theory*. To summarize, for a chemical reaction to take place, particles of the reactants must collide before any rearrangement of atoms or ions occurs; a certain minimum energy is required of the colliding particles; and a certain orientation is required of the colliding particles for a successful rearrangement of atoms or ions (Figure 4.3).

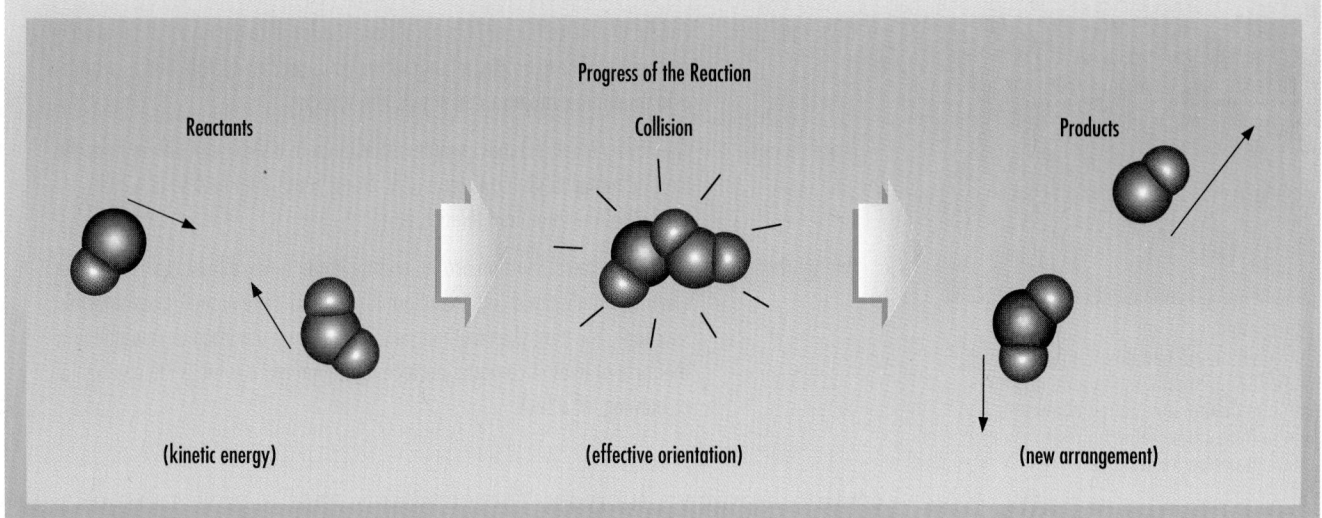

Progress of the Reaction

Reactants   Collision   Products

(kinetic energy)   (effective orientation)   (new arrangement)

Figure 4.3
Collision-reaction theory explains that chemical particles (entities) must collide with the correct orientation in order to react.

Exercise

1. Provide two examples each of physical, chemical, and nuclear changes.

2. How do physical, chemical, and nuclear changes compare in terms of the quantity of energy exchanged?

3. What different entities are rearranged in physical, chemical, and nuclear changes?

4. According to the collision-reaction theory, what are the requirements for a chemical reaction to take place?

5. Why are classification systems created?

6. There is debate among chemists as to whether dissolving is a physical change or a chemical change. What does this tell you about classification systems?

4.2 CHEMICAL REACTIONS

Chemical changes are also called chemical reactions. (Recall that in chemical reactions, new substances are produced.) How do you know if an unfamiliar change is a chemical reaction? Certain characteristic evidence is associated with chemical reactions (Table 4.2, page 138).

Since physical changes, as well as chemical changes, involve changes in state and energy, it is not always easy to interpret changes in matter. It is sometimes impossible to distinguish between a physical change and a chemical change by means of a simple observation. A chemical analysis of the mixture may be required to show that a new substance has been produced. Diagnostic tests that are specific to

Figure 4.4
A solution of sodium hypochlorite, NaClO(aq), sold as bleach, reacts with colored dyes and destroys the dyes. The color change is evidence of a chemical reaction.

Table 4.2

EVIDENCE OF CHEMICAL REACTIONS	
Evidence	**Description and Example**
Color change	The final product(s) may have a different color than the colors of the starting material(s) (Figure 4.4).
Odor change	The final material(s) may have a different odor than the odors of the starting material(s). For example, mixing solutions of sodium acetate and hydrochloric acid produces a mixture that smells like vinegar.
State change	The final material(s) may include a substance in a state that differs from the starting material(s). Most commonly, either a gas (Figure 4.5) or a solid (precipitate) is produced (Figure 4.6).
Energy change	When a chemical reaction occurs, energy in the form of heat, light, sound, or electricity is absorbed (an endothermic reaction) or released (an exothermic reaction). For most chemical reactions, the energy absorbed or released is in the form of heat. A common example of an energy change is the combustion, or burning, of a fuel.

certain chemicals increase the certainty that a new substance has formed in a chemical reaction. Appendix C, C.3 describes diagnostic tests for chemicals such as hydrogen and oxygen. If the diagnostic test entails a single step for a specific chemical, you may find it convenient to summarize this test using the format, "If [procedure] and [evidence], then [analysis]." An example of a diagnostic test is shown in Figure 4.8 on page 141.

Conservation of Mass in Chemical Changes

In chemical changes, the total mass of matter present before the change is always the same as the total mass present after the change, no matter how different the new substances appear. This was one of the

Figure 4.5
Vinegar, CH₃COOH(aq), added to a baking soda solution, NaHCO₃(aq), produces gas bubbles. There is also a change in odor. The changes in state and odor are interpreted as evidence of a chemical reaction.

Figure 4.6
A silver nitrate solution added to a sample of tap water produces a cloudy mixture containing a white solid (precipitate) that slowly settles to the bottom of the container. The change in state and color can be interpreted as evidence of a chemical reaction.

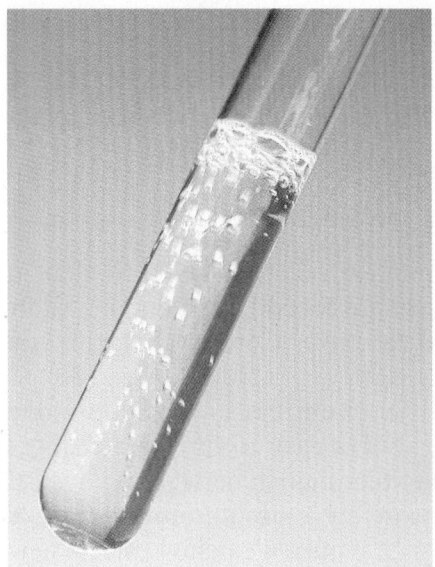

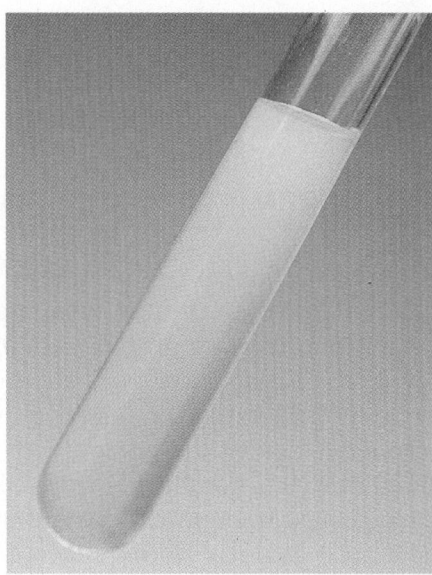

compelling reasons why scientists accepted the atomic theory of matter. If a chemical change is thought of as a rearrangement of particles at the molecular level, then it is simple to argue that the mass must be constant. The individual particles do not change, except in the ways they are associated with each other. Any system used to represent chemical reactions must follow the law of conservation of mass.

Communicating Chemical Reactions

A **balanced chemical equation** is one in which *the total number of each kind of atom or ion in the reactants is equal to the total number of the same kind of atom or ion in the products.*

The balanced chemical equation and molecular models in Figure 4.7 both represent the reaction of nitrogen dioxide gas and water to produce nitric acid and nitrogen monoxide gas. By studying the molecular models, you can see that the number of nitrogen atoms is three on both the reactant and the product sides of the equation arrow. Likewise, the number of oxygen atoms is seven on both sides of the equation arrow, and the number of hydrogen atoms is two.

If more than one molecule is involved (for example, three molecules of nitrogen dioxide in Figure 4.7), then a number called a **coefficient** is placed in front of the chemical formula. In this example, *three* molecules of nitrogen dioxide and *one* molecule of water react to produce *two* molecules of nitric acid and *one* molecule of nitrogen monoxide. Coefficients should not be confused with formula subscripts, which are part of the chemical formula for a substance.

Another subscript is used to show a substance's state of matter. It is not part of the theoretical description given by the molecular models. Chemical formulas showing state of matter subscripts provide both a theoretical and an empirical description of a substance.

Figure 4.7
Molecular models for the reactants and products in the chemical reaction of nitrogen dioxide and water. Models such as these help us to visualize non-observable processes.

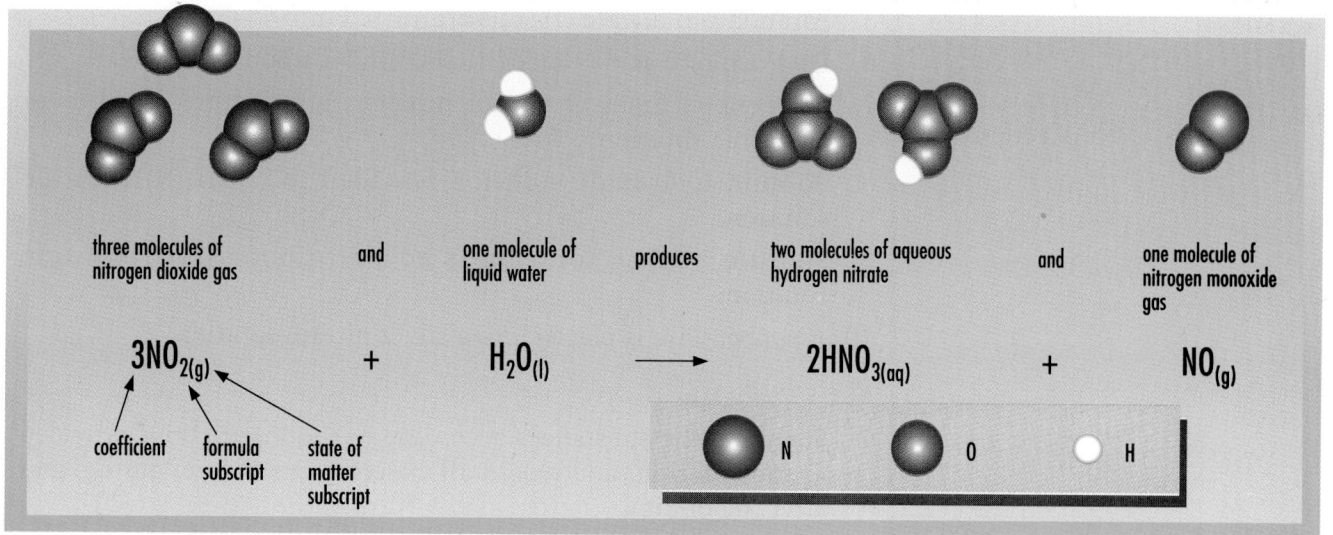

three molecules of nitrogen dioxide gas   and   one molecule of liquid water   produces   two molecules of aqueous hydrogen nitrate   and   one molecule of nitrogen monoxide gas

$$3NO_{2(g)} \quad + \quad H_2O_{(l)} \quad \longrightarrow \quad 2HNO_{3(aq)} \quad + \quad NO_{(g)}$$

coefficient   formula subscript   state of matter subscript

N   O   H

SUMMARY: CHEMICAL REACTION EQUATIONS

- A chemical reaction is communicated by a balanced chemical equation in which the same number of each kind of atom or ion appears on the reactant and product sides of the equation.

- A *coefficient* in front of a chemical formula in a chemical equation communicates the number of molecules or formula units that are involved in the reaction.

- Within formulas, a numerical subscript communicates the number of atoms or ions present in one molecule or formula unit of a substance.

- A state of matter subscript is used to communicate the physical state of the substance at SATP.

- [ ] Problem
- [ ] Prediction
- [ ] Design
- [ ] Materials
- [ ] Procedure
- [x] Evidence
- [x] Analysis
- [ ] Evaluation
- [ ] Synthesis

INVESTIGATION

4.1 Evidence of Chemical Reactions

The purpose of this investigation is to observe reactions that provide evidence of chemical changes. For each combination of substances given, assume that if any chemical change occurs, it happens rapidly at room temperature.

Problem

What evidence, if any, shows that a chemical change has occurred when the following substances are mixed?

1. A zinc strip is momentarily placed into a hydrochloric acid solution.
2. A couple of drops of blue bromothymol blue solution are added to hydrochloric acid.
3. A few drops of silver nitrate solution are added to hydrochloric acid.
4. Hydrochloric acid is added to sodium acetate solution.
5. Ammonium nitrate crystals are stirred into water.
6. Hydrochloric acid is added to sodium bicarbonate solution.
7. A couple of drops of phenolphthalein solution are added to an ammonia solution.
8. Sodium hydroxide solution is added to cobalt(II) chloride solution.
9. Sodium nitrate solution is added to potassium chloride solution.
10. A copper wire is placed into a silver nitrate solution.

Procedure

1. Combine the substances according to the instructions provided at each station, and record all observations before, during, and after the mixing.

2. Clean all apparatus and the laboratory bench area, and dispose of used chemicals as directed by your teacher, before proceeding to the next station.

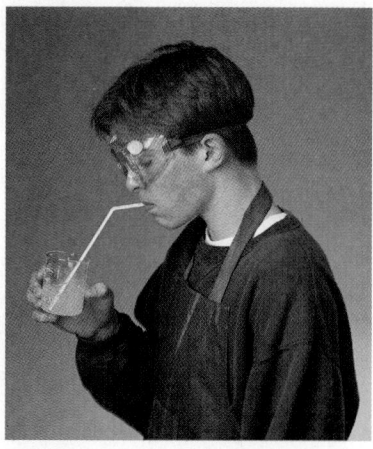

Figure 4.8
If an unknown gas is bubbled through a limewater solution, and the mixture becomes cloudy, then the gas most likely contains carbon dioxide. The limewater diagnostic test provides evidence for carbon dioxide gas in the breath you exhale.

Exercise

7. List four changes that can be used as evidence for chemical reactions.
8. Provide two examples from everyday life of each of the four types of changes listed in your answer to the previous question.
9. Use an "If [procedure] and [evidence], then [analysis]" format to write diagnostic tests for an acid and for hydrogen.
10. What scientific law led John Dalton to create the theory that atoms are conserved in a chemical reaction?
11. Which of the following chemical equations is balanced correctly?

(a) $H_{2(g)} + O_{2(g)} \rightarrow H_2O_{(g)}$

(b) $2\ NaOH_{(aq)} + Cu(ClO_3)_{2(aq)} \rightarrow Cu(OH)_{2(s)} + 2\ NaClO_{3(aq)}$

(c) $Pb_{(s)} + AgNO_{3(aq)} \rightarrow Ag_{(s)} + Pb(NO_3)_{2(aq)}$

(d) $2\ NaHCO_{3(s)} \rightarrow Na_2CO_{3(s)} + CO_{2(g)} + H_2O_{(l)}$

4.3 BALANCING CHEMICAL REACTION EQUATIONS

Models are used to represent individual molecules and the atomic rearrangements believed to occur in chemical reactions. Since atoms, ions, and molecules are much too small to see, observable changes in a chemical reaction must involve extremely large numbers of particles. In order to represent the changes observed during reactions, a convenient way to communicate the large numbers of particles is required.

You are already familiar with some terms used to define convenient numbers (Table 4.3). For example, a dozen is a convenient number referring to items such as eggs or donuts. Since atoms, ions, and molecules are extremely small particles, a convenient number for them must be much greater than a dozen. A convenient amount used by chemists is called the **mole** (SI symbol, mol). Modern methods of estimating this number of particles (entities) have led to the value 6.02×10^{23}. This value is called **Avogadro's number**, named after the Italian chemist, Amadeo Avogadro (1776–1856). (Avogadro did not determine the number but he created the research idea.) A **mole** is the amount of substance with the number of particles (entities) corresponding to Avogadro's number. For example,

- one mole of sodium is 6.02×10^{23} Na atoms;
- one mole of chlorine is 6.02×10^{23} Cl_2 molecules;
- one mole of sodium chloride is 6.02×10^{23} NaCl formula units.

Essentially, a mole represents a number (6.02×10^{23}, Avogadro's number), just as a dozen represents the number 12.

Table 4.3

CONVENIENT NUMBERS		
Quantity	Number	Example
pair	2	shoes
dozen	12	eggs
gross	144	pencils
ream	500	paper
mole	6.02×10^{23}	molecules

Figure 4.9
These amounts of carbon, table salt, and sugar each contain about a mole of entities (atoms, formula units, molecules) of the substance. The mole represents a convenient and specific quantity of a chemical.

Although the mole represents an extraordinarily large number, a mole of a substance is an observable quantity that is convenient to measure and handle. Figure 4.9 shows a mole of each of three common substances: one element, one ionic compound, and one molecular compound. In each case, a mole of entities is a sample size that is convenient for lab work. Later in this chapter you will learn how to determine the mass of a mole of any pure substance, once you know its formula. The eventual goal is to use the mole as the unit of stoichiometry to determine the quantities of chemicals that react and are produced in a chemical reaction.

Translating Balanced Chemical Equations

A balanced chemical equation can be interpreted theoretically in terms of individual atoms, ions, or molecules, or groups of them. Consider the reaction equation for the industrial production of the fertilizer ammonia.

$N_{2(g)}$	+	$3 H_{2(g)}$	$\rightarrow$	$2 NH_{3(g)}$
1 molecule		3 molecules		2 molecules
1 dozen molecules		3 dozen molecules		2 dozen molecules
1 mol nitrogen		3 mol hydrogen		2 mol ammonia
6.02×10^{23} molecules		$3(6.02 \times 10^{23})$ molecules		$2(6.02 \times 10^{23})$ molecules
6.02×10^{23} molecules		18.06×10^{23} molecules		12.04×10^{23} molecules

Note that the numbers in each row are *in the same ratio (1:3:2)* whether individual molecules, large numbers of molecules, or moles are considered. When moles are used to express the coefficients in the balanced equation, the ratio of reacting amounts is called the **mole ratio**.

A complete translation of the balanced chemical equation for the formation of ammonia is: "One mole of nitrogen gas and three moles of hydrogen gas react to form two moles of ammonia gas." This translation includes all the symbols in the equation, including coefficients and states of matter.

EXAMPLE

Translate the following chemical equation into an English sentence.

$6 CO_{2(g)} + 6 H_2O_{(l)} \rightarrow C_6H_{12}O_{(aq)} + 6 O_{2(g)}$

Six moles of carbon dioxide gas react with six moles of liquid water to produce one mole of aqueous glucose and six moles of oxygen gas.

Exercise

12. Translate the following English sentences into internationally understood balanced chemical equations.
    (a) Two moles of solid aluminum and three moles of aqueous copper(II) chloride react to form three moles of solid copper

and two moles of aqueous aluminum chloride. (This reaction does not always produce the expected products.)

(b) One mole of solid copper reacts with two moles of hydrochloric acid to produce one mole of hydrogen gas and one mole of copper(II) chloride. (When tested in the laboratory, this prediction is falsified.)

(c) Two moles of solid mercury(II) oxide decomposes to produce two moles of liquid mercury and one mole of oxygen gas. (This decomposition reaction is a historical but dangerous method of producing oxygen.)

(d) Methanol (antifreeze and fuel) is produced from natural gas in world-scale quantities by the following reaction series.

(i) One mole of methane gas reacts with one mole of steam to produce one mole of carbon monoxide gas and three moles of hydrogen gas.

(ii) One mole of carbon monoxide gas reacts with two moles of hydrogen gas to produce one mole of liquid methanol.

13. Translate each of the following chemical equations into an English sentence including the mole ratio for all the substances involved.

(a) Fire-starter for camp fires often involves the following reaction.

$$2 \ CH_3OH_{(l)} + 3 \ O_{2(g)} \rightarrow 2 \ CO_{2(g)} + 4 \ H_2O_{(g)}$$

(b) Phosphoric acid for fertilizer production can be produced from rock phosphorus.

$$Ca_3(PO_4)_{2(s)} + 3 \ H_2SO_{4(aq)} \rightarrow 2 \ H_3PO_{4(aq)} + 3 \ CaSO_{4(s)}$$

(c) The reaction of sodium with water is potentially dangerous.

$$2 \ Na_{(s)} + 2 \ HOH_{(l)} \rightarrow H_{2(g)} + 2 \ NaOH_{(aq)}$$

(d) Sulfuric acid can be used as a catalyst to dehydrate sugar.

$$C_{12}H_{22}O_{11(s)} \rightarrow 12 \ C_{(s)} + 11 \ H_2O_{(g)}$$

Although a chemical equation is international, its translation depends on the language used. For example, the French translation for the formation of ammonia is: "Une mole d'azote gazeux et trois moles d'hydrogène gazeux réagit pour produire deux moles d'ammoniac gazeux."

Balancing Chemical Equations

A chemical equation is a simple, precise, logical, and international method of communicating the experimental evidence of a reaction. The evidence used when writing a chemical equation is often obtained in stages. First are some general observations that a chemical change has occurred. These are likely followed by a series of diagnostic tests to identify the products of the reaction. At this stage an unbalanced chemical equation can be written, and then the theory of conservation of atoms can be used to predict the coefficients necessary to balance the reaction equation. In most cases, trial and error, as well as intuition and experience, play an important role in successfully balancing chemical equations. The following summary outlines a systematic approach to balancing equations. Use it as a guide as you study the example that follows the summary.

SUMMARY: BALANCING CHEMICAL EQUATIONS

Step 1: Write the chemical formula for each reactant and product, including the state of matter for each one.

Step 2: Try balancing the atom or ion present in the greatest number. Find the lowest common multiple to obtain coefficients to balance this particular atom or ion.

Step 3: Repeat step 2 to balance each of the remaining atoms and ions.

Step 4: Check the final reaction equation to ensure that all atoms and ions are balanced.

Tips

- Do not make the common mistake of creating chemical formulas to balance equations. The chemical formula subscripts are fixed, only the coefficients should be changed.
- If an atom/ion appears in more than two places in the chemical equation, balance it last.
- Balance elements last.
- If possible, balance polyatomic ions as a group.

EXAMPLE

A simple technology for recycling silver is to trickle waste solutions containing silver ions over scrap copper. For example, copper metal reacts with aqueous silver nitrate to produce silver metal and aqueous copper(II) nitrate, as shown in Figure 4.10. Write the balanced chemical equation.

Step 1: $?Cu_{(s)} + ?AgNO_{3(aq)} \rightarrow ?Ag_{(s)} + ?Cu(NO_3)_{2(aq)}$

Step 2: Oxygen atoms are present in the greatest number, so balance them first. (Always balance elements last.) Balance the nitrate ion as a group.

$?Cu_{(s)} + 2\,AgNO_{3(aq)} \rightarrow ?Ag_{(s)} + 1\,Cu(NO_3)_{2(aq)}$

Step 3: Balance Ag and Cu atoms.

$Cu_{(s)} + 2\,AgNO_{3(aq)} \rightarrow 2\,Ag_{(s)} + Cu(NO_3)_{2(aq)}$

Step 4: The amounts in moles of copper, silver, and nitrate are one, two, and two on both the reactant and the product sides of the equation arrow. (This is a mental check and no statement is required.)

Figure 4.10

A piece of copper before it is placed in a beaker of aqueous silver nitrate (left), during the reaction (center), and after the reaction (right). The chemical equation must represent this evidence.

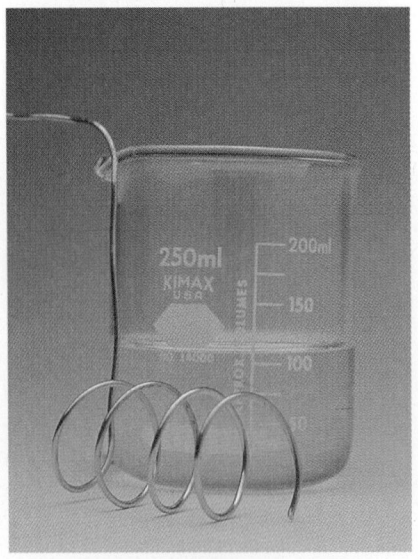
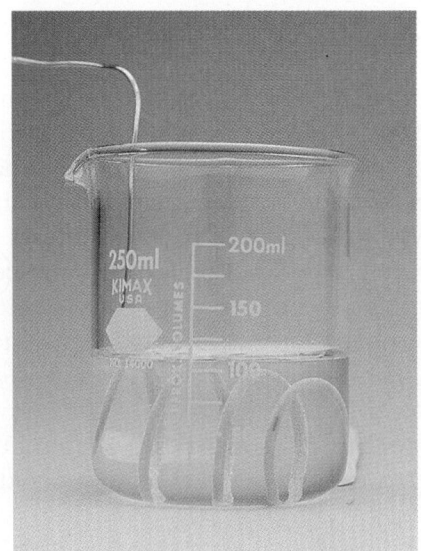
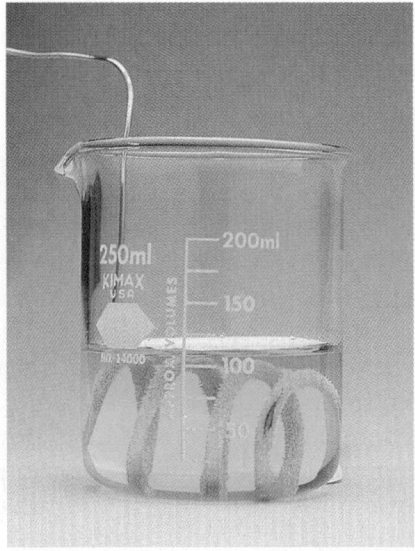

Use the following techniques for balancing chemical reactions:

- Persevere and realize that, like solving puzzles, several attempts may be necessary for more complicated chemical equations.

- The most common student error is to use incorrect chemical formulas to balance the chemical equation. *Always write correct chemical formulas first and then balance the equation as a separate step.*

- If polyatomic ions remain intact, balance them as a single unit.

- Delay balancing any atom that is present in more than two substances in the chemical equation until all other atoms or ions are balanced. (Oxygen is a common example.)

- Balance elements last.

- If a fractional coefficient is required to balance an atom, multiply all coefficients by the denominator of the fraction to obtain integer values.

For example, in balancing the following reaction equation, hydrogen atoms are balanced first, then nitrogen, and oxygen is balanced last. This requires 7 mol of oxygen atoms.

$$2\,NH_{3(g)} + ?\,O_{2(g)} \rightarrow 3\,H_2O_{(g)} + 2\,NO_{2(g)}$$

The only number that can balance the oxygen atoms is $\frac{7}{2}$. By doubling all coefficients, the reaction equation can then be balanced using only integers.

$$4\,NH_{3(g)} + 7\,O_{2(g)} \rightarrow 6\,H_2O_{(g)} + 4\,NO_{2(g)}$$

Exercise

14. Balance the following equations that communicate reactions that occur before, during and after the formation of acid rain.

(a) $C_{(s)} + O_{2(g)} \rightarrow CO_{2(g)}$

(b) $S_{8(s)} + O_{2(g)} \rightarrow SO_{2(g)}$

(c) $Cu(OH)_{2(s)} + H_2SO_{4(aq)} \rightarrow HOH_{(l)} + CuSO_{4(aq)}$

(d) $CaSiO_{3(s)} + H_2SO_{3(aq)} \rightarrow H_2SiO_{3(aq)} + CaSO_{3(s)}$

(e) $CaCO_{3(s)} + HNO_3 \rightarrow H_2CO_{3(aq)} + Ca(NO_3)_{2(aq)}$

(f) $Al_{(s)} + H_2SO_{4(aq)} \rightarrow H_{2(g)} + Al_2(SO_4)_{3(aq)}$

(g) $SO_{2(g)} + H_2O_{(l)} \rightarrow H_2SO_{3(aq)}$

(h) $Fe_{(s)} + H_2SO_{3(aq)} \rightarrow H_{2(g)} + Fe_2(SO_3)_{3(s)}$

(i) $N_{2(g)} + O_{2(g)} \rightarrow NO_{(g)}$

(j) $CO_{2(g)} + H_2O_{(l)} \rightarrow H_2CO_{3(aq)}$

(k) $CH_{4(g)} + O_{2(g)} \rightarrow CO_{2(g)} + H_2O_{(g)}$

(l) $C_4H_{10(g)} + O_{2(g)} \rightarrow CO_{2(g)} + H_2O_{(g)}$

(m) $FeS_{(s)} + O_{2(g)} \rightarrow FeO_{(s)} + SO_{2(g)}$

(n) $H_2S_{(g)} + O_{2(g)} \rightarrow H_2O_{(g)} + SO_{2(g)}$

(o) $CaCO_{3(s)} + SO_{2(g)} + O_{2(g)} \rightarrow CaSO_{4(s)} + CO_{2(g)}$

Write a balanced chemical equation for each of the following reactions in questions 15 to 19. Assume that substances are pure and at SATP unless the states of matter are given. Also classify the primary perspective presented in the accompanying statements.

15. Research indicates that sulfur dioxide gas reacts with oxygen in the atmosphere to produce sulfur trioxide gas.

16. Sulfur trioxide gas traveling across international boundaries causes disagreements between governments.

    sulfur trioxide + water → sulfuric acid

17. The means exist for industry to reduce sulfur dioxide emissions; for example, by treatment with lime.

    calcium oxide + sulfur dioxide + oxygen → calcium sulfate

18. Restoring acid lakes to normal is expensive; for example, adding lime to lakes from the air.

    calcium oxide + sulfurous acid → water + calcium sulfite

19. Fish in acidic lakes may die from mineral poisoning due to the leaching of minerals from lake bottoms.

    solid aluminum silicate + sulfuric acid →

    aqueous hydrogen silicate + aqueous aluminum sulfate

■ Problem
✔ **Prediction**
■ Design
■ Materials
■ Procedure
✔ **Evidence**
✔ **Analysis**
✔ **Evaluation**
■ Synthesis

▌Technological Applications
You will find it useful to memorize the technological applications of the chemicals and reactions indicated with an asterisk.

INVESTIGATION

4.2 Balancing Chemical Reaction Equations

The purpose of this investigation is to test the balancing of reaction equations by using molecular models.

Problem

What are the balanced equations for the following chemical equations?

Experimental Design

Molecular model kits are used to construct models of the reactants and products of these chemical reactions.

1. Hydrogen is used as fuel for the Space Shuttle.★

   $H_{2(g)} + O_{2(g)} \rightarrow H_2O_{(g)}$

2. Ammonia fertilizer is produced for agricultural use.

   $N_{2(g)} + H_{2(g)} \rightarrow NH_{3(g)}$

3. Natural gas burns for home heating.★

   $CH_{4(g)} + O_{2(g)} \rightarrow CO_{2(g)} + H_2O_{(g)}$

4. Hydrogen chloride gas is produced and then dissolved in water to make hydrochloric acid for use as a rust remover.

   $H_{2(g)} + Cl_{2(g)} \rightarrow HCl_{(g)}$

5. Methanol is used as fuel in a fondue burner.

   $CH_3OH_{(l)} + O_{2(g)} \rightarrow CO_{2(g)} + H_2O_{(g)}$

Materials

molecular model kit

Procedure

1. Construct models of the reactant molecules for reaction 1 as communicated by the predicted balanced chemical equation.

2. Draw a structural formula diagram (a sketch showing the arrangement of the bonded atoms) for each reactant (Figure 4.11), and list the total number of each kind of atom represented.

3. Disassemble the reactant molecules and use only the same atoms to assemble models of the product molecules for reaction 1. (No atoms should be left over or added.)

4. Draw a structural formula diagram for each product, and list the total number of each kind of atom represented.

5. Write a balanced chemical equation from your structural formula diagrams.

6. Repeat steps 1 to 5 for each of the other four reaction equations.

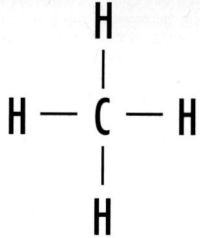

Figure 4.11
Communicate your model as a structural formula diagram presented below the photo.

4.4 METRIC UNITS AND CALCULATIONS

The mole is a number that has gained acceptance because of its convenience for communicating and calculating quantities of chemicals that react and are produced. The mole does not have much application outside of describing reacting quantities — but reactions are primarily what chemistry is about. With this in mind, the mole is introduced above within the context of writing and balancing chemical equations. The major application (and the reason for the creation of the mole as a concept) is for calculating reacting amounts. Although the full application of the mole for this purpose is delayed until Chapter 7 *Chemical Reaction Calculations*, you are introduced to the conversion of measured values into and out of amount in moles at this point. Although you may not see much purpose in these conversions here, you will find a very extensive purpose in Chapter 7 and beyond. You start off below with a quick review of metric (communicating and converting) before moving into some calculation basics — accuracy, precision, rounding, certainty, and significant digits.

Communicating in Metric

Communication systems require that a wide assortment of conventions be accepted by a large population. As outlined on page 55, for communication systems to be widely accepted they need to be more *international, precise, logical,* and *simple* than competing systems of communication. The International System of Units, SI, is used by the vast majority of countries in the world as the preferred system of units. For scientific communication all scientists work

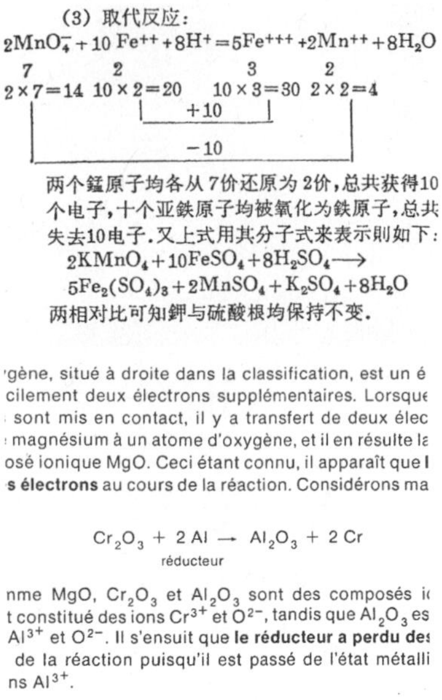

(3) 取代反应：

$$2MnO_4^- + 10\ Fe^{++} + 8H^+ = 5Fe^{+++} + 2Mn^{++} + 8H_2O$$

7   2   3   2

$2 \times 7 = 14 \quad 10 \times 2 = 20 \quad 10 \times 3 = 30 \quad 2 \times 2 = 4$

+10
−10

两个锰原子均各从7价还原为2价,总共获得10个电子,十个亚铁原子均被氧化为铁原子,总共失去10电子。又上式用其分子式来表示则如下：

$$2KMnO_4 + 10FeSO_4 + 8H_2SO_4 \longrightarrow$$
$$5Fe_2(SO_4)_3 + 2MnSO_4 + K_2SO_4 + 8H_2O$$

两相对比可知钾与硫酸根均保持不变。

'gène, situé à droite dans la classification, est un é
cilement deux électrons supplémentaires. Lorsque
sont mis en contact, il y a transfert de deux élec
magnésium à un atome d'oxygène, et il en résulte la
osé ionique MgO. Ceci étant connu, il apparaît que l
s électrons au cours de la réaction. Considérons ma

$$Cr_2O_3 + 2\ Al \longrightarrow Al_2O_3 + 2\ Cr$$
réducteur

nme MgO, Cr_2O_3 et Al_2O_3 sont des composés ic
t constitué des ions Cr^{3+} et O^{2-}, tandis que Al_2O_3 es
Al^{3+} et O^{2-}. Il s'ensuit que le réducteur a perdu des
de la réaction puisqu'il est passé de l'état métalli
ns Al^{3+}.

Figure 4.12
No matter what national language is being used for scientific communication, SI and IUPAC rules for communicating quantities and chemical formulas are followed internationally. This commitment by chemists speeds the evolution of scientific knowledge for the eventual benefit of all life on our planet.

and/or communicate with SI, otherwise the transfer of knowledge within the scientific community would be slowed.

The quantities and base units and their respective symbols for SI are listed in Appendix E, E.2. Use this section of Appendix E for future reference. Some of the common SI quantities, units, and symbols are listed in Table 4.4 on this page. If you have not learned these already, you will have to learn them before finishing your chemistry education.

Table 4.4

SOME SI QUANTITIES AND UNITS			
Quantity	Symbol	Unit	Symbol
length	l	metre	m
time	t	second	s
mass	m	gram	g
molar mass	M	grams per mole	g/mol
volume	v	litre	L
amount	n	mole	mol
temperature	t	degree Celsius	°C
energy	E	joule	J

SI Prefixes

SI prefixes and scientific notation are two internationally accepted methods for communicating very large and very small values or of communicating values with only a certain number of digits. SI prefixes are commonly used in multiples of 10^3, although there are a few common exceptions such as deci (10^{-1}) and centi (10^{-2}). See Appendix E, E.2 and Table 4.5 on this page. You should at least memorize mega, kilo, milli, and micro.

Rule of a Thousand

The SI prefixes and scientific notation can be used for the rule of a thousand and for converting from one SI value to another. **The rule of a thousand** says that you should express your final answer in a question as a value between 0.1 and 1000; e.g., 1250 g should be expressed as 1.250 kg. In this textbook, all answers follow the rule of a thousand. Your teacher may indicate that its use is optional. The precision and certainty rules presented in the next section also make it necessary for you to change the SI prefix and/or to use scientific notation.

Converting with SI Prefixes

SI prefixes make it very convenient to convert from one value to another. These conversions are necessary to follow the certainty and precision rules presented below and the rule of a thousand presented above. For the majority of these conversions the decimal place is moved by three places left or right. You can use a formal systematic approach by writing down a conversion factor (as shown below), or you can more efficiently move the decimal place and then mentally check

Table 4.5

SOME SI PREFIXES		
Prefix	Symbol	Factor
giga	G	10^9
mega*	M	10^6
kilo*	k	10^3
deci	d	10^{-1}
centi	c	10^{-2}
milli*	m	10^{-3}
micro*	μ	10^{-6}
nano	n	10^{-9}

*Memorize these.

the answer to see if it makes sense. (Follow the technique that your teacher wants you to use.)

For example, if your answer for a question calculated out to be 1583 g of lead and you wanted to express the answer using the rule of a thousand, you could do the following.

First select a conversion factor; in this case it is 1000 g/kg.

Then use the conversion factor such that the units work out the way that you want them.

$$m_{Pb} = 1583 \cancel{g} \times \frac{1 \text{ kg}}{1000 \cancel{g}} = 1.583 \text{ kg}$$

Note that you decide on whether to use the conversion factor as is or as its reciprocal by checking to see if the units cancel (divide) out to get what you want.

Alternately, do the conversion in your head by moving the decimal place three places.

$$m_{Pb} = 1583 \text{ g} = 1.583 \text{ kg}$$

If you are converting merely to follow the rule of a thousand, provide both answers—that way if you make an error in the conversion, you have grounds for an appeal for full or part marks.

EXAMPLE

Convert 0.0842 mol of zinc to follow the rule of a thousand.

$$n_{Zn} = 0.0842 \cancel{\text{mol}} \times \frac{1000 \text{ mmol}}{1 \cancel{\text{mol}}} = 84.2 \text{ mmol}$$

Scientific Notation

Scientific notation is a method suited to communicating very large or small values, but it can also be used to communicate values that can be communicated with SI prefixes. In general SI prefixes are more efficient to use than scientific notation. **Scientific notation** is a convenient method for expressing a value between 1 and 10 multiplied by a power of 10. For example, the values in Table 4.7 are expressed in regular notation and in scientific notation.

To enter a value with scientific notation into your calculator or computer you do not enter, for example, 4.98×10^{-2}. You usually enter

Table 4.6

USING THE RULE OF A THOUSAND	
Value	**SI Prefix**
1250 g	1.250 kg
13 456 mL	13.456 L
0.034 s	34 ms
0.006 52 m	6.52 mm
5982 kmol	5.982 Mmol

Rule of a Thousand Examples
Here are some common examples of the use of the rule of a thousand.

candy bar	49 g
refined sugar	4 kg
soft drinks	300 mL
gasoline	48.3 L
pain relief tablets	325 mg
vitamin capsules	200 mg
bulk fertilizer	25 t
concrete	7.5 m^3
carpet	12.4 m^2

Table 4.7

REGULAR AND SCIENTIFIC NOTATIONS		
Regular Notation	**Scientific Notation**	**SI Prefix**
1602 L	1.602×10^3 L	1.602 kL
0.000 002 3 mol	2.3×10^{-6} mol	2.3 μmol
24 326 g	$2.432\ 6 \times 10^4$ g	24.326 kg
0.0498 m	4.98×10^{-2} m	49.8 mm

something like [4] [•] [9] [8] [EXP] [2] [+/-] (See Appendix E, E.2), or whatever your calculator instructions indicate.

If you have to enter a value such as 24.3 kg into your calculator as grams in scientific notation, you enter it as a value of 24.3×10^3 g. If necessary, 32.3 μmol is entered in your calculator most conveniently as 32.3×10^{-6} mol. (You never report an answer in this format, but you may enter it into your calculator this way.)

Exercise

20. Provide the quantity symbol for the following quantities.
    (a) time
    (b) temperature
    (c) mass
    (d) amount
    (e) volume
    (f) molar mass

21. Provide the SI unit symbol for the following units.
    (a) litre
    (b) gram
    (c) second
    (d) degree Celsius
    (e) mole
    (f) grams per mole

22. Convert the following abbreviations into SI symbols. (Note that abbreviations should always have periods but SI symbols never have periods.)
    (a) 24 hrs.
    (b) 368 sec.
    (c) 50 Km. per hr.
    (d) 78 ml.
    (e) 100 gms.
    (f) 0.69 g/mole

23. When expressing a value, an SI rule requires you to either use words or symbols, but not a mix of the two. Provide the SI unit symbol complete with SI prefix symbol for the following values.
    e.g., two hundred and fifty grams is 250 g
    (a) thirty-two decimal six millilitres
    (b) zero decimal zero three two six litres
    (c) twenty-five degrees Celsius
    (d) one hundred and fifty-two decimal four seven grams per mole
    (e) zero decimal seven eight four moles
    (f) seven hundred and four millimoles
    (g) one hundred kilograms
    (h) zero decimal one zero zero megagrams
    (i) six thousand and fifty-two milliseconds
    (j) six decimal zero five two seconds
    (k) five hundred kilojoules
    (l) zero decimal five zero zero megajoules

24. Convert the following values to obey the rule of a thousand.
    (a) 2051 g
    (b) 0.0067 kg
    (c) 54 891 mL
    (d) 0.0985 L
    (e) 1996 ms
    (f) 21 060 mol
    (g) 0.023 mol
    (h) 0.0040 kmol

25. Express the following values in scientific notation without any SI prefix in the final units.
    (a) 2988 mol
    (b) 2.988 mmol
    (c) 3765 kg
    (d) 3765 g
    (e) 0.0067 L
    (f) 0.0067 mL

26. (Discussion) Read the following statement out loud — "The kilometer is an odometer that reads distances in kilometres. Just like a micrometer is a measuring device to measure lengths in micrometres." Why are there so many people who say kilometre wrong?

Accuracy, Precision, and Certainty

In the scientific community, there are a number of specific definitions for various terms — just as you find for most words in the dictionary. In this section you consider accuracy, precision, and certainty. Each of these terms is necessary for general scientific discussions as well as specific calculations. In general, accuracy is used to express how close your measurements or answers are to an accepted value; precision is used to express how close your measurements or answers are to each other; and certainty is used to express a combination of how large and how precise your measurement or answer is.

Accuracy

The **accuracy** of a value is an expression of how close the experimental value is to the accepted or predicted value. The comparison of the two values (experimental and predicted) is often measured in terms of a *percent difference*. This is *not a percent error*, because there may be very little error associated with a large percent difference, if the prediction is validly falsified by the evidence gathered. (Another similar concept *percent yield* is covered in Chapter 7, page 261.)

$$\% \text{ difference} = \frac{|\text{experimental value} - \text{predicted value}|}{|\text{predicted value}|} \times 100$$

Note that the percent difference is expressed most simply as an absolute value (|value|) without any positive or negative sign. This expression of accuracy is often used in the Evaluation section of investigation reports. The percent error associated with a percent difference is assumed to be a maximum of 10%. Therefore, if the experimental design, procedure, and skills are adequate, any error above 10% is assumed to be due to an unacceptable authority used to make the prediction.

EXAMPLE

Judge the acceptability of an authority in an investigation where an experimental value of 32.34 g and a predicted value of 35.47g are obtained.

$$\% \text{ difference} = \frac{\overset{3.13\text{ g}}{|32.34\text{ g} - 35.47\text{ g}|}}{35.47\text{ g}} \times 100 = 8.82\%$$

The prediction is verified because the percent difference is less than the allowed percent error of 10%; i.e., 8.82%. The [authority] is judged to be acceptable because the prediction is verified.

(a)

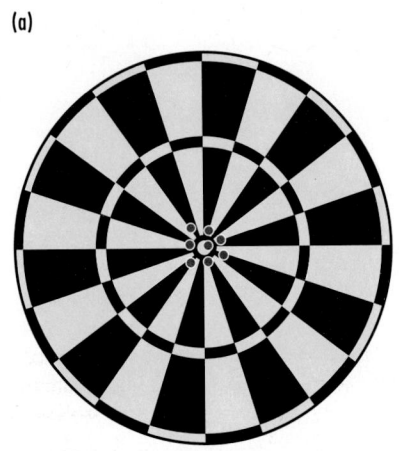

(b)

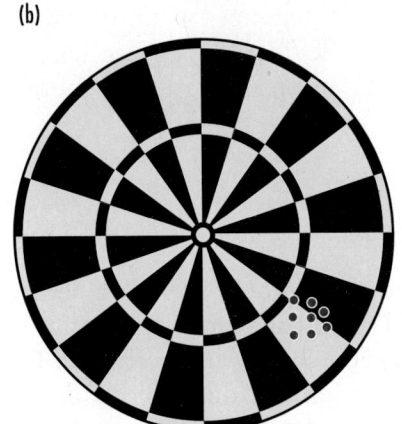

(c)

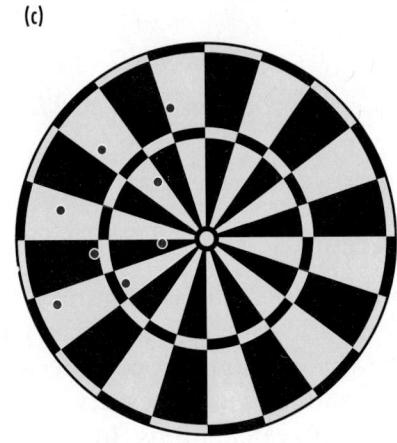

Figure 4.13
The positions of the darts in each of these figures are analogous to measured or calculated results in a laboratory setting. The results in (a) are precise and accurate, in (b) they are precise but not accurate, and in (c) they are neither precise nor accurate.

Precision

The **precision** of a measured or calculated value is the *place value of the last measurable digit and is determined by the measuring instrument.* A mass of 17.13 g is more precise than 17.1 g. The precision is determined by the particular system or instrument used; for example, a centigram balance versus a decigram balance.

Accuracy is an expression of how close a value is to the accepted, expected, or predicted value, whereas precision is a measure of the reproducibility or consistency of a result (Figure 4.13). Accuracy is generally attributed to an error in the system *(a systematic error)*; precision is associated with a *random error* of measurement. For example, if you used a balance without zeroing it, you might obtain measurements that have high precision (reproducibility) but low accuracy. The same is true for calibrating a pH meter at pH 7.00 and then making a measurement of a very high or low pH. The systematic error might be high (low accuracy), even though the random error of the measurement is low (high precision).

You may not know how uncertain the last measured digit is. On a centigram balance, the error of measurement in the last digit is usually considered to be ±0.01 g. Measurements such as 12.39 g, 12.40 g, and 12.41 g all have the same precision (hundredths), and may all be equally correct masses for the same object. The precision with which you read a thermometer might be ±0.2°C (for example, 12.0°C, 121.2°C, or 21.4°C) and a ruler might be read to ±0.5 mm; you must decide, for example, whether to record 11.0 mm, 11.5 mm, or 12.0 mm.

EXAMPLE

What is the inferred precision for the following sets of measurements?

1. 22.6°C, 45.8°C, and 87.2°C   *Answer: ±0.2°C*
2. 65 mm, 120 mm, 205 mm, and 325 mm   *Answer: ±5 mm*
3. 1.02 g, 0.99 g, 1.00 g, and 0.97 g   *Answer: ±0.01 g*

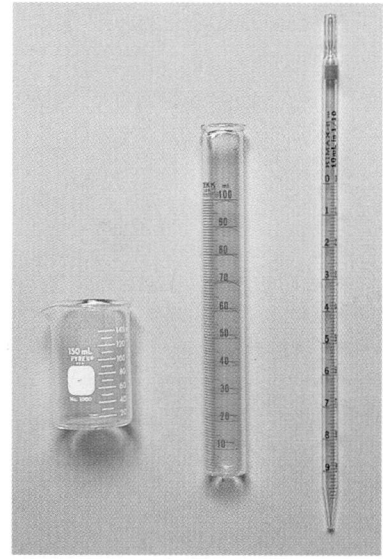

Figure 4.14
The precision of glassware for measuring volume increases from the beaker (±10 mL), to the graduated cylinder (±1 mL), and to the pipet (±0.1 mL).

Precision Rules for Calculations

According to the **precision rule**, a result obtained by adding or subtracting measured or calculated volumes is rounded to the same precision (number of decimal places) as the least precise value used in the calculation. For example, 12.6 g + 2.07 g + 0.142 g totals to 14.812 g on your calculator. This value is rounded to one-tenth of a gram and reported as 14.8 g because the first measurement limits the precision of the final result to tenths of a gram, and the rounding rule suggests leaving 8 as is. The final result is reported to the least number of decimal places in the values added or subtracted.

Rounding

Calculations are usually based on measurements (for example, in the Analysis section of a report). To report a calculated result correctly, you must know the place value at which the result becomes uncertain and a method for rounding the answer. *Rounding* means checking the first digit following the digit that will be rounded. If this digit is less than 5, it and all the following digits are discarded. If this digit is 5 or greater, it and all the following digits are discarded, and the preceding digit is increased by one.

> **Precision Rule**
> A result obtained by adding or subtracting measured values is rounded to the same precision (number of decimal places) as the least precise value used in the calculation. Multiplying or dividing by an exact number follows the precision rule. Multiplying by an exact number is the same as adding the value to itself that many times. Any value without an attached unit is assumed to be an exact value.

EXAMPLE

What is the answer for each of the following measured or calculated values with varying precision?

1. 40.08 g + 12.01 g + 48.00 g =   *Answer: 100.09 g*
2. 12.3 mL + 99 mL + 3.65 mL =
   Answer: 114.95 mL rounds to 115 mL
3. 14.8 s + 105 s + 22.6 s =   *Answer: 142.4 s rounds to 142 s*
4. 73.6°C − 4.35°C =   *Answer: 69.25°C rounds to 69.3°C*
5. 4.67 g × 3 =   *Answer: 14.01 g*
6. 1.25 L − 106 mL =
   Answer: 1.25 L − 0.106 L = 1.144 L rounds to 1.14 L

Certainty

How certain you are about a measurement depends on two factors — the precision of the instrument and the value of the measured quantity. More precise instruments give more certain values; for example, 15°C as opposed to 15.215°C. Consider two measurements with the same precision: 0.4 g and 12.8 g. If the balance used is precise to ±0.2 g, the value 0.4 g could vary by as much as 50%. However, 0.2 g of 12.8 g is a variation of less than 2%. For both factors — precision of instrument and value of the measured quantity — the more digits in a measurement, the more certain you are about the measurement. The **certainty** of any measurement is communicated by the number of significant digits in the measurement. In a measured or calculated value, **significant digits** are all those digits that are certain plus one estimated (uncertain) digit. Significant digits include all digits correctly reported from a measurement, except leading zeros. Leading zeros are the zeros at the beginning of a decimal fraction and are written only to locate the decimal point. For example, 6.20 mL (3 significant digits) has the same number of significant digits as 0.00620 L.

EXAMPLE

For each of the following measurements, the certainty (number of significant digits) is stated beside the measured or calculated value.

22.07 g	a certainty of 4 significant digits
0.41 mL	a certainty of 2 significant digits
700 mol	a certainty of 3 significant digits
0.020 50 km	a certainty of 4 significant digits
2×10^{40} m	a certainty of 1 significant digit

Certainty Rule for Calculations

According to the **certainty rule**, a result obtained by multiplying or dividing measured values is rounded to the same certainty (number of significant digits) as the *least certain value* used in the calculation. Significant digits are primarily used to determine the certainty of a result obtained from calculations using several measured values. For example, 0.002 489 mol × 6.94 g/mol is displayed as 0.1727366 g on a calculator. This is correctly reported as 0.173 g or as 173 mg because the second measured value used (6.94) limits the final result to a certainty of three significant digits.

EXAMPLE

Determine the certainty of the following calculated values and round the answer to the correct number of significant digits and correct units.

1. $n_{H_2} = 31.67 \text{ g} \times \dfrac{1 \text{ mol}}{2.02 \text{ g}} = 15.7 \text{ mol}$

2. $m_{CO_2} = 0.23 \text{ mol} \times \dfrac{44.01 \text{ g}}{1 \text{ mol}} = 10 \text{ g}$

Exercise

27. What criteria are used for judging a system of communication accepted by the scientific community?

28. If accuracy can be expressed as a percent difference, how can precision and certainty be expressed?

29. Communicate the accuracy of the following values as a percent difference.
    (a) experimental value; 212 mL; predicted value: 223 mL
    (b) experimental value; 15.1 g; predicted value: 14.8 g
    (c) experimental value; 67 s; predicted value: 50 s

30. Communicate the inferred precision of the following sets of measurements as a ± value.
    (a) 0.3 mL, 12.6 mL, 24.5 mL, 36.0 mL
    (b) 2.5 mm, 11.0 mm, 23.5 mm, 54.5 mm
    (c) 20 mg, 25 mg, 30 mg, 35 mg
    (d) 96°C, 94°C, 98°C

31. Communicate the certainty of the following measured or calculated values as a number of significant digits.
    (a) 16.05 g
    (b) 7.0 mL
    (c) 10 cm^2
    (d) 0.563 kg
    (e) 0.000 5 L
    (f) 90.00 g/mol

32. Round the following calculator values to a certainty of three significant digits.
    (a) 34.568 672 L
    (b) 87.44 g
    (c) 313.49 mL
    (d) 6.344 9 kg
    (e) 299.5 s
    (f) 0.012 55 h

33. Round the following calculator values to a certainty of two significant digits. Provide two answers — one with SI prefixes following the rule of a thousand and one with scientific notation.
    (a) 3490 m
    (b) 555 L
    (c) 0.006 449 kg
    (d) 0.0799 km
    (e) 48 234.87 kg
    (f) 5989 s

34. Perform the following calculations and express the answer to the correct precision.
    (a) 87.6 g – 4.36 g =
    (b) 2.01 mm + 12.6 mm + 87 mm =
    (c) 0.58 kg + 120 g + 85 g =
    (d) 1.06 m – 124 mm =

Figure 4.15
A high certainty (large number of significant digits) is hard to obtain. For example, a bathroom scale (top) has a higher capacity but lower precision than a centigram (middle) or milligram balance (bottom). High certainty (as expressed in significant digits) is obtained from measuring devices with the best combination of capacity and precision.

35. Perform the following calculations and express the answer to the correct certainty (number of significant digits).

(a) $n_{Cu} = 7.46 \text{ g} \times \dfrac{1 \text{ mol}}{63.55 \text{ g}} =$

(b) $m_{C} = 2.0 \text{ mol} \times \dfrac{12.01 \text{ g}}{1 \text{ mol}} =$

(c) $n_{CuSO_4} = 100.0 \text{ mL} \times \dfrac{0.500 \text{ mol}}{1 \text{ L}} =$

(d) $m_{CuSO_4 \cdot 5H_2O} = 0.0500 \text{ mol} \times \dfrac{249.71 \text{ g}}{1 \text{ mol}} =$

36. Perform the following more advanced calculations by using the precision and/or certainty rule where appropriate.

(a) $m_{NH_3} = 101 \text{ mol} \times \dfrac{17.04 \text{ g}}{1 \text{ mol}} =$

(b) $n_{CuSO_4} = 250.0 \text{ mL} \times \dfrac{5.00 \text{ mol}}{1 \text{ L}} =$

(c) $C_{NaCl} = \dfrac{15.5 \text{ mmol}}{10.00 \text{ mL}} =$

(d) $v_{avg} = \dfrac{13.6 \text{ mL} + 13.5 \text{ mL} + 13.6 \text{ mL}}{3} =$

(e) % difference $= \dfrac{|3.67 \text{ g} - 3.61 \text{ g}|}{3.61 \text{ g}} \times 100 =$

(f) $q = 50.0 \text{ g} \times 4.19 \text{ J/(g} \bullet \text{°C)} \times (34.2 - 25.4)\text{°C} =$

Figure 4.16
The one-kilogram prototype kept at the International Bureau of Weights and Measures in Sèvres, near Paris, France, is the only exact mass on our planet. All other masses are measured (uncertain) masses relative to this prototype.

4.5 THE MOLE

The mole was created by chemists as a unit of stoichiometry for determining the quantity of chemicals that react and are produced in a chemical reaction. You have seen that the mole is a useful unit of measure for communicating reacting amounts in a balanced chemical equation. Now you learn how to measure amount in moles, indirectly, by measuring mass directly.

Relative Atomic Mass

The relative mass of an entity is communicated by comparing the mass of the entity to the mass of a reference or standard mass. For example, all masses on Earth are compared with the mass of the international prototype (model) of one kilogram kept at the International Bureau of Weights and Measures (Figure 4.16). The concept of relative mass was used in the 19th century to assign relative atomic masses to the elements. By comparing the mass of each element that would react with a fixed mass, say sixteen grams, of oxygen, a **relative atomic mass** could be assigned to the elements. For purposes of this exercise, you start by assuming that all elements are monatomic. (After that the process becomes more complicated, but the general principle is the same.) For example,

$$Mg_{(s)} \quad + \quad O_{(g)} \rightarrow MgO_{(s)}$$
$$24 \text{ g} \qquad\quad 16 \text{ g}$$

$$Ca_{(s)} \quad + \quad O_{(g)} \rightarrow CaO_{(s)}$$
$$40 \text{ g} \qquad\quad 16 \text{ g}$$

$$2H_{(g)} \quad + \quad O_{(g)} \rightarrow H_2O_{(g)}$$
$$2 \text{ g} \qquad\quad 16 \text{ g}$$

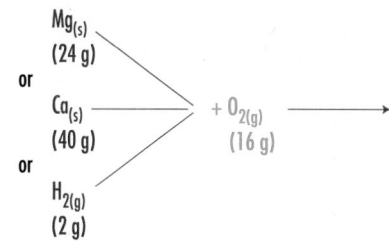

Figure 4.17
The relative atomic masses of the elements can be determined by reacting the elements with a common element, e.g., oxygen.

From this evidence, the relative atomic masses on a scale where oxygen is given a relative atomic mass of 16 are: magnesium 24, calcium 40, and hydrogen 1 (2 divided by 2). The reason for choosing 16 g as the relative reacting mass for oxygen is to give the lightest element (hydrogen) a relative atomic mass of 1. After years of research (and using further research results as to whether elements were monatomic, diatomic, or otherwise, plus using the formula of the resulting compound to determine the mole ratio of the elements reacting), a valid and reliable scale of relative atomic masses of the elements was accepted by the scientific community.

There were several changes over the years as to what the reference element should be. Some proposed scales of relative atomic mass were based on oxygen (as above), whereas others were based on hydrogen (1) and carbon (12). The modern scale of relative atomic masses is based on carbon, and, more specifically, the carbon-12 ($^{12}_{6}C$) isotope. One **atomic mass unit (1 amu)** is defined as exactly 1/12th of the mass of a carbon-12 atom. Because not all carbon atoms are ^{12}C (some, for example, are ^{14}C), the relative atomic mass of naturally occurring carbon is not exactly 12 — it is 12.01. (See atomic masses on the periodic table of the elements.) The relative atomic mass can be thought of as an average of the mass numbers of the common isotopes of the elements (although this is not entirely accurate). For example, a naturally occurring sample of chlorine is composed of about 75% chlorine-35 and 25% chlorine-37. The average mass number is close to the relative atomic mass of 35.45 for chlorine.

75% chlorine-35 yields: $\qquad \dfrac{75.53}{100} \times 35 = 26.44$

25% chlorine-37 yields: $\qquad \dfrac{24.47}{100} \times 37 = 9.05$

natural chlorine yields: $\qquad 26.44 + 9.05 = 35.49$ versus 35.45

You obtain a fairly accurate value — close to the accepted relative atomic mass of 35.45.

All masses of chemical entities, whether we are talking about protons, isotopes, atoms, or molecules, are communicated as a value relative to the mass of the carbon-12 isotope — which is defined as having a mass of exactly 12. This is an example of the international cooperation between scientists that makes it possible for them to communicate effectively with one another. Many conventions of communication are used constantly by chemists — this is just another example of what IUPAC and SI have accomplished to facilitate international communication.

Exercise

37. What is the definition of an amu?

38. What three elements have been used, at one time or another, as the reference element for communicating relative atomic masses?

39. What isotope is currently used as the reference for communicating relative atomic masses?

40. What is the composition of a carbon-12 isotope?

41. Why is the relative atomic mass of carbon, as referenced on the periodic table, not exactly 12?

Molar Mass

The **molar mass** of a substance is the mass of one mole of the substance and is expressed in units of grams per mole (g/mol) (Figure 4.18). Each substance has a different molar mass, which can be calculated as follows.

1. Write the correct chemical formula for the substance.
2. Determine the amount in moles of each atom (or monatomic ion) in one formula unit of the chemical.
3. Use the atomic molar masses from the periodic table and the amounts in moles to determine the molar mass of the chemical.
4. Communicate the molar mass in units of grams per mole, precise to two decimal places; for example, 78.50 g/mol.

You may also think of the molar mass as a ratio of the mass of a particular chemical to the amount of the chemical in moles. Molar mass is a convenient factor to use when converting between mass and amount in moles. In the following example, M represents the molar mass, and the numbers and atomic molar masses are written out; with practice you will eventually be able to determine the molar mass with your calculator and then write only the result.

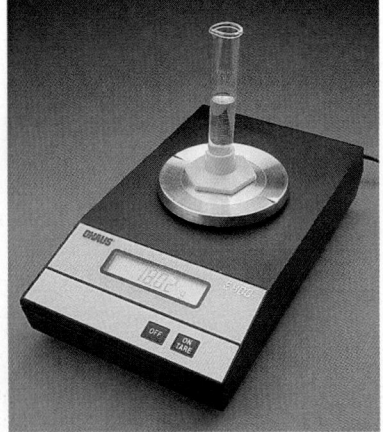

Figure 4.18
One molar mass of water is 18.02 g of H_2O, containing 6.02×10^{23} molecules. A molar mass of H_2O is contained in 22.4 L of water vapor at STP, 18.0 mL of liquid water, or 18.02 g of ice.

EXAMPLE

What is the molar mass of ammonium phosphate, $(NH_4)_3PO_4$?

$$M_{(NH_4)_3PO_4} = 3\,N + 12\,H + P + 4\,O$$
$$= (3 \times 14.01) + (12 \times 1.01) + (1 \times 30.97) + (4 \times 16.00)$$
$$= 149.12$$

The molar mass of ammonium phosphate is 149.12 g/mol.

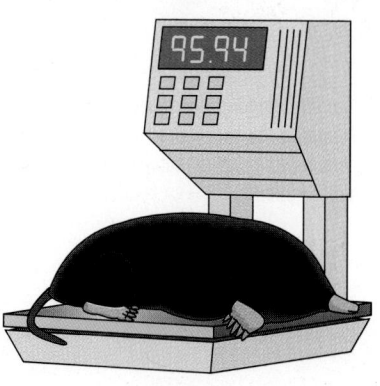

"molar mass"

Mass-Amount Conversions

In order to use the mole ratio from the balanced equation to determine the masses of reactants and products in chemical reactions, you must be able to convert a mass to an amount in moles and vice versa. To do this, you use either the molar mass as a conversion factor in grams per mole (g/mol), or the reciprocal of the molar mass, in moles per gram (mol/g), and cancel the units. Examples of each conversion follow; n represents amount in moles and m represents mass in grams.

EXAMPLE

Convert a mass of 1.5 kg of calcium carbonate to an amount in moles.

Molar mass of $CaCO_3$

$$= (1 \times 40.08) + (1 \times 12.01) + (3 \times 16.00) = 100.09 \text{ g/mol}$$

$$n_{CaCO_3} = 1.5 \text{ kg} \times \frac{1 \text{ mol}}{100.09 \text{ g}} = 0.015 \text{ kmol} = 15 \text{ mol}$$

$$\text{or} \quad 0.015 \text{ kmol} \times \frac{1000 \text{ mol}}{1 \text{ kmol}} = 15 \text{ mol}$$

or

$$n_{CaCO_3} = 1.5 \text{ kg} \times \frac{1000 \text{ g}}{1 \text{ kg}} \times \frac{1 \text{ mol}}{100.09 \text{ g}} = 15 \text{ mol}$$

(Note that the reciprocal of the molar mass is chosen as the appropriate conversion factor.)

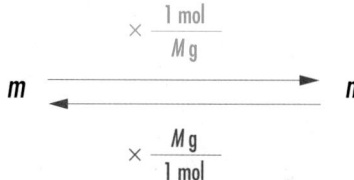

Figure 4.19
Molar mass was created to convert between mass and amount (in moles).

EXAMPLE

Convert a reacting amount of 3.46 mmol of sodium sulfate into mass in grams.

Molar mass of $Na_2SO_4 = 142.04$ g/mol

$$m_{Na_2SO_4} = 3.46 \text{ mmol} \times \frac{142.04 \text{ g}}{1 \text{ mol}} = 491 \text{ mg}$$

$$\text{or} \quad 491 \text{ mg} \times \frac{1 \text{ g}}{1000 \text{ mg}} = 0.491 \text{ g}$$

or

$$m_{Na_2SO_4} = 3.46 \text{ mmol} \times \frac{1 \text{ mol}}{1000 \text{ mmol}} \times \frac{142.04 \text{ g}}{1 \text{ mol}} = 0.491 \text{ g}$$

(Note that molar mass is calculated separately and inserted where needed.)

Exercise

42. Calculate the molar mass of each of the following substances. The molar masses of water and of carbon dioxide should be memorized for efficient work.
    (a) $H_2O_{(l)}$ (water)*
    (b) $CO_{2(g)}$ (respiration product)*
    (c) $NaCl_{(s)}$ (pickling salt, sodium chloride)*
    (d) $C_{12}H_{22}O_{11(s)}$ (table sugar, sucrose)*
    (e) $(NH_4)_2Cr_2O_{7(s)}$ (ammonium dichromate)

43. Calculate the amount of pure substance present (in moles) in each of the following samples of pure substances.
    (a) A 10.00 kg bag of table sugar*
    (b) A 500 g box of pickling salt*
    (c) 40.0 g of propane, $C_3H_{8(l)}$, in a camp stove cylinder
    (d) 325 mg of acetylsalicylic acid (Aspirin), $C_6H_4CH_3COOCOOH_{(s)}$, in a headache relief tablet
    (e) 150 g of isopropanol (rubbing alcohol), $CH_3CH_2OHCH_{3(l)}$, from a pharmacy

44. Calculate the mass of each of the following specified amounts of pure substances
    (a) 4.22 mol of ammonia in a window-cleaning solution*
    (b) 0.224 mol of sodium hydroxide (lye) in a drain-cleaning solution*
    (c) 57.3 mmol of water vapor produced by a laboratory burner
    (d) 9.44 kmol of potassium permanganate fungicide
    (e) 0.77 mol of ammonium sulfate fertilizer

45. Calculate the mass of each reactant and product from the amount in moles shown in the following equations and show how your calculations agree with the law of conservation of mass.
    (a) $H_{2(g)} + Cl_{2(g)} \rightarrow 2\ HCl_{(g)}$
    (b) $2\ CH_3OH_{(l)} + 3\ O_{2(g)} \rightarrow 2\ CO_{2(g)} + 4\ H_2O_{(g)}$

Avogadro's Constant

Avogadro's constant is defined as Avogadro's number of entities per mole, i.e., 6.02×10^{23}/mol. (Entities in a chemistry context refer to, for example, atoms, molecules, electrons, and formula units.) The constant is a convenient conversion factor created by chemists to convert between number of entities and amount in moles. Although the numbers that you deal with are unfathomable to the human mind, working with the numbers gives you a bit of an appreciation for how small atoms and molecules really are. The questions that you deal with below mainly involve two conversion factors, Avogadro's constant and molar mass. Both of these conversion factors were invented and retained by chemists to help us convert back and forth between amount in moles. The amount in moles is necessary to your understanding of the balanced equation and the quantities of chemicals that react and are produced. (See Chapter 7.)

For example, if you wanted to find the number of copper atoms (N_{Cu}) in one penny, you would have to start out by measuring the mass of a penny and then use the conversion factors that the units dictate until you get your answer. The basic strategy is calculate what you can until you get what you want. Let the units be your guide.

Suppose you measured the mass of 150 pennies to be 395 g. You could first calculate the mass of one penny by dividing 395 g by 150 pennies. Next multiply the mass of one penny by the reciprocal of the molar mass of copper (because the units tell you so), and then multiply by Avogadro's constant (to cancel moles and get the number of copper atoms).

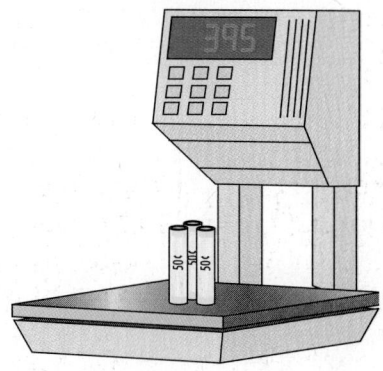

Figure 4.20
The number of copper atoms in one penny is 2.49 × 10²². Unfortunately, there is not an amountmeter or molemeter to measure the number of atoms directly.

$$N_{Cu} = \frac{395\ \cancel{g}}{150\ \text{penny}} \times \frac{1\ \cancel{mol}}{63.55\ \cancel{g}} \times \frac{6.02 \times 10^{23}}{1\ \cancel{mol}} = 2.49 \times 10^{22}/\text{penny}$$

Note that the communication of the answer should be kept as international as possible, with as few words as possible. Unfortunately, there is no international symbol for a penny.

The number of copper atoms in a penny is 2.49×10^{22}.

EXAMPLE ───────────────────────────────

Determine the number of formula units of sodium chloride in 0.563 mol of table salt.

$$N_{NaCl} = 0.56\ \text{mol} \times \frac{6.02 \times 10^{23}}{1\ \text{mol}} = 3.4 \times 10^{23}$$

EXAMPLE ───────────────────────────────

How many sugar (sucrose) molecules are there in a one-kilogram bag (assume 1000 g) of sugar?

$$N_{C_{12}H_{22}O_{11}} = 1000\ \text{g} \times \frac{1\ \text{mol}}{342.34\ \text{g}} \times \frac{6.02 \times 10^{23}}{1\ \text{mol}} = 1.76 \times 10^{24}$$

The number of sucrose molecules in a one-kilogram bag of sugar is 1.76×10^{24}.

EXAMPLE ───────────────────────────────

What is the mass of one water molecule?

$$m_{H_2O} = \frac{18.02\ \text{g}}{1\ \text{mol}} \times \frac{1\ \text{mol}}{6.02 \times 10^{23}} = 2.99 \times 10^{-23}\ \text{g}$$

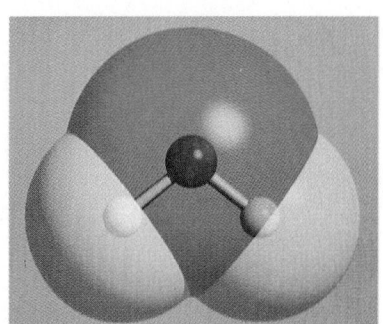

Figure 4.21
The mass of one water molecule is an unimaginable, 2.99 × 10⁻²³ g.

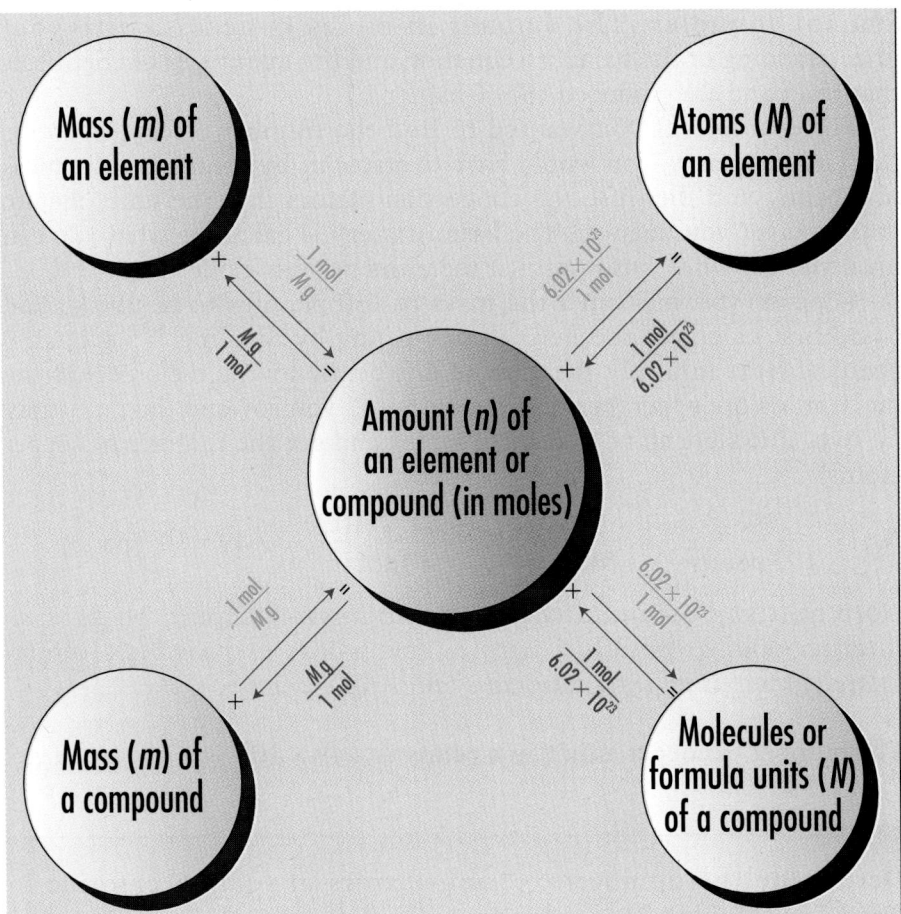

Figure 4.22
The mole is central to determining reacting amounts of chemicals. Molar mass (M) and Avogadro's constant (N_A) are conversion factors used to convert among mass (m), number (N), and amount (n).

Exercise

46. Predict the number of entities in each of the following chemical samples.
    (a) 15 mol of solid carbon dioxide — in dry ice *
    (b) 15 g of ammonia gas — in household cleaners *
    (c) 15 g of hydrogen chloride gas — in hydrochloric acid
    (d) 15 g of sodium chloride — in table salt *

47. Predict the mass of the characteristic entity in each of the following chemical samples.
    (a) water from respiration *
    (b) carbon dioxide from respiration *
    (c) glucose from photosynthesis *
    (d) oxygen from photosynthesis *

48. If necessary, use density as a conversion factor to answer the following questions.
    (a) If the density of pure ethanol is 0.789 g/mL, how many ethanol molecules are there in a 17 mL sample of ethanol — the approximate quantity of ethanol in a bottle of beer? *
    (b) How many nickel atoms are there in 0.72 cm³ of a nickel sample — the approximate volume of a Canadian quarter.

(The density of elements can be referenced on the *Nelson Chemistry* periodic table.)*
    (c) How many water molecules are there in a 100 mL sample of pure water? (You should have the density of water memorized.)*
49. (Enrichment) Calculate the volume of one characteristic entity (molecule or atom) in each of the following chemical samples.
    (a) water from a spring
    (b) gold in a ring

4.6 CLASSIFYING CHEMICAL REACTIONS

By analyzing the evidence obtained from many chemical reactions, it is possible to distinguish patterns. On the basis of these patterns, certain generalizations about reactions can be formulated. The generalizations in Table 4.8 are based on extensive evidence and provide an empirical classification of most, but not all, common chemical reactions. The five types of reactions are described in the sections that follow. (For now, any reactions that do not fit these categories are classified as "other.")

Table 4.8

CHEMICAL REACTIONS	
Reaction Type	**Generalization**
formation	elements → compound
simple decomposition	compound → elements
complete combustion	substance + oxygen → most common oxides
single replacement	element + compound → element + compound
double replacement	compound + compound → compound + compound

Formation Reactions

A **formation reaction** is the reaction of two or more elements to form either an ionic compound (from a metal and a nonmetal) or a molecular compound (from two or more nonmetals). An example that you have already seen is the reaction of magnesium and oxygen shown in Figure 4.23.

chemical equation: $2\,Mg_{(s)} + O_{2(g)} \rightarrow 2\,MgO_{(s)}$

word equation: magnesium + oxygen → magnesium oxide

The only molecular products that you will be able to predict at this time are those whose formulas were memorized from Tables 3.5 and 3.6 (pages 116, 118); for example, H_2O.

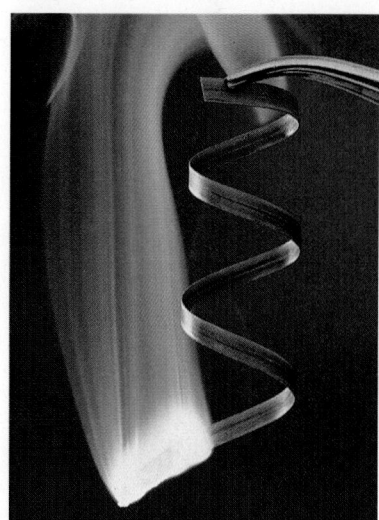

Figure 4.23
The formation of magnesium oxide by reacting the elements magnesium and oxygen.

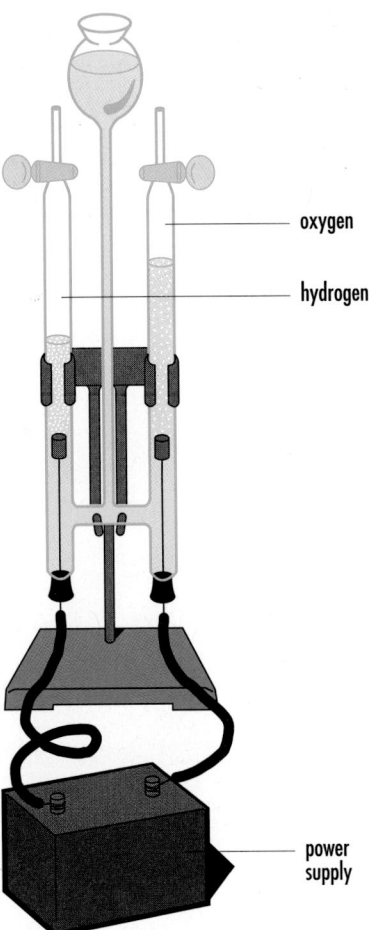

Simple Decomposition Reactions

A **simple decomposition reaction** is the breakdown of a compound into its component elements, that is, the reverse of a formation reaction. Simple decomposition reactions are important historically since they were used to determine chemical formulas. They remain important today in the industrial production of some elements from compounds available in the natural environment. A well-known example that is easy to demonstrate is the simple decomposition of water. (Figure 4.24).

chemical equation: $2 H_2O_{(l)} \rightarrow 2 H_{2(g)} + O_{2(g)}$

word equation: water $\rightarrow$ hydrogen + oxygen

Figure 4.24
The simple decomposition of water is accomplished in a Hoffman apparatus.

Exercise

50. For formation and simple decomposition reactions, write the generalizations that can be used to classify the reactions and predict the products.

51. Rewrite each of the following reactions as a word or balanced chemical equation, and classify each reaction as formation or simple decomposition. Assume the SATP states of matter unless otherwise indicated.
   E.g., $2 Na_{(s)} + Cl_{2(g)} \rightarrow 2 NaCl_{(s)}$ (formation)
   sodium + chlorine $\rightarrow$ sodium chloride
   (a) lithium oxide $\rightarrow$ lithium + oxygen
   (b) $2 KBr_{(s)} \rightarrow 2 K_{(s)} + Br_{2(l)}$
   (c) $6 K_{(s)} + N_{2(g)} \rightarrow 2 K_3N_{(s)}$
   (d) magnesium oxide $\rightarrow$ magnesium + oxygen
   (e) $16 Al_{(s)} + 3 S_{8(s)} \rightarrow 8 Al_2S_{3(s)}$
   (f) Two moles of solid calcium react with two moles of solid carbon and three moles of oxygen gas to produce two moles of solid calcium carbonate.

52. For each of the following reactions, classify the reaction type as formation or simple decomposition, predict the product(s) of the reaction, and complete and balance the chemical equation or complete the word equation. Assume the most common ion charges and that the products are at SATP.
   (a) Very reactive sodium metal reacts with the poisonous gas chlorine to produce an inert, edible chemical.
   $Na_{(s)} + Cl_{2(g)} \rightarrow$
   (b) Since the Bronze Age (about 3000 B.C.), copper has been produced by heating the ore, $CuO_{(s)}$.
   copper(II) oxide $\rightarrow$
   (c) A frequent technological problem associated with the operation of swimming pools is that copper pipes react with aqueous chlorine.
   $Cu_{(s)} + Cl_{2(aq)} \rightarrow$

(d) A major scientific breakthrough occurred in 1807 when Humphry Davy isolated potassium by passing electricity through molten (melted) potassium oxide.

$K_2O_{(l)} \rightarrow$

(e) When aluminum reacts with air, a tough protective coating forms. This coating prevents acidic substances, such as soft drinks (Figure 4.25), from corroding the aluminum.

$Al_s + O_{2(g)} \rightarrow$

(f) Sodium hydroxide can be decomposed into its elements by melting it and adding electricity.

$NaOH_{(l)} \rightarrow$

(g) When zinc is exposed to oxygen, a protective coating forms on the surface of the metal. This reaction makes zinc-coating of metals (galvanizing) a desirable process for resisting corrosion.

zinc + oxygen $\rightarrow$

(h) Translate the last equation above into an English sentence. Include the amounts in moles.

(i) What is the mole ratio for zinc oxide to oxygen in equation (g)?

Figure 4.25
Aluminum does not corrode in air because of a strongly adhering oxide coating.

OTTO MAASS (1890 – 1961)

After receiving his doctorate from Harvard University, Otto Maass joined the staff of McGill University in Montreal, where he taught in the chemistry department. He was chair of the department for 18 years. A dedicated teacher for 35 years, he directed the work of 137 graduate students at McGill, and also influenced the chemistry programs of other Canadian universities. At one time, every university chemistry department in Canada had one of his former students either as its head or among its senior professors.

In his research, Maass worked in many diverse areas, using mainly his own original equipment. Maass was known for the simplicity of his experimental designs and apparatus. He carried out sophisticated research with minimal materials. Maass and his students made important contributions in both pure and applied chemistry. In 1940, he became the Director General of the Pulp and Paper Research Institute, an enterprise involving the co-operative efforts of government, industry, and academia. Under his direction, the Institute became a model for such co-operation around the world.

With the outbreak of World War II, Maass was appointed Director of the Chemical Warfare and Smoke Division of the Canadian Department of National Defence. He tackled his duties with characteristic energy. His work was facilitated by the interest and assistance of former students who occupied positions in industries and at universities. If he needed information or assistance, he would call one of his former students on the telephone, and get what he needed the next day. This direct approach was somewhat startling to the bureaucracy, but such was the nature of this man. From 1939 to 1945, research in the McGill Chemistry Department was devoted to the war effort, and Maass persuaded many other Canadian universities to assist as well. Research included the development of the explosive RDX, and defenses against the possible use of poisonous gases. Military historians credit the defensive techniques and offensive resources developed by Maass with having prevented the use of poison gas during World War II. A brilliant teacher, researcher, and organizer, Otto Maass was one of Canada's great scientists.

- *What are some of Otto Maass's achievements?*
- *How did his research prevent the use of poisonous gas during World War II?*

Combustion Reactions

Figure 4.26
A trained fire-breather illustrates an unusual example of a combustion reaction. The flammable liquid is vaporized and ignited simultaneously by means of a careful, safe procedure. Only trained people should perform this act.

A **complete combustion reaction** is the burning of a substance with sufficient oxygen available to produce the most common oxides of the elements making up the substance that is burned. Some combustions, like those in a burning candle or an untuned automobile engine, are incomplete, and also produce the less common oxides such as carbon monoxide. Combustion reactions (Figure 4.26) are exothermic; these reactions provide the major source of energy for technological use in our society.

In order to successfully predict the products of a complete combustion reaction, you must know the composition of the most common oxides. If the substance being burned contains

- carbon, then $CO_{2(g)}$ is produced.
- hydrogen, then $H_2O_{(g)}$ is produced.
- sulfur, then $SO_{2(g)}$ is produced.
- nitrogen, then assume $NO_{2(g)}$ is produced.
- a metal, then the oxide of the metal with the most common ion charge is produced (Figure 4.28, page 168).

A typical example of a complete combustion reaction is the burning of butane (Investigation 4.3).

chemical equation: $2\,C_4H_{10(g)} + 13\,O_{2(g)} \rightarrow 8\,CO_{2(g)} + 10\,H_2O_{(g)}$

word equation: butane + oxygen $\rightarrow$ carbon dioxide + water

Exercise

53. For the complete combustion reaction, write the generalization that can be used to classify the reaction and predict the products of the reaction.

54. What are the most common oxides of carbon, hydrogen, sulfur, nitrogen, and iron?

55. For each of the following complete combustion reactions, predict the product(s) of the reaction, and complete and balance the chemical equation or complete the word equation. Assume the pure state of matter at SATP unless otherwise indicated.
    (a) In Canada, many homes are heated by the combustion of natural gas (assume methane).
    $CH_{4(g)} + O_{2(g)} \rightarrow$
    (b) Nitromethane, $CH_3NO_{2(l)}$, is a fuel commonly burned in drag-racing vehicles.
    $CH_3NO_{2(l)} + O_{2(g)} \rightarrow$
    (c) Most automobiles in the world currently burn gasoline (assume octane) as a fuel.
    octane + oxygen $\rightarrow$
    (d) Mercaptans (assume $C_2H_5SH_{(g)}$) are added to natural gas to give it a distinct odor. The mercaptan burns with the natural gas.
    $C_2H_5SH_{(g)} + O_{2(g)} \rightarrow$

INVESTIGATION

4.3 The Combustion of Butane

The purpose of this investigation is to test the generalization for a complete combustion reaction.

Problem

What are the products of the combustion of butane?

Experimental Design

A sample of butane, $C_4H_{10(g)}$, from a lighter is burned under an Erlenmeyer flask filled with air (Figure 4.27). In addition to direct observations, diagnostic tests for water and carbon dioxide gas are conducted before and after the reaction.

Materials

lab apron
safety glasses
insulated mitt or tongs
butane lighter
2 dry 250 mL Erlenmeyer flasks

stopper for Erlenmeyer flask
limewater solution
tweezers
cobalt(II) chloride paper

Procedure

1. Record observations of a dry strip of $CoCl_2$ paper in a dry stoppered flask.

2. Remove the $CoCl_2$ paper from the flask.

3. Add a few millilitres of limewater to the flask. Stopper, shake, and observe.

4. Hold the second dry Erlenmeyer flask upside down, using an insulated mitt or tongs.

5. Light the lighter and hold it underneath the opening with most of the flame inside the opening. Let the lighter burn for about 20 s (Figure 4.27).

6. Immediately insert the rubber stopper in the opening of the flask.

☐ Problem
✔ Prediction
☐ Design
☐ Materials
☐ Procedure
✔ Evidence
✔ Analysis
✔ Evaluation
☐ Synthesis

 CAUTION

The neck of the Erlenmeyer flask may get quite hot after holding the flame inside the opening. Hold the flask by the bottom, or with insulated mitts or tongs.

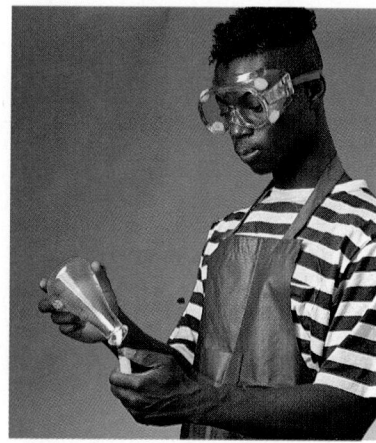

Figure 4.27
The reactants and products of this reaction can be determined by diagnostic tests.

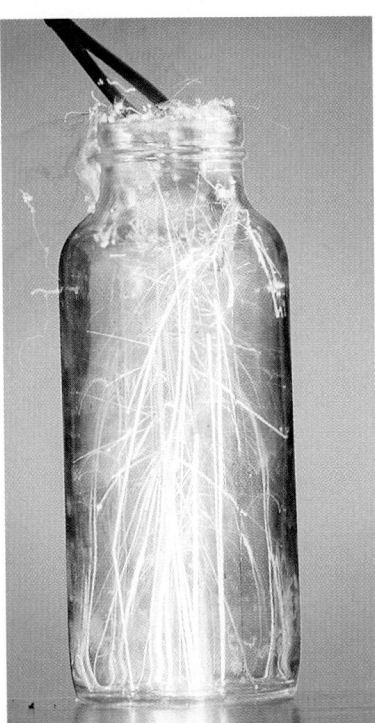

Figure 4.28
A spectacular combustion of a metal is the burning of steel wool in pure oxygen. This reaction is used in fireworks; note that it is also a formation reaction for iron(III) oxide.
$4\,Fe_{(s)} + 3\,O_{2(g)} \rightarrow 2\,Fe_2O_{3(s)}$

7. Use tweezers to obtain a dry, blue strip of $CoCl_2$ paper. Quickly, remove the stopper, drop the strip into the flask, and replace the stopper.

8. Rotate the flask so that the $CoCl_2$ strip slides around the inside wall of the flask. Record observations of the color of the strip.

9. Turn the flask right-side up, add a few millilitres of limewater solution, and replace the stopper.

10. Shake the flask and record the evidence.

Exercise

57. Rewrite each of the following reactions as a word equation or a balanced chemical equation, and classify each reaction as formation, simple decomposition, or complete combustion. (Some reactions may have two classifications.)
    (a) Homes may be heated by this reaction.
    $$C_3H_{8(g)} + 5\,O_{2(g)} \rightarrow 3\,CO_{2(g)} + 4\,H_2O_{(g)}$$
    (b) This reaction by itself is empirically impossible.
    $$C_3H_{8(g)} \rightarrow 3\,C_{(s)} + 4\,H_{2(g)}$$
    (c) This equation represents one of many reaction products.
    $$3\,C_{(s)} + 4\,H_{2(g)} \rightarrow C_3H_{8(g)}$$
    (d) Coal or barbecue briquettes burn.
    carbon + oxygen → carbon dioxide
    (e) Gasoline antifreeze burns in an automobile engine.
    methanol + oxygen → carbon dioxide + water
    (f) Poisonous hydrogen sulfide from natural gas is eventually converted to elemental sulfur using this reaction as a first step.
    $$2\,H_2S_{(g)} + 3\,O_{2(g)} \rightarrow 2\,SO_{2(g)} + 2\,H_2O_{(g)}$$
    (g) Hydrogen gas may be the fuel of the future.
    hydrogen + oxygen → water
    (h) Steel wool may burn as shown in Figure 4.28.
    $$4\,Fe_{(s)} + 3\,O_{2(g)} \rightarrow 2\,Fe_2O_{3(s)}$$
    (i) Toxic hydrogen cyanide gas can be destroyed in a waste treatment plant.
    Four moles of hydrogen cyanide gas react with nine moles of oxygen gas to produce four moles of carbon dioxide gas, two moles of water vapor, and four moles of nitrogen dioxide gas.

58. Classify the following reactions as formation, simple decomposition, or complete combustion. Predict the products of the reactions, write the formulas and states of matter, and balance the reaction equations.
    (a) $Al_{(s)} + F_{2(g)} \rightarrow$
    (b) $NaCl_{(s)} \rightarrow$
    (c) $S_{8(g)} + O_{2(g)} \rightarrow$
    (d) methane + oxygen →

(e) aluminum oxide $\rightarrow$

(f) propane burns

(g) $Hg_{(l)} + O_{2(g)} \rightarrow$

(h) iron(III) bromide $\rightarrow$

(i) $C_4H_{10(g)} + O_{2(g)} \rightarrow$

59. Write a consumer, commercial, or industrial context for two of the chemical reactions in question 58.

60. Write a description of reactants and products for two of the chemical reactions in question 58.

Chemical Reactions in Solution

The reactions discussed so far involve pure substances. The remaining two reaction types, single and double replacements, usually occur in aqueous solutions. As you know from Chapter 3, substances dissolved in water are indicated by the subscript (aq). In order to predict products (with states of matter) of single and double replacement reactions, you need to understand the nature of solutions and learn a method of determining whether a substance dissolves in water to an appreciable extent.

A solution is a homogeneous mixture (page 40) of a **solute** (the substance dissolved) and a **solvent** (the substance, usually a liquid, that does the dissolving). Figure 4.29 shows a common example involving table salt and water. The **solubility** of a substance, which is discussed in more detail in Chapter 5, is the maximum quantity of the substance that will dissolve in a solvent at a given temperature. For example, if you continue to add salt to water, more and more salt will dissolve until a maximum is reached. After you reach this point, any salt added will simply settle to the bottom and remain in the solid state. For substances like sodium chloride (in table salt), the maximum quantity that dissolves in certain solvents is large compared with other solutes. Such solutes are said to have a *high solubility*. When high-solubility substances are formed as products in a single or double replacement reaction, the maximum quantity of solute that can dissolve is rarely reached; thus, the solute remains in solution, and an (aq) subscript is appropriate. Other substances, such as calcium carbonate (in limestone and chalk), have very low solubilities. When these substances are formed in a chemical reaction, the maximum quantity that can dissolve is usually reached and the substance settles to the bottom as a solid. Solid substances formed from reactions in solution are known as **precipitates** (Figure 4.30, page 170).

A *solubility chart* outlines solubility generalizations for a large number of ionic compounds; see Table 4.9, page 170. A major purpose of this chart is to predict the state of matter for ionic compounds formed as products in chemical reactions in solution. This summary of solubility evidence is listed in two categories — *high solubility (aq)* (for example, sodium chloride) and *low solubility (s)* (for example, calcium carbonate). The following example demonstrates how to use the chart. Suppose iron(III) phosphate is predicted as a product in a chemical reaction. What is the solubility of $FePO_4$?

Figure 4.29
Table salt (the solute) is being dissolved in water (the solvent) to make a solution.

Table 4.9

SOLUBILITY OF IONIC COMPOUNDS AT SATP – GENERALIZATIONS							
Anion	**Cl⁻, Br⁻, I⁻**	**S²⁻**	**OH⁻**	**SO₄²⁻**	**CO₃²⁻, PO₄³⁻, SO₃²⁻**	**CH₃COO⁻**	**NO₃⁻**
High Solubility (aq) ≥ 0.1 mol/L (at SATP)	most	Group 1, NH_4^+ Group 2	Group 1, NH_4^+ Sr^{2+}, Ba^{2+}, Tl^+	most	Group 1, NH_4^+	most	all
Low Solubility (s) < 0.1 mol/L (at SATP)	$Ag^+, Pb^{2+}, Tl^+,$ $Hg_2^{2+}(Hg^+), Cu^+$	most	most	$Ag^+, Pb^{2+}, Ca^{2+},$ $Ba^{2+}, Sr^{2+}, Ra^{2+}$	most	Ag^+	none

All Group 1 compounds, including acids, and all ammonium compounds, are assumed to have high solubility in water.

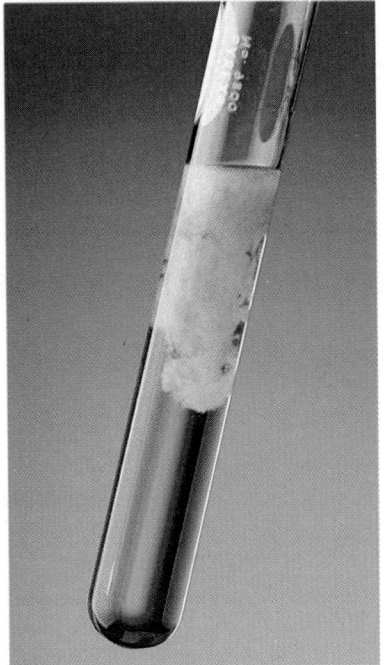

Figure 4.30
When an iron(III) nitrate solution is added to a sodium phosphate solution, a yellow precipitate forms immediately. Diagnostic tests indicate that the low solubility product is iron(III) phosphate, as predicted.

You might find it useful to memorize the uses of the chemicals indicated with an asterisk.

1. In the top row of the chart, locate the column containing the negative ion, PO_4^{3-}.

2. Look at the two boxes below this anion to determine in which category the positive ion Fe^{3+} belongs. (A process of elimination may be necessary.)

3. Since Fe^{3+} is not in Group 1, it must belong in the *low solubility* category.

If iron(III) phosphate is predicted as a product in a chemical reaction in solution, then the prediction is that a precipitate of $FePO_{4(s)}$ forms.

The solubility of ionic compounds can be predicted from the solubility chart. At this point you will not be expected to predict the solubility of molecular compounds in water, but you should memorize the examples in Table 4.10. Most elements have a low solubility in water.

Exercise

61. Which group of monatomic ions forms compounds all having high solubility in water?

62. Which positive polyatomic ion forms compounds all having high solubility in water?

63. Use Table 4.9 to predict the solubility of the following ionic compounds. Indicate high solubility by writing (aq) after the chemical formula (e.g., $NaCl_{(aq)}$) and low solubility by writing (s) (e.g., $CaCO_{3(s)}$).
    (a) KCl (fertilizer)*
    (b) Ca(NO₃)₂ (fireworks)
    (c) Na₂SO₄ (Glauber's salt)
    (d) AgCH₃COO (oxidizing agent)
    (e) ammonium bromide (fireproofing)
    (f) barium sulfide (vulcanizing)
    (g) lead(II) iodide (photography)
    (h) calcium hydroxide (slaked lime)*
    (i) iron(III) hydroxide (rust)
    (j) lead(II) sulfate (car battery)
    (k) calcium phosphate (rock phosphorus)
    (l) potassium permanganate (fungicide)
    (m) sodium tripolyphosphate (detergents)

(n) ammonium nitrate (fertilizer)

(o) cobalt(II) chloride (humidistat)

(p) calcium carbonate (limestone)

64. The following chemical reactions occur in a water environment (e.g., in water in a beaker). Write the balanced chemical equation, including the states of matter (e.g., $_{(s)}$ or $_{(aq)}$) for each reaction.

(a) lead(II) nitrate + lithium chloride →
    lead(II) chloride + lithium nitrate

(b) iron(III) chloride + sodium hydroxide →
    iron(III) hydroxide + sodium chloride

(c) ammonium iodide + silver nitrate →
    silver iodide + ammonium nitrate

(d) The net reaction during the discharge cycle of a car battery is one mole of lead and one mole of solid lead(IV) oxide reacting with two moles of sulfuric acid to produce two moles of water and two moles of lead(II) sulfate.

65. Use the ionic solubility table (Table 4.9), the generalization that all elements (except chlorine) have low solubility in water, and the molecular solubility in Table 4.10 to predict the solubility of the following chemicals in water. Write the chemical formula with $_{(aq)}$ to show high solubility and with the pure state of matter, $_{(s)}$, $_{(l)}$, or $_{(g)}$, to show low solubility.

(a) Zn (dry cell)

(b) P_4 (white phosphorus)

(c) $C_{12}H_{22}O_{11}$ (sugar)*

(d) methanol (antifreeze)*

(e) methane (fuel)*

(f) octane (gasoline)*

(g) barium sulfate (gastric X rays)

(h) sodium hydroxide (drain cleaner)*

(i) ammonia (cleaner)*

(j) hydrogen fluoride (etches glass)

(k) iodine (disinfectant)

(l) iron (steel)*

66. The following chemical reactions occur in a water environment, e.g., in water in a beaker. Write the balanced chemical equation, including SATP states of matter, e.g., $_{(s)}$, $_{(l)}$, $_{(g)}$, or $_{(aq)}$, for each reaction.

(a) Copper can be extracted from solution by reusing cans which contain iron.
    iron + copper(II) sulfate → copper + iron(III) sulfate

(b) Water can be clarified by producing a gelatinous precipitate.
    aluminum sulfate + calcium hydroxide →
    aluminum hydroxide + calcium sulfate

(c) Chlorine can be used to extract bromine from sea water.
    Cl_2 + NaBr → Br_2 + NaCl

(d) During photosynthesis in a plant, carbon dioxide reacts with water to produce glucose and oxygen.
    carbon dioxide + water → glucose + oxygen

67. Does low solubility mean zero solubility? Use your answer for question 66(d) to assist in answering this question.

Table 4.10

SOLUBILITY OF SELECTED MOLECULAR COMPOUNDS	
Solubility	Examples
high	$NH_{3(aq)}$, $H_2S_{(aq)}$, $H_2O_{2(aq)}$, $CH_3OH_{(aq)}$, $C_2H_5OH_{(aq)}$, $C_{12}H_{22}O_{11(aq)}$, $C_6H_{12}O_{6(aq)}$
low	$CH_{4(g)}$, $C_3H_{8(g)}$, $C_8H_{18(l)}$

Solubility
• The solubility of ionic compounds is referenced from a solubility chart.
• The solubility of molecular compounds is given or memorized from Table 4.10.
• The solubility of elements is memorized as being low.

Single Replacement Reactions

A **single replacement reaction** is the reaction of an element with a compound to produce a new element and an ionic compound. This reaction usually occurs in aqueous solutions. For example, silver can be produced from copper and a solution of silver ions (Figure 4.31).

$$Cu_{(s)} \quad + \quad 2\,AgNO_{3(aq)} \quad \rightarrow \quad 2\,Ag_{(s)} \quad + \quad Cu(NO_3)_{2(aq)}$$

copper + silver nitrate → silver + copper(II) nitrate
(metal)   (compound)   (metal)   (compound)

Iodine can be produced from chlorine and aqueous sodium iodide.

$$Cl_{2(g)} \quad + \quad 2\,NaI_{(aq)} \quad \rightarrow \quad I_{2(s)} \quad + \quad 2\,NaCl_{(aq)}$$

chlorine + sodium iodide → iodine + sodium chloride
(nonmetal) (compound)   (nonmetal) (compound)

The predicted high solubility (aq) states for the two ionic products, copper(II) nitrate and sodium chloride, are obtained from the solubility chart (Table 4.9, page 170). Empirical evidence shows that *a metal replaces a metal ion to liberate a different metal as a product* (as in the first preceding example) and *a nonmetal replaces a nonmetal ion to liberate a different nonmetal as a product* (as in the second example). Reactive metals such as those in Groups 1 and 2 react with water to replace the hydrogen, forming hydrogen gas and a hydroxide compound. (In these reactions, hydrogen acts like a metal.)

Double Replacement Reactions

A **double replacement reaction** can occur between two ionic compounds in solution. In the reaction, the ions "change partners" to form the products. If one of the products has low solubility, it may form a precipitate, as shown in Figure 4.32. As the term implies, **precipitation** is a double replacement reaction in which a precipitate forms. For example,

$$CaCl_{2(aq)} \quad + \quad Na_2CO_{3(aq)} \quad \rightarrow \quad CaCO_{3(s)} \quad + \quad 2\,NaCl_{(aq)}$$

compound + compound → compound + compound

In another kind of double replacement reaction, an acid reacts with a base, producing water and an ionic compound. This kind of double replacement reaction is known as **neutralization**. The reaction between hydrochloric acid and potassium hydroxide is an example.

$$HCl_{(aq)} \quad + \quad KOH_{(aq)} \quad \rightarrow \quad HOH_{(l)} \quad + \quad KCl_{(aq)}$$

acid   +   base   →   water + ionic compound (a salt)

When writing chemical equations for both precipitation and neutralization reactions, consult the solubility chart (Table 4.9, page 170) to determine the state of matter of the ionic products.

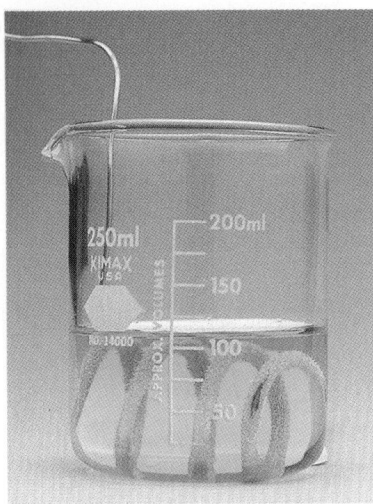

Figure 4.31
The blue solution verifies the prediction that the most common ion of copper, $Cu^{2+}_{(aq)}$, is formed.

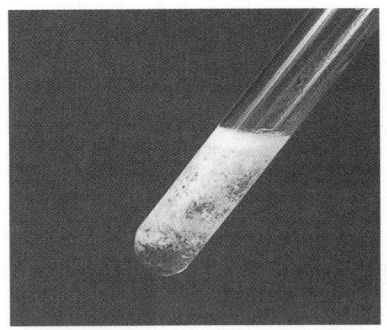

Figure 4.32
A white precipitate is formed when aqueous solutions of calcium chloride and sodium carbonate are mixed.

Exercise

68. For single and double replacement reactions, write the generalizations that can be used to classify the reaction and predict the products.

69. Rewrite each of the following reactions as a word or balanced chemical equation and classify each reaction as single or double replacement. (Assume the most common ion charge and SATP state of matter if not indicated otherwise.)

 (a) $Cu_{(s)} + 2\ AgNO_{3(aq)} \rightarrow 2\ Ag_{(s)} + Cu(NO_3)_{2(aq)}$

 (b) $KI_{(aq)} + AgNO_{3(aq)} \rightarrow AgI_{(s)} + KNO_{3(aq)}$

 (c) chlorine + aqueous sodium bromide →

 bromine + aqueous sodium chloride

 (d) sulfuric acid + aqueous sodium hydroxide →

 water + aqueous sodium sulfate

 (e) aqueous calcium nitrate + aqueous sodium phosphate →

 solid calcium phosphate + aqueous sodium nitrate

 (f) Two moles of solid aluminum react with one mole of solid iron(III) oxide to produce two moles of solid iron and one mole of solid aluminum oxide.

70. If a product in a water mixture has low solubility, the product is produced as a solid and is called a precipitate. Indicate the solubility in water of each of the following reaction products by writing the chemical formula along with a state of matter subscript, e.g., $_{(s)}$ or $_{(aq)}$.

 (a) potassium nitrate (saltpeter)

 (b) calcium chloride (desiccant)

 (c) magnesium hydroxide (antacid)

 (d) aluminum sulfate (water treatment)

 (e) lead(II) iodide (yellow pigment)

 (f) calcium phosphate (rock phosphorus)

 (g) hydrogen chloride (a gas)

 (h) ammonium carbonate (smelling salts)

 (i) silver acetate (lustrous needles)

 (j) iodine (disinfectant)

71. For each of the following reactions, classify the reaction type as single or double replacement, predict the products of the reaction, and complete and balance the chemical equation or complete the word equation. (Assume the most common ion charge and SATP state of matter if not indicated otherwise.)

 (a) Silver metal is recovered in a laboratory by placing aluminum foil in aqueous silver nitrate.

 $Al_{(s)} + AgNO_{3(aq)} \rightarrow$

 (b) Bromine is mined from the ocean by bubbling chlorine gas through ocean water containing sodium bromide (Figure 4.33).

 $Cl_{2(g)} + NaBr_{(aq)} \rightarrow$

 (c) A traditional laboratory method of producing hydrogen gas is to react zinc metal with sulfuric acid.

 zinc + sulfuric acid →

 (d) The presence of the chloride ion in an environmental sample

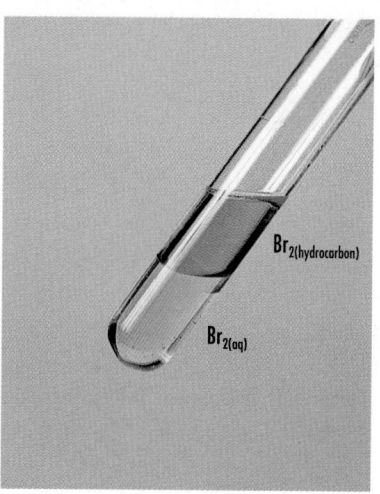

Figure 4.33
The orange color in the aqueous and in the hydrocarbon solutions is evidence for the production of bromine.

Figure 4.34
Would an alchemist have been able to fool you into thinking that copper-plated iron is gold?

is indicated by the formation of a white precipitate when aqueous silver nitrate is added to the sample.

aqueous silver nitrate + aqueous magnesium chloride →

(e) An analytical chemist uses sodium oxalate to precipitate calcium ions in a water sample from an acid lake.

$Na_2OOCCOO_{(aq)} + CaCl_{2(aq)} →$

(f) Sodium metal reacts vigorously with water to produce a flammable gas and a basic (hydroxide) solution.

sodium + water →

(g) When aqueous potassium hydroxide is added to a well-water sample, the formation of a rusty-brown precipitate indicates the presence of an iron(III) compound in the water.

$KOH_{(aq)} + FeCl_{3(aq)} →$

(h) A chemist in a consumer-protection laboratory adds aqueous sodium hydroxide to determine the concentration of acetic acid in a vinegar sample. Write the balanced chemical equation for the reaction.

(i) A dishonest 16th-century alchemist, who tried to fool people into believing that iron could be changed into gold, dipped an iron bar into aqueous copper(II) sulfate (Figure 4.34). Write the balanced chemical equation for the reaction.

(j) Translate equation (i) into an English sentence. Include the amounts in moles.

(k) What is the mole ratio for iron to copper in equation (i)?

72. (Discussion) What standard process does the scientific community use to check on the empirical claims of individual members of the community?

Problem 4A Testing Reaction Generalizations

Complete the Prediction and diagnostic tests of the investigation report. Write up the diagnostic tests, including any controls, as part of the Prediction or the Experimental Design.

Problem

What are the products of the reaction of sodium metal and water?

Experimental Design

A small piece of sodium metal is placed in distilled water and some diagnostic tests are carried out to identify the products.

Problem 4B Single and Double Replacement Reactions

The purpose of this problem is to predict and analyze single and double replacement reactions in preparation for an evaluation of these reaction types as part of Investigation 4.4. Complete the Prediction (including possible diagnostic tests) and Analysis (including reaction types) in your investigation report.

Problem

What reaction products are formed when the following substances are mixed?

(a) Aqueous chlorine is added to a potassium iodide solution.

(b) Solutions of magnesium chloride and sodium hydroxide are mixed.

(c) Solutions of aluminum nitrate and sodium phosphate are mixed.

(d) Nickel metal is added to hydrochloric acid.

(e) Sodium hydroxide solution is added to a chromium(III) chloride solution.

(f) Lithium metal is placed in water.

(g) A clean cobalt strip is placed in a silver nitrate solution.

(h) Nitric acid is added to an ammonium acetate solution.

Experimental Design

Diagnostic test information such as evidence of chemical reactions (Table 4.2, page 138), ion colors and solubilities (reference tables, inside back cover), and specific tests for products (Appendix C, C.3) are predicted, for convenience, along with the balanced chemical equations. The general plan is to observe the substances before and after mixing and conduct the appropriate diagnostic tests.

Evidence

SINGLE AND DOUBLE REPLACEMENT REACTIONS	
Reaction	**Observations**
(a)	• The colorless solutions produced a yellow-brown color when mixed. • A violet color appeared in the chlorinated hydrocarbon layer.
(b)	• The colorless solutions produced a white precipitate when mixed.
(c)	• The colorless solutions produced a white precipitate when mixed.
(d)	• The silvery solid added to the colorless solution produced gas bubbles and a green solution. • The gas produced a pop sound when ignited.
(e)	• The colorless sodium hydroxide and green chromium(III) chloride solutions produced a dark precipitate and a colorless solution.
(f)	• The soft, silvery solid and colorless liquid produced gas bubbles and a colorless solution. • The gas produced a pop sound when ignited. • Red litmus turned blue in the final solution. • The final solution produced a bright red flame color.
(g)	• The silvery solid and colorless solution produced a pink solution and silvery needles.
(h)	• The colorless solutions remained colorless when mixed. • A vinegar odor was produced.

Certainty
The certainty with which you defend your empirically based answer in the Analysis section of your report depends upon the quality and quantity of the evidence gathered from the diagnostic tests.

Now that you have studied each major type of reaction in some detail, you should be able to make certain predictions about types of reactions that will occur. You should also be able to write the correct chemical equation for the reaction. To do this, you need to use your knowledge of chemical formulas from Chapter 3, and your knowledge of balancing equations from this chapter.

1. Use the reaction generalizations to classify the reaction.

2. Use the reaction generalizations to predict the products of the chemical reaction and write the chemical equation.
   (a) Predict the chemical formulas from theory for ionic compounds and write the formulas from memory for molecular compounds and elements.
   (b) Include states of matter, using previously stated rules and generalizations.

3. Balance the equation without changing the chemical formulas.

INVESTIGATION

4.4 Testing Replacement Reaction Generalizations

The purpose of this investigation is to evaluate the single and double replacement generalizations by testing predictions for a number of reactions. For each combination of reactants given, assume that the reaction is *spontaneous*; that is, it occurs when reactants are mixed.

Problem

What reaction products result when the following substances are mixed?

1. Aqueous solutions of barium hydroxide and sulfuric acid are mixed.
2. Aqueous chlorine is added to a sodium bromide solution.
3. A clean zinc strip is placed in a copper(II) sulfate solution.
4. Solutions of calcium chloride and sodium carbonate are mixed.
5. Solutions of cobalt(II) chloride and sodium hydroxide are mixed.
6. Calcium metal is placed in water.
7. Hydrochloric acid is added to a sodium acetate solution.
8. A magnesium strip is placed in hydrochloric acid.

Experimental Design

Predictions of possible products are made, including any diagnostic test information such as evidence of chemical reactions (Table 4.2, page 138) and specific tests for products (Appendix C, C.3). The general plan is to observe the reactants before and after mixing, and to note any evidence supporting or contradicting the predictions.

- Problem
- ✔ Prediction
- Design
- Materials
- Procedure
- ✔ Evidence
- ✔ Analysis
- ✔ Evaluation
- Synthesis

CAUTION

Because of the many corrosive substances used in this investigation, wear eye protection at all times. Some of the substances are also toxic.

Procedure

1. Record observations of the reactant chemicals at one of the laboratory stations.

2. Perform the reaction according to the directions given at the station (for example, Figure 4.35) and record observations before, during, and after the reaction.

3. Clean all apparatus and the laboratory bench before proceeding to the next station

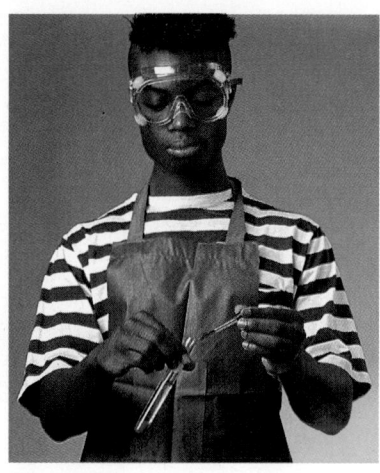

Figure 4.35
Solutions may be conveniently combined using a dropper.

Exercise

73. Rewrite each of the following reactions as a word or balanced chemical equation and classify each of the reactions. Predict the product(s) where necessary. (Assume the most common ion charges and SATP states of matter if not indicated otherwise.)

   (a) $2 HgO_{(s)} \rightarrow 2 Hg_{(l)} + O_{2(g)}$
   (b) glucose + oxygen → carbon dioxide + water (respiration)
   (c) zinc + aqueous lead(II) nitrate → lead + aqueous zinc nitrate
   (d) potassium + oxygen → potassium oxide
   (e) $Na_2S_{(aq)} + Zn(NO_3)_{2(aq)} \rightarrow$
   (f) propane + oxygen →
       (fuel for homes, automobiles, and barbecues)
   (g) potassium + phosphorus →
   (h) One mole of aqueous cobalt(II) nitrate reacts with two moles of aqueous sodium hydroxide to produce one mole of solid cobalt(II) hydroxide and two moles of aqueous sodium nitrate.

74. For each of the following reactions, classify the reaction, predict the product(s) of the reaction, and write the balanced chemical equation and word equation. (Assume the most common ion charges and SATP states of matter if not indicated otherwise.)

   (a) A pollution problem is caused when nitrogen reacts with oxygen in the combustion cylinder of an automobile engine.
   (b) A fuel is produced when water is electrolytically decomposed into its elements.
   (c) Used tin cans (made of iron) may be reused to extract copper from a copper(II) nitrate solution.
   (d) If a house with aluminum siding burns at a high enough temperature, the aluminum can catch fire.
   (e) To test for toxic mercury(I) sulfate in an aqueous sample, add aqueous sodium chloride and determine whether a precipitate forms.
   (f) Sulfurous acid in an acid lake is neutralized by adding solid slaked lime $(Ca(OH)_{2(s)})$.
   (g) Spilled gasoline is burned as the chosen alternative for cleaning up the spill.
   (h) Lithium metal is added to water to test the ability of the single replacement reaction generalization to predict the products of the reaction.

Mercury(I) Ion
The mercury(I) ion can be written as a monatomic ion, Hg^+, or, more acceptably, as a polyatomic ion, Hg_2^{2+}.

(i) Laboratory oxygen can be produced by heating potassium chlorate to produce potassium chloride and oxygen.

(j) Write the diagnostic tests used to test for the products predicted in parts (g), (h), and (i).

75. The following questions relate to the *Perspectives on Acid Rain* feature. Read the acid rain feature and then classify the perspective in each question below as one of the five presented in the feature. Classify the reactions as one of the five you have studied or as "other." Predict the reaction products and write balanced chemical equations complete with states of matter.

(a) Research indicates that natural acid rain is caused by carbon dioxide produced when wood (assume carbon) burns in a forest fire.

(b) One industrial solution to sulfur emissions starts with burning unwanted hydrogen sulfide.

(c) Fish and plants die in an acid lake if slaked lime (calcium hydroxide) is not used to neutralize sulfuric acid in the lake.

(d) Millions of dollars are lost each year due to damage caused by acid rain; e.g., nitric acid rain reacts with steel (assume iron) structures.

(e) Pressure by voters has resulted in legislation requiring potential sulfur dioxide gas emissions to be decomposed to inert sulfur and beneficial oxygen.

76. Normal rain is rain that has not been affected by acid-forming compounds either manufactured by human activities or produced by significant natural events, such as volcanic eruptions.

(a) Is normal rain neutral (pH 7)? If not, what is its approximate pH?

(b) What is the cause of the slight acidity of normal rain? Include a balanced chemical equation in your answer.

(c) Which oxides are responsible for the formation of acids in the atmosphere?

(d) List both the natural and manufactured sources of the oxides you listed in (c).

(e) Write a balanced chemical equation showing the production of sulfurous acid from sulfur dioxide and water.

(f) Write a balanced chemical equation to show how sulfuric acid might be formed in the atmosphere.

77. Write an experimental design to monitor the acidity of precipitation over an extended period of time.

78. Using your present knowledge and values, what action, if any, should be taken against an industry that emits sulfur dioxide into the atmosphere?

79. Suppose the industry mentioned in question 78 is a smelter. The smelter is the main industry in the town where you live, and most of the residents of the town, including some members of your family, work at this industry. The smelter will have to close if the owners are forced to make expensive renovations. Does this change your answer to the previous question? Elaborate on your answer.

What exactly is acid rain? We recognize that it is a serious problem, but why don't we solve it? Like so many other issues related to science and technology, there are a variety of perspectives. Here is a brief outline of five different perspectives to help you organize your knowledge of acid rain.

A Scientific Perspective

Acid rain is any form of natural precipitation that is noticeably more acidic than *normal rain*. A pH of less than 5.6 is usually considered to indicate *acid rain*. The high acidity of acid rain is due to sulfur oxides and nitrogen oxides reacting with water in the atmosphere. Oxides of sulfur and nitrogen are released from sources such as automobiles and coal-burning power plants. The slight acidity of normal rain is the result of natural carbon dioxide dissolving in atmospheric moisture to form carbonic acid. Nitrogen oxides from lightning strikes and plant decay, and sulfur oxides from volcanic eruptions, are other natural sources of acids in rain.

A Technological Perspective

Technologies are now available for the development and use of alternative fuels, the removal of sulfur from fossil fuels, and the recovery of oxides from exhaust gases. For example, some industries have added sulfur oxide recovery units to smokestacks at large smelters. In addition to reducing sulfur dioxide emissions by over 50 percent at such smelters, sulfuric acid is produced as a valuable by-product.

An Ecological Perspective

Hundreds of lakes in Eastern Canada are now devoid of aquatic plant and animal life due to their high acidity from acid rain. Some organisms are more susceptible than others to changes in acidity, but eventually the increased acidity (lower pH) leads to the "death" of a lake. Forests have also been destroyed by acid

rain, both in Canada and abroad. In British Columbia, however, where acid rain has not been detected at any significant level, these ecological problems have not developed.

An Economic Perspective

Use of alternative energy sources or implementation of pollution-reducing technologies means spending money that consumers and industries may feel they can't afford. If an industry shuts down, people lose jobs and the cost of social assistance escalates. From the same perspective, future costs of doing nothing are likely to be staggering.

A Political Perspective

Political pressures and opinions have resulted in legislation limiting the production of sulfur oxides and nitrogen oxides. Some people argue that there should be even stiffer legislation to regulate industries, but others argue that there should be less government interference.

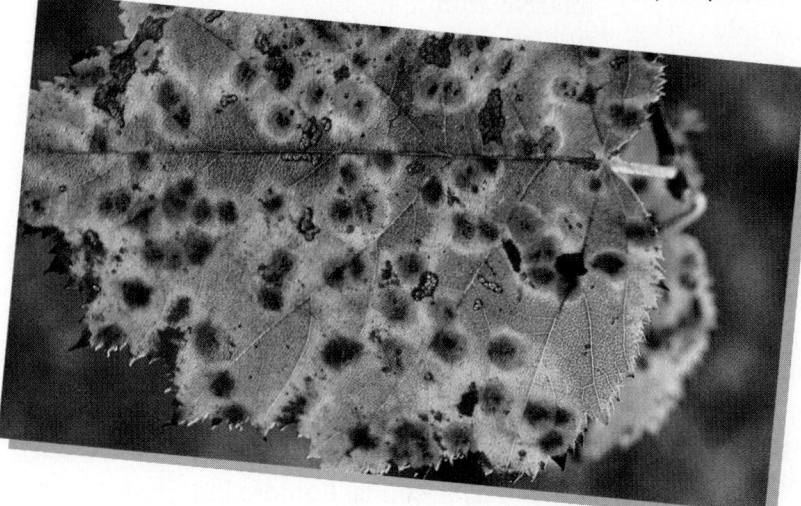

OVERVIEW

Chemical Reactions

Summary

- The key idea of the kinetic molecular theory is that the smallest particles of a substance are in continuous, random motion. This theory forms the basis of the model for the states of matter and the collision-reaction theory of chemical reactions.

- Empirical evidence for chemical reactions may include changes in color, odor, state, or energy.

- Chemical reactions and conservation of mass are explained by means of molecular models, and are communicated in the form of balanced chemical equations.

- A mole of entities, a convenient number for expressing the amount of a chemical, is equal to Avogadro's number, 6.02×10^{23}.

- SI is a metric system of measurement that is international, precise, logical, and simple.

- Accuracy, precision, and certainty are concepts used when communicating quantitatively.

- Molar mass and Avogadro's constant are used as conversion factors among mass, amount, and number of entities.

- Chemical reactions may be classified as formation, simple decomposition, complete combustion, single replacement, or double replacement.

- The states of matter of the products of single and double replacement reactions in solution are predicted using solubility generalizations.

Key Words

accuracy
atomic mass unit
Avogadro's constant
Avogadro's number
balanced chemical equation
certainty
certainty rule
chemical change
coefficient
complete combustion reaction
double replacement reaction
formation reaction
kinetic molecular theory
least certain value
molar mass
mole
mole ratio
neutralization
nuclear change
percent difference
physical change
precipitate
precipitation
precision
precision rule
relative atomic mass
rule of a thousand
scientific notation
significant digit
simple decomposition reaction
single replacement reaction
solubility
solute
solvent

Review

1. What is the central idea of the kinetic molecular theory?

2. Describe solids, liquids, and gases in terms of theoretical types of molecular motion.

3. List the three parts of the collision-reaction theory.

4. How many particles (entities) are there in one mole of a substance?

5. Use a convenient format for communicating a diagnostic test for these chemicals.
(a) hydrogen   (b) carbon dioxide

6. What evidence supports the theoretical statement that atoms are conserved in chemical reactions?

7. Distinguish between the meaning of coefficients in chemical equations and formula subscripts.

8. What steps should be followed in order to balance a chemical equation?

9. What four criteria are used to judge whether a communication system is acceptable to the international scientific community?

10. What are the SI prefixes for one million, one thousand, one thousandth, and one millionth?

11. Convert the following measurements to follow the rule of a thousand.
    (a) 47 624 g
    (b) 0.028 L
    (c) 2000 mol
    (d) 5026 mL
    (e) 0.004 73 mol
    (f) 0.097 s
    (g) 0.002 kg
    (h) 13 429 mmol

12. What is the quantity symbol for each of the following quantities?
    (a) mass
    (b) amount
    (c) molar mass

13. What is the SI unit symbol for each of the following units?
    (a) gram
    (b) millimole
    (c) gram per mol
    (d) micromole
    (e) millilitre
    (f) kilogram

14. Convert the following into SI unit symbols.
    (a) 15 gm./mol
    (b) 25.2 sec.
    (c) 50 Km/hr.

15. Express the following values in scientific notation without SI prefixes.
    (a) 7342 mol
    (b) 0.0368 L
    (c) 8150.4 g

16. If in the laboratory a student determined the molar mass of sulfuric acid to be 95.1 g/mol, what is the accuracy (communicated by percent difference) of this value?

17. Perform the following calculations and communicate the answers with the correct precision or certainty, as appropriate.
    (a) $m = 40.08 \text{ g} + 12.01 \text{ g} + 48.00 \text{ g} =$
    (b) $n = 1.64 \text{ mol} + 7.2 \text{ mol} + 0.108 \text{ mol} =$
    (c) $m = 1.50 \text{ mol} \times \dfrac{79.64 \text{ g}}{1 \text{ mol}} =$
    (d) $n = 10 \text{ g} \times \dfrac{1 \text{ mol}}{2.02 \text{ g}} =$

18. What is the definition of an amu?

19. What is the molar mass of each of the following chemicals?
    (a) oxygen (respiration gas)
    (b) calcium carbonate (limestone)
    (c) dinitrogen tetraoxide (pollutant)
    (d) calcium phosphate (rock phosphorus)
    (e) ethanol (grain alcohol)
    (f) sodium carbonate decahydrate (washing soda)

20. Provide the chemical formula, SATP state, and solubility in water for each of the following common names.
    (a) table salt   (b) table sugar
    (c) baking soda   (d) ammonia
    (e) methanol   (f) propane
    (g) glucose   (h) pencil lead

21. What is the solubility of each of the following chemicals in water?
    (a) copper
    (b) potassium carbonate
    (c) natural gas
    (d) silver chloride
    (e) lead(II) sulfate
    (f) barium nitrate
    (g) hydrogen chloride
    (h) ammonium sulfate

22. Write a general word equation for each of the five classes of reactions studied in this chapter.

23. On what basis do you predict whether a metal or a nonmetal will be the product of a single replacement reaction?

24. Write the chemical formula for the most common oxide of each of the following.
    (a) carbon   (b) hydrogen
    (c) sulfur   (d) iron

25. What is the solubility generalization for elements in water?

Applications

26. Convert the following masses into amounts (in moles).
    (a) 1.000 kg of table salt
    (b) 1.000 kg of table sugar
    (c) 1.000 kg of dry ice
    (d) 1.000 kg of water

27. Convert the following amounts into masses.
    (a) 1.50 mol of liquid oxygen
    (b) 1.50 mol of liquid propane
    (c) 1.50 mmol of liquid mercury
    (d) 1.50 kmol of liquid bromine

28. How many molecules are there in each of the following masses or amounts?
    (a) 0.42 mol of acetic acid (vinegar)
    (b) 7.6×10^{-4} mol of carbon monoxide (poisonous gas)
    (c) 100 g of carbon tetrachloride (poisonous fluid)
    (d) 100 g of hydrogen sulfide (rotten egg gas)

29. Write a sentence to describe each of the following balanced chemical equations, including coefficients (in moles) and states of matter. State the mole ratio for the complete reaction equation.
    (a) $2\,NiS_{(s)} + 3\,O_{2(g)} \rightarrow 2\,NiO_{(s)} + 2\,SO_{2(g)}$
    (b) $2\,Al_{(s)} + 3\,CuCl_{2(aq)} \rightarrow 2\,AlCl_{3(aq)} + 3\,Cu_{(s)}$
    (c) $2\,H_2O_{2(l)} \rightarrow 2\,H_2O_{(l)} + O_{2(g)}$

30. For each of the following reactions, classify the reaction and balance the equation.
    (a) $NaCl_{(s)} \rightarrow Na_{(s)} + Cl_{2(g)}$
    (b) $Na_{(s)} + O_{2(g)} \rightarrow Na_2O_{(s)}$
    (c) $Na_{(s)} + HOH_{(l)} \rightarrow H_{2(g)} + NaOH_{(aq)}$
    (d) $AlCl_{3(aq)} + NaOH_{(aq)} \rightarrow Al(OH)_{3(s)} + NaCl_{(aq)}$
    (e) $Al_{(s)} + H_2SO_{4(aq)} \rightarrow H_{2(g)} + Al_2(SO_4)_{3(aq)}$
    (f) $C_8H_{18(l)} + O_{2(g)} \rightarrow CO_{2(g)} + H_2O_{(g)}$

31. For each pair of reactants, classify the reaction type, complete the chemical equation, and balance the equation. Also, state the mole ratio for each equation.
    (a) $Ni_{(s)} + S_{8(s)} \rightarrow$
    (b) $C_6H_{6(l)} + O_{2(g)} \rightarrow$
    (c) $K_{(s)} + HOH_{(l)} \rightarrow$

32. Chlorine gas is bubbled into a potassium iodide solution and a color change is observed. Write the balanced chemical equation for this reaction and describe a diagnostic test for one of the products.

33. For each of the following reactions, translate the information into a balanced reaction equation. Then classify the main perspective — scientific, technological, ecological, economic, or political — suggested by the introductory statement.
    (a) Oxyacetylene torches are used to produce high temperatures for cutting and welding metals such as steel. This involves burning acetylene, $C_2H_{2(g)}$, in pure oxygen.
    (b) In chemical research conducted in 1808, Sir Humphry Davy produced magnesium metal by decomposing molten magnesium chloride using electricity.
    (c) An inexpensive application of single replacement reactions uses scrap iron to produce copper metal from waste copper(II) sulfate solutions.
    (d) The emission of sulfur dioxide into the atmosphere creates problems across international borders. Sulfur dioxide is produced when zinc sulfide is roasted in a combustion-like reaction in a zinc smelter.
    (e) Burning leaded gasoline added toxic lead compounds to the environment, which damaged both plants and animals. Leaded gasoline contained tetraethyl lead, $Pb(C_2H_5)_{4(l)}$, which undergoes a complete combustion reaction in a car engine.

Extensions

34. Smog is a major problem in many large cities. Find out what chemicals contribute to this problem and write chemical reaction equations for the production and further reactions of two of the chemicals. What are the ecological implications of the presence of smog? What might be done to help solve this problem? In your answer, consider a variety of perspectives.

35. Chemical industries provide many useful and essential products and processes. The manufacture of sulfuric acid in the contact process yields an annual worldwide production of about one trillion tonnes, making it the most commonly used acid in the world. Research the main chemical reactions involved in the contact process and list some by-products of processes involving sulfuric acid. What precautions are necessary when handling concentrated sulfuric acid?

Problem 4C Testing Reaction Generalizations

Complete the Prediction and Experimental Design of the investigation report. Include four diagnostic tests in the Experimental Design to determine whether the predicted reaction has taken place and the predicted products have formed.

Problem

What are the products of the reaction of aqueous copper(II) chloride and sodium hydroxide solutions?

UNIT III

THE MEASURE
OF
CHEMISTRY

Scientists select an aspect of the world around them and define systems that interest them. A system can be as simple as a solution in a beaker, a gas sealed in its container, or a chemical reaction in a flask. Generally, qualitative observations come first, but ultimately scientists seek to obtain and explain quantitative measures of the system. In both scientific and technological work, quantitative descriptions based on the best available measurements are the most highly valued.

Other systems comprise more than chemicals and their reactions. Social, political, economic, and ecological systems are closely related to science and technology; for informed decisions on STS issues, interdisciplinary knowledge is essential. New manufacturing processes, for example, are extremely complex; we must consider not only the underlying science and technology, but also how these processes affect individuals, politics, the economy, and the environment.

5 Solutions

"Water, water, everywhere,
Nor any drop to drink."

In Samuel Taylor Coleridge's classic poem "The Rime of the Ancient Mariner," written in 1798, an old seaman describes the desperation of becalmed sailors driven mad by thirst. Today, sea-going ships carry distillation equipment to convert salt water into drinking water.

In Canada, we depend on fresh water from lakes and rivers for drinking, cooking, irrigation, electric power generation, and recreation. Water is one example of a *solution* that is necessary for life; even the purest spring water contains dissolved minerals and gases. So many substances dissolve in water that it has been called "the universal solvent." Many household products, including soft drinks, fruit juices, vinegar, cleaners, and medicines, are aqueous solutions. ("Aqueous" comes from the Latin *aqua* for "water.") Our blood plasma is mostly water, and many substances essential to life are dissolved in it, including oxygen and carbon dioxide.

The ability of so many materials to dissolve in water also has some negative implications. Human activities have introduced thousands of unwanted substances into water supplies. These substances include paints, cleaners, industrial waste, insecticides, fertilizers, salt from highways, and other contaminants. Rain may become acidic if it contains dissolved gases produced when fossil fuels are burned. Learning about aqueous solutions will help you understand science-related social issues such as water quality and acid rain.

Solutions are homogeneous mixtures of substances composed of at least one solute and one solvent (Figure 5.1). Both solutes and solvents may be gases, liquids, or solids. The chemical formula representing a solution highlights the solute by using its chemical formula and communicates the identity of the solvent using a state of matter subscript. For example,

$NH_{3(aq)}$ ammonia gas dissolved in water (solvent)

$NaCl_{(aq)}$ solid sodium chloride dissolved in water (solvent)

$I_{2(al)}$ solid iodine dissolved in alcohol (solvent)

$C_2H_5OH_{(aq)}$ liquid ethanol dissolved in water (solvent)

In metal alloys, such as brass or the mercury amalgam used in tooth fillings, the solution is used in solid form after the dissolving has taken place in liquid form. Common liquid solutions that have a solvent other than water include varnish, furniture polish, and gasoline (Figure 5.2). However, by far the most numerous and versatile solutions are those in which water is the solvent (Figure 5.3, page 188). This chapter deals primarily with aqueous solutions.

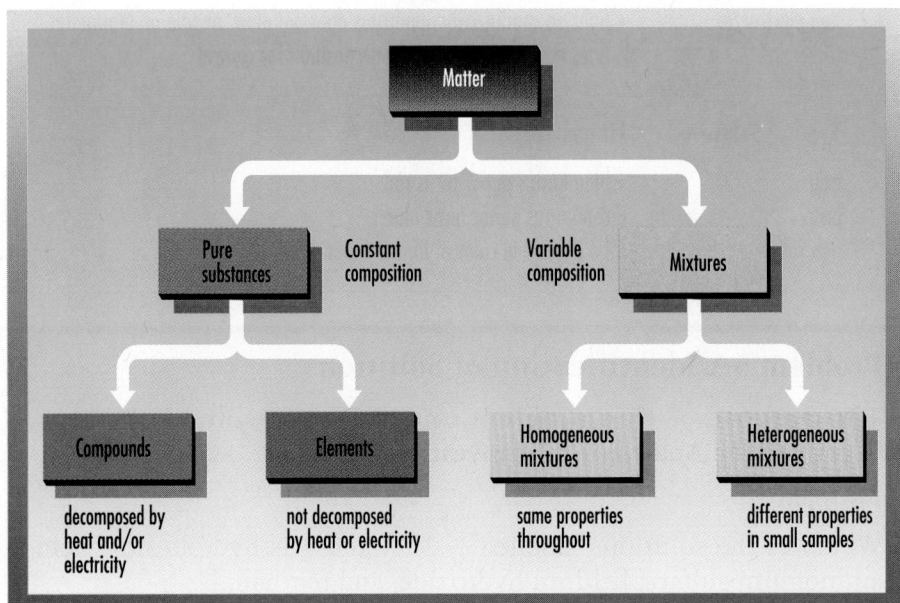

Figure 5.1
Matter is classified according to physical and chemical properties. This classification helps chemists to organize and communicate large quantities of knowledge about substances.

Figure 5.2
Gasoline is a non-aqueous solution containing many different liquids.

Solutions of Electrolytes and Non-Electrolytes

All aqueous solutions are clear or transparent. Opaque or translucent (cloudy) mixtures, such as milk, contain undissolved particles large enough to block or scatter light waves; these mixtures are considered to be heterogeneous. Compounds are **electrolytes** if their aqueous solutions conduct electricity. Compounds are **non-electrolytes** if their aqueous solutions do not conduct electricity. Most household aqueous solutions, such as fruit juices and cleaning solutions, contain electrolytes. The conductivity of a solution is easily tested with a simple conductivity apparatus (Figure 3.3, page 98) or an ohmmeter

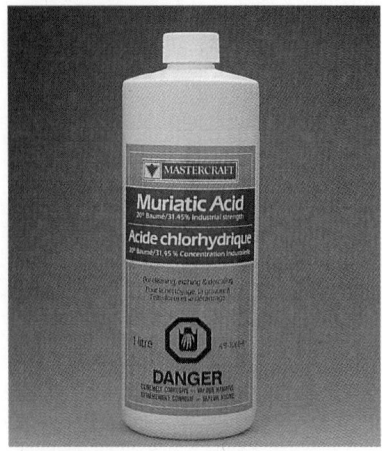

Figure 5.3
Hydrochloric acid (often sold under its archaic name, muriatic acid) contains hydrogen chloride gas dissolved in water. It is used to etch concrete before painting it, clean rusted metal, and adjust acidity in swimming pools.

Figure 5.4
An ohmmeter can indicate the conductivity of a solution.

(Figure 5.4). This evidence also provides a diagnostic test to determine the class of a solute — electrolyte or non-electrolyte. This very broad classification of compounds into electrolyte and non-electrolyte categories can be related to the main types of compounds classified in Chapter 3. Electrolytes are high solubility ionic compounds, including bases as ionic hydroxides, and acids. Non-electrolytes are generally molecular compounds.

Acidic, Basic, and Neutral Solutions

Another empirical method of classifying solutions uses litmus paper as a diagnostic test to classify solutes as acids, bases, or neutral substances (pages 101–102). This classification also relates to the main types of compounds classified in Chapter 3. Acids form acidic solutions, bases (ionic hydroxides) form basic solutions, and most ionic and molecular compounds form neutral solutions. The evidence provided by both the conductivity and litmus tests for different classes of solutes is summarized below.

Table 5.1

Type of Solute	Conductivity Test
electrolyte	• light on conductivity apparatus glows; needle on ohmmeter moves
non-electrolyte	• light on conductivity apparatus does not glow; needle on ohmmeter does not move compared with position for control

Type of Solute	Litmus Test
acid	• blue litmus paper turns red
base	• red litmus paper turns blue
neutral	• no change in color of litmus paper

Problem 5A Identification of Solutions

The purpose of this problem is to identify some solutions. Complete the Analysis of the investigation report.

Problem

Which of the solutions labelled 1, 2, 3, and 4 is hydrobromic acid, ammonium sulfate, lithium hydroxide, and methanol?

Experimental Design

Each solution, at the same temperature and concentration, is tested with litmus paper and with a conductivity apparatus.

Evidence

Solution	Litmus	Conductivity
1	no change	none
2	blue to red	high
3	no change	high
4	red to blue	high

Like all such tests, the conductivity test and the litmus test require control of other variables. For example, when either of these diagnostic tests is done, the temperature of the solution and the quantity of dissolved solute are kept the same for all substances tested.

INVESTIGATION

5.1 Chemical Analysis

The purpose of this investigation is to perform an introductory chemical analysis of several unknown white solid solutes, using the diagnostic tests discussed so far.

Problem

Which of the white solids labelled 1, 2, 3, and 4 is calcium chloride, citric acid, glucose, and calcium hydroxide?

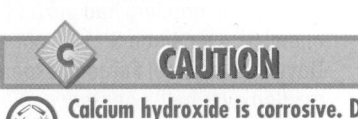

- ☐ Problem
- ☐ Prediction
- ☑ Design
- ☑ Materials
- ☑ Procedure
- ☑ Evidence
- ☑ Analysis
- ☐ Evaluation
- ☐ Synthesis

C CAUTION

Calcium hydroxide is corrosive. Do not touch any of the solids.

5.2 UNDERSTANDING SOLUTIONS

The year 1887 saw the proposal of a radical new theory by Svante Arrhenius. He hypothesized that particles of a substance, when dissolving, separate from each other and disperse into the solution. Non-electrolytes disperse electrically neutral particles throughout the solution. As Figure 5.5 shows, for example, molecules of sucrose

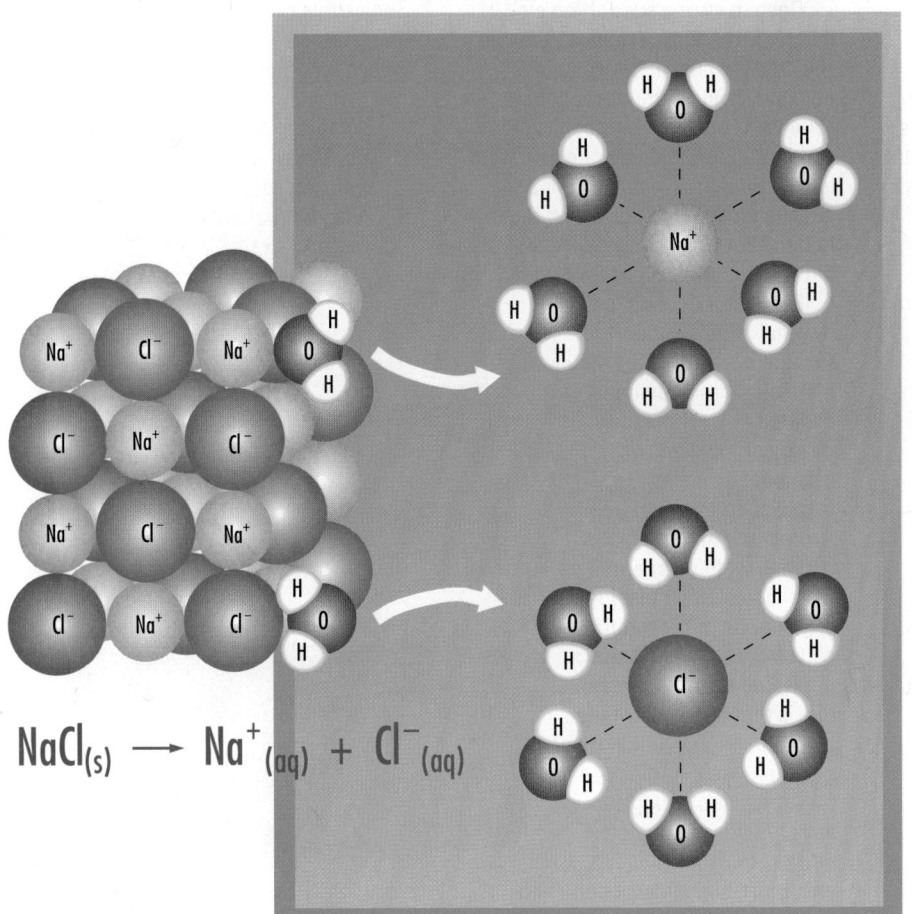

$$NaCl_{(s)} \longrightarrow Na^+_{(aq)} + Cl^-_{(aq)}$$

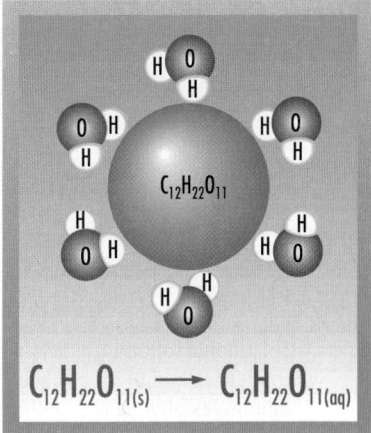

$$C_{12}H_{22}O_{11(s)} \longrightarrow C_{12}H_{22}O_{11(aq)}$$

Figure 5.5
This model illustrates sucrose dissolved in water. The model, showing electrically neutral particles in solution, agrees with the evidence that a sucrose solution does not conduct electricity.

Figure 5.6
This model represents the dissociation of sodium chloride.

separate from each other and disperse in an aqueous solution as individual molecules of sucrose surrounded by water molecules.

Arrhenius's explanation of the conductivity of solutions of electrolytes was the radical part of his proposal. He agreed with the accepted theory that electric current involves the movement of electric charge. Ionic compounds form conducting solutions. Therefore, according to Arrhenius, electrically charged particles must be present in the solutions. For example, when a compound such as table salt dissolves, it dissociates into individual aqueous ions (Figure 5.6).

SVANTE ARRHENIUS

Svante Arrhenius was born in Wijk, Sweden, in 1859. While attending the University of Uppsala near his home, he became intrigued by the problem of how and why some aqueous solutions conduct electricity, but others do not. This problem had puzzled chemists ever since Sir Humphry Davy and Michael Faraday experimented over half a century earlier by passing electric currents through chemical substances.

Faraday believed that an electric current produces new charged particles in a solution. He called these electric particles *ions* (the Greek word for "wanderer"). He could not explain what ions were, or why they did not form in solutions of substances such as sugar or alcohol dissolved in water.

As a university student, Arrhenius noticed that conducting solutions differed from non-conducting solutions in terms of another important property. The freezing point of any aqueous solution is lower than the freezing point of pure water; the more solute that is dissolved in the water, the more the freezing point is depressed. We apply this property in automobile antifreeze solutions, and when we spread salt on roads and sidewalks in winter to melt ice.

Arrhenius found that the freezing point depression of electrolytes in solution was always two or three times lower than that of non-electrolytes in solutions of the same concentration. He concluded that when a compound such as pure salt, NaCl, dissolves, it does not separate into NaCl molecules in solution, but rather into two types of particles. Since the NaCl solution also conducts electricity, he reasoned that the particles must be electrically charged. In Arrhenius's view, the

conductivity and freezing point evidence indicated that pure substances that form electrolytes were composed of *ions*, not neutral atoms. These ions appeared to be electrically charged atoms and not, as Faraday thought, particles of electricity that were somehow produced by the electric current. The stage was now set for a scientific controversy. Faraday was an established, respected scientist and his explanation agreed with Dalton's model of indivisible, neutral atoms. Arrhenius was an unknown university student and his theory contradicted Dalton's model.

Despite strong supporting evidence, Arrhenius's creative idea was rejected by most of the scientific community, including his teachers. When Arrhenius presented his theory and its supporting evidence as part of his doctoral thesis, the examiners questioned him for a gruelling four hours. They grudgingly passed him, but with the lowest possible mark.

For over a decade, only a few individuals supported Arrhenius's theory. Gradually, more supporting evidence accumulated. J. J. Thomson's discovery of the electron in 1897 dramatically upset established thinking. Soon, Arrhenius's theory of ions became widely accepted as the simplest and most logical explanation of the nature of electrolytes. In 1903 he won the Nobel Prize for the same thesis that had nearly failed him in his Ph.D. examination nineteen years earlier.

Arrhenius's struggle to have his ideas accepted is not very unusual. Ideally, scientists are completely open-minded,

but in reality, science is an activity practiced by people, and many people tend to resist change. Also, scientists attempt to explain new evidence in terms of existing or slightly revised models. The scientific community may be reluctant to accept new ideas that conflict very radically with familiar ones.

After receiving his Nobel Prize, Arrhenius focused his creative energies on other scientific mysteries. For example, he studied the role of carbon dioxide in the atmosphere and suggested that changes in carbon dioxide concentration could dramatically affect the Earth's climate. Today, this continuing discussion is referred to as the "greenhouse effect" or "global warming."

- *What two empirical properties did Arrhenius use to develop his theory of ions?*
- *Why did Arrhenius almost fail his doctoral exam?*

Dissociation is the separation of ions that occurs when an ionic compound dissolves in water. A dissociation equation is used to communicate this separation of ions. Two examples of dissociation equations for ionic compounds dissolving in water are shown below.

$$NaCl_{(s)} \rightarrow Na^+_{(aq)} + Cl^-_{(aq)}$$

$$(NH_4)_2SO_{4(s)} \rightarrow 2\,NH_4^+_{(aq)} + SO_4^{2-}_{(aq)}$$

Notice that the formula for the solvent, $H_2O_{(l)}$, does not appear as a reactant in the equation. Although water is necessary for the process of dissociation, it is not consumed and hence is not a reactant.

Arrhenius eventually extended his theory to explain some of the properties of acids and bases (see Chapter 4). According to Arrhenius, **bases** are ionic hydroxide compounds that dissociate into individual positive ions and negative hydroxide ions in solution. He believed the hydroxide ion was responsible for the properties of basic solutions; for example, turning red litmus paper blue. The dissociation of bases is similar to that of any other ionic compound, as shown in the following dissociation equation for barium hydroxide.

$$Ba(OH)_{2(s)} \rightarrow Ba^{2+}_{(aq)} + 2\,OH^-_{(aq)}$$

Acid solutes are electrolytes, but as pure substances most are molecular compounds. The properties of acids appear only when these substances, such as $HCl_{(g)}$ and $H_2SO_{4(l)}$, dissolve in water. Since acids are electrolytes, the accepted theory is that acid solutions must contain ions. However, the pure solute is molecular, so only neutral molecules are present. This unique behavior requires an explanation other than dissociation. According to Arrhenius, **acids** separate into individual molecules and *ionize* into positive hydrogen ions and negative ions.

Ionization is the reaction of neutral atoms or molecules to form charged ions. In the case of acids, Arrhenius assumed that the water solvent somehow causes the acid molecules to ionize, but he didn't propose an explanation for this. The aqueous hydrogen ions are believed to be responsible for changing the color of litmus in an acidic solution. Hydrogen chloride gas dissolving in water to form

Table 5.2

ACIDS, BASES, AND NEUTRAL SUBSTANCES		
Type of Substance	**Empirical Definition**	**Theoretical Definition**
acids	• turn blue litmus red and are electrolytes • neutralize bases	• some hydrogen compounds ionize to produce $H^+_{(aq)}$ ions • $H^+_{(aq)}$ ions react with $OH^-_{(aq)}$ ions to produce water
bases	• turn red litmus blue and are electrolytes • neutralize acids	• ionic hydroxides dissociate to produce $OH^-_{(aq)}$ ions • $OH^-_{(aq)}$ ions react with $H^+_{(aq)}$ ions to produce water
neutral substances	• do not affect litmus • some are electrolytes • some are non-electrolytes	• no $H^+_{(aq)}$ or $OH^-_{(aq)}$ ions are formed • some are ions in solution • some are molecules in solution

hydrochloric acid is a typical example of this category of acid. An ionization equation, shown below, is used to communicate this process.

$$HCl_{(g)} \rightarrow H^+_{(aq)} + Cl^-_{(aq)}$$

Arrhenius's theory was a major advance in understanding chemical substances and solutions. Arrhenius also provided the first comprehensive theory of acids and bases. The empirical and theoretical definitions of acids and bases are summarized in Table 5.2, page 191.

Substances in Water

Not all substances dissolve in water to an appreciable extent. For example, a piece of chalk (containing calcium carbonate and calcium sulfate) dropped into a glass of water remains a solid in the water. Solid calcium chloride, however, will dissolve and disappear, forming a solution that will conduct electricity. The solubility of ionic compounds (including bases as ionic hydroxides) can be predicted from a solubility chart (inside back cover of this book).

Acidic solutions vary in their electrical conductivity. Acids that are extremely good conductors are called *strong acids*. Sulfuric acid, nitric acid, and hydrochloric acid are examples of strong acids that are almost completely ionized when in solution. (These strong acids are listed on the inside back cover of this book under "Concentrated Reagents.") Most other common acids are *weak acids*. The conductivity of acidic solutions varies a great deal. The accepted explanation is that the degree of ionization of acids varies.

The solubility of molecular compounds in water varies and cannot easily be predicted until these compounds are studied in further detail. For now, the examples in Table 4.10 (page 171) should be memorized.

To understand the properties of aqueous solutions and the reactions that take place in solutions, it is necessary to know the major entities present when any substance is in a water environment. Table 5.3 summarizes this information. The information is based on the solubility and electrical conductivity of substances as determined in the laboratory. Your initial work in chemistry will deal mainly with strong acids and other highly soluble compounds.

Blood Plasma

Human blood plasma contains the following ions: $Na^+_{(aq)}$, $K^+_{(aq)}$, $Ca^{2+}_{(aq)}$, $Mg^{2+}_{(aq)}$, $HCO_3^-_{(aq)}$, $Cl^-_{(aq)}$, $HPO_4^{2-}_{(aq)}$, and $SO_4^{2-}_{(aq)}$, as well as many complex acid and protein molecules. A physician can gain information about the state of your health by testing for the quantity of these substances present in a blood sample.

Table 5.3

MAJOR ENTITIES PRESENT IN A WATER ENVIRONMENT			
Type of Substance	**Solubility in Water**	**Typical Pure Substance**	**Major Entities Present When Substance Is Placed in Water**
ionic compounds	high	$NaCl_{(s)}$	$Na^+_{(aq)}$, $Cl^-_{(aq)}$, $H_2O_{(l)}$
	low	$CaCO_{3(s)}$	$CaCO_{3(s)}$, $H_2O_{(l)}$
bases	high	$NaOH_{(s)}$	$Na^+_{(aq)}$, $OH^-_{(aq)}$, $H_2O_{(l)}$
	low	$Ca(OH)_{2(s)}$	$Ca(OH)_{2(s)}$, $H_2O_{(l)}$
molecular substances	high	$C_{12}H_{22}O_{11(s)}$	$C_{12}H_{22}O_{11(aq)}$, $H_2O_{(l)}$
	low	$C_8H_{18(l)}$	$C_8H_{18(l)}$, $H_2O_{(l)}$
strong acids	high	$HCl_{(g)}$	$H^+_{(aq)}$, $Cl^-_{(aq)}$, $H_2O_{(l)}$
weak acids	high	$CH_3COOH_{(l)}$	$CH_3COOH_{(aq)}$, $H_2O_{(l)}$
elements	low	$Cu_{(s)}$	$Cu_{(s)}$, $H_2O_{(l)}$
	low	$N_{2(g)}$	$N_{2(g)}$, $H_2O_{(l)}$

Exercise

1. Write an empirical definition and a theoretical definition of an electrolyte. What types of solutes are electrolytes?

2. How is a solution different from a pure substance? How is it similar?

3. Write equations to represent the dissociation or ionization of the following chemicals when they are placed in water.
   - (a) sodium fluoride
   - (b) sodium phosphate
   - (c) hydrogen nitrate
   - (d) aluminum sulfate
   - (e) ammonium hydrogen phosphate
   - (f) cobalt(II) chloride-6-water

4. Write a definition of an acid according to Arrhenius's theory.

5. Each of the following substances is either placed in water, or is produced in an aqueous chemical reaction. For each mixture, classify the chemicals using Table 5.3, and list the formulas of the major entities present in the water environment.
   - (a) zinc
   - (b) sodium bromide
   - (c) oxygen
   - (d) nitric acid
   - (e) calcium phosphate
   - (f) methanol
   - (g) aluminum sulfate
   - (h) potassium dichromate
   - (i) acetic acid
   - (j) sulfur
   - (k) copper(II) sulfate
   - (l) silver chloride
   - (m) paraffin wax, $C_{25}H_{52(s)}$

Problem 5B Qualitative Analysis

Complete the Analysis and Evaluation of the investigation report. In your evaluation, suggest improvements to the design, using your knowledge of chemicals and Table 5.4.

Problem

Which of the chemicals numbered 1 to 7 is $KCl_{(s)}$, $Ba(OH)_{2(s)}$, $Zn_{(s)}$, $C_6H_5COOH_{(s)}$, $Ca_3(PO_4)_{2(s)}$, $C_{25}H_{52(s)}$ (paraffin wax), and $C_{12}H_{22}O_{11(s)}$?

Experimental Design

The chemicals are tested for solubility, conductivity, and effect on litmus paper. Equal amounts of each chemical are added to equal volumes of water.

Evidence

Chemical	Solubility in Water	Conductivity of Solution	Effect of Solution on Litmus Paper
1	high	none	no change
2	high	high	no change
3	none	none	no change
4	high	high	red to blue
5	none	none	no change
6	none	none	no change
7	low	low	blue to red

Table 5.4

ELECTRICAL CONDUCTIVITY			
Class	Solid	Liquid	Aqueous
metal	✔	✔	–
nonmetal	X	X	–
ionic	X	✔	✔
molecular	X	X	X
acid	X	X	✔

Water is a useful solvent because of its unique properties:
- Water is plentiful and inexpensive.
- Water is a liquid at room temperature.
- Water has a very wide temperature range compared to other liquids.
- Water is capable of dissolving many different types of substances.

The reactions of baking powder, Drano®, or Eno® begin only when the compounds in these mixtures dissolve in water. Medication administered as part of an intravenous solution acts much faster than a solid pill that is swallowed, as the solid must first dissolve before it can act. Sulfur oxides and nitrogen oxides released into the atmosphere as gases become pollutants when they dissolve in moisture to form acids. In a laboratory, dry lead(II) nitrate solid mixed with sodium iodide solid produces only a slight color change, compared with the instantaneous yellow precipitation that is produced when the reactants are mixed as solutions (see Figure 5.7, page 195).

- Problem
- Prediction
- ✔ Design
- Materials
- Procedure
- ✔ Evidence
- ✔ Analysis
- Evaluation
- Synthesis

CAUTION

Solutions may be toxic, irritant, or corrosive; avoid eye and skin contact.

5.3 REACTIONS IN SOLUTION

In technological applications both in homes and at worksites, many chemicals are more easily handled when they are in solution. Transporting, loading, and storing chemicals are often more convenient and efficient when the chemicals are in solution. Also, performing a reaction in solution can change the rate (speed), the extent (completeness), and the type (kind of product) of the chemical reaction. For all systems, solutions make it easy to

- *handle chemicals*, for example, ammonia gas is dissolved in water for use as a household cleaner;

- *react chemicals*, for example, baking powder is dissolved in water to initiate the reaction; and

- *control reactions*, for example, the rate, extent, and type of reactions are easily controlled in solution.

According to the collision-reaction theory, all chemical reactions involve collisions among atoms, ions, or molecules. In a mixture of a solid and a gas, collisions can occur only on the relatively small surface area of the solid. If the solid is finely crushed it has far more surface area, so more collisions will occur and the substances will react more quickly. Dissolving a solid in a solution is the ultimate "breaking up" of the solid, since the substance is reduced to the smallest possible particles of an element or compound — separate atoms, ions, or molecules. If both reactants exist in dissociated form, the greatest number of collisions can occur. This situation is only possible in gas or liquid solutions.

INVESTIGATION

5.2 The Iodine Clock Reaction

Technological problem solving often involves a systematic trial-and-error approach that is guided by knowledge and experience. Usually one variable at a time is manipulated, while all other variables are controlled. Variables that may be manipulated include concentration, volume, and temperature. The purpose of this investigation is to find a method for getting a reaction to occur in a specified time period.

Problem

What technological process can be employed to have solution A react with solution B in a time of 20 ± 1 s?

Net Ionic Equations

A student mixed solutions of lead(II) nitrate and sodium iodide and observed the formation of a bright yellow precipitate (Figure 5.7). Another student recorded the same observation after mixing solutions of lead(II) acetate and magnesium iodide. Are these different reactions? The balanced chemical equations for these two reactions show some similarities and some differences.

(1) $Pb(NO_3)_{2(aq)} + 2\ NaI_{(aq)} \rightarrow PbI_{2(s)} + 2\ NaNO_{3(aq)}$

(2) $Pb(CH_3COO)_{2(aq)} + MgI_{2(aq)} \rightarrow PbI_{2(s)} + Mg(CH_3COO)_{2(aq)}$

Using the Arrhenius theory of dissociation, these reactions can be described more precisely. Each of the high solubility ionic compounds are believed to exist as separate ions. For reaction (1),

$$Pb^{2+}_{(aq)} + 2\ NO_3^-{}_{(aq)} + 2\ Na^+{}_{(aq)} + 2\ I^-{}_{(aq)} \rightarrow PbI_{2(s)} + 2\ Na^+{}_{(aq)} + 2\ NO_3^-{}_{(aq)}$$

It is apparent that some reactant ions — sodium and nitrate ions — are unchanged in this reaction. Ignoring these ions, you can write a **net ionic equation**, which shows only those entities that change in a chemical reaction.

$$Pb^{2+}_{(aq)} + 2\ \cancel{NO_3^-{}_{(aq)}} + 2\ \cancel{Na^+{}_{(aq)}} + 2\ I^-{}_{(aq)} \rightarrow PbI_{2(s)} + 2\ \cancel{Na^+{}_{(aq)}} + 2\ \cancel{NO_3^-{}_{(aq)}}$$

$$Pb^{2+}_{(aq)} + 2\ I^-{}_{(aq)} \rightarrow PbI_{2(s)}\ \text{(net ionic equation)}$$

Applying the same procedure to reaction (2),

$$Pb^{2+}_{(aq)} + 2\ \cancel{CH_3COO^-{}_{(aq)}} + \cancel{Mg^{2+}{}_{(aq)}} + 2\ I^-{}_{(aq)} \rightarrow PbI_{2(s)} + \cancel{Mg^{2+}{}_{(aq)}} + 2\ \cancel{CH_3COO^-{}_{(aq)}}$$

$$Pb^{2+}_{(aq)} + 2\ I^-{}_{(aq)} \rightarrow PbI_{2(s)}\ \text{(net ionic equation)}$$

The same observation made from apparently different chemical reactions gives the same net ionic equations. The ions that do not change during a chemical reaction are called **spectator ions**. Like spectators at a basketball game, spectator ions are present but not part of the action, that is, the chemical reaction.

Figure 5.7
A solution of sodium iodide reacts immediately when mixed with a lead(II) nitrate solution.

SUMMARY: WRITING NET IONIC EQUATIONS

Step 1. Write the usual (non-ionic) balanced chemical equation.
Step 2. Dissociate all high solubility ionic compounds and ionize all strong acids to show the complete ionic equation.
Step 3. Cancel identical entities appearing on both reactant and product sides.
Step 4. Write the net ionic equation, reducing coefficients if necessary.

EXAMPLE

Write the net ionic equation for the reaction of aqueous barium chloride and aqueous sodium sulfate.

$$BaCl_{2(aq)} + Na_2SO_{4(aq)} \rightarrow BaSO_{4(s)} + 2\ NaCl_{(aq)}$$

$$Ba^{2+}_{(aq)} + 2\ \cancel{Cl^-{}_{(aq)}} + 2\ \cancel{Na^+{}_{(aq)}} + SO_4^{2-}{}_{(aq)} \rightarrow BaSO_{4(s)} + 2\ \cancel{Na^+{}_{(aq)}} + 2\ \cancel{Cl^-{}_{(aq)}}$$

$$Ba^{2+}_{(aq)} + SO_4^{2-}{}_{(aq)} \rightarrow BaSO_{4(s)}$$

Net ionic equations are useful in communicating reactions other than precipitation reactions, or double replacement reactions. The following example is a good illustration.

EXAMPLE

Write the net ionic equation for the reaction of zinc metal and aqueous copper(II) sulfate.

$$Zn_{(s)} + CuSO_{4(aq)} \rightarrow Cu_{(s)} + ZnSO_{4(aq)}$$

$$Zn_{(s)} + Cu^{2+}_{(aq)} + \cancel{SO_4^{2-}}_{(aq)} \rightarrow Cu_{(s)} + Zn^{2+}_{(aq)} + \cancel{SO_4^{2-}}_{(aq)}$$

$$Zn_{(s)} + Cu^{2+}_{(aq)} \rightarrow Cu_{(s)} + Zn^{2+}_{(aq)}$$

EXAMPLE

Write the net ionic equation for the reaction of hydrochloric acid and barium hydroxide solution.

$$2\ HCl_{(aq)} + Ba(OH)_{2(aq)} \rightarrow BaCl_{2(aq)} + 2\ HOH_{(l)}$$

$$2\ H^+_{(aq)} + \cancel{2\ Cl^-}_{(aq)} + \cancel{Ba^{2+}}_{(aq)} + 2\ OH^-_{(aq)} \rightarrow \cancel{Ba^{2+}}_{(aq)} + \cancel{2\ Cl^-}_{(aq)} + 2\ HOH_{(l)}$$

$$H^+_{(aq)} + OH^-_{(aq)} \rightarrow HOH_{(l)}\ \text{(coefficients reduced to 1)}$$

Exercise

6. An acceptable method for the treatment of soluble lead waste is to precipitate the lead as a low solubility lead(II) silicate. Write the net ionic equation for the reaction of aqueous lead(II) nitrate and aqueous sodium silicate.

7. One industrial method of producing bromine is to react sea water, containing sodium bromide, with chlorine gas. Write the net ionic equation for this reaction.

8. Strontium compounds are often used in flares because their flame color is bright red. One industrial example is the reaction of aqueous solutions of strontium nitrate and sodium carbonate. Write the next ionic equation for this reaction.

9. Silver may be recovered from a solution by adding copper metal. Write the net ionic equation for the reaction of copper with aqueous silver nitrate.

10. In a hard water analysis, sodium oxalate solution reacts with calcium hydrogen carbonate present in the hard water to precipitate a calcium compound. Write the net ionic equation for this reaction.

Qualitative Chemical Analysis

Chemical analysis of an unknown sample includes **qualitative analysis**, the identification of specific substances present, and **quantitative analysis**, the measurement of the quantity of a substance present. A typical analysis is that done on a blood sample to determine if ethanol is present, and if so, how much is present. As another example, drinking water is frequently analyzed for the presence of a wide variety of dissolved substances, some potentially harmful, some beneficial.

An example of a qualitative analysis that you have already seen is the color reaction of litmus paper to identify the presence of hydrogen ions (indicating an acid) or hydroxide ions (indicating a base) in a

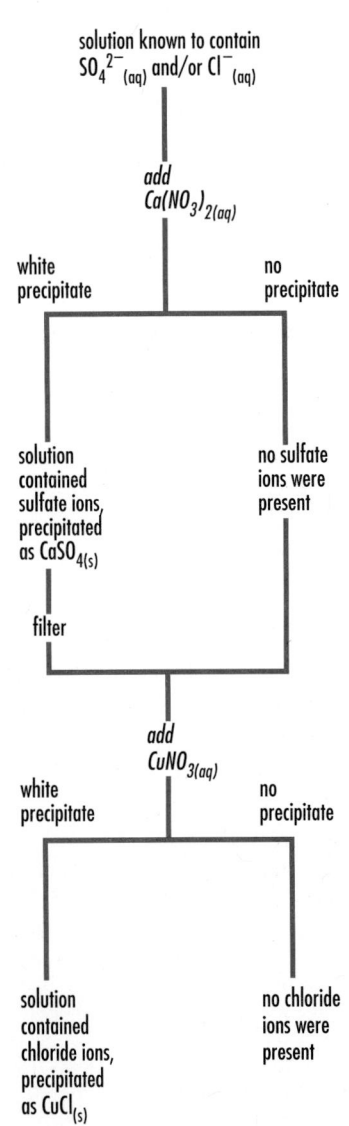

Figure 5.8
Analysis of a solution for sulfate and chloride ions. In this example, the two tests could be done in reverse order. This is not always the case in sequential analysis.

solution. Many other ions can be identified by means of selective precipitation; you can refer to a solubility table to predict precipitates formed in chemical reactions.

Suppose you were given a solution that contained either sulfate ions, chloride ions, or both ions. How could you determine which ions were present? To answer this question, the experimental design involves two diagnostic tests.

- *Test for the presence of sulfate ions.*
  Calcium ions form a low solubility compound with sulfate ions, so if an excess of calcium nitrate solution is added to a sample of the test solution and a precipitate forms, then sulfate ions are probably present. Refer to the solubility table on this book's inside back cover to verify that chloride ions do not precipitate with calcium ions, but that sulfate ions do. If $SO_4^{2-}{}_{(aq)}$ is present, then

$$Ca^{2+}{}_{(aq)} + SO_4^{2-}{}_{(aq)} \rightarrow CaSO_{4(s)}$$

- *Test for the presence of chloride ions.*
  Unlike copper(II) chloride, a compound containing copper(I) ions and chloride ions would have low solubility. If a copper(I) nitrate solution is added to the filtrate from the previous test and a precipitate forms, then chloride ions are probably present. Copper(I) sulfate would not precipitate. If $Cl^-{}_{(aq)}$ is present, then

$$Cu^+{}_{(aq)} + Cl^-{}_{(aq)} \rightarrow CuCl_{(s)}$$

With this experimental design as a guide, you can organize the materials and the procedure and answer the question (Figure 5.8).

In the preceding example, both the given solution and the diagnostic test solutions contain dissociated electrolytes. If a precipitate is observed, the collision-reaction theory suggests that collisions occurred between two kinds of ions to form a low solubility solid. The nitrate ions present in the diagnostic test solutions are spectator ions. By planning a careful experimental design, you can do a sequence of diagnostic tests to detect many different ions, beginning with only one sample of a solution.

INVESTIGATION

5.3 Sequential Qualitative Analysis

The purpose of this investigation is to do a qualitative chemical analysis of an unknown solution using precipitation reactions.

Problem

Are there any lead(II) ions and/or strontium ions present in a sample solution?

Experimental Design

A sodium chloride solution is used as a diagnostic test for the presence of lead(II) ions. The solution is known to contain only lead and/or strontium ions. If a precipitate forms, the solution is filtered to remove the precipitate from the test solution. A sodium sulfate solution is then added to the filtrate as a diagnostic test for strontium ions.

▢	Problem
▢	Prediction
▢	Design
✔	Materials
✔	Procedure
✔	Evidence
✔	Analysis
▢	Evaluation
▢	Synthesis

 CAUTION

Lead and strontium ions are toxic. Dispose of all materials as directed by your teacher.

The Aurora Borealis

The aurora borealis, also called the northern lights, is a natural light display that shimmers from red through green. The farther north you live, the more often you can observe this beautiful display at night. The aurora borealis originates with solar flares, which expel charged particles known as the "solar wind." Some of these particles become trapped in the Earth's magnetic field. Near the poles, these particles spiral downward into the atmosphere and strike, energize, and ionize molecules in the air. In both northern lights and flame tests, high energy ions lose energy in a form that is seen as colored light.

Figure 5.9
Copper(II) ions usually impart a green color to a flame. The green color of the flame and the blue color of solutions of copper(II) ions can be used as diagnostic tests for copper(II) ions.

- ▢ Problem
- ▢ Prediction
- ▢ Design
- ▢ Materials
- ▢ Procedure
- ✔ Evidence
- ✔ Analysis
- ▢ Evaluation
- ▢ Synthesis

Qualitative Analysis by Color

Some ions impart a specific color to a solution, a flame, or a gas discharge tube. For example, copper(II) ions produce a blue aqueous solution and usually a green flame. A *flame test*, a test for the presence of metal ions such as copper(II), is conducted by dipping a clean platinum or nichrome wire into a solution and then into a flame (Figure 5.9). The initial flame must be nearly colorless and the wire, when dipped in water, must not produce a color in the flame. Usually, the wire is cleaned by dipping it alternately into hydrochloric acid and then into the flame, until very little color is produced. The colors of some common ions in aqueous solutions and in flames are listed on the inside back cover of this book.

Exercise

11. List two household substances that are purchased as solids but are then made into solutions before use.

12. What color are the following ions in an aqueous solution?
    (a) iron(III)   (c) $Cu^{2+}_{(aq)}$
    (b) sodium   (d) $Ni^{2+}_{(aq)}$

13. What color are the following ions in a flame test?
    (a) calcium   (c) Na^+
    (b) copper(II)   (d) K^+

14. Design a diagnostic test for carbonate ions using a reactant that would not precipitate sulfide ions in the sample. Include a net ionic equation.

15. Design an experiment to determine if acetate and/or carbonate ions are present in a solution. Include net ionic equations.

16. Design an experiment to determine if potassium and/or strontium ions are present in a solution. Include net ionic equations.

17. (Enrichment) Design an experiment to analyze a single sample of a solution for any or all of the $Tl^+_{(aq)}$, $Ba^{2+}_{(aq)}$, and $Ca^{2+}_{(aq)}$ ions.

INVESTIGATION

5.4 Qualitative Analysis by Color

The purpose of this investigation is to do qualitative chemical analyses using solution and flame colors.

Problem

Which of the solutions labelled 1, 2, 3, and 4 contains one of the metal ions listed in the ion colors chart (see inside back cover of this book)?

Experimental Design

The solution and flame color of several known solution samples are observed, and then the unknown solutions are identified.

Materials

lab apron
safety glasses
well plate or 14 small test tubes with rack
platinum wire with holder/handle
laboratory burner
dropper bottles of $LiCl_{(aq)}$, $NaCl_{(aq)}$, $KCl_{(aq)}$, $CaCl_{2(aq)}$, $SrCl_{2(aq)}$, $BaCl_{2(aq)}$, $CuCl_{2(aq)}$, $CoCl_{2(aq)}$, $FeCl_{3(aq)}$, $HCl_{(aq)}$
(optional) masking tape

Procedure

1. Using a strip of masking tape or paper, label each of 10 wells or test tubes with the formula of one of the known solutions provided.

2. Add 10 to 15 drops of each solution into its labelled container.

3. Clean the platinum wire by alternately dipping the wire into the $HCl_{(aq)}$ and then heating it in the hottest part of the burner flame until the wire glows but shows no other color.

4. Observe the solution color of one of the known solutions.

5. Dip the platinum wire into the solution, hold in the hottest part of the burner flame, and determine the flame color.

6. Repeat steps 3 to 5 for each of the remaining known solutions.

7. Obtain 10 to 15 drops of one of the unknown solutions, and repeat steps 3 to 5.

8. Repeat step 7 for each of the remaining three unknown solutions.

9. Dispose of the HCl as instructed by your teacher, taking the necessary precautions when neutralizing the acid in case of splattering.

5.4 CONCENTRATION OF A SOLUTION

Most solutions are colorless and aqueous. Because of this similarity, and because of the need for numerical values for quantitative analysis, solutions are commonly described in terms of a numerical ratio that compares the quantity of solute to the quantity of the solution. Such a numerical ratio is called the solution's **concentration**. Chemists describe a solution of a given substance as *dilute* if it has a relatively small quantity of solute per unit volume of solution (Figure 5.10). A *concentrated* solution, on the other hand, has a relatively large quantity of solute per unit volume.

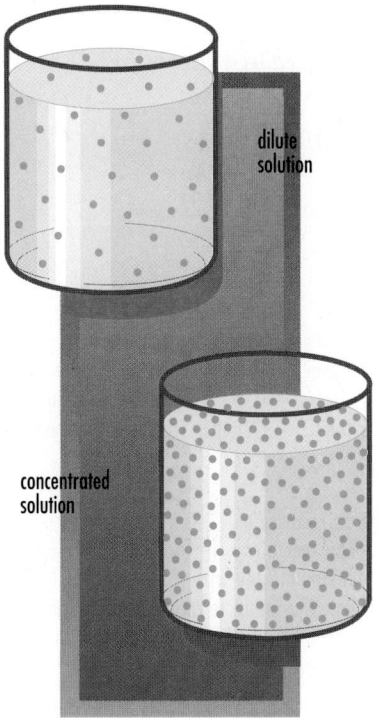

Figure 5.10
The model of the dilute solution shows fewer particles per unit volume compared with the model of the concentrated solution.

Communicating Concentration Ratios

In general, the concentration of a solution is expressed by the ratio

$$\text{concentration} = \frac{\text{quantity of solute}}{\text{quantity of solution}}$$

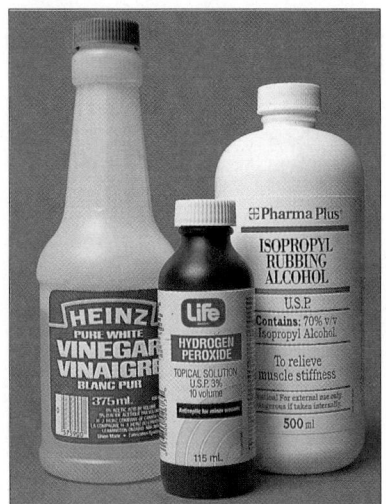

Figure 5.11
The concentrations of different consumer products depend on the product and sometimes the brand name. Concentrations are usually expressed as a percentage.

---

1 ppm = 1 g/10⁶ mL
     = 1 g/1000 L
     = 1 mg/L
     = 1 mg/kg
     = 1 μg/g

Parts per Million, Billion, and Trillion

1 ppm	1 drop in a full bathtub
1 ppb	1 drop in a full swimming pool
1 ppt	1 drop in 1000 swimming pools

Certainty
The certainty of an answer may be communicated with significant digits. The answer is no more certain (i.e., has no more significant digits) than the least certain value. See Appendix E, *Communication Skills*.

Measurement	Significant Digits
250 mL	3
0.24 mg/L	2
0.060 mg	2
500 mL	3
5%	1

For calculations dealing with molar concentration, you can use the following relationship.

$$C = \frac{n}{v}$$

where *C* is the molar concentration in moles per litre, *n* is the amount of solute in moles, and *v* is the volume of solution in litres.

For example, many consumer products, such as vinegar (acetic acid), are conveniently labelled with their concentration ratios expressed as percentages (Figure 5.11). A vinegar label listing "5% acetic acid (by volume)" means that there are 5 mL of pure acetic acid dissolved in every 100 mL of the vinegar solution. This type of concentration is often designated as % V/V or % by volume.

$$5\% \text{ V/V} = \frac{5 \text{ mL}}{100 \text{ mL}}$$

Another common concentration ratio used for consumer products is "percent weight volume" or % W/V. (In consumer and commercial applications, "weight" is used instead of "mass," which explains the W in the W/V label.) For example, a hydrogen peroxide topical solution used as an antiseptic is 3% W/V (Figure 5.11). This means that 3 g of hydrogen peroxide is in every 100 mL of solution.

$$3\% \text{ W/V} = \frac{3 \text{ g}}{100 \text{ mL}}$$

For very dilute solutions, a concentration unit is chosen to give reasonable numbers for very small quantities of solute. For example, the concentration of toxic substances in the environment or of chlorine in a swimming pool is usually expressed as parts per million (ppm) or even smaller ratios. One part per million of chlorine in a swimming pool corresponds to one gram of chlorine in a million millilitres of pool water, which is equivalent to one milligram of chlorine per litre of water.

Chemistry is primarily about chemical reactions, which are communicated using balanced chemical equations. The coefficients in these equations represent amounts of chemicals in units of moles. Therefore, concentration in the study and practice of chemistry is communicated using **molar concentration**, also known as **molarity**. Molar concentration or molarity, *C*, is the amount of solute in moles dissolved in one litre of solution.

$$\text{molar concentration} = \frac{\text{amount of solute (in moles)}}{\text{volume of solution (in litres)}}$$

The units of molar concentration (mol/L) come directly from this ratio. Although not an SI unit, it is still common among chemists to use the unit **molar**, M, with 1 M = 1 mol/L, to express molarity or molar concentration (Figure 5.12).

Calculations Involving Concentration

A common calculation involves the quantity of solute, volume of solution, and concentration of solution. When two values are known, the other can be calculated. Because concentration is a ratio, a simple procedure is to use the concentration ratio as part of a proportion statement. For example, a box of apple juice has sugar concentration of 12 g/100 mL (12% W/V). What mass of sugar is present in a 175 mL glass of juice? Set the concentration ratio equal to the ratio for the sample size.

$$\frac{m}{175 \text{ mL}} = \frac{12 \text{ g}}{100 \text{ mL}}$$

(Read as "*m* is to one hundred and seventy-five millilitres as twelve grams is to one hundred millilitres.")

CAREER

ENVIRONMENTAL CHEMIST

In Scotland, where she grew up, Dr. Margaret-Ann Armour trained as a physical organic chemist. After working for several years as a research chemist in the paper-making industry, where she became interested in the effects of chemicals on the environment, Dr. Armour diversified her work.

According to Armour, good researchers must have special attributes in addition to their fund of knowledge. Perseverance and optimism are essential qualities. Results may be slow to appear, and without optimism it is difficult to persevere. Above all, a researcher must be a good communicator. "A scientist isn't a lone individual working in isolation in the laboratory," she says. Because science has become increasingly specialized, "it's essential to be able to communicate with people in other disciplines." Research groups are often composed of people with diverse interests, training, and talents. Science needs people with a broad range of experience. "What scientists mostly do is ask questions and people need to ask new questions," Armour observes, adding that people from different backgrounds and cultures provide a varied supply of creative questions. Good questions arise as well from original thinking. A solid scientific education should encourage students to look beyond the facts to the underlying ideas.

As the Assistant Chair of the Department of Chemistry at the University of Alberta, Armour has to juggle a heavy load of administration, research, and interaction with students and the public. Her research group develops and tests methods for recycling and disposing of waste and surplus chemicals. As research into hazardous wastes has commanded the attention of government and business, Armour has learned to communicate effectively with non-scientists as well as with other scientists. She has devised strategies to explain complex ideas so that a lay audience can understand them.

Armour is troubled by the public's poor grasp of some pollution issues. Information imparted by the media is often simplistic or misleading, and it can be distorted by political or business influences. For example, reports on PCBs (polychlorinated biphenyls) have often appeared in the news. These compounds were used as transformer oils in the electrical industry and they're usually described in the media as deadly. "In fact," says Armour, "they are fairly inert; while they should be kept out of food, they can remain in the soil without causing harm. Only under certain highly specific conditions that cause the PCBs to decompose, do they become extremely toxic." On the other hand, some chemicals in the environment, such as chromium compounds, are very hazardous, but these are rarely mentioned in the popular media. The public needs to be protected from exposure to serious hazards as well as from the fear of infamous but less dangerous chemicals.

As an advocate of women's participation in science, Armour helped establish a committee called WISEST (Women in Scholarship, Engineering, Science, and Technology) at the University of Alberta. WISEST encourages young women to aspire to careers in science, partly by means of a summer program that gives students a chance to work with scientists and technologists.

According to Armour, there is a looming shortage of chemistry teachers in universities. Chemistry enrolments in undergraduate programs are down, so the future supply of teachers will be limited. Because professors will be retiring in large numbers, the demand for instructors will increase. Small supply and large demand spell opportunities for young people looking forward to a career in chemistry.

- *According to Dr. Armour, what are some attributes of good researchers?*
- *Why is it important to be sceptical about chemical information in the media?*

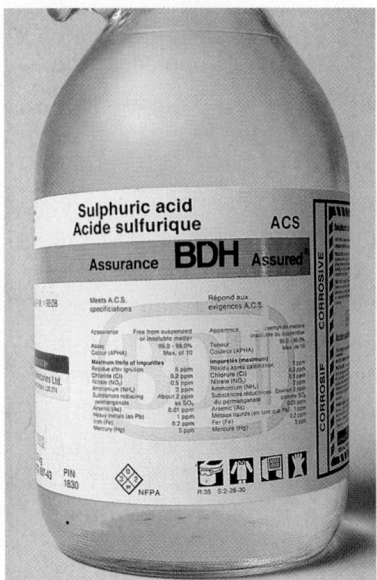

Figure 5.12
A bottle of concentrated sulfuric acid has a molar concentration of 17.8 mol/L, or a molarity of 17.8 M.

Solving this equation gives a value of 21 g for *m*, the mass of sugar in 175 mL of apple juice. When using a proportion, it is important that the two ratios are in the same order (for example, sugar to solution) on both sides of the equality.

Alternatively, the concentration ratio can be used as a conversion factor. You are already familiar with this procedure although you likely have done the calculation automatically. For example, if you work 4 h at a pay rate of $6/h, you expect to be paid $24. The $6/h is the conversion factor is necessary to convert the time worked into your wages.

$$\text{wage} = 4 \text{ h} \times \frac{\$6}{1 \text{ h}} = \$24$$

Similarly, the mass of sugar in a 175 mL glass of apple juice that has a concentration of 12 g/100 mL is calculated as follows.

$$m_{\text{sugar}} = 175 \text{ mL} \times \frac{12 \text{ g}}{100 \text{ mL}} = 21 \text{ g}$$

Sometimes it may be necessary to use the reciprocal of a ratio or conversion factor. For example, how long will you have to work at $6/h to earn $20 for a music CD?

$$\frac{t}{\$20} = \frac{1 \text{ h}}{\$6} \quad \text{or} \quad t = \$20 \times \frac{1 \text{ h}}{\$6}$$

$$t = 3.3 \text{ h}$$

All of these methods are equivalent, and you should choose the one that is best for you.

EXAMPLE

A sample of well water contains 0.24 g/L of dissolved iron. What mass of iron is present in 1.2 L of water in a kettle?

$$\frac{m}{1.2 \text{ L}} = \frac{0.24 \text{ mg}}{1 \text{ L}} \quad m_{\text{Fe}} = 29 \text{ mg}$$

or

$$m_{\text{Fe}} = 1.2 \text{ L} \times \frac{0.24 \text{ mg}}{1 \text{ L}} = 0.29 \text{ mg}$$

EXAMPLE

A sample of laboratory ammonia solution has a molar concentration of 14.8 mol/L. What amount of ammonia is present in a 2.5 L bottle?

$$\frac{n}{2.5 \text{ L}} = \frac{14.8 \text{ mol}}{1 \text{ L}} \quad n_{\text{NH}_3} = 37 \text{ mol}$$

or

$$n_{\text{NH}_3} = 2.5 \text{ L} \times \frac{14.8 \text{ mol}}{1 \text{ L}} = 37 \text{ mol}$$

What volume of 0.25 mol/L salt solution contains 0.10 mol of sodium chloride?

$$\frac{V}{0.10 \text{ mol}} = \frac{1 \text{ L}}{0.25 \text{ mol}} \qquad V_{\text{NaCl}} = 0.40 \text{ L}$$

or

$$V_{\text{NaCl}} = 0.10 \text{ mol} \times \frac{1 \text{ L}}{0.25 \text{ mol}} = 0.40 \text{ L}$$

In chemistry, the mole is a very important unit; however, measurements in a laboratory are usually of mass (in grams) and of volume (in millilitres). A common chemistry calculation involves the mass of a substance, the volume of solution, and the molar concentration of solution. This type of calculation requires the use of two ratios — molar mass and molar concentration. For example, a chemical analysis requires 2.00 L of 0.150 mol/L $AgNO_{3(aq)}$. What mass of silver nitrate solid is required to prepare this solution? You will need to determine the amount of silver nitrate in moles.

$$n_{\text{AgNO}_3} = 2.00 \text{ L} \times \frac{0.150 \text{ mol}}{1 \text{ L}} = 0.300 \text{ mol}$$

You can then convert this amount into a mass of silver nitrate using its molar mass. (See Chapter 4, page 159.)

$$m_{\text{AgNO}_3} = 0.300 \text{ mol} \times \frac{169.88 \text{ g}}{1 \text{ mol}} = 51.0 \text{ g}$$

Each of these steps can be done using a proportion. Another alternative that can be used is to combine the two steps shown above into one calculation.

$$m_{\text{AgNO}_3} = 2.00 \text{ L} \times \frac{0.150 \text{ mol}}{1 \text{ L}} \times \frac{169.88 \text{ g}}{1 \text{ mol}} = 51.0 \text{ g}$$

In order to successfully use this approach, you need to pay particular attention to the units in the calculation. Cancel units to check your procedure.

A student dissolves 5.00 g of solid sodium carbonate to make 250 mL of solution. What is the molarity of the solution?

$$n_{\text{NaCO}_3} = 5.00 \text{ g} \times \frac{1 \text{ mol}}{105.99 \text{ g}} = 0.0472 \text{ mol}$$

$$C_{\text{NaCO}_3} = \frac{0.0472 \text{ mol}}{0.250 \text{ L}} = 0.189 \text{ M}$$

18. What concentration ratio is often found on the labels of consumer products? Why do you think this unit is used instead of mol/L?

19. If the average concentration of PCBs (polychlorinated biphenyls) in the body tissue of a human is 4.00 mg/kg (4.00 ppm), what mass of PCBs is present in a 64.0 kg person?

20. The maximum concentration of salt in water at 0°C is 31.6 g/100 mL. What mass of salt can be dissolved in 250 mL of solution?

21. A household ammonia solution has a concentration of 1.24 mol/L. What volume of this solution would contain 0.500 mol of NH_3?

22. To prepare for an experiment using flame tests, a student requires 100 mL of 0.10 mol/L solutions of each of the following substances. Calculate the required mass of each solid.
(a) $NaCl_{(s)}$ (b) $KCl_{(s)}$ (c) $CaCl_{2(s)}$

23. An experiment is planned to study the chemistry of a home water-softening process. The brine used in this process has a concentration of 25 g/100 mL. What is the molar concentration of this solution?

24. What volume of 0.055 mol/L glucose solution contains 2.0 g of glucose?

Figure 5.13
In many school laboratories, electronic balances that measure masses to within 0.01 g or 0.0001 g have replaced mechanical balances. These balances are efficient and produce reliable and reproducible measurements of mass.

Preparation of Standard Solutions from a Solid

Solutions with precisely known concentrations, called **standard solutions**, are routinely prepared in both scientific research laboratories and industrial processes. They are used in chemical analysis as well as for the precise control of chemical reactions. To prepare a standard solution, precision equipment is required to measure the mass of solute and volume of solution. Electronic balances are used for precise and efficient measurement of mass (Figure 5.13). For measuring a precise volume of the final solution, a container called a *volumetric flask* is used (Figure 5.14).

Figure 5.14
Volumetric glassware comes in a variety of shapes and sizes. The Erlenmeyer flask on the far left has only approximate volume markings, as does the beaker. The graduated cylinders have much better precision, but for high precision a volumetric flask (on the right) is used. The volumetric flask shown here, when filled to the line, contains 100.0 mL ±0.16 mL at 20°C. This means that a volume measured in this flask is uncertain by less than 0.2 mL at the specified temperature.

INVESTIGATION

5.5 Demonstration: A Standard Solution from a Solid

The purpose of this demonstration is to illustrate the skills required to prepare a standard solution from a pure solid. You will need these skills in many investigations in this book. No problem is stated, since no specific analysis is being done — your task is to learn the procedure.

Materials

lab apron	stirring rod
safety glasses	wash bottle of pure water (deionized or distilled)
$CuSO_4 \cdot 5H_2O_{(s)}$	100 mL volumetric flask with stopper
150 mL beaker	small funnel
centigram balance	medicine dropper
laboratory scoop	meniscus finder

Procedure

Refer to Appendix C, C.3 for a summary of preparing standard solutions from pure solutes.

1. (Pre-lab) Calculate the mass of solid copper(II) sulfate-5-water needed to prepare 100.0 mL of a 0.5000 mol/L solution.

2. Obtain the calculated mass of copper(II) sulfate-5-water in a clean, dry 150 mL beaker.

3. Dissolve the solid in 40 mL to 50 mL of pure water.

4. Transfer the solution into a 100 mL volumetric flask. Rinse the beaker two or three times with small quantities of pure water, transferring the rinsings into the volumetric flask.

5. Add pure water until the volume is 100.0 mL.

6. Stopper the flask and mix the contents thoroughly by repeatedly inverting the flask.

- Problem
- Prediction
- Design
- Materials
- Procedure
- Evidence
- Analysis
- Evaluation
- Synthesis

C CAUTION

Copper(II) sulfate powder can be irritant and the solution is slightly corrosive.

Pure Water

The term *pure water* is often used to refer to deionized or distilled water. Of course it is not absolutely pure — this is not possible even with the best equipment. The degree of water purity required depends on the application, whether for boating, swimming, cooking, or drinking. What criteria would you apply in judging the purity of drinking water and laboratory water?

Exercise

25. A high school laboratory technician prepares 2.00 L of a 0.100 mmol/L aqueous solution of cobalt(II) chloride-2-water for an experiment. Show your work for the pre-lab calculation and write a complete specific procedure for preparing this solution, as in Investigation 5.5.

26. To test the hardness of water (Figure 5.15), a chemical analysis is done using 100.0 mL of a 0.250 mol/L solution of ammonium oxalate. What mass of ammonium oxalate is needed to make the standard solution?

27. Calculate the mass of solid lye (sodium hydroxide) needed to make 500 mL of a 10.0 mol/L cleaning solution (Figure 5.16).

28. A technician prepares 500.0 mL of a 75.0 mmol/L solution of potassium permanganate as part of a quality control analysis in the manufacture of hydrogen peroxide. Calculate the mass of potassium permanganate required to prepare the solution.

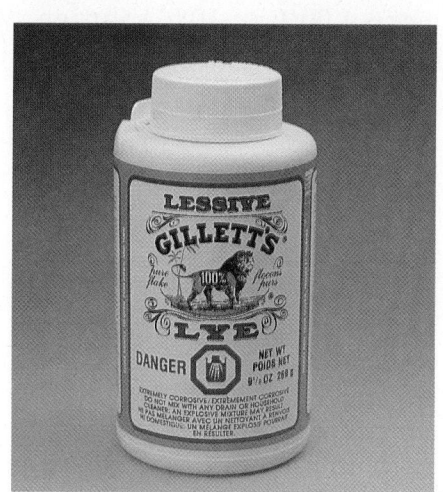

Figure 5.15
Water hardness is caused by calcium and magnesium compounds. The photograph shows the effect that hard-water deposits such as calcium carbonate can have in a water pipe.

Figure 5.16
Solutions of sodium hydroxide in very high concentration are sold as cleaners for clogged drains. The same solution can be made less expensively by dissolving solid lye (a commercial name for sodium hydroxide) in water. The pure chemical is very caustic and the label on the lye container recommends rubber gloves and eye protection.

Preparation of Standard Solutions by Dilution

Classifying Solutions
Solutions with a molar concentration of less than 0.1 mol/L may be described as *dilute*, while solutions with a concentration of greater than 1 mol/L may be referred to as *concentrated*. There are no accepted definitions.

Dilution is the process of decreasing the concentration of a solution, usually by adding more solvent. You apply this process when you add water to a concentrated fruit juice, fabric softener, or cleaning product. Because dilution is a simple, quick procedure, normal scientific practice is to begin with a more concentrated starting solution, called a **stock solution**, and to add solvent (usually water) to decrease the concentration to the desired level.

Calculating the new concentration after a dilution is straightforward, because the quantity of solute is not changed by adding more solvent. This means that the change in concentration is inversely related to the change in the solution's volume. For example, if water is added to 6% hydrogen peroxide disinfectant until the total volume is doubled, the concentration becomes one-half the original value, or 3%. This relationship can be expressed as an equation, in which symbols are used to represent initial concentration (c_i), final concentration (c_f), initial volume before dilution (v_i), and final volume after dilution (v_f). Molar concentration is expressed as an uppercase C.

$$v_i c_i = v_f c_f \quad \text{or} \quad v_i C_i = v_f C_f$$

Any one of the values expressed may be calculated for the dilution of a solution, provided the other values are known.

EXAMPLE ———————————————————————————

Water is added to 200 mL of 2.40 mol/L $NH_{3(aq)}$ cleaning solution, until the final volume is 1.000 L. Find the molar concentration of the final, diluted solution.

$$
\begin{aligned}
v_i C_i &= v_f C_f \\
200 \text{ mL} \times 2.40 \text{ mol/L} &= 1000 \text{ mL} \times C_f \\
C_f &= 0.480 \text{ mol/L} \\
&\ \ NH_3
\end{aligned}
$$

EXAMPLE ———————————————————————————

A student is instructed to dilute some concentrated $HCl_{(aq)}$ (36%) to make 4.00 L of 10% solution. What volume of hydrochloric acid solution should the student initially take to do this?

$$
\begin{aligned}
v_i c_i &= v_f c_f \\
v_i \times 36\% &= 4.00 \text{ L} \times 10\% \\
v_i &= 1.1 \text{ L} \\
&\ HCl
\end{aligned}
$$

The dilution technique is especially important in manipulating the concentration of a solution. For example, in scientific or technological research, a reaction that proceeds too rapidly or too violently with a concentrated solution may be controlled by lowering the concentration. In the medical and pharmaceutical industries, where low concentrations are common, prescriptions require not only minute quantities, but also extremely precise measurement. Precise dilution can produce solutions of much more accurate concentration compared with dilute solutions prepared directly from the pure solute.

The preparation of standard solutions by dilution requires a means of transferring precise volumes of solution. You know how to use graduated cylinders to measure volumes of solution, but graduated cylinders are not precise enough when working with small volumes. To deliver a very precise, small volume of solution, a laboratory device called a *pipet* is used. A 10 mL *graduated pipet* has graduation marks every tenth of a millilitre. (See Figure C10 of Appendix C, C.2.) This type of pipet can transfer any volume from 0.1 mL to 10.0 mL, and is typically precise to ±0.1 mL. A *volumetric pipet* transfers only one specific volume, but has a very high precision. (See Figure C9 of Appendix C, C.2.) For example, a 10 mL volumetric pipet is designed to transfer 10.00 mL of solution with a precision of ±0.02 mL. Sometimes called a *delivery pipet*, the volumetric pipet is often inscribed with *TD* to indicate that it is calibrated *to deliver* a particular volume with a specified precision. Both kinds of pipet come in a range of sizes and are used with a pipet bulb (Appendix C, C.2).

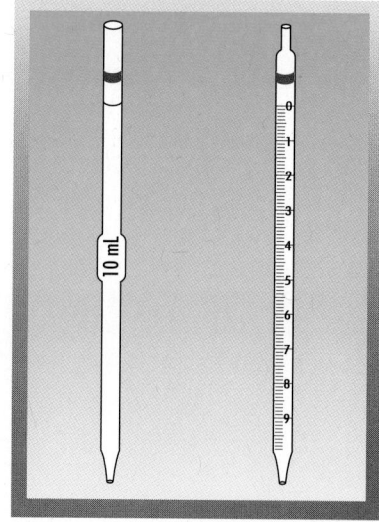

Figure 5.17
A volumetric pipet (a) transfers a fixed volume, and a graduated pipet (b) measures a range of volumes.

INVESTIGATION

5.6 Demonstration: A Standard Solution by Dilution

The purpose of this demonstration is to illustrate the procedure and skills for precisely diluting a stock solution in order to prepare a standard solution (Figure 5.18, page 208). As in Investigation 5.5, no problem is stated since no specific chemical analysis is being done. A typical problem statement would be in the context of a chemical analysis and the preparation of the standard solution would be part of an overall procedure.

Materials

lab apron
safety glasses
0.5000 mol/L $CuSO_{4(aq)}$
150 mL beaker
10 mL volumetric pipet
pipet bulb
wash bottle of pure water
100 mL volumetric flask with stopper
small funnel
medicine dropper
meniscus finder

Procedure

Refer to Appendix C, C.2 and C.3 for information on using a pipet and on preparing standard solutions by dilution.

1. (Pre-lab) Calculate the volume of 0.5000 mol/L stock solution of $CuSO_{4(aq)}$ required to prepare 100.0 mL of 0.05000 mol/L solution.

- Problem
- Prediction
- Design
- Materials
- Procedure
- Evidence
- Analysis
- Evaluation
- Synthesis

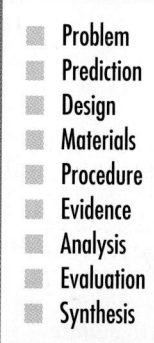

CAUTION
Copper(II) sulfate solution is slightly corrosive.

2. Add 40 mL to 50 mL of pure water to a clean 100 mL volumetric flask.

3. Measure 10.00 mL of the stock solution using a 10 mL volumetric pipet.

4. Transfer the 10.00 mL of solution into the 100 mL volumetric flask (Figure 5.18(a)).

5. Add pure water until the final volume is reached (Figure 5.18(b)).

6. Stopper the flask and mix the solution thoroughly (Figure 5.18(c)).

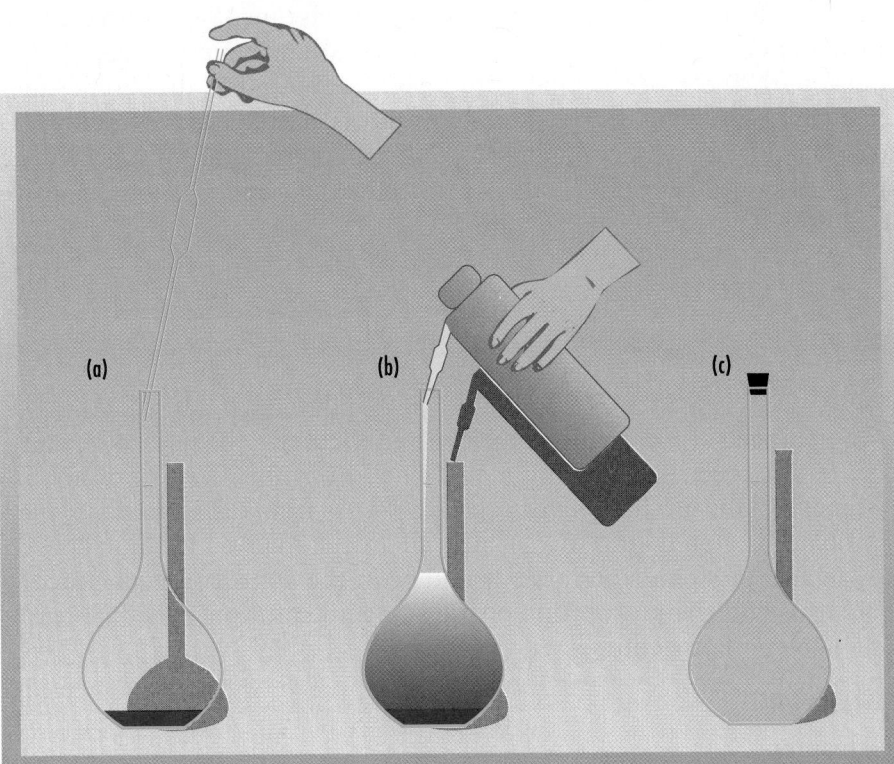

(a) (b) (c)

Figure 5.18
(a) 10.00 mL of $CuSO_{4(aq)}$ is transferred to a volumetric flask.
(b) The initial amount of copper(II) sulfate solute is not changed by adding water to the flask, since water is the solvent and not the solute.
(c) In the final dilute solution, the 25.0 mmol of copper(II) sulfate is still present, but it is distributed throughout a larger volume; in other words, it is diluted.

| Certainty
The certainty of an answer may be communicated with significant digits. The answer is no more certain (i.e., has no more significant digits) than the least certain value. See Appendix E, *Communication Skills*.

Exercise

29. What volume of concentrated 17.8 mol/L sulfuric acid would a laboratory technician need to make 2.00 L of 0.200 mol/L solution by dilution of the original, concentrated solution?

30. A 1.00 L bottle of concentrated acetic acid is diluted to prepare a 0.400 mol/L solution. Find the volume of diluted solution that is prepared. (Refer to the list of concentrated reagents on the inside back cover of this book for the molar concentration of concentrated acetic acid.)

31. A 10.00 mL sample of a test solution is diluted in an environmental laboratory to a final volume of 250.0 mL. The concentration of the diluted solution is found to be 0.274 g/L. What was the concentration of the original test solution?

32. A laboratory technician needs 1.00 L of 0.125 mol/L sulfuric acid solution for a quantitative analysis experiment. A commercial 5.00 mol/L sulfuric acid solution is available from a chemical supply company. Write a complete, specific procedure for preparing the solution (Figure 5.19). Include all necessary calculations.

33. In a study of reaction rates, you need to dilute the copper(II) sulfate solution prepared in Investigation 5.6. You take 5.00 mL of 0.05000 mol/L $CuSO_{4(aq)}$ and dilute this to a final volume of 100.0 mL.
    (a) What is the final concentration of the dilute solution?
    (b) What mass of solute is present in 10.0 mL of the final dilute solution?
    (c) Can this final dilute solution be prepared directly using the pure solid? Defend your answer.

34. A student tries a reaction and finds that the volume of solution that reacts is too small to be measured precisely. She takes a 10.00 mL volume of the solution with a pipet, transfers it into a clean 250 mL volumetric flask, adds pure water to increase the volume to 250.0 mL, and mixes the solution thoroughly.
    (a) Compare the concentration of the dilute solution to the original solution.
    (b) Compare the volume that will react now to the volume that reacted initially.
    (c) Predict the speed or rate of the reaction using the diluted solution compared with the original solution. Explain your answer.

35. (Discussion) For many years the adage "The solution to pollution is dilution" was used by individuals, industries, and governments. They did not realize at that time that chemicals, diluted by water in a river or by air in the atmosphere, could be concentrated in another chemical system later. Identify and describe a system in which pollutants can become concentrated.

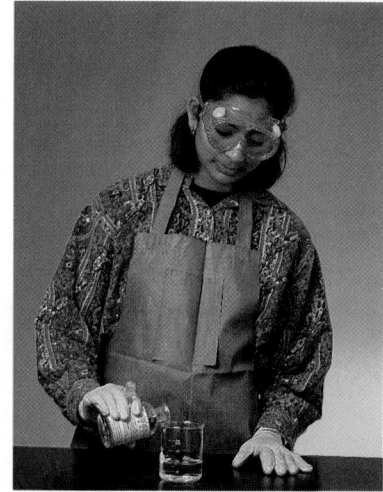

Figure 5.19
When diluting all concentrated reagents, especially acids, always add the concentrated reagent to water. This allows the reagent that is more dense to mix with the solvent, distributing the heat of solution. If water were added to concentrated sulfuric acid, it could form a surface layer and enough heat could be produced between the layers of liquid to boil the water and splatter acid out of the container.

Figure 5.20
Large amounts of salt can be crystallized from ocean water.

Solubility

Suppose you add some pickling salt (pure sodium chloride) to a jar of water and shake the jar until the salt dissolves. What happens if you continue this process? Eventually some solid salt remains at the bottom of the jar, despite your efforts to make it dissolve. A **saturated solution** is a solution at maximum concentration, in which no more solute will dissolve. If the container is sealed, and there are no temperature changes, no further changes will occur in the concentration of the solution or in the quantity of undissolved solute. The **solubility** of a substance is the concentration of a saturated solution of that solute in a particular solvent at a particular temperature; solubility is a specific maximum concentration. For example, the solubility of sodium sulfate in water at 0°C is 4.76 g/100 mL. If more solute is present, it does not dissolve under ordinary conditions. There are several experimental designs that can be used to determine the solubility of a solid; for

Table 5.5

SOLUBILITY OF SODIUM CHLORIDE IN WATER	
Temperature (°C)	Solubility (g/100 mL)
0	31.6
70	33.0
100	33.6

example, **crystallization** involves the removal of the solvent from the solution, leaving a solid (Figure 5.20, page 209).

5.7 The Solubility of Sodium Chloride in Water

The purpose of this investigation is to test the solubility data (Table 5.5, page 209) by using a graph of this data to predict the solubility of sodium chloride in water at a particular temperature. (See graphing notes in Appendix E.) The predicted value is then compared with the value determined experimentally by crystallization from a saturated solution.

Problem

What is the solubility of sodium chloride at room temperature?

Experimental Design

A precisely measured volume of a saturated $NaCl_{(aq)}$ solution at room temperature is heated to crystallize the salt.

Materials

lab apron	10 mL pipet
safety glasses	pipet bulb
face shield	distilled water
saturated $NaCl_{(aq)}$ solution	gas burner or hot plate
centigram balance	matches or striker
thermometer	laboratory stand
100 mL beaker	iron ring
evaporating dish	wire gauze

Procedure

1. Measure and record the mass of an evaporating dish to a precision of 0.01 g.

2. Obtain 40 mL to 50 mL of saturated $NaCl_{(aq)}$ in a 100 mL beaker.

3. Measure and record the temperature of the solution to a precision of 0.2°C.

4. Pipet a 10.00 mL sample of the saturated solution into the evaporating dish. (See Appendix C, C.2 for instructions on using a pipet.)

5. Using a burner, heat the solution until the water boils away, and dry crystalline $NaCl_{(s)}$ remains (Figure 5.21).

6. Allow the evaporating dish to cool.

7. Measure and record the mass of the evaporating dish and its contents.

8. Reheat the evaporating dish and the residue and repeat steps 6 and 7. If the mass remains constant, this confirms that the sample is dry.

■ Problem
✔ Prediction
■ Design
■ Materials
■ Procedure
✔ Evidence
✔ Analysis
✔ Evaluation
■ Synthesis

CAUTION

Always wear safety glasses when handling or heating chemicals. A face shield is advisable, in case the solution splatters. Be careful when handling hot objects.

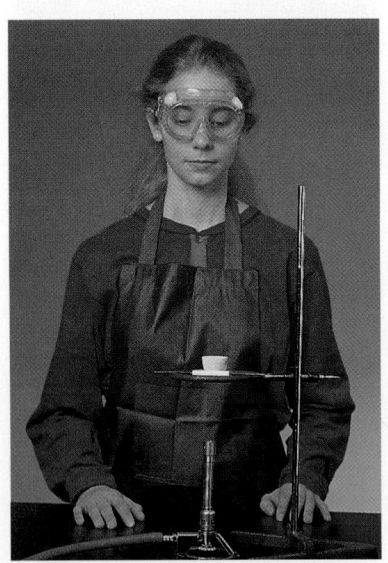

Figure 5.21
A saturated sodium chloride solution is heated to crystallize the salt. (See laboratory burner notes in Appendix C.)

Solubility Rules and Examples

Based on a large number of experiments, several generalizations can be made about the solubility of substances in water.

- Solids usually have higher solubility in water at higher temperatures.
- Gases always have higher solubility in water at lower temperatures.
- Gases always have higher solubility in water at higher pressures.
- Some liquids, such as mineral oil, do not dissolve in water at all, but form a separate layer. Such liquids are said to be **immiscible** with water.
- Some liquids, such as methanol, dissolve in water in any proportion and have no maximum concentration. Such liquids are said to be **miscible** with water.
- Elements generally have low solubility in water, but the halogens and oxygen dissolve sufficiently in water to be important in some solution reactions. (A few elements react with water to form solutions — for example, Group I and II metals.)

There are many common examples of saturated solutions that undergo changes. When water evaporates from a saturated solution, solids crystallize out of the solution. This occurs naturally in the formation of stalactites and stalagmites (Figure 5.22). You might see another example of crystallization in the kitchen if you leave the top off a container of syrup (a sugar solution). When you open a bottle or a can containing a carbonated beverage, you lower the gas pressure inside the container. The solubility of the carbon dioxide decreases and some of the dissolved gas comes out of solution. If you allow a glass of cold tap water to stand for a while, bubbles form on the inside of the glass, because the dissolved air becomes less soluble as the water warms to room temperature.

A solubility table of ionic compounds is best understood by assuming that most substances dissolve in water to some extent. The solubilities of various ionic compounds range from high solubility, like that of table salt, to negligible solubility, like that of silver chloride. The classification of compounds into high and low solubility categories allows you to predict the state of a compound formed in a reaction in aqueous solution. The cutoff point between high and low solubility is arbitrary. A solubility of 0.1 mol/L is commonly used in chemistry as this cutoff point because most ionic compounds have solubilities significantly greater or less than this value, which is a typical concentration for laboratory work. Of course, some compounds with intermediate solubility seem to be exceptions to the rule. Calcium sulfate, for example, has intermediate solubility, but enough of it will dissolve in water that the solution noticeably conducts electricity.

Figure 5.22
Stalactites and stalagmites form in caves when calcium carbonate crystallizes from groundwater solutions.

"The Bends"
When diving underwater using air tanks, a diver breathes air at the same pressure as the surroundings. The increased pressure underwater forces more air to dissolve in the diver's bloodstream. If a diver comes up too quickly, the solubility of air (mostly nitrogen) decreases as the pressure decreases, and nitrogen bubbles form in the blood vessels. These nitrogen bubbles are the cause of a diving danger known as "the bends" (so named because divers typically bend over in agony as they try to relieve the pain). Nitrogen bubbles are especially dangerous if they form in the brain or spinal cord. The bends may be avoided by ascending very slowly or corrected by using a decompression chamber, as shown in the photograph.

Exercise

36. Give examples of two liquids that are immiscible and two that are miscible with water.

37. Can more oxygen dissolve in a litre of water in a cold stream or a litre of water in a warm lake? Include your reasoning.

38. State why you think clothes might be easier to clean in hot water.

39. Why do carbonated beverages go "flat" when opened and left at room temperature and pressure?

Potash (potassium chloride) is used in the production of fertilizers. It is mined in Saskatchewan and New Brunswick as a red ore, which contains a mixture of potassium chloride and sodium chloride, as well as reddish particles of clay (mainly quartz, iron oxides, and calcium sulfides). To separate the valuable potash from the ore, miners rely on solubility. The crushed potash ore is placed in a series of solutions. These solutions contain a blend of potassium chloride and sodium chloride in water, maintained at the saturation point at all times. The solutions also contain two chemicals that act to "collect" the potassium chloride before it can dissolve. One, an 18-carbon amine (see Table 9.1) from beef fat, is hydrophobic (repels water) and is slightly attracted to potassium chloride. It acts to "grab" the potash. The second chemical is a "frother," a substance that causes gas bubbles to form in the solution. These bubbles push the "grabbed" potassium chloride to the surface, where it can be skimmed off. The clay particles settle to the bottom of the solution. The final purification of the potassium chloride involves washing it in fresh water. Any remaining sodium chloride dissolves more quickly than the potassium chloride and can be removed.

Problem 5C Solubility and Temperature

The purpose of this problem is to test the generalization about the effect of temperature on the solubility of an ionic compound. Complete the Prediction, Analysis, and Evaluation of the investigation report.

Problem

How does the temperature of a saturated mixture affect the solubility of potassium nitrate?

Experimental Design

Potassium nitrate is added to four flasks of pure water until no more potassium nitrate will dissolve and there is excess solid in each beaker. Each mixture is sealed and stirred at a different temperature until no further changes occur. The same volume of each solution is removed and evaporated to crystallize the solid. The specific relationship of temperature to the solubility of potassium nitrate is determined by graphical analysis. The temperature is the manipulated variable and the solubility is the responding variable.

Evidence

	SOLUBILITY OF POTASSIUM NITRATE AT VARIOUS TEMPERATURES		
Temperature (°C)	Volume of Solution (mL)	Mass of Empty Beaker (g)	Mass of Beaker Plus Solid (g)
0.0	10.0	92.74	93.99
12.5	10.0	91.75	93.95
23.0	10.0	98.43	101.71
41.5	10.0	93.37	100.15

5.5 CONCENTRATION OF IONS

In solutions of ionic compounds and strong acids, the electrical conductivity suggests the presence of ions in the solution. When these solutes produce aqueous ions, expressing the concentration of individual ions in moles per litre (mol/L) is important. As you have already seen in the discussion of net ionic equations (page 194) and qualitative analysis (page 196), precipitation reactions involve only one ion in a mixture (for example, chloride ions) reacting with another reagent. The molar concentrations of the ions in a solution depend on the relative numbers of ions making up the compound; for example, Cl^- ions in $NaCl_{(aq)}$ and $CaCl_{2(aq)}$.

These dissociation equations for ionic compounds or strong acids allow you to determine the molar concentration of either the ions or the compounds in solution. The ion concentration is always a whole number multiple of the compound concentration. For convenience, square brackets are commonly placed around formulas to indicate the

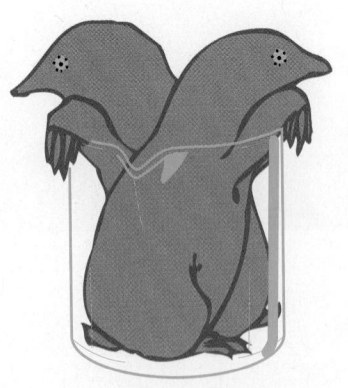
Two moles per litre.

molar concentration of the substance within the brackets. For example, $[NH_{3(aq)}]$, $[SO_4{}^{2-}{}_{(aq)}]$, $[NaOH_{(aq)}]$, and $[H^+{}_{(aq)}]$ indicate the molar concentrations of aqueous ammonia, sulfate ions, sodium hydroxide, and hydrogen ions respectively.

When sodium chloride dissolves in water, each mole of sodium chloride produces one mole of sodium ions and one mole of chloride ions.

$$NaCl_{(aq)} \quad \rightarrow \quad Na^+{}_{(aq)} \quad + \quad Cl^-{}_{(aq)}$$
$$\text{1 mol} \qquad\qquad \text{1 mol} \qquad\qquad \text{1 mol}$$

If $[NaCl_{(aq)}] = 1$ mol/L, then $[Na^+{}_{(aq)}] = 1$ mol/L and $[Cl^-{}_{(aq)}] = 1$ mol/L.

Calcium chloride dissociates in water to produce individual calcium and chloride ions. Each mole of calcium chloride produces one mole of calcium ions and two moles of chloride ions.

$$CaCl_{2(aq)} \quad \rightarrow \quad Ca^{2+}{}_{(aq)} \quad + \quad 2\,Cl^-{}_{(aq)}$$
$$\text{1 mol} \qquad\qquad \text{1 mol} \qquad\qquad \text{2 mol}$$

If $[CaCl_{2(aq)}] = 1$ mol/L, then $[Ca^{2+}{}_{(aq)}] = 1$ mol/L and $[Cl^-{}_{(aq)}] = 2$ mol/L. Notice that the individual ion concentrations can easily be predicted from the concentration of the compound and the subscripts of the ions in the formula of the compound. Even so, it is good practice to write the dissociation equation prior to calculating concentrations. This practice will help you avoid errors and is good preparation for your later study of equilibrium.

EXAMPLE ⎯⎯⎯⎯⎯⎯⎯⎯⎯⎯⎯⎯⎯⎯⎯⎯⎯⎯⎯⎯⎯⎯⎯⎯⎯⎯

What is the molar concentration of aluminum ions and sulfate ions in a 0.40 mol/L solution of $Al_2(SO_4)_{3(aq)}$?

$$Al_2(SO_4)_{3(aq)} \;\rightarrow\; 2\,Al^{3+}{}_{(aq)} + 3\,SO_4{}^{2-}{}_{(aq)}$$

$$[Al^{3+}{}_{(aq)}] = 0.40 \text{ mol/L} \times 2 = 0.80 \text{ mol/L}$$

$$[SO_4{}^{2-}{}_{(aq)}] = 0.40 \text{ mol/L} \times 3 = 1.20 \text{ mol/L}$$

EXAMPLE ⎯⎯⎯⎯⎯⎯⎯⎯⎯⎯⎯⎯⎯⎯⎯⎯⎯⎯⎯⎯⎯⎯⎯⎯⎯⎯

Determine the molar concentration of barium and hydroxide ions in a solution made by dissolving 5.48 g of barium hydroxide to make a volume of 250 mL.

$$Ba(OH)_{2(aq)} \;\rightarrow\; Ba^{2+}{}_{(aq)} + 2OH^-{}_{(aq)}$$

$$n_{Ba(OH)2} = 5.48 \text{ g} \times \frac{1 \text{ mol}}{171.35 \text{ g}} = 0.0320 \text{ mol}$$

$$[Ba(OH)_{2(aq)}] = \frac{0.0320 \text{ mol}}{0.250 \text{ L}} = 0.128 \text{ mol/L}$$

$$[Ba^{2+}{}_{(aq)}] = 0.128 \text{ mol/L} \times 1 = 0.128 \text{ mol/L}$$

$$[OH^-{}_{(aq)}] = 0.128 \text{ mol/L} \times 2 = 0.256 \text{ mol/L}$$

_____

What mass of magnesium bromide must be dissolved to make 1.50 L of solution with a bromide ion concentration of 0.30 mol/L?

$$MgBr_{2(aq)} \rightarrow Mg^{2+}_{(aq)} + 2\,Br^-_{(aq)}$$

$$[MgBr_{2(aq)}] = 0.30\text{ mol/L} \times \frac{1}{2} = 0.15\text{ mol/L}$$

$$n_{MgBr_2} = 1.50\text{ L} \times \frac{0.15\text{ mol}}{1\text{ L}} = 0.23\text{ mol}$$

$$m_{MgBr_2} = 0.23\text{ mol} \times \frac{184.11\text{ g}}{1\text{ mol}} = 41\text{ g}$$

In a chemistry laboratory, an industrial process, or the natural environment, a solution may be diluted by the addition of more solvent or may be mixed with another solution without any chemical reaction occurring. In both cases, the concentrations of the entities present will change. If a simple dilution with more solvent occurs, the usual dilution calculation (page 206) is used to determine the final concentration of each molecule or ion. If two solutions, each containing ions, are mixed, the amount of each ion present is determined from the volume and concentration of the original solution. The final concentration of each ion is then determined by taking the total amount of each ion and dividing by the total volume of the final mixture. This procedure will apply only if no chemical reaction has taken place.

EXAMPLE _____

In a lead waste container, 50 mL of 0.10 mol/L lead(II) nitrate and 40 mL of 0.20 mol/L lead(II) acetate are combined. Assuming no chemical reaction occurs, what is the final concentration of each ion in the final solution?

$$Pb(NO_3)_{2(aq)} \rightarrow Pb^{2+}_{(aq)} + 2\,NO_3^-_{(aq)}$$

$$[Pb(NO_3)_{2(aq)}] = 0.10\text{ mol/L}$$

$$[Pb^{2+}_{(aq)}] = 0.10\text{ mol/L} \times 1 = 0.10\text{ mol/L}$$

$$n_{Pb^{2+}} = 50\text{ mL} \times \frac{0.10\text{ mol}}{1\text{ L}} = 5.0\text{ mmol}$$

$$[NO_3^-_{(aq)}] = 0.10\text{ mol/L} \times 2 = 0.20\text{ mol/L}$$

$$n_{NO_3^-} = 50\text{ mL} \times \frac{0.20\text{ mol}}{1\text{ L}} = 10\text{ mmol}$$

$$Pb(CH_3COO)_{2(aq)} \rightarrow Pb^{2+}_{(aq)} + 2\,CH_3COO^-_{(aq)}$$

$$[Pb(CH_3COO)_{2(aq)}] = 0.20\text{ mol/L}$$

$$[Pb^{2+}_{(aq)}] = 0.20\text{ mol/L} \times 1 = 0.20\text{ mol/L}$$

$$n_{Pb^{2+}} = 40\text{ mL} \times \frac{0.20\text{ mol}}{1\text{ L}} = 8.0\text{ mmol}$$

$$[CH_3COO^-_{(aq)}] = 0.20 \text{ mol/L} \times 2 = 0.40 \text{ mol/L}$$

$$n_{CH_3COO^-} = 40 \text{ mL} \times \frac{0.40 \text{ mol}}{1 \text{ L}} = 16 \text{ mmol}$$

$$[Pb^{2+}_{(aq)}] = \frac{5.0 \text{ mmol} + 8.0 \text{ mmol}}{50 \text{ mL} + 40 \text{ mL}} = \frac{13.0 \text{ mmol}}{90 \text{ mL}} = 0.14 \text{ mol/L}$$

$$[NO_3^-_{(aq)}] = \frac{10 \text{ mmol} + 0 \text{ mmol}}{50 \text{ mL} + 40 \text{ mL}} = \frac{10 \text{ mmol}}{90 \text{ mL}} = 0.11 \text{ mol/L}$$

$$[CH_3COO^-_{(aq)}] = \frac{0 \text{ mmol} + 16 \text{ mmol}}{50 \text{ mL} + 40 \text{ mL}} = \frac{16 \text{ mmol}}{90 \text{ mL}} = 0.18 \text{ mol/L}$$

Exercise

40. Find the concentration of each ion in the following solutions.
    (a) 0.41 mol/L $Na_2S_{(aq)}$
    (b) 1.2 mol/L $Sr(NO_3)_{2(aq)}$
    (c) 0.13 mol/L $(NH_4)_3PO_{4(aq)}$

41. A 250 mL solution is prepared by dissolving 2.01 g of iron(III) chloride in water. What is the concentration of each ion in the solution?

42. In order to prepare for a chemical analysis, a lab technician requires 500 mL of each of the following solutions. Calculate the mass of solid required for each solution.
    (a) $[Cl^-_{(aq)}] = 0.400$ mol/L from $CaCl_{2(s)}$
    (b) $[CO_3^{2-}_{(aq)}] = 0.35$ mol/L from $Na_2CO_3 \cdot 10H_2O_{(s)}$

43. For both economic and environmental reasons, silver ion solutions are often collected for future recycling. In a container, 250 mL of 0.023 mol/L silver nitrate and 105 mL of 0.11 mol/L silver nitrate solutions are mixed. What is the concentration of each ion in the final solution?

44. Assuming no chemical reaction occurs, what is the concentration of each ion in the final solution obtained from each of the following mixtures?
    (a) 1.5 L of 0.60 mol/L $NiCl_{2(aq)}$ and 1.0 L of 0.40 mol/L $KBr_{(aq)}$
    (b) 200 mL of 0.30 mol/L $Al(NO_3)_{3(aq)}$ and 100 mL of 0.20 mol/L $Mg(NO_3)_{2(aq)}$
    (c) 6.20 g of $NaI_{(s)}$ and 8.75 g of $KI_{(s)}$ in a total volume of 500 mL

Communicating Hydrogen Ion Concentration

The molar concentration of hydrogen ions is of critical importance in chemistry. According to Arrhenius's theory, hydrogen ions are responsible for the properties of acids, and the higher the concentration of hydrogen ions, the more *acidic* a solution will be. Similarly, the higher the concentration of hydroxide ions, the more *basic* a solution will be. You might not expect a neutral solution or pure water to contain any hydrogen or hydroxide ions at all. However, careful testing yields evidence that water always contains tiny amounts of both hydrogen and hydroxide ions, due to slight ionization. In a sample of pure water, about two of every billion molecules have ionized to form hydrogen and hydroxide ions.

$$H_2O_{(l)} \rightarrow H^+_{(aq)} + OH^-_{(aq)}$$

In pure water at SATP, the hydrogen ion concentration is very low, about 1×10^{-7} mol/L. This value is often negligible; for example, a conductivity test will show no conductivity for pure water unless the equipment is extremely sensitive.

Aqueous solutions exhibit a phenomenally wide range of hydrogen ion concentrations — from more than 10 mol/L for a concentrated hydrochloric acid solution, to less than 10^{-15} mol/L for a concentrated sodium hydroxide solution. Any aqueous solution can be classified as acidic, neutral, or basic, using a scale based on the hydrogen ion concentration.

- In a neutral solution, $[H^+_{(aq)}]$ is equal to 1×10^{-7} mol/L.
- In an acidic solution, $[H^+_{(aq)}]$ is greater than 1×10^{-7} mol/L.
- In a basic solution, $[H^+_{(aq)}]$ is less than 1×10^{-7} mol/L.

The extremely wide range of hydrogen ion concentration led to a convenient shorthand method of communicating these concentrations. In 1909, Danish chemist Sören Sörenson introduced the term pH or "power of hydrogen." The **pH** of a solution is defined as the negative of the exponent to the base ten of the hydrogen ion concentration (expressed as moles per litre). For example, a concentration of 10^{-7} mol/L has a pH of 7 (neutral), and a pH of 2 corresponds to a hydrogen ion concentration of 10^{-2} mol/L (acidic). The pH is specified on the labels of consumer products such as shampoos; in water-quality tests for pools and aquariums; in environmental studies of acid rain; and in laboratory investigations of acids and bases. Since each pH unit corresponds to a factor of 10 in the concentration, the huge $[H^+_{(aq)}]$ range can now be communicated by a much simpler set of positive numbers (Figure 5.23). A *neutralization* reaction is a reaction between an acid and a base. In a neutralization reaction, the pH changes, becoming closer to 7.

Figure 5.23
The pH scale can communicate a broad range of hydrogen ion concentrations, in a wide variety of substances.

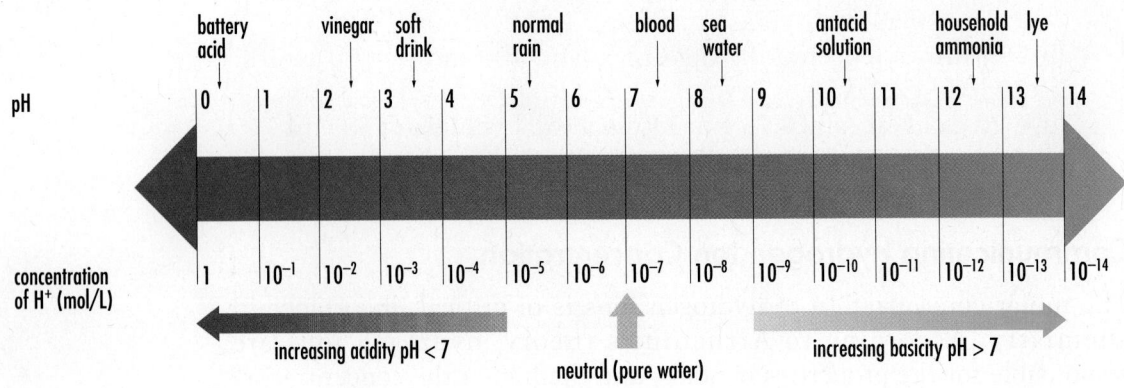

45. Write a dissociation equation to explain the electrical conductivity of each of the following chemicals.
    (a) potash: potassium chloride
    (b) Glauber's salt: sodium sulfate
    (c) TSP: sodium phosphate

46. What is the molar concentration of the cation and the anion in a 0.14 mol/L solution of each of the following chemicals?
    (a) saltpeter: KNO_3
    (b) road salt: calcium chloride
    (c) fertilizer: ammonium phosphate

BIOMAGNIFICATION

Many of us think that mammals are the dominant life form on Earth, but in many ways this is not true. Insects far outnumber all other forms of animal life, and fossil evidence shows that they have been around much longer than mammals. Today, one-third of all food grown or stored for human use is consumed by insects, and half of all human deaths and deformities are due to diseases traceable to insects. It's no wonder that insecticides have been considered important for human health and crop protection and thus have been widely used.

One of the best-known insecticides is DDT (dichlorodiphenyltrichloro-ethane). The

insecticidal properties of this compound were discovered in 1939, and over the next twenty years it was widely used in North America. However, the serious side effects of DDT have resulted in restrictions on its use. DDT accumulates in animals and humans because its solubility in fat (100 000 ppm) is so much higher than its solubility in water (0.0012 ppm). This means that animals and humans absorb DDT in their fatty tissues rather than in their blood, from which it could be eliminated.

The most publicized, although not necessarily the most serious, effect of DDT is associated with declines in the reproductive success of predatory bird species. Eggs produced by mature females are thin-shelled and have high concentrations of DDT residues. In British Columbia, the peregrine falcon population was severely affected by DDT use in Okanagan orchards and the Fraser Valley. Since the ban of DDT, however, the falcon has returned to the Fraser Valley. In the Okanagan, the DDT contamination in the soils is still too high to support a return of these birds. (The slower rate of DDT breakdown in the Okanagan soils is attributed to the type of earthworms present.)

The predatory birds accumulate harmful levels of DDT as a result of *biomagnification*. This is the increase in concentration of a chemical from

its background level in the environment as it moves through a food chain (in other words, the chemical becomes most concentrated in predatory animals). The table shows the evidence gathered from one research study on increased concentration of DDT.

BIOMAGNIFICATION OF DDT	
Location of DDT	Concentration (ppm)
water in Lake Michigan	0.000 002
amphipods (tiny crustaceans)	0.410
fish	3 – 6
gulls	99

Biomagnification of DDT occurs because of the differences in solubility of DDT in fat and water. Problems caused by such phenomena highlight the need for international cooperation. For example, DDT use is banned in Canada and in the United States, but not in Mexico, where many peregrine falcons spend the winter.

- *Why does DDT accumulate in humans and other animals?*
- *What is biomagnification?*

47. Measurements of pH can provide a quick estimate of the hydrogen ion concentration in an aqueous solution. What is the hydrogen ion concentration in the following solutions?
    (a) pure water: pH = 7
    (b) household ammonia: pH = 11
    (c) vinegar: pH = 2
    (d) soda pop: pH = 4
    (e) drain cleaner: pH = 14

48. Hydrogen ion concentration is a theoretical concept used to explain the behavior of acids. Express the following concentrations as pH values.
    (a) grapefruit juice: $[H^+_{(aq)}] = 10^{-3}$ mol/L
    (b) rainwater: $[H^+_{(aq)}] = 10^{-5}$ mol/L
    (c) milk: $[H^+_{(aq)}] = 10^{-7}$ mol/L
    (d) soap: $[H^+_{(aq)}] = 10^{-10}$ mol/L

49. If a water sample test shows a pH of 5, by what factor would the hydrogen ion concentration have to be decreased to neutralize the sample?

50. What amount of hydrogen ions, in moles, is present in 100 L of the following solutions? Do what the units tell you to do.
    (a) wine: $[H^+_{(aq)}] = 1 \times 10^{-3}$ mol/L
    (b) sea water: pH = 8.0
    (c) stomach acid: $[HCl_{(aq)}] = 0.05$ mol/L

51. (Discussion) Many chemicals that are potentially toxic or harmful to the environment have maximum permissible concentration levels set by government legislation.
    (a) If the chemical is dangerous, should the limit be zero?
    (b) Is a zero level theoretically possible?
    (c) Is a zero level empirically measurable?
    (d) If a non-zero limit is set, how do you think this limit should be determined?

52. (Discussion) What are some benefits and risks of using acidic and basic substances in your home? Provide some examples where you consider the benefits to exceed the risks and some where you consider the risks to exceed the benefits.

Problem 5D Qualitative Analysis

Complete the Analysis of the investigation report and evaluate the experimental design.

Problem

Which of the 0.1 mol/L solutions labelled 1, 2, 3, 4, and 5 is $KCl_{(aq)}$, $CaBr_{2(aq)}$, $HCl_{(aq)}$, $CH_3OH_{(aq)}$, and $NaOH_{(aq)}$?

Experimental Design

The solutions are prepared so that they all have the same concentration and temperature. A sample of each solution is observed to determine its color and then tested for pH, conductivity, and effect on litmus paper. Each solution is tested in an identical way.

Evidence

	PROPERTIES OF THE SOLUTIONS TESTED				
Solution	Effect on Red Litmus Paper	Effect on Blue Litmus Paper	Color	Conductivity	pH
1	changed to blue	no change	none	high	13
2	no change	no change	none	high	7
3	no change	no change	none	very high	7
4	no change	no change	none	none	7
5	no change	changed to red	none	high	1

HOUSEHOLD CHEMICAL SOLUTIONS

An amazing number of solutions are available for household use at your local drugstore, hardware store, and supermarket. There are so many that you may feel you're encountering a bewildering array of names, instructions, warnings, and concentration labels.

For several reasons, knowledge of chemistry can be advantageous when you make decisions about buying and using these products. For example, reading the information on household product labels is important for safety. Hazard symbols and safety warnings on labels are pointless if they go unnoticed, or are not understood. Every year people are injured because they are unaware that bleach (sodium hypochlorite solution) should never be mixed with acids such as vinegar. Although both solutions are effective cleaners for certain stains, when they are combined they react to produce the highly toxic gas, chlorine. Trying to use both at once — for example, in cleaning a toilet — has been known to transform a bathroom into a death trap. Also, bleaches should not be used with products containing liquid ammonia. The gases given off may cause eye irritation, coughing, and nausea.

Another household chemical, isopropyl alcohol, is sold in its pure form as a disinfectant, and in 70% concentration as rubbing alcohol.

The label on each bottle states that the solution should not be taken internally. This is not a casual comment; the alcohol is toxic when swallowed, and cannot be used as a substitute for the ethyl alcohol used in alcoholic beverages.

Many products are sold under common or classical names. As you have learned previously, hydrochloric acid is sometimes sold as muriatic acid. Sodium hydroxide, called *lye* as a pure solid, has a variety of names when sold as a concentrated solution for use in cleaning plugged drains. Generic or "no-name" products often contain the same kind and quantity of active ingredients as brand name products. You can save time, trouble, and money by knowing that, in most cases, the chemical names of compounds used for home products must be given on the label. If you discover that your favorite brand of rust remover is an acetic acid solution, you can substitute vinegar to do the same job less expensively.

Concentration is another factor to consider when buying solutions. Hydrogen peroxide disinfectant is sold as aqueous solutions with concentrations of 3% or 6%. Vinegar is packaged as 5% or 7% acetic acid (by volume).

Cough syrups contain widely varying concentrations of medication. Consumers expect the purchase price of a product to reflect the quantity of active ingredients the solution contains, but that is often not the case.

As you can see, a little knowledge applied to your purchase and use of solutions can make a real difference, in terms of economy, efficiency, and personal safety.

- *How can a knowledge of chemistry be useful when buying or using consumer products?*
- *State two examples of a pair of consumer products that should not be mixed together.*

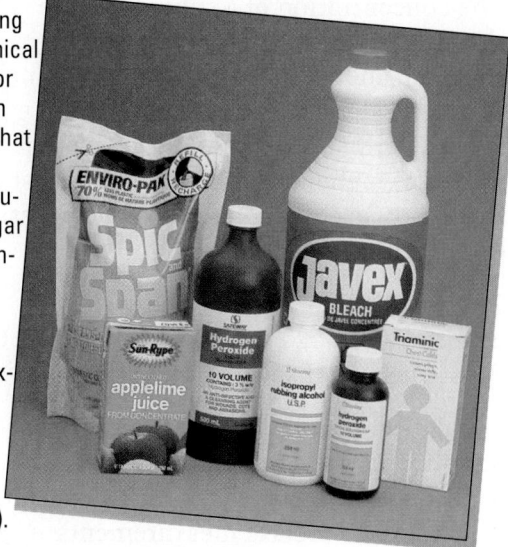

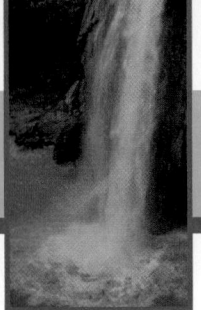

OVERVIEW

Solutions

Summary

- Solutions are homogeneous mixtures of a solute and a solvent. Based on solution properties, solutes may be classified as electrolytes or non-electrolytes, and also as acids, bases, or neutral substances.

- According to Arrhenius's theory, non-electrolytes dissolve in solution into separate, electrically neutral molecules, and electrolytes dissociate to produce positive and negative ions. In this chapter, acids are defined as hydrogen compounds that ionize to produce hydrogen ions when dissolved, and bases are defined as ionic hydroxides that dissociate to produce hydroxide ions when dissolved.

- Solutions are especially important in chemistry because the dissolving of a reactant is often necessary for reactions to occur.

- Precipitation reactions, color of solutions, and flame tests are used as qualitative tests for specific ions in solution.

- The concentration of a solution is the ratio of the quantity of solute to the quantity of the solution, and is the basis of a number of solution calculations, including ion concentration. Molar concentration, or molarity, is the way that chemists most frequently use to express the concentration of a solution.

- The solubility of a substance is the concentration of a saturated solution of that substance under specific conditions.

- Standard solutions may be prepared by dissolving a solid or by diluting a stock solution. Volumetric flasks and pipets are used to make precise measurements.

- The pH of a solution is the negative of the exponent to the base ten of the hydrogen ion concentration and describes the acidity of a solution.

Key Words

acid
base
concentration
crystallization
dilution
dissociation
electrolyte
immiscible
ionization
miscible
molar
molar concentration
molarity
net ionic equation
non-electrolyte
pH
qualitative analysis
quantitative analysis
saturated solution
solubility
solution
spectator ion
standard solution
stock solution

Review

1. List two diagnostic tests used to classify solutes in aqueous solutions.

2. Write the chemical name for the solute and the solvent in each of the following solutions.
   (a) $CaCl_{2(aq)}$ (b) $NH_{3(aq)}$

3. What classes of compounds are
   (a) electrolytes?
   (b) non-electrolytes?

4. What two properties of solutions did Arrhenius study to develop his theory of dissociation of electrolytes?

5. Which classes of compounds dissociate and which ionize?

6. How can you predict the solubility of ionic and molecular compounds?

7. According to Arrhenius's theory, which ions are responsible for the acidic and the basic properties of a solution?

8. Why are solutions important in the study of chemistry?

9. State four examples of solutions used by consumers and describe how each one is used.

10. How is a net ionic equation different from a non-ionic equation?

11. What are some advantages of writing a chemical equation in the net ionic form?

12. What are two main types of chemical analysis, and how are they different?

13. State three examples of qualitative tests for specific ions in solution.

14. Distinguish between the terms "dilute" and "concentrated."

15. Give one example in which a high concentration of a solute is beneficial and one example in which it is harmful.

16. What do all concentration units have in common?

17. What quantity (or conversion factor) is used to convert between the following?
    (a) mass and amount of a substance
    (b) volume of solution and amount of dissolved substance

18. Convert the following.
    (a) 10.0 g of sodium hydroxide into an amount in moles
    (b) 0.35 mol of copper(II) nitrate into a mass
    (c) 2.65 g equivalent to 0.109 mol into the molar mass of a substance

19. What is a standard solution, and why is such a solution necessary?

20. What are two methods used to prepare standard solutions?

21. What specific volumetric equipment is required to
    (a) contain the solution in the final steps of preparing a standard solution?
    (b) deliver precisely 7.8 mL of a solution in a dilution procedure?
    (c) deliver precisely 10.00 mL of a sample to be analyzed in a titration?

22. What are the generalizations for the change in solubility of most solids and gases in water as the temperature of the solution drops?

23. Cooking oil and water are immiscible. What does this mean?

24. On what two factors does an ion concentration depend?

25. What is the pH of a solution for each of the following hydrogen ion concentrations?
    (a) 10^{-15} mol/L
    (b) 10^{-4} mol/L
    (c) 10 mol/L

Applications

26. Describe a diagnostic test or a simple procedure that would distinguish between the following.
    (a) a solution of an ionic compound and a solution of a molecular compound
    (b) a solution of an acid and a solution of a base
    (c) a solution of a molecular compound and pure water

27. Write a balanced equation for each of the following pure substances dissolving in water.
    (a) solid strontium hydroxide
    (b) solid potassium phosphate
    (c) hydrogen bromide gas
    (d) solid magnesium acetate

28. Using the theory of Arrhenius, list all of the entities (atoms, ions, or molecules) believed to be present when each of the following substances is present in water.
    (a) calcium chloride
    (b) ethanol
    (c) ammonium carbonate
    (d) copper
    (e) lead(II) hydroxide
    (f) hydrogen sulfate
    (g) aluminum sulfate
    (h) sulfur

29. Write the net ionic equation for each of the following reactions. All solutions are assumed to be at least 0.10 mol/L in concentration.
    (a) sodium hydroxide and cobalt(II) chloride solutions
    (b) silver nitrate and calcium iodide solutions
    (c) silver nitrate solution and zinc metal
    (d) hydrochloric acid and solid calcium hydroxide
    (e) the precipitation of aluminum hydroxide in a qualitative analysis

30. Design an experiment to identify six solutions, where each solution contains one of the following aqueous ions: sodium, lithium, calcium, nickel(II), copper(II), and iron(III). Include a table showing what evidence is expected.

31. A solution is suspected to contain chloride ions and sulfide ions. Assuming that the solution does not contain any interfering ions, design an experiment to test for the presence of these two ions.

32. An unknown solution conducts electricity, turns red litmus paper blue, and forms a precipitate when sodium sulfate solution is added. What is one possible chemical formula for the solute present in the original solution?

33. A brine (sodium chloride) solution is prepared by dissolving 3.13 g of sodium chloride to make 20.0 mL of solution. What is the concentration of this solution expressed in g/100 mL?

34. Convert the following.
    (a) 0.35 mol of $NaCl_{(aq)}$ in 1.5 L of solution into a molar concentration
    (b) 25 mL of 0.80 mol/L $Mg(NO_3)_{2(aq)}$ into an amount in moles
    (c) 0.246 mol of $NH_{3(aq)}$ in a 2.40 mol/L solution into a volume of solution
    (d) 25.00 g of $CuCl_{2(s)}$ in 1.20 L of solution into a molar concentration
    (e) 50.0 mL of 0.228 mol/L $Na_2CO_{3(aq)}$ into a mass of $Na_2CO_{3(s)}$

35. The label on a bottle of sparkling spring water lists 440 mg/L of total dissolved minerals. What mass of minerals is present in a 1.50 L bottle of the spring water?

36. Standard solutions of sodium oxalate are used in a variety of chemical analyses. What mass of sodium oxalate is required to prepare 250.0 mL of a 0.375 mol/L solution?

37. Standard solutions of potassium hydrogen tartrate, $KHC_4H_4O_6$, are used in chemical analyses to determine the concentration of solutions of bases such as sodium hydroxide.
    (a) Calculate the mass of potassium hydrogen tartrate that is measured to prepare 100.0 mL of a 0.150 mol/L standard solution.
    (b) Write a complete procedure for the preparation of this standard solution, including specific quantities and equipment.

38. Calculate the volume of concentrated phosphoric acid (14.6 mol/L) that must be diluted to prepare 500 mL of a 1.25 mol/L solution.

39. It is desirable in chemical analyses to dilute a stock solution to produce a standard solution.
    (a) What volume of a 0.400 mol/L stock solution of potassium dichromate is required to produce 100.0 mL of a 0.100 mol/L solution?
    (b) Write a complete procedure for the preparation of this standard solution, including specific quantities and equipment.

40. What initial volume of concentrated laboratory hydrochloric acid should be diluted to prepare 5.00 L of 0.125 mol/L solution for an experiment?

41. If water is added to a 25.0 mL sample of 2.70 g/L $NaOH_{(aq)}$ until the volume becomes 4.00 L, find the concentration of the final solution.

42. A 25.0 mL sample of a saturated potassium chlorate solution is evaporated to form 2.16 g of crystals. What is the solubility of potassium chlorate in g/100 mL?

43. The solubility of oxygen gas in water at 25°C is 42 mg/L. What mass of oxygen is dissolved in the water in a 25.0 L home aquarium, assuming a saturated solution at 25°C?

44. Consider the following experimental problem: "What changes occur when a

saturated solution of sodium carbonate at room temperature is cooled in an ice bath?" Write a prediction using the format, "According to...."

45. Write dissociation equations and calculate the molar concentration of the cations and the anions in each of the following solutions.
    (a) 2.24 mol/L $Na_2S_{(aq)}$
    (b) 0.44 mol/L $Fe(NO_3)_{2(aq)}$
    (c) 0.175 mol/L $K_3PO_{4(aq)}$
    (d) 8.75 g of cobalt(III) sulfate in 0.500 L of solution

46. Calculate the mass of the solid required to obtain the ion concentration specified.
    (a) $Zn(NO_3)_{2(s)}$ to obtain 175 mL of 0.100 mol/L $NO_3^-{}_{(aq)}$
    (b) $Na_3PO_{4(s)}$ to obtain 0.65 L of 1.2 mol/L $Na^+{}_{(aq)}$

47. Determine the concentration of each ion in the final solutions listed below.
    (a) 75 mL of 0.50 mol/L $CuSO_{4(aq)}$ and 50 mL of 0.40 mol/L $Cu(NO_3)_{2(aq)}$
    (b) 28.6 g of $AlCl_{3(s)}$ and 15.5 g of $CaCl_{2(s)}$ in 1.50 L of solution

48. The average Canadian uses 200 L of water per day. The carbonate ion concentration in a sample of well water is tested and found to be 225 ppm. Determine the mass of carbonate ions that is present in 200 L of well water.

49. Sea water contains 4×10^{-6} ppm of dissolved gold. What volume of sea water contains 1 g of gold?

Extensions

50. If a solvent evaporates from a saturated solution, some solute usually crystallizes, especially if one or more crystals are already present. This fact can be utilized to grow large, single crystals of a solid. A small, single crystal is hung from a thread in a saturated solution, and the solvent is allowed to slowly evaporate. Using this design, try growing large crystals of salt, alum, and/or bluestone. Paint some lacquer on your final product to protect it.

51. The solubility of a gas always decreases as the temperature increases; for example, the solubility of carbon dioxide in water is 0.335 g/100 mL at 0°C and 0.169 g/100 mL at 20°C. You may have noticed the greater "fizzing" of warm soft drinks versus cold ones. However, a cold ice cube placed in a warm drink often releases gas bubbles. Suggest a hypothesis for the release of gas by the cold ice cube. Write an experimental design in terms of specific variables and test your hypothesis.

52. Prepare to debate the positive or negative side of the resolution, "Chemicals are a benefit to our society." Be ready to make a five-minute oral presentation from a variety of perspectives.

Problem 5E Cation Analysis

Complete the Analysis of the investigation report. Draw a flowchart to accompany your analysis and, if necessary, consult a reference book such as *The CRC Handbook of Chemistry and Physics* or *The Merck Index*. In the Evaluation section of your report, suggest an additional procedure step to identify the fifth cation present in the solution.

Problem

What four of five cations are present in the solution sample provided?

Experimental Design

A design involving diagnostic tests of solution color, flame color, and solubility is employed to identify four of the five cations present. A series of tests is done on one sample. Whenever a precipitate forms, the mixture is filtered and further tests are carried out on the filtrate.

Evidence

Diagnostic Test	Observation
solution color	green
litmus test	blue to red
$KCH_3COO_{(aq)}$ added	white precipitate
$KCl_{(aq)}$ added to green filtrate	white precipitate
$KOH_{(aq)}$ added to colorless filtrate	no precipitate
$K_2SO_{4(aq)}$ added	white precipitate
flame test on filtrate	yellow flame

Gases

The photograph shows a dramatic example of how a gas can save human lives. In a car crash, an airbag can protect a driver from serious injury. Upon collision, the bag inflates automatically, activated by sensors in the steering column and in the bumper. After cushioning the impact, it gradually deflates as gas escapes through the permeable bag. Following a trip to the automobile body shop rather than to the hospital, the driver can have the airbag mechanism recharged and the triggering devices reset.

Gases play other major roles in the operation of automobiles. For example, tires and shock absorbers are inflated with pressurized air to provide a safe and comfortable ride. Air enters through the car's vents to keep passengers cool in summer and warm in winter. Inside the combustion cylinders of the engine, a gasoline and oxygen explosion produces a large amount of gas at high temperature. This moves a piston, converting chemical energy into motion. The gases emitted by automobile exhausts, such as carbon and nitrogen oxides, diffuse into the atmosphere as pollutants.

Gases play a large part in both technology and in our natural environment. You can sometimes hear and feel gases — for example, when a strong wind blows. You may detect the odor or the taste of some gases. Because many gases are invisible, their study requires some imagination.

Some gases such as fluorine, chlorine, and oxygen are very reactive, others such as nitrogen are slightly reactive, and others such as the noble gases are extremely inert. Although they have different chemical properties, gases have remarkably similar physical properties.

- Gases always fill their containers. They have neither a shape nor a volume of their own. (Recall that liquids have a volume of their own but no particular shape; solids have both a shape and a volume of their own.)

- Gases are highly *compressible*. Unlike the volumes of liquids and solids, gas volumes become significantly smaller when the pressure on a sample is increased, and significantly larger when the pressure is reduced. Actions such as pumping air into a bicycle tire or spraying an aerosol depend on this property.

- Gases *diffuse*, or move spontaneously throughout any available space. The fragrance of an air freshener or of perfume lingering in the air is a common example of this property.

- Temperature affects either the volume or the pressure of a gas, or both. Experiments show that if a gas sample is free to expand, its volume increases as its temperature increases. For example, when the air in a hot air balloon is heated, the volume of the air increases, and some of the gas is expelled. Since the density of hot air left in the balloon is less than that of the colder air outside, the balloon rises. The temperature-pressure relationship explains why throwing an aerosol can into a fire is so hazardous — heat causes the pressure of the contained gases to increase so much that the can may explode.

A **gas** is empirically defined as a substance that fills and assumes the shape of its container, diffuses rapidly, and mixes readily with other gases. Gases increase in volume, pressure, or both when heated, and decrease in volume when pressure is applied.

Pressure and Volume: Boyle's Law

Pressure is force per unit area. As you stand on the ground, gravity exerts a downward force on you. Whatever you have on your feet, the resulting force that you exert on the ground is the same. However, when you wear snowshoes, for example, the force is distributed over a larger area, so you exert less pressure on the ground directly below your feet. Wearing snowshoes, you can walk over snow instead of sinking into it.

Scientists have agreed, internationally, on units, symbols, and standard values for pressure. The SI unit for pressure is the *kilopascal* (kPa), which represents a force of 1000 N (newtons) on an area of 1 m²; 1 kPa = 1000 N/m² or 1 kPa = 1 kN/m². According to kinetic molecular theory, gases exert pressure due to the forces exerted by gas particles colliding with objects in their path. *Atmospheric pressure* is the pressure exerted by the air. At sea level, average atmospheric pressure is about 101 kPa. Scientists used this value as a basis to define *one*

In 1643, Evangelista Torricelli (1608 – 1647), following up on a suggestion from Galileo, accidentally invented a way of measuring atmospheric pressure. He was investigating Aristotle's notion that nature abhors a vacuum. His experimental design involved inverting a glass tube filled with mercury and placing it in a tub of mercury. Noticing that the mercury level changed from day to day, he realized that his device, which came to be called a *mercury barometer*, was a means of measuring atmospheric pressure. In Torricelli's honor, standard pressure was at one time defined as 760 mm Hg, or 760 Torr. (Mercury vapor is toxic; in modern mercury barometers, a thin film of water or oil is added to prevent the evaporation of mercury.)

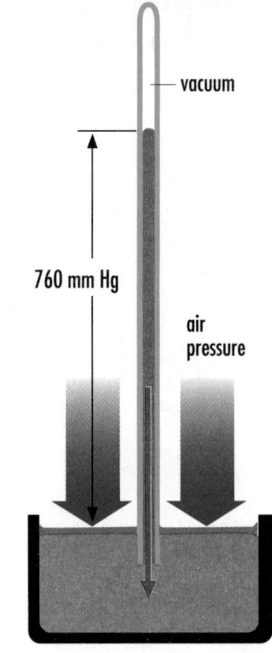

standard atmosphere (1 atm), or standard pressure, as exactly 101.325 kPa. For more convenience, standard ambient pressure has been more recently defined as exactly 100 kPa.

Standard conditions for work with gases are a temperature of 0°C and a pressure of 1 atm (101.325 kPa); these conditions are known as standard temperature and pressure (**STP**). Standard ambient temperature and pressure (**SATP**) are defined as 25°C and 100 kPa.

	Problem
	Prediction
✔	Design
	Materials
	Procedure
✔	Evidence
✔	Analysis
	Evaluation
	Synthesis

Enrichment Analysis

Instead of using the number of books or masses added as the unit of measurement to represent the pressure, calculate the pressure in SI units of kilopascals (kN/m²). To do this, multiply the mass in kilograms by 9.81 N/kg and divide by the cross-sectional area of the cylinder in square metres. The number you get will be in pascals, which can easily be converted to kilopascals. The total pressure on the sample of air can now be found by adding your calculated pressure to the uncorrected air pressure in your location.

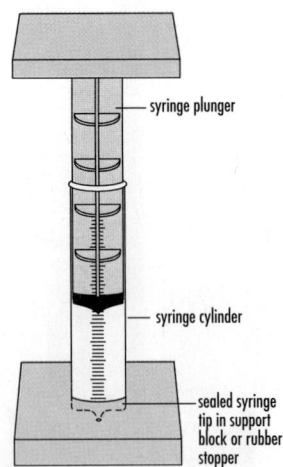

Figure 6.1
Set-up of Boyle's law apparatus for Investigation 6.1.

INVESTIGATION

6.1 Pressure and Volume of a Gas

The purpose of this investigation is to determine the general relationship between the pressure and volume of a gas. Include a graph and a word statement describing the relationship as part of your analysis. (For more accurate results, it will be necessary to correct the measured pressure for the vapor pressure of water. See Table 6.3 on page 245.)

Problem

What effect does increasing the pressure have on the volume of a gas?

Materials

Boyle's law apparatus or
35 mL plastic syringe
large rubber stopper
cork borer
5 textbooks or equal masses (1 kg)

Procedure

1. Pull out the syringe plunger so that 30 mL of air is inside the cylinder.

2. If a syringe cap is not provided, bore a small hole deep enough in the rubber stopper so that the tip of the syringe will be inside the stopper. This should be a tight fit. Make sure the tip of the syringe does not leak.

3. Hold the syringe barrel vertical and measure the initial volume.

4. While holding the syringe securely, carefully place one textbook or mass on the end of the plunger (Figure 6.1). (Your partner should balance the mass and be prepared to catch it if it starts to tilt.)

5. Record the mass and new volume of air.

6. Repeat steps 4 and 5 for a total of 4 or 5 books or masses.

Analysis of the evidence from Investigation 6.1 suggests an inverse variation between the pressure and the volume of a gas; that is, as the pressure increases, the volume decreases (Figure 6.2). To determine the mathematical nature of this relationship requires a further analysis of the pressure and volume evidence. Using the evidence given in SI units from Table 6.1, you can see that when the pressure is doubled (100 kPa

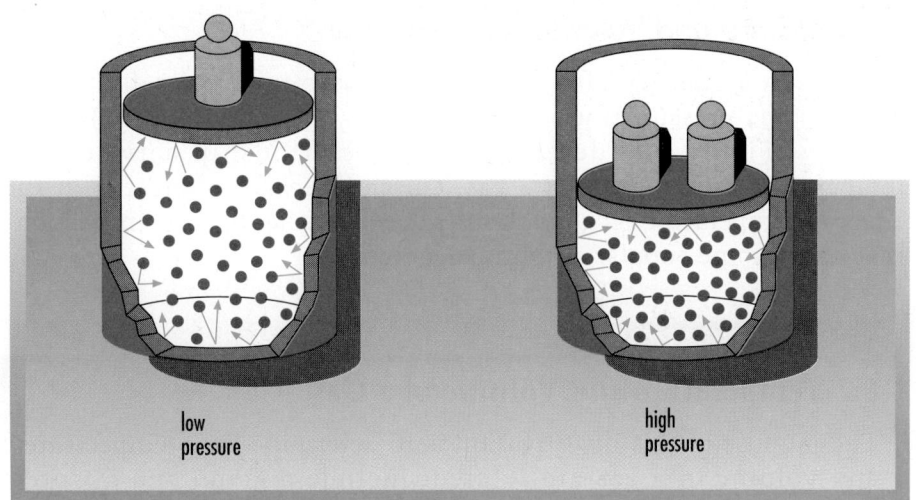

low pressure

high pressure

to 200 kPa), the volume is halved (3.00 L to 1.50 L). If the pressure is tripled, the volume is reduced to one-third. Check the other values to see similar results.

If p_1 and v_1 represent the initial conditions, then some of the other conditions may be stated as follows.

$$(p_1, v_1) \quad (2p_1, \frac{1}{2}v_1) \quad (3p_1, \frac{1}{3}v_1) \quad (4p_1, \frac{1}{4}v_1)$$

For all conditions listed above, note that the product of the pressure and volume is a constant, p_1v_1 or mathematically, the relationship is represented as $pv = k$.

This simple relationship was first determined by Robert Boyle in 1662 (Figure 6.3). **Boyle's law** states that *as the pressure on a gas increases, the volume of the gas decreases proportionally, provided that the temperature and amount of gas remain constant.* Boyle's law can be conveniently written comparing any two sets of pressure and volume measurements.

$$p_1v_1 = p_2v_2 \quad (Boyle's\ law)$$

Table 6.1

PRESSURE AND VOLUME OF A GAS SAMPLE	
Pressure (kPa)	Volume (L)
100	3.00
200	1.50
300	1.00
400	0.75
500	0.60

Exercise

1. Give some examples of gases that are useful and some that are harmful.

2. What are three ways of reducing the volume of a gas in a shock absorber (cylinder and piston) of an automobile?

3. Define atmospheric pressure. Why does atmospheric pressure depend on your location or vary over time at your location?

4. What is the value of the average atmospheric pressure in SI units at sea level? What is this value in three other, non-SI units?

5. A bicycle pump contains 0.65 L of air at 101 kPa. If the tube is closed, what pressure is required to change the volume to 0.25 L?

6. A weather balloon containing 35 L of helium at 98 kPa is released and rises through the atmosphere. Assuming the temperature is constant, what is the volume of the balloon when the atmospheric pressure is 25 kPa?

Figure 6.3
English chemist Robert Boyle (1627 – 1691) determined the effect of pressure on the volume of a gas in quantitative terms: "We have shown that the strengths required to compress air are in reciprocal proportions, or thereabouts, to the spaces comprehending the same portion of the air."

Temperature and Volume: Charles' Law

More than a century after Boyle had determined the relationship between the pressure and volume of a gas, French physicist Jacques Charles (Figure 6.4) determined the relationship between the temperature and volume of a gas. Charles became interested in the effect of temperature on gas volume after observing the hot air balloons that had become popular as flying machines.

Figure 6.4
Jacques Charles (1746 – 1823) designed and flew the first hydrogen balloon in 1783. Applying Archimedes' concept of buoyancy, Henry Cavendish's calculations for the density of hydrogen, and his own observations, he invented the hydrogen balloon. Later, his experiences and experiments led to the formulation of Charles' law.

▢	Problem
▢	Prediction
▢	Design
▢	Materials
▢	Procedure
✔	Evidence
✔	Analysis
✔	Evaluation
▢	Synthesis

 **CAUTION**

Heat the water slowly and ensure that the tested gas in the syringe does not eject the syringe plunger.

INVESTIGATION

6.2 Temperature and Volume of a Gas

The purpose of this investigation is to determine how temperature and volume of a gas are related. Include a graph and a word statement describing the relationship as part of your analysis. In your evaluation, pay particular attention to the sources of experimental uncertainties.

Problem

What effect does increasing the temperature have on the volume of a gas?

Experimental Design

A volume of air is sealed inside a syringe, which is then placed in a water bath. As the temperature of the water is manipulated, the volume of air is measured as the responding variable. The amount of gas inside the syringe and the pressure on the gas are two controlled variables.

Materials

lab apron
safety glasses
plastic syringe (35–60 mL)
cap or stopper for the syringe tip
buret clamp
thermometer and clamp
600 mL beaker
ring stand
wire gauze
plastic stirring rod
laboratory burner and striker

Procedure

1. Set the syringe plunger to about 15–20 mL of air.
2. Seal the tip of the syringe with a cap or stopper.
3. Set up the ring stand with the 600 mL beaker on the wire gauze above the burner (Figure 6.5).
4. Use the buret clamp to hold the syringe as far as possible into the beaker without touching the sides or bottom.
5. Clamp the thermometer so that the bulb is beside the end of the plunger but not touching the syringe.

6. Add water at room temperature to about 1 cm from the top of the beaker.

7. After a few minutes, record the temperature and volume of air.

8. Light the burner and adjust to obtain a small blue flame.

9. Heat the water slowly, stirring occasionally.

10. Record the gas volume and temperature about every 10°C until about 90°C. (It may be necessary to tap or twist the plunger occasionally to make sure it is not stuck.)

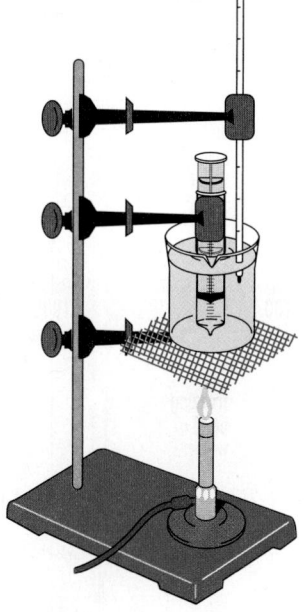

Figure 6.5
Set-up of apparatus for Investigation 6.2.

Kelvin Temperature Scale

The mathematical equation describing the relationship between temperature and volume may not be apparent from your graph; however, if the two variables are graphed as in Figure 6.6(a), a straight line is obtained, so a simple relationship does exist. When the line is extrapolated downward it meets the horizontal axis at –273°C. It appears that if the gas did not liquefy, its volume would become zero at –273°C. If this experiment is repeated with different quantities of gas or with samples of different gases, straight-line relationships between

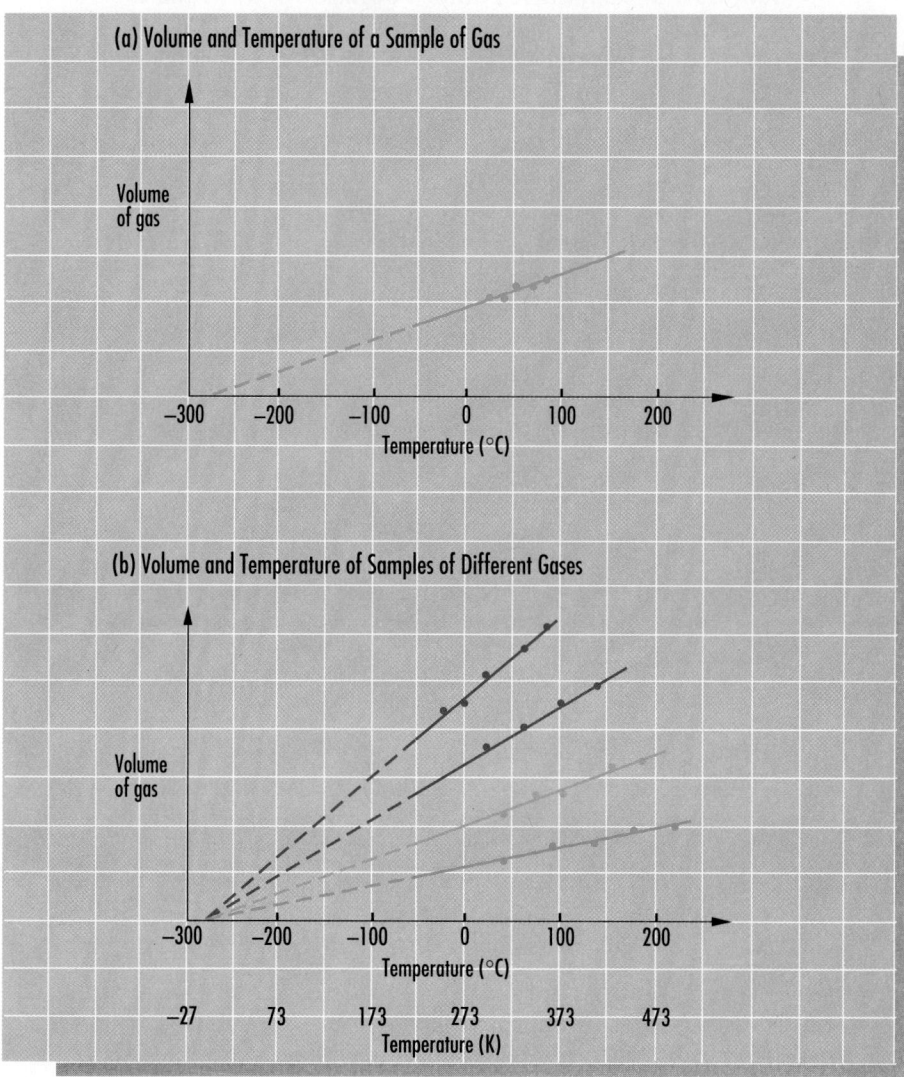

STP and SATP are each defined by two exact values with infinite significant digits (i.e., STP is 273.15 K and 101.325 kPa and SATP is 298.15 K and 100 kPa). For convenience, however, use STP as 273 K and 101 kPa, and SATP as 298 K and 100 kPa.

Although 0°C is defined as exactly 273.15 K, use 0°C as equal to 273 K for convenience.
  0°C = 273 K
  30°C = 303 K
  –20°C = 253 K

In Chapter 5 you studied another example of an inverse relationship — that between volume and concentration when a sample of solution is diluted: $v_i C_i = v_f C_f$.

Figure 6.6
When the graphs of several careful volume-temperature experiments are extrapolated, all the lines meet at absolute zero, –273°C or 0 K.

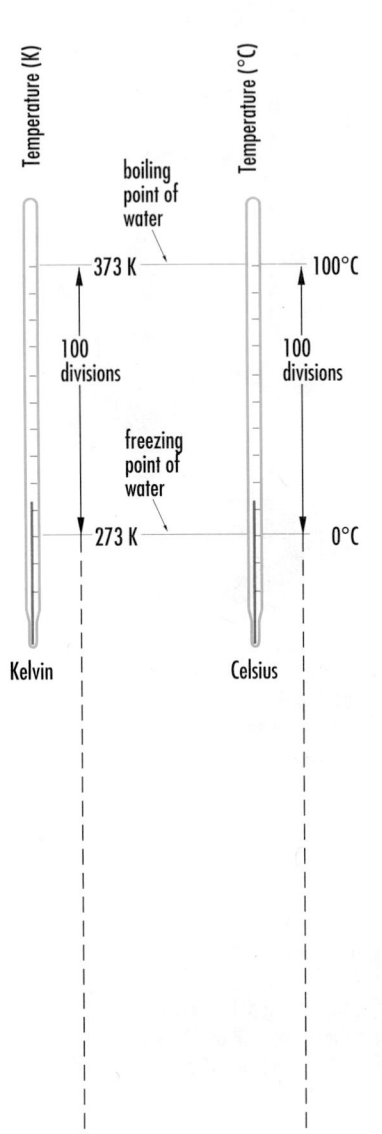

Temperature (K)   Temperature (°C)

boiling
point of
water

373 K ————— 100°C

100   100
divisions   divisions

freezing
point of
water

273 K ————— 0°C

Kelvin   Celsius

absolute
zero

0 K ————— –273°C

Figure 6.7
*Jacques Charles predicted –273°C
to be the temperature at which the
volume of a gas would become zero,
if the gas could remain a gas at that
low temperature. Scottish physicist
and mathematician Lord Kelvin
(1824 – 1907) considered –273°C
to be the temperature at which the
kinetic energy of all particles of
solids, liquids, or gases would
become zero.*

Figure 6.8
*The volume of a gas in a container
with a movable piston increases as
the temperature of the gas
increases.*

temperature and volume are also observed. When the lines are extrapolated, they all meet at –273°C, as shown in Figure 6.6(b). This temperature, called **absolute zero**, is thought to represent the lowest temperature that can be obtained.

Absolute zero is the basis of another temperature scale, called the absolute or **Kelvin temperature scale**. On the Kelvin scale, absolute zero (–273°C) is zero kelvins (0 K) as shown in Figure 6.6(b). (Note that no degree symbol is used for kelvin.) To convert degrees Celsius to kelvins, add 273 (Figure 6.7).

Charles' Law

The relationship between volume and temperature in kelvins of a gas is shown in Figure 6.6(b). This relationship is described as a direct variation; that is, as the temperature increases, the volume increases. Mathematically, this relationship is represented as

$$v = kT$$

Table 6.2

ANALYSIS OF TEMPERATURE AND VOLUME OF A GAS SAMPLE			
Temperature (°C)	Temperature T (K)	Volume v (L)	Constant v/T (L/K)
25	298	5.00	0.0168
50	323	5.42	0.0168
75	348	5.84	0.0168
100	373	6.26	0.0168
125	398	6.68	0.0168

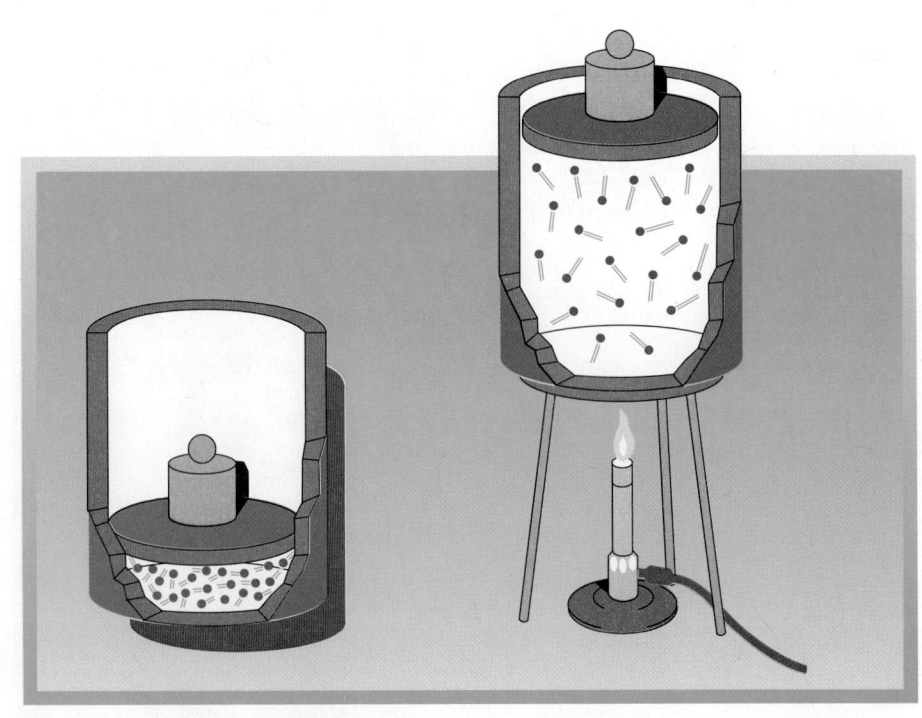

This means that the quotient of the two variables (v/T) has a constant value (k), which is the slope of the straight line graph (Figure 6.6(b)). A constant value is clearly shown by the analysis in Table 6.2.

The relationship between volume and absolute temperature is known as **Charles' law**, which states that *as the temperature of a gas increases, the volume increases proportionally, provided that the pressure and amount of gas remain constant* (Figure 6.8). Charles' law can be conveniently written comparing any two sets of volume and temperature measurements.

$$\frac{v_1}{T_1} = k \quad \text{and} \quad \frac{v_2}{T_2} = k$$

$$\text{therefore,} \quad \frac{v_1}{T_1} = \frac{v_2}{T_2} \quad (Charles'\ law)$$

The Combined Gas Law

When Charles' law and Boyle's law are combined, the resulting **combined gas law** states the relationships among the volume, temperature, and pressure of any fixed quantity of gas. This law states that the product of the pressure and volume of a gas sample is proportional to its absolute temperature.

$$pv = kT$$

$$\frac{pv}{T} = k$$

This equation is valid when the amount of gas is constant. The relationship can be expressed in a convenient form for calculations involving changes in volume, temperature, or pressure for a particular gas sample.

$$\frac{p_1v_1}{T_1} = \frac{p_2v_2}{T_2} \quad (combined\ gas\ law)$$

The combined gas law is a useful starting point for all cases involving pressure, volume, and temperature even if one of these variables is constant (as in Boyle's and Charles' laws). For example, a steel cylinder with a fixed volume contains a gas at a pressure of 652 kPa and a temperature of 25°C. If the cylinder is heated to 150°C, what will be the new pressure? Because the volume is constant, we can cancel v_1 and v_2 from the combined gas law equation.

$$\frac{p_1\cancel{v_1}}{T_1} = \frac{p_2\cancel{v_2}}{T_2}$$

We can now substitute the pressures and temperatures (after converting to kelvins).

$$\frac{652\ \text{kPa}}{298\ \text{K}} = \frac{p_2}{423\ \text{K}}$$

$$p_2 = 925\ \text{kPa}$$

Assuming that the steel walls are sufficiently strong, the gas will have a pressure of 925 kPa inside the cylinder.

Scuba Diving

Diving is a popular activity on the West Coast, and scuba (self-contained underwater breathing apparatus) makes this sport exciting. A scuba tank containing compressed air is attached to a regulator that releases the air at the same pressure as the underwater surroundings. The pressure underwater can be quite substantial. Every 10 m of depth adds about 1 atm (100 kPa) of pressure to the normal air pressure. At a depth of 20 m, the total pressure is about 300 kPa. Breathing pressured air is necessary to balance the internal and external pressures on the chest to allow divers to inflate their lungs. However, this also creates problems if divers ascend to normal pressure too quickly or while holding their breath. According to Boyle's law, if the pressure is decreased from 300 kPa to 100 kPa, the volume of air in the lungs would increase three times, rupturing the diver's lungs. This is one reason why a person needs an understanding of gases and gas laws in order to obtain a scuba diving license and to dive safely.

A balloon containing hydrogen gas at 20°C and a pressure of 100 kPa has a volume of 7.50 L. Balloons are free to expand so that the gas pressure within them remains equal to the air pressure outside. Calculate the volume of the balloon after it rises 10 km into the upper atmosphere, where the temperature is –36°C and the outside air pressure is 28 kPa. Assume that no hydrogen gas escapes.

Initial Conditions	Final Conditions
p_1 = 100 kPa	p_2 = 28 kPa
v_1 = 7.50 L	v_2 = ?
T_1 = 20°C = 293 K	T_2 = –36°C = 237 K

$$\frac{p_1 v_1}{T_1} = \frac{p_2 v_2}{T_2}$$

$$v_2 = \frac{p_1 v_1 T_2}{p_2 T_1}$$

$$= \frac{100 \text{ kPa} \times 7.50 \text{ L} \times 237 \text{ K}}{28 \text{ kPa} \times 293 \text{ K}}$$

$$= 22 \text{ L}$$

or

$$v_2 = 7.50 \text{ L} \times \frac{100 \text{ kPa}}{28 \text{ kPa}} \times \frac{237 \text{ K}}{293 \text{ K}} = 22 \text{ L}$$

The lightness of baked goods such as bread and cakes is a result of gas bubbles trapped in the dough or batter when it is heated. The leavening, or production of gas bubbles, can be due to vaporization of water, expansion of gases already in the dough or batter, or leavening agents such as yeast or baking powder. Yeasts are living organisms that feed on sugar, producing carbon dioxide and either water or ethanol; baking powder is a mixture of sodium hydrogen carbonate and a solid acid that react together to produce carbon dioxide; the bubbles of gas are part of the light and delectable baked goods that result from kitchen chemistry.

Exercise

7. Carbon dioxide produced by yeast in bread dough causes the dough to rise, even before baking. During baking, the carbon dioxide expands. Predict the final volume of 0.10 L of carbon dioxide in bread dough that is heated from 25°C to 190°C at constant pressure.

8. An automobile tire has a volume of 27 L at 225 kPa and 18°C.

   (a) What volume would the air inside the tire occupy if the pressure was 98 kPa and the temperature remained the same?

   (b) How many times larger is the new volume compared with the original volume? How does this compare with the change in pressure?

9. A balloon has a volume of 5.00 L at 20°C and 100 kPa. What is its volume at 35°C and 90 kPa?

10. A storage tank is designed to hold a fixed volume of butane gas at 150 kPa and 35°C. To prevent dangerous pressure buildup, the tank has a relief valve that opens at 250 kPa. At what (Celsius) temperature does the valve open?

11. In a cylinder of a diesel engine, 500 mL of air at 40.0°C and 1.00 atm is powerfully compressed just before the diesel fuel is injected. The resulting pressure is 35.0 atm. If the final volume is 23.0 mL, what is the final temperature in the cylinder?

12. A cylinder of helium gas has a volume of 1.0 L. The gas in the cylinder exerts a pressure of 800 kPa at 30°C. What volume would this gas occupy at SATP?

13. For any of the calculations in the previous questions, does the result depend on the identity of the gas? Explain briefly.

14. What assumption was made in all of the previous calculations?

6.2 AVOGADRO'S THEORY AND MOLAR VOLUME

The early study of gases was strictly empirical. Boyle's and Charles' laws were developed before Dalton's atomic theory was published in 1803. The kinetic molecular theory (page 136) was refined by about 1860, thanks to the development of mathematical tools such as statistical analysis.

Before any theory is accepted by the scientific community, supporting evidence must be available. The more phenomena that a theory can explain, the more widely accepted it becomes. The kinetic molecular theory is strongly supported by experimental evidence.

- The kinetic molecular theory explains why gases, unlike solids and liquids, are compressible. If most of the volume of a gas sample is empty space, it should be possible to force the particles closer together.

- The kinetic molecular theory explains the concept of gas pressure. Pressure is considered to be the result of gas particles colliding with objects, for example, the walls of a container. The pressure exerted by a gas sample is the total force of these collisions distributed over an area of the container wall; in other words, force per unit area.

- The kinetic molecular theory explains Boyle's law. If the volume of a container is reduced, gas particles will move a shorter distance before colliding with the walls of the container. They will collide with the walls more frequently, resulting in increased pressure on the container.

- The kinetic molecular theory explains Charles' law. According to kinetic theory, an increase in temperature represents an increase in the average speed of particle motion. In a container in which the pressure can be kept constant (for example, in a cylinder with a piston or in a flexible-walled container such as a weather balloon), faster-moving molecules will collide more frequently with the container walls. They will also collide with more force, causing the walls to move outward. Thus, the volume of a gas sample increases with increasing temperature.

The Law of Combining Volumes and Avogadro's Theory

In 1808, Joseph Gay-Lussac, a French scientist and a colleague of Jacques Charles, measured the relative volumes of gases involved in chemical reactions. His observations led to the **law of combining volumes**, which states that *when measured at the same temperature and pressure, volumes of gaseous reactants and products of chemical reactions are always in simple ratios of whole numbers*. An example of this is the decomposition of water, in which the volumes of hydrogen and oxygen produced are always in a ratio of 2:1.

Two years after this law was formulated, the Italian scientist Amadeo Avogadro proposed an explanation. Avogadro was intrigued by the fact that reacting volumes of gases were in whole-number ratios, just like the coefficients in a balanced equation. (Remember that this was only about eight years after Dalton had presented his atomic theory of matter.) Suggesting a relationship between the volume ratios and coefficient ratios, he proposed that *equal volumes of gases at the same temperature and pressure contain equal numbers of molecules*, a statement that is best called **Avogadro's theory**.

This theoretical concept explains the law of combining volumes. For example, if a reaction occurs between two volumes of one gas and one volume of another at the same temperature and pressure, the theory indicates that two molecules of the first substance react with one molecule of the second. Another example is the reaction of nitrogen and hydrogen, in which ammonia is produced (Figure 6.9).

Avogadro's initial idea was a hypothesis. Although it is still sometimes referred to as a hypothesis, the idea is no longer tentative but is firmly established. Therefore, Avogadro's idea has the status of a theory.

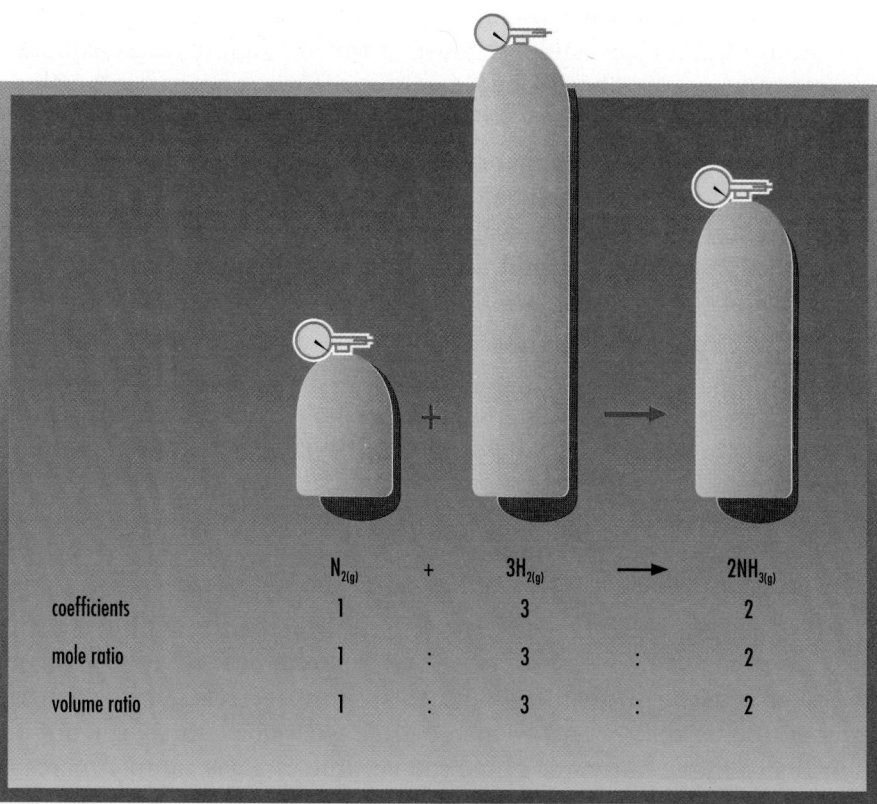

	$N_{2(g)}$	+	$3H_{2(g)}$	$\longrightarrow$	$2NH_{3(g)}$
coefficients	1		3		2
mole ratio	1	:	3	:	2
volume ratio	1	:	3	:	2

Figure 6.9
One volume of nitrogen reacts with three volumes of hydrogen, producing two volumes of ammonia.

The Law of Combining Gas Volumes

When all gases are at the same temperature and pressure, the law of combining gas volumes provides an efficient way of predicting the volumes of gases involved in a chemical reaction. As explained by Avogadro's theory, the mole ratios provided by the balanced equation are also the volume ratios.

For example,

$2 C_4H_{10(g)}$	$+$	$13 O_{2(g)}$	$\rightarrow$	$8 CO_{2(g)}$	$+$	$10 H_2O_{(g)}$
2 mol		13 mol		8 mol		10 mol
2 vol		13 vol		8 vol		10 vol
4 L		26 L		16 L		20 L

EXAMPLE

Predict the volume of oxygen required for the complete combustion of 120 mL of butane from a lighter. The oxygen and butane are measured at the same temperature and pressure.

$$2 C_4H_{10(g)} + 13 O_{2(g)} \rightarrow 8 CO_{2(g)} + 10 H_2O_{(g)}$$
$$120 \text{ mL} \qquad v$$

$$\frac{v_{O_2}}{120 \text{ mL}} = \frac{13}{2}$$

$$v_{O_2} = 780 \text{ mL}$$

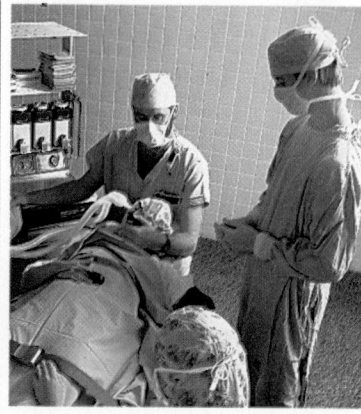

"I have procured air [oxygen] ... between five and six times as good as the best common air that I have ever met with." In 1774, English chemist Joseph Priestley (1733 – 1804) decomposed mercury(II) oxide into mercury and oxygen by heating the red powdery compound using solar radiation focused through a lens. The result was the first reported pure oxygen gas sample. Today, oxygen is produced for medical use and for combustion reactions such as oxy-acetylene welding torches.

Exercise

15. The production of sulfuric acid is a very important chemical industry in any developing or developed country (Figure 6.10). The following questions are in the context of producing sulfuric acid.

    (a) One technology for removing sulfur from sour natural gas involves reacting sulfur dioxide with hydrogen sulfide. Use the reaction equation to predict the volume of sulfur dioxide needed to react completely with 248 kL of hydrogen sulfide. The gases are measured at 350°C and 250 kPa.

    $$16 H_2S_{(g)} + 8 SO_{2(g)} \rightarrow 3 S_{8(s)} + 16 H_2O_{(g)}$$

    (b) A step in the manufacture of sulfuric acid involves burning sulfur, $S_{8(s)}$. Predict the volume of oxygen required to produce 250 L of sulfur dioxide with all gases at 450°C and 200 kPa.

    (c) In the presence of the catalyst, $V_2O_{5(s)}$, sulfur dioxide reacts with oxygen to form sulfur trioxide.

    $$2 SO_{2(g)} + O_{2(g)} \rightarrow 2 SO_{3(g)}$$

    Predict the volumes of sulfur dioxide and oxygen needed to produce 325 L of sulfur trioxide when all gases are measured at the same temperature and pressure.

16. The production of nitric acid is important to the fertilizers and the explosives industries.

    (a) The production of nitric acid by the Ostwald process begins with the combustion of ammonia.

Figure 6.10
More sulfuric acid is manufactured in North America than any other chemical.

$$4\,NH_{3(g)} + 5\,O_{2(g)} \rightarrow 4\,NO_{(g)} + 6\,H_2O_{(g)}$$

Predict the volume of oxygen required to react with 100 L of ammonia, as well as the volumes of nitrogen oxide and water vapor produced. All gases are measured at 800°C and 200 kPa.

(b) In another step of the Ostwald process, nitrogen monoxide reacts with oxygen to form nitrogen dioxide. Predict the volume of oxygen at 800°C and 200 kPa required to produce 750 L of nitrogen dioxide at the same temperature and pressure.

(c) Nitric acid is produced by reacting nitrogen dioxide with water.

$$3\,NO_{2(g)} + H_2O_{(l)} \rightarrow 2\,HNO_{3(l)} + NO_{(g)}$$

Predict the volume of nitrogen monoxide produced by the reaction of 100 L of nitrogen dioxide with excess water. Both gases are measured at the same temperature and pressure.

(d) A high-nitrogen fertilizer is made by reacting ammonia gas with nitric acid to produce aqueous ammonium nitrate. Can the law of combining volumes be used to predict the volume of ammonia gas required to react with 100 L of nitric acid? Justify your answer.

Molar Volume of Gases

The evolution of scientific knowledge often involves integrating two or more concepts. For example, Avogadro's idea and the mole concept (Chapter 4) can be integrated. According to Avogadro's theory, equal volumes of any gas at the same temperature and pressure contain an equal number of particles. A *mole* is a specific number of particles. Therefore, for all gases at each specific pressure and temperature, there must be a certain volume that contains exactly one mole of particles. The volume that one mole of a gas occupies at a specified temperature and pressure is called its **molar volume**. The molar volume is the same for all gases at the same temperature and pressure. For scientific work, the most useful specific pressure and temperature conditions are either

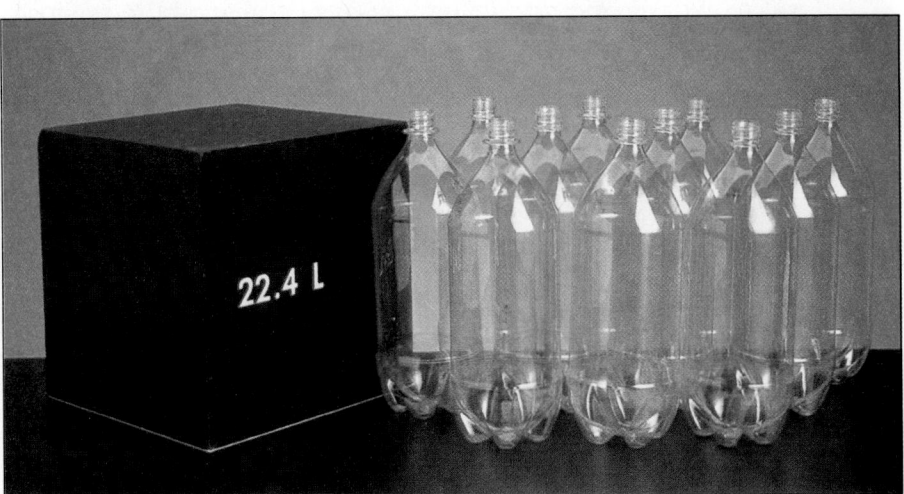

Figure 6.11
At STP, 1 mol of gas is contained in 11 "empty" 2 L pop bottles.

SATP or STP. It has been determined empirically that the molar volume of a gas at SATP is 24.8 L/mol. The molar volume of a gas at STP is 22.4 L/mol (Figure 6.11).

Knowing the molar volume of gases allows scientists to work with easily measured volumes of gases when specific masses of gases are needed. Measuring the volume of a gas is much more convenient than measuring its mass. Imagine trapping a gas in a container and trying to measure its mass on a balance — and then making corrections for the buoyant force of the surrounding air. Also, working with gas volumes is more precise, as the process involves measuring relatively large volumes rather than relatively small masses. Molar volume can be used as a conversion factor to convert amount in moles to volume, and vice versa, as shown in the following examples. Notice how the units cancel and how the results are expressed with appropriate certainties. (For a review of expressing the certainty in significant digits, see "Quantitative Precision and Certainty" in Appendix E, E.3.)

EXAMPLE

What amount in moles of oxygen is available for a combustion reaction in a volume of 5.6 L at STP?

$$n_{O_2} = 5.6 \, \cancel{L} \times \frac{1 \text{ mol}}{22.4 \, \cancel{L}} = 0.25 \text{ mol}$$

EXAMPLE

What volume is occupied by 0.024 mol of carbon dioxide gas at SATP?

$$v_{CO_2} = 0.024 \, \cancel{\text{mol}} \times \frac{24.8 \text{ L}}{1 \, \cancel{\text{mol}}} = 0.60 \text{ L}$$

In SI symbols, the relationship of amount (n), volume (v), and molar volume (V) is expressed as $n = \dfrac{v}{V}$ or $v = nV$

Exercise

17. Use the kinetic molecular theory to explain the following observed properties of gases.
    (a) Gas pressure increases when the volume of the gas is kept constant and the temperature increases.
    (b) Gas pressure increases when the temperature is kept constant and the volume of the gas decreases.
    (c) The fragrance of an open bottle of perfume is evident throughout a room.
    (d) At SATP, the average speed of air (oxygen and nitrogen) molecules is about 450 m/s, which is approximately the speed of a bullet fired from a rifle. Nevertheless, it takes several minutes for the odor of a perfume to diffuse throughout a room.

Serendipity is the good fortune of making an accidental discovery of something valuable. Perhaps the earliest recorded case of serendipity in science involves Archimedes, who accidentally discovered how to determine the volume of irregular objects by noticing the overflow of water while he was bathing.

Torricelli's invention of the barometer (page 225) appears to have been the result of serendipity. Another example is the discovery of x-rays by Wilhelm Röntgen in 1895, as he conducted an electric tube experiment that caused a chemical to glow more than one metre away. Serendipity also played a part in the discoveries of Ernest Rutherford. While attempting to verify J. J. Thomson's model of the atom, Rutherford obtained aberrant results, but instead of discounting his observations he investigated them further, and developed his own contribution to atomic theory. Some recent examples of serendipity include the invention of Velcro fasteners, Teflon, Ivory soap, corn flakes, and Post-it notes.

Velcro fasteners, shown magnified in the two photographs on the right below, were devised in the early 1950s by George de Mestral. One day, after walking in the Swiss countryside, he had to pull cockleburs off his trousers when he returned home. Curious about what made these burrs stick so well, he discovered the hook structures that you can see in the magnified photograph on the left below. The hook-and-loop structure, duplicated in plastic fasteners, is just one of many technologies that copy natural objects.

Teflon was discovered in 1938 by Roy Plunkett, who worked for the Du Pont Chemical Company. He tried to release some gas from a cylinder of tetrafluoroethylene, but although the valve was clear and the mass of the cylinder indicated it was not empty, he couldn't get any out. Instead of discarding the cylinder and substituting a new one, the inquisitive chemist sawed the cylinder in half so he could see inside. He found the incredibly inert and slippery plastic polytetrafluoroethylene, now known as Teflon. Its initial application was in gaskets for equipment used to manufacture the first atomic bomb in the early 1940s. Today Teflon is commonly used as a non-stick coating on cookware, and as a chemically resistant stopcock in burets used in chemistry labs.

Post-it notes were developed in 1974 by Art Fry, an employee of 3M Corporation. He marked so many places in his books with slips of paper that the slips always fell out. He remembered that, years before, a colleague had developed a weak adhesive that wasn't permanent; it had been rejected as having no practical value. Fry put two and two together and the rest, as they say, is history.

An important aspect of serendipity is illustrated by two quotations. According to French chemist and microbiologist Louis Pasteur, "In the fields of observation, chance favors only the prepared mind." And Hungarian biochemist Albert von Szent-Györgyi said, "Discovery consists of seeing what everybody has seen, and thinking what nobody has thought." Horace Walpole, the English scholar who coined the term "serendipity" in the 18th century, said that curiosity and natural talent play a role in accidental discoveries, but that training in a broad range of subjects and in flexible thinking are probably more important.

- *What is an example of a technology that copies a natural object?*
- *In your opinion, what is the main message promoted in this article?*

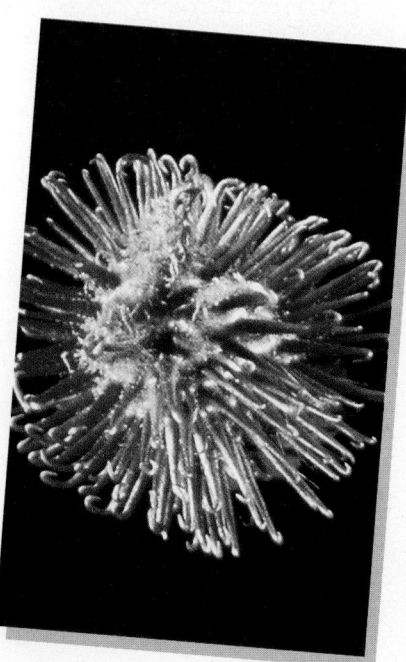

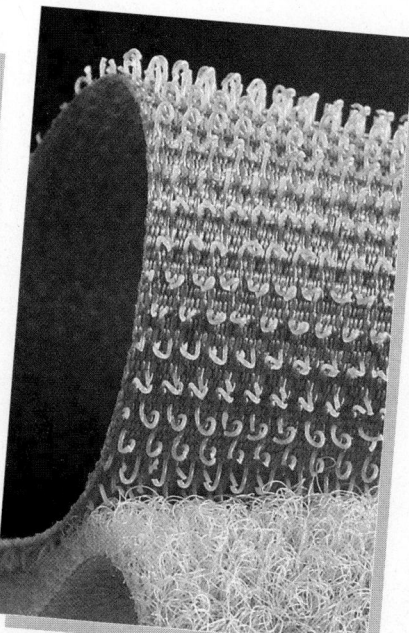

18. Weather balloons filled with hydrogen gas are occasionally reported as UFOs. They can reach altitudes of about 40 km. What volume does 7.50 mol of hydrogen gas in a weather balloon occupy at SATP?

19. Sulfur dioxide gas is emitted from marshes, volcanoes, and refineries that process crude oil and natural gas. What amount of sulfur dioxide is contained in 50 mL of the gas at SATP?

20. Neon gas under low pressure emits the red light that glows in advertising signs (Figure 6.12). What volume does 2.25 mol of neon gas occupy at STP before being added to neon tubes in a sign?

21. Oxygen is released by plants during photosynthesis and is used by plants and animals during respiration. What amount in moles of oxygen is present in 20.0 L of air at STP? Assume that air is 20% oxygen (by volume).

Figure 6.12
When an electric current is passed through a glass tube containing the noble gas neon, the gas glows a characteristic red color.

Molar Volume and Molar Mass

Gases such as oxygen and nitrogen are often liquefied for storage and transportation, then allowed to vaporize for use in a technological application. Helium is stored and transported as a compressed gas. Both liquefied and compressed gases are sold by mass. Molar volume and molar mass can be combined to calculate the volume of gas that is available from a known mass of a substance.

For example, helium-filled balloons (Figure 6.13) are often used for party decorations. Because these balloons are less dense than air, they stay aloft and will keep rising unless tied down by a string. What volume does 3.50 g of helium gas occupy at SATP? To answer this question, we need to first convert the mass into an amount in moles.

$$n_{He} = 3.50\ g \times \frac{1\ mol}{4.00\ g} = 0.875\ mol$$

Now we can convert this amount into a volume at SATP using the molar volume constant.

$$v_{He} = 0.875\ mol \times \frac{24.8\ L}{1\ mol} = 21.7\ L$$

Once these two steps are clearly understood, they can be combined into a single calculation as shown below.

$$v_{He} = 3.50\ g \times \frac{1\ mol}{4.00\ g} \times \frac{24.8\ L}{1\ mol} = 21.7\ L$$

Molar mass and molar volume can also be combined to determine the density of gases under standard conditions. A density in grams per litre (g/L) is obtained by using the molar volume to obtain the amount of gas in 1 L and then converting this amount into a mass using the molar mass. For example, at SATP, the amount in moles of helium in 1 L can be calculated as follows.

Figure 6.13
Helium-filled balloons are popular items for parties and store promotions.

$$n_{He} = 1.00 \cancel{L} \times \frac{1 \text{ mol}}{24.8 \cancel{L}} = 0.0403 \text{ mol}$$

This amount of helium in 1 L is then converted to a mass.

$$m_{He} = 0.0403 \cancel{\text{mol}} \times \frac{4.00 \text{ g}}{1 \cancel{\text{mol}}} = 0.161 \text{ g}$$

Therefore, the density of helium at SATP is 0.161 g in 1.00 L or 0.161 g/L. Once you have understood these two steps, they can be combined into one calculation.

$$m_{He} = 1.00 \cancel{L} \times \frac{1 \cancel{\text{mol}}}{24.8 \cancel{L}} \times \frac{4.00 \text{ g}}{1 \cancel{\text{mol}}} = 0.161 \text{ g}$$

Figure 6.14
For years, the fuel of choice for automobiles has been gasoline.

Exercise

22. Volatile liquids vaporize if left in open containers. To keep these vapors from the air, consumers are responsible for keeping containers tightly sealed when not in use. Convert the following masses of vaporized liquids into amounts in moles.
    (a) 50.0g of $C_8H_{18(g)}$ (gasoline vapor)
    (b) 70.0 g of methanol (vapors from windshield washer antifreeze)
    (c) 575 mg of chlorine (found in household bleach)

23. One gram of baking powder or 0.25 g of baking soda produces about 0.13 g of carbon dioxide. What volume is occupied by 0.13 g of carbon dioxide gas at SATP?

24. Millions of tonnes of nitrogen dioxide are dumped as smog into the atmosphere each year by automobiles. What is the volume of 1.00 t (1.00 Mg) of nitrogen dioxide at SATP?

25. Human beings exhale millions of tonnes of carbon dioxide into the atmosphere each year. Determine the volume occupied by 1.00 t of carbon dioxide at STP.

26. To completely burn 1.0 L of gasoline in an automobile engine requires about 1.9 kL of oxygen at SATP (Figure 6.14). What mass of oxygen gas is consumed by burning 1.0 L of gasoline?

27. Water vapor plays an important role in the weather patterns on Earth. What mass of water must vaporize to produce 1.00 L of water vapor at SATP?

28. What is the density of carbon dioxide at SATP?

29. Dry air is approximately 80% nitrogen and 20% oxygen. Calculate the average density of dry air.

30. (a) Using your answers to questions 28 and 29, suggest one reason why carbon dioxide is used as a fire extinguisher.

    (b) What is another important characteristic of carbon dioxide that makes it suitable for use as a fire extinguisher?

INVESTIGATION

6.3 The Molar Mass of a Volatile Liquid

Before the invention of modern technologies, such as mass spectrometers, the molar mass of a substance was determined by a number of different physical and chemical methods. In this investigation, the molar mass of a liquid that is easily vaporized is determined from the measured volume of the vapor (converted to STP conditions), the molar volume of a gas at STP, and the mass of the condensed vapor. If the identity of the liquid is known, evaluate the design of this experiment using the percentage difference between your experimental value of the molar mass and the known molar mass.

Problem

What is the molar mass of a volatile liquid?

Experimental Design

Excess liquid is slowly vaporized inside a flask with a small opening. The pressure, volume, and temperature of the vapor are measured and the mass of the vapor is determined by condensing the vapor back into its liquid state.

Materials

lab apron
safety glasses
125 mL Erlenmeyer flask
10 mL graduated cylinder
100 mL or 250 mL graduated cylinder
600 mL beaker
buret clamp
hot plate (if liquid is flammable) or
    wire gauze and laboratory burner
ring stand
plastic stirring rod
thermometer
small elastic band
aluminum foil
volatile liquid
pin
barometer
boiling chips (optional)

Procedure

1. Measure the mass of a 5 cm by 5 cm square of aluminum foil, elastic band, and 125 mL Erlenmeyer flask.

2. Measure 3 mL of the unknown volatile liquid and pour it into the flask.

3. Fold the aluminum foil tightly over the flask opening and secure, just under the lip, with the elastic band. Use a pin to make a *tiny* hole, as small as possible, in the center of the foil.

Problem
Prediction
Design
Materials
Procedure
✔ Evidence
✔ Analysis
✔ Evaluation
Synthesis

GASES **241**

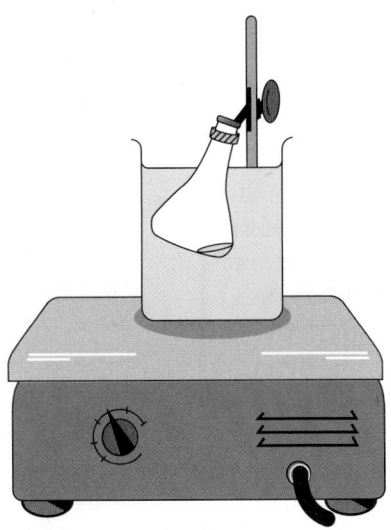

Figure 6.15
The flask should be almost completely surrounded by water and the water heated slowly to the maximum temperature specified by your teacher.

4. Clamp the flask to the stand so that the flask is slightly tilted and as far into the beaker as possible (Figure 6.15).

5. Add water to the beaker so that the flask is almost completely surrounded by water. Add a few boiling chips if directed by your teacher.

6. Heat the beaker of water slowly, with occasional stirring, until the last traces of the liquid have evaporated. Do not exceed the maximum temperature specified by your teacher.

7. When all of the liquid in the flask has evaporated, immediately remove the flask from the beaker and measure the temperature of the water.

8. Record the atmospheric pressure.

9. After the flask has cooled to room temperature, wipe the flask and cap carefully with a paper towel to remove all the water. Be sure to remove any water trapped between the ends of the aluminum foil and the flask.

10. Determine the final mass of the flask, cap, and condensed liquid.

11. Remove the cap from the flask and fill the flask completely with tap water. Use a large graduated cylinder to measure the volume of water in the flask.

Air Quality in the Lower Fraser Valley

Clean air is important for healthy living, and most areas of British Columbia have good air quality. The exception is the Lower Fraser Valley, which has 54% of the province's population. Here, mountains to the north and south trap polluted air from Greater Vancouver and funnel it up the valley, resulting in some of the worst air quality in Canada. Especially at risk are children, the elderly, and people with respiratory ailments such as asthma.

Motor vehicles are the primary source of this air pollution. Data for 1990 indicate that the Greater Vancouver region introduced over 600 kt of pollutants into the air basin. Over half of this was carbon monoxide (CO), with lesser amounts of other pollutants:

- volatile organic compounds such as solvent and fuel vapors
- nitrogen oxides (NO_x)
- sulfur oxides (SO_x)
- particulates such as soot, fly ash, lead, and dust

The province now requires most vehicles in the region to pass an annual inspection of emissions of carbon monoxide, hydrocarbons, and nitrogen oxides in order to be licenced. To help reduce the harmful effects of air pollution, the province introduced in 1995 the toughest auto emission standards in Canada. By 2001, all new cars sold in the province will have to meet new low-emission vehicle standards that will reduce today's emission levels by 70%. Cleaner gasoline fuels and alternative fuels like natural gas (methane) and propane are also part of these initiatives. It is expected that zero-emission vehicles will be introduced early in the next century. Inner-city vehicles may be powered by storage batteries or hydrogen-oxygen fuel cells (see Chapter 16). All of these measures should help to improve the air quality in the Lower Fraser Valley.

6.3 THE IDEAL GAS LAW

The gas laws discussed so far are exact only for an ideal gas. An **ideal gas** is a hypothetical gas that obeys all the gas laws perfectly under all conditions; that is, it does not condense into a liquid when cooled, and graphs of its volume and temperature and of its pressure and temperature are perfectly straight lines. Theoretically, an ideal gas is composed of particles of zero size that have no attraction to each other.

Real gases *do* condense and *do* have particles that attract each other, and therefore real gases do not follow the gas laws exactly. Real gases deviate most from ideal gas behavior at lower temperatures and higher pressures. They behave more like ideal gases as temperature increases and pressure decreases. The accepted theoretical interpretation of these findings is that the farther apart the molecules of a gas are, and the faster they are moving, the less attraction there is between molecules, and the more closely the gas approaches ideal behavior. The size of the molecules also appears to affect the deviation from ideal behavior. For example, the molar volume of an ideal gas at STP is 22.414 L/mol; for helium, the value is 22.426 L/mol; for oxygen, 22.392 L/mol; and for chlorine, 22.063 L/mol. The smaller the molecules, the more closely the gas resembles an ideal gas.

In this book, all gases are dealt with as if they were ideal. A single, ideal-gas equation describes the interrelationship of pressure, temperature, volume, and amount of matter — the four variables that define a gaseous system.

- According to Boyle's law, the volume of a gas is inversely proportional to the pressure: $v \propto 1/p$.
- According to Charles' law, the volume of a gas is directly proportional to the Kelvin temperature: $v \propto T$.
- According to Avogadro's theory, the volume of a gas is directly proportional to the amount of matter: $v \propto n$.

Combining these three statements produces the following relationship:

$$v \propto \frac{1}{p} \times T \times n$$

Another way of stating this is:

$$v = (\text{a constant}, R) \times \frac{1}{p} \times T \times n$$

$$v = \frac{nRT}{p}$$

$$pv = nRT$$

This last equation is known as the **ideal gas law**; the constant R is known as the **universal gas constant**. The value for the universal gas constant can be obtained by substituting STP (or SATP) conditions for one mole of an ideal gas into the ideal gas law and solving for R.

The apparatus shown below consists of a metal sphere to which a pressure gauge is attached. Because the gas inside the sphere cannot expand, the relationship between temperature and pressure of a gas can be determined. If we were to plot a graph of pressure against absolute temperature, a straight line would be obtained. This means that the pressure of a gas is directly related to its absolute temperature; that is, as the temperature increases, the pressure increases proportionally. This relationship is shown below.

$$\frac{p_1}{T_1} = \frac{p_2}{T_2}$$

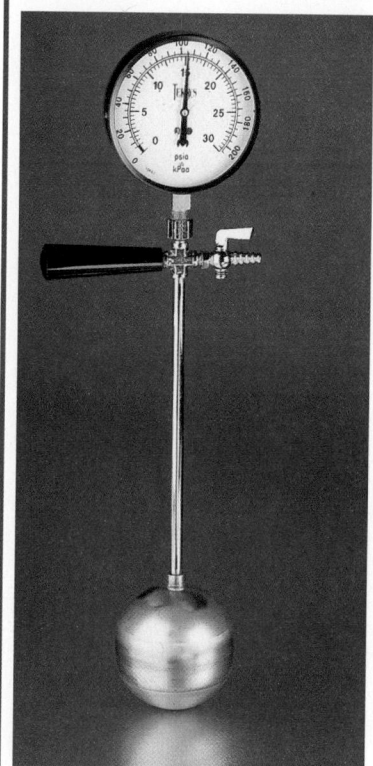

$$R = \frac{pv}{nT}$$

$$= \frac{101.3 \text{ kPa} \times 22.4 \text{ L}}{1.00 \text{ mol} \times 273 \text{ K}}$$

$$= \frac{8.31 \text{ kPa} \cdot \text{L}}{\text{mol} \cdot \text{K}}$$

The value of the universal gas constant depends on the units chosen to measure volume, pressure, and temperature. If any three of the four variables in the ideal gas law are known, the fourth can be calculated by means of this equation. However, often the mass of a gas is a known quantity. In this case, a two-step calculation is required.

For example, 0.78 g of hydrogen at 22°C and 125 kPa is produced. What volume of hydrogen would be expected? To use the ideal gas law, you first need to convert the mass into an amount in moles of hydrogen.

$$n_{H_2} = 0.78 \text{ g} \times \frac{1 \text{ mol}}{2.02 \text{ g}} = 0.39 \text{ mol}$$

Now you can use the ideal gas law to determine the volume of hydrogen at the conditions specified.

$$pv = nRT$$

$$v_{H_2} = \frac{nRT}{p}$$

$$= \frac{0.39 \text{ mol} \times 8.31 \text{ kPa} \cdot \text{L} \times 295 \text{ K}}{125 \text{ kPa} \times 1 \text{ mol} \cdot \text{K}}$$

$$= 7.6 \text{ L}$$

EXAMPLE _____

What mass of neon gas should be introduced into an evacuated 0.88 L tube to produce a pressure of 90 kPa at 30°C?

$$pv = nRT$$

$$n_{Ne} = \frac{pv}{RT}$$

$$= \frac{90 \text{ kPa} \times 0.88 \text{ L} \cdot \text{mol} \cdot \text{K}}{8.31 \text{ kPa} \cdot \text{L} \times 303 \text{ K}}$$

$$= 0.031 \text{ mol}$$

$$m_{Ne} = 0.031 \text{ mol} \times \frac{20.18 \text{ g}}{1 \text{ mol}} = 0.63 \text{ g}$$

Exercise

31. What amount of methane gas is present in a sample that has a volume of 500 mL at 35.0°C and 210 kPa?

32. Determine the pressure in a 50 L compressed air cylinder if 30 mol of air is present in the container, which is heated to 40°C.

33. What volume does 50 kg of oxygen gas occupy at a pressure of 150 kPa and a temperature of 125°C?

34. At what temperature does 10.5 g of ammonia gas exert a pressure of 85.0 kPa in a 30.0 L container?

35. Starting with the ideal gas law, derive a formula to calculate the molar mass M of a gas, given the mass and volume of the gas at a specific pressure and temperature. Recall that $n = m/M$; that is, the amount in moles is equal to the mass divided by the molar mass.

36. Using the formula derived in question 35, calculate the molar mass of 1.00 L of gas that has a mass of 1.25 g and that exerts a pressure of 100 kPa at 0°C.

37. What is the density, in grams per litre, of propane at 22°C and 96.7 kPa?

38. A 1.49 g sample of a pure gas occupies a volume of 981 mL at 42.0°C and 117 kPa.
    (a) Determine the molar mass of the compound.
    (b) If the chemical formula is known to be XH_3, identify the element "X."

■ Problem
✔ Prediction
■ Design
■ Materials
■ Procedure
✔ Evidence
✔ Analysis
✔ Evaluation
■ Synthesis

C CAUTION

🔥 Butane is flammable. Do not conduct this experiment near an open flame. Good ventilation in the laboratory is required.

INVESTIGATION

6.4 Determining the Molar Mass of a Gas

In this investigation you will determine the molar mass of a sample of gas with the purpose of evaluating the experimental design used to obtain the result. Butane, $C_4H_{10(g)}$, is suggested but you may substitute another gas.

Problem

What is the molar mass of butane?

Experimental Design

A sample of butane gas from a lighter is collected in a graduated cylinder by downward displacement of water. The volume, temperature, and pressure of the gas are measured, along with the change in mass of the butane lighter. To obtain a more accurate molar mass of the experimental yield, you can correct the measured pressure for the vapor pressure of water. To do this, subtract the appropriate value of the vapor pressure of water (see Table 6.3) from

Table 6.3

VAPOR PRESSURE OF WATER	
Temperature (°C)	Vapor Pressure (kPa)
17	1.94
18	2.06
19	2.20
20	2.34
21	2.49
22	2.64
23	2.81
24	2.98
25	3.17
26	3.36
27	3.57
28	3.78
29	4.01
30	4.24

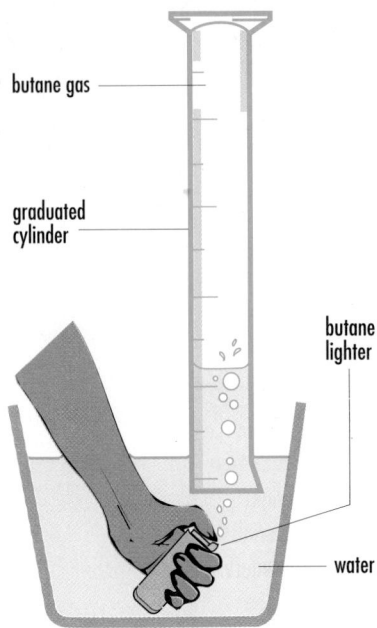

butane gas

graduated
cylinder

butane
lighter

water

Figure 6.16
*The gas from a butane lighter can
be collected by downward
displacement of water. This
apparatus can be used to determine
the molar mass of butane.*

your measured pressure. The design is evaluated on the basis of the accuracy of the experimental value for the molar mass of butane, which is compared with the accepted value.

Materials

lab apron
safety glasses
butane lighter (with flint removed) or butane cylinder with tubing
plastic bucket
500 mL graduated cylinder or 600 mL graduated beaker
balance
thermometer
barometer

Procedure

1. Determine the initial mass of the butane lighter.

2. Pour water into the bucket until it is two-thirds full and then completely fill the graduated cylinder with water and invert it in the bucket (Figure 6.16). Ensure that no air has been trapped in the cylinder.

3. Hold the butane lighter in the water and under the cylinder and release the gas until you have collected 400 mL to 500 mL of gas. Make sure all the bubbles enter the cylinder.

4. Equalize the pressures inside and outside the cylinder by adjusting the position of the cylinder until the water levels inside and outside the cylinder are the same.

5. Read the measurement on the cylinder and record the volume of gas collected.

6. Record the ambient (room) temperature and pressure.

7. Dry the butane lighter and determine its final mass.

8. Release the butane gas from the cylinder in a fume hood or outdoors.

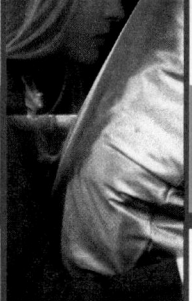

OVERVIEW

Gases

Summary

- Gases diffuse rapidly, are easily compressed, and, when heated, expand significantly.
- Volume and pressure are inversely related for gases. This relationship is called Boyle's law, expressed as

$$p_1v_1 = p_2v_2$$

- Volume and absolute temperature are directly related for gases. This relationship is called Charles' law, expressed as

$$\frac{v_1}{T_1} = \frac{v_2}{T_2}$$

- Volumes of gaseous reactants and products in a chemical reaction are in the same simple whole-number ratios as the coefficients in the reaction equation. This relationship is called the law of combining volumes.
- Avogadro's theory states that equal volumes of gases contain equal numbers of molecules, provided the pressure and temperature are the same.
- One mole of a gas occupies a specific volume called its molar volume, 22.4 L at STP or 24.8 L at SATP.
- The ideal gas law, $pv = nRT$, generalizes Boyle's law and Charles' law, and specifies the relationship among the four quantities that define a gas sample — pressure, volume, amount, and temperature.

Key Words

absolute zero
Avogadro's theory
Boyle's law
Charles' law
combined gas law
gas
ideal gas
ideal gas law
Kelvin temperature scale
law of combining volumes
molar volume
pressure
SATP
STP
universal gas constant

Review

1. State each of the following laws in a sentence beginning, "The volume of a gas sample...."
   (a) Boyle's law
   (b) Charles' law
   (c) the law of combining volumes
2. Compare the physical and chemical properties of the different gaseous elements.
3. State the ideal gas law in words.
4. Under what conditions is a real gas most similar to an ideal gas?
5. What is the major difference between a law and a theory? Use Charles' law and the kinetic molecular theory to support your answer.
6. Avogadro's idea is sometimes called a principle, a hypothesis, a law, or a theory. Is Avogadro's idea empirical or theoretical? Explain your answer.
7. Convert each of the following temperatures into absolute temperatures.
   (a) freezing point of water
   (b) 21°C room temperature
   (c) 37°C body temperature
   (d) absolute zero
8. Determine the molar mass of each of the following gases.
   (a) dinitrogen oxide (laughing gas)
   (b) propane (alternative automobile fuel)
9. Convert each of the following masses into an amount in moles.
   (a) 14 g of neon (used in neon signs)
   (b) 598 mg of uranium hexafluoride (separates uranium isotopes)
   (c) 29.8 kg of sulfur dioxide (produces acid rain)
10. Calculate the mass of each of the following.
   (a) 26 mol of bromine (an orange-red gas)

(b) 8.34 μmol of krypton (discovered in 1898)

(c) 2.7 kmol of sulfur trioxide (produces sulfuric acid)

11. Convert each of the following gas volumes into an amount in moles.
    (a) 5.1 L of carbon monoxide gas at SATP
    (b) 20.7 mL of fluorine gas at STP
    (c) 90 kL of nitrogen dioxide gas at SATP

12. Calculate the volumes at SATP of the following amounts of gas.
    (a) 500 mol of hydrogen (most common element in the universe)
    (b) 56 kmol of hydrogen sulfide (found in sour natural gas)

Applications

13. Argon gas is an inert carrier gas that moves other gases through a research or industrial system. What is the volume occupied by 4.2 kg of argon gas at SATP?

14. Freon gas is a chlorofluorocarbon (CFC) used as a coolant in air conditioners and refrigerators. If 500 mL of freon at 1.50 atm and 24°C is compressed to 250 mL at 3.50 atm, what is the final temperature of the gas?

15. The pressure of atmospheric air systems affects our weather. Describe two kinds of weather systems.

16. One of the most common uses of carbon dioxide gas is carbonating beverages such as soft drinks.
    (a) What is the new volume of a 300 L sample of carbon dioxide gas when the pressure doubles?
    (b) What is the new volume of a 300 L sample of carbon dioxide gas when the temperature increases from 30°C to 60°C?
    (c) What is the molar volume of carbon dioxide gas at 22°C and 94.0 kPa?
    (d) Design an experiment to determine the volume of carbon dioxide dissolved in a soft drink.

17. Pressurized hydrogen gas is used to fuel some prototype automobiles. What is the new volume of a 28.8 L sample of hydrogen in which the pressure is increased from 100 kPa to 350 kPa?

18. In Vancouver, a helium balloon containing 4.0 L of helium at 101 kPa was purchased. The balloon was taken on a trip to Banff where the atmospheric pressure was only 91 kPa.
    (a) Why is the atmospheric pressure in Banff generally lower than in Vancouver?
    (b) What will be the new volume of the balloon in Banff?

(c) What assumption is made in this calculation?

19. A glass container can hold an internal pressure of only 195 kPa before breaking. The container is filled with a gas at 19.5°C and 96.7 kPa and then heated. At what temperature will the container break?

20. Electrical power plants and ships commonly use steam to drive turbines, producing mechanical energy from the pressure of the steam. The rotating turbine is connected to a generator that produces electricity. Steam enters a turbine at a high temperature and pressure, and exits, still a gas, at a lower temperature and pressure. Determine the final pressure of steam that is converted from 10.0 kL at 600 kPa and 150°C to 18.0 kL at 110°C.

21. The Industrial Revolution that occurred in Europe during the mid-18th to the mid-19th centuries followed a technological advance: the development of a steam engine that could produce mechanical energy from heat. The pressure of steam converts heat into mechanical energy. What is the pressure increase in the boiler of a steam engine when the temperature is increased from 100°C to 200°C? (Express the increase as a percent.)

22. Yeast cells in bread dough convert sugar into either carbon dioxide and water, or carbon dioxide and ethanol, as shown in the following chemical equations.

$$C_6H_{12}O_{6(s)} + 6\,O_{2(g)} \rightarrow 6\,CO_{2(g)} + 6\,H_2O_{(g)}$$

$$C_6H_{12}O_{6(s)} \rightarrow 2\,CO_{2(g)} + 2\,C_2H_5OH_{(g)}$$

(a) Use the law of combining volumes to predict the volume of carbon dioxide produced when 50 mL of oxygen gas reacts with glucose.
(b) When the baking cycle is complete, which of the two reactions will produce the greater degree of leavening? Justify your answer.

23. What is the volume occupied by 1.0 g of carbon dioxide gas trapped in bread dough at SATP?

24. Steam production during baking is a secondary reason why bread and cakes rise. What volume of water vapor is produced inside a cake when 1.0 g of water is vaporized at 190°C and 103 kPa?

25. Large quantities of chlorine gas are produced from salt to make bleach and for water treatment. What is the volume of 26.5 kmol of chlorine gas at 400 kPa and 35°C?

26. Bromine is produced by reacting chlorine with bromide ions in sea water. What amount of

bromine is present in an 18.8 L sample of gas at 60 kPa and 140°C?

27. A student is trying to identify a pure gas sample. She decides to determine the molar mass of the gas, and obtains the following evidence.

    mass of evacuated container = 7.02 g
    mass of container plus gas = 9.31 g
    volume of container = 1.25 L
    temperature of gas = 23.4°C
    pressure of gas = 102.2 kPa

(a) From the evidence gathered, what is the molar mass of this common gas?

(b) What is a possible identity of the gas? Can you be certain of this? Briefly explain your reasoning.

28. "Standard ambient temperature and pressure" is a convention established by scientists to suit conditions on Earth. Suppose scientists were to establish standard conditions on the planet Venus as 800°C and 7500 kPa. What is the molar volume of Venus's mainly carbon dioxide atmosphere under these standard conditions?

29. Uranium hexafluoride is a very dense gas used to separate isotopes of uranium for nuclear applications. What is the density of this gas at SATP?

30. Suppose you were trapped in a room in which there was a slow natural gas leak (assume pure methane). In order to breathe as little natural gas as possible, should you be near the ceiling or the floor? Justify your answer.

31. A typical passenger hot air balloon contains 5.7 ML of air at an average temperature of 100°C. Show that the density of the air in the balloon is noticeably less than the surrounding air at SATP.

Extensions

32. Draw a concept map of the predominant ideas in this chapter, starting with gas volume.

33. As air passes over mountains, the gases become cooler when they expand and become warmer when they are compressed. Use the kinetic molecular theory to explain this.

34. Chinook winds cause rapid changes in weather. Calculate the final volume of a cubic metre (1.00 m³) of air at –23°C and 102 kPa when the temperature and pressure change to 12°C and 96 kPa during a chinook.

35. A temperature inversion is a weather pattern that can trap polluted air near ground level. Describe the circumstances and the process by which the polluted air becomes trapped.

36. To illustrate ideal behavior and real behavior of a gas, sketch a graph for (a) and (b). You will need to consider the influence of intermolecular forces at low temperatures and high pressure.
(a) volume and pressure of a gas
(b) volume and temperature of a gas

37. Design an experiment to test one of the gas laws. Assume that you have only everyday materials available to you, such as a pump, a pressure gauge, a balloon, a pail, hot and cold water, a measuring cup, a tape measure, and an outdoor alcohol thermometer.

38. Identify and evaluate the benefits and risks of using compressed gases such as methane, propane, and hydrogen, as fuels for vehicles.

Problem 6A Analyzing Gas Samples

Sulfur dioxide gas analysis is an important technique for monitoring emissions from gas plants and oil refineries. Complete the Analysis of the investigation report on the quantity of sulfur dioxide gas in a sample of air.

Problem

What mass of sulfur dioxide gas is present in a 20.00 L sample of air?

Experimental Design

The air sample at SATP is bubbled through a sodium hydroxide solution to remove the sulfur dioxide gas. The temperature and pressure are kept constant throughout the experiment. The new volume is measured as the sample emerges from the solution, and the mass of sulfur dioxide is calculated from the evidence gathered.

Evidence

initial volume of air = 20.00 L
final volume of air = 19.74 L
initial temperature = 25°C
final temperature = 25°C
initial pressure = 100 kPa
final pressure = 100 kPa

7 Chemical Reaction Calculations

One of the earliest chemical technologies is the control of fermentation — the production of alcohol from sugar. Alcohol, in the human body, induces chemical reactions that affect the co-ordination and judgment of the drinker and can lead to serious accidents. This is why Canada has laws stipulating limits to the concentration of alcohol allowed in the blood of a motorist.

When a driver is asked to breathe into a breathalyzer, the device measures the alcohol content in the exhaled air and indicates the result as a concentration of alcohol in the blood. For example, a reading of 0.08 on a breathalyzer means that the blood alcohol content is 0.08%, or 80 mg of alcohol in 100 mL of blood. A police officer takes a breath sample for on-the-spot analysis; because of the possibility of challenges in court, the officer must be prepared to defend the reliability and accuracy of the reading. If the breathalyzer test indicates an alcohol concentration above the legal limit, a second or a third sample may be analyzed more precisely in a laboratory using a technique called titration, which you will learn about in this chapter.

Chemical analysis involves knowledge of chemical reactions, understanding of diverse experimental designs, and practical skills to apply this knowledge and understanding. In this chapter, you will have opportunities to develop all of these.

QUANTITATIVE ANALYSIS

Complete chemical analysis of a substance usually begins with qualitative analysis (page 195) and is often followed by quantitative analysis (Figure 7.1). *Quantitative chemical analysis* involves the scientific concepts and technological skills needed to determine the quantity of a substance in a sample. Many techniques are available, but problem-solving skills, experience with analytical procedures, and general empirical knowledge are all important aspects of quantitative analysis. The chemistry and the technology of quantitative analysis are closely related; knowledge and skills in both areas are essential for chemical technologists in medicine, agriculture, and industry.

In one type of chemical analysis, precipitation is part of the experimental design. The sample under investigation is combined with an excess quantity of another reactant to make sure that all of the sample reacts. The reactant in the sample that is completely consumed is called the **limiting reagent**. The reactant that is present in more than the required amount is called the **excess reagent**.

Figure 7.1
Successful quantitative analysis depends on careful and precise work. It is also important for technologists to have attitudes that include honesty, perseverance, respect for evidence, and openness to unexpected results.

Problem 7A Chemical Analysis Using a Graph

Lab technicians sometimes perform the same chemical analysis on many samples every day. For example, in a medical laboratory, blood and urine samples are routinely analyzed for specific chemicals such as cholesterol and sugar. In many industrial and commercial laboratories, technicians read the required quantity of a chemical from a graph that has been prepared in advance. To illustrate this practice, complete the Analysis of the investigation report.

Problem

What mass of lead(II) nitrate is present in 20.0 mL of a solution?

Experimental Design

Samples of two different lead(II) nitrate solutions are used. Each sample is reacted with an excess quantity of a potassium iodide solution, producing lead(II) iodide, which has a low solubility and settles to the bottom of the beaker (Figure 7.2). After the contents of the beaker are filtered and dried, the mass of lead(II) iodide is determined. The reference data supplied in Table 7.1, relating the mass of $Pb(NO_3)_2$ to the mass of PbI_2 for this reaction, are graphed. The analysis is completed by reading from the graph the mass of lead(II) nitrate present in each solution.

Evidence

A bright yellow precipitate formed.

	Solution 1	Solution 2
Volume used (mL)	20.0	20.0
Mass of filter paper (g)	0.99	1.02
Mass of dried paper plus precipitate (g)	5.39	8.57

Figure 7.2
When lead(II) nitrate reacts with potassium iodide, a bright yellow precipitate forms.

Table 7.1

REFERENCE DATA: REACTION OF LEAD(II) NITRATE AND POTASSIUM IODIDE	
Mass of PbI$_2$ Produced (g)	**Mass of Pb(NO$_3$)$_2$ Reacting (g)**
1.39	1.00
2.78	2.00
4.18	3.00
5.57	4.00
6.96	5.00

Figure 7.3
Once the precipitate settles and the top layer becomes clear, you can test for the completeness of the reaction. Carefully run a drop or two of the excess reagent down the side of the beaker and watch for additional precipitation, which indicates that some of the limiting reagent remains in the solution.

▨	Problem
▨	Prediction
▨	Design
▨	Materials
▨	Procedure
✔	Evidence
✔	Analysis
✔	Evaluation
▨	Synthesis

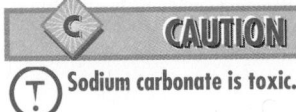

⬦ Sodium carbonate is toxic.

Table 7.2

REFERENCE DATA: REACTION OF SODIUM CARBONATE AND CALCIUM CHLORIDE	
Mass of CaCO₃ Produced (g)	**Mass of Na₂CO₃ Reacting (g)**
0.47	0.50
0.94	1.00
1.42	1.50
1.89	2.00
2.36	2.50

Practical Considerations

In a chemical analysis, it may not be possible to predict the quantity of excess reagent to use, so a procedure involving trial and error may be required. For precipitation reactions, you may need to use the following procedure to verify that a sample has completely reacted.

- Precisely measure a convenient volume of the limiting reagent solution.
- Slowly add an approximately equal volume of the excess reagent solution.
- Allow the precipitate to settle enough so that you can see a clear solution at the top of the mixture.
- Add a few more drops of the excess reagent (Figure 7.3).
- If any cloudiness is visible, repeat the procedure, using volumes that you judge to be appropriate.

An alternative procedure primarily used for verification is to filter the mixture and carry out a diagnostic test on the filtrate to determine if any unreacted chemical remains; for example, precipitation or a flame test might be used to test for any unreacted limited reagent that remains.

INVESTIGATION

7.1 Analysis of Sodium Carbonate Solution

The purpose of this investigation is to practice the skills used by chemical technicians in the quantitative analysis of a substance by precipitation. For a description of the method for filtering a precipitate, see Appendix C, C.3.

Problem

What mass of sodium carbonate is present in a 50.0 mL sample of a solution?

Experimental Design

The mass of sodium carbonate present in the sample solution is determined by having it react with an excess quantity of a calcium chloride solution. The mass of calcium carbonate precipitate produced is used to determine the mass of sodium carbonate that reacted; the figure is read from a graph of the reference data in Table 7.2.

Materials

lab apron
safety glasses
$Na_2CO_{3(aq)}$
$CaCl_{2(aq)}$
wash bottle of pure water
50 mL or 100 mL graduated cylinder
100 mL beaker
250 mL beaker
400 mL beaker
stirring rod
medicine dropper

filter paper
filter funnel, rack, and stand
centigram balance

Procedure

1. Measure 50.0 mL of $Na_2CO_{3(aq)}$ in the graduated cylinder and pour into a clean 250 mL beaker.

2. Obtain about 60 mL of $CaCl_{2(aq)}$ in a clean 100 mL beaker.

3. Slowly add about 50 mL of $CaCl_{2(aq)}$, with stirring, to the $Na_2CO_{3(aq)}$.

4. Allow the mixture to settle. When the top layer of the mixture becomes clear, add a few extra drops of $CaCl_{2(aq)}$ using the dropper.

5. If any cloudiness is visible, then repeat steps 2 to 4, adding as much $CaCl_{2(aq)}$ as necessary.

6. Measure the mass of a piece of filter paper.

7. Filter the mixture and discard the filtrate in the sink.

8. Dry the precipitate and filter paper overnight on a folded paper towel.

9. Measure the mass of the dried filter paper plus precipitate.

Quantitative Predictions

Chemical engineers design and control the chemical technology in a processing plant. For example, soda ash plants use limestone (calcium carbonate) and salt (sodium chloride) to produce soda ash (sodium carbonate). This technology requires quantitative predictions of the quantities of limestone and salt needed to make a specific quantity of soda ash. Quantitative predictions made to ensure that industrial processes work well are based largely on an understanding of the relative quantities of reactants consumed and products produced in a chemical system. This understanding can be entirely empirical, but more often, it is related to a knowledge of the balanced chemical equation. The main objective of this chapter is to use balanced chemical equations and quantitative relationships (from Chapters 4, 5, and 6) to develop and test methods of predicting relative quantities of reactants and products in a chemical reaction.

| INVESTIGATION

7.2 Decomposing Malachite

The chemical in this investigation is known by several names. The systematic or IUPAC name for it is *copper(II) hydroxide carbonate*. From this name the formula of the substance can be determined. Geologists, however, refer to this substance as *malachite*. It also has a common name that appears in chemical supply catalogues — *basic copper carbonate*. Copper(II) hydroxide carbonate is a green

- Problem
- ✔ Prediction
- Design
- Materials
- ✔ Procedure
- ✔ Evidence
- ✔ Analysis
- ✔ Evaluation
- Synthesis

Figure 7.4
Use the glass stirring rod to break up lumps of malachite and to mix the contents of the dish while they are being heated. Large lumps may decompose on the outside but not on the inside.

double salt with the chemical formula $Cu(OH)_2 \cdot CuCO_{3(s)}$. This double salt decomposes completely when heated to 200°C, forming copper(II) oxide, carbon dioxide, and water vapor. The purpose of this investigation is to test the use of a balanced chemical equation in predicting the relative amounts of chemicals in a chemical reaction.

Problem

How is the amount in moles of copper(II) oxide produced related to the amount in moles of malachite reacted in the decomposition of malachite?

Experimental Design

A known mass of malachite is heated strongly until the color changes completely from green to black (Figure 7.4). The mass of black product is determined. The results from several laboratory groups are combined in a graph to answer the Problem question. The mass of malachite used is the manipulated (independent) variable and the mass of copper(II) oxide produced is the responding (dependent) variable. (If possible, use a computer spreadsheet program to graph and analyze the best fit line.)

Materials

lab apron
safety glasses
porcelain dish (or crucible and clay triangle)
small ring stand
hot plate or laboratory burner (Figure 1.11 (c), page 44)
glass stirring rod
sample of malachite (not more than 3 g)
centigram balance
laboratory scoop or plastic spoon

Exercise

1. What is the purpose of a quantitative chemical analysis?

2. How is a graph used in a quantitative chemical analysis?

3. In order to prepare tables of reference data, experiments are done to obtain the evidence. In the experiments to prepare Table 7.1 on page 251, which is the manipulated (independent) variable and which is the responding (dependent) variable?

4. In a quantitative chemical analysis of a substance, why is it desirable to use an excess of one reactant?

5. Using the analysis from Investigation 7.2, complete the following table comparing amounts of malachite and copper(II) oxide in the decomposition reaction.

Amount of Malachite (mol)	Amount of Copper(II) Oxide (mol)
10	
2.6	
	10
	6.4

6. What amount in moles of copper(II) oxide would be produced by decomposing 75 g of malachite?

7. What amount in moles of malachite would be consumed to produce 36 g of copper(II) oxide?

GRAVIMETRIC STOICHIOMETRY: CALCULATING MASSES

Analysis of the evidence from Investigation 7.2 indicates that when malachite is decomposed, the ratio of the amounts in moles of copper(II) oxide and malachite is a simple ratio of 2:1. This is the same ratio given by the coefficients of these substances in the balanced chemical equation. Two moles of copper(II) oxide are produced from one mole of malachite.

$$Cu(OH)_2 \cdot CuCO_{3(s)} \rightarrow 2\ CuO_{(s)} + CO_{2(g)} + H_2O_{(g)}$$

Unfortunately, there is no instrument that measures amounts in moles directly. Therefore, some measurable quantity such as mass is required to predict and analyze the quantities of reactants and products in a chemical reaction. However, the key to any relationship between two substances in a chemical reaction depends on the mole ratio from the balanced chemical equation.

The procedure for calculating the masses of reactants or products in a chemical reaction is called **gravimetric stoichiometry**. Although this expression is quite a mouthful, these two words can be understood by looking at the root words. *Gravimetric* refers to mass measurement. *Stoichiometry* comes from Greek and Old English words meaning a series of steps for measuring something. In general, stoichiometry is the procedure used to calculate the quantities of chemicals in chemical reactions. Gravimetric stoichiometry is restricted to determining the masses of chemicals that are involved in a chemical reaction, either as reactants or products.

Suppose that you decomposed 1.00 g of malachite. What mass of copper(II) oxide would be formed? To answer this question, start by writing the balanced chemical equation. Underneath the balanced equation, write the mass that is given and the symbol m for the mass to be calculated, along with the conversion factors. In this example, one mass is given and the conversion factors (that is, the molar masses) are calculated from the chemical formulas and the information in the

Underwater Construction
Bridges across waterways as well as many docks in British Columbia rely on concrete for strength. The concrete construction in these structures is done partly on land and partly underwater. This is possible because concrete doesn't have to "dry"; it goes from liquid to solid through a series of chemical reactions with water, called "curing." Concrete is a mixture of sand (and/or stone) and cement. The active compounds in cement (calcium silicates and calcium aluminates) are unstable and react with water. The reaction products crystallize over time, binding together the sand or stone of the concrete mix into a rock-like substance capable of withstanding heavy traffic or severe weather conditions.

periodic table. The balanced chemical equation provides the mole ratio of the two chemicals.

$$Cu(OH)_2 \cdot CuCO_{3(s)} \rightarrow 2\,CuO_{(s)} + CO_{2(g)} + H_2O_{(g)}$$

1.00 g m
221.13 g/mol   79.55 g/mol

In the second step, the given mass of malachite is converted to an amount in moles.

$$n_{Cu(OH)_2 \cdot CuCO_3} = 1.00\ g \times \frac{1\ mol}{221.13\ g} = 0.004\,52\ mol$$

The third step is to calculate, using the mole ratio 1:2, the amount of copper(II) oxide that will be produced.

$$\frac{n_{CuO}}{n_{Cu(OH)_2 \cdot CuCO_3}} = \frac{2}{1}$$

$$\frac{n_{CuO}}{0.004\,52\ mol} = \frac{2}{1}$$

$$n_{CuO} = 0.004\,52\ mol \times \frac{2}{1} = 0.009\,04\ mol$$

In the final step, calculate the mass represented by this amount of CuO.

$$m_{CuO} = 0.009\,04\ mol \times \frac{79.55\ g}{1\ mol} = 0.719\ g$$

The three calculation steps described above can be combined into a single calculation as shown below.

$$m_{CuO} = 1.00\ \text{g malac} \times \frac{1\ \text{mol malac}}{221.13\ \text{g malac}} \times \frac{2\ \text{mol CuO}}{1\ \text{mol malac}} \times \frac{79.55\ g\ CuO}{1\ \text{mol CuO}}$$
$$= 0.719\ g$$

According to the stoichiometric method, the mass of copper(II) oxide produced by the decomposition of 1.00 g of malachite is 0.719 g.

The certainty here, three significant digits, is determined by the least certain value, 1.00 g. Note that the mass of copper(II) oxide has been obtained by knowing only the balanced chemical equation and the molar masses. The actual experiment was not necessary. The following example illustrates how to communicate the stoichiometric procedure.

EXAMPLE

Iron is the most widely used metal in North America (Figure 7.5). It may be produced by the reaction of iron(III) oxide, from iron ore, with carbon monoxide to produce iron metal and carbon dioxide. What mass of iron(III) oxide is required to produce 100.0 g of iron?

$$Fe_2O_{3(s)} + 3\,CO_{(g)} \rightarrow 2\,Fe_{(s)} + 3\,CO_{2(g)}$$

m 100.0 g
159.70 g/mol   55.85 g/mol

$$n_{Fe} = 100.0\ g \times \frac{1\ mol}{55.85\ g} = 1.791\ mol$$

Figure 7.5
Wrought iron is a very pure form of iron. The ornate gates on Parliament Hill in Ottawa are made of wrought iron.

$$n_{Fe_2O_3} = 1.791 \text{ mol} \times \frac{1}{2} = 0.8953 \text{ mol}$$

$$m_{Fe_2O_3} = 0.8953 \text{ mol} \times \frac{159.70 \text{ g}}{1 \text{ mol}} = 143.0 \text{ g}$$

or $m_{Fe_2O_3} = 100.0 \text{ g Fe} \times \dfrac{1 \text{ mol Fe}}{55.85 \text{ g Fe}} \times \dfrac{1 \text{ mol Fe}_2\text{O}_3}{2 \text{ mol Fe}} \times \dfrac{159.70 \text{ g Fe}_2\text{O}_3}{1 \text{ mol Fe}_2\text{O}_3}$

$$= 143.0 \text{ g}$$

This stoichiometric procedure can also be extended to determine the number of molecules *(N)* that have reacted or are produced. For example, if 20 g of butane, $C_4H_{10(g)}$, are completely burned in a lighter, how many molecules of carbon dioxide are produced? The chemical equation for this combustion is shown below. In all stoichiometry problems it is useful to list the measurements and conversion factors underneath the appropriate chemical formulas.

$$2 \text{ C}_4\text{H}_{10(g)} \quad + \quad 13 \text{ O}_{2(g)} \quad \rightarrow \quad 8 \text{ CO}_{2(g)} \quad + \quad 10 \text{ H}_2\text{O}_{(g)}$$

20 g N

58.14 g/mol 6.02×10^{23}/mol

Coefficients in a balanced chemical equation represent reacting amounts in moles. Therefore, the key to all stoichiometry is to work in units of moles. This means that you first have to convert the information about the known substance into an amount in moles,

$$n_{C_4H_{10}} = 20 \text{ g} \times \frac{1 \text{ mol}}{58.14 \text{ g}} = 0.344 \text{ mol}$$

Now you can "jump" to the required substance, carbon dioxide, using the mole ratio given by the coefficients.

$$n_{CO_2} = 0.344 \text{ mol} \times \frac{8}{2} = 1.38 \text{ mol}$$

Finally, this amount is converted into whatever quantity is requested in the question. In this example, you need to convert the amount in moles to the number of molecules, using the Avogadro constant.

$$N_{CO_2} = 1.38 \text{ mol} \times \frac{6.02 \times 10^{23}}{1 \text{ mol}} = 8.3 \times 10^{23}$$

Again, once you understand the steps involved in a stoichiometry calculation, they can be combined into a single calculation.

$$N_{CO_2} = 20 \text{ g C}_4\text{H}_{10} \times \frac{1 \text{ mol C}_4\text{H}_{10}}{58.14 \text{ g C}_4\text{H}_{10}} \times \frac{8 \text{ mol CO}_2}{2 \text{ mol C}_4\text{H}_{10}} \times \frac{6.02 \times 10^{23} \text{ CO}_2}{1 \text{ mol CO}_2}$$

$$= 8.3 \times 10^{23}$$

Exercise

8. Powdered zinc metal reacts violently with sulfur (S_8) when heated to produce zinc sulfide (Figure 7.6). Predict the mass of sulfur required to react with 25 g of zinc.

Figure 7.6
The reaction of powdered zinc and sulfur is rapid and highly exothermic. Because of the numerous safety precautions that would be necessary, it is not usually carried out in school laboratories.

9. Bauxite ore contains aluminum oxide, which is decomposed using electricity to produce aluminum metal. What mass of aluminum metal can be produced from 125 g of aluminum oxide?

10. Determine the mass of oxygen required to completely burn 10.0 g of propane.

11. Calculate the mass of lead(II) chloride precipitate produced when 2.57 g of sodium chloride in solution reacts in a double replacement reaction with excess aqueous lead(II) nitrate.

12. Predict the mass of hydrogen gas produced when 2.73 g of aluminum reacts in a single replacement reaction with excess sulfuric acid.

13. What mass of copper(II) hydroxide precipitate is produced by the reaction in solution of 2.67 g of potassium hydroxide with excess aqueous copper(II) nitrate?

14. How many molecules of oxygen are produced from the simple decomposition of 25.0 g of water?

Testing the Stoichiometric Method

The most rigorous test of any scientific concept is whether or not it can be used to make predictions. If the prediction is shown to be valid, then the concept is judged to be acceptable. The prediction is falsified if the percent difference between the actual and the predicted values is considered to be too great; for example, more than 10%. The concept may then be judged unacceptable. (See "Evaluation" in Appendix B, B.2.) Percent difference between an experimental value and predicted value is the primary criterion for the evaluation of an accepted value (for example, a constant) or an accepted method (for example, stoichiometry). It is assumed that reagents are pure and skills are adequate for the experiment that is conducted.

As in Investigation 7.1, filtration is a common experimental design for testing stoichiometric predictions. Stoichiometry is used to *predict* the mass of precipitate that will be produced, and filtration is used to *analyze* the mass of precipitate actually produced in a reaction. Problems 7B and 7C and Investigation 7.3 test the validity of the stoichiometric method. In all examples, an excess of one reactant is used to ensure the complete reaction of the limiting reagent.

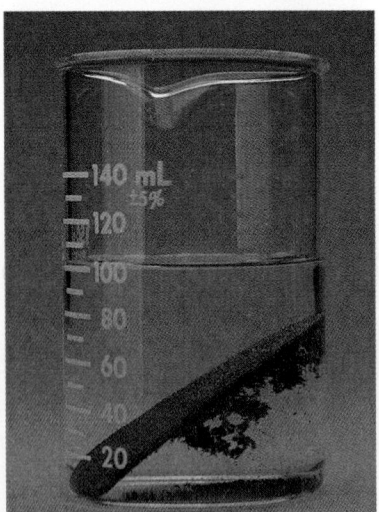

Figure 7.7
Zinc reacts with a solution of lead(II) nitrate.

Problem 7B Testing Gravimetric Stoichiometry

The purpose of this problem is to test the stoichiometric method. Complete the Prediction, Analysis, and Evaluation of the investigation report. Evaluate the experimental design, the prediction, and the method of stoichiometry.

Problem

What mass of lead is produced by the reaction of 2.13 g of zinc with an excess of lead(II) nitrate in solution (Figure 7.7)?

Experimental Design

A known mass of zinc is placed in a beaker with an excess of lead(II) nitrate solution. The lead produced in the reaction is separated by filtration and dried. The mass of the lead is determined.

Evidence

In the beaker, crystals of a shiny black solid were produced and all of the zinc reacted.

mass of filter paper = 0.92 g

mass of dried filter paper plus lead = 7.60 g

Limiting and Excess Reagents

Using the stoichiometric procedure it is now possible to calculate an excess of a chemical used or recognize how much is in excess. This calculation is necessary to develop an efficient design of an experiment to test the stoichiometric method (as in Investigation 7.3). As a general guideline, an excess of 10% should be sufficient to ensure that the limiting reagent is completely consumed.

Suppose you decided to test the method of stoichiometry using the reaction of 2.00 g of copper(II) sulfate in solution with an excess of sodium hydroxide in solution. What would be a reasonable mass of sodium hydroxide to use? To answer this question, you should calculate the minimum mass required and then add 10% to this mass. The first part of this plan follows the usual steps of stoichiometry.

$$CuSO_{4(aq)} \quad + \quad 2\,NaOH_{(aq)} \quad \rightarrow \quad Cu(OH)_{2(s)} \quad + \quad Na_2SO_{4(aq)}$$

2.00 g $\qquad\quad$ m

159.71 g/mol $\quad$ 40.00 g/mol

$$n_{CuSO_4} = 2.00\ g \times \frac{1\ mol}{159.71\ g} = 0.0125\ mol$$

$$n_{NaOH} = 0.0125\ mol \times \frac{2}{1} = 0.0250\ mol$$

$$m_{NaOH} = 0.0250\ mol \times \frac{40.00\ g}{1\ mol} = 1.00\ g$$

or

$$m_{NaOH} = 2.00\ g\ \text{CuSO}_4 \times \frac{1\ mol\ \text{CuSO}_4}{159.71\ g\ \text{CuSO}_4} \times \frac{2\ mol\ \text{NaOH}}{1\ mol\ \text{CuSO}_4} \times \frac{40.00\ g\ NaOH}{1\ mol\ \text{NaOH}}$$

$$= 1.00\ g$$

According to the stoichiometry, 1.00 g of $NaOH_{(s)}$ is required to react with 2.00 g of $CuSO_{4(s)}$. Therefore, 1.10 g of $NaOH_{(s)}$ (required mass plus 10%) would be a reasonable value to use in the experiment.

Problem 7C Designing and Testing Stoichiometry

The purpose of this problem is to test the method of stoichiometry and to prepare you for Investigation 7.3. Complete the Prediction, Experimental Design (including the mass of excess reagent to be used), Materials, Procedure, Analysis, and Evaluation of the investigation report.

FORENSIC CHEMIST

Carolyn Krausher is a chemist with an RCMP forensic laboratory. She works with police forces, customs officers, and fire department investigators to analyze materials that are related to a criminal offense. The types of materials vary widely, but they typically include paint-smeared clothing of hit-and-run victims or chips of paint found at the scene of an automobile accident. Krausher has analyzed paint smears on a pry bar that showed it had been used as a break-and-enter tool. She has identified particles on a suspect's clothing as being from the packing inserted between the inner and outer walls of a safe that was damaged in a burglary attempt.

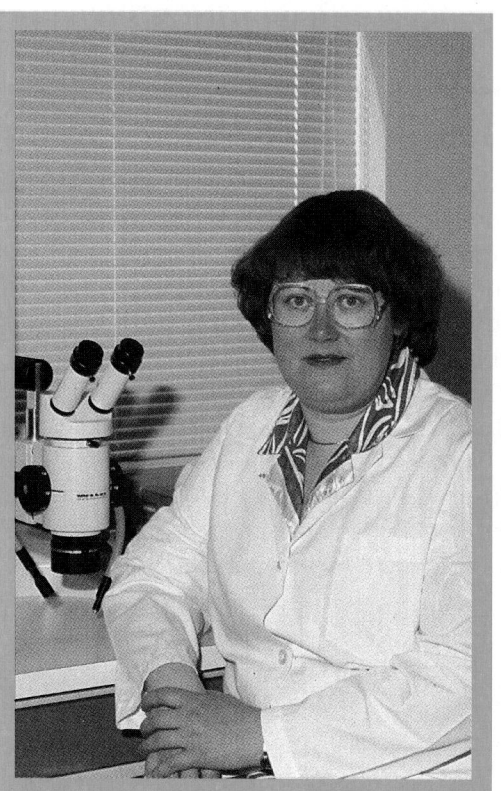

To identify materials, Krausher relies on a variety of technologies, including microscopy, gas chromatography, infrared spectroscopy, and X-ray diffraction. An understanding of statistics is also important in determining the significance of an analysis.

Krausher decided to become a forensic chemist during her fourth year of a B.Sc. program specializing in chemistry. In high school, Krausher enrolled in core academic subjects, achieving her highest grades in mathematics, physics, and chemistry. She chose chemistry as her area of specialization because she thought it offered her the widest range of career options.

Krausher comments on her profession: "I like the case work, which is the core of the job — working with court exhibits. You get an impact smear on the clothing of a hit-and-run victim with a bit of red or blue paint, or whatever. You can see it, but you know it's too smeared to actually work with. To sort through the debris and find a paint chip that you can work with, for example, is really rewarding. In other situations, if you find several kinds of paint at the scene, as well as glass and other materials, you might have enough to start to bring the picture together. You strengthen the evidence as you consider all the bits and pieces."

A forensic chemist prepares reports for investigators; these reports become part of the case for or against an accused person. Krausher often appears in court as an expert witness, answering questions about the analysis of the material on exhibit, even though fewer than 10% of her analyses actually lead to court cases.

One attractive aspect of this job is the variety. There is no typical day. In a crime laboratory, a forensic chemist must be self-motivated and must assign priorities to tasks so that the most pressing work is completed first.

A civilian member of the RCMP, Krausher has had to meet the same security requirements as regular RCMP personnel. For a forensic chemist, though, the physical requirements are less stringent. If accepted into an RCMP crime laboratory, a chemist must be willing to transfer to another location, although this is not as common as for the regular force. RCMP crime laboratories are located in Vancouver, Edmonton, Regina, Winnipeg, Ottawa, Sackville (New Brunswick), and Halifax.

Krausher enjoys her job and recommends this career to students interested in chemistry. Her advice to high school students is to strive to do well in chemistry, physics, and mathematics, especially statistics. Computer studies would also be a valuable option.

- *In general, what does a forensic chemist do?*
- *What high school courses should you take to prepare yourself for a career as a forensic chemist?*

Problem

What is the mass of precipitate formed when 2.98 g of sodium phosphate in solution reacts with an excess of aqueous calcium nitrate?

Evidence

A white precipitate formed when the sodium phosphate and calcium nitrate solutions were mixed.

mass of filter paper = 0.93 g
mass of dried filter paper plus precipitate = 3.82 g

In the diagnostic test for excess calcium nitrate, a sodium carbonate solution was added to the filtrate; a white precipitate formed.

INVESTIGATION

7.3 Testing the Stoichiometric Method

The purpose of this investigation is to test the stoichiometric method. Part of your design should include a calculation of the mass of excess reagent you will use.

Problem

What is the mass of precipitate produced by the reaction of 2.00 g of strontium nitrate in 75 mL of water with excess (2.50–3.00 g) copper(II) sulfate in 75 mL of water?

Applications of Stoichiometry

Once you have tested a scientific concept sufficiently, you can use it as a tool in other scientific and technological work. You have probably found that all of the tests of stoichiometry to this point have shown the method to be acceptable. Now you can consider the stoichiometric approach to be valid and you can use it as an analytical tool.

Stoichiometry can be used to test experimental designs, technological skills, purity of chemicals (Figure 7.8), and the quantitative nature of a particular reaction. In all cases, an evaluation can be done using a **percent yield**. This is the ratio of the actual or experimental quantity of product obtained (actual yield) to the maximum possible quantity of product (predicted or theoretical yield) obtained from a stoichiometry calculation.

$$\text{percent yield} = \frac{\text{actual yield}}{\text{predicted yield}} \times 100$$

Another application of stoichiometry is the identification of limiting and excess reagents in a chemical reaction. If two known quantities of chemicals are reacted, one will usually be completely consumed (the limiting reagent) and the other will be in excess. This is determined using the same stoichiometric principles that you have already employed. Like all stoichiometry problems, the mole ratio from the balanced chemical equation is the key part of the solution. Consider the following reaction and starting quantities.

- Problem
- ✔ Prediction
- ✔ Design
- ✔ Materials
- ✔ Procedure
- ✔ Evidence
- ✔ Analysis
- ✔ Evaluation
- Synthesis

CAUTION

Strontium nitrate is moderately toxic; there is risk of fire when it is in contact with organic chemicals and it may explode when bumped or heated. Copper(II) sulfate is a strong irritant and is toxic if ingested.

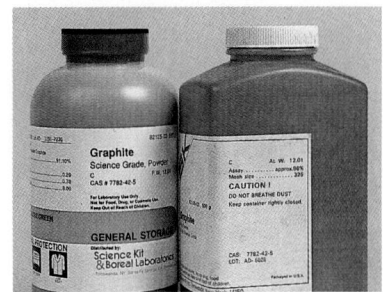

Figure 7.8
Chemicals come in a wide variety of grades (purities). Some low-purity or technical grades may only be 80–90% pure, whereas high-purity or reagent grades may be better than 99% pure. The purity of a chemical can significantly affect experimental results when studying chemical reactions.

$$Cu_{(s)} \quad + \quad 2\,AgNO_{3(aq)} \quad \rightarrow \quad 2\,Ag_{(s)} \quad + \quad Cu(NO_3)_{2(aq)}$$

10.0 g   20.0 g

63.55 g/mol   169.88 g/mol

According to the balanced equation, one mole of copper reacts completely with two moles of silver nitrate. To determine which is limiting (and therefore which is in excess), you need to convert the quantities given into amounts in moles.

$$n_{Cu} = 10.0\,g \times \frac{1\,mol}{63.55\,g} = 0.157\,mol$$

$$n_{AgNO_3} = 20.0\,g \times \frac{1\,mol}{169.88\,g} = 0.118\,mol$$

You now need to test one of these values using the mole ratio from the chemical equation. In other words, assume one chemical is completely used up and see if a sufficient amount of the second chemical is available. If copper is the limiting reagent, then the amount in moles of silver nitrate that is required is shown below.

$$n_{AgNO_3} = 0.157\,mol \times \frac{2}{1} = 0.315\,mol$$

Obviously, this value (0.315 mol) is much greater than the amount we actually have available (0.118 mol). Therefore, the assumption is incorrect — copper cannot be the limiting reagent; it must be present in excess. Notice that it does not matter which chemical you initially assume is limiting. You will be able to identify limiting and excess reagents no matter which chemical you first assume is the limiting reagent.

Once the limiting and excess reagents have been identified, a number of questions can immediately be answered. How much of the excess reagent will remain after the reaction? How much product will be obtained? It is important to note that *all predictions made from a balanced chemical equation must be based on the limiting reagent*. To find the excess quantity of copper, find the mass of copper required to react with the 20.0 g of silver nitrate and compare that with the starting mass of copper. You know that the 20.0 g of silver nitrate is equivalent to 0.118 mol, and the chemical equation shows that the mole ratio of copper to silver nitrate is 1:2. Therefore, the amount in moles of copper that reacts can be calculated as follows:

$$n_{Cu} = 0.118\,mol \times \frac{1}{2} = 0.059\,mol$$

This can now be converted, using the molar mass of copper, to a mass of copper:

$$m_{Cu} = 0.059\,mol \times \frac{63.55\,g}{1\,mol} = 3.7\,g$$

The excess quantity of copper is 10.0 g – 3.7 g, or 6.3 g.

To find the yield of silver product expected, use the amount in moles of the limiting reagent, $AgNO_3$, to predict the mass of this product.

$$n_{Ag} = 0.118\,mol \times \frac{2}{2} = 0.118\,mol$$

$$m_{Ag} = 0.118\,mol \times \frac{107.87\,g}{1\,mol} = 12.7\,g$$

$A + B \rightarrow C + D$

Assuming A is limiting, is sufficient B available according to the mole ratio?
• Yes, then A is limiting and B is in excess.
• No, then A is in excess and B is limiting.

In an experiment, 26.8 g of iron(III) chloride in solution is combined with 21.5 g of sodium hydroxide in solution. Which reactant is in excess and by how much? What mass of each product will be obtained?

$$FeCl_{3(aq)} \quad + \quad 3\,NaOH_{(aq)} \quad \rightarrow \quad Fe(OH)_{3(s)} \quad + \quad 3\,NaCl_{(aq)}$$

26.8 g   21.5 g m m

162.20 g/mol   40.00 g/mol   106.88 g/mol   58.44 g/mol

$$n_{FeCl_3} = 26.8\ g \times \frac{1\ mol}{162.20\ g} = 0.165\ mol$$

$$n_{NaOH} = 21.5\ g \times \frac{1\ mol}{40.00\ g} = 0.538\ mol$$

If $FeCl_3$ is the limiting reagent, the amount of NaOH required is

$$n_{NaOH} = 0.165\ mol \times \frac{3}{1} = 0.496\ mol$$

Therefore, NaOH is in excess, and the excess amount and mass are

$$n_{NaOH} = 0.538\ mol - 0.496\ mol = 0.042\ mol$$

$$m_{NaOH} = 0.042\ mol \times \frac{40.00\ g}{1\ mol} = 1.7\ g$$

The mass of the two products can also be calculated.

$$m_{Fe(OH)_3} = 0.165\ mol\ FeCl_3 \times \frac{1\ mol\ Fe(OH)_3}{1\ mol\ FeCl_3} \times \frac{106.88\ g\ Fe(OH)_3}{1\ mol\ Fe(OH)_3}$$
$$= 17.7\ g$$

$$m_{NaCl} = 0.165\ mol\ FeCl_3 \times \frac{3\ mol\ NaCl}{1\ mol\ FeCl_3} \times \frac{58.44\ g\ NaCl}{1\ mol\ NaCl}$$
$$= 29.0\ g$$

Exercise

15. How is a scientific concept tested?

16. When testing gravimetric stoichiometry using an experiment, should you combine the two reactants in the same ratio as they appear in the balanced chemical equation? Justify your answer.

17. When calculating a percent yield, where does the value of the actual yield come from? Where does the predicted yield come from?

18. A quick, inexpensive source of hydrogen gas is the reaction of zinc with hydrochloric acid. If 0.35 mol of zinc is placed in 0.60 mol of hydrochloric acid,
    (a) which reactant will be completely consumed?
    (b) what mass of the other reactant will remain after the reaction is complete?

Figure 7.9
Electroplating produces a thin metal coating on objects such as the car door handles shown in the photograph. Chromium plating is used for esthetic as well as technical reasons because it creates a shiny surface and also prevents corrosion. However, environmental damage may result if the toxic solutions used in electroplating are dumped as waste. Treating toxic wastes to transform them into safe materials is sometimes prohibitively expensive.

19. A chemical technician is planning to react 3.50 g of lead(II) nitrate with excess potassium bromide in solution.
   (a) What would be a reasonable mass of potassium bromide to use in this reaction?
   (b) Predict the mass of precipitate expected.
   (c) What assumptions must be made about the chemicals and reaction?

20. In a chemical analysis, 3.00 g of silver nitrate in solution was reacted with excess sodium chromate to produce 2.81 g of precipitate. What is the percent yield?

21. A solution containing 9.8 g of barium chloride is mixed with a solution containing 5.1 g of sodium sulfate.
   (a) Which reactant is in excess?
   (b) Determine the excess mass.
   (c) Predict the mass of precipitate.

22. A solution containing 18.6 g of chromium(III) chloride is reacted with a 15.0 g piece of zinc to produce chromium metal (Figure 7.9).
   (a) Which reactant is in excess?
   (b) Determine the excess mass.
   (c) If 5.1 g of chromium metal are formed, what is the percent yield?

Problem 7D Testing a Chemical Process

The purpose of this problem is to perform a quality control test on a chemical process. A sample of a solution used in the process is taken to the laboratory to determine the mass of sodium silicate in a specific volume of the solution. Complete the Analysis and Evaluation of the investigation report. In the Evaluation, suggest at least one way to improve the efficiency of this laboratory analysis.

Problem

What is the mass of sodium silicate in a 25.0 mL sample of the solution used in a chemical process?

Prediction

If the process is operating as expected, the mass of sodium silicate in a 25.0 mL sample should be between 6.40 g and 6.49 g.

Experimental Design

An excess quantity of iron(III) nitrate is added to the sodium silicate sample and the precipitate is separated by filtration. After the precipitate has dried, its mass is determined.

Evidence

mass of filter paper = 0.98 g
mass of dried filter paper plus precipitate = 9.45 g
The color of the filtrate was yellow-orange.

Many chemical reactions involve gases. One common consumer example is the combustion of propane in a home gas barbecue (Figure 7.10). The reaction of chlorine in a water treatment plant is a commercial example. An important industrial application of a chemical reaction involving gases is the production of the fertilizer ammonia from nitrogen and hydrogen gases. These technological examples feature gases as either valuable products, such as ammonia, or as part of an essential process, such as water treatment.

Studies of chemical reactions involving gases (for example, the law of combining volumes, page 234) have helped scientists to develop ideas about molecules and explanations for chemical reactions, such as the collision-reaction theory (page 137). In both technological applications and scientific studies of gases, it is necessary to accurately calculate quantities of gaseous reactants and products.

The method of stoichiometry applies to all chemical reactions. This section extends stoichiometry to gases, using gas volume and molar volume (page 233). For example, if 300 g of propane burns in a gas barbecue, what volume of oxygen measured at the standard conditions of STP is required for the reaction? To answer this question, write a balanced chemical equation to relate the propane to the oxygen. List the given and the required measurements and the conversion factors for each chemical, just as you did in the previous section.

Figure 7.10
Propane gas barbecues have become very popular. Charcoal barbecues are now banned in parts of California because they produce five times as much pollution (nitrogen oxides, hydrocarbons, and particulates) as gas barbecues.

$$C_3H_{8(g)} \quad + \quad 5\,O_{2(g)} \quad \rightarrow \quad 3\,CO_{2(g)} \quad + \quad 4\,H_2O_{(g)}$$

$C_3H_{8(g)}$	$5\,O_{2(g)}$
300 g	v (STP)
44.11 g/mol	22.4 L/mol

Since propane and oxygen are related by their mole ratio, you must convert the mass of propane to an amount in moles.

$$n_{C_3H_8} = 300\text{ g} \times \frac{1\text{ mol}}{44.11\text{ g}} = 6.80\text{ mol}$$

The balanced equation indicates that one mole of propane reacts with five moles of oxygen. Use this mole ratio to calculate the amount of oxygen required, in moles. (This step is common to all stoichiometry calculations.)

$$n_{O_2} = 6.80\text{ mol} \times \frac{5}{1} = 34.0\text{ mol}$$

The final step involves converting the amount of oxygen to the required quantity; in this case, volume.

$$v_{O_2} = 34.0\text{ mol} \times \frac{22.4\text{ L}}{1\text{ mol}} = 762\text{ L}$$

Note that the final step uses the molar volume at STP as a conversion factor, in the same way that molar mass is used in gravimetric stoichiometry.

Once you understand the logic of the individual steps, the calculation may be done as a single step.

$$v_{O_2} = 300 \text{ g C}_3\text{H}_8 \times \frac{1 \text{ mol C}_3\text{H}_8}{44.11 \text{ g C}_3\text{H}_8} \times \frac{5 \text{ mol O}_2}{1 \text{ mol C}_3\text{H}_8} \times \frac{22.4 \text{ L O}_2}{1 \text{ mol O}_2}$$

$$= 762 \text{ L}$$

The following example illustrates how to communicate solutions to stoichiometric problems involving gases.

EXAMPLE _____

Hydrogen gas is produced when sodium metal is added to water. What mass of sodium is necessary to produce 20.0 L of hydrogen at SATP?

$$2 \text{ Na}_{(s)} + 2 \text{ HOH}_{(l)} \longrightarrow \text{H}_{2(g)} + 2 \text{ NaOH}_{(aq)}$$

m	20.0 L
22.99 g/mol	24.8 L/mol

$$n_{H_2} = 20.0 \text{ L} \times \frac{1 \text{ mol}}{24.8 \text{ L}} = 0.806 \text{ mol}$$

$$n_{Na} = 0.806 \text{ mol} \times \frac{2}{1} = 1.61 \text{ mol}$$

$$m_{Na} = 1.61 \text{ mol} \times \frac{22.99 \text{ g}}{1 \text{ mol}} = 37.1 \text{ g}$$

or $m_{Na} = 20.0 \text{ L H}_2 \times \frac{1 \text{ mol H}_2}{24.8 \text{ L H}_2} \times \frac{2 \text{ mol Na}}{1 \text{ mol H}_2} \times \frac{22.99 \text{ g Na}}{1 \text{ mol Na}}$

$$= 37.1 \text{ g}$$

Note that the general steps of the stoichiometry calculation are the same for both solids and gases. Changes from mass to amount or from volume to amount, or vice versa, are done using the molar mass or the molar volume, respectively, of the substance. Although the molar mass depends on the chemical involved, the molar volume of a gas depends only on temperature and pressure. If the conditions are not standard (that is, STP or SATP) then the ideal gas law ($pv = nRT$, page 243), rather than the molar volume, is used to find the amount or volume of a gas, as in the following example.

EXAMPLE _____

In an industrial application known as the Haber process (Chapter 13), ammonia to be used as fertilizer results from the reaction of nitrogen and hydrogen. What volume of ammonia at 450 kPa pressure and 80°C can be obtained from the complete reaction of 7.5 kg of hydrogen?

$$\text{N}_{2(g)} + 3 \text{ H}_{2(g)} \longrightarrow 2 \text{ NH}_{3(g)}$$

7.5 kg	v
2.02 g/mol	450 kPa, 80°C

$$n_{H_2} = 7.5 \text{ kg} \times \frac{1 \text{ mol}}{2.02 \text{ g}} = 3.7 \text{ kmol}$$

$$n_{NH_3} = 3.7 \text{ kmol} \times \frac{2}{3} = 2.5 \text{ kmol}$$

$$v_{NH_3} = \frac{nRT}{p} = \frac{2.5 \text{ kmol} \times \dfrac{8.31 \text{ kPa·L}}{1 \text{ mol·K}} \times 353 \text{ K}}{450 \text{ kPa}} = 16 \text{ kL}$$

$$\text{or} \quad v_{NH_3} = 7.5 \text{ kg}_{H_2} \times \frac{1 \text{ mol}_{H_2}}{2.02 \text{ g}_{H_2}} \times \frac{2 \text{ mol}_{NH_3}}{3 \text{ mol}_{H_2}} \times \frac{8.31 \text{ kPa·L}}{1 \text{ mol·K}} \times \frac{353 \text{ K}}{450 \text{ kPa}}$$
$$= 16 \text{ kL}$$

Exercise

23. What volume of oxygen at STP is needed to completely burn 15 g of methanol in a fondue burner (Figure 7.11)?

24. Hydrogen gas is burned in pollution-free vehicles in which pure hydrogen and oxygen gases react to produce water vapor. What volume of hydrogen at 40°C and 150 kPa can be burned using 300 L of oxygen gas measured at the same conditions? (Recall the law of combining volumes, page 234.)

25. A Down's Cell is used in the industrial production of sodium from the decomposition of molten sodium chloride. A major advantage of this process compared with earlier technologies is the production of the valuable by-product, chlorine. What volume of chlorine gas is produced (measured at SATP), along with 100 kg of sodium metal, from the decomposition of sodium chloride?

26. Most combustion reactions use oxygen from the air (20% oxygen). What mass of propane from a tank can be burned using 125 L of air at SATP?

27. A typical Canadian home heated with natural gas consumes 2.00 ML of natural gas during the month of December. What volume of oxygen at SATP is required to burn 2.00 ML of methane measured at 0°C and 120 kPa?

Figure 7.11
In alcohol burners like this one, the pale blue color indicates a very clean flame. This is quite different from the yellow flame of a candle, which contains incomplete combustion products such as soot.

INVESTIGATION

7.4 Percent Yield of Hydrogen

The purpose of this investigation is to determine the percent yield of a gas in a chemical reaction. There are several possible approaches that can be used in the design and analysis. The suggested method is to predict the volume of gas at STP and, in your analysis, convert the measured volume to STP conditions using the combined gas law or the ideal gas law. (For more accurate

- ☐ Problem
- ✔ Prediction
- ☐ Design
- ☐ Materials
- ☐ Procedure
- ✔ Evidence
- ✔ Analysis
- ✔ Evaluation
- ☐ Synthesis

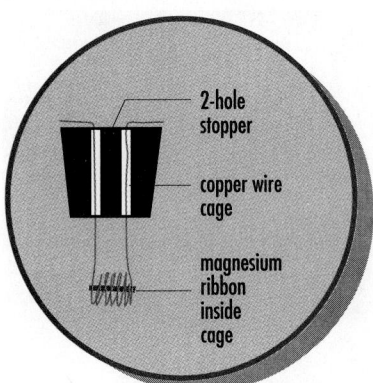

2-hole stopper

copper wire cage

magnesium ribbon inside cage

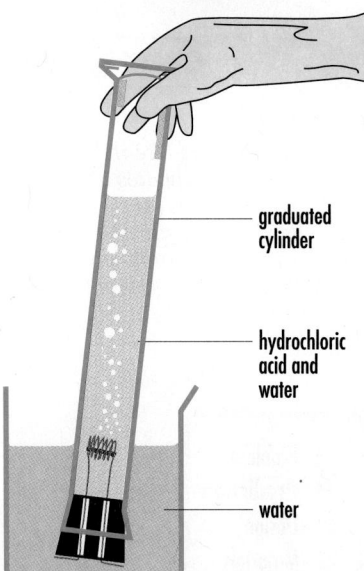

graduated cylinder

hydrochloric acid and water

water

Figure 7.12
While holding the cylinder so it does not tip, rest it on the bottom of the beaker. The acid, which is more dense than water, will flow down toward the stopper and react with the magnesium. The hydrogen produced should remain trapped in the graduated cylinder.

results of the actual yield, it will be necessary to correct the measured pressure for the vapor pressure of water. See Table 6.3 on page 246.) In your evaluation, comment on the possible reasons for the percentage obtained.

Problem

What is the percent yield of hydrogen from the reaction of magnesium with excess hydrochloric acid?

Experimental Design

A known mass of magnesium ribbon reacts with excess hydrochloric acid. The temperature, pressure, and volume of the hydrogen gas that is produced are measured.

Materials

lab apron
safety glasses
disposable plastic gloves
100 mL graduated cylinder
2-hole stopper to fit cylinder
thermometer
barometer
large beaker (600 mL or 1000 mL)
magnesium ribbon (60 mm to 70 mm)
6 mol/L hydrochloric acid
piece of fine copper wire (100 mm to 150 mm)
centigram or analytical balance

Procedure

1. Obtain a strip of magnesium ribbon about 60 mm to 70 mm long.

2. Measure and record the mass of the magnesium.

3. Fold the magnesium ribbon to make a small compact bundle, no larger than a pencil eraser.

4. Wrap the fine copper wire all around the magnesium, making a cage to hold it but leaving 30 mm to 50 mm of the wire free for a handle.

5. Carefully pour 10 mL to 15 mL of 6 mol/L hydrochloric acid into the graduated cylinder.

6. Slowly fill the graduated cylinder to the brim with water from a beaker. As you fill the cylinder, pour slowly down the side of the cylinder to minimize mixing of the water with the acid at the bottom. In this way, the liquid at the top of the cylinder is relatively pure water and the acid remains at the bottom.

7. Half-fill the large beaker with water.

8. Bend the copper-wire handle through the holes in the stopper so that the cage holding the magnesium is positioned about 10 mm below the bottom of the stopper.

9. Insert the stopper into the graduated cylinder — the liquid in the cylinder will overflow a little. Cover the holes in the stopper with your finger. Working quickly, invert the cylinder, and immediately lower it so that the stopper is below the surface of the water in the beaker before you remove your finger from the stopper holes (Figure 7.12).

10. Observe the reaction, then wait about 5 min after the bubbling stops to allow the contents of the graduated cylinder to reach room temperature.

11. Raise or lower the graduated cylinder so that the level of liquid inside the beaker is the same as the level of liquid in the graduated cylinder. (This equalizes the gas pressure in the cylinder with the pressure of the air in the room.)

12. Measure and record the volume of gas in the graduated cylinder.

13. Record the laboratory (ambient) temperature and pressure.

14. The liquids in this investigation may be poured down the sink, but rinse the sink with lots of water.

CAUTION

Rinse your hands well after step 9 in case you got any dilute acid on your skin.

Hydrogen gas, produced in the reaction of hydrochloric acid and magnesium, is flammable. Ensure that there is adequate ventilation and that there are no open flames in the classroom.

7.4 SOLUTION STOICHIOMETRY

You have already seen the usefulness of gravimetric stoichiometry and gas stoichiometry for both predictions and analyses. However, the majority of stoichiometric work in research and in industry involves solutions, particularly aqueous solutions. Solutions are easy to handle and reactions in solution are relatively easy to control.

Stoichiometric calculations of substances in solution involve the molar concentration and the volume of a solution, rather than the molar mass and the mass of a substance. However, the general stoichiometric method remains the same. The major difference is that molar concentration (page 200), rather than molar mass or molar volume, is used as a conversion factor to convert solution volume to amount in moles. Another difference in solution stoichiometry compared with gravimetric stoichiometry is that calculations are usually based on equivalent reacting amounts of two reactants as determined by the mole ratio in the balanced equation. In other words, excess reagents are not usually employed in stoichiometry problems based solely on solutions. This is particularly true for the quantitative chemical analysis discussed in section 7.6.

Consider the following example. Solutions of ammonia and phosphoric acid are used to produce ammonium hydrogen phosphate fertilizer (Figure 7.13). What volume of 14.8 mol/L $NH_{3(aq)}$ is needed for the ammonia to react completely with 1.00 kL of 12.9 mol/L $H_3PO_{4(aq)}$ to produce fertilizer?

The first step of the stoichiometric method involves writing a balanced chemical equation so that a relationship between the amount of ammonia (in moles) and the amount of phosphoric acid (in moles) can be established. Beneath the equation, list both the given and the required measurements and the conversion factors.

$$2\,NH_{3(aq)} \quad + \quad H_3PO_{4(aq)} \quad \rightarrow \quad (NH_4)_2HPO_{4(aq)}$$

v 1.00 kL

14.8 mol/L   12.9 mol/L

In the second step, the information given for phosphoric acid is converted to an amount in moles.

$$n_{H_3PO_4} = 1.00\ kL \times \frac{12.9\ mol}{1\ L} = 12.9\ kmol$$

In the third step, the mole ratio is used to calculate the amount of the required substance, ammonia. According to the balanced chemical equation, two moles of ammonia react for every one mole of phosphoric acid.

$$n_{NH_3} = 12.9\ kmol \times \frac{2}{1} = 25.8\ kmol$$

In the final step, the amount of ammonia is converted to the quantity requested in the question. To obtain the volume of ammonia, the molar concentration is used to convert the amount in moles to the solution volume.

$$v_{NH_3} = 25.8\ kmol \times \frac{1\ L}{14.8\ mol} = 1.74\ kL$$

The three calculation steps listed above can be summarized into one step as shown below.

$$v_{NH_3} = 1.00\ \cancel{kL\ H_3PO_4} \times \frac{12.9\ \cancel{mol\ H_3PO_4}}{1\ \cancel{L\ H_3PO_4}} \times \frac{2\ \cancel{mol\ NH_3}}{1\ \cancel{mol\ H_3PO_4}} \times \frac{1\ L\ NH_3}{14.8\ \cancel{mol\ NH_3}}$$

$$= 1.74\ kL$$

According to the stoichiometric method, the required volume of ammonia solution is 1.74 kL. The following example shows how to communicate a solution to a stoichiometric problem involving solutions.

EXAMPLE

As part of a chemical analysis, a technician determines the concentration of a sulfuric acid solution. In the experiment, a 10.00 mL sample of sulfuric acid reacts completely with 15.9 mL of 0.150 mol/L potassium hydroxide solution. Calculate the molar concentration of the sulfuric acid.

$$H_2SO_{4(aq)} \quad + \quad 2\,KOH_{(aq)} \quad \rightarrow \quad 2\,HOH_{(1)} + K_2SO_{4(aq)}$$

10.00 mL   15.9 mL

C 0.150 mol/L

$$n_{KOH} = 15.9\ mL \times \frac{0.150\ mol}{1\ L} = 2.39\ mmol$$

$$n_{H_2SO_4} = 2.39\ mmol \times \frac{1}{2} = 1.19\ mmol$$

$$C_{H_2SO_4} = \frac{1.19\ mmol}{10.00\ mL} = 0.119\ mol/L$$

or $C_{H_2SO_4} = 15.9\ \cancel{mL\ KOH} \times \dfrac{0.150\ \cancel{mol\ KOH}}{1\ \cancel{L\ KOH}} \times \dfrac{1\ mol\ H_2SO_4}{2\ \cancel{mol\ KOH}} \times \dfrac{1}{10.00\ \cancel{mL}}$

$$= 0.119\ mol/L$$

Figure 7.13
Fertilizers can have a dramatic effect on plant growth. The plants on the left were fertilized with an ammonium hydrogen phosphate fertilizer.

SUMMARY: GRAVIMETRIC, GAS, AND SOLUTION STOICHIOMETRY

Step 1: Write a balanced chemical equation and list the measurements and conversion factors for the given substance and the one to be calculated.

Step 2: Convert the given measurement to an amount in moles by using the appropriate conversion factor.

Step 3: Calculate the amount of the other substance by using the mole ratio from the balanced equation.

Step 4: Convert the calculated amount to the final quantity requested by using the appropriate conversion factor.

Exercise

28. Ammonium sulfate fertilizer is manufactured by having sulfuric acid react with ammonia. In a laboratory study of this process, 50.0 mL of sulfuric acid reacts with 24.4 mL of a 2.20 mol/L ammonia solution to produce the ammonium sulfate solution. From this evidence, calculate the molar concentration of the sulfuric acid at this stage in the process.

29. Slaked lime can be added to an aluminum sulfate solution in a water treatment plant to clarify the water. Fine particles in the water stick to the precipitate produced. Calculate the volume of 0.0250 mol/L calcium hydroxide solution required to react completely with 25.0 mL of 0.125 mol/L aluminum sulfate solution.

30. In designing an experiment similar to Investigation 7.1, a chemistry teacher wants 75.0 mL of 0.200 mol/L iron(III) chloride solution to react completely with an excess quantity of 0.250 mol/L sodium carbonate solution.
    (a) What is the minimum volume of sodium carbonate solution needed?
    (b) What would be a reasonable volume of sodium carbonate to use in this experiment?

Problem 7E Testing Solution Stoichiometry

You have already tested the stoichiometric method for gravimetric and gas stoichiometry, but the testing of a scientific concept is never finished. Scientists keep looking for new experimental designs and new ways of testing a scientific concept. A test of stoichiometry using solutions is presented below. Complete the Prediction, Analysis, and Evaluation (of the Prediction) of the investigation report. Pay particular attention to the evaluation of the experimental design.

Problem

What mass of precipitate is produced by the reaction of 20.0 mL of 0.210 mol/L sodium sulfide with an excess quantity of aluminum nitrate solution?

Experimental Design

The two solutions provided react with each other and the resulting precipitate is separated by filtration and dried. The mass of the dried precipitate is determined.

Evidence

A yellow precipitate resembling aluminum sulfide was formed.
mass of filter paper = 0.97 g
mass of dried filter paper plus precipitate = 1.17 g
A few additional drops of the sodium sulfide solution added to the filtrate produced a precipitate.

Problem 7F Determining a Solution Concentration

Once a scientific concept has passed several tests, it can be used in industry. Suppose you are a technician in an industry that needs to determine the molar concentration of a silver nitrate solution which, due to its cost, is being recycled. Complete the Analysis of the investigation report.

Problem

What is the molar concentration of silver nitrate in the solution to be recycled?

Experimental Design

A sample of the silver nitrate solution to be recycled reacts with an excess quantity of sodium sulfate in solution. The precipitate formed is filtered and the mass of dried precipitate is measured.

Evidence

A white precipitate was formed in the reaction. A similar precipitate was formed when a few drops of silver nitrate were added to the filtrate.
volume of silver nitrate solution = 100 mL
mass of filter paper = 1.27 g
mass of dried filter paper plus precipitate = 6.74 g

INVESTIGATION

- Problem
- ✔ Prediction
- Design
- ✔ Materials
- ✔ Procedure
- ✔ Evidence
- ✔ Analysis
- ✔ Evaluation
- Synthesis

7.5 Percent Yield of Barium Sulfate

Barium sulfate is a white, odorless, tasteless powder that has a variety of different uses. Barium sulfate is used as a weighting mud in oil-drilling, is used in the manufacture of paper, paints, and inks, and is taken internally for gastrointestinal X-ray analysis. The reaction studied in this investigation is similar to the one used in the manufacture of barium sulfate. The purpose of this investigation is to determine the suitability of this reaction by determining the percent yield of product.

Problem

What is the percent yield of barium sulfate in the reaction of barium chloride and sodium sulfate solutions?

Experimental Design

A 40.0 mL sample of 0.15 mol/L sodium sulfate solution is mixed with 50.0 mL of 0.100 mol/L barium chloride solution. A diagnostic test is used to test the filtrate for excess reagent. The actual yield is compared with the predicted yield to assess the completeness of this reaction.

CAUTION

Soluble barium compounds are toxic.

7.5 CHEMICAL TECHNOLOGY: THE SOLVAY PROCESS

Science and technology are different activities, but they are mutually dependent. Science, an *international* discipline, is involved with *natural* products and processes, and is more *theoretical* in its approach, emphasizing *ideas* over practical applications. Scientific ideas are evaluated according to how well they can predict and explain natural phenomena. Compared with science, technology is more *localized*, is involved with processes and manufactured products, and is more *empirical*, often involving a trial-and-error approach and emphasizing *methods and materials*. Technological products and processes are evaluated on the basis of their simplicity, reliability, efficiency, and cost.

Technologies can be classified, by scale, into three classes. *Industrial technologies* usually involve large-scale production of substances from natural raw materials. Examples include mining, oil refining, and the production of chemicals such as sodium carbonate. *Commerical technologies* are smaller scale processes involved in the production of goods such as computers, home appliances, cars, food, and clothing. *Consumer technologies* involve the use by individuals of products or processes such as refrigerators, sewing machines, shampoos, and shrink-wrap packaging.

"A review of the history of civilization clearly shows that science and technology have played a key role in shaping our culture. Various technological innovations have acted as springboards...in an ever-widening range of pursuits and activities. Increasing awareness of this effect, particularly in contemporary society, has created the need for better understanding the nature of technology."
James Burke, 1987

Exercise

31. List four differences between science and technology.
32. Classify each of the following questions as requiring scientific or technological activities to find the answers. (Do not answer the questions.)
    (a) What coating on a nail can prevent corrosion?
    (b) Which chemical reactions are involved in the corrosion of iron?
    (c) What is the accepted explanation for the chemical formula of water?
    (d) What process produces nylon thread continuously?
    (e) Why is a copper(II) sulfate solution blue?
    (f) How can automobiles be redesigned to achieve safer operation?

The Solvay Process

The worldwide production of glass and soap requires huge amounts of sodium carbonate, or soda ash. Until the late 1700s, the main source of sodium carbonate was burned plants; the ashes were mixed with water and the soda ash was extracted. During the 19th century in France, an industrial method called the LeBlanc process was developed for producing soda ash, but this process required burning a lot of coal, which was expensive. (It also caused considerable air pollution, but at the time this was not considered to be a problem.) In 1865 Ernest Solvay, a Belgian chemist, began to perfect the ammonia-soda process for the production of soda ash, and in 1867 Solvay's process was installed for the first time in his small factory in Belgium.

Since the LeBlanc process was so firmly established, the new Solvay process did not gain immediate acceptance. But the cost of the new process was one-third the cost of the old LeBlanc process, so Solvay processing plants were eventually built in every major industrialized country (Figure 7.14). The wide use of his invention brought Solvay a great deal of money, much of which he channelled into charitable work in Brussels.

The overall, or net, reaction in the Solvay process, involving calcium carbonate and sodium chloride, is one that does not occur spontaneously at room temperature.

$$CaCO_{3(s)} + 2\,NaCl_{(aq)} \rightarrow Na_2CO_{3(aq)} + CaCl_{2(aq)}$$

Imagine adding chalk to a salt solution — no reaction occurs. How then can this reaction be implemented industrially to produce large quantities of soda ash? Solvay's design involved an indirect route with a series of intermediate reactions. His major breakthrough involves a reaction between ammonium hydrogen carbonate and sodium chloride that at first glance seems improbable.

$$NH_4HCO_{3(aq)} + NaCl_{(aq)} \rightarrow NH_4Cl_{(aq)} + NaHCO_{3(s)}$$

What Solvay discovered by experimentation is that in cold water ammonium chloride has a higher solubility than sodium hydrogen carbonate. As a result, sodium hydrogen carbonate can be separated out of the solution by crystallization (page 39). This separation by solubility allows the $NaHCO_3$ to be separated and sold as baking soda, or to be decomposed into Na_2CO_3 as washing soda.

The ingenuity of Solvay's design becomes apparent when you write out the reactions and see that all of the intermediate products are recycled as reactants in other reactions. Nothing is left as a by-product except calcium chloride, which today is sold as road salt and as a drying agent (desiccant). See for yourself in the following questions.

Figure 7.14
The Solvay process is used in this soda ash plant. The saleable products are washing soda, baking soda, and road salt. Canada's only soda ash plant is located in Amherstburg, Ontario.

Solution Mining
Solution mining is the extraction of soluble ionic compounds from underground deposits. In this process, hot water or steam is pumped into the deposit, forming a concentrated solution which is then pumped to the surface. Sodium chloride is often recovered by means of this technology. Using this process, some chemical companies create caverns in underground salt layers about 2 km below the Earth's surface. Each cavern is large enough to hold over one hundred million litres (100 ML), a volume equal to about 1500 railway tank cars, or about the size of a large hockey arena. The companies can then use the caverns to store petrochemicals after removing the salt.

Exercise

33. Write and balance the reaction equations for the Solvay process from the word equations below.
    (a) Limestone, $CaCO_{3(s)}$, is decomposed by heat to form calcium oxide (lime) and carbon dioxide.

(b) Carbon dioxide reacts with aqueous ammonia and water to form aqueous ammonium hydrogen carbonate.

(c) In the same vessel, the aqueous ammonium hydrogen carbonate reacts with brine, $NaCl_{(aq)}$, to produce aqueous ammonium chloride and solid baking soda, $NaHCO_{3(s)}$.

(d) Heating the separated baking soda decomposes it into solid washing soda, water vapor, and carbon dioxide.

(e) The first of two recycling reactions involves the reaction of lime $(CaO_{(s)})$ with water to produce slaked lime $(Ca(OH)_{2(s)})$.

(f) Next, the slaked lime is added to the aqueous ammonium chloride (an intermediate product) to produce ammonia, aqueous calcium chloride, and water.

(g) Write the net (overall) reaction for the Solvay process (page 274).

34. *Intermediates* are produced part way through a process and become reactants in a later reaction. Cross out all intermediate products in the reaction equations you have written. Do not be concerned about quantities of reactants and products. What you have left should combine to give you the net unbalanced reaction for the Solvay process.

35. *Raw materials* are the materials that are consumed in the net reaction. What are the raw materials for the Solvay process? Where are these raw materials obtained? What makes these materials suitable for a large-scale chemical process?

36. The *primary products* and the *by-products* of a chemical process depend on how marketable the products are. What are the primary product and by-product of the Solvay process?

37. What intermediate in the Solvay process is highly marketable? What are some consequences of removing this intermediate from the system of reactions?

38. Resources other than chemical and technological resources are required for most chemical processes. What additional natural resources are needed for the Solvay process?

39. (Discussion) What makes the Solvay process so economical?

40. The net reaction in the Solvay process has brine (sodium chloride solution) and limestone (calcium carbonate) reacting to produce soda ash (sodium carbonate) and calcium chloride.

(a) Calculate the mass of soda ash that can be produced in a Solvay plant from the reaction of 10.0 kL of 5.40 mol/L brine solution and excess limestone.

(b) What volume of 5.40 mol/L brine solution is required when 500 kg of limestone reacts during the Solvay process?

(c) Find the volume of 3.00 mol/L calcium chloride solution that the Solvay process can produce from the reaction of 350 L of 2.00 mol/L brine solution and excess limestone.

41. Baking soda is a by-product in the Solvay process. Solid crystals of sodium hydrogen carbonate are produced by the reaction of $NH_4HCO_{3(aq)}$ and $NaCl_{(aq)}$.

(a) Calculate the volume of 0.700 mol/L ammonium hydrogen carbonate solution required in the Solvay process to react completely with 4.00 kL of a 3.00 mol/L brine solution.

(b) (Discussion) In the solubility table on this book's inside back cover, all sodium compounds are listed as having a high solubility, but the $NaHCO_3$ forms as a solid in the Solvay process reaction. How can you account for this discrepancy?

42. (Discussion) The large masses and volumes in questions 40 and 41 represent industrial quantities. Even larger quantities are used in newer processing plants that are built on what is called a "world scale" to produce quantities for international distribution. Why do you think chemical plants are being built on an increasingly large scale?

7.6 CHEMICAL ANALYSIS BY TITRATION

Titration is a common experimental design used to determine the concentration of substances in solution (see Appendix C, C.3). **Titration** involves the carefully measured and controlled addition of a solution from a buret into a measured volume of a sample solution, usually in an Erlenmeyer flask (Figure 7.15) until the reaction is judged to be complete. Titration is a good example of a chemical technology that is reliable, efficient, economical, and simple to use.

A *buret* is a precisely marked glass cylinder with a stopcock at one end. It measures a volume of reacting solution. A typical titration is the chemical analysis of acetic acid in a sample of vinegar, using a sodium hydroxide solution in a buret. A chemical analysis by titration typically involves a number of volumetric techniques: the preparation of the standard solution, the use of a pipet to transfer portions of the sample for analysis, and the technique of titration itself. The solution in the buret (called the **titrant**) is added to the sample until the reaction is complete.

A titration analysis of an unknown concentration requires that a chemical reaction between a reagent and the unknown sample be spontaneous, fast, quantitative, and stoichiometric. In order to obtain precise and reliable results, the concentration of the reactant used to analyze the unknown must be accurately known; that is, you must be able to prepare a standard solution. If this is not possible, for example, with sodium hydroxide or hydrochloric acid solutions, then a primary standard is used to determine the concentration of the reactant before a chemical analysis of the unknown takes place. This process is known as *standardization* and uses a **primary standard** whose solid form has a relatively high molar mass and purity, shows stability on drying, is readily available, and is known to react quantitatively with the substance being standardized.

When doing a titration, there will be a point at which the reaction is complete. In other words, chemically equivalent amounts of reactants, as determined by the mole ratio, have been combined. The point at which the exact theoretical amount of titrant has been added to the sample is called the **equivalence point**. To measure this point experimentally, we look for a sudden change in some observable property of the solution, such as color. The point during a titration

Titration Requirements
A chemical reaction must be
- spontaneous — chemicals react on their own without a continuous addition of energy
- fast — chemicals react instantaneously when mixed
- quantitative — the reaction is more than 99% complete
- stoichiometric — there is a simple, whole number ratio of moles of reactants and products

when this sudden change is observed is called the **endpoint**. At the endpoint the titration is stopped and the volume of titrant is recorded. Ideally, the empirical endpoint and the theoretical equivalence point coincide.

A titration analysis should involve several trials, using different samples of the unknown solution to improve the reliability of the answer. A typical requirement is to repeat measurements until three trials result in volumes within 0.1 mL to 0.2 mL. These three results are then averaged before carrying out the volumetric stoichiometry calculation. Refer to Appendix C, C.3, for information on titration using a buret.

An example of a titration analysis is the determination of the concentration of hydrochloric acid in a product that is used to treat concrete prior to painting. A sodium carbonate solution can be used as the reagent to analyze the hydrochloric acid. Suppose 1.59 g of anhydrous sodium carbonate, $Na_2CO_{3(s)}$, was dissolved to make 100.0 mL of solution. Samples (10.00 mL) of this standard solution were then taken and titrated with the $HCl_{(aq)}$ product, which had been diluted by a factor of 10. The titration evidence collected is shown in Table 7.3.

Table 7.3

TITRATION OF 10.00 mL OF $Na_2CO_{3(aq)}$ WITH DILUTED $HCl_{(aq)}$				
Trial	1	2	3	4
Final buret reading (mL)	13.3	26.0	38.8	13.4
Initial buret reading (mL)	0.2	13.3	26.0	0.6
Volume of $HCl_{(aq)}$ added (mL)	13.1	12.7	12.8	12.8

To analyze this evidence you first need to calculate the molar concentration of the sodium carbonate solution.

$$n_{Na_2CO_3} = 1.59 \text{ g} \times \frac{1 \text{ mol}}{105.99 \text{ g}} = 0.0150 \text{ mol}$$

$$[Na_2CO_{3(aq)}] = \frac{0.0150 \text{ mol}}{0.1000 \text{ L}} = 0.150 \text{ mol/L}$$

Now you can start the stoichiometry procedure by writing the balanced chemical equation.

$$2 \text{ HCl}_{(aq)} \quad + \quad Na_2CO_{3(aq)} \quad \rightarrow \quad H_2CO_{3(aq)} \quad + \quad 2 \text{ NaCl}_{(aq)}$$

12.8 mL   10.00 mL
C 0.150 mol/L

Notice in Table 7.3 that four trials were done, and the volume in the first trial is significantly higher than in the others. The volume you use for $HCl_{(aq)}$ should be your best average, typically three results within ±0.1 mL. The value, 12.8 mL, shown below the chemical equation is the average of trials 2, 3, and 4. (Keep the unrounded value in your calculator as usual.) The rest of the stoichiometry procedure follows the usual steps already discussed.

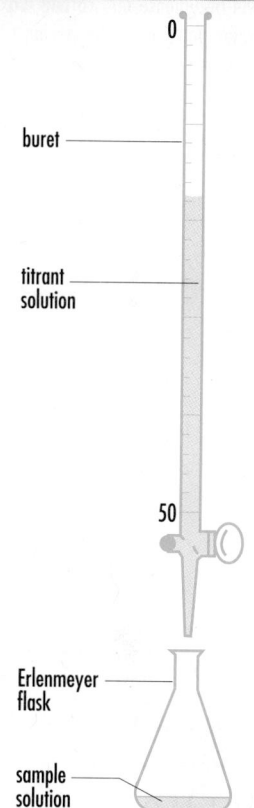

Figure 7.15
An initial reading of volume is made on the buret before any titrant is added to the sample solution. Then titrant is added until the reaction is complete; that is, when a drop of titrant changes the color of the sample. The final buret reading is then taken. The difference in buret readings is the volume of titrant added.

Any property of a solution, such as color or conductivity, that changes abruptly can be used as an endpoint. However, some changes may not be very sharp or may be difficult to measure accurately. This introduces an error into the experiment. The difference between the empirical endpoint and the theoretical equivalence point is known as the titration error.

$$n_{Na_2CO_3} = 10.00 \text{ mL} \times \frac{0.150 \text{ mol}}{1 \text{ L}} = 1.50 \text{ mmol}$$

$$n_{HCl} = 1.50 \text{ mmol} \times \frac{2}{1} = 3.00 \text{ mmol}$$

$$[HCl_{(aq)}] = \frac{3.00 \text{ mmol}}{12.8 \text{ mL}} = 0.235 \text{ mol/L}$$

Alternatively, these steps may be combined into one calculation as shown below.

$$[HCl_{(aq)}] = 10.00 \text{ mL Na}_2\text{CO}_3 \times \frac{0.150 \text{ mol Na}_2\text{CO}_3}{1 \text{ L Na}_2\text{CO}_3} \times \frac{2 \text{ mol HCl}}{1 \text{ mol Na}_2\text{CO}_3} \times \frac{1}{12.8 \text{ mL}}$$
$$= 0.235 \text{ mol/L}$$

Since the sample of concrete cleaner had been diluted by a factor of 10, the original concentration of hydrochloric acid must be 10 times greater, or 2.35 mol/L.

Carbonic acid, $H_2CO_{3(aq)}$, is relatively unstable and will decompose into carbon dioxide gas and water. You see this happening when you open a bottle of a carbonated drink. In this titration example, the decomposition of carbonic acid will be slow and not noticeable because of the low concentrations.

INVESTIGATION

7.6 Titration Analysis of ASA

Acetylsalicylic acid, known as ASA, is the most commonly used drug — with over 10 000 t manufactured in North America every year. ASA, $C_8H_7O_2COOH$, is an organic acid like acetic acid and reacts with strong bases such as sodium hydroxide in the same way as acetic acid. Sodium hydroxide is not a primary standard because the solid always contains some water, and a specific mass cannot be accurately measured. For this reason, the sodium hydroxide must first be standardized with a primary standard such as potassium hydrogen phthalate, $KC_7H_4O_2COOH$. Potassium hydrogen phthalate (KHP) reacts with sodium hydroxide according to the following reaction equation.

$$KC_7H_4O_2COOH_{(aq)} + NaOH_{(aq)} \rightarrow KNaC_7H_4O_2COO_{(aq)} + HOH_{(l)}$$

The purpose of this investigation is to accurately determine the ASA content of a consumer product.

Problem

What is the mass of ASA in a consumer tablet?

Experimental Design

A standard solution of KHP is prepared. This solution is used to standardize a given solution of sodium hydroxide by titrating samples of KHP with sodium hydroxide using phenolphthalein as the indicator. An ASA tablet is dissolved in methanol and then titrated with the standardized sodium hydroxide using phenolphthalein as the indicator.

Materials

lab apron	50 mL buret
safety glasses	10 mL volumetric pipet
$KC_7H_4O_2COOH_{(s)}$	pipet bulb
$NaOH_{(aq)}$	ring stand
ASA tablets	buret clamp
phenolphthalein	centigram balance
methanol	laboratory scoop

- Problem
- ✔ Prediction
- Design
- Materials
- Procedure
- ✔ Evidence
- ✔ Analysis
- ✔ Evaluation
- Synthesis

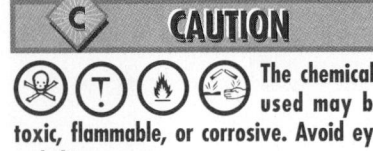

CAUTION

The chemicals used may be toxic, flammable, or corrosive. Avoid eye and skin contact.

wash bottle of pure water
(2) 100 mL or 150 mL beakers
250 mL beaker
100 mL volumetric flask with stopper
(2) 250 mL Erlenmeyer flasks

stirring rod
small funnel
medicine dropper
meniscus finder

Procedure

Standardization

1. Prepare a 100 mL standard solution of 0.150 mol/L KHP. (See Appendix C, C.3)

2. Obtain about 70 mL of $NaOH_{(aq)}$ in a clean, dry, labelled 150 mL beaker.

3. Set up the buret with $NaOH_{(aq)}$ following the accepted procedure for rinsing and clearing the air bubble. (See Appendix C, C.3.)

4. Pipet a 10.00 mL sample of KHP into a clean Erlenmeyer flask and add 1 or 2 drops of phenolphthalein indicator.

5. Record the initial buret reading to the nearest 0.1 mL.

6. Titrate the KHP sample with $NaOH_{(aq)}$ until a single drop produces a permanent change from colorless to faint pink.

7. Record the final buret reading to the nearest 0.1 mL.

8. Repeat steps 4 to 7 until three consistent results are obtained.

ASA Analysis

9. Add about 30 mL of methanol to one ASA tablet in a clean Erlenmeyer flask.

10. Stir and crush the tablet until the solid has mostly dissolved. (The final mixture may be slightly cloudy because of the presence of inert ingredients.)

11. Add 1 or 2 drops of phenolphthalein indicator.

12. Record the initial buret reading to the nearest 0.1 mL.

13. Titrate the ASA sample with $NaOH_{(aq)}$ until a single drop produces a permanent change from colorless to pink.

14. Record the final buret reading to the nearest 0.1 mL.

15. Repeat steps 9 to 14 until three consistent results are obtained.

16. Dispose of all solutions into the sink and flush with lots of water.

INVESTIGATION

7.7 Titration Analysis of Vinegar

Consumer products are required by law to have the minimum quantity of the active ingredient listed on the product label. Companies who produce chemical products usually employ analytical chemists and technicians to monitor the characteristics of the final product in a process known as quality control. Nevertheless, government consumer affairs departments also use chemists and technicians to check products, particularly in

- Problem
- Prediction
- Design
- ✔ Materials
- ✔ Procedure
- ✔ Evidence
- ✔ Analysis
- ✔ Evaluation
- Synthesis

CAUTION

Chemicals used may be flammable, or corrosive.

response to consumer complaints. The purpose of this investigation is to test the manufacturer's claim of the concentration of acetic acid in a consumer sample of vinegar.

Problem

What is the molar concentration of acetic acid in a sample of vinegar?

Prediction

According to the label, the manufacturer claims that the vinegar contains 5% acetic acid, which translates into a 0.87 mol/L concentration of acetic acid.

Experimental Design

A sample of vinegar from the school cafeteria, fast-food chain, or food store is diluted by a factor of five to make a 100.0 mL final solution. The diluted solution is titrated with a standardized sodium hydroxide solution (from Investigation 7.6) using phenolphthalein as the indicator.

7.7 EXPERIMENTAL DESIGNS FOR ANALYSIS

Chemists and chemical technologists use many different techniques and experimental designs for chemical analysis. The choice of an experimental design depends upon the time and equipment available and the degree of accuracy that is required. In this section you will have an opportunity to choose from and evaluate several experimental designs (Table 7.4). You will investigate many other quantitative and qualitative techniques in further studies of chemistry.

Experience with the experimental designs in Table 7.4 gives you insight into the alternatives available to chemists, whether they work in research, industry, or environmental protection. If several designs can provide an answer to an experimental problem, alternative designs are evaluated by considering cost, simplicity, efficiency, accuracy, safety, and environmental factors.

Table 7.4

EXPERIMENTAL DESIGNS FOR CHEMICAL ANALYSIS	
Procedure	**Description**
crystallization	The solvent is vaporized from a solution, with or without heating, leaving a solid whose mass is measured.
filtration	A low solubility solid, formed as a product of a single or double replacement reaction, is separated by means of a filter, and its mass is measured.
gas collection	A gas formed as a product of a reaction is collected, and its volume, temperature, and pressure are measured.
titration	A titrant in a buret is progressively added to a measured volume of a solution in an Erlenmeyer flask. The volume of titrant at the endpoint is measured.

The following series of problems all pose the same problem: "What is the molar concentration of a hydrochloric acid solution in a solution of kettle-scale remover?" The experimental designs include filtration, gas collection, and titration. Be prepared to suggest other experimental designs that could also solve the problem. Complete the Materials and Analysis of each investigation report and, where possible, evaluate the experimental designs.

Problem 7G Filtration Analysis

Problem

What is the molar concentration of the hydrochloric acid in a solution of kettle-scale remover?

Experimental Design

The hydrochloric acid in a solution of kettle-scale remover reacts with an excess quantity of a 1.05 mol/L lead(II) nitrate solution to form a precipitate, which is filtered and dried.

Evidence

A white precipitate formed when the solutions were mixed.
volume of $HCl_{(aq)}$ = 25.0 mL
mass of filter paper = 0.89 g
mass of dried filter paper plus dried precipitate = 9.71 g
Several drops of a potassium iodide solution added to the filtrate produced a yellow precipitate.

Problem 7H Gas Volume Analysis

Problem

What is the molar concentration of the hydrochloric acid in a solution of kettle-scale remover?

Experimental Design

The hydrochloric acid in a solution of kettle-scale remover reacts with an excess quantity of zinc. The gas that is generated is collected and its volume at STP is determined.

Evidence

A colorless gas and a colorless solution formed when the reactants were mixed. Zinc was left over after the reaction stopped.

volume of $HCl_{(aq)}$ = 25.0 mL
volume of gas collected = 749 mL at STP

Problem 7I Titration Analysis

Problem

What is the molar concentration of the hydrochloric acid in a solution of kettle-scale remover?

> **Analytical Techniques**
> Quantitative chemical analysis techniques can be classified as gravimetric, volumetric, optical, or electrical. Gravimetric and volumetric analyses are discussed in this chapter. Optical analysis can be used, for example, to determine the concentration of a colored solution by the percentage of light absorbed. Electrical analysis can involve conductivity, electroplating, or voltage measurements (see Chapter 16). Qualitative analysis techniques include, among many others, all of the diagnostic tests listed in Appendix C.

Experimental Design

The hydrochloric acid in a solution of kettle-scale remover is titrated with a standardized solution of barium hydroxide. The color change of bromothymol blue indicator to green indicates the equivalence point.

Evidence

TITRATION OF 10.00 mL SAMPLES OF $HCl_{(aq)}$ WITH 0.974 mol/L $Ba(OH)_{2(aq)}$				
Trial	1	2	3	4
Final buret reading (mL)	15.6	29.3	43.0	14.8
Initial buret reading (mL)	0.6	15.6	29.3	1.2
Volume of $Ba(OH)_{2(aq)}$ added (mL)				
Color at endpoint	blue	green	green	green

Evaluating Experimental Designs

Evaluating procedures and experimental designs is a useful skill. Whether you are assessing the information in a scientific report (see Appendix B), a newspaper, or a sales presentation, you need to evaluate the framework on which the information is based. Experience in evaluating experimental designs will prepare you for this. Following are some criteria used to evaluate experimental designs in the scientific community.

- In order to conclude that an effect has been caused by a variable, only one variable should be manipulated at a time.
- The experimental design should, if possible, include a control group or procedure that provides a comparison.
- All variables other than the manipulated variable and responding variable should be kept constant.
- The responding variable that is selected should be observed reliably, consistently, and simply.
- The experimental design should produce results that are reproducible time after time by skilled experimenters.
- The experimental design should be time-efficient while allowing several trials or repetitions.
- The cost of performing the experiment should be as low as possible to achieve the desired results.
- The design should have as little adverse effect on the environment (including people) as possible.
- The experimental design should be the simplest design available to achieve the desired result.

An acceptable experimental design should satisfy as many of these criteria as possible. Trade-offs are always necessary, but they should not jeopardize the ethics and validity of the experimental design.

Exercise

43. (a) Which of the three experimental designs in Problems 7G, 7H, and 7I was best suited to determine the molar concentration of the hydrochloric acid? Justify your opinion by stating your criteria and the significance of the criteria.

   (b) Suggest an alternative experimental design that uses zinc and hydrochloric acid but is different from Problem 7H.

44. Each of the problems listed below is followed by a brief plan that is part of a proposed experimental design. Identify the flaw in each plan and include your reasoning.

   (a) What is the concentration of the ammonia solution?
      The ammonia solution is boiled to crystallize the solute.

   (b) What is the chemical formula for water?
      Tap water is heated in a beaker with a laboratory burner to collect the gases.

   (c) What is the concentration of the sodium phosphate solution?
      A sample of the sodium phosphate solution is added to an excess quantity of copper(II) sulfate and the water is boiled away to determine the mass of the precipitate formed.

Create experimental designs to test the following claims.
- A lecturer on mind-over-matter claims that, after sufficient training, people can walk on hot coals without burning their feet. (For hints, see page 35.)
- A salesperson claims that the Ultra-Superfilter vacuum cleaner cleans your carpet better than your current vacuum cleaner.
- A psychic claims that three out of ten people in a particular group have auras that only the psychic can see.
- Some astrologers claim that when they interview people they can tell which astrological sign they were born under.

MEDICAL LABORATORY TESTING

You may have had a blood or urine sample taken for testing at a medical laboratory. Once in the laboratory, a sample can undergo many different tests and analyses. For example, the chemical components of blood and other body fluids can be measured to determine the level of glucose, cholesterol, and hormones or to detect the presence of drugs. Technologists working in microbiology study bacteria, fungi, viruses, and parasites that invade the body. Those working in hematology look for disease in blood cells and the clotting mechanisms of the blood. Technologists working in histotechnology prepare body tissues for examination under a microscope.

Many of the routine tests are done by automated analysis machines. With blood samples, for example, automated instruments can determine a white cell count, red cell count, and hemoglobin level. A medical laboratory technologist oversees these tests and may do further analyses on samples to provide the doctor with specific information such as the different types of white cells present. Other tests — for example, some of those on urine and stool samples —

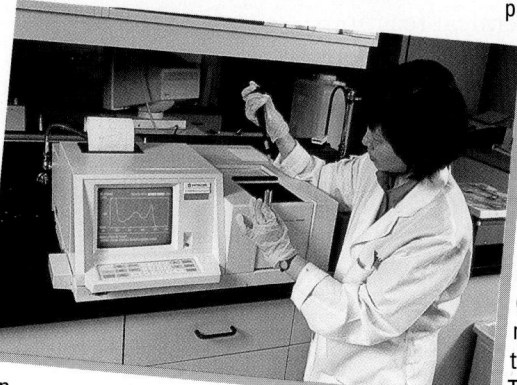

require the technologist to set up the analysis manually.

Even though the analysis process is highly automated, a medical laboratory technologist needs a knowledge of basic chemistry and biology to follow the many laboratory procedures. The ability to make various solutions and dilutions, use a pipet, perform titrations, use a microscope, and practice laboratory safety are all important skills.

In British Columbia, grade 12 students who would like to work in this field can enter a two-year program at a post-secondary level followed by a one-year internship, usually in a hospital laboratory. Students who complete this program can then become certified as medical laboratory technologists by the Canadian Society of Laboratory Technologists.

OVERVIEW

Chemical Reaction Calculations

Summary

- Quantitative chemical analysis is routinely done by means of graphs, which can also be used to predict quantities that react or are produced in chemical reactions.

- Many chemical reactions, particularly gravimetric chemical analysis, are done using an excess of one reactant so that the other reactant (the limiting reagent) being analyzed is completely consumed.

- Stoichiometry is a procedure for calculating the quantities of reactants and products in a chemical reaction and may be used in either the prediction or analysis sections of an investigation report.

- Gravimetric, solution, and gas stoichiometry all involve the same basic steps: write a balanced chemical equation to obtain a mole ratio; convert a given measurement to an amount in moles; calculate the required amount from the mole ratio; and convert the required amount in moles into the desired quantity.

- The quantitative study of chemical reactions is used to test the principles of stoichiometry (evaluated by a percent difference). Once shown to be acceptable, stoichiometry is then used to test designs, skills, purity, and the quantitative nature of reactions (evaluated by a percent yield).

- Science deals with natural products and processes, emphasizes ideas, is evaluated according to the validity of its predictions and explanations, and is international in scope.

- Technology deals with manufactured products and manufacturing processes; it emphasizes methods and materials, is evaluated according to its economy, efficiency, and reliability, and is simple in its application.

- A titration is a common experimental design for the quantitative analysis of solutions. Titration is the addition of a titrant in a buret to a sample in an Erlenmeyer flask until the endpoint of the reaction is reached.

- Crystallization, filtration, gas collection, and titration are experimental techniques for quantitative analysis.

Key Words

endpoint
equivalence point
excess reagent
gravimetric
limiting reagent
percent yield
primary standard
stoichiometry
titrant
titration

Review

1. Consider a quantitative analysis in which a sample reacts with another chemical to produce a precipitate.
   (a) Which substance is the limiting reagent?
   (b) What is the purpose of using an excess quantity of the second reactant?

2. (a) How is a graph used in a quantitative analysis?
   (b) What are some advantages of using a graph?

3. In all stoichiometry, why is it necessary to always convert to or convert from amounts in moles?

4. (a) List three types of stoichiometry calculations.
   (b) State the type of quantity determined in each type of calculation.

(c) For each quantity listed, what conversion factor is used to convert to and from an amount in moles?

5. In which sections of a lab report are stoichiometry calculations likely to be found?

6. Why is a balanced chemical equation necessary when doing stoichiometry calculations?

7. (a) What is a percent yield?
   (b) List some reasons why the percentage yield of product in a chemical reaction is generally less than 100%.

8. (a) In a gravimetric analysis, how do you know experimentally when the reaction is complete?
   (b) In a solution analysis using a titration, how do you know experimentally when the reaction is complete?

9. State four experimental designs commonly involved in a chemical analysis.

10. Other than the choice of variables and controls, what are some criteria used to evaluate experimental designs? Make a list of key terms summarizing the criteria.

11. How is chemical science different from chemical technology in terms of emphasis and scope?

12. According to what four criteria is a technology evaluated?

13. List some careers related to chemical analysis.

Applications

14. Calcium ions present in hard water can be analyzed by adding sodium oxalate to precipitate calcium oxalate. Identify the limiting reagent and excess reagent in this analysis.

15. Baking soda can be used as a fire extinguisher for small fires. When heated, the baking soda decomposes into sodium carbonate solid, carbon dioxide, and water vapor.
    (a) If 2.4 mol of baking soda decomposes, what amount in moles of each of the products is formed?

(b) What mass of solid product will remain after the complete decomposition of the contents of a 1.00 kg box of baking soda?
(c) Suggest a reason why baking soda can be used as a fire extinguisher.

16. Identify the limiting and excess reagents for each of the following pairs of reactants. How much excess is present in each case?
    (a) $Zn_{(s)}$ + $CuSO_{4(aq)}$ →
        0.42 mol   0.22 mol
    (b) $Cl_{2(aq)}$ + $NaI_{(aq)}$ →
        10 mmol   10 mmol
    (c) $AlCl_{3(aq)}$ + $NaOH_{(aq)}$ →
        20 g   19 g

17. After malachite is decomposed, the next step in the production of copper metal is the reaction of copper(II) oxide and carbon to produce copper metal and carbon dioxide. Determine the mass of carbon required to react with 50.0 kg of copper(II) oxide.

18. Isooctane, $C_8H_{18(l)}$, is one of the main constituents of gasoline. Calculate the mass of carbon dioxide gas produced by the complete combustion of 692 g of isooctane.

19. Silver-plated tableware is popular because it is less expensive than sterling silver. Silver nitrate solution is used by an electroplating business like that shown below to replate silver tableware for their customers. To test the purity of the solution, a technician adds 10.00 mL of 0.500 mol/L silver nitrate to an excess quantity of 0.480 mol/L NaOH solution. From the reaction, 0.612 g of precipitate is obtained.

(a) What would be a reasonable volume of excess sodium hydroxide to add?

(b) State a specific diagnostic test that could be done to verify that an excess had been added.

(c) Calculate the predicted yield of precipitate.

(d) What is the percent yield? What does this tell you about the purity of the solution?

20. Some antacid products contain aluminum hydroxide to neutralize excess stomach acid. Determine the volume of 0.10 mol/L stomach acid (assumed to be $HCl_{(aq)}$) that can be neutralized by 912 mg of aluminum hydroxide in an antacid tablet.

21. Sulfuric acid is produced on a large scale from readily available raw materials. One step in the industrial production of sulfuric acid is the reaction of sulfur trioxide with water. Calculate the molar concentration of sulfuric acid produced by the reaction of 10.0 Mg of sulfur trioxide with an excess quantity of water to produce 7.00 kL of acid.

22. Analysis shows that 9.44 mL of 50.6 mmol/L $KOH_{(aq)}$ is needed for the titration of 10.00 mL of water from an acidic lake. Determine the molar concentration of acid in the lake water, assuming that the acid is sulfuric acid.

23. A convenient source of oxygen in a laboratory is the decomposition of aqueous hydrogen peroxide to produce water and oxygen. What volume of 0.88 mol/L hydrogen peroxide is required to produce 500 mL of oxygen gas at SATP?

24. Hydrogen gas is produced industrially from the reaction of methane with steam to produce hydrogen and carbon dioxide gases. What volume of hydrogen gas, measured at STP, can be produced from 1.0 t of steam?

25. Hydrogen gas can be produced by the electrolytic decomposition of water. What volume of hydrogen gas is produced, along with 52 kL of oxygen gas, at 25°C and 120 kPa?

26. What volume of hydrogen gas is produced by the reaction of a 25.0 g strip of zinc with 100 mL of a 0.197 mol/L nitric acid solution at SATP? (Hint: First find which is the limiting reagent.)

27. A 10.00 mL sample of oxalic acid, $HOOCCOOH_{(aq)}$, is boiled until dry to crystallize the oxalic acid. Use the following evidence to calculate the molar concentration of the oxalic acid solution.

mass of empty evaporating dish = 84.56 g
mass of dish plus solid oxalic acid = 85.97 g

28. Write three experimental designs that use stoichiometry to determine the concentration of oxalic acid, $HOOCCOOH_{(aq)}$.

29. Write two different experimental designs to answer this question: "What is the concentration of sodium hydroxide in an unknown solution?" State the chemicals and techniques that you intend to use in your designs.

30. Evaluate the following experimental or industrial designs.

(a) Litmus indicator is used to determine the equivalence point of a reaction between solutions of lead(II) nitrate and sodium bromide.

(b) Silver nitrate is used to precipitate sulfate ions from an industrial solution.

(c) Excess lead(II) nitrate is used in an industrial process to remove sulfite ions from a solution.

(d) A concentrated solution of hydrochloric acid is diluted and used as a standard solution to determine the concentration of a potassium hydroxide solution.

31. Make a list of theories, laws, generalizations, and rules that you must know to answer a stoichiometry question. Then construct a concept map showing how all this knowledge is related.

Extensions

32. Chemical technicians in a water treatment plant perform a number of routine analyses on water samples. Find out what is analyzed and what techniques and equipment are used for the quantitative analysis. Describe some examples of large-scale stoichiometry involved in treating a municipal water supply.

33. Do a search on the internet to look for chemical calculation computer programs. Download any free "demo" versions and

demonstrate these to your teacher and classmates.

34. In the late 1970s some psychics claimed that they could bend spoons with the psycho-kinetic powers of their minds. Write an experimental design to test this claim at a show where you have offered a $10 000 prize if someone can bend a spoon under the set conditions.

35. Research the training and requirements involved in becoming an analytical chemist. Include a list of workplaces where analytical chemists are employed. If possible, interview someone to learn more about this occupation.

Problem 7J A Chemical Analysis

Oxalic acid is a common acid with many technological applications. From a scientific perspective, oxalic acid has the characteristic of having two –COOH groups and, like $H_2SO_{4(aq)}$, reacts in a 1:2 mole ratio with sodium hydroxide. Complete the Analysis and Evaluation of the investigation report.

Problem

What is the concentration of oxalic acid, $HOOCCOOH_{(aq)}$, in a rust-removing solution?

Prediction

According to the manufacturer's label, the concentration of oxalic acid in the rust-removing solution is 10% W/V or 1.11 mol/L.

Experimental Design

The original oxalic acid solution (rust remover) is diluted by a factor of 100; that is, 10.00 mL to 1000 mL. The concentration of dilute oxalic acid solution is determined by titration with a sodium hydroxide solution.

Evidence

VOLUME OF 0.0161 mol/L SODIUM HYDROXIDE REQUIRED TO NEUTRALIZE 10.00 mL OF DILUTED OXALIC ACID			
Trial	1	2	3
Final buret reading (mL)	14.3	27.8	41.1
Initial buret reading (mL)	0.2	14.3	27.8
Volume of $NaOH_{(aq)}$ used (mL)			

"Joy in looking and comprehending is

nature's most beautiful gift."

Albert Einstein, Swiss/American physicist

(1879 – 1955)

UNIT IV

CHEMICAL BONDING
IN
MOLECULAR SYSTEMS

The diversity of matter in our world is amazing. Chemists are continually finding and analyzing substances, mostly molecular substances, never known before. Sometimes these chemicals are discovered in the laboratory, and sometimes they are found in nature — for example, in a fungus growing on a tree.

Often a chemical found in nature is available in only small quantities. If the chemical has beneficial applications — for example, in medicine — chemists may study the chemical structure and bonding and try to duplicate the natural product. Or they may try to alter it slightly, producing new molecules. There is no known limit to the number and variety of molecular compounds, especially organic substances.

Experimental work of this kind involves understanding the differences in structure and electron configuration that distinguish one compound from another. In particular, this research involves investigating chemical bonds in compounds. The empirical and theoretical study of the bonding in molecular substances is the focus of this unit.

8 Chemical Bonding

What holds substances together? The layers of materials in a ski are glued together, but what atomic or molecular "glue" holds the water molecules in a snowflake? This "glue" is not a substance at all, but a natural force of attraction — similar to the force that attracts a falling object to Earth or to the force that attracts a magnet to a refrigerator door. Electrical forces among atoms, ions, and molecules are the "glue" that keeps solid objects in the world around us, from rocks to people, from falling apart.

The beautiful, complex structure of a snowflake and the simple, cubic shape of a salt crystal are clues to the attractions and the arrangement of the particles within these substances. These attractions and arrangements of particles, in turn, explain the properties and reactions of substances. Chemical researchers try to understand what holds particles together, why certain structures are formed, and how and why substances react. The amazing diversity of living and non-living things makes this job a fascinating and challenging task.

The question of what holds substances together is important, since a great deal of chemistry involves taking substances apart and putting them together in new ways. In this chapter, you will investigate chemical bonding — the forces that hold atoms and molecules together — and the relationship of chemical bonding to the structure and properties of matter.

Much of our knowledge of matter comes from studies of chemical reactions. Dalton's atomic theory was based on evidence from formation and decomposition reactions. Mendeleyev's development of the periodic table was based primarily on reactions of elements that form compounds with similar chemical formulas (Table 8.1). The unreactive character of the noble gases has been pivotal in the development of theories of atoms, ions, and molecules. By comparing the chemical reactivity of elements and compounds, you can establish patterns, such as an order of reactivity, or **activity series**, for elements in different families and periods. Scientists use patterns such as these to develop or refine concepts of matter.

Table 8.1

SIMILARITIES IN CHEMICAL FORMULAS WITHIN FAMILIES			
Group 1	**Group 2**	**Group 16**	**Group 17**
$LiCl$	BeO	H_2O	CF_4
$NaCl$	MgO	H_2S	CCl_4
KCl	CaO	H_2Se	CBr_4
$RbCl$	SrO	H_2Te	CI_4

Problem 8A Reactivity of Elements

The purpose of this investigation is to develop a generalization about trends in reactivity within the periodic table. You will examine the relative reactivity of the alkali metals with water and the relative reactivity of the halogens with sodium metal. Complete the Analysis of the investigation report.

Problem

How does reactivity change within groups of the periodic table?

Experimental Design

Alkali metals are placed in water and their reactivity is observed. Samples of sodium metal are placed in each of the halogens and their reactivity is observed.

Evidence

REACTIVITY OF GROUP 1 AND GROUP 17 ELEMENTS		
Element	**Reaction with Water**	**Reaction with Sodium**
cesium	instant explosion	
lithium	colorless gas slowly produced, doesn't ignite	
potassium	colorless gas rapidly produced, explodes in 5 s	
rubidium	colorless gas rapidly produced, explodes in 2 s	
sodium	colorless gas burns as produced	
bromine		rapid reaction, forms white solid
chlorine		very rapid reaction, forms white solid
fluorine		instant reaction, forms white solid
iodine		reacts on heating, forms white solid

There are many observable patterns in the chemical properties of elements. Within a period, chemical reactivity tends to be high in Group 1 metals, lower in elements toward the middle of the table, and increasing to a maximum in the Group 17 nonmetals. Metals react differently from nonmetals. Within a family of representative metals, reactivity increases moving down the group; within a family of nonmetals, reactivity increases moving up the group (Figure 8.1). Two patterns, introduced in Chapter 3, will be discussed in more detail in this chapter. These patterns are:

- Nonmetals react with other nonmetals to form molecular compounds. These compounds can be solid, liquid, or gaseous at SATP and they do not conduct electricity, no matter what their state or form. The melting points of molecular solids are relatively low compared with ionic solids.

- Metals react with nonmetals to form ionic compounds. Ionic compounds are non-conducting crystalline solids at SATP. These compounds have relatively high melting points and conduct electricity in their liquid and aqueous states.

Reactivity of Elements

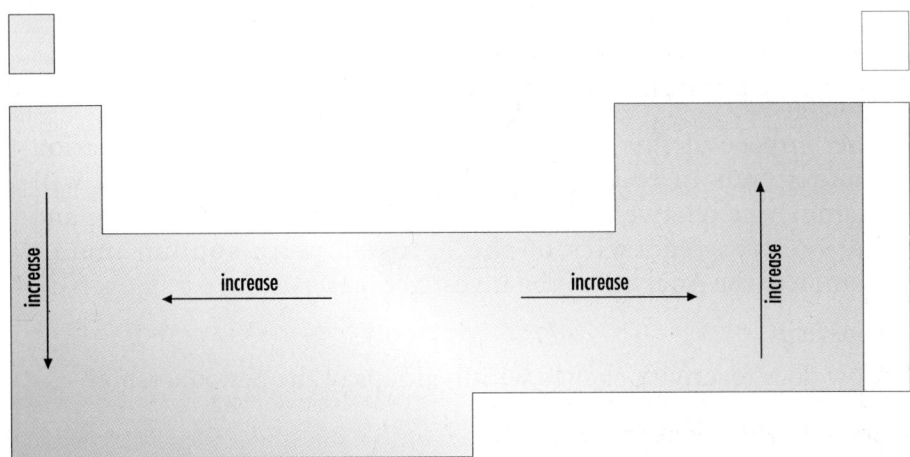

Figure 8.1
Trends in reactivity within the periodic table.

Exercise

1. Name one basis for the classification of elements into families in the periodic table.

2. By examining the periodic tables shown in Figure 8.1 and inside the front cover, which element would you expect to be
   (a) the most reactive metal?
   (b) the most reactive nonmetal?

3. In terms of chemical reactivity, why are the noble gases classified as a chemical family?

4. What is the significance of the "staircase line" in the periodic table?

5. When a metal reacts with a nonmetal, what class of compound is formed? What common evidence supports this classification?

6. (Enrichment) Which metals occur naturally in their element form (i.e., are noble metals)? What does this tell you about their position in an activity series?

Understanding Chemical Periodicity

The similarity in the chemical formulas within families and the empirical trend in the reactivity of elements demonstrate that certain chemical properties of elements are periodic in nature. As we develop a theoretical explanation for this observed periodicity, the natural place to look is at the accepted theories of atomic structure and at the measurable properties of the atom. Three such measurable properties are atomic radius, ionic radius, and ionization energy.

Atomic and Ionic Radii

Atomic radius can be defined in different ways, as illustrated in Figure 8.2. The **van der Waals radius** is defined *as one-half the distance of closest approach between the nuclei of two non-bonded atoms.* In metals such measurements can be made using the technique of X-ray diffraction. The **covalent bond radius** is defined as *one-half the distance between the nuclei of the two atoms in a molecule.* **Ionic radius** is defined as *the radius of an anion or a cation in an ionic crystal.*

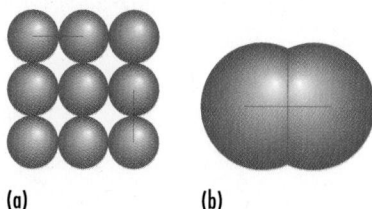

(a)   (b)

Figure 8.2
These drawings illustrate (a) the van der Waals radius in non-bonded atoms and (b) the covalent bond radius in a molecule.

Figure 8.3 shows the atomic and ionic radii of elements in the periodic table. For the atomic radii in Figure 8.3, the van der Waals values are used for metals and the covalent bond values are used for nonmetals. The values for the noble gases are estimated, since data from bonded atoms are not available. The trends in atomic radius are explained theoretically in terms of nuclear charge and energy levels. The gradual decrease in atomic radius as one moves from left to right in a given row of the periodic table is believed to be caused by the increase in nuclear charge with increasing atomic number. The increase from top to bottom in a given group of the periodic table is believed to be the result of the increase in the number of energy levels. Each additional energy level places electrons a greater distance from the nucleus and shields the other electrons from the attractive force of the nucleus, resulting in a decreased attractive force and a larger radius.

Evidence indicates that when an atom forms an anion, its radius increases, as shown in Figure 8.3. The generally accepted theoretical explanation is that the nuclear charge remains the same and the repulsion resulting from the additional electrons enlarges the electron cloud. When an atom forms a cation, its radius decreases. It is believed that while the nuclear charge remains the same, removing an electron reduces electron-electron repulsion and causes the electron cloud to shrink.

Ionization Energy

Ionization energy is defined as the minimum energy required to remove an electron from a gaseous atom in its lowest energy state. For example,

$$X_{(g)} + \text{energy} \rightarrow X^{+}_{(g)} + e^{-} \qquad \text{(first ionization)}$$

One way to detect trends in the properties of elements and atoms is to plot each property against the element's atomic number. A periodic property yields a graph with a visible repeating pattern.

Atomic and Ionic Radii of Selected Elements (in picometres)

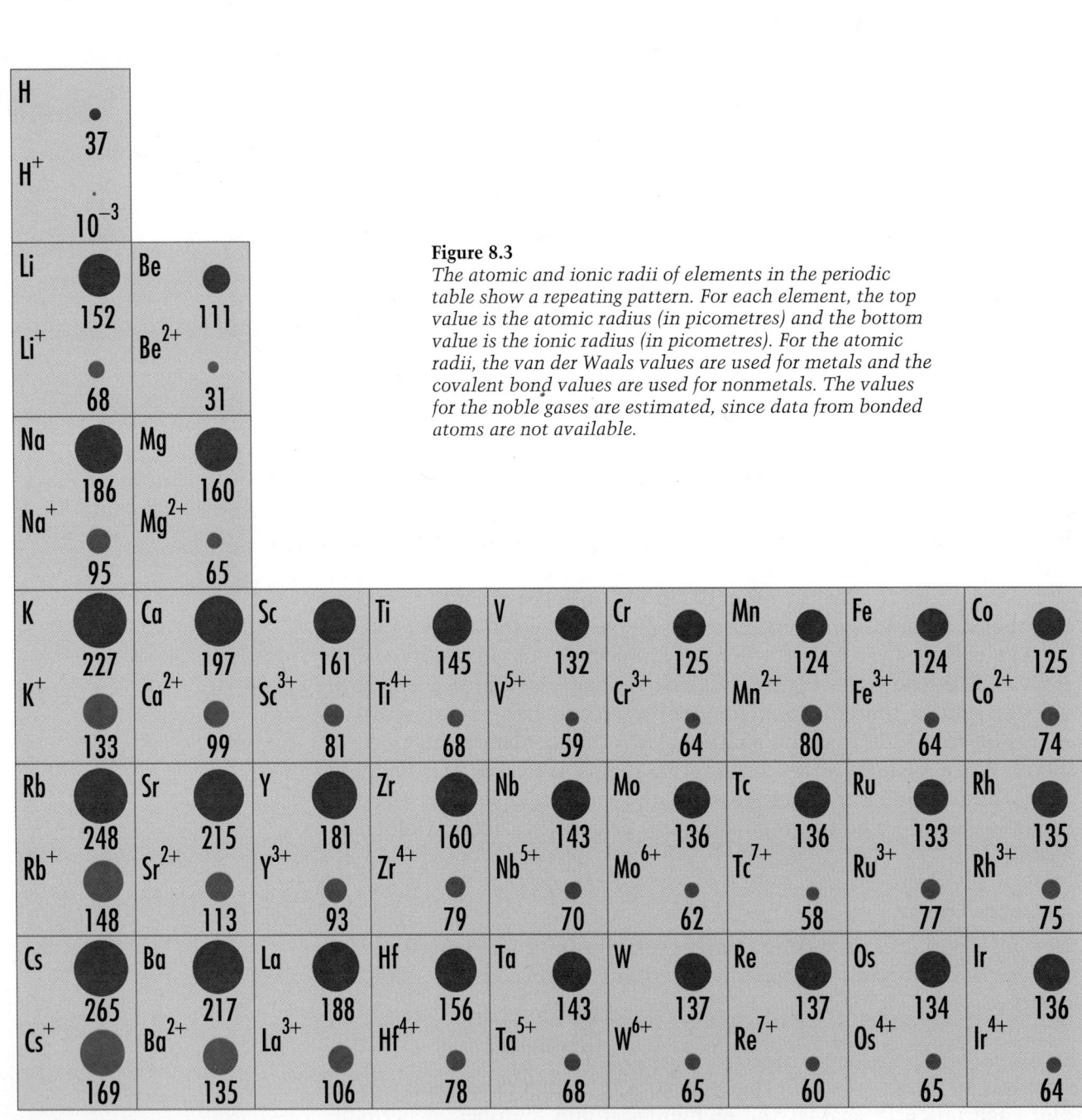

Figure 8.3
The atomic and ionic radii of elements in the periodic table show a repeating pattern. For each element, the top value is the atomic radius (in picometres) and the bottom value is the ionic radius (in picometres). For the atomic radii, the van der Waals values are used for metals and the covalent bond values are used for nonmetals. The values for the noble gases are estimated, since data from bonded atoms are not available.

10	11	12	13	14	15	16	17	18
							H 37; H^- 208	He 50
			B 88	C 77	N 70; N^{3-} 171	O 66; O^{2-} 140	F 64; F^- 136	Ne 62
			Al 143; Al^{3+} 50	Si 117	P 110; P^{3-} 212	S 104; S^{2-} 184	Cl 99; Cl^- 181	Ar 95
Ni 124; Ni^{2+} 72	Cu 128; Cu^{2+} 72	Zn 133; Zn^{2+} 74	Ga 122; Ga^{3+} 62	Ge 123; Ge^{4+} 53	As 121; As^{3-} 222	Se 117; Se^{2-} 198	Br 114; Br^- 196	Kr 112
Pd 138; Pd^{2+} 86	Ag 144; Ag^+ 126	Cd 149; Cd^{2+} 97	In 163; In^{3+} 81	Sn 140; Sn^{4+} 71	Sb 141; Sb^{3+} 90	Te 137; Te^{2-} 221	I 133; I^- 216	Xe 130
Pt 138; Pt^{4+} 70	Au 144; Au^{3+} 91	Hg 160; Hg^{2+} 110	Tl 170; Tl^+ 144	Pb 175; Pb^{2+} 120	Bi 155; Bi^{3+} 96	Po 167; Po^{4+} 65	At 142; At^- 227	Rn 140

Problem 8B Ionization Energy

The purpose of this problem is to identify a pattern in the relationship between the ionization energy of an element and its atomic number. Complete the Analysis of the investigation report. This is an excellent opportunity to use computer graphing programs where they are available.

Problem

What is the pattern in the graph when the first ionization energy of an element is plotted against its atomic number?

Experimental Design

The available ionization energies of the first 100 elements are plotted against the element's atomic number, and the pattern of the graph is examined.

Evidence

FIRST IONIZATION ENERGIES OF ELEMENTS (kJ/mol)											
H	1	1312	Fe	26	759	Sb	51	834	Os	76	839
He	2	2372	Co	27	758	Te	52	869	Ir	77	878
Li	3	520	Ni	28	737	I	53	1008	Pt	78	868
Be	4	899	Cu	29	745	Xe	54	1170	Au	79	890
B	5	801	Zn	30	906	Cs	55	376	Hg	80	1007
C	6	1086	Ga	31	579	Ba	56	503	Tl	81	589
N	7	1402	Ge	32	762	La	57	538	Pb	82	716
O	8	1314	As	33	947	Ce	58	528	Bi	83	703
F	9	1681	Se	34	941	Pr	59	523	Po	84	812
Ne	10	2081	Br	35	1140	Nd	60	530	At	85	—
Na	11	496	Kr	36	1351	Pm	61	535	Rn	86	1037
Mg	12	738	Rb	37	403	Sm	62	543	Fr	87	—
Al	13	578	Sr	38	549	Eu	63	547	Ra	88	509
Si	14	786	Y	39	616	Gd	64	592	Ac	89	499
P	15	1012	Zr	40	660	Tb	65	564	Th	90	587
S	16	1000	Nb	41	664	Dy	66	572	Pa	91	568
Cl	17	1251	Mo	42	685	Ho	67	581	U	92	584
Ar	18	1521	Tc	43	702	Er	68	589	Np	93	597
K	19	419	Ru	44	711	Tm	69	597	Pu	94	585
Ca	20	590	Rh	45	720	Yb	70	603	Am	95	578
Sc	21	631	Pd	46	805	Lu	71	524	Cm	96	581
Ti	22	658	Ag	47	731	Hf	72	642	Bk	97	601
V	23	650	Cd	48	868	Ta	73	761	Cf	98	608
Cr	24	653	In	49	558	W	74	770	Es	99	619
Mn	25	717	Sn	50	709	Re	75	760	Fm	100	627

The graph from Problem 8B shows that as atomic number increases, the first ionization energy of the elements rises and falls in a regular pattern. In each period the alkali metal has the lowest ionization energy and the noble gas has the highest ionization energy. Recall that within a given period the outermost electrons are in the same energy level. The evidence indicates that if other factors are constant, the attraction for the outer-level electron increases as the nuclear charge increases. The theoretical explanation is that as the force of attraction between the nucleus and electrons increases, more energy is needed to remove the electron from the atom.

The graph from Problem 8B also shows that within a given group the ionization energy decreases as the atomic number increases. The theoretical explanation is that because atomic radius increases and additional inner layers of electrons reduce the effect of the nuclear charge, there is less attractive force between the nucleus and outer-level electrons so that less energy is required to remove them from the atom.

Finally, the graph from Problem 8B reveals that ionization energy does not increase uniformly across a period, suggesting the existence of energy sublevels within the main energy levels. Stronger evidence for sublevels within the main energy levels came from research as spectroscopic technology improved and it was discovered that single spectral lines are really groups of very fine lines. The sublevels were initially classified according to the characteristics of the fine lines observed in the spectra of atoms. The sublevel symbols are shown in Table 8.2. The relative energies of the sublevels within the principal energy levels are $s<p<d<f$. The letter abbreviations of the energy sublevels are used to describe the electrons in atoms.

Table 8.2

CLASSIFICATION OF ENERGY SUBLEVELS		
Sublevel Name	Sublevel Symbol	Number of Orbitals
sharp	s	1
principal	p	3
diffuse	d	5
fundamental	f	7

Atomic Orbitals

Electrons in atoms are classified according to their energies and are believed to exist in regions of space around the nucleus, known as **orbitals**. According to the theory of quantum mechanics (page 82), an orbital is defined as a region in space in which there is a 95% probability of finding an electron with a given energy. Electrons are thought to occupy a region in space in somewhat the same way that clouds occupy regions of the atmosphere. The size and shape of orbitals depends on the energy of the electron.

Orbital diagrams are an extension of the energy level diagrams discussed in Chapter 2, which represent the number of electrons in the principal energy levels. An orbital diagram is a one-dimensional energy graph depicting the relative energies of orbitals in the principal energy levels and sublevels. In order to describe the distribution of electrons in an atom the procedure will be restricted to atoms in their lowest energy state, assuming isolated gaseous atoms. In Figure 8.4, a circle is used to represent an electron orbital in which two electrons can exist. Empirical evidence supports **Hund's rule** which states that no electron pairing takes place in p, d, or f orbitals until each orbital of the given set contains one electron.

Figure 8.4
Diagram of relative energies of orbitals in sublevels. Each circle represents an orbital capable of holding two electrons. In multi-electron atoms the lowest energy orbitals are filled first.

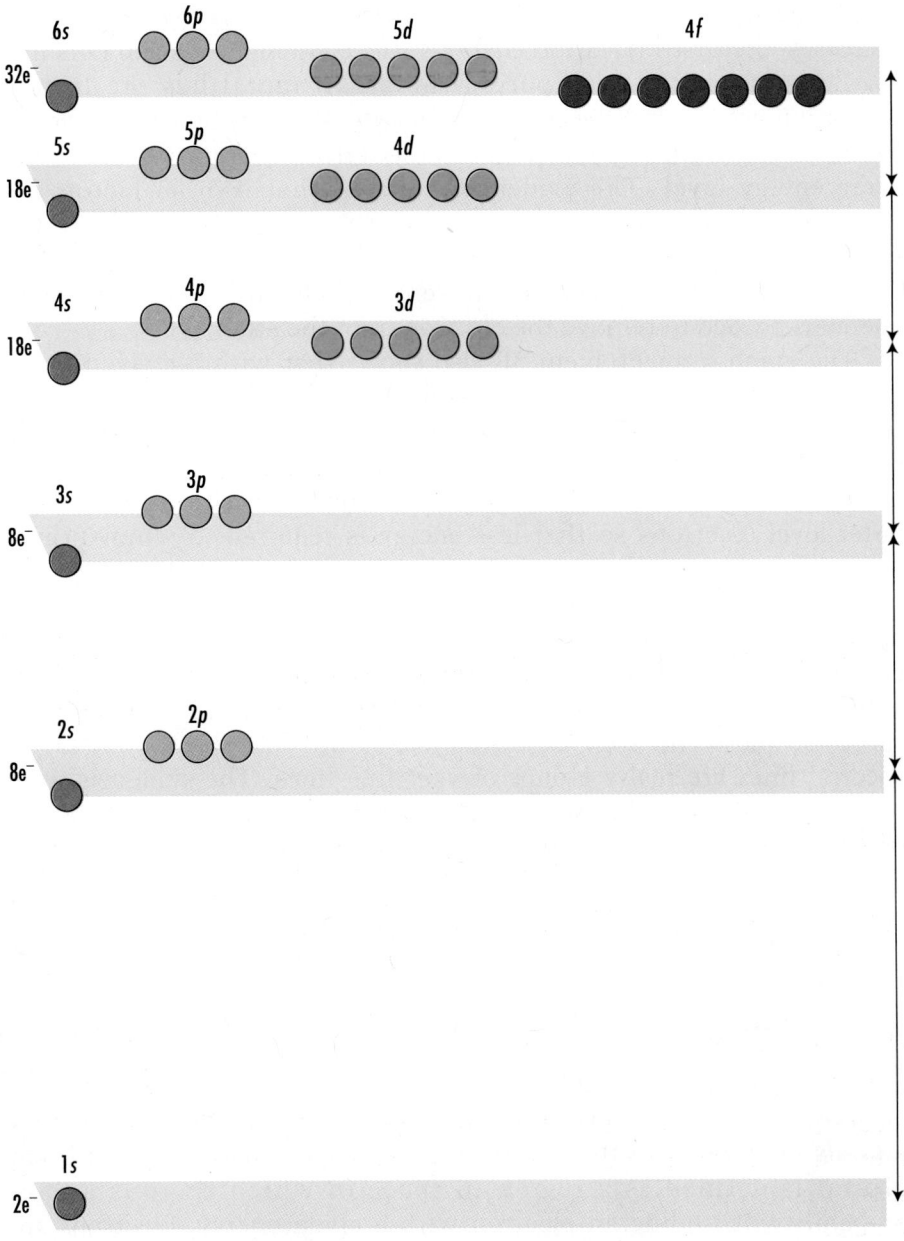

SUMMARY: PROCEDURE FOR ORBITAL DIAGRAMS

Step 1: Determine the total number of electrons from the atomic number of the element.

Step 2: Place electrons in orbitals in order of increasing energy according to the orbital diagram in Figure 8.4. Use an arrow pointing up to represent the first electron placed in an orbital, and an arrow pointing down to represent the second electron.

Step 3: Beginning at the lowest energy level, place electrons singly in the *s*, *p*, *d*, and *f* orbitals until each orbital of the same energy contains one electron. Additional electrons are placed in the half-filled orbitals until the total number of electrons has been represented.

Draw an orbital diagram for oxygen.

1. Oxygen has an atomic number of 8, and therefore has 8 electrons.

2. The first two electrons will occupy the 1s orbital. ⇅

   The next two electrons will occupy the 2s orbital. ⇅

3. The next three electrons are placed singly in each of the three 2p orbitals. ↑ ↑ ↑

   The last (eighth) electron must be paired with one of the electrons in any of the 2p orbitals. ⇅ ↑ ↑

   The final diagram would be drawn as

   $2s$ ⇅ $2p$ ⇅ ↑ ↑   or $2s$ ↑↓ $2p$ ↑↓ ↑ ↑
   $1s$ ⇅ $1s$ ↑↓

Exercise

7. Complete orbital diagrams for phosphorus, iron, tin, and tungsten.

8. Complete orbital diagrams for potassium and fluoride ions. Which noble gas atoms have the same orbital diagrams as these ions?

9. Complete orbital diagrams for barium and sulfide ions. Which noble gas atoms have the same orbital diagrams as these ions?

Electron Configurations

Orbital diagrams are a good way of visualizing the energies of the electrons in an atom, but they are rather cumbersome to draw. The method of electron configurations provides the same information in a more concise format. An electron configuration is a listing of the number and kinds of electrons in order of increasing energy, written in a single line.

SUMMARY: PROCEDURE FOR WRITING ELECTRON CONFIGURATION

Step 1: Determine the total number of electrons in the atom or simple ion.

Step 2: Start assigning electrons in increasing order of main energy levels and sublevels.

Step 3: Continue assigning electrons by filling each sublevel before going to the next sublevel until the total number of electrons has been assigned. Figure 8.5 provides two simple means of remembering the order of the sublevels.

Figure 8.5

(a) Aufbau diagram. To determine the ground state electron configuration of a multi-electron atom, start with the 1s orbital and move downward following the direction of the arrows. The order goes as follows: 1s → 2s → 2p → 3s → 3p → 4s → 3d → 4p → 5s . . .

(b) Classification of elements by sublevels being filled. Groups of elements in the periodic table can be classified on the basis of the type of orbital being filled with electrons. To determine the ground state electron configuration of a multi-electron atom, start with the hydrogen atom and move from left to right across each period. The order goes as follows: 1s → 2s → 2p → 3s → 3p → 4s → 3d → 4p → 5s ...

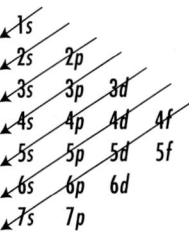

(a) Aufbau diagram

(b) Classification of elements by sublevels being filled

EXAMPLE

What is the electron configuration of (a) oxygen and (b) sodium in their ground state?

(a) Oxygen has 8 electrons. According to Figure 8.5, the electron configuration of an oxygen atom in its ground state is O: $1s^2\ 2s^2\ 2p^4$.

(b) Sodium has 11 electrons. According to Figure 8.5, the electron configuration for a sodium atom in its ground state is summarized as Na: $1s^2\ 2s^2\ 2p^6\ 3s^1$.

The noble gas family is unique not only empirically but also theoretically, because electron configurations in this family all involve completely filled energy levels. The electron configuration of an element can be abbreviated by using, as a starting point, the electron configuration of the noble gas that precedes it in the periodic table. This method is called the "core" or "kernel" method because the noble gas symbol represents the nucleus and inner electrons (core or kernel) of the atom. For example, [He] represents $1s^2$, and [Ne] represents $1s^2\ 2s^2\ 2p^6$. With the core or kernel method, the electron configuration of oxygen is written as O: [He] $2s^2\ 2p^4$, and sodium is written as Na: [Ne] $3s^1$. When atoms of the representative elements form ions, they usually gain or lose electrons to acquire the same number of electrons as the nearest noble gas on the periodic table. Consequently, ions of the representative elements are often **isoelectronic** with one of the noble gases.

In predicting the electron configurations of the atoms and ions of the transition metals, there are two important exceptions to the rules outlined above. The first exception involves elements in Groups 6 and 11. For example, the electron configuration of Cr is [Ar] $4s^1\ 3d^5$ rather than [Ar] $4s^2\ 3d^4$, and the electron configuration of Cu is [Ar] $4s^1\ 3d^{10}$ rather than [Ar] $4s^2\ 3d^9$. Apparently electron configurations with the d orbitals either half filled or completely filled are more stable, even though they involve promoting electrons from s orbitals to the slightly higher energy d orbitals. The second exception involves the electron configuration of transition metal ions. For example, the electron configuration of the Mn^{2+} is [Ar] $3d^5$ rather than [Ar] $4s^2 3d^3$. Empirical work indicates that when transition metal ions form, the electrons are removed from the ns orbital before the $(n-1)d$ orbital. Apparently the shape of the s orbital makes it easier to remove the electrons even though they have lower energy than the d electrons.

Exercise

10. Write full electron configurations for each of the period 3 elements.

11. Write full electron configurations for each of the halogens. Describe how the halogen configurations are similar. Does this general similarity apply to other families?

12. Write full electron configurations for a fluoride ion and a sodium ion. With which noble gas are these ions isoelectronic?

13. Identify the elements whose atoms have the following electron configurations.
    (a) $1s^2 2s^2$
    (b) $1s^2 2s^2 2p^5$
    (c) $1s^2 2s^2 2p^6 3s^1$
    (d) $1s^2 2s^2 2p^6 3s^2 3p^4$

14. The last electron represented in an electron configuration is related to the position of the element in the periodic table (see Figure 8.5). For each of the following sections of the periodic table, indicate the sublevel (s,p,d,f) of the last electron.
    (a) Groups 1 and 2
    (b) Groups 3 to 12 (transition metals)
    (c) Groups 13 to 18
    (d) lanthanides and actinides

15. Use the kernel method to write electron configurations for each of the Group 11 metals, copper, silver, and gold.

Orbitals and Bonds

Modern atomic theory has developed an explanation of the reactivity of elements in the periodic table (Chapter 2). According to Bohr's atomic model, periodicity of the chemical properties of elements is related to the number of valence electrons in the atoms. (Recall from Chapter 2 that valence electrons are those occupying the highest electron energy level of an atom.) For example, the alkali metals form similar compounds with chlorine because all the alkali metal atoms have one valence electron. In quantum mechanics, this idea is refined to include a classification and description of valence electrons. This refinement of the concept of valence electrons enables us to explain and predict chemical formulas for molecular and ionic substances.

Since chemical reactions are thought to involve only the outer or valence electrons, in explanations of reactions and bonding the discussion of orbitals is usually restricted to the valence orbitals of an atom. According to the theory of quantum mechanics, the number and

Simplicity
Scientists generally agree that simplicity is a characteristic of an acceptable theory. Einstein never fully accepted the theory of quantum mechanics, partly because of its complexity. In spite of his suspicions, he admired the ability of quantum mechanics to explain and predict observations.

occupancy of valence orbitals in the representative elements are determined by the following theoretical rules, plus the descriptions in Table 8.3 and Figure 8.6.

- There are four valence orbitals in the valence level of atoms of representative elements. Hydrogen, which has a very simple structure, is an exception to this and many other rules; it has only one valence orbital.

- An orbital may contain 0, 1, or 2 electrons. This means that two (but not more than two) electrons may share the same region of space at the same time.

- Electrons occupy any empty valence orbitals within an s, p, d, or f sublevel before forming electron pairs.

- A maximum of eight electrons can occupy s plus p orbitals in the valence level of an atom. This is known as the **octet rule**. (The noble gases have this valence electronic structure; their lack of reactivity indicates that eight electrons filling an s plus p valence level is a very stable structure.)

$3e^- \rightarrow 1e^-\ 1e^-\ 1e^-$ $6e^- \rightarrow 2e^-\ 2e^-\ 1e^-\ 1e^-$
$8e^- \rightarrow 2e^-\ 2e^-\ 2e^-\ 2e^-$ $8e^- \rightarrow 2e^-\ 2e^-\ 2e^-\ 2e^-$
$2e^- \rightarrow 2e^-$ $2e^- \rightarrow 2e^-$
$13p^+$ $16p^+$

Al atom   S atom

Figure 8.6
To explain bonding, valence levels above the first level are considered to have four orbitals, each of which may contain 0, 1, or 2 electrons.

Table 8.3

THEORETICAL DEFINITIONS OF ORBITALS			
Orbital	Number of Electrons in the Orbital	Description of Electrons	Type of Electrons
empty	0	—	—
half-filled	1	unpaired	bonding
filled	2	lone pair	non-bonding

Valence electrons are classified in terms of orbital occupancy. A **bonding electron** is a single electron occupying an orbital. A **lone pair** is a pair of electrons occupying a filled orbital.

Lewis Models

In 1916, American theoretical chemist G.N. Lewis created a simple model of the arrangement of electrons in atoms that explains and predicts empirical formulas. An **electron dot diagram** that represents the **Lewis model** for an element includes the chemical symbol and dots that represent the valence electrons.

Figure 8.7 shows the relationship of the restricted quantum mechanics model and the Lewis model for an oxygen atom. The purpose of drawing electron dot diagrams is to determine how many of the valence electrons are bonding electrons. Figure 8.8 shows the electron dot diagrams for the elements in period 2. To draw electron dot diagrams of atoms:

- Write the element symbol to represent the nucleus and any filled energy levels of the atom.

- Use a dot to represent each valence electron.

- Start by placing a single valence electron into each of four valence orbitals (represented by the four sides of the element symbol).

$6e^-$
$2e^-$
$8p^+$

$:\ddot{O}\cdot$

Electron energy level diagram of oxygen atom

Electron dot diagram of oxygen atom

Figure 8.7
An energy level diagram and an electron dot diagram represent the quantum mechanics model and the Lewis model of an oxygen atom. These models help to explain the chemical properties of oxygen.

- If additional locations are required for electrons, start filling the four orbitals with a second electron until up to eight positions for valence electrons have been occupied.

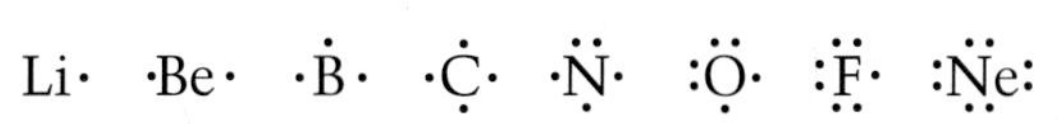

$$\text{Li}\cdot \quad \cdot\text{Be}\cdot \quad \cdot\overset{\displaystyle\cdot}{\text{B}}\cdot \quad \cdot\overset{\displaystyle\cdot}{\text{C}}\cdot \quad \cdot\overset{\displaystyle\cdot}{\text{N}}\cdot \quad :\overset{\displaystyle\cdot}{\text{O}}\cdot \quad :\overset{\displaystyle\cdot}{\text{F}}\cdot \quad :\overset{\displaystyle\cdot}{\underset{\displaystyle\cdot}{\text{N}}}\text{e}:$$

Figure 8.8
Electron dot diagrams of period 2 elements.

Electronegativities

On theoretical grounds, chemists believe that atoms have different abilities to attract electrons. For example, the farther away from the nucleus electrons are, the weaker is their attraction to the nucleus. Also, inner electrons (those closer to the nucleus) shield the valence electrons from the attraction of the positive nucleus.

Chemists use the term **electronegativity** to describe the relative ability of an atom to attract a pair of bonding electrons in its valence level. Electronegativity is usually assigned on a scale developed by Linus Pauling (Figure 8.9), who won a Nobel Prize for his work on bonding theory. Empirically, Pauling based his scale on energy changes in chemical reactions. According to the scale, fluorine has the highest electronegativity, 4.0, and cesium has the lowest electronegativity, 0.7. Note that these are the most reactive nonmetal and metal, respectively. Metals tend to have low electronegativities and nonmetals tend to have high electronegativities. A high electronegativity is interpreted as a high attraction for electrons. Electronegativity is a periodic property of atoms and is closely related to ionization energy. See the periodic table on the inside front cover of this book for data on electronegativities of atoms.

Figure 8.9
Linus Pauling (1901 – 1994) was a dual winner of the Nobel Prize. In 1954 he won the prize in chemistry for his work on molecular structure, and in 1962 he received the Nobel Peace Prize for campaigning against the nuclear bomb.

Chemical Bonding

Imagine that two atoms, each with an orbital containing one bonding electron, collide in such a way that these half-filled orbitals overlap. As the two atoms collide, the nucleus of each atom attracts and attempts to "capture" the bonding electron of the other atom. A "tug-of-war" over the bonding electrons occurs. Which atom wins? Comparing the electronegativities of the two atoms can predict the result of the contest. If the electronegativities of both atoms are relatively high, neither atom may win and the pair of bonding electrons may be shared between the two atoms. The simultaneous attraction of two nuclei for a shared pair of bonding electrons is known as a **covalent bond**. This is the type of bond formed between two nonmetals.

If the electronegativities of two colliding atoms are quite different, the atom with the stronger attraction for electrons may succeed in removing the bonding electron from the other atom. An *electron transfer* then occurs, and positive and negative ions are formed. An **ionic bond** results from the simultaneous attraction of positive ions and negative ions in a three-dimensional array. This type of bond, formed between a metal and a nonmetal, creates a definite, repeating pattern, or *crystal* (Figure 8.10).

The formation of a chemical bond involves a competition for bonding electrons in unfilled orbitals. If the two atoms have equal electronegativities, then the bonding electrons are shared equally. If they have unequal electronegativities, the electrons may be unequally

Figure 8.10
The regular geometric shape of ionic crystals is evidence for an orderly array of positive and negative ions.

shared or they may be nearly completely transferred. Chemical bonds form a continuum from equal sharing to almost complete electron transfer. This explanation of the formation of chemical bonds is *simple*, *logical*, and *consistent* with other scientific ideas, such as atomic theory and collision-reaction theory. This explanation also accounts for the observation that metals combine with nonmetals to form ionic compounds, whereas nonmetals combine with nonmetals to form molecular compounds.

Exercise

16. Describe in terms of electrons and orbitals: bonding electron, lone pair.
17. Write a theoretical definition of electronegativity, covalent bond, and ionic bond, and describe the bonds in terms of a difference in electron activity.
18. Prepare a table with the headings: element symbol, electronegativity, group number, number of valence electrons, electron dot diagram, number of bonding electrons, and number of lone pairs of electrons. Fill in the table with the elements in period 3.
19. How do the electron dot diagrams that represent metal atoms differ from those of nonmetals?
20. Using the electronegativity data in the periodic table on this book's inside front cover, describe the variation in electronegativities within a group and a period.
21. (Discussion) Which element is an exception to almost every rule or generalization about elements? What is unique about this element compared with all other elements in the table?

Metals

The category of metals includes both representative and transition metals and is characterized by elements that are shiny, silvery, flexible, malleable solids with good electrical conductivity. Further evidence from the analysis of X-ray diffraction patterns shows that all metals have a continuous and very compact crystalline structure. With few exceptions, all metals have closely packed structures.

An acceptable theory for metals must explain the characteristic metallic properties, provide testable predictions, and be as simple as possible. According to current beliefs, the properties of metals are the result of the nondirectional nature of the bonding as a result of loosely held, mobile electrons. As illustrated in Figure 8.11, the valence electrons are believed to fill the spaces between the positive centers. This simple model, known as the *electron sea* model, incorporates the ideas of

- low ionization energy of metal atoms to explain loosely held electrons
- empty valence orbitals to explain electron mobility
- electrostatic attractions of positive centers and the negatively charged electron "sea" to explain the strong, nondirectional bonding

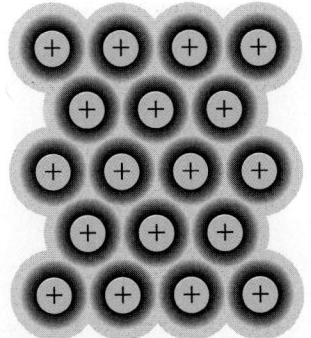

Figure 8.11
In this model of metallic bonding, each circled positive charge represents the nucleus and inner electrons of a metal atom, surrounded by a mobile sea of electrons.

Figure 8.11 shows a cross-section of the crystal structure of a metal. Each circled positive charge represents the nucleus and inner electrons (kernel) of a metal atom. The shaded area surrounding the circled positive charges represents the mobile sea of electrons. The electron sea model may be used to explain the empirical properties of metals. Because the electrons are believed to be quite mobile, it is easy to use a battery to force additional electrons onto one end of a metal sample and cause other electrons to leave the sample from the other end. A sea of electrons implies no specific or directional bonding. Because the electrons essentially provide the "electrostatic glue" holding the atomic centers together, structures should be continuous and closely packed. The flexibility and malleability of metals are thought to be the result of the nondirectional nature of the bonds and the ability of planes of atoms to slide over each other with no increase in electrostatic repulsion. Finally, the shiny appearance and silver color of metals are believed to be caused by the ability of their valence electrons to absorb and re-emit the energy from all wavelengths of visible light.

Metalloids

The "staircase line" in the periodic table, separating metals and nonmetals, is helpful in predicting types and properties of compounds. However, the staircase line does not represent a sharp division between metals and nonmetals. Elements near this line have some properties of both metals and nonmetals, as well as some properties unique to this region. These elements, known as **metalloids** (Figure 8.12), are very hard, have high melting points, and are either non-conductors or semiconductors of electricity. For example, silicon is a lustrous, silver-gray solid like many metals, but it is brittle and conducts electricity only slightly.

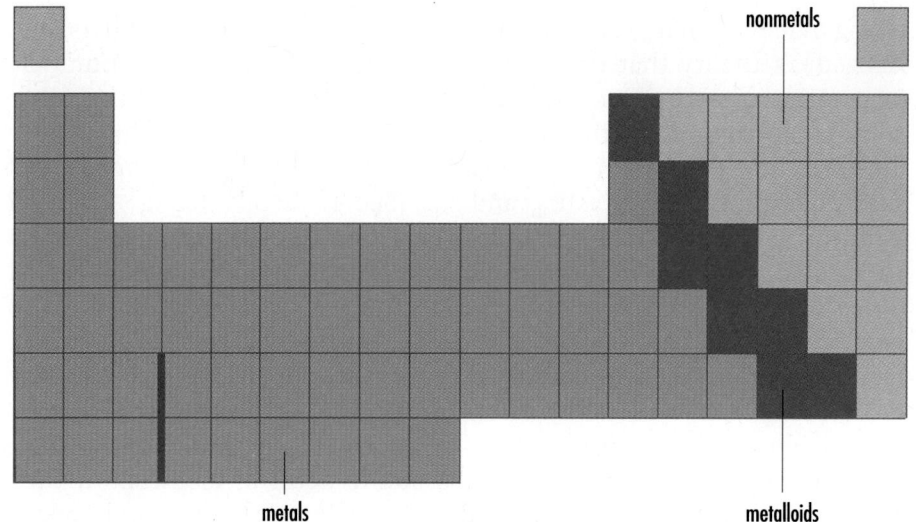

Figure 8.12
Metalloids occupy the region near the staircase line of the periodic table. This class of elements represents a revision of the classification of elements as either metals or nonmetals.

According to quantum mechanics as illustrated in electron dot diagrams, metalloids are unique because they have many bonding electrons compared with both metals and nonmetals. The concept of a network of covalent bonds explains the properties of metalloids (Figure 8.13). The hardness and high melting points of metalloids are explained by strong, directional, covalent bonds; their poor electrical conductivity is explained by the covalent bonding of all the valence electrons; and their crystal shapes are explained by the regular, three-dimensional arrays of covalently bonded atoms.

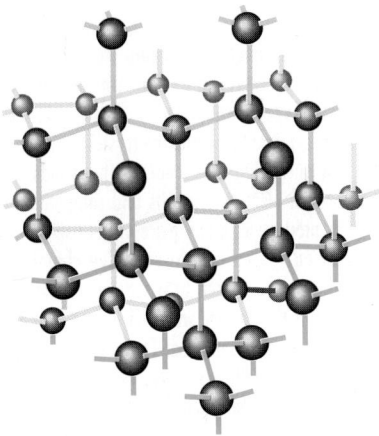

Figure 8.13
Metalloids like silicon have a diamond-like structure. The structure of diamond, shown here, is theoretically described as a network of covalently bonded atoms forming a macromolecule the size of the crystal itself.

Molecular Elements
Many molecular elements are diatomic and some are polyatomic.

hydrogen	$H_{2(g)}$
nitrogen	$N_{2(g)}$
oxygen	$O_{2(g)}$
fluorine	$F_{2(g)}$
chlorine	$Cl_{2(g)}$
iodine	$I_{2(s)}$
bromine	$Br_{2(l)}$
phosphorus	$P_{4(s)}$
sulfur	$S_{8(s)}$

8.2 MOLECULAR SUBSTANCES

The theory of chemical bonding developed in this chapter paves the way for a more complete understanding of molecular elements and compounds. In this section, molecular substances are discussed in terms of their properties and their chemical bonding.

Molecular Elements

Using evidence that gases react in simple ratios of whole numbers, and Avogadro's theory that equal volumes of gases contain equal numbers of particles, early scientists were able to determine that the most common forms of hydrogen, oxygen, nitrogen, and the halogens are diatomic molecules. Modern evidence shows that phosphorus and sulfur commonly occur as $P_{4(s)}$ and $S_{8(s)}$. (See Table 3.5 on page 116 for a summary of the chemical formulas of molecular elements.) An acceptable theory of molecular elements must provide an explanation for evidence such as these empirical formulas. Recall that according to atomic theory, an atom such as chlorine, with seven valence electrons, requires one electron to complete the stable octet. This electron may be obtained from a metal by electron transfer, or the required electron may be obtained by sharing a valence electron with another atom. Two chlorine atoms could each obtain a stable octet of electrons if they shared a pair of electrons in a covalent bond. A *covalent bond* between the atoms results from the simultaneous attraction of two nuclei for a shared pair of electrons, and this explains why chlorine molecules are diatomic.

$$:\!\overset{..}{\underset{..}{Cl}}\!\cdot \ + \ \cdot\overset{..}{\underset{..}{Cl}}\!: \ \rightarrow \ :\!\overset{..}{\underset{..}{Cl}}\!:\!\overset{..}{\underset{..}{Cl}}\!:$$

According to atomic theory, oxygen atoms have six valence electrons and evidence shows that the element is diatomic (Figure 8.14). Sharing a pair of bonding electrons would leave both oxygen atoms with less than a stable octet. Initially, molecular theory could not explain the diatomic character of oxygen. Instead of replacing the theory, scientists revised it by introducing the idea of a *double bond*. If two oxygen atoms can share a pair of electrons, perhaps they can share two pairs of electrons at once to form a double covalent bond.

$$:\ddot{O}\cdot \; + \; \cdot\ddot{O}: \; \rightarrow \; :\ddot{O}::\ddot{O}:$$

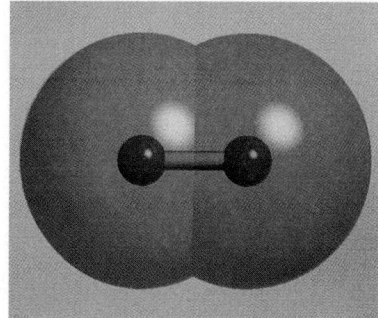

Figure 8.14
The oxygen molecule is represented here by a computer-generated combination of space-filling and ball-and-stick models.

$$Cl - Cl \qquad O = O$$

Figure 8.15
Structural diagrams for chlorine and oxygen molecules. A single line represents a single covalent bond and a double line represents a double covalent bond.

This arrangement is consistent with accepted theory, since stable octets of electrons result. This idea explains many empirically known molecular formulas, such as $O_{2(g)}$, in a simple way, without the necessity of changing most of the previous assumptions.

Lewis models, based on the ideas of quantum mechanics and the concept of covalent bonds, are a form of electron "bookkeeping" to account for valence electrons. They do not show what orbitals look like or where electrons may actually be at any instant — they simply keep track of which electrons are involved in bonds. Once this is understood, an even simpler and more efficient model, known as a **structural diagram**, can be used to represent bonding in molecules (Figure 8.15). In structural diagrams, lone pairs of electrons are not indicated, and shared pairs of electrons are represented by lines.

Molecular Compounds

Molecular compounds cannot be represented by a simplest ratio formula in the way that ionic compounds can. Simplest ratio formulas indicate only the relative numbers of atoms in a compound, but give no clue about the actual number or arrangement of atoms. For example, the simplest ratio formula CH represents a compound containing equal numbers of carbon and hydrogen atoms. Empirical work indicates that several compounds, such as acetylene ($C_2H_{2(g)}$) and benzene ($C_6H_{6(l)}$), can be described by means of the simplest ratio formula CH. To distinguish between these compounds, it is necessary to represent them with molecular formulas.

Exercise

22. List some examples of molecular elements and compounds.

23. Draw electron dot diagrams for the elements in the halogen family. How are these diagrams consistent with the concept of a chemical family?

24. The electron dot diagram of a hydrogen molecule is an exception to the octet rule.　　H:H

    Considering the positions of hydrogen and helium in the periodic table, how is the electron dot diagram for hydrogen a good explanation of its empirical formula?

25. Use an electron dot diagram to explain the empirical formula for nitrogen, N_2.

26. According to atomic theory, a sulfur atom has six valence electrons, including two bonding electrons. Therefore, any molecule containing sulfur would include two covalent bonds from each sulfur atom.

    (a) Use an electron dot diagram to predict the simplest chemical formula for sulfur.

    (b) Suggest an explanation for the empirical formula of sulfur, S_8.

27. (Enrichment) Explain the empirical formula for phosphorus, P_4. Check your answer against the known structure in a reference and evaluate your explanation.

DOROTHY CROWFOOT HODGKIN (1910 –)

Biochemist Dorothy Mary Crowfoot Hodgkin was born in Egypt of English parents. Her father was an archeologist and the family travelled with him on his research expeditions. Dorothy Hodgkin enrolled at Somerville College, Oxford, England, where she began to study large, complex molecules using X-ray diffraction techniques (page 330).

In Ph.D. studies at Cambridge, Hodgkin worked with John Desmond Bernal. An inspiring teacher and creative scientist, Bernal pioneered the use of X-ray diffraction analysis in the study of biologically important molecules such as enzymes. Hodgkin, a talented and determined young scientist, worked with Bernal to refine the techniques of X-ray diffraction analysis. Her efforts led to the first successful applications of this technique. For her Cambridge Ph.D. dissertation, Hodgkin studied the X-ray diffraction of pepsin, a digestive enzyme.

Upon returning to Oxford to teach chemistry, Hodgkin continued to apply X-ray diffraction techniques to determine the structure of complex organic molecules. From 1942 to 1949 she studied the structure of penicillin, using a computer to analyze X-ray diffraction photographs and to discover the arrangement of the atoms in the molecule. This was the first direct use of a computer to solve a biochemical problem. The later sophistication of computers enhanced the power of X-ray diffraction by speeding up the long and tedious calculations. Even with the assistance of computers, however, working out molecular structures took years to complete. By 1955 Dr. Hodgkin had determined the structure of vitamin B_{12}, a compound that is essential for the production of human red blood cells. (Crystals of vitamin B_{12} are shown above.) For her research on vitamin B_{12}, Dorothy Hodgkin was awarded the 1964 Nobel Prize in chemistry.

- *What two technologies are combined in the study of large, biologically important molecules?*
- *Why did Dorothy Hodgkin receive a Nobel Prize in chemistry?*

Empirical Molecular Formulas

In order to determine molecular formulas empirically, we need evidence of both percentage composition and molar mass. Percentage composition can be measured by a number of different technologies. Chemists in the 19th century, using the relatively simple technologies available, were able to determine empirical formulas of molecular substances. Today, chemists employ sophisticated technologies to gather evidence for the empirical formulas of molecular compounds. *Combustion analyzers*, such as that shown in Figure 8.16, make very

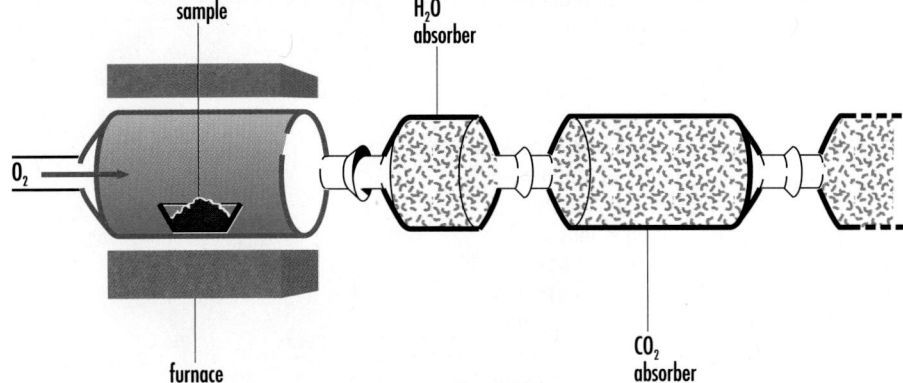

Figure 8.16
A substance burned in a combustion analyzer produces oxides that are captured by absorbers in chemical traps. The initial and final masses of each trap indicate the masses of the oxides produced. These masses are then used in the calculation of the percentage composition of the substance burned.

precise measurements of the relative masses of elements such as carbon, hydrogen, nitrogen, oxygen, and sulfur in compounds. Several milligrams of a compound are burned inside the combustion chamber and the quantities of combustion products, such as carbon dioxide, water vapor, and sulfur dioxide, are measured. Computer analysis provides a printout listing the percent by mass of each element detected in the compound. From this information, the simplest ratio formula is obtained.

To determine the molecular formula for a compound, a molar mass for the substance must be determined as well. Although a number of laboratory methods can determine molar mass, chemists most often rely on a *mass spectrometer* (Figure 8.17).

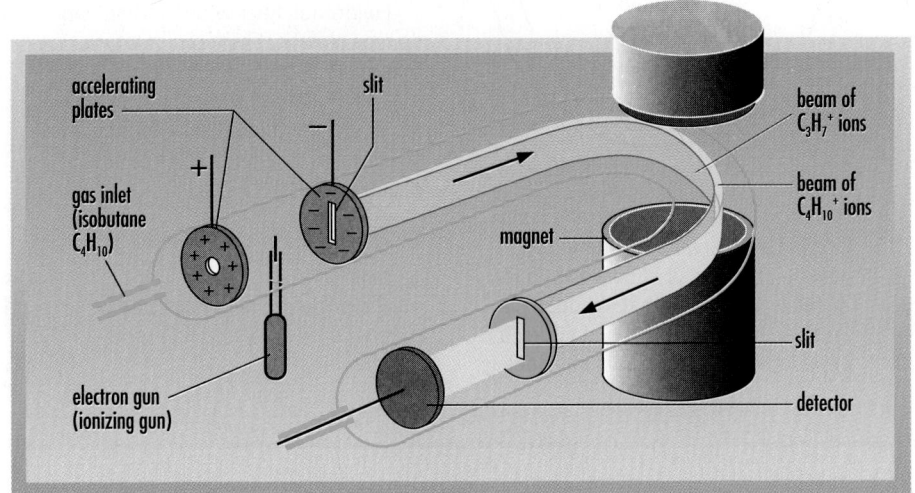

Figure 8.17
A mass spectrometer is used to determine the masses of ionized particles by measuring the amount of deflection in the path of the particles as they pass through a magnetic field.

For some time now, scientists have been aware that pure carbon exists in two forms known as network solids: diamond and graphite. In a diamond the carbon atoms form a

sturdy, three-dimensional network in which each atom is rigidly fixed by covalent bonds to its four closest neighbors (Figure 8.13, page 306). In graphite, the carbon atoms form two-dimensional sheets in which the carbon atoms are located at the corners of hexagons. The discovery of a third structural form of pure carbon is causing great excitement in the scientific community. This third form, C_{60}, is a round molecule that possesses extraordinary stability. Because the structure of this hollow, cagelike molecule has the symmetry of the geodesic dome invented by American architect R. Buckminster Fuller, scientists have named it *buckminsterfullerene*.

Chemists at Rice University in Texas first suggested the existence of an entirely new class of carbon molecules — including C_{60} — which they called *fullerenes*. Along with colleagues at the University of Sussex in England, they had proposed an unusual soccer-ball shape

for the 60-atom carbon cluster buckminsterfullerene; like a geodesic dome, a soccer ball consists of a network of 12 pentagons and 20 hexagons. In the course of their experiments, the researchers at Rice discovered a whole class of carbon molecules with the geodesic-dome structure. What intrigued them was the remarkable stability of all these molecules.

Because fullerenes are readily soluble and vaporizable molecules that remain stable in air, they are well suited to a wide range of applications, such as lubricants, catalysts, and medicines. British scientists have reported generating macroscopic quantities of fully fluorinated "teflon balls" ($C_{60}F_{60}$) that could work as an excellent lubricant, because their freely spinning spherical shape resembles that of minuscule ball bearings. It may also be possible to link fullerenes in bulk to form a framework with regular spaces between the spheres, to carry catalysts to reaction sites.

The most intriguing properties of fullerenes are electrical. In their vari-

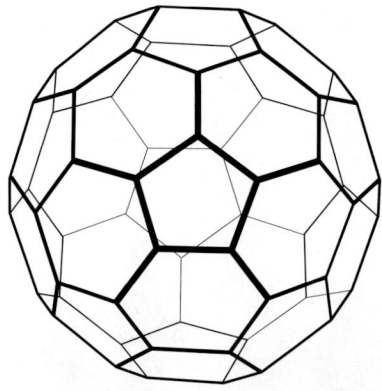

ous compound forms, fullerenes can function as insulators, conductors, semiconductors, or superconductors. Research indicates that although pure crystals of C_{60} are semiconductors, a conductor can be produced by doping

C_{60} with potassium to form crystals of K_3C_{60}. When K_3C_{60} is cooled below 18 K, it becomes a superconductor. If a greater amount of potassium is used in the doping reaction, the

resulting crystal is an insulator. The ability to absorb and subsequently give up electrons might even make fullerenes the basis of a new class of battery, lighter and more efficient than the lead-acid batteries now in use. Since 1985, inexpensive ways of producing fullerenes have been developed, including a process that uses an electric arc struck between two carbon rod electrodes in a helium atmosphere. This method may soon produce C_{60} as inexpensively as aluminum.

Fullerenes form when carbon condenses slowly at high temperature. Scientists speculate that they may be among the oldest and most common molecules in the universe, formed in the first generation of stars about 15 billion years ago.

- *Describe the structure of the C_{60} molecule.*
- *Describe the technological applications of two fullerene compounds.*

In a mass spectrometer, a small gaseous sample is bombarded by a beam of electrons, which causes the molecules to break up into charged fragments. (For example, the two main fragments for a particular compound are shown in Figure 8.17. These fragments have been accelerated by an electric field and then deflected by a magnetic field. The amount of deflection depends on the mass and the charge of the fragment. Thus, from the amount of deflection, the molar mass of the original sample can be determined from the molar mass of the largest fragment. The mass spectrograph (the printout from a mass spectrometer) of the compound shown in Figure 8.18 (page 312) shows several fragments. The following example shows how evidence from a combustion analyzer and a mass spectrometer is used to determine a molecular formula.

EXAMPLE

Complete the Analysis of the investigation report.

Problem

What is the molecular formula of the fluid in a butane lighter?

Evidence

From combustion analysis: percent by mass of carbon = 82.5%
                           percent by mass of hydrogen = 17.5%
From mass spectrometry: molar mass = 58 g/mol
(See the mass spectrometer in Figure 8.17 and the mass spectrograph in Figure 8.18, page 312.)

Analysis

Assume one mole (58 g) of the compound is analyzed.

$$m_C = \frac{82.5}{100} \times 58 \text{ g} = 48 \text{ g}$$

$$n_C = 48 \text{ g} \times \frac{1 \text{ mol}}{12.01 \text{ g}} = 4.0 \text{ mol}$$

$$m_H = \frac{17.5}{100} \times 58 \text{ g} = 10 \text{ g}$$

$$n_H = 10 \text{ g} \times \frac{1 \text{ mol}}{1.01 \text{ g}} = 10 \text{ mol}$$

or $$n_C = \frac{82.5}{100} \times 58 \text{ g} \times \frac{1 \text{ mol}}{12.01 \text{ g}} = 4.0 \text{ mol}$$

$$n_H = \frac{17.5}{100} \times 58 \text{ g} \times \frac{1 \text{ mol}}{1.01 \text{ g}} = 10 \text{ mol}$$

- Find the mass of each element in one mole by multiplying the percentage by the mass of one mole.
- Find the amount in moles of each element by multiplying the mass in one mole by the reciprocal of the atomic molar mass.
- Find the whole-number ratio of atoms in the molecule.

The mole ratio of carbon atoms to hydrogen atoms in the compound analyzed is 4:10.

According to the evidence gathered in this investigation, the empirical molecular formula of the fluid in a butane lighter is C_4H_{10}.

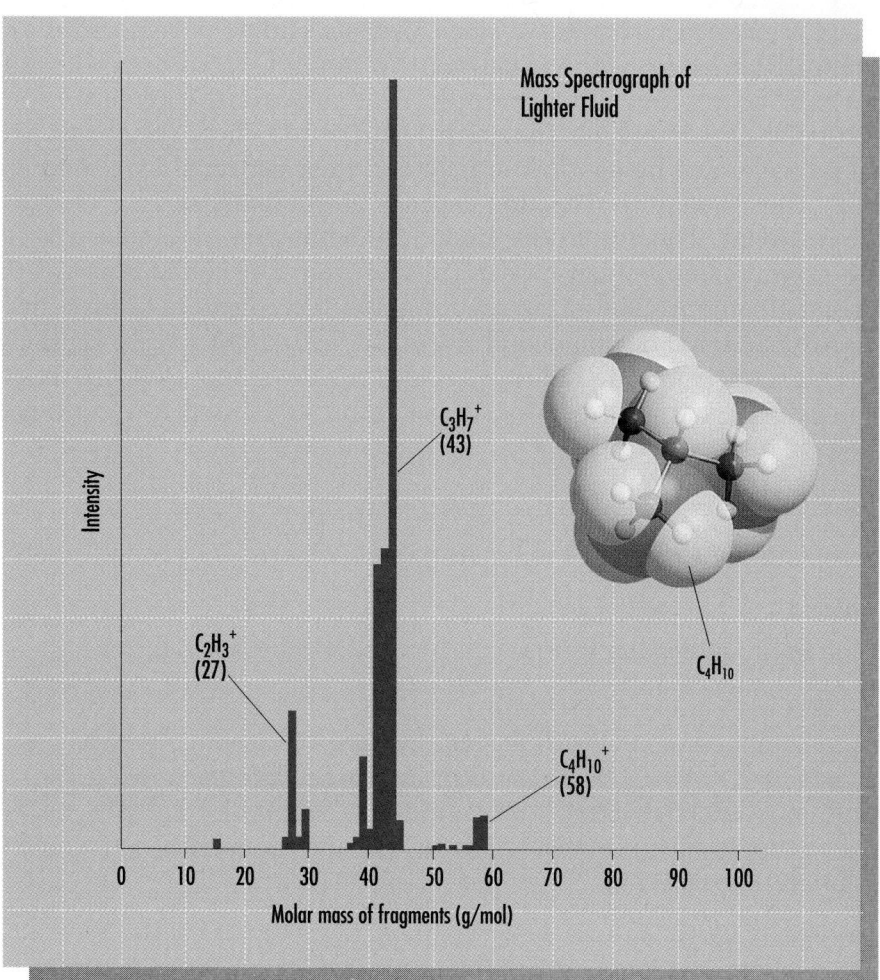

Figure 8.18
A mass spectrograph of lighter fluid is used to determine the molar mass of the chemical in the fluid. Expert analysis of the mass spectrograph provides the basis for the molecular model shown here.

Exercise

28. Analysis of an air pollutant indicates that the compound is 30.4% nitrogen and 69.6% oxygen. The mass spectrograph for the pollutant shows that its molar mass is 92.0 g/mol. Determine the molecular formula and the chemical name of the polluting compound.

29. An important raw material for the petrochemical industry is ethane, which is extracted from natural gas. Determine the molecular formula for ethane using the following evidence.

    molar mass = 30.1 g/mol

    percent by mass of carbon = 79.8%

    percent by mass of hydrogen = 20.2%

30. (Discussion) What ways of knowing the chemical formula of molecular compounds are you currently able to employ?

Problem 8C Chemical Analysis

Carbohydrates are an important source of food energy. A food chemist extracts a carbohydrate from honey and submits a sample to a spectroscopy lab for analysis. Complete the Analysis of the investigation report.

Problem

What is the molecular formula of the unknown carbohydrate?

Experimental Design

A sample of the carbohydrate is burned in a combustion analyzer to determine the percent by mass of each element in the compound. Another sample is analyzed by a mass spectrometer to determine the molar mass of the carbohydrate.

Evidence

molar mass = 180.2 g/mol
percent by mass of carbon = 40.0%
percent by mass of hydrogen = 6.8%
percent by mass of oxygen (remainder of sample) = 53.2%

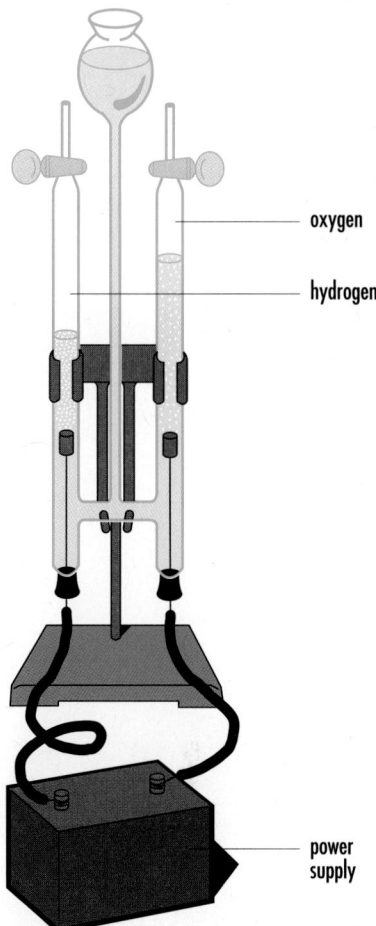

oxygen

hydrogen

power supply

Figure 8.19
When water is decomposed by electricity, the volume of hydrogen gas produced is twice that of the oxygen gas. This evidence led to the acceptance of H_2O as the empirical formula for water.

Explaining Molecular Formulas

Explanations of molecular formulas are based on the same ideas used to explain molecular elements. The rules of the restricted quantum mechanics theory, the idea of overlapping half-filled orbitals, and a consideration of differences in electronegativity all work together to produce a logical, consistent, simple explanation of experimentally determined molecular formulas.

A covalent bond in molecular compounds, just as in molecular elements, is a strong, directional force within a complete structural unit: the molecule. The theoretical interpretation of the empirical formula for water, H_2O, is that a single molecule contains two hydrogen atoms and one oxygen atom held together by covalent bonds (Figures 8.19 and 8.20). The purpose of explaining a molecular formula is to show the arrangement of the atoms that are bonded together. As shown below, an oxygen atom requires two electrons to complete a stable octet. These two electrons are thought to be supplied by the bonding electrons of two hydrogen atoms. In this way, oxygen achieves a stable octet and all atoms complete their unfilled energy levels.

$$:\overset{..}{\underset{..}{O}}\cdot \; + \; \overset{\cdot H}{\cdot H} \; \rightarrow \; :\overset{..}{\underset{H}{O}}:H \quad \text{or} \quad \underset{H}{\overset{O-H}{|}}$$

The idea of a *double covalent bond* explains empirically known molecular formulas such as $O_{2(g)}$ and $C_2H_{4(g)}$. The atoms involved must share more than one bonding electron. A double bond involves the sharing of *two* pairs of electrons between two atoms. Similarly, a *triple covalent bond* involves two atoms sharing *three* pairs of electrons. There is no empirical evidence for the formation of a bond involving

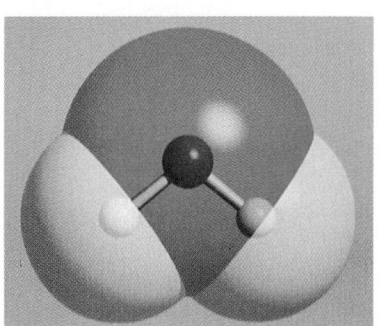

Figure 8.20
The empirical formula for water is described theoretically by this computer-generated model.

more than three pairs of electrons. According to accepted rules and models, there are many atoms that can form more than one kind of covalent bond. For example, carbon, nitrogen, and oxygen, which are three of the most important elements in molecules in living organisms, can all form more than one kind of covalent bond. The maximum number of single covalent bonds that an atom can form is known as its **bonding capacity**, which is determined by its number of bonding electrons (Table 8.4). For example, nitrogen, with a covalent bonding capacity of three, can form three single bonds, one single bond and one double bond, or one triple bond.

A covalent bond in which one of the atoms donates both electrons is called a **coordinate covalent** bond. The concept is useful in explaining the structure of many molecules and polyatomic ions. The properties of coordinate covalent bonds do not differ from those of a normal covalent bond because all electrons are alike, regardless of source. The following equation shows the formation of a coordinate covalent bond.

$$H-\underset{\underset{H}{|}}{\overset{\overset{H}{|}}{N}}: \ + \ \underset{\underset{H}{|}}{\overset{\overset{H}{|}}{B}}-H \ \rightarrow \ H-\underset{\underset{H}{|}}{\overset{\overset{H}{|}}{N}}-\underset{\underset{H}{|}}{\overset{\overset{H}{|}}{B}}-H \quad \text{or} \quad H:\underset{\underset{\ddot{H}}{}}{\overset{\overset{H}{}}{\ddot{N}}}: \ + \ \underset{\underset{H}{}}{\overset{\overset{H}{}}{\ddot{B}}}:H \ \rightarrow \ H:\underset{\underset{H}{}}{\overset{\overset{H}{}}{\ddot{N}}}:\underset{\underset{H}{}}{\overset{\overset{H}{}}{\ddot{B}}}:H$$

Table 8.4

BONDING CAPACITIES OF SOME COMMON ATOMS			
Atom	Number of Valence Electrons	Number of Bonding Electrons	Bonding Capacity
carbon	4	4	4
nitrogen	5	3	3
oxygen	6	2	2
halogens	7	1	1
hydrogen	1	1	1

In the discussion presented in this textbook, the sharing of electrons has been restricted to the bonding electrons. This restricted theory also requires that all valence electrons in molecules be paired. Of course, no theory in science is absolute, and there are exceptions to this theory. Some molecules, such as nitrogen monoxide, appear to have unpaired electrons or appear not to follow the octet rule. Rather than developing a more detailed theory, this textbook specifically notes such cases as exceptions.

SUMMARY: DRAWING ELECTRON DOT DIAGRAMS AND STRUCTURAL DIAGRAMS FOR MOLECULAR COMPOUNDS

Step 1: Draw the electron dot diagram of the atom that has the highest bonding capacity.

Step 2: Form shared pairs of bonding electrons with the remaining atoms.

Step 3: If any bonding electrons remain on adjacent atoms, form a double or triple bond.

Step 4: In the finished electron dot diagram, all atoms (except hydrogen) should contain a stable octet, counting lone pairs plus shared pairs of electrons.

Step 5: Draw the structural diagram for the molecule.

Although the molecular formulas of some compounds provide clues about which atoms are bonded, trial and error may be necessary to obtain a reasonable model. For example, this approach produces a model of the bonding implicit in the empirically determined molecular formula for formaldehyde, H_2CO.

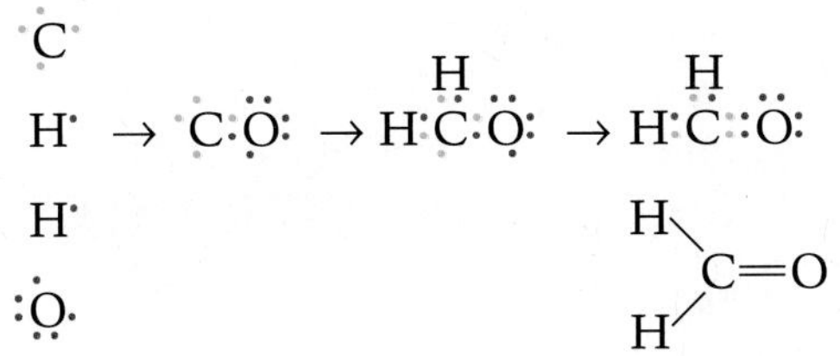

When molecules have more than one possible arrangement of atoms, the molecular formula may be written differently to indicate groups of atoms that are bonded together. For example, the molecular formula for ethanol, C_2H_5OH, clearly shows that one hydrogen atom is bonded to the oxygen atom. Dimethyl ether, which has the same number and kind of atoms but very different chemical and physical properties, is written as CH_3OCH_3 or $(CH_3)_2O$ to distinguish it from ethanol.

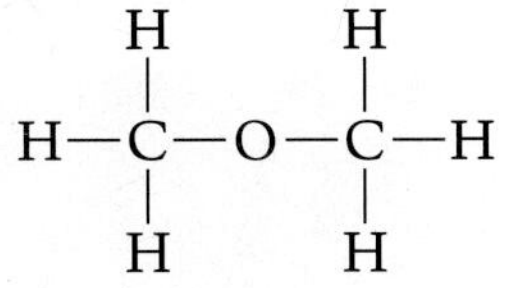

dimethyl ether

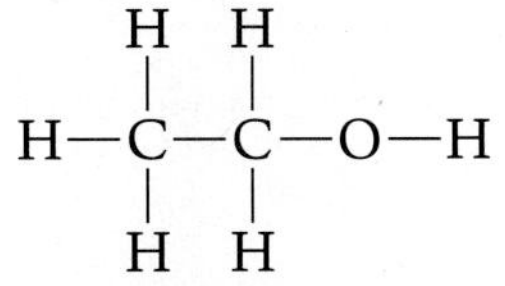

ethanol

Exercise

31. Why is it incorrect for the structural diagram of H_2S to be written as H—H—S?

32. Why do you think the molecular formula for methanol is usually written as CH_3OH instead of CH_4O?

33. For each of the following molecular compounds and ions, name the compound/ion and explain the empirically determined formula by drawing an electron dot diagram and a structural diagram of the molecule/ion.
    (a) HCl
    (b) NH_3
    (c) H_2S
    (d) CO_2
    (e) NH_4^+

34. Use the bonding capacities listed in Table 8.4 to draw a structural diagram of each molecule.
    (a) H_2O_2
    (b) C_2H_4
    (c) HCN
    (d) C_2H_5OH
    (e) CH_3OCH_3
    (f) CH_3NH_2

35. (Discussion) What criteria can be used to assess the explanatory power of a new scientific theory?

- [ ] Problem
- [ ] Prediction
- [ ] Design
- [ ] Materials
- [ ] Procedure
- [x] Evidence
- [x] Analysis
- [x] Evaluation
- [ ] Synthesis

INVESTIGATION

8.1 Molecular Models

Chemists use molecular models to explain and predict molecular structure, relating structure to the properties and reactions of substances. The purpose of this investigation is to explain some known chemical reactions by using molecular models. Evaluate your explanation by assessing whether it is logical and consistent, and as simple as possible.

Problem

How can theory, represented by molecular models, explain the following series of chemical reactions that have occurred in a laboratory?

(a) $CH_4 + Cl_2 \rightarrow CH_3Cl + HCl$
(b) $C_2H_4 + Cl_2 \rightarrow C_2H_4Cl_2$
(c) $N_2H_4 + O_2 \rightarrow N_2 + 2 H_2O$
(d) $CH_3CH_2CH_2OH \rightarrow CH_3CHCH_2 + H_2O$
(e) $HCOOH + CH_3OH \rightarrow HCOOCH_3 + H_2O$

Experimental Design

Molecular model kits, designed to show the bonding capacities of atoms, test the ability of theory to explain the specified chemical reactions.

Materials

molecular model kit (Figure 8.21)

Procedure

1. For the first reaction, construct a structural model for each reactant and record the structures in your notebook.

2. Study your models to determine the minimum rearrangement that would produce structural models of the products.

3. Rearrange the models to form the products, and record the structures in your notebook.

4. Repeat steps 1 to 3 for each chemical equation.

Figure 8.21
You can use molecular model kits to test the explanatory power of bonding theories.

- [ ] Problem
- [ ] Prediction
- [ ] Design
- [ ] Materials
- [ ] Procedure
- [x] Evidence
- [x] Analysis
- [ ] Evaluation
- [ ] Synthesis

INVESTIGATION

8.2 Evidence for Double Bonds

The purpose of this investigation is to obtain empirical evidence for the existence of double bonds. Two compounds, cyclohexane and cyclohexene (Figure 8.22), are believed to be almost identical, except for the presence of a double bond between two carbon atoms in cyclohexene. These compounds illustrate a relationship between structure and reactivity — cyclohexene reacts rapidly with bromine but cyclohexane does not. The reaction is indicated by the disappearance of the color of the bromine.

Problem

Which of the common substances supplied contain molecules with double covalent bonds between carbon atoms?

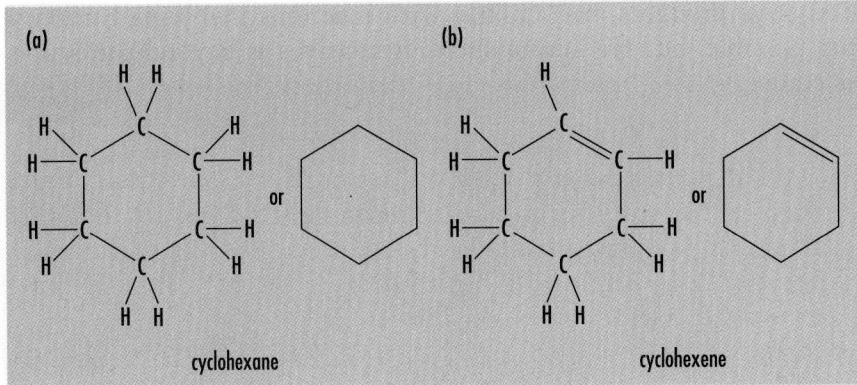

cyclohexane   cyclohexene

Experimental Design

The unknown samples and two controls (cyclohexane and cyclohexene) are tested by adding a few drops of a solution containing bromine in trichlorotrifluoroethane (or bromine in water). After each sample is mixed with the bromine solution, any evidence of a chemical reaction is noted.

Materials

lab apron
safety glasses
small test tubes with stoppers
test tube rack
waste container, with lid, for organic substances
bromine solution in a dropper bottle
cyclohexane in dropper bottle
cyclohexene in dropper bottle
common substances, such as mineral oil, paint thinner, liquid
 paraffin, soybean oil, corn oil, peanut oil

Procedure

1. Add 10 drops of cyclohexane to a clean test tube.

2. Add 1 drop of bromine solution to the test tube. Shake the test tube gently. Repeat this procedure with up to 4 drops of bromine solution.

3. Dispose of all materials into the labelled waste container.

4. Repeat steps 1 to 3 using cyclohexene. Use a clean test tube.

5. Repeat steps 1 to 3 using the samples provided. Use a clean test tube each time.

Figure 8.22
In (a), the structural diagrams of cyclohexane show that all bonds are single bonds. In (b), the cyclohexene structures indicate one carbon-carbon double bond. The second structure in diagrams (a) and (b) represents the same molecule with a line diagram where each corner of the structure represents a carbon atom, and any bonds not shown are assumed to join hydrogen atoms to carbon atoms.

Predicting Molecular Formulas

The rules and ideas presented so far (to explain known chemical formulas) can be adapted to predict the molecular formulas of compounds formed from two nonmetallic elements. To simplify the predictions, use the fewest possible number of atoms and the fewest number of double or triple bonds to predict the simplest product. Alternative products, particularly with reactions involving carbon, are often possible, but predicting such alternatives is beyond the scope of this course.

Exercise

36. Predict the simplest molecular formula and write the chemical name for a product of each of the following reactions. Show your reasoning by including both an electron dot diagram and a structural diagram of the product.
    (a) $I_{2(s)} + Br_{2(l)} \rightarrow$
    (b) $P_{4(s)} + Cl_{2(g)} \rightarrow$
    (c) $O_{2(g)} + Cl_{2(g)} \rightarrow$
    (d) $C_{(s)} + S_{8(s)} \rightarrow$
    (e) $S_8 + O_{2(g)} \rightarrow$

37. (Extension) Compare your predictions from question 36 with the empirical evidence presented in a reference such as *The CRC Handbook of Chemistry and Physics*.

38. (Discussion) Evaluate the theory used to predict molecular formulas.

Problem 8D Predicting Molecular Formulas

The purpose of this problem is to test the predictive power of accepted theories and models for molecular formulas. Complete the Prediction, Analysis, and Evaluation of the investigation report.

Problem

What are the molecular formula and the chemical name of the simplest compound formed when oxygen reacts with fluorine?

Evidence

From combustion analysis: percent by mass of oxygen = 29.5%

percent by mass of fluorine = 70.5%

From mass spectrometry: molar mass = 54.0 g/mol

8.3 ENERGY CHANGES

Exothermic and endothermic reactions are defined in Table 4.2 on page 138.

Energy transfer is an important factor in all chemical changes. Exothermic reactions such as the combustion of gasoline in a car engine or the metabolism of fats and carbohydrates in a human body (Figure 8.23) release energy into the surroundings. Endothermic

Figure 8.23
In the human body, exothermic chemical reactions occur as fats and carbohydrates are metabolized.

reactions, such as photosynthesis (Figure 8.24) or the decomposition of water into hydrogen and oxygen (Figure 8.19, page 313) remove energy from the surroundings. Knowledge of energy and energy changes is important to society and to industry, and the study of energy changes provides chemists with important information about chemical bonds.

Just as glue holds objects together, electrical forces hold atoms together. In order to pull apart objects that are glued together, you have to supply some energy. Similarly, if atoms or ions are bonded together, energy is required to separate them. Separated atoms or ions release energy when they bond together again.

bonded particles + energy → separated particles (endothermic process)

separated particles → bonded particles + energy (exothermic process)

The stronger the bond holding the particles together, the greater is the energy required to separate them. **Bond energy** is the energy required to break a chemical bond. It is also the energy released when a bond is formed. Even the simplest of chemical reactions may involve the breaking and forming of several individual bonds. The terms *exothermic* and *endothermic* are empirical descriptions of overall changes that can be explained by knowledge of bond changes. Consider the decomposition of water, for example:

Figure 8.24
Plants use energy from the sun in a series of endothermic reactions called photosynthesis.

$$2\,H_2O_{(l)} \rightarrow 2\,H_{2(g)} + O_{2(g)}$$

In this reaction, hydrogen-oxygen bonds in the water molecules must be broken before the hydrogen-hydrogen and oxygen-oxygen bonds can be formed.

Since the overall change is endothermic, the energy required to break the O—H bonds must be greater than the energy released when the H—H and O=O bonds form. In any endothermic reaction, more energy is needed to break bonds in the reactants than is released by bonds formed in the products. For exothermic reactions, the opposite is true (Figure 8.25, page 320).

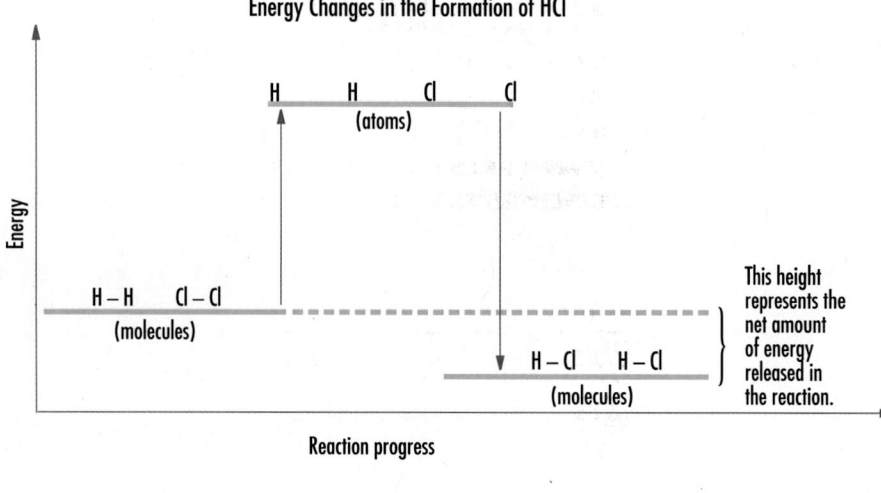

Energy Changes in the Formation of HCl

H   H   Cl   Cl
(atoms)

Energy

H – H   Cl – Cl
(molecules)

H – Cl   H – Cl
(molecules)

This height represents the net amount of energy released in the reaction.

Reaction progress

$$H_{2(g)} + Cl_{2(g)} \longrightarrow 2 HCl_{(g)} + energy$$

Figure 8.25
Energy is absorbed in order to break the H—H and Cl—Cl bonds, but more energy is released when the H—Cl bonds form. The overall result is an exothermic reaction.

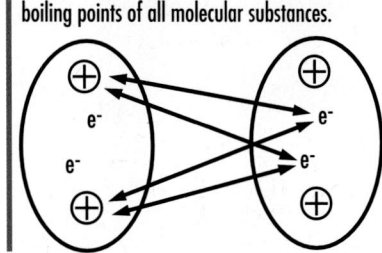
Relative Strength of Chemical Bonds

According to the kinetic molecular theory, there is little space between the particles of solids and liquids. The particles in gases are widely spaced and, as a result, the forces of attraction between gas particles are considered to be negligible. When a liquid boils, particles break free of the bonds holding them together and separate as particles of a gas. The higher the boiling point, the greater the energy required to separate the molecules. Thus, boiling point temperature provides an indirect measure of the forces or bonds that hold particles together.

Ionic compounds generally have high boiling points, whereas molecular compounds generally have low boiling points. This evidence suggests that bonds among the ions in a crystal are significantly stronger than bonds among the molecules in a molecular substance. However, the covalent bonds within molecules must be strong because heat does not generally cause molecular substances to decompose.

Intermolecular Forces

The weak forces or bonds among molecules are known as **intermolecular forces**. Many different observations, such as surface tension, changes of state, and heats of vaporization provide evidence that there are three kinds of intermolecular forces, discussed in the following pages.

Two of these forces, London forces and dipole-dipole forces, are collectively known as **van der Waals** forces.

London Forces

Weak attractive forces were first described by Fritz London in 1930. **London forces**, also known as *dispersion forces*, are the weak attractive forces that result when electrons in one molecule are attracted by the positive nuclei of atoms in nearby molecules. Inside a molecule, the distances between electrons and nuclei are small, and therefore the attractions between electrons and nuclei are strong. These strong attractive forces result in ionic or covalent bonds. Between molecules, London forces are comparatively weak because the electron-nuclei attractions occur over much greater distances. For similar molecules, boiling points are an indirect measure of the strength of these attractions. As shown in Table 8.5, boiling points increase as the total number of electrons in the molecule increases; this is interpreted to mean that the strength of the London forces is greater in the larger molecules.

Table 8.5

EVIDENCE FOR LONDON FORCES IN THE HALOGEN FAMILY			
Halogen	Number of Electrons per Molecule	Boiling Point (°C)	
fluorine, F_2	18	−188	increasing
chlorine, Cl_2	34	−34.6	strength of
bromine, Br_2	70	58.8	London
iodine, I_2	106	184	forces

Dipole-Dipole Forces

A chemical bond may involve equal sharing of an electron pair in a covalent bond, the complete transfer of an electron in an ionic bond, or the unequal sharing of a pair of electrons between two atoms. A covalent bond resulting from unequal sharing of a pair of electrons is known as a **polar covalent bond**. The unequal sharing, caused by unequal attractions for the bonding electrons, results in an uneven distribution of electrons within the bond. For example, in the HCl molecule, the bonding electron pair is pulled more strongly toward the chlorine (electronegativity 3.0) than toward the hydrogen (electronegativity 2.1). This results in a slight buildup of negative charge at the chlorine end of the molecule, leaving the hydrogen end slightly positive. These partial charges are indicated by the Greek letter "delta," δ (Figure 8.26); for example, the hydrogen end of the HCl molecule carries a δ+ charge and the chlorine end a δ– charge. If polar bonds cause the molecule as a whole to have oppositely charged ends, then the molecule is called a **polar molecule**.

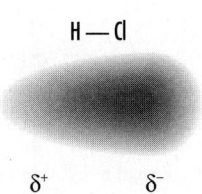

Figure 8.26
In an HCl molecule, the electrons are pulled more strongly to the chlorine end, resulting in a polar covalent bond. In diatomic molecules such as this, a polar bond causes the molecule itself to be polar.

The Greek letter delta has two meanings as a symbol in science. The lowercase delta, δ, means "partial," and the uppercase delta, Δ, means "change in."

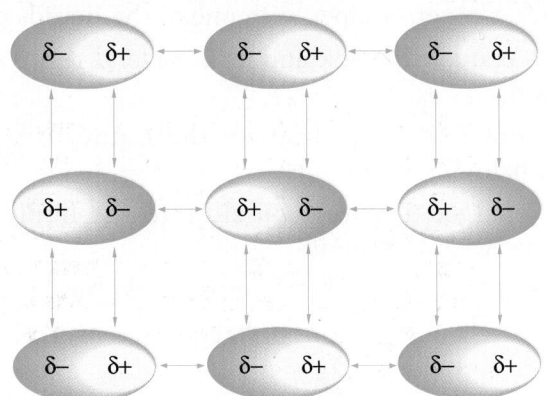

Figure 8.27
A polar molecule is simultaneously attracted to all the other polar molecules around it.

Polar molecules such as HCl tend to line up so that the slightly positive end is near the slightly negative end of a nearby molecule (Figure 8.27). This attraction between oppositely charged ends of polar molecules is known as a **dipole-dipole force**. Both experimental evidence and theoretical calculations indicate that, for most molecules, the dipole-dipole forces are much weaker than London forces. Also, dipole-dipole forces usually have only small effects on properties of substances composed of polar molecules. For solubility, however, it has been found that polar solutes dissolve in polar solvents and non-polar solutes dissolve in non-polar solvents; that is, *like dissolves like.*

Problem
Prediction
Design
Materials
Procedure
✔ Evidence
✔ Analysis
Evaluation
Synthesis

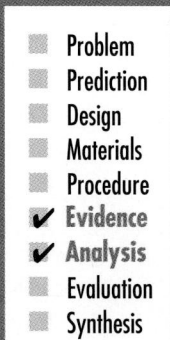

CAUTION

All unknown substances are potentially hazardous. Avoid inhaling vapors and avoid skin contact.

Empirical Rules for Polar and Non-polar Molecules

The following rules can be used to predict, from the molecular formula, whether a molecule is polar or non-polar.

Polar Molecules

Type	Examples
AB	$HCl_{(g)}$, $CO_{(g)}$
N_xA_y	$NH_{3(g)}$, $NF_{3(g)}$
O_xA_y	$H_2O_{(l)}$, $OCl_{2(g)}$
$C_xA_yB_z$	$CHCl_{3(l)}$, $C_2H_5OH_{(l)}$

Non-polar Molecules

Type	Examples
A_x	$Cl_{2(g)}$, $N_{2(g)}$
C_xA_y	$CO_{2(g)}$, $CH_{4(g)}$

INVESTIGATION

8.3 Evidence for Polar Molecules

The purpose of this investigation is to test for the presence of polar molecules in a variety of pure chemical substances.

Problem

Which of various molecular substances contain polar molecules?

Experimental Design

A thin stream of each liquid is tested by holding a positively charged acetate strip or a negatively charged vinyl strip near the liquid (Figure 8.28).

Materials

lab apron
safety glasses
samples of various liquids
50 mL buret, or 10 mL pipet, or medicine dropper
buret clamp and stand for buret (if used)
buret funnel for buret (if used)
400 mL beaker or pan
acetate strip
vinyl strip
paper towel

Procedure

1. Fill the buret/pipet/dropper with one of the liquids provided.

2. Rub the acetate strip back and forth several times in a piece of paper towel.

3. Allow a thin stream or drops of the liquid to pour into the container below.

4. Hold the charged acetate strip close to the liquid stream and observe the stream of liquid.

5. Repeat steps 1 to 4 with the charged vinyl strip.

6. Rotate to the next station and repeat steps 1 to 5 using the other liquids provided.

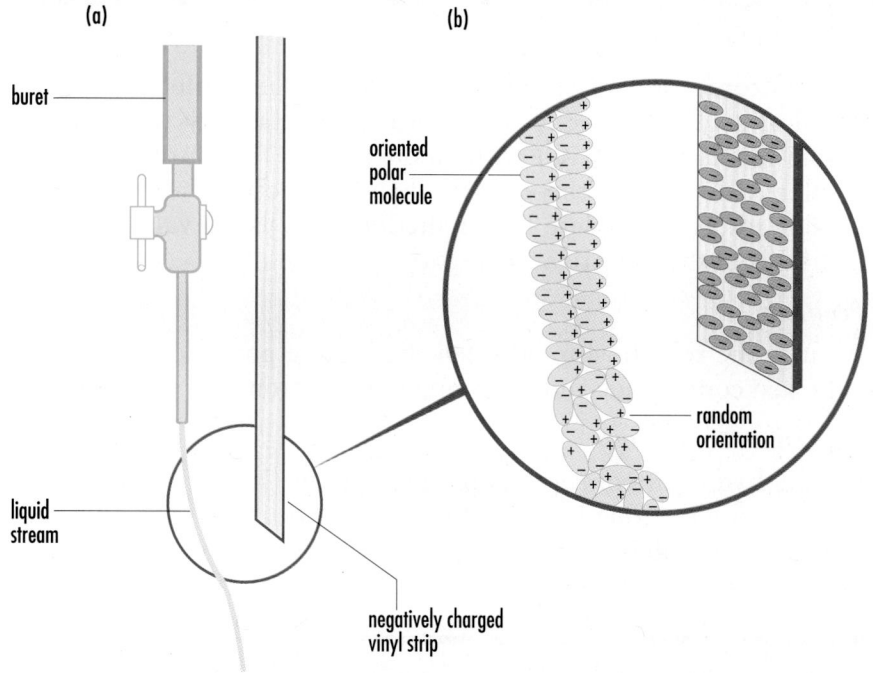

Labels on figure:
(a)

buret

liquid
stream

negatively charged
vinyl strip

(b)

oriented
polar
molecule

random
orientation

Figure 8.28
(a) Testing a liquid with a charged strip provides evidence for the existence of polar molecules in a substance.
(b) Polar molecules in a liquid become oriented so that their positive poles are closer to a negatively charged material. Near a positively charged material they become oriented in the opposite direction. Polar molecules are thus attracted by either kind of charge.

Problem 8E Polarity of Molecules

Empirical rules can be used to predict whether a substance will be polar or non-polar, based on its molecular formula. Polarity of a substance determines its solubility in water. Complete the Prediction, Analysis, and Evaluation of the investigation report.

Problem

Which of various molecular substances contain polar molecules?

Experimental Design

A thin stream of each liquid is tested by holding an electrically charged object near the liquid stream.

Evidence

THE EFFECT OF ELECTRIC CHARGE ON MOLECULAR LIQUIDS		
Substance	**Positive Charge**	**Negative Charge**
$Br_{2(l)}$	unaffected	unaffected
$CCl_{4(l)}$	unaffected	unaffected
$CH_3OH_{(l)}$	attracted	attracted
$CH_3OCH_{3(l)}$	attracted	attracted
$C_5H_{12(l)}$	unaffected	unaffected
$CS_{2(l)}$	unaffected	unaffected
$CH_3COCH_{3(l)}$	attracted	attracted
$H_2O_{2(l)}$	attracted	attracted
$NCl_{3(l)}$	attracted	attracted
$N_2H_{4(l)}$	attracted	attracted

Table 8.6

Group	Hydrogen Compound	Boiling Point (°C)
14	CH_4	–164
	SiH_4	–112
	GeH_4	–89
	SnH_4	–52
15	NH_3	–33
	PH_3	–87
	AsH_3	–55
	SbH_3	–17
16	H_2O	100
	H_2S	–61
	H_2Se	–42
	H_2Te	–2
17	HF	20
	HCl	–85
	HBr	–67
	HI	–36

Table 8.6 title: BOILING POINTS OF THE HYDROGEN COMPOUNDS OF ELEMENTS IN GROUPS 14 TO 17

Problem 8F London Forces

London forces are believed to exist among all molecules in the liquid or solid state. Polar molecules exert relatively weak dipole-dipole forces in addition to London forces. The purpose of this problem is to test the theoretical rule for predicting the strengths of London forces. Complete the Prediction, Analysis, Evaluation, and Synthesis of the investigation report.

Problem

What is the relationship between the boiling points of a family of hydrogen compounds and the number of electrons per molecule?

Experimental Design

For the hydrogen compounds of elements in Groups 14 to 17, the numbers of electrons in the various elements are determined from the atomic numbers. The boiling points are listed in Table 8.6. The evidence is graphed for analysis.

Hydrogen Bonds

When a hydrogen atom is bonded to a very electronegative atom such as fluorine (4.0), oxygen (3.5), or nitrogen (3.0), some unusual properties result. In order to explain evidence that cannot be explained by London forces and dipole-dipole forces (such as the evidence in Problem 8F), chemists created a concept to describe a third kind of intermolecular force. **Hydrogen bonds** are special, relatively strong dipole-dipole forces between molecules containing F—H, O—H, and N—H bonds. Some of the unusual properties of H_2O — such as the lower density of ice compared with water, the very high capacity for absorbing heat, the higher than expected boiling point, and the powerful action of water as a solvent — can be attributed to hydrogen bonds among water molecules.

According to current theory, there are two parts to the explanation for this special intermolecular force. First, the large difference in electronegativities between hydrogen and either fluorine, oxygen, or nitrogen produces highly polar bonds. Second, the small size of the

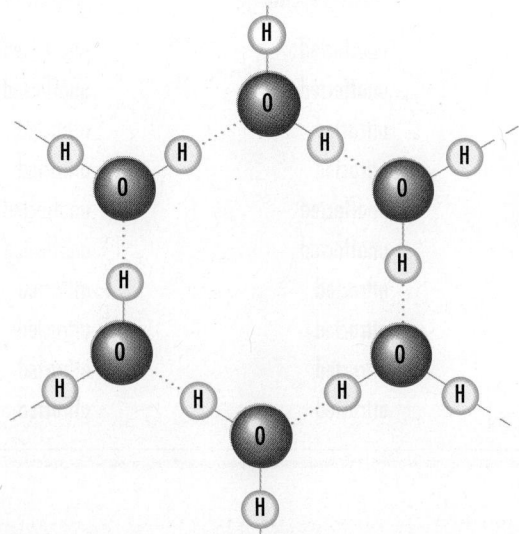

Figure 8.29
In ice, hydrogen bonds between the molecules result in a regular hexagonal crystal structure. The · · · H— represents a hydrogen (proton) being shared between two pairs of electrons.

hydrogen atom means that the positive pole is highly concentrated, so that it exerts a strong attraction on the negative pole of a nearby molecule (Figure 8.29). Another way of thinking about the hydrogen bond is to consider that a hydrogen atom stripped of its electron is a proton. A hydrogen bond can therefore be considered as the simultaneous attraction of a proton by two pairs of electrons.

INVESTIGATION

8.4 Hydrogen Bond Formation

You have learned that energy is required to break chemical bonds and that energy is released when new bonds are formed. The purpose of this investigation is to test these ideas in relation to hydrogen bonding.

Problem

Are additional hydrogen bonds formed when water and glycerol are mixed?

Experimental Design

Water, $H_2O_{(l)}$, and glycerol, $C_3H_5(OH)_{3(l)}$, liquids that contain O—H bonds (Figure 8.30), are mixed, and any change in energy is measured using a thermometer. If a significant temperature increase is noted, then additional hydrogen bonds are probably present.

Materials

lab apron
safety glasses
10 mL of water
10 mL of glycerol
(2) 50 mL graduated cylinders

nested pair of polystyrene cups
cup lid with center hole
thermometer
250 mL beaker (for support)

- [ ] Problem
- [x] Prediction
- [ ] Design
- [ ] Materials
- [x] Procedure
- [x] Evidence
- [x] Analysis
- [x] Evaluation
- [ ] Synthesis

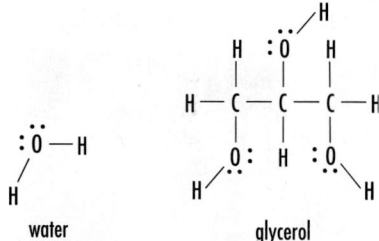

Figure 8.30
The structures of water and glycerol suggest the possibility of more hydrogen bonds after these two substances form a solution.

HYDROGEN BONDS IN BIOCHEMISTRY

Hydrogen bonds appear to have a marked effect on the shape of large, biologically important molecules such as proteins and DNA (deoxyribonucleic acid). Proteins are made up of long chains of amino acids, which fold into specific three-dimensional structures because of the attractions among different parts of the chain. The shapes of the various kinds of protein molecules are essential to their functions as enzymes, hormones, antibodies, and struc-

tural materials. A DNA molecule stores the genetic information in a cell, and is made up of two chains of compounds called nucleotides. Hydrogen bonds between the two

chains of DNA, broken and re-formed when the genetic information is copied, are an integral part of the passing of information from one generation to the next.

hydrogen bonds

thymine   adenine   cytosine   guanine

Dr. Ronald Gillespie

Dr. Ronald James Gillespie was born in England, where he studied chemistry. In 1950 Gillespie became a lecturer at University College where he began working with Dr. R. Nyholm, a professor of chemistry. Their studies led them to question the molecular shape theories then in use. Together they developed the VSEPR theory in an effort to explain molecular shape without having to use mathematical models.

In 1958, Gillespie joined McMaster University, where he continued to develop and refine the VSEPR theory. Gillespie has received 18 scientific awards since 1949. He is the author of the textbook *Molecular Geometry* and has written numerous articles for chemistry and chemical education journals. Gillespie has shown a commitment to education as chairman of the Chemical Education Division of the Chemical Institute of Canada (CIC) and of the International Conference on New Directions in the Chemistry Curriculum for high school and first-year university chemistry.

VSEPR Theory

Evidence indicates that molecules have definite three-dimensional shapes. These shapes appear to be important in determining the physical and chemical properties of molecular substances. A simple and remarkably effective theory for predicting and explaining molecular shapes was developed in part by the Canadian chemist, Ronald Gillespie of McMaster University in Ontario. According to the valence shell electron pair repulsion theory, known as the **VSEPR theory**, the shape of a molecule is determined by the number of valence electron pairs that surround the central atom (the atom with the highest bonding capacity). The valence electron pairs include both shared (bonding) and unshared (lone pairs). According to VSEPR (pronounced "vesper") theory these electron pairs repel each other and arrange themselves as far apart as possible, to minimize the repulsion. For example, if an atom has two shared pairs of electrons and no lone pairs, a linear arrangement is expected.

$$X:A:X \quad \text{or} \quad X—A—X$$

This arrangement keeps the electron pairs on the opposite sides of the central atom (A) and as far away as possible from each other.

Note that in Table 8.7, "A" represents a nonmetal central atom and "X" represents nonmetal atoms bonded to the central atom. Only arrangements around central atoms are predicted. If more than one central atom is present, the shape around each atom is predicted separately. In our restricted VSEPR theory, lone pairs around "X" atoms do not significantly affect the shape of the molecule.

Table 8.7

USING VSEPR THEORY TO PREDICT MOLECULAR SHAPE					
Class	**Shape**	**Model**	**Polarity***	**Example**	
AX_2	linear	X—A—X	non-polar	CS_2, CO_2	⎫
AX_3	trigonal planar	X / A \ X X	non-polar	BF_3, BH_3	⎬ no lone pairs around the central atom
AX_4	tetrahedral	X | A / X X X	non-polar	CH_4, $SiCl_4$	⎭
$:AX_3$	pyramidal	A / X X X	polar	NH_3, PCl_3	⎫
$\ddot{A}X_2$	V-shaped	A / X X	polar	H_2O, OCl_2	⎬ lone pairs around the central atom
$\ddot{A}X$	linear	A—X	polar	HCl, BrF	⎭

*when X represents identical atoms

SUMMARY: GUIDELINES FOR USING VSEPR THEORY TO PREDICT MOLECULAR SHAPE

Step 1: Write the Lewis structure for the molecule, considering only the electron pairs around the central atom.

Step 2: Count the number of electron pairs around the central atom (bonding and lone pairs). Treat double and triple bonds as though they were single bonds.

Step 3: Refer to Table 8.7 and use the molecule's classification type to predict its shape and polarity.

EXAMPLE

Predict the shape and polarity of an H_2S molecule.

1. The Lewis structure is $: \overset{..}{\underset{H}{S}} : H$

2. The central atom has two lone pairs and two bonding pairs.
3. According to Table 8.7, the H_2S molecule is V-shaped and polar.

Exercise

39. What valence orbital and valence electron condition must exist in order for a covalent bond to form between two approaching atoms?

40. Draw diagrams of the Lewis models for each of the following molecular substances. Use VSEPR theory to predict the shape around the central atom and the polarity of the molecule.
    (a) CH_4 (d) CO_2
    (b) NH_3 (e) CCl_4
    (c) H_2O (f) HCN

41. Predict whether the following substances will dissolve in water.
    (a) $C_5H_{12(l)}$ (c) $C_3H_5(OH)_{3(l)}$
    (b) $CH_3OH_{(l)}$

42. Predict the type(s) of chemical bonds present in samples of the following pure substances.
    (a) $C_8H_{18(l)}$ (e) $Si_{(s)}$
    (b) $Cu_{(s)}$ (f) $Al_2O_{3(s)}$
    (c) $MgCl_{2(s)}$ (g) $C_2H_4(OH)_{2(l)}$
    (d) $C_2H_3Cl_{3(l)}$ (h) $NH_4NO_{3(s)}$

43. Predict which of the following molecular compounds are composed of polar molecules.
    (a) $C_8H_{18(l)}$ (e) $BrCl_{(l)}$
    (b) $S_{8(s)}$ (f) $C_{10}H_{22(l)}$
    (c) $C_2Cl_{4(l)}$ (g) $C_2H_4Br_{2(l)}$
    (d) $CHCl_{3(l)}$

44. Predict which of the following molecular compounds contain hydrogen bonds. Show the hydrogen bonds using structural models of at least two molecules of the substance.
    (a) $H_2O_{2(l)}$ (d) $CH_3OCH_{3(l)}$
    (b) $CH_3F_{(l)}$ (e) $CH_3OH_{(l)}$
    (c) $NH_2OH_{(l)}$

Ionic compounds are abundant in nature. Soluble ionic compounds are present in both fresh water and salt water. Ionic compounds with low solubility make up most rocks and minerals. Relatively pure deposits of sodium chloride occur in Alberta, and Saskatchewan has the world's largest deposits of potassium chloride and sodium sulfate.

You can also find ionic compounds at home. Iodized table salt consists of sodium chloride with a little potassium iodide added. Antacids contain a variety of compounds, such as magnesium hydroxide and calcium carbonate. Many home cleaning products contain sodium hydroxide. Other examples of ionic compounds are rust (iron(III) hydroxide) that forms on the steel bodies of cars and tarnish (silver sulfide) that forms on silver. Ionic compounds have low chemical reactivity compared with the elements from which they are formed. Some ionic compounds, such as lime (calcium oxide) and potash (potassium chloride), are so stable that they were classified as elements until the early 1800s. (At that time elements were defined as pure substances that could not be decomposed by strong heating.)

Reactive metals such as sodium, magnesium, and calcium are not found as elements in nature, but occur instead in ionic compounds. The least active metals, such as silver, gold, platinum, and mercury, do not react readily to form compounds and may therefore be found uncombined in nature.

For thousands of years, metallurgists have used ionic compounds to extract metals from naturally occurring compounds. In these processes an ionic compound such as hematite (Fe_2O_3) in iron ore is reduced to a pure metal, which can then be used to make tools, weapons, and machines. Iron, the main constituent of steel, is the most widely used metal. Unfortunately, iron is reactive and readily *corrodes*, or reacts with substances in the environment, re-forming to an ionic compound (Figure 8.31). A lot of time and money are spent trying to prevent or slow the corrosion of iron and other metals — for example, by having cars rust-proofed at automotive centers.

Figure 8.31
Rusting is a common and expensive problem. In an industrial society much time and money are spent repairing damage caused by corrosion.

Formation of Ionic Compounds

Ionic compounds are formed in many ways. *Binary ionic compounds* are the simplest ionic compounds; they may be formed in the reaction of a metal with a nonmetal. For example,

$$2\,Na_{(s)} \;+\; Cl_{2(g)} \;\rightarrow\; 2\,NaCl_{(s)}$$

The reaction of ammonia and hydrogen chloride gases produces the ionic compound ammonium chloride, which appears as a white smoke of tiny solid particles (Figure 8.32).

$$NH_{3(g)} \;+\; HCl_{(g)} \;\rightarrow\; NH_4Cl_{(s)}$$

The conductivity of molten ionic compounds and aqueous solutions of ionic compounds suggests that charged particles are present. According to the restricted quantum mechanics theory, the stability of ionic compounds suggests that their electronic structure is similar to that of the noble gases, which have filled energy levels. By tying in these ideas with the collision-reaction theory, scientists explain the formation of an ionic compound as a collision between a metal and a nonmetal atom that results in a transfer of electrons, forming positive and negative ions that have filled energy levels. Scientists consider that the electron transfer is encouraged by the large difference in electronegativity of metal and nonmetal atoms. Electron dot diagrams represent these ideas. For example, a theoretical description of the formation of sodium fluoride shows how a stable octet structure is formed.

$$Na^{\cdot} + \cdot \ddot{\underset{\cdot\cdot}{F}}: \rightarrow Na^{+}[:\ddot{\underset{\cdot\cdot}{F}}:]^{-}$$

The overall process in the formation of an ionic compound such as sodium fluoride involves a loss of electrons by a metal and a gain of electrons by a nonmetal. Loss of electrons is called **oxidation**, which can be represented in an **oxidation half-reaction**; for example,

$$Na_{(s)} \rightarrow Na^{+}_{(s)} + e^{-}$$

The other half of the process involves a fluorine atom gaining the electron lost by the sodium atom. Gain of electrons is called **reduction**. This gain can be represented by a **reduction half-reaction**; for example,

$$F_{2(g)} + 2\,e^{-} \rightarrow 2\,F^{-}_{(s)}$$

Note that fluorine, like other halogens, occurs as diatomic molecules. The two fluorine atoms require a total of two electrons in the reduction half-reaction.

Two sodium atoms are required to supply the two electrons. In order to balance the two half-reactions (that is, to make electrons lost equal electrons gained), the entire half-reaction for sodium is doubled. Then the two half-reactions are added together to give a net or overall reaction.

$$2\,[Na_{(s)} \rightarrow Na^{+}_{(s)} + e^{-}], \text{ which means: } \quad 2\,Na_{(s)} \rightarrow 2\,Na^{+}_{(s)} + 2\,e^{-}$$

$$\frac{F_{2(g)} + 2\,e^{-} \rightarrow 2\,F^{-}_{(s)}}{2\,Na_{(s)} + F_{2(g)} \rightarrow 2\,NaF_{(s)}}$$

A Model for Ionic Compounds

To be acceptable, a theory of bonding must be able to explain the properties of ionic compounds — why they are hard solids with high melting and boiling points, and why they are conductors in their molten and aqueous states. All ionic compounds are hard solids at SATP, so the ions must be held together or bonded very strongly in a rigid structure. In the model for ionic compounds (Figure 3.7, page 104), ions are considered to be spheres arranged in a regular pattern.

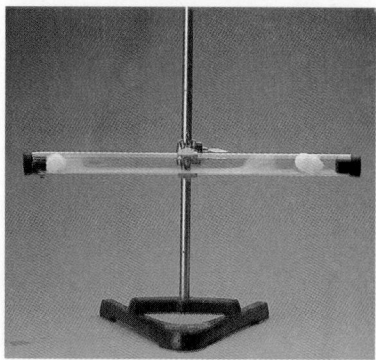

Figure 8.32
The solid ionic compound ammonium chloride forms from the reaction of ammonia and hydrogen chloride gases when small quantities of concentrated solutions are placed into the ends of a tube such as this.

Sodium fluoride is added to many toothpastes; it acts to harden enamel in teeth, so that cavities are less likely to form.

Oxidation and Reduction
The term "reduction" derives from metallurgy. When a metal is extracted from its ore, a very large volume of ore produces only a small volume of metal. A large volume is *reduced* to a small one. The term "oxidation" is relatively recent. Originally, oxidation meant reaction with oxygen; later, the term was expanded to mean reaction with nonmetals. Oxidation and reduction reactions, as presently defined, are the most common classes of chemical reactions.

Figure 8.33
The cubic shape of table salt crystals provides a clue about the internal structure of sodium chloride.

X-Rays and Three-Dimensional Models

Three-dimensional models of crystals are determined experimentally by X-ray diffraction. When a beam of X-rays is reflected from the top layers of an ionic crystal, a regular pattern is obtained. The wavelength of the X-rays determines the way they are reflected from the spaces between ions. (This is similar to the way in which the wavelength of light determines how it is reflected from the grooves on the surface of a compact disk.) Scientists can use the pattern of the reflected or transmitted X-rays and their knowledge of the wavelength to infer not only the arrangement but also the size and separation of the ions.

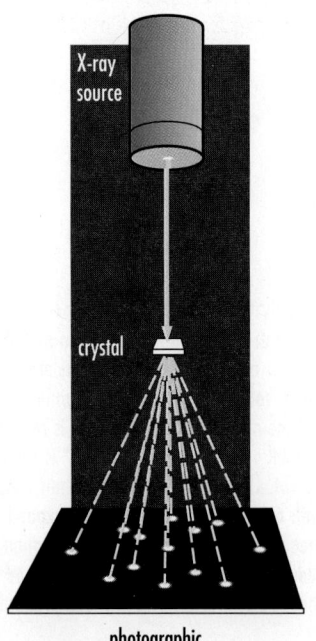

Depending on the sizes and charges of the ions, different arrangements are possible, but whatever the pattern, it will allow the greatest number of oppositely charged ions to approach each other closely while preventing the close approach of ions having the same charge. In all cases, any ion will be surrounded by ions of opposite charge. This creates strong attractions, and explains why ionic compounds are hard solids with high melting and boiling points. The arrangement of ions for a given compound is called its **crystal lattice**. The model also explains why ionic compounds are brittle — the ions cannot be rearranged without the addition of a lot of energy to break the crystal lattice apart.

The model for ionic compounds leads to representations of ionic crystals such as the model of sodium chloride in Figure 3.7 (page 104). The diagram shows each ion surrounded by six ions of opposite charge, held firmly within the crystal by strong electrostatic attractions. The observable shape of sodium chloride crystals (Figure 8.33) supports this model. Although all ionic compounds have hard and brittle crystalline forms, and high melting and boiling points, these properties vary in degree depending on the nature of the ions forming the compound.

According to laboratory evidence and the ion model, ion attractions are non-directional — all positive ions attract all nearby negative ions. There are no distinct neutral molecules in ionic compounds. The chemical formula shows only a formula unit expressing the simplest whole number ratio of ions. For example, a crystal of sodium chloride, $NaCl_{(s)}$, contains equal numbers of sodium ions and chloride ions, and a crystal of calcium fluoride, $CaF_{2(s)}$, contains one calcium ion for every two fluoride ions; however, there are no distinct molecules of sodium chloride or molecules of calcium fluoride.

Exercise

45. Why are ionic compounds abundant in nature?

46. Write a brief explanation for the formation of a binary ionic compound from its constituent elements.

47. What evidence suggests that ionic bonds are strong?

48. Potassium chloride is a substitute for table salt for people who need to reduce their intake of sodium ions. Use electron dot diagrams to represent the formation of potassium chloride from its elements. Show the electronegativities of the reactant atoms.

49. Use electron dot diagrams to represent the reaction of calcium and oxygen atoms. Name the ionic product.

50. The empirically determined chemical formula for magnesium chloride is $MgCl_2$. Write reduction and oxidation half-reaction equations, as well as the net reaction equation, to explain the empirical formula of magnesium chloride.

51. Write reduction and oxidation half-reaction equations, as well as the net reaction equation, to predict the chemical formula of the product of the reaction of aluminum and oxygen.

52. Based only on differences in electronegativity, what compound would you expect to be the most strongly ionic of all binary compounds?

53. What problem arises in trying to predict the type of compound formed by the reaction of gold and selenium? What does this problem indicate about the completeness of the rules and theories for ionic compounds?

54. What is the difference between the information expressed by the chemical formula of an ionic compound such as NaCl, and the molecular formula of a substance such as H_2O?

55. (Enrichment) Prepare a list of ionic compounds from the labels of some products found at home. Include the product name and as many chemical formulas as possible.

MICHAEL SMITH

One morning in October 1993, Dr. Michael Smith, biochemist at the University of British Columbia, woke to the news that he had just been awarded the 1993 Nobel Prize in Chemistry. The scientific work that the Royal Swedish Academy recognized as deserving of the prize was performed by Smith and his colleagues in their UBC laboratory between 1977 and 1982—although back then the work had little effect on the wider scientific community. What Smith's team developed was a technique called "site-specific mutagenesis," through which they were able to alter the genetic code in individual segments of DNA.

To appreciate Smith's work, it helps to understand life's biochemical hierarchy. Genes are the parts of DNA that prescribe specific outward traits in an organism. Each gene is made up of several thousand nucleotide pairs (organic acids). Three nucleotide pairs in a row specify an amino acid. In turn, amino acids build proteins, proteins build cells, and cells build organisms like fruit flies and humans.

When the chemical composition of genes undergoes random changes, the result is mutation. Much scientific effort has gone into studying how and why mutations take place, and how various proteins and genes relate to the function of cells. With the help of Smith's

technique, researchers can now alter specific gene sites and change any amino acids that are part of the sequence making up a protein. In this manner, scientists are able to mimic the way evolution causes mutation, but under controlled conditions. The role of each amino acid can be explored using this technique.

Site-specific mutagenesis is providing scientists with a much better understanding of how biological systems function. It is also proving to have several exciting practical applications in the area of biological engineering. For instance, a company co-founded by Smith is currently using the technique to produce insulin from yeast cells implanted with human genes.

George Hunter, executive director of the British Columbia Biotechnology Alliance, spoke about Dr. Michael Smith's work in an interview: "The information he developed is fundamental to biotechnology in the same way that a hammer is fundamental to building a house."

- *List life's biological hierarchy, from large to small.*
- *What is "site-specific mutagenesis"?*

OVERVIEW

Chemical Bonding

Summary

- Chemical reactivity is summarized by patterns within the periodic table — patterns such as chemical families, activity series, and the types of compounds formed.

- Periodicity in atomic radius and ionization energy of the elements supports and helps explain the observed patterns in reactivity.

- Lewis models and electronegativity are used to explain chemical bonding as a continuum ranging from equal sharing of electrons (molecular elements) to unequal sharing of electrons (polar molecular compounds) to almost complete electron transfer (ionic compounds).

- All chemical bonds — covalent, ionic, and intermolecular — result from a simultaneous attraction of oppositely charged particles.

- Electron dot diagrams and structural diagrams show covalent bonds to explain and predict formulas of molecular elements and compounds.

- The endothermic or exothermic nature of chemical reactions can be explained by the concept of bond energy.

- Intermolecular forces — London forces, dipole-dipole forces, and hydrogen bonds — are relatively weak compared with covalent bonds and ionic bonds.

- The formation and properties of ionic compounds are explained by the collision of metal and nonmetal atoms, which results in electron transfer (oxidation and reduction), forming ions with stable octets.

Key Words

activity series
bond energy
bonding capacity
bonding electron
coordinate covalent bond
covalent bond
covalent bond radius
crystal lattice
dipole-dipole forces
electron dot diagram
electronegativity
Hund's rule
hydrogen bond
intermolecular forces
ionic bond
ionic radius
isoelectronic
Lewis model
London forces
lone pair
metalloids
octet rule
orbital
oxidation
oxidation half-reaction
polar covalent bond
polar molecule
reduction
reduction half-reaction
structural diagram
van der Waals forces
van der Waals radius
VSEPR theory

Review

1. How does the chemical reactivity vary
   (a) among the elements in Groups 1 and 2 of the periodic table?
   (b) among the elements in Groups 16 and 17?
   (c) within period 3?
   (d) within Group 18?

2. Describe the trends in the periodic table for each of the following atomic properties:
   (a) atomic radius
   (b) ionization energy
   (c) electronegativity

3. How are the positions of two reacting elements in the periodic table related to the type of compound and bond formed? State two generalizations.

4. What is the maximum number of electrons in the valence level of an atom of a representative element?

5. How do the electronegativities of representative metals compare with representative nonmetals?

6. Draw an electron dot diagram for each of the following atoms. In each, identify the number of bonding electrons and the lone pairs.
   (a) Ca
   (b) Al
   (c) Ge
   (d) N
   (e) S
   (f) Br
   (g) Ne

7. Draw an electron dot diagram and a structural diagram for each molecule in the following reactions.
   (a) $N_{2(g)} + I_{2(s)} \rightarrow NI_{3(s)}$
   (b) $H_2O_{2(aq)} \rightarrow H_2O_{(l)} + O_{2(g)}$

8. Why did scientists propose the idea of double and triple covalent bonds?

9. What empirical evidence is there for double and triple bonds?

10. List three criteria used to evaluate a theoretical explanation of an observation.

11. List two examples each for exothermic and endothermic chemical changes.

12. What information does a boiling point provide about intermolecular forces?

13. List three types of intermolecular forces and give an example of a molecular substance having each type of force.

14. Write an empirical definition of an ionic compound.

15. Summarize the theoretical structure of ionic compounds.

16. Use electron dot diagrams to explain the electron rearrangement in the following chemical reactions.
    (a) magnesium atoms + sulfur atoms → magnesium sulfide
    (b) aluminum atoms + chlorine atoms → aluminum chloride

17. What is the difference in the meaning of the numbers in a molecular formula and in an ionic formula?

18. List the types of bonds that according to our current theories, are believed to be present in each substance.
    (a) $C_6H_{14(l)}$
    (b) $Fe_{(s)}$
    (c) $CaCl_{2(s)}$
    (d) $C_2H_3Cl_{3(l)}$
    (e) $C_{(s)}$
    (f) $Fe_2O_{3(s)}$
    (g) $C_2H_5OH_{(l)}$
    (h) $C_{12}H_{22}O_{11(s)}$

19. Using VSEPR theory, predict the shape around the central atom of each of the following molecules.
    (a) $HF_{(g)}$
    (b) $BCl_{3(g)}$
    (c) $SiH_{4(g)}$
    (d) $CCl_{4(l)}$
    (e) $HCN_{(g)}$
    (f) $OCl_{2(g)}$
    (g) $NCl_{3(g)}$
    (h) $H_2O_{2(l)}$

20. Draw diagrams of models to explain the following types of bonding.
    (a) London forces
    (b) dipole-dipole forces
    (c) hydrogen bonding
    (d) metallic bonding
    (e) ionic bonding
    (f) bonding in metalloids

21. Metals are shiny, malleable conductors of heat and electricity.
    (a) Write full electron configurations for these metals: magnesium, potassium, chromium, silver, and mercury.
    (b) How do the electron configurations of metals explain their conductivity, malleability, and shininess?

22. Metalloids are characterized by hardness, high melting points, and poor electrical conductivity.

(a) Write full electron configurations for these metalloids: boron, silicon, germanium, arsenic, and antimony.
(b) How do the electron configurations of metalloids explain their properties?

23. Nonmetals have relatively low melting points and are non-conductors in any state.
(a) Write full electron configurations for these nonmetals: nitrogen, sulfur, bromine, xenon, and iodine.
(b) Draw Lewis diagrams for these nonmetals: N_2, S_8, Br_2, Xe, and I_2.
(c) How do the Lewis diagrams of nonmetals explain their relatively low melting points and non-conductivity?

24. Aluminum is a strong, light, easily shaped metal that is widely used in aircraft construction.
(a) Write an equation to represent the oxidation of aluminum atoms to aluminum ions.
(b) Compare the radius of an aluminum atom with an aluminum ion, and explain the difference.

25. Chlorine is a reactive nonmetal that is used as an industrial and domestic bleach.
(a) Write an equation to represent the reduction of chlorine molecules to chloride ions.
(b) Compare the radius of a chlorine atom with a chloride ion, and explain the difference.

Applications

26. All chemical bonds are thought to be the result of simultaneous attractions between oppositely charged particles. For each chemical bond listed below, indicate which types of particles are involved.
(a) covalent bond
(b) London forces
(c) dipole-dipole forces
(d) hydrogen bonds
(e) ionic bond

27. An activity series is a list of substances in order of chemical reactivity under specified conditions.
(a) List the nonmetals of Group 17 from most reactive to least reactive.

(b) What type of half-reaction do these non-metals undergo?
(c) Explain your order in (a) in terms of the tendency of the nonmetals to gain or lose electrons.
(d) How consistent is the order in your activity series with the order of electronegativities given in the periodic table on this book's inside front cover?

28. The most common oxides of period 2 elements are as follows:
Na_2O, MgO, Al_2O_3, SiO_2, P_2O_5, SO_2, Cl_2O
(a) Which oxides are classified as ionic and which are classified as molecular?
(b) Calculate the difference in electronegativity between the two elements in each oxide.
(c) How is the difference in electronegativity related to the properties of the compound?

29. Determine the molecular formula for nicotine from the following evidence.
molar mass = 162.24 g/mol
percent by mass of carbon = 74.0%
percent by mass of hydrogen = 8.7%
percent by mass of nitrogen = 17.3%

30. Predict the structural model, chemical formula, and name for the simplest product in each of the following chemical reactions. Indicate any products that violate the octet rule.
(a) $H_{2(g)}$ + $P_{4(s)}$ →
(b) $Si_{(s)}$ + $Cl_{2(g)}$ →
(c) $C_{(s)}$ + $O_{2(g)}$ →
(d) $B_{(s)}$ + $F_{2(g)}$ →

31. How are intermolecular forces similar to covalent bonds, and how are they different?

32. Write oxidation and reduction half-reaction equations and the net reaction equation to explain the following chemical reactions.
(a) $2 K_{(s)}$ + $Br_{2(l)}$ → $2 KBr_{(s)}$
(b) $2 Sr_{(s)}$ + $O_{2(g)}$ → $2 SrO_{(s)}$

33. Explain, using the theory of chemical bonds, the high melting and boiling points of ionic compounds.

Extensions

34. Modern technologies allow scientists to measure indirectly the length of chemical

bonds. Use the information in Table 8.8 to determine the effect of bond type on bond length.

Table 8.8

COVALENT BOND LENGTHS		
Typical Compound	Covalent Bond Type	Bond Length (nm)
CH_4	C—H	0.109
C_2H_6	C—C	0.154
C_2H_4	C=C	0.134
C_2H_2	C≡C	0.120
CH_3OH	C—O	0.143
CH_3COCH_3	C=O	0.123
CH_3NH_2	C—N	0.147
CH_3CN	C≡N	0.116

35. Use the concept of bond energy to briefly explain the energy changes in Figure 8.25 on page 320.

36. Use electron dot diagrams and the idea of a coordinate covalent bond to explain the following chemical reactions.
    (a) $NH_{3(g)} + HCl_{(g)} \rightarrow NH_4Cl_{(s)}$
    (b) $H_2O_{(l)} + H_2O_{(l)} \rightarrow H_3O^+_{(aq)} + OH^-_{(aq)}$
    (c) $NH_{3(g)} + BF_{3(g)} \rightarrow NH_3BF_{3(s)}$

37. Some medical professionals are concerned about the level of saturated and unsaturated fats in the foods we eat. Find out how the terms "saturated fats" and "unsaturated fats" are related to the concept of single and double bonds. Locate products such as margarine at home or in a grocery store, and list any information printed on the labels or packaging that describes the products' saturated and unsaturated fat content.

38. Knowing the molecular formula does not always tell you the structural formula of a compound. For example, the C_4H_{10} compound found in a butane lighter may be one of two different molecules. Draw structural diagrams to represent these two molecules. Do some research to determine how mass spectrometry, infrared spectrometry, and nuclear magnetic resonance (NMR) can be used to determine which of the two molecules is present in the lighter fluid.

Problem 8G Bonding Theory

The purpose of this problem is to test the predictive power of the theory of chemical bonds. Complete the Prediction, Analysis, and Evaluation of the investigation report.

Problem

What compound forms in the reaction of phosphorus and fluorine?

Evidence

From mass spectrometry:
    molar mass = 126 g/mol
From combustion analysis:
    percent by mass of phosphorus = 24.5%
    percent by mass of fluorine = 75.5%

9 Organic Chemistry

As the 19th century dawned, John Dalton was attempting to convince the scientific community that all matter consists of atoms. By 1872, Dmitri Mendeleyev had organized the known elements into a periodic table, but no theory existed to explain the table. As we near the 21st century, atomic theory enables scientists to predict and then explain the properties of new compounds, and to design molecules for specific purposes. Of the more than 10 million compounds that have been discovered, at least 90% are molecular compounds of the element carbon. More than one-quarter million new compounds are synthesized in laboratories each year, and almost all of these are molecular compounds of carbon as well.

In the natural world, plants and animals synthesize millions of carbon compounds. Understanding the properties of such compounds is a major part of chemistry. Many manufactured chemicals are copies of natural products. After isolating and identifying chemicals from natural products, chemists and engineers invent processes to synthesize these or similar chemicals for some technological application or social purpose. Synthetically produced chemicals discussed in this chapter include gasoline, solvents, polyesters, synthetic sweeteners, artificial flavorings, and medicines.

THE CHEMISTRY OF CARBON COMPOUNDS

In the early 19th century, Swedish chemist Jöns Jacob Berzelius classified compounds into two categories: those obtained from living organisms, which he called *organic*, and those obtained from mineral sources, which he called *inorganic*. At that time, most chemists believed that organic chemicals could be synthesized only in living systems. A theory known as "vitalism" proposed that the laws of nature are somehow different for living and non-living systems, and that the synthesis of organic compounds involved a "vital force."

This theory was shown to be unacceptable in 1828 by German chemist Friedrich Wöhler (1800 – 1882). Wöhler performed a revolutionary laboratory experiment in which he used the inorganic compound ammonium cyanate, $NH_4OCN_{(s)}$, to synthesize urea, $H_2NCONH_{2(s)}$, a well-known organic compound produced by many living organisms. In the years following Wöhler's experiment, chemists synthesized many other organic compounds. For example, acetic acid, $CH_3COOH_{(l)}$, a relatively simple molecule, was synthesized in 1845. Sucrose, $C_{12}H_{22}O_{11(s)}$ (Figure 9.1), has a more complex structure and so it was not synthesized until 1953 — by Canadian chemist Raymond Lemieux.

Figure 9.1
Sucrose occurs naturally in sugar beets and sugar cane. No sugar cane is grown in Canada, although it is refined in New Brunswick, Quebec, and Ontario. Sugar beets are grown and refined in Quebec, Manitoba, and Alberta. High-fructose corn syrup is refined in Ontario.

Today, **organic chemistry** is defined as the study of the molecular compounds of carbon. The properties of organic compounds are a result of the covalent bonds within their molecules. The oxides of carbon and the compounds of carbonate, bicarbonate, cyanide, cyanate, and thiocyanate ions are not considered organic compounds. Inorganic compounds such as these contain ionic bonds.

Compounds of Carbon

Animals, plants, and fossil fuels contain a remarkable variety of carbon compounds. Early chemical technology was developed to extract compounds from living systems, such as ethanol produced from sugar undergoing fermentation by yeast cells. Technology was also developed to mine coal and other fossil fuels; these inexpensive resources required little or no processing. With increasingly sophisticated technology, new uses for carbon compounds have developed. Technological research and development have produced not only better fuels, but also many new compounds.

The number of known compounds of carbon far exceeds the number of compounds of all other elements combined. Carbon atoms can form four bonds, like atoms of some other elements such as silicon. Carbon atoms have the special property that they can bond together to form chains, rings, spheres, sheets, and tubes of almost any size. Another unique property is carbon's ability to form combinations of single, double, and triple covalent bonds. No other element can do this.

Structural Models and Diagrams

As we have seen, molecular formulas are useful for communicating the relative numbers of atoms present in a molecule, and, in some cases, to suggest a molecule's structure. As the number of atoms in a molecule increases, the molecular formula must be expanded in order to communicate the structure of the molecule. One simple alternative is to cluster groups of atoms, such as $CH_3CH_2CH_2CH_2CH_3$ to represent C_5H_{12}. This *expanded molecular formula* is actually just one of three possible structures for this compound (Figure 9.2). Substances with the same molecular formula but different structures are called **isomers**. As the number of carbon atoms in a molecule increases, the number of possible isomers increases dramatically. For example, $C_{10}H_{22}$ has 75 possible isomers; $C_{20}H_{42}$ has 366 319; and $C_{30}H_{62}$ has 4 111 846 763. Considering all the possible compounds with double and triple bonds, as well as other kinds of atoms besides carbon and hydrogen, the number of possible carbon compounds is enormous.

Chemists have invented other ways to communicate the structures of these compounds. *Ball-and-stick models* and *space-filling models* such as those shown in Figures 9.2 and 9.3 help us visualize the structures of molecules.

Models known as structural diagrams also communicate molecular structure. A *complete structural diagram*, as in Figure 9.4 (a), shows all

Figure 9.2
Each of the three isomers of C_5H_{12} has different physical and chemical properties.

Figure 9.3
Different kinds of models are used to represent different aspects of molecules. This is a space-filling model of pentane, used to show the shape of the molecule. Ball-and-stick models, shown in Figure 9.2, are particularly effective in showing types of covalent bonds and the angles between the bonds.

atoms and bonds; a *condensed structural diagram*, Figure 9.4 (b), omits the C—H bonds but shows the carbon-carbon bonds. A *line structural diagram*, Figure 9.4 (c), is an efficient way to represent long chains of carbon atoms; the end of each line segment represents a carbon atom, and hydrogen atoms are not shown.

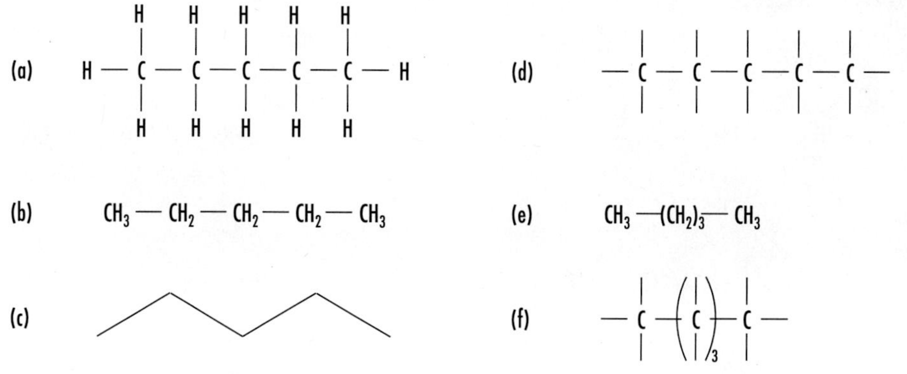

(a)
$$H - C - C - C - C - C - H$$
with H atoms above and below each carbon

(b) $CH_3 - CH_2 - CH_2 - CH_2 - CH_3$

(c)   /\/\/

(d) $- C - C - C - C - C -$

(e) $CH_3 -(CH_2)_3- CH_3$

(f) $- C -(C)_3- C -$

Figure 9.4
These structural diagrams represent the same isomer of C_5H_{12}.

Exercise

1. How does the modern definition of organic chemistry compare with the original definition?

2. State two unique features of the covalent bonding of carbon atoms.

3. Most carbon compounds contain hydrogen. In addition, they often contain oxygen, sulfur, phosphorus, nitrogen, and/or halogen atoms. What is the bonding capacity of each of these other atoms?

4. Using Figure 9.4 as reference, draw a complete structural diagram, a condensed structural diagram, and a line structural diagram for the three isomers of C_5H_{12}.

9.1 Models of Organic Compounds

The purpose of this investigation is to examine the structure of some isomers of organic compounds and to practice drawing structural diagrams.

Problem

What are the structures of the isomers of C_4H_{10}, $C_2H_3Cl_3$, C_2H_6O, and C_2H_7N?

Materials

molecular model kit (Figure 9.5)

Procedure

1. Assemble two different isomeric models of C_4H_{10} and record three different structural diagrams for each model.

2. Assemble two different models for each of the other molecular formulas and record their complete and condensed structural diagrams.

Problem
Prediction
Design
Materials
Procedure
Evidence
✔ **Analysis**
Evaluation
Synthesis

Figure 9.5
Various kits are available for constructing models of molecules. Each kit has advantages and disadvantages, but all of them help you visualize the theoretical structure of compounds.

Figure 9.6
This image shows the ball-and-stick model of methane superimposed on its space-filling model, as generated by a sophisticated computer program. This model most accurately represents scientists' empirical and theoretical knowledge of the methane molecule.

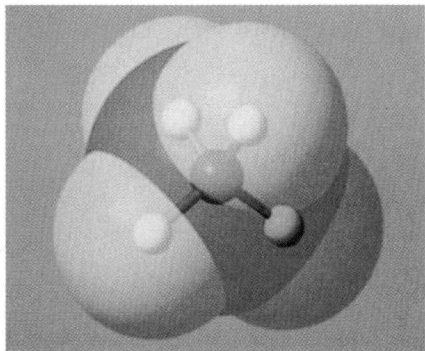

Families of Organic Compounds

In order to cope with the huge number of organic substances, chemists classify them into families based on the characteristic structures and bonds believed to exist within the molecules. **Functional groups** are characteristic arrangements of atoms within a molecule that are believed to be largely responsible for properties of the compound. For example, evidence indicates that the physical and chemical properties of ethanol

$$CH_3—CH_2—OH$$

are largely determined by the presence of the –OH group of atoms, which is known as the **hydroxyl** functional group. In Investigation 9.2 you will examine structural diagrams and identify possible functional groups of several organic compounds.

9.2 Classifying Organic Compounds

The purpose of this investigation is to provide practice in classification. You will also learn about some organic compounds found in various commercial and consumer products.

Problem

How can selected organic compounds be classified according to the functional groups in their molecular structures?

Experimental Design

Information from chemical references and empty containers of commercial and consumer products containing one or more organic substances are investigated to determine the name, toxicity, and structure of selected organic compounds. The compounds are then classified according to similar functional groups. (Some compounds may contain more than one functional group and thus fit more than one classification.)

Procedure

1. Observe one of the samples provided and read the names of compounds listed on the product label.

2. Using the information sheet provided, record the product's commercial name and its use. For the selected organic compound contained in the product, record the toxicity rating, the IUPAC name, and a structural diagram.

3. Repeat steps 1 and 2 for the other samples provided.

- ▦ Problem
- ▦ Prediction
- ▦ Design
- ▦ Materials
- ▦ Procedure
- ✔ Evidence
- ✔ Analysis
- ▦ Evaluation
- ▦ Synthesis

**Chemicals Aren't Toxic**
Toxicologists have argued that it is not chemicals that are toxic, it is the quantity or concentration that makes chemicals toxic. For example, the LD_{50} oral dose of sodium chloride in rats is 3.75 g/kg (moderately toxic).

TOXICITY RATINGS

Detailed information on the safety of compounds is found in references such as *Clinical Toxicology of Commercial Products* and Material Safety Data Sheets (MSDS). A typical classification of toxicity is the LD_{50}, which is the quantity of a substance that researchers estimate would be a lethal dose for 50% of a particular species exposed to that quantity of the substance. The table provides toxicity ratings and LD_{50} values for human beings.

Toxicity Rating		LD_{50} Oral Dose (/kg)	LD_{50} for 70 kg Human
6	extremely toxic	less than 5 mg	a taste (less than 7 drops)
5	very toxic	5 to 50 mg	7 drops to 5 mL
4	quite toxic	50 to 500 mg	5 to 25 mL
3	moderately toxic	0.5 to 5 g	30 to 300 mL
2	slightly toxic	5 to 15 g	300 mL to 1 L
1	almost non-toxic	above 15 g	more than 1 L

Classifying Organic Compounds

Organic chemists divide carbon compounds into families, classifying them according to functional groups. These groups, the sites where chemists believe reactions usually take place, help to explain many of the chemical properties of organic compounds. Table 9.1 lists families of organic compounds, each of which you will study in this chapter. In the general formulas, R represents any chain of carbon and hydrogen atoms. R(H) indicates that the substituent may be a chain or a single hydrogen atom. X represents a halogen atom.

Table 9.1

FAMILIES OF ORGANIC COMPOUNDS		
Family Name	**General Formula**	**Example**
alkanes	$-\overset{\mid}{\underset{\mid}{C}}-\overset{\mid}{\underset{\mid}{C}}-$	propane, $CH_3-CH_2-CH_3$
alkenes	$-\overset{\mid}{C}=\overset{\mid}{C}-$	propene (propylene), $CH_2=CH-CH_3$
alkynes	$-C\equiv C-$	propyne, $CH\equiv C-CH_3$
aromatics	(benzene ring)	toluene, (benzene ring)$-CH_3$
organic halides	$R-X$	chloropropane, $CH_3-CH_2-CH_2-Cl$
alcohols	$R-OH$	propanol, $CH_3-CH_2-CH_2-OH$
ethers	R_1-O-R_2	methoxyethane, $CH_3-O-CH_2-CH_3$ (methyl ethyl ether)
carboxylic acids	$R(H)-\overset{O}{\overset{\|}{C}}-OH$	propanoic acid, $CH_3-CH_2-\overset{O}{\overset{\|}{C}}-OH$
aldehydes	$R(H)-\overset{O}{\overset{\|}{C}}-H$	propanal, $CH_3-CH_2-\overset{O}{\overset{\|}{C}}-H$
ketones	$R_1-\overset{O}{\overset{\|}{C}}-R_2$	propanone (acetone), $CH_3-\overset{O}{\overset{\|}{C}}-CH_3$
esters	$R_1(H)-\overset{O}{\overset{\|}{C}}-O-R_2$	methyl ethanoate (methyl acetate), $CH_3-\overset{O}{\overset{\|}{C}}-O-CH_3$
amines	$R_1-\overset{R_2(H)}{\overset{\mid}{N}}-R_3(H)$	propylamine, $CH_3-CH_2-CH_2-\overset{H}{\overset{\mid}{N}}-H$
amides	$R_1(H)-\overset{O}{\overset{\|}{C}}-\overset{R_2(H)}{\overset{\mid}{N}}-R_3(H)$	propanamide, $CH_3-CH_2-\overset{O}{\overset{\|}{C}}-\overset{H}{\overset{\mid}{N}}-H$

9.2 HYDROCARBONS

Coal, crude oil, oil sands, heavy oil, and natural gas are non-renewable sources of fuels. They are also the primary sources of **hydrocarbons** — compounds containing only carbon and hydrogen atoms. Hydrocarbons are the starting points in the synthesis of thousands of products including fuels, plastics, and synthetic fibres. Some hydrocarbons are obtained directly by physical separation from petroleum and natural gas, whereas others come from oil and gas refining (Figure 9.7).

Refining is the technology that includes separating complex mixtures into purified components. The refining of coal and natural gas involves physical processes; for example, coal may be crushed and

Figure 9.7
On February 13, 1947, after 132 dry holes, Leduc Number 1 became the first Imperial Oil well to produce oil in western Canada. Today, drilling for oil and gas is a sophisticated operation involving computerized drilling rigs and computer analysis of geological data.

RAYMOND LEMIEUX (1920 –   )

"I wanted to play hockey, but at 125 pounds I didn't have a chance!" With one career denied him, Raymond Lemieux turned to another, and hockey's loss was chemistry's gain. A professor at the University of Alberta, Dr. Lemieux has earned international recognition for his work on the chemistry of carbohydrates. Besides writing over 200 scientific publications, Dr. Lemieux holds more than 30 patents and is the founder of several research and chemical companies.

While working at the National Research Council's Prairie Regional Laboratory in Saskatoon in 1953, Lemieux became the first scientist to synthesize sucrose, a sugar known as a *disaccharide*. (The molecules of disaccharides consist of two *monosaccharides* chemically combined; a monosaccharide has six carbon atoms in its basic molecular

formula, a disaccharide has twelve.) This feat, described as "the Mount Everest of organic chemistry," was followed by another — the synthesis of a second disaccharide, maltose.

In the 1960s Lemieux studied the structures of *trisaccharides* that occur on the surfaces of cells. The structural differences in the trisaccharides on human red blood cells are believed to determine an individual's blood type. Differences in these trisaccharide structures are also factors in the rejection that often occurs when an organ is transplanted from one individual to another. In 1975, Lemieux synthesized three blood group trisaccharides and eliminated the need to use whole blood for typing newly donated blood at blood banks.

Even in retirement, Dr. Lemieux often works seven days a week in the

laboratory. He has always enjoyed his work, and somehow every project leads to another idea, and to yet another project.

treated with solvents. Components of natural gas are separated either by solvent absorption or by condensation and distillation. Petroleum refining is more complex than coal or gas refining, but many more products are obtained from crude oil.

Petroleum Refining

Petroleum is a complex mixture of hundreds of thousands of compounds. Some of these compounds boil at temperatures as low as 20°C. The least volatile components of crude oil, however, boil at temperatures above 400°C. The differences in boiling points of the compounds making up petroleum enable the separation of these compounds in a process called *fractional distillation*, or **fractionation**.

When crude oil is heated to 500°C in the absence of air, most of its constituent compounds vaporize. The compounds with boiling points higher than 500°C remain as mixtures called asphalts and tars. The vaporized components of the petroleum rise and gradually cool in a metal tower (Figure 9.8). Where the temperature in the higher parts of the tower is below the boiling points of the vaporized compounds, the

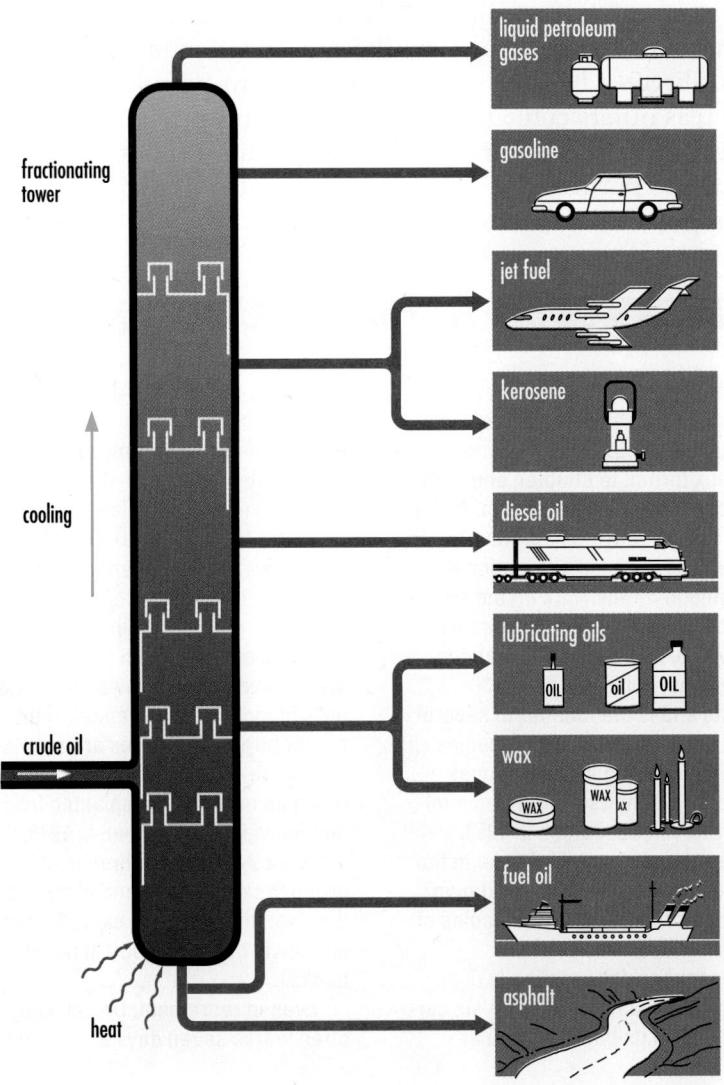

Figure 9.8
A fractional distillation tower contains trays positioned at various levels. Heated crude oil enters near the bottom of the tower. The bottom of the tower is kept hot, and the temperature gradually decreases toward the top of the tower. The lower the boiling point of a fraction, the higher the tray on which it condenses.

substances in the vapor begin to condense. Those substances with high boiling points condense in the lower, hotter parts of the tower, whereas those with lower boiling points condense near the cooler top of the tower. At various levels in the tower, trays collect mixtures of substances as they condense, each mixture containing compounds with similar boiling points. These mixtures are called petroleum *fractions*.

The fractions with the lowest boiling points contain the smallest molecules. The low boiling points are due to the fact that small molecules have fewer electrons and weaker London forces, compared with large molecules (page 320). The fractions with higher boiling points contain much larger molecules. Some typical fractions are shown in Table 9.2. The physical process of fractionation is followed by chemical processes in which the fractions are converted into valuable products (Figure 9.9, page 346).

Table 9.2

FRACTIONAL DISTILLATION OF PETROLEUM			
Boiling Point Range of Fraction (°C)	Carbon Atoms per Molecule	Fraction (Intermediate Product)	Applications
below 30	1 to 5	gases	gaseous fuels for cooking and for heating homes
30 to 90	5 to 6	petroleum ether	dry cleaning, solvents, naphtha gas, camping fuel
30 to 200	5 to 12	straight-run gasoline	automotive gasoline
175 to 275	12 to 16	kerosene	fuel for diesel and jet engines and for kerosene heaters; cracking stock (raw materials for fuel and petrochemical industries)
250 to 375	15 to 18	light gas or fuel oil	furnace oil; cracking stock
over 350	16 to 22	heavy gas oil	lubricating oils; cracking stock
over 400	18 and up	greases	lubricating greases; cracking stock
over 450	20 and up	paraffin waxes	candles, waxed paper, cosmetics, polishes; cracking stock
over 500	26 and up	unvaporized residues	asphalts and tars for roofing and paving

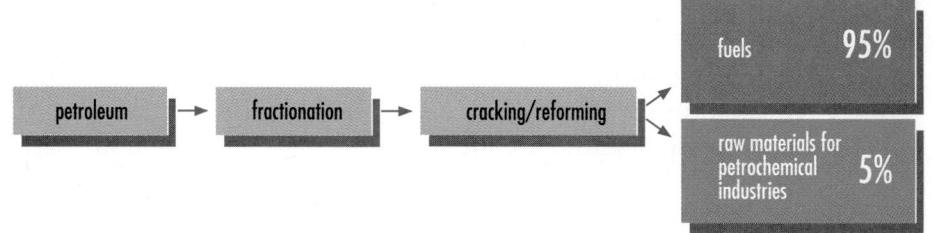

Figure 9.9
Almost all of the products of petroleum refining are burned as fuel for heating and transportation. Only 5% of the original mass of petroleum is used as starting chemicals (feedstock) in the manufacture of solvents, greases, plastics, synthetic fibres, and pharmaceuticals.

Cracking, Reforming, and Combustion Reactions

Straight fractional distillation of petroleum does not produce enough hydrocarbons in the gasoline fraction (called *straight-run gasoline*) to meet the demand for gasoline. Other fractions are chemically altered to produce more gasoline hydrocarbons with 5 to 12 carbon atoms per molecule. Hydrocarbons are broken into smaller fragments in a technological process called **cracking**, which occurs in the absence of air. For example, hydrocarbons of large molar mass (C_{15} to C_{18}) are converted into gasoline hydrocarbons (C_5 to C_{12}).

$$C_{17}H_{36(l)} \rightarrow C_9H_{20(l)} + C_8H_{16(l)}$$

Originally, only high temperatures caused these reactions in an industrial process called *thermal cracking*. Today, a catalyst speeds up the reactions in a process called *catalytic cracking*.

The opposite of cracking is a **reforming reaction**. In catalytic reforming, large molecules are formed from smaller ones. For example,

$$C_5H_{12(l)} + C_5H_{12(l)} \rightarrow C_{10}H_{22(l)} + H_{2(g)}$$

Reforming reactions most commonly convert low-grade gasolines into higher grades, and make larger hydrocarbon molecules for synthetic lubricants and petrochemicals (Figure 9.10).

Figure 9.10
In some of the towers of oil refineries, crude oil is distilled in order to separate its components. Other towers and equipment are designed for catalytic cracking and reforming reactions.

In addition to cracking and reforming, combustion is a very common hydrocarbon reaction. Ninety-five percent of petroleum ends up being used as fuels in combustion reactions to produce energy. For example,

$$2\,C_8H_{18(l)} + 25\,O_{2(g)} \rightarrow 16\,CO_{2(g)} + 18\,H_2O_{(g)} + energy$$

Alkanes

Although hydrocarbons can be classified according to empirical properties, a more common classification is based upon empirical formulas. Hydrocarbons whose empirical formulas indicate only single carbon-to-carbon bonds are called **alkanes**. The simplest member of the alkane series is methane, $CH_{4(g)}$, which is the main constituent of the natural gas sold for home heating. The molecular formulas of the smallest alkanes are shown in Table 9.3. Each formula in the series has one more CH_2 group than the one preceding it. Derived from empirical formulas and from bonding capacity, the general formula for all alkanes is C_nH_{2n+2}; that is, a series of CH_2 units plus two terminal hydrogen atoms.

The first syllable in the name of an alkane is a prefix that indicates the number of carbon atoms in the molecule (Figure 9.11). The prefixes shown in Table 9.3 are used in naming all organic compounds. The same prefixes identify groups of atoms that form branches on the structures of larger molecules. A *branch* is any group of atoms that is not part of the main structure of the molecule. For example, a hydrocarbon branch is called an **alkyl branch**. In the names of alkyl branches, the prefixes are followed by a *-yl* suffix (Table 9.4).

Names and Structures of Branched Alkanes

When there are branches on a carbon chain, the name of the compound indicates this. For example, consider the three isomers of C_5H_{12} shown in Figure 9.2 (page 338). The unbranched isomer is named pentane. The numbers on the following structural diagram show how the carbon atoms are identified.

pentane

In the second isomer, there is a continuous chain of four carbon atoms with a methyl group on the second carbon atom. To name this structure, identify the *parent chain* — the longest continuous chain of carbon atoms. Here, the four carbons indicate that the parent chain is butane. The carbon atoms of this parent chain are numbered from the end closest to the branch, so this isomer is called 2-methylbutane.

2-methylbutane

In the third isomer of pentane, two methyl groups are attached to a three-carbon (propane) parent chain. This third pentane isomer is

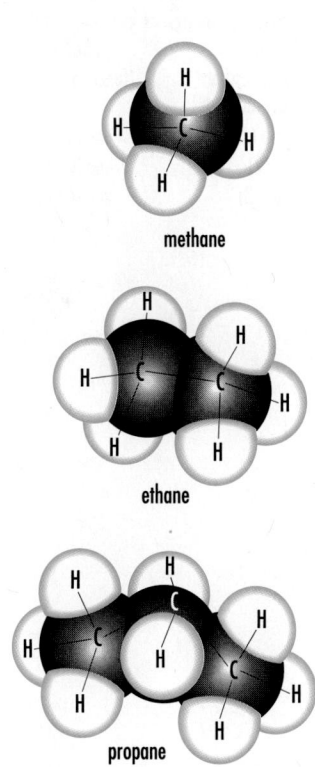

Table 9.3

THE ALKANE FAMILY OF ORGANIC COMPOUNDS	
IUPAC Name	**Formula**
methane	$CH_{4(g)}$
ethane	$C_2H_{6(g)}$
propane	$C_3H_{8(g)}$
butane	$C_4H_{10(g)}$
pentane	$C_5H_{12(l)}$
hexane	$C_6H_{14(l)}$
heptane	$C_7H_{16(l)}$
octane	$C_8H_{18(l)}$
nonane	$C_9H_{20(l)}$
decane	$C_{10}H_{22(l)}$
–ane	C_nH_{2n+2}

Memorize the prefixes indicating one to ten carbon atoms.

methane

ethane

propane

Figure 9.11
The prefix meth- indicates one carbon atom, eth- signifies two carbon atoms, and prop- signifies three carbon atoms. The ending -ane indicates a chain of carbon atoms with single bonds only.

Table 9.4

EXAMPLES OF ALKYL BRANCHES	
Branch	**Name**
$-CH_3$	methyl
$-C_2H_5$ ($-CH_2CH_3$)	ethyl
$-C_3H_7$ ($-CH_2CH_2CH_3$)	propyl

Nomenclature Rules

The two general rules used throughout organic nomenclature are:

- use the longest continuous chain containing the primary functional group
- use the lowest set of numbers to indicate where substituents or functional groups exist in the molecule

General Construct of Names

Prefix	**+Suffix**
(number of carbon atoms per molecule)	(functional group)
e.g., meth (1)	ane (C−C)
eth (2)	ene (C=C)
prop (3)	yne (C≡C)

named 2,2-dimethylpropane. Note the use of the comma, the hyphen, and single words when naming isomers of branched alkanes.

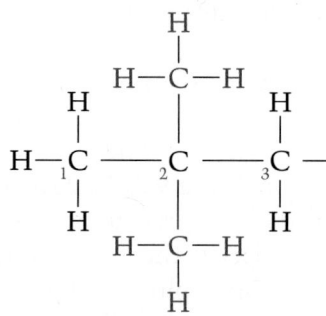

2,2-dimethylpropane

SUMMARY: NAMING BRANCHED ALKANE STRUCTURES

Step 1: Identify the longest continuous chain of carbon atoms — the parent chain — in the structural diagram. Number the carbon atoms, starting from the end closest to the branch(es).

Step 2: Identify any branches and their location number on the parent chain.

Step 3: Write the complete IUPAC name, following this format: *(number of location)-(branch name)(parent chain)*.

When writing the name of the alkane the branches are listed (a) in alphabetical order or (b) in order of complexity.

EXAMPLE

Write the IUPAC name corresponding to the following structural diagram.

$$CH_3 \underset{1}{\text{---}} \underset{2}{CH} \text{---} \underset{3}{CH} \text{---} \underset{4}{CH_2} \text{---} \underset{5}{CH_2} \text{---} \underset{6}{Cl}$$

with CH_3 on carbon 2 and $CH_2\text{---}CH_3$ on carbon 3

Step 1: The longest continuous chain has six carbon atoms. Therefore, the name of the parent chain is *hexane*.

Step 2: There is a *methyl* group branch at the second carbon atom, and an *ethyl* group branch at the third carbon atom of the parent chain.

Step 3: With the branches named in alphabetical order, the compound is *3-ethyl-2-methylhexane*. With the branches named in order of complexity, the compound is *2-methyl-3-ethylhexane*.

A structural diagram can illustrate an IUPAC name. For example, 3-ethyl-2,4-dimethylpentane is a gasoline molecule with a pentane parent chain consisting of five carbon atoms joined by single covalent bonds.

$$\underset{1}{\text{---}C} \text{---} \underset{2}{C} \text{---} \underset{3}{C} \text{---} \underset{4}{C} \text{---} \underset{5}{C} \text{---}$$

Numbering this straight chain from left to right establishes the location of the branches. An ethyl branch is attached to the third carbon atom and a methyl branch is attached to each of the second and fourth carbon atoms.

$$
\begin{array}{c}
CH_3 \\
| \\
CH_2 \\
| \\
-\overset{1}{C}-\overset{2}{C}-\overset{3}{C}-\overset{4}{C}-\overset{5}{C}- \\
| \quad | \quad | \\
CH_3 \quad CH_3
\end{array}
$$

In the following complete structural diagrams, hydrogen atoms are shown at any of the four bonds around each carbon atom that are left after the branches have been located.

$$
\begin{array}{c}
CH_3 \\
| \\
H \quad H \quad CH_2 \quad H \quad H \\
| \quad | \quad | \quad | \quad | \\
H-C-C-C-C-C-H \\
| \quad | \quad | \quad | \quad | \\
H \quad CH_3 \quad H \quad CH_3 \quad H
\end{array}
\quad \text{or} \quad
\begin{array}{c}
CH_3 \\
| \\
CH_2 \\
| \\
CH_3-CH-CH-CH-CH_3 \\
| \quad\quad | \\
CH_3 \quad CH_3
\end{array}
$$

2,4-dimethyl-3-ethylpentane

Exercise

6. Naphtha, commonly used as a camping fuel, is a mixture of alkanes with five or six carbon atoms per molecule.
   (a) Draw structural diagrams and write the IUPAC names for all the isomers of C_5H_{12}.
   (b) Draw structural diagrams and write the IUPAC names for all the isomers of C_6H_{14}.

7. Automotive gasoline is largely composed of alkanes containing five to twelve carbon atoms per molecule. Write IUPAC names for the following components of straight-run gasoline.

   (a)
$$
\begin{array}{c}
CH_3-CH-CH_2-CH_2-CH_2-CH_2-CH_3 \\
| \\
CH_3
\end{array}
$$

Cycloalkanes

On the evidence of empirical formulas and chemical properties, chemists believe that organic carbon compounds sometimes take the form of **cyclic hydrocarbons** — hydrocarbons with a closed ring. When all the carbon-carbon bonds in a cyclic hydrocarbon are single bonds, the compound is called a **cycloalkane**. For example, cyclopropane and cyclobutane (Figure 9.12) are the two simplest cycloalkanes. Cyclic hydrocarbons are usually represented by line structural diagrams.

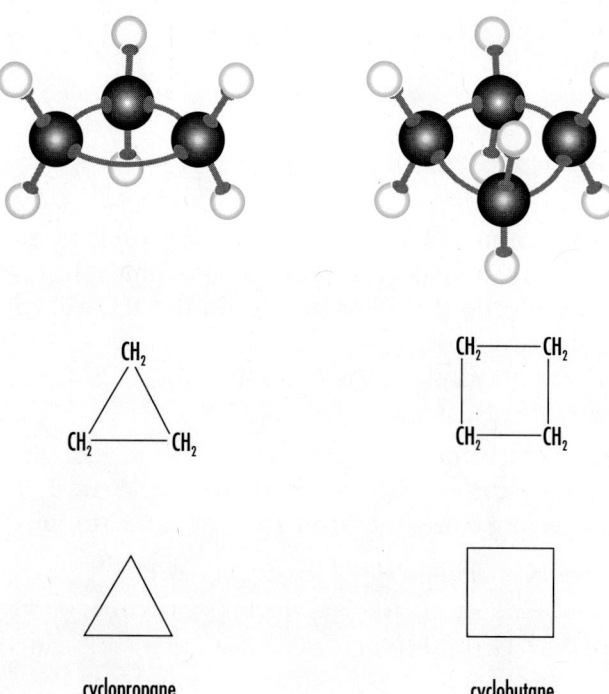

Figure 9.12
Cycloalkanes such as cyclopropane and cyclobutane are similar to alkanes, except that the two ends of the molecule are joined to form a ring of atoms. These models show approximate orientations of the atoms. Condensed diagrams and line structural diagrams are drawn in the shape of regular polygons.

cyclopropane

cyclobutane

Exercise

10. Since petroleum contains many large alkanes, cracking reactions are common in the first stage of oil refining. For each of the following word equations, draw a complete structural diagram of each reactant and product.
    (a) hexane + hydrogen → ethane + butane
    (b) 2-methylpentane + hydrogen → propane + propane
    (c) 2,2-dimethylbutane + hydrogen → ethane + methylpropane

11. Reforming reactions increase the yield of desirable products, such as compounds whose molecules have longer chains or more branches. For each of the following word equations, draw structural diagrams when IUPAC names are given and write IUPAC names when structural diagrams are given.
    (a) $CH_3-CH_2-CH_3$ + $CH_3-CH_2-CH_2-CH_2-CH_3$ →
    $$CH_3-(CH_2)_6-CH_3 + H-H$$
    (b) cyclohexane + ethane → ethylcyclohexane + hydrogen
    (c)

    $$CH_3-CH_2-CH_2-CH_2-CH_2-CH_3 \rightarrow CH_3-\overset{\overset{\displaystyle CH_3}{|}}{\underset{\underset{\displaystyle CH_3}{|}}{C}}-CH_2-CH_3$$

    (d) Draw structural diagrams and write the IUPAC names for two other isomers of the product given in (c).

12. Most of the products of cracking and reforming reactions end up in fuel mixtures such as gasoline. Complete the following equations for complete combustion, including structures and IUPAC names. Recall that complete combustion involves a reaction with oxygen to produce the most common oxides.
    (a) 2,2,4-trimethylpentane + oxygen →
    (b)

    $$CH_3-\overset{\overset{\displaystyle CH_3}{|}}{CH}-CH_2-CH_3 + O{=}O \rightarrow$$

    (c) ⬠ + O=O →

13. Classify and write structural formula equations for the following organic reactions. Do not balance the equations.
    (a) methane + butane → pentane + hydrogen
    (b) propane + pentane → octane + hydrogen
    (c) butane + oxygen →
    (d) decane + hydrogen → heptane + propane
    (e) 3-ethyl-5-methylheptane + hydrogen →
    ethane + propane + methylbutane

Alkenes and Alkynes

Analysis reveals that hydrocarbons containing double or triple covalent bonds are minor constituents in natural gas and petroleum. However, these compounds are often formed during cracking reactions and are valuable components of gasoline. Hydrocarbons containing double or triple bonds are important in the petrochemical industry because they

Cracking Processes
In oil refineries, cracking is accomplished using two different process designs: catalytic cracking and thermal cracking. A catalyst increases the rate of the reaction at lower temperatures, while thermal cracking requires higher temperatures.

Margarine
Margarine containing vegetable oils whose molecules have many double bonds is said to be *polyunsaturated*. The molecules of *saturated* fats, in animal products such as butter, are fully hydrogenated.

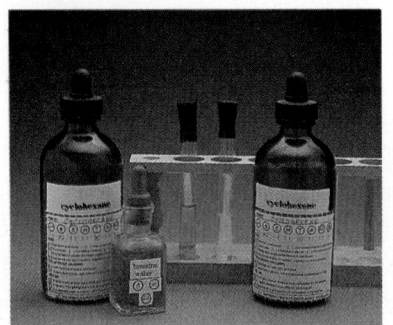

Figure 9.13
Bromine water is used in a diagnostic test for unsaturated organic compounds. When an equal amount of bromine is added simultaneously to cyclohexane and cyclohexene, the unsaturated cyclohexene reacts instantaneously, decolorizing the bromine. In the cyclohexane, which is saturated, there is no noticeable reaction.

Table 9.5

THE ALKENE FAMILY OF ORGANIC COMPOUNDS	
IUPAC Name (common name)	**Formula**
ethene (ethylene)	$C_2H_{4(g)}$
propene (propylene)	$C_3H_{6(g)}$
butene (butylene)	$C_4H_{8(g)}$
pentene	$C_5H_{10(1)}$
hexene	$C_6H_{12(1)}$
—ene	C_nH_{2n}

Figure 9.14
Ethene and propene are the simplest members of the alkene family.

Table 9.6

THE ALKYNE FAMILY OF ORGANIC COMPOUNDS	
IUPAC Name (common name)	**Formula**
ethyne (acetylene)	$C_2H_{2(g)}$
propyne	$C_3H_{4(g)}$
butyne	$C_4H_{6(g)}$
pentyne	$C_5H_{8(1)}$
hexyne	$C_6H_{10(1)}$
—yne	C_nH_{2n-2}

are the starting materials for the manufacture of many derivatives, including plastics.

A double or a triple bond between two carbon atoms in a molecule affects the chemical properties of the molecule. For example, hydrocarbons with double bonds react quickly with bromine, compared with alkanes, which react very slowly (Figure 9.13). Organic compounds with carbon-carbon double bonds are said to be *unsaturated*, because fewer hydrogen atoms are attached to the carbon atom framework compared with the number of hydrogen atoms that would be attached if all the bonds were single. Unsaturated hydrocarbons react readily with small diatomic molecules, such as bromine and hydrogen. This type of reaction is an **addition reaction**. Addition of a sufficient quantity of hydrogen, called **hydrogenation**, converts unsaturated hydrocarbons to saturated ones.

Hydrocarbons with carbon-carbon double bonds are members of the **alkene** family (Figure 9.14). The names of alkenes with only one double bond feature the same prefixes as in the names of alkanes, together with the suffix *-ene* (Table 9.5). (Ethene is the starting material for a huge variety of consumer, commercial, and industrial products, some of which are listed on page 361.)

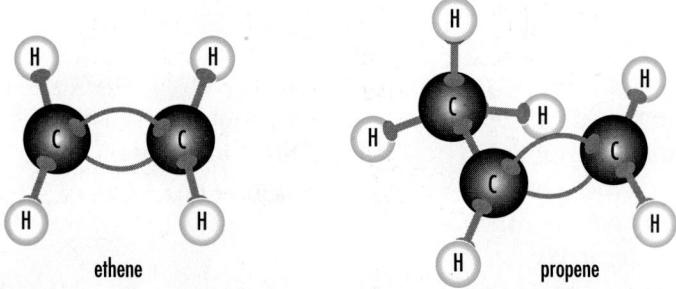

ethene   propene

The **alkyne** family has chemical properties that can be explained only by the presence of a triple bond between carbon atoms (Figure 9.15). Like alkenes, alkynes are unsaturated and react immediately with small molecules such as hydrogen or bromine in an addition reaction. Alkynes are named like alkenes, except for the *-yne* suffix. The simplest alkyne, ethyne or acetylene, is used as a fuel (Figure 9.16). Table 9.6 lists the first five members of the alkyne family. Isomers exist for all alkynes larger than propyne.

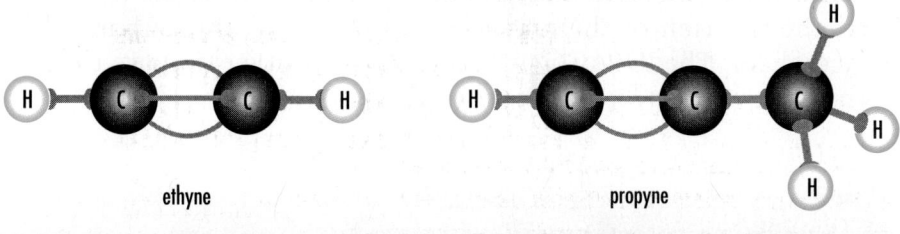

ethyne   propyne

Figure 9.15
The triple covalent bonds of ethyne and propyne are the shortest, strongest, and most reactive of all carbon-carbon bonds.

Figure 9.16
The flame of an oxyacetylene torch is hot enough to melt most metals.

Steroids

Steroids are unsaturated compounds based on a structure of four rings of carbon atoms. The best known and most abundant steroid is cholesterol, which is an essential constituent of cell walls, but which has also been associated with diseases of the cardiovascular system. Cholesterol that coats the interior surfaces of arteries contributes to health problems such as high blood pressure. Other steroids include the male and female sex hormones, and anti-inflammatory agents such as cortisone. Oral contraceptives include two synthetic steroids. Some athletes have used anabolic steroids to enhance muscle development and physical performance, but such use may cause permanent damage.

cortisone

cholesterol

Naming Alkenes and Alkynes

Since the location of a multiple bond affects the chemical and physical properties of a compound, an effective naming system should specify the multiple bond location. Alkenes and alkynes are named much like alkanes, with two additional points to consider.

- The longest or parent chain of carbon atoms must contain the multiple bond, and the chain is numbered from the end closest to the multiple bond.
- The name of the compound's parent chain is preceded by a number that indicates the position of the multiple bond on the parent chain.

For example, there are two possible butene isomers, 1-butene and 2-butene.

$$\underset{1}{CH_2} = \underset{2}{CH} - \underset{3}{CH_2} - \underset{4}{CH_3} \qquad \underset{1}{CH_3} - \underset{2}{CH} = \underset{3}{CH} - \underset{4}{CH_3}$$

1-butene   2-butene

In the following branched alkyne structure, the parent chain is pentyne.

$$\underset{1}{CH_3} - \underset{2}{C} \equiv \underset{3}{C} - \underset{4}{\overset{\overset{\displaystyle CH_3}{|}}{CH}} - \underset{5}{CH_3}$$

4-methyl-2-pentyne

The location of the multiple bond in an alkyne takes precedence over the location of the branches in numbering the carbons of the parent chain. The IUPAC name, 4-methyl-2-pentyne, follows the same format as that used for alkanes (page 347).

(page 347)

Dienes

Dienes are alkenes with two double bonds. The nomenclature of dienes is similar to alkenes with one double bond, except that you must communicate the position of the two double bonds. Dienes are isomers of alkynes and have a general formula of C_nH_{2n-2}. The nomenclature of dienes involves communicating the length of the hydrocarbon chain and the positions of the double bonds. For example, 1, 3-pentadiene has a double bond between carbon 1 and 2 and between carbon 3 and 4 in a five-carbon chain. The *ane* of the alkane name is changed to *adiene*.

Exercise

14. All cycloalkanes have isomers that are alkenes. Draw structural diagrams and write the IUPAC names for all the isomers of C_4H_8.

15. Alkenes with two double bonds are called *dienes*. All dienes have isomers that are alkynes. Draw structural diagrams and write the IUPAC names for all the isomers of C_4H_6.

16. Alkenes and alkynes are the starting materials in the manufacture of a wide variety of organic compounds. Draw structural diagrams for the starting materials for the products in parentheses.

    (a) propene (polypropylene)

    (b) methylpropene (synthetic rubber)

    (c) 2-methyl-1,3-butadiene (natural rubber)

Geometric Isomers

In alkanes, the rotation of attached groups about the carbon-carbon single bond is quite free. The situation is different for alkenes, where rotation about the carbon-carbon double bond is not possible without breaking the bond. Molecular models are quite useful in simulating this difference in rotation ability. Consequently, alkenes can have *geometric* isomers, which differ from each other with respect to the position of attached groups relative to the double bond. The term *cis* means that two attached groups are on the same side of the double bond, and *trans* means that two attached groups are across from each other. For example, consider the two geometrical isomers of 2-butene, $CH_3CH{=}CHCH_3$.

cis–2–butene   *trans*-2–butene

In general, *cis* isomers are polar and *trans* isomers are not, so the melting points, boiling points, and solubility of *cis* and *trans* isomers can be quite different.

9.3 Structures and Properties of Isomers

The purpose of this investigation is to examine the structures and physical properties of some isomers of unsaturated hydrocarbons.

Problem

What are the structures and physical properties of the isomers of C_4H_8 and C_4H_6?

Experimental Design

Structures of possible isomers are determined by means of a molecular model kit. Once each structure is named, the boiling and melting points are obtained from a reference such as *The CRC Handbook of Chemistry and Physics* or *The Merck Index*.

Materials

molecular model kits
chemical reference

Procedure

1. Use the required "atoms" to make a model of C_4H_8.

2. Draw a structural diagram of the model and write the IUPAC name for the structure.

3. By rearranging bonds, produce models for all other isomers of C_4H_8, including cyclic structures. Draw structural diagrams and write the IUPAC name for each structure before disassembling the models.

4. If you construct a model which contains a double C=C bond, test the restricted rotation of groups about the bond axis.

5. Repeat steps 1 to 4 for C_4H_6.

6. In a reference, find the melting point and the boiling point of each of the compounds identified.

- Problem
- Prediction
- Design
- Materials
- Procedure
- ✔ Evidence
- ✔ Analysis
- Evaluation
- Synthesis

Aromatics

Historically, organic compounds with an aroma or odor were called *aromatic compounds*. Today, chemists define **aromatics** as benzene, $C_6H_{6(l)}$, and all other carbon compounds that have benzene-like structures and properties. The molecular structure of benzene intrigued chemists for many years because the properties of this compound, which are listed below, could not be explained by the accepted theories of bonding and reactivity.

- The molecular formula of benzene, based on its percentage composition and molar mass, is C_6H_6.

- The melting point of benzene is 5.5°C, the boiling point is 80.1°C, and tests show that the molecules are non-polar.

Commercial Aromatics
A large number of aromatic compounds have commercial uses. Some of the most widely known aromatics are acetylsalicylic acid (ASA or aspirin), benzocaine (a local anesthetic), methyl salicylate (known as oil of wintergreen, applied externally to aching muscles), and ephedrine (a nasal decongestant). Other aromatics include amphetamines, which stimulate the central nervous system, adrenaline, a hormone that also stimulates the central nervous system, and vanillin, a flavoring agent. As well, various small aromatic compounds are added to gasoline to improve the burning quality of the mixture.

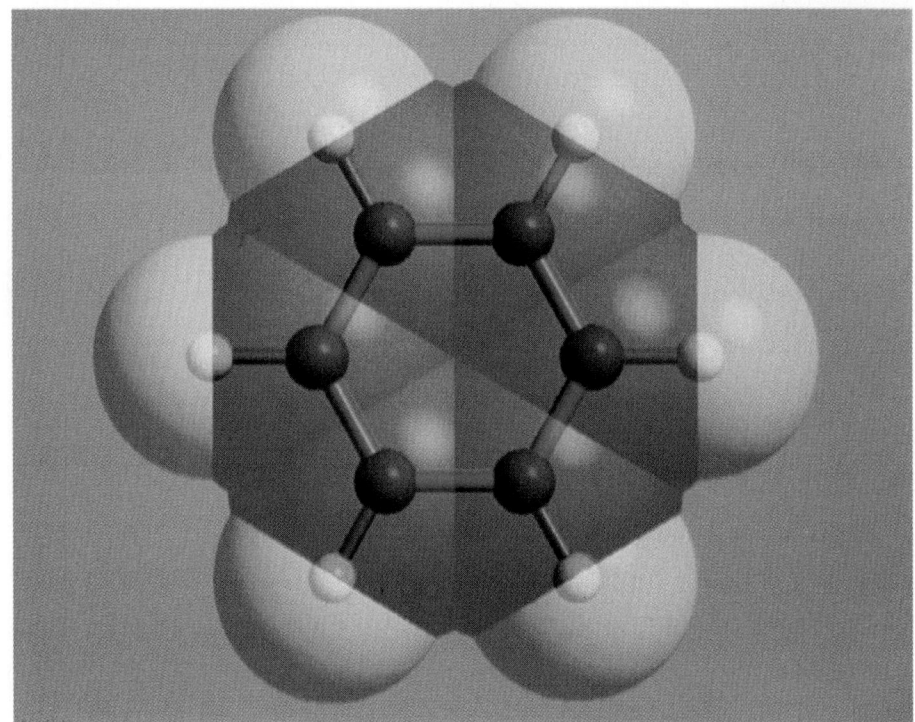

Figure 9.17
Apparently, Kekulé conceived the idea of the ring structure of benzene in a daydream: "I fell half asleep. Again the atoms gambolled before my eyes. Smaller groups this time kept modestly to the background. My mind's eyes, trained by repeated visions of a similar kind, could now distinguish larger structures, of various shapes; long rows, sometimes more closely fitted together; all twining and twisting in snakelike motion. But look! What was that? One of the snakes grabbed its own tail, and the form whirled mockingly before my eyes. As if struck by lightning I awoke; ...I spent the rest of the night [working] out the consequences.... If we learn to dream we shall perhaps discover the truth."

aspirin
(ASA, acetylsalicylic acid)

benzocaine   vanillin

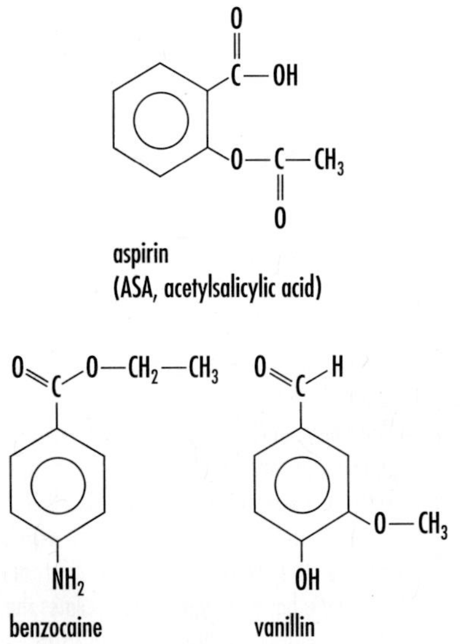

Figure 9.18
Common aromatic compounds include aspirin, benzocaine, and vanillin.

- There is no empirical support for the idea that there are double or triple bonds in benzene. For example, it is very unreactive with bromine.

- X-ray diffraction indicates that all the carbon-carbon bonds in benzene are the same length.

- Evidence from chemical reactions indicates that all carbons in benzene are identical and that each carbon is bonded to one hydrogen.

Even after the empirical formula for benzene was determined in 1825 by English scientist Michael Faraday, visualizing a model of the benzene molecule that followed accepted bonding rules proved difficult. Finally, in 1865 German architect and chemist August Kekulé (1829 – 1896), who popularized the use of structural models, proposed a cyclic structure for benzene. Since evidence indicates that all bonds between the carbon atoms in benzene are identical in length and in strength, an acceptable model requires the even distribution of valence electrons around the entire molecule. A model of benzene is shown in Figure 9.17. Consider this molecule as having 18 valence electrons (three for each carbon atom), distributed around a 6-carbon atom ring, forming a strong hexagonal structure. This structure is particularly stable, and the reactions of benzene are similar to those of alkanes. Structures of all aromatic compounds include bonding similar to that in the benzene ring (Figure 9.18). To distinguish them from aromatics, organic compounds with chain or cyclic structures of single, double, or triple bonds are classified as **aliphatic compounds**.

Naming Aromatics

Simple aromatics are usually named as relatives of benzene. If an alkyl group is bonded to a benzene ring, it is named as an alkylbenzene (Figure 9.19). The alkyl group is considered a substitute for a hydrogen atom. Since all of the carbon atoms of benzene are equivalent to each other, no number is required in the names of compounds of benzene that contain one substituent.

When two hydrogen atoms of the benzene ring have been substituted, three isomers are possible. These isomers are named as alkylbenzenes, using the lowest possible pair of numbers to indicate the location of the two alkyl groups on the benzene ring. The numbering starts at one of the substituents and goes clockwise or counterclockwise to obtain the lowest possible pair of numbers.

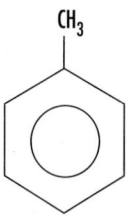

Figure 9.19
Methylbenzene, commonly known as toluene, is a solvent used in glues and lacquers. It is toxic to humans but is preferred to benzene as a solvent, because benzene is both toxic and carcinogenic.

1,2-diethylbenzene   1,3-diethylbenzene   1,4-diethylbenzene

For some larger molecules, it is more convenient to consider the benzene ring as a branch. In such molecules, the benzene ring is called a **phenyl group**, $-C_6H_5$. For example, the following compound is named 2-phenylbutane, according to the naming system for branched alkanes (page 347).

$$CH_3 - CH - CH_2 - CH_3$$

SUMMARY: ORGANIC REACTIONS OF HYDROCARBONS

Cracking

$$\text{large molecule} \underset{\text{heat}}{\overset{\text{catalyst}}{\rightarrow}} \text{smaller molecules}$$

Reforming

$$\text{small molecule(s)} \underset{\text{heat}}{\overset{\text{catalyst}}{\rightarrow}} \text{larger molecules or one with more branches}$$

Complete Combustion

$$\text{compound} + O_{2(g)} \rightarrow CO_{2(g)} + H_2O_{(g)}$$

Addition (Hydrogenation)

$$\text{alkene or alkyne} + H_{2(g)} \rightarrow \text{alkane}$$

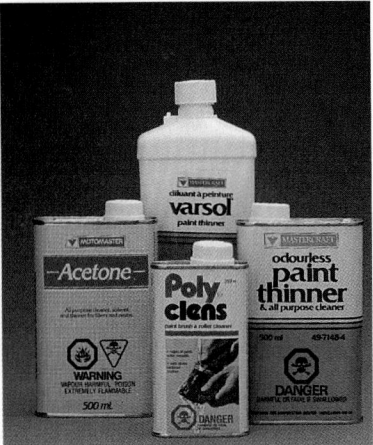

Figure 9.20
Isomers of dimethylbenzene, known as xylenes, are used as solvents. Acetone, which is also a solvent, is not a xylene.

Exercise

17. Write IUPAC names for the following hydrocarbons. Draw and name any geometric isomers formed by these compounds.
    (a) CH₃—CH₂—CH₂—CH₂—CH₃
    (b) CH₃—CH=CH—CH₂—CH₃
    (c) CH≡C—CH₂—CH₂—CH₃
    (d) CH₂=CH—CH₂—CH₃
    (e) CH₃—CH—CH=CH—CH₃
             |
            CH₃

18. Why is the following not an acceptable structural formula for benzene?

19. The isomers of dimethylbenzene, commonly called xylenes, are used as solvents for adhesives (Figure 9.20). Write alternative IUPAC names and draw structural diagrams for the three xylene isomers.
    (a) *o*-xylene
    (b) *m*-xylene
    (c) *p*-xylene

20. Draw structural diagrams for the following hydrocarbons.
    (a) *trans*-diphenylethene
    (b) *cis*-3-hexene
    (c) 1,2,4-trimethylbenzene
    (d) 1-ethyl-2-methylbenzene

21. In addition to alkanes, cracking reactions may also involve alkenes, alkynes, and aromatics. For each of the following reactions, draw a structural diagram equation. Include all reactants and products.
    (a) 1-butene → ethyne + ethane
    (b) 3-methylheptane → 2-butene + butane
    (c) 3-methylheptane →
                     propene + 2-methyl-1-butene + hydrogen
    (d) propylbenzene → methylbenzene + ethene

22. For each of the following reforming reactions, draw a structural diagram or write the IUPAC name for each reactant and product.
    (a) 2-methyl-1-pentene → 2,3-dimethyl-1-butene
    (b)

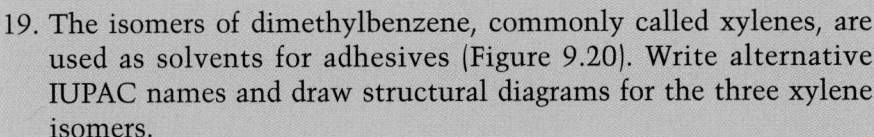

23. All hydrocarbons undergo combustion reactions. Complete the following combustion reaction equations, including both the structural equation and the word equation. Assume a complete combustion reaction.

(a) $CH \equiv CH + O = O \rightarrow$

(b)

$$\underset{}{\text{CH}_3} \text{—} \bigodot \;\; + \;\; O = O \;\; \rightarrow$$

24. Classify each of the following reactions as one of the four types summarized on page 357. Write the names and the structures for all reactants and products.

(a) methyl-2-butene + hydrogen →

(b) ethylbenzene → phenylethene + hydrogen

(c) $CH_3 - C \equiv C - CH_3 + H - H \text{ (excess)} \rightarrow$

(d)

$$\bigodot \;\; + \;\; CH_2 = CH_2 \;\; \rightarrow \;\; \bigodot\text{—}C_2H_5$$

(e)

$$CH_3 - CH = \underset{\underset{\text{CH}_3}{|}}{\overset{\overset{\text{C}_2\text{H}_5}{|}}{C}} - CH - CH_3 \;\; + \;\; O = O \;\; \rightarrow$$

25. Make a concept map or flow chart connecting the various classes of hydrocarbons.

26. Classify and write structural formula equations for the following organic reactions. Do not balance the equations.

(a) ethane → ethene + hydrogen

(b) 2-butene + hydrogen → butane

(c) 4,4-dimethyl-2-pentyne + hydrogen →
propene + methylpropane

(d) methylbenzene + oxygen →

(e) ethene + 2-butene → 3-methylpentane

27. For environmental problems, there are no clearcut answers (Figure 9.21). Consider, for example, the following resolution for a debate. "Fossil fuels are our best energy resource and we should maximize their exploitation in the future." Identify various perspectives (Appendix D, D.4), consult reference materials, brainstorm with classmates, refer to the point-counterpoint example on page 360, and use a decision-making model (Appendix D, D.4) to develop a thesis. Then, present your research, analysis, and evaluation as a report or as part of a class debate.

Figure 9.21
When large quantities of crude oil are moved from their source to refineries, accidents will inevitably occur. Oil spills have harmful effects on waterfowl and on other animals and plants inhabiting the environment where a spill occurs. What trade-offs should be considered in proposing a solution to this environmental problem?

Problem 9A Molecular Structure of Unknown Liquid

Complete the Analysis of the investigation report.

Problem

What is the molecular structure and name of an unknown liquid that is known to be organic?

Experimental Design

A sample of the unknown liquid is tested for saturation, using the bromine diagnostic test. Additional samples are analyzed using a combustion analyzer and a mass spectrometer.

Evidence

No immediate reaction with bromine was observed.
percent by mass of carbon = 91.1%
percent by mass of hydrogen = 8.9%
molar mass = 92.2 g/mol

The Fossil Fuel Debate

For every issue there are various perspectives. For every point made from a particular perspective there will usually be counterpoints. The example below presents point-counterpoint arguments using economic, ecological, and social perspectives.

EXAMPLE

Point-Counterpoint on the Use of Fossil Fuels

Point	Counterpoint
Alternative energy sources, such as solar, are too expensive as a replacement for fossil fuels. Fossil fuel equipment is already purchased and here for our use.	Solar energy is free and renewable. The equipment costs would dramatically decrease with mass production. Installation costs would be recovered from energy savings.
Mining industries have developed and implemented extensive environmental controls and recovery of land when coal mining is finished.	Fossil fuel production, such as coal strip mining and tar sands mining, irrevocably destroys the natural habitat of many plants and animals.
Fossil fuels will not be needed for petrochemical use for future generations — new methods and materials will be used to supply their needs.	Fossil fuels are precious, finite sources for petrochemicals, needed for the health and happiness of future generations.

"Minds are like parachutes. They only function when they are open." — Sir James Dewar (1842 – 1923), Scottish chemist and physicist

76 L of petroleum can provide the gasoline to drive a vehicle 485 km, or it can produce 24 shirts, 2 automobile tires, 5 m² of carpeting material, 30 m of 1.3 cm-diameter rope, 12 windbreakers, 4 sleeping bags, 2 tents, 6 duffel bags, 4 sweaters, 1 blanket, and 15 parkas.

PERSPECTIVES ON THE USE OF FOSSIL FUELS

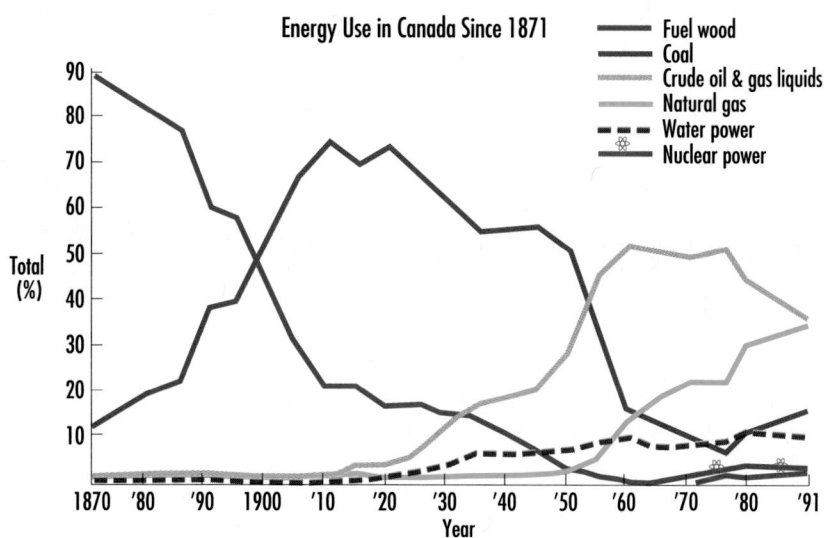

Energy Use in Canada Since 1871

Legend:
— Fuel wood
— Coal
— Crude oil & gas liquids
— Natural gas
- - - Water power
⚛ Nuclear power

Total (%) axis: 10, 20, 30, 40, 50, 60, 70, 80, 90

Year axis: 1870, '80, '90, 1900, '10, '20, '30, '40, '50, '60, '70, '80, '91

Canada's high energy use is due partly to the climate and the large area over which the population is distributed. Attitudes toward energy use and conservation are also factors in this high consumption.

Less than 5% of our fossil fuels are used to produce petrochemicals. The economic importance of petrochemicals lies in the fact that basic raw materials are processed and reprocessed many times. For example, the numbers of jobs in various industries that rely on petrochemicals are shown below.

Ethylene is one of the most important petrochemicals. For every 11 jobs involved in the manufacture of ethylene, 116 jobs are created in manufacturing vinyl chloride (chloroethene), 600 jobs in manufacturing polyvinyl chloride (PVC), and about 6000 jobs in manufacturing other commercial and consumer products such as pipes and tiles.

The graph above shows the sources of energy consumed by Canadians between 1870 and 1980. By 1900, fossil fuels had replaced wood as the main energy source. By 1950, fossil fuels such as oil and natural gas had replaced coal.

By 1985, fossil fuels accounted for about 87% of total energy use in Canada. Energy from hydroelectricity accounted for 11% and energy from nuclear reactors, 2%. This dependence on fossil fuels for energy is likely to continue in the 21st century.

As you can see in the graph below, Canadians are the world's largest per capita consumers of energy. Approximately 45% of total energy production in Canada ends up as waste in the form of heat lost in the generation and transmission of electricity. The amount of energy lost is more than that available to many developing countries to support their populations and economies.

- *Describe how the use of fossil fuels in Canada has changed since 1950.*
- *Name five commercial products that are made from ethylene.*

Per Capita Energy Consumption by Country (1986)

Mass of coal equivalent (t): 1–10

Countries: Canada, United States, Australia, United Kingdom, Sweden, Spain, Argentina, Brazil, China, Thailand

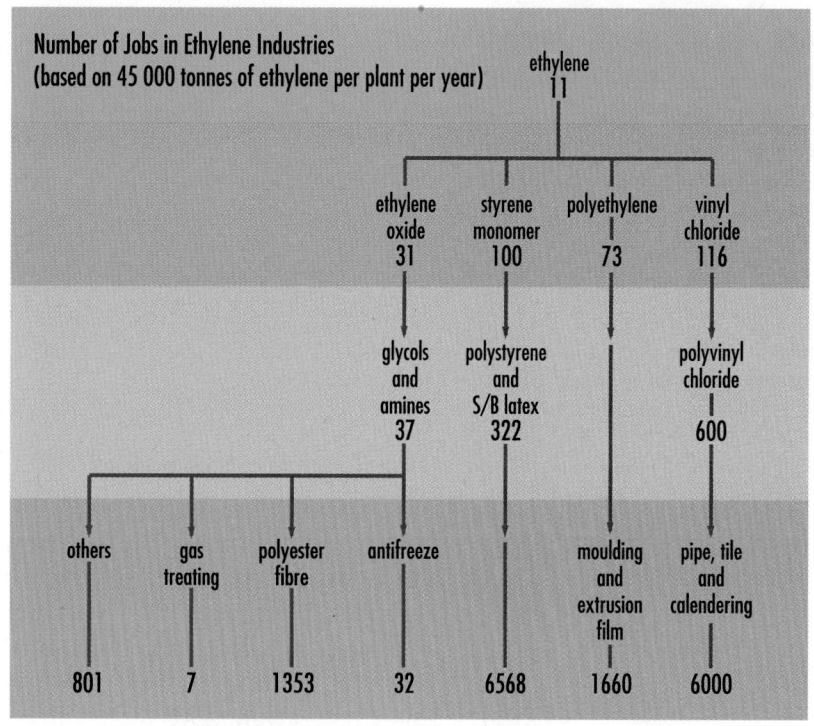

Number of Jobs in Ethylene Industries
(based on 45 000 tonnes of ethylene per plant per year)

ethylene 11

- ethylene oxide 31
- styrene monomer 100
- polyethylene 73
- vinyl chloride 116

- glycols and amines 37
- polystyrene and S/B latex 322
- polyvinyl chloride 600

- others 801
- gas treating 7
- polyester fibre 1353
- antifreeze 32
- moulding and extrusion film 6568
- 1660
- pipe, tile and calendering 6000

In 1982, a 30% decrease in the ozone layer — a decrease called an ozone "hole" — was noticed for the first time by a team of British researchers working in Halley Bay, Antarctica. The British team's results surprised American researchers who had been measuring ozone levels by weather satellite since 1978. American satellite data are transmitted to Earth and are automatically processed by computers before scientists examine them. The Americans had not noticed the decrease in ozone levels because their computers were programmed to reject low measurements as invalid anomalies and to reset these values arbitrarily. The British scientists had also been monitoring atmospheric concentrations of chlorofluorocarbons (CFCs) and they raised the possibility that the decreasing ozone levels and the increasing CFC concentrations in the atmosphere were related. Since 1982, American computers processing total ozone mapping spectrophotometer (TOMS) data no longer reject low values, and alarming depletions of 60% to 70% in ozone levels over Antarctica have been detected.

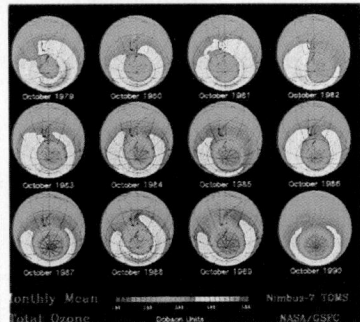

These NASA satellite photos show ozone levels over Antarctica as measured by the ozone-detecting device called TOMS. Look for recent photos on the Internet.

■ Organic Halides
The general formula for an organic halide is R — X.

9.3 HYDROCARBON DERIVATIVES

Organic compounds are divided, for convenience, into two main classes: hydrocarbons and hydrocarbon derivatives. **Hydrocarbon derivatives** are molecular compounds of carbon and at least one other element that is not hydrogen. (See the list of organic compound families in Table 9.1 on page 342.) Most, but not all, hydrocarbon derivatives also contain hydrogen. For ease of classification, such compounds are named as if they had been produced by the modification of a hydrocarbon molecule.

Organic Halides

Organic halides are organic compounds in which one or more hydrogen atoms have been replaced by halogen atoms. These compounds include many common products such as freons (chlorofluorocarbons) used in refrigerators and air conditioners, and Teflon (polytetrafluoroethylene) used in cookware and labware.

Many organic halides are toxic and many are also carcinogenic, so their benefits must be balanced against potential hazards. Two such compounds, the insecticide DDT (dichlorodiphenyltrichloroethane) and the PCBs (polychlorinated biphenyls) used in electrical transformers, have been banned because of public concern about toxicity.

IUPAC nomenclature for halides follows the same format as that for branched-chain hydrocarbons. The branch is named by shortening the halogen name to *fluoro-*, *chloro-*, *bromo-*, or *iodo-*. For example, CH_3Cl is chloromethane and C_2H_5Br is bromoethane.

When translating IUPAC names for organic halides into structural diagrams, draw the parent chain and add branches at locations specified in the name. For example, 1,2-dichloroethane indicates that this compound has a two carbon (eth-), single bonded parent chain (-ane), with one chlorine atom on each carbon (1,2-dichloro-).

$$
\begin{array}{ccc}
Cl & & Cl \\
| & & | \\
H-C & - & C-H \\
| & & | \\
H & & H
\end{array}
$$

1,2-dichloroethane

Organic Reactions of Halides

As illustrated in Figure 9.13 on page 352, reactions of unsaturated hydrocarbons with bromine occur rapidly. Alkenes and alkynes also add hydrogen to their multiple bonds in an addition reaction called hydrogenation (page 352). It seems logical that the addition of halogen or hydrogen halide molecules to the carbons of a double or triple bond would be a common method of preparing halides. Experiment supports this expectation. The rapid rate of these reactions is explained by the idea that no strong covalent bond is broken — the electron rearrangement does not involve separation of the carbon atoms. For example, ethene reacts with chlorine, producing 1,2-dichloroethane.

$$H-\underset{\underset{H}{|}}{C}=\underset{\underset{H}{|}}{C}-H \ + \ Cl-Cl \ \rightarrow \ H-\underset{\underset{H}{|}}{\overset{\overset{H}{|}}{C}}-\underset{\underset{Cl}{|}}{\overset{\overset{H}{|}}{C}}-H$$

$$\text{ethene} \qquad + \qquad \text{chlorine} \qquad \rightarrow \qquad \text{1,2-dichloroethane}$$

The addition of halogens to alkynes results in alkenes or alkanes. For example, in the initial reaction of ethyne with bromine, 1,2-dibromoethene is produced.

$$H-C\equiv C-H \ + \ Br-Br \ \rightarrow \ H-\underset{\underset{Br}{|}}{C}=\underset{\underset{Br}{|}}{C}-H$$

$$\text{ethyne} \qquad + \qquad \text{bromine} \qquad \rightarrow \qquad \text{1,2-dibromoethene}$$

Since addition reactions involving multiple bonds are very rapid, the alkene product, 1,2-dibromoethene, can easily undergo a second addition step to produce 1,1,2,2-tetrabromoethane.

$$H-\underset{\underset{Br}{|}}{C}=\underset{\underset{Br}{|}}{C}-H \ + \ Br-Br \ \rightarrow \ H-\underset{\underset{Br}{|}}{\overset{\overset{Br}{|}}{C}}-\underset{\underset{Br}{|}}{\overset{\overset{Br}{|}}{C}}-H$$

$$\text{1,2-dibromoethene} \quad + \quad \text{bromine} \quad \rightarrow \quad \text{1,1,2,2-tetrabromoethane}$$

The addition of hydrogen halides (HF, HCl, HBr, or HI) to unsaturated compounds will produce isomers, since the hydrogen halide molecules can add in two different orientations.

$$H-\underset{\underset{H}{|}}{C}=\underset{}{C}-\underset{}{C}-H \ + \ H-Cl \ \rightarrow \ H-\underset{\underset{H}{|}}{\overset{\overset{H}{|}}{C}}-\underset{\underset{Cl}{|}}{\overset{\overset{H}{|}}{C}}-\underset{\underset{H}{|}}{\overset{\overset{H}{|}}{C}}-H \ + \ H-\underset{\underset{Cl}{|}}{\overset{\overset{H}{|}}{C}}-\underset{\underset{H}{|}}{\overset{\overset{H}{|}}{C}}-\underset{\underset{H}{|}}{\overset{\overset{H}{|}}{C}}-H$$

$$\text{propene} \qquad + \text{ hydrogen chloride} \rightarrow \quad \text{2-chloropropane} \quad + \quad \text{1-chloropropane}$$

Another reaction that produces halides is a **substitution reaction**, which involves the breaking of a carbon-hydrogen bond in an alkane or aromatic ring and the replacement of the hydrogen atom with another atom or group of atoms. These reactions often occur slowly at room temperature, indicating that very few of the molecular collisions at room temperature are energetic enough to break carbon-hydrogen bonds. Light energy may be necessary for the substitution reaction to proceed at a noticeable rate. Consider the following example, the reaction of propane with bromine vapor.

$$C_3H_{8(g)} \ + \ Br_{2(g)} \ \overset{\text{light}}{\rightarrow} \ C_3H_7Br_{(l)} \ + \ HBr_{(g)}$$

In this reaction, a hydrogen atom of the propane molecule is substituted with a bromine atom. Propane contains hydrogen atoms bonded in two different locations, those on an end-carbon atom and those on the middle-carbon atom, so two different products are formed.

$$
\begin{array}{c}
\text{H}\ \ \text{H}\ \ \text{H} \\
|\ \ \ \ |\ \ \ \ | \\
\text{H}-\text{C}-\text{C}-\text{C}-\text{H}\ +\ \text{Br}-\text{Br}\ \rightarrow\ \text{H}-\text{C}-\text{C}-\text{C}-\text{H}\ +\ \text{H}-\text{C}-\text{C}-\text{C}-\text{H}\ +\ \text{H}-\text{Br} \\
|\ \ \ \ |\ \ \ \ | \\
\text{H}\ \ \text{H}\ \ \text{H}
\end{array}
$$

propane   +   bromine   →   1-bromopropane   +   2-bromopropane   +   hydrogen bromide
                                      (b.p. 71°C)   (b.p. 59°C)

Benzene rings are stable structures and, like alkanes, react slowly with halogens. For example, the reaction of benzene with chlorine produces chlorobenzene and hydrogen chloride. As with alkanes, further substitution can occur in benzene rings until all hydrogen atoms are replaced by halogen atoms.

Evidence Causes Theory
The evidence that the reaction of benzene and chlorine is slow and produces HCl (which turns blue litmus red) causes us to believe that this reaction is a substitution reaction.

benzene   +   chlorine   →   chlorobenzene   +   hydrogen chloride

In another organic reaction known as **elimination**, an alkyl halide reacts with a hydroxide ion to produce an alkene by removing a hydrogen and a halide ion from the molecule. Elimination of alkyl halides is one of the most common methods of preparing alkenes. The following reaction is an example.

$$
\begin{array}{c}
\text{H}\ \ \text{H}\ \ \text{H} \\
|\ \ \ \ |\ \ \ \ | \\
\text{H}-\text{C}-\text{C}-\text{C}-\text{H}\ +\ \text{OH}^-\ \rightarrow\ \text{H}-\text{C}=\text{C}-\text{C}-\text{H}\ +\ \text{H}-\text{O}\ +\ \text{Br}^- \\
|\ \ \ \ |\ \ \ \ | \\
\text{H}\ \ \text{Br}\ \ \text{H}
\end{array}
$$

2-bromopropane   + hydroxide ion →   propene   +   water   +   bromide ion

Table 9.7

HALIDE AND NITRO FUNCTIONAL GROUPS	
—F	fluoro
—Cl	chloro
—Br	bromo
—I	iodo
—NO$_2$	nitro

Percent Yield
The percent yield of isomers such as those of dichlorobenzene can be explained by the relative number of reaction sites available and the relative size of the substituents. For example, explain why 1,3-dichlorobenzene is the highest percent yield of the three isomers.

Exercise

28. The halide and nitro derivatives of hydrocarbons form a wide variety of useful products. Draw structural formulas for the following examples.
    (a) 1,2-dichloro-1,1,2,2-tetrafluoroethane (refrigerant)
    (b) 1,2-dimethylbenzene (solvent)
    (c) 1-methyl-2,4,6-trinitrobenzene (explosive)
    (d) 1,4-dichlorobenzene (moth repellant)
    (e) tetrafluoroethene (Teflon monomer)
    (f) 1,2-dichloroethane (solvent for rubber)
    (g) *cis*-1,2-dichloroethene (solvent for fats)

29. Write IUPAC names for the formulas given.
    (a) CHI$_3$ (antiseptic)
    (b) CH$_2$=C—CH$_2$Cl (insecticide)
                 |
                 CH$_3$
    (c) CH$_2$Cl$_2$ (paint remover)

(d) CH_3NO_2 (rocket fuel)

(e) $CH_2Br — CHBr — CH_2Br$ (soil fumigant)

30. Classify the following as substitution or addition reactions. Predict all possible products for only the initial reaction. Complete the word equation and the structural diagram equation in each case. You need not balance the equations.

(a) trichloromethane + chlorine →

(b) propene + bromine →

(c) ethylene + hydrogen iodide →

(d) ethane + chlorine →

(e) $Cl — C \equiv C — Cl$ + F — F (excess) →

(f)
$$H—\overset{\displaystyle H}{\underset{}{C}}=\overset{\displaystyle H}{\underset{}{C}}-\overset{\displaystyle H}{\underset{\displaystyle H}{C}}-\overset{\displaystyle H}{\underset{\displaystyle H}{C}}-H \quad + \quad H—Cl \quad \rightarrow$$

(g)
  + Cl—Cl →

31. A major use of alkyl halides is in the preparation of unsaturated compounds. Predict all possible initial products of the following elimination reactions. Write word equations and structural diagram equations. Do not balance the equations.

(a)
$$H—\overset{\displaystyle H}{\underset{\displaystyle H}{C}}-\overset{\displaystyle H}{\underset{\displaystyle Cl}{C}}-H \quad + \quad OH^- \quad \rightarrow$$

(b)
$$H—\overset{\displaystyle H}{\underset{\displaystyle H}{C}}-\overset{\displaystyle H}{\underset{\displaystyle Cl}{C}}-\overset{\displaystyle H}{\underset{\displaystyle H}{C}}-\overset{\displaystyle H}{\underset{\displaystyle H}{C}}-H \quad + \quad OH^- \quad \rightarrow$$

32. Classify and write structural formula equations for the following organic reactions. Do not balance the equations.

(a) propane + chlorine →
           1-chloropropane + 2-chloropropane + hydrogen chloride

(b) propene + bromine → 1,2-dibromopropane

(c) benzene + iodine → iodobenzene + hydrogen iodide

(d) 2-butene + hydrogen chloride → 2-chlorobutane

(e) bromobenzene + chlorine →

33. The synthesis of an organic compound typically involves a series of reactions, for example, some substitutions and some additions.

(a) Design an experiment beginning with a hydrocarbon to prepare 1,1,2-trichloroethane.

(b) (Discussion) What experimental complications might arise in attempting the reactions suggested in part (a)?

Alcohol-Blended Fuels

Researchers developed unleaded gasoline in response to contamination of the environment by leaded gasoline. Oil companies began to produce and sell alcohol-blended fuels to improve their unleaded gasoline. A 5% methanol-blended gasoline and a 3% ethanol-blended gasoline are sold in some parts of Canada and a 9.5% methanol/ethanol-blended gasoline is sold in other areas.

Methanol is an efficient fuel and is relatively inexpensive to produce. However, because methanol does not mix well with gasoline and tends to corrode engine parts, ethanol is used as a co-solvent to neutralize the corrosive reaction. Together they provide an additive that burns well, is environmentally friendly, acts as a gas-line antifreeze, and ensures a smooth-running gasoline engine.

Alcohols

Alcohols
The general formula for an alcohol is R—OH

Alcohols have certain characteristic properties that can be explained by the presence of a hydroxyl (–OH) functional group attached to a hydrocarbon chain. Alcohols boil at much higher temperatures than do hydrocarbons of comparable molar mass. Chemists explain that alcohol molecules, because of the –OH functional group, form hydrogen bonds (page 324) and thus liquid alcohols are less volatile. Shorter-chain alcohols are very soluble in water, apparently because they form hydrogen bonds with water molecules.

Because the hydrocarbon portion of the molecule of long-chain alcohols is non-polar, larger alcohols are good solvents for non-polar molecular compounds as well. (See the generalization about solubility, "like dissolves like," on page 322.) Alcohols are frequently used as solvents in organic reactions because they are effective for both polar and non-polar compounds. Alcohols are also used as starting materials in the synthesis of other organic compounds.

Simple alcohols are named from the alkane of the parent chain. The -e is dropped from the end of the alkane name and is replaced with *-ol*. For example, the simplest alcohol, with one carbon atom, has the IUPAC name "methanol." Methanol is sometimes called wood alcohol because it was once made by heating wood shavings in the absence of air. The modern method of preparing methanol combines carbon monoxide and hydrogen at high temperature and pressure in the presence of a catalyst.

$$CO_{(g)} + 2\,H_{2(g)} \rightarrow CH_3OH_{(l)}$$

Methanol
Methanol, sold as methyl hydrate, is used throughout Canada as gasline antifreeze and windshield washer fluid.

Methanol is toxic to humans. Drinking even small amounts of it or inhaling the vapor for prolonged periods can lead to blindness or death.

Ethanol, $C_2H_5OH_{(l)}$, can be prepared by the fermentation of sugars. In the fermentation process, enzymes produced by yeast cells act as catalysts in the breakdown of sugar molecules.

$$C_6H_{12}O_{6(s)} \rightarrow 2\,CO_{2(g)} + 2\,C_2H_5OH_{(l)}$$

Alkanols
Alcohols include subgroups such as alkanols, alkanediols, alkenols, alkenediols, and phenols.

In terms of industrial applications, ethanol is the most important synthetic organic chemical. It is a solvent in lacquers, varnishes, perfumes, and flavorings, and is a raw material in the synthesis of other organic compounds.

When naming alcohols with more than two carbon atoms, the position of the hydroxyl group is indicated. For example, there are two isomers of propanol, C_3H_7OH: 1-propanol is used as a solvent for lacquers and waxes, as a brake fluid, and in the manufacture of propanoic acid; 2-propanol, or isopropanol, is sold as rubbing alcohol and is used to manufacture oils, gums, and acetone. Both isomers of propanol are toxic to humans if taken internally. Alcohols that contain more than one hydroxyl group are called *polyalcohols*; their names indicate the positions of the hydroxyl groups. For example, 1,2-ethanediol (ethylene glycol) is used as antifreeze for car radiators. 1,2,3-propanetriol (glycerine) is a base material in many cosmetics and functions as a moisturizer in foods such as chocolates. Glycerine, also called glycerol, is sold in drugstores.

1°, 2°, and 3° Alcohols
Structural models of alcohols with four or more carbon atoms suggest that three structural types of alcohols exist.
• *primary alcohols*, in which the carbon atom carrying the –OH group is bonded to one other carbon atom, as in $CH_3CH_2CH_2CH_2OH_{(l)}$, 1-butanol.
• *secondary alcohols*, in which the carbon atom carrying the –OH group is bonded to two other carbon atoms, as in $CH_3CHOHCH_2CH_{3(l)}$, 2-butanol.
• *tertiary alcohols*, in which the carbon atom carrying the –OH group is bonded to three other carbon atoms, as in $(CH_3)_3COH_{(l)}$, 2-methyl-2-propanol.

$$\underset{\text{1,2-ethanediol}}{\begin{array}{c} \quad\;\;\text{H}\;\;\;\text{H} \\ \quad\;\;| \quad\; | \\ \text{H}-\text{C}-\text{C}-\text{H} \\ \quad\;\;| \quad\; | \\ \quad\;\;\text{OH}\;\;\text{OH} \end{array}} \qquad \underset{\text{1,2,3-propanetriol}}{\begin{array}{c} \quad\;\;\text{H}\;\;\;\text{H}\;\;\;\text{H} \\ \quad\;\;| \quad\; | \quad\; | \\ \text{H}-\text{C}-\text{C}-\text{C}-\text{H} \\ \quad\;\;| \quad\; | \quad\;| \\ \quad\;\;\text{OH}\;\;\text{OH}\;\;\text{OH} \end{array}}$$

Elimination Reactions

Like organic halides, alcohols undergo elimination reactions to produce alkenes (Figure 9.22). This type of reaction is catalyzed by concentrated sulfuric acid, which removes or eliminates a hydrogen atom and a hydroxyl group, as shown in the following equation.

$$\underset{\text{ethanol}}{\begin{array}{c} \quad\;\;\text{H}\;\;\;\text{H} \\ \quad\;\;| \quad\; | \\ \text{H}-\text{C}-\text{C}-\text{H} \\ \quad\;\;| \quad\; | \\ \quad\;\;\text{H}\;\;\;\text{OH} \end{array}} \xrightarrow{\text{acid}} \underset{\text{ethene}}{\begin{array}{c} \quad\;\;\text{H}\;\;\;\text{H} \\ \quad\;\;| \quad\; | \\ \text{H}-\text{C}=\text{C}-\text{H} \\ \end{array}} + \underset{\text{water}}{\begin{array}{c} \\ \text{H}-\text{O} \\ | \\ \text{H} \end{array}}$$

Ethers

The family of organic compounds known as **ethers** contain an oxygen atom bonded between two hydrocarbon groups, and have the general formula, **R$_1$—O—R$_2$**. Ethers are named by adding *oxy* to the prefix for the smaller hydrocarbon group and joining it to the alkane name of the larger hydrocarbon group. Hence, CH_3—O—C_2H_5 is called *methoxyethane*. In contrast to alcohols, ethers demonstrate low solubility in water, relatively low boiling points, and no evidence of hydrogen bonding. The most well known ether is ethoxyethane, CH_3—CH_2—O—CH_2—CH_3, also known as diethyl ether. A volatile, highly flammable liquid, ethoxyethane was once used extensively as an anesthetic, but is now mainly used as a solvent for fats and oils.

Other than flammability, ethers are relatively unreactive compounds. They undergo chemical change only when treated with powerful reagents under vigorous conditions. Ethers are formed by the condensation reaction of alcohols. A **condensation reaction** is characterized by the joining of two molecules and the elimination of a small molecule, usually water. When concentrated sulfuric acid is used as a catalyst, methoxymethane is formed from methanol, as shown by the following equation.

$$\overset{H^+_{(aq)}}{CH_3OH_{(l)} + CH_3OH_{(l)} \rightarrow CH_3OCH_{3(l)} + HOH_{(l)}}$$

Exercise

34. Write IUPAC names for the following compounds, and determine if any two are isomers.

   (a)$\underset{\quad\;\;\;\;\;\;\;\;\;\;\;\;|}{CH_3-CH-CH_2-CH_3}$
   $\qquad\qquad\;\; OH$

Figure 9.22
An ethylene gas generator produces ethylene, which speeds up the ripening of fruits such as bananas. The reactant, ethanol, undergoes an elimination reaction in the presence of an acid catalyst.

(b) CH_3—CH—CH_2—CH_2—CH_2
　　　　　|　　　　　　　　　|
　　　　　OH　　　　　　　　OH

(c) CH_3—CH_2—O—CH_3

(d) CH_3—O—CH_2—CH_2—CH_3

35. The major disadvantages of using ethoxyethane as an anesthetic are its irritating effects on the respiratory system and the occurrence of post-anesthetic nausea and vomiting. For this reason it has been largely replaced by methoxypropane, which is relatively free of side effects.

　(a) Draw structural formulas of ethoxyethane and methoxypropane, and determine if they are isomers.

　(b) Write an equation to show the formation of ethoxyethane from ethanol.

36. Alcohols can be made by addition reactions. Write IUPAC names or draw structural diagrams to represent each of the following reactions.

　(a) 2-butene + water → 2-butanol

　(b) CH_2 = CH_2　+　H—O—Cl　→　CH_2 — CH_2
　　　　　　　　　　　　　hydrogen　　　　　|　　　　|
　　　　　　　　　　　　　hypochlorite　　OH　　Cl

37. Elimination reactions of alcohols are generally slow, and require an acid catalyst and heating. For each of the following reactions, write names and draw structural diagrams, as required.

　(a) 1-propanol → propene + water

　(b) CH_3 — CH_2 — CH_2 — CH_2 — OH →

38. Only a few of the simpler alcohols are used in combustion reactions. Alcohol-gasoline mixtures, known as gasohol, are the most common examples. Write a balanced chemical equation, using molecular formulas, for the complete combustion of the following alcohols.

　(a) ethanol (in gasohol)

　(b) 2-propanol (in gas-line antifreeze)

39. Classify and write structural formula equations for the following organic reactions. Do not balance the equations.

　(a) ethene + water → ethanol

　(b) 2-butanol → 1-butene + 2-butene + water

　(c) ethoxyethane + oxygen →

　(d) ethene + hypochlorous acid ($HOCl_{(aq)}$) → 2-chloroethanol

　(e) methanol + oxygen →

Aldehydes and Ketones

Evidence indicates that two families of organic compounds, called aldehydes and ketones, contain the **carbonyl** functional group, –CO–. This group consists of a carbon atom with a double covalent bond to an oxygen atom (see the following structural diagram). In **aldehydes**, the carbonyl group is on the terminal carbon atom of a chain. Aldehydes are named by replacing the final *-e* of the name of the corresponding alkane with the suffix *-al*. The simplest aldehydes are methanal, commonly called formaldehyde, and ethanal, commonly called acetaldehyde.

$$\overset{\displaystyle O}{\underset{\displaystyle |}{\underset{\text{carbonyl group}}{-C-}}} \qquad \overset{\displaystyle O}{\underset{\displaystyle |}{\underset{\text{methanal}}{H-C-H}}} \qquad \overset{\displaystyle O}{\underset{\displaystyle |}{\underset{\text{ethanal}}{CH_3-C-H}}}$$

Aldehydes
An aldehyde is represented by the general formula $$\overset{\displaystyle O}{\underset{\displaystyle |}{R(H)-C-H}}$$

The smaller aldehyde molecules have sharp, irritating odors. The larger ones have flowery odors and are diluted to make perfumes. Methanal is a starting material in the manufacture of Bakelite plastics. Its strong, pungent odor was a familiar one in biology laboratories. Now, because its toxicity to humans has been established, biological specimens are no longer left soaking in methanal. Ethanal is used primarily in the synthesis of other organic compounds such as acetic acid.

ETHANOL: DEALING WITH A SOCIAL PROBLEM

Ethanol is the alcohol present in alcoholic beverages. For this purpose, it is obtained by the fermentation of carbohydrates from a wide variety of sources (e.g., rye, corn, grapes). Since ancient times, people have known that drinking fermented fruit juice produces a euphoric state. During the Middle Ages, alchemists discovered that they could concentrate the active ingredient in these fermented juices by distillation, and since then many types of distilled spirits have been developed. The concentration of alcohol in alcoholic beverages is expressed in percent by volume and varies from 3.9% (light beer) to 75% (overproof rum). In Canada, the alcohol content of alcoholic beverages is typically 5.0% for beer, 12% for wine, and 40% for distilled liquor.

Because of its use both as a highly taxed beverage and an important industrial chemical, ethanol is routinely provided to industry in a form that is unfit to drink. Rendering it unfit to drink is accomplished by adding a *denaturant*, a toxic substance such as methanol.

Like other short-chain alcohols, ethanol is poisonous, and a concentration of only 0.40% in the blood can cause death. According to Canadian law, individuals whose blood alcohol levels are greater than 0.08% are considered impaired. Driving while

impaired is a serious threat to the driver, the passengers, and anyone else on the road.

Excessive use of alcohol is a serious social problem in Canada and many other parts of the world. One means of combating alcoholism is a drug, disulfiram, commonly called "antabuse." In the human body, ethanol is converted to ethanal and then to ethanoic acid. Ethanal, an aldehyde, is believed to be the main cause of a hangover following excessive consumption of alcohol. Ethanal can cause headaches, gastric upset, vertigo, and other negative sensations. In the treatment of alcoholics, these unpleasant effects can be turned to advantage by administering the drug disulfiram. The drug has no

effect in the absence of alcohol. However, when a patient consumes alcohol, disulfiram slows the rate at which ethanal is converted to ethanoic acid, and the amount of ethanal in the body increases. This buildup causes symptoms such as heart palpitations, flushed skin, and nausea that can discourage drinking.

People with a strong craving for ethanol have sometimes consumed methanol (methyl hydrate) when their supply of ethanol ran out. Consumption of even small amounts of methanol is life-threatening. In the body, methanol is converted first to methanal (formaldehyde) and then to methanoic (formic) acid which destroys the myelin sheath around nerves, causing blindness and death.

- *Calculate the volume of ethanol in each of the following drinks.*
  (a) 355 mL of beer (5.0% ethanol by volume)
  (b) 150 mL of wine (12% ethanol by volume)
  (c) 45 mL of rum (40% ethanol by volume)
- *Why is industrial ethanol denatured? Include an economic and a societal perspective in your answer.*
- *Draw structural formulas for methanol, methanal, ethanol, ethanal, 2-propanol, and propanone.*

A **ketone** differs from an aldehyde only in the position of the carbonyl (–CO–) group. In a ketone, the carbonyl group can be present anywhere in a carbon chain *except* at the end of the chain (Figure 9.23). This difference in the position of the carbonyl group affects the chemical reactivity of the molecule, and makes it possible to distinguish aldehydes from ketones empirically (Figure 9.24).

A ketone is named by replacing the *-e* ending of the name of the corresponding alkane with *-one*. The simplest ketone is propanone, CH_3COCH_3, commonly known as *acetone*. Acetone is an effective solvent found in many nail polish removers, plastic cements, resins, and varnishes. Supermarkets and hardware stores sell it as a cleaner as well. Because acetone is both volatile and flammable, use it only in well-ventilated areas.

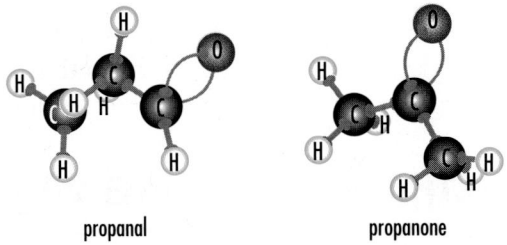

propanal   propanone

Figure 9.23
Aldehydes and ketones containing the same number of carbon atoms are isomers of each other, but have different properties.

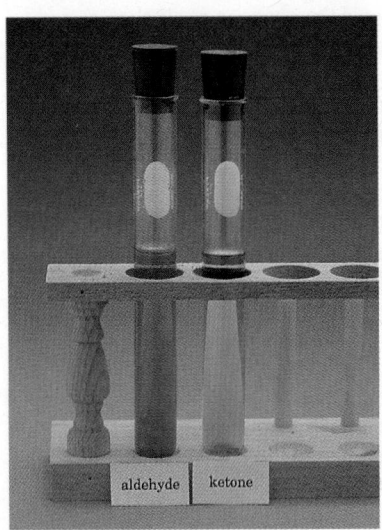

Figure 9.24
In a diagnostic test, Fehling's solution distinguishes aldehydes from ketones. An aldehyde converts the blue copper(II) ion in the Fehling's solution to a red precipitate of copper(I) oxide. A ketone does not react with Fehling's solution.

Carboxylic Acids

The family of organic compounds known as **carboxylic acids** contain the **carboxyl** functional group, –COOH, which includes both the carbonyl and hydroxyl groups.

$$\begin{array}{c} O \\ \parallel \\ -C-OH \end{array}$$
carboxyl group

The characteristic properties of carboxylic acids are explained by the presence of this group. Carboxylic acids are found in citrus fruits, crabapples, rhubarb, and other foods characterized by a sour, tangy taste. Carboxylic acids also have distinctive odors (Figure 9.25).

As one would predict from the structure of carboxylic acids, the molecules of these compounds are polar and form hydrogen bonds both with each other and with water molecules. These acids exhibit the same solubility behavior as alcohols; that is, the smaller members (one to four carbon atoms) of the acid series are miscible with water, whereas larger ones are virtually insoluble. Carboxylic acids have the properties of acids; a litmus test can distinguish these compounds from other hydrocarbon derivatives.

Carboxylic acids are named by replacing the *-e* ending of the corresponding alkane name with *-oic*, followed by the word "acid." The first member of the carboxylic acid family is methanoic acid, HCOOH, commonly called formic acid (Figure 9.26). Methanoic acid is used in removing hair from hides and in coagulating and recycling rubber.

Figure 9.25
Tracking dogs, with their acute sense of smell, are trained to follow odors such as the characteristic blend of carboxylic acids in the sweat from a person's feet.

Ethanoic acid, commonly called acetic acid, is the compound that makes vinegar taste sour. Wine vinegar and cider vinegar are produced naturally when sugar in fruit juices is fermented first to alcohol, then to ethanoic acid. This acid is employed extensively as a textile dye and as a solvent for other organic compounds.

glucose → ethanol → ethanoic acid (acetic acid)

fruit juice → wine → vinegar

Some acids contain two or three carboxyl groups. For example, oxalic acid, used in commercial rust removers and in copper and brass cleaners, consists of two carboxyl groups bonded together. Tartaric acid occurs in grapes, and citric acid in citrus fruits.

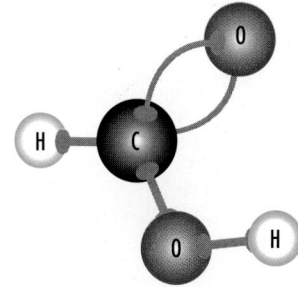

Figure 9.26
Methanoic acid, HCOOH, is the irritating component in the sting of bees and other insects. In fact, the traditional name for this acid, formic acid, is derived from formica, the Latin word for ant.

$$
\begin{array}{ccc}
& & \mathrm{CH_2-COOH} \\
& & | \\
\mathrm{COOH} & \mathrm{HO-CH-COOH} & \mathrm{HO-C-COOH} \\
| & | & | \\
\mathrm{COOH} & \mathrm{HO-CH-COOH} & \mathrm{CH_2-COOH} \\
\text{oxalic acid} & \text{tartaric acid} & \text{citric acid}
\end{array}
$$

Organic Reactions of Carboxylic Acids

Carboxylic acids react as other acids do, in neutralization reactions, for example, and they also undergo a variety of organic reactions. In a **condensation reaction**, a carboxylic acid combines with another reactant, forming two products — an organic compound and a compound such as water. For example, a carboxylic acid can react with an alcohol, forming an ester and water. This condensation reaction is known as **esterification**.

Esters
The general formula for an ester is

$$
\mathrm{R_1(H)-\overset{\displaystyle O}{\overset{\displaystyle \|}{C}}-O-R_2}
$$

where R_2 is the alkyl branch from the alcohol and R_1 (or H) is from the acid.

$$CH_3-\overset{\displaystyle O}{\overset{\|}{C}}-OH \quad + \quad HO-CH_3 \quad \rightarrow \quad CH_3-\overset{\displaystyle O}{\overset{\|}{C}}-O-CH_3 \quad + \quad HOH$$

| carboxylic acid | + | alcohol | → | an ester | + | water |

The **ester** functional group is similar to that of an acid, except that the hydrogen atom of the carboxyl group has been replaced by a hydrocarbon branch. Esters occur naturally in many plants (Figure 9.27) and are responsible for the odors of fruits and flowers. Esters are often added to foods to enhance aroma and taste. Other commercial applications include cosmetics, perfumes, synthetic fibres, and solvents.

The name of an ester has two parts. The first part is the name of the alkyl group from the alcohol used in the esterification reaction. The second part comes from the acid. The ending of the acid name is changed from -*oic acid* to -*oate*. For example, in the reaction of methanol and ethanoic acid represented above, the ester formed is methyl ethanoate, which is used in the manufacture of artificial leather.

The flavor of food is strongly influenced by its odor. Artificial flavorings are made by mixing synthetic esters to give the approximate odor (e.g., raspberry or banana) of the natural substance. For artificial fruit flavors, organic acids are usually added to give the sharp taste characteristic of fruit. Artificial flavors can only approximate the real thing, because it would be too costly to include all the components of the complex mixture of compounds present in the natural fruit or spice. Table 9.8 shows the main ester used to create the odors of certain artificial flavors.

Figure 9.27
Edible oils such as vegetable oils are liquid glycerol esters of unsaturated fatty acids. Fats such as shortening are solid glycerol esters of saturated fatty acids. Adding hydrogen to the double bonds of the unsaturated oil converts the oil to a saturated fat. Most saturated fats are solids at room temperature.

Table 9.8

THE ODORS OF SELECTED ESTERS		
Odor	**Name**	**Formula**
apple	methyl butanoate	$CH_3CH_2CH_2COOCH_3$
apricot	pentyl butanoate	$CH_3CH_2CH_2COOCH_2CH_2CH_2CH_2CH_3$
banana	3-methylbutyl ethanoate	$CH_3COOCH_2CH_2\overset{\displaystyle CH_3}{\overset{\|}{C}HCH_3}$
cherry	ethyl benzoate	$C_6H_5COOC_2H_5$
orange	octyl ethanoate	$CH_3COOCH_2CH_2CH_2CH_2CH_2CH_2CH_2CH_3$
pineapple	ethyl butanoate	$CH_3CH_2CH_2COOCH_2CH_3$
red grape	ethyl heptanoate	$CH_3CH_2CH_2CH_2CH_2CH_2COOCH_2CH_3$
rum	ethyl methanoate	$HCOOCH_2CH_3$
wintergreen	methyl salicylate	(benzene ring with OH group and $-\overset{\displaystyle}{\underset{\displaystyle O}{\overset{\|}{C}}}-O-CH_3$)

40. Write complete structural diagram equations and word equations for the formation of the following esters. Refer to Table 9.8 and identify the odor of each ester formed.

   (a) ethyl methanoate

   (b) ethyl benzoate

   (c) methyl butanoate

   (d) 3-methylbutyl ethanoate

41. Name the following esters, and the acids and alcohols from which they could be prepared. Determine if any two of the esters are isomers.

   (a) $CH_3CH_2COOCH_2CH_3$

   (b) $CH_3CH_2CH_2COOCH_3$

   (c) $HCOOCH_2CH_2CH_2CH_3$

   (d) $CH_3COOCH_2CH_2CH_3$

INVESTIGATION

9.4 Synthesizing an Ester

The purpose of this investigation is to synthesize and observe the properties of two esters.

Problem

What are some physical properties of ethyl ethanoate (ethyl acetate) and methyl salicylate?

Experimental Design

The esters are produced by the reaction of appropriate alcohols and acids, using sulfuric acid as a catalyst. The solubility and the odor of the esters are observed.

Materials

lab apron
safety glasses
dropper bottles of ethanol, methanol, glacial ethanoic (acetic) acid, and concentrated sulfuric acid
vial of salicylic acid (2-hydroxybenzoic acid)
(2) 25 × 250 mm test tubes
250 mL beaker or styrofoam cup
(2) 50 mL beakers
(2) 10 mL graduated cylinders
laboratory scoop
balance
hot plate or hot tap water
thermometer
ring stand with test tube clamp

- Problem
- Prediction
- Design
- Materials
- Procedure
- ✔ Evidence
- ✔ Analysis
- Evaluation
- Synthesis

 **CAUTION**

Both ethanoic and sulfuric acids are dangerously corrosive. Protect your eyes, and do not allow the acids to come into contact with skin, clothes, or lab desks. Both methanol and ethanol are flammable; do not use near an open flame.

 CAUTION

Excessive inhalation of the products may cause headaches or dizziness. Use your hand to waft the odor from the beaker toward your nose. The laboratory should be well ventilated.

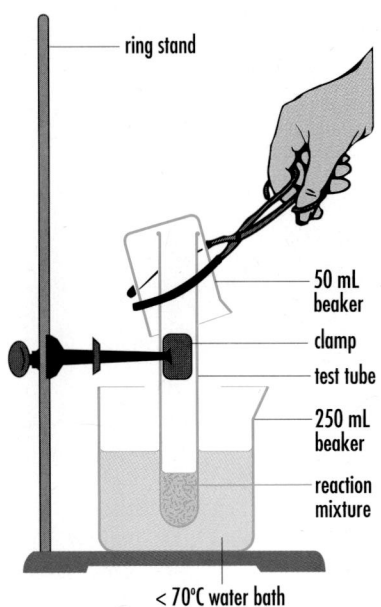

Figure 9.28
Set-up for Investigation 9.4.

ring stand
50 mL beaker
clamp
test tube
250 mL beaker
reaction mixture
< 70°C water bath

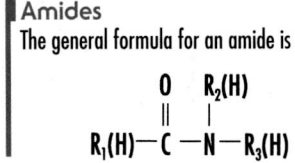

Amides
The general formula for an amide is

$$\underset{\text{R}_1(\text{H})}{}-\overset{\overset{\text{O}}{\parallel}}{\text{C}}-\overset{\overset{\text{R}_2(\text{H})}{|}}{\text{N}}-\text{R}_3(\text{H})$$

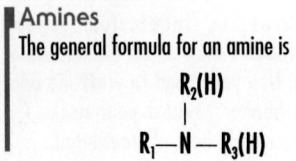

Amines
The general formula for an amine is

$$\text{R}_1-\overset{\overset{\text{R}_2(\text{H})}{|}}{\text{N}}-\text{R}_3(\text{H})$$

Procedure

1. Add about 5 mL of ethanol and 6 mL of ethanoic acid to one of the test tubes.
2. Have your teacher add 8 to 10 drops of concentrated sulfuric acid to the mixture. (The acid will act as a catalyst.)
3. Set up a hot water bath using the 250 mL beaker. (The temperature of the water should not exceed 70°C.)
4. Clamp the test tube so that it is immersed in hot water to the depth of the mixture.
5. As a safety precaution to block any eruption of the volatile mixture, invert a 50 mL beaker above the end of the test tube and heat the mixture for about 10 min (Figure 9.28).
6. After heating the mixture, rinse the 50 mL beaker with cold tap water and add about 30 mL of cold water to the beaker.
7. Cool the test tube with cold tap water and pour the contents of the test tube into the cold water in the beaker. Observe and smell the mixture carefully, using the correct technique for smelling chemicals.
8. Repeat steps 1 to 7, using 3.0 g of salicylic acid, 10 mL of methanol, and 20 drops of sulfuric acid.

Amides

An **amide** has a functional group consisting of a carbonyl group bonded to a nitrogen atom. Amides, as well as esters, can be formed in condensation reactions. The following structural diagram equation shows the reaction of ammonia with a carboxylic acid to form an amide and water.

$$\underset{\text{ethanoic acid}}{CH_3-\overset{\overset{O}{\parallel}}{C}-OH} \;+\; \underset{\text{ammonia}}{H-\overset{\overset{H}{|}}{N}-H} \;\rightarrow\; \underset{\text{ethanamide}}{CH_3-\overset{\overset{O}{\parallel}}{C}-NH_2} \;+\; \underset{\text{water}}{HOH}$$

Amide functional groups occur in proteins, the large molecules found in all living organisms. Amide linkages, also called peptide bonds, join amino acids together in proteins.

Amide names consist of the name of the alkane with the same number of carbon atoms, with the final -*e* replaced by the suffix -*amide*. The same name results if you change the suffix of the carboxylic acid reactant from -*oic acid* to -*amide*. For example, the amide shown in the condensation reaction above is called ethanamide, commonly called acetamide.

Amines

Amines consist of one or more hydrocarbon groups bonded to a nitrogen atom. X-ray diffraction analysis reveals that the functional group of an amine is a nitrogen atom bonded by single covalent bonds to one, two, or three carbon atoms. Amines are polar substances that are extremely soluble in water, as they form strong hydrogen bonds both to each other and to water. Many amines have peculiar, disagreeable odors (Figure 9.29).

The names of amines include the names of the alkyl groups attached to the nitrogen atom, followed by the suffix -*amine*; for example, methylamine, $CH_3NH_{2(g)}$, and ethylmethylamine, $C_2H_5NHCH_{3(l)}$.

Amines with one, two, or three hydrocarbon groups attached to the central nitrogen atom are referred to, respectively, as *primary*, *secondary*, and *tertiary*. Chemically, amines are similar to ammonia in forming basic solutions and reacting with strong acids.

Amines are used extensively in the synthesis of medicines. A group of amines found in many plants are called *alkaloids*. Many alkaloids can influence the function of the central nervous systems of animals. For example, the opium alkaloids are a group of over twenty amines found in the residue after the evaporation of the juice from the opium poppy. These alkaloids can cause death resulting from respiratory depression. The opiates are powerful painkillers, but unfortunately they are highly addictive.

1°, 2°, and 3° Amines

H
|
R — N — H
primary amine

R
|
R — N — H
secondary amine

R
|
R — N — R
tertiary amine

Primary or secondary amines react with carboxylic acids in a condensation reaction to produce amides and water.

Figure 9.29
The smell of rotting fish is due partly to amines such as dimethylamine. Amines are products of the decomposition of amino acids by bacteria.

Exercise

42. Draw structural formulas for three isomers of C_3H_9N, and classify them as primary, secondary, or tertiary amines. Write IUPAC names for each isomer.

43. Write structural formula equations to represent the formation of the following amides.
    (a) methanamide
    (b) propanamide

44. Short-chain alcohols, amides, and amines are very soluble in water. Use bonding theory to explain why these compounds are particularly soluble in water.

ORGANIC CHEMISTRY IN OUTER SPACE

For centuries humans have wondered if there is other life in the universe. The presence of organic compounds in outer space indicates that living organisms might exist elsewhere in the universe. During the past two decades the search for organic compounds has intensified through space probes to Earth's nearest neighbors. The search for extraterrestrial organic compounds also includes the study of interstellar gas, the thin wisps of gas between the stars.

This area of space research involves absorption spectroscopy. When visible light or other electromagnetic radiation passes through a gas, certain specific wavelengths are absorbed by the molecules of the gas. When the transmitted radiation is analyzed, it forms an *absorption spectrum*, a continuous spectrum broken by dark lines (see the photograph below). The dark lines correspond to the wavelengths absorbed by the gas through which the radiation passed.

Since every compound absorbs a unique range of wavelengths,

absorption spectra can be used to determine the chemical composition of the gas through which the radiation travelled. On average, the density of matter in interstellar space is about 10^{-24} g/cm^3, but over vast distances there is enough gas to absorb detectable amounts of light. Like human fingerprints, the absorption spectra of atoms and molecules identify them precisely. The identity of

molecules in interstellar space is determined from the absorption lines these molecules produce in the spectra of distant stars. More than 40 kinds of molecules have been detected in interstellar space. These include methanal (HCHO), methanoic acid (HCOOH), methanol (CH$_3$OH), methanamide (NH$_2$CHO), propyne (CH$_3$CCH), and ethanol (C$_2$H$_5$OH).

Knowledge of the composition of interstellar matter is essential to understanding the formation of stars and the origin of life in the universe. The simple molecules that make up the complex substances in living things exist in the space between the stars. Perhaps the first life forms on Earth developed from these simple molecules. While scientists know that space contains a host of organic compounds, evidence that space contains the complex chemistry associated with life is inconclusive.

• *Describe absorption spectroscopy.*
• *Draw structural diagrams for five organic compounds that have been detected in interstellar space.*

Exercise

45. Prepare a table with the headings: Family, General Formula, and Naming System. Complete the table for the following organic families.
    (a) alcohols
    (b) ethers
    (c) aldehydes
    (d) ketones
    (e) carboxylic acids
    (f) esters
    (g) amines
    (h) amides

46. Many organic compounds have more than one functional group in a molecule. Copy the following structural diagrams. Circle and label the functional groups: hydroxyl, carboxyl, carbonyl, ester, amine, and amide. Suggest either a source or a use for each of these substances.

(a)

vanillin

(b)

acetylsalicylic acid (ASA)

(c)

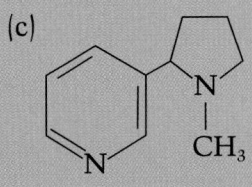
nicotine

(d)

$$H_2N - \overset{\overset{\displaystyle O}{\|}}{C} - NH_2$$
urea

(e)

$$HO - \overset{\overset{\displaystyle O}{\|}}{C} - CH_2 - \overset{\overset{\displaystyle OH}{|}}{\underset{\underset{\displaystyle OH}{|}}{\underset{\displaystyle C=O}{C}}} - CH_2 - \overset{\overset{\displaystyle O}{\|}}{C} - OH$$
citric acid

Hydrogen Bonding
Evidence from boiling points and solubility indicates that the following classes of organic compounds exhibit hydrogen bonding.

alcohols	R—OH
carboxylic acids	R(H)—COOH
amides	R(H)—CONR(H)$_2$
amines	R—NR(H)$_2$

47. Carboxylic acids, like inorganic acids, can be neutralized by bases. However, carboxylic acids also undergo organic reactions. Classify the following reactions as neutralization or esterification. Write the complete structural formulas and word equations for the reactions.

(a)

$$CH_3-\overset{\overset{\large O}{\|}}{C}-OH \quad + \quad NaOH \quad \rightarrow$$

(b)

$$CH_3-CH_2-\overset{\overset{\large O}{\|}}{C}-OH \quad + \quad CH_3-OH \quad \rightarrow$$

(c) benzoic acid + potassium hydroxide →

(d) ethanol + methanoic acid →

48. Fats and oils are naturally occurring esters that store chemical energy in plants and animals. Fatty acids, such as octadecanoic acid (also known as stearic acid), typically combine with 1,2,3-propanetriol, known as glycerol, to form fat, a triester. Complete the following chemical equation by predicting the structural diagram for the ester product.

$$\begin{array}{l} CH_2OH \\ \quad | \\ CHOH \\ \quad | \\ CH_2OH \end{array} \quad + \quad CH_3(CH_2)_{16}COOH \quad \rightarrow$$

glycerol + stearic acid

49. Classify the chemicals and write a complete structural diagram equation for each of the following organic reactions. Where possible, classify the reactions and name the chemicals as well.
(a) $C_2H_6 + Br_2 \rightarrow C_2H_5Br + HBr$
(b) $C_3H_6 + Cl_2 \rightarrow C_3H_6Cl_2$
(c) $C_6H_6 + I_2 \rightarrow C_6H_5I + HI$
(d) $CH_3CH_2CH_2CH_2Cl + OH^- \rightarrow CH_3CH_2CHCH_2 + H_2O + Cl^-$
(e) $C_3H_7COOH + CH_3OH \rightarrow C_3H_7COOCH_3 + HOH$
(f) $C_2H_5OH \rightarrow C_2H_4 + H_2O$
(g) $C_6H_5CH_3 + O_2 \rightarrow CO_2 + H_2O$
(h) $C_2H_5OH \rightarrow CH_3CHO \rightarrow CH_3COOH$
(i) $CH_3CHOHCH_3 \rightarrow CH_3COCH_3$
(j) $NH_3 + C_4H_9COOH \rightarrow C_4H_9CONH_2 + H_2O$
(k) $NH_3 + CH_4 \rightarrow CH_3NH_2 + H_2$

Common Organic Compounds

The technological applications of the following chemicals should be memorized as common knowledge.

$CH_{4(g)}$	natural gas
$C_3H_{8(g)}$	barbecue fuel
$C_4H_{10(g)}$	lighter fluid
$C_8H_{18(l)}$	in gasoline
$C_2H_{4(g)}$	polyethylene monomer
$C_2H_{2(g)}$	cutting torch
$C_6H_{6(l)}$	in gasoline
$CH_3OH_{(l)}$	gasoline antifreeze
$C_2H_5OH_{(l)}$	beverage alcohol
$C_2H_4(OH)_{2(l)}$	radiator antifreeze
$HCHO_{(l)}$	specimen preservative
$(CH_3)_2CO_{(l)}$	nail polish remover
$CH_3COOH_{(l)}$	vinegar

Problem 9B Chemical Analysis of an Organic Compound

Complete the Analysis and Evaluation of the investigation report.

Problem

What are the molecular formula and structure of an unknown organic substance?

Experimental Design

A sample of an unknown gas is analyzed using a combustion analyzer, and the mass of a specific volume of the unknown gas at SATP is measured.

9.4 SYNTHESIZING ORGANIC COMPOUNDS

New organic substances are synthesized as part of research or to demonstrate a new type of reaction. Others are synthesized if a chemist needs a compound with specific chemical and physical properties. Large amounts of some synthetic compounds are routinely produced chemically (Figure 9.30) and large amounts of others are extracted from living systems (Figure 9.31).

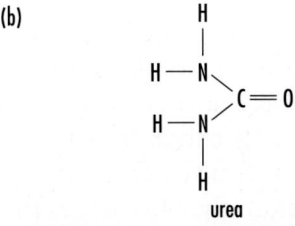

(a)

ASA

(b)

urea

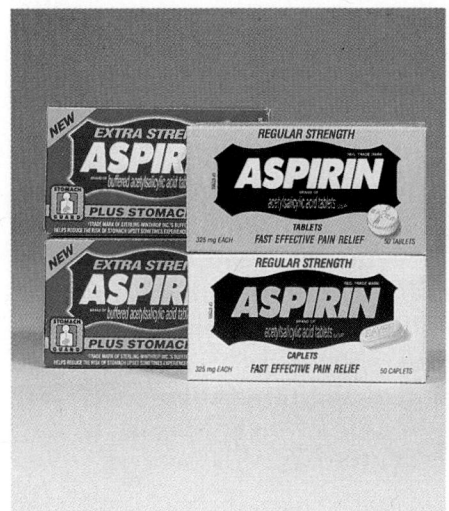

Figure 9.30
(a) More than ten million kilograms of Aspirin, or acetylsalicylic acid (ASA), are produced in North America annually. Because of this compound's ability to reduce pain and inflammation, it is the most prevalent drug in our society.
(b) Urea, the organic compound first synthesized by Wöhler in 1828 (page 337), is produced in even larger quantities than ASA. Urea is used primarily as plant fertilizer and as an animal feed additive.

Polymers

Polymerization is the formation of very large molecules (polymers) from many small molecular units (**monomers**). **Polymers** are substances whose molecules are made up of many similar small molecules linked together in long chains. These compounds have long existed in nature, but were only synthesized by technological processes in the twentieth century. They have molar masses up to millions of grams per mole.

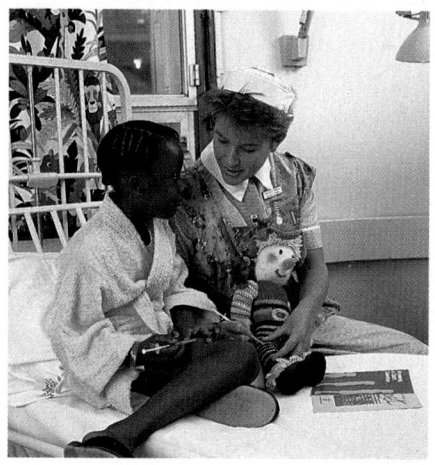

Figure 9.31
(a) Table sugar is extracted from sugar cane or sugar beets.
(b) Insulin for people with diabetes is extracted from the pancreatic fluids of animals such as pigs and is also produced by genetically altered bacteria. To control diabetes, insulin is injected into the bloodstream.

Addition Polymers

Many plastics are produced by the polymerization of alkenes. For example, polyethylene (polyethene) is made by polymerizing ethene molecules in a reaction known as **addition polymerization**. Polyethylene is used to make plastic insulation for wires and containers such as plastic milk bottles, refrigerator dishes, and laboratory wash bottles. Addition polymers are formed when monomer units join each other in a process that involves the rearranging of electrons in double or triple bonds in the monomer. In addition polymerization, the polymer is the only product formed.

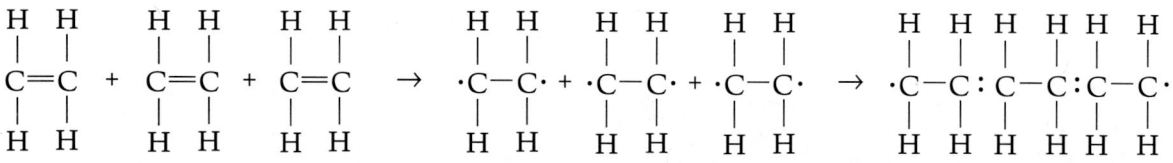

ethylene

part of polyethylene

Using tetrafluoroethene instead of ethene in an addition polymerization reaction produces the substance polytetrafluoroethene, commonly known as Teflon (page 362). Teflon has properties similar to polyethylene, such as a slippery surface and an unreactive nature. But Teflon has a much higher melting point than polyethylene, so it is used to coat cooking utensils. Polypropylene, polyvinyl chloride, Plexiglas, polystyrene, and natural rubber are also addition polymers (Figure 9.32).

Condensation Polymers

Condensation polymerization involves the formation of a small molecule (such as H_2O, NH_3, or HCl) from the functional groups of two different monomer molecules. The small molecule is said to be "condensed out" of the reaction. The monomer molecules bond at the site where atoms are removed from their functional groups. To form a condensation polymer, the monomer molecules must each have at least two functional groups; that is, they must be *bifunctional*.

A *polyester* is a polymer formed from the reaction of a bifunctional acid monomer with a bifunctional alcohol monomer. The resulting ester has a free hydroxyl group at one end and a free carboxyl group at

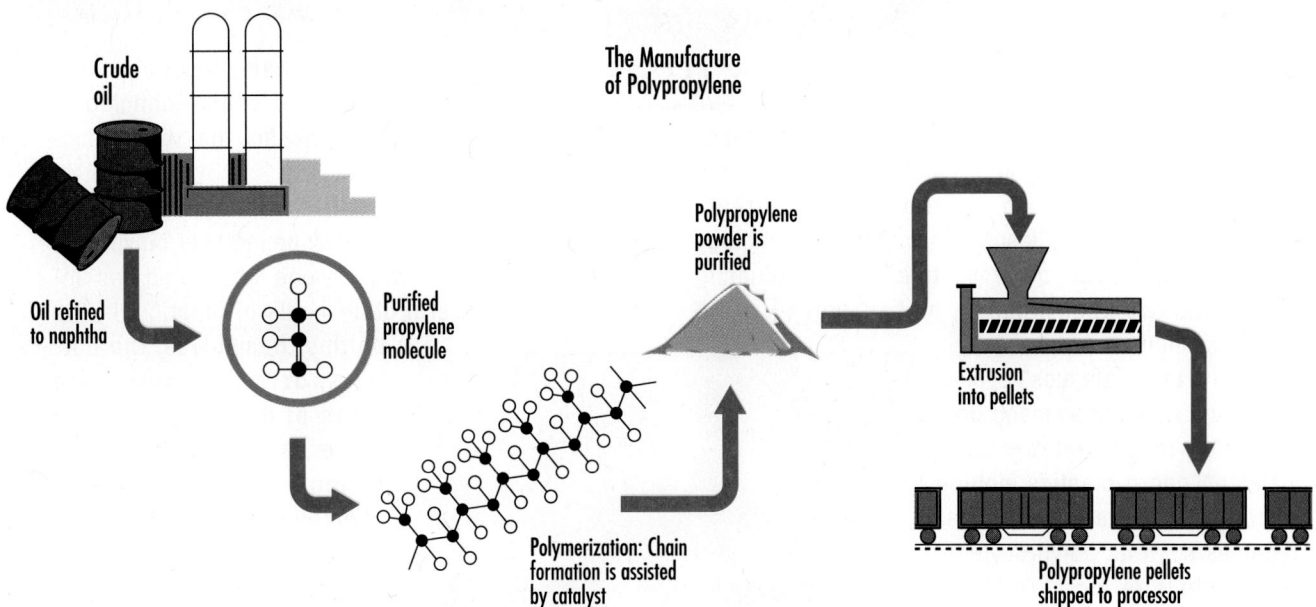

Crude oil

The Manufacture of Polypropylene

Oil refined to naphtha

Purified propylene molecule

Polymerization: Chain formation is assisted by catalyst

Polypropylene powder is purified

Extrusion into pellets

Polypropylene pellets shipped to processor

Figure 9.32
Polypropylene is one of many chemicals derived from crude oil.

the other end. As a result, further reactions can occur at both ends of the ester. The model below represents the production of the polyester known as Dacron, also called Fortrel and Terylene.

1,4-benzenedioic acid + 1,2-ethanediol → part of polyester + water
Dacron

A *polyamide* is a large molecule formed when a bifunctional acid monomer reacts with a bifunctional amine monomer. For example, Kevlar is a polyamide that has a breaking strength considerably greater than that of steel; it is used in cord to reinforce tires and in bulletproof vests. Nylon is also a polyamide.

1,6-hexanedioic acid + 1,6-diaminohexane → part of the Nylon 6-6 polymer chain + water

ORGANIC CHEMISTRY **381**

PLASTICS TECHNOLOGIST

When Tracy DesLaurier graduated from high school in 1981, he was interested in science, computers, and technology, but he had no idea what career to pursue. After a year of working at various "pay-the-rent" jobs, he enrolled in Plastics Engineering Technology at a Canadian technology institute. Since graduating from this two-year program, he has worked in the plastics industry.

The term "plastics" applies to many substances made from a variety of different raw materials. Modern plastics include many materials with familiar names such as polypropylene, polyethylene, polyurethane, polyvinyl chloride (PVC), and polystyrene. All these substances are synthetic polymers made up of chains of smaller molecules.

DesLaurier has worked mainly with a plastic called polyethylene terephthalate (PET). This plastic is used to make soft drink bottles, strapping tape for fastening boxes, "blisters" in commercial packaging, and fibres for carpets, clothing, and insulation.

During the 1980s, about the time when DesLaurier was preparing to re-enter the work force, people were becoming concerned about the amount of waste produced from objects that were used once and thrown away. New technologies for recycling processes — for paper, metals, and many kinds of plastics — were being developed. A resourceful self-starter, DesLaurier was hired as one of five employees in a new company set up to provide an alternative to the annual dumping of millions of pounds of post-consumer PET in Canada. Besides the owner, the company consisted of a mechanical engineer, a millwright (industrial mechanic), a secretary, and two plastics technologists. DesLaurier's job was to establish methods of reclaiming PET from soft drink bottles, and his colleague's job was to establish a method for extrusion of post-consumer PET. Since joining this firm, DesLaurier has been involved in designing the company's plant, specifying equipment, and exchanging information about recycling technologies with other companies in the plastics and waste management industries.

A bale of crushed soft-drink bottles contains more than just PET. There may be metal or polypropylene caps, paper or polyethylene labels, and cap liners made of several other kinds of plastics. In the bottles themselves, the bottoms, or "base cups," are made of polyethylene, and there is also soft drink residue, and dirt. All these contaminants must be separated from the PET for the recycled material to be reusable. The bottles are ground up, the labels and other dry contaminants are removed, and the mixture is washed. Metal and other plastics can be removed using a density separation process, producing pure, clean, post-consumer PET. The flakes of PET are then melted under heat and pressure and manufactured into new products. At his home, for example, DesLaurier has a dense plush carpet made from 40% nylon and 60% recycled PET. Other types of carpet are manufactured using up to 100% recycled PET.

The company DesLaurier works for grew from 6 to 30 individuals within a few years. As the company grew and changed, so did DesLaurier's job. In recent years, he has been spending more time than he would like in airports, as he travels across the continent to confer with other firms involved in plastics recycling.

To become a plastics technologist, you need a background in chemistry and physics and a willingness to continue learning on the job. "The field of plastics technology offers an intellectually rewarding and well-paying career in an expanding field," says DesLaurier.

- *Name five commercial products made from polyethylene terephthalate (PET).*
- *In the recycling process, how are metals and other plastics separated from PET?*

Natural Polymers

Proteins are a basic structural material in plants and animals. Scientists estimate that there are more than ten billion different proteins in Earth's living organisms. Remarkably, all of these proteins are constructed from only about 20 *amino acids*. Amino acids are structurally bifunctional, containing both an amine group ($-NH_2$) and a carboxyl group ($-COOH$). The condensation reaction of the carboxyl group of one molecule with the amine group of another forms amide (peptide) links. This process, repeated from one molecule to the next, produces the very long molecular structures of proteins. The reaction of the amino acids glycine and alanine illustrates the formation of a peptide bond.

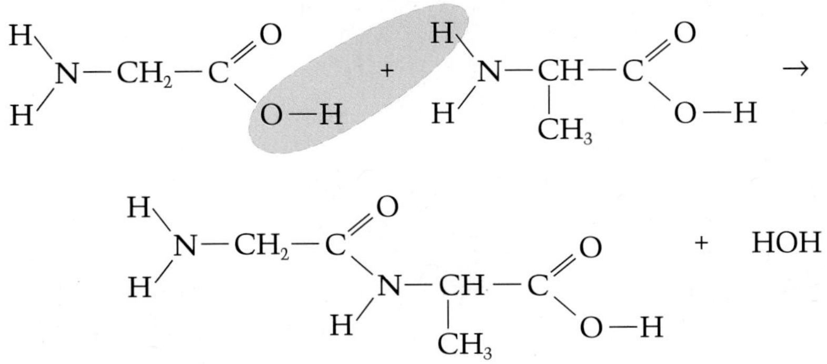

glycine   +   alanine   →   a dipeptide   +   water

Among **carbohydrates** — compounds with the general formula $C_x(H_2O)_y$ — polymerization occurs as well. Simple sugar molecules are the monomers; they undergo a condensation polymerization reaction, in which a water molecule is formed and the monomers join together to form a larger molecule. For example, the sugars glucose and fructose can form sucrose and water.

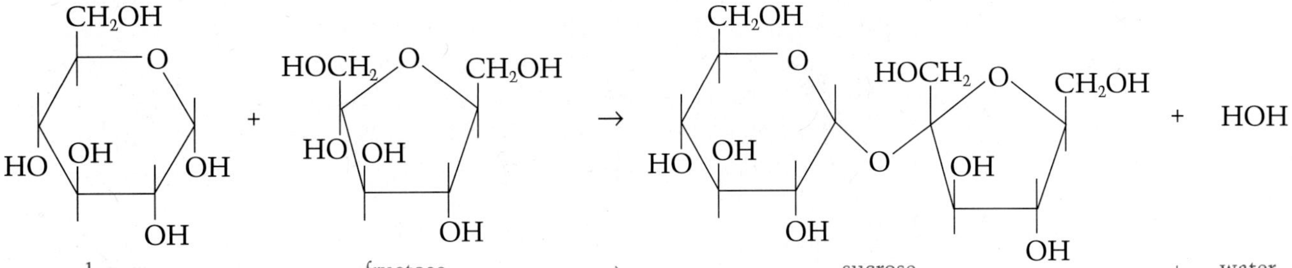

glucose   +   fructose   →   sucrose   +   water

Both starch and cellulose consist of long chains of glucose molecules. The bonds in starch molecules are slightly different from the bonds in cellulose. Humans can digest starch, breaking apart the molecules in the digestive process and releasing glucose, which is then used as a source of energy. Humans cannot digest cellulose (Figure 9.33).

Aspartame

Aspartame, shown below, is formed from the condensation reaction of two amino acids. This substance tastes sweet but supplies little energy. In the sweetener NutraSweet, this compound is added to calorie-reduced food products and carbonated beverages.

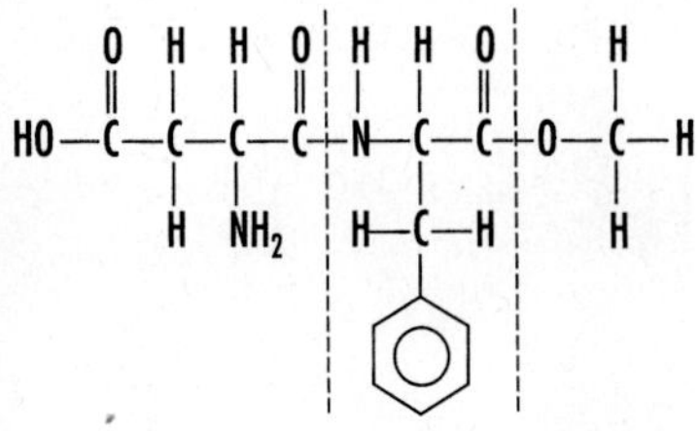

portion from aspartic acid   portion from phenylalanine   portion from methanol

Insulin

In 1922 Frederick Banting and Charles Best showed that, in diabetics, insulin is the chemical that is deficient. Insulin, now known to be a protein consisting of 48 linked amino acids with a molar mass of 5733 g/mol, helps to regulate the body's use of glucose, a carbohydrate.

Figure 9.33
Cellulose occurs in wood and many other plant materials. Although humans cannot digest cellulose, many microorganisms can break it down. As the breakdown occurs, wood "rots."

Figure 9.34
The carbon cycle is a unique illustration of the interrelationship of all living things with the environment — a key connection is the bonding of the carbon atom.

The Carbon Cycle

The chemistry of carbon compounds is interconnected with almost every aspect of our lives, involving science, technology, and social issues in one way or another. Atoms of carbon move throughout the biosphere of our planet in a cycle known as the *carbon cycle* (Figure 9.34).

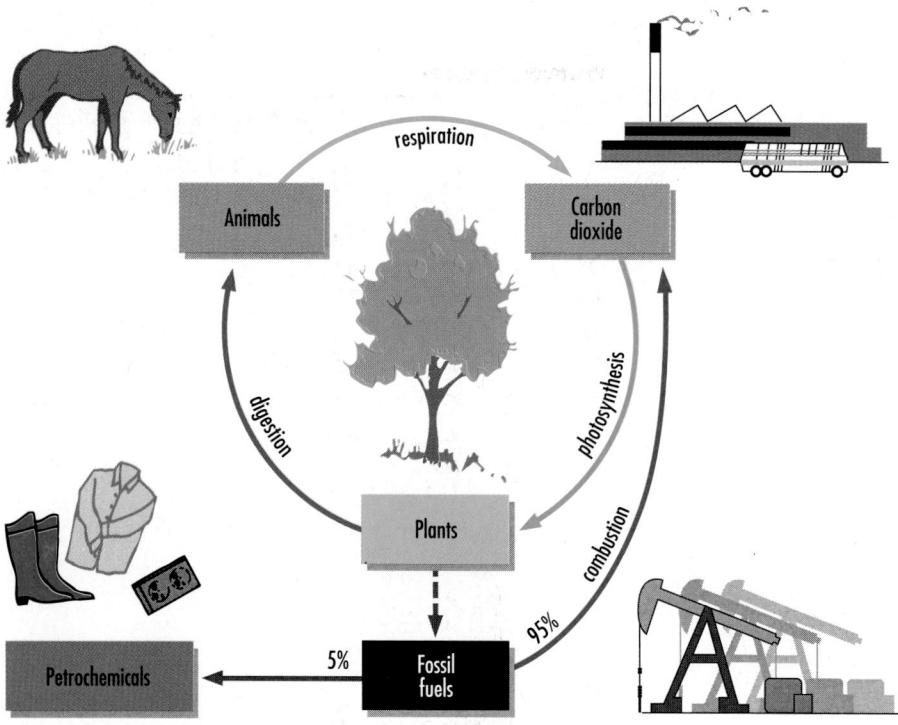

Exercise

50. What is the monomer from which polypropylene is made?

51. What other product results when nylon is made?

52. Suggest a reason why the particular type of nylon shown on page 381 is called "Nylon 6-6."

53. Teflon, made from tetrafluoroethene monomer units, is a polymer that provides a non-stick surface on cooking utensils. Write a structural diagram equation to represent the formation of polytetrafluoroethene.

54. Polyvinyl chloride, or PVC plastic, has numerous applications. Write a structural equation to represent the polymerization of chloroethene (vinyl chloride).

55. Alkyd resins used in paints are polyesters. Using structural diagrams, write a chemical equation to represent the first step in the reaction of 1,2,3-propanetriol with 1,2-benzenedioic acid. Note that many possible structures can form as a result of a three-dimensional growth of the polymer.

56. (Discussion) As with most consumer products, the use of polyethylene has benefits and problems. What are some beneficial uses of polyethylene and what problems result from these uses? Suggest alternative substances for each application.

9.5 Some Properties and Reactions of Organic Compounds

The purpose of this investigation is to observe some properties and reactions of organic compounds.

Problem

What observations can be made about the properties and reactions of some organic compounds?

Experimental Design

Part I

Physical and chemical properties of isomers of $C_4H_{10}O$ are tabulated and compared. Solubilities, melting points, and boiling points of the isomers are found in a reference, such as *The CRC Handbook of Chemistry and Physics* or *The Merck Index*. Evidence for the reaction of the alcohol isomers with a potassium permanganate solution, and with concentrated hydrochloric acid

▨	Problem
▨	Prediction
▨	Design
▨	Materials
▨	Procedure
✔	Evidence
✔	Analysis
✔	Evaluation
▨	Synthesis

◼ MAUD LEONORA MENTEN (1879–1960) ◼

In 50 years of brilliant and intense effort, Maud Menten made discoveries related to enzymes, blood sugar, hemoglobin, kidney function, and the treatment of cancer.

Born in Port Lambton, Ontario, Menten spent much of her early life in British Columbia, attending high school in Chilliwack. She graduated with a bachelor of medicine degree from the University of Toronto in 1907, and four years later she became one of the first Canadian women to receive a medical doctorate. In 1916, Maud Menten earned a Ph.D. in biochemistry from the University of Chicago.

While a graduate student, Menten turned her attention to the study of enzymes, catalysts in living organisms. In collaboration with Leonor Michaelis in Berlin, she studied a reaction called the inversion of sucrose (table sugar). If kept sterile (that is, if kept free of all living organisms), a solution of sucrose remains stable for a long time. However, the addition of a small amount of the enzyme invertase causes a rapid conversion of sucrose to the simpler

sugars glucose and fructose. The invertase acts as a catalyst, so it is not consumed in the reaction.

In 1913, Menten and Michaelis proposed an equation relating the rate of an enzyme-catalyzed reaction to the concentrations of the enzyme and the reacting substance. They developed the theory that an enzyme functions by forming a "complex" made up of the enzyme and the reacting substance. The "complex" then breaks apart into the products

and releases the enzyme molecule for another cycle. The work of Menten and Michaelis unravelled some of the mysteries of enzymes, and established that this type of biologically produced organic compound could be studied in the same ways that simpler chemicals are studied. Although knowledge of enzymes has expanded since 1913, the Michaelis-Menten equation is still considered as important today as when it was first published.

Menten continued her brilliant career at the University of Pittsburgh, retiring at age 71 as head of the pathology department. She had been a respected teacher who demanded excellence from her students. Following retirement, she spent two years doing research in Vancouver before ill health curtailed her efforts. Dr. Menten's accomplishments showed great versatility. In her spare time, she studied astronomy and languages, played the clarinet, and painted. She loved the outdoors, and enjoyed swimming in the Pacific, exploring ocean beaches, and camping and climbing in the Rockies.

solution, are obtained. A teacher demonstration provides evidence for the reaction of the three alcohol compounds with metallic sodium.

Part II

The reactions of cyclohexane and cyclohexene with a basic solution of potassium permanganate are investigated. Carbon compounds containing double and triple bonds react rapidly with bromine (page 352), so it seems reasonable that they should react quickly with other reactive substances such as potassium permanganate.

Part III

The chemical and physical properties of benzoic acid are examined, using sodium hydroxide and hydrochloric acid. Carboxylic acids are acidic because a hydrogen ion is easily removed from the –COOH group. Like inorganic acids, organic acids should undergo simple double replacement reactions. The solubility in water of organic acids should depend on the size of the non-polar part of the molecule; small molecules such as ethanoic (acetic) acid, $CH_3COOH_{(l)}$, are known to be soluble.

Materials

lab apron
safety glasses
safety screen (for teacher demonstration)
overhead projector (optional for teacher demonstration)
sodium metal (for teacher demonstration only)
3 small beakers (for teacher demonstration)
knife (for teacher demonstration)
forceps (for teacher demonstration)
dropper bottle of 1-butanol
dropper bottle of 2-butanol
dropper bottle of 2-methyl-2-propanol
dropper bottle of concentrated (12 mol/L) $HCl_{(aq)}$
dropper bottle of 6.0 mol/L $NaOH_{(aq)}$
dropper bottle of 0.010 mol/L $KMnO_{4(aq)}$
dropper bottle of cyclohexane, $C_6H_{12(l)}$
dropper bottle of cyclohexene, $C_6H_{10(l)}$
vial of benzoic acid, $C_6H_5COOH_{(s)}$
litmus paper
laboratory scoop
stirring rod
6 small test tubes with stoppers
test tube rack
(3) 50 mL beakers

Procedure

Part I *Teacher Demonstration*

1. Prepare the safety screen. Pour approximately 20 mL of 1-butanol, 2-butanol, and 2-methyl-2-propanol into labelled 50 mL beakers.

2. Add a *small* piece of sodium metal to each beaker. Observe and record evidence of reaction.

3. Add water to each beaker *slowly, with stirring,* so that any remaining sodium metal reacts completely.

Part I *Student Investigation*

1. (a) Pour about 2 mL of 1-butanol, 2-butanol, and 2-methyl-2-propanol into labelled test tubes and place them in a test tube rack.
   (b) Add approximately 2 mL of potassium permanganate solution to each of the test tubes.
   (c) Stopper and shake each test tube using the accepted technique for this procedure. Remove the stoppers after shaking the test tubes and replace each test tube in the rack.
   (d) Observe and record any evidence of change that becomes apparent over the next 5 min.

2. (a) Add approximately 2 mL of 1-butanol, 2-butanol, and 2-methyl-2-propanol to labelled 50 mL beakers.
   (b) Add approximately 10 mL of concentrated hydrochloric acid to each beaker.
   (c) Stir *carefully* to mix the contents of the beaker. After 1 min, look for cloudiness in the water layer — an indication of the formation of an alkyl halide with low solubility.

Part II

1. Pour about 1 mL of cyclohexane and cyclohexene into separate labelled test tubes, and place the test tubes in a test tube rack.

2. Add about 2 mL of potassium permanganate solution to each test tube.

3. Add about 2 mL of sodium hydroxide solution to each test tube.

4. Stopper and shake the tubes, and observe any changes. Immediately remove the stoppers. After 5 min, observe any further changes.

Part III

1. Add benzoic acid into a test tube to a depth of about 1 cm.

2. Add about 10 mL of water to the test tube; stopper and shake it. Immediately remove the stopper.

3. Test the liquid with litmus paper.

4. Add sodium hydroxide drop by drop to the test tube, mixing after each drop, until no further change occurs.

5. Add hydrochloric acid drop by drop to the test tube, mixing after each drop, until no further change occurs.

CAUTION

Sodium metal must be used in small pieces. Avoid contact with skin and eyes, as it is a caustic irritant which causes severe burns. Contact with water liberates flammable gases.

1-butanol, 2-butanol, 2-methyl-2-propanol, cyclohexane, and cyclohexene are flammable and produce harmful vapor which may irritate the skin, eyes, and respiratory system. Avoid breathing the vapor.

Concentrated hydrochloric acid and sodium hydroxide are corrosive. Avoid skin and eye contact. If any is splashed in the eyes, rinse for at least 15 min and get medical attention.

DISPOSAL TIP

Pour the contents of the test tubes and the beakers into waste containers marked with the names of the chemicals.

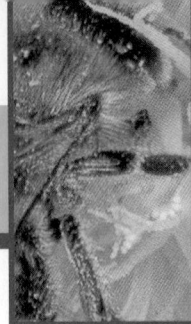

OVERVIEW

Organic Chemistry

Summary

- Organic chemistry is the study of molecular compounds of carbon, including fossil fuels, petrochemicals, and biologically important molecules.

- The huge diversity of organic compounds is explained in terms of the unique ability of carbon atoms to bond to other carbon atoms, and in terms of the number and variety of covalent bonds possible between carbon atoms.

- IUPAC nomenclature for organic compounds is based on functional groups that classify organic substances into families; for example, four hydrocarbon families — alkanes, alkenes, alkynes, and aromatics — and hydrocarbon derivatives — halides, alcohols, aldehydes, ketones, carboxylic acids, esters, amines, and amides.

- Most organic compounds undergo combustion reactions, some of which produce energy for industrialized societies. Other organic reactions, such as addition, substitution, elimination, esterification, and polymerization, are characteristic of different functional groups.

- Very large organic molecules, known as polymers, have important technological applications, such as the manufacture of plastics. Natural polymers such as proteins are biologically important molecules.

Key Words

addition
addition polymerization
alcohol
aldehyde
aliphatic compound
alkane
alkene
alkyl branch
alkyne
amide
amine
aromatic
carbohydrate
carbonyl
carboxyl
carboxylic acid
condensation polymerization
condensation reaction
cracking
cyclic hydrocarbon
cycloalkane
elimination
ester
esterification
ether
fractionation
functional group
hydrocarbon
hydrocarbon derivative
hydrogenation
hydroxyl
isomer
ketone
monomer
organic chemistry
organic halide
phenyl group
polymer
polymerization
protein
refining
reforming
substitution

Review

1. What are two general technological uses of hydrocarbons?

2. Why does carbon form so many different compounds?

3. When space probes are sent to the moon or to the planets, soil samples are collected and analyzed for organic compounds. Why do scientists search for the presence of organic compounds?

4. Hydrocarbons can be classified as saturated or unsaturated. In the context of organic chemistry, what is the meaning of the terms *saturated* and *unsaturated*? Classify the following hydrocarbons as saturated or unsaturated.
   (a) C_5H_{10}
   (b) C_5H_{12}
   (c) C_6H_{10}
   (d) hexane
   (e) heptene

5. Using examples that contain three carbon atoms, draw structural formulas and write IUPAC names for examples of the following organic families. Where isomers exist, draw and name each isomer.
   (a) alcohols
   (b) ethers
   (c) aldehydes
   (d) ketones
   (e) carboxylic acids
   (f) esters
   (g) amines
   (h) amides

6. For each of the following organic families, provide the general formula and some information about a common example.
   (a) alkanes
   (b) alkenes
   (c) alkynes
   (d) aromatics
   (e) organic halides
   (f) alcohols
   (g) ethers
   (h) aldehydes
   (i) ketones
   (j) carboxylic acids
   (k) esters

7. Structural models are used in the study of organic chemistry, since structure is the basis of both nomenclature and the study of reactions. Draw a structural diagram for each of the following compounds. Identify the organic family to which each one belongs.
   (a) 2-methylbutane
   (b) ethylbenzene
   (c) 3-hexyne
   (d) 2,3-dimethyl-2-pentene
   (e) 2-propanol
   (f) methylamine
   (g) methyl ethanoate

8. Write the IUPAC name for each structural diagram shown below. Identify the organic family for each compound.

   (a)
   $$\begin{array}{ccc} Cl & & Cl \\ | & & | \\ CH_2 & \!\!\!\!-\!\!\!\! & CH_2 \end{array}$$

   (b)
   $$\begin{array}{cc} & CH_3 \\ & | \\ CH_2\!=\!C & \!-CH_3 \end{array}$$

   (c)

   (d)
   $$\begin{array}{c} O \\ \| \\ CH_3-C-H \end{array}$$

   (e)
   $$\begin{array}{c} O \\ \| \\ CH_3-C-CH_2-CH_3 \end{array}$$

   (f)
   $$\begin{array}{c} O \\ \| \\ CH_3-C-NH_2 \end{array}$$

9. How do polymer molecules differ from other molecules? Give an example of a polymer in a living system and one in a non-living system.

Applications

10. Draw structural diagrams and write the IUPAC names for all the isomers of $C_2H_2Cl_2$. Predict the polarity of each isomer.

11. Comparing the boiling points of ethers with their alcohol isomers provides an indication of the comparative strength of the intermolecular forces in these compounds.
    (a) Write IUPAC names for each of the following formulas.

Formula	Boiling Point (°C)
CH_3—O—CH_3	−25
CH_3—CH_2—OH	78
CH_3—O—CH_2—CH_3	11
CH_3—CH_2—CH_2—OH	97
CH_3—CH_2—O—CH_2—CH_3	35
CH_3—CH_2—CH_2—CH_2—OH	117

(b) Describe and explain the difference between the boiling points of the ethers and their alcohol isomers.

12. Classify each of the following organic reactions. Write IUPAC names and draw structural diagrams where required for all reactants and products, assuming only a single-step reaction.
   (a) ethane + 2-butene → 3-methylpentane
   (b) 2,4-dimethylhexane + hydrogen →
            butane + methylpropane
   (c) 1-ethyl-2-methylbenzene + oxygen →
   (d) cyclohexene + chlorine →
   (e) C_3H_8 + $CH_3(CH_2)_3CH_3$ →
            $CH_3CHCH_3(CH_2)_4CH_3$ + H_2
   (f) ⬠ + Br—Br →
   (g) $CH_3(CH_2)_2COOH$ + $CH_3(CH_2)_2OH$ →
   (h)
            Cl
            |
   CH_3—CH—CH_3 + OH^- →

13. Classify each of the following reactions. Write IUPAC names and draw structural diagrams for all reactants and products. Assume only a single-step reaction and do not balance the equations.
   (a) 1,2-dibromobenzene + bromine →
   (b) 1-butene + water →
   (c) $CH_3CCCH_2CH_3$ + HI (excess) →
   (d) $CH_3CHCl(CH_2)_3CH_3$ + OH^- →

14. Suggest a reaction or a sequence of reactions to synthesize each of the following compounds. Write chemical equations using structural diagrams.
   (a) ethyl ethanoate (fingernail polish solvent)
   (b) ethanol (common solvent, alcoholic beverages, gasohol)
   (c) propene (monomer for polypropylene plastic)
   (d)
       / F   F \
         | |
      —C—C—   (Teflon polymer)
         | |
       \ F   F /n
   (e) H—C≡C—H (oxyacetylene welding)
   (f)
          Cl
          |
      Cl—C—F   (CFC-12, a refrigerant)
          |
          F

Extensions

15. Chlorofluorocarbons (CFCs) are stable gases that are ordinarily unreactive. What are some benefits of CFCs? If they are unreactive, why is their release into the atmosphere cause for concern? Should governments ban CFCs? Why or why not?

16. A mass spectrometer is used to determine the molar mass of a substance and, often, its identity. When a sample is placed into the spectrometer, high-energy electrons bombard the gas molecules of the sample and break up the molecules into a variety of charged fragments (ions). These charged fragments are then separated according to their masses, and the number of ions is counted. A mass spectrograph, such as the one below, shows the relative numbers of ions with various ion masses. The peak in the graph with the highest mass usually corresponds to a singly charged, but complete, molecule of the sample. The peak at 46 corresponds to $C_2H_5OH^+$. How a molecule breaks up into fragments, such as CH_3^+, is related to its molecular structure. Identify possible fragments for the peaks labelled a, b, c, d, and e.

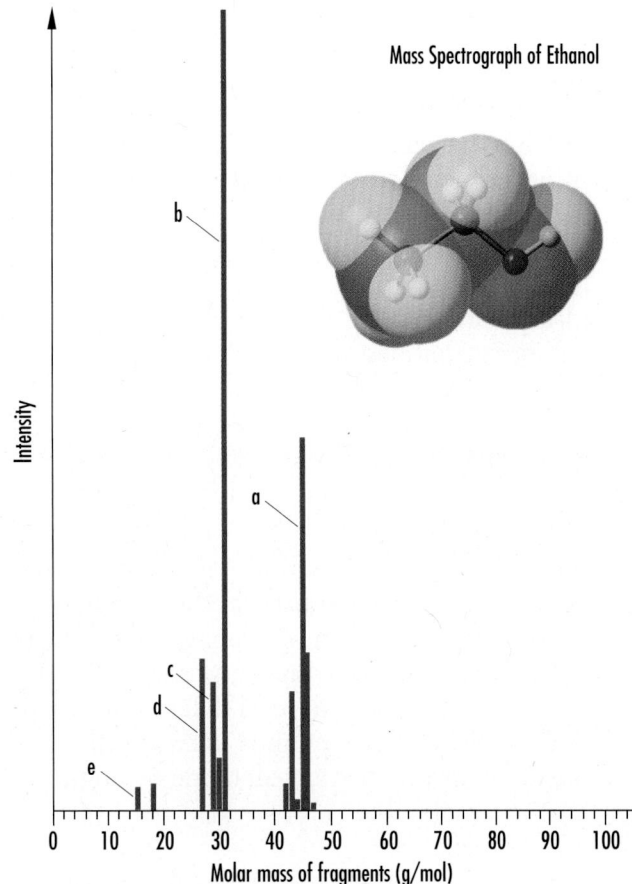

Mass Spectrograph of Ethanol

Intensity

Molar mass of fragments (g/mol)

Problem 9C Determining Structure

Complete the Analysis of the investigation report.

Problem

What are the structure and the IUPAC name of C_3H_4O?

Evidence

molar mass = 56 g/mol
bromine test: color disappeared immediately
Fehling's test: red precipitate formed

Problem 9D Determining Percent Yield

Side reactions and by-products are common for organic reactions. The yield of a desired product is often expressed as a percent. Complete the Analysis of the investigation report.

Problem

What is the percent yield in the initial substitution reaction between methane and chlorine?

Experimental Design

A quantity of methane reacts with chlorine gas. The products of the reaction are separated by condensation of the gaseous products into separate fractions.

Evidence

mass of methane reacted = 1.00 kg
mass of chloromethane produced = 2.46 kg

"People must know the past to

understand the present and

to face the future."

Nellie L. McClung, pioneer author,

Clearing in the West (1935)

UNIT V

ENERGY CHANGES IN CHEMICAL SYSTEMS

Energy transformations are the basis for all activities. The most important energy changes for us are the ones that occur within our own internal chemical factories and keep us alive. Beyond these chemical systems are other essential systems that control the flow of energy in our environment.

In the chemical process called photosynthesis, radiant energy from the sun is stored within the molecules of glucose. This stored chemical energy provides the energy for life. Fossils fuels also contain stored chemical energy waiting to be transformed into other forms of energy. In all chemical systems, energy is classified, described, and explained based on the changes that it undergoes.

Energy is a major factor in social change on our planet. Technologies that consume energy are created for a social purpose, but they often have drawbacks. The control and use of our present sources of energy, as well as the development of new sources, will continue to have far-reaching environmental, economic, social, and political effects for many years.

10 Energy Changes

The term "energy" is used in everyday language in many different ways. Sometimes we say that we have "no energy" to do chores or that an active child is "full of energy." We get "energy" from breakfast cereals. Our society is preoccupied by energy — its availability, management, benefits, and future sources. North Americans consume more than one-quarter of the world's energy output. Because of our preference for disposable products, and our reliance on non-renewable resources, a fundamental change in thinking is needed to alter this statistic. Increasingly, those nations that do not have sufficient energy are demanding a larger share of the world's energy supply.

Scientists define energy as the ability to do work. Energy is classified in terms of its primary sources — chemical, nuclear, solar, and geothermal — and also in terms of its useful forms, such as heat and electricity. Our knowledge of energy derives from the changes it undergoes. Our society depends on heat and on mechanical and electrical energy. However, as the old saying goes, "You can't get something for nothing." These valued forms of energy must come from other energy forms. The source of heat and mechanical and electrical energy for technological uses is almost entirely chemical energy from fossil fuels.

In this chapter, you will study some interchangeable forms of energy, as well as heat and phase changes. The following chapter focuses on chemical energy.

10.1 CLASSIFYING ENERGY CHANGES

Energy is essential to life. Breathing, digestion, the beating of our hearts, and the production of our food, shelter, and clothing all depend on energy. Apart from these essential needs, other energy uses, for example, electrical appliances and motor vehicle transportation, are determined by the society in which we live. Where does this energy come from? An *energy resource* is a natural substance or process that provides a useful form of energy from one of four basic energy sources (Table 10.1). For example, oil and natural gas are two natural resources whose chemical energy we convert into heat and other useful forms of energy. Nuclear energy stored in uranium atoms, for example, is the energy source for nuclear reactors that produce heat and electricity. Both radiant energy from the sun (Figure 10.1) and geothermal energy from the Earth (Figure 10.2) are utilized as energy sources to a limited extent. Unfortunately, these two energy sources are often too variable, produce too little energy, or are too far from populated areas. The most significant solar energy resource is the water in rivers, which is used by hydroelectric power generating stations (Figure 10.3, page 396). The use of flowing water is classified as a solar energy resource because the water cycle in the biosphere is powered by solar energy.

As you learned in previous studies, energy can be converted from one form to another. In fact, almost all of our scientific understanding of energy comes from studying changes in the forms of energy.

"The extra calories [energy] needed for one hour of intense mental effort would be completely met by the eating of one oyster cracker or one half of a salted peanut."
— Francis Benedict, American chemist (1870 – 1957)

Figure 10.1
More than two thousand mirrors in this solar electric generating station reflect sunlight to a boiler at the top of the central tower, where steam is formed to turn electric turbines.

Figure 10.2
Geysers are the result of water heated deep inside the Earth. The water is converted to steam, which escapes through cracks in the Earth's crust. In some countries, such as Iceland and Italy, steam from geysers provides heat for entire towns.

Table 10.1

ENERGY SOURCES, RESOURCES, AND FORMS		
Energy Sources	**Natural Resources**	**Technologically Useful Energy Forms**
• chemical	• fossil fuels, plants	heat,
• nuclear	• uranium, hydrogen	electrical energy,
• solar	• direct radiant energy from sun; wind; water	mechanical energy,
• geothermal	• geysers, hot springs	light, sound

Figure 10.3
British Columbia's climate and geography provide ideal conditions for the production of hydroelectric power. British Columbia has the second largest potential among all provinces for electric power generation.

Exercise

1. List two energy-consuming devices that you use every day that are essential; two that are practical, efficient, or convenient; and two that are non-essential.

2. For each example in question 1, identify the technologically useful form of energy (Table 10.1) that best describes the end use of the energy.

3. For each example in question 1, identify the energy source and the natural resource (Table 10.1) from which the energy was obtained.

4. List three examples of energy-conserving strategies or products that you, or someone you know, could employ.

5. What are some advantages of solar and geothermal energy sources, compared with sources of chemical and nuclear energy?

Heat

Most familiar forms of energy are eventually converted to **thermal energy** — the energy available from a substance as a result of the motion of its molecules. Thermal energy is a by-product of any energy conversion and also the final product after useful work has been done. For example, the chemical energy from the combustion of gasoline in a car engine is converted mostly to thermal energy and partly to energy of motion of the car, which in turn is also converted to thermal energy by frictional forces. The electrical energy used to operate a TV set is converted mostly to thermal energy and to some light and sound energy. The light and sound energy are then absorbed by materials in the surroundings and converted to thermal energy. Humans produce energy from food and lose energy as heat (Figure 10.4).

Thermal energy is directly related to the motion of molecules in a substance. **Heat**, q, is energy transferred between substances. (An object possesses thermal energy but cannot possess heat.) When heat

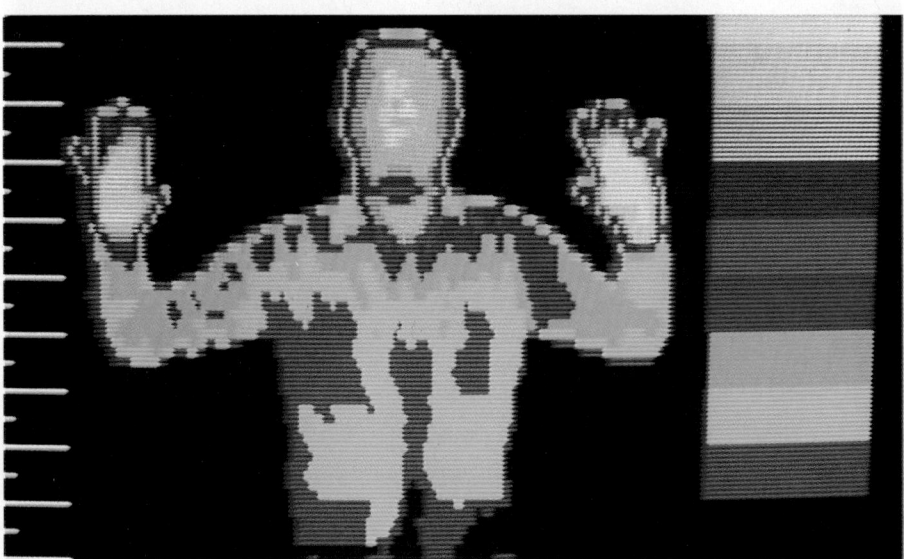

Figure 10.4
As shown by this thermogram, humans radiate heat.

transfers between a system and its surroundings, evidence obtained from measurements of the temperature of the surroundings is used to classify the change as exothermic or endothermic (Figure 10.5). *Exothermic* changes usually involve an increase in the temperature of the surroundings, and *endothermic* changes usually involve a decrease in the temperature of the surroundings. Analyzing this evidence requires a clear understanding of the relationship among heat, thermal energy, and temperature.

According to the kinetic molecular theory (page 136), solids, liquids, and gases are composed of particles that are continually moving and colliding with other particles. The physical properties of the three states of matter are explained by three types of motion. As Table 10.2 and Figure 10.6 show, these are translational (straight line), rotational (spinning), and vibrational (back-and-forth). Any moving object has energy called **kinetic energy**. A moving car, bird, or molecule all have kinetic energy. In all examples, the faster the motion of an object, the greater its kinetic energy. Because the molecules of a substance are always colliding, at any instant some molecules will be moving faster than others. Therefore, in a large group of molecules there will be a range of kinetic energies from very low to very high values.

Figure 10.5

In an exothermic change, energy exits the system, increasing the thermal energy of the surroundings. In an endothermic change, energy enters the system, decreasing the thermal energy of the surroundings.

Table 10.2

EMPIRICAL AND THEORETICAL DESCRIPTIONS OF THE STATES OF MATTER		
State	**Empirical Properties**	**Molecular Motion**
solids	• definite shape and volume • virtually incompressible • do not flow readily	mainly vibrational
liquids	• assume shape of container but have a definite volume • virtually incompressible • flow readily	some vibrational, rotational, and translational
gases	• assume shape and volume of container • highly compressible • flow readily	mainly translational

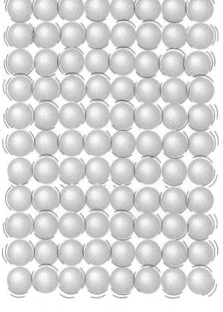

solid
(vibrational)

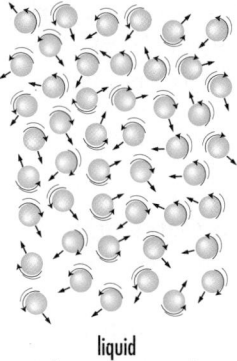

liquid
(vibrational, rotational,
and translational)

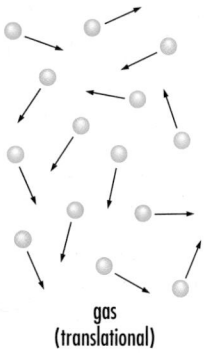

gas
(translational)

Figure 10.6

According to the kinetic molecular theory, the motion of particles is different in solids, liquids, and gases. Particles in solids have primarily vibrational motion; particles in liquids have vibrational, rotational, and translational motion; and particles in gases have mainly translational motion (kinetic energy).

The **temperature** of a substance is a measure of the average kinetic energy of its particles. The thermal energy available from a substance is related to the total kinetic energy of all of the molecules. As long as there are no phase changes, the transfer of heat to a substance causes faster molecular motion increasing both the temperature (average kinetic energy) and the thermal energy (total kinetic energy) of the substance. Therefore, a change in the temperature of the substance, as measured with a thermometer, is explained theoretically as a change in the average kinetic energy ΔE_k of the particles in the substance. The Greek letter Δ pronounced "delta," represents "change in." For example, Δt is translated as "change in temperature" and ΔE_k is translated as "change in kinetic energy." If the temperature of the substance has decreased and there are no phase changes, then heat has flowed out of the substance.

Thus, the temperature change Δt of a substance in a system or its surroundings depends on the quantity of heat q flowing into or out of the substance. The temperature change also depends on the quantity of the substance and on the heat capacity of the substance. The *heat capacity* is the heat required to change the temperature of a unit quantity of the substance; a substance is said to have a large heat capacity when a relatively large quantity of heat must flow to produce a given temperature change. There are two types of heat capacity, defined as follows. (Recall that the SI unit for energy is the **joule, J**.)

- The **specific heat capacity**, c, is the quantity of heat required to raise the temperature of a unit mass (e.g., one gram) of a substance by one degree Celsius. For example, the specific heat capacity of water is 4.19 J/(g•°C).

- The **volumetric heat capacity**, c, is the quantity of heat required to raise the temperature of a unit volume (e.g., one cubic metre) of a substance by one degree Celsius. This quantity is especially useful for liquids and gases. For example, the volumetric heat capacity of air is 1.2 kJ/(m³•°C). Since the density of water at SATP is almost exactly 1 g/mL or 1 kg/L, the values of the specific and volumetric heat capacities for water are numerically the same.

$$c_{H_2O} = 4.19 \ \frac{J}{g•°C} = 4.19 \ \frac{kJ}{L•°C} = 4.19 \ \frac{MJ}{m^3•°C}$$

The quantity of heat q that flows varies directly with the quantity of substance (mass m or volume v), the specific or volumetric heat capacity c, and the temperature change Δt.

$$q = mc\Delta t \ \text{ or } \ q = vc\Delta t$$

Specific and volumetric heat capacities vary for different substances and for different states of matter (Table 10.3). The tables on the inside back cover of this book list values for the three states of water and for some other chemicals. When calculating the quantity of heat that flows into or out of a substance, the heat capacity constant must correspond to the state of matter of the substance and to the measured quantity (mass or volume) of the substance. For example, calculation of the quantity of heat flowing into a measured mass of ice requires the specific heat capacity for $H_2O_{(s)}$, 2.01 J/(g•°C). Cancelling units in your calculation will help ensure that you use the correct constant. In this book, quantities of heat transferred are calculated as absolute values by subtracting the lower temperature from the higher temperature.

EXAMPLE

Many hot water heaters use the combustion of natural gas to heat the water in the tank (Figure 10.7). When 150 L of water at 10°C is heated to 65°C, how much energy flows into the water?

$$q = vc\Delta t$$
$$= 150 \text{ L} \times 4.19 \ \frac{\text{kJ}}{\text{L}\bullet\text{°C}} \times (65 - 10)\text{°C}$$
$$= 35 \text{ MJ}$$

or

$$q = mc\Delta t$$
$$= 150 \text{ kg} \times 4.19 \ \frac{\text{J}}{\text{g}\bullet\text{°C}} \times (65 - 10)\text{°C}$$
$$= 35 \text{ MJ}$$

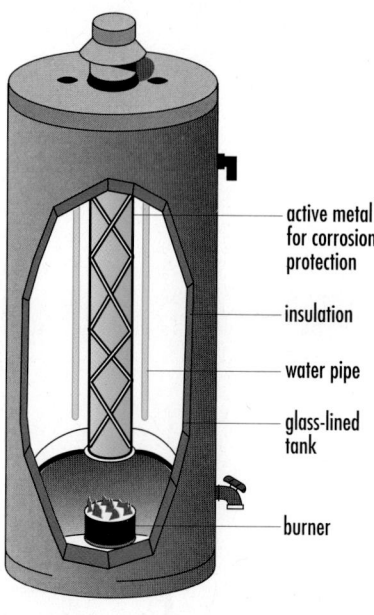

Figure 10.7
The chemical energy from the combustion of natural gas or oil provides the heat that flows into the water in the insulated, glass-lined tank above the burner.

active metal for corrosion protection
insulation
water pipe
glass-lined tank
burner

Exercise

6. Table 10.3 gives the specific heat capacities for ice, water, and steam. For each of these states of matter, describe a situation in which its specific heat capacity is used.

7. Calculate the quantity of heat that flows into 1.50 L of water at 18.0°C that is heated in an electric kettle to 98.7°C.

8. In an industrial plant, 100 kg of steam is heated from 100°C to 210°C. Calculate the quantity of heat that flows into the steam.

9. Some North American native peoples use rocks heated in fire pits to produce steam in "sweat lodges" for purification rites (Figure 10.8). If the specific heat capacity of rock is 0.86 J/(g•°C), what quantity of heat is released by a 2.5 kg rock cooled from 350°C to 15°C?

10. Aqueous ethylene glycol is commonly used in car radiators as an antifreeze and coolant. A 50% ethylene glycol solution in a radiator has a volumetric heat capacity of 3.7 kJ/(L•°C). What volume of this aqueous ethylene glycol is required to absorb 250 kJ for a temperature change of 10°C?

11. Solar energy can preheat cold water for domestic hot water tanks.
    (a) What quantity of heat is obtained from solar energy if 100 L of water is preheated from 10°C to 45°C?
    (b) If natural gas costs 0.351¢/MJ, calculate the money saved if the volume of water in part (a) is heated 1500 times per year.

Table 10.3

SPECIFIC HEAT CAPACITIES FOR H_2O	
ice	$H_2O_{(s)}$ 2.01 J/(g•°C)
water	$H_2O_{(l)}$ 4.19 J/(g•°C)
steam	$H_2O_{(g)}$ 2.01 J/(g•°C)

Figure 10.8
In the sweat lodges of some North American native peoples, rocks are heated to high temperatures before water is sprinkled on them to produce steam.

12. The solar-heated water in question 11 might be heated to the final temperature in a natural gas water heater.
    (a) What quantity of heat flows into 100 L of water heated from 45°C to 70°C?
    (b) At 0.351¢/MJ, what is the cost of heating 100 L of water 1500 times per year?
    (c) (Discussion) Would you invest in a solar energy water preheater if the technology were available in your area?
13. What type of hot water heater is used in your home? Is the natural resource that provides the heat renewable or non-renewable?

■ Problem
✔ Prediction
■ Design
✔ Materials
■ Procedure
✔ Evidence
✔ Analysis
✔ Evaluation
■ Synthesis

CAUTION

Do not use flammable materials to construct your water heater.

INVESTIGATION

10.1 Designing and Evaluating a Water Heater

A water heater is an insulated container with an energy source to heat the water. The insulation is not perfect, so the water tends to cool as heat flows from the container to the surroundings. In this investigation, two criteria, *specific energy* and *cooling rate* are used to evaluate a water heater. Specific energy is the heat that flows into the water (for the temperature change from room temperature to 60°C) per unit mass of the total container. A well-designed water heater has a high specific energy in joules per kilogram. Cooling rate is the heat flowing out of the water (q) per minute. Based on this criterion, a well-designed water heater has a low cooling rate in joules per minute.

Problem

What is the best design for a simple water heater?

Procedure

1. Obtain your teacher's approval of the safety of your design, then construct your water heater, including a measured volume of water at room temperature.

2. Measure the total mass of the water heater (without the energy source).

3. Measure the initial temperature of the water.

4. Heat the water to 60°C while stirring it constantly.

5. Remove the heat source and let the water heater sit for 10 min.

6. Stir well and measure the final temperature of the water.

7. Repeat steps 1 to 6 using either the same design or a modified design.

Enthalpy Changes

When studying energy changes, scientists find it convenient to distinguish between the substance or group of substances undergoing a change, called a *chemical system*, and the system's environment, called the *surroundings*. A chemical reaction in a beaker is an **open system**, since both energy and matter can flow into or out of the system. The surroundings include the surface on which the beaker rests and the air around the beaker. Figure 10.9 shows another example of an open system. A **closed system**, for example, a sealed flask containing reactants, allows the transfer of energy in or out, but it does not allow the transfer of matter. Most important changes in chemistry occur in open systems under constant pressure, the pressure of the atmosphere. The total kinetic and potential energy of a system under constant pressure is called the **enthalpy** of the system. When a system changes, under constant pressure, energy is often absorbed or released to the surroundings and the system is said to undergo an **enthalpy change**. An enthalpy change is given the symbol ΔH, pronounced "delta H." It is impossible to measure an individual enthalpy or the total energy of a system; however, enthalpy changes can be determined from the energy changes of the surroundings (Figure 10.5, page 397). Enthalpy changes occur, for example, during phase changes, chemical reactions, and nuclear reactions (Figure 10.10).

A **phase change** is a change in the state of matter without any change in the chemical composition of the system. When water freezes on a skating rink, or wax melts and drips from a candle, or clouds form in the atmosphere, a phase change is occurring. *Phase changes always involve energy changes but they never involve temperature changes.* In fact, the constant temperature at which a phase change occurs (for example, the melting point or boiling point) is a characteristic physical property of a pure substance (Figure 10.11). A graph of temperature versus time as heat is transferred to a system is called a **heating curve**. When heat is transferred from a system, the temperature-time graph is called a **cooling curve**.

Figure 10.9
The sugar in a marshmallow burns in an open system.

Figure 10.10
A burning candle is an open system and illustrates several enthalpy changes — combustion reaction, and the melting and subsequent solidification of wax.

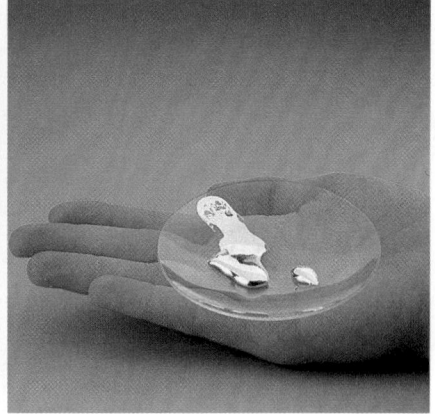

Figure 10.11
The quantity of energy needed to melt gallium metal is low. Its melting point is also so low that the metal melts below body temperature (37°C).

10.2 Heating and Cooling Curves

- Problem
- Prediction
- Design
- Materials
- Procedure
- ✔ Evidence
- ✔ Analysis
- Evaluation
- Synthesis

CAUTION

Handle all heated materials with great care.

Heating curves and cooling curves provide a visual representation of the temperature changes and enthalpy changes that occur when a pure substance is heated or cooled. The purpose of this investigation is to experimentally determine the heating and cooling curves for a pure substance.

Problem

How does the temperature and state of matter change as a pure substance is heated or cooled?

Experimental Design

A sample of a pure solid is heated quickly to liquefy the sample. It is then cooled slowly, and the temperature is recorded at regular time intervals until the sample is completely solidified. The solid is then heated slowly while recording the temperature at regular time intervals. For both the heating and cooling processes, a temperature-time graph is drawn.

Materials

lab apron
safety glasses
stoppered test tube with a pure solid
laboratory burner and striker
wire gauze
(2) ring stands
ring clamp
buret clamp
(2) thermometers
(2) 250 mL beakers
plastic stirring rod
stopwatch or clock with a second hand

Procedure

Cooling the Molten Substance

1. Remove the stopper from the test tube.
2. Set up the ring stand, ring clamp, wire gauze, and laboratory burner.
3. Add about 200 mL of water to a 250 mL beaker, place on the wire gauze and start heating.
4. Use a buret clamp to clamp the test tube in the water.
5. Heat the beaker until the water temperature is about 90°C and the solid is completely melted. Shut off the burner.
6. Place a dry thermometer carefully into the test tube.
7. Add about 200 mL of room temperature water to the second

250 mL beaker and set this on the second ring stand.

8. Carefully loosen the buret clamp from the first stand and transfer the clamp with the test tube to a second ring stand. Lower the test tube into the beaker of water so that the liquid inside the tube is below the level of the water.

9. Start the temperature readings immediately, and continue to take a reading every 30 s thereafter.

10. Between temperature readings, gently stir the liquid sample with the thermometer and stir the water with the stirring rod.

11. Record any other observations along with the temperature and time measurements until the temperature has dropped 10–15°C below the freezing point. (Allow the thermometer to freeze into the solid.)

Heating the Solid

12. Light the burner and heat the water in the beaker left on the first stand to a temperature of about 90°C. Shut off the burner.

13. Loosen the clamp holding the test tube and transfer the test tube into the hot water. Quickly clamp into place.

14. Start the temperature readings immediately, and continue to take a reading every 30 s thereafter.

15. Between temperature readings, gently stir the water with the stirring rod and, when possible, stir the liquid sample with the thermometer.

16. Record any other observations along with the temperature and time measurements until the temperature has risen 10°C to 15°C above the melting point.

17. When the sample is still molten, remove the thermometer and quickly wipe it with a paper towel.

18. Loosen the buret clamp and lift the clamp and test tube out of the hot water. When the test tube has cooled, replace the stopper and return the test tube to its original location.

Phase Changes

Both heating and cooling curves clearly show horizontal sections during phase changes (Figure 10.12). A horizontal section of the graph means that the temperature remains constant in spite of the fact that heat is being removed or added to the substance. Since the temperature remains constant during a phase change, no change in the average kinetic energy of the molecules occurs. According to chemical bonding theory (Chapter 8), energy is required to overcome the forces or bonds that hold particles together. The heat flowing from the surroundings does work to separate the bonded particles. This increases the potential energy E_p of the separated particles. During the opposite phase changes, freezing and condensing, the stored potential energy is released as particles rearrange to form bonds. In the cases of freezing and condensing, the chemical potential energy is released as heat, which flows to the surroundings during the phase change. Theoretically, the enthalpy change during any phase change is explained as a change in the chemical potential energy of the system.

In this book we use enthalpy changes (ΔH) to represent energy changes of a system at constant pressure and at the same initial and final temperatures. We use q to represent heat transferred when there are temperature changes, for example, during heating or cooling of a substance.

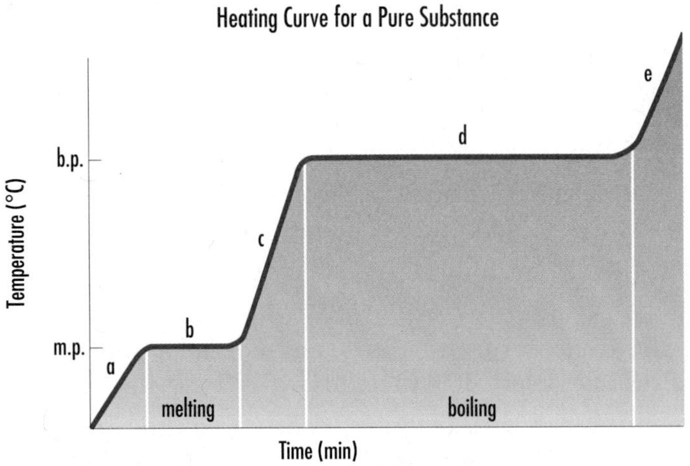

Heating Curve for a Pure Substance
Temperature (°C)
b.p.
m.p.
a
b
c
d
e
melting
boiling
Time (min)

Figure 10.12
As a pure substance is heated, the following changes can occur: **a** *– an increase in the temperature of the solid (q);* **b** *– a solid to liquid phase change (ΔH);* **c** *– an increase in the temperature of the liquid (q);* **d** *– a liquid to gas phase change (ΔH); and* **e** *– an increase in temperature of the gas (q).*

Chemical Changes

In a chemical reaction, there is a change in the composition of the system as reactants are converted to products. Enthalpy changes occur during all the chemical reactions you have seen, such as the combustion of gas in a burner or the precipitation of a solid in a double replacement reaction. In order to control variables and allow comparisons, energy changes of chemical reactions are measured at the same conditions of temperature and pressure, such as SATP, before and after the reaction. Under these conditions, the enthalpy change of a chemical system is the change in the chemical potential energy of the system. All chemical reactions are either exothermic or endothermic. If a reaction is exothermic, energy is released to the surroundings (temperature increases), and, by inference, the chemical potential energy of the system decreases (Figure 10.13). If energy is absorbed by a system from its surroundings (temperature decreases) in an endothermic reaction, then the chemical potential energy of the system increases.

Figure 10.13
A person at rest gives off energy to the environment at about the same rate as a candle. Exothermic reactions reduce the chemical potential energy of the chemical systems (the person and candle); the energy transferred to the surroundings, represented by the arrows, increases the temperature of the surroundings.

Although most chemical changes involve ten to a hundred times more energy than phase changes, a similar theoretical explanation may be used for both types of changes. The greater change in enthalpy is explained by the stronger ionic and covalent bonds involved in chemical changes compared with the intermolecular bonds involved in phase changes. Energy changes during chemical reactions are discussed in greater detail in Chapter 11.

Nuclear Changes

The energy from nuclear reactions is also explained in terms of changes in enthalpy. Nuclear theory suggests that a nuclear reaction is a change in the nucleus of an atom to form a different kind of atom. The energy produced by the sun and by nuclear reactors is the result of nuclear reactions. These exothermic nuclear reactions are also explained in terms of changes in potential energy as bonds break and form, but these bonds are among particles (protons and neutrons) in the nuclei of atoms. Since nuclear reactions produce the largest quantities of energy of any types of reactions, they must involve the strongest bonds. What we know about the theoretical processes in a reaction system is inferred from what we observe in the surroundings. These inferences must be consistent with theories about bonding built from other observations.

According to the law of conservation of energy, for all phase changes, chemical reactions, and nuclear reactions that start and finish at the same conditions, the energy absorbed from the surroundings or released to the surroundings is equal to the change in the potential energy of the system (Table 10.4).

Table 10.4

COMPARISON OF ENERGY CHANGES		
Energy Change	**Empirical Evidence**	**Theoretical Explanation**
flow of heat, q	a temperature change with no change in state or chemicals	ΔE_k as a result of an increase or decrease in the speed of the particles
phase change, ΔH	exothermic or endothermic change forming a new state of matter	ΔE_P as a result of changes in the intermolecular bonds among particles
chemical reaction, ΔH	exothermic or endothermic change forming new chemical substances	ΔE_P as a result of changes in the ionic or covalent bonds among ions or atoms
nuclear reaction, ΔH	exothermic or endothermic change forming new elements or subatomic particles	ΔE_P as a result of changes in the nuclear bonds among nuclear particles (nucleons)

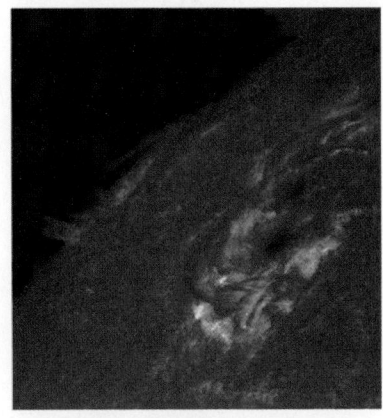

Figure 10.14
As a result of high temperatures, the sun is a swirling ball of charged particles. This fourth state of matter, a gas that has been completely broken down into charged particles, is called a plasma.

10.2 PHASE CHANGES OF A SYSTEM

Three states of matter, solid, liquid, and gas, are familiar to all of us, since we encounter these states in our daily lives (Figure 10.14). These states can be changed from one to another in a phase change (Figure 10.15, page 406). Melting or **fusion**, the change from a solid to a liquid, is endothermic. Also endothermic is boiling or vaporizing, the change from a liquid to a gas. The opposite processes, condensing a gas and freezing or solidifying a liquid, are exothermic. **Sublimation** is a phase change in which a solid changes directly to or from a gas with no intermediate liquid state (Figure 10.16, page 406).

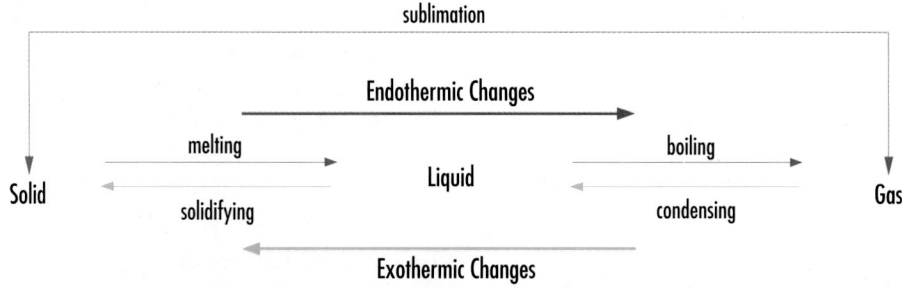

sublimation

Endothermic Changes

melting → boiling →

Solid   Liquid   Gas

solidifying   condensing

Exothermic Changes

Figure 10.15
A solid absorbs energy as it changes to a liquid. A liquid in turn absorbs additional energy to change to a gas. If the direction of the phase changes is reversed, energy is released.

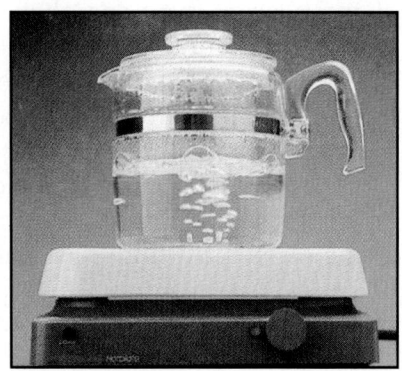

Figure 10.16
Dry ice (solid carbon dioxide) sublimes when heated. The white cloud is formed by water vapor in the air condensing when it is cooled by the invisible CO_2 gas.

Figure 10.17
Moisture condenses and freezes inside greenhouse windows, releasing energy which keeps the inside temperature well above the outside temperature.

All phase changes of pure substances have a specific value for the enthalpy change. For example, one mole (18.02 g) of ice at 0°C absorbs 6.03 kJ of energy as it changes to water. One mole of water at 100°C absorbs 40.8 kJ of energy as it changes to steam. The same quantities of energy per mole are released to the surroundings during the opposite phase changes of freezing and condensing (Figure 10.17). The enthalpy change per mole of a substance undergoing a change is called the **molar enthalpy** and is represented by the symbol H. It is customary to include a subscript on the molar enthalpy change to indicate the type of change occurring; for example, H_{vap} represents the molar enthalpy of vaporization. Molar enthalpy values are obtained empirically and are listed in reference books in tables such as Table 10.5.

Table 10.5

MOLAR ENTHALPIES OF PHASE CHANGES FOR SELECTED SUBSTANCES			
Chemical Name	Formula	Molar Enthalpy of Fusion (kJ/mol)	Molar Enthalpy of Vaporization (kJ/mol)
sodium	Na	2.6	101
chlorine	Cl_2	6.40	20.4
sodium chloride	NaCl	28	171
water	H_2O	6.03	40.8
ammonia	NH_3	–	1.37
freon-12	CCl_2F_2	–	34.99
methanol	CH_3OH	–	39.23
ethylene glycol	$C_2H_4(OH)_2$	–	58.8
sodium sulfate-10-water	$Na_2SO_4 \cdot 10H_2O$	78.0*	–

* This value represents molar enthalpy of solution (see page 414).

The term "greenhouse effect" refers to the heating of the Earth in a process similar to the heating of a greenhouse when the sun shines on it. Greenhouse gases play the role of the glass panels of the greenhouse. The greenhouse effect has made the Earth habitable. Without this effect, the Earth would be much cooler, probably without life as we know it. However, human activities now appear to be increasing the greenhouse effect.

The major greenhouse gases are carbon dioxide, chlorofluorocarbons (CFCs), and methane. Scientists estimate that 50% to 55% of the greenhouse effect is caused by carbon dioxide, 20% to 25% by CFCs, and 20% to 25% by methane. Smaller contributions are made by dinitrogen oxide and other gases. A large proportion of the human-produced carbon dioxide in the atmosphere is the result of combustion processes. All of the CFCs are thought to be a result of human activities. However, an estimated 40% of atmospheric methane comes from natural sources such as swamps, marshes, lakes, and oceans. Sources of methane from human activities include livestock (15%), rice paddies (20%), coal mining (6%), and oil and natural gas production (6%).

Opinions vary as to the rate and probable extent of the warming. Will the warming occur so slowly that people and natural systems can adapt? Or might conditions shift suddenly, wreaking unimaginable havoc? Based on current emissions of greenhouse gases, computer models suggest an increase of 1.5°C in the average temperature of the Earth by the year 2050, and an increase of more than 3.0°C by the year 2100.

Temperature increases of this magnitude are predicted to cause a rise in sea levels of up to one metre over current levels, an extension of frost-free seasons by up to two months at high latitudes, and increased probability of prairie droughts. If the predicted droughts and changes in sea levels occur, much of the best agricultural land in the world will become unproductive and a vast area of populated land will become uninhabitable.

Because the certainty of predictions is low, many people, including some scientists, believe that the threat of the greenhouse effect is minimal. The Earth's temperature has fluctuated in the past, for example, during the ice ages. It is possible that, independent of human interference, the temperature of the Earth is increasing naturally. It is also possible that the Earth is in the midst of a cooling trend and that the human-generated greenhouse effect is preventing another ice age. Models of the atmosphere are complex but inadequate for making precise predictions. The capacity of the oceans to absorb higher levels of carbon dioxide is not known, nor is the effect of an increased concentration of atmospheric carbon dioxide on plant growth understood.

Although the rate and extent of global warming are difficult to predict, it seems reasonable to reduce the production of greenhouse gases to avoid upsetting the delicate balance of the biosphere. We can create technologies to switch from high carbon fuels to low carbon fuels and use conventional fuels more efficiently. We can also practice energy conservation and exploit energy sources that do not produce carbon dioxide, such as solar energy, wind power, fuel cells, and photovoltaic cells.

- *How is the Earth like a greenhouse?*
- *Many people question the predicted temperature increases and have little confidence in the computer models (the authority upon which the predictions are made). Should we just wait and see what happens in the future?*

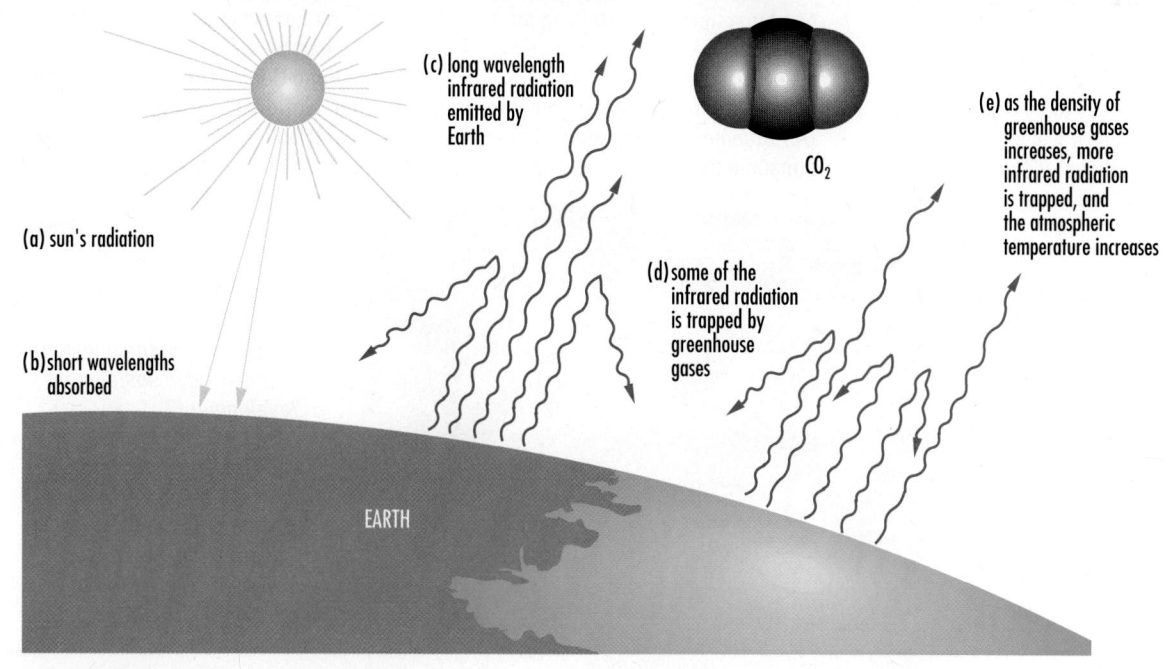

(a) sun's radiation

(b) short wavelengths absorbed

(c) long wavelength infrared radiation emitted by Earth

CO_2

(d) some of the infrared radiation is trapped by greenhouse gases

(e) as the density of greenhouse gases increases, more infrared radiation is trapped, and the atmospheric temperature increases

EARTH

To calculate an enthalpy change, the molar enthalpy value must be obtained from a reference source and the amount in moles of the substance undergoing the phase change must be determined. For example, a freon such as $CCl_2F_{2(g)}$ that is used as a refrigerant is alternately vaporized in tubes inside a refrigerator, absorbing heat, and condensed in tubes outside the refrigerator, releasing heat. This results in energy being transferred from the inside to the outside of the refrigerator. The boiling point of $CCl_2F_{2(l)}$ is –29.8°C and its molar enthalpy of vaporization is 34.99 kJ/mol (Table 10.5). If 500 g of $CCl_2F_{2(l)}$ vaporized at SATP, the expected enthalpy change ΔH_{vap} can be calculated as follows.

$$\Delta H_{vap} = nH_{vap}$$
$$\text{CCl}_2\text{F}_2$$
$$= 500 \text{ g} \times \frac{1 \text{ mol}}{120.91 \text{ g}} \times \frac{34.99 \text{ kJ}}{1 \text{ mol}}$$

$$= 145 \text{ kJ}$$

SUSTAINABLE DEVELOPMENT

Global warming. Acid rain. Depletion of the ozone layer. Destruction of forest lands. Disappearing fish stocks. Problems like these over the past twenty years have triggered a growing realization that the planet is facing an environmental assault on several fronts.

In 1983, the United Nations General Assembly created the World Commission on Environment and Development to examine and report on the crisis. The commission, headed by Dr. Gro Harlem Bruntland of Norway, held hearings and sponsored studies around the world. After five years, the Bruntland Commission released its final report. *Our Common Future* delivered a strong message: people must change their way of thinking if uncontrolled industrial development and environmental degradation — opposite sides of the same coin — are to be checked. All nations, said the commission, must make sustainable development a priority.

What is *sustainable development?* It means development that balances the use of the environment and natural resources with their conservation and renewal. It is development carried out by present generations to meet their needs, while conserving the resources for future genera-

tions to use in meeting their needs. Protecting the environment in which our human activity takes place and from which our resources are drawn is, therefore, essential. Growing consumption of fossil fuels, for example, has been directly linked to increases in atmospheric pollutants as well as to changes in atmospheric chemical cycles (such as those of carbon, nitrogen, and sulfur). It is important to seek alternative, renewable energy resources and to develop their use on a large scale.

The Bruntland Commission emphasized that we all depend on one biosphere for sustaining our lives. Yet each community and each nation often pursues its own activities with little regard for the impact on others. *Our Common Future* noted: "Some consume the Earth's resources at a

rate that would leave little for future generations. Others, many more in number, consume far too little and live with the prospect of hunger, squalor, disease, and early death."

In British Columbia, public demand for sustainable development has generally focused on the province's forests. Years of forestry practices that concentrated on efficient timber extraction had many costs, including loss of wildlife habitats, destruction of fish spawning streams, lack of effective replanting, and damage to landscapes often as a result of poor road-building practices. Today this attitude has changed, prompted in large part by the work of the Bruntland Commission and its promotion of worldwide sustainable development. Government and industry in British Columbia are now working together to ensure that they can sustain timber productivity while protecting other resources for use and enjoyment in the years ahead.

- *What two perspectives are emphasized in sustainable development?*
- *Use the internet or other data base to search for information on sustainable development as it relates to development and the environment on an issue of your choice.*

Exercise

14. Which phase changes are exothermic? Which are endothermic?

15. Calculate the enthalpy change ΔH_{vap} for the vaporization of 100 g of water at 100°C.

16. Ethylene glycol, $C_2H_4(OH)_{2(l)}$, is used in car radiator antifreeze. The melting point of ethylene glycol is –11.5°C, the boiling point is 198°C, and the molar enthalpy of vaporization H_{vap} is given in Table 10.5.
    (a) Sketch a heating curve of ethylene glycol from –50.0°C to 250°C.
    (b) What is the enthalpy change needed to completely vaporize 500 g of ethylene glycol at 198°C with no temperature change occurring?

17. Under certain atmospheric conditions, the temperature of the surrounding air rises as a snowfall begins, because of the energy released to the atmosphere as water changes to snow. What is the enthalpy change ΔH_{fr} for the freezing of 1.00 t of water at 0°C to 1.00 t of snow at 0°C?

18. Geothermal energy is obtained by pumping water down to the hot rock in the mantle of the Earth. The heat in the rock causes the water to boil and the resulting steam is collected. Many places in Canada have an abundant, untapped supply of geothermal energy. Calculate the enthalpy change ΔH_{cond} for the condensation of 100 kg of steam obtained from geothermal wells.

19. During sunny days, chemicals can store solar energy in homes for later release. Certain hydrated salts dissolve in their water of hydration when heated and release heat when they solidify. For example, Glauber's salt, $Na_2SO_4 \cdot 10\,H_2O_{(s)}$, solidifies at 32°C, releasing 78.0 kJ/mol of salt. What is the enthalpy change for the solidification of 1.00 kg of Glauber's salt used to supply energy to a home?

20. What happens to the energy released by an exothermic enthalpy change?

Since an enthalpy change in a chemical system is often observed as a change in temperature of the surroundings, ΔH is sometimes referred to as a heat change rather than an enthalpy change; for example, the terms *heat of vaporization* and *heat of combustion* are sometimes used rather than the more accurate terms, *enthalpy of vaporization* and *enthalpy of combustion*.

Total Energy Changes of a System

In many situations involving a transfer of energy, a system may undergo a sequence of phase and temperature changes. For example, one type of household vaporizer uses electrical energy to heat water to its boiling point and then to convert the water into steam. Measurements of the temperature of the water in a household vaporizer would result in a heating curve similar to the one shown in Figure 10.18.

Suppose 2.5 L (2.5 kg) of water at 12°C is completely changed to steam at 100°C. First, the water at 12°C must be heated to the boiling point. The heat required for this process can be determined from the volume of water, its volumetric heat capacity (page 398), and the temperature change.

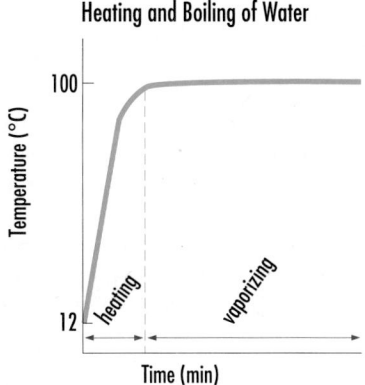

Figure 10.18
The heating curve for water in a vaporizer shows two distinct regions, heating and vaporizing.

$$q = vc\Delta t$$

$$= 2.5 \text{ L} \times 4.19 \ \frac{\text{kJ}}{\text{L}\cdot°\text{C}} \times \overbrace{(100 - 12)}^{88°\text{C}}°\text{C}$$

$$= 0.92 \text{ MJ}$$

or

$$q = mc\Delta t$$

$$= 2.5 \text{ kg} \times 4.19 \ \frac{\text{J}}{\text{g}\cdot°\text{C}} \times \overbrace{(100 - 12)}^{88°\text{C}}°\text{C}$$

$$= 0.92 \text{ MJ}$$

When the water has been heated to the boiling point, rapid vaporization begins. The enthalpy change for the vaporization is calculated from the amount of water and its molar enthalpy of vaporization (Table 10.5, page 406).

$$\Delta H_{vap} = nH_{vap}$$

$$= 2.5 \text{ kg} \times \frac{1 \text{ mol}}{18.02 \text{ g}} \times \frac{40.8 \text{ kJ}}{1 \text{ mol}}$$

$$= 5.7 \text{ MJ}$$

The total energy change of the system equals the heat absorbed by the water as it is heated to the boiling point plus the enthalpy change as a result of the phase change from water to steam.

$$\Delta E_{total} = q + \Delta H_{vap}$$
$$= 0.92 \text{ MJ} + 5.7 \text{ MJ}$$
$$= 6.6 \text{ MJ}$$

EXAMPLE

Calculate the total energy change for 1000 g of molten iron at 1700°C that changes to solid iron at 80°C (Figures 10.19 and 10.20). The specific heat capacity of liquid iron is 0.82 J/(g•°C) and that of solid iron is 0.52 J/(g•°C). The molar enthalpy of solidification for iron is 15 kJ/mol at 1535°C.

$$\Delta E_{total} = \underset{\text{(liquid cooling)}}{q} + \underset{\text{(freezing)}}{\Delta H_{fr}} + \underset{\text{(solid cooling)}}{q}$$

$$= mc\Delta t + nH_{fr} + mc\Delta t$$

$$= 1000 \text{ g} \times 0.82 \ \frac{\text{J}}{\text{g}\cdot°\text{C}} \times (1700 - 1535)°\text{C}$$

$$+ 1000 \text{ g} \times \frac{1 \text{ mol}}{55.85 \text{ g}} \times \frac{15 \text{ kJ}}{1 \text{ mol}}$$

$$+ 1000 \text{ g} \times 0.52 \ \frac{\text{J}}{\text{g}\cdot°\text{C}} \times (1535 - 80)°\text{C}$$

$$= 1.16 \text{ MJ}$$

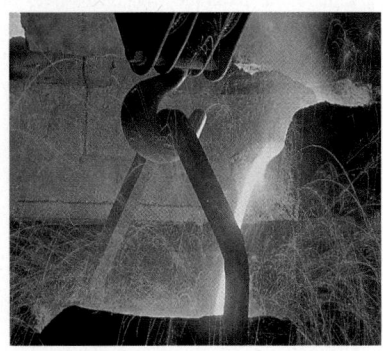

Figure 10.19
Molten iron is poured into a form in which it cools and solidifies, forming a cast-iron object.

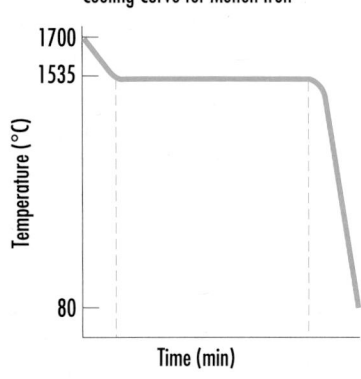

Figure 10.20
A cooling curve for molten iron shows the three distinct regions required for a total energy change calculation.

▌**Using a Calculator**
Determining the total energy change of a system involves careful, multi-step calculations. You should round the result for each stage to the appropriate certainty on paper, but you should also add the full calculator display in each stage to the memory of your calculator. Make sure that all results added to memory have the same units. Usually the certainty of the final result is determined by the least certain measurement, but when calculating totals you should be careful to use both the certainty rule for multiplication and the precision rule for addition and subtraction (Appendix E, E.3).

Exercise

21. On a winter day at –15°C, a camper puts 750 g of snow into a pot over an open fire and heats the water to 37°C (Figure 10.21).
    (a) Sketch a heating curve for the system and indicate the phase or phase change for each section of the graph.
    (b) Describe each part of the heating curve theoretically, by labelling each section with ΔE_k or ΔE_p.
    (c) Label each section with the formulas that would be used to calculate the total energy change of the system. Then list the formulas in order of occurrence.
    (d) Assuming pure water and standard conditions, calculate the total energy needed to change 750 g of snow at –15°C to water at 37°C.
    (e) Survival experts recommend that you do *not* eat snow even if you are stranded without water. Eating snow greatly increases the risk of hypothermia, the lowering of the body's core temperature. Use your calculations from part (d) to provide a scientific basis for this advice.

22. Electrical energy is often produced by using steam to turn turbines connected to electric generators (Figure 10.22). The steam is produced from water, using energy from the chemical combustion of fossil fuels, the nuclear fission of uranium, or other sources.
    (a) Geothermal energy is a renewable energy source that can supply steam to make electricity. Calculate the total energy change when 100 kg of injected water is heated from 10°C to steam at 100°C at standard pressure.
    (b) In a nuclear power plant the thermal energy needed to produce steam is produced by the fission of uranium rather than by the more conventional method of burning fossil fuels. What is the total energy change required to heat 27 t of liquid water from 70°C to 100°C, convert the water to steam, and then heat the steam from 100°C to 260°C?
    (c) (Discussion) Many different technologies, including chemical, nuclear, hydro, geothermal, solar, wind, tidal, and ocean thermal, are used to produce electrical energy. The choice of technology always involves trade-offs of competing values; for example, considerations of safety, cost, and environmental factors. List several kinds of electric power generating plants and list one advantage and one disadvantage of each. You may include a variety of perspectives.

Figure 10.21
Melting snow to produce warm drinking water is much safer than eating snow.

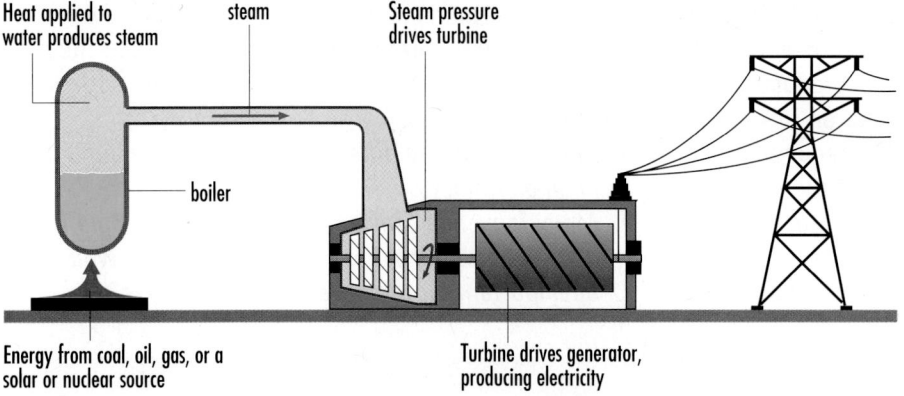

Heat applied to water produces steam

steam

Steam pressure drives turbine

boiler

Energy from coal, oil, gas, or a solar or nuclear source

Turbine drives generator, producing electricity

Figure 10.22
Most electric power generating stations operate on similar principles. An energy source (for example, coal, oil, gas, the sun, or nuclear fission) provides the heat to change the water in the boiler to steam. The steam pressure rotates turbine blades connected to an electric generator. The energy conversion, from chemical or nuclear energy to electrical energy, is only about 30% efficient.

CHEMICAL PROCESS ENGINEER

"To be a chemical engineer you have to be able to visualize things in your head," says Mark Archer, a chemical engineer in an oil refinery. "You have to love problem solving and looking for new ways of doing things." Archer has worked for nine years at a Petro-Canada refinery as a process engineer. He is currently in charge of the fluidized catalytic cracking ("Cat") unit of the refinery. When Archer makes a decision about how to make the cracking process more effective, he has to consider many variables and how they affect one another. By changing one variable (such as temperature or flow rate), he might affect the performance of another variable (for example, pressure). Computers are used to transform the volume of data into information. This information is required to make decisions in optimizing the unit.

In high school, Archer liked the logic and the precision of his science courses. For students considering chemical engineering as a career, Archer recommends studying math (including calculus), chemistry, physics, and computer science. At university, Archer majored in chemical engineering but he selected options in electrical engineering, as well as extra courses in computer science. This diverse background has been a great asset in his career. For a process engineer, understanding how the chemistry is affected by the computerized control of valves and other equipment is a definite plus.

At the oil refinery, Archer works as part of a team. People in the marketing division research the types and quantities of petrochemical products to be sold. This information is then passed on to Archer, who manipulates variables so that the required yields will result. For example, 55% to 60% of the crude oil that is refined is usually converted into gasoline because of the high demand for this product. Archer works closely with an operations coordinator who has the practical know-how to operate the "Cat" unit.

The products of the "Cat" are monitored by laboratory technicians and chemists. There are over 4000 different pieces of data collected throughout the refinery in this monitoring procedure, including temperatures, pressures, flows, and chemical compositions. Some data are collected every minute, some daily, and some weekly. Workers with formal academic training and with practical training are needed to carry out the monitoring. Chemical engineers make decisions about how to optimize the process.

In an oil refinery, men and women are employed in several other units as well, including the crude, isomer, reformer, coker, hydro-treater, butamer, alkylation, and utilities units. The units are managed by a team of engineers, operators, technicians, and maintenance people. Environmental chemists and environmental engineers are also employed at an oil refinery. Liquid and gas emissions are monitored to make sure that environmental standards are met.

After nine years on the job, Archer still enjoys his work. When not on the job, Archer, a husband and father, relaxes by playing golf, doing woodworking projects, and pursuing amateur astronomy. Archer is still looking to the future; he is currently taking university courses toward an MBA (Master of Business Administration) so that he will be able to move into management one day. Archer advises students to think carefully about the courses they choose in order to "position themselves for a changing world."

- *Make a list of the different occupations found in an oil refinery.*
- *Many jobs, like Mark Archer's, require people to work as a team. What skills and attitudes would be important for team members to possess?*

Communication of Enthalpy Constants and Changes

The value of a ΔH depends on the quantity of a substance that undergoes a change. For example, one mole of ice as it melts has an enthalpy change of 6.03 kJ, whereas the enthalpy change for two moles of ice is 12.06 kJ. When enthalpy changes appear in a reference source, they are always reported per unit quantity of the substance, either as a molar enthalpy (in J/mol) or as a specific enthalpy (in J/g). Molar enthalpy constants may represent endothermic or exothermic changes. By convention, *endothermic enthalpy changes are reported as positive values and exothermic enthalpy changes are reported as negative values.* This sign convention is based on whether the system loses energy to the surroundings or gains energy from the surroundings. For example, when ice melts, the system gains energy from the surroundings and the molar enthalpy constant is reported as a positive quantity to indicate an endothermic change. For ice at 0°C,

$$H_{melting} = +6.03 \text{ kJ/mol}$$

The law of conservation of energy implies that the reverse process has an equal and opposite energy change. For liquid water at 0°C,

$$H_{freezing} = -6.03 \text{ kJ/mol}$$

This sign convention is used internationally to communicate molar enthalpy constants in reference sources. Another method of communicating changes in enthalpy is presented below, and two more methods will be discussed in Chapter 11 (page 427).

The sign convention represents the change from the perspective of the chemical system itself, not from that of the surroundings. An increase in the temperature of the surroundings implies a negative change in the enthalpy of the chemical system.

Potential Energy Diagrams

A **potential energy diagram** is a theoretical description of an enthalpy change. The energy transferred during a phase change results from changes in the chemical potential energy of the particles as bonds are broken or formed. Potential energy is stored or released as the positions of the particles change, much like stretching a spring and then releasing it. The enthalpy change measured during any transition from one state to another is described theoretically as an increase or decrease in potential energy ΔE_p.

$$\underset{\text{(molecules)}}{\Delta E_p} = \underset{\text{(system)}}{\Delta H}$$

In potential energy diagrams, an increase in potential energy of the molecules describes an endothermic process in a system (Figure 10.23). A decrease in potential energy of the molecules describes an exothermic process in a system (Figure 10.24). For our purposes, no numbers need be placed on the y-axis; only the change in potential energy (change in enthalpy) of the system is shown in the diagrams.

Since an exothermic change involves a decrease in enthalpy, the direction of this change is communicated as a negative value by $\Delta H < 0$. The direction of an endothermic change is communicated as a positive value by $\Delta H > 0$.

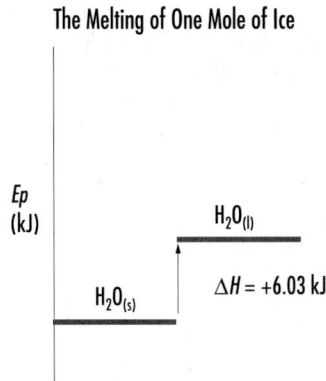

Figure 10.23
A potential energy diagram for the melting of one mole of ice shows an increase in potential energy; this explains the positive sign for ΔH in this endothermic change.

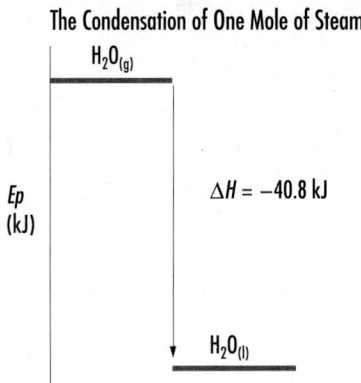

Figure 10.24
A potential energy diagram for the condensation of one mole of steam shows a decrease in potential energy; this explains the negative sign for ΔH in this exothermic change.

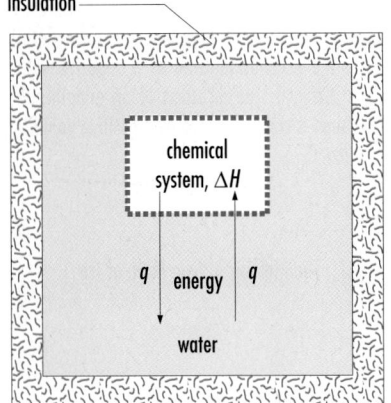

Figure 10.25
The system inside the calorimeter undergoes either a phase change such as fusion, or a chemical change such as a double replacement reaction. Energy is either absorbed from the water or released to the water. An increase in the temperature of the water indicates an exothermic change ($\Delta H < 0$) of the system and a decrease in the temperature of the water indicates an endothermic change ($\Delta H > 0$).

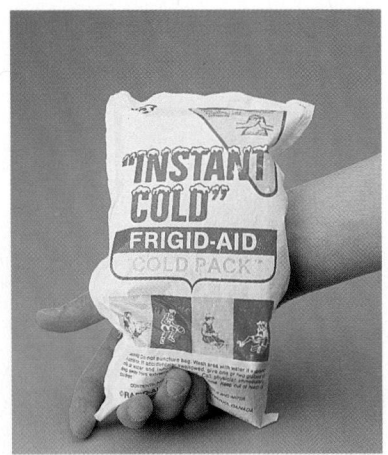

Figure 10.26
Cold packs contain an ionic compound such as ammonium nitrate and a separate pouch of water. To initiate dissolving, the pouch of water is broken by squeezing it.

Exercise

23. Communicate the following molar enthalpies using the conventional symbols of chemistry.
    (a) the molar enthalpy of freezing of $H_2O_{(l)}$
    (b) the molar enthalpy of vaporization of $H_2O_{(l)}$

24. Communicate the freezing and vaporization of water as chemical potential energy diagrams, including the enthalpy change ΔH.

25. The following equation describes the sublimation of ice.

$$H_2O_{(s)} \rightarrow H_2O_{(g)}$$

    (a) What is the value for the molar enthalpy of sublimation of ice?
    (b) Draw a potential energy diagram to describe this sublimation.

26. Using the information in Table 10.5 on page 406, sketch a potential energy diagram, including the value of ΔH, to describe each of the following phase changes.
    (a) the vaporization of 1.00 mol of liquid ammonia
    (b) the condensation of 1.000 mol of freon-12 gas
    (c) the solidification of 1.00 mol of Glauber's salt, sodium sulfate-10-water, from the aqueous state

10.3 CALORIMETRY OF PHYSICAL AND CHEMICAL SYSTEMS

According to the law of conservation of energy, energy is neither created nor destroyed in any physical or chemical change. In other words, energy is only converted from one form to another. Study of energy changes requires an **isolated system**, that is, one in which neither matter nor energy can move in or out. Carefully designed experiments and precise measurements are also needed. **Calorimetry** is the technological process of measuring energy changes using an isolated system called a **calorimeter** (Figure 10.27, page 416). The chemical system being studied is surrounded by a known quantity of water inside the calorimeter. Energy is transferred between the chemical system and the water (Figure 10.25). Water is used because it is readily available and inexpensive and it has one of the highest specific heat capacities. For example, to determine the molar enthalpy of ice melting, the ice may be placed in water inside a calorimeter. As the ice melts, heat transfers from the water to the ice. The total energy gained by the ice is equal to the energy lost by the calorimeter water, as long as both the ice (system) and the water (surroundings) are part of an isolated system. In other words, for measurement to be accurate, no energy may be transferred between the inside of the calorimeter and the environment outside the calorimeter.

The specific and volumetric heat capacities listed on the inside back cover of this book were determined by means of calorimetry. A variety of physical changes, such as phase changes, dissolving, and dilution, can be studied using calorimeters. The dissolving of substances in water may involve noticeable energy changes. The cold packs used in sports to treat sprains and bruises are a practical application of this fact. They contain a chemical that has a very high endothermic molar enthalpy of solution (Figure 10.26). This means that the system absorbs heat from the surroundings, which in this case are the injured tissues of the body. The result is that the injured part of the body feels cold.

Analysis of calorimetric evidence is based on the law of conservation of energy and on several assumptions. The law of conservation of energy may be expressed in several ways, for example, "The total energy change of the chemical system is equal to the total energy change of the calorimeter surroundings." Using this method, both the enthalpy change and the quantity of heat are calculated as absolute values, without a positive or negative sign.

$$\underset{\text{(system)}}{\Delta H} \quad = \quad \underset{\text{(calorimeter)}}{q}$$

The main assumption is that no heat is transferred between the calorimeter and the outside environment. A simplifying assumption is that any heat absorbed or released by the calorimeter materials, such as the container, is negligible. Also, a dilute aqueous solution is assumed to have a density and specific heat capacity equal to that of pure water.

For example, in a calorimetry experiment, 4.24 g of lithium chloride is dissolved in 100 mL (100 g) of water at an initial temperature of 16.3°C. The final temperature of the solution is 25.1°C. To calculate the molar enthalpy of solution H_s for lithium chloride, the first step is to use the law of conservation of energy.

$$\underset{\text{(LiCl dissolving)}}{\Delta H_s} \quad = \quad \underset{\text{(calorimeter water)}}{q}$$

The enthalpy change of lithium chloride dissolving to form a solution is determined by the same mathematical formulas used earlier in this chapter.

$$nH_s = vc\Delta t$$
or
$$nH_s = mc\Delta t$$

Assuming that the dilute solution has the same physical properties as pure water, the molar enthalpy of solution can now be obtained by substituting the given information and the appropriate constants into this equation.

$$nH_s = mc\Delta t$$

$$4.24 \text{ g} \times \frac{1 \text{ mol}}{42.39 \text{ g}} \times H_s = 100 \text{ g} \times 4.19 \frac{\text{J}}{\text{g} \cdot °\text{C}} \times \overbrace{(25.1 - 16.3)}^{8.8°\text{C}}°\text{C}$$

$$\underset{\text{LiCl}}{H_s} = 37 \text{ kJ/mol}$$

Since the temperature of the water in the calorimeter increases, the dissolving of lithium chloride is exothermic. Therefore, the molar enthalpy of solution for lithium chloride is reported as –37 kJ/mol. Note that the certainty of the final answer (two significant digits) is determined by the temperature change of 8.8°C.

10.3 Molar Enthalpy of Solution: Hot and Cold Packs

Hot and cold packs (Figure 10.26) usually contain a chemical and water. The hot pack chemical has an exothermic molar enthalpy of solution, and the cold pack chemical has an endothermic molar enthalpy of solution. The purpose of this investigation is to investigate the energy changes for hot and cold pack chemicals.

Problem

What is the molar enthalpy of solution of calcium chloride and of ammonium nitrate?

Experimental Design

For each compound, the energy change upon dissolving is qualitatively observed. The approximate mass of the compound required to make 50 mL of a 1.00 mol/L solution is calculated and then is measured precisely. **Use the MSDS to determine the hazards associated with the compound and take necessary precautions.** The temperature change is measured as the compound dissolves in the water in a calorimeter (Figure 10. 27).

Materials

lab apron
safety glasses
$CaCl_{2(s)}$
$NH_4NO_{3(s)}$
small self-sealing plastic bag
distilled water bottle
plastic spoon
weighing boat or paper
centigram balance
50 mL graduated cylinder
medicine dropper
calorimeter apparatus (Figure 10.27)

Procedure

1. Place approximately 5 mL of calcium chloride solid in the plastic bag.

2. Pour about 20 mL of water into the bag, close the top, and mix to dissolve the solid.

3. Hold the bottom of the bag in your hand and note the energy change.

4. Empty the contents into the sink, and rinse and dry the inside of the plastic bag.

5. Repeat steps 2 to 5 using ammonium nitrate solid.

6. Measure 50 mL of water in a graduated cylinder and place it in the calorimeter.

7. Obtain the required mass of calcium chloride in a suitable container.

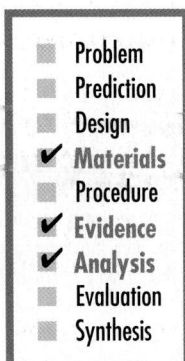

Problem
Prediction
Design
✔ Materials
Procedure
✔ Evidence
✔ Analysis
Evaluation
Synthesis

CAUTION

Handle all heated materials with great care.

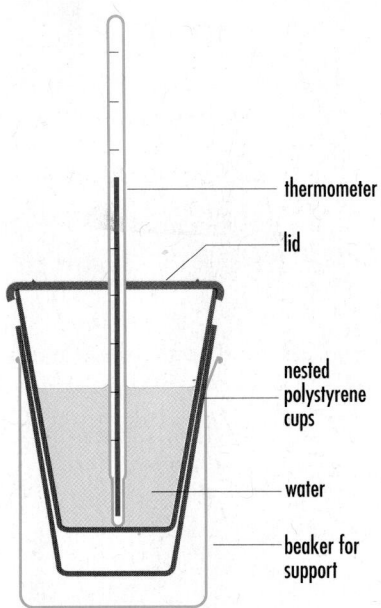

Figure 10.27
A simple laboratory calorimeter consists of nested polystyrene cups as the insulated container, a measured quantity of water, and a thermometer. The chemical system that will undergo an enthalpy change is placed in or dissolved in the water of the calorimeter. Energy transfers between the chemical system and the surrounding water are monitored by measuring changes in the temperature of the water.

thermometer

lid

nested polystyrene cups

water

beaker for support

8. Record the initial temperature of the water.

9. Add the calcium chloride to the water.

10. Cover the calorimeter and stir with the thermometer until a maximum temperature change is obtained.

11. Record the final temperature of the water.

12. Dispose of the contents of the calorimeter into the sink. Rinse and dry the inside of the calorimeter.

13. Repeat steps 7 to 12 using ammonium nitrate.

Exercise

27. List three assumptions made in student investigations involving simple calorimeters.

28. In a calorimetry experiment such as Investigation 10.3, which measurements limit the certainty of the experimental result?

29. In a chemistry experiment, 10 g of urea, $NH_2CONH_{2(s)}$, is dissolved in 150 mL of water in a simple calorimeter. A temperature change from 20.4°C to 16.7°C is measured. Calculate the molar enthalpy of solution for the fertilizer urea.

30. A laboratory technician initially adds 43.1 mL of concentrated, 11.6 mol/L hydrochloric acid to water to form 500 mL of dilute solution. The temperature of the solution changes from 19.2°C to 21.8°C. Calculate the molar enthalpy of dilution of hydrochloric acid.

    Note that caution and suitable safety procedures are required when diluting strong, concentrated acids. Localized heating, without sufficient mixing, may cause the solution to boil and splatter corrosive liquid.

31. A 10.0 g sample of liquid gallium metal, at its melting point, is added to 50 g of water in a polystyrene calorimeter. The temperature of the water changes from 24.0°C to 27.8°C as the gallium solidifies. Calculate the molar enthalpy of solidification for gallium. List any assumptions.

Problem10A Designing a Calorimetry Lab

A chemistry teacher designs a calorimetry lab in which students prepare a 250 mL solution of ammonium nitrate whose molar enthalpy of solution is reported in a reference source as +25 kJ/mol. Complete the Prediction of the investigation report.

Problem

What mass of ammonium nitrate should be dissolved to produce a temperature decrease of 5.0°C?

Problem 10B Molar Enthalpy of a Phase Change

When determining molar enthalpies of phase changes, the *total* energy change of a system is usually required. Complete the Prediction, Analysis, and Evaluation of the investigation report.

Problem

What is the molar enthalpy of fusion of ice?

Experimental Design

The prediction is tested by adding a measured quantity of ice, dried with a paper towel, to water in a polystyrene calorimeter. The final temperature is measured when the calorimeter and ice water mixture have reached the lowest temperature.

Evidence

mass of calorimeter = 3.76 g
mass of calorimeter + water = 103.26 g
mass of calorimeter + water + melted ice = 120.59 g
initial temperature of ice = 0.0°C
initial temperature of water = 32.4°C
final temperature of water = 15.6°C

Calorimetry of Chemical Changes

Chemical reactions that occur in aqueous solutions can be studied using a polystyrene calorimeter like the one shown in Figure 10.27, page 416. The chemical system usually involves reactant solutions that are considered to be equivalent to calorimeter water. The analysis is identical to the analysis of energy changes during phase changes and dissolving.

Problem
Prediction
✔ Design
✔ Materials
✔ Procedure
✔ Evidence
✔ Analysis
✔ Evaluation
Synthesis

CAUTION

Wear safety glasses. Both sodium hydroxide and sulfuric acid are corrosive chemicals. Rinse with lots of cold water if these chemicals contact your skin.

INVESTIGATION

10.4 Molar Enthalpy of Reaction

Evaluating experimental designs and estimating the certainty of empirically determined values are important skills in interpreting scientific statements. The purpose of this investigation is to test the calorimeter design and calorimetry procedure by verifying a widely accepted value for the molar enthalpy of a neutralization reaction. The accuracy (percent difference) obtained in this investigation is used to evaluate the calorimeter and the assumptions made in the analysis, not to evaluate the prediction and its authority, *The CRC Handbook of Chemistry and Physics*. The ultimate authority in this experiment is considered to be the reference value used in the prediction.

Problem

What is the molar enthalpy of neutralization for sodium hydroxide when 50 mL of aqueous 1.0 mol/L sodium hydroxide reacts with an excess quantity of 1.0 mol/L sulfuric acid?

Prediction

According to *The CRC Handbook of Chemistry and Physics*, the molar enthalpy of neutralization for sodium hydroxide with sulfuric acid is –57 kJ/mol.

Exercise

32. It is commonly assumed in calorimetry labs with polystyrene calorimeters that a negligible quantity of heat is absorbed or released by the solid calorimeter materials such as the cup, stirring rod, and thermometer. Use the empirical data in Table 10.6 to evaluate this assumption.
    (a) For a temperature change of 5.0°C, calculate the energy change of the water only.
    (b) For a temperature change of 5.0°C, calculate the total energy change of the water, polystyrene cups, stirring rod, and thermometer.
    (c) Calculate the percent error introduced by using only the energy change of the water.
    (d) Evaluate the assumption of negligible heat transfer to the solid calorimeter materials.

Table 10.6

TYPICAL QUANTITIES FOR MATERIALS IN A SIMPLE CALORIMETER		
Material	Specific Heat Capacity (J/(g•°C))	Mass (g)
water	4.19	100.00
polystyrene cups	0.30	3.58
glass stirring rod	0.84	9.45
thermometer	0.87	7.67

Bomb Calorimeters and Heat Capacity

Many chemical reactions do not take place in aqueous solutions. From a technological perspective, one of the most important exothermic reactions is combustion. A polystyrene calorimeter cannot be used to study the energy changes of combustion reactions, as it has a low melting point and burns readily.

A design for the study of reactions such as combustion, which cannot occur in aqueous solution, is to carry out the reaction in a container inside a bomb calorimeter (Figure 10.28, page 420). The inner reaction compartment is called a *bomb* because in early models it often exploded. Modern bomb calorimeters are strong enough to withstand explosive reactions. The enthalpy change of the reaction is determined from the temperature change of the calorimeter.

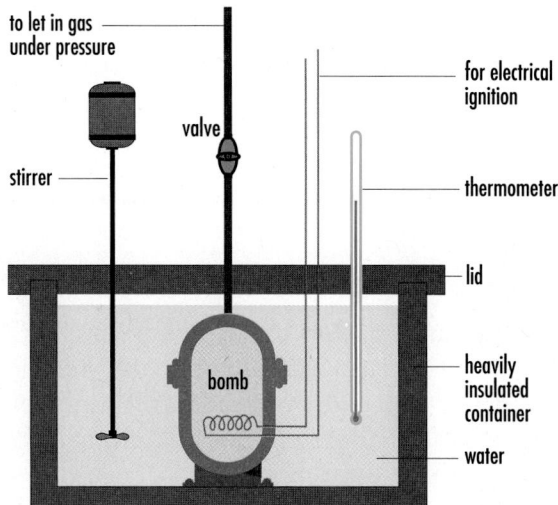

Figure 10.28
The reactants are placed inside the calorimeter's bomb, which is surrounded by the calorimeter water. Once the calorimeter is sealed and the initial temperature measured, the combustion reaction is initiated by an electric heater or spark. Stirring is essential in order to obtain a uniform final temperature for the water.

Bomb calorimeters are used in research to measure enthalpy changes of combustion of fuels, oil, foodstuffs, forage crops, and explosives. But calorimeters that are larger and more sophisticated than polystyrene cups usually have a noticeable heat transfer to or from the calorimeter materials. These modern calorimeters have fixed components, including the volume of water used. The total energy change of the calorimeter is the sum of the energy changes of all of the components.

$$\underset{\text{(calorimeter)}}{\Delta E_{total}} = \underset{\text{(water)}}{vc\Delta t} + \underset{\text{(containers)}}{m_1c_1\Delta t} + \underset{\text{(stirrer)}}{m_2c_2\Delta t} + \underset{\text{(thermometer)}}{m_3c_3\Delta t}$$

Because the temperature change is identical for all components and the same components are used over and over again, this total energy calculation can be simplified. The different constants in the equation can be replaced by a single constant C, the heat capacity of the particular calorimeter. The **heat capacity** of a calorimeter is the total energy absorbed or released per degree Celsius for the calorimeter and its contents.

$$\Delta E_{total} = (vc + m_1c_1 + m_2c_2 + m_3c_3)\, \Delta t$$

$$\Delta E_{total} = C\Delta t, \text{ where } C \text{ is the heat capacity of the calorimeter}$$

Manufacturers may provide a value for the heat capacity of the calorimeter, or the calorimeter may be calibrated by the user with a well-known standard before it is used for calorimetric analysis. Suppose a 1.50 g sample of sucrose, together with excess oxygen gas, is placed in the bomb of a calorimeter whose heat capacity is specified by the manufacturer as 8.57 kJ/°C. The temperature changes from 25.00°C to 27.88°C. According to the law of conservation of energy, the enthalpy of combustion ΔH_c equals the energy change q of the calorimeter.

$$\underset{\text{(sucrose)}}{\Delta H_c} = \underset{\text{(calorimeter)}}{q}$$

The terms *vc* and *mc* both have units of energy per degree Celsius, as shown below using unit values.

$$vc = 1\,\mathcal{k} \times 1\,\frac{kJ}{\mathcal{k}\cdot°C} = 1\,\frac{kJ}{°C}$$

$$mc = 1\,\mathcal{g} \times 1\,\frac{J}{\mathcal{g}\cdot°C} = 1\,\frac{J}{°C}$$

The enthalpy change is defined as nH_c and the total energy change of the calorimeter as $C\Delta t$.

$$nH_c = C\Delta t$$

$$1.50 \text{ g} \times \frac{1 \text{ mol}}{342.34 \text{ g}} \times H_c = 8.57 \frac{\text{kJ}}{°\text{C}} \times 2.88°\text{C}$$

$$H_c = 5.63 \text{ MJ/mol}$$
$$C_{12}H_{22}O_{11}$$

Since the temperature of the calorimeter has increased, the combustion is exothermic and the molar enthalpy of combustion of sucrose is reported as –5.63 MJ/mol.

The names and units for different heat quantities can be confusing. You have to be careful to distinguish among the following:

specific heat capacity	J/(g•°C)
heat capacity	J/°C
specific heat	J/g

Exercise

33. An oxygen bomb calorimeter has a heat capacity of 6.49 kJ/°C. The complete combustion of 1.12 g of acetylene produces a temperature change from 18.60°C to 27.15°C. Calculate the molar enthalpy of combustion H_c for acetylene, $C_2H_{2(g)}$.

34. Canadian inventors have developed zeolite, a natural aluminum silicate mineral, as a storage medium for solar heat. Zeolite releases heat when hydrated with water. In a test, zeolite is used to heat water in a tank that has a heat capacity of 157 kJ/°C. What is the enthalpy change of hydration (ΔH_h) for zeolite if the temperature of the water increases from 27°C to 73°C?

35. Besides the molar enthalpy of combustion as determined in a bomb calorimeter, what other properties or factors are involved in evaluating alternative automobile fuels such as propane, ethanol, and hydrogen?

Problem 10C Calibrating a Bomb Calorimeter

Before molar or specific enthalpies of reaction can be determined, a bomb calorimeter must be calibrated using a primary standard of precisely known molar enthalpy. Complete the Analysis of the investigation report.

Problem

What is the heat capacity of a newly assembled oxygen bomb calorimeter?

Experimental Design

An oxygen bomb calorimeter is assembled, and several samples of the primary standard, benzoic acid, are burned using a constant pressure of excess oxygen. The evidence that is collected determines the heat capacity of the calorimeter for future experiments.

Evidence

In *The CRC Handbook of Chemistry and Physics*, the molar enthalpy of combustion for benzoic acid is reported as

$$H_c = -3231 \text{ kJ/mol}$$
$$_{C_6H_5COOH}$$

Averaging
In general, you will always be correct to calculate first and then average the results.

CALORIMETRIC EVIDENCE FOR THE BURNING OF BENZOIC ACID			
Trial	1	2	3
Mass of $C_6H_5COOH_{(s)}$ (g)	1.024	1.043	1.035
Initial temperature (°C)	24.96	25.02	25.00
Final temperature (°C)	27.99	28.10	28.06

Problem 10D Energy Content of Foods

Bomb calorimeters can be used in the determination of the energy content of foods by combustion analysis. Complete the Analysis and Evaluation of the investigation report.

Problem

Which substance, fat or sugar, has the higher energy content in kilojoules per mole?

Experimental Design

A sample of one component of fat (stearic acid, $C_{18}H_{36}O_2$) is completely burned in a bomb calorimeter. The molar enthalpy of combustion is determined and compared with the previously determined value for sucrose (page 421).

Evidence

mass of stearic acid = 1.14 g
heat capacity of calorimeter = 8.57 kJ/°C
initial temperature (°C) = 25.00°C
final temperature (°C) = 30.28°C

Figure 10.29
An experimental dummy covered with sensors is torched into a fireball by researchers studying burns and fire-retardant clothing.

Molar Enthalpies of Phase and Chemical Changes

Calorimetry is the main source of experimental evidence for molar enthalpies. Molar enthalpies of substances in chemical reactions are typically larger (10^2 to 10^4 kJ/mol) than molar enthalpies in phase changes (10^0 to 10^2 kJ/mol). However, it is often not possible to predict whether a chemical change will be exothermic or endothermic. Apart from the observation that combustion reactions are exothermic (Figure 10.29), few generalizations exist that help predict whether a chemical reaction will absorb or release heat.

OVERVIEW

Energy Changes

Summary

- Energy may be classified in terms of sources, natural resources, and technologically useful forms.

- A scientific perspective on energy includes empirical descriptions of heat and enthalpy changes and corresponding theoretical descriptions (kinetic and potential energy changes).

- The transfer of heat q appears as temperature changes of a system. Theoretically, this corresponds to changes in the speeds of particles present. Heat quantities are calculated using a variety of constants: $q = mc\Delta t$ or $q = vc\Delta t$ or $q = C\Delta t$.

- Enthalpy changes occur during phase, chemical, or nuclear changes and are explained in terms of changes in potential energy resulting from changes in bonding. Enthalpy changes are calculated using a molar enthalpy constant: $\Delta H = nH$.

- Calorimetry is the technique most commonly used to determine molar enthalpies. Energy transfers between a chemical system and a calorimeter are based on the law of conservation of energy.

Key Words

calorimeter
calorimetry
closed system
cooling curve
enthalpy
enthalpy change
fusion (melting)
heat
heat capacity
heating curve
isolated system
joule
molar enthalpy
open system
phase change
potential energy diagram
specific heat capacity
sublimation
temperature
thermal energy
volumetric heat capacity

Review

Several energy constants are listed on the inside back cover of this book. Use these tables as a reference when a particular constant is required but not given.

1. Name some examples of ways in which you rely on energy from chemical reactions.

2. Our society depends primarily on energy from chemical sources such as fossil fuels. What are some alternative energy sources and natural resources?

3. What is the relationship between heat and temperature?

4. What quantity of energy is necessary to heat 2.57 L of water from 3.0°C to 95.0°C using a propane camp stove?

5. List three major classes of enthalpy changes in matter and the evidence that distinguishes each one from the others.

6. Which phase changes are endothermic and which are exothermic? Compare the magnitudes of the energy changes.

7. Describe phase changes, chemical reactions, and nuclear reactions in terms of kinetic and potential energy.

8. How does the molar enthalpy of a phase change compare with the molar enthalpy of a chemical change? Include approximate values.

9. Which types of energy change produce the largest quantities of energy? Why?

10. What quantity of energy is required to change 9.53 g of ice at 0.0°C to water at 0.0°C on an automobile windshield?

11. What is the sign convention used to report endothermic and exothermic molar

enthalpies? Provide a rationale for this convention.

12. All calorimeters have several characteristic components.
    (a) List the components common to calorimeters.
    (b) What scientific law is used in the analysis of calorimetric evidence?
    (c) List the main assumptions made when using a simple laboratory calorimeter.

13. For each quantity listed below, state the quantity symbol and SI unit symbol(s): specific heat capacity, heat capacity, temperature change, heat, change in potential energy, amount of a substance, molar enthalpy, and enthalpy change.

14. Sketch general potential energy diagrams for endothermic and exothermic enthalpy changes.

Applications

15. Bricks in a fireplace will absorb heat and release it long after the fire has gone out. A student conducted an experiment to determine the specific heat capacity of a brick. Based on the evidence obtained in this experiment, 16 kJ of energy was transferred to a 938 g brick as the temperature of the brick changed from 19.5°C to 35.0°C. Calculate the specific heat capacity of the brick.

16. A solar floor is typically made of concrete, which absorbs energy when exposed to direct sunlight. How much energy, obtained from the sun, is absorbed if the temperature of an 8.0 m × 4.0 m × 0.10 m insulated concrete solar floor is raised from 18.0°C to 30.0°C?

17. Many houses are referred to as "sieves" by conservationists because air travels in and out easily through cracks and openings. About one-quarter of the heating bill of a typical house is a result of this movement of air.
    (a) What quantity of heat must a furnace provide to warm the air in a 10.0 m × 11.0 m × 2.40 m house from –25.0°C to 20.5°C?
    (b) What is a simple, inexpensive way to improve the energy efficiency of a typical house?

18. A burn caused by 2.50 g of steam at 100°C is more severe than a burn caused by 2.50 g of water at 100°C. Assuming a final temperature of 35°C in both cases, what is the difference? Include relevant calculations in your answer.

19. In an investigation, ice at –25°C is converted to steam at 115°C at standard pressure.
    (a) Draw a heating curve for the conversion of ice at –25°C to steam at 115°C.
    (b) Label regions of the graph corresponding to temperature or phase changes.

20. In a study of the properties of chlorine, chlorine gas at 25°C and standard pressure is cooled to a temperature of –150°C.
    (a) Sketch a cooling curve, including appropriate phase transition temperatures obtained from the periodic table.
    (b) Label each section of the graph in part (a) with one of the labels q or ΔH.
    (c) Label regions on your graph corresponding to kinetic and potential energy changes.

21. A hiker fills a pot with 2.39 kg of snow at –12.4°C and heats it over an open fire until it melts and is heated to 97.8°C. Calculate the total energy change for converting the snow to hot water.

22. Nuclear fusion reactors that produce energy from reactions similar to those that occur in the sun are still at an experimental stage. Calculate the total energy change required to change 1.00 t of water in a nuclear fusion reactor from 85°C to steam at 250°C in a closed system at standard pressure.

23. In early experimental versions of solar-power towers (Figure 10.1, page 395), water was heated under pressure by solar energy reflected and focused by hundreds of mirrors surrounding the tower. Assume that the energy constants for pressurized water are the same as for water under standard pressure.
    (a) Water under pressure at 85°C is heated to its boiling point at 120°C, converted to steam, and then heated to 150°C. Sketch a heating curve for this system.
    (b) Determine the total energy change as 120 kg of pressurized water is heated in this solar power system.

24. Artificial ice for indoor skating rinks is prepared by circulating a saturated calcium chloride solution through pipes beneath the ice. The solution is cooled by the "ice plant," which usually employs an ammonia heat pump. Calculate the enthalpy change for the vaporization of 1.00 kg of ammonia used as a refrigerant in producing artificial ice. The boiling point of ammonia is –33.35°C and its molar enthalpy of vaporization is 1.37 kJ/mol.

25. Ethane in natural gas is used in the production of ethylene (ethene) for plastics. The boiling point of ethane is –89°C and its molar enthalpy of vaporization is 15.65 kJ/mol.
    (a) What quantity of energy is released when 100 kg of ethane gas is condensed from natural gas?
    (b) What conditions of temperature and pressure are needed to condense ethane?
    (c) What amount of ethane requires 1.00 MJ of energy for vaporization?
    (d) What volume of air can be cooled from 29°C to 19°C by the vaporization of 1.0 kg of ethane used as a refrigerant?

26. The molar enthalpy of combustion of natural gas is –802 kJ/mol. Assuming 100% efficiency and assuming that natural gas consists only of methane, what is the minimum mass of natural gas that must be burned in a laboratory burner to heat 3.77 L of water from 16.8°C to 98.6°C?

27. The molar enthalpy of combustion for a gasoline assumed to be octane is –1.3 MJ/mol. A particular engine has a heat capacity of 105 kJ/°C. Assuming 100% efficiency, and assuming that gasoline consists only of octane, what is the minimum mass of gasoline that must be burned to change the temperature of an engine from 18°C to 120°C?

28. A 77.5 g piece of brass is heated to 98.7°C in a boiling water bath. The brass is quickly transferred to a calorimeter containing 102.76 g of water at 18.5°C. The final temperature of the calorimeter and the brass is 23.5°C. Calculate the specific heat capacity of the brass.

29. The energy content of foodstuffs is determined by combustion in a bomb calorimeter that has a heat capacity of 9.22 kJ/°C. When 3.00 g of butter is burned in excess oxygen, the temperature of the bomb calorimeter changes from 19.62°C to 31.89°C. Calculate the specific enthalpy of combustion of butter in units of kJ/g.

Extensions

30. (a) A pure liquid is suspected to be ethanol. Using the energy concepts from this chapter, list as many experimental designs as possible to confirm or refute the suspected identity of the liquid.
    (b) Describe some other experimental designs that could be used to determine if the unknown liquid is ethanol.

31. (Internet) Prepare a report about one aspect of sustainable development, such as forestry or the use of "waste" materials. A good starting point is the web site maintained by the Center for Renewable Energy and Sustainable Technology (CREST) at the following location: http://solstice.crest.org/online/aeguidc/index.html. Follow the links under "Sustainable Living."

32. Electrical energy used by a small appliance, such as a kettle, is determined by the power consumed and the length of time it is used. This relationship is expressed as $\Delta E = Pt$ where P is the power in watts (joules per second) and t is the time in seconds.
    (a) How efficient is your kettle? Design and conduct an experiment to compare the energy input (electrical) with the energy output (flow of heat to the water).

    $$\text{Percent efficiency} = \frac{\text{output}}{\text{input}} \times 100$$

    (b) How would you design a more efficient kettle? Outline your improved design. Evaluate the need for an improved design in terms of the efficiency calculated in part (a).

Problem 10E Solidification of Wax

Complete the Analysis of the investigation report and evaluate the Experimental Design.

Problem

What is the molar enthalpy of solidification of paraffin wax?

Experimental Design

Liquid paraffin wax, $C_{25}H_{52}$, at its melting point is placed in a polystyrene calorimeter. The final temperature is recorded when the wax just solidifies.

Evidence

volume of water in calorimeter = 150 mL
mass of paraffin wax per trial = 25.00 g

Trial	1	2	3
Final temperature (°C)	27.1	27.7	27.5
Initial temperature (°C)	20.4	21.2	20.9

11 Reaction Enthalpies

A fireworks display is a beautiful example of the energy changes that may take place during chemical reactions. Although we don't regularly see fireworks, our everyday lives depend on many important, though less spectacular, transformations of chemical energy. Our bodies regulate, nourish, and renew themselves through complex series of chemical reactions. In order to adapt comfortably to a northern climate, we rely heavily on the heat produced by chemical reactions. For example, burning natural gas in furnaces keeps us warm. We rely on gasoline burned in car engines to travel from one place to another. Electrical utility companies burn natural gas, oil, or coal and convert the released heat into electricity.

In previous studies, you have not included energy in chemical equations even in reactions such as combustion where there is obviously a change in energy. In this chapter you will investigate the energy changes in chemical reactions, learning how to measure energy that is released or absorbed. As well, you will learn how to communicate these changes. An understanding of energy in chemical reactions is necessary for informed debate on the need for and selection of alternative energy sources.

11.1 COMMUNICATING ENTHALPY CHANGES

All chemical reactions involve energy changes. The enthalpy change of a reaction is sometimes referred to as the *heat of reaction* or the *change in heat content*. Likewise, molar enthalpies of reaction are also called *molar heats of reaction*. However, in this chapter the preferred term for the energy change in a chemical system is *enthalpy change*. Most information about energy changes comes from the experimental method of calorimetry (page 414). The molar enthalpies obtained from these studies can communicate the energy changes of chemical reactions in several different ways: by stating the molar enthalpy of a specific reaction; by stating the enthalpy change for a balanced reaction equation; by including an energy value as a term in a balanced reaction equation; or by drawing a chemical potential energy diagram.

All four of these methods of expressing energy changes are equivalent. The first three are closer to empirical descriptions and the fourth method is a theoretical description similar to the potential energy diagrams drawn in Chapter 10 to describe phase changes. Each of these methods of communicating energy changes in chemical reactions is described in the following sections.

Method 1: Molar Enthalpies of Reaction

The *molar enthalpy of reaction* for a substance is the quantity of heat released or absorbed by the chemical reaction of one mole of the substance at constant pressure. Molar enthalpies are usually measured by calorimetry (Figure 11.1). To communicate a molar enthalpy, both the substance and the reaction must be specified. The substance is conveniently specified by its chemical formula. Some chemical reactions are well known and specific enough to be identified by name only. For instance, reference books often list molar enthalpies of formation (H_f) and combustion (H_c). (See the table listing molar enthalpies of formation in Appendix F.) No chemical equation is necessary, since these two types of reaction are readily understood by chemists. For example, the molar enthalpy of formation for methanol at SATP is communicated internationally as

$$H_{f_{CH_3OH}} = -239.1 \text{ kJ/mol}$$

This means that 239.1 kJ of energy is released to the surroundings when one mole of methanol is formed from its elements. The following chemical equation communicates the formation reaction assumed to occur.

$$C_{(s)} + 2 H_{2(g)} + \tfrac{1}{2} O_{2(g)} \rightarrow CH_3OH_{(l)}$$

A molar enthalpy that is determined when the initial and final conditions of the chemical system are SATP is called a **standard molar enthalpy** of reaction. The symbol $H°$ distinguishes standard molar enthalpies from molar enthalpies H, which are measured at other conditions of temperature and pressure. Standard molar enthalpies allow chemists to create tables to compare enthalpy values and to increase the precision of frequently used values by careful calorimetry.

Since an exothermic change involves a decrease in enthalpy, the direction of this change is communicated as a negative value by $\Delta H < 0$. The direction of an endothermic change is communicated as a positive value by $\Delta H > 0$.

(a)   **Exothermic Reaction**

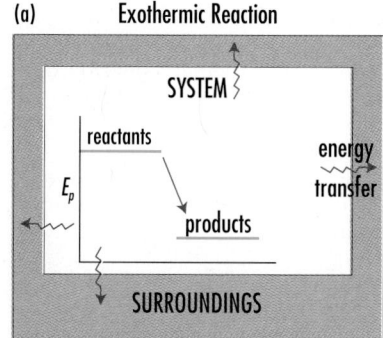

Surroundings are warmed as chemical system releases energy.

(b)   **Endothermic Reaction**

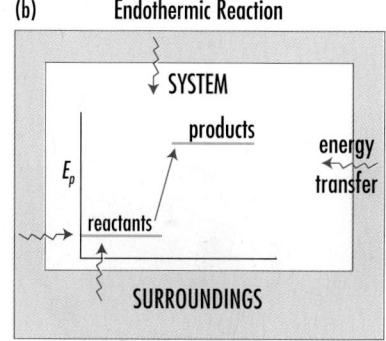

Surroundings are cooled as chemical system absorbs energy.

Figure 11.1
A calorimeter provides controlled surroundings by which the experimenter can monitor the energy changes of a chemical system. Chemists infer that during an exothermic reaction, illustrated in (a), the decrease in the system's enthalpy is transferred as heat to the surroundings. This is indicated by a temperature increase of the surroundings. During an endothermic reaction, illustrated in (b), heat flows from the surroundings to the chemical system. This is indicated by a decrease in temperature of the surroundings; this corresponds to an increase in the enthalpy of the chemical system.

Figure 11.2
Methanol burns more completely than gasoline and produces lower levels of some pollutants. The technology of methanol-burning vehicles was originally developed for racing cars because methanol burns faster than gasoline. However, its energy content is lower so that it takes twice as much methanol as gasoline to drive a given distance.

For an exothermic reaction, the standard molar enthalpy is measured by taking into account all the energy required to change the reaction system from SATP in order to initiate the reaction *and* all the energy released following the reaction, as the products are cooled to SATP. For example, the standard molar enthalpy of combustion of methanol is

$$H_c^\circ \underset{CH_3OH}{} = -638.0 \text{ kJ/mol}$$

This means that the complete combustion of one mole of methanol (Figure 11.2) releases 638.0 kJ of energy according to the following balanced equation.

$$CH_3OH_{(l)} + \tfrac{3}{2} O_{2(g)} \rightarrow CO_{2(g)} + 2 H_2O_{(g)}$$

For a *standard* value, the initial and final conditions of the chemical system must be SATP. In this case, the carbon dioxide and water vapor are produced at a high temperature. They would be allowed to cool to SATP before the final measurement of the energy produced.

If a chemical reaction is not well known or if the equation for the reaction is not obvious, then the chemical equation must be stated along with the molar enthalpy. For example, methanol is produced industrially by the high-pressure reaction of carbon monoxide and hydrogen gases.

$$CO_{(g)} + 2 H_{2(g)} \rightarrow CH_3OH_{(l)}$$

Chemists have determined the standard molar enthalpy for methanol in this reaction, H_r°, to be –128.6 kJ/mol. The symbol for molar enthalpy of reaction uses the subscript "*r*" to refer to the reaction given. Note that this is not a formation reaction since the reactants are not elements.

Method 2: Enthalpy Changes, ΔH

Molar enthalpies can be used to calculate the enthalpy change during a chemical reaction; a molar enthalpy and a balanced chemical equation are required. The enthalpy change is calculated using the empirical definition presented in Chapter 10,

$$\Delta H_r = nH_r$$

where *n* is the amount in moles of the substance whose molar enthalpy is known.

For example, sulfur dioxide and oxygen react to form sulfur trioxide (Figure 11.3). The standard molar enthalpy in terms of sulfur dioxide in this reaction is –98.9 kJ/mol. To calculate the enthalpy change for this reaction, first write the balanced chemical equation.

$$2 SO_{2(g)} + O_{2(g)} \rightarrow 2 SO_{3(g)}$$

Then obtain the amount of sulfur dioxide from the balanced equation and use $\Delta H_c^\circ = nH_c^\circ$.

$$\Delta H_c^\circ = nH_c^\circ$$
$$\Delta H_c^\circ = 2 \text{ mol} \times \frac{-98.9 \text{ kJ}}{1 \text{ mol}} = -197.8 \text{ kJ}$$

Figure 11.3
Most sulfuric acid is produced in plants like this by the contact process, which includes two exothermic combustion reactions. Sulfur reacts with oxygen, forming sulfur dioxide; sulfur dioxide, in contact with a catalyst, reacts with oxygen, forming sulfur trioxide.

Report the enthalpy change for the reaction by writing it next to the balanced equation, as follows:

$$2\,SO_{2(g)} + O_{2(g)} \rightarrow 2\,SO_{3(g)} \qquad\qquad \Delta H_c^\circ = -197.8 \text{ kJ}$$

The enthalpy change depends on the actual amount in moles of reactants and products in the chemical reaction. Therefore, if the balanced equation for the reaction is written differently, the enthalpy change should be reported differently. For example,

$$SO_{2(g)} + \tfrac{1}{2}\,O_{2(g)} \rightarrow SO_{3(g)} \qquad\qquad \Delta H_c^\circ = -98.9 \text{ kJ}$$

$$2\,SO_{2(g)} + O_{2(g)} \rightarrow 2\,SO_{3(g)} \qquad\qquad \Delta H_c^\circ = -197.8 \text{ kJ}$$

Both chemical equations agree with the empirically determined molar enthalpy for sulfur dioxide in this reaction.

$$H_c^\circ{}_{SO_2} = \frac{-197.8 \text{ kJ}}{2 \text{ mol}} = \frac{-98.9 \text{ kJ}}{1 \text{ mol}} = -98.9 \text{ kJ/mol}$$

Unlike molar enthalpies of formation or combustion, the enthalpy changes for most reactions must be accompanied by a balanced chemical equation.

Method 3: Energy Terms in Balanced Equations

Another way to report the enthalpy change in a chemical reaction is to include it as a term in a balanced equation. If a reaction is endothermic, it requires a certain quantity of energy for the reactants to continuously react. This energy (like the reactants) is transformed as the reaction progresses and is listed along with the reactants. For example,

$$H_2O_{(l)} + 285.8 \text{ kJ} \rightarrow H_{2(g)} + \tfrac{1}{2}\,O_{2(g)}$$

If a reaction is exothermic, energy is released as the reaction proceeds (Figure 11.4) and is listed along with the products. For example,

$$Mg_{(s)} + \tfrac{1}{2}\,O_{2(g)} \rightarrow MgO_{(s)} + 601.6 \text{ kJ}$$

In order to specify the initial and final conditions for measuring the enthalpy change of the reaction, the temperature and pressure may be specified at the end of the equation.

$$Mg_{(s)} + \tfrac{1}{2}\,O_{2(g)} \rightarrow MgO_{(s)} + 601.6 \text{ kJ} \qquad \text{(at SATP)}$$

Method 4: Potential Energy Diagrams

To explain observed energy changes, chemists theorize that changes in chemical potential energy occur during a reaction. This energy is a stored form of energy that is related to the relative positions of particles and the strengths of the bonds between them. As bonds break and re-form and the positions of atoms are altered, changes in potential energy occur. Evidence of a change in enthalpy of a chemical system is provided by a temperature change of its surroundings.

Figure 11.4
Combustion reactions are the most familiar exothermic reactions. The searing heat produced by a burning building is a formidable obstacle facing firefighters.

A *potential energy diagram* shows the potential energy of the reactants and the products of a chemical reaction (Figures 11.5, 11.6, and 11.7). The difference between the initial and final energies in a potential energy diagram is the enthalpy change, obtained from calorimetry by measuring the temperature change of the calorimeter. A temperature change is caused by a flow of heat into or out of the chemical system.

Each of the four methods of communicating the molar enthalpy or change in enthalpy of a chemical reaction has advantages and disadvantages. To best understand energy changes in chemical reactions, you should learn all four methods. Figure 11.8 illustrates these methods for an exothermic and an endothermic reaction.

Figure 11.5
This figure shows potential energy diagrams for (a) exothermic and (b) endothermic chemical changes. A potential energy diagram represents a balanced chemical equation with the reactants and products positioned at different values on the vertical energy scale. The horizontal axis represents the progress of the reaction.

Figure 11.6
The standard molar enthalpy of formation for magnesium oxide is obtained from the data table in Appendix F. Since this formation is observed to be exothermic, the reactants must have a higher potential energy than the product.

Figure 11.7
The standard molar enthalpy of the decomposition of water is obtained by reversing the sign of the standard molar enthalpy of formation of water. Since this reaction is endothermic, the reactant (water) must have a lower potential energy than the products (hydrogen and oxygen).

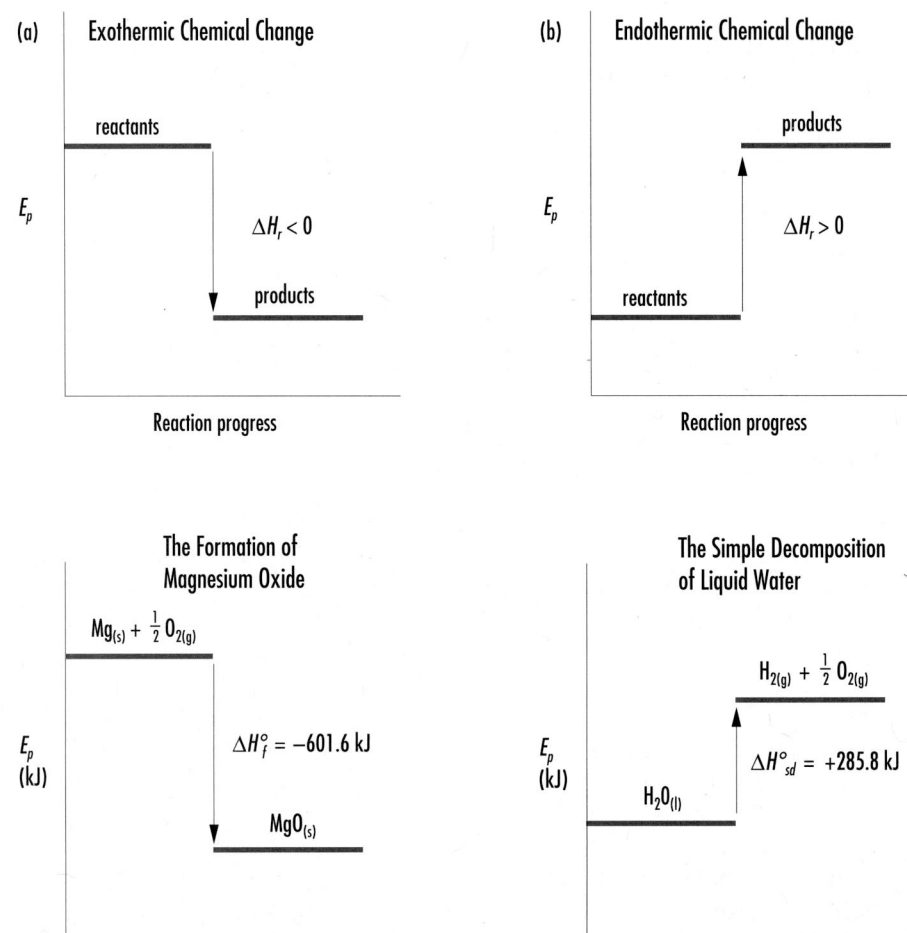

SUMMARY: FOUR WAYS OF COMMUNICATING ENERGY CHANGES

	Exothermic Changes	Endothermic Changes
1. Molar Enthalpy	$H < 0$	$H > 0$
2. Enthalpy Change	reactants $\rightarrow$ products; $\Delta H < 0$	reactants $\rightarrow$ products; $\Delta H > 0$
3. Term in a Balanced Equation	reactants $\rightarrow$ products + energy	reactants + energy $\rightarrow$ products
4. Potential Energy Diagram	E_p (reactants) > E_p (products)	E_p (reactants) < E_p (products)

1. Molar enthalpy for cellular respiration:

$H°_{respiration}$ = −2802.7 kJ/mol

$C_6H_{12}O_6$

Molar enthalpy for photosynthesis:

$H°_{photosynthesis}$ = +2802.7 kJ/mol

$C_6H_{12}O_6$

2. $C_6H_{12}O_{6(s)} + 6\,O_{2(g)} \longrightarrow 6\,CO_{2(g)} + 6\,H_2O_{(l)} \quad \Delta H° = -2802.7\ kJ$

$6\,CO_{2(g)} + 6\,H_2O_{(l)} \longrightarrow C_6H_{12}O_{6(s)} + 6\,O_{2(g)} \quad \Delta H° = +2802.7\ kJ$

3. $C_6H_{12}O_{6(s)} + 6\,O_{2(g)} \longrightarrow 6\,CO_{2(g)} + 6\,H_2O_{(l)} + 2802.7\ kJ$

$6\,CO_{2(g)} + 6\,H_2O_{(l)} + 2802.7\ kJ \longrightarrow C_6H_{12}O_{6(s)} + 6\,O_{2(g)}$

4. Potential Energy Diagram for **Cellular Respiration**

Potential Energy Diagram for **Photosynthesis**

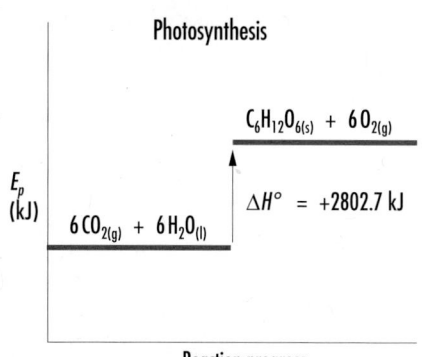

Figure 11.8
Energy is transformed in cellular respiration and in photosynthesis. Cellular respiration, a series of exothermic reactions, is the breakdown of foodstuffs, such as glucose, that takes place within cells. Photosynthesis, a series of endothermic reactions, is the process by which green plants use light energy to make glucose from carbon dioxide and water.

Exercise

1. Translate the empirical molar enthalpies given below into a balanced chemical equation, including the enthalpy change (ΔH).
   (a) The standard molar enthalpy of combustion for methanol is −638.0 kJ/mol.
   (b) The standard molar enthalpy of formation for liquid carbon disulfide is 89.0 kJ/mol.
   (c) The standard molar enthalpy of roasting (combustion) for zinc sulfide is −441.3 kJ/mol.
   (d) The standard molar enthalpy of simple decomposition, $H°_{sd}$, for iron(III) oxide is 824.2 kJ/mol.

2. For each of the following balanced chemical equations and enthalpy changes, write the symbol and calculate the molar enthalpy of combustion for the substance that reacts with oxygen.
   (a) $2\,H_{2(g)} + O_{2(g)} \rightarrow 2\,H_2O_{(g)}$ $\qquad\qquad\qquad \Delta H°_c = -483.6\ kJ$
   (b) $4\,NH_{3(g)} + 7\,O_{2(g)} \rightarrow 4\,NO_{2(g)} + 6\,H_2O_{(g)} + 1134.4\ kJ$
   (c) $2\,N_{2(g)} + O_{2(g)} + 163.2\ kJ \rightarrow 2\,N_2O_{(g)}$
   (d) $3\,Fe_{(s)} + 2\,O_{2(g)} \rightarrow Fe_3O_{4(s)}$ $\qquad\qquad\qquad \Delta H°_c = -1118.4\ kJ$

3. For each of the following reactions, translate the given molar enthalpy into a balanced chemical equation using the ΔH_r notation and then rewrite the equation, including the energy as a term in the equation.
   (a) Propane obtained from natural gas is used as a fuel in barbecues and vehicles (Figure 11.9, page 432). The standard molar enthalpy of combustion for propane, as determined by calorimetry, is −2.04 MJ/mol.

Remember that the unit of *H* is the kJ/mol and the unit of ΔH is the kJ, and that when multiplying by an exact number you use the precision rule.

Figure 11.9
Propane-fueled vehicles are not allowed to park in underground parking lots. Propane is denser than air, and a dangerous quantity of propane could accumulate in the event of a leak.

(b) Nitrogen monoxide forms at the high temperatures inside an automobile engine. The standard molar enthalpy of formation for nitrogen monoxide is 90.2 kJ/mol.

(c) Some advocates of alternative fuels have suggested that cars could run on ethanol. The standard molar enthalpy of combustion for ethanol is –1.28 MJ/mol.

4. In Investigation 10.4 (page 418) you studied the energy changes in the neutralization of a strong acid and a strong base.

$$H_2SO_{4(aq)} + 2\,NaOH_{(aq)} \rightarrow Na_2SO_{4(aq)} + 2\,H_2O_{(l)} + 114\ kJ$$

(a) Write this chemical equation using the ΔH_r notation.

(b) Calculate the molar enthalpy of neutralization for sulfuric acid.

(c) Calculate the molar enthalpy of neutralization for sodium hydroxide.

5. The standard molar enthalpy of combustion for hydrogen is –241.8 kJ/mol. The standard molar enthalpy of decomposition for water vapor is 241.8 kJ/mol.

(a) Write both chemical equations using the ΔH_r° notation.

(b) How does the enthalpy change for the combustion of hydrogen compare with the enthalpy change for the simple decomposition of water vapor? Suggest a generalization to include all pairs of chemical equations that are the reverse of one another.

6. Communicate the energy change by using four different methods for each of the following chemical reactions. Assume standard conditions (SATP) for the measurements of initial and final states and consult Appendix F to obtain the standard molar enthalpies.

(a) the formation of acetylene (ethyne) fuel

(b) the simple decomposition of aluminum oxide

(c) the complete combustion of carbon fuel

11.2 PREDICTING ENTHALPY CHANGES

Calorimetry is the basis for most information about energy changes. However, not every reaction of interest to scientists and engineers can be studied by means of a calorimetric experiment. For example, the rusting of iron is extremely slow and therefore results in temperature changes too small to be measured using a conventional calorimeter. The energy of formation of carbon monoxide is impossible to measure with a calorimeter because the combustion of carbon produces carbon dioxide and carbon monoxide simultaneously. Chemists have devised a number of methods to predict an enthalpy change for reactions that are inconvenient to study experimentally. All of the methods are based on the experimentally established principle that *net changes in some properties of a system are independent of the way the system changes from the initial state to the final state.* A temperature change is an example of a property that satisfies this principle. A net temperature change $(t_f - t_i)$ does not depend on whether the temperature changed slowly, quickly, or rose and fell several times between the initial temperature and the final temperature. This same principle applies to enthalpy changes. If several reactions occur in different ways but the initial reactants and final products are the same, the net enthalpy change is the same as long as the reactions have the same initial and final conditions (Figure 11.10).

Predicting ΔH_r: Hess's Law

Based on experimental measurements of enthalpy changes, Swiss chemist G. H. Hess suggested in 1840 that *the addition of chemical equations yields a net chemical equation whose enthalpy change is the sum of the individual positive and negative enthalpy changes.* This generalization has been tested in many experiments and is now accepted as the law of additivity of enthalpies of reaction, also known as **Hess's law** of heat summation. Hess's law can be written as an equation using the uppercase Greek letter Σ (pronounced "sigma") to mean "the sum of."

$$\Delta H_{net} = \Delta H_1 + \Delta H_2 + \Delta H_3 + \dots$$

or

$$\Delta H_{net} = \Sigma \Delta H_r$$

Hess's discovery allowed the determination of the enthalpy change of a reaction without direct calorimetry, using two rules for chemical equations and enthalpy changes that you already know.

- If a chemical equation is reversed, then the sign of ΔH_r changes.
- If the coefficients of a chemical equation are altered by multiplying or dividing by a constant factor, then the ΔH_r is altered in the same way.

For example, consider the enthalpy change for the formation of carbon monoxide.

$$C_{(s)} + \tfrac{1}{2} O_{2(g)} \rightarrow CO_{(g)} \qquad \Delta H_f^\circ = ?$$

This reaction cannot be studied calorimetrically since the combustion of carbon produces carbon dioxide as well as carbon monoxide. However, the enthalpy of complete combustion for carbon and for carbon monoxide can be measured by calorimetry and the enthalpy of formation for carbon monoxide can be determined using Hess's law, as follows:

(1) $C_{(s)} + O_{2(g)} \rightarrow CO_{2(g)}$ $\qquad\qquad \Delta H_c^\circ = -393.5 \text{ kJ}$
(2) $2\,CO_{(g)} + O_{2(g)} \rightarrow 2\,CO_{2(g)}$ $\qquad\qquad \Delta H_c^\circ = -566.0 \text{ kJ}$

Rearrange these two equations and then add them together to obtain the chemical equation for the formation of carbon monoxide. The first term in the formation equation for carbon monoxide is one mole of solid carbon. Therefore, leave equation (1) unaltered so that $C_{(s)}$ will

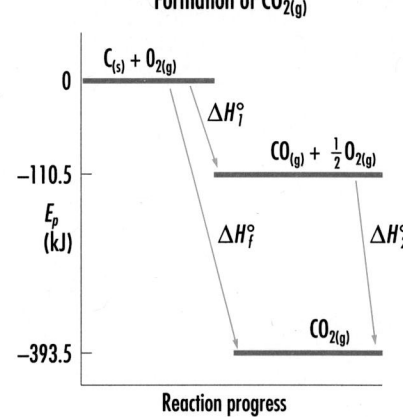

Figure 11.10
Carbon and oxygen react, forming carbon dioxide. The enthalpy change is –393.5 kJ. Carbon and oxygen react to form carbon monoxide (ΔH_1° = –110.5 kJ), which reacts to form carbon dioxide (ΔH_2° = –283.0 kJ). The net enthalpy change of the two-step reaction is –110.5 kJ + (–283.0 kJ) = –393.5 kJ which is identical to that of the overall reaction.

appear on the reactant side when we add the equations. However, we want 1 mol of $CO_{(g)}$ to appear as a product, so reverse equation (2) and divide each of its terms (including the enthalpy change) by 2.

$$C_{(s)} + O_{2(g)} \rightarrow CO_{2(g)} \qquad\qquad \Delta H° = -393.5 \text{ kJ}$$

$$CO_{2(g)} \rightarrow CO_{(g)} + \tfrac{1}{2}O_{2(g)} \qquad\qquad \Delta H° = +283.0 \text{ kJ}$$

Note that the sign of the enthalpy change in equation (2) has changed, since the equation has been reversed. Now add the reactants, products, and enthalpy changes to get a net reaction equation. Note that $CO_{2(g)}$ can be cancelled because it appears on both sides of the net equation. Similarly, $\tfrac{1}{2}O_2$ can be cancelled from each side of the equation, resulting in:

$$C_{(s)} + O_{2(g)} \rightarrow CO_{2(g)} \qquad\qquad \Delta H° = -393.5 \text{ kJ}$$

$$CO_{2(g)} \rightarrow CO_{(g)} + \tfrac{1}{2}O_{2(g)} \qquad\qquad \Delta H° = +283.0 \text{ kJ}$$

$$C_{(s)} + \tfrac{1}{2}O_{2(g)} \rightarrow CO_{(g)} \qquad\qquad \Delta H_f° = -110.5 \text{ kJ}$$

While manipulating equations (1) and (2), you should check the desired equation and plan ahead to ensure that the substances end up on the correct sides and in the correct amounts.

SUMMARY: ENTHALPY OF REACTION AND HESS'S LAW

To determine an enthalpy change of a reaction by using Hess's law, follow these steps:

1. Write the net reaction equation, if it is not given.
2. Manipulate the given equations so they will add to yield the net equation.
3. Cancel and add the remaining reactants and products.
4. Add the component enthalpy changes to obtain the net enthalpy change.
5. Determine the molar enthalpy, if required.

EXAMPLE

Problem

What is the standard molar enthalpy of formation of butane?

Experimental Design

Since the formation of butane cannot be determined calorimetrically, Hess's law is chosen as the method to obtain the value of the standard molar enthalpy of formation.

Evidence

The following values were determined by calorimetry.

(1) $C_4H_{10(g)} + \tfrac{13}{2}O_{2(g)} \rightarrow 4CO_{2(g)} + 5H_2O_{(g)} \qquad \Delta H_c° = -2657.4 \text{ kJ}$

(2) $C_{(s)} + O_{2(g)} \rightarrow CO_{2(g)} \qquad\qquad\qquad\qquad \Delta H_f° = -393.5 \text{ kJ}$

(3) $2H_{2(g)} + O_{2(g)} \rightarrow 2H_2O_{(g)} \qquad\qquad\qquad \Delta H_f° = -483.6 \text{ kJ}$

Analysis

$$4\,C_{(s)} + 5\,H_{2(g)} \rightarrow C_4H_{10(g)}$$

$$4\,CO_{2(g)} + 5\,H_2O_{(g)} \rightarrow C_4H_{10(g)} + \tfrac{13}{2}\,O_{2(g)} \qquad \Delta H^\circ = +2657.4 \text{ kJ}$$

$$4\,C_{(s)} + 4\,O_{2(g)} \rightarrow 4\,CO_{2(g)} \qquad \Delta H^\circ = -1574.0 \text{ kJ}$$

$$5\,H_{2(g)} + \tfrac{5}{2}\,O_{2(g)} \rightarrow 5\,H_2O_{(g)} \qquad \Delta H^\circ = -1209.0 \text{ kJ}$$

$$\text{Net} \quad 4\,C_{(s)} + 5\,H_{2(g)} \rightarrow C_4H_{10(g)} \qquad \Delta H^\circ = -125.6 \text{ kJ}$$

$$H^\circ_{f\,C_4H_{10}} = \frac{\Delta H^\circ_f}{n} = \frac{-125.6 \text{ kJ}}{1 \text{ mol}} = -125.6 \text{ kJ/mol}$$

According to the evidence gathered and Hess's law, the standard molar enthalpy of formation of butane is –125.6 kJ/mol.

To obtain the formation equation and its enthalpy change:
• reverse equation (1) and change the sign of the ΔH
• multiply equation (2) and its ΔH by 4
• multiply equation (3) and its ΔH by $\tfrac{5}{2}$

Exercise

7. The standard enthalpy changes for the formation of aluminum oxide and iron(III) oxide are

$$2\,Al_{(s)} + \tfrac{3}{2}\,O_{2(g)} \rightarrow Al_2O_{3(s)} \qquad \Delta H^\circ_f = -1675.7 \text{ kJ}$$

$$2\,Fe_{(s)} + \tfrac{3}{2}\,O_{2(g)} \rightarrow Fe_2O_{3(s)} \qquad \Delta H^\circ_f = -824.2 \text{ kJ}$$

Calculate the standard enthalpy change for the following reaction.

$$Fe_2O_{3(s)} + 2\,Al_{(s)} \rightarrow Al_2O_{3(s)} + 2\,Fe_{(s)} \qquad \Delta H^\circ_r = ?$$

8. Coal gasification converts coal into a combustible mixture of carbon monoxide and hydrogen, called *coal gas* (Figure 11.11, page 436), in a gasifier.

$$H_2O_{(g)} + C_{(s)} \rightarrow CO_{(g)} + H_{2(g)} \qquad \Delta H^\circ_r = ?$$

Calculate the standard enthalpy change for this reaction from the following chemical equations and standard enthalpy changes.

$$2\,C_{(s)} + O_{2(g)} \rightarrow 2\,CO_{(g)} \qquad \Delta H^\circ_f = -221.0 \text{ kJ}$$

$$2\,H_{2(g)} + O_{2(g)} \rightarrow 2\,H_2O_{(g)} \qquad \Delta H^\circ_f = -483.6 \text{ kJ}$$

9. The coal gas described in question 8 can be used as a fuel, for example, in a combustion turbine (Figure 11.11, page 436).

$$CO_{(g)} + H_{2(g)} + O_{2(g)} \rightarrow CO_{2(g)} + H_2O_{(g)} \qquad \Delta H^\circ_c = ?$$

Predict the change in enthalpy for this combustion reaction from the following information.

$$2\,C_{(s)} + O_{2(g)} \rightarrow 2\,CO_{(g)} \qquad \Delta H^\circ_f = -221.0 \text{ kJ}$$

$$C_{(s)} + O_{2(g)} \rightarrow CO_{2(g)} \qquad \Delta H^\circ_f = -393.5 \text{ kJ}$$

$$2\,H_{2(g)} + O_{2(g)} \rightarrow 2\,H_2O_{(g)} \qquad \Delta H^\circ_f = -483.6 \text{ kJ}$$

10. As an alternative to combustion, coal gas can undergo a process called *methanation*.

$$3\,H_{2(g)} + CO_{(g)} \rightarrow CH_{4(g)} + H_2O_{(g)} \qquad \Delta H^\circ_r = ?$$

The reaction of powdered aluminum and iron(III) oxide is known as the "thermite" reaction and is very exothermic. (See Figure 15.8.)

Determine the standard enthalpy change for this methanation reaction using the following chemical equations and the values for the standard enthalpy changes.

$$2\,H_{2(g)} + O_{2(g)} \rightarrow 2\,H_2O_{(g)} \qquad \Delta H_f^\circ = -483.6 \text{ kJ}$$

$$2\,C_{(s)} + O_{2(g)} \rightarrow 2\,CO_{(g)} \qquad \Delta H_f^\circ = -221.0 \text{ kJ}$$

$$CH_{4(g)} + 2\,O_{2(g)} \rightarrow CO_{2(g)} + 2\,H_2O_{(g)} \qquad \Delta H_c^\circ = -802.7 \text{ kJ}$$

$$C_{(s)} + O_{2(g)} \rightarrow CO_{2(g)} \qquad \Delta H_f^\circ = -393.5 \text{ kJ}$$

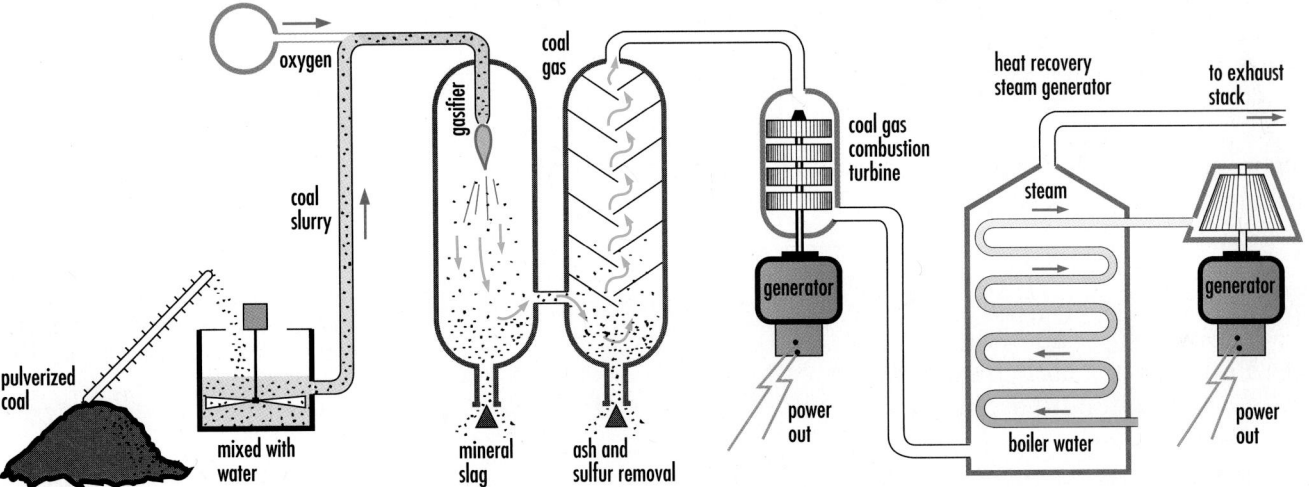

Figure 11.11
Electric power generating stations that use coal as a fuel are only 30% to 40% efficient. Coal gasification and combustion of the coal gas provide one alternative to burning coal. Efficiency is improved by using both a combustion turbine and a steam turbine to produce electricity.

Problem 11A Analysis Using Hess's Law

Most natural gas is burned as fuel to provide heat. However, natural gas is also a source of hydrogen gas for producing ammonia-based fertilizers. The purpose of this problem is to provide practice in the analysis of evidence related to Hess's law, using the production of hydrogen from methane and steam. Complete one possible Analysis for the investigation report.

Problem

What is the standard enthalpy change for the production of hydrogen from methane and steam?

$$CH_{4(g)} + H_2O_{(g)} \rightarrow CO_{(g)} + 3\,H_{2(g)} \qquad \Delta H_r^\circ = ?$$

Evidence

$$2\,C_{(s)} + O_{2(g)} \rightarrow 2\,CO_{(g)} \qquad \Delta H_c^\circ = -221.0 \text{ kJ}$$

$$CH_{4(g)} + 2\,O_{2(g)} \rightarrow CO_{2(g)} + 2\,H_2O_{(g)} \qquad \Delta H_c^\circ = -802.7 \text{ kJ}$$

$$CO_{(g)} + H_2O_{(g)} \rightarrow CO_{2(g)} + H_{2(g)} \qquad \Delta H_r^\circ = -41.2 \text{ kJ}$$

$$2\,H_{2(g)} + O_{2(g)} \rightarrow 2\,H_2O_{(g)} \qquad \Delta H_c^\circ = -483.6 \text{ kJ}$$

$$C_{(s)} + 2\,H_{2(g)} \rightarrow CH_{4(g)} \qquad \Delta H_f^\circ = -74.4 \text{ kJ}$$

$$C_{(s)} + H_2O_{(g)} \rightarrow CO_{(g)} + H_{2(g)} \qquad \Delta H_r^\circ = +131.3 \text{ kJ}$$

$$2\,CO_{(g)} + O_{2(g)} \rightarrow 2\,CO_{2(g)} \qquad \Delta H_c^\circ = -566.0 \text{ kJ}$$

$$CO_{(g)} + H_{2(g)} + O_{2(g)} \rightarrow CO_{2(g)} + H_2O_{(g)} \qquad \Delta H_r^\circ = -524.8 \text{ kJ}$$

Problem 11B Testing Hess's Law

The following data are from a test of Hess's law using a calorimeter. Use these data in your prediction, assuming that the water of combustion is a liquid in the bomb calorimeter.

$$5\,C_{(s)} + 6\,H_{2(g)} \rightarrow C_5H_{12(l)} \qquad\qquad \Delta H_f^\circ = -173.5 \text{ kJ}$$

$$C_{(s)} + O_{2(g)} \rightarrow CO_{2(g)} \qquad\qquad \Delta H_f^\circ = -393.5 \text{ kJ}$$

$$H_{2(g)} + \tfrac{1}{2}\,O_{2(g)} \rightarrow H_2O_{(g)} \qquad\qquad \Delta H_f^\circ = -241.8 \text{ kJ}$$

$$H_2O_{(l)} \rightarrow H_2O_{(g)} \qquad\qquad \Delta H_{vap}^\circ = +44.0 \text{ kJ}$$

Then complete the Prediction, Analysis, and Evaluation of the investigation report.

Problem

What is the standard molar enthalpy of combustion of pentane?

Experimental Design

Hess's law is used to predict the standard molar enthalpy of combustion of pentane. To test the prediction and the acceptability of the law, the standard molar enthalpy of combustion of pentane is determined calorimetrically.

Evidence

mass of pentane reacted = 2.15 g
volume of water equivalent to calorimeter = 1.24 L
initial temperature of calorimeter and contents = 18.4°C
final temperature of calorimeter and contents = 37.6°C

INVESTIGATION

11.1 Applying Hess's Law

Magnesium burns rapidly, releasing heat and light (Figure 1.5, page 32).

$$Mg_{(s)} + \tfrac{1}{2}\,O_{2(g)} \rightarrow MgO_{(s)}$$

The enthalpy change of this reaction can be measured using a bomb calorimeter, but not a polystyrene cup calorimeter. The enthalpy change for the combustion of magnesium can be determined by applying Hess's law to the following three chemical equations.

$$MgO_{(s)} + 2\,HCl_{(aq)} \rightarrow MgCl_{2(aq)} + H_2O_{(l)}$$

$$Mg_{(s)} + 2\,HCl_{(aq)} \rightarrow MgCl_{2(aq)} + H_{2(g)}$$

$$H_{2(g)} + \tfrac{1}{2}\,O_{2(g)} \rightarrow H_2O_{(l)} \qquad\qquad \Delta H_f^\circ = -285.8 \text{ kJ}$$

- Problem
- Prediction
- Design
- Materials
- ✔ Procedure
- ✔ Evidence
- ✔ Analysis
- ✔ Evaluation
- Synthesis

Problem

What is the molar enthalpy of combustion for magnesium?

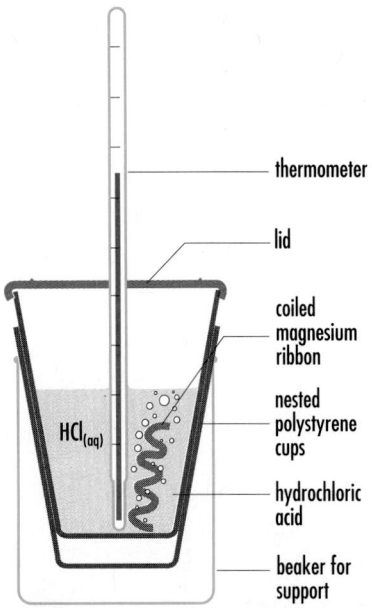

Figure 11.12
Magnesium ribbon reacts rapidly in dilute hydrochloric acid. With nested polystyrene cups, the enthalpy change can be determined by measuring the temperature change of the HCl solution.

Prediction

According to the table of standard molar enthalpies of formation (Appendix F), the standard molar enthalpy of combustion for magnesium is –601.6 kJ/mol. The molar enthalpy of combustion of magnesium is the same as the molar enthalpy of formation of magnesium oxide because both processes have the same chemical equation.

$$H^\circ_{c\ \text{Mg}} = H^\circ_{f\ \text{MgO}} = -601.6 \text{ kJ/mol}$$

$$Mg_{(s)} + \tfrac{1}{2} O_{2(g)} \rightarrow MgO_{(s)} \qquad \Delta H^\circ_c = \Delta H^\circ_f = -601.6 \text{ kJ}$$

Experimental Design

The enthalpy changes for the first two reactions with hydrochloric acid are determined empirically using a polystyrene calorimeter (Figure 11.12). The three ΔH° values are used, along with Hess's law, to obtain the molar enthalpy of combustion for magnesium.

Materials

lab apron
safety glasses
magnesium ribbon (maximum 15 cm strip)
magnesium oxide powder (maximum 1.00 g sample)
1.00 mol/L hydrochloric acid (use 50 mL each time)
polystyrene calorimeter with lid
50 mL or 100 mL graduated cylinder
laboratory scoop or plastic spoon
steel wool
weighing boat or paper
centigram balance
ruler

Predicting ΔH, Using Formation Reactions

Chemists rely on conventions to simplify explanations and communication. For example, SATP is a set of internationally accepted conditions that defines a *standard state*. Since elements are the building blocks of compounds and since absolute potential energies cannot be measured, it is convenient to set at zero the value for the potential energy of elements in their most stable form at SATP. This convention, defining elements as the reference point at which the potential energy is zero, is the **reference energy state**. This convention does not mean that the potential energy of an element is *always* considered to be zero; in another situation, a different convention might be more convenient. (Similarly, the Celsius temperature scale sets 0°C at the freezing point of pure water. This is a convenient reference point but it does not mean that water molecules have zero kinetic energy at that temperature.)

The enthalpy change measured in a formation reaction can now be theoretically described as a change in potential energy from zero (the potential energy of the elements) to some final value determined by the enthalpy change. For example,

$$H_{2(g)} + \tfrac{1}{2} O_{2(g)} \rightarrow H_2O_{(l)} \qquad \Delta H_f^\circ = -285.8 \text{ kJ}$$

E_p (kJ)	0	0	−285.8

The potential energy decreases from 0 kJ for the reactants to −285.8 kJ for the product. In other words, the reactants are at a higher chemical potential energy than the product. This decrease in potential energy is transferred to the surroundings and appears as heat or other forms of energy. Suppose you were seated on a bicycle at the top of a hill and coasted downhill. Your potential energy at the top of the hill is converted into kinetic energy as you move from a point of higher potential energy (top) to lower potential energy (bottom) (Figure 11.13). Of course, if you want to return to the top of the hill, you must supply the energy to move from a lower to a higher potential energy. Similarly, to convert the water back into hydrogen and oxygen requires that energy be added, specifically 285.8 kJ/mol of water.

Figure 11.13
As a cyclist coasts downhill, his potential energy decreases — it is converted to kinetic energy.

Tables of standard molar enthalpies of formation (Appendix F) can be used to compare the stabilities of compounds. Most compounds are formed exothermically from their elements, and the molar enthalpies of formation are negative. This means that the compounds are energetically more stable than their elements (defined at 0 kJ/mol). **Thermal stability** is the tendency of a compound to resist decomposition when heated. The more heating required to decompose a compound, the more stable the compound. In other words, *the more exothermic the formation, the more stable the compound is relative to its elements.* For example, the standard molar enthalpies of formation for tin(II) oxide and tin(IV) oxide can be obtained in Appendix F .

$$H_f^\circ = -280.7 \text{ kJ/mol} \qquad\qquad H_f^\circ = -577.6 \text{ kJ/mol}$$
$$\text{SnO} \qquad\qquad\qquad\qquad\qquad \text{SnO}_2$$

Tin(IV) oxide is more stable than tin(II) oxide because tin(IV) oxide has a more negative molar enthalpy of formation.

Note the position of zero potential energy in Figure 11.10 (page 433). In this example, one side of the chemical equation has only elements in their natural states at SATP. If a chemical equation is expressed as a sum of formation reactions only, the calculation of the enthalpy change is simpler than if a variety of reaction equations is used. For example, the slaking of lime, calcium oxide, is represented by the following chemical equation (Figure 11.14).

$$CaO_{(s)} + H_2O_{(l)} \rightarrow Ca(OH)_{2(s)} \qquad \Delta H_r^\circ = \text{?}$$

To find the standard enthalpy change for this reaction, write the formation equation and corresponding standard enthalpy change (Appendix F) for each compound in the given equation.

Figure 11.14
Adding lime to a lake can help neutralize the effects of acid rain. However, the restoration of a lake requires more than just neutralizing the excess acidity of the water.

$$Ca_{(s)} + \tfrac{1}{2}O_{2(g)} \rightarrow CaO_{(s)} \qquad \Delta H_f^\circ = 1 \text{ mol} \times -634.9 \text{ kJ/mol}$$

$$H_{2(g)} + \tfrac{1}{2}O_{2(g)} \rightarrow H_2O_{(l)} \qquad \Delta H_f^\circ = 1 \text{ mol} \times -285.8 \text{ kJ/mol}$$

$$Ca_{(s)} + O_{2(g)} + H_{2(g)} \rightarrow Ca(OH)_{2(s)} \qquad \Delta H_f^\circ = 1 \text{ mol} \times -986.1 \text{ kJ/mol}$$

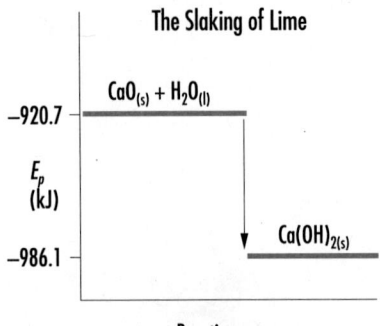

The Slaking of Lime

-920.7 — $CaO_{(s)} + H_2O_{(l)}$

E_p
(kJ)

-986.1 — $Ca(OH)_{2(s)}$

Reaction progress

Figure 11.15
Potential energy diagram for the slaking of lime. The two summation (Σ) terms in the mathematical formula become the positions of the reactants and the products on the potential energy scale.

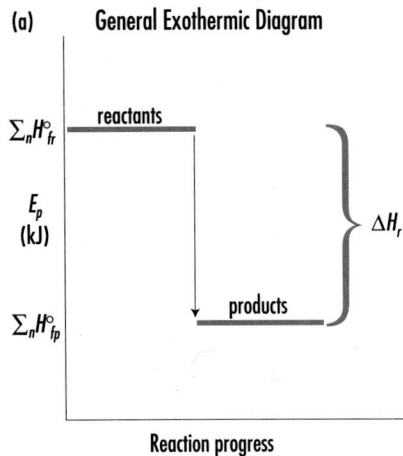

(a)   General Exothermic Diagram

$\Sigma_n H^\circ_{fr}$ ———— reactants

E_p
(kJ)

$\Sigma_n H^\circ_{fp}$ ———— products

$\Big\} \Delta H_r$

Reaction progress

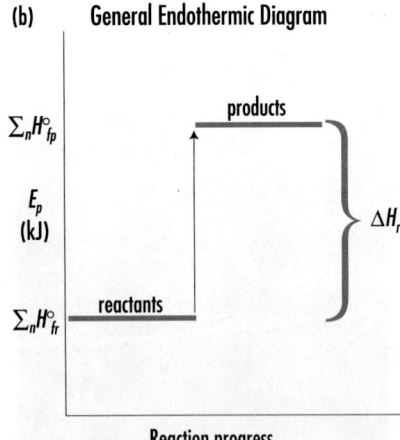

(b)   General Endothermic Diagram

$\Sigma_n H^\circ_{fp}$ ———— products

E_p
(kJ)

$\Big\} \Delta H_r$

$\Sigma_n H^\circ_{fr}$ ———— reactants

Reaction progress

Figure 11.16
The two summation terms in the net enthalpy change formula provide the location of the reactants and products on potential energy diagrams.

By adding the third equation to the reverse of the first two equations, the chemical equation required for the slaking of lime is obtained.

$$Ca_{(s)} + O_{2(g)} + H_{2(g)} \rightarrow Ca(OH)_{2(g)} \qquad \Delta H^\circ_f$$

$$CaO_{(s)} \rightarrow Ca_{(s)} + \tfrac{1}{2}O_{2(g)} \qquad -\Delta H^\circ_f$$

$$H_2O_{(1)} \rightarrow H_{2(g)} + \tfrac{1}{2}O_{2(g)} \qquad -\Delta H^\circ_f$$

Applying Hess's law gives the following equation.

$$\Delta H^\circ_r = \underset{Ca(OH)_2}{\Delta H^\circ_f} + \underset{CaO}{(-\Delta H^\circ_f)} + \underset{H_2O}{(-\Delta H^\circ_f)}$$

Notice that the net enthalpy change is equal to the enthalpy of formation for the product minus the enthalpies of formation of the reactants.

$$\Delta H^\circ_r = \underset{Ca(OH)_2}{\Delta H^\circ_f} - \underset{CaO \quad H_2O}{(\Delta H^\circ_f + \Delta H^\circ_f)}$$

Substituting the definition $\Delta H = nH$ and combining terms result in the following formula, where $\Sigma n H^\circ_{fp}$ is the standard enthalpy change of the products, and $\Sigma n H^\circ_{fr}$ is the standard enthalpy change of the reactants.

$$\Delta H^\circ_r = \underset{Ca(OH)_2}{\Sigma n H^\circ_{fp}} - \underset{CaO + H_2O}{\Sigma n H^\circ_{fr}}$$

$$= \underset{Ca(OH)_2}{n H^\circ_f} - \underset{CaO \quad H_2O}{(n H^\circ_f + n H^\circ_f)}$$

$$= -986.1 \text{ kJ} - (-920.7 \text{ kJ})$$

$$= -65.4 \text{ kJ}$$

According to Hess's law and empirically determined molar enthalpies of formation, the standard enthalpy change for the slaking of lime is reported as follows.

$$CaO_{(s)} + H_2O_{(1)} \rightarrow Ca(OH)_{2(s)} \qquad \Delta H^\circ_r = -65.4 \text{ kJ}$$

Therefore, the H°_r for $Ca(OH)_2$ in this reaction is -65.4 kJ/mol.

The enthalpy change of the reaction in this example can be theoretically described by a potential energy diagram (Figure 11.15). Note that the derived formula provides a negative enthalpy change for an exothermic reaction, consistent with the accepted convention. Also note that the value of $\Sigma n H^\circ_{fp}$ appears on the diagram as the chemical potential energy of the product, and the value of $\Sigma n H^\circ_{fr}$ is the potential energy of the reactants.

SUMMARY: USING ENTHALPIES OF FORMATION TO PREDICT ΔH_r

According to Hess's law, the net enthalpy change for a chemical reaction is equal to the sum of the enthalpies of formation of the products minus the sum of the enthalpies of formation of the reactants (Figure 11.16).

$$\Delta H_r = \Sigma n H_{fp} - \Sigma n H_{fr}$$

EXAMPLE

What is the standard molar enthalpy of combustion of methane fuel?

$$CH_{4(g)} + 2\,O_{2(g)} \rightarrow CO_{2(g)} + 2\,H_2O_{(g)}$$

$$\Delta H_c^{\circ} = \Sigma n H_{fp}^{\circ} - \Sigma n H_{fr}^{\circ}$$

$$= \left(1\ mol\ \times \frac{-393.5\ kJ}{1\ mol} + 2\ mol\ \times \frac{-241.8\ kJ}{1\ mol}\right)$$

$$- \left(1\ mol\ \times \frac{-74.4\ kJ}{1\ mol} + 2\ mol\ \times \frac{0\ kJ}{1\ mol}\right)$$

$$= -877.1\ kJ - (-74.4\ kJ)$$

$$= -802.7\ kJ$$

$$H_{c_{CH_4}}^{\circ} = \frac{\Delta H_c^{\circ}}{n} = \frac{-802.7\ kJ}{1\ mol} = -802.7\ kJ/mol$$

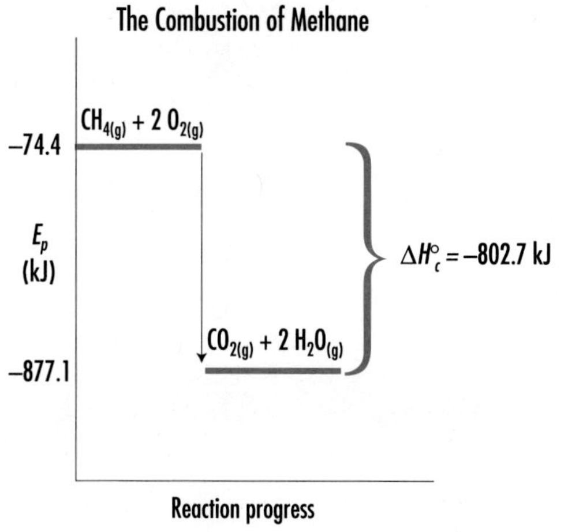

The Combustion of Methane

$CH_{4(g)} + 2\,O_{2(g)}$ at -74.4

E_p (kJ)

$CO_{2(g)} + 2\,H_2O_{(g)}$ at -877.1

$\Delta H_c^{\circ} = -802.7\ kJ$

Reaction progress

Figure 11.17
The combustion of methane can also be communicated using a potential energy diagram.

The combustion of methane is communicated in the potential energy diagram (Figure 11.17).

Exercise

11. Methane, the major component of natural gas, is used as a source of hydrogen gas to produce ammonia. Ammonia is used as a fertilizer and a refrigerant, and is used to manufacture fertilizers, plastics, cleaning agents, and prescription drugs. The following questions refer to some of the chemical reactions of these processes.

    (a) The first step in the production of ammonia is the reaction of methane with steam using a nickel catalyst. Predict the ΔH_r° for the following reaction.

    $$CH_{4(g)} + H_2O_{(g)} \rightarrow CO_{(g)} + 3\,H_{2(g)}$$

    (b) The second step of this process is the further reaction of carbon monoxide to produce more hydrogen. Both iron and zinc-copper catalysts are used. Predict the ΔH_r°.

    $$CO_{(g)} + H_2O_{(g)} \rightarrow CO_{2(g)} + H_{2(g)}$$

    (c) After the carbon dioxide gas is removed by dissolving it in water, the hydrogen reacts with nitrogen obtained from the air. Predict the ΔH_f° to form two moles of ammonia.

12. Nitric acid, required in the production of nitrate fertilizers, is produced from ammonia by the Ostwald process (Figure 11.18). Predict the standard enthalpy change for each reaction in the process, as written, and then predict the standard molar enthalpy of reaction for the first reactant listed in each equation.

    (a) $4\,NH_{3(g)} + 5\,O_{2(g)} \rightarrow 4\,NO_{(g)} + 6\,H_2O_{(g)}$

    (b) $2\,NO_{(g)} + O_{2(g)} \rightarrow 2\,NO_{2(g)}$

    (c) $3\,NO_{2(g)} + H_2O_{(l)} \rightarrow 2\,HNO_{3(l)} + NO_{(g)}$

Figure 11.18
An Ostwald process plant converts ammonia to nitric acid cleanly and efficiently. Unreacted gases and energy from the exothermic reactions are recycled. Catalytic combustors burn noxious fumes to minimize environmental effects and to supply additional energy to operate the plant.

Figure 11.19
*Fertilizers such as ammonium
nitrate have had a dramatic impact
on crop yields. Since the 19th
century, average crop yields per acre
have increased almost five-fold for
corn and eight-fold for wheat.
However, run-off from fertilized
fields is a source of water pollution.
Also, the high cost of chemical
fertilizers has driven some farmers
into debt.*

13. Ammonium nitrate fertilizer is produced by the reaction of ammonia with the nitric acid resulting from the series of reactions given in question 12. Ammonium nitrate is one of the most important fertilizers for increasing crop yields (Figure 11.19).
    (a) Predict the standard enthalpy change of the reaction used to produce ammonium nitrate.

    $$NH_{3(g)} + HNO_{3(l)} \rightarrow NH_4NO_{3(s)}$$

    (b) Sketch a potential energy diagram for the reaction of ammonia and nitric acid.

14. (Discussion) Evaluate the technology outlined in questions 11, 12, and 13 from at least five perspectives.

Problem 11C Testing ΔH_r° from Formation Data

The purpose of this problem is to test the use of molar enthalpies of formation as a method of predicting the enthalpy change of a reaction. Complete the Prediction, Analysis, and Evaluation of the investigation report.

Problem

What is the molar enthalpy of combustion of methanol?

Experimental Design

Methanol is burned in excess oxygen in a bomb calorimeter whose heat capacity is 10.9 kJ/°C. Assume that water is produced in the form of a liquid.

Evidence

mass of methanol reacted = 4.38 g
heat capacity of bomb calorimeter = 10.9 kJ/°C
initial temperature of calorimeter = 20.4°C
final temperature of calorimeter = 27.9°C

Problem 11D Determining Standard Molar Enthalpies of Formation

The prediction of enthalpy changes using standard molar enthalpies of formation depends entirely on the availability of tables of standard molar enthalpies of formation. Many of these H_f° values can be initially determined from H_c° values by using the formation method equation. In this problem you will use a known H_c° value to calculate a corresponding H_f° value. Complete the Analysis of the investigation report. Work out your own problem-solving approach here, using the hint provided in this paragraph.

Problem

What is the standard molar enthalpy of formation for hexane, $C_6H_{14(l)}$?

Table 11.1

COMPARISON OF ENTHALPY CHANGES	
Type	**Molar Enthalpies**
phase	$10^0 - 10^2$ kJ/mol
chemical	$10^2 - 10^4$ kJ/mol
nuclear	$10^6 - 10^{10}$ kJ/mol

11.3 NUCLEAR REACTIONS

Of all energy changes, nuclear reactions involve the greatest quantities of energy (Table 11.1). The nuclear reactions that occur in the sun are important to us because they supply the energy that sustains life on Earth (Figure 11.20).

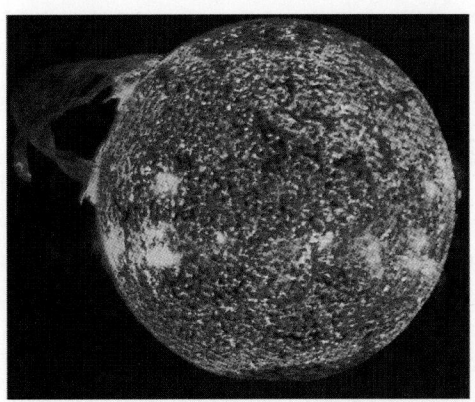

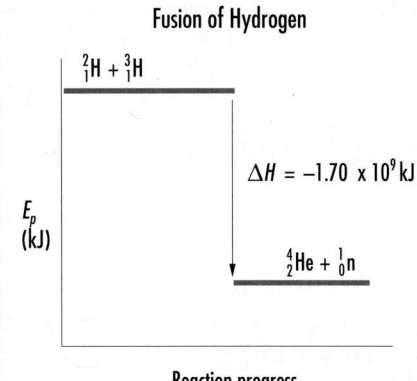

Fusion of Hydrogen

$^2_1H + ^3_1H$

E_p (kJ)

$\Delta H = -1.70 \times 10^9$ kJ

$^4_2He + ^1_0n$

Reaction progress

Figure 11.20
Direct solar radiation provides the energy required for green plants to produce food and oxygen daily. Indirectly, solar energy is also the source of energy from winds, water, and fossil fuels. According to current theory, fossil fuels are the remains of plants and animals that originally depended on sunlight for energy. Fossil fuels are therefore considered a stored form of solar energy.

Figure 11.21
A potential energy diagram of a nuclear fusion reaction that is used in the research and development of nuclear fusion reactors.

Enthalpy changes in nuclear reactions are a result of potential energy changes as bonds among the particles of the nucleus — protons and neutrons — are broken or formed. There are many different nuclear reactions taking place in the sun, as in other stars; in one of the main reactions, four hydrogen atoms fuse, producing one helium atom. Scientists and engineers think that using a similar reaction, the fusion of two isotopes of hydrogen, is a promising possibility in the development of nuclear fusion reactors on Earth. In this reaction a helium atom (^4_2He), a neutron (^1_0n), and a large quantity of energy are produced. This nuclear reaction is communicated by the equation below. (See page 76 for an explanation of the symbols in the following equation.)

$$^2_1H + ^3_1H \rightarrow ^4_2He + ^1_0n \qquad \Delta H = -1.70 \times 10^9 \text{ kJ}$$

$$^2_1H + ^3_1H \rightarrow ^4_2He + ^1_0n + 1.70 \times 10^9 \text{ kJ}$$

This means that 1.7×10^9 kJ of energy is released for every mole of helium produced. A potential energy diagram for this nuclear reaction is similar to that for an exothermic chemical reaction (Figure 11.21).

For convenience when comparing enthalpy changes, scientific notation is combined with the SI prefix, k, in kJ.

Another important nuclear reaction is the fission or splitting of uranium into two smaller nuclei. Nuclear fission reactions provide the energy for nuclear power generating stations; they have molar enthalpies on the order of 10^{10} kJ/mol.

$$\ce{^{235}_{92}U} + \ce{^{1}_{0}n} \rightarrow \ce{^{141}_{56}Ba} + \ce{^{92}_{36}Kr} + 3\ce{^{1}_{0}n} + 1.9 \times 10^{10} \text{ kJ}$$

Problem 11E Calorimetry of a Nuclear Reaction

There are many types of nuclear reactions. Some nuclear reactions, such as nuclear fission and fusion, are difficult to study in a laboratory. However, nuclear decay reactions are easier to study. For example, a radioactive isotope undergoing an exothermic nuclear decay reaction releases enough energy to boil water in a beaker. Use this information to design an experiment to answer the following question. Complete the Experimental Design of the investigation report.

Problem

What is the molar enthalpy of nuclear decay for a radioactive isotope?

11.4 ENERGY AND SOCIETY

Our society has become very dependent on fossil fuels for energy. This is both a blessing and a curse. Inexpensive fossil fuels have contributed to our high standard of living. However, we may be paying dearly for this good fortune. Environmental problems such as global warming, rising costs of scarce resources, and shortages of raw materials for the petrochemical industry are some of the possible disadvantages of this dependency. There are three major demands for energy from fossil fuels — heating, transportation, and industry. What are some alternatives to fossil fuels? Options include both the use of different fuels and more economical management of fossil fuels (Table 11.2).

Figure 11.22
A well-insulated home with the majority of windows facing south is the main requirement for obtaining heat from direct sunlight. Solar heating and retaining the heat generated by people and appliances can reduce heating bills by as much as 90%.

Table 11.2

ALTERNATIVES TO CURRENT FOSSIL FUEL USES	
Energy Demands	**Alternative Energy Sources and Practices**
heating	• solar heating (Figure 11.22), heat pumps, geothermal energy, biomass gas, and electricity from hydro and nuclear plants • improved building insulation and design
transportation	• alcohol/gasohol and hydrogen fuels (Figure 11.23), and electric vehicles (powered by batteries and fuel cells) • mass transit, bicycles, and walking
industry	• solar energy, nuclear energy, and hydroelectricity • improved efficiency and waste heat recovery (Figure 11.24)

Figure 11.23
This experimental car burns hydrogen as a fuel, producing water vapor as an exhaust, but no carbon dioxide. The hydrogen is stored in a tank as a metal hydride.

Figure 11.24
A great deal of energy is wasted by motors, compressors, and exhaust emissions. Recovering this energy is one way in which industries improve their energy efficiency. The greenhouse in the photograph uses heat from the compressors on the Trans-Canada Pipeline.

MARIE SKLODOWSKA CURIE (1867 – 1934)

Marie Sklodowska was born in Poland, which was at that time under Russian domination. She worked as a governess until she saved enough money to move to Paris. In 1891 she began to study science at the Sorbonne and graduated two years later at the top of her class. Her marriage in 1895 to Pierre Curie initiated a partnership that soon achieved worldwide significance.

Searching for a topic for her doctoral thesis, Marie Curie became intrigued by a recent discovery by Henri Becquerel. In 1896 Becquerel had discovered that compounds of uranium spontaneously emitted rays that exposed photographic plates. She decided to investigate the emissions from uranium and to find out if the property discovered in uranium was exhibited by other elements. Curie coined the term "radioactivity"

to describe the rays. She discovered that thorium also produced rays. In her studies of uranium and thorium, she noted that the mineral pitchblende was more radioactive than pure uranium. Together, Marie and Pierre Curie set out to isolate the extra source of the radioactivity, which led them to discover two new elements, polonium and radium. To obtain sufficient radium to study its chemical properties thoroughly, they undertook the arduous processing of eight tonnes of pitchblende to get one gram of radium.

The Curies and Becquerel shared the 1903 Nobel Prize in physics — Becquerel for the discovery of radioactivity and the Curies for their investigations into the nature of radioactivity. In 1911 Marie Curie was awarded the Nobel Prize in chemistry for the discovery of radium and polonium and for the isolation of pure radium.

Besides her talent as a researcher, Marie Curie was a respected teacher. In 1900 she was appointed lecturer in physics at a girls' school, where she introduced a method of teaching science based on experimental demonstrations. When her husband was killed in a traffic accident in 1906, Dr. Curie was appointed to the professorship that he had held, becoming the first woman to teach at the Sorbonne.

During World War I (1914 – 1918) Curie drove an ambulance and, with her daughter Irene Joliot-Curie,

worked on the application of X rays to the treatment of wounds. After the war, Curie became involved in the supervision of the Paris Institute of Radium, which became a major center for the study of nuclear physics and chemistry. In 1934 Dr. Marie Curie died of leukemia, in all likelihood caused by excessive exposure to radiation.

Irene Curie and her husband, Frederic Joliot, continued the family tradition of researching radioactivity. The Joliot-Curies were awarded the 1935 Nobel Prize in chemistry for the synthesis of new radioactive isotopes of nitrogen, phosphorus, and silicon.

- *What well-known term did Marie Curie invent? What does it mean?*
- *What are some positive and negative aspects of radiation?*

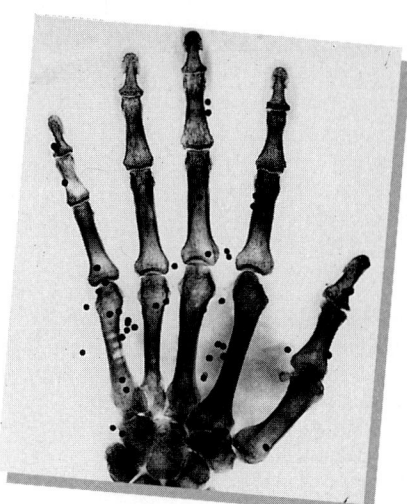

Exercise

15. A number of energy sources are available for heating: oil, gas, coal, wood, solar, geothermal, and nuclear. These sources produce heat directly. Indirect sources of heat include electric motors and lights.
    (a) Which two energy sources are the most common for heating?
    (b) Which two energy sources do you think will be most common by the year 2030?
    (c) Which two principal energy sources would you choose by the year 2030?

16. Suppose you represent a large utility company applying to a government regulatory body for permission to build a new power station to satisfy the electricity demand in your area. Your choices are coal, natural gas, hydro, solar, or nuclear power generating stations (Figure 11.25). Which type of station would you recommend and where should it be built? What are the alternatives to building a new power station? Be prepared to defend your decision from a variety of perspectives.

17. (Research) Energy consumers were briefly euphoric when successful "cold fusion" was announced by two scientists in 1989. Successful cold fusion would represent an inexpensive, clean, readily available source of energy. Do some library research or search the internet and find out what this was all about.
    (a) Write a report on the current prospects of cold fusion.
    (b) Explain why the initial announcement of the cold fusion "breakthrough" was so controversial.

18. (Discussion) Refer to Table 11.2 on page 444. List at least five more alternative energy sources or practices in each of the three categories. Then refer to books in your library to learn more about any one of them. Devise strategies for integrating your selected energy source or practice into your home, school, or community.

Figure 11.25
There is a variety of energy sources for generating electricity.
(a) In a hydroelectric power station, water collected behind a dam is released through a pipe to the turbine.
(b) Water in a boiler is heated in one of several ways:
• chemical energy from the combustion of fossil fuels in a thermal electric power plant
• nuclear energy from the fission of uranium in a nuclear power plant
• direct radiant solar energy reflected from many mirrors onto the boiler in a solar power plant
• geothermal energy from the interior of the Earth in a geothermal power plant

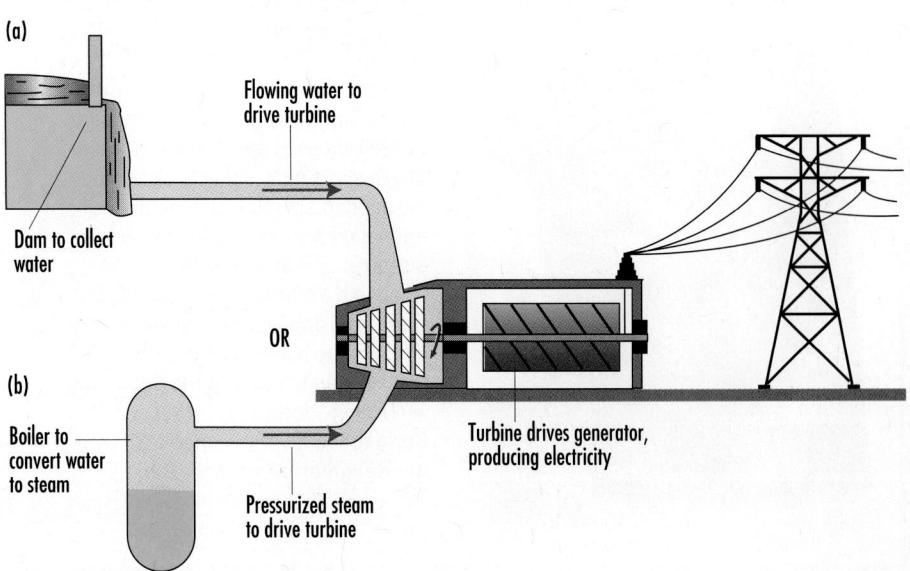

THE NUCLEAR POWER ALTERNATIVE

Nuclear power stations have much in common with conventional power stations fired by fossil fuels. In both, heat is used to boil water and the resulting steam drives a turbine. The spinning turbines, in turn, drive generators that produce electricity. In a conventional power station, burning natural gas, oil, or coal supplies the necessary heat; in a nuclear power station, nuclear fission provides the heat. When struck by a slow-moving (thermal) neutron, the nucleus of the uranium-235 isotope splits into two smaller nuclei. More neutrons are ejected, which may produce a chain reaction, as shown in the illustration on the right.

Uranium is a very concentrated energy source. For example, when placed in a CANDU reactor, a fuel bundle 50 cm long and 10 cm in diameter, with a mass of 22 kg, can produce as much energy as 400 t of coal or 2000 barrels of oil. At present, approximately 16% of the world's electricity is generated by nuclear power stations like the one shown in the photograph.

Canadian nuclear reactors use natural uranium containing about 0.7% uranium-235 and 99.3% uranium-238. The energy is produced by the fission of uranium-235 inside the reactor. Because of the vast quantities of energy released in fission, it has great potential as a commercial power source. Advocates of nuclear energy point out advantages such as its low fuel costs. Also, nuclear power does not contribute to global warming and acid rain because it does not produce any carbon dioxide or sulfur dioxide. There are, however, disadvantages to nuclear energy that must be weighed against its advantages. These disadvantages include the possible release of radioactive materials in a reactor malfunction; the difficulty of disposing of the highly toxic radioactive wastes; the large capital costs of building nuclear reactors; the short lifetime and the de-commissioning expense of nuclear reactors; and the risk that nuclear weapons will be manufactured from the plutonium produced during a nuclear reaction.

In the 1950s, people had high expectations of endless, inexpensive nuclear energy. Few would have predicted that concerns about reactor safety and radioactive wastes would severely dampen the enthusiasm for nuclear energy. But public attitudes changed after nuclear accidents such as the one at Chernobyl in the Ukraine, where a serious accident in a nuclear reactor on April 28, 1986, spewed a deadly, steam-driven cloud of radioactive plutonium, cesium, and uranium dioxide into the atmosphere. A nuclear reactor cannot explode like a nuclear bomb, as only steam explosions can occur.

In 1979, a reactor malfunctioned at Three Mile Island in Pennsylvania, but fortunately, the containment structure worked well. There have been no major nuclear accidents in Canada; in fact, Canadian reactors have the best safety and energy performance records in the world.

Less dramatic than a reactor malfunction, but also serious, is the continuing problem of radioactive waste disposal. Scientists and engineers have worked to devise a safe and economically feasible method of disposing of radioactive waste for many years, but the public remains skeptical. Burial in arid regions or in granite layers to avoid contaminating ground water are two possibilities. Chemists have been developing suitable materials for encasing radioactive substances to prevent their escape into the environment. Lead-iron-phosphate glass is a promising material, since the nuclear waste can be chemically incorporated into a stable glass and then buried in a safe place.

We must evaluate the nuclear energy alternative carefully. On one hand, it may be difficult to meet future energy needs and environmental demands without increased use of

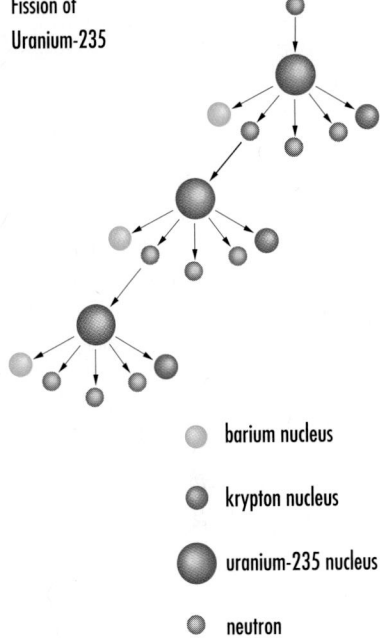

Fission of Uranium-235

- barium nucleus
- krypton nucleus
- uranium-235 nucleus
- neutron

nuclear power. On the other hand, an increased role for nuclear energy appears unlikely in the near future, given the wide range of its disadvantages and current public opinion.

- *What two nuclear accidents have dampened the public's interest in nuclear power plants?*
- *Why is the fission of uranium described as a chain reaction?*

Multi-Step Energy Calculations

In practice, energy calculations rarely involve only a single-step calculation of heat or enthalpy change. In most practical situations, several energy calculations might be required. These calculations may involve a combination of energy change definitions such as,

- heat flows, $q = mc\Delta t$ or $q = vc\Delta t$ or $q = C\Delta t$
- enthalpy changes, $\Delta H = nH$
- Hess's law, $\Delta H_{net} = \Sigma \Delta H$

$$\Delta H_r = \Sigma n H_{fp} - \Sigma n H_{fr}$$

For example, if the enthalpy change of a reaction and the quantity of reactant or product are known, a prediction of energy absorbed or released can be made. In the Solvay process for the production of sodium carbonate (page 274), one step is the endothermic decomposition of sodium hydrogen carbonate.

$$2\,NaHCO_{3(s)} + 129.2\ kJ \rightarrow Na_2CO_{3(s)} + CO_{2(g)} + H_2O_{(g)}$$

What quantity of chemical energy ΔH_r° is required to decompose 100 kg of sodium hydrogen carbonate? To answer this question, you first need to know the energy absorbed per mole of $NaHCO_3$, in other words, its standard molar enthalpy.

$$H_{r\ NaHCO_3}^\circ = \frac{\Delta H_r^\circ}{n} = \frac{129.2\ kJ}{2\ mol} = 64.6\ kJ/mol$$

This means that 64.6 kJ of energy is required for every mole of $NaHCO_3$ decomposed. Converting 100 kg to an amount in moles and multiplying by the standard molar enthalpy will give us the required ΔH_r°.

$$\Delta H_r^\circ = nH_r^\circ$$
$$= 100\ kg \times \frac{1\ mol}{84.01\ g} \times \frac{64.6\ kJ}{1\ mol}$$
$$= 76.9\ MJ$$

Therefore, 76.9 MJ is required to decompose 100 kg of sodium hydrogen carbonate.

In many cases, the enthalpy change of a particular reaction may not be given. The usual procedure is to determine the ΔH using Hess's law and then to proceed as in the previous example. The following example illustrates this method.

When the enthalpy change of a reaction is stated as a term in the equation, assume that the initial and final conditions under which the energy change is measured are SATP. If the conditions are other than SATP, they should be stated at the end of the equation.

EXAMPLE

What quantity of energy can be obtained from the roasting of 50.0 kg of zinc sulfide ore?

$$ZnS_{(s)} + \tfrac{3}{2}O_{2(g)} \rightarrow ZnO_{(s)} + SO_{2(g)}$$

$$\Delta H_c^\circ = \Sigma n H_{fp}^\circ - \Sigma n H_{fr}^\circ$$

$$= (1 \text{ mol} \times \frac{-350.5 \text{ kJ}}{1 \text{ mol}} + 1 \text{ mol} \times \frac{-296.8 \text{ kJ}}{1 \text{ mol}})$$

$$- (1 \text{ mol} \times \frac{-206.0 \text{ kJ}}{1 \text{ mol}} + \tfrac{3}{2} \text{ mol} \times \frac{0 \text{ kJ}}{1 \text{ mol}})$$

$$= -647.3 \text{ kJ} - (-206.0 \text{ kJ})$$

$$= -441.3 \text{ kJ}$$

$$H_c^\circ_{\text{ZnS}} = \frac{\Delta H_c^\circ}{n}$$

$$= \frac{-441.3 \text{ kJ}}{1 \text{ mol}} = -441.3 \text{ kJ/mol}$$

$$\Delta H_c^\circ_{\text{ZnS}} = n H_c^\circ$$

$$= 50.0 \text{ kg} \times \frac{1 \text{ mol}}{97.44 \text{ g}} \times \frac{-441.3 \text{ kJ}}{1 \text{ mol}}$$

$$= -226 \text{ MJ}$$

According to the formation method of Hess's law, 226 MJ of energy can be obtained.

A multi-step energy calculation is shown in the following example. The energy produced by a chemical reaction is used to heat another substance. The key step in the procedure is based on the law of conservation of energy. Note the similarity to calorimetry calculations.

$$\text{total enthalpy change} = \text{quantity of heat}$$
$$\Delta H_r^\circ = q$$

EXAMPLE

What mass of octane is completely burned during the heating of 20 L of aqueous ethylene glycol automobile coolant from –10°C to 70°C? The volumetric heat capacity of the aqueous ethylene glycol is 3.7 kJ/(L•°C).

$$2 \, C_8H_{18(l)} + 25 \, O_{2(g)} \rightarrow 16 \, CO_{2(g)} + 18 \, H_2O_{(g)}$$

$$\Delta H_c^\circ = \Sigma n H_{fp}^\circ - \Sigma n H_{fr}^\circ$$

$$= (16 \text{ mol} \times \frac{-393.5 \text{ kJ}}{1 \text{ mol}} + 18 \text{ mol} \times \frac{-241.8 \text{ kJ}}{1 \text{ mol}})$$

$$- (2 \text{ mol} \times \frac{-250.1 \text{ kJ}}{1 \text{ mol}} + 25 \text{ mol} \times \frac{0 \text{ kJ}}{1 \text{ mol}})$$

$$= -10 \ 648.4 \text{ kJ} - (-500.2 \text{ kJ})$$

$$= -10 \ 148.2 \text{ kJ}$$

$$H_c^\circ_{C_8H_{18}} = \frac{-10 \ 148.2 \text{ kJ}}{2 \text{ mol}} = -5074.1 \text{ kJ/mol}$$

$\Delta H^\circ_c = q$ is based on the assumption that the heat released by the combustion equals the heat gained by the glycol.

$$\Delta H^\circ_c = q$$
$$\text{(octane)} \quad \text{(glycol)}$$
$$nH^\circ_c = vc\Delta t$$

$$n \times 5074.1 \text{ kJ/mol} = 20 \text{ L} \times 3.7\frac{\text{kJ}}{\text{L}\cdot°\text{C}} \times 80°\text{C}$$

$$n_{\text{C}_8\text{H}_{18}} = 1.2 \text{ mol}$$

$$m_{\text{C}_8\text{H}_{18}} = 1.2 \text{ mol} \times \frac{114.26 \text{ g}}{1 \text{ mol}} = 0.13 \text{ kg}$$

According to the molar enthalpy of formation method and the law of conservation of energy, the mass of octane required is 0.13 kg.

Fire!

As the cost of oil and gas has risen, more people have turned to wood as a cheaper alternative for fuel. It is estimated that 6% of Canadians now depend solely on wood for home heating, while 14% use it as a supplement. Unfortunately, as more people use wood, the number of house fires has increased dramatically. The problem is creosote. Creosote is a complex mixture of various phenols and phenolic ethers produced during the incomplete combustion of wood. As the hot gases released by a wood fire rise up through the chimney, any creosote oil in those gases will condense and coat the inside surface. If the chimney is not regularly cleaned, this creosote may build into a thick layer. Sparks rising up the chimney may ignite the creosote, causing a fire that can quickly spread to the entire home. Such fires can be prevented by regular maintenance of the chimney and stove or fireplace. Newer, high-efficiency wood-burning appliances don't release creosote. However, these appliances cost more and many people are unwilling to pay the extra cost.

Exercise

19. Coal is a major energy source for electricity, of which industry is the largest user (Figure 11.26). Anthracite coal is a high-molar-mass carbon compound with a composition of about 95% carbon by mass. A typical simplest-ratio formula for anthracite coal is $C_{52}H_{16}O_{(s)}$. What is the quantity of energy available from burning 100 kg of anthracite coal in a thermal electric power plant, according to the following chemical equation?

$$2\,C_{52}H_{16}O_{(s)} + 111\,O_{2(g)} \rightarrow 104\,CO_{2(g)} + 16\,H_2O_{(g)}$$
$$\Delta H^\circ_c = -44.0 \text{ MJ}$$

20. What are some alternatives to non-renewable fossil fuels such as coal for electric power generation? Write a short report discussing the advantages and disadvantages of one alternative, or list the advantages and disadvantages in a table.

21. Transportation accounts for about 30% of energy use in Canada. Most of this energy is supplied by burning gasoline. Using the following typical gasoline combustion equation, calculate the energy produced per kilogram of octane burned.

$$2\,C_8H_{18(l)} + 25\,O_{2(g)} \rightarrow 16\,CO_{2(g)} + 18\,H_2O_{(g)}$$
$$\Delta H^\circ_c = -10\,148.2 \text{ kJ}$$

22. Alternative transportation fuels include methanol and hydrogen.
    (a) Calculate the energy produced per kilogram of methanol burned.
    (b) Calculate the energy produced per kilogram of hydrogen burned.
    (c) In terms of energy content, how do these two alternative fuels compare with gasoline (octane) in question 21?

(d) What factors other than energy content are important when comparing different automobile fuels? Include several perspectives.

23. In a typical household, about one-quarter of the energy consumed is used to heat water.
   (a) What amount of methane undergoing complete combustion is required to heat 100 L of water from 5°C to 70°C?
   (b) How might we heat water more efficiently?
   (c) What alternative energy resources are available for heating water?

24. Canadian (CANDU) nuclear reactors produce energy by nuclear fission of uranium-235, as displayed in the following equation. If the molar enthalpy of fission for uranium-235 is 1.9×10^{10} kJ/mol, how much energy can be obtained from the fission of 1.00 kg (4.26 mol) of uranium-235?

$$^{235}_{92}\text{U} + {}^{1}_{0}\text{n} \rightarrow {}^{141}_{56}\text{Ba} + {}^{92}_{36}\text{Kr} + 3{}^{1}_{0}\text{n} + 1.9 \times 10^{10} \text{ kJ}$$

Figure 11.26
Strip mining of coal and reclamation of the land occur simultaneously at the site shown in the photograph. The coal mined here is burned at a nearby power generating station.

Problem 11F Molar Enthalpy of Formation

During the catalytic reforming process used in an oil refinery, cyclohexanes are converted into aromatic hydrocarbons. Because they are excellent fuels, aromatic compounds are used in high-test gasolines for racing cars. The purpose of this problem is to practice calculations based on calorimetric evidence. Note that in a bomb calorimeter, combustion produces liquid water. Complete the Analysis of the investigation report.

Problem

What is the molar enthalpy of formation for cyclohexane?

Experimental Design

A small quantity of cyclohexane, $C_6H_{12(l)}$, is burned in a bomb calorimeter. Initial and final temperatures of the water are recorded.

Evidence

mass of cyclohexane = 1.43 g
heat capacity of calorimeter = 10.5 kJ/°C
initial temperature of calorimeter = 20.32°C
final temperature of calorimeter = 26.67°C

OVERVIEW

Reaction Enthalpies

Summary

- Enthalpy changes of chemical reactions may be communicated by specifying the standard molar enthalpy $H°$ for a chemical in the reaction or the standard enthalpy change $\Delta H°$ of the reaction, by including an energy term in a balanced chemical equation, or by drawing a potential energy diagram.

- Enthalpy changes are determined calorimetrically by using Hess's law with a list of known chemical equations or by using Hess's law with a table of standard molar enthalpies of formation.

  $$\Delta H_{net} = \Sigma \Delta H$$

  $$\Delta H_r = \Sigma n H_{fp} - \Sigma n H_{fr}$$

- Nuclear reactions involve changes in the bonding of particles in the nucleus; these reactions involve the greatest enthalpy changes. The enthalpy changes of nuclear reactions can be communicated in the same four ways as for chemical reactions.

- Multi-step energy calculations involve a combination of energy concepts, including calorimetry and Hess's law.

Key Words

Hess's law
potential energy diagram
reference energy state
standard molar enthalpy
thermal stability

Review

1. List four ways to communicate an enthalpy change.

2. How does the sign of a ΔH correspond to the location (reactant or product side) of the energy term in a chemical equation?

3. You are given the following information for butane: $H_c° = -2657$ kJ/mol.
   (a) Describe in words what this information conveys.
   (b) Translate this information into a balanced chemical equation with a $\Delta H_c°$.
   (c) Rewrite the chemical equation, including the enthalpy change as a term in the equation.
   (d) What is the value of $H_f°$ for $C_4H_{10(g)}$ and why is this value different from the information given above?

4. What is the reference zero point for chemical potential energy diagrams of all chemical reactions?

5. In general, how does the change in chemical potential energy of an endothermic reaction compare with the change in an exothermic reaction?

6. Calorimetry can be used to determine a molar enthalpy empirically. List two ways to *predict* enthalpy changes of chemical reactions.

7. What is the general principle underlying all methods of predicting $\Delta H_r°$?

8. When applying Hess's law, what are two rules for manipulating the values of ΔH for chemical equations?

9. An efficient way to predict an enthalpy change uses Hess's law and a table of standard molar enthalpies of formation.
   (a) What is the mathematical formula for this method?
   (b) Describe in words what this formula means.
   (c) How do the two summation terms (Σ) relate to a chemical potential energy diagram?

10. How do nuclear reactions compare with chemical reactions in terms of
    (a) the magnitude of the energy change?
    (b) the sign of the energy change?

11. For each of our society's three major energy demands,
    (a) list the best alternative, in your opinion, to fossil fuels.
    (b) list ways of reducing our consumption of fossil fuels other than switching to alternatives.

Applications

12. In a calorimetry experiment, the standard molar enthalpy of combustion for propane is determined to be –2.25 MJ/mol. Translate this information into a balanced chemical equation.

13. Baking soda can extinguish small grease fires in a kitchen. When thrown on a fire, the baking soda absorbs energy and decomposes to produce carbon dioxide, which helps smother the fire.

    $$2\,NaHCO_{3(s)} + 129\ kJ \rightarrow Na_2CO_{3(s)} + H_2O_{(g)} + CO_{2(g)}$$

    (a) What is the standard molar enthalpy for carbon dioxide in this reaction?
    (b) Rewrite this equation using the ΔH_r° notation.
    (c) Draw a potential energy diagram to communicate this information.

14. The following reaction is important in catalytic converters in automobiles.

    $$2\,CO_{(g)} + 2\,NO_{(g)} \rightarrow N_{2(g)} + 2\,CO_{2(g)} + 746\ kJ$$

    (a) Rewrite this equation using the ΔH_r° notation.
    (b) What is the standard molar enthalpy of reaction for nitrogen monoxide?
    (c) What quantity of energy is released by the chemical system into the environment when 500 g of nitrogen monoxide reacts?

15. Hydrazine, used as a rocket fuel, undergoes the following combustion.

    $$N_2H_{4(l)} + 3\,O_{2(g)} \rightarrow 2\,NO_{2(g)} + 2\,H_2O_{(g)} + 400\ kJ$$

    (a) Given the information above, report the standard molar enthalpy of combustion for hydrazine.
    (b) What is the standard molar enthalpy of reaction for oxygen in this chemical equation?

    (c) How much energy is released to the surroundings if 8.00 g of oxygen is consumed?

16. When copper(II) sulfide is roasted in air, copper(II) oxide and sulfur dioxide are formed.
    (a) Calculate the ΔH_r° for the reaction of one mole of $CuS_{(s)}$.
    (b) Draw and label a potential energy diagram.

17. In cellular respiration, glucose and oxygen combine to form carbon dioxide, liquid water, and energy.
    (a) Calculate the ΔH_r° for this reaction.
    (b) Compare this reaction with the combustion of glucose. How are they similar and how are they different?

18. Ammonia forms the basis of a large fertilizer industry. Laboratory research has shown that nitrogen from the air reacts with water, using sunlight and a catalyst to produce ammonia and oxygen. This research, if technologically feasible on a large scale, may lower the cost of ammonia fertilizer.
    (a) Determine the ΔH_r° of the reaction using a chemical equation balanced with whole number coefficients.
    (b) Calculate the quantity of solar energy needed to produce 1.00 kg of ammonia.
    (c) If 3.60 MJ of solar energy is available per square metre each day, what area of solar collectors would provide the energy to produce 1.00 kg of ammonia in one day?
    (d) What assumption is implied in the previous calculation?

19. Calculate the enthalpy change for
    (a) the condensing of 1.00 mol of steam to water at 100°C.
    (b) the formation of 1.00 mol of water from its elements.
    (c) the formation of 1.00 mol of helium-4 in a fusion reaction (page 443).
    (d) (Discussion) If the preceding enthalpy changes were represented on a graph with a scale of 1 cm to 100 kJ, calculate the distance for each enthalpy change in parts (a), (b), and (c). How many pages (28 cm per page) would be needed to represent the nuclear enthalpy change?

20. Most ethane extracted from natural gas is converted into ethene (ethylene) by the following cracking reaction. The ethylene is used to produce hundreds of consumer products.

$$C_2H_{6(g)} \rightarrow C_2H_{4(g)} + H_{2(g)} \qquad \Delta H_r^\circ = \text{?}$$

(a) What is the standard enthalpy change for the cracking of ethane?

(b) Draw a potential energy diagram to represent the cracking of ethane.

21. Acetylene is commonly used in oxyacetylene welding. What is the standard molar enthalpy of combustion for acetylene?

22. Arrange the following compounds in order of increasing stability: ethane, ethylene, acetylene.

23. Polyvinyl chloride is a polymer produced from ethylene. Ethylene first reacts with chlorine to produce the vinyl chloride (chloroethene) monomer and other products. Because of the mixture of by-products, a direct calorimetric determination of the enthalpy change of the ethylene and chlorine reaction is impossible. Use the following calorimetrically determined enthalpy changes and Hess's law to predict the standard enthalpy change for the reaction of ethylene with chlorine.

$$C_2H_{4(g)} + Cl_{2(g)} \rightarrow C_2H_3Cl_{(g)} + HCl_{(g)}$$
$$\Delta H_r^\circ = \text{?}$$

$$H_{2(g)} + Cl_{2(g)} \rightarrow 2\,HCl_{(g)} \qquad \Delta H_f^\circ = -184.6 \text{ kJ}$$

$$C_2H_{4(g)} + HCl_{(g)} \rightarrow C_2H_5Cl_{(l)} \quad \Delta H_r^\circ = -65.0 \text{ kJ}$$

$$C_2H_3Cl_{(g)} + H_{2(g)} \rightarrow C_2H_5Cl_{(l)}$$
$$\Delta H_r^\circ = -138.9 \text{ kJ}$$

24. Ethylene glycol (1,2-ethanediol) is a petrochemical produced in large quantities from ethylene, using the following two-step process. Ethylene glycol is used as antifreeze, as hydraulic fluid, and as a raw material in the manufacture of polyesters.

$$C_2H_{4(g)} + \tfrac{1}{2}O_{2(g)} \rightarrow C_2H_4O_{(g)}$$

$$C_2H_4O_{(g)} + H_2O_{(1)} \rightarrow C_2H_4(OH)_{2(1)}$$

(a) Write the net chemical equation for the production of ethylene glycol from ethylene.

(b) Calculate the ΔH_r° for the net equation.

(c) Draw a potential energy diagram for the net reaction.

(d) What is the enthalpy change involved in the manufacture of 1.00 t of ethylene glycol?

25. Chloroethane, $C_2H_5Cl_{(1)}$, is a petrochemical used to produce tetraethyl lead, an anti-knock additive in gasoline. In the reaction of ethylene with hydrogen chloride in a bomb calorimeter, chloroethane is produced. The following evidence was gathered in an experimental determination of the molar enthalpy of reaction for chloroethane.

mass of chloroethane = 15.78 g
heat capacity of bomb calorimeter = 4.06 kJ/°C
initial temperature = 19.03°C
final temperature = 22.92°C

(a) According to this evidence, what is the molar enthalpy for chloroethane in this reaction?

(b) Write a balanced chemical equation, including the enthalpy change for this reaction.

(c) Calculate the molar enthalpy of formation of chloroethane.

26. What mass of propane must be burned to heat 2.50 L of water from 10°C to 80°C?

27. Design at least three experiments to identify a pure liquid as methanol. Rank the experiments in terms of certainty of results.

Extensions

28. (Internet or CD-ROM encyclopedia; search using "automobile fuel alternative.") Choose an alternative fuel for automobiles. What are the advantages and disadvantages of this substance as a replacement for gasoline? Include a scientific perspective (enthalpy changes), a technological perspective (octane ratings and any other technical considerations), an ecological perspective (environmental impact), an economic perspective (relative costs), and any other perspectives important to this issue.

29. Cold packs are used to treat sports injuries (Figure 10.26 on page 414). These packages contain an ionic compound that has an

endothermic heat of solution. Design a new product to replace or compete with cold packs. Investigate the use of an endothermic chemical reaction as a cold pack. Use two common materials — citric acid and baking soda. In your research, determine and compare enthalpy changes. Design your new product based on the technological criteria of reliability, economy, and simplicity.

30. Determining standard enthalpies of reaction requires that the initial and final conditions for the reactants and products, respectively, must be at SATP. Design a calorimeter capable of making this measurement without the calorimeter water affecting the final conditions. Assume you are burning a sample of propane in the calorimeter.

Problem 11G H_f for Calcium Oxide

Complete the Prediction and the Analysis of the investigation report.

Problem

What is the enthalpy change for the reaction?

$$Ca_{(s)} + \tfrac{1}{2} O_{2(g)} \rightarrow CaO_{(s)}$$

Experimental Design

Calcium metal reacts with hydrochloric acid in a calorimeter and the enthalpy change is determined. Similarly, the enthalpy change for the reaction of calcium oxide with hydrochloric acid is determined. These two chemical equations are combined with the formation equation for water to determine the required enthalpy change.

Evidence

concentration of $HCl_{(aq)}$ = 1.0 mol/L
volume of $HCl_{(aq)}$ = 100 mL

Reactant	Mass (g)	Initial Temperature (°C)	Final Temperature (°C)
$Ca_{(s)}$	0.52	21.3	34.5
$CaO_{(s)}$	1.47	21.1	28.0

Problem 11H H_c by Four Methods

The molar enthalpy of combustion of a substance can be determined experimentally by calorimetry. It can also be predicted from standard molar enthalpies of formation, from Hess's law using formation reactions, or from bond energies obtained from a reference. Predict the molar enthalpy of combustion of lighter fluid, methylpropane, by all three methods and complete the Analysis of the investigation report. Evaluate the methods used to determine the enthalpy of combustion.

Problem

What is the molar enthalpy of combustion of methylpropane?

Experimental Design

The molar enthalpy of combustion is predicted by three different methods, and is also determined in a bomb calorimeter.

Evidence

heat capacity of bomb calorimeter = 9.35 kJ/°C
mass of methylpropane burned = 1.52 g
initial temperature of calorimeter = 20.21°C
final temperature of calorimeter = 28.25°C

REACTION KINETICS
AND
EQUILIBRIUM

"We can never achieve absolute truth

but we can live hopefully by a system

of calculated probabilities."

Agnes Meyer, American writer and

social worker (1887 – 1970)

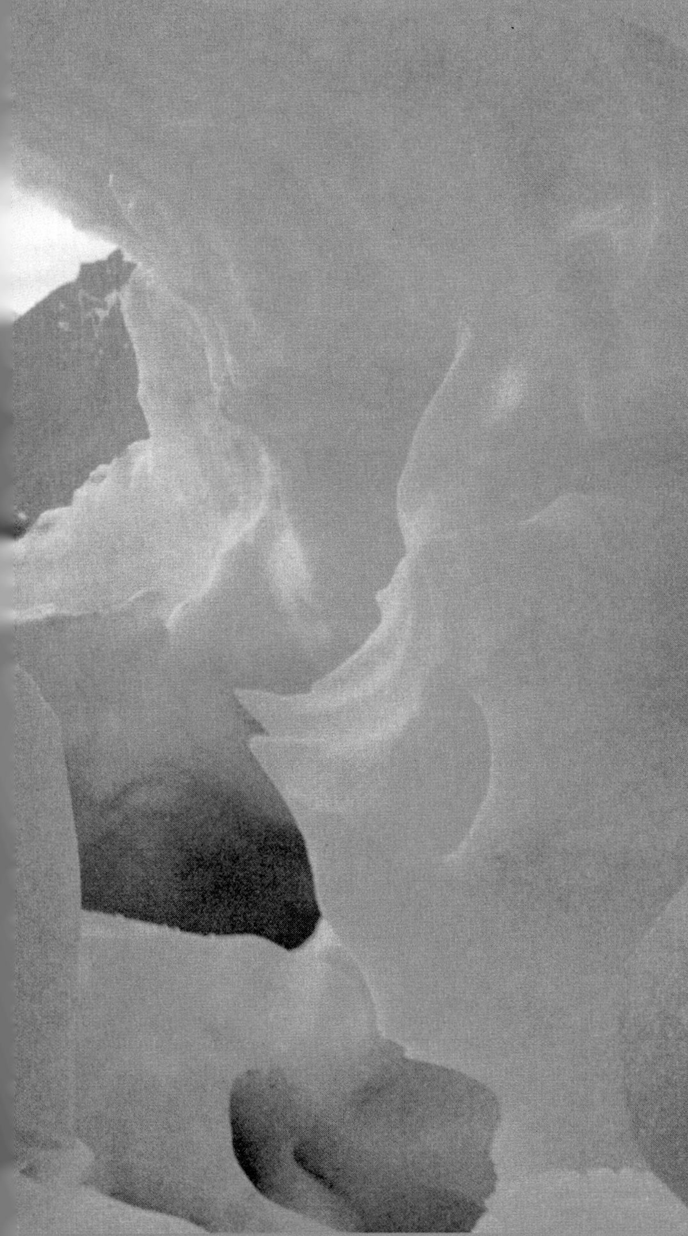

Equilibrium represents a balance, either a static balance of forces as in the case of snow on a mountainside or a dynamic balance of the rates of two opposing reactions. Any balance may be momentarily disrupted as evidenced by an avalanche or a precipitation in a chemical reaction. This always leads to a new balance or equilibrium. Understanding the dynamic equilibrium of a chemical system requires a clear understanding of reaction kinetics.

This unit of study will deepen your knowledge of the nature of science. As you revise your concepts about rates of reaction and equilibrium, especially in acid-base systems, your ideas about the nature of the scientific endeavor will be challenged and refined. Addressing issues such as acid deposition in the environment provides opportunities for STS decision making. The integration of knowledge for problem solving — in the context of science, technology, and society — is the ultimate challenge in your study of chemistry in high school.

12 Chemical Kinetics

So-called cold-blooded animals such as this garter snake have metabolic rates of reaction that increase or decrease in response to changes in the temperature of their surroundings. Scientists who study rates are involved in chemical kinetics — the study of particle interactions at a molecular level during a chemical reaction.

Chemical kinetics is a fundamental area of chemistry that crosses over into many other areas of science and engineering. Biologists are interested in the changes in the rate of some of your bodily reactions when you have a fever, as well as in the progress of reactions involved in areas such as growth and bone regeneration. Automotive engineers want to increase the rate at which noxious pollutants are oxidized in car exhaust systems and to decrease the rate of rusting of the metal structure of the car. Agriculture specialists are concerned with the progress of reactions involved in spoilage and decay and with the ripening rate of many foodstuffs.

For most of your study so far, discussion has been limited to reactant and product conditions. Conditions have been described and predicted before and after the chemical reaction, with the progress of the reaction itself represented only by a single arrow in a chemical equation. As you study particle mechanics during chemical reactions, you will increase the complexity and detail of your knowledge. Because changes at the molecular level are not directly observable, the topics in this chapter and those following are predominantly theoretical.

12.1 RATE OF REACTION

From a scientific perspective, **chemical kinetics** starts with empirical research of reaction times and progressive changes in measurable properties. Chemical kinetics also includes the development of theoretical concepts and descriptions to explain and predict observed rates of reaction. As is often the case in science, theoretical knowledge of chemical kinetics has lagged behind empirical knowledge. This is particularly true for rates of reactions in solution and most heterogeneous systems of interest. The scientific community feels truly confident only about theoretical explanations for some relatively simple homogeneous gas phase reactions.

In the past few years, technological advance in instrumentation and computers has greatly aided researchers in the field of reaction rates, especially for very high speed reactions like explosions. Faster and more powerful computers aid theories of chemical kinetics by allowing more detailed development of theoretical reaction models.

Chemical kinetics is an area of chemistry that still presents good research opportunities for future chemists. The scientific community uses the tactful phrase "not well understood" to describe areas of knowledge like chemical kinetics where theoretical explanations are very incomplete.

Measurement of Reaction Rates

A **reaction rate** is usually obtained by somehow measuring the rate at which a product is formed or the rate at which a reactant is consumed. Properties such as mass, color, conductivity, volume, and pressure may be measured, depending on the nature of the reaction investigated. Rates of reaction are expressed mathematically in terms of a change in property of a reactant or product per unit of time. A variety of units may be used, such as in 1.2 mmol/min or 1.5 mL/s. For the sake of convenience and consistency — and to allow easy comparison of many reaction types — reaction rates are often expressed as a change in concentration per unit of time, as mol/(L•s), for example. Symbolically, where r is the average reaction rate, ΔC is the change in concentration, and Δt is the elapsed time, the expression

$$\text{average reaction rate} = \frac{\text{change in concentration}}{\text{elapsed time}}$$

becomes

$$r = \frac{\Delta C}{\Delta t}$$

Concentration, calculated from empirical (measured) data, is the most useful variable to use because reaction rate does *not* depend on the size of the sample but *is* normally dependent on the concentrations of the reactants. Evidence shows that, for most reactions, the concentration changes are more rapid near the beginning of the reaction and the rate decreases with the time elapsed.

Typically for a reaction such as

$$CH_{4(g)} + Cl_{2(g)} \rightarrow CH_3Cl_{(g)} + HCl_{(g)}$$

the concentration of $Cl_{2(g)}$ during the progress of the reaction would plot as a curve with a negative slope, as shown in Figure 12.1. The rate of reaction at any given point in time is the *absolute value* of the slope of the curve at that point. Figure 12.2 shows lines drawn tangent to the curve at two times during the progress of the reaction. Later in the reaction, the slope is less steep, meaning the rate is slower.

Figure 12.1
The concentration of a reactant during the progress of a reaction decreases continuously.

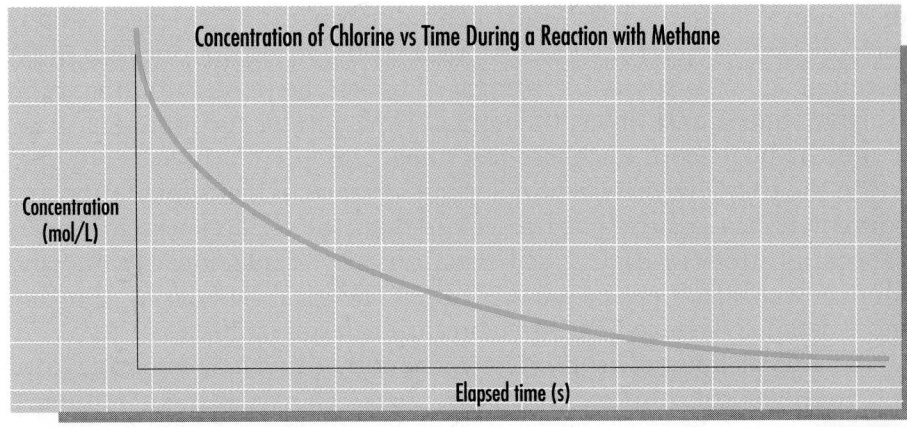

Concentration of Chlorine vs Time During a Reaction with Methane

Concentration (mol/L)

Elapsed time (s)

Figure 12.2
The rate of reaction of a particular reactant during the progress of a reaction, measured at times A and B.

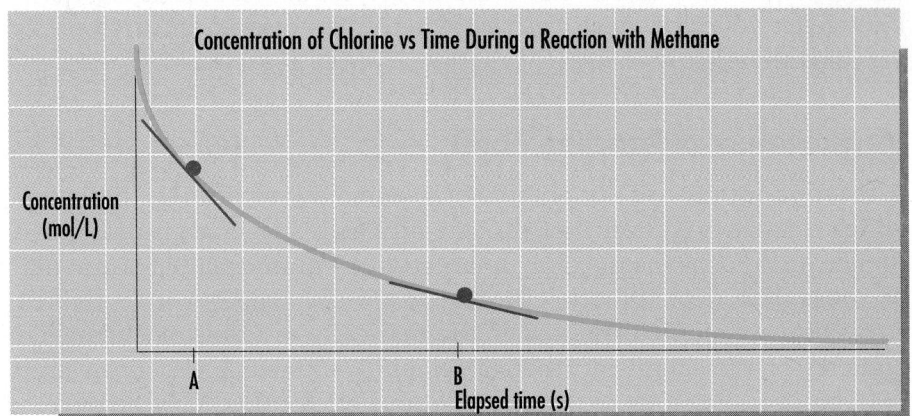

Concentration of Chlorine vs Time During a Reaction with Methane

Concentration (mol/L)

A
B
Elapsed time (s)

Figure 12.3
The rate of production of a low solubility gaseous product, measured by water displacement volume.

Many methods are available to measure a reaction rate. The method chosen depends on the kinds of substances involved in the reaction and the nature and characteristics of the reaction. An ideal method would allow a direct measurement of a reactant or product without disturbing the progress of the reaction itself, which is possible for some reactions. If a reaction produces a gas, such as the reaction of zinc and hydrochloric acid in aqueous solution, the gas may be collected and its volume and/or pressure may be measured as the reaction proceeds (Figure 12.3). The reaction progress is not affected because the gas has very low solubility and bubbles up out of the solution, with the result that the gas leaves the reaction zone anyway.

For some reactions, a property such as conductivity changes in the reaction vessel as the reaction proceeds. For example, the hydrolysis of alkyl halides in aqueous ethanol solution starts with neutral molecules as reactants and produces charged ions, so that as the reaction proceeds and more ions form, the conductivity increases (Figure 12.4). This conductivity can be measured and plotted graphically as a function of time.

$$(CH_3)_3CCl_{(aq)} + H_2O_{(l)} \rightarrow (CH_3)_3COH_{(aq)} + H^+_{(aq)} + Cl^-_{(aq)}$$

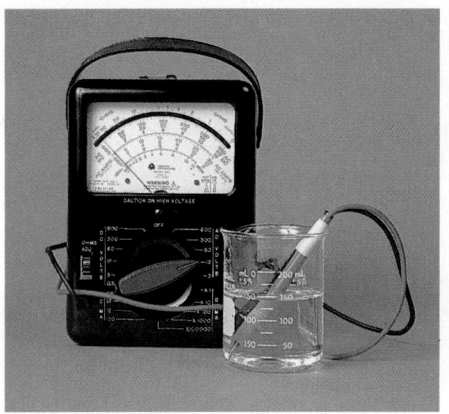
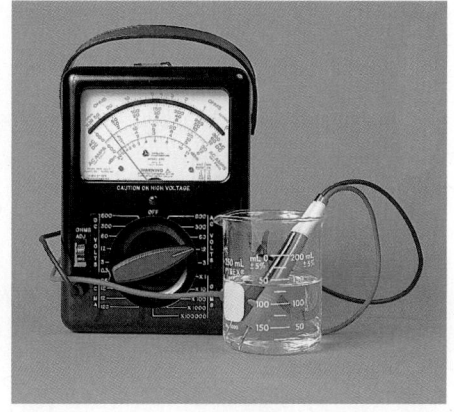

Figure 12.4
The rate of production of an ionic product, measured by conductivity.

For some reactions, especially in solution, measurement of a colored reactant or product may be made by measuring the intensity of its color. For precise measurements of color intensity, a spectrophotometer is used, which adds the advantage that the machine can accurately measure light wavelengths outside the human visible spectrum (Figure 12.5). The following chemical equation represents an example where color may be used to measure concentration of a reactant or product.

$$ClO^-_{(aq)} + I^-_{(aq)} \rightarrow IO^-_{(aq)} + Cl^-_{(aq)}$$
$$\text{colorless} \quad\quad \text{colorless} \quad\quad \text{yellow} \quad\quad \text{colorless}$$

For this reaction, the yellow color of the aqueous hypoiodite ions appears initially and becomes more evident as the reaction progresses and more $IO^-_{(aq)}$ ions form.

Figure 12.5
The rate of production of a colored product, measured by light absorbency.

INVESTIGATION

12.1 The Rate of a Chemical Reaction

The purpose of this experiment is to apply a standard empirical trial-and-error technique to begin a study of reaction rates. By determining what reaction conditions affect the rate of a common reaction, you gather evidence to use later to develop a useful theory to explain reaction rates in general.

Problem

How does altering reaction conditions affect the rate of a reaction?

Experimental Design

In this experiment an excess of hydrochloric acid is reacted with magnesium metal.

$$Mg_{(s)} + 2\ HCl_{(aq)} \rightarrow H_{2(g)} + MgCl_{2(aq)}$$

The acid concentration and reaction temperature are changed in separate trials to determine whether these variables affect the rate of reaction.

- Problem
- Prediction
- Design
- ✔ Materials
- ✔ Procedure
- ✔ Evidence
- ✔ Analysis
- Evaluation
- Synthesis

 **CAUTION**

Hydrochloric acid is corrosive. Avoid contact with skin, eyes, clothing, or the desk. If you spill this acid on your skin, wash immediately with lots of cool water.

Hydrogen gas, produced in the reaction of hydrochloric acid and magnesium, is flammable. Ensure that there is adequate ventilation and that there are no open flames in the classroom.

Exercise

1. State three examples of properties, directly related to reactants or products, that could be used to measure a reaction rate.

2. In an experiment, 1.00 g of zinc metal reacts completely in 2.00 L of (initially) 1.00 mol/L hydrochloric acid over a 15.0 s time interval.
   (a) Express the average reaction rate of the zinc in units of mol/s.
   (b) Find the final concentration of hydrochloric acid after 15 s.
   (c) Express the average rate of reaction of the hydrochloric acid in units of mol/(L·s).
   (d) Express the average rate of production of aqueous zinc chloride in units of mol/(L·s).
   (e) What units could be used to express the rate of formation of hydrogen?

3. In a complete sentence, give an example of a reaction rate that concerns many Canadians — one that involves cars, cold mornings, and cold batteries.

4. Sketch the graph of Figure 12.1, page 460 and add possible plot lines for the molar concentrations of methane, $CH_{4(g)}$, and for hydrogen chloride, $HCl_{(g)}$, as the reaction progresses. Assume that the initial concentration of methane is double the initial concentration of chlorine.

5. Suggest how the progress of a slow acid-base reaction in solution might be measured.

12.2 FACTORS AFFECTING REACTION RATES

INVESTIGATION

Problem
Prediction
Design
Materials
✔ Procedure
✔ Evidence
✔ Analysis
Evaluation
Synthesis

CAUTION

Sulfurous acid, sulfuric acid, and hydrogen peroxide are corrosive; handle the solutions with care. Avoid contact with skin, eyes, clothing, and the desk. If you spill these on your skin, wash immediately with lots of cool water.

12.2 Demonstration: Variables Affecting Reaction Rates

The purpose of this demonstration is to provide further evidence for the effect of reaction variables on the rate of three different chemical reactions. The concepts of variable control and variable manipulation are also clearly illustrated by the procedure used.

Problem

What variables affect the rates of chemical reactions?

Experimental Design

The iodine clock reaction is performed several times, varying the concentration of one of the solutions, or varying the temperature of the reaction, or adding a catalyst each time.[1] Cornstarch, a complex carbohydrate, is placed in a burner flame in the form of a solid lump and in the form of a finely divided powder.[2] Hydrogen peroxide solution is allowed to decompose in an open container, with and without the presence of a catalyst.[3]

Empirical Effect of the Nature of Reactants

For similar reactions, the nature of the reactant chosen affects the reaction rate. For example, the metals zinc, iron, and lead all react with hydrochloric acid to produce hydrogen gas as one of the products. When conditions of temperature, concentration, amount, and physical shape are the same for these reactions, the rates are quite different, as shown in Figure 12.6. It has become traditional to speak of the *activity series* for common metals based on exactly this comparison — the rate of reaction with simple acids. Historically, a concern with metals has always been their rate of corrosion, and scientists have searched for methods that would prevent or slow the corrosion rate.

Figure 12.6
The different rates of reaction of zinc (left), iron (center), and lead (right) with equally concentrated samples of hydrochloric acid.

In homogeneous chemical systems, such as reactions in solution, most reactions of monatomic ions — for example, silver ions and chloride ions — are extremely fast. Reactions of the more complicated polyatomic ions or of molecular substances are often much slower. For example, glucose molecules and iron(II) ions both react with acidified permanganate ions in a highly visible way. The permanganate solution is a deep purple color, and the disappearance of this color visibly indicates the point at which all of the permanganate ions have reacted. The reaction rate of covalently bonded glucose molecules with permanganate ions is very slow compared with the rate of reaction with monatomic iron(II) ions, as shown in Figure 12.7.

The old mnemonic — Please Send Charlie McCarthy A Zebra In That Long Hog Crate Marked Striped Printed Goods — is a mental trick for remembering that the relative reactivity of metals (and hydrogen) with aqueous hydrogen ions (acids) follows this order: potassium, sodium, calcium, magnesium, aluminum, zinc, iron, tin, lead, *hydrogen*, copper, mercury, silver, platinum, and gold. The metals listed from copper to gold will not react with common acids.

Figure 12.7
The series of photos shows the different reaction rates of glucose molecules and of iron(II) ions with acidified permanganate ions, showing the reaction at initial mixing (left), after 5 s (center), and after 100 s (right).

Figure 12.8
The photograph shows the reaction of solid iron and of iron fibers (steel wool) heated in air. The rate is much faster with the finely divided metal because of the increased surface exposure.

Empirical Effect of Concentration and Surface Area

Empirical evidence indicates that if the concentration of a reactant is increased then the reaction rate generally increases. This is supported by the evidence gathered so far in this chapter, as well. For example, a higher concentration of hydrochloric acid noticeably increases the rate of hydrogen gas production in Investigation 12.1. This is such a common effect that everyday language routinely uses the word "concentrated" to imply faster and more effective, as we see in advertising for cleaning compounds and fabric softeners.

Where a reaction system is heterogeneous, the reaction occurs at the interface of two different phases present, and the exposed surface area where the two reacting phases are in contact affects the reaction rate. In Investigation 12.1, the reaction takes place only where the magnesium metal surface is in contact with the hydrochloric acid solution. A reaction with finely divided iron fibers is much faster than with one solid piece of equal mass because the small pieces have a much greater overall surface area, as shown in Figure 12.8. In general, if the surface area increases, reaction rate increases proportionally. Heterogeneous reactions are very common, including any reaction of a gas with a solid or liquid and any reaction of a solid with a solution.

The classical example illustrating the effect of surface area is to consider the difficulty of lighting a solid block of wood with a match. If the same wood block is then whittled into fine shavings, the wood lights easily and burns rapidly. If the same mass of wood is made into very fine sawdust and blown into the air, it may burn so rapidly that it becomes an explosion hazard.

Empirical Effect of Temperature

Based on the empirical evidence from this chapter, if the temperature of the system increases then the reaction rate generally increases. This is commonly evident in the drying of paint and the setting of glues (really the completion of reactions) and especially in that very common manipulation of reaction rate variables we call cooking. Chemists have long known that as a rule of thumb, around SATP, a 10°C rise in temperature often doubles or triples the rate of a chemical reaction.

Most of the reactions making up human physiology are operating at the preferred rate at about 37°C, but illness can necessitate the speeding up of some reactions — notably the ones that produce disease-fighting agents. When you have an abnormally high temperature, the fever is a way of increasing the rate of these immune system reactions — at the expense of also speeding up others that can be harmful.

Lowering the temperature of the system can decrease the rate of reaction, which is useful for food storage (Figure 12.9). Insects such as this butterfly (Figure 12.10) are active only in warm sunlight because their metabolic rates are slowed by cool temperatures. Humans apply cold to burned areas to slow — and thereby minimize the effect of — unwanted physiological reactions.

Empirical Effect of Catalysis

Catalysis refers to the effect on reactions of a **catalyst** — a substance that significantly increases the rate of a chemical reaction without itself being consumed or changed. The chemical composition and amount of a catalyst are identical at the start and at the end of a reaction. In green plants, for example, the process of photosynthesis can take place only in the presence of the catalyst chlorophyll (Figure 12.11, page 466). Most catalysts accelerate reactions very significantly, even when present in very tiny amounts compared with the amount of reactants present.

The action of catalysts seriously perplexed early chemists, who had problems with the concept of something obviously being involved in a chemical reaction but not being changed by that reaction. Effective catalysts for reactions have almost all been discovered by purely empirical methods — trying everything to see what worked. Chemists learned early that finely divided metals catalyze many reactions, including the one shown in Figure 12.12, page 466. Perhaps the most common consumer example of catalysis today is the use of finely divided platinum and palladium in catalytic converters in car exhaust systems (Figure 12.13, page 466). These catalysts speed the combustion of the exhaust gases so that a higher proportion of the exhaust will be the relatively harmless, completely oxidized products. Highly toxic CO gas, for example, is not nearly as serious a problem if it can be oxidized to CO_2 before being emitted to the atmosphere.

Enzymes in human body processes are normally extremely complex molecular substances (proteins) that act as catalysts, so that a great number of human physiological reactions are actually controlled by the amount of enzyme present. Enzymes are also of great importance for catalyzing reactions in the food, beverage, cleaner, and pharmaceutical industries.

The success of any chemical industry will depend on the control of chemical reactions by selecting conditions that will optimize yield of the desired product and minimize production of undesired substances. For many industrial processes, the use of catalysts (see Table 12.1, page 466) makes the difference between success and failure by making the reaction rate fast enough to be profitable. (For an example of this, see the feature in Chapter 13 on the Haber process, page 505.)

Figure 12.9
A refrigerator is a common technological device used to decrease reaction rates.

Figure 12.10
Butterflies, like other insects, have metabolic rates that increase and decrease in response to temperature changes in their surroundings.

Catalysts
A catalyst is a substance that increases the rate of a reaction but is not consumed by the reaction. Many commercial and industrial catalysts are heterogeneous, which means that they provide a solid surface upon which the reactants can be adsorbed and reacted. The catalytic converter in a car is a good example of a heterogeneous catalyst.

A catalyst reduces the energy required for the reactants to form products, but does not alter the net enthalpy change, ΔH, of the overall reaction.

Inhibitors
Some substances are known to slow or stop reaction rates if present in the reaction system. We believe that these inhibitor substances act differently from catalysts, controlling the reaction rate in a completely different way. A major commercial use of inhibitors is in foodstuffs and medical products to delay deterioration, where the inhibitors are called preservatives.

Figure 12.11
Green plants contain chlorophyll, which acts as a catalyst during photosynthesis to help convert carbon dioxide and water into glucose and oxygen.

Table 12.1

MAJOR INDUSTRIAL PRODUCTS, PROCESSES, AND CATALYSTS		
Product	**Process**	**Catalyst**
sulfuric acid	contact	vanadium
ammonia	Haber	iron
gasoline	catalytic reforming	Mo or Pt on alumina
polyethylene	chain reaction (addition polymerization)	Al and Ti complexes (Ziegler-Natta catalysts)

Figure 12.12
Hydrogen peroxide solution decomposes to produce oxygen. This happens much more rapidly in the presence of a contact lens disinfectant disk. The plastic disk is coated with finely divided platinum metal, which acts as a catalyst for the reaction.

SUMMARY: FACTORS AFFECTING REACTION RATES

- The rate of any reaction depends on the nature of the chemical substances reacting. As a rule, reactions involving ionic bond changes are usually very rapid, while reactions involving covalent bond changes are often much slower.

- An increase in reactant concentration or in reactant surface area increases the rate of a reaction.

- An increase in temperature increases the rate of a reaction.

- A catalyst increases the rate of a reaction, without itself being consumed.

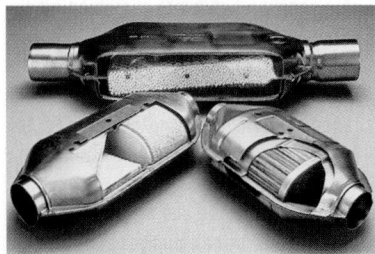

Figure 12.13
An automobile catalytic converter increases the rate of oxidation of exhaust gases: $NO_{(g)}$ is converted into $N_{2(g)}$ and $O_{2(g)}$; $CO_{(g)}$ is converted to $CO_{2(g)}$; and unburned hydrocarbons are converted to $CO_{2(g)}$ and $H_2O_{(g)}$.

Exercise

6. Identify four different factors that are likely to affect the rate of a reaction. Where possible state a generalization for each factor.

7. Give one technological example of each of the following.
   (a) a relatively slow reaction
   (b) a relatively fast but controlled reaction
   (c) a very fast, uncontrolled reaction

8. Enzymes in your body are generally present in extremely small quantities, but any substances that affect your enzymes are almost always very toxic and dangerous. Explain why this should be so, using reaction rates in your explanation.

9. (Discussion) Many boxed dry cereals contain BHT (butylated hydroxytoluene) in the packaging material. This compound is also called 2,6-di-tert-butyl-p-cresol. Use a reference such as a dictionary of chemistry to find out what BHT has to do with reaction rates.

The quantitative dependence of a reaction rate on the various factors that have an effect on it cannot be predicted from theory but must be determined empirically — from analysis of experimental evidence. The specific relationship of reaction rate to concentration of a particular reactant present cannot be reliably predicted from the balanced chemical equation, although the equation does provide a starting point for examining the relationship. Once discovered, the relationship often turns out not to be a simple direct or inverse proportion, and considerable detail and theory are required to provide a satisfactory scientific explanation. Nevertheless, certain generalizations for many reactions can be made.

Empirical Determination of Rate Laws

In 1863, two Norwegian chemists, Cato Guldberg and Peter Waage, announced that the rate of a reaction is exponentially proportional to the product of the concentrations of the reactants. This generalization, initially obtained from the study of decomposition of esters, has been verified for many chemical reactions and is now known as the **rate law**. For a stylized reaction equation where coefficients and formulas for the chemical reactants are symbolized as follows,

$$a\,X + b\,Y \rightarrow (\text{products})$$

according to the rate law, the rate will always be proportional to the product of the concentrations of the reactants, where these concentrations are raised to some exponential values. This can be expressed as

$$r \propto [X]^m[Y]^n$$

where m and n, determined empirically, may have any real number value, including fractions or a possible value of zero.

Written as an equation

$$r = k\,[X]^m[Y]^n$$

where k is a specific **rate constant**, determined empirically, that is valid only for a specific reaction at a given temperature.

All theories and laws in science are restricted to some degree, and the rate law is no exception. The following restrictions apply to this law:

- The concentration exponent, referred to often as the order of the reaction for a particular substance, must be determined empirically by laboratory investigation as it cannot be predicted from theory.

- The rate equation is only valid for the initial concentrations at a specified temperature, and in the course of most reactions the concentrations and temperature change as soon as the reaction begins.

> **Coefficients and Order**
> Coefficients in a balanced chemical equation relate to the mass balance. Exponents in a rate law (each called the order of the substance) are empirically determined. The order of a substance is generally not related to the coefficient of that substance in the balanced reaction equation.

For example, what is the rate law for the "iodine clock" reaction (Investigation 12.2) which produces the following evidence when the reactant concentrations are varied and the rates are measured? It is found experimentally that the rate is proportional to the concentrations of iodate ion and of bisulfite ion.

$$r = k\,[IO_{3\ (aq)}^-]^m\,[HSO_{3\ (aq)}^-]^n$$

Trial	Initial $[IO_{3\ (aq)}^-]$ (mmol/L)	Initial $[HSO_{3\ (aq)}^-]$ (mmol/L)	Initial Rate of I_2 Production (mmol/(L·s))
1	4.0	6.0	1.60
2	2.0	6.0	0.80
3	2.0	3.0	0.20

Comparing the first two trials shows that when the concentration of $IO_{3\ (aq)}^-$ is halved, the rate drops by a factor of $\frac{0.80}{1.60}$ or $\frac{1}{2}$. This is in direct proportion to the change in concentration of $IO_{3\ (aq)}^-$, so the exponent "m" in the rate law is 1.

Comparing the first and the third trials shows that when the concentration of $HSO_{3\ (aq)}^-$ is halved, the rate drops by a factor of $\frac{0.20}{0.80}$ or $\frac{1}{4}$. This is a direct square proportion to the change in concentration. Since $(\frac{1}{2})^2 = \frac{1}{4}$, so the exponent "$n$" in the rate law is 2.

Entering the values from trial 1 (any trial will do) in the rate law, with the concentrations expressed in mol/L, and then solving for k, gives

$$r = k\,[IO_{3\ (aq)}^-]\,[HSO_{3\ (aq)}^-]^2$$

$$0.001\,60 \text{ mol/(L·s)} = k \times 0.0040 \text{ mol/L} \times (0.0060 \text{ mol/L})^2$$

$$k = \frac{0.001\,60 \text{ mol/(L·s)}}{0.0040 \text{ mol/L} \times (0.0060 \text{ mol/L})^2} = 1.1 \times 10^4 \text{ L}^2/(\text{mol}^2 \cdot \text{s})$$

The rate law is $\quad r = 1.1 \times 10^4 \text{ L}^2/(\text{mol}^2 \cdot \text{s})\,[IO_{3\ (aq)}^-]\,[HSO_{3\ (aq)}^-]^2$

Exercise

10. For a reaction where the rate law is $\ r = k\,[NH_{4\ (aq)}^+]\,[NO_{2\ (aq)}^-]$,
    (a) calculate k at temperature T_1, if the rate, r, is 2.40×10^{-7} mol/(L·s) when $[NH_{4\ (aq)}^+]$ is 0.200 mol/L and $[NO_{2\ (aq)}^-]$ is 0.00500 mol/L.
    (b) calculate r at temperature T_2, if the rate constant, k, is 3.20×10^{-4} L^2/(mol^2·s) when $[NH_{4\ (aq)}^+]$ is 0.100 mol/L and $[NO_{2\ (aq)}^-]$ is 0.0150 mol/L.

11. For a reaction where the rate law is $\ r = k\,[Cl_{2(g)}]\,[NO_{(g)}]^2$,
    (a) calculate k at a temperature where the rate, r, is 0.0242 mol/(L·s) when $[Cl_{2(g)}]$ is 0.20 mol/L and $[NO_{(g)}]$ is 0.20 mol/L.
    (b) calculate r at a temperature where the rate constant, k, is 3.00 when $[Cl_{2(g)}]$ is 0.44 mol/L and $[NO_{(g)}]$ is 0.025 mol/L.

12. State the effect on the value of a specific reaction rate constant, k, in a given rate law expression when
    (a) the temperature of the reaction is increased.
    (b) the initial concentration of any reactant is decreased.

Problem 12A The Empirical Determination of a Rate Law Constant

A typical method for determining a rate law expression relationship involves measuring all variables under controlled conditions and then changing a reactant concentration by some simple numerical factor in order to determine what effect this change will have on the overall reaction rate. Mixing an acidic solution containing $IO_3^-{}_{(aq)}$ ions with another solution containing $I^-{}_{(aq)}$ ions begins a reaction that proceeds, in several reaction steps, to finally produce molecular iodine as one of the products. The Analysis of an investigation report has been partly done based on the evidence given. Complete this Analysis.

Problem

What is the rate law constant for the following reaction?

$$IO_3^-{}_{(aq)} + 5\,I^-{}_{(aq)} + 6\,H^+{}_{(aq)} \rightarrow 3\,I_{2(aq)} + 3\,H_2O_{(l)}$$

Experimental Design

A series of reaction trials are performed, with the concentration of one reactant doubled each time after the initial control trial. The initial rate of formation of molecular iodine product is measured for each trial.

Evidence

Initial $[IO_3^-]$ (mol/L)	Initial $[I^-]$ (mol/L)	Initial $[H^+]$ (mol/L)	Initial Rate of Production of I_2 (mol/(L•s))
0.10	0.10	0.10	5.0×10^{-4}
0.20	0.10	0.10	1.0×10^{-3}
0.10	0.20	0.10	1.0×10^{-3}
0.10	0.10	0.20	2.0×10^{-3}

Analysis

According to the rate law,

$$r = k[IO_3^-]^a[I^-]^b[H^+]^c.$$

Comparing the first two trials shows that the effect of doubling the concentration of iodate ions is a doubling of the reaction rate. This means that the exponent "a" in the rate law expression has a value of one — a simple direct relationship exists. The reaction rate is said to be first order in $IO_3^-{}_{(aq)}$.

Comparing the first and third trials shows that doubling the concentration of iodide ions doubles the reaction rate. This means that the exponent "b" in the rate law expression has a value of one — once again, a simple direct relationship exists. The reaction rate is said to be first order in $I^-{}_{(aq)}$.

Comparing the first and fourth trials shows that doubling the concentration of hydrogen ions quadruples the reaction rate. This means that the exponent "c" in the rate law expression has a value of two — a simple direct squared relationship exists. The reaction rate is said to be second order in $H^+{}_{(aq)}$.

Although empirical knowledge is the foundation of science, the ultimate aim of science is to understand processes by inventing consistent, logical, and simple theories. Theoretical knowledge is developed and communicated by way of models and analogies, and the resultant knowledge is tested by its ability to explain and to predict. It is not unusual for theories to be more successful at explanations than at predictions. The collision-reaction theory is an excellent example of this type of theory. Explanations of reaction rates and factors affecting rates are much easier to obtain than theoretical predictions of specific reaction rates.

Kinetic Collision Theory

Elastic collisions are ones in which kinetic energy is conserved.

The main ideas of the collision-reaction theory are:

- A chemical sample consists of particles (atoms, ions, or molecules) that are in constant random motion at various speeds, rebounding elastically from collisions with each other. The average kinetic energy of the particles is perceived as the temperature of the sample. Figure 12.14 shows the distribution of kinetic energies among particles in a sample at two different temperatures.

- A chemical reaction must involve collisions of reactant particles.

- An effective collision requires sufficient energy and correct orientation (positioning) of the colliding particles so that bonds can be broken and new bonds formed.

- Ineffective collisions involve particles that rebound from the collision, essentially unchanged in nature.

- The rate of a given reaction depends on two factors related to collisions: the frequency of collisions and the fraction of those collisions that are effective.

Figure 12.14
Temperature is a measure of the average kinetic energy of the particles. This graph shows how the distribution of kinetic energies changes when a substance is heated or cooled. At any temperature in any substance there are some particles with low kinetic energy and some with high kinetic energy. The higher the temperature, the more particles there are with higher kinetic energies.

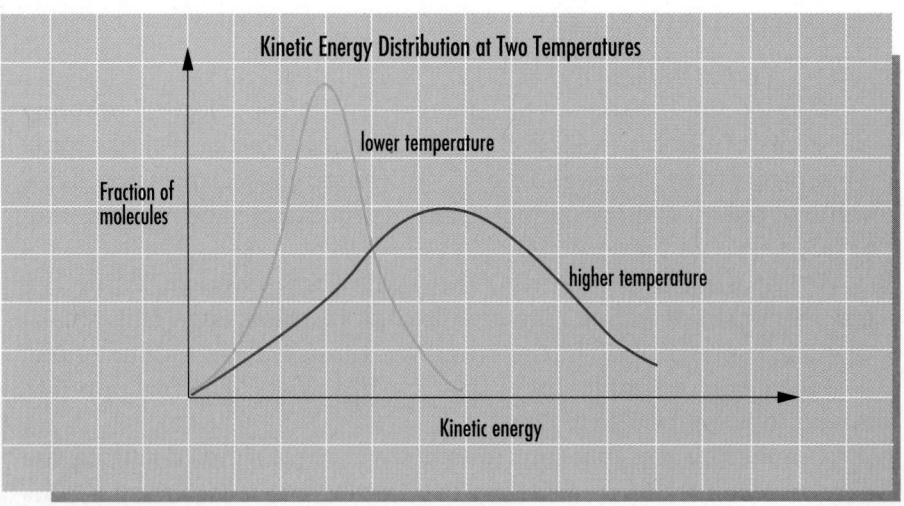

Activation Energy Theory and Diagrams

Consider an empty soft drink can. In the nitrogen-oxygen mixture in the can, an estimated 10^{30} molecular collisions occur every second. This is an absolutely enormous number — and visualizing one thousand billion billion billion events per second is not the sort of thing the human brain is designed to do. Gases are by far the least dense of the states of matter, so in the condensed phases of matter (liquids and solids), the number of collisions per second must be even greater. In liquids and solids, the particles are moving at the same speeds as they are in gases (if the temperature is the same), but they are very much closer together. If each collision produced a reaction when reacting substances were combined, the rate of any reaction would be extremely rapid, appearing essentially instantaneous. The nitrogen and oxygen molecules in air, for example, would react completely to form (toxic) nitrogen dioxide, $NO_{2(g)}$ in about 5×10^{-9} s (five billionths of a second). Since empirically measured reaction rates are often relatively slow, and in many cases too slow to be measurable (as in the nitrogen-oxygen reaction of air), we are forced to interpret the evidence as showing that normally only an extremely tiny fraction of the collisions actually produce new substances. The theoretical explanation for this empirical evidence involves the concept of **activation energy** — the minimum energy with which particles must collide before they can rearrange in structure, resulting in an *effective* collision.

The concept of chemical activation energy, E_{act}, can be illustrated by an analogy with gravitational potential energy. Consider a billiard ball rolling on a smooth track shaped as shown in Figure 12.15. When a ball leaves point A moving right, as it rises on the uphill portion of the track it slows down — as kinetic energy converts to potential energy. The ball can only successfully overcome the rise of the track and proceed to point B if it has enough initial speed (kinetic energy). We could call this situation an *effective* trip. The minimum kinetic energy required is analogous to the *activation energy* for a reaction. If the ball doesn't have enough kinetic energy it will not reach the top of the track and will just roll back to point A. This is like two molecules colliding without enough energy to rearrange their bonds — they just rebound.

Note that a ball that returns to point A will have the same energy it began with, but a ball that makes it to point B will have more energy

An Analogy for Activation Energy

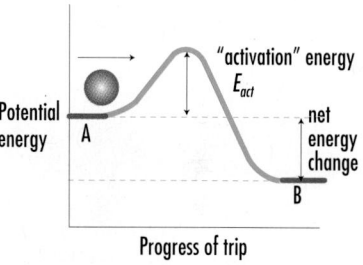

Figure 12.15
On a trip from A to B there is a net decrease in overall (net) energy, but there must be an initial increase in potential energy (activation energy) for the trip to be possible.

Figure 12.16
The self-sustaining exothermic reaction of hydrogen and oxygen requires only a single spark to begin. This is useful when desired, as in the operation of the space shuttle main engines (left), and disastrous in other cases, as in the explosion of the German airship Hindenburg *in Lakehurst, New Jersey, in 1937 (right).*

because it will be moving faster. The example above is also analogous to energy change, for an exothermic reaction. The heat of reaction (net energy change) is immediately released to the surroundings; in kinetic theory terms, this means that the energy is released to any other nearby molecules. These other molecules then move faster, collide with more energy, and become more likely to react. The energy released when the first few molecules of a hydrogen-oxygen mixture react (initiated by a spark or flame) is quickly transferred to other molecules, allowing the reaction to proceed unaided by external sources of energy. The reaction, once begun, is self-sustaining as long as enough molecules remain to make collisions likely. Exothermic reactions often drive themselves in this way once begun, as shown in Figure 12.16, page 471.

Consider the reaction of carbon monoxide with nitrogen dioxide, plotted as potential energy of the molecules versus progress of reaction; that is, the progress over time of the particle activity that constitutes the reaction (Figure 12.17).

$$CO_{(g)} + NO_{2(g)} \rightarrow CO_{2(g)} + NO_{(g)}$$

Figure 12.17
Over the progress of this exothermic reaction depicting an effective collision between molecules in the gas phase, the potential energy increases to a maximum at the point of closest approach, then decreases to a final value lower than the initial energy. The potential energy lost by the molecules shows up as kinetic energy produced. Formation of products would raise the temperature of the system.

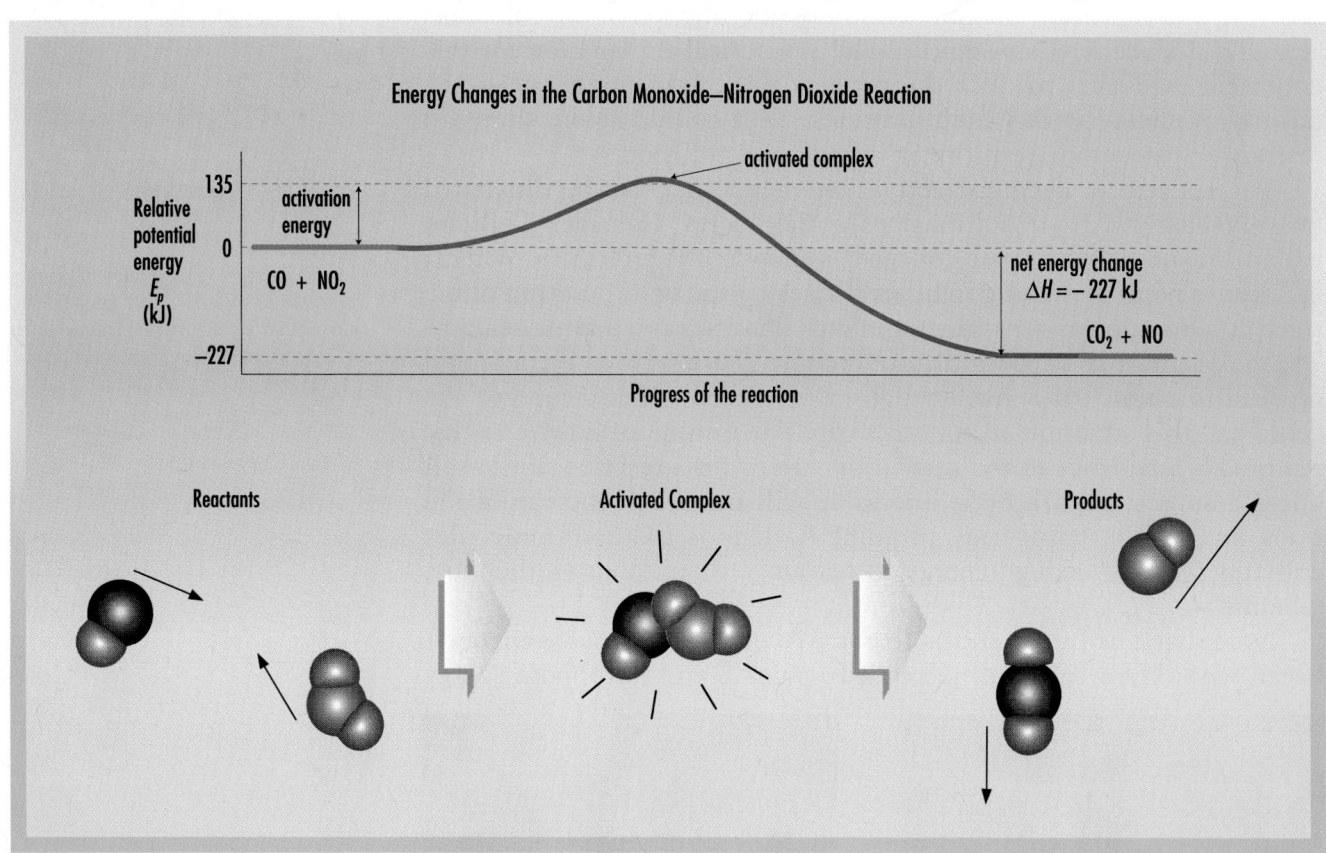

We can discuss the progress of the reaction shown in Figure 12.17 in terms of molecular collisions, by moving along the plot from left to right. Along the flat region to the left, the molecules are moving toward each other, but are still distant from each other. As the molecules approach more closely, they are affected by repulsion forces and begin to slow down, as some of their kinetic energy is changed to potential energy (stored as a repelling electric field between them). If the molecules have enough kinetic energy — which means more energy than is required to get to the energy level of the activated complex — they can approach closely enough for their bond structure to rearrange. **Activated complex** is the chemical species containing partially broken and partially formed bonds representing the maximum potential energy point in the change.

In this example, the energy initially absorbed to break the nitrogen-oxygen bond is less than the energy released when a new carbon-oxygen bond forms. Repulsion forces push the product molecules apart, converting potential energy to kinetic energy. If the overall energy change is measured by comparing reactants and products at the same temperature, the difference is observed as a net release of heat to the surroundings. If the energy difference is measured at constant pressure (the usual situation for a reaction open to the atmosphere), it is called the enthalpy of reaction, ΔH, which is –227 kJ for this particular (exothermic) example.

In other types of reactions the final potential energy may be greater than the initial potential energy, meaning that a net input of energy has to take place for such a reaction to happen. In these cases the enthalpy or molar enthalpy of reaction is an increase, so it has a positive value, for example, H_f° = +26.5 kJ/mol HI for the formation reaction of hydrogen iodide:

$$H_{2(g)} + I_{2(g)} \rightarrow 2\,HI_{(g)} \qquad \Delta H^\circ = +53.0\ \text{kJ}$$

During such reactions the temperature of the surroundings tends to drop, as heat is absorbed in the progress of the reaction. We call such reactions endothermic (see Figure 12.18).

Figure 12.18
Over the progress of this endothermic reaction depicting an effective collision between molecules in the gas phase, the potential energy increases to a maximum at the point of closest approach, then decreases to a final value higher than the initial energy. The potential energy gain of the molecules comes at the expense of kinetic energy. Formation of products would lower the temperature of the system.

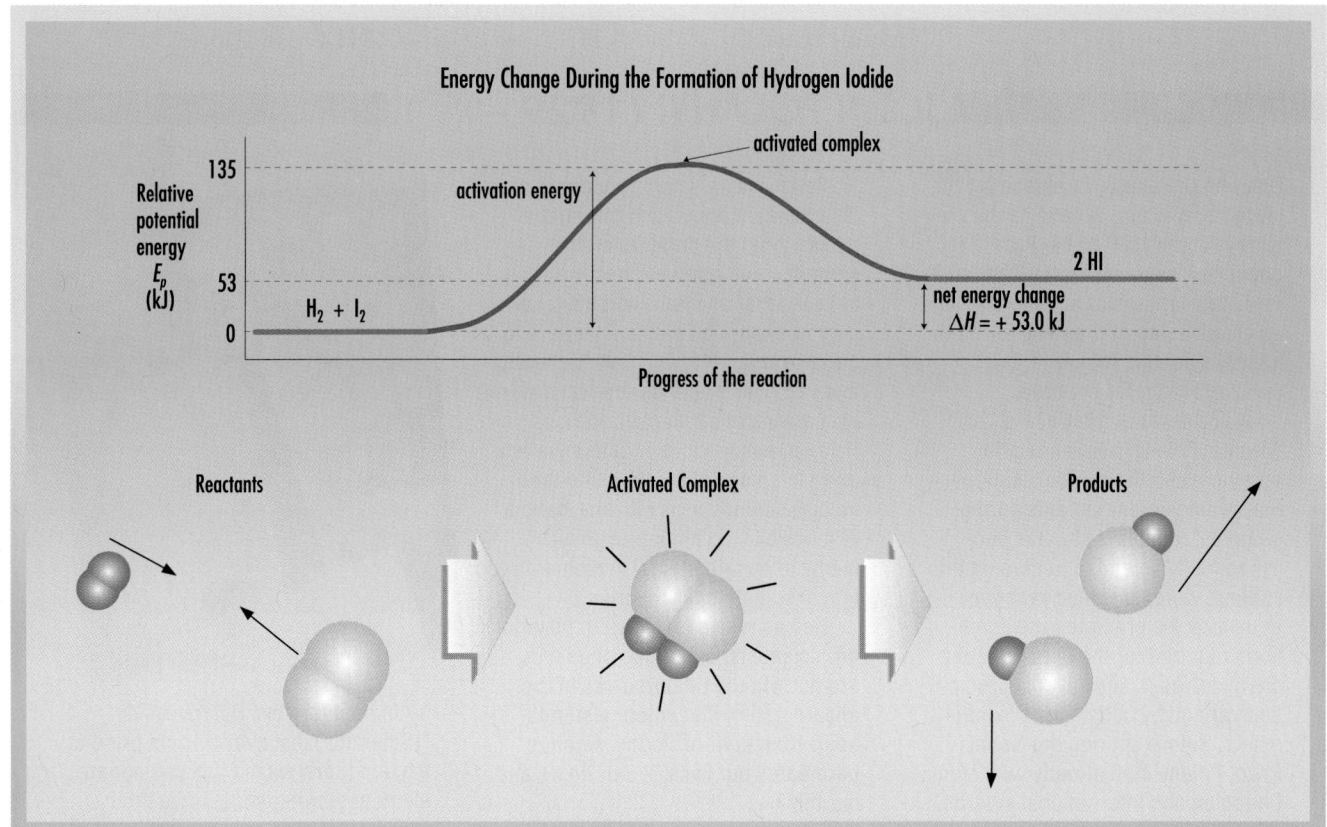

Reaction Mechanism Theory

Collisions between particles at the molecular level last for extremely short time periods. Calculations show that collisions of three particles simultaneously must be much less frequent than two-particle collisions and that any collision involving four or more particles is extremely unlikely indeed. Scientists believe most chemical reactions actually occur as a sequence of individual steps that involve collisions of two particles. A classic example is oxidation of hydrogen bromide, which is rapid between 400°C and 600°C. This reaction has been studied extensively, and illustrates a **reaction mechanism** — the individual steps during the progress of a reaction. All substances are simple molecules in the gas phase.

$$4\,HBr_{(g)} \;+\; O_{2(g)} \;\rightarrow\; 2\,H_2O_{(g)} \;+\; 2\,Br_{2(g)}$$

Empirically, the evidence shows that increasing the concentration of oxygen increases the reaction rate, just as we would expect. But it turns out that increasing the concentration of hydrogen bromide has the same degree of effect! Since four molecules of HBr are involved for every molecule of O_2, it seems logical to expect a change in HBr concentration to have a much greater effect on the rate, but measurement shows this is not the case.

We explain by theorizing that the reaction likely occurs in the following discrete steps, or elementary processes, each of which involves a two-particle collision occurring at a different rate.

These steps
sum to give the
equation for the
overall reaction.

$HBr_{(g)} \;+\; O_{2(g)} \;\rightarrow\; HOOBr_{(g)}$	(slow)
$HOOBr_{(g)} \;+\; HBr_{(g)} \;\rightarrow\; 2\,HOBr_{(g)}$	(fast)
$2[HOBr_{(g)} \;+\; HBr_{(g)} \;\rightarrow\; H_2O_{(g)} \;+\; Br_{2(g)}]$	(fast)
$4\,HBr_{(g)} \;+\; O_{2(g)} \;\rightarrow\; 2\,H_2O_{(g)} \;+\; 2\,Br_{2(g)}$	

JOHN POLANYI (1929 –)

Dr. John C. Polanyi, Canadian Nobel laureate, is widely known as the co-winner of the 1986 Nobel Prize in chemistry, as an educator who publicizes the dangers of nuclear war, and as an advocate of the need for fundamental scientific research that uncovers the laws of nature.

A professor at the University of Toronto, Polanyi has refined the empirical and theoretical descriptions of molecular motions during chemical reactions and has published many scientific papers on this subject, which is called *reaction dynamics*. He predicted the conditions required for the operation of a chemical laser, and saw his prediction verified by subsequent experiments. Before sharing the Nobel Prize, Polanyi had already won many Canadian and international awards.

Polanyi is an active proponent of disarmament, striving to inform the public about the dangers of the spread of nuclear weapons. He believes that an informed public can exert pressure on political leaders to curtail armament and move the world away from its dependence on warfare as a means of settling differences. Polanyi reasons that because we are living in an age of science and technology, scientists, as citizens, have a responsibility to influence governments to ensure ethical implementation of scientific discoveries.

He has written, "Those of us who are scientists, and those of us who are not, should rid ourselves of the absurd notion that science stands apart from culture. Today, science permeates our lives — our doing and our thinking."

In an interview, Polanyi said, "[Scientific] discovery would grind to a halt if there wasn't this passionate element where people have their vision of what truth looks like."

The theoretical interpretation is that the first reaction step is relatively slow because it has a fairly high activation energy. The rate of the overall reaction is basically controlled by this step. The second step cannot react HOOBr any faster than the first step can produce it, so the rate of the reaction overall is the same as the rate of the slowest step — in this case, the first. The slowest reaction step in any reaction mechanism is called the **rate-determining step**. Substances such as HOBr and HOOBr — which are formed during the reaction but immediately react again and are not present when the reaction is complete — are called **intermediate products** or reaction intermediates. A potential energy diagram would look like Figure 12.19.

Exercise

Consider the potential energy diagram for the hypothetical reaction in Figure 12.20. This reaction is reversible as under certain conditions products will react to re-form the reactant molecules.

13. (a) What is the activation energy for the following net forward reaction?
   $$4\,A + 2\,B \rightarrow 2\,C + 3\,D$$
   (b) What is the activation energy for the following net reverse reaction?
   $$2\,C + 3\,D \rightarrow 4\,A + 2\,B$$
   (c) What is the reaction enthalpy (net energy change) for the net forward reaction?
   (d) What is the reaction enthalpy (net energy change) for the net reverse reaction?
   (e) What is the rate-determining step for the forward reaction?
   (f) What is the rate-determining step for the reverse reaction?
   (g) Which reaction (forward or reverse) is exothermic?
   (h) Which letters represent activated complexes?
   (i) Which letters represent intermediate products?

14. Explain what you would expect to occur if the original collision of particles in the first step of the forward reaction has a total available kinetic energy equivalent to 55 kJ.

15. Explain what you would expect to occur if the original collision of particles in the first step of the reverse reaction has a total available kinetic energy equivalent to 55 kJ.

Oxidation of $HBr_{(g)}$

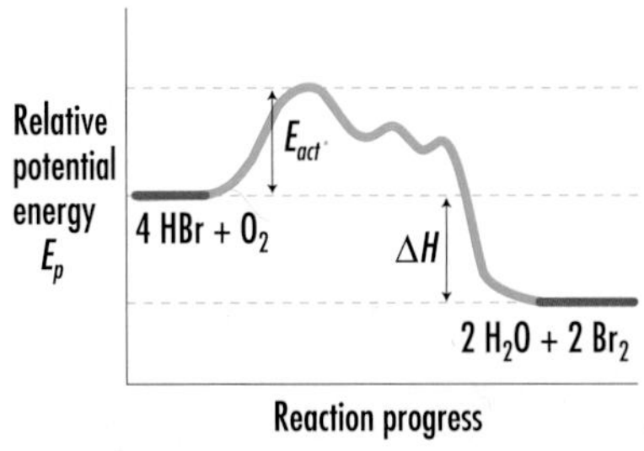

Figure 12.19
Over the progress of this reaction, the potential energy increase necessary to reach the first activated complex stage is the greatest increase required, so this is the rate-determining (slowest) step. Energy released as kinetic energy past this point is sufficient to quickly carry the reaction mechanism to completion. An overall decrease in energy means this reaction is exothermic.

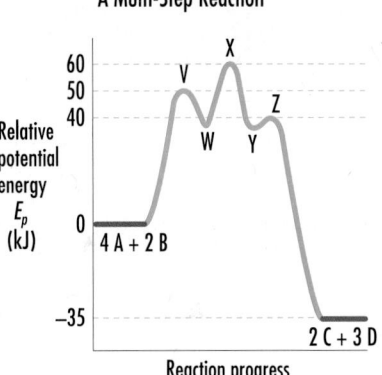

Figure 12.20
Potential energy diagram for questions 13 to 15.

12.5 EXPLAINING REACTION RATES

Evidence strongly indicates that only effective collisions result in reaction, and only a minuscule fraction of collisions are ordinarily effective. Much of what determines whether a collision will be effective must depend on the nature of the reacting entities.

Theoretical Effect of the Nature of Reactants

An effective collision not only requires a minimum energy but also an appropriate orientation during the collision. To better explain the collision-reaction theory, we need to consider other factors such as the nature of the bonds present and the type of reaction occurring. We believe that reactions involving molecular substances and/or complex ions are often slower than reactions of simple ions because any breaking of a covalent bond necessarily means a significantly high activation energy. Collision orientation can explain part of the effect of changing reactants for similar types of reactions. Consider a collision between a molecule of CO and one of NO_2 oriented spatially as shown in Figure 12.21. Compare the orientation with that shown in Figure 12.17 (page 472).

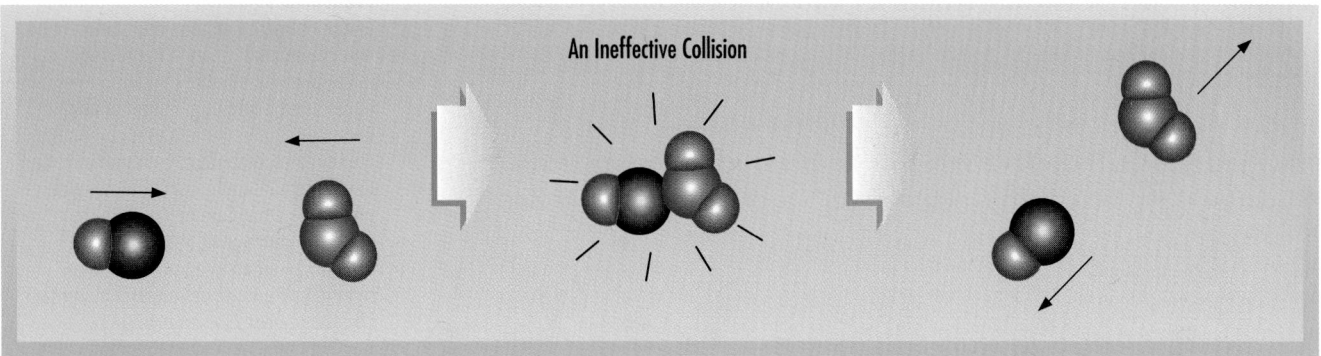

An Ineffective Collision

Figure 12.21
Over the progress of this collision between molecules in the gas phase, the orientation of the molecules in the collision is such that no rearrangement of bonds can occur.

Theoretical Effect of Concentration and Surface Area

If the concentration of a reactant increases, the reaction rate generally increases. A higher concentration of a reactant means a greater number of particles per unit volume. A greater number of particles are obviously more likely to collide as they move randomly within a fixed volume. If twice as many particles are present there should be twice the probability of an effective collision. Therefore, for simple reactions, the rate of reaction should be directly proportional to the concentration of a reactant. We find experimentally that this is correct for single-step reactions. However, many reactions are multi-step ones. In these cases, a reactant may not even be involved in the rate-determining step of the reaction mechanism, and many other factors are involved in determining the overall rate.

Surface area effects may be thought of most simply by considering that another substance can only have particle collisions with a liquid or solid phase at the surface where the substances are in contact. The number of particles per square millimetre of surface of a solid is fixed. However, the area of surface exposed for a given quantity depends on how finely divided the sample of solid is. We make use of this in foodstuffs by finely grinding pepper to increase flavor and by using icing (finely powdered) sugar and ground coffee to increase the rate at which they react. Dividing a solid into finer and finer pieces has a limit when you reach the elementary particles of which the solid is composed. Sugar cannot be divided more finely than into its individual molecules. A moment's reflection should convince you that this is what the dissolving process does — it divides a solid or liquid solute into the theoretical maximum number of separate particles, creating the maximum possible surface area. This is why so many reactions are only possible or are normally only performed in solution, including nearly all of the reactions of human physiology.

Theoretical Effect of Temperature

If the temperature of a system increases, the reaction rate generally increases. Theoretically, temperature is believed to be a measure of the average kinetic energy of the particles in a sample. Experimental evidence shows that a relatively small increase in temperature seems to have a disproportionate effect on reaction rate. An increase of about 10°C will often double the rate of a reaction. When you consider that a rise from 27°C to 37°C represents an absolute increase from 300 K to 310 K, it can be seen that a 3% increase in temperature seems to be causing a 100% increase in reaction rate.

The explanation for the temperature effect lies in the concept of activation energy. For a given activation energy, E_{act}, a much smaller fraction of molecules has the required kinetic energy at a lower temperature compared with a higher temperature. A temperature rise that is a small increase in overall energy may cause a very large increase in the number of particles that have energy exceeding the activation energy. An analogy can be made here with test score distributions — scaling all provincial exam scores upward by 5% could in theory easily double the number of marks higher than 80%. The term "threshold energy" is sometimes used; it is the minimum kinetic energy required to convert kinetic energy to potential energy during the formation of the activated complex. Note that on a kinetic energy distribution graph the energy for effective collision is measured horizontally (Figure 12.22), not vertically as on a potential energy diagram.

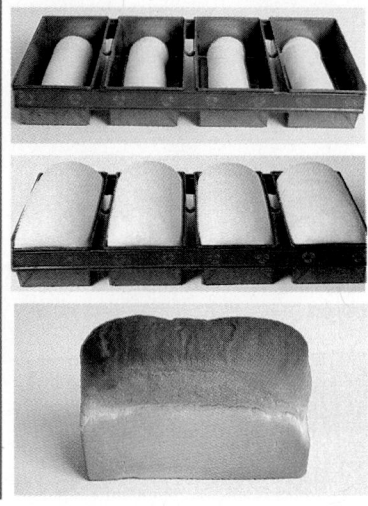

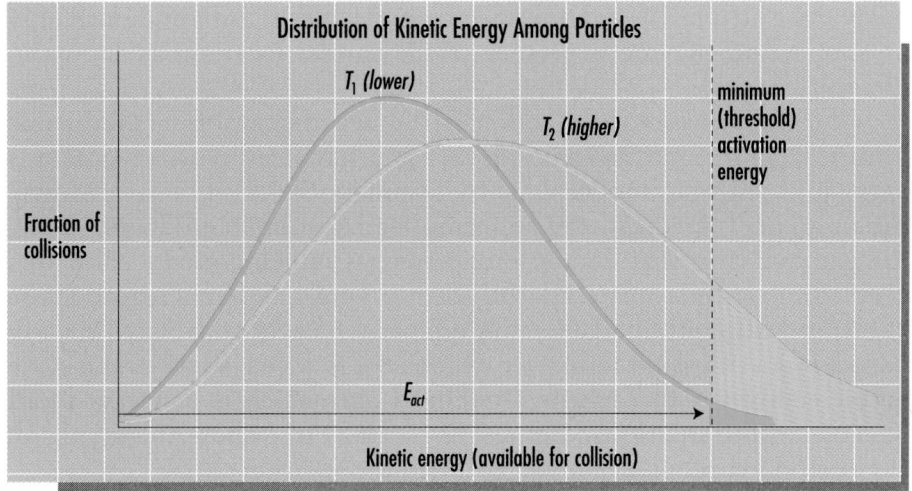

Figure 12.22
The molecules with enough energy to create a successful collision are represented by the area enclosed under the graph line and to the right of the (dashed) minimum energy level. Note the very large increase in the number of these molecules at the higher sample temperature.

Theoretical Effect of Catalysis

Theoretically, catalysts accelerate a reaction by providing an alternative lower energy pathway from reactants to products. That is, a catalyst allows the reaction to occur by a different mechanism, inserting different intermediate steps, but resulting in the same products overall. If the new pathway (mechanism) has a lower activation energy, a greater fraction of molecules possess the minimum required energy and the reaction rate increases. Since the activation energy is lowered by exactly the same amount for the reverse reaction, the rate of any reverse reaction increases as well (see Figure 12.23, page 478).

Figure 12.23
The reaction proceeds by a three-step mechanism when a catalyst is present, but nonetheless proceeds much faster than by the one-step uncatalyzed mechanism. The catalyzed mechanism theoretically has a much lower activation energy, so more collisions are successful.

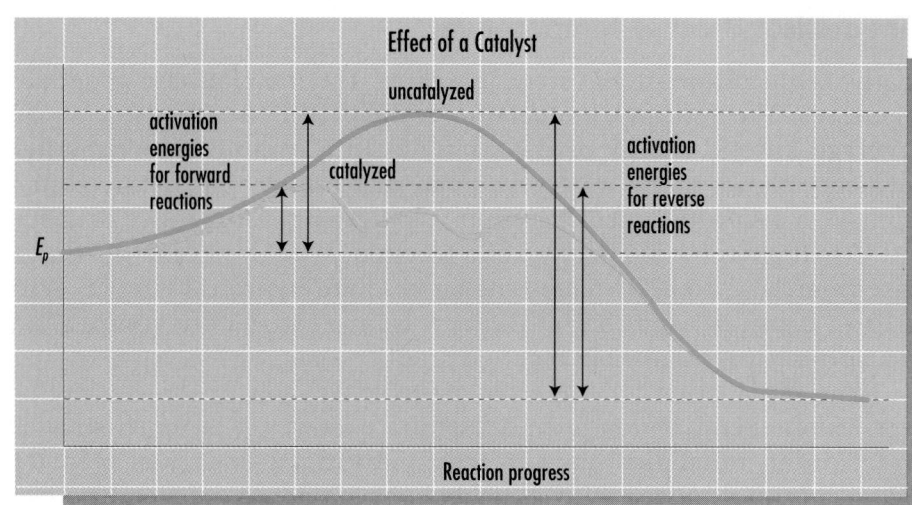

Effect of a Catalyst

uncatalyzed

activation energies for forward reactions

catalyzed

activation energies for reverse reactions

E_p

Reaction progress

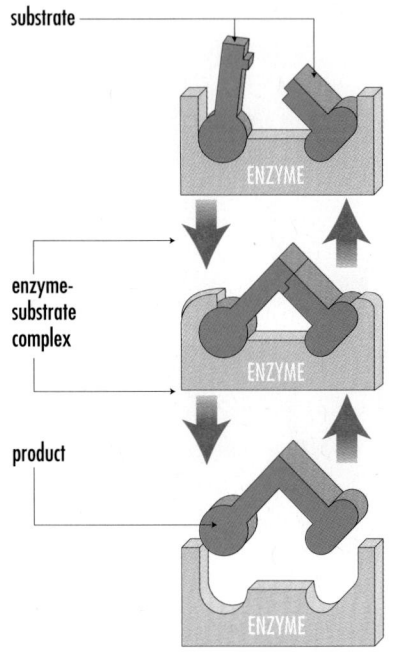

substrate

enzyme-substrate complex

product

ENZYME

Figure 12.24
Note that the enzyme shape is specific for the reactant (substrate) molecules. Almost all enzymes catalyze only one specific reaction.

Figure 12. 25
The photographs show aqueous formic acid solution at room temperature (left), the addition of sulfuric acid (center) to the formic acid, and the reaction of the acidified solution (right). This reaction must be performed in a fume hood, because $CO_{(g)}$ is extremely toxic to humans.

The actual mechanism by which catalysis occurs is not well understood for most reactions, and discovering acceptable catalysts has traditionally been a trial-and-error, empirical system. Catalysts are extremely important in chemical technology and industry because they allow the use of lower temperatures, thus preventing decomposition of reactants and products and also lessening unwanted side reactions. The result is an increase in the efficiency and economics of many industrial processes. A few catalyzed reactions have been studied in detail, so that we believe we understand the mechanism changes involved. Most of the catalysts (enzymes) for biological reactions work by shape and orientation. They fit substrate proteins into locations on the enzyme as a key fits into a lock, enabling specific molecules to link or detach only as long as the enzyme is present in the correct location, as shown in Figure 12.24.

One well-understood reaction is the decomposition of methanoic (formic) acid in aqueous solution. At room temperature this reaction is very slow, with no noticeable activity unless the solution is acidified. When an acid is added, the solution begins to bubble (Figure 12.25). Testing indicates that carbon monoxide gas is being produced. If this reaction proceeds until all the formic acid has been reacted, the solution still contains the same quantity of acid that was initially added. It doesn't seem to matter which acid is added, so we assume the acting catalyst is the aqueous hydrogen ion, which is common to all aqueous acidic solutions. Figures 12.26 and 12.27 show the graphs of the reaction without a catalyst and with the addition of one.

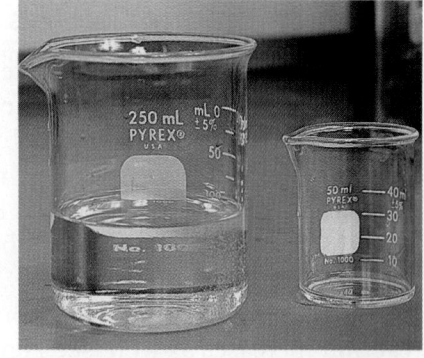

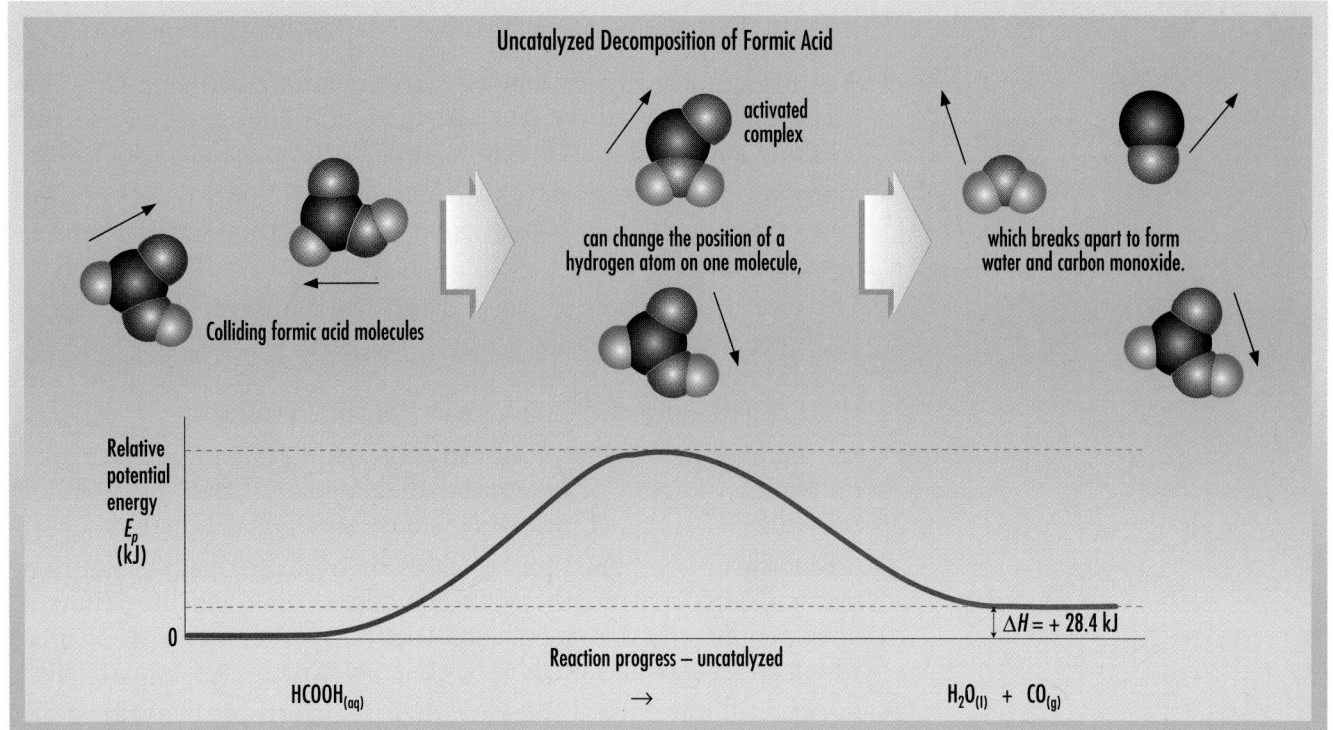

Figure 12.26
The uncatalyzed reaction proceeds too slowly to notice at room temperature. Note that this reaction is endothermic, requiring an overall increase in (input of) energy.

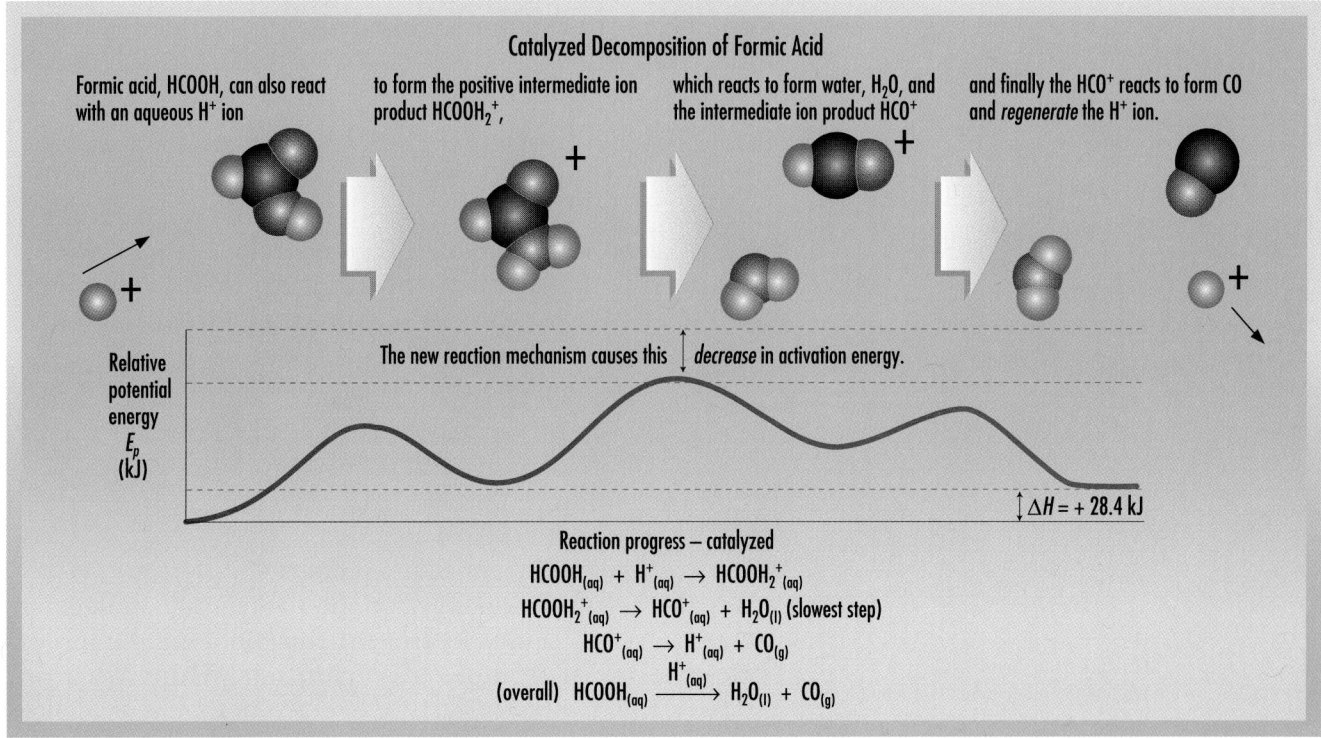

Figure 12.27
The catalyzed reaction proceeds rapidly enough to produce bubbles of gaseous carbon monoxide gas at room temperature. A higher fraction of the molecular collisions become effective because the activation energy is appreciably decreased. In a reaction mechanism, catalysts are present as a reactant in an early step and as a product in a later step. Catalysts are normally written above the reaction arrow in a reaction equation.

- A minimum *activation* energy is required for a collision between particles to be *effective*, that is, energetic enough to cause the breaking and forming of bonds to produce new particle structures.

- Many reactions occur as a sequence of *elementary* processes, or two particle-collision steps, which make up the overall reaction *mechanism*.

- The rate of any reaction depends on the nature of the chemical substances reacting because both the strength of bond(s) to be broken and the location of the bond(s) in the particle structure affect the likelihood that any given collision is effective.

- An increase in reactant concentration or in reactant surface area increases the rate of a reaction because the total number of collisions possible per unit time is increased proportionately.

- An increase in temperature increases the rate of a reaction for two reasons. The total number of collisions possible per unit time is increased slightly, but more importantly, the fraction of collisions which are energetic enough to be effective is increased dramatically.

- A catalyst increases the rate of a reaction by providing an alternative pathway to the same product formation that has a lower activation energy. A much larger fraction of collisions is effective following the changed reaction mechanism. Catalysts are involved in the reaction mechanism at some point, but are *regenerated* before the reaction is complete.

Exercise

16. The reaction of hydrogen with chlorine at room temperature is so slow as to be undetectable if the container is completely dark, but is explosively fast if sunlight is allowed to fall on the reactants. The following reaction mechanism has been suggested for this reaction.

$$Cl_{2(g)} + \text{light energy} \rightarrow Cl_{(g)} + Cl_{(g)}$$
$$Cl_{(g)} + H_{2(g)} \rightarrow HCl_{(g)} + H_{(g)}$$
$$H_{(g)} + Cl_{2(g)} \rightarrow HCl_{(g)} + Cl_{(g)}$$
$$Cl_{(g)} + Cl_{(g)} \rightarrow Cl_{2(g)}$$

(a) Write the net overall reaction equation.
(b) Identify the reaction intermediate products.
(c) Discuss the activation energy for the collision of molecular chlorine with molecular hydrogen, and for the collision of atomic chlorine with molecular hydrogen. Which reaction must have the greater activation energy, and what evidence can be used to support your argument?

17. The reaction of hydrogen and oxygen (see Figure 12.16, page 471) is exothermic and self-sustaining. Write the equation for this reaction, and provide a reason why it is not likely that the reaction occurs as a single step.

OVERVIEW

Chemical Kinetics

Summary

- Reactions occur at rates that vary from explosively fast to undetectably slow. Rates are often measured as a change in property of one substance per unit of time.

- Factors that affect reaction rates include the nature of reacting substances, the concentration of reactants, the exposed surface area of condensed phase reactants, the temperature of the system, and the presence of catalysts.

- According to the rate law, the rate of a reaction depends on concentration. The concentration exponent values for this law must be determined experimentally — they cannot be predicted from theory.

- Reaction rates are explained using kinetic molecular theory, assuming that most particle collisions do not result in reaction because a minimum activation energy is required, as well as an effective particle orientation.

- Reactions are generally assumed to occur in a series of reaction steps, the slowest of which determines the rate of the overall reaction. The sequence is called the reaction mechanism, and includes intermediate products which are produced in the course of the reaction but immediately reacted again, so they do not appear in the overall equation.

- Catalysts, which are present before and after a reaction, are assumed to speed up reactions by changing the reaction mechanism to a different one with a lower activation energy, resulting in a higher fraction of successful collisions.

Key Words

activated complex
activation energy
catalyst
chemical kinetics
enzyme
intermediate product
rate constant
rate-determining step
rate law
reaction mechanism
reaction rate

Review

1. List four factors that affect the reaction rate of a homogeneous reaction.

2. What is the generalization for the effect of surface area on reaction rate in a heterogeneous system? Provide a simple explanation of the generalization using collision-reaction theory.

3. Use the following chemical reaction equation to predict the effect of the changes stated on the rate of the reaction of marble chips in hydrochloric acid.

$$2\ HCl_{(aq)} + CaCO_{3(s)} \rightarrow$$
$$CO_{2(g)} + H_2O_{(l)} + CaCl_{2(aq)} + heat\ energy$$

(a) The concentration of hydrochloric acid is increased.
(b) The reaction mixture is cooled.
(c) Finely ground calcium carbonate is used instead of large chips of calcium carbonate.
(d) The partial pressure of $CO_{2(g)}$ in the container is increased.

4. Use collision-reaction theory to provide an explanation of the following observations:
   (a) Zinc metal reacts much more rapidly in concentrated $HCl_{(aq)}$ than in dilute $HCl_{(aq)}$.
   (b) Paints and stains often have instructions that they should not be applied below 10°C.
   (c) Food spoils more rapidly on a counter than in a refrigerator.
   (d) A natural gas furnace requires a pilot light or electronic igniter in order to operate.
   (e) Dust in grain elevators has been blamed for several violent explosions, which have completely demolished the elevators.
   (f) For people who have contact lenses, a sterilizing kit is available containing hydrogen peroxide and a platinum-coated disk. Oxygen gas is rapidly released from the decomposition of hydrogen peroxide only when the disk is placed in the solution.
   (g) Baking powder, containing a mixture of a solid acid and a solid base, does not react when dry but reacts rapidly when dissolved in water.

Applications

5. Sketch a potential energy diagram for the endothermic formation reaction of nitrogen and oxygen to produce nitrogen dioxide. Using appropriate symbols, label the activation energy and enthalpy of reaction on the diagram.

6. A piece of sodium metal cuts easily with a steel knife, and the freshly exposed shiny surface immediately reacts with air to form a dull oxide coating. What can be stated about the activation energy for this reaction?

7. The Haber process for producing ammonia — a very important industrial chemical — is usually carried out at a temperature of about 500°C. At this temperature only about a third of the hydrogen and nitrogen reactants change to form ammonia product. The reactants will change almost completely to product at 200°C, but the reaction is never done at this lower temperature. Industrially, this reaction is done as a continuous procedure, which means that $N_{2(g)}$ and $H_{2(g)}$ are continuously added to the reaction vessel and $NH_{3(g)}$ is continuously removed. Explain why high temperature is the logical choice for a reaction condition.

8. Is it likely that the following equation represents the reaction mechanism for the combustion of propane? Explain.
   $$C_3H_{8(g)} + 5\,O_{2(g)} \rightarrow 3\,CO_{2(g)} + 4\,H_2O_{(g)}$$

9. At 25°C a catalyzed solution of formic acid produces 44.2 mL of carbon monoxide gas in 30.0 s. Calculate the rate of reaction. What can you state about how long you would expect the production of the same volume to take at 30°C? What can you state about how long it would take at 25°C without the catalyst?

10. Write definitions of *reactant*, *product*, *catalyst*, and *intermediate product* in terms of whether the substance is present at the beginning, at the end, and/or during the progress of a reaction.

11. Use the following diagram to answer the questions below:

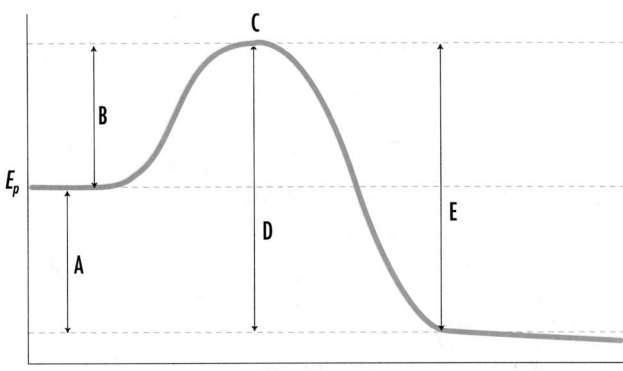

Reaction progress

   (a) The energy change, ΔH, for the reverse reaction is represented by what letter?
   (b) The activation energy for the forward reaction is represented by what letter?
   (c) The activated complex for the reaction is represented by what letter?

12. Consider the following hypothetical reaction:
   $$2\,A_{(g)} + B_{(g)} \rightarrow 2\,C_{(g)}$$
   For the forward reaction, $\Delta H = +50$ kJ, and for the reverse reaction $E_{act} = +35$ kJ.
   (a) Draw a potential energy diagram for this reaction.
   (b) What is E_{act} for the forward reaction?
   (c) What is ΔH for the reverse reaction?

Extensions

13. A cube of solid reactant with sides of $1.00\ cm^2$ is submerged in a liquid and reacts to form 20 mL of a gas product per second. The solid-liquid interface is $6.00\ cm^2$ of surface area. If you sliced this cube (like a block of cheese) into 10 slices, and then replaced it in the liquid, what would you expect the rate of reaction to be?

14. Consider the reaction of $S_2O_8^{2-}$ ions with I^- ions, for which a suggested mechanism has been determined by supporting empirical evidence, as shown. Assume that the overall reaction is slightly exothermic.

$$2\ Cu^{2+}_{(aq)} + 2\ I^-_{(aq)} \rightarrow 2\ Cu^+_{(aq)} + I_{2(aq)} \quad \text{(fast)}$$

$$Cu^+_{(aq)} + S_2O_8^{2-}_{(aq)} \rightarrow CuSO_4^+_{(aq)} + SO_4^{2-}_{(aq)} \text{(slow)}$$

$$Cu^+_{(aq)} + CuSO_4^+_{(aq)} \rightarrow 2\ Cu^{2+}_{(aq)} + SO_4^{2-}_{(aq)} \text{(fast)}$$

(a) Identify the catalyst and the intermediate products in this reaction.
(b) Identify the reactants and products, and write the overall reaction equation.
(c) Sketch a potential energy vs reaction progress graph for this reaction.
(d) Explain what effect increasing the concentration of I^- ions would have on the overall rate.
(e) Explain what effect increasing the concentration of $S_2O_8^{2-}$ ions would have.

15. The following reaction occurs in the combustion cylinder of an automobile.

$$N_{2(g)} + O_{2(g)} + energy \rightleftharpoons 2\ NO_{(g)}$$

What conditions or technological innovations could be applied that would decrease the production of nitrogen monoxide?

Problem 12B The Empirical Determination of a Rate Law

The typical method for determining a relationship is to measure all variables under controlled conditions and then to change one (a reactant concentration, which becomes the independent variable) by some simple numerical factor to determine what the effect will be on the rate of reaction, which becomes the dependent variable. For the hypothetical reaction following, assume that the exponential values for concentrations in the reaction rate expression are integers. Complete the Analysis of an investigation report based on the evidence given.

Problem

What is the rate law for the following hypothetical chemical reaction?

$$a\ W + b\ X + c\ Y \rightarrow d\ Z$$

Experimental Design

A series of reaction trials are performed, with the concentration of one reactant doubled each time after the initial control trial. The initial rate of formation of product "Z" is measured for each trial.

Evidence

Initial [W] (mol/L)	Initial [X] (mol/L)	Initial [Y] (mol/L)	Initial Rate of Formation of Z (mol/(L•s))
0.100	0.100	0.100	12.0
0.200	0.100	0.100	12.0
0.100	0.200	0.100	24.0
0.100	0.100	0.200	48.0

13 Equilibrium in Chemical Systems

An expert juggler in performance is similar to a chemical system at equilibrium. As in a closed chemical system, nothing enters or leaves. During the performance, there is no net observable change. Although there is no net change, there is internal movement. This is a *dynamic equilibrium*, with some balls moving upward and some moving downward at any given moment. There is no net change because the rate of the upward movement is equal to the rate of the downward movement. What could disturb this equilibrium? If you threw the juggler more balls, or if a mosquito flew close to his face, or if a sudden noise startled him, some balls might leave the system, disrupting the equilibrium.

There are many examples of dynamic equilibrium in chemical systems. Although they are not exactly like the situation described here, there are similarities. Chemical systems at equilibrium have constant properties; nothing appears to be happening. Disturbing dynamic equilibria in industrial processes is one task carried out by chemical engineers, who try to encourage or discourage particular reactions by manipulating the conditions under which the reactions occur. Some general concepts apply to all chemical systems at equilibrium; these are a focus of this chapter.

Scientists describe chemical systems in terms of empirical properties such as temperature, pressure, composition, and amounts of substances present. Chemical systems are simpler to study when made separate from their surroundings by a definite boundary, allowing control over the system so that no matter can enter or leave. Such a physical arrangement is called a **closed system**. A solution in a test tube or a beaker can be considered a closed system, as long as no gas is used or produced in the reaction. Systems involving gases must be closed on all sides. The separation of a system from its surroundings means that relevant empirical properties and conditions can be described and changes occurring within the system can be studied. The use of controlled systems is an integral part of scientific study.

One example of a chemical system at equilibrium is a soft drink in a closed bottle. Nothing appears to change, until the bottle is opened. Removing the bottle cap and reducing the pressure alters the equilibrium state, as the carbon dioxide is allowed to leave the system (Figure 13.1). Carbonated drinks that have gone "flat" because of the decomposition of carbonic acid can be carbonated again by the addition of pressurized carbon dioxide to the solution to reverse the reaction and restore the original equilibrium.

Collision-reaction theory is fundamental to the study of chemical systems. As originally introduced in this text to provide a basis for stoichiometric calculations, this theory assumed that reactions are always spontaneous, rapid, quantitative, and stoichiometric. Common experience, however, shows that this assumption is not always the case. Not all reactions are rapid; for example, corrosion of a car body may take years. A study of oxidation-reduction reactions soon provides evidence that many reactions are not spontaneous. This chapter will examine and test one of these assumptions in detail and thereby significantly increase your understanding of chemical systems.

Figure 13.1
When the pressure on this equilibrium system changes, the equilibrium is disturbed.

INVESTIGATION

13.1 Extent of a Chemical Reaction

Evidence supporting the assumption that reactions are quantitative was obtained in Chapter 7 with stoichiometry experiments that produced precipitates. In a quantitative reaction, the *limiting reagent* is completely consumed. To identify the limiting reagent you can test the final reaction mixture for the presence of the original reactants. For example, in a diagnostic test you might try to precipitate ions from the final reaction mixture that were present in the original reactants.

The purpose of this investigation is to test the validity of the assumption that chemical reactions are quantitative.

Problem

What are the limiting and excess reagents in the chemical reaction of selected quantities of aqueous sodium sulfate and aqueous calcium chloride?

- ☐ Problem
- ☑ Prediction
- ☐ Design
- ☐ Materials
- ☑ Procedure
- ☑ Evidence
- ☑ Analysis
- ☑ Evaluation
- ☐ Synthesis

Experimental Design

Samples of sodium sulfate solution and calcium chloride solution are mixed in different proportions and the final mixture is filtered. Samples of the filtrate are tested for the presence of excess reagents, using the following diagnostic tests.

- If a few drops of $Ba(NO_3)_{2(aq)}$ are added to the filtrate and a precipitate forms, then excess sulfate ions are present.

  $$Ba^{2+}_{(aq)} + SO_4^{2-}_{(aq)} \rightarrow BaSO_{4(s)}$$

- If a few drops of $Na_2CO_{3(aq)}$ are added to the filtrate and a precipitate forms, then excess calcium ions are present.

  $$Ca^{2+}_{(aq)} + CO_3^{2-}_{(aq)} \rightarrow CaCO_{3(s)}$$

Materials

lab apron
safety glasses
25 mL of 0.50 mol/L $CaCl_{2(aq)}$
25 mL of 0.50 mol/L $Na_2SO_{4(aq)}$
1.0 mol/L $Na_2CO_{3(aq)}$ in dropper bottle
saturated $Ba(NO_3)_{2(aq)}$ in dropper bottle
(2) 50 mL or 100 mL beakers
two small test tubes
10 mL or 25 mL graduated cylinder
filtration apparatus
filter paper
wash bottle
stirring rod

Anomalies

Anomalies, or discrepant events, play an important role in the acquisition and development of scientific knowledge. Sometimes these events have been ignored, discredited, or elaborately explained away by scientists who do not wish to question or reconsider accepted laws and theories. However, anomalies sometimes lead to the restriction, revision, or replacement of scientific laws and theories.

Exercise

1. The evidence gathered in Investigation 13.1 may be classified as an anomaly — an unexpected result that contradicts previous rules or experience.
   (a) Write the balanced formula equation for the double replacement reaction of sodium sulfate and calcium chloride solutions.
   (b) Write a statement describing the anomaly that occurred, using chemical names from the formula equation.
   (c) Write the net ionic equation for the reaction.
   (d) Use chemical names from the net ionic equation to write a statement about the anomaly.
   (e) Which of the previous statements more accurately describes the chemical system, according to reaction kinetic theory?

2. When scientists first encounter an apparent anomaly they carefully evaluate the experimental design, procedure, and technological skills involved in an investigation. One important consideration is the reproducibility of the evidence. Compare your evidence in Investigation 13.1 with the evidence collected by other groups. Is there support for the reproducibility of this evidence?

13.2 EQUILIBRIUM IN CHEMICAL SYSTEMS

Evidence obtained from many reactions contradicts the assumption that reactions are always quantitative. In some reactions, there is direct evidence for the presence of both reactants after the reaction appears to have stopped. This apparent anomaly can be explained consistently in terms of the collision-reaction theory, by the idea that a reverse reaction can occur. That is, the products, calcium sulfate and sodium chloride, can react to re-form the original reactants. The final state of this chemical system can be explained as a competition between collisions of reactants to form products and collisions of products to re-form reactants.

$$Na_2SO_{4(aq)} + CaCl_{2(aq)} \overset{\text{forward}}{\underset{\text{reverse}}{\rightleftharpoons}} CaSO_{4(s)} + 2\,NaCl_{(aq)}$$

This competition requires that the system be closed so that reactants and products cannot escape from the reaction container. The chemical system in Investigation 13.1 can be considered a closed system, bounded by the volume of the liquid phase.

We assume that any closed chemical system with constant macroscopic properties (no observable change occurring) is in a state of **equilibrium**, usually classified, for convenience, as one of three types. **Phase equilibrium** involves a single chemical substance undergoing a phase change in a closed system (Figure 13.2). **Solubility equilibrium** involves a single chemical solute interacting with a solvent substance (Figure 13.3). A **chemical reaction equilibrium** involves several substances — the reactants and products of a chemical reaction. All three types of equilibrium are explained by a theory of **dynamic equilibrium** — a balance between two opposite processes occurring at the same *rate*.

The terms *forward* and *reverse* are used to identify which process is being referred to, and are specific to a written equilibrium equation. When any equation is written with arrows to show that the change occurs both ways, the left-to-right change direction is called the **forward reaction**, and the right-to-left change direction is called the **reverse reaction**.

Phase Equilibrium

In a closed system, a phase change may establish an equilibrium, such as the evaporation/condensation equilibrium of Figure 13.2. We explain this *establishment* of equilibrium by using kinetic theory. We assume that when a liquid is placed in a closed container, initially only evaporation occurs: some molecules gain enough energy in collisions to leave the surface of the liquid phase and move into the gas phase. As the number of molecules in the gas phase increases, however, increasingly more of them collide with the liquid surface and lose enough energy to join the condensed phase. In time, the *rates* at which molecules are evaporating and condensing will become equal, so that while both processes are still occurring, no change in the system is observable. The concentration of substance in the gas phase remains

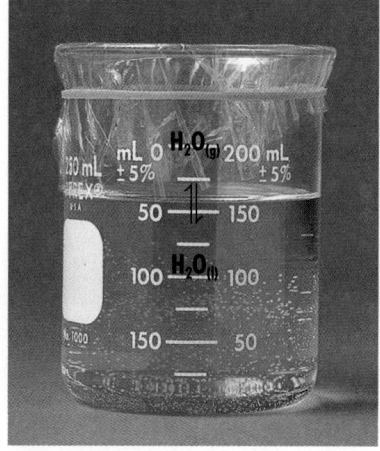

Figure 13.2
Water placed in a sealed container, like this beaker sealed with plastic wrap, evaporates until the vapor pressure inside the container becomes constant. According to the theory of dynamic equilibrium, when the vapor pressure is constant the rate of evaporation is equal to the rate of condensation.

$$H_2O_{(l)} \rightleftharpoons H_2O_{(g)}$$

Figure 13.3
In a saturated solution of iodine, the concentration of the dissolved solute is constant. According to the theory of dynamic equilibrium, the rate of the dissolving process is equal to the rate of the crystallizing process.

$$I_{2(s)} \rightleftharpoons I_{2(aq)}$$

constant. Note that in an *open* container no equilibrium can be established — for molecules that change to the gas phase and escape the system, no reverse process can occur. Another important point to note is the role played by *temperature* in the establishment of the equilibrium. The tendency of any liquid to evaporate increases at higher temperatures, so the concentration (and thus the pressure) of the vapor is greater if the equilibrium is established at a higher temperature. (See the water vapor pressure table on the inside back cover.)

The other common phase equilibrium is the one normally called melting/freezing. Chemists do not think of these as two separate changes, but rather as one change referred to by the direction of interest. Solid/liquid phase equilibrium can normally establish only at one temperature, the melting/freezing point, and this is useful in temperature control. If a large amount of crushed ice is placed in water, for example, initially the ice melts much more rapidly than the water freezes, so you observe melting as the net change. As the temperature drops, the rate of melting decreases and the rate of freezing increases, until at a temperature of 0°C the rates become equal. Chemists know that in a well-stirred ice/water mixture the temperature must be precisely 0°C, so an ice/water slush can be used to control temperatures for experiments. This situation is illustrated by an equilibrium equation.

$$H_2O_{(s)} \rightleftharpoons H_2O_{(l)} \qquad t = 0°C$$

Solubility Equilibrium

Most substances dissolve in a solvent to a certain extent, and then dissolving appears to stop. If the solution is in a *closed system*, one in which no substance can enter or leave, then observable properties become constant, or are *in equilibrium.*

According to the kinetic molecular theory, particles are always moving and collisions are always occurring in a system, even if no changes are observed. The initial dissolving of sodium chloride in water is thought to be the result of collisions between water molecules and ions that make up the crystals. At equilibrium, water molecules still collide with the ions at the crystal surface. Chemists assume that dissolving of the solid sodium chloride is still occurring at equilibrium. Some of the dissolved sodium and chloride ions must, therefore, be colliding and crystallizing out of the solution to maintain a balance. If both dissolving and crystallizing take place at the same rate, no observable changes would occur in either the concentration of the solution or in the quantity of solid present. The balance that exists when two opposing processes occur at the same rate is known as *dynamic equilibrium* (Figure 13.4).

Testing the Theory of Dynamic Equilibrium

You can try a simple experiment to illustrate dynamic equilibrium. Dissolve pickling (coarse) salt to make a saturated solution with excess solid in a small jar. Ensure that the lid is firmly in place, then shake the jar and record the time it takes for the contents to settle so that the solution is clear. Repeat this process once a day for two weeks.

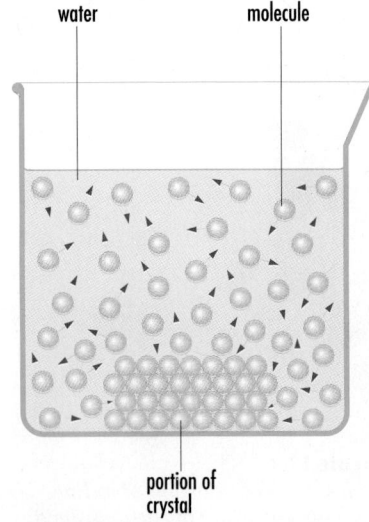

water   molecule

portion of
crystal

Figure 13.4
Dynamic equilibrium — in a saturated solution such as this, with excess solute present, dissolving and crystallizing occur at the same rate.

Although the same quantity of undissolved salt is present each day, the settling becomes much faster over time. This happens because the solid particles in the jar become fewer in number, but larger in size. In terms of the tiniest particles, this occurs because dissolving occurs just slightly faster than crystallizing, whereas the reverse is true for the largest particles. Eventually, the smallest particles disappear, and the largest ones grow in size. Chemists usually allow precipitates to sit for a while before filtering them, because larger particles filter more quickly. They call this step letting the precipitate "digest." This evidence supports the idea that both dissolving and crystallizing are occurring simultaneously.

The theory of dynamic equilibrium can be tested by using a saturated solution of iodine in water. Radioactive iodine is used as a marker to follow the movements of some of the molecules in the mixture. To one sample of a saturated solution containing an excess of solid normal iodine, a few crystals of radioactive iodine are added. To a similar second sample, a few millilitres of a saturated solution of radioactive iodine are added (Figure 13.5). The radioactive iodine emits radiation which can be detected by a Geiger counter to show the location of the radioactive iodine. After a few minutes, the solution and the solid in both samples clearly show increased radioactivity over the average background readings. Assuming the radioactive iodine molecules are chemically identical to normal iodine, the experimental evidence supports the idea of simultaneous dissolving and crystallizing of iodine molecules in a saturated system.

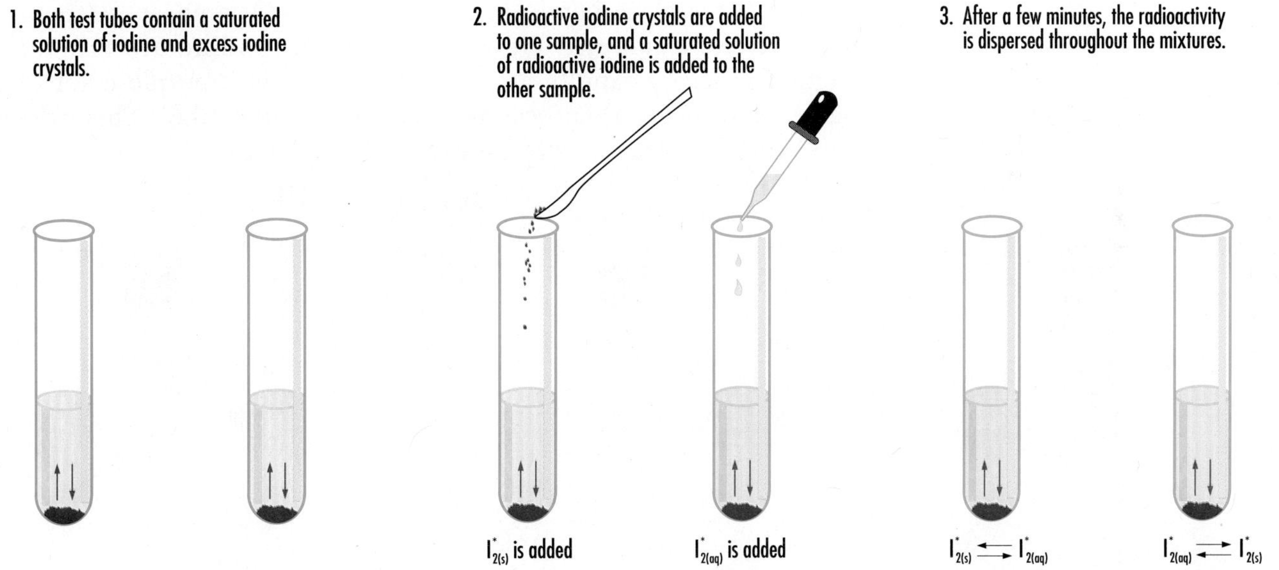

1. Both test tubes contain a saturated solution of iodine and excess iodine crystals.

2. Radioactive iodine crystals are added to one sample, and a saturated solution of radioactive iodine is added to the other sample.

3. After a few minutes, the radioactivity is dispersed throughout the mixtures.

$I^*_{2(s)}$ is added $I^*_{2(aq)}$ is added

$I^*_{2(s)} \rightleftharpoons I^*_{2(aq)}$ $I^*_{2(aq)} \rightleftharpoons I^*_{2(s)}$

Figure 13.5
Radioactive iodine (I_2^*), added to a saturated solution of normal iodine (I_2), is eventually distributed throughout the mixture.

A solubility equilibrium must contain both dissolved and undissolved solute at the same time. This state can be established by starting with a solute and adding it to a solvent. Consider adding calcium sulfate to water in a large enough quantity that not all will dissolve. We say we have added *excess* solute. A net ionic equilibrium equation can be written for a saturated solution established this way.

$$CaSO_{4(s)} \rightleftharpoons Ca^{2+}_{(aq)} + SO_4^{2-}_{(aq)}$$

Now consider a situation where two solutions, containing very high concentrations of calcium and sulfate ions respectively, are mixed. In this situation, the initial rate at which ions combine to form solid crystals is much greater than the rate at which those crystals dissolve, so we observe precipitation until the rates become equal and equilibrium is established. See the equation that follows.

$$Ca^{2+}_{(aq)} + SO_4^{2-}_{(aq)} \rightleftharpoons CaSO_{4(s)}$$

The important point is that at equilibrium the rates of change are equal. *How* the equilibrium is established is not a factor. Viewed this way, the chemical reaction in Investigation 13.1, and in fact most ionic compound precipitation reactions, are really just examples of solubility equilibrium considered in reverse.

Chemical Reaction Equilibrium

Chemical reaction equilibria are more complex than phase or solubility equilibria, due to the variety of possible chemical reactions and the greater number of substances involved. An explanation of chemical equilibrium systems requires a synthesis of ideas from kinetic molecular theory, collision-reaction theory, and the concepts of reversibility and dynamic equilibrium. Although this synthesis is successful as an explanation, it has only limited application in predicting quantitative properties of an equilibrium system.

Chemists have studied the reaction of hydrogen gas and iodine gas extensively, because the molecules are relatively simple and the reaction takes place between molecules in the gas phase, without the necessity of a solvent. Once hydrogen and iodine are mixed, the reaction proceeds rapidly at first. The initial dark purple color of the iodine vapor fades, then becomes constant (Figure 13.6). This evidence is theoretically described by an equilibrium equation.

$$H_{2(g)} + I_{2(g)} \rightleftharpoons 2\,HI_{(g)}$$

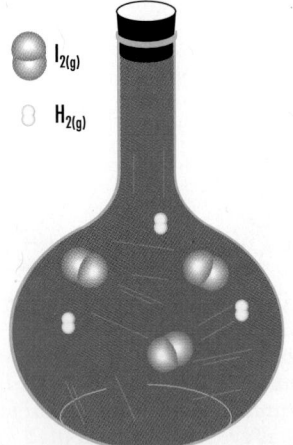

Figure 13.6
Initially, hydrogen and iodine are added to the system. The color of iodine vapor is the only easily observable property.

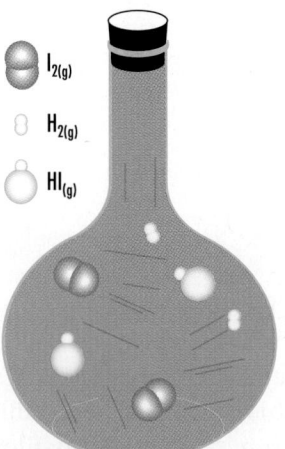

Early in the reaction, hydrogen and iodine form hydrogen iodide faster than hydrogen iodide forms hydrogen and iodine. Overall, the amount of iodine decreases, so its color appears to lighten. Both hydrogen and hydrogen iodide are colorless gases.

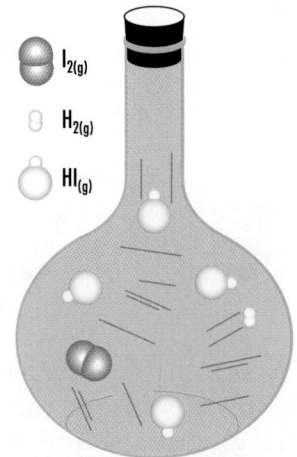

At equilibrium, the system contains all three substances. The purple color shows that some iodine remains. The constancy of the color is evidence that equilibrium exists. Forward and reverse reactions are occurring at equal rates.

Table 13.1 contains data from three experiments with the hydrogen-iodine system: one in which hydrogen and iodine are mixed; one in which hydrogen, iodine, and hydrogen iodide are mixed; and one in which only hydrogen iodide is present initially. At a temperature of 448°C, the system quickly reaches an observable equilibrium each time. Chemists use evidence such as that in Table 13.1 to describe a state of equilibrium in two ways: in terms of percent reaction and in terms of an equilibrium constant.

Table 13.1

	THE HYDROGEN-IODINE SYSTEM AT 448°C					
System	Initial System Concentrations (mmol/L)			Equilibrium System Concentrations (mmol/L)		
	$H_{2(g)}$	$I_{2(g)}$	$HI_{(g)}$	$H_{2(g)}$	$I_{2(g)}$	$HI_{(g)}$
1	1.00	1.00	0	0.22	0.22	1.56
2	0.50	0.50	1.70	0.30	0.30	2.10
3	0	0	3.20	0.35	0.35	2.50

Percent Reaction at Chemical Equilibrium

A **percent reaction** or **percent yield** is defined as the yield of product measured at equilibrium compared with the maximum possible yield of product. In other words, percent reaction is one way of communicating the *position of an equilibrium*. The maximum possible yield of product is calculated using the method of stoichiometry, assuming a quantitative forward reaction with no reverse reaction. Percent reaction provides an easily understood way to discuss amounts of chemicals present in equilibrium systems. For example, analysis of the evidence in Table 13.1 shows that at 448°C the hydrogen-iodine system reaches an equilibrium with a percent reaction of 78% (Table 13.2). A percent reaction is a single value that can be used to describe and compare chemical reaction equilibria.

Table 13.2

	PERCENT REACTION OF THE HYDROGEN-IODINE SYSTEM AT 448°C		
System	Equilibrium [HI]* (mmol/L)	Maximum Possible [HI]* (mmol/L)	Percent Reaction (%)
1	1.56	2.00	78.0
2	2.10	2.70	77.8
3	2.50	3.20	78.1
*Square brackets [ ] indicate molar concentration.			

To communicate that an equilibrium exists, equilibrium arrows ($\rightleftharpoons$) are used. To communicate the extent of a reaction, the percent reaction is written above the equilibrium arrows in a chemical equation. The following equation describes the position of the hydrogen-iodine equilibrium system (Table 13.2).

$$H_{2(g)} + I_{2(g)} \overset{78\%}{\rightleftharpoons} 2\,HI_{(g)} \qquad t = 448°C$$

All chemical reactions are now thought of as occurring in both forward and reverse directions. Any reaction falls loosely into one of four categories. Reactions that favor reactants very strongly — much less than 1% — are observed as being non-spontaneous. In these reactions, mixing reactants has no observable result. Reactions that have noticeable equilibrium conditions may be reacted less or more than 50%, favoring reactants or products respectively. Significant amounts of both reactants and products are always present. Finally, reactions that favor products very strongly, much more than 99%, are observed to be complete (*quantitative*). These are generally written with a single arrow to indicate that the effect of the reverse reaction is negligible. Table 13.3 shows how percent reaction is used to classify equilibrium systems and how the classification is communicated in reaction equations.

Table 13.3

CLASSES OF CHEMICAL REACTION EQUILIBRIA		
Percent Reaction	**Description of Equilibrium**	**Position of Equilibrium**
<50%	reactants favored	<50% $\rightleftharpoons$
>50%	products favored	>50% $\rightleftharpoons$
>99%	quantitative	>99% or $\rightarrow$ $\rightleftharpoons$

When there is no limiting reagent for a reaction, and when one cannot assume complete reaction, stoichiometric calculations require a little more thought. Such calculations may conveniently be set up as an ICE table, meaning that the *initial, change,* and *equilibrium* values are arranged in tabular form.

EXAMPLE

Consider the previous reaction equation for the formation of hydrogen iodide at 448°C. Assume the reaction is begun with 1.00 mmol/L concentrations of $H_{2(g)}$ and $I_{2(g)}$ (see Table 13.1, page 491). The equilibrium concentration of $I_{2(g)}$ (determined by color intensity) is 0.22 mmol/L. The ICE table is set up as follows.

THE $H_{2(g)}$ + $I_{2(g)}$ $\rightleftharpoons$ 2 $HI_{(g)}$ EQUILIBRIUM			
Concentration	$[H_{2(g)}]$ (mmol/L)	$[I_{2(g)}]$ (mmol/L)	$[HI_{(g)}]$ (mmol/L)
Initial	1.00	1.00	0.00
Change			
Equilibrium		0.22	

Begin by letting ΔC stand for the change in concentration of the iodine from its initial value to its (lower) equilibrium value.

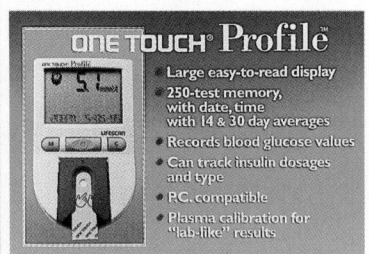

$$\Delta C = C_{eq} - C_i = (0.22 - 1.00)\ \text{mmol/L} = -0.78\ \text{mmol/L}$$

According to the stoichiometric ratios from the balanced equation, the value of ΔC for hydrogen consumed must also be –0.78 mmol/L, and must be 2×0.78 mmol/L or +1.56 mmol/L for the hydrogen iodide produced. Continuing, we get

THE $H_{2(g)} + I_{2(g)} \rightleftharpoons$ 2 $HI_{(g)}$ EQUILIBRIUM			
Concentration	$[H_{2(g)}]$ (mmol/L)	$[I_{2(g)}]$ (mmol/L)	$[HI_{(g)}]$ (mmol/L)
Initial, C_i	1.00	1.00	0.00
Change, ΔC	–0.78	–0.78	+1.56
Equilibrium, C_{eq}	0.22	0.22	1.56

The stoichiometric calculation can use concentrations directly, rather than amounts in moles, because the volume must be the same for every gaseous substance in a closed container. The volume is a common factor in the calculation.

▌ICE Tables

It is often convenient to create an ICE (initial, change, and equilibrium) table to describe what happens to the concentrations as an equilibrium is established or re-established. For example, for the reaction

$$A_{(aq)} + B_{(aq)} \rightleftharpoons 2\ G_{(aq)}$$

C	$A_{(aq)}$	$B_{(aq)}$	$G_{(aq)}$
C_i	C_i	C_i	C_i
ΔC	$C_{eq} - C_i$	$C_{eq} - C_i$	$C_{eq} - C_i$
C_{eq}	C_{eq}	C_{eq}	C_{eq}

$\Delta C = C_{eq} - C_i$
$C_{eq} = C_i + \Delta C$
$\Delta C_A = \Delta C_B$ (in this case)
$\Delta C_G = 2\ \Delta C_A$ (in this case)

Note that ΔC is negative for a decrease in concentration and positive for an increase in concentration.

Exercise

3. For a chemical system at equilibrium:
   (a) What are the observable characteristics?
   (b) Why is the equilibrium considered "dynamic"?
   (c) What is considered "equal" about the system?

4. In a gaseous reaction system, 2.00 mol of methane, $CH_{4(g)}$, is initially added to 10.00 mol of chlorine, $Cl_{2(g)}$. At equilibrium the system contains 1.40 mol of chloromethane, $CH_3Cl_{(g)}$, and some hydrogen chloride, $HCl_{(g)}$.
   (a) Write a balanced reaction equation for this equilibrium and calculate the maximum possible yield of chloromethane product.
   (b) Calculate the percent reaction of this equilibrium and state whether products or reactants are favored.

5. Combustion reactions, such as the burning of methane, obviously favor products so strongly they are normally written with a single arrow. Assuming the forward reaction has a very low activation energy (Chapter 12) and the reverse reaction has a very high activation energy (to account for the difference in the tendency to occur), sketch a possible potential energy diagram representing the progress of such a reaction.

6. After 4.0 mol of $C_2H_{4(g)}$ and 2.50 mol of $Br_{2(g)}$ are placed in a sealed container, the reaction

$$C_2H_{4(g)} + Br_{2(g)} \rightleftharpoons C_2H_4Br_{2(g)}$$

establishes an equilibrium. The diagram shows the concentration of $C_2H_{4(g)}$ as it changes over time at a fixed temperature until equilibrium is reached.

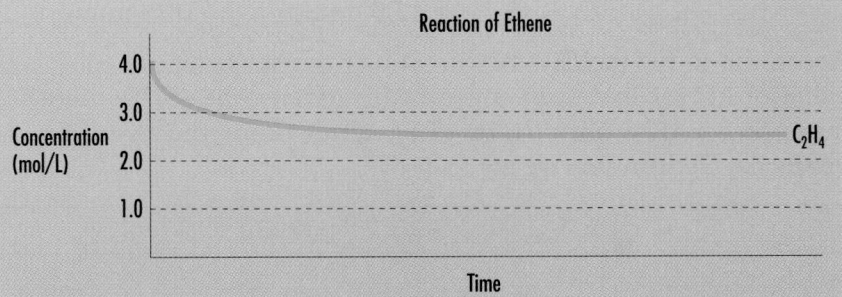

Reaction of Ethene

Concentration (mol/L)

Time

(a) Sketch this graph. Draw lines on your copy to show how the concentration of each of the other two substances changes.
(b) Create an ICE table.
(c) Calculate the percent reaction at equilibrium.

7. For each of the following, write the chemical reaction equation with appropriate equilibrium arrows, as shown in Table 13.3 (page 492).
   (a) The Haber process is used to manufacture ammonia fertilizer from hydrogen and nitrogen gases. Under less than desirable conditions, only an 11% yield of ammonia is obtained at equilibrium.
   (b) A mixture of carbon monoxide and hydrogen, known as water gas, is used as a supplementary fuel in many large industries. At high temperatures, the reaction of coke and steam forms an equilibrium mixture in which the products (carbon monoxide and hydrogen gases) are favored. (Assume that coke is pure carbon.)
   (c) Because of the cost of silver, many high school science departments recover silver metal from waste solutions containing silver compounds or silver ions. A quantitative reaction of waste silver ion solutions with copper metal results in the production of silver metal and copper(II) ions.
   (d) One step in the industrial process used to manufacture sulfuric acid is the production of sulfur trioxide from sulfur dioxide and oxygen gases. Under certain conditions the reaction produces a 65% yield of products.

8. The empirical and theoretical concepts of equilibrium can be applied to many different chemical reaction systems. Use the generalizations from your study of organic chemistry to predict the position of equilibrium for bromine placed in a reaction container with ethylene at a high temperature.

9. Interpret the graph shown to answer the questions about the reaction.

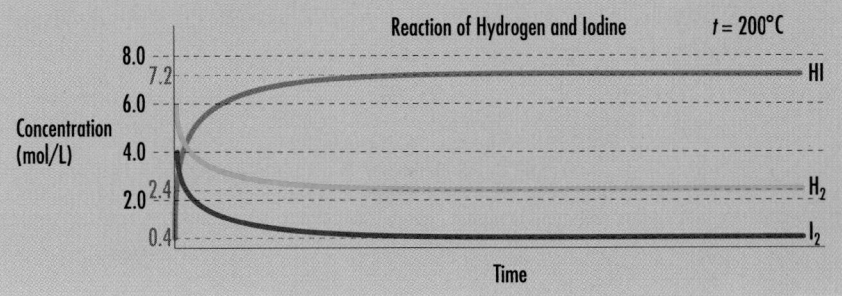

Reaction of Hydrogen and Iodine   t = 200°C

Concentration (mol/L)

Time

Coal Gas

Coal can be mined from deep deposits by drilling to, and then pumping very hot steam into, a coal seam. The steam and coal react to yield coal gas — carbon monoxide and hydrogen. The coal gas is retrieved through a second concentric tube. The process produces a lower yield but is safer than sending miners into deep mine shafts. The product is a cleaner fuel to handle and to burn.

(a) All three substances are gases. If the container has a volume of 2.00 L, what amount (in moles) of each substance was present initially?

(b) What amount (in moles) of hydrogen iodide had formed at equilibrium? (Create an ICE table.)

(c) What is the percent reaction at equilibrium?

(d) In Chapter 12, the reaction rate was described as the slope of the concentration curve of a given reactant. Does this mean that the rate of reaction of all three substances becomes zero when their concentrations become constant at equilibrium? Explain.

Problem 13A The Synthesis of an Equilibrium Law

The following chemical equation represents a chemical equilibrium.

$$Fe^{3+}_{(aq)} + SCN^{-}_{(aq)} \rightleftharpoons FeSCN^{2+}_{(aq)}$$

This equilibrium is convenient to study because the color of the system characterizes the state of the system (Figure 13.7). The purpose of this problem is the synthesis of an equilibrium law. Complete the Analysis of the investigation report.

Problem

What mathematical formula, using equilibrium concentrations of reactants and products, gives a constant for the iron(III)-thiocyanate reaction system?

Experimental Design

Reactions are performed using various initial concentrations of iron(III) nitrate and potassium thiocyanate solutions. The equilibrium concentrations of the reactants and the product are determined from the measurement and analysis of the intensity of the color. Possible mathematical relationships among the concentrations are tried, and then are analyzed to determine if the mathematical formula gives a constant value.

Evidence

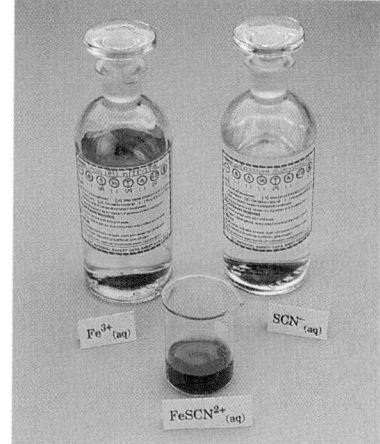

Figure 13.7
The two reactants combine to form a dark red equilibrium mixture. The red color of the solution is the color of the aqueous thiocyanato iron(III) product, $FeSCN^{2+}_{(aq)}$.

Computers
Scientists often use computers to analyze numerical evidence in order to establish mathematical relationships among experimental variables. The mathematical formulas derived are useful in understanding chemical processes and in applying these processes to technology.

	IRON(III)-THIOCYANATE EQUILIBRIUM AT SATP		
Trial	$[Fe^{3+}_{(aq)}]$ (mol/L)	$[SCN^{-}_{(aq)}]$ (mol/L)	$[FeSCN^{2+}_{(aq)}]$ (mol/L)
1	3.91×10^{-2}	8.02×10^{-5}	9.22×10^{-4}
2	1.48×10^{-2}	1.91×10^{-4}	8.28×10^{-4}
3	6.27×10^{-3}	3.65×10^{-4}	6.58×10^{-4}
4	2.14×10^{-3}	5.41×10^{-4}	3.55×10^{-4}
5	1.78×10^{-3}	6.13×10^{-4}	3.23×10^{-4}

Analysis

Test the following mathematical relationships for constancy.

1. $[Fe^{3+}_{(aq)}][SCN^-_{(aq)}][FeSCN^{2+}_{(aq)}]$

2. $[Fe^{3+}_{(aq)}] + [SCN^-_{(aq)}] + [FeSCN^{2+}_{(aq)}]$

3. $\dfrac{[FeSCN^{2+}_{(aq)}]}{[Fe^{3+}_{(aq)}][SCN^-_{(aq)}]}$

4. $\dfrac{[Fe^{3+}_{(aq)}]}{[FeSCN^{2+}_{(aq)}]}$

5. $\dfrac{[SCN^-_{(aq)}]}{[FeSCN^{2+}_{(aq)}]}$

The Equilibrium Constant *K*

Analysis of the evidence from experiments such as those in Problem 13A reveals a mathematical relationship that provides a constant value for a chemical system over a range of concentrations. This constant value is called the **equilibrium constant *K*** for the reaction system. Evidence and analysis of many equilibrium systems have resulted in the following **equilibrium law**.

For the reaction $a\,A + b\,B \rightleftharpoons c\,C + d\,D$

the equilibrium law is $K = \dfrac{[C]^c[D]^d}{[A]^a[B]^b}$

In this mathematical expression, A, B, C, and D represent chemical entities and *a, b, c,* and *d* represent their coefficients in the balanced chemical equation.

EXAMPLE _____

Write the equilibrium law for the reaction of nitrogen monoxide gas with oxygen gas to form nitrogen dioxide gas.

$$2\,NO_{(g)} + O_{2(g)} \rightleftharpoons 2\,NO_{2(g)}$$

$$K = \dfrac{[NO_{2(g)}]^2}{[NO_{(g)}]^2[O_{2(g)}]}$$

A balanced chemical equation with whole number coefficients is used to write the mathematical expression of the equilibrium law, as the coefficients of the balanced equation become the exponents of the concentrations. The higher the numerical value of the equilibrium constant, the greater the tendency of the system to favor the forward direction. That is, the greater the equilibrium constant, the greater the percent reaction and the more the products are favored at equilibrium. The reciprocal of the equilibrium law also gives a constant, which would be the reciprocal of the equilibrium constant shown above. However, using the *products over reactants* convention results in a *relationship* between the numerical value of K and the forward extent of the equilibrium that is easier to visualize; scientists agree that this is easier to understand.

Both methods of expressing the position of an equilibrium — the equilibrium constant and the percent reaction — have restricted application. The value of the equilibrium constant is found by experiment to depend on temperature, and the value is also affected by large changes in the equilibrium concentration of a reactant or a product. A moderate change in the concentration of any one of the reactants or products results in a change in the other concentrations, so that the equilibrium constant remains the same. The equilibrium constant provides only a measure of the equilibrium position of the reaction; it does not provide any information on the rate of the reaction. Although the percent reaction expression is easily understood, the value is different for every change in the initial concentration of reactants. Like equilibrium constant expressions, percent reaction expressions give no information about rate of reaction, and are dependent on temperature. Because they hold for a significant range of different concentrations, equilibrium constant expressions have been found to be more useful, and K values for reactions are in common use throughout the scientific community.

Equilibrium constants are usually adjusted to reflect the fact that pure substances in solid or liquid (condensed) states have concentrations that are essentially fixed — the number of moles per unit volume is a *constant* value. For example, a litre of liquid water at SATP has a mass of 1.00 kg (an amount of 55.5 mol) and thus a fixed concentration of 55.5 mol/L. The concentration of condensed states is not generally included in the expression — it is assumed that their constant values become part of the expressed equilibrium constant. Substances in a gaseous or dissolved state have *variable* concentrations, which are always shown in an equilibrium expression.

Equilibrium Constant Units
It is common practice to ignore units and list only the numerical value of an equilibrium constant. The units for the equilibrium constant vary, since they depend on the coefficients and the mathematical expression of the equilibrium law. In the nitrogen oxide-nitrogen dioxide example, the unit for the equilibrium constant is 1/(mol /L) or L/mol. For some equilibrium constants, all the units cancel.

EXAMPLE

Write the equilibrium expression for the decomposition of solid ammonium chloride to gaseous ammonia and gaseous hydrogen chloride.

$$NH_4Cl_{(s)} \rightleftharpoons NH_{3(g)} + HCl_{(g)}$$

$$K = [NH_{3(g)}][HCl_{(g)}]$$

K Involving Solids
The concentration of a solid, such as $NH_4Cl_{(s)}$, is omitted from the equilibrium expression; more correctly, it is a constant value which is included in the value of the equilibrium constant, K.

The role of temperature in equilibrium constant expressions is critical, although the temperature is not written in the expression directly. The value of the equilibrium constant, K, always depends on the temperature. Any stated numerical value for an equilibrium constant, or any calculation using an equilibrium constant expression, must specify a temperature.

Since equilibrium depends on the concentrations of reacting substances, these substances must be represented in the expression as they actually exist — meaning that ions in solution must be represented as individual entities. Equilibrium constant expressions are always written from the *net ionic* form of reaction equations, balanced with simplest whole number (integral) coefficient values.

EXAMPLE

Write the equilibrium law for the reaction of zinc in copper(II) chloride solution.

$$Zn_{(s)} + Cu^{2+}_{(aq)} \rightleftharpoons Cu_{(s)} + Zn^{2+}_{(aq)}$$

$$K = \frac{[Zn^{2+}_{(aq)}]}{[Cu^{2+}_{(aq)}]}$$

K Involving Solids
Again, note that the constant concentrations of the solids, as well as the constant concentration of the spectator ions (the chloride ions in this example), are omitted from the equilibrium expression.

Problem 13B Determining an Equilibrium Constant

Complete the Analysis of the investigation report.

Problem

What is the value of the equilibrium constant for the decomposition of phosphorus pentachloride gas to phosphorus trichloride gas and chlorine gas?

Evidence

equilibrium temperature = 200°C
$[PCl_{3(g)}] = [Cl_{2(g)}] = 0.014$ mol/L
$[PCl_{5(g)}] = 4.3 \times 10^{-4}$ mol/L

Exercise

10. Write a balanced equation with integer coefficients and the expression of the equilibrium law for each of the following reaction systems.
    (a) Hydrogen gas reacts with chlorine gas to produce hydrogen chloride gas in the industrial process that eventually produces hydrochloric acid.
    (b) In the Haber process, nitrogen reacts with hydrogen to produce ammonia gas.
    (c) At some time in the future, industry and consumers may make more extensive use of the combustion of hydrogen as an energy source.
    (d) When aqueous ammonia is added to an aqueous nickel(II) ion solution, the $Ni(NH_3)_6^{2+}_{(aq)}$ complex ion is formed (Figure 13.8).
    (e) In the Solvay process for making washing soda, one reaction involves heating solid calcium carbonate (limestone) to produce solid calcium oxide (quicklime) and carbon dioxide.

Figure 13.8
A $Ni^{2+}_{(aq)}$ solution is green. Ammonia reacts with the nickel(II) ion to form the intensely blue hexaamminenickel(II) ion, $Ni(NH_3)_6^{2+}_{(aq)}$.

(f) In Investigation 13.1, aqueous solutions of sodium sulfate and calcium chloride are mixed. (Remember to use a net ionic equation.)

(g) In a sealed can of soda, carbonic acid, $H_2CO_{3(aq)}$, decomposes to liquid water and carbon dioxide gas.

(h) In an insulated cup, ice cubes added to water establish a phase equilibrium. (Explain why there is only one condition for this equilibrium.)

11. Write the expression of the equilibrium law for the hydrogen-iodine-hydrogen iodide system at 448°C. Using the evidence in Table 13.1 on page 491, calculate the value of the equilibrium constant.

12. In the Haber process for synthesizing ammonia gas from nitrogen and hydrogen, the value of K is 6.0×10^{-2} for the reaction at 500°C. In a sealed container at equilibrium at 500°C, the concentrations of $H_{2(g)}$ and of $N_{2(g)}$ are measured to be 0.50 mol/L and 1.50 mol/L respectively. Write the equilibrium constant expression and calculate the equilibrium concentration of $NH_{3(g)}$.

13. Liquid butane in lighters escapes in the gas phase when the lighter is opened. Butane lighters do not work well outdoors in very cold weather. Use the phase equilibrium expression for butane to explain why this is true, and to predict whether the pressure (concentration) of butane gas in the lighter is related to the quantity of liquid butane present.

14. At a certain constant (very high) temperature, 1.00 mol of $HBr_{(g)}$ is introduced into a 2.00 L container. Decomposition of this gas to hydrogen and bromine gases quickly establishes an equilibrium, at which point the molar concentration of $HBr_{(g)}$ is measured to be 0.100 mol/L.

(a) Write a balanced equation for the reaction.

(b) Write the equilibrium constant expression.

(c) Calculate the amount (in moles) of $HBr_{(g)}$ present at equilibrium.

(d) Calculate the amount of $HBr_{(g)}$ that has been reacted to form $H_{2(g)}$ and $Br_{2(g)}$ products when equilibrium is established.

(e) Calculate the amounts of $H_{2(g)}$ and $Br_{2(g)}$ that have been produced, and thus are present, when equilibrium is established.

(f) Calculate the concentration of all substances present at equilibrium.

(g) Calculate K for this reaction at this temperature.

The Solubility Product Constant K_{sp}

A special case of equilibrium involves any situation where excess solute is in equilibrium with its solution. As explained earlier, such an equilibrium can be established by starting with excess solute and dissolving it until the solution is saturated, or by mixing solutions of two substances that result in a product precipitating. It is simpler to think of this type of equilibrium in terms of the solubility of a substance, and in fact the equilibrium characteristics are generally defined in this way. Assuming a water solvent (other solvents are similar), the solutes fall generally into two types: those that ionize when dissolving, and those that do not. For molecular compounds the expression is typically very simple. Consider the expression for sucrose dissolving in water.

$$C_{12}H_{22}O_{11(s)} \rightleftharpoons C_{12}H_{22}O_{11(aq)}$$

The equilibrium expression is $K = [C_{12}H_{22}O_{11(aq)}]$ at t °C

In simple cases such as this, the equilibrium constant is equal to the concentration of the solution, and the value varies only with temperature. This is equivalent to saying that the solubility depends only on the temperature of the solution.

When a dissolved substance dissociates into ions, however, the situation is very different. Consider the expression for a compound with low solubility, copper(I) chloride.

$$CuCl_{(s)} \rightleftharpoons Cu^{+}_{(aq)} + Cl^{-}_{(aq)}$$

The equilibrium expression is $K_{sp} = [Cu^{+}_{(aq)}][Cl^{-}_{(aq)}]$

with the value of the constant $K_{sp} = 1.7 \times 10^{-7}$ at 25°C.

For any ionizing solute the equilibrium constant is the *product* of the concentrations of the ions in solution, and for that reason is often called the **solubility product constant** of the substance, symbolized as K_{sp}. For ionic substances with a more complex formula, like calcium phosphate, the K_{sp} expression is also more complex.

$$Ca_3(PO_4)_{2(s)} \rightleftharpoons 3\ Ca^{2+}_{(aq)} + 2\ PO_4^{3-}_{(aq)}$$

The equilibrium expression is $K_{sp} = [Ca^{2+}_{(aq)}]^3[PO_4^{3-}_{(aq)}]^2$

with the value of the constant $K_{sp} = 2.1 \times 10^{-33}$ at 25°C.

K_{sp} values are listed routinely in chemistry reference materials (see Table 13.4 and Appendix F). Tables of K_{sp} values normally list only ionic compounds with low solubility, because under ordinary laboratory conditions highly soluble ionic compounds do not form precipitates. Their solutions, as commonly used, are not saturated — no equilibrium is established. References typically list solubilities of highly soluble substances as mol/L or g/100 mL values rather than as K_{sp} values.

Table 13.4

SOLUBILITY PRODUCT CONSTANTS AT 25°C		
Name	**Formula**	**K_{sp}**
cobalt(II) hydroxide	$Co(OH)_2$	1.1×10^{-15}
lithium carbonate	Li_2CO_3	8.2×10^{-4}
mercury(I) chloride	Hg_2Cl_2	1.5×10^{-18}
nickel(II) carbonate	$NiCO_3$	1.4×10^{-7}
tin(II) sulfide	SnS	3.2×10^{-28}
zinc hydroxide	$Zn(OH)_2$	7.7×10^{-17}
Reference — *CRC Handbook of Chemistry and Physics* (76th Ed.)		

A straightforward calculation will convert a solubility value to (or from) a K_{sp} value.

EXAMPLE

Find K_{sp} for magnesium fluoride at 25°C, given a solubility of 0.001 72 g/100 mL.

$$MgF_{2(s)} \rightleftharpoons Mg^{2+}_{(aq)} + 2\,F^-_{(aq)} \qquad\qquad K_{sp} = [Mg^{2+}_{(aq)}][F^-_{(aq)}]^2$$

$$[Mg^{2+}_{(aq)}] = [MgF_{2(aq)}]$$
$$= \frac{0.001\ 72\ g}{100\ mL} \times \frac{1\ mol}{62.31\ g} \times \frac{1000\ mL}{1\ L}$$
$$= 2.7 \times 10^{-4}\ mol/L$$

$$[F^-_{(aq)}] = 2[Mg^{2+}_{(aq)}]$$
$$= 2 \times 2.7 \times 10^{-4}\ mol/L$$
$$= 5.5 \times 10^{-4}\ mol/L$$

$$K_{sp} = [Mg^{2+}_{(aq)}][F^-_{(aq)}]^2$$
$$= (2.7 \times 10^{-4})(5.5 \times 10^{-4})^2$$
$$= 8.1 \times 10^{-11}$$

K_{sp} Units
The units for K_{sp} vary from mol/L to (mol/L)^2 to (mol/L)^3, etc. The communication convention in the scientific community is to omit the units for K_{sp} values.

EXAMPLE

Find the molar solubility of zinc hydroxide at 25°C, where K_{sp} is 7.7×10^{-17}.

$$Zn(OH)_{2(s)} \rightleftharpoons Zn^{2+}_{(aq)} + 2\,OH^-_{(aq)}$$
$$K_{sp} = [Zn^{2+}_{(aq)}]\,[OH^-_{(aq)}]^2 = 7.7 \times 10^{-17}$$
$$[OH^-_{(aq)}] = 2\,[Zn^{2+}_{(aq)}]$$
$$K_{sp} = [Zn^{2+}_{(aq)}]\,(2\,[Zn^{2+}_{(aq)}])^2 = 7.7 \times 10^{-17}$$
$$4\,[Zn^{2+}_{(aq)}]^3 = 7.7 \times 10^{-17}$$
$$[Zn^{2+}_{(aq)}] = \sqrt[3]{\frac{7.7 \times 10^{-17}}{4}}$$
$$= 2.7 \times 10^{-6}\ mol/L$$

$$[Zn(OH)_{2(aq)}] = [Zn^{2+}_{(aq)}]$$
$$= 2.7 \times 10^{-6}\ mol/L$$

The molar solubility of zinc hydroxide is 2.7×10^{-6} mol/L.

Exercise

(Refer to Appendix F or Table 13.4 for required K_{sp} values.)

15. Calculate the molar solubility of barium sulfate at 25°C.

16. Calculate the solubility at 25°C of silver bromide, in g/100 mL.

17. Calculate the molar concentration of fluoride ions in a saturated solution of strontium fluoride at 25°C.

18. Calculate the K_{sp} of thallium(I) chloride at boiling water temperature where the concentration of a saturated solution is 2.4 g/100 mL.

19. Calculate the K_{sp} of calcium fluoride at 20°C where the concentration of a saturated solution is determined by evaporating a 100 mL sample of saturated solution to dryness. A mass of 0.0016 g of solid remains after evaporation.

20. Mercury(I) chloride dissolves as shown by the equation

$$Hg_2Cl_{2(s)} \rightleftharpoons Hg_2^{2+}{}_{(aq)} + 2\,Cl^-{}_{(aq)}$$

Calculate the mass of compound required to make 500 mL of a saturated solution of mercury(I) chloride at 25°C.

13.3 QUALITATIVE CHANGES IN EQUILIBRIUM SYSTEMS

Observing the effects of varying system properties on the equilibrium of systems contributes greatly to our understanding of the equilibrium state. From a technological perspective, controlling the extent of equilibrium by manipulating properties is very desirable, because control leads to more efficient and economic processes. From a scientific perspective, observing equilibrium leads to improved theories that explain the nature of equilibrium, thus increasing our understanding.

Equilibrium is an area of study where, historically, technology has led science. Manipulation of reactions was first used in response to some human need, and was not explained until much later by successive theories of increasing validity. This section of the chapter will deal with equilibrium manipulation in the same way, with empirical descriptions of equilibrium manipulation given first, followed by theoretical explanations of the observed results.

Le Châtelier's Principle

According to **Le Châtelier's principle**, when a chemical system at equilibrium is disturbed by a change in a property of the system, the system adjusts in a way that opposes the change to reach a new equilibrium (Figure 13.9). The application of Le Châtelier's principle involves a three-stage process: an initial equilibrium state, a shifting non-equilibrium state, and a new equilibrium state.

Le Châtelier's principle provides a method of predicting the response of a chemical system to an imposed change. Using this simple and completely empirical approach, chemical engineers could produce more of the desired products, making technological processes more efficient and more economical. For example, Fritz Haber used Le Châtelier's principle to devise a process for the economical production of ammonia from atmospheric nitrogen. (See the Haber process, page 505.)

Henri Louis Le Châtelier (1850 – 1936)
Le Châtelier, French chemist and engineer, worked in chemical industries. To maximize the yield of products, Le Châtelier used systematic trial and error. After measuring properties of equilibrium states in chemical systems, he discovered a pattern and stated it as a generalization. This generalization has been supported extensively by evidence and is now considered a scientific law. By convention, it is known as Le Châtelier's principle.

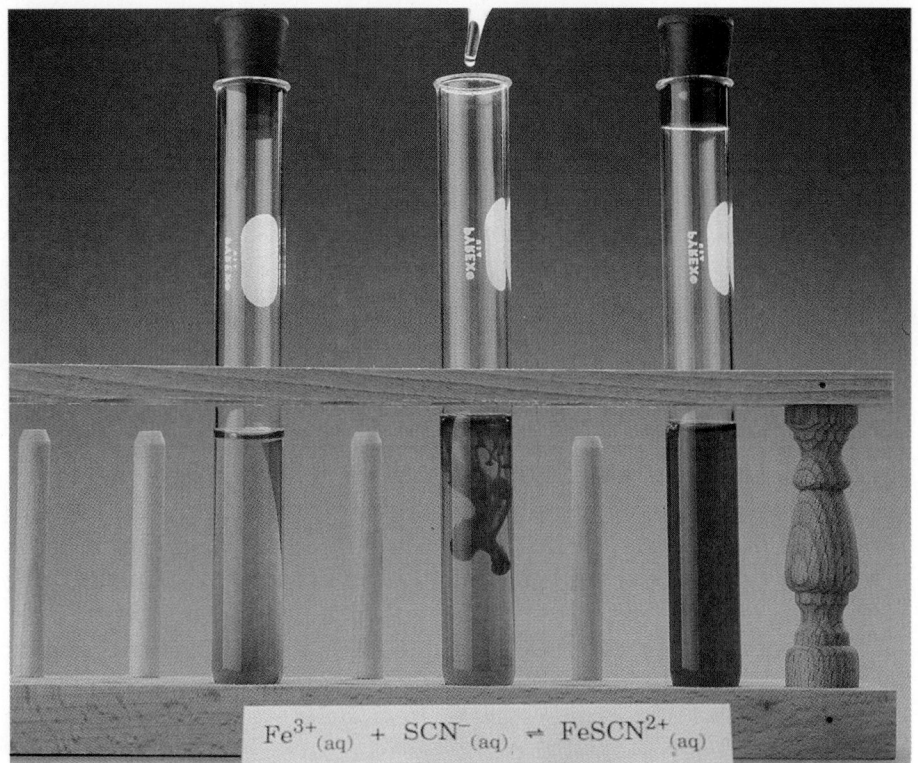

Le Châtelier's Principle and Concentration Changes

Le Châtelier's principle predicts that if the addition of a reactant to a system at equilibrium changes the concentration of that substance, then that system will undergo an **equilibrium shift** forward (to the right). The effect of the shift is that temporarily we observe the reactant concentration decreasing, as some of the added reactant changes to products. This period of change ends with the establishment of a *new* equilibrium state, in which concentrations are usually different from their original values, and once again there are no observable changes. The system has changed in such a way as to *oppose* the change introduced. For example, the production of freon-12, a CFC refrigerant, involves the following equilibrium reaction.

$$CCl_{4(l)} + 2\,HF_{(g)} \rightleftharpoons \underset{\text{freon-12}}{CCl_2F_{2(g)}} + 2\,HCl_{(g)}$$

To improve the yield of the primary product, freon-12, more hydrogen fluoride is added to the initial equilibrium system. The additional amount of reactant disturbs the equilibrium state and the system shifts to the right, consuming some of the added hydrogen fluoride by reaction with carbon tetrachloride. As a result, more freon-12 is produced and a new equilibrium state is obtained. In chemical reaction equilibrium shifts, the imposed concentration change is normally only *partially* counteracted, and the final equilibrium state concentrations of the reactants and products are usually different from the values at the original equilibrium state. See Figure 13.10 for a graphic interpretation of the freon-12 equilibrium shift.

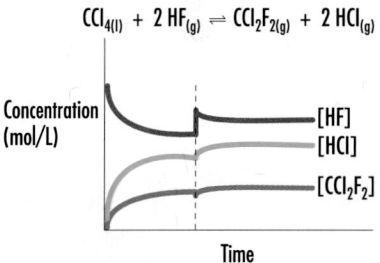

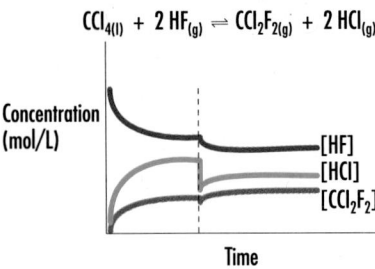

$$CCl_{4(l)} + 2\,HF_{(g)} \rightleftharpoons CCl_2F_{2(g)} + 2\,HCl_{(g)}$$

Concentration (mol/L)

[HF]
[HCl]
[CCl$_2$F$_2$]

Time

Figure 13.11
The reaction establishes an equilibrium which is disturbed (at the time indicated by the vertical dotted line) by the removal of $HCl_{(g)}$. The equilibrium shifts forward, increasing the concentration of both products while decreasing HF concentration, until a new equilibrium is established. The initial K value and the final K value are the same.

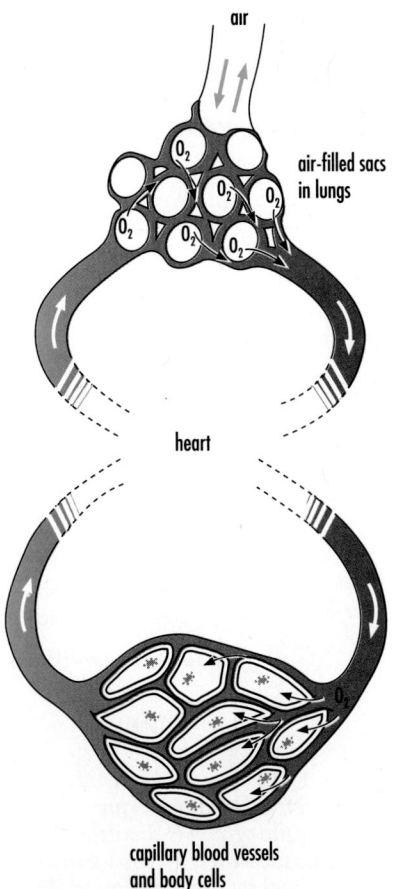

air

O_2

air-filled sacs
in lungs

heart

O_2

capillary blood vessels
and body cells

Figure 13.12
Oxygenated blood from the lungs is pumped by the heart to body tissues. The deoxygenated blood returns to the heart and is pumped to the lungs. Shifts in equilibrium occur over and over again as oxygen is picked up in the lungs and released throughout the body.

The removal of a product (if the removal decreases concentration as well as amount) will also shift an equilibrium forward, to produce more product to counteract the change imposed. The freon-12 reaction can be shifted forward by removing either gaseous product, since decreasing the amount of a gas lowers its concentration in any reaction container of fixed size. See Figure 13.11.

Note that adding more carbon tetrachloride, $CCl_{4(l)}$, would have no effect on the equilibrium state in the container. This reactant is (and stays) in liquid form, so its concentration is constant — it would not be increased by increasing the amount of $CCl_{4(l)}$ present.

Adjusting an equilibrium state by adding and/or removing a substance is by far the most common application of Le Châtelier's principle. For industrial chemical reactions, engineers strive to design processes where reactants are added continuously and products are continuously removed, so that an equilibrium is never allowed to establish. If the reaction is always shifting forward, the process is always making product (and presumably, the industry is always making money).

Another example, the final step in the production of nitric acid, is represented by the following system.

$$3\,NO_{2(g)} + H_2O_{(l)} \rightleftharpoons 2\,HNO_{3(aq)} + NO_{(g)}$$

In this industrial process, nitrogen monoxide gas is removed from the chemical system by a reaction with oxygen gas. The removal of the nitrogen monoxide causes the system to shift to the right — some nitrogen dioxide and water react, replacing some of the removed nitrogen monoxide. As the system shifts, more of the desired product, nitric acid, is produced.

A vitally important biological equilibrium is that of hemoglobin (a protein in red blood cells), oxygen, and oxygenated hemoglobin.

$$Hb + O_2 \rightleftharpoons HbO_2$$

As blood circulates to the lungs, the high concentration of oxygen shifts the equilibrium to the right and the blood becomes oxygenated (Figure 13.12). As the blood circulates throughout the body, cell reactions consume oxygen. This removal of oxygen shifts the equilibrium to the left and more oxygen is released.

Rate Theory and Concentration Changes

Kinetic theory provides a simple explanation of the equilibrium shift that occurs when a reactant concentration is increased. We assume that when reactant is added, with more reactant particles present per unit volume, collisions are suddenly much more frequent for the forward reaction. This *increases* the rate significantly. Since the reverse reaction rate is not changed, the rates are no longer equal, and for a time the difference in rates results in an observed increase of products.

Of course, as the concentration of products increases, so does the reverse reaction rate. At the same time the new forward rate decreases as reactant is consumed, until eventually the two rates become equal to each other again. The rates at the new equilibrium state are faster than those at the original state, because the system now contains a larger number of particles (and therefore a higher concentration) in dynamic equilibrium. If a substance is removed causing an equilibrium shift, the explanation is similar except that the initial effect is to suddenly *decrease* one of the equilibrium rates by decreasing the concentration.

Addition or removal of a substance in solid or liquid state does not change the *concentration* of that substance. The reaction of condensed phases takes place only at an exposed surface — and if the surface area exposed is changed it is always exactly the same change in available area for both forward and reverse reaction collisions. The forward and reverse rates must change by exactly the same amount if they change at all, so equilibrium is not disturbed and no shift occurs.

THE HABER PROCESS: A CASE STUDY IN TECHNOLOGY

One example of the use of Le Châtelier's principle is the Haber process for producing ammonia, a process invented in 1913 by German chemist Fritz Haber. Germany needed a new source of nitrogen compounds for the production of explosives, because supplies of nitrates from Chile were cut off by the British blockade during World War I. Haber developed a method of synthesizing ammonia from hydrogen (from the electrolysis of brine) and nitrogen (obtained from the atmosphere).

$$N_{2(g)} + 3H_{2(g)} \rightleftharpoons 2NH_{3(g)} + energy$$

As in many profitable industrial chemical processes, the process does not involve the reaction of a single batch of chemicals. The addition of nitrogen and hydrogen and the removal of ammonia gas continue without interruption. According to Le Châtelier's principle, removing ammonia, keeping the temperature low, and keeping the total pressure

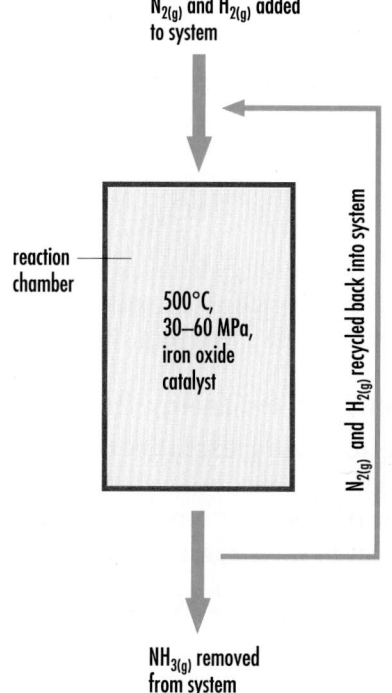

$N_{2(g)}$ and $H_{2(g)}$ added to system

reaction chamber

500°C, 30–60 MPa, iron oxide catalyst

$N_{2(g)}$ and $H_{2(g)}$ recycled back into system

$NH_{3(g)}$ removed from system

high all shift the system to the right and increase the yield of ammonia. Unfortunately, the reaction of nitrogen and hydrogen at low temperatures is so slow that the process becomes uneconomical. Adding heat increases the rate of the reaction, which is important in any continuous process such as this. The relationship between percent yield and temperature is shown in the graph below. Haber discovered that using an iron oxide catalyst eliminates the need for excessively high temperatures. An industrial plant using a

modification of the Haber process might operate with a temperature of about 500°C and a pressure of 50 MPa. Under these conditions, the yield of ammonia is about 40%.

Today, the Haber process and modifications of it are used to produce large quantities of ammonia for use as a chemical fertilizer. As shown below, left, ammonia fertilizer may be added directly to the soil. The ammonia dissolves in moisture present in the soil and, if the soil is slightly acidic, ammonia is converted to the ammonium ion.

Haber Process Conditions

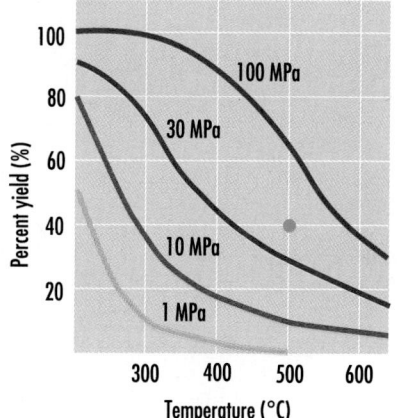

- *Why is a low temperature, which gives a higher percent yield of ammonia, not used in the Haber process?*
- *What role does iron oxide play in the Haber process?*

Le Châtelier's Principle and Temperature Changes

The energy in a chemical equilibrium equation is treated as though it were a reactant or a product.

$$\text{reactants} + \text{energy} \rightleftharpoons \text{products}$$

$$\text{reactants} \rightleftharpoons \text{products} + \text{energy}$$

Energy can be added to or removed from a system by heating or cooling the container. In either situation, the equilibrium shifts to minimize the change. If the system is cooled, the equilibrium shifts so that more heat is produced. If heat is added, the equilibrium shifts in the direction in which energy is absorbed.

For example, in the salt-sulfuric acid process used in the production of hydrochloric acid, the system is heated in order to increase the percent yield of hydrogen chloride gas.

$$2\,NaCl_{(s)} + H_2SO_{4(l)} + \text{energy} \rightleftharpoons 2\,HCl_{(g)} + Na_2SO_{4(s)}$$

The energy that is added causes the system to shift to the right, absorbing some of the added energy.

In the production of sulfuric acid by the contact process, the product is favored by keeping the system at a low temperature.

$$2\,SO_{2(g)} + O_{2(g)} \rightleftharpoons 2\,SO_{3(g)} + \text{energy}$$

Removing energy causes the system to shift to the right. This shift yields more sulfur trioxide while at the same time partially replacing the energy that was removed.

Rate Theory and Energy Changes

Kinetic theory explains the equilibrium shift that occurs when the energy of a system at equilibrium is increased or decreased as a result of an imbalance of reaction rates. Consider the previously mentioned contact process reaction equation — a typical exothermic reaction.

$$2\,SO_{2(g)} + O_{2(g)} \rightleftharpoons 2\,SO_{3(g)} + 198\ kJ$$

Rate theory explains the result of cooling the system by assuming that *both* forward and reverse reaction rates are slower at lower temperatures, but that the reverse rate decreases *more* than the forward rate. While the rates remain unequal the observed result is the production of more product and more energy. The shift causes concentration changes which will increase the reverse rate and decrease the forward rate until they become equal again, at a new, lower temperature. See Figure 13.13.

Le Châtelier's Principle and Gas Volume Changes

According to Boyle's law, the concentration of a gas in a container is directly related to the pressure of the gas. Decreasing the volume by half doubles the concentration of every gas in the container. Thus, changing the volume of any equilibrium system involving gases may cause a shift in the equilibrium. To predict whether a change in pressure will affect a system's equilibrium, you must consider the total amount in moles of gas reactants and the total amount in moles

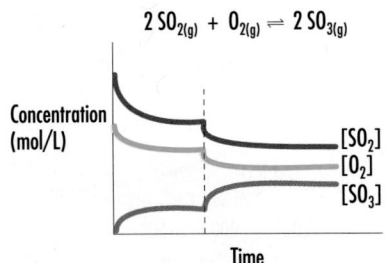

Figure 13.13
The reaction establishes an equilibrium which is disturbed (at the time indicated by the vertical dotted line) by a decrease in temperature. The equilibrium shifts forward, increasing the concentration of SO$_3$ product while decreasing the concentration of both reactants, until a new equilibrium is established.

of gas products. For example, in the equilibrium reaction of sulfur dioxide and oxygen, three moles of gaseous reactants produce two moles of gaseous products.

$$2\,SO_{2(g)} + O_{2(g)} \rightleftharpoons 2\,SO_{3(g)}$$

If the volume is decreased, the overall pressure is increased; this causes a shift to the right, which decreases the number of gas molecules (three moles to two moles) and reduces the pressure. If the volume is increased, the pressure is decreased, and the shift is in the opposite direction. A system with equal numbers of gas molecules on each side of the equation, such as the equilibrium reaction between hydrogen and iodine (page 490), is not shifted by a change in volume. Similarly, systems involving only liquids or solids are not affected by changes in pressure.

Rate Theory and Gas Volume Changes

When a system involving gaseous reactants and products is changed in volume, the resulting equilibrium shift is again explained as an imbalance of reaction rates.

$$2\,SO_{2(g)} + O_{2(g)} \rightleftharpoons 2\,SO_{3(g)} + 198\ kJ$$

Rate theory explains the result of decreasing the volume of this system by assuming that *both* forward and reverse reaction rates become faster because the concentrations (partial pressures) of reactants and products both increase. For this example, however, the forward rate increases *more* than the reverse rate, because there are more particles involved in the forward reaction. This means the increase in the total number of collisions is greater for the forward reaction process. Again, while the rates remain unequal, the observed result is the production of more product. The shift causes concentration changes which increase the reverse rate and decrease the forward rate until they become equal again. See Figure 13.14.

Catalysts and Equilibrium Systems

Catalysts are used in most industrial chemical systems. A catalyst decreases the time required to reach an equilibrium position, but does not affect the final position of equilibrium. The presence of a catalyst in a chemical reaction system lowers the activation energy for both forward and reverse reactions by an equal amount, so the equilibrium establishes much more rapidly but at the *same position* as it would without the catalyst present. Forward and reverse rates are increased equally. The value of catalysts in industrial processes is to decrease the time required for equilibrium shifts created by manipulating other variables, allowing a more rapid overall production of the desired product.

Enthalpy, Entropy, and Equilibrium Systems

The establishment of dynamic equilibrium states in chemical systems can be viewed theoretically as a balance between two fundamental processes operating in our universe. One of these is explained as the tendency of all matter to occupy a state of lowest possible energy. Using this theory, we explain phenomena such as the flow of heat

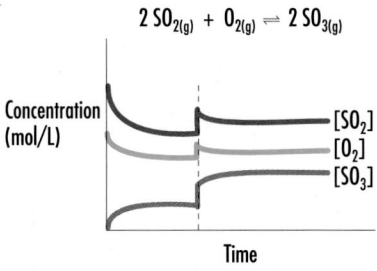

$$2\,SO_{2(g)} + O_{2(g)} \rightleftharpoons 2\,SO_{3(g)}$$

Figure 13.14
The reaction equilibrium is disturbed by a decrease in container volume (at the time indicated by the vertical dotted line). The equilibrium shifts forward, increasing the concentration of SO_3 while decreasing the concentration of reactants, until a new equilibrium is established.

Catalysts
A catalyst does not affect the final position of an equilibrium because the rates of the forward and reverse reactions are increased by the same factor. The final equilibrium concentrations are reached in a shorter time compared with the same, but uncatalyzed, reaction.

energy from higher to lower temperature regions. The basic principle of chemical bonding is explained by stating that bonded particles have lower energy than non-bonded particles, usually expressed as heat content of the system at constant pressure, called enthalpy.

The other universal principle is a little more difficult to discuss, because it is more abstract. We say that all particle systems tend to occupy a state of highest possible **entropy**. One simplistic way to define entropy is to say that it is the quantity of *randomness* or freedom of movement of particles in a system. The more orderly and organized particles are, the less entropy they have. The tendency to maximum entropy explains why, when you toss a deck of playing cards in the air, they don't come down in a neat pile arranged by value and suit. The universe tends naturally to a state of maximum disorder. By this reasoning, human beings are glaring examples of low entropy — consisting of an enormous number of extremely well-ordered molecular and ionic systems.

A competition exists between the tendencies of energy to decrease and of entropy to increase. Humans remain organized systems only if they absorb several thousand kilojoules daily. When a solid melts or a liquid vaporizes, the particles become more randomly arranged and entropy increases, but at the same time the energy present must increase. Whenever the tendencies of energy and entropy oppose each other in a system, that system will establish a balance between these two processes at some point — a dynamic equilibrium.

Consider a previous example.

$$2\ SO_{2(g)}\ +\ O_{2(g)}\ \rightleftharpoons\ 2\ SO_{3(g)}\ +\ 198\ kJ$$

From an energy versus entropy point of view, the forward reaction favors the tendency to minimum energy. However, the reverse reaction converts two molecules to three — and when the same amount of matter is in more separate particles the system is considered more disordered. (Think of breaking a glass.) So the reverse reaction favors the tendency to maximum entropy. Because the processes oppose each other, a dynamic equilibrium will be established in a closed system.

An interesting exception involves the mixing of water and ethanol in their liquid states. The tendency to maximum entropy favors the mixing, because the molecules are more random when intermixed than when surrounded only by identical molecules. In this case, however, the tendency to minimum energy also favors the mixing, which is exothermic. Since both natural tendencies favor the same direction, this is a system which cannot establish equilibrium — water and ethanol mix completely in any proportion at all, and are said to be completely *miscible*.

SUMMARY: VARIABLES AFFECTING CHEMICAL EQUILIBRIA

Variables	Direction of Change	Response of System
concentration	increase	shifts to consume some of the added reactant or product
	decrease	shifts to replace some of the removed reactant or product

Variables	Direction of Change	Response of System
temperature	increase	shifts to consume some of the added heat
	decrease	shifts to replace some of the removed heat
volume (overall pressure)	increase (decrease in pressure)	shifts toward the side with the larger total amount of gaseous entities
	decrease (increase in pressure)	shifts toward the side with the smaller total amount of gaseous entities

INVESTIGATION

13.2 Demonstration: Equilibrium Shifts

The purpose of this demonstration is to test Le Châtelier's principle by studying two chemical equilibrium systems: the equilibrium between two oxides of nitrogen (Figure 13.15), and the equilibrium of carbon dioxide gas and carbonic acid.

$$N_2O_{4(g)} + \text{energy} \rightleftharpoons 2\,NO_{2(g)}$$
colorless   reddish brown

$$CO_{2(g)} + H_2O_{(l)} \rightleftharpoons H^+_{(aq)} + HCO_3^-_{(aq)}$$

The second equilibrium system, produced by the reaction of carbon dioxide gas and water, is commonly found in the human body and in carbonated drinks. A diagnostic test is necessary to detect shifts in this equilibrium. Bromothymol blue, an acid-base indicator, can detect an increase or decrease in the hydrogen ion concentration in this system. Bromothymol blue turns blue when the hydrogen ion concentration decreases and yellow when the hydrogen ion concentration increases.

Problem

How does a change in temperature affect the nitrogen dioxide-dinitrogen tetraoxide equilibrium system? How does a change in pressure affect the carbon dioxide–carbonic acid equilibrium system?

Materials

lab apron
safety glasses
(2) $NO_{2(g)}/N_2O_{4(g)}$ sealed flasks
carbon dioxide-bicarbonate equilibrium mixture (pH = 7)
bromothymol blue indicator in dropper bottle
small syringe with needle removed (5 to 50 mL)
solid rubber stopper to seal end of syringe
beaker of ice-water mixture
beaker of hot water

- [ ] Problem
- [x] Prediction
- [ ] Design
- [ ] Materials
- [ ] Procedure
- [x] Evidence
- [x] Analysis
- [x] Evaluation
- [ ] Synthesis

hot water $NO_{2(g)} \rightleftharpoons N_2O_{4(g)}$ ice water

Figure 13.15
Each of these flasks contains an equilibrium mixture of dinitrogen tetraoxide and nitrogen dioxide. Shifts in equilibrium can be seen when one of the flasks is heated or cooled.

Procedure

1. Place the sealed $NO_{2(g)}/N_2O_{4(g)}$ flasks in hot and cold water baths and record your observations.

2. Place two or three drops of bromothymol blue indicator in the carbon dioxide-bicarbonate equilibrium mixture.

3. Draw some of the carbon dioxide-bicarbonate equilibrium mixture into the syringe, then block the end with a rubber stopper.

4. Slowly move the syringe plunger and record your observations.

INVESTIGATION

13.3 Testing Le Châtelier's Principle

The purpose of this investigation is to test Le Châtelier's principle by applying stress to five different chemical equilibria. In order to complete the Prediction section of the report, you must read the Experimental Design, Materials, and Procedure.

- ▨ Problem
- ✔ Prediction
- ▨ Design
- ▨ Materials
- ▨ Procedure
- ✔ Evidence
- ✔ Analysis
- ✔ Evaluation
- ▨ Synthesis

Problem

How does applying stresses to particular chemical equilibria affect the systems?

Part 1: $CoCl_4^{2-}{}_{(aq)} + 6\ H_2O_{(l)} \rightleftharpoons Co(H_2O)_6^{2+}{}_{(aq)} + 4\ Cl^-{}_{(aq)} + $ energy

Part 2: $H_2Tb_{(aq)} \rightleftharpoons H^+{}_{(aq)} + HTb^-{}_{(aq)}$

$\quad\quad HTb^-{}_{(aq)} \rightleftharpoons H^+{}_{(aq)} + Tb^{2-}{}_{(aq)}$

Part 3: $Fe^{3+}{}_{(aq)} + SCN^-{}_{(aq)} \rightleftharpoons FeSCN^{2+}{}_{(aq)}$

Part 4: $Cu(H_2O)_4^{2+}{}_{(aq)} + 4\ NH_{3(aq)} \rightleftharpoons Cu(NH_3)_4^{2+}{}_{(aq)} + 4\ H_2O_{(l)}$

Part 5: $CrO_4^{2-}{}_{(aq)} + 2\ H^+{}_{(aq)} \rightleftharpoons Cr_2O_7^{2-}{}_{(aq)} + H_2O_{(l)}$

Experimental Design

Stresses are applied to five chemical equilibrium systems and evidence is gathered to test predictions made using Le Châtelier's principle. Control samples are used in all cases.

Part I Cobalt(II) Complexes
Water, saturated silver nitrate, and heat are added to, and heat is removed from, samples of the provided equilibrium mixture.

Part II Thymol Blue Indicator
Hydrochloric acid and sodium hydroxide are added to samples of the provided equilibrium mixture.

Part III Iron(III)-Thiocyanate Equilibrium
Iron(III) nitrate, potassium thiocyanate, and sodium hydroxide are added to samples of the provided equilibrium system.

Part IV Copper(II) Complexes
Aqueous ammonia and hydrochloric acid are added to samples of the provided equilibrium mixture.

Part V Chromate-Dichromate Equilibrium

Aqueous sodium hydroxide, hydrochloric acid, and then aqueous barium nitrate are added to a sample of the provided equilibrium mixture.

Materials

lab apron
safety glasses
100 mL beaker
large waste beaker
6 to 12 small test tubes
test-tube rack
distilled water
crushed ice
hot water bath
cobalt(II) chloride equilibrium mixture in ethanol
0.2 mol/L $AgNO_{3(aq)}$ in dropper bottle
blue thymol blue indicator in dropper bottle
0.1 mol/L $HCl_{(aq)}$ in dropper bottle
0.1 mol/L $NaOH_{(aq)}$ in dropper bottle
iron(III) thiocyanate equilibrium mixture
0.2 mol/L $Fe(NO_3)_{3(aq)}$ in dropper bottle
0.2 mol/L $KSCN_{(aq)}$ in dropper bottle
6.0 mol/L $NaOH_{(aq)}$ in dropper bottle
0.1 mol/L $CuSO_{4(aq)}$
1.0 mol/L $NH_{3(aq)}$ in dropper bottle
1.0 mol/L $HCl_{(aq)}$ in dropper bottle
chromate-dichromate equilibrium mixture
0.1 mol/L $Ba(NO_3)_{2(aq)}$ in dropper bottle

Procedure

Dispose of the chemicals as directed by your teacher. Most may be washed down the drain with large amounts of water.

Part I Cobalt(II) Complexes

1. Obtain 25 mL of the equilibrium mixture with the cobalt(II) chloride complex ions.

2. Place a small amount of the mixture into each of five small test tubes.

3. Use the fifth test tube as a control for comparison purposes.

4. Add drops of water to one test tube until a change is evident. Record the evidence.

5. Add drops of 0.2 mol/L silver nitrate to another test tube and record the evidence.

6. Heat another equilibrium mixture in a water bath and record the evidence.

7. Cool an equilibrium mixture in an ice bath and record the evidence.

Diagnostic Tests
The equilibria chosen for this investigation involve chemicals that provide color solutions. The predictions based upon the Le Châtelier's principle are tested by observing color changes. For example, if the solution colors are observed, and the color changes from yellow to orange, then the equilibrium has shifted from chromate to dichromate ions.

Solution Colors
The following colors are used for diagnostic tests.

$CoCl_4^{2-}{}_{(aq)}$	blue
$Co(H_2O)_6^{2+}{}_{(aq)}$	pink
$H_2Tb_{(aq)}$	red
$HTb^-{}_{(aq)}$	yellow
$Tb^{2-}{}_{(aq)}$	blue
$Fe^{3+}{}_{(aq)}$	yellow
$SCN^-(aq)$	colorless
$FeSCN^{2+}{}_{(aq)}$	red
$Cu(H_2O)_4^{2+}{}_{(aq)}$	pale blue
$Cu(NH_3)_4^{2+}{}_{(aq)}$	deep blue
$CrO_4^{2-}{}_{(aq)}$	yellow
$Cr_2O_7^{2-}{}_{(aq)}$	orange

Part II Thymol Blue Indicator

1. Add about 5 mL of distilled water to each of two small test tubes.

2. Add 1 to 3 drops of thymol blue indicator to the water in each test tube to obtain a noticeable color.

3. Use one test tube of solution as a control.

4. Add drops of 0.1 mol/L $HCl_{(aq)}$ to the experimental test tube to test for the predicted color changes.

5. Add drops of 0.1 mol/L $NaOH_{(aq)}$ to the same tube to test for the predicted color changes.

Part III Iron(III)-Thiocyanate Equilibrium

1. Obtain about 20 mL of the iron(III)-thiocyanate equilibrium solution.

2. Place about 5 mL of the equilibrium solution in each of three test tubes.

3. Use one test tube as a control.

4. Add drops of $Fe(NO_3)_{3(aq)}$ to an equilibrium mixture until a change is evident.

5. Add drops of 6.0 mol/L $NaOH_{(aq)}$ to this new equilibrium mixture until a change occurs. (Iron(III) hydroxide has a low solubility.)

6. Add drops of $KSCN_{(aq)}$ to another equilibrium mixture until a change is evident.

Part IV Copper(II) Complexes

1. Obtain 2 mL of 0.1 mol/L $CuSO_{4(aq)}$ in a small test tube.

2. Add three drops of 1.0 mol/L $NH_{3(aq)}$ to establish the equilibrium mixture.

3. Add more 1.0 mol/L $NH_{3(aq)}$ to the above equilibrium mixture and record the results.

4. Add 1.0 mol/L $HCl_{(aq)}$ to the equilibrium mixture from step 3 and record the results.

Part V Chromate-Dichromate Equilibrium

1. Obtain 15 mL of the chromate-dichromate equilibrium mixture.

2. Place 5 mL samples of the equilibrium mixture into each of three small test tubes.

3. Add 0.1 mol/L $HCl_{(aq)}$ drop by drop to one sample or to 0.1 mol/L $K_2CrO_{4(aq)}$ and record the evidence.

4. Add 0.1 mol/L $NaOH_{(aq)}$ drop by drop to another sample (or the previous $HCl_{(aq)}$ sample or 0.1 mol/L $K_2Cr_2O_{7(aq)}$) and record the evidence.

5. Add 0.1 mol/L $Ba(NO_3)_{2(aq)}$ drop by drop to a third sample and record the evidence. (Barium chromate has low solubility.)

Ensure that all equipment and surfaces are left clean and that your hands are washed thoroughly before leaving the laboratory.

Controls

Whether stated in the Procedure or not, you are to use controls as often as possible. For example, before adding sodium hydroxide to a new equilibrium solution, split the solution into two samples in order to have a control sample for color comparison.

Exercise

21. What three types of changes shift the position of a chemical equilibrium?

22. For each of the following chemical systems at equilibrium, use Le Châtelier's principle to predict the effect of the change imposed on the chemical system. Indicate the direction in which the equilibrium is expected to shift. For each example, sketch the graph of concentrations versus reaction progress, plotted from *just before* the change to the established equilibrium. Also state for each example whether the tendency to maximum entropy favors the forward or reverse reaction.

    (a) $H_2O_{(l)}$ + energy $\rightleftharpoons$ $H_2O_{(g)}$

    The container is heated.

    (b) $H_2O_{(l)}$ $\rightleftharpoons$ $H^+_{(aq)}$ + $OH^-_{(aq)}$

    A few crystals of $NaOH_{(s)}$ are added to the container.

    (c) $CaCO_{3(s)}$ + energy $\rightleftharpoons$ $CaO_{(s)}$ + $CO_{2(g)}$

    $CO_{2(g)}$ is removed from the container.

    (d) $CH_3COOH_{(aq)}$ $\rightleftharpoons$ $H^+_{(aq)}$ + $CH_3COO^-_{(aq)}$

    A few drops of pure $CH_3COOH_{(l)}$ are added to the system.

23. The following equation represents part of the industrial production of nitric acid. Predict the direction of the equilibrium shift for each of the following changes. Explain any shift in terms of the changes in forward and reverse reaction rates.

    $$4\,NH_{3(g)} + 5\,O_{2(g)} \rightleftharpoons 4\,NO_{(g)} + 6\,H_2O_{(g)} + \text{energy}$$

    (a) $O_{2(g)}$ is added to the system.
    (b) The temperature of the system is increased.
    (c) $NO_{(g)}$ is removed from the system.
    (d) The pressure of the system is increased by decreasing the volume.

24. The following chemical equilibrium system is part of the Haber process for the production of ammonia.

    $$N_{2(g)} + 3\,H_{2(g)} \rightleftharpoons 2\,NH_{3(g)} + \text{energy}$$

    Suppose you are a chemical process engineer. Use Le Châtelier's principle to predict five specific changes that you might impose on the equilibrium system to increase the yield of ammonia.

25. In a solution of copper(II) chloride, the following equilibrium exists.

    $$CuCl_4^{2-}{}_{(aq)} + 4\,H_2O_{(l)} \rightleftharpoons Cu(H_2O)_4^{2+}{}_{(aq)} + 4\,Cl^-{}_{(aq)}$$
    dark green   blue

    For the following stresses put on the equilibrium, predict the shift in the equilibrium and draw a graph of concentration versus time to communicate the shift.
    (a) Hydrochloric acid is added.
    (b) Silver nitrate is added.

26. Identify the nature of the changes imposed on the following equilibrium system at the five times indicated by coordinates A, B, C, D, and E.

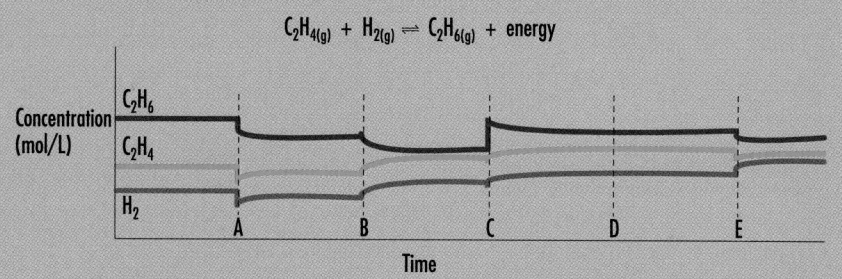

$$C_2H_{4(g)} + H_{2(g)} \rightleftharpoons C_2H_{6(g)} + energy$$

Concentration (mol/L) — C_2H_6, C_2H_4, H_2

A   B   C   D   E

Time

27. In which of the following cases does the tendency to maximum entropy not oppose the tendency to minimum enthalpy?

(a) $H_2O_{(l)} \rightarrow H_2O_{(g)}$

(b) $N_{2(g)} + 3\,H_{2(g)} \rightarrow 2\,NH_{3(g)}$ $\Delta H = -91$ kJ

(c) $KOH_{(s)} \rightarrow K^+_{(aq)} + OH^-_{(aq)} + heat$

(d) $2\,C_{(s)} + 2\,H_{2(g)} \rightarrow C_2H_{4(g)}$ $\Delta H = +53$ kJ

Problem 13C The Nitrogen Dioxide Equilibrium

The purpose of this problem is to test Le Châtelier's principle. Complete the Prediction, Analysis, and Evaluation sections of the report. (Evaluate the Design, Prediction, and authority only.)

Problem

How does increasing the pressure affect the nitrogen dioxide–dinitrogen tetraoxide equilibrium?

Experimental Design

A sample of nitrogen dioxide gas is compressed in a syringe and the intensity of the color is used as evidence to test the prediction.

Evidence

The orange-brown nitrogen dioxide gas color increases in intensity and then decreases in intensity when the plunger on the syringe is depressed (Figure 13.16).

Figure 13.16
$2\,NO_{2(g)} \rightleftharpoons N_2O_{4(g)}$
An increase in pressure on the nitrogen dioxide–dinitrogen tetraoxide equilibrium in the closed system results initially in a more intense color followed by a less intense color.

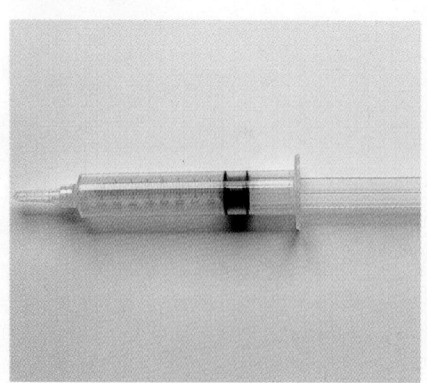
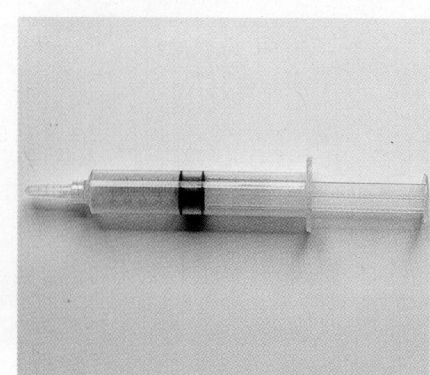
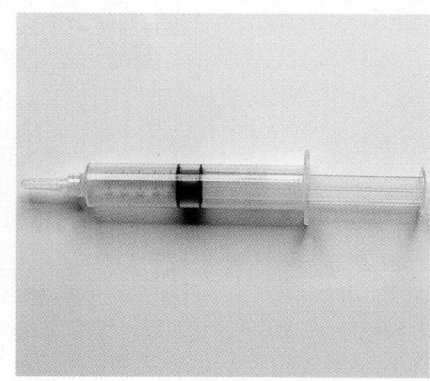

13.4 QUANTITATIVE CHANGES IN EQUILIBRIUM SYSTEMS

Knowledge of the equilibrium law and of stoichiometric calculation technique often make it possible to do quantitative predictions of amounts and concentrations of substances reacting in closed systems. Quantities that are involved in establishing an initial equilibrium, beginning with only reactants present, can be calculated and predicted, as well as changes of quantities in an established equilibrium (shifts).

Recall that the convention used for the equilibrium expression — the ratio of product to reactant concentrations — means that the size of K varies with the extent (position) of the equilibrium. All things being equal, the larger the value of K, the more the reaction as written favors products. All things are *not* equal, of course, if the reactions being compared do not have parallel forms of the equilibrium constant expression. Consider the two following examples.

$$2\ HI_{(g)} \rightleftharpoons H_{2(g)} + I_{2(g)} \qquad\qquad K = 50 \text{ at } 450°C$$

$$CO_{2(g)} + H_{2(g)} \rightleftharpoons CO_{(g)} + H_2O_{(g)} \qquad\qquad K = 1.1 \text{ at } 900°C$$

Both of these reaction equations involve two gas molecules on each side, and the equilibrium law expressions each have two concentrations expressed in the numerator and two in the denominator. The larger value of K for the first reaction makes it possible to state that the first reaction proceeds farther to completion before equilibrium is established.

Now consider these two examples.

$$SrF_{2(s)} \rightleftharpoons Sr^{2+}_{(aq)} + 2\ F^-_{(aq)}$$
$$K_{sp} = [Sr^{2+}_{(aq)}][F^-_{(aq)}]^2 = 4 \times 10^{-9} \text{ at } 25°C$$

$$CaCO_{3(s)} \rightleftharpoons Ca^{2+}_{(aq)} + CO_3^{2-}_{(aq)}$$
$$K_{sp} = [Ca^{2+}_{(aq)}][CO_3^{2-}_{(aq)}] = 5 \times 10^{-9} \text{ at } 25°C$$

Although the two have nearly equivalent K values, the expression for calcium carbonate has only two concentration values compared with three for the strontium fluoride. A quick calculation of the sort already performed in this chapter will show that the molar solubility of the $CaCO_3$ is significantly higher. Generally then, it is only possible to state that a very large K value means a reaction favoring products, and a very small K value means a reaction favoring reactants. Also, in general, if K is *significantly* larger for one reaction than another, the reaction with the larger value is more complete.

Temperature must also be considered in any quantitative calculation involving K values. The only factor that changes the value of K for a given reaction is temperature: changes in concentration, unless extreme, do not have a significant effect on the numerical value of K, and neither does the presence or absence of a catalyst. A temperature change that *increases* the value of K for a reaction shifts equilibrium to the right (more complete), and a temperature change that *decreases* the value of K for a reaction shifts equilibrium to the left (less complete).

The Chemical System Trial Reaction Quotient

If the concentrations of the substances in any closed chemical system are known, it can be useful to determine whether or not the system is at equilibrium — or, if not, in what direction the system will shift to reach equilibrium. If reactants and products are both present, determining shift direction is not often easy or obvious. In such a case, the concentrations can be substituted into the equilibrium law expression to produce a trial value that is called a **reaction quotient**, usually symbolized as **Q**. The result of such a trial calculation must be one of three possible situations.

- Q is equal to K, and the system is at equilibrium.
- Q is greater than K, and the system is shifting left to reach equilibrium, because the product to reactant ratio is too high.
- Q is less than K, and the system is shifting right to reach equilibrium, because the product to reactant ratio is too low.

EXAMPLE

In a container at 450°C, nitrogen and hydrogen are reacted to produce ammonia. The equilibrium constant is 0.064 and at a specific time the substance concentrations are $[N_{2(g)}] = 4.0$ mol/L, $[H_{2(g)}] = 0.020$ mol/L, and $[NH_{3(g)}] = 2.2 \times 10^{-4}$ mol/L. Predict the direction of reaction shift, if any.

$$N_{2(g)} + 3\,H_{2(g)} \rightleftharpoons 2\,NH_{3(g)}$$

Calculate the value of $Q = \dfrac{[NH_{3(g)}]^2}{[N_{2(g)}][H_{2(g)}]^3} = \dfrac{(2.2 \times 10^{-4})^2}{(4.0)(0.020)^3} = 1.5 \times 10^{-3}$

$Q = 1.5 \times 10^{-3} = 0.0015$, which is lower than the value of K, so the reaction will be shifting right at this time, producing ammonia faster than it decomposes.

A stoichiometric calculation can also be used to determine final equilibrium concentrations for a system, if initial reactant concentrations are known. Consider, for example, the following equilibrium.

$$CO_{(g)} + H_2O_{(g)} \rightleftharpoons CO_{2(g)} + H_{2(g)} \qquad\qquad K = 4.20 \text{ at } 900°C$$

In a container, carbon monoxide and water vapor are reacted to produce carbon dioxide and hydrogen. 2.00 mol of each reactant are placed in a 500 mL reaction flask. Initial concentrations are $[CO_{(g)}] = 4.00$ mol/L, and $[H_2O_{(g)}] = 4.00$ mol/L.

Shift calculations may conveniently be set up as an ICE table for clarity. Begin by letting ΔC stand for the change in concentration of the carbon monoxide from its initial value to its (lower) equilibrium value.

Concentration	$[CO_{(g)}]$ (mol/L)	$[H_2O_{(g)}]$ (mol/L)	$[H_{2(g)}]$ (mol/L)	$[CO_{2(g)}]$ (mol/L)
Initial	4.00	4.00	0.00	0.00
Change	ΔC	ΔC	ΔC	ΔC
Equilibrium	$4.00 - \Delta C$	$4.00 - \Delta C$	ΔC	ΔC

The 1:1:1:1 stoichiometric ratio from the balanced equation allows the calculation that the concentrations of both reactants will decrease by the same amount, ΔC, and the concentrations of both products will increase by that same amount.

$$K = \frac{[CO_{2(g)}][H_{2(g)}]}{[CO_{(g)}][H_2O_{(g)}]} = \frac{(\Delta C)(\Delta C)}{(4.00 - \Delta C)(4.00 - \Delta C)} = 4.20$$

$$\frac{\Delta C^2}{(4.00 - \Delta C)^2} = 4.20$$

Taking the square root of both sides and then multiplying both sides by $(4.00 - \Delta C)$ gives

$$\Delta C = (2.05)(4.00 - \Delta C)$$

Solve for ΔC.

$$\Delta C = 2.69 \text{ mol/L}$$

Concentration	$[CO_{(g)}]$ (mol/L)	$[H_2O_{(g)}]$ (mol/L)	$[H_{2(g)}]$ (mol/L)	$[CO_{2(g)}]$ (mol/L)
Initial	4.00	4.00	0.00	0.00
Change	2.69	2.69	2.69	2.69
Equilibrium	1.31	1.31	2.69	2.69

At equilibrium

$$[CO_{(g)}] = [H_2O_{(g)}] = 1.31 \text{ mol/L}$$
$$[CO_{2(g)}] = [H_{2(g)}] = 2.69 \text{ mol/L}$$

Note: *If the initial reactant concentrations are not identical, use of the quadratic formula is required to solve questions of this type, as the left side of the equation to calculate ΔC is not a perfect square.*

The Trial Ion Product: Predicting Precipitation

A very important situation for using the reaction quotient occurs when mixing solutions that contain ions which combine to form a low solubility compound. After mixing, if these ions are present in too high a concentration, a precipitate must form to allow establishment of an equilibrium. In this situation, the reaction quotient may be called the *trial ion product*.

For example, 100 mL of 0.100 mol/L $CaCl_{2(aq)}$ and 100 mL of 0.0400 mol/L $Na_2SO_{4(aq)}$ are mixed at 20°C. Determine whether a precipitate will form. For $CaSO_4$ at 20°C, $K_{sp} = 3.6 \times 10^{-5}$.

$$CaSO_{4(s)} \rightleftharpoons Ca^{2+}_{(aq)} + SO_4^{2-}_{(aq)} \qquad Q = [Ca^{2+}_{(aq)}][SO_4^{2-}_{(aq)}]$$

According to the balanced equation

$$CaCl_{2(s)} \rightleftharpoons Ca^{2+}_{(aq)} + 2 Cl^-_{(aq)}$$

$[Ca^{2+}_{(aq)}] = [CaCl_{2(aq)}] = 0.100$ mol/L before the solutions are mixed.

Similarly, $[SO_4^{2-}_{(aq)}] = [Na_2SO_{4(aq)}] = 0.0400$ mol/L before mixing.

Note: *Mixing two solutions always increases the overall volume, so the initial concentration of ions in both solutions is always decreased by the act of mixing them, in proportion to the initial/final volume ratio.*

In this instance, after mixing,

$$[Ca^{2+}_{(aq)}] = 0.100 \text{ mol/L} \times \frac{100 \text{ mL}}{200 \text{ mL}} = 0.0500 \text{ mol/L}$$

Similarly, after mixing,

$$[SO_4^{2-}_{(aq)}] = 0.0400 \text{ mol/L} \times \frac{100 \text{ mL}}{200 \text{ mL}} = 0.0200 \text{ mol/L}$$

$Q = (0.0500)(0.0200) = 1.00 \times 10^{-3}$, which is much greater than the K_{sp}.

The reaction must shift to the left, meaning a precipitate will form.

Solutions of ionic substances exhibit the **common ion effect**. When a reaction equilibrium involving ions exists in a solution, the equilibrium can be shifted by dissolving into the solution any other compound that adds a common ion, or any compound that also reacts with a present ion. Consider the following equilibrium system.

$$Fe^{3+}_{(aq)} + SCN^-_{(aq)} \rightleftharpoons FeSCN^{2+}_{(aq)}$$

nearly colorless   colorless   deep red color

Addition of (soluble) $Fe(NO_3)_{3(s)}$ to this reaction system produces more $Fe^{3+}_{(aq)}$ ions as the compound dissolves. Increasing the concentration of $Fe^{3+}_{(aq)}$ ions will shift the original equilibrium to the right, darkening the color of the solution. See Figure 13.9, page 503.

The reverse of this effect can occur if an ion is introduced which reacts with and removes an ion present in the original equilibrium. If (soluble) $NaOH_{(s)}$ is added to the solution above, the $OH^-_{(aq)}$ ions produced react with the $Fe^{3+}_{(aq)}$ ions present to produce a precipitate of (low solubility) $Fe(OH)_{3(s)}$. Decreasing the concentration of $Fe^{3+}_{(aq)}$ ions shifts the original equilibrium to the left, observed as a lightening of the color of the solution.

Exercise

28. Which of the following compounds is least soluble?
    (a) TlBr $K_{sp} = 3.6 \times 10^{-6}$
    (b) CuBr $K_{sp} = 5.9 \times 10^{-9}$
    (c) AgBr $K_{sp} = 5.0 \times 10^{-13}$
    (d) AgCl $K_{sp} = 1.7 \times 10^{-10}$

29. For an exothermic reaction at equilibrium,
    $$A + B \rightleftharpoons C + D + \text{energy}$$
    If all substances are molecular gases, explain the effect on the value of the equilibrium constant of:
    (a) increasing the amount of A.
    (b) decreasing the container volume.
    (c) raising the temperature.
    (d) adding a catalyst.

30. In a closed container, nitrogen and hydrogen are reacted to produce ammonia. The equilibrium constant is 0.050. At a

specific time in the reaction process, substance concentrations are $[N_{2(g)}] = 2.0 \times 10^{-4}$ mol/L, $[H_{2(g)}] = 4.0 \times 10^{-3}$ mol/L, and $[NH_{3(g)}] = 2.2 \times 10^{-4}$ mol/L.

Predict the direction of reaction shift.

$$N_{2(g)} + 3\ H_{2(g)} \rightleftharpoons 2\ NH_{3(g)}$$

31. For each mixture, use a Q value (trial ion product) to predict whether a precipitate will form. Refer to Appendix F for K_{sp} values.
    (a) 50 mL of 0.040 mol/L $Ca(NO_3)_{2(aq)}$ plus 150 mL of 0.080 mol/L $(NH_4)_2SO_{4(aq)}$
    (b) 50 mL of 2.2×10^{-9} mol/L $AgNO_{3(aq)}$ plus 50 mL of 0.050 mol/L $NH_4Cl_{(aq)}$
    (c) 100 mL of 2.1×10^{-3} mol/L $Pb(NO_3)_{2(aq)}$ plus 50 mL of 0.0060 mol/L $NaI_{(aq)}$

32. Consider the system
    $$CO_{2(g)} + H_{2(g)} \rightleftharpoons CO_{(g)} + H_2O_{(g)}$$
    Initially, 0.25 mol of water and 0.20 mol of carbon monoxide are placed in the reaction vessel. At equilibrium, spectroscopic evidence shows that 0.10 mol of carbon dioxide is present. Find K for this system.

33. Consider the system
    $$2\ HBr_{(g)} \rightleftharpoons H_{2(g)} + Br_{2(g)}$$
    Initially, 0.25 mol of hydrogen and 0.25 mol of bromine are placed into a 500 mL electrically heated reaction vessel. K for the reaction at the temperature used is 0.020.
    (a) Find the substance concentrations at equilibrium.
    (b) Calculate the amount (in moles) of each substance present at equilibrium.
    (c) Calculate the reaction extent as a percent reaction.

13.5 WATER EQUILIBRIUM

Pure water has a very slight conductivity that is only observable if measurements are made with very sensitive instruments (Figure 13.17). According to Arrhenius's theory, conductivity is due to the presence of ions. Therefore, the conductivity observed in pure water must be the result of ions produced by the ionization of some water molecules into hydrogen ions and hydroxide ions. Because the conductivity is so slight, the equilibrium at SATP must greatly favor the water molecules.

$$H_2O_{(l)} \overset{<10^{-6}\%}{\rightleftharpoons} H^+_{(aq)} + OH^-_{(aq)}$$

$$K = \frac{[H^+_{(aq)}][OH^-_{(aq)}]}{[H_2O_{(l)}]} = \text{a very small number}$$

Evidence indicates that fewer than two water molecules in one billion ionize at SATP. Because the concentration of water in pure water and in dilute aqueous solutions is essentially constant, a new constant, which incorporates both the constant concentration of $H_2O_{(l)}$ and the equilibrium constant, can be calculated. This new constant is called the ion product or **ionization constant for water, K_w.**

$$K_w = [H^+_{(aq)}][OH^-_{(aq)}] = 1.0 \times 10^{-14} \, (\text{mol/L})^2 \text{ at SATP}$$

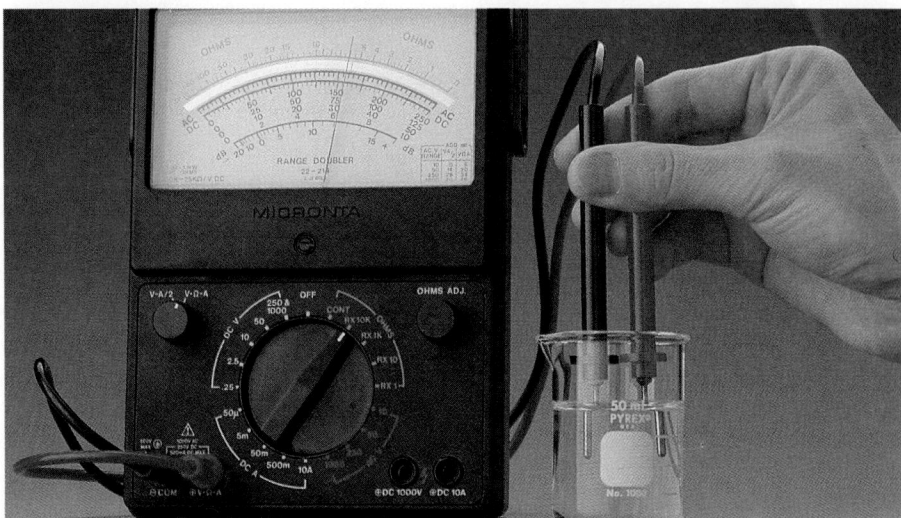

Figure 13.17
A sensitive meter shows the electrical conductivity of distilled water in a laboratory. Successive distillations to increase purity will lower but never eliminate the conductivity of water as measured by increasingly sensitive instruments.

The equilibrium equation for the ionization of water shows that hydrogen ions and hydroxide ions are formed in a 1:1 ratio. Therefore, the concentration of hydrogen ions and hydroxide ions in pure water and neutral solutions must be equal. Using the mathematical expression for K_w and the value of K_w at SATP, the concentrations of $H^+_{(aq)}$ and $OH^-_{(aq)}$ can be calculated.

$$[H^+_{(aq)}] = [OH^-_{(aq)}] = 1.0 \times 10^{-7} \text{ mol/L in neutral solution}$$

The ionization of water is especially important in the empirical and theoretical study of acidic and basic solutions. According to Arrhenius's theory, an acid is a substance that ionizes in water to produce hydrogen ions. The additional hydrogen ions provided by the acid increase the hydrogen ion concentration in the water; the concentration will be greater than 10^{-7} mol/L, so the solution is *acidic*. A *basic* solution is one in which the hydroxide ion concentration is greater than 10^{-7} mol/L; a basic solution is produced, for example, by the dissociation in water of an ionic hydroxide such as sodium hydroxide. One important observation is that the ionization constant, K_w, applies to all aqueous solutions. Another important point is that the numerical value of K_w is valid at SATP, but *not* at temperatures that are much higher or lower. Recall that the value of the equilibrium K depends on the temperature. For higher temperatures, K_w is a greater value, so products are more favored. This means a few more water molecules become ionized in aqueous systems when the molecular collisions are faster and more frequent.

K_w may be used to calculate either the hydrogen ion concentration or the hydroxide ion concentration in an aqueous solution, if the other concentration is known.

$$\text{Since } [H^+_{(aq)}][OH^-_{(aq)}] = K_w$$

$$\text{then} \qquad [\text{H}^+_{(aq)}] = \frac{K_w}{[\text{OH}^-_{(aq)}]}$$

$$\text{and} \qquad [\text{OH}^-_{(aq)}] = \frac{K_w}{[\text{H}^+_{(aq)}]}$$

EXAMPLE

A 0.15 mol/L solution of hydrochloric acid at 25°C is found to have a hydrogen ion concentration of 0.15 mol/L. Calculate the concentration of the hydroxide ions.

$$\text{HCl}_{(aq)} \rightarrow \text{H}^+_{(aq)} + \text{Cl}^-_{(aq)}$$

$$[\text{H}^+_{(aq)}] = [\text{HCl}_{(aq)}] = 0.15 \text{ mol/L}$$

$$[\text{OH}^-_{(aq)}] = \frac{K_w}{[\text{H}^+_{(aq)}]}$$

$$= \frac{1.0 \times 10^{-14} \text{ (mol/L)}^2}{0.15 \text{ mol/L}}$$

$$= 6.7 \times 10^{-14} \text{ mol/L}$$

EXAMPLE

Calculate the hydrogen ion concentration in a 0.25 mol/L solution of barium hydroxide.

$$\text{Ba(OH)}_{2(s)} \rightarrow \text{Ba}^{2+}_{(aq)} + 2\,\text{OH}^-_{(aq)}$$

$$[\text{OH}^-_{(aq)}] = 2 \times [\text{Ba(OH)}_{2(aq)}] = 2 \times 0.25 \text{ mol/L} = 0.50 \text{ mol/L}$$

$$[\text{H}^+_{(aq)}] = \frac{K_w}{[\text{OH}^-_{(aq)}]}$$

$$= \frac{1.0 \times 10^{-14} \text{ (mol/L)}^2}{0.50 \text{ mol/L}}$$

$$= 2.0 \times 10^{-14} \text{ mol/L}$$

EXAMPLE

Determine the hydrogen ion and hydroxide ion concentrations in 500 mL of an aqueous solution containing 2.6 g of dissolved sodium hydroxide.

$$n_{\text{NaOH}} = 2.6 \text{ g} \times \frac{1 \text{ mol}}{40.00 \text{ g}} = 0.065 \text{ mol}$$

$$[\text{NaOH}_{(aq)}] = \frac{0.065 \text{ mol}}{0.500 \text{ L}} = 0.13 \text{ mol/L}$$

$$\text{NaOH}_{(s)} \rightarrow \text{Na}^+_{(aq)} + \text{OH}^-_{(aq)}$$

$$[\text{OH}^-_{(aq)}] = [\text{NaOH}_{(aq)}] = 0.13 \text{ mol/L}$$

$$[\text{H}^+_{(aq)}] = \frac{K_w}{[\text{OH}^-_{(aq)}]}$$

$$= \frac{1.0 \times 10^{-14} \text{ (mol/L)}^2}{0.13 \text{ mol/L}}$$

$$= 7.7 \times 10^{-14} \text{ mol/L}$$

34. The hydrogen ion concentration in an industrial effluent is 4.40 mmol/L (4.40×10^{-3} mol/L). Determine the concentration of hydroxide ions in the effluent.

35. The hydroxide ion concentration in a household cleaning solution is 0.299 mmol/L. Calculate the hydrogen ion concentration in the cleaning solution.

36. Calculate the hydroxide ion concentration in a solution prepared by dissolving 0.37 g of hydrogen chloride in 250 mL of water.

37. Calculate the hydrogen ion concentration in a saturated solution of calcium hydroxide (limewater) that has a solubility of 6.9 mmol/L.

38. What is the hydrogen ion concentration in a solution made by dissolving 20.0 g of potassium hydroxide in water to form 500 mL of solution?

39. (Enrichment) Calculate the percent ionization of water at SATP. Recall that 1.000 L of water has a mass of 1000 g.

Figure 13.18
Chromate ions, $CrO_4^{2-}{}_{(aq)}$, are yellow; dichromate ions, $Cr_2O_7^{2-}{}_{(aq)}$, have an orange color.

Problem 13D The Chromate-Dichromate Equilibrium

In an aqueous solution, chromate ions are in equilibrium with dichromate ions (Figure 13.18).

$$2\,CrO_4^{2-}{}_{(aq)} + 2\,H^+{}_{(aq)} \rightleftharpoons Cr_2O_7^{2-}{}_{(aq)} + H_2O_{(l)}$$

The position of this equilibrium depends on the acidity of the solution. Complete the Prediction and Experimental Design (including diagnostic tests) of the investigation report.

Problem

How does changing the hydrogen ion concentration affect the chromate-dichromate equilibrium?

Communicating Concentrations: pH and pOH

A concentrated acid solution may have a hydrogen ion concentration exceeding 10 mol/L. A concentrated base solution may have a hydrogen ion concentration of 10^{-15} mol/L or less. A similarly wide range of hydroxide ion concentrations occurs. Because of the tremendous range of hydrogen ion and hydroxide ion concentrations and the lengthy English translation of a concentration value such as 4.72×10^{-11} mol/L, scientists rely on a simple system for communicating concentrations. This system, called the **pH** scale, was developed in 1909 by Danish chemist Sören Sörenson. Expressed as a numerical value without units, the pH of a solution is the negative of the logarithm to the base ten of the hydrogen ion concentration. This is illustrated in Figure 5.23 on page 216.

$$pH = -\log[H^+_{(aq)}]$$

Values of pH can be calculated from the hydrogen ion concentration, as shown in the following example. The digits preceding the decimal point in a pH value are determined by the digits in the exponent of the given hydrogen ion concentration. These digits serve to locate the position of the decimal point in the concentration value and have no connection with the certainty of the value. However, *the number of digits following the decimal point in the pH value is equal to the number of significant digits in the hydrogen ion concentration.* For example, a hydrogen ion concentration of 2.7×10^{-3} mol/L corresponds to a pH of 2.57.

EXAMPLE

Communicate a hydrogen ion concentration of 4.7×10^{-11} mol/L as a pH value.

$$
\begin{aligned}
pH &= -\log[H^+_{(aq)}] \\
&= -\log(4.7 \times 10^{-11}) \quad \text{(two significant digits)} \\
&= 10.33 \quad\quad\quad\quad\quad \text{(two digits following the decimal point)}
\end{aligned}
$$

On many calculators, $-\log(4.7 \times 10^{-11})$ may be entered by pushing the following sequence of keys.

| 4 | • | 7 | EXP |

| 1 | 1 | +/- | log | +/- |

If pH is measured in an acid-base analysis, a conversion from pH to the molar concentration of hydrogen ions may be necessary. This conversion is based on the mathematical concept that a base ten logarithm represents an exponent.

$$[H^+_{(aq)}] = 10^{-pH}$$

The method of calculating the hydrogen ion concentration from the pH value is shown in the following example.

EXAMPLE

Communicate a pH of 10.33 as a hydrogen ion concentration.

$$
\begin{aligned}
[H^+_{(aq)}] &= 10^{-pH} \\
&= 10^{-10.33} \text{ mol/L} \quad \text{(two digits following the decimal point)} \\
&= 4.7 \times 10^{-11} \text{ mol/L} \quad \text{(two significant digits)}
\end{aligned}
$$

On many calculators, $10^{-10.33}$ may be entered by pushing the following sequence of keys.

| 1 | 0 | • | 3 | 3 |

| +/- | either | INV | log |

or | 2nd | log |

Hydroxide Ion Concentration, pOH

Although pH is used in most applications, in some applications it may be convenient to describe hydroxide ion concentrations in a similar way. The definition of **pOH** follows the same format and the same certainty rule as for pH.

$$pOH = -\log[OH^-_{(aq)}] \quad \text{and} \quad [OH^-_{(aq)}] = 10^{-pOH}$$

The mathematics of logarithms allows us to express a simple relationship between pH and pOH.

$$pH + pOH = 14.00 \text{ (at SATP)}$$

This relationship enables a quick conversion between pH and pOH.

SUMMARY: pH AND pOH

$$K_w = [H^+_{(aq)}][OH^-_{(aq)}]$$

$$pH = -\log[H^+_{(aq)}] \qquad pOH = -\log[OH^-_{(aq)}]$$

$$[H^+_{(aq)}] = 10^{-pH} \qquad [OH^-_{(aq)}] = 10^{-pOH}$$

$$pH + pOH = 14.00 \text{ (at SATP)}$$

Exercise

40. Food scientists and dieticians measure the pH of foods when they devise recipes and special diets.

    (a) Complete the following table.

	ACIDITY OF FOODS			
Food	$[H^+_{(aq)}]$ (mol/L)	$[OH^-_{(aq)}]$ (mol/L)	pH	pOH
oranges	5.5×10^{-3}			
asparagus				5.6
olives		2.0×10^{-11}		
blackberries				10.60

    (b) Based on pH only, predict which of the foods would taste most sour.

41. To clean a clogged drain, 26 g of sodium hydroxide is added to water to make 150 mL of solution. What are the pH and pOH values for the solution?

42. What mass of potassium hydroxide is contained in 500 mL of solution that has a pH of 11.5?

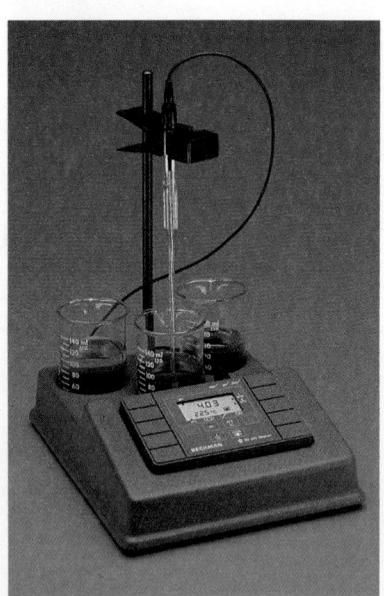

Figure 13.19
Arnold Beckman invented the pH meter in 1935, 26 years after Sören Sörenson had developed the concept of pH for communicating hydrogen ion concentration.

The pH Meter

A pH meter measures the voltage between electrodes in a solution and displays this measurement as a pH value (Figure 13.19). A potential difference is generated between a reference half-cell that has a constant reduction potential and the other half-cell that is in contact with an external solution of unknown hydrogen ion concentration. For convenience, the two electrodes are usually combined into a single electrode. The combination electrode contains one reference half-cell (electrode and electrolyte) and the electrode of a second half-cell. When the combination electrode is immersed in an aqueous solution, the solution acts as an electrolyte, completing the voltaic cell. Since the reference half-cell has a constant reduction potential, the net cell potential depends only on the reduction potential of the second half-cell, which in turn is dependent on the hydrogen ion concentration. The voltage of the complete cell is converted and displayed as a pH value by the meter.

INVESTIGATION

13.4 pH of Common Substances

One reason for the wide acceptance of the pH scale is the availability of a convenient, rapid, and precise measuring instrument. The purpose of this investigation is to show the technological advantages of a pH meter.

Problem

What generalizations can be made about the pH of foods and cleaning agents?

Experimental Design

The pH of a variety of solutions is measured. An attempt is made to develop generalizations concerning the pH of foods and cleaning agents.

Materials

lab apron
safety glasses
pH meter and pH 7 buffer solution
wash bottle of distilled water
400 mL waste beaker
several 100 mL beakers
various cleaning agents, such as ammonia, drain cleaner, and shampoo (Figure 13.20)
various food products, such as juices, pop, vinegar, and milk

Procedure

Substances must be dissloved in water before measuring the pH.

1. Rinse the electrode of the pH meter with distilled water.
2. Place the pH meter electrode in a standard buffer solution and calibrate the instrument by adjusting the meter to read the pH of the buffer.
3. Rinse the pH meter electrode with distilled water.
4. Place the electrode in a beaker containing a sample and record the pH reading.
5. Rinse the pH meter electrode with distilled water.
6. Repeat steps 4 and 5 with each sample provided.

	Problem
	Prediction
	Design
	Materials
	Procedure
✔	Evidence
✔	Analysis
	Evaluation
	Synthesis

CAUTION

 Some of the materials being tested are very corrosive. Do not allow them to come into contact with eyes, skin, or clothing. If there is any skin or eye contact, immediately rinse with plenty of water. The eyes should be flushed for at least 15 min and the teacher informed.

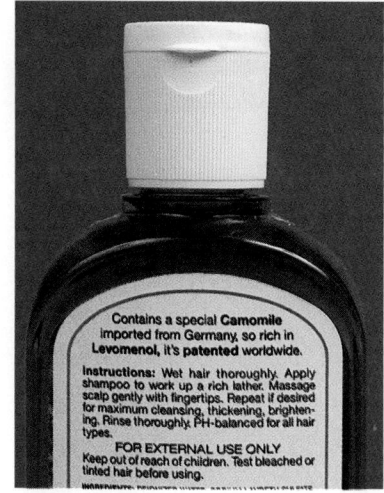

Figure 13.20
Some shampoos are pH balanced, that is they have a pH similar to the natural pH of your scalp/hair (about 6.5). The question arises as to whether pH balanced shampoos clean your hair the same as basic shampoos do.

Acids and bases can be distinguished by means of a variety of properties (Table 13.5). Some properties of acids and bases are defining properties, some of which are used as diagnostic tests, such as the litmus test (Figure 13.21).

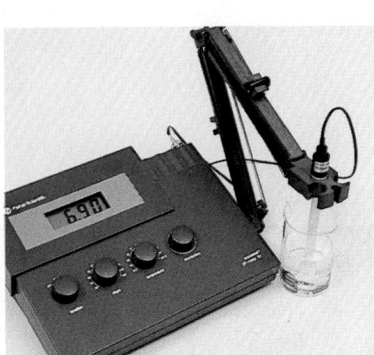

Figure 13.21
A pH meter measures the voltage generated by a pH dependent voltaic cell and the scale converts the millivolt reading into a pH reading.

Table 13.5

EMPIRICAL PROPERTIES OF ACIDS AND BASES	
Acids	**Bases**
taste sour	taste bitter and feel slippery
turn blue litmus red	turn red litmus blue
have pH less than 7	have pH greater than 7
neutralize bases	neutralize acids
react with active metals to produce hydrogen gas	
react with carbonates to produce carbon dioxide	

Problem 13E Strengths of Acids

According to Arrhenius's theory, acids ionize in solution to produce hydrogen ions. The purpose of this problem is to compare the acidity of several acids. Complete the Experimental Design (including a list of the variables) and the Analysis of the investigation report.

Problem

What is the order of several common acids in terms of decreasing acidity?

Evidence

ACIDITY OF 0.10 mol/L ACIDS		
Acid Solution	**Formula**	**pH**
hydrochloric acid	$HCl_{(aq)}$	1.00
acetic (ethanoic) acid	$CH_3COOH_{(aq)}$	2.89
hydrofluoric acid	$HF_{(aq)}$	2.11
methanoic acid	$HCOOH_{(aq)}$	2.38
nitric acid	$HNO_{3(aq)}$	1.00
hydrocyanic acid	$HCN_{(aq)}$	5.15

Strong and Weak Acids

Acidic solutions of different substances at the same concentration do not possess acid properties to the same degree. The pH of an acid may be only slightly less than 7, or it may be as low as 1. Other properties can also vary. For example, acetic acid does not conduct an electric current as well as hydrochloric acid of equal concentration

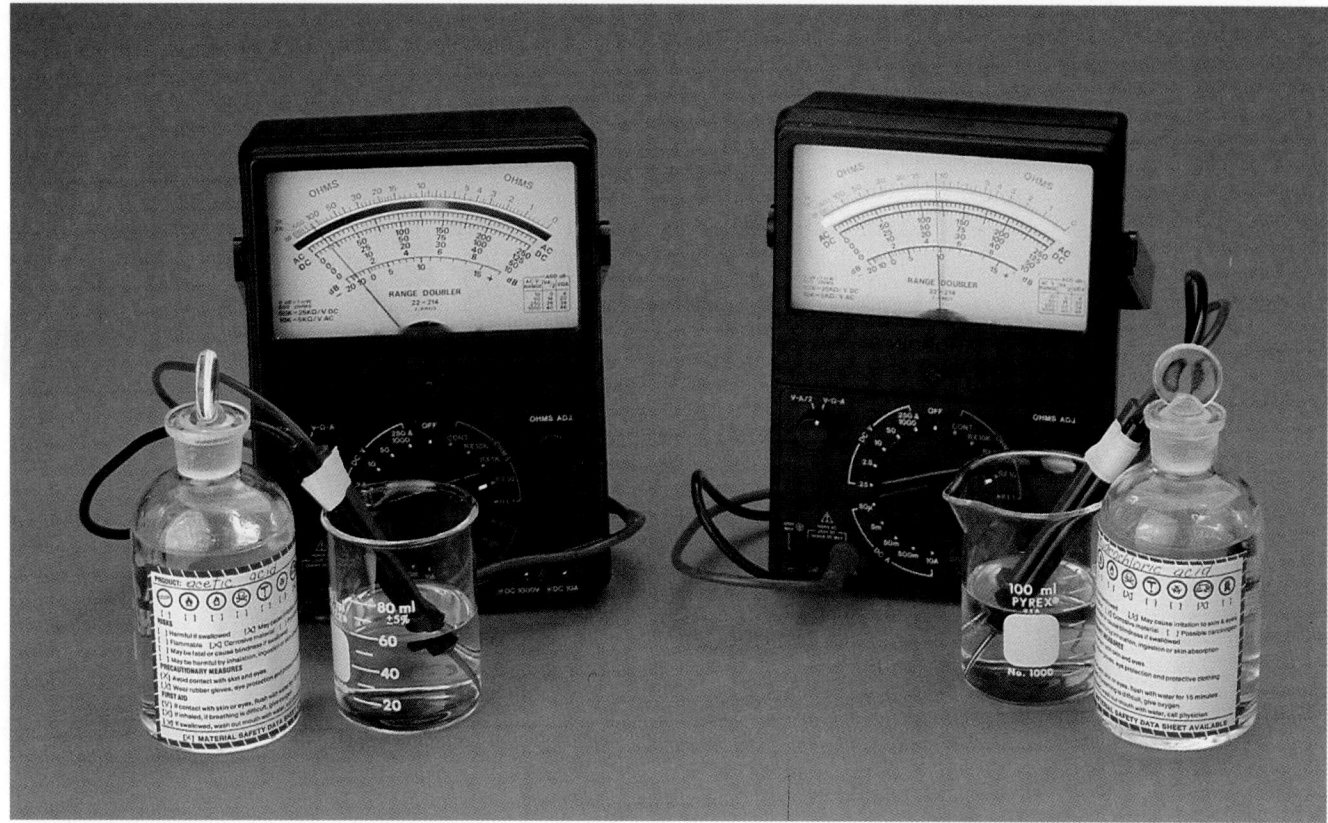

Figure 13.22
In solutions of equal concentration, weak acid such as acetic acid conducts electricity to a lesser extent than does a strong acid such as hydrochloric acid.

(Figure 13.22). When chemical reactions of these acids are observed, it is apparent that acetic acid, although it reacts in the same manner and amount as hydrochloric acid, does not react as quickly. The concepts of strong and weak acids were developed to describe and explain these differences in properties of acids.

An acid is described as *weak* if its characteristic properties are less than those of a common strong acid, such as hydrochloric acid. Weak acids are weak electrolytes and react at a slower rate than strong acids do; the pH of solutions of weak acids are closer to 7 than the pH of strong acids of equal concentration. There are relatively few strong acids; hydrochloric, sulfuric, and nitric acids are the most common. Most common acids are weak (Figure 13.23).

The empirical distinction between strong and weak acids can be explained by combining Arrhenius's theory and equilibrium principles. A **strong acid** is an acid that ionizes quantitatively in water to form hydrogen ions. For example, hydrogen chloride ionizes completely in water.

$$HCl_{(aq)} \rightarrow H^+_{(aq)} + Cl^-_{(aq)}$$

A **weak acid** is an acid that ionizes partially in water to form hydrogen ions. Measurements of pH indicate that most weak acids ionize less than 50%. Acetic acid, a common weak acid, ionizes only 1.3% in solution at 25°C and 0.10 mol/L concentration.

$$CH_3COOH_{(aq)} \overset{1.3\%}{\rightleftharpoons} H^+_{(aq)} + CH_3COO^-_{(aq)}$$

Figure 13.23
Many naturally occurring acids are weak carboxylic acids. Methanoic acid is found in the stingers of certain ants, butanoic acid in rancid butter, citric acid in citrus fruits, oxalic acid in tomatoes, and long-chain fatty acids, such as stearic acid, in animal fats.

According to Arrhenius's theory, weak acids have weaker acidic properties than strong acids of equal concentration because there are fewer hydrogen ions present in the weak acid solution. The hydrogen ion concentration of any acid solution can be calculated by multiplying the percent ionization (as a fraction) by the molar concentration of the acid solute. (For percent ionization, refer to the table of acids and bases in Appendix F.) For example, in a 0.10 mol/L $HCl_{(aq)}$ solution, 100% of the HCl molecules ionize.

$$HCl_{(aq)} \rightarrow H^+_{(aq)} + Cl^-_{(aq)}$$

$$[H^+_{(aq)}] = \frac{100}{100} \times 0.10 \text{ mol/L}$$

$$= 0.10 \text{ mol/L}$$

In a 0.10 mol/L solution of acetic acid, only 1.3% of the CH_3COOH molecules ionize to form hydrogen ions.

$$CH_3COOH_{(aq)} \overset{1.3\%}{\rightleftharpoons} H^+_{(aq)} + CH_3COO^-_{(aq)}$$

$$[H^+_{(aq)}] = \frac{1.3}{100} \times 0.10 \text{ mol/L}$$

$$= 1.3 \times 10^{-3} \text{ mol/L}$$

Relative strengths of acids are determined from the measured pH of solutions of the same concentration. The strength or the percent ionization is calculated from the pH as illustrated in the following example. (This is the method used to obtain reference values for percent ionization, such as the values in the table of acids and bases in Appendix F.)

$$[H^+_{(aq)}] = \frac{p}{100} \times [HA_{(aq)}]$$

where p = percent ionization and $[HA_{(aq)}]$ = concentration of acid

EXAMPLE

The pH of a 0.10 mol/L methanoic acid solution is 2.38. Calculate the percent ionization of methanoic acid.

$$[H^+_{(aq)}] = 10^{-pH}$$

$$= 10^{-2.38} \text{ mol/L}$$

$$= 4.2 \times 10^{-3} \text{ mol/L}$$

$$p = \frac{[H^+_{(aq)}]}{[HCOOH_{(aq)}]} \times 100$$

$$= \frac{4.2 \times 10^{-3} \text{ mol/L}}{0.10 \text{ mol/L}} \times 100$$

$$= 4.2\%$$

The table of acids and bases in Appendix F can be prepared by measuring the pH of 0.10 mol/L solutions, converting to a percent ionization, and ranking the acids in order of decreasing strength (percent ionization).

Ionization Constants for Acids

The strength of an acid can also be communicated using an equilibrium constant that expresses the extent of ion formation by the acid. Acids placed in water reach a state of equilibrium. For example, concentrated acetic acid dissolved in water ionizes in the water to an extent of 0.42% in a 1.0 mol/L solution at SATP.

$$CH_3COOH_{(aq)} \rightleftharpoons H^+_{(aq)} + CH_3COO^-_{(aq)}$$

For a better description of this equilibrium, chemists use the equilibrium law to calculate the equilibrium constant, known as the **acid ionization constant**, K_a. The ionization constant for acetic acid can be found in the table of acids and bases in Appendix F.

$$K_a = \frac{[H^+_{(aq)}]\,[CH_3COO^-_{(aq)}]}{[CH_3COOH_{(aq)}]} = 1.8 \times 10^{-5}\ \text{mol/L}$$

One advantage of the K_a value is that it can be used over a range of concentrations of an acid to predict the hydrogen ion concentration. The percent ionization, although simpler to use, only predicts the $[H^+_{(aq)}]$ accurately within a narrow range around the specified concentration for the acid, for example, 0.10 mol/L. The use of the K_a value for predicting the hydrogen ion concentration and the pH of a 1.0 mol/L acetic acid solution is illustrated below.

As mentioned before, percent ionization depends on the concentration of the acid. Notice that 0.10 mol/L acetic acid has a much greater percent ionization (1.3%) than 1.0 mol/L acetic acid (0.42%) at SATP.

$$CH_3COOH_{(aq)} \rightleftharpoons H^+_{(aq)} + CH_3COO^-_{(aq)} \qquad K_a = 1.8 \times 10^{-5}\ \text{mol/L}$$

Since the hydrogen ion and acetate ion concentrations are equal, the equilibrium expression can be solved for the concentration of the hydrogen ion, *provided* the assumption can be made that at equilibrium the concentration of the acid has been decreased by a negligible amount. This assumption is *only* valid for weak acids that ionize less than about 5% in aqueous solution, when calculating to two significant digits. In solving such a question, this assumption should be *stated* clearly.

At equilibrium,

$$[H^+_{(aq)}] = [CH_3COO^-_{(aq)}] \text{ and}$$
$$[CH_3COOH_{(aq)}] = 1.0\ \text{mol/L} - (\text{very small value})$$
$$\doteq 1.0\ \text{mol/L}$$
$$[H^+_{(aq)}] = [CH_3COO^-_{(aq)}]$$

$$\text{therefore, } K_a = \frac{[H^+_{(aq)}]^2}{[CH_3COOH_{(aq)}]}$$

$$[H^+_{(aq)}] = \sqrt{1.8 \times 10^{-5}\ \text{mol/L} \times [CH_3COOH_{(aq)}]}$$
$$= \sqrt{1.8 \times 10^{-5}\ \text{mol/L} \times 1.0\ \text{mol/L}}$$
$$= 4.2 \times 10^{-3}\ \text{mol/L}$$

The pH can then be calculated from the hydrogen ion concentration.

$$pH = -\log [H^+_{(aq)}]$$
$$= -\log (4.2 \times 10^{-3})$$
$$= 2.37$$

This method of calculating hydrogen ion concentrations is restricted to those cases for which the initial concentration of the acid is numerically much larger than the K_a value (for example, 100 times larger). Since most K_a values are small, the examples presented here usually provide an accurate prediction. (For cases where this restriction is not met, it is necessary to consult a chemistry reference book.) This process can be reversed for calculating a K_a value from the pH of an acidic solution. The example below communicates the procedure for this type of calculation. The K_a value obtained can then be used to calculate the hydrogen ion concentration or pH of carbonic acid solutions over a range of concentrations.

Although units are normally omitted from equilibrium constants, where they have a bewildering variety of expressions, units are always mol/L for acid ionization constants (K_a), and are frequently used with K_a values for that reason.

EXAMPLE

Suppose you measured the pH of a 0.25 mol/L carbonic acid solution to be 3.48. What is the K_a for carbonic acid?

$$[H^+_{(aq)}] = 10^{-pH}$$
$$= 10^{-3.48} \text{ mol/L}$$
$$= 3.3 \times 10^{-4} \text{ mol/L}$$

$$H_2CO_{3(aq)} \rightleftharpoons H^+_{(aq)} + HCO_3^-_{(aq)}$$

$$K_a = \frac{[H^+_{(aq)}][HCO_3^-_{(aq)}]}{[H_2CO_{3(aq)}]}$$

$$= \frac{(3.3 \times 10^{-4} \text{ mol/L})^2}{0.25 \text{ mol/L}}$$

$$= 4.4 \times 10^{-7} \text{ mol/L}$$

For purposes of easy comparison, the K_a values of different acids are sometimes expressed as pK values. A pK value is the negative logarithm of the K_a. A reported pK value of 12 would represent an acid with a K_a of 1×10^{-12}, a very weak acid.

Strong Bases

According to Arrhenius, a *base* is a substance that increases the hydroxide ion concentration of a solution. Ionic hydroxides have varying solubility in water, but all are **strong bases** because ionic hydroxides dissociate completely when they dissolve in water. The basic properties of ionic hydroxides vary only with their concentration in solution.

$$NaOH_{(aq)} \rightarrow Na^+_{(aq)} + OH^-_{(aq)}$$

$$Ba(OH)_{2(aq)} \rightarrow Ba^{2+}_{(aq)} + 2 OH^-_{(aq)}$$

The pH and the conductivity of a $Ba(OH)_{2(aq)}$ solution are found to be higher than those of a $NaOH_{(aq)}$ solution of equal concentration. The barium hydroxide solution is more basic because barium hydroxide dissociates to yield two hydroxide ions per formula unit.

Problem 13F Qualitative Analysis

Complete the Analysis of the investigation report.

Problem

Which of the unknown solutions provided is $HBr_{(aq)}$, $CH_3COOH_{(aq)}$, $NaCl_{(aq)}$, $C_{12}H_{22}O_{11(aq)}$, $Ba(OH)_{2(aq)}$, and $KOH_{(aq)}$?

Experimental Design

The solutions, which have been prepared with equal concentrations, are each tested with a conductivity apparatus and with both red and blue litmus.

Evidence

LITMUS AND CONDUCTIVITY TESTS ON UNKNOWN SOLUTIONS

Solution	Red Litmus	Blue Litmus	Conductivity
1	blue	no change	very high
2	no change	red	low
3	no change	no change	none
4	no change	red	high
5	no change	no change	high
6	blue	no change	high

Exercise

43. Propose an alternative experimental design to answer the problem in Problem 13F. In other words, what design would be suitable if litmus paper and/or a conductivity apparatus were not available?

44. Write a theoretical definition for the strength of an acid. What empirical properties provide evidence for differing acid strengths?

45. Refer to the percent ionization given in the table of acids and bases in Appendix F.
    (a) What is the hydrogen ion concentration of a 0.10 mol/L solution of hydrofluoric acid?
    (b) What is the hydrogen ion concentration of a 2.3 mmol/L solution of nitric acid?
    (c) What is the hydrogen ion concentration of a 0.10 mol/L solution of hydrocyanic acid?
    (d) Which of the solutions in (a) to (c) is most acidic?

46. The hydrogen ion concentration in a 0.100 mol/L solution of propanoic acid is determined to be 1.16×10^{-3} mol/L. Calculate the percent ionization of propanoic acid in water.

47. A 0.10 mol/L solution of lactic acid, found in sour milk, has a pH of 2.43. Calculate the percent ionization of lactic acid in water.

48. Unlike the rest of the hydrogen halides, hydrogen fluoride is a weak acid. However, hydrofluoric acid has the special property

Figure 13.24
Hydrofluoric acid is a weak acid but it etches glass. The etching is obviously not due to the hydrogen/hydronium ion.

of etching glass, a property that is used in the production of frosted effects on glass (Figure 13.24). Write the K_a expression for hydrofluoric acid and calculate the hydrogen and fluoride ion concentrations in a 2.0 mol/L solution of this acid at 25°C.

49. Phosphoric acid is used in rust-remover solutions. Use the first ionization constant to predict the hydrogen ion concentration and pH of a 1.0 mol/L solution of phosphoric acid at 25°C.

50. Ascorbic acid is the chemical ingredient of Vitamin C. A student prepares a 0.20 mol/L aqueous solution of ascorbic acid, measures its pH, and finds it to be 2.40. Based on this evidence, what is the K_a for ascorbic acid?

Deriving K_b

Consider the ammonia equilibrium reaction stated to the right.

$$K = \frac{[OH^-_{(aq)}][NH_4^+_{(aq)}]}{[H_2O_{(l)}][NH_{3(aq)}]}$$

Since $[H_2O_{(l)}]$ is a constant for most aqueous solutions, this value is incorporated into the K_b just as it was into K_w and K_a.

$$K[H_2O_{(l)}] = \frac{[OH^-_{(aq)}][NH_4^+_{(aq)}]}{[NH_{3(aq)}]}$$

$$K_b = \frac{[OH^-_{(aq)}][NH_4^+_{(aq)}]}{[NH_{3(aq)}]}$$

Since the balancing entity generally has the same concentration as the hydroxide ion, the base ionization constant for any weak base (WB) can be defined using the equilibrium concentrations of the entities present.

$$K_b = \frac{[OH^-_{(aq)}][\text{balancing entity}]}{[WB]} = \frac{[OH^-_{(aq)}]^2}{[WB]}$$

Weak Bases

According to Arrhenius's theory (page 191), bases are soluble ionic hydroxides that dissociate in water into positive ions and negative hydroxide ions. According to Le Châtelier's principle, the hydroxide ions added to water cause a shift in the water ionization equilibrium, decreasing the hydrogen ion concentration and producing a pH greater than seven. However, some molecular and ionic compounds, other than hydroxides, also dissolve in water to produce basic solutions. These solutions are not as basic as equivalent ionic hydroxide solutions, and therefore the compounds dissolving to make these solutions are called **weak bases**. Soluble ionic hydroxide compounds in aqueous solutions are the only common examples of strong bases.

The Arrhenius definition is too restricted to explain weak bases that are not hydroxide compounds. We cannot discard the Arrhenius idea completely, though, without seriously affecting our understanding of acidic and basic solutions in terms of the water equilibrium. However, we can revise the definition of a base. A simple revision of the Arrhenius definition, so that it still includes hydroxide ions, is the idea that *weak bases react non-quantitatively with water to form an equilibrium that includes aqueous hydroxide ions.*

$$\text{base} + H_2O_{(l)} \rightleftharpoons OH^-_{(aq)} + \text{balancing entity}$$

As illustrated in Investigation 13.4, cleaning solutions, such as sodium phosphate and ammonia, are generally basic. The basic properties of these solutions are explained by the formation of hydroxide ions.

E.g., $\quad NH_{3(aq)} + H_2O_{(l)} \rightleftharpoons OH^-_{(aq)} + NH_4^+_{(aq)}$

The equilibrium constant for the reaction of a weak base with water is defined according to the equilibrium law and then modified to produce the **base ionization constant, K_b,** for a weak base (WB). (See the derivation in margin.)

$$K_b = \frac{[OH^-_{(aq)}]^2}{[WB]}$$

Notice the parallel between the mathematical expression for K_b and the one for K_a (page 529). The same restrictions and approximations used with the acid ionization constant, K_a, also apply to calculations using the base ionization constant.

EXAMPLE

The pH of a 0.100 mol/L sodium carbonate solution is measured to be 11.66. What is the K_b for the carbonate ion?

$$pOH = 14.00 - 11.66 = 2.34$$

$$[OH^-_{(aq)}] = 10^{-2.34} \text{ mol/L} = 4.6 \times 10^{-3} \text{ mol/L}$$

$$CO^{2-}_{3\,(aq)} + H_2O_{(l)} \rightleftharpoons OH^-_{(aq)} + HCO^-_{3\,(aq)}$$

$$K_b = \frac{[OH^-_{(aq)}]^2}{[CO^{2-}_{3\,(aq)}]} = \frac{(4.6 \times 10^{-3} \text{ mol/L})^2}{0.100 \text{ mol}}$$

$$= 2.1 \times 10^{-4} \text{ mol/L}$$

CANADIAN NOBEL PRIZE WINNERS

In 1971, Gerhard Herzberg (below, left) became the first Canadian to win the Nobel Prize in chemistry. Herzberg was born on Christmas Day, 1904, in Hamburg, Germany. As a boy he dreamed of becoming an astronomer. At university, he specialized in spectroscopy, the study of the light emitted or absorbed by molecules. His research in this field at the University of Saskatchewan and at the National Research Council in Ottawa earned him an international reputation and the Nobel Prize. In announcing the prize, the Swedish Academy stated that Herzberg's ideas and discoveries "stimulated the whole modern development of chemistry from chemical kinetics to cosmochemistry."

Herzberg has always enjoyed passionate debate, whether the topic is molecular spectroscopy, freedom of speech, world hunger, or science funding in Canada.

Henry Taube (page 620) was born in Saskatchewan in 1915 and studied at the University of Saskatchewan. While at the University of Chicago, he began to research electron transfer reactions, which earned him the 1983 Nobel Prize in chemistry.

John Polanyi (page 474) was the co-winner of the 1986 Nobel Prize in chemistry. Polanyi's scientific career began in England, where his family had emigrated after leaving Germany in 1934. Since 1956, Polanyi, a chemistry professor at the University of Toronto, has made significant breakthroughs in the empirical and theoretical descriptions of molecular motions in chemical reactions. Polanyi has been outspoken in his views on science and society, especially regarding the dangers of nuclear war and the obligation of scientists to publicize the promises and perils of science and technology.

Like Henry Taube, Sidney Altman was born in Canada but moved to the United States to pursue his career. As a high school student in Montreal, Altman knew he wanted to be a scientist. He moved to the United States after high school and is now professor of chemistry at Yale University in New Haven, Connecticut. Altman (below, right) received the 1989 Nobel Prize in chemistry for his work on the ability of RNA (ribonucleic acid) to catalyze chemical reactions in cells. His work disproved a 75-year-old belief that all catalysts in cells are proteins.

Drs. Herzberg, Taube, Polanyi, and Altman are not the only Nobel Prize winners associated with Canada. Ernest Rutherford (1908) and James Sumner (1946) won prizes in chemistry for their work in radiation and the nuclear atom. Sir Frederick Banting and John MacLeod won the prize in medicine in 1923 for their discovery of insulin. Lester Pearson, Prime Minister of Canada from 1963 to 1968, won the Nobel Peace Prize for his role in the Suez crisis. Richard Taylor (page 85) was the co-winner of the 1990 Nobel Prize in physics for gathering evidence for the existence of quarks. In 1992, Rudolph Marcus, a graduate of McGill University, was awarded the prize in chemistry for his work in electrochemistry. Michael Smith (page 331), who works on biochemistry and genetic engineering at the University of British Columbia, was awarded the prize in chemistry in 1993. And McMaster University Professor Bertram Brockhouse won the 1994 prize in physics.

- *How many Nobel Prize winners are mentioned in this article?*

- *What was Lester Pearson's job when he won the Nobel prize? Why was he awarded the prize?*

Relationship between K_a and K_b

Acetic acid, present in vinegar, is a common weak acid. A sodium acetate solution is slightly basic; therefore, acetate ions are a weak base. The equilibrium chemical reactions and corresponding equilibrium constants are shown below.

$$CH_3COOH_{(aq)} \rightleftharpoons H^+_{(aq)} + CH_3COO^-_{(aq)}$$

$$K_a = \frac{[H^+_{(aq)}][CH_3COO^-_{(aq)}]}{[CH_3COOH_{(aq)}]}$$

$$CH_3COO^-_{(aq)} + H_2O_{(l)} \rightleftharpoons OH^-_{(aq)} + CH_3COOH_{(aq)}$$

$$K_b = \frac{[OH^-_{(aq)}][CH_3COOH_{(aq)}]}{[CH_3COO^-_{(aq)}]}$$

Notice that most of the terms in these two equilibrium constant expressions are the same. Suppose that the two expressions are multiplied and identical terms cancelled.

$$K_a \times K_b = \frac{[H^+_{(aq)}]\cancel{[CH_3COO^-_{(aq)}]}}{\cancel{[CH_3COOH_{(aq)}]}} \times \frac{[OH^-_{(aq)}]\cancel{[CH_3COOH_{(aq)}]}}{\cancel{[CH_3COO^-_{(aq)}]}}$$

$$= [H^+_{(aq)}][OH^-_{(aq)}]$$

$$= K_w$$

For acids and bases whose chemical formulas differ only by a hydrogen, i.e., conjugate acid-base pairs,

$$K_a K_b = K_w \quad \text{or} \quad K_b = \frac{K_w}{K_a}$$

This expression is particularly useful for calculating K_b, using K_a values given on an acid-base table (Appendix F). Conjugate acid-base pairs appear opposite each other on the acid-base table. The K_b can be calculated as shown in the example in the margin.

The 1:1 ratio of hydroxide ions to bicarbonate ions in the chemical equation means that their concentrations are equal. This calculation is simplified by assuming that an insignificant amount of carbonate ion reacts with water. The equilibrium concentration of carbonate ion would be closer to (0.100 mol/L − 4.6 × 10⁻³ mol/L) or 0.0954 mol/L. Using 0.0954 mol/L in the calculation and rounding the answer give a value of 2.2×10^{-4} mol/L for K_b, which to two significant digits (with the last digit uncertain) is not "significantly" different from the value given in the example.

■ **Example**
What is the K_b for acetate ions?

$$K_b = \frac{K_w}{K_a} = \frac{1.0 \times 10^{-14} \ (mol/L)^2}{1.8 \times 10^{-5} \ mol/L}$$

$$= 5.6 \times 10^{-10} \ mol/L$$

■ **Technological Applications**
In the questions in the Exercise, the chemicals have technological applications as listed below.

- $CN^-_{(aq)}$ in fumigants and for extracting gold and silver from ores
- $SO_4^{2-}_{(aq)}$ in fertilizers
- $C_2H_5COO^-_{(aq)}$ in fungicides
- $C_6H_5NH_2_{(aq)}$ in dyes
- $NO_2^-_{(aq)}$ as preservative
- codeine in medicines

Exercise

51. List some empirical properties that would be useful when distinguishing strong bases from weak bases.

52. For each of the following weak bases, write the chemical equilibrium equation and the mathematical expression for K_b.
    (a) $CN^-_{(aq)}$ (b) $SO_4^{2-}_{(aq)}$

53. The hydroxide ion concentration in a 0.157 mol/L solution of sodium propanoate, $NaC_2H_5COO_{(aq)}$, is found to be 1.1×10^{-5} mol/L. Calculate the base ionization constant for the propanoate ion.

54. Aniline, $C_6H_5NH_2$, is closely related to ammonia and is also a weak base. If the pH of a 0.10 mol/L aniline solution was found to be 8.81, what is its K_b?

55. Using the acid-base table, determine the K_b for nitrite ions.

56. Codeine (use Cod as a chemical symbol) has a K_b of 1.73×10^{-6} mol/L. Calculate the pH of a 0.020 mol/L codeine solution.

57. What would be the pH of a 0.18 mol/L cyanide ion solution?

OVERVIEW

Equilibrium in Chemical Systems

Summary

- Studies of closed chemical systems show that all empirical properties eventually become constant. This is explained by the theory of dynamic equilibrium — a balance between the rates of forward and reverse reactions of a system.

- The position of a chemical reaction equilibrium system is specified by either a percent reaction or an equilibrium constant.

- Chemical reaction equilibria can be classified as ones in which the reactants are favored (<50%), the products are favored (>50%), or the reaction is quantitative (>99%).

- When a property such as temperature, concentration, or pressure (as a result of a volume change) is altered in a chemical system at equilibrium, the system shifts toward the side of the equilibrium that will counteract the change.

- The equilibrium of saturated solutions may be expressed as a solubility product constant, K_{sp}, which is convenient for predicting solution and/or ion concentrations and formation of precipitates.

- The evidence of slight conductivity of pure water at SATP indicates an ionization equilibrium, described by the equilibrium expression $K_w = [H^+_{(aq)}][OH^-_{(aq)}]$.

- $pH = -\log[H^+_{(aq)}]$ and $pOH = -\log[OH^-_{(aq)}]$ are convenient descriptions of the hydrogen ion and hydroxide ion concentrations in aqueous solutions.

- The different acidic properties of various acids indicate that a few acids are strong, but many acids are weak. The percent ionization is the basis of calculations of hydrogen ion concentration of different acids.

Key Words

acid ionization constant K_a
base ionization constant K_b
chemical reaction equilibrium
closed system
common ion effect
dynamic equilibrium
entropy
equilibrium
equilibrium constant K
equilibrium law
equilibrium shift
forward reaction
ionization constant for water K_w
Le Châtelier's principle
percent reaction
pH
phase equilibrium
pK
pOH
reaction quotient Q
reverse reaction
solubility equilibrium
solubility product constant K_{sp}
strong acid
strong base
weak acid
weak base

Review

1. Write an empirical definition of chemical equilibrium.

2. What main idea explains chemical equilibrium?

3. What are two ways to describe the relative amounts of reactants and products present in a chemical reaction at equilibrium?

4. Describe and explain a situation in which a soft drink is in
   (a) a non-equilibrium state.
   (b) an equilibrium state.

5. Write a statement of Le Châtelier's principle.

6. What variables are commonly manipulated when a chemical equilibrium system is shifted?

7. How does a change in volume of a closed system containing gases affect the pressure of the system?

8. How does the hydrogen ion concentration compare with the hydroxide ion concentration if a solution is
   (a) neutral?
   (b) acidic?
   (c) basic?

9. In many processes in industry, engineers try to maximize the yield of a product. In what two ways can concentration be manipulated in order to increase the yield of a product?

10. Does a catalyst affect a state of equilibrium? What does it do?

11. What two diagnostic tests can distinguish a weak acid from a strong acid?

12. According to Arrhenius's theory, what do all bases have in common?

13. Make a list of all mathematical formulas introduced in this chapter.

Applications

14. For each of the following descriptions, write a chemical equation for the system at equilibrium. Communicate the position of the equilibrium with equilibrium arrows. Then write a mathematical expression of the equilibrium law for each chemical system.
    (a) A combination of low pressure and high temperature provides a percent yield of less than 10% for the formation of ammonia in the Haber process.
    (b) At high temperatures, the formation of water vapor from hydrogen and oxygen is quantitative.
    (c) The reaction of carbon monoxide with water vapor to produce carbon dioxide and hydrogen has a percent yield of 67% at 500°C.

15. In a sealed container, nitrogen dioxide is in equilibrium with dinitrogen tetraoxide.
    $$2\,NO_{2(g)} \rightleftharpoons N_2O_{4(g)}$$
    $$K = 1.15 \text{ L/mol}, \; t = 55°C$$
    (a) Write the mathematical expression for the equilibrium law applied to this chemical system.
    (b) If the equilibrium concentration of nitrogen dioxide is 0.050 mol/L, predict the concentration of dinitrogen tetraoxide.
    (c) Write a prediction for the shift in equilibrium that occurs when the concentration of nitrogen dioxide is increased.

16. Scientists and technologists are particularly interested in the use of hydrogen as a fuel. What interpretation can be made about the relative proportions of reactants and products in this system at equilibrium?
    $$2\,H_{2(g)} + O_{2(g)} \rightleftharpoons 2\,H_2O_{(g)}$$
    $$K = 1 \times 10^{80} \text{ L/mol at SATP}$$

17. Predict the shift in the following equilibrium system resulting from each of the following changes.
    $$4\,HCl_{(g)} + O_{2(g)} \rightleftharpoons$$
    $$2\,H_2O_{(g)} + 2\,Cl_{2(g)} + 113 \text{ kJ}$$
    (a) an increase in the temperature of the system
    (b) a decrease in the system's total pressure due to an increase in the volume of the container
    (c) an increase in the concentration of oxygen
    (d) the addition of a catalyst

18. Chemical engineers use Le Châtelier's principle to predict shifts in chemical systems at equilibrium resulting from changes in the reaction conditions. Predict the changes necessary to maximize the yield of product in each of the following industrial chemical systems.
    (a) the production of ethene (ethylene)
    $$C_2H_{6(g)} + energy \rightleftharpoons C_2H_{4(g)} + H_{2(g)}$$
    (b) the production of methanol
    $$CO_{(g)} + 2\,H_{2(g)} \rightleftharpoons CH_3OH_{(g)} + energy$$

19. Apply Le Châtelier's principle to predict whether, and in which direction, the following established equilibrium would be shifted by the change imposed.
    $$2\,CO_{(g)} + O_{2(g)} \rightleftharpoons 2\,CO_{2(g)} + heat energy$$

(a) temperature is increased
(b) vessel volume is increased
(c) oxygen is added
(d) platinum catalyst is added
(e) carbon dioxide is removed

20. For each example, predict whether and in which direction an established equilibrium would be shifted by the change imposed. Explain any shift in terms of changes in forward and reverse reaction rates.
   (a) $Cu^{2+}_{(aq)} + 4NH_{3(g)} \rightleftharpoons Cu(NH_3)_4^{2+}_{(aq)}$
      $CuSO_{4(s)}$ is added
   (b) $CaCO_{3(s)} + energy \rightleftharpoons CaO_{(s)} + CO_{2(g)}$
      temperature is decreased
   (c) $Na_2CO_{3(s)} + energy \rightleftharpoons Na_2O_{(s)} + CO_{2(g)}$
      sodium carbonate added
   (d) $H_2CO_{3(aq)} + energy \rightleftharpoons CO_{2(g)} + H_2O_{(l)}$
      vessel volume decreased
   (e) $KCl_{(s)} \rightleftharpoons K^+_{(aq)} + Cl^-_{(aq)}$ $AgNO_{3(s)}$ is added
   (f) $CO_{2(g)} + NO_{(g)} \rightleftharpoons CO_{(g)} + NO_{2(g)}$
      vessel volume increased
   (g) $Fe^{3+}_{(aq)} + SCN^-_{(aq)} \rightleftharpoons FeSCN^{2+}_{(aq)}$
      $Fe(NO_3)_{3(s)}$ added

21. Find K_{sp} for calcium fluoride at 25°C, given that a 1.00 L sample of the saturated solution, evaporated to dryness, produced 26.76 mg of solid CaF_2.

22. Find the molar solubility of calcium oxalate at 25°C, where K_{sp} is 2.3×10^{-9}.

23. In which of the following cases does the tendency to maximum entropy favor the forward reaction?
   (a) $Br_{2(l)} \rightarrow Br_{2(g)}$
   (b) $N_{2(g)} + 3 H_{2(g)} \rightarrow 2 NH_{3(g)}$ ΔH is negative
   (c) $LiCl_{(s)} \rightarrow Li^+_{(aq)} + Cl^-_{(aq)} + heat$
   (d) $6 C_{(s)} + 3 H_{2(g)} \rightarrow C_6H_{6(l)}$ $\Delta H = +49$ kJ
   (e) $CaCO_{3(s)} + energy \rightarrow CaO_{(s)} + CO_{2(g)}$

24. In a container at high temperature, ethyne (acetylene) and hydrogen are reacted to produce ethene (ethylene). The equilibrium constant is 0.072. At a specific time the substance concentrations are $[C_2H_2]$ = 0.40 mol/L, $[H_2]$ = 0.020 mol/L, and $[C_2H_4]$ = 3.2×10^{-4} mol/L. Predict the direction of the reaction shift.
   $$C_2H_{2(g)} + H_{2(g)} \rightleftharpoons C_2H_{4(g)}$$

25.(a) $H_{2(g)} + Br_{2(g)} \rightleftharpoons 2 HBr_{(g)}$ $K = 12.0$ at t°C
      8.00 mol of hydrogen and 8.00 mol of bromine are added to a 2.00 L reaction container. Predict the concentrations at equilibrium.

(b) $H_{2(g)} + Br_{2(g)} \rightleftharpoons 2 HBr_{(g)}$ $K = 12.0$ at t°C
   12.0 mol of hydrogen and 12.0 mol of bromine are added to a 2.00 L reaction container. Predict the concentrations at equilibrium.

26. Hydrocyanic acid is a very weak acid.
   (a) Write an equilibrium reaction equation for the ionization of 0.10 mol/L $HCN_{(aq)}$. Include the percent ionization at SATP.
   (b) Calculate the hydrogen ion concentration and the pH of a 0.10 mol/L solution of $HCN_{(aq)}$.

27. At 25°C, the hydrogen ion concentration in vinegar is 1.3 mmol/L. Calculate the hydroxide ion concentration.

28. At 25°C, the hydroxide ion concentration in normal human blood is 2.5×10^{-7} mol/L. Calculate the hydrogen ion concentration and the pH of blood.

29. Acid rain has a pH less than that of normal rain. The presence of dissolved carbon dioxide, which forms carbonic acid, gives normal rain a pH of 5.6. What is the hydrogen ion concentration in normal rain?

30. If the pH of a solution changes by 3 pH units as a result of adding a weak acid, by how much does the hydrogen ion concentration change?

31. If 8.50 g of sodium hydroxide is dissolved to make 500 mL of cleaning solution, determine the pOH of the solution.

32. What mass of hydrogen chloride gas is required to produce 250 mL of a hydrochloric acid solution with a pH of 1.57?

33. Determine the pH of a 0.10 mol/L hypochlorous acid solution.

34. Calculate the pH and pOH of a hydrochloric acid solution prepared by dissolving 30.5 kg of hydrogen chloride gas in 806 L of water. What assumptions are made when doing this calculation?

35. Acetic (ethanoic) acid is the most common weak acid used in industry. Determine the pH and pOH of an acetic acid solution prepared by dissolving 60.0 kg of pure, liquid acetic acid to make 1.25 kL of solution.

36. Determine the mass of sodium hydroxide that must be dissolved to make 2.00 L of a solution with a pH of 10.35.

37. Write an experimental design for the identification of four colorless solutions: a strong acid solution, a weak acid solution, a neutral molecular solution, and a neutral ionic solution. Write sentences, create a flow chart, or design a table to describe the required diagnostic tests.

38. Sketch a flow chart or concept map that summarizes the conversion of $[H^+_{(aq)}]$ to and from $[OH^-_{(aq)}]$, pH, and concentration of solute. Make your flow chart large enough that you can write the procedure between the quantity symbols in the diagram.

39. Acetylsalicylic acid (ASA) is a painkiller used in many headache tablets. This drug forms an acidic solution that attacks the digestive system lining. *The Merck Index* lists its K_a at 25°C to be 3.27×10^{-4} mol/L. Predict the pH of a saturated 0.018 mol/L solution of acetylsalicylic acid, $C_6H_4COOCH_3COOH_{(aq)}$. How might the pH change as the temperature changes to 37°C?

40. Boric acid is used for weatherproofing wood and fireproofing fabrics. Assuming that only one hydrogen ion is released per molecule of hydrogen borate that ionizes, what do you predict for the pH of a 0.50 mol/L solution of boric acid?

41. Salicylic acid, $C_6H_4OHCOOH$, is an active ingredient of solutions, such as Clearasil, that are used to treat acne. Since the K_a for this acid was not listed in any convenient references, a student tried to determine the value experimentally. If the pH of a saturated (1 g/460 mL) solution of salicylic acid was found to be 2.4 at 25°C, calculate the ionization constant for this acid.

42. Sodium ascorbate is the sodium salt of ascorbic acid and is used as an antioxidant in food products. The pH of a 0.15 mol/L solution of the ascorbate ion, $HC_6H_6O_6^-{}_{(aq)}$, is 8.65. Calculate the K_b of the ascorbate ion.

43. Sodium hypochlorite is a strong oxidizing agent that is a fire hazard when in contact with organic materials. Solutions of sodium hypochlorite are used as bleach and disinfectant. Determine the hydroxide ion concentration of a 4% sodium hypochlorite solution sold as household bleach.

Extensions

44. A halogen light bulb contains a tungsten (wolfram) filament, $W_{(s)}$, in a mixed atmosphere of a noble gas and a halogen; for example, $Ar_{(g)}$ and $I_{2(g)}$ (see the photograph below). The operation of a halogen lamp depends, in part, on the equilibrium system,

$$W_{(s)} + I_{2(g)} \rightleftharpoons WI_{2(g)}$$

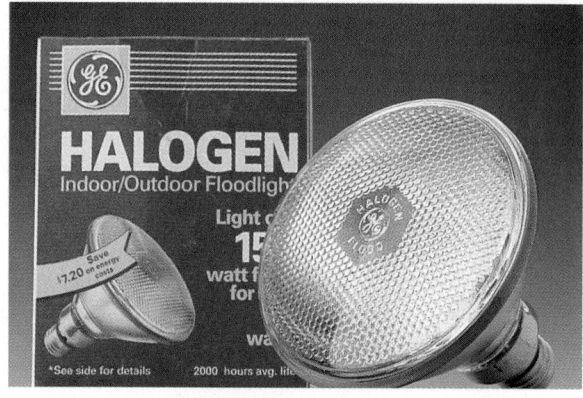

Find out the role of temperature in the operation of a halogen lamp. For example, how is it possible for a halogen lamp to operate with the filament at 2700°C when the tungsten normally would not last very long at this high temperature? Why is such a high temperature desirable?

45. When the Olympic Games were held in Mexico in 1968, many athletes arrived early to train in the higher altitude (2.3 km) and lower atmospheric pressure of Mexico City. Exertion at high altitudes, for people who are not acclimatized, may make them dizzy or "lightheaded" from lack of oxygen. Use the theory of dynamic equilibrium and Le Châtelier's principle to explain this observation. How are people who normally live at high altitudes physiologically adapted to their reduced-pressure environment?

46. $H_{2(g)} + Br_{2(g)} \rightleftharpoons 2\ HBr_{(g)}$ $K = 12.0$ at t°C
12.0 mol of hydrogen and 8.00 mol of bromine are added to a 2.00 L reaction container. Predict the concentrations at equilibrium.

47. $CO_{(g)} + H_2O_{(g)} \rightleftharpoons CO_2 + H_{2(g)}$

$K = 4.00$ at 900°C

In a container, carbon monoxide and water vapor are reacting to produce carbon dioxide and hydrogen. The concentrations are $[CO_{(g)}] = 4.00$ mol/L, $[H_2O_{(g)}] = 2.00$ mol/L, $[CO_{2(g)}] = 4.00$ mol/L, and $[H_{2(g)}] = 2.00$ mol/L. Predict the direction of the reaction shift, and the concentrations at equilibrium.

INVESTIGATION

13.5 Studying a Chemical Equilibrium System

The purpose of this investigation is to solve a problem concerning the effect of an energy change on the following equilibrium system.

$$Fe^{3+}_{(aq)} + SCN^{-}_{(aq)} \rightleftharpoons FeSCN^{2+}_{(aq)}$$

almost colorless   colorless   red

Write a problem statement and then design and carry out an investigation to determine the role of energy in this equilibrium system.

- ✔ Problem
- ✔ Prediction
- ✔ Design
- ✔ Materials
- ✔ Procedure
- ✔ Evidence
- ✔ Analysis
- ✔ Evaluation
- ☐ Synthesis

 CAUTION

Iron(III) compounds are irritant. Potassium thiocyanate is toxic. Avoid skin and eye contact. If there is any skin or eye contact, immediately rinse with plenty of water. The eyes should be flushed for at least 15 min and the teacher informed.

14 Acid-Base Reactions

Acid indigestion, commercial antacid remedies for indigestion, pH-balanced shampoos — you don't have to look far in a magazine or a newspaper to find a reference to acids or acidity. Many people think that all acids are corrosive, and therefore dangerous, because strong acids react with many substances. This can make the use of boric acid as an eyewash, for example, seem like a very risky procedure.

Popular references to acids and bases offer no insight into what these substances are or what they do. In fact, such references usually emphasize only one perspective, such as the environmental damage caused by an acid or the cleaning power of a base. As a result, popular ideas are often confusing; an amateur gardener who has just read an article condemning the destruction of forests by acid rain may be puzzled by instructions on an evergreen fertilizer stating that evergreens are acid-loving plants.

This chapter presents evidence and develops concepts about the substances we call acids and bases. Another key idea developed in this chapter, using acid-base theories as an example, is the increasing power of a theory to explain and predict, as the theory becomes more complete and less restrictive. The development of theories over time is a characteristic of science; few topics show this development as well as a study of acids and bases.

14.1 CHANGING IDEAS ON ACIDS AND BASES

Historically, the empirical properties of substances are often known to chemists long before a theory is developed to explain and predict their behavior. For example, several of the distinguishing properties of acids and bases were known by the middle of the 17th century (page 101). Additional properties, such as pH and the nature of acid-base reactions (page 522), were discovered by the early 20th century.

Early attempts at an acid-base theory tended to focus on acids and to ignore bases. Over time, several theories travelled through cycles of formulation, testing, acceptance, further testing, and eventual rejection. Following is a brief historical summary of acid-base theories and the evidence that led to their revision.

- Antoine Lavoisier (1743 – 1794) assumed that oxygen was responsible for acid properties and that acids were combinations of oxides and water. For example, sulfuric acid, H_2SO_4, was described as hydrated sulfur trioxide, $SO_3 \cdot H_2O$. There were immediate problems with this theory because some oxide solutions, such as CaO, are basic, and several acids, such as HCl, are not formed from oxides. This evidence led to the rejection of the oxygen theory, although we retain the generalization that nonmetallic oxides form acidic solutions.

- Sir Humphry Davy (1778 – 1829) advanced a theory that the presence of hydrogen gave a compound acidic properties. Justus von Liebig (1803 – 1873) later expanded this theory to include the idea that acids are salts of hydrogen. This meant that acids could be thought of as ionic compounds in which hydrogen had replaced the metal ion. However, this theory did not explain why many compounds containing hydrogen have neutral properties (for example, CH_4) or basic properties (for example, NH_3).

- Svante Arrhenius (1859 – 1927) developed a theory in 1887 that provided the first useful theoretical definition of acids and bases (page 189). *Acids were described as substances that ionize in aqueous solution to form hydrogen ions, and bases as substances that dissociate to form hydroxide ions in solution.* This theory explained the process of neutralization by assuming that H^+ and OH^- ions combine to form H_2O. The various strengths of acids were explained in terms of the degree of ionization.

In science, it is unwise to assume that any scientific concept is complete (Figure 14.1). Whenever scientists assume that they understand a subject, two things usually happen.

- First, conceptual knowledge tends to remain static for a while, because little conflicting evidence exists or because any conflicting evidence is ignored.

- Second, when enough conflicting evidence accumulates, a revolution in thinking occurs within the scientific community in which the current theory is drastically revised or entirely replaced. Often, the revolutionary discoveries occur as a result of a flash of insight following a period of hard work.

Acid Formulas
A carry-over from the 19th century idea that acids are salts of hydrogen is the practice of writing hydrogen first in the formulas of substances known to form acidic solutions.

Figure 14.1
In science, no theory can be proven. Well-established, accepted theories have a substantial quantity of supporting evidence. On the other hand, a theory can be disproven by a single, significant, reproducible observation.

Scientific Concepts
"For the first time, I saw a medley of haphazard facts fall into line and order. All the jumbles and recipes and hotchpotch of the inorganic chemistry of my boyhood seemed to fit themselves into the scheme before my eyes — as though one were standing beside a jungle and it suddenly transformed itself into a Dutch garden. 'But it's true,' I said to myself. 'It's very beautiful. And it's true.'" — C.P. Snow (1905 – 1980), English writer, physicist, and diplomat

14.1 Testing Arrhenius's Acid-Base Definitions

The purpose of this investigation is to test Arrhenius's definitions of acid and base. A number of common substances in solution are identified as acid, base, or neutral, using one or more diagnostic tests. In your experimental design, be sure to identify all variables, including any controls.

Problem

Which of the substances tested may be classified as acid, base, or neutral?

Materials

lab apron
safety glasses
aqueous 0.10 mol/L solutions of:
    hydrogen chloride (a gas in solution)
    hydrogen acetate (vinegar)
    sodium hydroxide (lye, caustic soda)
    calcium hydroxide (slaked lime)
    ammonia (cleaning agent)
    sodium carbonate (washing soda, soda ash)
    sodium hydrogen carbonate (baking soda)
    sodium hydrogen sulfate (bowl cleaner)
    calcium oxide (lime)
    carbon dioxide (carbonated beverages)
    aluminum nitrate (salt solution)
    sodium nitrate (fertilizer)
conductivity apparatus, blue litmus paper, red litmus paper, and
    any other materials necessary for diagnostic tests

- ▢ Problem
- ✔ Prediction
- ✔ Design
- ▢ Materials
- ✔ Procedure
- ✔ Evidence
- ✔ Analysis
- ✔ Evaluation
- ▢ Synthesis

CAUTION

Chemicals used include toxic, corrosive, and irritant materials. Avoid eye and skin contact. If you spill any of the chemical solutions on your skin, immediately rinse the area with lots of cool water. In the unlikely situation of getting some of the chemicals in your eye, immediately rinse your eye for at least 15 min and inform your teacher.

Prediction

The prediction is not to be based on experience or a personal hypothesis, but only on Arrhenius's theoretical definitions.
acid $\rightarrow$ H$^+_{(aq)}$ + anion
base $\rightarrow$ cation + OH$^-_{(aq)}$
ionic $\rightarrow$ no H$^+_{(aq)}$ or OH$^-_{(aq)}$
Assume that the Arrhenius concept restricts dissociation and ionization into only two ions.

Revision of Arrhenius's Definitions

Evidence from Investigation 14.1 clearly indicates the limited ability of Arrhenius's definitions to predict acidic or basic properties of a substance in aqueous solution. Only five predictions that would reasonably be made using Arrhenius's definitions are verified: the acids, $HCl_{(aq)}$ and $CH_3COOH_{(aq)}$, the bases, $NaOH_{(aq)}$ and $Ca(OH)_{2(aq)}$, and neutral $NaNO_{3(aq)}$. Seven predictions are falsified. There were problems predicting the properties of solutions of compounds of hydrogen polyatomic anions, such as $NaHCO_{3(aq)}$ and $NaHSO_{4(aq)}$; oxides of metals and nonmetals, such as $CaO_{(aq)}$ or $CO_{2(g)}$; basic compounds that are neither oxides nor hydroxides, such as $NH_{3(aq)}$ and $Na_2CO_{3(aq)}$; and acidic compounds such as $Al(NO_3)_{3(aq)}$. Each of these substances, except for $NaNO_{3(aq)}$, fails to produce a neutral solution, as Arrhenius's definition would predict. Therefore, the theoretical definitions of acid and base need to be revised or replaced.

    The ability of a theoretical concept to explain evidence is not valued as much as its ability to predict the results of new experiments.

Good theoretical progress is made when theories not only explain what is known but enable correct predictions about new situations. Revising Arrhenius's acid-base definitions to explain the results of Investigation 14.1 involves two key ideas: collisions with water molecules and the nature of the hydrogen ion. Since all substances tested are in aqueous solution, then particles will constantly be colliding with, and may also react with, the water molecules present.

It is highly unlikely that the particle we call an aqueous hydrogen ion, $H^+_{(aq)}$, is really an aqueous version of a hydrogen atom stripped of its only electron. The hydrogen ion is a proton, a tiny particle with a highly concentrated positive charge (high-charge density). If such a particle comes near polar water molecules, it is likely to bond strongly to one or more of the molecules (Figure 14.2); that is, it is likely to be hydrated. There is no evidence for unhydrated hydrogen ions in aqueous solution. However, experiments have provided clear evidence for the existence of hydrated protons (Figure 14.3). The simplest representation of a hydrated proton is $H_3O^+_{(aq)}$, commonly called the **hydronium ion** (Figure 14.4).

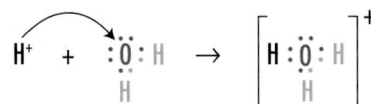

Figure 14.2
The Lewis model for a hydrogen ion has no electrons. A water molecule is believed to have two lone pairs of electrons, as shown in its electron dot model. The hydrogen ion (proton) is believed to bond to one of these lone pairs of electrons to produce the H_3O^+ ion.

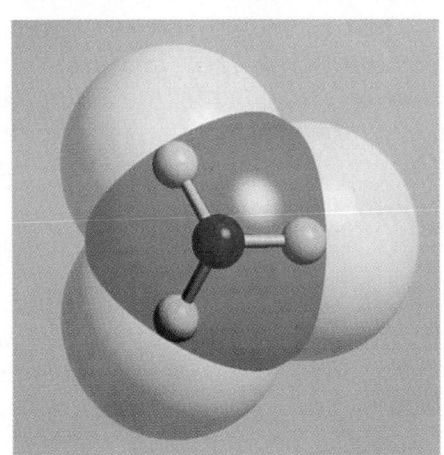

Figure 14.3
By passing infrared light through solutions of acids, Paul Giguère of the Université Laval, Quebec, obtained clear evidence for the existence of hydronium ions in solution.

Figure 14.4
The hydronium ion has a pyramidal structure. The oxygen atom is the apex and the three identical hydrogen atoms form the base of the pyramid.

The formation of acidic solutions by HCl may now be explained as a reaction with water, with hydronium ions being formed (Figure 14.5).

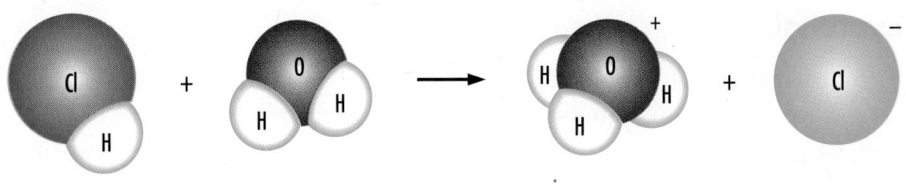

$$HCl_{(aq)} + H_2O_{(l)} \rightarrow H_3O^+_{(aq)} + Cl^-_{(aq)}$$

Figure 14.5
When gaseous hydrogen chloride dissolves in water, the HCl molecules are thought to collide and react with water molecules to form hydronium and chloride ions.

A strong acid, such as HCl (above) is considered to react completely, or nearly so, with water. In Chapter 13, we saw that weak molecular acids, such as hydrofluoric and acetic acids, ionized to less than 50%. We may now explain the equilibrium resulting from the reaction of such acids with water in terms of the formation of hydronium ions. The double arrows represent the equilibrium that is present in these acid solutions, and the predominant entities present are the acid molecules. (The K_a for these acids can be obtained from the acids and bases table in Appendix F.)

$$HF_{(aq)} + H_2O_{(l)} \rightleftharpoons H_3O^+_{(aq)} + F^-_{(aq)} \qquad K_a = 6.6 \times 10^{-4} \text{ mol/L}$$

$$CH_3COOH_{(aq)} + H_2O_{(l)} \rightleftharpoons H_3O^+_{(aq)} + CH_3COO^-_{(aq)}$$
$$K_a = 1.8 \times 10^{-5} \text{ mol/L}$$

In general, *acidic solutions form when substances react with water to form hydronium ions.*

The concept of acids reacting with water to produce hydronium ions is a small adjustment in thinking when explaining or predicting the behavior of molecular acids (e.g., $HCl_{(aq)}$). In fact, either ionization to produce hydrogen ions or reaction with water to produce hydronium ions is acceptable in many contexts. However, in order to explain the acid-base behaviors of weak bases and ions, the reaction with water concept is necessary. Since consistent and unified theories are valued by the scientific community, the concept of acids and bases reacting with water to form acidic and basic solutions is often applied, even when there is no need, e.g., in the explanation of molecular acids.

Just as the properties of molecular acids may be explained by a reaction with water concept, the production of weak-base solutions may be explained by a reaction with water of molecules or ions. The key to the explanation is the production of hydroxide ions. Recall from Chapter 13 that ionic hydroxides produce strong-base solutions by simple dissociation.

$$Ba(OH)_{2(s)} \rightarrow Ba^{2+}_{(aq)} + 2\,OH^-_{(aq)}$$

Here there is no need to invoke a reaction with water for ionic hydroxides because hydroxide ions are present as a result of the dissociation of the ionic compound, $Ba(OH)_2$. However, as you will see below in the section on hydrolysis, entities other than ionic hydroxides require a reaction with water to explain the production of a basic solution.

In general, *basic solutions form when substances dissociate or react with water to form hydroxide ions.*

Hydrolysis

In Investigation 14.1, evidence was obtained that ammonia and sodium carbonate form basic aqueous solutions. Arrhenius's theory did not help you to predict this; nor does it explain this evidence. The modified-Arrhenius theory — the reaction-with-water or hydrolysis theory — can explain the evidence, as illustrated in the chemical equations below.

$$NH_{3(aq)} + H_2O_{(l)} \rightleftharpoons OH^-_{(aq)} + NH_4^+_{(aq)}$$

Sodium carbonate is an ionic compound with high solubility and, according to the Arrhenius theory, dissociates in water to provide aqueous ions of sodium and carbonate.

$$Na_2CO_{3(s)} \rightarrow 2\,Na^+_{(aq)} + CO_3^{2-}_{(aq)}$$

The sodium ion is not listed on the acid-base chart and is eliminated as a possible contributor because many sodium compounds (e.g., $NaCl_{(aq)}$)

— like all group 1 ions — form neutral solutions. The basic character of carbonate solutions can be explained by the reaction-with-water theory.

$$CO_3{}^{2-}{}_{(aq)} + H_2O_{(l)} \rightleftharpoons OH^-{}_{(aq)} + HCO_3{}^-{}_{(aq)}$$

According to the modified-Arrhenius theory, weak bases are substances that react with water to increase the hydroxide ion concentration. If you look at the acid-base table in the Appendix, all bases other than the amide ion ($NH_2{}^-$), the oxide ion (O^{2-}), and the hydroxide ion (OH^-) are weak bases. Since amides and oxides have low solubility in water and react with water to produce hydroxide ions, the hydroxide ion is the only strong base in an aqueous solution (see note on leveling effect).

The only neutral anions are $Cl^-{}_{(aq)}$, $Br^-{}_{(aq)}$, $I^-{}_{(aq)}$, $NO_3{}^-{}_{(aq)}$, and $ClO_4{}^-{}_{(aq)}$.

SIR KARL POPPER (1902 – 1994)

Born in Austria, Karl Raimund Popper studied science and philosophy at the University of Vienna. As a student he was active in politics, music, and social work with children. After university, he earned his living as a secondary school teacher of mathematics and physics, while continuing to pursue his interest in philosophy.

Popper's first book, *Logik der Forschung*, was published in 1934. It

presented a revolutionary view of the nature of scientific knowledge. Published in English in 1959 as *The Logic of Scientific Discovery*, the book addresses a central weakness of scientific induction, the process by which a general statement is derived from individual observations. Popper pointed out that although statements derived in this way can never be proven, they can be disproven by a single authenticated negative example. In Popper's view, hypotheses should not be tested by trying to prove them right by means of many verified predictions, but by trying to prove them wrong by means of one falsified prediction. For example, he stated that we can never prove, inductively, that "all swans are white"; however, a single observation of a black swan can disprove the statement. This principle of testing by attempted falsification is central to Popper's view of science.

Popper emigrated to New Zealand in 1937, just before Austria was annexed by Nazi Germany. If he had remained in Austria he might well have died in a concentration camp because of his Jewish ancestry. Throughout World War II, he lectured in philosophy at the University of New Zealand and worked on his second book, *The Open Society and Its Enemies*. In this book, published in 1945, Popper applies the principle of falsification to the social sciences. In Popper's view, a social or political system is equivalent to a highly complex scientific theory. Such a system should be tested, not to determine where it is succeeding, but to determine where it is failing. Popper believes that the only type of society in which errors can be eliminated from the system is one in which people are free to criticize the actions of the existing government and, if necessary, to replace it.

Popper emphasizes that criticism should be actively encouraged and gratefully welcomed. It is only in having one's work, actions, and opinions criticized that one can identify a need for improvement or modification.

Popper moved to Britain in 1946, where he held the post of Professor of Logic and Scientific Method at the London School of Economics until his retirement in 1969. He was knighted in 1965, and is now widely acclaimed as one of the world's greatest philosophers of science.

- *What did Popper feel you should try to do when testing a theory?*

- *Did Popper feel his principle applies only to scientific studies?*

In a test of the above hydrolysis theory, a student tested the pH of a sodium cyanide solution and found it to have a pH greater than 7. Can the concept of hydrolysis explain this evidence? Provide your reasoning.

$$NaCN_{(s)} \rightarrow Na^+_{(aq)} + CN^-_{(aq)}$$

$$CN^-_{(aq)} + H_2O_{(l)} \rightleftharpoons OH^-_{(aq)} + HCN_{(aq)}$$

Yes, the hydrolysis theory is acceptable for explaining the basic character of a sodium cyanide solution.

Hydrolysis of Amphoteric Ions

When an entity (usually restricted to ions) reacts with water to produce an acidic or basic solution, **hydrolysis** is said to have occurred. Hydrolysis is a concept that was created to explain many of the anomalies that were discovered in Investigation 14.1. To test the concept of hydrolysis further, let us try to explain the basic character of sodium hydrogen carbonate ($NaHCO_3$) and the acidic character of sodium hydrogen sulfate ($NaHSO_4$). We could explain the acidic solution formed by $NaHSO_{4(aq)}$ by invoking a secondary ionization theory, but that would not be consistent with the explanations required throughout this section. Again, we are looking for a logical and consistent theory that will explain all of the anomalies found in Investigation 14.1.

When $NaHCO_{3(s)}$ dissolves in water to form a conducting solution, sodium and hydrogen carbonate ions are believed to be present.

$$NaHCO_{3(s)} \rightarrow Na^+_{(aq)} + HCO_3^-{}_{(aq)}$$

Again, the sodium ion is not listed on the acid-base chart and is eliminated as a possible contributor. The basic character of hydrogen carbonate solutions can be explained by the reaction-with-water theory.

$$HCO_3^-{}_{(aq)} + H_2O_{(l)} \rightleftharpoons OH^-_{(aq)} + H_2CO_{3(aq)}$$

The explanation for the acidic character of sodium hydrogen sulfate is explained by the dissociation and subsequent hydrolysis of the hydrogen sulfate ion.

$$NaHSO_{4(s)} \rightarrow Na^+_{(aq)} + HSO_4^-{}_{(aq)}$$

$$HSO_4^-{}_{(aq)} + H_2O_{(l)} \rightleftharpoons H_3O^+_{(aq)} + SO_4^{2-}{}_{(aq)}$$

Of course, we could have explained why the hydrogen carbonate ion formed an acidic solution (which it did not do), or have explained why a hydrogen sulfate ion formed a basic solution (which it did not do). Ions that could, theoretically, form either an acidic or a basic solution are called **amphoteric**. All polyatomic ions that begin with hydrogen (e.g., HCO_3^- and HSO_4^-) are amphoteric. The term *amphiprotic* is now generally preferred over amphoteric to describe such entities because they can either donate or accept a hydrogen ion (proton).

In a test of the above hydrolysis theory, a student tested the pH of a sodium dihydrogen borate solution and found it to have a pH greater than 7. Can the concept of hydrolysis explain this evidence? Provide your reasoning.

$$NaH_2BO_{3(s)} \rightarrow Na^+_{(aq)} + H_2BO_3^-_{(aq)}$$

$$H_2BO_3^-_{(aq)} + H_2O_{(l)} \rightleftharpoons OH^-_{(aq)} + H_3BO_{3(aq)}$$

Yes, hydrolysis theory is acceptable for explaining the basic character of a sodium dihydrogen borate solution. Hydrolysis theory appears to have good explanatory power — it passes the test.

Hydrolysis of Metal and Nonmetal Oxides

In Investigation 14.1, calcium oxide produced a basic solution and carbon dioxide produced an acidic solution. Let us test the explanatory power of the hydrolysis theory by trying to explain this evidence. First, we must realize that calcium oxide and carbon dioxide have low solubility in water. Therefore, we show the pure substance reacting with water rather than dissolving in water, although dissolving and then reacting is a plausible alternative.

$$CaO_{(s)} + H_2O_{(l)} \rightarrow Ca^{2+}_{(aq)} + 2\,OH^-_{(aq)}$$

or

$$O^{2-}_{(s)} + H_2O_{(l)} \rightarrow 2\,OH^-_{(aq)}$$

Most metal oxides have low solubility in water, but the accepted theory is that the solid state oxide ions are converted to aqueous hydroxide ions by the reaction with water to form a basic solution. Note on your acid-base table that the oxide ion is an even stronger base than hydroxide ions.

Now let us try to explain the acidic character of carbon dioxide in water.

$$CO_{2(g)} + 2\,H_2O_{(l)} \rightleftharpoons H_3O^+_{(aq)} + HCO_3^-_{(aq)}$$

Note that it takes two water molecules to produce a hydronium ion and a balancing entity. This passes the test of being a logical and consistent explanation of the evidence. An alternative explanation is the two-step process presented below.

$$CO_{2(g)} + H_2O_{(l)} \rightleftharpoons H_2CO_{3(aq)}$$

$$\underline{H_2CO_{3(aq)} + H_2O_{(l)} \rightleftharpoons H_3O^+_{(aq)} + HCO_3^-_{(aq)}}$$

$$CO_{2(g)} + 2\,H_2O_{(l)} \rightleftharpoons H_3O^+_{(aq)} + HCO_3^-_{(aq)}$$

Chemists have done numerous tests on metal oxides and nonmetal oxides to determine the acidic and basic character of the solutions formed. A synthesis of the evidence from these tests has produced the following generalizations.

Metal oxides react with water to produce basic solutions.
Nonmetal oxides react with water to produce acidic solutions.

Oxide Ions in Water
Physical analysis shows no evidence for the existence of aqueous oxide ions. It is most convenient to show the ion in solid state, reacting with water.

Carbon Dioxide in Water
Physical analysis shows evidence that carbon dioxide exists primarily as $CO_{2(aq)}$ molecules in water. However, it has become common to represent this solution as $H_2CO_{3(aq)}$, allowing the use of a formula similar to other weak acids. Sulfur dioxide solution is a parallel case.

EXAMPLE

Magnesium oxide and sulfur dioxide are tested and found to form basic and acidic solutions, respectively. Explain these diagnostic test results and judge the hydrolysis concept.

$$MgO: \quad O^{2-}_{(s)} + H_2O_{(l)} \rightleftharpoons 2\,OH^-_{(aq)}$$

$$SO_2: \quad SO_{2(g)} + 2\,H_2O_{(l)} \rightleftharpoons H_3O^+_{(aq)} + HSO_3^-_{(aq)}$$

Based on these successful logical and consistent explanations, the hydrolysis concept is judged to be acceptable.

Hydrolysis of Metal Ions

In Investigation 14.1, you found that an aluminum nitrate solution is acidic. Further evidence indicates that solutions like sodium nitrate and calcium nitrate are neutral. This leads to the hypothesis that the aluminum ion is responsible for the acidity of the solution. Further tests with aluminum solutions show them to be acidic. The research indicates that group 1 and 2 metal ions (except for Be^{2+}) do not produce acidic solutions, but that highly charged small ions do form acidic solutions (see Table 14.1 and the relative strengths of acids and bases table in Appendix F). These ions are said to have high charge density — a large amount of charge in a small volume. The hydrolysis concept needs some revision in order to explain the acidity of these metal ions. Chemists use the concept of complex ions (in this case, hydrated ions) to maintain the logical consistency of the hydrolysis theory as presented below.

$$Al^{3+}_{(s)} + 6\,H_2O_{(l)} \rightleftharpoons Al(H_2O)_6^{3+}_{(aq)}$$

$$Al(H_2O)_6^{3+}_{(aq)} + H_2O_{(l)} \rightleftharpoons H_3O^+_{(aq)} + Al(H_2O)_5(OH)^{2+}_{(aq)}$$

Notice that this acid and its K_a are on your relative strengths of acids and bases table. Again, with revisions, the hydrolysis concept has sufficient explanatory power to be judged acceptable by the scientific community.

EXAMPLE

Explain why a copper(II) chloride solution is acidic. Assuming that the copper(II) complex ion is similar to that of the other metal ions on the acid-base table:

$$Cu(H_2O)_6^{2+}_{(aq)} + H_2O_{(l)} \rightarrow H_3O^+_{(aq)} + Cu(H_2O)_5(OH)^+_{(aq)}$$

As mentioned above, predicting is much more difficult than explaining evidence. Scientists regard a theory as acceptable only if it can predict results in new situations. The revisions we have made to Arrhenius's theoretical definitions do not offer reasons for these reactions nor do they supply predictions in many situations. Consider a solution formed by dissolving sodium hydrogen phosphate in water. Will such a solution be acidic, basic, or neutral? Valid equations can be

---

Table 14.1

K_a FOR SOME METAL IONS AT SATP

Metal Ion*	K_a (mol/L)
$Zr^{4+}_{(aq)}$	2.1
$Sn^{2+}_{(aq)}$	2.0×10^{-2}
$Fe^{3+}_{(aq)}$	1.5×10^{-3}
$Cr^{3+}_{(aq)}$	1.0×10^{-4}
$Al^{3+}_{(aq)}$	9.8×10^{-6}
$Be^{2+}_{(aq)}$	3.2×10^{-7}
$Fe^{2+}_{(aq)}$	1.8×10^{-7}
$Pb^{2+}_{(aq)}$	1.6×10^{-8}
$Cu^{2+}_{(aq)}$	1.0×10^{-8}

*The aqueous metal ion is a hydrated complex ion; e.g., $Cu(H_2O)_4^{2+}_{(aq)}$

Metal Ion Formulas

Aqueous ions of transition metals are usually written in a simplified form, without showing the number of water molecules present in the actual hydrated complex ion, as shown in Table 14.1.

written to predict that either hydronium ions or hydroxide ions will form when hydrogen phosphate ions react with water.

$$HPO_4^{2-}{}_{(aq)} + H_2O_{(l)} \rightleftharpoons H_3O^+{}_{(aq)} + PO_4^{3-}{}_{(aq)}$$

$$HPO_4^{2-}{}_{(aq)} + H_2O_{(l)} \rightleftharpoons OH^-{}_{(aq)} + H_2PO_4^-{}_{(aq)}$$

Nothing that you have studied so far in this book enables you to *predict* whether either of these reactions predominates. However, if the solution turns red litmus blue, then you can select one of the equations to *explain* the evidence.

The following section discusses a more advanced theory of acids and bases that better satisfies the objections of the last paragraph.

Exercise

1. How well does the modified-Arrhenius theory explain the results of Investigation 14.1?

2. Test the explanatory power of the revised Arrhenius definitions by explaining the following evidence. Write an ionization or dissociation equation where appropriate, and then write a net equation showing reactions with water to produce either hydronium or hydroxide ions (consistent with the evidence).
   (a) $HBr_{(g)}$ in solution shows a pH of 2 on pH paper.
   (b) $Na_3PO_{4(s)}$ forms a solution with a pH of 8.
   (c) $FeCl_{3(s)}$ in solution provides a pH meter reading of 5.
   (d) $NaHSO_{3(s)}$ in solution turns blue litmus red.
   (e) $Na_2HPO_{4(s)}$ in solution turns red litmus blue.
   (f) $Na_2O_{(s)}$ in solution turns red litmus blue.
   (g) $SO_{3(g)}$ in solution turns blue litmus red.
   (h) $KOH_{(s)}$ yields a solution with a pH = 12.

3. Make a list of all amphiprotic (amphoteric) ions from your acid-base table.

4. What test do you put a theory through after testing its explanatory power?

Problem 14A Metal and Nonmetal Oxides

The purpose of this problem is to test the generalizations that metal and nonmetal oxides react with water to form basic and acidic solutions. Complete the Prediction, Analysis, and Evaluation of the investigation report.

Problem

What kind of solutions (acidic, basic, or neutral) do the period 3 oxides form when placed in water?

Experimental Design

Oxides of elements in period 3 are tested to determine their acidic or basic nature. Soluble oxides are tested in water using litmus paper. To all the oxides, a strong acid (hydrochloric acid) and a strong base (sodium hydroxide) are added to determine if a neutralization reaction occurs.

Evidence

LITMUS AND NEUTRALIZATION TESTS ON OXIDES			
Oxide	Litmus Test	$HCl_{(aq)}$ Test	$NaOH_{(aq)}$ Test
$Na_2O_{(s)}$	red to blue	neutralizes	no reaction
$MgO_{(s)}$	red to blue	neutralizes	no reaction
$Al_2O_{3(s)}$	(insoluble)	neutralizes	neutralizes
$SiO_{2(s)}$	(insoluble)	no reaction	neutralizes
$P_2O_{3(s)}$	blue to red	no reaction	neutralizes
$SO_{3(g)}$	blue to red	no reaction	neutralizes
$Cl_2O_{(g)}$	blue to red	no reaction	neutralizes

Predicting the Results of Hydrolysis

Every acid and base, regardless if it is of the standard variety or involves hydrolysis, can be described quantitatively with a percent reaction or by an equilibrium (ionization) constant, K. The K_a of an acid can be determined empirically from a pH value or can be referenced on a table of acids and bases. The larger the K_a value, the stronger the acid. The K_b of a base can also be determined empirically from pH or pOH values or it can be calculated from a referenced K_a. (Most acid-base tables do not list K_b values, although they could.) Recall from Chapter 13, page 533 that $K_b = K_w/K_a$. Remember that a K_b is found by dividing K_w by the K_a of the conjugate acid (of the base). Therefore, the K_b of $NH_{3(aq)}$ below is calculated from the K_a of $NH_4^+{}_{(aq)}$. For example, the reactions in the previous section can be written with K_a or K_b values.

$$HF_{(aq)} + H_2O_{(l)} \rightleftharpoons H_3O^+{}_{(aq)} + F^-{}_{(aq)} \qquad K_a = 6.6 \times 10^{-4} \text{ mol/L}$$

$$NH_{3(aq)} + H_2O_{(l)} \rightleftharpoons OH^-{}_{(aq)} + NH_4^+{}_{(aq)} \qquad K_b = 1.7 \times 10^{-5} \text{ mol/L}$$

From the K_a or K_b value, the $[H_3O^+{}_{(aq)}]$, $[OH^-{}_{(aq)}]$, pH or pOH can also be calculated, as described and practiced in Chapter 13.

A problem occurs with hydrolysis when both the cation and the anion of a salt hydrolyze. The cation tends to increase the hydronium ion concentration while the anion tends to increase the hydroxide ion concentration. The extent of acidity or basicity is dependent on the relative size of the K_a or K_b. For example, an aqueous solution of aluminum sulfate contains the acidic hexaaquoaluminum(III) ion and the basic sulfate ion.

$$Al(H_2O)_6^{3+}{}_{(aq)} + H_2O_{(l)} \rightleftharpoons H_3O^+{}_{(aq)} + Al(H_2O)_5(OH)^{2+}{}_{(aq)}$$

$$K_a = 9.8 \times 10^{-6} \text{ mol/L}$$

$$SO_4^{2-}{}_{(aq)} + H_2O_{(l)} \rightleftharpoons OH^-{}_{(aq)} + HSO_4^-{}_{(aq)}$$

$$K_b = 1.0 \times 10^{-12} \text{ mol/L}$$

Since the K_a is larger than the K_b, the solution is predicted to be acidic, and evidence gathered in the laboratory would verify this prediction. (Sometimes you can just look at the relative position of the cation and anion on the table and make a decision. If the acid is relatively strong compared with the strength of the base, then the solution is acidic, and vice versa.)

Predicting the results of the hydrolysis of amphiprotic (amphoteric) ions is another problem area. The ability of the hydrolysis theory to explain but not predict the acidity/basicity of amphoteric ions is documented in the previous section. However, we should now be able to make this prediction by comparing the K_a and K_b values of the amphoteric ion.

EXAMPLE ───

Predict whether a baking soda solution is acidic or basic.

$NaHCO_{3(s)} \rightarrow Na^+_{(aq)} + HCO_3^-{}_{(aq)}$

$HCO_3^-{}_{(aq)} + H_2O_{(l)} \rightleftharpoons H_3O^+_{(aq)} + CO_3^{2-}{}_{(aq)} \qquad K_a = 4.7 \times 10^{-11} \text{ mol/L}$

$HCO_3^-{}_{(aq)} + H_2O_{(l)} \rightleftharpoons OH^-_{(aq)} + H_2CO_{3(aq)} \qquad K_b = 2.7 \times 10^{-8} \text{ mol/L}$

According to these relative K-values, the baking soda solution should be basic because the K_b is larger than the K_a. The hydrogen carbonate ion is a stronger base than it is an acid.

───

Predicting whether double hydrolysis of a salt produces an acidic or basic solution is often possible without calculating the K_b value, by looking at an acid-base table. *If the salt is composed of a stronger acid than a base, then the salt forms an acidic solution. If the salt is composed of a stronger base than an acid, then the salt forms a basic solution.* The analysis is based on the relative distance of the acid and base from the top or bottom of the relative strengths of acids and bases table.

> **Bi-Ions**
> Anions containing hydrogen in the formula are referred to here as bi-ions, for convenience. The common names of such ions often begin with the bi prefix, as in bicarbonate and bisulfate.

> **K_b of Amphiprotic Bi-Ions**
> The K_b for $HCO_3^-{}_{(aq)}$ is not $K_{b\,HCO_3^-} = K_w/K_{a\,HCO_3^-}$ but is $K_w/K_{a\,H2CO3}$. To find the K_b of bi-ions, such as $HCO_3^-{}_{(aq)}$, use the K_a of the corresponding conjugate acid (e.g., $H_2CO_{3(aq)}$), not the K_a of the bi-ion (e.g., $HCO_3^-{}_{(aq)}$) as an acid.

Exercise

5. Predict whether the following solutions are acidic, basic, or neutral.
   (a) table salt (saline or brine) solution
   (b) aluminum chloride (antiperspirant)
   (c) $Na_2CO_{3(aq)}$ (washing soda)
   (d) $NaHSO_{4(aq)}$ (Sani-Flush)
   (e) ammonium phosphate (fertilizer)
   (f) carbonated beverage (pop, wine, and beer)
   (g) ammonium dihydrogen phosphate (fertilizer)
   (h) baking soda (baking)
   (i) ammonium sulfate (fertilizer)
   (j) magnesium oxide (milk of magnesia)

6. What is the K_b for a sodium phosphate (rust cleaning) solution?

7. Predict the pH of a 0.10 mol/L sodium sulfite (developer) solution?

8. Predict the pH of a 0.30 mol/L ammonium nitrate (fertilizer) solution?

9. Estimate the pH of a 0.25 mol/L ammonium carbonate (baking powder) solution?

10. What is the strongest possible acid in an aqueous solution, and what is the strongest possible base in an aqueous solution?

11. What kind of fertilizers would be appropriate for acid-loving plants like evergreens?

12. What are the two major tests for theories that are acceptable to the scientific community? Does the concept of hydrolysis pass these tests?

INVESTIGATION

14.2 Testing the Concept of Hydrolysis

The scientific purpose of this investigation is to test the concept of hydrolysis. The pH of a wide assortment of salts is predicted and then tested.

Problem

What is the pH for the salts tested?

Materials

lab apron
safety glasses
pure water
0.10 mol/L aqueous solutions of:
    sodium carbonate
    sodium phosphate
    aluminum sulfate
    aluminum chloride
    sodium chloride
    ammonium chloride
    ammonium oxalate
    ammonium acetate
    ammonium carbonate
    ammonium sulfate
    potassium sulfate
    copper(II) sulfate
    iron(III) sulfate
    iron(III) chloride
    sodium hydrogen carbonate
    sodium hydrogen sulfate
pH paper, pH meter, and/or universal indicator
containers (small beakers, test tubes, or spot plates)
waste beakers

Problem
Prediction
✔ Design
Materials
✔ Procedure
✔ Evidence
✔ Analysis
✔ Evaluation
Synthesis

CAUTION

Chemicals used include toxic, corrosive, and irritant materials. Avoid eye and skin contact. If you spill any of the chemical solutions on your skin, immediately rinse the area with lots of cool water. In the unlikely situation of getting some of the chemicals in your eye, immediately rinse your eye for at least 15 min and inform your teacher.

THE BRØNSTED-LOWRY ACID-BASE CONCEPT

Acid and base definitions, revised to include the ideas of the hydronium ion and reaction with water, are more effective in describing, explaining, and predicting than the original definitions proposed by Arrhenius. However, these revised definitions are still too restrictive. Reactions of acids and bases do not always involve water. Also, evidence indicates that some entities that form basic solutions (such as $HCO_3^-{}_{(aq)}$) can actually neutralize stronger bases. A broader concept is needed to describe, explain, and predict these properties of acids and bases.

New theories in science usually result from looking at the evidence in a way that has not occurred to other observers. A new approach to acids and bases was adopted in 1923 by Johannes Brønsted (1879 – 1947) of Denmark and independently by Thomas Lowry (1874 – 1936) of England. These scientists focused on the role of an acid and a base in a reaction rather than on the acidic or basic properties of their aqueous solutions. An acid, such as hydrogen chloride, functions in a way opposite to a base, such as ammonia. According to the Brønsted-Lowry idea, hydrogen chloride donates a proton to a water molecule,

$$\overset{\overset{\displaystyle H^+}{\frown}}{HCl_{(aq)}} + H_2O_{(l)} \rightarrow H_3O^+{}_{(aq)} + Cl^-{}_{(aq)}$$

and ammonia accepts a proton from a water molecule.

$$\overset{\overset{\displaystyle H^+}{\frown}}{NH_{3(aq)}} + H_2O_{(l)} \rightleftharpoons OH^-{}_{(aq)} + NH_4^+{}_{(aq)}$$

Water does not have to be one of the reactants. For example, the hydronium ions present in a hydrochloric acid solution can react directly with dissolved ammonia molecules.

$$\overset{\overset{\displaystyle H^+}{\frown}}{H_3O^+{}_{(aq)}} + \underset{base}{NH_{3(aq)}} \rightarrow H_2O_{(l)} + NH_4^+{}_{(aq)}$$
$$\underset{acid}{}$$

We can describe this reaction as NH_3 molecules removing protons from H_3O^+ ions. Hydronium ions act as the acid, and ammonia molecules act as the base. Water is present as the solvent but not as a primary reactant. In fact, water does not even have to be present, as evidenced by the reaction of hydrogen chloride and ammonia gases (Figure 14.6).

$$\overset{\overset{\displaystyle H^+}{\frown}}{HCl_{(g)}} + \underset{base}{NH_{3(g)}} \rightarrow NH_4Cl_{(s)}$$
$$\underset{acid}{}$$

According to the Brønsted-Lowry concept, a **Brønsted-Lowry acid** is a proton donor and a **Brønsted-Lowry base** is a proton acceptor. A **Brønsted-Lowry neutralization** is a competition for protons that results

J. Brønsted, pictured above, independently created new theoretical definitions for acids and bases based upon proton transfer.

Figure 14.6
One hazard of handling concentrated solutions of ammonia and hydrochloric acid is gas fumes. In the photograph, ammonia gas and hydrogen chloride gas, escaping from the open bottles, react and form a white cloud of very tiny crystals of $NH_4Cl_{(s)}$.

Figure 14.7
Baking soda is a common household chemical, but it requires an uncommonly sophisticated theory to explain or predict its properties.

in a proton transfer from the strongest acid present to the strongest base present.

A substance can only be classified as a Brønsted-Lowry acid or base for a specific reaction. This point is important — protons may be gained in a reaction with one substance, but lost in a reaction with another substance. (For example, in the reaction of HCl with water shown above, water acts as the base; whereas, in the reaction of NH_3 with water, water acts as the acid.) A substance that appears to act as Brønsted-Lowry acid in some reactions and as Brønsted-Lowry base in other reactions is called *amphiprotic*. The hydrogen carbonate ion in baking soda (Figure 14.7), like every other hydrogen polyatomic ion (bi-ion), is amphiprotic, as shown by the following reactions.

$$HCO_{3\ (aq)}^{-} + H_2O_{(l)} \rightleftharpoons OH_{(aq)}^{-} + H_2CO_{3(aq)} \qquad K_a = 2.1 \times 10^{-4}\ mol/L$$
base   acid

$$HCO_{3\ (aq)}^{-} + H_2O_{(l)} \rightleftharpoons H_3O_{(aq)}^{+} + CO_{3\ (aq)}^{2-} \qquad K_a = 2.7 \times 10^{-8}\ mol/L$$
acid   base

$$HCO_{3\ (aq)}^{-} + H_3O_{(aq)}^{+} \rightleftharpoons H_2CO_{3(aq)} + H_2O_{(l)} \qquad \text{(neutralizes strong acid)}$$
base   acid

$$HCO_{3\ (aq)}^{-} + OH_{(aq)}^{-} \rightleftharpoons CO_{3\ (aq)}^{2-} + H_2O_{(l)} \qquad \text{(neutralizes strong base)}$$
acid   base

Since the Brønsted-Lowry definitions do not explain *why* a proton is donated or accepted, they fall short of being a comprehensive theory. *The advantage of the Brønsted-Lowry definitions is that they enable us to define acids and bases in terms of chemical reactions rather than simply as substances that form acidic and basic aqueous solutions.* A definition of acids and bases in terms of chemical reactions allows us to describe, explain, and predict many reactions in aqueous solution, non-aqueous solution, or pure states.

Exercise

13. Theories in science develop over a period of time. Illustrate this development by writing a theoretical definition of an acid, using the following concepts. Begin your answer with, "According to [authority], acids are substances that...."
    (a) the Arrhenius concept
    (b) the revised Arrhenius concept
    (c) the Brønsted-Lowry concept

14. How does the definition of a base according to Arrhenius compare with the Brønsted-Lowry definition?

15. Classify each reactant in the following equations as a Brønsted-Lowry acid or base.
    (a) $HF_{(aq)} + SO_{3\ (aq)}^{2-} \rightleftharpoons F_{(aq)}^{-} + HSO_{3\ (aq)}^{-}$
    (b) $CO_{3\ (aq)}^{2-} + CH_3COOH_{(aq)} \rightleftharpoons CH_3COO_{(aq)}^{-} + HCO_{3\ (aq)}^{-}$
    (c) $H_3PO_{4(aq)} + OCl_{(aq)}^{-} \rightleftharpoons H_2PO_{4\ (aq)}^{-} + HOCl_{(aq)}$
    (d) $HCO_{3\ (aq)}^{-} + HSO_{4\ (aq)}^{-} \rightleftharpoons SO_{4\ (aq)}^{2-} + H_2CO_{3(aq)}$

Conjugate Acids and Bases

According to the Brønsted-Lowry concept, acid-base reactions involve the transfer of a proton. These reactions are universally reversible and result in an acid-base equilibrium.

In a proton transfer reaction at equilibrium, both forward and reverse reactions involve Brønsted-Lowry acids and bases. For example, in an acetic acid solution (Figure 14.8), the forward reaction is explained as a proton transfer from acetic acid to water molecules and the reverse reaction is a proton transfer from hydronium to acetate ions.

$$\underset{\text{acid}}{CH_3COOH_{(aq)}} + \underset{\text{base}}{H_2O_{(l)}} \rightleftharpoons \underset{\text{base}}{CH_3COO^-_{(aq)}} + \underset{\text{acid}}{H_3O^+_{(aq)}}$$

This equilibrium is typical of all acid-base reactions. There will always be two acids (in the above example CH_3COOH and H_3O^+) and two bases (in the above example H_2O and CH_3COO^-) in any acid-base reaction equilibrium. Furthermore, the base on the right (CH_3COO^-) is formed by removal of a proton from the acid on the left (CH_3COOH). The acid on the right (H_3O^+) is formed by the addition of a proton to the base on the left (H_2O). A pair of substances that differ only by a proton is called a **conjugate acid-base pair**. An acetic acid molecule and an acetate ion are a conjugate acid-base pair. Acetic acid is the conjugate acid of the acetate ion and the acetate ion is the conjugate base of acetic acid. The hydronium ion and water are the second conjugate acid-base pair in this equilibrium. Conjugate acid-base pairs appear opposite each other in a table of acids and bases such as that in Appendix F.

Figure 14.8
Ethanoic (acetic) acid in vinegar is a weak acid, i.e., it has low electrical conductivity. This property is explained by a low reaction with water. Therefore, few free ions are in solution.

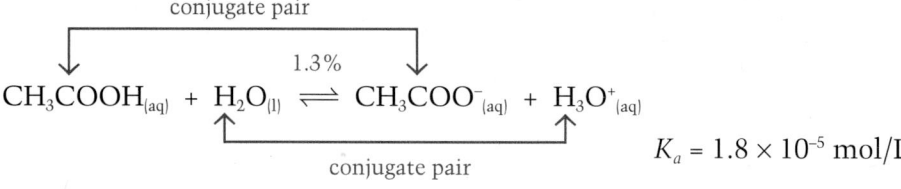

$$\overset{\text{conjugate pair}}{\underset{\text{conjugate pair}}{CH_3COOH_{(aq)} + H_2O_{(l)} \overset{1.3\%}{\rightleftharpoons} CH_3COO^-_{(aq)} + H_3O^+_{(aq)}}} \qquad K_a = 1.8 \times 10^{-5} \text{ mol/L}$$

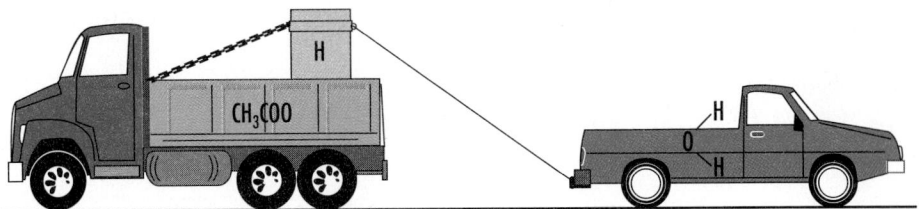

At equilibrium, only 1.3% of the CH_3COOH molecules have reacted with water in a 0.10 mol/L solution at SATP. It appears that the ability of the CH_3COO^- part of the acetic acid molecule to keep its proton (H^+) is much greater than the ability of H_2O to attract the proton away (Figure 14.9). This means that CH_3COO^- is a stronger base (that is, it has a greater attraction for protons) than H_2O. When HCl molecules react with water (Figure 14.5, page 543), the Cl has a much weaker attraction for its hydrogen than the H_2O does. The water molecule *wins* and the H of the HCl is completely transferred as a proton (H^+).

The terms *strong* acid and *weak* acid (page 527) can be explained by the Brønsted-Lowry concept and also by comparing the reactions of acids with the same base — for example, water. Using HA as the general symbol for any acid and A^- as its conjugate base, the empirically derived table of acids and bases (Appendix F) lists the position of equilibrium of acids reacting with water.

$$HA_{(aq)} + H_2O_{(l)} \rightleftharpoons H_3O^+_{(aq)} + A^-_{(aq)}$$

The extent of the proton transfer between HA and H_2O determines the strength of $HA_{(aq)}$. In Brønsted-Lowry terms, when a strong acid reacts with water, an almost complete transfer of protons results for the forward reaction and almost no transfer of protons for the reverse reaction; a nearly 100% reaction with water or a large equilibrium constant is found. Theoretically, a strong acid has a very low attraction for its proton and easily donates a proton to any base, even weak bases such as water. This leads to the interpretation that the conjugate base, A^-, of a strong acid must have a very weak attraction for protons. A useful generalization regarding the relative strengths of a conjugate acid-base pair is: *the stronger an acid, the weaker its conjugate base*; and conversely, *the weaker an acid, the stronger its conjugate base*. (See the acid-base table in Appendix F.)

No simple explanation, in terms of forces or bonds, can be given for the differing abilities of acids to donate protons or of bases to accept them. The inability to predict acid and base strengths for a substance not included in an empirically determined table of acids and bases is a major deficiency of all acid-base theories.

Strong Acids

A strong acid has a:
- high electrical conductivity
- low pH
- large K_a
- large percent reaction with water
- very few molecules in solution
- weak conjugate base
- weak attraction for protons

Exercise

19. Use the Brønsted-Lowry definitions to identify the two conjugate acid-base pairs in each of the following acid-base reactions.

    (a) $HCO_3^-{}_{(aq)} + S^{2-}{}_{(aq)} \rightleftharpoons HS^-{}_{(aq)} + CO_3^{2-}{}_{(aq)}$

    (b) $H_2CO_{3(aq)} + OH^-{}_{(aq)} \rightleftharpoons HCO_3^-{}_{(aq)} + H_2O_{(1)}$

    (c) $HSO_4^-{}_{(aq)} + HPO_4^{2-}{}_{(aq)} \rightleftharpoons H_2PO_4^-{}_{(aq)} + SO_4^{2-}{}_{(aq)}$

    (d) $H_2O_{(1)} + H_2O_{(1)} \rightleftharpoons H_3O^+{}_{(aq)} + OH^-{}_{(aq)}$

20. Some ions can form more than one conjugate acid-base pair. List the two conjugate acid-base pairs involving a hydrogen carbonate ion in the reactions in question 19.

Acid-Base Indicators

Many plant compounds and synthetic dyes change color when mixed with an acid or a base (Figures 14.10 and 14.11). Substances that change color when reacted with acids or bases are known as **acid-base indicators**. A very common indicator used in school laboratories is litmus, which is obtained from a lichen. Litmus paper

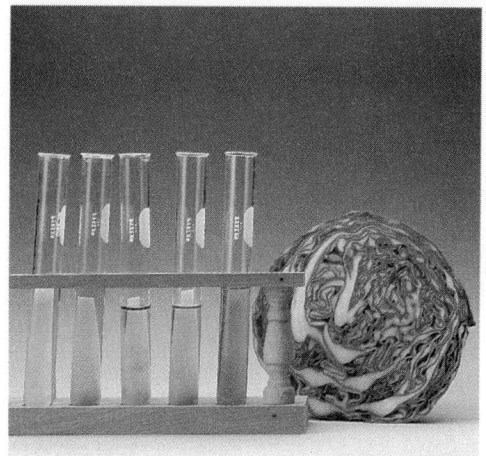

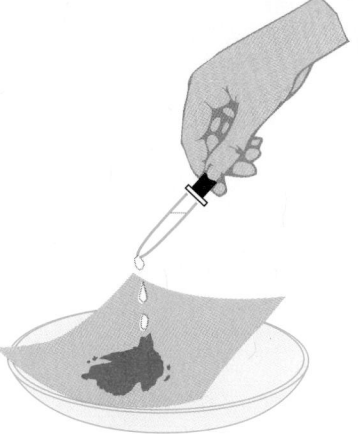

Figure 14.10
Purple cabbage boiled in water produces an extract that changes color in different solutions. The test tubes show the color of the cabbage juice in a strong acid (pH 1), a weak acid (pH 4),a neutral solution (pH 7), a weak base (pH 9) and a strong base (pH 14). All concentrations are 0.10 mol/L.

Figure 14.11
A few drops of dilute sodium hydroxide solution have been dropped onto goldenrod-colored paper. How do you think the red color can be reversed to the goldenrod color?

Figure 14.12
Sodium hydroxide solution has been added to hydrochloric acid containing phenolphthalein indicator, which is colorless in acids and red in bases. The red color indicates the temporary presence of some unreacted sodium hydroxide.

is prepared by soaking absorbent paper with litmus solution and then drying it. As you know, red and blue are the two colors of the litmus dye. Phenolphthalein, another indicator that you have used, is either colorless or red (Figure 14.12). Most indicators exist in one of two conjugate forms that are reversible and distinctly different in color.

The explanation of the behavior of acid-base indicators depends, in part, on both the Brønsted-Lowry concept and the equilibrium concept. *An indicator is a conjugate weak acid-base pair formed when an indicator dye dissolves in water.* Using $HIn_{(aq)}$ to represent the acid form and $In^-_{(aq)}$ to represent the base form of any indicator, the following equilibrium can be written. (Litmus colors are given below the equation as an example.)

$$\overset{\text{conjugate pair}}{\underset{\substack{\text{acid} \\ \text{red} \quad \text{(litmus color)}}}{HIn_{(aq)}} + H_2O_{(l)} \rightleftharpoons \underset{\substack{\text{base} \\ \text{blue}}}{In^-_{(aq)}} + H_3O^+_{(aq)}}$$

According to Le Châtelier's principle, an increase in the hydronium ion concentration will shift the equilibrium to the left. Then more indicator will change to the color of the acid form ($HIn_{(aq)}$). This happens, for example, when litmus is added to an acidic solution. Similarly, in basic solutions the hydroxide ions remove hydronium ions with the result that the equilibrium shifts to the right. Then the base color of the indicator (In^-) predominates. Since different indicators have different acid strengths, the acidity or pH of the solution at which an indicator changes color varies (Figures 14.13 and 14.14). These pH values have been measured and are reported in the table of acid-base indicators on the inside back cover of this book.

Figure 14.13
A few common acid-base indicators are shown here. Each indicator has its own pH range over which it changes color from the acid form (HIn) at the lower pH value to the base form (In⁻) at the higher pH value. Material Safety Data Sheets are available for these chemicals.

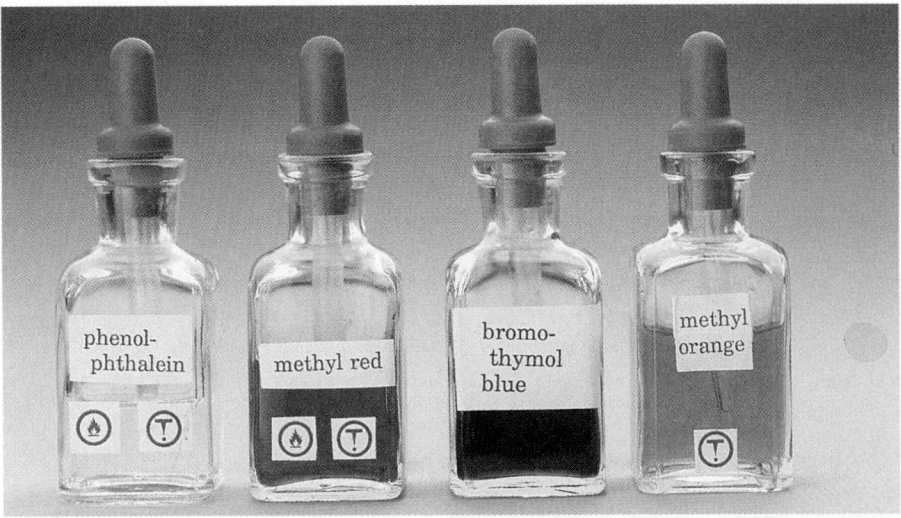

$pH < 4.8$

$pH > 6.0$

Figure 14.14
Methyl red exists predominantly in its red (acid) form at pH values less than 4.8, and in its yellow (base) form at pH values greater than 6.0. Between pH values of 4.8 and 6.0, intermediate orange colors occur, as both forms of the indicator are present in detectable quantities.

pH Test Strips

Litmus is not the only indicator paper available. Bromothymol blue paper is sold to test aquarium water for pH values between 6.0 and 7.6, to a precision of about 0.1 pH unit. Other test strips contain several different indicators and show different colors at different pH values. These give a composite color that can measure pH from 0 to 14 to within 1 to 2 pH units (Figure 14.15).

Figure 14.15
The test strips shown on the right have been dipped into the solution. Comparing the color of a strip with the color scale on the container gives an approximate pH. The inexpensive pH paper on the right bleeds (the indicator runs off the paper).

Exercise

21. According to the table of acid-base indicators on this book's inside back cover, what is the color of each of the following indicators in the solutions of given pH?
    (a) phenolphthalein in a solution with a pH of 11.7
    (b) bromothymol blue in a solution with a pH of 2.8
    (c) litmus in a solution with a pH of 8.2
    (d) methyl orange in a solution with a pH of 3.9

22. Complete the analysis for each of the following diagnostic tests. *If* [the specified indicator] is added to a solution, *and* the color of the solution turns [the given color], *then* the solution pH is _____.
    (a) methyl red (red)
    (b) alizarin yellow (red)
    (c) bromocresol green (blue)
    (d) bromothymol blue (green)

23. Separate samples of an unknown solution turned both methyl orange and bromothymol blue to yellow, and turned bromocresol green to blue.
    (a) Estimate the pH of the unknown solution.
    (b) Calculate the approximate hydronium ion concentration.

Pool Water Acidity

The pH of swimming pool water has to be monitored and adjusted to a pH of 7.2–7.6. A few drops of phenol red indicator is added to a sample of the pool water to determine the pH. Sodium carbonate (soda ash) is added to the water to increase the pH when necessary. Sodium bisulfate may be added to decrease the pH.

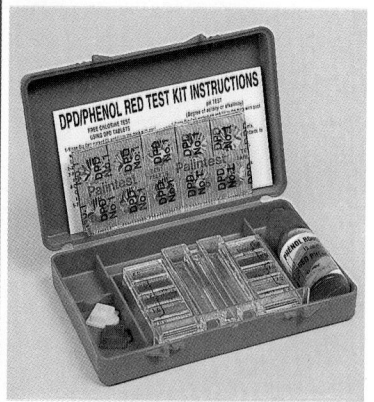

Problem 14B Using Indicators to Determine pH

One experimental design for determining the pH of a solution is testing the solution with indicators. Complete the Analysis of the investigation report. Include a table of indicators and pH.

Problem

What is the approximate pH of three unknown solutions?

Experimental Design

The unknown solutions were labelled A, B, and C. Each solution was tested with a different series of indicators.

Evidence

Solution A: After addition to the solution, methyl violet was blue, methyl orange was yellow, methyl red was red, and phenolphthalein was colorless.

Solution B: After addition to the solution, indigo carmine was blue, phenol red was yellow, bromocresol green was blue, and methyl red was yellow.

Solution C: After addition to the solution, phenolphthalein was colorless, thymol blue was yellow, bromocresol green was yellow, and methyl orange was orange.

Problem 14C Designing an Indicator Experiment

Solving puzzles is a common feature of the scientific enterprise. Science and technology olympics often employ puzzles similar to this one: Design an experiment that uses indicators to identify which of three unknown solutions labelled X, Y, and Z have pH values of 3.5, 5.8, and 7.8. There are several acceptable designs for this problem.

Problem 14D Testing a Unifying Theory

The purpose of this problem is to test an indicator-based experimental design for creating a table of relative strengths of acids and bases. Acid-base tables can be built by using a pH meter or paper (Problem 13E, page 526) or by relative percent yield (Problem 14G, page 566). Complete the Experimental Design, Analysis, and Evaluation sections of the report. (Reverse the Evaluation section sequence suggested in Appendix B, B.2 in order to finish the report with evaluating the experimental design.)

Problem

What are the relative strengths of the acids provided?

Prediction

According to the experimental design using pH, the relative strengths of the acids provided are in order $HCl_{(aq)}$ (strongest), $HNO_{3(aq)}$, $HF_{(aq)}$, $HCOOH_{(aq)}$, $CH_3COOH_{(aq)}$, and $HCN_{(aq)}$.

Evidence

INDICATOR COLORS WITH THE ACIDS						
Acid	HMv	H₂Tb	HOr	HBp	HCr	HCh
hydrofluoric HF$_{(aq)}$	blue	orange	orange	violet	blue	yellow
acetic CH₃COOH$_{(aq)}$	blue	yellow	yellow	purple	blue	yellow
nitric HNO₃$_{(aq)}$	green	red	red	violet	blue	yellow
hydrocyanic HCN$_{(aq)}$	blue	yellow	yellow	red	red	yellow
methanoic HCOOH$_{(aq)}$	blue	orange	orange	purple	blue	yellow
hydrochloric HCl$_{(aq)}$	green	red	red	violet	blue	yellow

INVESTIGATION

14.3 Indicators and Acid Strength

The purpose of this investigation is to test the experimental design of using indicators to build an acid-base table. There are several experimental designs used to build an acid-base table — pH (Problem 13E, page 526), indicators (this investigation), and relative percent yield (Problem 14G, page 566). Predict by using your acid-base table (Appendix F). Choose indicators from those provided to yield an efficient and valid experimental design. Communicate your answer in the Analysis section as an acid-base table. Evaluate your experimental design by reversing the Evaluation section sequence suggested in Appendix B, B.2.

Problem

What are the relative strengths of the provided acids and bases as expressed in an acid-base table?

Materials

lab apron
safety glasses
0.10 mol/L HCl$_{(aq)}$
0.10 mol/L NaHSO₄$_{(aq)}$
0.10 mol/L CH₃COOH$_{(aq)}$
0.10 mol/L NaHSO₃$_{(aq)}$
0.10 mol/L Na₂CO₃$_{(aq)}$
0.10 mol/L NaOH$_{(aq)}$
indicators, for example: methyl orange, methyl violet, bromothymol blue, phenolphthalein
(6) 13 × 100 mm test tubes for each indicator or spot plate (microchem)
test-tube rack

- ☐ Problem
- ✔ Prediction
- ✔ Design
- ☐ Materials
- ✔ Procedure
- ☐ Evidence
- ✔ Analysis
- ✔ Evaluation
- ☐ Synthesis

CAUTION

Chemicals used include toxic, corrosive, and irritant materials. Avoid eye and skin contact. If you spill any of the chemical solutions on your skin, immediately rinse the area with lots of cool water. In the unlikely situation of getting some of the chemicals in your eye, immediately rinse your eye for at least 15 min and inform your teacher.

Acid-Base Table

The acid-base table presents the acids and their conjugate bases in terms of decreasing strength, from the strongest acids (SA) to the weakest acids (WA).

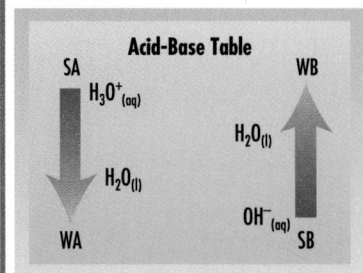

Predicting Acid-Base Equilibria

The failure of acid-base theory to predict strengths of acids and bases means that the theory cannot be used to predict the position of acid-base equilibria. In science, when a theory is unable to predict, an empirical generalization that will allow accurate prediction is often sought. The table of acids and bases in Appendix F was developed empirically, based on measured strengths of acids in water; the acids are listed in order of decreasing strength. Opposite each acid in the table is its conjugate base, whose strength is related to the acid — the weaker the acid, the stronger its conjugate base. This relationship results in an ordered list of acids and bases in which the strongest acid is at the upper left and the strongest base is at the lower right.

Problem 14E Position of Acid-Base Equilibria

The purpose of this problem is to develop a generalization for predicting the position of acid-base equilibria. Use the table of acids and bases in Appendix F and the evidence of position of equilibrium to complete the Analysis of the investigation report.

Problem

How do the positions of the reactant acid and base in the acid-base table relate to the position of equilibrium?

Evidence

1. $CH_3COOH_{(aq)} + H_2O_{(l)} \overset{<50\%}{\rightleftharpoons} H_3O^+_{(aq)} + CH_3COO^-_{(aq)}$

2. $HCl_{(aq)} + H_2O_{(l)} \overset{>99\%}{\rightleftharpoons} H_3O^+_{(aq)} + Cl^-_{(aq)}$

3. $CH_3COO^-_{(aq)} + H_2O_{(l)} \overset{<50\%}{\rightleftharpoons} CH_3COOH_{(aq)} + OH^-_{(aq)}$

4. $H_3PO_{4(aq)} + NH_{3(aq)} \overset{>50\%}{\rightleftharpoons} H_2PO_4^-_{(aq)} + NH_4^+_{(aq)}$

5. $HCO_3^-_{(aq)} + SO_3^{2-}_{(aq)} \overset{<50\%}{\rightleftharpoons} HSO_3^-_{(aq)} + CO_3^{2-}_{(aq)}$

6. $H_3O^+_{(aq)} + OH^-_{(aq)} \rightarrow H_2O_{(l)} + H_2O_{(l)}$

Predicting an Acid-Base Reaction

When making complex predictions, scientists often combine a variety of empirical and theoretical concepts. Prediction of the products and the extent of acid-base reactions requires a combination of concepts. According to the collision-reaction theory, a proton transfer may result from a collision between an acid and a base. In a reactant mixture, there are countless random collisions of all the entities present. Using a synthesis of the collision-reaction theory and the Brønsted-Lowry concept of acids and bases makes an explanation for the predominant reaction possible. It seems probable that in the competition for proton transfer, the substance that has the greatest attraction for protons (that is, the strongest base) and the substance that gives up its proton most easily (that is, the strongest acid) will react. This explanation assumes that only one proton is transferred per collision.

Table 14.2

ENTITIES IN AN AQUEOUS SOLUTION	
Substance Dissolved	**Predominant Entities (Ions, Atoms, or Molecules)**
high solubility ionic compound	separate ions; e.g., $K^+_{(aq)}$, $NO_3^-{}_{(aq)}$
strong acid	$H_3O^+{}_{(aq)}$, conjugate base; e.g., $H_3O^+{}_{(aq)}$, $Cl^-_{(aq)}$
weak acid	molecules; e.g., $HF_{(aq)}$, $CH_3COOH_{(aq)}$

The first step in predicting the products of an acid-base reaction is to list all entities as they exist in the initial aqueous mixture (Table 5.3, page 192 and Table 14.2). For example, suppose some spilled oven cleaner containing aqueous sodium hydroxide is neutralized with vinegar. The initial list includes the following entities.

$$Na^+_{(aq)}, OH^-_{(aq)}, CH_3COOH_{(aq)}, H_2O_{(l)}$$

Brønsted-Lowry definitions and the table of acids and bases are used to label all possible acids and bases as **A** or **B**, as shown below. Both possibilities are labelled for amphiprotic substances. Metal cations, except for the aquo ions that undergo hydrolysis (e.g., $Al(H_2O)_6^{3+}{}_{(aq)}$), are assumed to be spectator ions.

$$\begin{array}{cccc} & & \text{SA} & \text{A} \\ Na^+_{(aq)}, & OH^-_{(aq)}, & CH_3COOH_{(aq)}, & H_2O_{(l)} \\ & \text{SB} & & \text{B} \end{array}$$

The order of acids and bases in the acid-base table is used to identify the strongest acid (**SA**) and the strongest base (**SB**) among those labelled. The predominant proton transfer reaction occurs between the strongest acid and the strongest base. Arrange these reactants in an equation and transfer the proton to predict the products. Note that charge and mass must be balanced in the predicted equation.

$$\overset{H^+}{\overbrace{}}$$
$$CH_3COOH_{(aq)} + OH^-_{(aq)} \rightleftharpoons CH_3COO^-_{(aq)} + H_2O_{(l)}$$

In the final step, the position of equilibrium is predicted using the empirical rule developed in Problem 14E. *Products are favored* (>50% reaction) if the strongest acid is listed higher in the table of acids and bases than the strongest base. *Reactants are favored* (<50% reaction) if the strongest acid is listed lower on the acid-base table than the strongest base. Since CH_3COOH is higher than OH^- in the acid-base table, the products are favored for this reaction equilibrium.

$$\overset{>50\%}{CH_3COOH_{(aq)} + OH^-_{(aq)} \rightleftharpoons CH_3COO^-_{(aq)} + H_2O_{(l)}}$$

Position of Equilibrium
The relative positions of the strongest acid and the strongest base on an acid-base table can be used to determine the position of an acid-base equilibrium.

Products Favored

$$\begin{array}{ll} \text{SA} & >50\% \\ + & \rightleftharpoons \\ \quad\text{SB} & \end{array}$$

Reactants Favored

$$\begin{array}{ll} \quad\text{SB} & <50\% \\ + & \rightleftharpoons \\ \text{SA} & \end{array}$$

The position of equilibrium is communicated by the percent reaction, the percent yield, or the K value. Generally speaking, a $K > 1$ means >50% and a $K < 1$ means <50% reaction.

The prediction of the position of equilibrium is restricted to the categories of more than 50% or less than 50%. Studies of acid-base reactions show that many reactions involving relatively strong acids and bases (especially hydronium or hydroxide ions) are quantitative (that is, more than 99% complete). Quantitative reactions, other than $H_3O^+_{(aq)}$ plus $OH^-_{(aq)}$, cannot be predicted using concepts developed in this book. Throughout this chapter, quantitative reactions (>99% complete) other than strong acid-strong base reactions are identified for you.

EXAMPLE

Ammonium nitrate fertilizer is produced by the quantitative reaction of aqueous ammonia (Figure 14.16) with nitric acid. Write a balanced acid-base equilibrium equation.

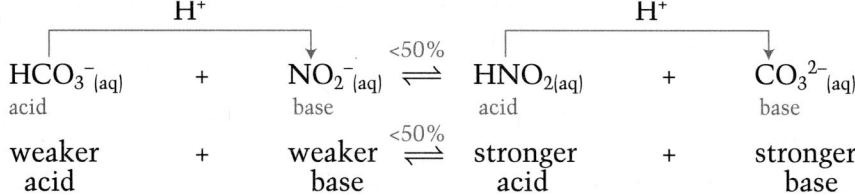

$$NH_{3(aq)} + H_3O^+_{(aq)} \rightarrow NH_4^+_{(aq)} + H_2O_{(l)}$$

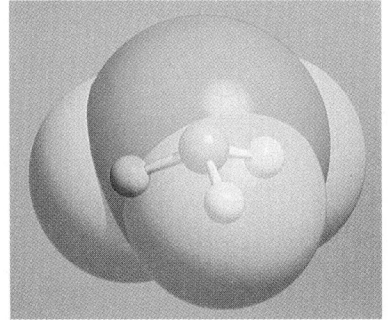

Figure 14.16
Evidence indicates that the ammonia molecule, NH_3, modeled in the figure, and the hydronium ion are both pyramidal.

An interesting consequence of the application of the Brønsted-Lowry concept leads to another useful generalization. In any acid-base equilibrium, the formulas of the stronger acid and stronger base are always on the same side of the reaction equation.

$$\underset{\text{acid}}{HCO_3^-_{(aq)}} + \underset{\text{base}}{NO_2^-_{(aq)}} \overset{<50\%}{\rightleftharpoons} \underset{\text{acid}}{HNO_{2(aq)}} + \underset{\text{base}}{CO_3^{2-}_{(aq)}}$$

weaker acid + weaker base $\overset{<50\%}{\rightleftharpoons}$ stronger acid + stronger base

Consulting the relative strengths of acids and bases table, we see that nitrous acid is a stronger acid than hydrogen carbonate ion, and that carbonate ion is a stronger base than nitrite ion. Because of these relative strengths, we predict that for this equilibrium, the reactants (left-side chemicals) are favored. *In general, in a Brønsted-Lowry acid-base equilibrium, the weaker acid and base are always favored at equilibrium.*

SUMMARY: A FIVE-STEP METHOD OF PREDICTING ACID-BASE REACTIONS

1. List all entities (ions, atoms, or molecules including $H_2O_{(l)}$) initially present as they exist in a water environment. (Refer to Table 5.3, page 192.)
2. Identify all possible acids and bases, using the Brønsted-Lowry definitions.
3. Identify the strongest acid and the strongest base present, using the table of acids and bases (Appendix F).
4. Transfer one proton from the acid to the base and predict the conjugate base and the conjugate acid as the products.
5. Predict the position of the equilibrium, using the generalization developed on page 563 and the table of acids and bases (Appendix F).

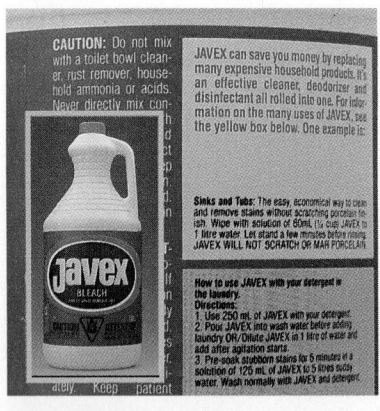

Figure 14.17
Bottles of household bleach display a warning against mixing the bleach (aqueous sodium hypochlorite) with acids. Does your prediction of the reaction between vinegar and hypochlorite ions provide any clues about the reason for the warning?

Exercise

Use the five-step method to make predictions for the predominant reactions in the following chemical systems.

24. Hydrofluoric acid and an aqueous solution of sodium sulfate are mixed to test the five-step method of predicting acid-base reactions.

25. Strong acids, such as perchloric acid, have been shown to react quantitatively with strong bases, such as sodium hydroxide.

26. Predict the acid-base reaction of bleach with vinegar (Figure 14.17).

27. Methanoic acid is added to an aqueous solution of sodium hydrogen sulfide.

28. A student mixes solutions of ammonium chloride and sodium nitrite in a chemistry laboratory.

29. Empirical work has shown that nitric acid reacts quantitatively with a sodium acetate solution.

30. A consumer attempts to neutralize an aqueous sodium hydrogen sulfate cleaner with a solution of lye. (See Appendix G, if you do not remember what lye is.)

31. Can ammonium nitrate fertilizer, added to water, be used to neutralize a muriatic acid (hydrochloric acid) spill?

Problem 14F Testing the Five-Step Method

Complete the Prediction, Analysis, and Evaluation of the investigation report.

Problem

What are the products and position of the equilibrium for sodium hydrogen carbonate (Figure 14.18) with stomach acid, vinegar, household ammonia, and lye, respectively?

Experimental Design

Each of the chemicals is prepared as a solution with a concentration between 0.1 mol/L and 1.0 mol/L. Evidence is gathered to test the predicted products and the position of the equilibrium .

Evidence

THE ADDITION OF BAKING SODA TO VARIOUS SOLUTIONS			
Reactant	Bubbles	Odor	pH
$HCl_{(aq)}$	yes	none	increases
$CH_3COOH_{(aq)}$	yes	disappears	increases
$NH_{3(aq)}$	no	remains	decreases
$NaOH_{(aq)}$	no	none	decreases

Figure 14.18
The versatility of baking soda is demonstrated by its use in extinguishing fires, in baking biscuits, and in neutralizing excess stomach acid. It is also used as a medium for local anesthetics — apparently, baking soda reduces stinging sensations by neutralizing the acidity of the anesthetic, with the result that the speed and efficiency of the anesthetic are improved. The broad range of uses for baking soda results, in part, from its amphiprotic character.

14.4 Testing Brønsted-Lowry Predictions

The purpose of this investigation is to test the Brønsted-Lowry concept of acids and bases and the five-step method for predicting acid-base reactions.

Problem

What reactions occur when the following substances are mixed? (Hints for diagnostic tests are in parentheses.)

1. ammonium chloride and sodium hydroxide solutions (odor)

2. hydrochloric acid and sodium acetate solutions (odor)

3. sodium benzoate and sodium hydrogen sulfate solutions (benzoic acid has low solubility)

4. hydrochloric acid and aqueous ammonium chloride (odor)

5. solid sodium chloride added to water (litmus)

6. solid aluminum sulfate added to water (litmus)

7. solid sodium phosphate added to water (litmus)

8. solid sodium hydrogen sulfate added to water (litmus)

9. solid sodium hydrogen carbonate added to hydrochloric acid (pH)

10. solid sodium hydrogen carbonate added to sodium hydroxide solution (pH)

11. solid sodium hydrogen carbonate added to sodium hydrogen sulfate solution (pH)

Experimental Design

Each prediction of a reaction using the established procedure is accompanied by a diagnostic test of the prediction. The certainty of the evaluation is increased by performing as many diagnostic tests as possible, complete with controls.

- ▪ Problem
- ✔ Prediction
- ▪ Design
- ✔ Materials
- ✔ Procedure
- ✔ Evidence
- ✔ Analysis
- ✔ Evaluation
- ▪ Synthesis

CAUTION

Chemicals used include toxic, corrosive, and irritant materials. Avoid eye and skin contact. If you spill any of the chemical solutions on your skin, immediately rinse the area with lots of cool water. In the unlikely situation of getting some of the chemicals in your eye, immediately rinse your eye for at least 15 min and inform your teacher. Remember to detect odors cautiously by wafting air toward your nose from the container.

▪Prediction

When predicting the products of the acid-base reactions in Investigation 14.4, at least one of the reactants is in an aqueous solution. So list all initial entities present as being in the aqueous state, even those added in the solid state.

▪Analysis

In the Analysis for Problem 14G, use the generalizations for >50% and <50% reactions to predict the relative positions of the acids and bases on an acid-base table. The evidence for reaction 1 is interpreted as

Acids   Bases

$CH_3COOH_{(aq)}$ >50%

                $C_3H_7COO^-_{(aq)}$

Problem 14G An Acid-Base Table

Complete the Analysis of the investigation report. Include a short table of acids and bases. This is the third design (after using pH and indicators earlier) that you can use to build an acid-base table.

Problem

What is the order of acid strength for the first four members of the carboxylic acid family?

$$CH_3COOH_{(aq)} + C_3H_7COO^-_{(aq)} \overset{>50\%}{\rightleftharpoons} CH_3COO^-_{(aq)} + C_3H_7COOH_{(aq)}$$

$$HCOOH_{(aq)} + CH_3COO^-_{(aq)} \overset{>50\%}{\rightleftharpoons} HCOO^-_{(aq)} + CH_3COOH_{(aq)}$$

$$C_2H_5COOH_{(aq)} + C_3H_7COO^-_{(aq)} \overset{<50\%}{\rightleftharpoons} C_2H_5COO^-_{(aq)} + C_3H_7COOH_{(aq)}$$

$$C_2H_5COOH_{(aq)} + HCOO^-_{(aq)} \overset{<50\%}{\rightleftharpoons} C_2H_5COO^-_{(aq)} + HCOOH_{(aq)}$$

14.3 pH CHANGES IN ACID-BASE REACTION SYSTEMS

For many acid-base reactions the appearance of the products resembles that of the reactants, so you cannot directly observe the progress of a reaction. Also, acids cannot easily be distinguished from bases except by measuring pH. The pH values and changes provide important information about the nature of acids and bases, the properties of conjugate acid-base pairs and indicators, and the stoichiometric relationships in acid-base reactions. A graph showing the continuous change of pH during an acid-base reaction is called a **pH curve** (titration curve) for the reaction.

THE NEXT STAGE: LEWIS ACID-BASE THEORY

"To restrict the group of acids to those substances that contain hydrogen interferes as seriously with the systematic understanding of chemistry as would the restriction of the term 'oxidizing agent' to substances containing oxygen." — Gilbert Lewis (1875 – 1946)

Increasingly less restricted views of nature are obtained as scientific concepts are developed. Acids had been restricted to compounds containing hydrogen since the early concepts of Humphry Davy. American chemist Gilbert N. Lewis, who developed the concept of covalent bonds, has referred to this association of acids with hydrogen as "the cult of the proton." Lewis defined acids as electron-pair acceptors and bases as electron-pair donors. The Lewis definitions incorporate all previous theories and definitions of acids and bases. In addition, the Lewis acid-base concept, freed from association with protons, is much broader and explains many more inorganic and organic reactions. In the diagram, each theory of acids is represented by an oval. The larger ovals represent theories that explain an increased number of observations.

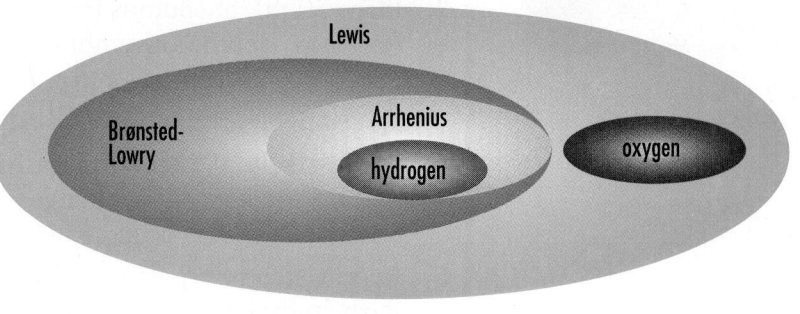

| Problem |
| Prediction |
| Design |
| Materials |
| Procedure |
| ✔ Evidence |
| ✔ Analysis |
| Evaluation |
| Synthesis |

CAUTION

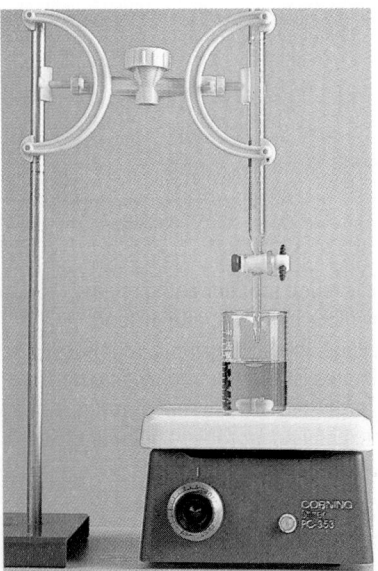

Acids and bases are corrosive and toxic. Avoid skin and eye contact. If you spill any of the chemical solutions on your skin, immediately rinse the area with lots of cool water. In the unlikely situation of getting some of the chemicals in your eye, immediately rinse your eye for at least 15 min and inform your teacher.

Figure 14.19
A buret and magnetic stirrer provide for a very efficient procedure.

INVESTIGATION

14.5 Demonstration: pH Curves

The purpose of this demonstration is to study pH curves and the function of an indicator in an acid-base reaction.

Problem

What are the shapes of the pH curves for the continuous addition of hydrochloric acid to a sample of a sodium hydroxide solution and to a sample of a sodium carbonate solution?

Experimental Design

Small volumes of hydrochloric acid are added continuously to a measured volume of a base. After each addition, the pH of the mixture is measured. The volume of hydrochloric acid is the manipulated variable and the pH of the mixture is the responding variable.

Materials

lab apron
safety glasses
0.10 mol/L $HCl_{(aq)}$
0.10 mol/L $NaOH_{(aq)}$
0.10 mol/L $Na_2CO_{3(aq)}$
bromothymol blue indicator
methyl orange indicator
pH 7 buffer solution for calibration of pH meter
distilled water
pH meter or pH probe or computer interface
magnetic stirrer
50 mL buret and funnel
150 mL beaker
(2) 250 mL beakers
(2) 50 mL graduated cylinders

Procedure

1. Set the temperature on the pH meter and calibrate it by adjusting it to indicate the pH of the known pH 7 buffer solution.

2. Place 50 mL of sodium hydroxide in a 150 mL beaker and add a few drops of bromothymol blue indicator.

3. Measure and record the pH of the 0.10 mol/L sodium hydroxide solution (Figure 14.19).

4. Successively add small quantities of $HCl_{(aq)}$, measuring the pH and noting any color changes after each addition, until about 80 mL of acid has been added.

5. Repeat steps 1 to 4 for 50 mL of 0.10 mol/L sodium carbonate with hydrochloric acid, using a 250 mL beaker and methyl orange indicator. Continue until 130 mL of $HCl_{(aq)}$ has been added.

Interpreting pH Curves

The pH curves for acid-base reactions have characteristic shapes. If the sample is a strong base, such as $NaOH_{(aq)}$, titrated with a strong acid, such as $HCl_{(aq)}$, the initial pH is high because the sample is a base and no acid has yet been added. The final pH is low because an excess of acid has been added (Figure 14.20). If the sample is a strong acid titrated with a strong base, the initial pH is low and the final pH is high because an excess of the strong base has been added (Figure 14.21). An experimentally determined initial pH, equivalence point, and the final pH from a titration may differ from the values obtained from stoichiometry calculations (see Figures 14.20 and 14.22). That is part of the unavoidable uncertainty in any experiment. The initial addition of the titrant (in the buret) to either an acid or base sample does not produce large changes in the pH of the solution. This relatively flat region of a pH curve is where a *buffering action* occurs. This means that the pH is relatively constant even though small amounts of a strong acid or base are being added. This occurs because the first amount of titrant is immediately consumed, leaving an excess of strong acid (or base), and the pH is changed very little.

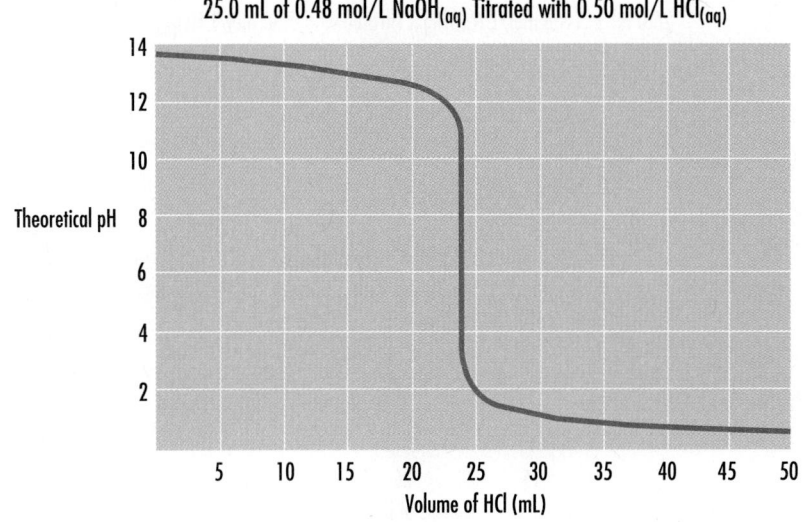

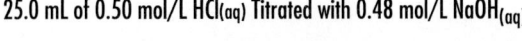

Figure 14.20
This calculated (theoretical) pH curve for the addition of 0.50 mol/L $HCl_{(aq)}$ to a 25.0 mL sample of 0.48 mol/L $NaOH_{(aq)}$ helps chemists to understand the nature of the acid-base reactions. The 0.48 mol/L concentration for $NaOH_{(aq)}$ illustrates an actual situation — sodium hydroxide is not a primary standard.

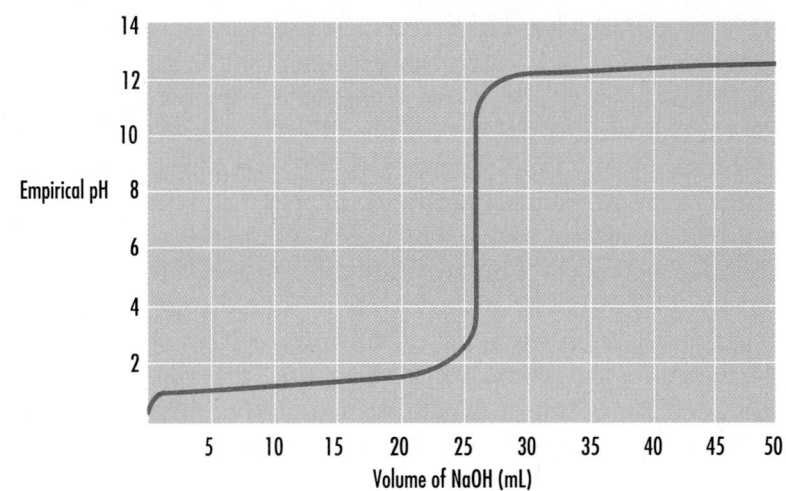

Figure 14.21
This experimentally determined (empirical) pH curve for the addition of 0.48 mol/L $NaOH_{(aq)}$ to a 25.0 mL sample of 0.50 mol/L $HCl_{(aq)}$ illustrates the typical shape of a strong acid/strong base pH curve.

Following this buffering region there is a very rapid change in pH for a very small additional volume of the titrant. For example, in the titration of NaOH$_{(aq)}$ with HCl$_{(aq)}$ (Figures 14.20 and 14.22), the midpoint of the sharp drop in pH occurs at a pH of 7. This is the *endpoint* of the titration (page 277). As you saw in Investigation 14.5, this pH endpoint should occur at the same time as an indicator endpoint, the abrupt change in the color of bromothymol blue. Either endpoint can be used to signal the end of the titration. If a suitable indicator is chosen, the titration can be done efficiently and inexpensively, using an indicator instead of a pH meter. The volume of the HCl titrant at the endpoint is about 24 mL (Figure 14.22); this volume is called the *equivalence point* (page 276). Theoretically, the *equivalence point* represents the stoichiometric quantity of titrant required by the balanced chemical equation. For example, 25 mL of 0.48 mol/L NaOH$_{(aq)}$ would require 24 mL of 0.50 mol/L HCl$_{(aq)}$ titrant for a complete reaction, according to the stoichiometry of this reaction.

$$OH^-_{(aq)} \quad + \quad H_3O^+_{(aq)} \quad \rightarrow \quad 2\,H_2O_{(l)}$$

25 mL × 0.48 mol/L	24 mL × 0.50 mol/L	
12 mmol	12 mmol	

Figure 14.22
Alizarin yellow is an unsuitable indicator for this titration because it changes color before the end of the reaction (pH 7). Orange IV is also unsuitable because it changes color after the end of the reaction. Bromothymol blue is suitable because its endpoint of pH 6.8 (assume the middle of its pH range) closely matches the endpoint of pH 7, and the color change is completely on the vertical portion of the pH curve, within the range where there is rapid change in pH. Note that this pH curve is the experimental (empirical) version of Figure 14.20.

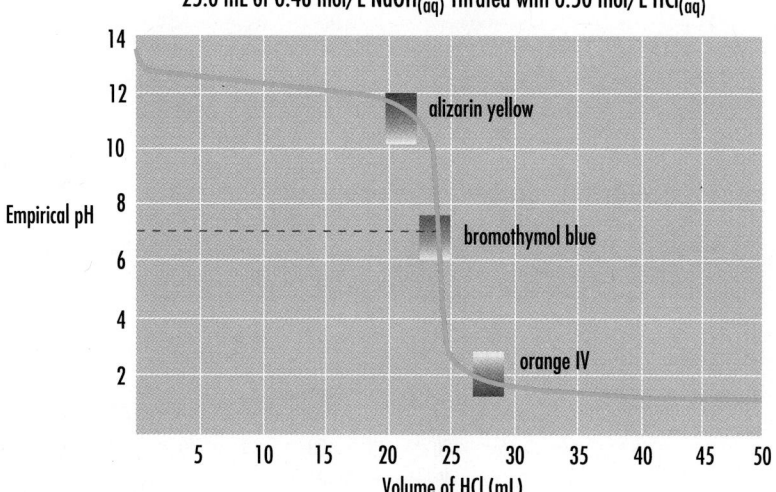

25.0 mL of 0.48 mol/L NaOH$_{(aq)}$ Titrated with 0.50 mol/L HCl$_{(aq)}$

Polyprotic Substances

The pH curve for the titration of sodium hydroxide with hydrochloric acid has only one observable endpoint. According to the Brønsted-Lowry concept, only one reaction has occurred. But the pH curve for the addition of HCl$_{(aq)}$ to Na$_2$CO$_{3(aq)}$ (Figure 14.23) displays two endpoints — two rapid changes in pH. pH curves such as this can be interpreted as indicating the number of quantitative reactions for polyprotic acids or bases. Here, for example, two successive reactions have occurred. The two endpoints in Figure 14.23 can be explained by two different proton transfer equations. First, protons transfer from hydronium ions to carbonate ions, which are the strongest base present in the initial mixture. The pH drops slightly as a CO$_3^{2-}$$_{(aq)}$/HCO$_3^-$$_{(aq)}$ buffer is established, and as carbonate ions are converted to less basic hydrogen carbonate ions.

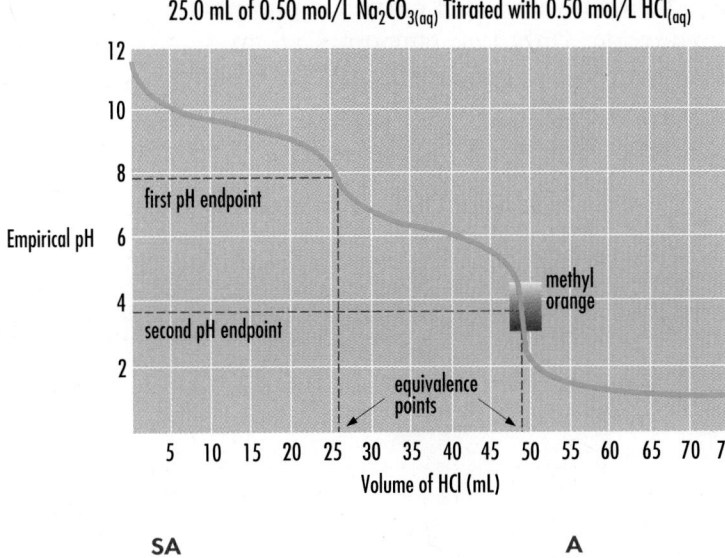

25.0 mL of 0.50 mol/L $Na_2CO_{3(aq)}$ Titrated with 0.50 mol/L $HCl_{(aq)}$

Figure 14.23
A pH curve for the addition of 0.50 mol/L $HCl_{(aq)}$ to a 25.0 mL sample of 0.50 mol/L $Na_2CO_{3(aq)}$ can be used to select an indicator for a titration.

$$\underset{B}{\underset{SA}{H_3O^+_{(aq)}},\ Cl^-_{(aq)},\ \underset{SB}{Na^+_{(aq)}},\ \underset{}{CO_3^{2-}_{(aq)}},\ \underset{B}{\overset{A}{H_2O_{(l)}}}}$$

$$H_3O^+_{(aq)}\ +\ CO_3^{2-}_{(aq)}\ \rightarrow\ H_2O_{(l)}\ +\ HCO_3^-_{(aq)}$$

Then in a second reaction protons transfer from additional hydronium ions to the hydrogen carbonate ions formed in the first reaction.

$$\underset{B}{\underset{SA}{H_3O^+_{(aq)}},\ Cl^-_{(aq)},\ Na^+_{(aq)},\ \underset{B}{\overset{A}{H_2O_{(l)}}},\ \underset{SB}{\overset{A}{HCO_3^-_{(aq)}}}}$$

$$H_3O^+_{(aq)}\ +\ HCO_3^-_{(aq)}\ \rightarrow\ H_2O_{(l)}\ +\ H_2CO_{3(aq)}$$

Notice from observing the pH curve that each reaction requires about 25 mL of hydrochloric acid to reach the endpoint and that the methyl orange color change marks the second endpoint of the titration.

Substances that may donate or accept more than one proton are called **polyprotic**. A carbonate ion is a *polyprotic base* because it can accept a total of two protons. Other polyprotic bases include sulfide ions and phosphate ions.

$$S^{2-}_{(aq)}\ \rightarrow\ HS^-_{(aq)}\ \rightarrow\ H_2S_{(aq)}$$

$$PO_4^{3-}_{(aq)}\ \rightarrow\ HPO_4^{2-}_{(aq)}\ \rightarrow\ H_2PO_4^-_{(aq)}\ \rightarrow\ H_3PO_{4(aq)}$$

Polyprotic acids that can donate more than one proton include oxalic acid and phosphoric acid.

$$HOOCCOOH_{(aq)}\ \rightarrow\ HOOCCOO^-_{(aq)}\ \rightarrow\ OOCCOO^{2-}_{(aq)}$$

$$H_3PO_{4(aq)}\ \rightarrow\ H_2PO_4^-_{(aq)}\ \rightarrow\ HPO_4^{2-}_{(aq)}\ \rightarrow\ PO_4^{3-}_{(aq)}$$

Evidence from pH measurements indicates that polyprotic substances become weaker acids or bases with every proton donated or accepted. This phenomenon is explained by the electrostatic theory that it is easier to donate an H$^+$ from neutral H_3PO_4 than from the negatively charged $H_2PO_4^-$ and the more negatively charged HPO_4^{2-}. According to Le Châtelier's principle, with each successive H$^+$ donation, there is more H$^+$ in solution pushing the reaction back toward the reactants.

Figure 14.24 shows the pH curve for phosphoric acid titrated with sodium hydroxide. Only two endpoints are present, corresponding to equivalence points of 26 mL and 53 mL. At the first equivalence point, equal amounts of $H_3PO_{4(aq)}$ and $OH^-_{(aq)}$ have been added.

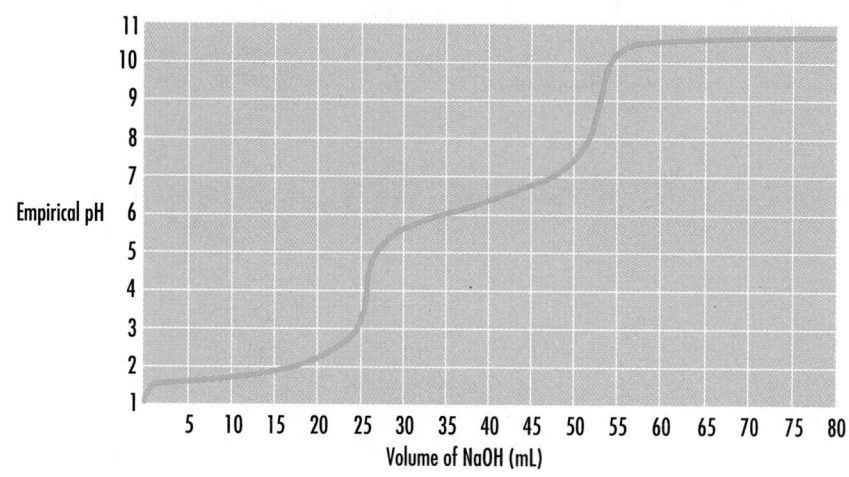

25.0 mL of 0.50 mol/L $H_3PO_{4(aq)}$ Titrated with 0.48 mol/L $NaOH_{(aq)}$

Empirical pH

Volume of NaOH (mL)

$$\begin{array}{cccc} & & \text{SA} & \text{A} \\ Na^+_{(aq)}, & OH^-_{(aq)}, & H_3PO_{4(aq)}, & H_2O_{(l)} \\ & \text{SB} & & \text{B} \end{array}$$

$$OH^-_{(aq)} + H_3PO_{4(aq)} \rightarrow H_2O_{(l)} + H_2PO_4^-_{(aq)}$$

pH Curves

pH curves provide a wealth of information:
- equivalence point(s)
- initial pH of solution
- pH when excess of titrant is added
- number of quantitative reactions
- pH endpoint(s)
- transition point(s) for selecting indicator(s)
- buffering potential

Since all the $H_3PO_{4(aq)}$ has reacted, the second plateau must represent the reaction of $OH^-_{(aq)}$ with $H_2PO_4^-_{(aq)}$. The second equivalence point corresponds to the completion of the reaction of $H_2PO_4^-_{(aq)}$ with an additional 25 mL of $OH^-_{(aq)}$ solution added. No $H_2PO_4^-_{(aq)}$ remains.

$$\begin{array}{ccccc} & & \text{SA} & \text{A} & \text{SA} \\ Na^+_{(aq)}, & OH^-_{(aq)}, & \cancel{H_3PO_{4(aq)}}, & H_2O_{(l)}, & H_2PO_4^-_{(aq)} \\ & \text{SB} & & \text{B} & \text{B} \end{array}$$

$$OH^-_{(aq)} + H_2PO_4^-_{(aq)} \rightarrow H_2O_{(l)} + HPO_4^{2-}_{(aq)}$$

No pH endpoint is apparent at 75 mL for the possible reaction of $HPO_4^{2-}_{(aq)}$ with $OH^-_{(aq)}$. A clue to this missing third endpoint can be obtained from the table of acids and bases, (Appendix F). The hydrogen phosphate ion is an extremely weak acid and apparently does not quantitatively lose its proton to $OH^-_{(aq)}$.

$$\overset{>50\%}{HPO_4^{2-}_{(aq)} + OH^-_{(aq)} \rightleftharpoons PO_4^{3-}_{(aq)} + H_2O_{(l)}}$$

As a general rule, *only quantitative reactions produce detectable endpoints in an acid-base titration.*

Figure 14.24
A pH curve for the addition of 0.48 mol/L $NaOH_{(aq)}$ to a 25.0 mL sample of 0.50 mol/L $H_3PO_{4(aq)}$ displays only two rapid changes in pH. This is interpreted as indicating that there are only two quantitative reactions for phosphoric acid with sodium hydroxide.

32. How is buffering action displayed on a pH curve?

33. How are quantitative reactions displayed on a pH curve?

34. How is a pH curve used to choose an indicator for a titration?

35. An acetic acid sample is titrated with sodium hydroxide (Figure 14.25.
    (a) Based on Figure 14.25, estimate the endpoint and the equivalence point.
    (b) Choose an appropriate indicator for this titration.
    (c) Write a Brønsted-Lowry equation for this reaction.

36. A sodium phosphate solution is titrated with hydrochloric acid (Figure 14.26).
    (a) Why are only two endpoints shown in Figure 14.26?
    (b) Write three Brønsted-Lowry equations for the pH curve in Figure 14.26. Communicate the position of each equilibrium.

37. Oxalic acid reacts quantitatively in a two-step reaction with a sodium hydroxide solution. Assuming that an excess of sodium hydroxide is added, sketch a pH curve (without any numbers) for all possible reactions.

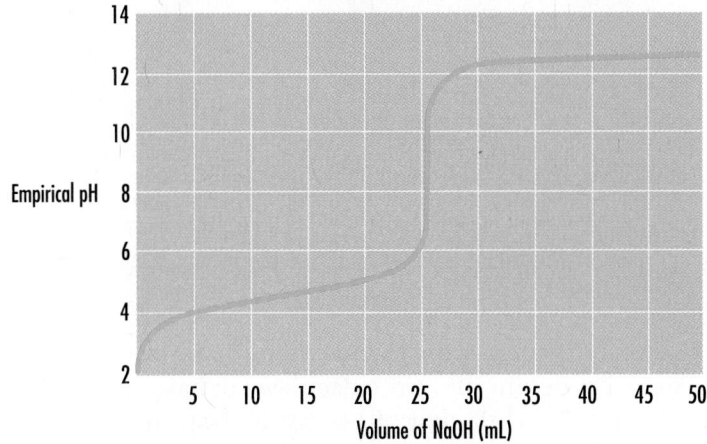

25.0 mL of 0.49 mol/L $CH_3COOH_{(aq)}$ Titrated with 0.48 mol/L $NaOH_{(aq)}$

Figure 14.25
The pH curve for the addition of 0.48 mol/L $NaOH_{(aq)}$ to 25.0 mL of 0.49 mol/L $CH_3COOH_{(aq)}$ illustrates pH changes during the reaction of a weak acid with a strong base.

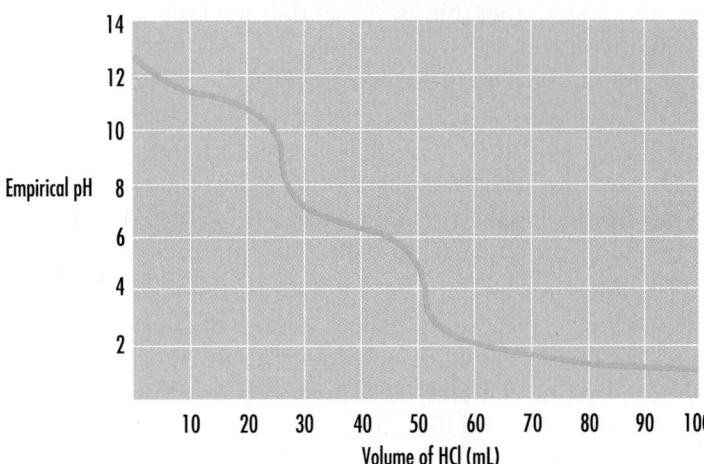

25.0 mL of 0.51 mol/L $Na_3PO_{4(aq)}$ Titrated with 0.50 mol/L $HCl_{(aq)}$

Figure 14.26
The pH curve for the addition of $HCl_{(aq)}$ to $Na_3PO_{4(aq)}$ can be interpreted using Brønsted-Lowry acid-base theory.

Weak Acid and Base pH Curves

Quantitative acid-base reactions generally involve the strongest acid in an aqueous solution ($H_3O^+_{(aq)}$) and the strongest base in an aqueous solution ($OH^-_{(aq)}$). The pH curves for these reactions are displayed in Figure 14.21 (page 569) and Figure 14.22 (page 570). The pH often changes by 6–10 pH units — a change in $[H^+_{(aq)}]$ by a factor of a remarkable one million (10^6) to ten billion (10^{10}). However, not all titrations involve the hydronium and the hydroxide ions.

pH curves show that some weak acids (e.g., acetic acid) with a $OH^-_{(aq)}$ solution produce a quantitative reaction (Figure 14.25, page 573). The reactions of some weak bases (e.g., $PO_4^{3-}_{(aq)}$ and $HPO_4^{2-}_{(aq)}$) with $H_3O^+_{(aq)}$ produce a quantitative reaction (Figure 14.26, page 573). The strongest of the weak acids also react quantitatively with the strongest of the weak bases, e.g., $HSO_4^-_{(aq)}$ with $CO_3^{2-}_{(aq)}$ is a quantitative reaction. The further apart the reacting acid and base are on the acids and bases table (i.e., the larger the difference in K_a), the more likely the reaction will be quantitative. The result of plotting many pH curves gives the general pH curves shown in Figure 14.27.

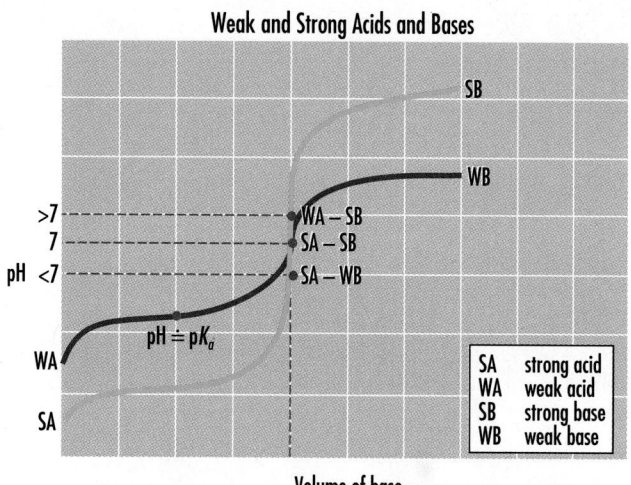

Figure 14.27
Laboratory evidence is generalized here to show the reactions of weak and strong acids and bases.

The pH endpoint for a strong acid–strong base (SA-SB) reaction is seven (7). The pH endpoint of a strong acid–weak base (SA-WB) reaction is less than seven (<7), while that of a weak acid–strong base (WA-SB) is greater than seven (>7). If weak acid–weak base reactions have a detectable endpoint, the pH is close to seven. The weaker the acid and/or base, the less detectable is the endpoint, and the less likely the reaction is quantitative.

In summary, a typical indicator chosen for a SA-SB reaction is bromothymol blue (pK_a = 7), while methyl orange (pK_a = 4) may be chosen for SA-WB reactions, and phenolphthalein (pK_a = 10) for WA-SB reactions. The initial and final pH values are approximately equal to the pK_a values of any weak acids and conjugate acids of any weak bases involved in the reactions. The pH at each of the endpoints in Figure 14.27 is determined only by the hydrolysis of the salt product at the equivalence point (where no excess reactants are present). The salt product that dominates the solution properties at the equivalence point can be predicted from the five-step Brønsted-Lowry method or from formula, complete ionic, and net ionic equations.

EXAMPLE

What is the relative pH at the equivalence points of SA-SB, SA-WB, and WA-SB reactions?

Strong Acid–Strong Base

    SA   **A**

$H_3O^+_{(aq)}$, $NO_3^-_{(aq)}$, $K^+_{(aq)}$, $OH^-_{(aq)}$, $H_2O_{(l)}$
        B   **SB**   **B**

$$H_3O^+_{(aq)} + OH^-_{(aq)} \rightarrow 2\,H_2O_{(l)}$$

or

$$HNO_{3(aq)} + KOH_{(aq)} \rightarrow H_2O_{(l)} + KNO_{3(aq)}$$

$$H^+_{(aq)} + \cancel{NO_{3\,(aq)}^-} + \cancel{K^+_{(aq)}} + OH^-_{(aq)} \rightarrow H_2O_{(l)} + \cancel{K^+_{(aq)}} + \cancel{NO_{3\,(aq)}^-}$$

$$H^+_{(aq)} + OH^-_{(aq)} \rightarrow H_2O_{(l)}$$

At the equivalence point, the hydrolysis of $KNO_{3(aq)}$ produces a neutral solution (pH = 7).

Strong Acid–Weak Base

    SA   **A**

$H_3O^+_{(aq)}$, $Cl^-_{(aq)}$, $NH_{3(aq)}$, $H_2O_{(l)}$
        B   **SB**   **B**

$$H_3O^+_{(aq)} + NH_{3(aq)} \rightarrow H_2O_{(l)} + NH_4^+_{(aq)}$$

or

$$HCl_{(aq)} + NH_{3(aq)} \rightarrow NH_4Cl_{(aq)}$$

$$H^+_{(aq)} + \cancel{Cl^-_{(aq)}} + NH_{3(aq)} \rightarrow NH_4^+_{(aq)} + \cancel{Cl^-_{(aq)}}$$

$$H^+_{(aq)} + NH_{3(aq)} \rightarrow NH_4^+_{(aq)}$$

At the equivalence point, the hydrolysis of $NH_4Cl_{(aq)}$ produces an acidic solution (pH <7).

Weak Acid–Strong Base

    SA   **A**

$CH_3COOH_{(aq)}$, $Ba^{2+}_{(aq)}$, $OH^-_{(aq)}$, $H_2O_{(l)}$
                SB   **B**

$$CH_3COOH_{(aq)} + OH^-_{(aq)} \rightarrow H_2O_{(l)} + CH_3COO^-_{(aq)}$$

or

$$2\,CH_3COOH_{(aq)} + Ba(OH)_{2(aq)} \rightarrow 2\,H_2O_{(l)} + Ba(CH_3COO)_{2(aq)}$$

$$2\,CH_3COOH_{(aq)} + \cancel{Ba^{2+}_{(aq)}} + 2\,OH^-_{(aq)} \rightarrow$$
$$2\,H_2O_{(l)} + \cancel{Ba^{2+}_{(aq)}} + 2\,CH_3COO^-_{(aq)}$$

$$CH_3COOH_{(aq)} + OH^-_{(aq)} \rightarrow H_2O_{(l)} + CH_3COO^-_{(aq)}$$

At the equivalence point, the hydrolysis of $Ba(CH_3COO)_{2(aq)}$ produces a basic solution (pH >7).

> **Net Ionic Equations**
> Entities common to both sides of the equation are cancelled out and the coefficients of the remaining entities are reduced to the simplest ratio in a net ionic equation.

Summary:

- Strong acid–strong base reactions are quantitative (100%) and have an endpoint with a pH = 7.
- pH curves provide evidence that many other reactions are quantitative.
- Strong acid–weak base quantitative reaction endpoints have a pH <7.
- Weak acid–strong base quantitative reaction endpoints have a pH >7.
- Weak acid–weak base quantitative reaction endpoints have a pH $\doteq$ 7.
- Hydrolysis determines the pH endpoints of strong/weak quantitative reactions.
- The five-step Brønsted-Lowry method or the formula and complete ionic method may be used to determine the net ionic equation for an acid-base reaction.

Exercise

38. Predict whether the pH endpoint is approximately 7 ($\doteq$7), greater than 7 (>7), or less than 7 (<7) for each of the following acid-base titrations (for the first quantitative reaction only).
    (a) hydroiodic acid + aqueous sodium hydrogen phosphate →
    (b) boric acid + aqueous sodium hydroxide →
    (c) aqueous sodium hydrogen sulfate + aqueous potassium hydroxide →
    (d) hydrochloric acid + solid magnesium hydroxide →
    (e) hydrosulfuric acid + aqueous sodium hydrogen carbonate →
    (f) sulfuric acid + aqueous ammonia →

39. Specific pH curves are the best method for choosing indicators for acid-base titrations. Use a combination of indicator K_as, hydrolysis, and/or general pH curves to choose methyl orange, bromothymol blue, or phenolphthalein for each of the following acid-base titrations (for the first quantitative reaction only).
    (a) $HBr_{(aq)} + Ca(OH)_{2(s)} \rightarrow$
    (b) $HNO_{3(aq)} + Na_2CO_{3(aq)} \rightarrow$
    (c) $KOH_{(aq)} + HNO_{2(aq)} \rightarrow$

40. Predict whether the pH endpoint is $\doteq$ 7, >7, or <7 for each of the following salts produced at the equivalence point.
    (a) $NH_4Cl_{(aq)}$
    (b) $Na_2S_{(aq)}$
    (c) $KNO_{3(aq)}$
    (d) $NaHSO_{4(aq)}$

41. Use the five-step Brønsted-Lowry method to predict the net ionic equation when the following chemicals are mixed.
    (a) solutions of perchloric acid and sodium carbonate. (A pH curve shows two protons are transferred quantitatively.)
    (b) solutions of nitrous acid and potassium hydroxide.
    (c) solutions of phosphoric acid and sodium hydroxide. (A pH curve shows two protons are transferred quantitatively.)

42. Write the formula, complete ionic, and net ionic equations for the reaction that occurs when the following chemicals are mixed.
   (a) Stomach acid is neutralized by solid magnesium hydroxide.
   (b) Aqueous ammonia is added to sulfurous acid.
   (c) A solution of sulfuric acid is neutralized by sodium hydroxide. (A pH curve indicates two quantitative reactions.)

43. Write a net ionic equation to represent the following acid-base reactions. (Use the method that you find convenient or that your teacher prefers.)
   (a) Oxalic acid is titrated with aqueous sodium hydroxide.
   (b) Sodium phosphate is titrated with hydrochloric acid. (The titration is stopped at the first equivalence point.)
   (c) Sodium hydrogen phosphate is titrated with hydrochloric acid. (An indicator is chosen for the first endpoint.)
   (d) Nitric acid is titrated with aqueous barium hydroxide.
   (e) A sulfuric acid spill is neutralized by adding excess lye.

44. Write a formula (non-ionic) equation to represent the following acid-base reactions.
   (a) Acetic acid is titrated with aqueous sodium hydroxide.
   (b) Nitrous acid is titrated with aqueous barium hydroxide.
   (c) Sodium carbonate is titrated with hydrobromic acid. (The titration is stopped after the first quantitative reaction.)
   (d) Carbonic acid is titrated with aqueous sodium hydroxide. (An indicator is used to stop the reaction with stoichiometric ratio of 1:1.)
   (e) A sulfuric acid spill is neutralized by adding excess lye (caustic soda).

Equations for Titrations
In an equation for a titration, a single arrow, $\rightarrow$, is used. The reaction represented by the equation must be quantitative so that stoichiometric calculations based on the equation are valid. In such cases, formula equations are often the most convenient form.

Indicator Transition Point, K_a, and pK_a

The pH at which an indicator changes color is called the **transition point**. The transition point is somewhere within the transition range referenced on the acid-base indicators table on the inside back cover of this text. This table tells you, for example, that the transition point for methyl orange is somewhere within the pH range of 3.2–4.4. At the transition point, the [In$^-$] = [HIn] and the color change is occurring.

According to the table of relative strengths of acids and bases (Appendix F), the K_a for methyl orange is approximately 10^{-4} mol/L. The pH at transition point is approximately 4. Similarly for bromothymol blue, the K_a is approximately 10^{-7} mol/L, and the transition point is approximately a pH of 7. Generally then, *the approximate pH at the transition point of an indicator is the exponential value of the K_a for the indicator.*

pK_a of Acids

It is often convenient to use a pK_a to communicate the K_a values of acids and categories of acids such as indicators. The pK_a is defined as the negative log of the K_a, not unlike the pH definition you learned in Chapter 13.

$$pH = -\log[H^+_{(aq)}]$$
$$pK_a = -\log K_a$$
$$\text{or } K_a = 10^{-pK_a}$$

The pK_a is useful for communicating generalizations such as those presented below.

- The pK_a of an acceptable indicator is equal to the pH of the indicator endpoint for a quantitative reaction (e.g., titration).
- The pH at the indicator endpoint is equal to the pK_a of the indicator (Figure 14.28).

Mathematical proofs for these generalizations are provided in post-secondary chemistry textbooks.

Another generalization and its corollary that are efficiently communicated by using pK_a are provided (without proof) below.

- The pK_a of an acid or base being titrated is equal to the pH on the mid-point of the buffering (reaction) plateau.
- The pH mid-point of a buffering (reaction) plateau during titration is equal to the pK_a of the acid or base being titrated (see Figure 14.28).

These generalizations make it easy to interpret and to predict pH curves for a wide variety of reactions.

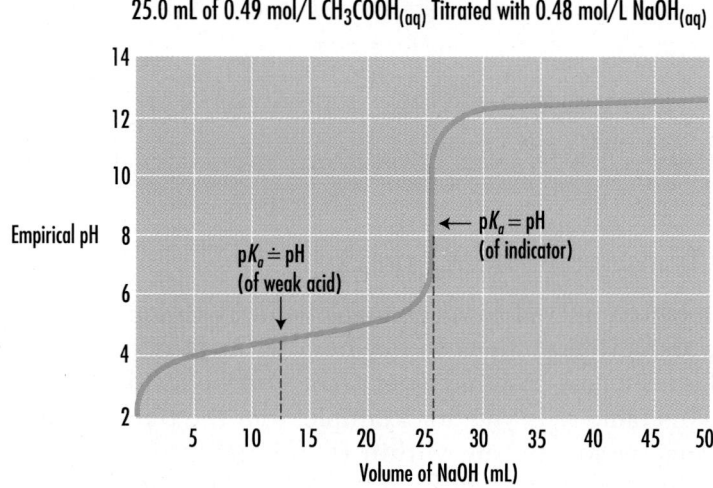

25.0 mL of 0.49 mol/L $CH_3COOH_{(aq)}$ Titrated with 0.48 mol/L $NaOH_{(aq)}$

Figure 14.28
The pH provides information about the strength of the initial reagent, the indicator, and the excess reagent.

Buffers

All pH curves involving a weak acid or weak base have at least one region where a buffering action occurs. The curves in these relatively constant pH regions are most nearly horizontal at a volume of titrant which is one-half the first equivalence point or halfway between successive equivalence points for polyprotic substances. The solution mixture present near these points has a special significance and is known as a **buffer solution** or **buffer**. For example, in the titration of acetic acid with sodium hydroxide (Figure 14.25, page 573), the pH is approximately 4.7 at a volume of 12.5 mL of sodium hydroxide. Since one-half of the equivalence volume has been added, one-half of the original acetic acid has reacted.

$$OH^-_{(aq)} + CH_3COOH_{(aq)} \rightarrow H_2O_{(l)} + CH_3COO^-_{(aq)}$$

The mixture in this buffering region contains approximately equal amounts of the unreacted weak acid, CH_3COOH, and its conjugate base, CH_3COO^-, produced in the reaction. A **buffer** is a mixture of a weak acid and its conjugate base. A buffer has the unique ability to maintain a nearly constant pH when small amounts of a strong acid or base are added.

Buffering action can be explained using Brønsted-Lowry equations. Suppose a small amount of $NaOH_{(aq)}$ is added to the acetic acid–acetate ion buffer described above. Using the five-step method for predicting acid-base reactions (page 564), the following equation is obtained.

<table>
<tr><td></td><td>SA</td><td></td><td>A</td><td></td></tr>
<tr><td>$Na^+_{(aq)}$, $OH^-_{(aq)}$,</td><td>$CH_3COOH_{(aq)}$,</td><td>$CH_3COO^-_{(aq)}$,</td><td>$H_2O_{(l)}$,</td></tr>
<tr><td>SB</td><td></td><td>B</td><td>B</td></tr>
</table>

$$OH^-_{(aq)} + CH_3COOH_{(aq)} \rightarrow H_2O_{(l)} + CH_3COO^-_{(aq)}$$

Figure 14.25 shows that this reaction is quantitative. A small amount of OH^- would convert a small amount of acetic acid to acetate ions. The overall effect is a small decrease in the ratio of acetic acid to acetate ions in the buffer and a slight increase in the pH (Figure 14.29(a)). This small change and the consumption of the added hydroxide ions in the process explains why the pH change is small. This buffer would work equally well if a small amount of a strong acid, such as $HCl_{(aq)}$, were added quantitatively to the buffer.

<table>
<tr><td>SA</td><td></td><td>A</td><td></td><td>A</td><td></td></tr>
<tr><td>$H_3O^+_{(aq)}$, $Cl^-_{(aq)}$,</td><td>$CH_3COOH_{(aq)}$,</td><td>$CH_3COO^-_{(aq)}$,</td><td>$H_2O_{(l)}$,</td></tr>
<tr><td>B</td><td></td><td>B</td><td>B</td></tr>
</table>

$$H_3O^+_{(aq)} + CH_3COO^-_{(aq)} \rightarrow H_2O_{(l)} + CH_3COOH_{(aq)}$$

The hydronium ion is consumed and the mixture now has a slightly higher ratio of acetic acid to acetate ions and a slightly lower pH (Figure 14.29(b)).

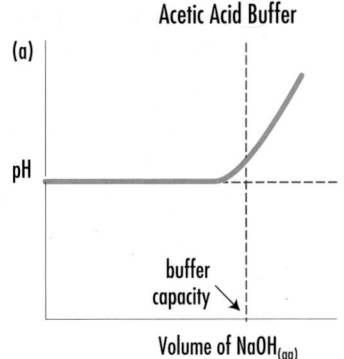

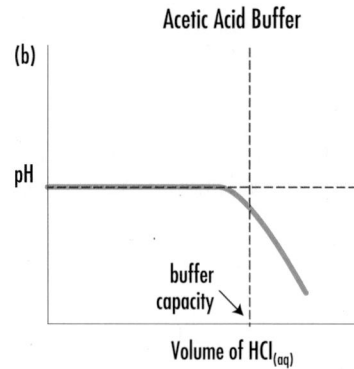

Figure 14.29
A batch amount of a buffer is eventually depleted and the pH changes quickly.

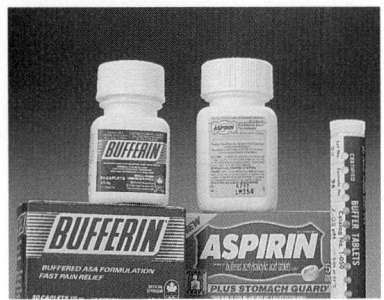

Figure 14.30
Many consumer and commercial products contain buffers. Buffered aspirin is a well-known example. Blood plasma and capsules for making buffer solutions (for example, to calibrate pH meters) are commercial examples of buffers.

The ability of buffers to maintain a relatively constant pH is important in many biological processes where certain chemical reactions occur at a specific pH value. Many aspects of cell functions and metabolism in living organisms are very sensitive to pH changes. For example, each enzyme carries out its function optimally over a small pH range. One important buffer within living cells is the conjugate acid-base pair, $H_2PO_4^-{}_{(aq)} - HPO_4^{2-}{}_{(aq)}$. The major buffer system in the blood and other body fluids is the conjugate acid-base pair, $H_2CO_3{}_{(aq)} - HCO_3^-{}_{(aq)}$. Blood plasma has a remarkable buffering ability, as shown by the empirical results in Table 14.3. Since the production of buffer is a *continuous* rather than a *batch* process, the buffer is not a limiting reagent.

Table 14.3

BUFFERING ACTION OF NEUTRAL SALINE SOLUTION AND OF BLOOD PLASMA		
Solution (1.0 L)	Initial pH of Mixture	Final pH after Adding 1 mL of 10 mol/L HCl
neutral saline	7.0	2.0
blood plasma	7.4	7.2

Human blood plasma normally has a pH of about 7.4. Any change of more than 0.4 pH units, induced by poisoning or disease, can be lethal. If the blood were not buffered, the acid absorbed from a glass of orange juice would probably be fatal.

Buffers are also important in many consumer, commercial, and industrial applications (Figure 14.30). Fermentation and the manufacture of antibiotics require buffering to optimize yields and to avoid undesirable side reactions. The production of various cheeses, yogurt, and sour cream are very dependent on controlling pH levels, since an optimum pH is needed to control the growth of micro-organisms and to allow enzymes to catalyze fermentation processes. Sodium nitrite and vinegar are widely used to preserve food; part of their function is to prevent the fermentation that takes place only at certain pH values.

The *CRC Handbook of Chemistry and Physics* provides recipes for preparing buffer solutions. For example, a buffer with a pH of 10.00 can be prepared by mixing 50 mL of 0.050 mol/L $NaHCO_3{}_{(aq)}$ with 10.7 mL of 0.10 mol/L $NaOH_{(aq)}$. The reaction to establish the buffer is described in the same way as all other acid-base reactions. The pH curve in Figure 14.23 (page 571) provides evidence that this reaction is quantitative.

Brønsted-Lowry Is a Unifying Concept

The five-step Brønsted-Lowry method to explain and predict acid-base reactions is a preferred, acceptable method because it unifies all quantitative and non-quantitative reactions.

- neutralization reactions
- hydrolysis reactions
- indicator reactions
- polyprotic reactions
- buffer reactions
- excess reactions
- titration reactions

$$\underset{\substack{\quad\quad B \quad\quad\quad SB \quad\quad B}}{Na^+{}_{(aq)}, \overset{SA}{HCO_3^-{}_{(aq)}}, \overset{}{OH^-{}_{(aq)}}, \overset{A}{H_2O_{(l)}}}$$

$$HCO_3^-{}_{(aq)} + OH^-{}_{(aq)} \rightarrow H_2O_{(l)} + CO_3^{2-}{}_{(aq)}$$

MEDICAL LABORATORY TECHNOLOGIST

"When I was in junior high I went to an open house at a technical college. The med lab displays were fascinating, and I knew then I wanted to be in this field."

Kathleen Kaminsky has been a medical laboratory technologist since 1984, but she discusses her work with an enthusiasm that makes it seem as though this is her first week on the job. She says that she had already decided in grade six that she would pursue medical laboratory work as a career. She was interested in math and science and found the science labs particularly intriguing. After graduating from high school, Kaminsky enrolled in post-secondary training at a Canadian technology institute. She took an intensive ten-month program, featuring courses with titles such as "Coagulation" and "Immunohistology," followed by twelve months of practical study in an affiliated training hospital. Completion of this program — both the coursework and the field placement — prepares a candidate to take examinations set by the Canadian Society of Laboratory Technologists in order to obtain national certification as a Registered Medical Laboratory Technologist.

Following her training, Kaminsky worked in private laboratories for two years, then moved to a major city hospital laboratory. Her responsibilities involve almost every type of chemical technology — from gas chromatograph analysis for alcohols and ketones, to phase-shift polarized microscope identification of antithyroid antibodies. Blood samples undergo a wide range of tests, from simple sugar analysis to the separation of individual proteins, a process called serum protein electrophoresis.

Kaminsky says, "The best part of my job is the unpredictability. Every day is different. I enjoy the independence of setting my own priorities, and the challenges, too. This is highly exacting work, with little margin for error, because the results are of critical importance to the people concerned." Another aspect of her work that Kaminsky appreciates is the development and change brought about by new technology, applied to both equipment and techniques. "In this job, you are always learning new things. We have been doing a lot of HDL (high density lipoprotein) analysis lately, for example, as doctors and patients become more aware of the role played by cholesterol in heart disease."

One new technology that is most welcome is equipment that continuously analyzes and oxygenates blood as it circulates in infants whose lungs are not sufficiently developed to function well on their own. This condition among newborns is called respiratory distress syndrome. Kaminsky says, "We used to lose many of these babies; now we save most of them."

Hospital medical laboratories operate continuously, so Kaminsky's job includes 12-hour shifts from 7 to 7, sometimes all day and sometimes all night. In this shift system, she works an average of 3.5 days each week, at various times, so that four- to six-day breaks are built into her schedule.

Usually, medical lab technologists enjoy standard job benefits: sick leave, long-term disability, medical insurance, and dental plans. Salaries are good — in the middle-income range — and pay increases with experience and greater responsibility. Kaminsky feels that if you have an interest in science, especially lab work, then medical lab technology can offer great personal satisfaction and rewards. "It's definitely demanding, both physically and mentally. You can be on your feet for most of a shift, but you still have to be alert and careful at all times. But even after a really hard day, I always go home knowing I've done something worthwhile.

- *What training and experience does Kathleen Kaminsky have?*
- *What difficulties and challenges does she face in her job?*

Since the sodium hydroxide is a limiting reagent, the resulting buffer solution contains only the weak acid–weak base conjugate pair, $HCO_3^-_{(aq)}$ — $CO_3^{2-}_{(aq)}$. From the perspective of hydrolysis, we can explain why this buffer is basic; the K_b for the strongest base ($CO_3^{2-}_{(aq)}$) is larger than the K_a for the strongest acid ($HCO_3^-_{(aq)}$).

A quick look at the acid-base table also tells you that the buffer is basic; the $HCO_3^-_{(aq)}$ — $CO_3^{2-}_{(aq)}$ conjugate pair is closer to the bottom (strong base) part of the table. Determining the actual pH of the buffer described above is beyond the scope of this textbook.

According to Le Châtelier's principle, the equilibrium between the $HCO_3^-_{(aq)}$ — $CO_3^{2-}_{(aq)}$ conjugate pair is shifted by the addition of $H_3O^+_{(aq)}$ or $OH^-_{(aq)}$.

$$HCO_3^-_{(aq)} \rightleftharpoons H^+_{(aq)} + CO_3^{2-}_{(aq)}$$

or $$HCO_3^-_{(aq)} + H_2O_{(l)} \rightleftharpoons H_3O^+_{(aq)} + CO_3^{2-}_{(aq)}$$

Adding $HCl_{(aq)}$ shifts the equilibrium to the left, converts $CO_3^{2-}_{(aq)}$ to $HCO_3^-_{(aq)}$, and decreases the pH — as expected, but only slightly. Adding $NaOH_{(aq)}$ to a sample of the same buffer shifts the equilibrium to the right, converts more $HCO_3^-_{(aq)}$ to $CO_3^{2-}_{(aq)}$, and increases the pH slightly (Figure 14.29). Note that Le Châtelier's principle may be used effectively as an alternative to the five-step, Brønsted-Lowry method for predicting buffer reactions.

Exercise

45. Give an empirical definition of a buffer.

46. List two buffers that help maintain a normal pH level in your body.

47. Use the five-step method to predict the quantitative reaction of a carbonic acid–hydrogen carbonate ion buffer
    (a) when a small amount of $HCl_{(aq)}$ is added.
    (b) when a small amount of $NaOH_{(aq)}$ is added.

48. What happens if a large amount of a strong acid or base is added to a buffer?

49. Use Le Châtelier's principle to predict what will happen to an acetic acid–acetate ion buffer
    (a) when a small amount of $HCl_{(aq)}$ is added.
    (b) when a small amount of $NaOH_{(aq)}$ is added.

50. Which of the following buffers are acidic, basic, or near neutral, and what is their order from lowest to highest pH?
    (a) hydrogen phosphate ion — phosphate ion
    (b) $HCOOH_{(aq)}$ — $HCOO^-_{(aq)}$
    (c) $H_2CO_{3(aq)}$ — $HCO_3^-_{(aq)}$
    (d) hydrogen sulfite ion — sulfite ion

51. Which of the following solution pairs will not form an effective buffer?
    (a) $HNO_{3(aq)}$ and $NaNO_{3(aq)}$
    (b) $NH_{3(aq)}$ and $NH_4Cl_{(aq)}$
    (c) $HCl_{(aq)}$ and $NaOH_{(aq)}$
    (d) $C_6H_5COOH_{(aq)}$ and $NaC_6H_5COO_{(aq)}$

INVESTIGATION

14.6 Buffers

The purpose of this investigation is to test our concept of buffers. Write the Experimental Design, Materials, and table of evidence to match the Procedure that is provided. The Materials list should include the size of the equipment used. The buffer is prepared by a reaction communicated by the following chemical equation.

$$H_2PO_4^-{}_{(aq)} + OH^-{}_{(aq)} \rightarrow HPO_4^{2-}{}_{(aq)} + H_2O_{(l)}$$

excess   limiting   (base part
(acid part   reagent   of buffer)
of buffer)

☐	Problem
✔	Prediction
✔	Design
✔	Materials
☐	Procedure
✔	Evidence
✔	Analysis
✔	Evaluation
☐	Synthesis

Problem

How does the pH change when a strong acid and a strong base is slowly added to a $H_2PO_4^-{}_{(aq)}$ — $HPO_4^{2-}{}_{(aq)}$ buffer?

Procedure

1. Obtain 50mL of 0.10 mol/L $KH_2PO_{4(aq)}$ and 29 mL of 0.10 mol/L $NaOH_{(aq)}$ in separate graduated cylinders.

2. Pour the $KH_2PO_{4(aq)}$ and then the $NaOH_{(aq)}$ into a beaker to prepare a buffer with a pH of 7.

3. Pour an equal amount of the buffer into two test tubes.

4. Add 0.10 mol/L $NaCl_{(aq)}$ as a control into a third and a fourth test tube.

5. Add two drops of bromocresol green to one buffer test tube and one control test tube.

6. Add and count drops of 0.10 mol/L $HCl_{(aq)}$ until the color changes.

7. Repeat steps 5 and 6 with phenolphthalein and 0.10 mol/L $NaOH_{(aq)}$.

8. Dispose of all solutions down the drain with running water.

CAUTION

Acids and bases are corrosive and toxic. Avoid skin and eye contact. If you spill any of the chemical solutions on your skin, immediately rinse the area with lots of cool water. In the unlikely situation of getting some of the chemicals in your eye, immediately rinse your eye for at least 15 min and inform your teacher.

CASE STUDY

Acid Deposition

For the past twenty years, acid deposition has received a great deal of attention in the media, where it is commonly called "acid rain" (see page 179). Acid deposition actually includes any form of acid precipitation (rain, snow, or hail) and condensation from acid fog, as well as acid dust from dry air. The study of acid deposition highlights some important aspects of the nature of science, the nature of technology, and the interaction of science, technology, and society.

Science of Acid Deposition

Acid deposition research illustrates both the empirical basis of scientific knowledge and the uncertainty associated with that knowledge.

Although natural emissions (from volcanos, lightning, and microbial action) contribute to acid deposition, it seems clear that its primary source is human activity. Empirical work indicates that the main causes of acid deposition in North America are sulfur dioxide, SO_2, and nitrogen oxides, NO_x. The major sources of SO_2 emissions in North America are coal-fired power generating stations and non-ferrous ore smelters. When coal is burned in power stations, sulfur in the coal is oxidized to SO_2. The roasting of sulfide ores in smelters also produces SO_2. In the atmosphere, SO_2 reacts with water to produce sulfurous acid, $H_2SO_{3(aq)}$, or is further oxidized to sulfuric acid, $H_2SO_{4(aq)}$. Because nitrogen oxides are produced whenever fuel is burned at high temperature, the main source of NO_x is motor vehicle emissions. At the high temperatures of combustion reactions, the nitrogen and oxygen present in the air combine to form a variety of nitrogen oxides, which produce nitrous and nitric acid when they react with atmospheric water and oxygen.

Sulfuric and sulfurous acids cause considerable environmental damage when they fall to Earth in the form of acid deposition. Experiments indicate that virtually anything that the acids contact (soil, water, plants, and structural materials) is affected to some degree (Figure 14.31). Scientists have repeatedly shown that acid deposition has increased the acidity of some lakes and streams to the point where aquatic life is depleted and waterfowl populations are threatened. Some environmental groups claim that 14 000 Canadian lakes have been damaged by acid deposition. Apparently, the greatest damage is done to lakes that are poorly buffered. When natural alkaline buffers such as limestone are present, they neutralize the acidic compounds from acid deposition. However, lakes lying on granitic strata are susceptible to immediate damage because acids cause metal ions to go into solution in a process called leaching. Especially harmful are cadmium, mercury,

Acid Rain
Normal rain is slightly acidic, due mostly to the presence of naturally occurring carbon dioxide in the atmosphere. The precipitation dissolves some of the CO_2, and the weak acid, H_2CO_3, is formed. With no other substances present from human activity, normal rainfall has a pH at or above 5.6.

Acid Deposition
Acid deposition is caused by nonmetal oxides and neutralized by metal oxides. The formula NO_x represents several oxides of nitrogen, including $N_2O_{(g)}$, $NO_{2(g)}$, and $N_2O_{4(g)}$. Likewise, SO_x represents sulfur oxides including $SO_{2(g)}$ and $SO_{3(g)}$, and CO_x represents gases like $CO_{2(g)}$ and $CO_{(g)}$.

Sulfuric acid is a component of acid rain.

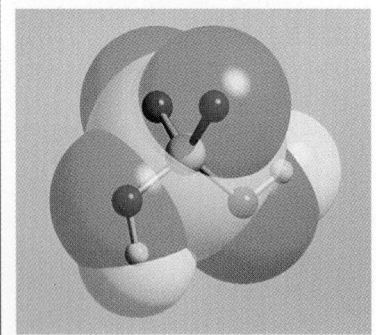

Figure 14.31
Acid rain has damaged this stone sculpture in Krakow, Poland. Krakow, a World Heritage Site, has been on the United Nations Immediate Attention List since 1973.

lead, arsenic, aluminum, and chromium, because they are toxic to living organisms. For example, aluminum is highly toxic to fish because it impairs gill function. Acid rain may also increase human exposure — through food and drinking water — to the metals listed above.

There is some controversy regarding the interpretation of acid deposition research. A 1988 U.S. federal task force on the effects of acid deposition concluded that relatively few lakes have been damaged and that further harm is unlikely to occur over the next few decades. This task force maintained that damage to human health, crops, and forests by acid deposition has yet to be proven. Critics of the report argue that the criteria used to define acid lakes were inadequate and that data were used selectively to support the predetermined conclusions of the report.

Acid deposition is also suspected as one of the causes of forest decline, particularly in forests at high altitudes and colder latitudes. Evidence continues to accumulate that acid deposition is causing serious harm to forests throughout the Northern hemisphere. The Black Forest of Germany has been particularly hard hit. Some observers contend that many forests receive as much as 30 times more acid than they would if rain fell through clean air. Damage to trees includes yellowing, premature loss of needles, and eventual death. Studies of tree rings reveal that tree growth is suppressed in regions that are prone to acid deposition. Some research indicates that, as the concentration of trace metal ions increases, ring growth decreases. Research also suggests that acid deposition damages the needles and leaves of trees, cutting down carbohydrate production.

Disputes over the effect of acid rain on forests highlight the wide range of interpretation of the empirical data. A small minority of scientists insists that there is currently no direct evidence linking acid deposition to elevated tree mortality rates and to decreases in the ring widths of tree trunks. They cite evidence suggesting that the reduction in the growth rate of trees at high altitudes and latitudes may be more directly related to a reduction in mean annual temperatures in those regions. These researchers point out that growth ring data correspond to fluctuations in mean annual temperatures over the last century.

Fish Warning

People of New Brunswick were recently warned to reduce the amount of freshwater fish they eat to between 6 and 22 meals per year. (Saltwater fish or those grown in aquaculture were not considered a risk to human health.) Pregnant women, nursing mothers, and children under eight years of age were advised not to eat any freshwater sport fish at all. The reason is that freshwater trout, bass, and land-locked salmon have been found to have higher than acceptable levels of mercury in their tissues. Scientists blame acid rain. They say that because of the increased acidity of lakewater, toxic metal salts in the lake sediments dissolved more readily. The fish then accumulated the metal as they passed the water through their gills. This was the first time that people in this area had been warned about possible contamination in fish, but it may not be the last. Nova Scotia was also testing the fish in its lakes and rivers to see if a similar warning was needed there.

Suspected damage from acid rain can be replicated in controlled studies in the laboratory.

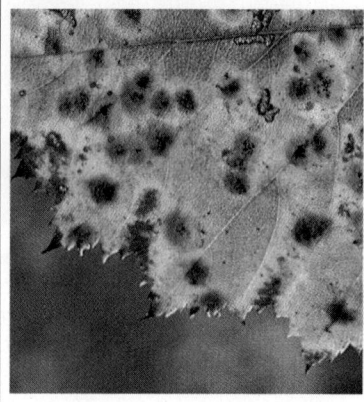

Some researchers report that acid deposition painted on seedlings in soil with inadequate nutrients actually has a beneficial effect on growth. Other research indicates that ground-level ozone is implicated in the extensive damage to Germany's Black Forest that is generally attributed to acid deposition. For example, lichens, which are adversely affected by SO_2, have been found growing abundantly on dying trees, which would not be an expected finding if sulfur dioxide emissions were the cause of the trees' death.

The complex nature of acid deposition makes disentangling the separate effects of pollution and climate change difficult. The effects of both these factors become more severe with increasing altitude, where trees are growing at the coldest temperatures they can normally tolerate. Under such conditions, any decrease in temperature moves the trees into a lethal climatic range. Scientists have also found that the moisture in clouds tends to be much more acidic than rainfall. The reasonable conclusion, supported by research, is that the acid deposition phenomenon elevates tree mortality most seriously at high altitudes where droplets from clouds are the main source of moisture.

Technology of Acid Deposition

Both technological attempts to deal with the causes and effects of acid rain and the instruments used in scientific research reflect some important aspects of the nature of technology.

Taller smoke stacks were introduced as an inexpensive technological fix for local air pollution problems. The thinking was that if the pollutants were released at higher altitudes, they would be diluted by air and would not reach harmful concentrations at ground level. The rationale for tall stacks was soon shown to be flawed. While local air quality did improve, the taller stacks spewed pollutants higher into the air than shorter ones could. High-tech instruments provide evidence that bands of smoke from tall stacks sometimes travelled hundreds of kilometres. Even after the visible smoke has dispersed, the invisible pollutants continued to travel thousands of kilometres from their source, often crossing international boundaries. Monitoring air chemistry with special instruments indicates that more than half the acid deposition in Eastern Canada originates as emissions from industries in the United States. Canadian emissions also contribute to acid deposition in the United States; between 10% and 25% of the deposition in the northeastern states apparently originates in Canada.

One technology for reducing acid deposition is the chemical scrubber, a device that processes the gases emitted by smelters and power plants, dissolving or precipitating the pollutants. Catalysts that reduce the nitrogen oxides produced by combustion reactions represent another technological response to the problem. For example, new automobiles are now outfitted with catalytic converters.

Technology also counterbalances the effects of acid rain. Adding basic materials (for example, lime and limestone) to lakes to neutralize the acid has had some success. Other research has found that certain types of bacteria can oxidize sulfur compounds, while other types can reduce sulfur. This finding suggests that micro-organisms might play a

Technology exists to reduce harmful emissions to a minimum. Environmental ethics, economics, and legal requirements are other perspectives that affect whether technology is used.

beneficial role in the control of lake acidification, particularly when the water remains in the lake for a long time. This research may lead to new technologies for using micro-organisms.

The various strategies for reducing acid rain may involve annual investments of billions of dollars. Because the costs are so high, it is essential that the atmospheric conditions involved in producing and transporting acidic precipitation be well understood. Much research on acid deposition has attempted to develop computer models to identify the source of the acid and the physical and chemical mechanisms by which it is transported to other locations. For example, development of sophisticated technology has made possible the tracking of airborne acidic material from smoke stacks. After a tracer compound is released in different parts of Canada and the United States, a sensitive detector samples the air downwind from the source. Although science and technology have been partners in causing the problem of acid deposition, they are now partners in the search for solutions.

Social Aspects of Acid Deposition

Acid deposition is a societal issue that reveals some important aspects of the interaction among science, technology, and society. The most common perspectives on this issue are environmental, economic, social, and political.

Acid deposition is causing serious environmental, economic, and social problems in Eastern Canada. The environmental problems described above generate enormous economic problems. Acid deposition is endangering fishing, tourism, agriculture, and forestry. The resource base at risk sustains approximately 8% of Canada's Gross National Product (GNP). It is estimated that acid deposition causes about one billion dollars' worth of damage in Canada annually.

Besides the social costs of economic losses, there are other human costs as well. Many respiratory problems are associated with exposure to air pollution. These problems range from aggravation of asthma cases and a consequent increase in hospital admissions, to eventual chronic lung disease. The acute effects of sulfuric acid on humans are usually more pronounced in asthmatics.

The economic and social problems caused by acid rain must be dealt with in the political arena. This entails mediating between the pro-development and pro-environment lobby groups. In addition, the fact that acid deposition is not subject to legislation regarding international boundaries makes it a contentious political issue between Canada and the United States. Although the acid deposition problem is not completely understood, both the Canadian and the American governments have passed legislation to reduce the discharge of sulfur and nitrogen oxides. Opponents of these measures argue that the regulations could severely hinder economic growth, because the cleanup effort will affect the coal-burning electric power plants that emit large amounts of SO_2. These power plants are perceived to be essential to the industrial growth, economic well-being, and social fabric of the Northeastern United States, the very region believed to be responsible for the acid deposition that is ravaging Eastern Canada.

Sulfur Recovery at Cominco's Zinc Operation

The recovery of sulfur at Cominco's zinc smelter in Trail, B.C., is a good example of how a potential environmental hazard can be turned into useful products. Sulfur dioxide and other waste gases produced during the smelting operation are processed in several stages to recover the sulfur. In an initial stage, the waste gases are fed into an acid plant that converts 95% of the SO_2 to sulfuric acid, which is sold as a product or used on-site in further sulfur recovery. The remaining SO_2 in the waste gas is removed by an ammonia scrubbing process. A by-product of this process, ammonium bisulfate, is then combined with sulfuric acid to make an ammonium sulfate solution for fertilizer production. Cominco has two fertilizer plants. In one, the ammonium sulfate solution is evaporated to produce a crystalline ammonium sulfate fertilizer. In the other, sulfuric acid from the acid plant is mixed with stripped electrolyte from the zinc processing and ammonia shipped from Alberta to produce granular ammonium sulfate fertilizer.

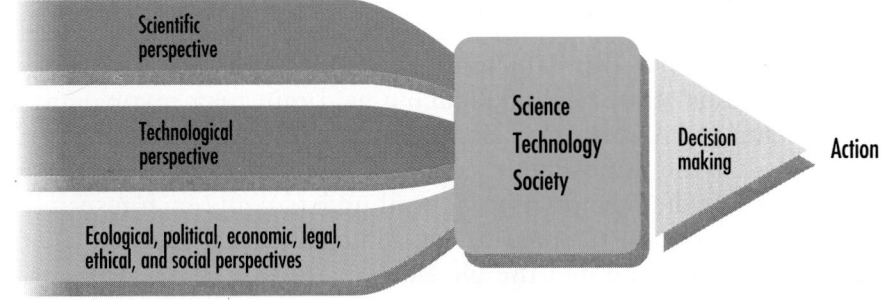

Many issues in society involve science and technology. Both the problems and the solutions involve complex interrelationships among these three categories. An example of an STS issue is the problem of acid rain.

Exercise

Statements about science-technology-society issues have been classified into at least 12 different perspectives (Appendix D, D.4). Although it is difficult to speak convincingly from only one perspective on an STS issue, a presentation that shows respect for a variety of perspectives can be effective.

52. What perspectives other than scientific, technological, ecological, economic, and political can be adopted on acid rain?

53. Make a table listing one or two statements from all possible perspectives, either for or against the resolution that emissions of precursors to acid deposition must be reduced immediately.

54. Assign a value (1 to 10) to each argument and make a decision for or against the argument, based upon your knowledge and values.

55. Within the context of the acid-deposition STS issue, provide an example of:
    (a) science assisting technology
    (b) technology assisting science
    (c) technology affecting society
    (d) society affecting science
    (e) society affecting technology

56. What role can science play in helping to find solutions to STS issues?

57. What problems do scientists sometimes encounter when trying to communicate with the general public or in a court of law? For example, what valued aspect of scientific communication becomes a drawback outside the scientific community?

58. Find examples of phrases that scientists use, when communicating with each other, that are not used in everyday conversation.

59. (Internet) Find some Internet sites dealing with acid rain and record some recent research information.

In Chapter 7 (page 276) and in Appendix C, C.3, you have been introduced to the preparation of standard solutions and to the process and skills of titration. In review, a *titration* is the progressive addition of a *titrant* from (for example) a buret into (for example) an Erlenmeyer flask containing the *sample* for the purpose of determining the concentration, molar mass, or quantity of a specified chemical in the sample. In the first stage, often a *primary standard* is used to standardize a titrant — determine an accurate value for the concentration of the titrant. A primary standard is a chemical for which an accurate concentration can be prepared, e.g., sodium carbonate $(Na_2CO_{3(s)})$ and potassium hydrogen phthalate $(KHOOCC_6H_4COO_{(s)})$. Hydrochloric acid and sodium hydroxide are not primary standards — hydrogen chloride gas vaporizes, especially from concentrated hydrochloric acid, and sodium hydroxide is hydroscopic — it gains mass by absorbing water from the air. Therefore, hydrochloric acid and sodium hydroxide must be standardized against, for example, sodium carbonate and potassium hydrogen phthalate, respectively, before being used as a titrant for determining the concentration or molar mass of another base or acid.

The typical titration process involves preparing a solution of a primary standard, titrating samples of this primary standard to standardize the titrant, and then titrating samples of the chemical of unknown mass, concentration, or molar mass against the standardized titrant. In each case, endpoints must be chosen to signal the equivalence points of the reactions. *Endpoints* are rapid changes in the properties of a chemical system and typically involve a rapid change in pH, electric potential, or color. For most acid-base titrations, a pH curve for the reaction is used to select an acid-base indicator that changes color at the chosen equivalence point. The chosen indicator is then added to the samples of the primary standard and the unknown before each of the titrations is started. The endpoint at the indicator *transition point* (color change) signals the chemist to stop the titration and record the volume of titrant used. The volume of titrant is the empirical measure of the *equivalence point* — the theoretical stoichiometric point of the reaction where the correct stoichiometric ratio of sample and titrant have reacted. Ideally, the endpoint (the chosen rapid change in a property of the chemical system) is at the stoichiometric point: however, all measurements have uncertainty and also affect the original system, e.g., it takes a small volume of titrant to react with the indicator.

The stoichiometric calculation usually involves the average of at least two consistent titration trials. Chemists demand high reproducibility from titration results. Equivalence points that are more than ±0.2 mL from a set of consistent results are recorded but not included in the average volume of titrant used. Titrations must involve reactions that obey the assumptions required of stoichiometric calculations — the reactions must be stoichiometric, spontaneous, and quantitative. In addition, the reactions used in titrations must be fast.

Figure 14.32
Sodium carbonate or soda ash is a common, inexpensive chemical. A nitric acid spill from railway cars is neutralized by blowing soda ash onto the spilled acid.

Reactions can be analyzed stoichiometrically most easily if they are quantitative (greater than 99% complete). When completing stoichiometric calculations from titration evidence, the number of quantitative reactions that occur can be determined initially from a pH curve. The number of protons completely transferred determines the mole ratio of reacting substances in the reaction equation. Either net ionic or formula (non-ionic) equations may be used to describe the reaction and to provide the mole ratio. In redox stoichiometry, net ionic equations from half-reaction equations are often more convenient to use. In acid-base reactions, formula (non-ionic) equations are often simpler and more convenient, as long as the concept of proton transfer is used to predict the products. The use of a formula equation requires that the number of protons transferred from polyprotic acids be carefully determined from a pH curve to establish the correct mole ratio.

In the neutralization of sodium hydroxide with hydrochloric acid, both reactants are monoprotic. The formula and net ionic equations are

$$HCl_{(aq)} + NaOH_{(aq)} \rightarrow H_2O_{(l)} + NaCl_{(aq)}$$
$$H_3O^+_{(aq)} + OH^-_{(aq)} \rightarrow 2\,H_2O_{(l)}$$

If sodium carbonate (Figure 14.32) is neutralized with a strong acid using the methyl orange endpoint, the pH curve indicates that two protons are transferred quantitatively (Figure 14.23, page 571).

$$2\,HCl_{(aq)} + Na_2CO_{3(aq)} \rightarrow H_2CO_{3(aq)} + 2\,NaCl_{(aq)}$$
$$2\,H_3O^+_{(aq)} + CO_3^{2-}_{(aq)} \rightarrow 2\,H_2O_{(l)} + H_2CO_{3(aq)}$$

Once either of these balanced equations is written, stoichiometry calculations can be done. Recall that for some polyprotic substances, all reactions may not be quantitative. For example, only one pH endpoint is obtained in the titration of sodium sulfite with hydrochloric acid (page 600). Therefore, only one proton is transferred.

$$HCl_{(aq)} + Na_2SO_{3(aq)} \rightarrow NaHSO_{3(aq)} + NaCl_{(aq)}$$
$$H_3O^+_{(aq)} + SO_3^{2-}_{(aq)} \rightarrow H_2O_{(l)} + HSO_3^-_{(aq)}$$

Oxalic acid is a common ingredient in solutions for removing rust. Suppose that an investigation is performed to determine the concentration of oxalic acid in a rust-removing solution. Three 10.0 mL samples of oxalic acid are titrated with a standardized 1.27 mol/L sodium hydroxide solution. Phenolphthalein color change is used as the second pH endpoint; the results are shown in Table 14.4.

Omitting Titration Trials
In higher level chemistry courses there are statistical tests to determine whether a titration trial should be used or not. At this post-secondary level, it is sufficient to use a simple rule such as, if a trial is more than ± 0.2 mL from the average of the most precise trials, omit it. Ethically, you may omit a trial in your calculations, but you must always report all trials.

Table 14.4

TITRATION OF 10.0 mL OF OXALIC ACID WITH SODIUM HYDROXIDE			
Trial	**1**	**2**	**3**
Final buret reading (mL)	12.1	23.5	34.9
Initial buret reading (mL)	0.3	12.1	23.5
Comment on endpoint	overshot	good	good
Decision	disregard	use	use

After omitting the first trial, the volume of 1.27 mol/L sodium hydroxide solution used in the calculations is 11.4 mL.

$$2\,NaOH_{(aq)} + HOOCCOOH_{(aq)} \rightarrow 2\,H_2O_{(l)} + Na_2OOCCOO_{(aq)}$$

11.4 mL   10.0 mL

1.27 mol/L C

$$n_{NaOH} = 11.4\ mL \times 1.27\ \frac{mol}{L} = 14.5\ mmol$$

$$n_{HOOCCOOH} = 14.5\ mmol \times \frac{1}{2} = 7.24\ mmol$$

$$C_{HOOCCOOH} = \frac{7.24\ mmol}{10.0\ mL} = 0.724\ mol/L$$

or $C_{HOOCCOOH} = 11.4\ mL\ NaOH \times \dfrac{1.27\ mol\ NaOH}{1\ L\ NaOH} \times$

$$\frac{1\ mol\ HOOCCOOH}{2\ mol\ NaOH} \times \frac{1}{10.0\ mL}$$

$$= 0.724\ mol/L$$

Exercise

60. In a chemical analysis, only one quantitative reaction could be detected in the pH curve for the reaction of sodium sulfite with hydrochloric acid. In a subsequent titration, 10.00 mL of sodium sulfite solution was titrated with 0.225 mol/L hydrochloric acid. The average volume required at the endpoint for this titration was 14.2 mL. What is the concentration of the sodium sulfite solution?

61. Chemical analysis of a stain remover containing oxalic acid was conducted by a commercial analytical chemistry firm. Oxalic acid solution was titrated with 0.485 mol/L potassium hydroxide to the second endpoint, using phenolphthalein (Table 14.5). Calculate the concentration of the oxalic acid in this brand of stain remover.

62. A 25.0 mL sample of a cleaning solution, containing sodium hydrogen sulfate, was titrated with 0.500 mol/L sodium hydroxide using phenolphthalein indicator. At the endpoint, one drop of $NaOH_{(aq)}$ was sufficient to change the phenolphthalein indicator from colorless to pink. At this point, a stoichiometrically equivalent 10.2 mL of $NaOH_{(aq)}$ had been added. What is the concentration of sodium hydrogen sulfate in the cleaning agent?

63. A titration of phosphoric acid used in a commercial rust-removing solution with 0.123 mol/L sodium hydroxide was completed to the end of the second quantitative reaction. The equivalence point values are obtained from Figure 14.33. What is the concentration of the phosphoric acid solution?

$$2\,NaOH_{(aq)} + H_3PO_{4(aq)} \rightarrow 2\,H_2O_{(l)} + Na_2HPO_{4(aq)}$$

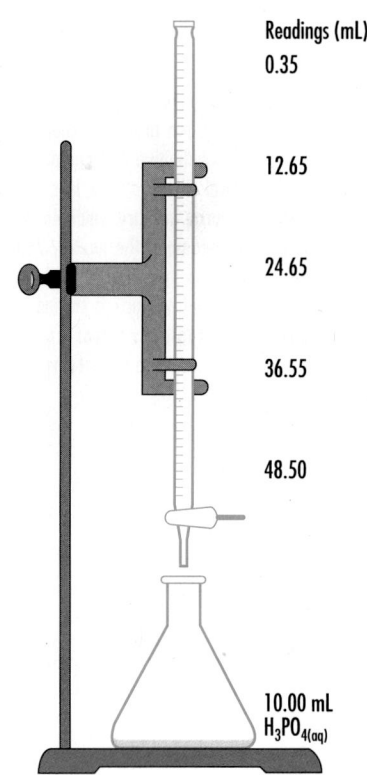

Readings (mL)

0.35

12.65

24.65

36.55

48.50

10.00 mL
$H_3PO_{4(aq)}$

Figure 14.33
Sodium hydroxide titrant is added in successive trials to a phosphoric acid sample.

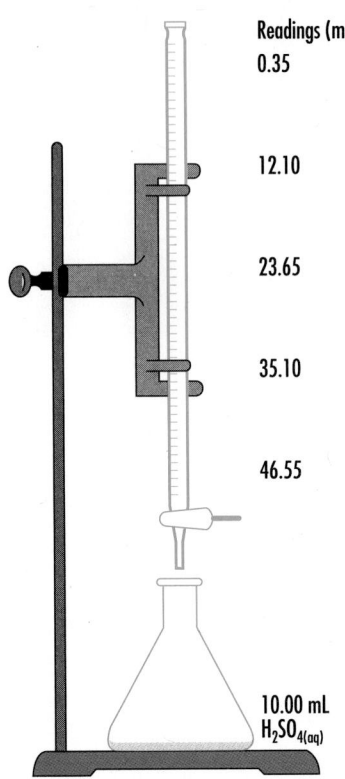

Readings (mL)
0.35

12.10

23.65

35.10

46.55

10.00 mL
H₂SO₄₍ₐq₎

Figure 14.34
Sodium hydroxide titrant is added to a sulfuric acid sample in successive trials.

▌ Chemical Analysis

Analytical chemistry is a branch of chemistry dealing with both qualitative and quantitative analyses. Analytical chemists work in diverse areas such as research, industry, medicine, and government (for example, the environment, consumer affairs, and law enforcement). The analysis of commercial products is carried out by industry for quality control and by government for consumer safety and information.

64. In a chemical analysis, 10.00 mL samples of sodium sulfide solution used in an industrial process were titrated with 0.150 mol/L hydrochloric acid to the end of the second quantitative reaction. An average of 16.8 mL of $HCl_{(aq)}$ was required. What is the concentration of the sodium sulfide solution?

65. A titration of sulfuric acid with 0.484 mol/L sodium hydroxide was completed to the second endpoint. The evidence is displayed in Figure 14.34. Evidence from pH curves indicates that the reaction of sulfuric acid with the sodium hydroxide involves two quantitative reactions. Calculate the concentration of the sulfuric acid solution.

Table 14.5

TITRATION OF 25.0 mL OXALIC ACID WITH POTASSIUM HYDROXIDE			
Trial	1	2	3
Final buret reading (mL)	17.1	32.7	48.3
Initial buret reading (mL)	1.4	17.1	32.7

Problem 14H Mass Percent of Sodium Phosphate

Sodium phosphate is used to clean paintbrushes and grease spills. Complete the Analysis of the investigation report.

Problem

What is the mass-by-volume percent of sodium phosphate in a cleaning solution?

Experimental Design

A mass of 1.36 g of sodium carbonate was used to make 100.0 mL of a primary standard solution. Using the methyl orange endpoint (second reaction step), samples of the sodium carbonate solution were titrated with hydrochloric acid solution to standardize the acid solution. The sodium phosphate solution was then titrated to the second endpoint with the standardized hydrochloric acid solution.

Evidence

TITRATION OF 10.00 mL OF SODIUM CARBONATE WITH $HCl_{(aq)}$				
Trial	1	2	3	4
Final buret reading (mL)	13.2	25.9	38.7	13.3
Initial buret reading (mL)	0.1	13.2	25.9	0.7

TITRATION OF 10.00 mL OF SODIUM PHOSPHATE WITH STANDARDIZED $HCl_{(aq)}$				
Trial	1	2	3	4
Final buret reading (mL)	19.2	37.6	19.7	38.1
Initial buret reading (mL)	0.4	19.2	1.2	19.7

Problem 14I Identifying an Unknown Acid

Complete the Analysis of the investigation report. Then write another experimental design or a series of designs to help identify the inorganic or organic acid with increased certainty.

Problem

What is the molar mass of an unknown acid?

Experimental Design

As part of the chemical analysis of an unknown white solid, a titration with a strong base was carried out, using a pH meter.

Evidence

mass of solid = 0.217 g
concentration of $NaOH_{(aq)}$ = 0.182 mol/L
volume of $NaOH_{(aq)}$ at second pH endpoint = 16.1 mL

INVESTIGATION

14.7 Ammonia Analysis

Ammonia is most often used by consumers as a household cleaner. In its simplest form, ammonia is sold as an aqueous solution with dilution instructions for various cleaning applications, such as washing laundry, cleaning glass, and removing wax. Ammonia is a relatively weak base and requires a strong acid to meet the quantitative reaction requirements for a titration. The pH curve for ammonia titrated with hydrochloric acid is shown in Figure 14.35, page 594.

   The purpose of this investigation is to test the molar concentration of a household ammonia solution (diluted by a factor of 10), using a titration with hydrochloric acid as the experimental design. Use $Na_2CO_{3(aq)}$ as a primary standard to standardize the HCl. You require 100.0 mL of 0.143 mol/L $Na_2CO_{3(aq)}$. The pH curve for this reaction is in Figure 14.36, page 594.

Problem

What is the molar concentration of the household ammonia sample provided?

Prediction

According to the literature, the concentration of a fresh household ammonia solution varies from 3.0% to 29% by mass. This concentration corresponds to a range of 1.8 mol/L to 17 mol/L.

- ▪ Problem
- ▪ Prediction
- ✔ Design
- ✔ Materials
- ✔ Procedure
- ✔ Evidence
- ✔ Analysis
- ✔ Evaluation
- ▪ Synthesis

CAUTION

Ammonia irritates skin and mucous membranes. Hydrochloric acid is a corrosive acid. Avoid eye and skin contact. If you spill any of the solutions on your skin, immediately wash the area with lots of cool water. In the unlikely situation of getting some of the chemicals in your eye, immediately rinse your eye for at least 15 min and inform your teacher.

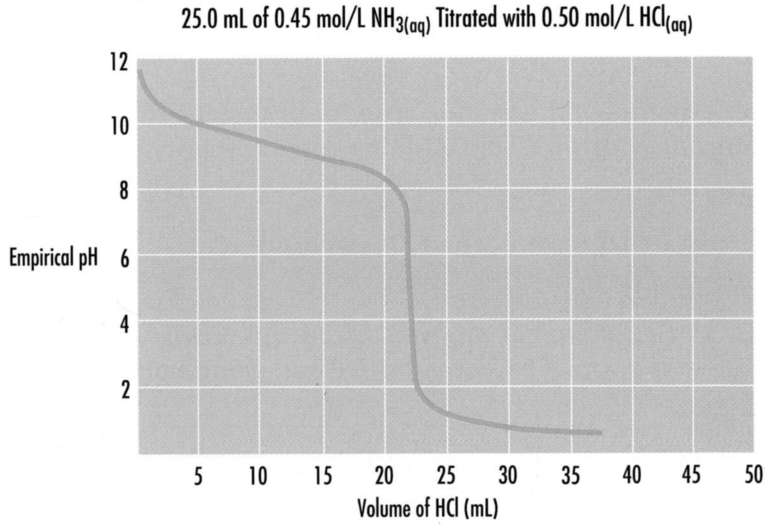

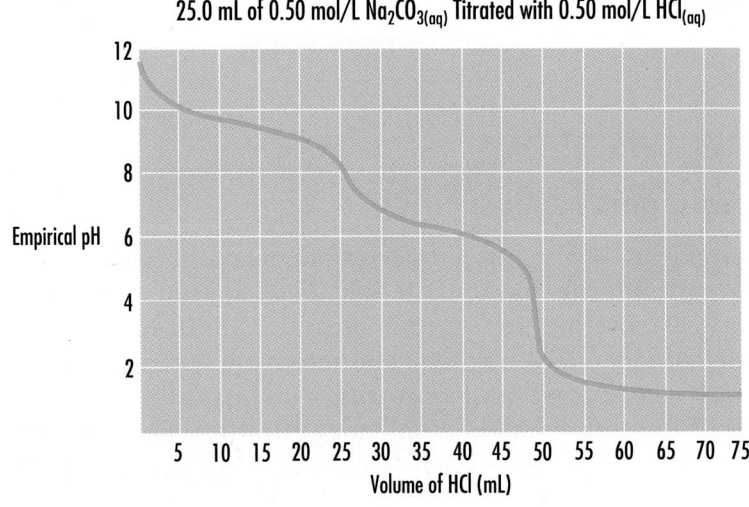

Figure 14.35
*pH curve for the addition of
0.50 mol/L HCl$_{(aq)}$ to 25.0 mL of
0.45 mol/L NH$_{3(aq)}$.*

Figure 14.36
*A pH curve for the addition of
0.50 mol/L HCl$_{(aq)}$ to a 25.0 mL
sample of 0.50 mol/L Na$_2$CO$_{3(aq)}$ can
be used to select an indicator for
titration.*

Recognize Excess Problems
You can recognize excess problems by noting
that you are given all of the four required
values.

Non-Equivalent Reacting Amounts

Titration is a common experimental technique to determine the
unknown concentration of a solution. In a titration we stop adding the
titrant when the indicator changes color. This procedure assumes that
the endpoint of the indicator accurately matches the pH endpoint of
the reaction. (See Figure 14.22, page 570.) Only then is the amount of
added titrant chemically equivalent to the amount of sample present.
Therefore the selection of an indicator is critical if the results are to be
accurate.

What happens when the indicator chosen changes color before or
after the equivalence point? In this situation, the amounts of titrant
and sample are no longer equivalent; one of them is in excess. For
example, consider the titration of 25.0 mL of 0.48 mol/L sodium
hydroxide with 0.50 mol/L hydrochloric acid, using orange IV as the
indicator. From Figure 14.22 on page 570, the equivalence point is
measured to be about 24 mL hydrochloric acid.

$$HCl_{(aq)} \quad + \quad NaOH_{(aq)} \rightarrow H_2O_{(l)} + NaCl_{(aq)}$$

28 mL   25.0 mL
0.50 mol/L   0.48 mol/L

The non-equivalent amounts are readily shown by the calculation of the amount of each reactant.

$$n_{NaOH} = 25.0 \text{ mL} \times \frac{0.48 \text{ mol}}{1 \text{ L}} = 12 \text{ mmol} \text{ (initially present)}$$

$$n_{HCl} = 28 \text{ mL} \times \frac{0.50 \text{ mol}}{1 \text{ L}} = 14 \text{ mmol} \text{ (added)}$$

Since the mole ratio is 1:1, the amount of $HCl_{(aq)}$ required to titrate the $NaOH_{(aq)}$ sample is 12 mmol.

$$n_{HCl} = 12 \text{ mmol} \times \frac{1}{1} = 12 \text{ mmol} \text{ (required amount)}$$

An excess of 2 mmol of $HCl_{(aq)}$ has been added. This means that the final solution is acidic instead of being neutral for this strong acid–strong base reaction (Figure 14.37).

$$\text{excess } n_{HCl} = 14 \text{ mmol} - 12 \text{ mmol} = 2 \text{ mmol}$$
$$v_f = 28 \text{ mL} + 25.0 \text{ mL} = 53 \text{ mL}$$
$$C_f \atop HCl = \frac{2 \text{ mmol}}{53 \text{ L}} = 0.04 \text{ mol/L}$$
$$pH = -\log(0.04) = 1.4$$

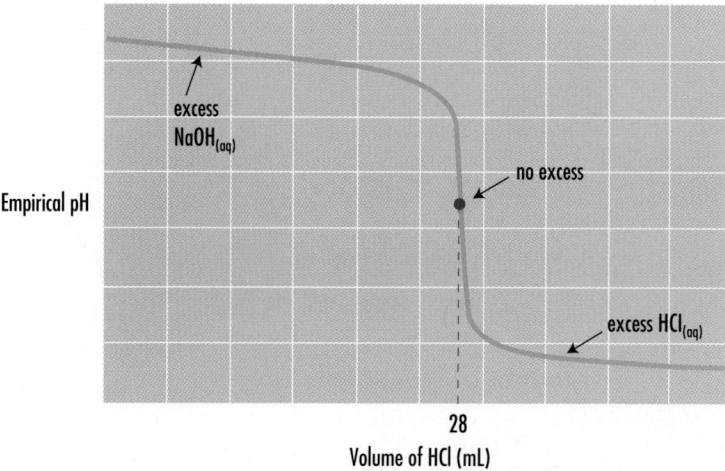

NaOH$_{(aq)}$ Titrated with HCl$_{(aq)}$

Empirical pH

excess
NaOH$_{(aq)}$

no excess

excess HCl$_{(aq)}$

28
Volume of HCl (mL)

Figure 14.37
To determine where the chemical system is on the pH curve, you must calculate the amount of each chemical and use the mole ratio to determine which chemical, if any, is in excess.

If the above example had been a chemical analysis of the concentration of the $NaOH_{(aq)}$ solution, the answer obtained from the typical stoichiometry calculation (assuming equivalent amounts) would be quite inaccurate.

Questions based on non-equivalent amounts or excess reagents are recognized when too much information is apparently provided, compared with typical acid-base stoichiometry questions. In excess-reagent questions, a solution volume and concentration are given for each reactant, and the questions require the calculation of some property of the final solution, for example, pH or ion concentration. A general procedure for solving excess-reagent questions is summarized below.

EXAMPLE

What is the pH of a solution produced by quantitatively reacting 20 mL of 0.10 mol/L aqueous hydrogen sulfide with 50 mL of 0.10 mol/L sodium hydroxide (Figure 14.38)?

$$H_2S_{(aq)} \quad + \quad 2\,NaOH_{(aq)} \quad \rightarrow \quad 2\,H_2O_{(l)} \quad + \quad Na_2S_{(aq)}$$

20.0 mL   50.0 mL

0.10 mol/L   0.10 mol/L

$$n_{H_2S} \; = \; 20\text{ mL} \times 0.10\text{ mol/L}$$
$$= \; 2.0\text{ mmol} \text{ (limiting reagent)}$$

$$n_{NaOH} \; = \; 50\text{ mL} \times 0.10\text{ mol/L}$$
$$= \; 5.0\text{ mmol} \text{ (excess reagent)}$$

$$n_{rea \atop NaOH} \; = \; 2.0\text{ mmol} \times \frac{2}{1}$$
$$= \; 4.0\text{ mmol} \text{ (required amount)}$$

$$n_{xs \atop NaOH} \; = \; 5.0\text{ mmol} - 4.0\text{ mmol}$$
$$= \; 1.0\text{ mmol} \text{ (excess amount)}$$

$$C_{f \atop NaOH} \; = \; \frac{1.0\text{ mmol}}{(50 + 20)\text{ mL}} = 0.014\text{ mol/L}$$

$$[OH^-_{(aq)}] \; = \; C_{NaOH} = 0.014\text{ mol/L}$$

$$[H_3O^+_{(aq)}] \; = \; \frac{1.0 \times 10^{-14}\ (\text{mol/L})^2}{0.014\text{ mol/L}}$$

$$= \; 7.0 \times 10^{-13}\text{ mol/L}$$

$$pH \; = \; -\log(7.0 \times 10^{-13})$$

$$= \; 12.16$$

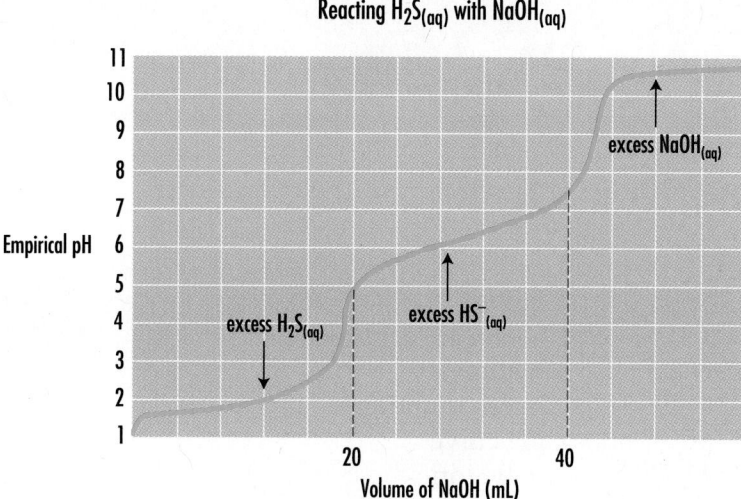

Reacting H₂S₍ₐq₎ with NaOH₍ₐq₎

excess NaOH₍ₐq₎

excess HS⁻₍ₐq₎

excess H₂S₍ₐq₎

Empirical pH

Volume of NaOH (mL)

Figure 14.38
A pH curve helps to visualize and understand excess problems.

Exercise

66. Predict the pH of a solution produced by mixing 100 mL of 0.25 mol/L nitric acid with 50 mL of 0.40 mol/L potassium hydroxide solution during an acid-base titration.

67. Calculate the hydrogen ion and hydroxide ion concentrations after 60 mL of 0.10 mol/L hydrochloric acid is used to "neutralize" 25 mL of 0.10 mol/L barium hydroxide.

68. Calculate the hydrogen ion and hydroxide ion concentrations after 100 mL of 0.20 mol/L sodium hydroxide solution is added to 40 mL of a 0.20 mol/L sulfuric acid solution during an experiment to determine the pH curve for the reaction.

69. A 10.0 mL sample of a hydrochloric acid solution of unknown concentration was titrated with a standardized 0.0759 mol/L sodium hydroxide solution. The endpoint of 11.5 mL was overshot by 1.5 mL of NaOH₍ₐq₎. According to the evidence, what is the pH of the solution after overshooting the endpoint? Sketch the pH curve for this reaction system.

OVERVIEW

Acid-Base Reactions

Summary

- Acid-base theory has developed and changed over time. Limitations in the predictive and explanatory power of a theory force scientists to expand the theory to a less restricted form.

- Both the role of water in acidic solutions and the evidence for hydronium ions lead to the Brønsted-Lowry concept of acids and bases as proton donors and proton acceptors in chemical reactions.

- The Brønsted-Lowry concept explains amphiprotic substances, hydrolysis, acidic and basic oxides, acid-base indicators, buffers, polyprotic substances, pH curves, and acid-base equilibria.

- Acid-base reactions can be predicted using a table of relative acid strengths and the Brønsted-Lowry concepts.

- Reaction progress can be recorded as a pH curve. This leads to a description of buffers, hydrolysis, quantitative reactions, strong and weak acid-base reactions, polyprotic substances, excess problems, and indicators in acid-base titrations.

- An endpoint indicates the equivalence point of a quantitative reaction in a titration used in acid-base stoichiometry calculations.

- An STS issue like acid rain can be described from a variety of perspectives including scientific, technological, ecological, economic, political, legal, ethical, and social.

Key Words

acid-base indicator
amphiprotic
amphoteric
Brønsted-Lowry acid
Brønsted-Lowry base
Brønsted-Lowry neutralization
buffer
conjugate acid-base pair
hydrolysis
hydronium ion
pH curve
polyprotic
transition point

Review

1. Formal concepts of acids have existed since the 18th century. State the main idea and the limitations of each of the following: the oxygen concept; the hydrogen concept; Arrhenius's concept; and the Brønsted-Lowry concept of acids.

2. What happens when scientists find a theory, such as Arrhenius's theory of acids, to be unacceptable?

3. State two main ways in which a theory or a theoretical definition may be tested.

4. In terms of modern evidence, what is the nature of a hydrogen ion in aqueous solution?

5. How has the theoretical definition of a base changed from Arrhenius's concept, to the revised Arrhenius's concept, and subsequently to the Brønsted-Lowry concept?

6. Aqueous solutions of nitric acid and nitrous acid of the same concentration are prepared.
(a) How do their pH values compare?
(b) Explain your answer using the Brønsted-Lowry concept.

7. Write at least three diagnostic tests to determine which of two provided solutions is acetic acid or hydrochloric acid. Use the *If* [procedure], *and* [evidence], *then* [analysis] format (Appendix C, C.3).

8. According to the Brønsted-Lowry concept, what determines the position of equilibrium in an acid-base reaction?

9. What generalization from the table of acids and bases (Appendix F) can be used to predict the position of an acid-base equilibrium?

10. State two examples of conjugate acid-base pairs, each involving the hydrogen sulfite ion.

11. If the pH of a solution is 6.8, what is the color of each of the following indicators in this solution?
    (a) methyl red
    (b) chlorophenol red
    (c) bromothymol blue
    (d) phenolphthalein
    (e) methyl orange

12. Draw a pH curve to illustrate the following information about acid-base reaction systems.
    (a) What is a buffering action?
    (b) Where does buffering action appear on a pH curve?
    (c) How are quantitative reactions represented on a pH curve?
    (d) Define pH endpoint and equivalence point.
    (e) How is a suitable indicator chosen for a titration?
    (f) Do non-quantitative reactions have an endpoint? Explain your answer briefly.

13. Sketch generalized pH curves on a single set of axes to illustrate the addition of a strong and a weak base to a strong and a weak acid.

14. Sketch a generalized pH curve for the addition of a strong base to a weak acid and label the approximate pH and pK_a of the weak acid and the required indicator.

15. Which of the following statements are always true when products are favored in an acid-base equilibrium?
    (a) The stronger base is a product.
    (b) The equilibrium constant is likely greater than one.
    (c) Minimum energy and maximum randomness favors products.
    (d) The stronger acid is a reactant.
    (e) The percent reaction is greater than 50%.
    (f) The pH of the final solution is greater than 7.
    (g) The reactant acid is above the reactant base in an acid-base table.
    (h) The forward reaction is likely exothermic.

16. State two different applications of buffers.

17. What is the pH of natural (normal) acid rain and what causes the acidity?

Applications

18. Predict, with reasoning, whether each of the following chemical systems will be acidic, basic, or neutral.
    (a) aqueous hydrogen bromide
    (b) aqueous potassium nitrite
    (c) aqueous ammonia
    (d) aqueous sodium hydrogen sulfate
    (e) carbonated beverages
    (f) limewater
    (g) vinegar

19. Write an experimental design to test the predictions in question 18.

20. Write two experimental designs to rank a group of bases in order of strength.

21. Write formula, complete ionic, and net ionic equations for the following reactions. (Reference Appendix G if necessary.)
    (a) Vinegar is used to neutralize an oven cleaner spill.
    (b) Baking soda solution is used as an antacid to neutralize excess stomach acid.
    (c) Household ammonia is used to stop potential staining by neutralizing some pop spilled on a carpet.

22. Compounds may be classified as ionic or molecular. Each of these classes can be subdivided into neutral substances, acids, or bases. Construct a flow chart that includes two examples for each of the six categories under the headings "Ionic" and "Molecular."

23. Identify all acids, bases, and conjugate pairs and predict the position of equilibrium in each of the following reactions.

    (a) $HCOOH_{(aq)} + CN^-_{(aq)} \rightleftharpoons$
    $$HCOO^-_{(aq)} + HCN_{(aq)}$$

    (b) $HPO_4^{2-}_{(aq)} + HCO_3^-_{(aq)} \rightleftharpoons$
    $$H_2PO_4^-_{(aq)} + CO_3^{2-}_{(aq)}$$

24. Separate samples of an unknown solution were tested with indicators. Congo red was red and chlorophenol red was yellow in the solution. Estimate the approximate pH and hydronium ion concentration of the solution.

25. The test of an acceptable theory is its ability to explain and predict a wide range of phenomena, i.e., be a unifying theory. Use the five-step procedure to write the chemical equations describing each of the following acid-base reactions. Write one diagnostic test for each prediction.
   (a) the addition of hydrofluoric acid to a solution of potassium sulfate
   (b) the addition of a solution of sodium hydrogen sulfate to a solution of sodium hydrogen sulfide
   (c) the titration of methanoic acid with sodium hydroxide solution
   (d) the addition of a small amount of a strong acid to a hydrogen phosphate ion–phosphate ion buffer solution
   (e) the addition of colorless phenolphthalein indicator $(HPh_{(aq)})$ to a strong base
   (f) the hydrolysis of the salt at the equivalence point when aqueous sodium bisulfate is reacted with aqueous sodium hydroxide
   (g) the addition of a blue solution of bromothymol blue to vinegar
   (h) sulfur dioxide forms sulfurous acid rain
   (i) limestone neutralizes sulfuric acid rain
   (j) the addition of aluminum chloride to water
   (k) the addition of washing soda to water
   (l) the addition of baking soda to water

26. Alternately, many acid-base phenomena can be described or predicted by using Le Châtelier's principle. Test the ability of the Le Châtelier principle to describe or predict the effect of the following reactions.
   (a) $NaOH_{(aq)}$ is added to the following buffer equilibrium.

   $$NH_{3(aq)} + H_2O_{(l)} \rightleftharpoons NH_4^+{}_{(aq)} + OH^-{}_{(aq)}$$
   $$\text{or } NH_{3(aq)} + H^+{}_{(aq)} \rightleftharpoons NH_4^+{}_{(aq)}$$

   (b) $NaOH_{(aq)}$ is added to a bromothymol blue indicator solution.

   $$HBb_{(aq)} + H_2O_{(l)} \rightleftharpoons H_3O^+{}_{(aq)} + Bb^-{}_{(aq)}$$
   $$\text{or } HBb_{(aq)} \rightleftharpoons H^+{}_{(aq)} + Bb^-{}_{(aq)}$$

   (c) $NaOH_{(aq)}$ titrant is added to a vinegar sample.

   $$CH_3COOH_{(aq)} + H_2O_{(l)} \rightleftharpoons$$
   $$H_3O^+{}_{(aq)} + CH_3COO^-{}_{(aq)}$$
   $$\text{or } CH_3COOH_{(aq)} \rightleftharpoons H^+{}_{(aq)} + CH_3COO^-{}_{(aq)}$$

   (d) $NaOH_{(aq)}$ is added to water.

   $$H_2O_{(l)} + H_2O_{(l)} \rightleftharpoons H_3O^+{}_{(aq)} + OH^-{}_{(aq)}$$
   $$\text{or } H_2O_{(l)} \rightleftharpoons H^+{}_{(aq)} + OH^-{}_{(aq)}$$

27. One way to evaluate a theory is to test predictions with new substances. Sodium methoxide, $NaCH_3O_{(s)}$, is dissolved in water. Will the final solution be acidic, basic, or neutral? Explain your answer using a net ionic equation.

28. In an experimental investigation of amphiprotic substances, samples of baking soda were added to a solution of sodium hydroxide and to a solution of hydrochloric acid. The pH of the sodium hydroxide changed from 13.0 to 9.5 after the addition of the baking soda. The pH of the hydrochloric acid changed from 1.0 to 4.5 after the addition of baking soda. Provide a theoretical explanation of these results including writing chemical equations to describe the reactions.

29. Each of seven unlabelled beakers was known to contain one of the following 0.10 mol/L solutions: $CH_3COOH_{(aq)}$, $Ba(OH)_{2(aq)}$, $NH_{3(aq)}$, $C_2H_4(OH)_{2(aq)}$, $H_2SO_{4(aq)}$, $HCl_{(aq)}$, and $NaOH_{(aq)}$. Describe diagnostic test(s) required to distinguish the solutions and label the beakers. Use the "If ——, and ——, then ——" format (Appendix C, C.3), a flow chart, or a table to communicate your answer.

30. Use the pH curve for the titration of sodium sulfite solution with hydrochloric acid (see the diagram below) to answer the following questions.

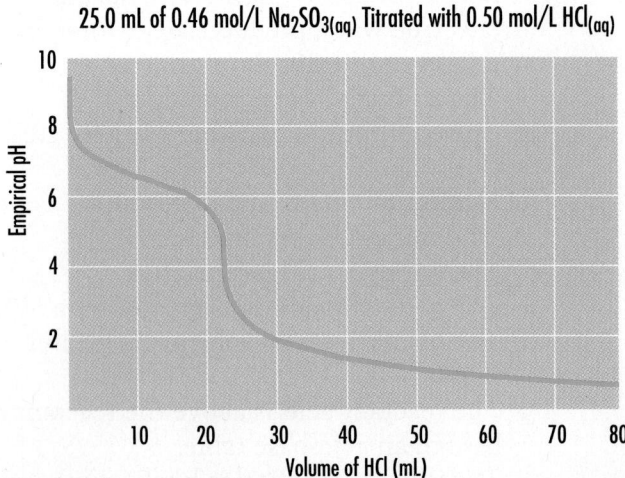

25.0 mL of 0.46 mol/L $Na_2SO_{3(aq)}$ Titrated with 0.50 mol/L $HCl_{(aq)}$

Empirical pH

Volume of HCl (mL)

(a) How many quantitative reactions have occurred?

(b) Write the chemical equation for each quantitative reaction.

(c) State the pH endpoint and the equivalence point for each reaction.

(d) Choose a suitable indicator to correspond to the pH endpoint(s).

(e) Identify the buffering region(s) and state the chemical formulas for the entities present in each region.

31. A pH meter was used to determine the endpoint of a titration of 10.00 mL samples of hypochlorous acid with 0.350 mol/L barium hydroxide solution. At the pH endpoint, a stoichiometrically equivalent volume of 12.6 mL of hydroxide solution was required. What is the molar concentration of the hypochlorous acid solution?

32. Sketch the pH (titration) curve for 15 mL of 0.10 mol/L $HCl_{(aq)}$ being added to 10 mL of 0.10 mol/L $NH_{3(aq)}$. Include the following information in your sketch. Show all calculations.

(a) the equivalence point of the reaction

(b) the initial pH of the ammonia reaction

(c) the pH after adding 5 mL of $HCl_{(aq)}$

(d) the entities present at the equivalence point

(e) the pH after adding 10 mL of $HCl_{(aq)}$

(f) the pH after adding 15 mL of $HCl_{(aq)}$

(g) an indicator for an endpoint

33. A 25.0 mL sample of a diluted rust-removing solution containing phosphoric acid was titrated to the second endpoint using 1.50 mol/L sodium hydroxide. The average equivalence point of the sodium hydroxide solution was 17.9 mL. What is the concentration of phosphoric acid in the rust-removing solution? Show your work.

34. A 50 mL volume of 0.56 mol/L hydrochloric acid was spilled on a counter. A student quickly decided to sprinkle calcium hydroxide (slaked lime) onto the spill. If 1.0 g of solid calcium hydroxide was used, would it completely neutralize the acid? Justify your answer by calculating the pH of the final mixture.

35. A series of experiments with a non-aqueous solvent determined that the products are highly favored in each of the following acid-base equilibria.

$$(C_6H_5)_3C^- + C_4H_4NH \rightleftharpoons$$
$$(C_6H_5)_3CH + C_4H_4N^-$$

$$CH_3COOH + HS^- \rightleftharpoons H_2S + CH_3COO^-$$

$$O^{2-} + (C_6H_5)_3CH \rightleftharpoons (C_6H_5)_3C^- + OH^-$$

$$C_4H_4N^- + H_2S \rightleftharpoons C_4H_4NH + HS^-$$

(a) Identify the Brønsted-Lowry acids, bases, and conjugate acid-base pairs in these chemical reactions.

(b) Arrange the acids in the four chemical reactions in order of decreasing acid strength; that is, prepare a table of acids and bases.

36. Critique the following experimental designs.

(a) Sodium hydroxide is titrated against a phosphoric acid solution to the third equivalence point using the bromothymol blue color change as the endpoint.

(b) The concentration of an acetic acid solution is determined by boiling the water away.

(c) The concentration of hydroxide ions in an ammonia solution is determined by precipitating the ions with a silver nitrate solution.

(d) Hydrochloric acid is used as a primary standard to determine the concentration of sodium sulfide solution.

(e) Litmus is used as a diagnostic test of the reaction between sodium hydrogen carbonate and sodium hydroxide.

(f) Cobalt chloride paper is used in a diagnostic test for the production of water in a strong acid–strong base reaction.

(g) The strength of an acid is determined by titration.

37. Create five or more different experimental designs to determine the concentration of a hydrochloric acid solution. At least two of your designs must use a concept not presented in this chapter.

38. Write three experimental designs for ordering some acids into a relative strengths of acids and bases table.

39. A titration curve is a graph of any solution property versus a volume of a titrant. An endpoint is the abrupt change in the solution property at the completion of a chemical reaction. Many properties, such as pH, color, rate of reaction, and conductivity, can be

used to construct a titration curve. A pH curve is the most familiar example of a titration curve. Write an experimental design to determine the concentration of a barium hydroxide solution, using a conductivity titration with sulfuric acid. Explain your method with appropriate equations and predict the titration curve.

40. All theories in science, including the Arrhenius, Brønsted-Lowry, and Lewis acid-base concepts, are restricted in some way. How do scientists decide which theory to use in a particular situation? For example, is the least restricted theory always the best one to use?

Extensions

41. Chloro-substituted acetic acids are used in organic synthesis, cleaners, and herbicides. These acids are prepared by the chlorination of acetic acid in the presence of small amounts of phosphorus. This is known as the Hell-Volhard-Zelinsky reaction. As is typical of organic syntheses, a mixture of products and some unreacted acetic acid are present at equilibrium. When these acids were studied separately, the data in Table 14.6 were obtained.
    (a) Calculate percent reactions with water or acid ionization constants. Write chemical equations to describe the reaction of these acids with water.
    (b) Suggest a theoretical explanation for the relative strengths of this series of acids.
    (c) A chemical technician is assigned the design of an acid-base titration to determine the amounts of each acid present in the mixture. She knows from experience that the titration of acetic acid with a strong base has a definite endpoint. Sketch a simplified pH curve for titration of a mixture of the four acids produced by chlorinating acetic acid. How could the technician determine relative amounts of the acids present in the mixture?

Table 14.6

COMPARISON OF 0.10 mol/L SOLUTIONS OF CHLOROACETIC ACIDS		
Substance	**Chemical Formula**	**pH of Solution**
acetic acid	$CH_3COOH_{(aq)}$	2.89
chloroacetic acid	$CH_2ClCOOH_{(aq)}$	1.94
dichloroacetic acid	$CHCl_2COOH_{(aq)}$	1.30
trichloroacetic acid	$CCl_3COOH_{(aq)}$	1.14

42. Prepare a concept map to describe scientific problem solving. In your map, include concepts such as certainty, falsification, and evaluation of experimental designs.

43. What role does science play in science-technology-society issues? Write a one-page essay that expresses an informed view. Use acid rain as an example.

44. Explain, using chemical equations, the formation of acids from nitrogen and sulfur oxides in the atmosphere. How are these acids deposited on Earth and what is the environmental impact of this deposition? How complete and certain is the scientific knowledge of acid deposition? Why are scientific and technological knowledge and skills necessary to inform the acid deposition debate?

45. Aristotle, Francis Bacon, Karl Popper, Thomas Kuhn, and others have interpreted and described the advance of scientific knowledge in their own ways. Create an essay, a work of art, or an experiment to portray the evolution of scientific knowledge in one of the chapters you have studied. Use nature of science concepts developed by Aristotle, Bacon, Popper, Kuhn, or other philosophers of science. Choose an approach that communicates your message, for example, prose, poetry, illustrations, drama, song, or a publication such as a pamphlet. Metaphors, analogies, models, and fiction are acceptable methods for communicating your interpretation of the nature of the scientific endeavor, as seen by philosophers of science.

46. Pure sulfuric acid, $H_2SO_{4(l)}$, reacts with solid potassium hydroxide. A complete transfer of protons releases 324.2 kJ of energy per mole of sulfuric acid.
    (a) Write the proton transfer reaction, assuming both protons are transferred.
    (b) What reaction conditions would favor a quantitative reaction?
    (c) If 100 kJ of energy is released, calculate the mass of potassium sulfate obtained.

47. Liquid ammonia can be used as a solvent for acid-base reactions.
    (a) What is the strongest acid species that could be present in this solvent? (Consider the reaction of a strong proton donor such as hydrogen chloride when it dissolves and reacts quantitatively in pure liquid ammonia.)
    (b) What is the strongest base that could be present in pure liquid ammonia?
    (c) The ionization equilibrium of ammonia as a solvent is similar to that of water as a solvent. Write the equilibrium equation for the ionization of the ammonia.
    (d) Sketch a titration curve for the addition of the strongest acid in ammonia to the strongest base. Instead of pH, what do you think would be used on the vertical axis of your graph?

48. Scientists develop theories by trying to explain empirical evidence. These ideas are usually tested by predicting new evidence and then evaluating these predictions experimentally.
    (a) According to the table of acids and bases, which is the stronger acid, H_2O or H_2S?
    (b) Why? Attempt to answer this question using electronegativities and Brønsted-Lowry concepts. Do these concepts provide a satisfactory explanation?
    (c) What other factor(s) might be important in explaining the relative strengths of H_2O and H_2S?
    (d) Test your hypothesis by predicting the relative strength of H_2Se and H_2Te as acids and checking your prediction in a reference book.

Problem 14J Interpretation of Results

The purpose of this problem is to test the effectiveness of concepts in predicting and/or explaining experimental results. Complete the Prediction, Analysis, Evaluation, and Synthesis of the investigation report. In the Evaluation, suggest several diagnostic tests that could be performed to increase the certainty of the interpretation.

Problem

What are the products of the reaction between aluminum and aqueous copper(II) chloride?

Experimental Design

The aluminum strip is placed in a copper(II) chloride solution. Diagnostic tests for copper and for any other product(s) observed are carried out.

Evidence

- An orange-brown solid formed on the aluminum strip.
- Gas bubbles were produced, especially early in the reaction.
- Oxygen and carbon dioxide tests on the gas were negative, but a hydrogen test was positive.

ELECTROCHEMICAL SYSTEMS

"A technological innovation is not necessarily the fruit of a new scientific discovery, but most often is an internal, intrinsic development of technology itself."

Jacques Ellul (1912 –   ),

French philosopher of technology

Of all chemical changes, electro-chemical reactions are among the most common, in both living and non-living systems. Photosynthesis, respiration, and metabolism are all electrochemical processes. Technological systems involving electrochemical reactions, such as the production of metal from their ores, have been used for thousands of years. In the development of these systems, technology has led science. More recently, however, a sophisticated understanding of modern chemistry is clearly needed before new technologies such as high-energy, non-polluting electrochemical cells can be developed. Science and technology nurture each other in a symbiotic relationship.

When you study electrochemistry, you realize the tremendous impact that technology continues to have on our society. You are challenged to understand not only new technological products and processes, but also the goals of technology and its approaches to problem solving.

15 Electrochemistry

The production of this book, or any other book, involves products and processes explained by **electrochemistry** — electron transfer in chemical reactions. Knowledge of electrochemistry will help you connect and clarify many seemingly unrelated reactions. For example, paper for this book is produced from trees that used photosynthesis reactions to grow. Harvesting trees requires machinery made from steel, which is produced by the electrochemical reduction of iron ore. The energy used to run the machines in a forest or in a pulp mill comes from the combustion of fossil fuels. Electrochemical reactions play a role in the production of paper from wood pulp. Photographs used in books may involve the reduction of silver ions to silver metal to form a negative image, which is printed using metal plates made by electrochemical reactions. All of the people involved, from those who harvest the trees to those who read the book, metabolize food to live and work. Electrochemical reactions are the most common type of chemical reaction and include those in photosynthesis, metallurgy, combustion, bleaching, metabolism, and respiration, as well as many others. An understanding of electrochemistry will give you a broader comprehension of many chemical reactions and their importance in both living and non-living systems.

15.1 OXIDATION AND REDUCTION

In prehistoric times, people learned to extract metals from rocks and minerals (Figure 15.1). This discovery initiated both the technology of *metallurgy* and humanity's progression from the Stone Age, through the Bronze Age and the Iron Age, to our increasingly technological modern age. Only a few metals, such as gold and silver, exist naturally in the form of a pure element. Most metals exist on Earth in a variety of compounds mixed with other substances in rocks called ores. The pure metals must be extracted, or refined, from the ores (Figure 15.2). For some metals, the basic procedures are quite simple and were developed early in human history; for others, more complex procedures have been developed more recently.

> "Technology is not defined as the sum of devices, machines, and inventions. It is the way we do things around here — it's a practice, a common practice." — Ursula Franklin, engineer and scholar

The Development of Metallurgical Processes

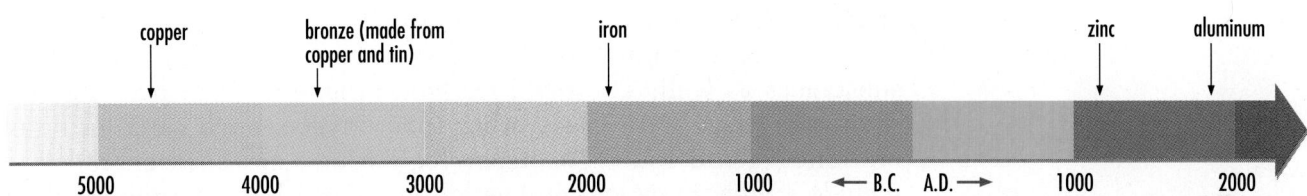

copper   bronze (made from copper and tin)   iron   zinc   aluminum

5000   4000   3000   2000   1000   ← B.C.   A.D. →   1000   2000

Figure 15.1
The technology of metallurgy has a long history, preceding by thousands of years the scientific understanding of the processes.

Although the technological processes of refining vary from one metal to another, the processes typically involve a large volume of ore that is *reduced* to a smaller volume of metal. From metallurgy, the term *reduction* came to be associated with producing metals from their compounds. For example, the production of iron, tin, and copper metals are typical examples of this reduction process.

$$Fe_2O_{3(s)} + 3\,CO_{(g)} \rightarrow 2\,Fe_{(s)} + 3\,CO_{2(g)}$$
$$SnO_{2(s)} + C_{(s)} \rightarrow Sn_{(s)} + CO_{2(g)}$$
$$CuS_{(s)} + H_{2(g)} \rightarrow Cu_{(s)} + H_2S_{(g)}$$

As you can see from these chemical equations, another substance, called a *reducing agent*, causes or promotes the reduction of a metal compound to an elemental metal. In the preceding examples, carbon monoxide is the reducing agent for the production of iron, carbon (charcoal) is the reducing agent for the production of tin, and hydrogen is the reducing agent for the production of copper. These are three of the most common reducing agents used in metallurgical processes.

Before humans discovered the technology of metal refining, they were routinely using fire, an even earlier technology. The technological use of fire did not require a detailed scientific understanding of the processes (Figure 15.3, page 608). Fire has been particularly crucial in the development of human cultures. Only relatively recently, in the 18th century, have we come to realize the role of oxygen in burning. Understanding the connection between corrosion and burning is an even more recent development. Corrosion such as rusting is now understood to be similar to combustion, although corrosion reactions occur more slowly. Reactions of substances with oxygen, whether they were the explosive combustion of gunpowder, the burning of wood, or the slow corrosion of iron (Figure 15.4), came to be called *oxidation*. As

Figure 15.2
The first stage of producing copper metal involves the mining of copper ore. The photograph shows the loading of ore at Island Copper, an open pit mine near Port Hardy, B.C..

Figure 15.3
Making steel is more complicated than making bronze and requires higher temperatures than the temperatures provided by a simple wood fire. Only a few cultures developed the technology to make steel early in their history. In Japan, steel was used in the crafting of samurai swords. At a time when there was no written language, the process of sword making was made into a ritual so that it could be more accurately passed on from one generation to the next.

Figure 15.4
The rusting of steel involves the oxidation of iron and is a major economic and technological problem in our society.

the study of chemistry developed, it became apparent that oxygen was not the only substance that could cause reactions with empirical characteristics similar to oxidation reactions. For example, metals can be converted to compounds by most nonmetals and by some other substances as well. The rapid reaction process we call burning may even take place with gases other than oxygen, such as chlorine or bromine (Figure 15.5). The term "oxidation" has been extended to include a wide range of combustion and corrosion reactions, such as the following.

$$2\,Mg_{(s)} + O_{2(g)} \rightarrow 2\,MgO_{(s)}$$
$$2\,Al_{(s)} + 3\,Cl_{2(g)} \rightarrow 2\,AlCl_{3(s)}$$
$$Cu_{(s)} + Br_{2(g)} \rightarrow CuBr_{2(s)}$$

A substance that causes or promotes the oxidation of a metal to produce a metal compound is called an *oxidizing agent*. In the reactions shown above, the oxidizing agents are oxygen, chlorine, and bromine.

Exercise

1. Write an empirical definition for each of the following terms.
   (a) reduction
   (b) oxidation
   (c) oxidizing agent
   (d) reducing agent
   (e) metallurgy
   (f) corrosion

2. For each of the following, identify the oxidizing agent or the reducing agent and classify the reaction of the metal or metal compound as reduction or oxidation.
   (a) $4\,Fe_{(s)} + 3\,O_{2(g)} \rightarrow 2\,Fe_2O_{3(s)}$
   (b) $2\,PbO_{(s)} + C_{(s)} \rightarrow 2\,Pb_{(s)} + CO_{2(g)}$
   (c) $NiO_{(s)} + H_{2(g)} \rightarrow Ni_{(s)} + H_2O_{(l)}$
   (d) $Sn_{(s)} + Br_{2(l)} \rightarrow SnBr_{2(s)}$
   (e) $Fe_2O_{3(s)} + 3\,CO_{(g)} \rightarrow 2\,Fe_{(s)} + 3\,CO_{2(g)}$
   (f) $Cu_{(s)} + 4\,HNO_{3(aq)} \rightarrow Cu(NO_3)_{2(aq)} + 2\,H_2O_{(l)} + 2\,NO_{2(g)}$

3. List three reducing agents used in metallurgy.

4. What class of elements serves as oxidizing agents for metals?

5. In the history of metallurgy, which came first, technological applications or scientific understanding? Elaborate on your answer.

$$Cu_{(s)} + Br_{2(g)} \rightarrow CuBr_{2(s)}$$

Figure 15.5
Copper metal is oxidized by reactive nonmetals such as bromine.

15.1 Single Replacement Reactions

The purpose of this investigation is to explain some single replacement reactions (page 172) in terms of oxidation and reduction. As part of the Experimental Design, include diagnostic tests (as in Appendix C, C.3) for the predicted products.

Problem

What are the products of the single replacement reactions for the following sets of reactants?

- copper and aqueous silver nitrate
- aqueous chlorine and aqueous sodium bromide
- magnesium and hydrochloric acid
- zinc and aqueous copper(II) sulfate
- aqueous chlorine and aqueous potassium iodide

Materials

lab apron	magnesium ribbon
safety glasses	zinc strip
five small test tubes	aqueous silver nitrate
two test tube stoppers	aqueous sodium bromide
test tube rack	aqueous copper(II) sulfate
steel wool	aqueous potassium iodide
wash bottle	hydrochloric acid
matches	chlorine water
copper strip	trichlorotrifluoroethane

Procedure

1. Set up five test tubes, each filled to a depth of 2–3 cm with one of the five aqueous solutions.

2. Add the element indicated to each test tube.

3. Perform diagnostic tests on each of the five mixtures. Record your evidence.

4. Dispose of the solutions as directed by your teacher.

- Problem
- ✔ Prediction
- ✔ Design
- Materials
- Procedure
- ✔ Evidence
- ✔ Analysis
- ✔ Evaluation
- Synthesis

Half-Reactions

A half-reaction represents what is happening to only one reactant in an overall reaction (page 329). It tells only part of the story. Another half-reaction is required to complete the description of the reaction. Splitting a chemical reaction equation into two parts not only makes the explanations simpler but also leads to some important applications, discussed in Chapter 16.

For example, when zinc metal is placed into a hydrochloric acid solution, gas bubbles form as the zinc slowly disappears. Diagnostic tests show that the gas is hydrogen and that zinc ions are present in the solution. What happens to the zinc and what happens to the hydrochloric acid? The half-reactions help to answer these questions. Zinc atoms are converted to zinc ions, and atomic theory requires that electrons be released, as shown by the following half-reaction equation.

$$Zn_{(s)} \quad \rightarrow \quad Zn^{2+}_{(aq)} \quad + \quad 2\,e^-$$

Simultaneously, hydrogen ions from the hydrochloric acid pick up electrons and are converted into hydrogen gas as shown below.

$$2\,H^+_{(aq)} \quad + \quad 2\,e^- \quad \rightarrow \quad H_{2(g)}$$

Notice that both of these half-reaction equations, or half-reactions, are balanced by mass (same number of element symbols on both sides) and by charge (same total charge on both sides).

Synthesis means "putting together." As a scientific process, synthesis involves combining knowledge (empirical, theoretical, or a mix of both) to obtain a broader and more general description or experimental design. Synthesis is a creative part of what scientists do. The purpose of this exercise is to combine the results of Investigation 15.1 with previous theoretical definitions of atoms and ions and with historical definitions of oxidation and reduction.

Gold Mining

In the gold rush that began in 1858, tens of thousands of miners flooded into British Columbia. They travelled by foot and packhorse up the Fraser River and into the Cariboo region of the interior, setting up tent cities and instant towns such as Barkerville. In deep creek gravels lay the most productive gold veins, and there the miners dug and panned for nuggets. Few became rich.

Gold remains important to British Columbia's economy even today, and the way it is mined still takes advantage of the metal's high density. The province's major gold producer is the Eskay Creek Mine located north of Stewart in northwestern B.C. So rich in gold is the ore that it only needs to be crushed and blended (to sink the dense gold for recovery) before it is shipped to market for processing and refining. Another gold producer is the nearby Snip Mine. There, gravity is used to separate the "free" gold (gold that is not trapped in other minerals) from crushed ore, after which the gold is melted down to make bars. The remaining material is then milled to produce a sulfide concentrate that contains finer particles of gold, which are then processed.

Exercise

6. The purpose of this question is to develop a theory to explain single replacement reactions. Refer to the chemical equation for the reaction of zinc metal and aqueous copper(II) sulfate in Investigation 15.1.
   (a) According to the chemical equation and atomic theory, what happens to the copper(II) ions as they react?
   (b) Write a half-reaction equation showing copper(II) ions converted to copper atoms. Balance this half-reaction equation with the appropriate number of electrons. (See page 329.)
   (c) According to the chemical equation and atomic theory, what happens to the zinc atoms as they react?
   (d) Write a half-reaction equation showing zinc atoms converted to zinc ions, including the appropriate number of electrons.
   (e) Which reactant gains electrons? Which loses electrons?
   (f) Does the sulfate ion change during this reaction? What is a substance called that does not change during a reaction?

7. Write a pair of balanced half-reaction equations — one showing a gain of electrons and one showing a loss of electrons — for each of the following pairs of reactants from Investigation 15.1.
   (a) copper and aqueous silver ions
   (b) magnesium and aqueous hydrogen ions

8. Chlorine and hydrogen, the nonmetals in the reactions in Investigation 15.1, have diatomic molecules but monatomic ions. The half-reaction equation, however, must be balanced for both atoms and charge. For example, chlorine molecules become choride ions.

$$Cl_{2(aq)} + 2\,e^- \rightarrow 2\,Cl^-_{(aq)}$$

   (a) In the reaction of chlorine with bromide ions, what happens to the bromide ions?
   (b) Write a balanced half-reaction equation showing this change.
   (c) Write a pair of balanced half-reaction equations for the reaction of chlorine with iodide ions.
   (d) What are the spectator ions in the reactions involving chlorine in Investigation 15.1?

Theoretical Definitions of Reduction and Oxidation

In a laboratory, single replacement reactions in aqueous solution are easier to study than the metallurgy or corrosion reactions discussed earlier in this chapter. However, all of these reactions share a common feature — ions are converted to atoms and atoms are converted to ions. For example, consider the reduction of aqueous silver nitrate to silver metal in the presence of solid copper (Figure 15.6). According to atomic theory, silver atoms are electrically neutral particles ($47p^+$, $47e^-$) and silver ions are charged particles ($47p^+$, $46e^-$). In this reaction, an electron is required to convert a silver ion into a silver atom. The following half-reaction equation explains the reduction of silver ions using the theoretical rules for atoms and ions. The gain of electrons is called **reduction**.

$$Ag^+_{(aq)} + e^- \rightarrow Ag^0_{(s)} \qquad \text{(reduction)}$$

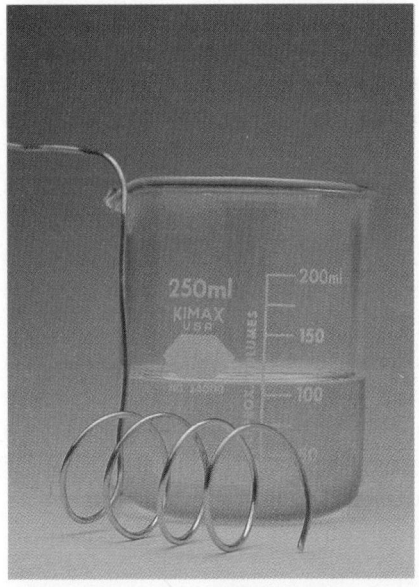
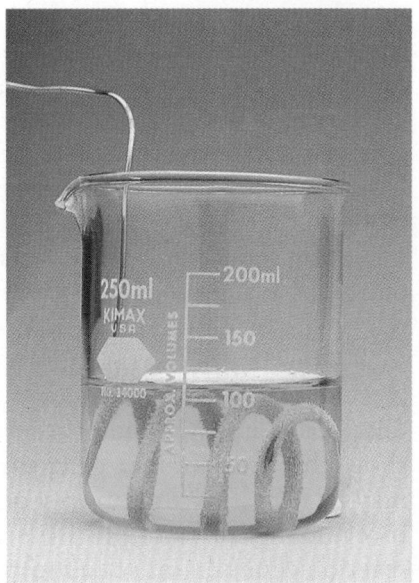

Figure 15.6
A piece of copper before it is placed into a beaker of silver nitrate solution (left). Note the changes after the reaction has occurred (right).

ASLIHAN YENER

Before becoming an archeologist, Aslihan Yener (1946 –   ) studied chemistry. Her analysis of bronze objects and the source of the metal used in them is based on comparing the ratios of the isotopes of lead that are present. If results are similar for a bronze artifact and for the metal from a particular mine, this is strong evidence that the metal came from that mine.

Bronze, a copper-tin alloy, was widely used in Europe and the Middle East from about 3000 B.C. to 1000 B.C. Weapons, tools, and many personal and religious objects were made from bronze. Archeologists know that copper was readily available in the Middle East, the center of bronze production and trading, but where did the metallurgists get their tin? Aslihan Yener, now an archeologist with the Smithsonian Institution in the United States, believes she has the answer. Her discoveries during expeditions to her homeland, Turkey, and her chemical analysis of bronze objects have established that the tin came from a large tin mine in the Taurus Mountains of Turkey.

Although this theoretical definition of reduction is in agreement with current atomic theory, it does not explain where the electrons come from. The explanation involves the function of the reducing agent. As crystals of silver metal are produced, the solution becomes blue, indicating that copper atoms are being converted to copper(II) ions. According to atomic theory, copper atoms (29p$^+$, 29e$^-$) must be losing electrons as they form copper(II) ions (29p$^+$, 27e$^-$). The loss of electrons is called **oxidation**.

$$Cu^0_{(s)} \rightarrow Cu^{2+}_{(aq)} + 2\,e^- \qquad \text{(oxidation)}$$

Evidence shows that the silver-colored solid and the blue color of the solution are simultaneously formed near the surface of the copper metal (Figure 15.6). Therefore, scientists believe that the electrons required by the silver ions are supplied by the copper atoms during collisions of the ions and atoms.

The theory of electrochemistry created to describe and explain reactions of this type suggests that electrons are gained by an oxidizing agent and lost by a reducing agent, and that the number of electrons gained in a reaction must equal the number of electrons lost. In contrast to the restricted technological definitions of oxidation and reduction as separate processes, a theoretical description requires oxidation and reduction to be simultaneous processes referred to as an oxidation-reduction reaction. For efficient communication, scientists often use a shortened version, **redox (red**uction-**ox**idation**) reaction**, to refer to an oxidation-reduction reaction.

Reduction and oxidation half-reaction equations and the overall (net) ionic equation summarize the electron transfer that is believed to take place during a redox reaction. In two half-reaction equations, to show that the number of electrons gained equals the number lost, it may be necessary to multiply one or both half-reaction equations by an integer. In the example illustrated in Figure 15.7, the silver half-reaction equation must be multiplied by 2.

$$2\,[Ag^+_{(aq)} + e^- \rightarrow Ag_{(s)}] \qquad \text{(two electrons gained)}$$
$$Cu_{(s)} \rightarrow Cu^{2+}_{(aq)} + 2\,e^- \qquad \text{(two electrons lost)}$$

Now, add the half-reaction equations and cancel terms that appear on both sides of the equation to obtain the net ionic equation.

$$2\,Ag^+_{(aq)} + \cancel{2\,e^-} + Cu_{(s)} \rightarrow 2\,Ag_{(s)} + Cu^{2+}_{(aq)} + \cancel{2\,e^-}$$

$$2\,Ag^+_{(aq)} + Cu_{(s)} \rightarrow 2\,Ag_{(s)} + Cu^{2+}_{(aq)}$$

Note that *reduction* and *oxidation* are processes; the **reducing agent (RA)** and the **oxidizing agent (OA)** are substances.

reduced to metal

$$2\,Ag^+_{(aq)} + Cu_{(s)} \rightarrow 2\,Ag_{(s)} + Cu^{2+}_{(aq)}$$

OA   RA

oxidized to metal ion

Silver ions are reduced to silver metal by reaction with copper metal, the reducing agent. Simultaneously, copper metal is oxidized to copper(II) ions by reaction with silver ions, the oxidizing agent.

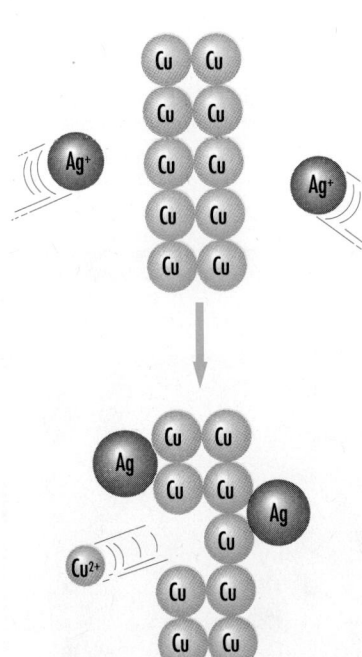

Figure 15.7
A model of the reaction of copper metal and silver nitrate solution illustrates aqueous silver ions reacting at the surface of a copper strip.

To evaluate this theory of oxidation and reduction you should look at its logical consistency with other accepted theories and definitions. The theoretical definitions of oxidation and reduction are consistent with the historical, empirical definitions presented earlier in this chapter; for example, a compound is reduced to a metal and a metal is oxidized to form a compound. Redox theory is also consistent with accepted atomic theory and the collision-reaction theory. Most importantly, redox theory explains the observations made by scientists.

> **SUMMARY: REDOX THEORY**
>
> - A redox reaction is a chemical reaction in which electrons are transferred between entities.
> - The number of electrons gained by one entity equals the number of electrons lost by another.
> - Reduction is a process in which electrons are gained.
> - Oxidation is a process in which electrons are lost.
> - A **reducing agent** promotes reduction by donating (losing) electrons in a redox reaction, and is itself oxidized.
> - An **oxidizing agent** causes oxidation by removing (gaining) electrons in a redox reaction, and is itself reduced.

"LEO the lion says GER" is a mnemonic to help remember that "Loss of Electrons is Oxidation" and "Gain of Electrons is Reduction."

Elephants never forget when they rely on mnemonics!

For most metals and metal ions, the oxidation and reduction half-reactions are relatively simple. According to atomic theory, metal cations gain electrons and are reduced to form the metal. The reverse process, a metal losing some of its electrons and being oxidized to a metal cation, can also occur.

$$\text{metal cation} \;+\; n\,e^- \underset{\text{oxidation}}{\overset{\text{reduction}}{\rightleftharpoons}} \text{metal}$$

Similarly, many nonmetals and simple, nonmetal anions also undergo a reduction and oxidation involving a transfer of electrons. This is also consistent with atomic theory because nonmetals tend to form negative ions (anions).

$$\text{nonmetal} \;+\; n\,e^- \underset{\text{oxidation}}{\overset{\text{reduction}}{\rightleftharpoons}} \text{nonmetal anion}$$

For example,

$$Cl_{2(aq)} \;+\; 2\,e^- \underset{\text{oxidation}}{\overset{\text{reduction}}{\rightleftharpoons}} 2\,Cl^-_{(aq)}$$

EXAMPLE

Use redox theory to write and label two balanced half-reaction equations to describe the reaction of zinc metal with aqueous gold(III) nitrate, as given by the following (unbalanced) chemical equation.

$$Zn_{(s)} + Au(NO_3)_{3(aq)} \rightarrow Au_{(s)} + Zn(NO_3)_{2(aq)}$$

$$Au^{3+}_{(aq)} + 3e^- \rightarrow Au_{(s)} \qquad \text{(reduction)}$$

$$Zn_{(s)} \rightarrow Zn^{2+}_{(aq)} + 2e^- \qquad \text{(oxidation)}$$

Exercise

9. Write a theoretical definition for each of the following terms.
   (a) redox reaction
   (b) reduction
   (c) oxidation
   (d) oxidizing agent
   (e) reducing agent

10. For each of the following, write the oxidation and reduction half-reaction equations. Balance each half-reaction equation by balancing atoms and charge.
    (a) $Sn_{(s)} + Cu(NO_3)_{2(aq)} \rightarrow Cu_{(s)} + Sn(NO_3)_{4(aq)}$
    (b) $Br_{2(l)} + KI_{(aq)} \rightarrow I_{2(s)} + KBr_{(aq)}$
    (c) $Ca_{(s)} + HNO_{3(aq)} \rightarrow H_{2(g)} + Ca(NO_3)_{2(aq)}$
    (d) $Al_{(s)} + Fe_2O_{3(s)} \rightarrow Fe_{(l)} + Al_2O_{3(s)}$ (see Figure 15.8)

11. Use your answers to question 10 to answer the following.
    (a) Classify each half-reaction equation as reduction or oxidation.
    (b) In each of the net chemical equations, label each reactant as a reducing agent or an oxidizing agent.

12. According to atomic theory, all atoms have an attraction for electrons. Which do you think is more consistent with atomic theory, to speak of oxidizing agents "pulling electrons away from reducing agents" or of reducing agents "giving electrons to oxidizing agents"? Justify your answer.

13. Ionic compounds can react in double replacement reactions.
    (a) Predict the balanced chemical equation for the reaction of aqueous solutions of iron(III) chloride and sodium hydroxide.
    (b) According to ideas discussed in this chapter, has a redox reaction taken place in the reaction above? Explain your answer.

Figure 15.8
The reduction of iron(III) oxide by aluminum powder is called the "thermite" reaction. Because this reaction is rapid and very exothermic, molten white-hot iron is produced. In this apparatus, widely used in the past, the thermite reaction occurs in the upper chamber and molten iron flows down into a mold between the ends of two rails, welding them together.

Writing Complex Half-Reaction Equations

Although most metals and nonmetals have relatively simple half-reaction equations, polyatomic ions and molecular compounds undergo more complicated oxidation and reduction processes. In most of these processes, the reaction takes place in an aqueous solution that very often is acidic or basic. Experimental evidence shows that water molecules, hydrogen ions, and hydroxide ions play an important role in these half-reactions. A method of writing half-reactions for polyatomic ions and molecular compounds requires that water molecules and hydrogen or hydroxide ions be included. This method, illustrated in the following example, is sometimes called the "half-reaction" or ion-electron" method.

Nitrous acid can be reduced in an acidic solution, with an appropriate reducing agent, to form nitrogen monoxide gas. What is the reduction half-reaction for nitrous acid? The first step is to write the reactants and products for the half-reaction equation to be determined.

$$HNO_{2(aq)} \rightarrow NO_{(g)}$$

If necessary, you should balance all atoms other than oxygen and hydrogen in this partial equation. In this example, there is only one nitrogen atom on each side. Next, add water molecules, present in an aqueous solution, to balance the oxygen atoms.

$$HNO_{2(aq)} \rightarrow NO_{(g)} + H_2O_{(l)}$$

Because the reaction takes place in an acidic solution, hydrogen ions are present, and these are used to balance the hydrogen on both sides of the equation.

$$H^+_{(aq)} + HNO_{2(aq)} \rightarrow O_{(g)} + H_2O_{(l)}$$

At this stage, all of the atoms should be balanced but the charge on both sides is not balanced. Add an appropriate number of electrons to balance the charge. Because electrons carry a negative charge, they are always added to the less negative, or more positive, side of the half-reaction.

$$e^- + H^+_{(aq)} + HNO_{2(aq)} \rightarrow NO_{(g)} + H_2O_{(l)}$$

This balanced half-reaction equation represents a gain of electrons — in other words, a reduction of the nitrous acid.

In a basic solution, the concentration of hydroxide ions greatly exceeds that of hydrogen ions. To develop a half-reaction in a basic solution, it would make sense to follow a similar procedure and introduce hydroxide ions instead of hydrogen ions as the half-reaction is constructed. Unfortunately, this approach does not always work easily. Therefore, for basic solutions, we will develop the half-reaction as if it occurred in an acidic solution and then convert the hydrogen ions into water molecules using hydroxide ions. This trick works because a hydrogen ion and a hydroxide ion react in a 1:1 ratio to form a water molecule. The following example illustrates the procedure for writing half-reaction equations that occur in basic solutions.

Using an appropriate oxidizing agent, copper metal can be oxidized in a basic solution to form copper(I) oxide. What is the half-reaction for this process? Following the same steps as before, write the formulas and balance the atoms, other than oxygen and hydrogen.

$$2\,Cu_{(s)} \rightarrow Cu_2O_{(s)}$$

Next, balance the oxygen using water molecules and balance the hydrogen using hydrogen ions, assuming an acidic solution. The charge is balanced using electrons.

$$H_2O_{(l)} + 2\,Cu_{(s)} \rightarrow Cu_2O_{(s)} + 2\,H^+_{(aq)} + 2\,e^-$$

Because the half-reaction occurs in a basic solution, to *both sides* of the equation, add the same number of hydroxide ions as there are hydrogen ions. This is done to maintain the balance of mass and charge.

$$2\ OH^-_{(aq)} + H_2O_{(l)} + 2\ Cu_{(s)} \rightarrow Cu_2O_{(s)} + 2\ H^+_{(aq)} + 2\ e^- + 2\ OH^-_{(aq)}$$

Combine equal numbers of hydrogen ions and hydroxide ions to form water molecules.

$$2\ OH^-_{(aq)} + H_2O_{(l)} + 2\ Cu_{(s)} \rightarrow Cu_2O_{(s)} + 2\ H_2O_{(l)} + 2\ e^-$$

Finally, cancel H_2O and anything else that is the same from both sides of the equation.

$$2\ OH^-_{(aq)} + 2\ Cu_{(s)} \rightarrow Cu_2O_{(s)} + H_2O_{(l)} + 2\ e^-$$

SUMMARY: WRITING COMPLEX HALF-REACTION EQUATIONS

Step 1: Write the chemical formulas for the reactants and products.
Step 2: Balance all atoms, other than O and H.
Step 3: Balance O by adding $H_2O_{(l)}$.
Step 4: Balance H by adding $H^+_{(aq)}$.
Step 5: Balance the charge on each side by adding e^- and cancel anything that is the same on both sides.

For basic solutions only

Step 6: Add $OH^-_{(aq)}$ to both sides equal in number to the $H^+_{(aq)}$ present.
Step 7: Combine $H^+_{(aq)}$ and $OH^-_{(aq)}$ on the same side to form $H_2O_{(l)}$.
Step 8: Cancel equal amounts of $H_2O_{(l)}$ and anything else that is the same on both sides.

EXAMPLE

Aqueous permanganate ions are reduced to solid manganese(IV) oxide in a basic solution. Write the reduction half-reaction equation.

$$4\ OH^-_{(aq)} + 3\ e^- + 4\ H^+_{(aq)} + MnO_4^-_{(aq)} \rightarrow MnO_{2(s)} + 2\ H_2O_{(l)} + 4\ OH^-_{(aq)}$$
$$4\ H_2O_{(l)} + 3\ e^- + MnO_4^-_{(aq)} \rightarrow MnO_{2(s)} + 2\ H_2O_{(l)} + 4\ OH^-_{(aq)}$$
$$MnO_4^-_{(aq)} + 2\ H_2O_{(l)} + 3\ e^- \rightarrow MnO_{2(s)} + 4\ OH^-_{(aq)}$$

Exercise

14. For each of the following, complete the half-reaction equation and classify it as an oxidation or a reduction.
    (a) dinitrogen oxide to nitrogen gas in an acidic solution
    (b) nitrite ions to nitrate ions in a basic solution
    (c) silver(I) oxide to silver metal in a basic solution
    (d) nitrate ions to nitrous acid in an acidic solution
    (e) hydrogen gas to water in a basic solution
    (f) chlorine to perchlorate ions in an acidic solution

15.2 Spontaneity of Redox Reactions

Until now in this textbook, it has been assumed that all chemical reactions are **spontaneous**; that is, they occur without a continuous addition of energy to the system. Spontaneous redox reactions in solution generally provide visible evidence of a reaction within a few minutes. The scientific purpose of this investigation is to test the assumption that all single replacement reactions are spontaneous.

Problem

Which combinations of copper, lead, silver, and zinc metals and their aqueous metal ion solutions produce spontaneous reactions?

Experimental Design

A drop of each solution is placed in separate locations on a clean area of each of the four metal strips.

Materials

lab apron
safety glasses
reusable strips of copper, lead, silver, and zinc metals
   (*Note that the lead strips bend much more easily than the zinc strips, which look similar.*)
0.10 mol/L solutions of copper(II) nitrate, lead(II) nitrate, silver nitrate, and zinc nitrate in dropper bottles
steel wool

- ☐ Problem
- ✔ Prediction
- ☐ Design
- ☐ Materials
- ✔ Procedure
- ✔ Evidence
- ✔ Analysis
- ✔ Evaluation
- ☐ Synthesis

C   CAUTION

These chemicals are toxic — especially the lead solution — and irritant. Avoid skin contact. Remember to wash your hands before leaving the laboratory.

D   DISPOSAL TIP

Rinse all of the metal strips thoroughly and return them so they can be used again.

Synthesis of a Reaction Spontaneity Rule

The evidence of products you obtained in Investigation 15.2 probably verified your predictions. However, the assumption that the reactions are spontaneous must be judged unacceptable, as only six of the combinations led to a reaction. The question that arises is, "How do you know when a chemical reaction will occur?" Based on the evidence collected in this investigation, the reactivity of the four metal ions can be compared by counting the number of spontaneous reactions observed.

Number of reactions that occurred	3	2	1	0
Ions	$Ag^+_{(aq)}$	$Cu^{2+}_{(aq)}$	$Pb^{2+}_{(aq)}$	$Zn^{2+}_{(aq)}$

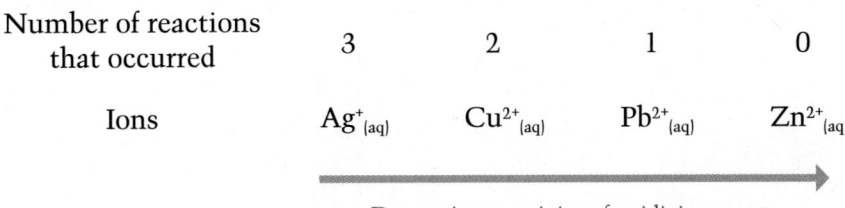

Decreasing reactivity of oxidizing agents

The order of reactivity of the four metals can be obtained in a similar way.

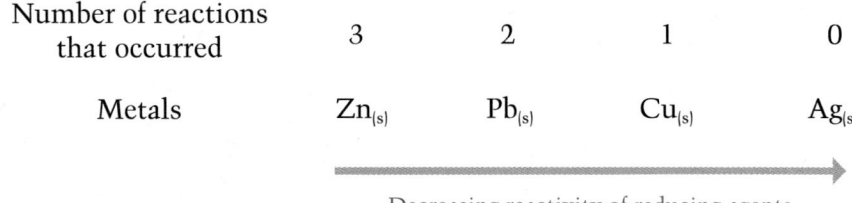

Number of reactions that occurred	3	2	1	0
Metals	$Zn_{(s)}$	$Pb_{(s)}$	$Cu_{(s)}$	$Ag_{(s)}$

Decreasing reactivity of reducing agents

In these reactions, the metal ions are the oxidizing agents and the silver ion is the strongest oxidizing agent (SOA) of the four ions. The metals are the reducing agents and the zinc metal is the strongest reducing agent (SRA). The two lists of reactivity can be combined as half-reaction equations, as shown in Table 15.1.

Table 15.1

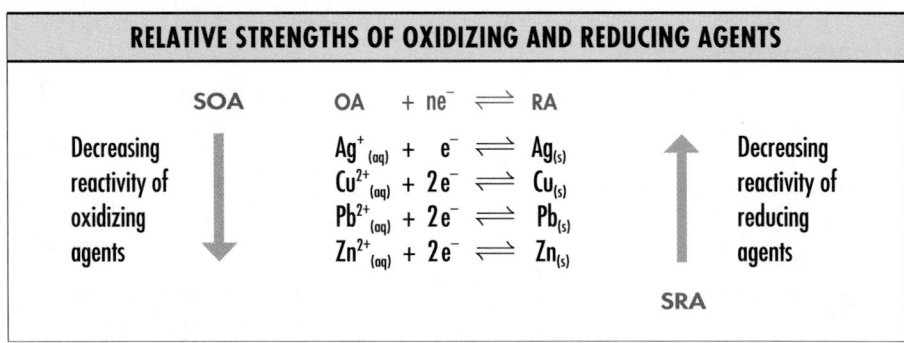

RELATIVE STRENGTHS OF OXIDIZING AND REDUCING AGENTS

$$SOA \qquad OA + ne^- \rightleftharpoons RA$$

Decreasing reactivity of oxidizing agents

$$Ag^+_{(aq)} + e^- \rightleftharpoons Ag_{(s)}$$
$$Cu^{2+}_{(aq)} + 2e^- \rightleftharpoons Cu_{(s)}$$
$$Pb^{2+}_{(aq)} + 2e^- \rightleftharpoons Pb_{(s)}$$
$$Zn^{2+}_{(aq)} + 2e^- \rightleftharpoons Zn_{(s)}$$

Decreasing reactivity of reducing agents

SRA

All the predicted reactions in Investigation 15.2 involve a metal ion and a metal atom. In Table 15.1, the metal ions are on the left side of the equations and the metal atoms are on the right side. For metal ions (the oxidizing agents), the half-reaction equations are read from left to right in the table. For metal atoms (the reducing agents), the half-reaction equations are read from right to left.

Exercise

15. Refer to your evidence from Investigation 15.2.
   (a) List the metal(s) that reacted spontaneously with a copper(II) ion solution.
   (b) Which metal(s) did not appear to react with a copper(II) ion solution?
   (c) Start with the position of $Cu^{2+}_{(aq)}$ in Table 15.1 and note the position of the metals that reacted and the metal(s) that did not react. For a metal that reacts spontaneously with $Cu^{2+}_{(aq)}$, where does the metal appear on a table of reduction half-reactions?
   (d) Repeat (a), (b), and (c) for the $Pb^{2+}_{(aq)}$ ion.
   (e) Your answer to parts (c) and (d) is an empirical hypothesis that can be tested by predicting the reaction evidence for the other metal ions in Investigation 15.2. Use the table to predict which are the other spontaneous reactions observed in this investigation. Is your hypothesis verified?

Evidence from many redox reactions, for which half-reactions have been listed in this way, has been used to establish a generalization, called the **redox spontaneity rule**. A spontaneous redox reaction occurs only if the oxidizing agent (OA) is above the reducing agent (RA) in a table of relative strengths of oxidizing and reducing agents. Figure 15.9 illustrates how you can use the rule, along with such a table of oxidizing and reducing agents, to predict whether or not a reaction is spontaneous.

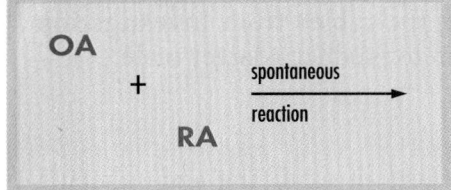
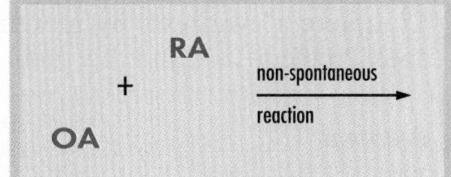

Figure 15.9
The redox spontaneity rule.

Problem 15A Spontaneity of Reactions

To develop a table of relative strengths of oxidizing and reducing agents and to check the consistency of the spontaneity rule for nonmetals, the following experiment was performed. Complete the Analysis of the investigation report (including a table like Table 15.1).

Problem

What is the relative strength of oxidizing agents among chlorine, bromine, and iodine?

Experimental Design

Each of the three halogens is separately added to solutions of the other three halide ions. A halogen test is conducted on each final mixture and, as a control, on each aqueous halogen.

Evidence

REACTIONS OF HALOGENS WITH SOLUTIONS OF HALIDES			
	$Br_{2(aq)}$	$Cl_{2(aq)}$	$I_{2(aq)}$
$Br^-_{(aq)}$	—	✔	—
$Cl^-_{(aq)}$	—	—	—
$I^-_{(aq)}$	✔	✔	—

✔ indicates color change in the halogen test compared with the control

Henry Taube

Henry Taube (1915 –   ) was born and raised in Saskatchewan, where he earned his M.Sc. degree from the University of Saskatchewan in Saskatoon. He completed his Ph.D. studies at the University of California in Berkeley in 1940, then worked briefly at several universities, including the University of Alberta. When he was unable to find a permanent position in Canada, he returned to the United States. In 1946 he joined the Department of Chemistry at the University of Chicago. Here, he began the research that led to a Nobel Prize in chemistry 37 years later, in 1983.

Henry Taube's research deals primarily with reaction rates and mechanisms of inorganic reactions in solution. Dr. Taube and his associates have substantially increased our understanding of oxidation-reduction reactions in solution. His work on electron-transfer reactions led to a synthesis of theoretical concepts of electron structure with empirical concepts of rates of reaction. This unity of theoretical and empirical concepts has provided many other scientists with a successful strategy for investigating redox reactions. Henry Taube is one of eight Nobel laureates in chemistry who were educated in Canada or who have lived and worked in Canada (see page 533).

Problem 15B Tables of Relative Strengths of Oxidizing and Reducing Agents — Design 1

A research team is developing a table of relative strengths of oxidizing and reducing agents. One team member had completed Investigation 15.2 and another had completed the investigation reported in Problem 15A. A third member used a combination of metals, nonmetals, and solutions.

Complete the analysis of the evidence gathered in the third experiment by constructing a table like Table 15.1, page 618. Complete a synthesis by merging the tables from Investigation 15.2, Problem 15A, and this problem to produce a larger table.

Problem

What is the table of relative strengths of oxidizing and reducing agents for copper, silver, bromine, and iodine?

Experimental Design

Each substance is placed in solutions of the ions of the other three substances.

Evidence

REACTIONS OF METALS AND NONMETALS WITH SOLUTIONS OF IONS

	$I_{2(aq)}$	$Cu^{2+}_{(aq)}$	$Ag^+_{(aq)}$	$Br_{2(aq)}$
$I^-_{(aq)}$	—	—	✔	✔
$Cu_{(s)}$	✔	—	✔	✔
$Ag_{(s)}$	—	—	—	✔
$Br^-_{(aq)}$	—	—	—	—

✔ indicates color change and/or precipitate

A Second Experimental Design for Building Tables of Relative Strengths of Oxidizing and Reducing Agents

Once a spontaneity rule is developed from experimental evidence, the rule may be used to generate half-reaction tables. The evidence to be analyzed in this case is a net ionic equation, accompanied by observations of spontaneity. In the following design, the spontaneity rule, rather than the number of reactions observed, is used to order the oxidizing and reducing agents to produce a table. The procedural knowledge for this type of analysis and synthesis is illustrated by the following example.

Three reactions among indium, cobalt, palladium, and copper were investigated. The reaction equations below indicate that two spontaneous reactions occurred. Using these equations, construct a table of relative strengths of oxidizing and reducing agents.

$$3\,Co^{2+}_{(aq)} + 2\,In_{(s)} \rightarrow 2\,In^{3+}_{(aq)} + 3\,Co_{(s)}$$
$$Cu^{2+}_{(aq)} + Co_{(s)} \rightarrow Co^{2+}_{(aq)} + Cu_{(s)}$$
$$Cu^{2+}_{(aq)} + Pd_{(s)} \rightarrow \text{no evidence of reaction}$$

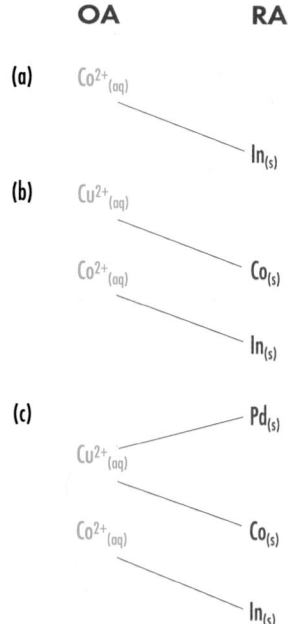

To construct a table from this information, work with one equation at a time. Identify the oxidizing and reducing agents for the first reaction, and arrange them in two columns using the spontaneity rule. For the first reaction, this step is shown in Figure 15.10(a). $Co^{2+}_{(aq)}$ is the oxidizing agent and $In_{(s)}$ is the reducing agent. Since the reaction is spontaneous, the oxidizing agent is above the reducing agent in the list. In the second reaction, $Cu^{2+}_{(aq)}$ is the oxidizing agent and $Co_{(s)}$ is the reducing agent. This reaction is also spontaneous; therefore $Cu^{2+}_{(aq)}$ is above $Co_{(s)}$ in the list. Since a metal appears on the same line as its ion in a half-reaction table, add $Co_{(s)}$ and extend the list as shown in Figure 15.10(b). No reaction occurs for the third pair of reagents. If a reaction had occurred, $Cu^{2+}_{(aq)}$ would be the oxidizing agent and $Pd_{(s)}$ would be the reducing agent. As this reaction is not spontaneous, the oxidizing agent appears below the reducing agent. Figure 15.10(c) shows the list extended to include $Pd_{(s)}$. To complete the table, write balanced half-reaction equations for each oxidizing/reducing agent pair.

SOA $\quad Pd^{2+}_{(aq)} + 2\,e^- \rightleftharpoons Pd_{(s)}$

$\quad\quad\quad Cu^{2+}_{(aq)} + 2\,e^- \rightleftharpoons Cu_{(s)}$

$\quad\quad\quad Co^{2+}_{(aq)} + 2\,e^- \rightleftharpoons Co_{(s)}$

$\quad\quad\quad In^{3+}_{(aq)} + 3\,e^- \rightleftharpoons In_{(s)} \quad$ SRA

Figure 15.10
The relative position of a pair of oxidizing and reducing agents indicates whether a reaction will be spontaneous or not.

Exercise

16. Two reactions were performed to obtain evidence that could be used to practice the spontaneity rule. Construct a table of relative strengths of oxidizing and reducing agents.

    $$Co^{2+}_{(aq)} + Zn_{(s)} \rightarrow Co_{(s)} + Zn^{2+}_{(aq)}$$
    $$Mg^{2+}_{(aq)} + Zn_{(s)} \rightarrow \text{no evidence of reaction}$$

17. In a school laboratory four metals were combined with each of four solutions. The evidence collected is represented by the following chemical equations. Construct a table of relative strengths of oxidizing and reducing agents.

    $$Be_{(s)} + Cd^{2+}_{(aq)} \rightarrow Be^{2+}_{(aq)} + Cd_{(s)}$$
    $$Cd_{(s)} + 2\,H^+_{(aq)} \rightarrow Cd^{2+}_{(aq)} + H_{2(g)}$$
    $$Ca^{2+}_{(aq)} + Be_{(s)} \rightarrow \text{no evidence of reaction}$$
    $$Cu_{(s)} + 2\,H^+_{(aq)} \rightarrow \text{no evidence of reaction}$$

18. Is the redox spontaneity rule (Figure 15.9, page 619) empirical or theoretical? Justify your answer.

19. Use the relative strengths of nonmetals and metals as oxidizing and reducing agents, as indicated in the following unbalanced equations, to construct a table of half-reactions.

    $$Ag_{(s)} + Br_{2(l)} \rightarrow AgBr_{(s)}$$
    $$Ag_{(s)} + I_{2(s)} \rightarrow \text{no evidence of reaction}$$
    $$Cu^{2+}_{(aq)} + I^-_{(aq)} \rightarrow \text{no redox reaction}$$
    $$Br_{2(l)} + Cl^-_{(aq)} \rightarrow \text{no evidence of reaction}$$

Non-spontaneity of a reaction is communicated in several ways: with the phrases "no evidence of reaction"; "non-spontaneous," or "no reaction"; or with "non-spont." or "ns" written over the equation arrow.

By convention, the strongest oxidizing agent is at the top left in a table of relative strengths of oxidizing and reducing agents and the strongest reducing agent is at the bottom right of the table.

Problem 15C Tables of Relative Strengths of Oxidizing and Reducing Agents — Design 2

In an experiment, the relative reactivity of four metals and hydrogen was tested. Complete the Analysis of the investigation report, including a table of relative strengths of oxidizing and reducing agents.

Problem

What is the relative strength of oxidizing and reducing agents for strontium, cerium, nickel, hydrogen, platinum, and their aqueous ions?

Experimental Design

A series of reactions was attempted and the results recorded.

Evidence

$$3\,Sr_{(s)} + 2\,Ce^{3+}_{(aq)} \rightarrow 3\,Sr^{2+}_{(aq)} + 2\,Ce_{(s)}$$
$$Ni_{(s)} + 2\,H^+_{(aq)} \rightarrow Ni^{2+}_{(aq)} + H_{2(g)}$$
$$2\,Ce^{3+}_{(aq)} + 3\,Ni_{(s)} \rightarrow \text{no evidence of reaction}$$
$$Pt_{(s)} + 2\,H^+_{(aq)} \rightarrow \text{no evidence of reaction (assume } Pt^{4+}_{(aq)})$$

To find the strongest oxidizing agent, start at the top left of a table of half-reactions, and read down. To find the strongest reducing agent, start at the bottom right and read up.

Evidence collected in many experiments, like Investigation 15.2 and Problems 15A, 15B, and 15C, has been analyzed to produce an extended table of oxidizing and reducing agents. Some of these are included in the table in Appendix F. You can use this table to compare oxidizing and reducing agents, and to predict spontaneous redox reactions.

Exercise

20. Use the table of relative strengths of oxidizing and reducing agents (Appendix F) to arrange the following metal ions in order of decreasing strength as oxidizing agents: lead(II) ions, silver ions, zinc ions, and copper(II) ions. How does this order compare with the evidence you gathered in Investigation 15.2?

21. According to the same table in question 20, what classes of substances usually behave as oxidizing agents?

22. According to the same table in question 20, what classes of substances usually behave as reducing agents?

23. Use the theory of quantum mechanics (in the restricted version for monatomic ions described on pages 84–85) to explain why nonmetals behave as oxidizing agents and metals behave as reducing agents. Is there logical consistency between the theory of quantum mechanics and the empirically determined table of oxidizing and reducing agents?

24. Trends in the reactivity of elements (Figure 8.2, page 293) show that fluorine is the most reactive nonmetal. How does this relate to the position of fluorine in the table of oxidizing and reducing agents? State one reason why this element is the most reactive

nonmetal. Why is your reason an explanation? (Keep asking a series of "why" questions until your theoretical knowledge is expended. Does your theory pass the test of being able to explain the empirically determined table?)

25. From your own knowledge, list two metals that are found as elements and two that are never found as elements in nature. Test your answer by referring to the position of these metals in the table of oxidizing and reducing agents.

URSULA FRANKLIN

Ursula Franklin (1921 –) is an internationally recognized Canadian scientist and scholar, whose career illustrates the vital connections among science, technology, and society. Born in Germany in 1921, she received her Ph.D. in experimental physics from the Technical University in Berlin in 1948. The following year she received a fellowship to study in Canada, where she did post-doctoral research at the University of Toronto. She was associated with the Ontario Research Foundation for fifteen years and then, in 1967, returned to the University of Toronto, where she became a professor in 1973.

Franklin specializes in the structures of metals and alloys. She pioneered the study of metallurgy in ancient cultures and has worked on the dating of copper, bronze, and ceramic artifacts used in prehistoric cultures in Canada. She is the director of the University of Toronto's Collegium Archeometrium, which brings together scholars from the university and the Royal Ontario Museum in interdisciplinary studies of the past.

Another of Franklin's interests is the social impact of technology. In her lectures, she strives to open the eyes of scientists and non-scientists to the impact of science and technology on the quality of human life. In her view, "There is no hierarchical relationship between science and technology. Science is not the mother of technology. Science and technology today have parallel or side-by-side relationships; they stimulate and utilize each other."

Franklin cites recent developments in telecommunications as an example of how technology can affect society. People now use the telephone instead of writing, or fax or e-mail documents instead of mailing them. She points out , "In addition to carrying out established tasks in a different manner, there are genuinely new activities possible now that could not have been accomplished with the old technologies."

Technology has developed means of overcoming the limitations of distance and time. Radio, television, film, and video create new realities with intense emotional impact. People feel like participants rather than observers. These developments have a far-reaching effect on the way we view the world and think about the future.

Franklin has helped develop Canadian policies on science and technology through her membership on both the Science Council of Canada and the Natural Science and Engineering Research Council. She has voiced serious concerns about our lack of support for research in science and technology. She points out that, because of distracting concerns over funding and academic acceptability, creative and imaginative thought is often stifled before it can be developed into research questions. She insists, "We need to have diversity, a seed bed for experimentation."

Besides her impressive scientific accomplishments, Franklin has done extensive work to better the lives of people. She is a tireless advocate for Science for Peace. She has worked to improve opportunities for women in scientific and academic positions and to make scientific information understandable and accessible to the public. Colleagues describe her as a "forthright, warm, supportive, concerned, and honest person" and as having a "delightful sense of humor."

Her achievements have resulted in numerous awards and honorary degrees. In 1982 she was named an Officer of the Order of Canada, and in 1984, she was appointed a University Professor at the University of Toronto, an honor acknowledging that her academic and scientific interests go far beyond a single discipline. Ursula Franklin is realistic, but not pessimistic, about social and scientific issues, and her enthusiasm about what can be done inspires others. As a scientist and as a Canadian, she exemplifies how scientific knowledge can be used to benefit society.

- *What pioneering work did Ursula Franklin do?*
- *What is her view of the relationship between science and technology?*

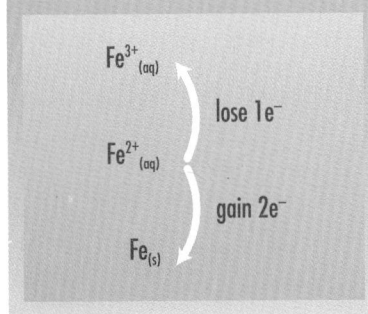

Figure 15.11
Iron(II) ions can either lose or gain electrons and therefore can act as either reducing agents or oxidizing agents.

26. Identify three oxidizing agents (other than $Fe^{2+}_{(aq)}$ shown in Figure 15.11) from the table of the relative strengths of oxidizing and reducing agents (Appendix F) that can also act as reducing agents. Try to explain this unique behaviour.

27. Use the empirically determined redox spontaneity rule (page 619) to predict whether or not the following mixtures would show evidence of a reaction; that is, predict whether the reactions are spontaneous. (Do not write the equations for the reaction.)
    (a) nickel metal in a solution of silver ions
    (b) zinc metal in a solution of aluminum ions
    (c) an aqueous mixture of copper(II) ions and iodide ions
    (d) chlorine gas bubbled into a bromide ion solution
    (e) an aqueous mixture of copper(II) ions and tin(II) ions
    (f) copper metal in nitric acid

28. Describe two experimental designs to collect evidence from which half-reaction tables can be built.

29. (Discussion) Of the two parallel ways of knowing, empirical and theoretical, which, to this point, has been the most useful to you in predicting the spontaneity of redox reactions?

15.2 PREDICTING REDOX REACTIONS

A redox reaction may be explained as a transfer of valence electrons from one substance to another. Evidence indicates that the majority of atoms, molecules, and ions are stable and do not readily release electrons. Since two entities are involved in an electron transfer, this transfer can be explained as a competition for electrons. As in a tug-of-war analogy, each entity exerts a pull on the electrons of the other. During a collision, if the pull is successful in transferring electrons, there is a spontaneous redox reaction (Figure 15.12); if the pull is unsuccessful, no reaction occurs (Figure 15.13). Empirical results, summarized in a table of oxidizing and reducing agents, support this idea.

> An empirical concept may be used to predict the products of a redox reaction and a theoretical concept may be used to describe and explain it. For redox reactions, no simple theory has been developed to predict products accurately.

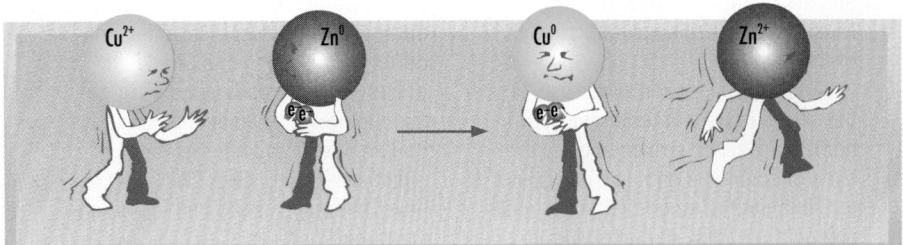

Figure 15.12
Copper has a stronger attraction for valence electrons than zinc does.

Predicting Redox Reactions Using a Table

Arrhenius's ideas about solutions (pages 189–192) provide an important starting point for predicting redox reactions. In solutions, molecules and ions act independently of each other. A first step in predicting redox reactions is to list all entities that are present. (Some helpful reminders are listed in Table 15.2.) For example, when copper metal is placed into an acidic potassium permanganate solution, copper atoms, potassium ions, permanganate ions, hydrogen ions, and water molecules are all present. Next, refer to the related table in Appendix F and label all possible oxidizing and reducing agents in the starting mixture. The permanganate ion is listed as an oxidizing agent only in an acidic solution. To indicate this combination, draw an arc between the permanganate and hydrogen ions as shown, and label the pair as an oxidizing agent.

$$\underset{\substack{\text{RA}}}{\overset{}{Cu_{(s)}}} \qquad \underset{}{\overset{\text{OA}}{K^+_{(aq)}}} \qquad \overset{\overset{\text{OA}}{\frown}}{\underset{}{MnO_4^-{}_{(aq)}}} \;\; \overset{\text{OA}}{H^+_{(aq)}} \qquad \underset{\substack{\text{RA}}}{\overset{\text{OA}}{H_2O_{(1)}}}$$

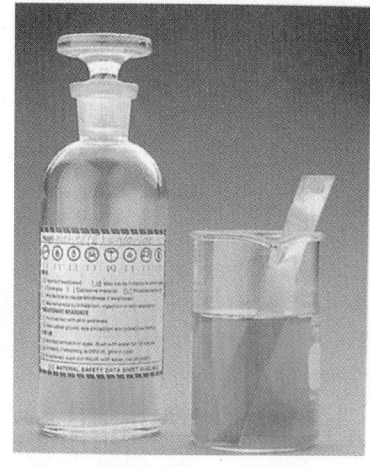

Figure 15.13
A piece of copper can be left sitting in a nickel(II) ion solution indefinitely with no evidence of a reaction. The green nickel (II) ion color remains and the copper metal is not affected. Collisions between copper atoms and nickel(II) ions apparently do not result in the transfer of electrons. This evidence suggests that copper atoms have a greater attraction for valence electrons than nickel(II) ions do.

Exercise

30. List all entities initially present in the following mixtures and identify all possible oxidizing and reducing agents.
    (a) A lead strip is placed in a copper(II) sulfate solution.
    (b) A gold coin is placed in a nitric acid solution.
    (c) A potassium dichromate solution is added to an acidic iron(II) nitrate solution.
    (d) An aqueous chlorine bleach solution is added to a phosphorous acid.
    (e) A potassium permanganate solution is mixed with an acidified tin(II) chloride solution.
    (f) Iodine solution is added to a basic mixture containing manganese (II) oxide.

Predicting Redox Reactions in Solution

Using a table to identify the strongest oxidizing and reducing agents, redox reactions can be predicted in a mixture containing several different entities. Assuming that collisions are completely random, the strongest oxidizing agent and the strongest reducing agent will react. (In some cases, further reactions may occur as well, but this book considers only the primary reaction.) Using both the empirical table and the theoretical concept of competition for electrons allows predictions, descriptions, and explanations of the most likely redox reaction in a chemical mixture.

Suppose a solution of potassium permanganate is slowly poured into an acidic iron(II) sulfate solution. Does a redox reaction occur and, if so, what is the reaction equation? To make a prediction, the entities initially present are identified and labelled as possible oxidizing agents, reducing agents, or both, as shown below.

$$\underset{}{\overset{\text{OA}}{K^+_{(aq)}}} \qquad \overset{\overset{\text{OA}}{\frown}}{\underset{}{MnO_4^-{}_{(aq)}}} \;\; \overset{\text{OA}}{H^+_{(aq)}} \qquad \underset{\substack{\text{RA}}}{\overset{\text{OA}}{Fe^{2+}_{(aq)}}} \qquad \overset{\overset{\text{OA}}{\frown}}{\underset{}{SO_4^{2-}{}_{(aq)}}} \;\; \underset{\substack{\text{RA}}}{\overset{\text{OA}}{H_2O_{(1)}}}$$

Use the related table in Appendix F to choose the strongest oxidizing agent from your list and to write its reduction half-reaction equation.

$$MnO_4^-{}_{(aq)} + 8\,H^+_{(aq)} + 5\,e^- \rightarrow Mn^{2+}_{(aq)} + 4\,H_2O_{(1)}$$

Table 15.2

HINTS FOR LISTING AND LABELLING ENTITIES
• Aqueous solutions contain $H_2O_{(1)}$ molecules.
• Acidic solutions contain $H^+_{(aq)}$ ions.
• Basic solutions contain $OH^-_{(aq)}$ ions.
• Some oxidizing or reducing agents are combinations; for example, the combination of $MnO_4^-{}_{(aq)}$ and $H^+_{(aq)}$.
• $H_2O_{(1)}$, $Fe^{2+}_{(aq)}$, $Cu^+_{(aq)}$, and $Sn^{2+}_{(aq)}$ may act as either oxidizing or reducing agents.

Figure 15.14
A solution of potassium permanganate is being added to an acidic solution of iron(II) ions. The dark purple color of $MnO_4^-{}_{(aq)}$ ions instantly disappears. The accepted interpretation is that they react to produce the almost colorless $Mn^{2+}{}_{(aq)}$ ions.

Repeat this for the strongest reducing agent and its oxidation half-reaction equation.

$$Fe^{2+}{}_{(aq)} \rightarrow Fe^{3+}{}_{(aq)} + e^-$$

Now, balance the number of electrons transferred by multiplying one or both half-reaction equations by an integer so that the number of electrons gained equals the number of electrons lost. Cancel any common terms from the two half-reaction equations before adding the equations to obtain the net ionic equation.

$$MnO_4^-{}_{(aq)} + 8\,H^+{}_{(aq)} + 5\,e^- \rightarrow Mn^{2+}{}_{(aq)} + 4\,H_2O_{(l)}$$

$$5\,[\,Fe^{2+}{}_{(aq)} \rightarrow Fe^{3+}{}_{(aq)} + e^-\,]$$

$$\overline{MnO_4^-{}_{(aq)} + 8\,H^+{}_{(aq)} + 5\,Fe^{2+}{}_{(aq)} \rightarrow 5\,Fe^{3+}{}_{(aq)} + Mn^{2+}{}_{(aq)} + 4\,H_2O_{(l)}}$$

Finally, use the spontaneity rule to predict whether or not the net ionic equation represents a spontaneous redox reaction. Indicate this by writing "spont." or "non-spont." over the equation arrow.

$$MnO_4^-{}_{(aq)} + 8\,H^+{}_{(aq)} + 5\,Fe^{2+}{}_{(aq)} \xrightarrow{\text{spont.}} 5\,Fe^{3+}{}_{(aq)} + Mn^{2+}{}_{(aq)} + 4\,H_2O_{(l)}$$

This prediction may be tested by mixing the solutions (Figure 15.14) and performing some diagnostic tests. If the solutions are mixed and the purple color of the permanganate ion disappears, then it is likely that the permanganate ion reacted. If the pH of the solution is tested before and after reaction, and the pH has increased, then the hydrogen ions likely reacted.

Disproportionation

A **disproportionation** is the reaction of a single substance with itself to produce two different substances. This type of reaction is often a redox reaction and occurs when a substance can act either as an oxidizing agent or as a reducing agent. For example, in an aqueous solution of iron(II) chloride, the iron(II) ion is both the strongest oxidizing agent and the strongest reducing agent present.

$$Fe^{2+}{}_{(aq)} + 2\,e^- \rightarrow Fe_{(s)}$$

$$2[Fe^{2+}{}_{(aq)} \rightarrow Fe^{3+}{}_{(aq)} + e^-]$$

$$\overline{3\,Fe^{2+}{}_{(aq)} \rightarrow Fe_{(s)} + 2\,Fe^{3+}{}_{(aq)}}$$

The same procedure for predicting the most likely redox reaction is used for disproportionation as is used for other redox reactions predicted from a table of relative strengths of oxidizing and reducing agents.

Disproportionation reactions are also referred to by the more descriptive terms of "self oxidation-reduction" or "autoxidation."

> ## SUMMARY: PREDICTING REDOX REACTIONS
>
> Step 1: List all entities present and classify each as a possible oxidizing agent, reducing agent, or both.
>
> Step 2: Choose the strongest oxidizing agent as indicated in the table of relative strengths of oxidizing and reducing agents, and write the reduction half-reaction equation.

CAREER

ANALYTICAL CHEMIST

Dr. Mary Fairhurst is a senior analytical chemist and research associate at a Canadian chemical company. She supervises several laboratories and is the leader of a research group investigating ways of optimizing processes in the chlor-alkali plant. In a chlor-alkali plant, the electrolysis of aqueous sodium chloride is used to produce chlorine, sodium hydroxide, and hydrogen. The chlorine is used primarily as a bleach at pulp and paper mills. The sodium chloride is obtained from salt deposits that lie more than a kilometre below the Earth's surface.

The caverns left by the solution mining process are often used as underground "storage tanks" for petrochemicals such as vinyl chloride. Employees at both the chlor-alkali plant and the vinyl chloride plant must understand chemistry in order to analyze samples from the process stream and to adjust the process for efficient operation.

Although she spends some time in the laboratory, Fairhurst more often works at a computer terminal, confers with colleagues by telephone, and attends meetings. Meetings are the core of industrial problem solving, where groups of professionals may brainstorm quick remedies to a production problem, improvements to the safety of plant workers, reduction in plant emissions, budgets, and a myriad of other matters that affect the operation of a plant.

In her present job, Fairhurst needs a broad understanding of computerized analytical techniques, such as gas and liquid chromatography, mass spectroscopy, infrared spectroscopy, nuclear magnetic resonance spectroscopy, differential scan calorimetry, and laser-based particle sizers. Some traditional chemical techniques are also important tools, including acid-base titrations and gravimetric analyses.

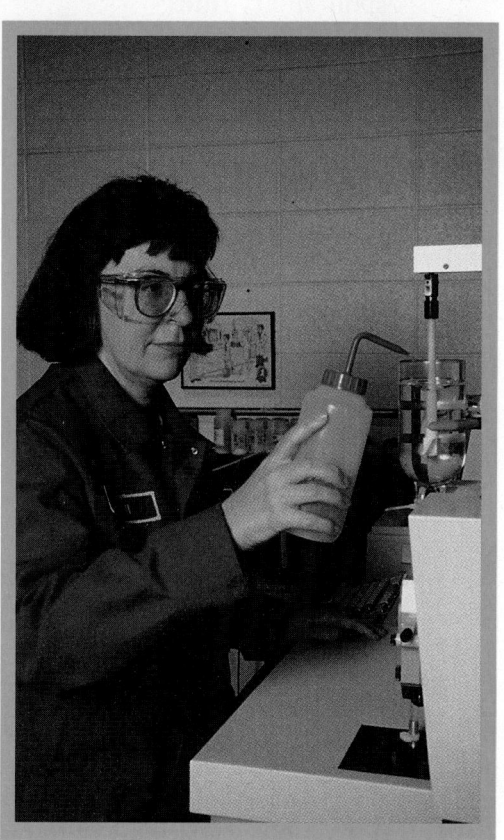

Fairhurst's groups may be asked questions such as whether or not a new technique can be successfully used in the plant, how to improve the size of crystals produced in a particular process, or why two different analytical techniques provide different results. Fairhurst has had to learn how to do statistical analysis of experimental results, a skill she did not learn in school but which is extremely important in her work.

When asked what she enjoys most about her career, Fairhurst replies, "What I've always liked is the variety of tasks. You may be in the lab collecting data; that's fun, using some sophisticated instruments. Then you may spend a couple of days at a computer terminal trying to figure out what the data really mean, or you may be meeting with other people who collect similar data. I like the variety — you never have enough time to get bored, and you never have enough time to feel that you know everything. I think the more I do in science, the more I see how little we do know. There is always a lot more to learn and many more experiments to do."

While she was in high school, Fairhurst studied all the sciences. At university she enrolled in Honors Chemistry after seeing the applications of chemistry and finding that she enjoyed laboratory work. After receiving her B.Sc., she enrolled in a Ph.D. program leading to her doctorate in analytical chemistry.

Fairhurst also gets involved with students studying science in junior and senior high schools. She has worked as a judge at the finals of the Canadian Science Fair, and she is often asked to visit high school classrooms to talk about chemistry.

- *State three examples of questions that Mary Fairhurst and her colleagues have to answer.*
- *In addition to laboratory work, what are some other parts of Mary Fairhurst's job?*

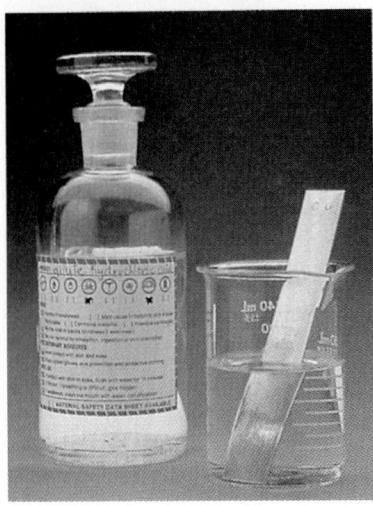

Figure 15.15
Copper metal and hydrochloric acid solution show no evidence of reaction.

Step 3: Choose the strongest reducing agent as indicated in the table, and write the oxidation half-reaction equation.

Step 4: Balance the number of electrons lost and gained in the half-reaction equations by multiplying one or both equations by a number. Then add the two balanced half-reaction equations to obtain a net ionic equation.

Step 5: Predict whether the net ionic equation represents a spontaneous or non-spontaneous redox reaction using the spontaneity rule.

EXAMPLE

In a chemical industry, could copper pipe be used to transport a hydrochloric acid solution? To answer this question,

(a) predict the redox reaction and its spontaneity, and

(b) describe two diagnostic tests that could be done to test your prediction.

(a)

$$\begin{array}{c} \text{SOA} \qquad\qquad \text{OA} \\ Cu_{(s)} \quad H^+_{(aq)} \quad Cl^-_{(aq)} \quad H_2O_{(l)} \\ \text{SRA} \qquad\qquad \text{RA} \searrow_{RA} \nearrow \text{RA} \end{array}$$

$$2\,H^+_{(aq)} + 2\,e^- \rightarrow H_{2(g)}$$

$$Cu_{(s)} \rightarrow Cu^{2+}_{(aq)} + 2\,e^-$$

$$2\,H^+_{(aq)} + Cu_{(s)} \xrightarrow{\text{non-spont.}} H_{2(g)} + Cu^{2+}_{(aq)}$$

Since the reaction is non-spontaneous, it should be possible to use a copper pipe to carry hydrochloric acid.

(b) *If* the mixture is observed, *and* no gas is produced, *then* it is likely that no hydrogen gas was produced (Figure 15.15). If the color of the solution is observed, and the color did not change to blue, then copper probably did not react to produce copper(II) ions. (If the solution is tested for pH before and after adding the copper, and the pH did not increase, then hydrogen ions probably did not react.)

Exercise

31. For the following, use the method in the example above to predict the most likely redox reaction. For any spontaneous reaction, describe one diagnostic test to identify a primary product.
    (a) During a chemistry education demonstration, zinc metal is placed in a hydrochloric acid solution.
    (b) In the industrial production of iodine, chlorine gas is bubbled into sea water containing iodide ions.
    (c) A gold ring is placed into a hydrochloric acid solution.
    (d) Nitric acid is painted onto a copper sheet to etch a design.

(e) The steel of an automobile fender is exposed to acid rain. (Assume that steel is made mainly of iron.)

32. As part of the design of a chemical analysis, a chemical technician prepares several solutions. Will each of the solutions listed below be stable if stored for a long period of time? Justify your answer.
(a) acidic tin(II) chloride
(b) copper(I) nitrate

33. In Chapter 4 (page 172), predictions of reactions were made according to the single replacement generalization assuming the formation of the most common ion.
(a) Use the generalization about single replacement reactions to predict the reaction of iron metal with a copper(II) sulfate solution.
(b) Use redox theory and a table showing half-reactions (Appendix F) to predict the most likely redox reaction of iron metal with a copper(II) sulfate solution.
(c) Write one qualitative and one quantitative experimental design to test the two different predictions for the reaction between iron metal and copper(II) sulfate.

34. Oxygen gas is bubbled into an aqueous solution of iron(II) iodide containing excess hydrochloric acid. Predict all spontaneous reactions, in the order in which they will occur.

35. An excess of cobalt metal is left sitting in an aqueous mixture containing silver ions, iron(III) ions, and copper(II) ions for an extended period of time. Write a balanced redox equation for every reaction that has occurred.

Problem 15D Testing Redox Concepts

Write a Prediction and an Experimental Design (including diagnostic tests) to complete the investigation report.

Problem

What are the products of the reaction of tin(II) chloride with an ammonium dichromate solution acidified with hydrochloric acid?

- ■ Problem
- ✔ Prediction
- ✔ Design
- ■ Materials
- ■ Procedure
- ✔ Evidence
- ✔ Analysis
- ✔ Evaluation
- ■ Synthesis

INVESTIGATION

15.3 Demonstration: Sodium Metal

The purpose of this demonstration is to test the five-step method for predicting redox reactions. As part of the Experimental Design, include a list of diagnostic tests using the "If [procedure], and [evidence], then [analysis]" format for every product predicted. (This format is described in Appendix C, C.3.)

Problem

What are the products of the reaction of sodium metal with water?

CAUTION

This reaction of sodium metal must be demonstrated with great care, because a great deal of heat is produced. Use only a piece the size of a small pea, use a safety screen, wear a lab apron, safety glasses, and face shield, and keep observers at least two metres away.

Predicting Redox Reactions by Constructing Half-Reactions

In the previous section you were able to predict a net ionic redox reaction using half-reaction equations provided on a table of relative strengths of oxidizing and reducing agents. What if half-reaction equations are not provided? In this case, you can use the half-reaction method (page 626) to balance the oxidation and reduction half-reactions and then balance electrons to obtain the net ionic equation.

For example, zinc metal reacts with nitrate ions in an acidic solution to produce zinc ions and ammonium ions. The first step is to start each of the two half-reaction equations by writing the formulas for the reactants and products.

$$Zn_{(s)} \rightarrow Zn^{2+}_{(aq)}$$
$$NO_3^-{}_{(aq)} \rightarrow NH_4^+{}_{(aq)}$$

Notice that all atoms, other than oxygen and hydrogen, are already balanced. Now add water molecules, where necessary, to balance any oxygen atoms.

$$Zn_{(s)} \rightarrow Zn^{2+}_{(aq)}$$
$$NO_3^-{}_{(aq)} \rightarrow NH_4^+{}_{(aq)} + 3 H_2O_{(l)}$$

Because the solution is acidic, hydrogen ions are then added, where necessary, to balance hydrogen atoms.

$$Zn_{(s)} \rightarrow Zn^{2+}_{(aq)}$$
$$10 H^+{}_{(aq)} + NO_3^-{}_{(aq)} \rightarrow NH_4^+{}_{(aq)} + 3 H_2O_{(l)}$$

Balance the charge by adding electrons to each half-reaction equation.

$$Zn_{(s)} \rightarrow Zn^{2+}_{(aq)} + 2 e^- \quad \text{(oxidation)}$$
$$8 e^- + 10 H^+{}_{(aq)} + NO_3^-{}_{(aq)} \rightarrow NH_4^+{}_{(aq)} + 3 H_2O_{(l)} \quad \text{(reduction)}$$

Zinc loses electrons and is the reducing agent, while the nitrate ion gains electrons and is the oxidizing agent.

The zinc half-reaction equation is now multiplied by a factor of 4 to equalize the electrons. In all methods of balancing redox equations, the number of electrons lost must equal the number of electrons gained.

$$4 Zn_{(s)} \rightarrow 4 Zn^{2+}_{(aq)} + 8 e^-$$
$$8 e^- + 10 H^+{}_{(aq)} + NO_3^-{}_{(aq)} \rightarrow NH_4^+{}_{(aq)} + 3 H_2O_{(l)}$$

Finally, add the two half-reaction equations and cancel electrons and anything else that is the same on both sides.

$$4 Zn_{(s)} + 10 H^+{}_{(aq)} + NO_3^-{}_{(aq)} \rightarrow 4 Zn^{2+}_{(aq)} + NH_4^+{}_{(aq)} + 3 H_2O_{(l)}$$

Once the reactants and products are separated into two processes, each half-reaction equation is balanced using the method discussed previously (page 626). If the starting mixture is basic, the same additional steps as given before are followed: assume an acidic solution first and obtain the balanced redox equation, then add hydroxide ions to convert the hydrogen ions into water and then cancel to obtain the net ionic equation. You can check your final answer by counting the number of each kind of atom on both sides of the equation and by checking the net charge on each side.

EXAMPLE

The following reactants and products have been determined in a basic solution. Write the balanced redox equation.

$$SO_3^{2-}{}_{(aq)} + CrO_4^{2-}{}_{(aq)} \rightarrow SO_4^{2-}{}_{(aq)} + CrO_2^{-}{}_{(aq)}$$

$$3[H_2O_{(l)} + SO_3^{2-}{}_{(aq)} \rightarrow SO_4^{2-}{}_{(aq)} + 2 H^+{}_{(aq)} + 2 e^-]$$

$$2[3 e^- + 4 H^+{}_{(aq)} + CrO_4^{2-}{}_{(aq)} \rightarrow CrO_2^{-}{}_{(aq)} + 2 H_2O_{(l)}]$$

$$2 H^+{}_{(aq)} + 3 SO_3^{2-}{}_{(aq)} + 2 CrO_4^{2-}{}_{(aq)} \rightarrow 3 SO_4^{2-}{}_{(aq)} + 2 CrO_2^{-}{}_{(aq)} + H_2O_{(l)}$$

$$2 OH^-{}_{(aq)} + 2 H^+{}_{(aq)} + 3 SO_3^{2-}{}_{(aq)} + 2 CrO_4^{2-}{}_{(aq)}$$
$$\rightarrow 3 SO_4^{2-}{}_{(aq)} + 2 CrO_2^{-}{}_{(aq)} + H_2O_{(l)} + 2 OH^-{}_{(aq)}$$

$$H_2O_{(l)} + 3 SO_3^{2-}{}_{(aq)} + 2 CrO_4^{2-}{}_{(aq)} \rightarrow 3 SO_4^{2-}{}_{(aq)} + 2 CrO_2^{-}{}_{(aq)} + 2 OH^-{}_{(aq)}$$

Final check:

	Left	Right
H	2	2
O	18	18
S	3	3
Cr	2	2
charge	10-	10-

Exercise

36. Construct your own half-reaction equations to predict the balanced net ionic redox reaction for each of the following mixtures.

(a) $OCl^-{}_{(aq)} + I^-{}_{(aq)} \rightarrow Cl^-{}_{(aq)} + I_3^-{}_{(aq)}$ (acidic)

(b) $Zn_{(s)} + SO_4^{2-}{}_{(aq)} \rightarrow SO_{2(g)} + Zn^{2+}{}_{(aq)}$ (acidic)

(c) $Sn_{(s)} + NO_3^-{}_{(aq)} \rightarrow SnO_{2(s)} + NO_{(g)}$ (acidic)

(d) $MnO_{2(s)} + SO_3^{2-}{}_{(aq)} \rightarrow Mn^{2+}{}_{(aq)} + S_2O_6^{2-}{}_{(aq)}$ (acidic)

(e) $MnO_4^-{}_{(aq)} + C_2O_4^{2-}{}_{(aq)} \rightarrow CO_{2(g)} + MnO_{2(s)}$ (basic)

(f) $Cl_{2(aq)} \rightarrow Cl^-{}_{(aq)} + ClO_3^-{}_{(aq)}$ (basic)

(g) $ClO_3^-{}_{(aq)} + N_2H_{4(aq)} \rightarrow NO_{(g)} + Cl^-{}_{(aq)}$ (basic)

(h) $Au_{(s)} + CN^-{}_{(aq)} + O_{2(g)} \rightarrow Au(CN)_4^-{}_{(aq)} + OH^-{}_{(aq)}$ (basic)

15.3 OXIDATION STATES

Historically, oxidation and reduction were considered to be separate processes, more of interest for technology than for science. With modern atomic theory came the idea of an electron transfer involving both a gain of electrons by one entity and a loss of electrons by another entity. This theory of redox reactions is most easily understood for atoms or monatomic ions. Metals and monatomic anions tend to lose electrons (become oxidized), whereas nonmetals and monatomic cations tend to gain electrons (become reduced). More complex redox reactions, such as the reduction of iron(III) oxide by carbon monoxide in the technological process of iron production, the oxidation of glucose in the biological process of respiration, or the use of dichromate ions as a strong oxidizing agent in chemical analysis are not adequately described or explained with simple redox theory. Using half-reactions from a table or ones developed using the procedure given in the previous section is one means of describing such reactions. However, this provides only a limited description and little explanation.

In order to describe oxidation and reduction of molecules and polyatomic ions, chemists have developed a method of "electron bookkeeping" to track the loss and gain of electrons. In this system, the **oxidation state** of an atom in an entity is defined as the *apparent* net electric charge that an atom would have if electron pairs in covalent bonds belonged entirely to the more electronegative atom. An oxidation state is a useful idea for keeping track of electrons but it does not usually represent an actual charge on an atom — oxidation states are arbitrary charges.

An **oxidation number** is a positive or negative number corresponding to the oxidation state assigned to an atom. In a covalently bonded molecule or polyatomic ion, the more electronegative atoms are considered to be negative and the less electronegative atoms are considered to be positive. For example, in a water molecule the oxygen atom (whose electronegativity is listed in the periodic table as 3.5) is assigned the bonding electron from each hydrogen atom (electronegativity 2.1). That is, the oxidation number of the oxygen atom is –2 and the oxidation number of each hydrogen atom is +1. In order to distinguish these numbers from actual electrical charges, oxidation numbers are written in this book as positive or negative numbers; that is, with the sign preceding the number. Oxidation numbers can be assigned to many common atoms and ions (Table 15.3) and they can then be used to determine the oxidation numbers of other atoms.

Table 15.3

COMMON OXIDATION NUMBERS		
Atom or Ion	**Oxidation Number**	**Examples**
all atoms in elements	0	Na is 0, Cl in Cl_2 is 0
hydrogen in compounds	+1	H in HCl is +1
except hydrides	–1	H in LiH is –1
oxygen in compounds	–2	O in H_2O is –2
except in peroxides	–1	O in H_2O_2 is –1
all monatomic ions	charge on the ion	Na^+ is +1, S^{2-} is –2

For example, the oxidation number of carbon in methane, CH_4, is determined using the oxidation number of hydrogen as +1 and the idea that a methane molecule is electrically neutral. The oxidation numbers of the one carbon atom (x) and the four hydrogen atoms (4 times +1) must equal zero.

$$x + 4(+1) = 0$$
$$x = -4$$

$$\overset{x\ +1}{CH_4} \quad \text{or} \quad \overset{-4\ +1}{CH_4}$$

Therefore, carbon in methane has an oxidation number of –4.

In a polyatomic ion, the total of the oxidation numbers of all atoms must equal the charge of the ion. The oxidation number of manganese in the permanganate ion, MnO_4^-, is determined using the oxidation number of oxygen as -2 and the knowledge that the charge on the ion is $1-$. The total of the oxidation numbers of the one manganese atom and the four oxygen atoms (4 times -2) must equal the charge on the ion ($1-$).

$$x + 4(-2) = -1$$
$$x = +7$$

$$\overset{x}{\underset{}{Mn}}\overset{-2}{\underset{}{O_4^-}} \quad \text{or} \quad \overset{+7}{\underset{}{Mn}}\overset{-2}{\underset{}{O_4^-}}$$

Therefore, the oxidation number of manganese in MnO_4^- is $+7$.

SUMMARY: DETERMINING OXIDATION NUMBERS

- Assign common oxidation numbers (Table 15.3).
- The total of the oxidation numbers of atoms in a molecule or ion equals the value of the net electric charge of the molecule or ion.
  (a) The oxidation number for a compound is zero.
  (b) The oxidation number for a polyatomic ion equals the charge on the ion.
- Any unknown oxidation number is determined algebraically from the sum of the known oxidation numbers and the net charge on the entity.

Exercise

37. Determine the oxidation number of
    (a) S in SO_2
    (b) Cl in $HClO_4$
    (c) S in SO_4^{2-}
    (d) Cr in $Cr_2O_7^{2-}$
    (e) I in MgI_2
    (f) H in CaH_2

38. Determine the oxidation number of nitrogen in
    (a) $N_2O_{(g)}$
    (b) $NO_{(g)}$
    (c) $NO_{2(g)}$
    (d) $NH_{3(g)}$
    (e) $N_2H_{4(g)}$
    (f) $NaNO_{3(s)}$
    (g) $N_{2(g)}$
    (h) $NH_4Cl_{(s)}$

39. Determine the oxidation number of carbon in
    (a) graphite (elemental carbon)
    (b) glucose
    (c) sodium carbonate
    (d) carbon monoxide

40. Carbon can be progressively oxidized in a series of organic reactions. Determine the oxidation number of carbon in each of the compounds in the following series of oxidations.

    methane $\rightarrow$ methanol $\rightarrow$ methanal $\rightarrow$

    methanoic acid $\rightarrow$ carbon dioxide

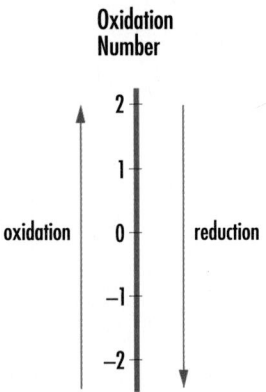

Oxidation Number

oxidation   0   reduction

Figure 15.16
In a redox reaction, both oxidation and reduction occur.

Redox Reactions in Living Organisms

The ability of carbon to take on different oxidation states is essential to life on Earth. Photosynthesis involves a series of reduction reactions in which the oxidation number of carbon changes from +4 in carbon dioxide to an average of 0 in sugars such as glucose. In cellular respiration, carbon undergoes a series of oxidations, after which the oxidation number of carbon is again +4 in carbon dioxide.

When carbon dioxide is released into the atmosphere, some of it reacts with water to form carbonic acid, and this accounts for some of the natural acidity of rain water.

$$CO_{2(g)} + H_2O_{(l)} \rightarrow H_2CO_{3(aq)}$$

When oxidation numbers are assigned to each atom, no change is found; therefore, this is not classified as a redox reaction.

Oxidation Numbers and Oxidation-Reduction Reactions

Although the concept of oxidation states is somewhat arbitrary, because it is based on assigned charges, it is self-consistent and allows predictions of apparent electron transfer. If the oxidation number of an atom or ion changes during a chemical reaction, then an electron transfer (that is, an oxidation-reduction reaction) is believed to occur. In this system, an increase in the oxidation number is defined as an **oxidation** and a decrease in oxidation number is a **reduction**. If oxidation numbers are listed as positive and negative numbers on a line as they are in Figure 15.16, then the process of oxidation involves a change to a more positive value ("up" on the number line) and reduction is a change to a more negative value ("down" on the number line). If the oxidation numbers do not change, this is interpreted as no transfer of electrons. A reaction in which all oxidation numbers remain the same is not classified as a redox reaction.

When natural gas burns in a furnace, carbon dioxide and water form. Carbon is oxidized from –4 in methane to +4 in carbon dioxide as it reacts with oxygen. Simultaneously, oxygen is reduced from 0 in oxygen gas to –2 in both products.

$$\underset{\text{oxidation}}{\overset{\text{reduction}}{\underset{-4\ +1}{CH_{4(g)}} + 2\ \underset{0}{O_{2(g)}} \rightarrow \underset{+4\ -2}{CO_{2(g)}} + 2\ \underset{+1\ -2}{H_2O_{(g)}}}}$$

Methane is the reducing agent because it causes the oxygen to be reduced, and oxygen is the oxidizing agent because it causes the carbon in methane to be oxidized. Combustion reactions are common examples of redox reactions.

EXAMPLE

The determination of blood alcohol content from a sample of breath or blood involves the reaction of the sample with acidic potassium dichromate solution. If ethanol is present, chromium(III) ions, water, and acetic acid are produced. Use oxidation numbers to show that this is an oxidation-reduction reaction and identify the oxidizing agent and the reducing agent.

$$\overset{\text{reduction}}{\underset{\text{oxidation}}{\underset{+6\ -2}{Cr_2O_7{}^{2-}}_{(aq)} + \underset{+1}{H^+}_{(aq)} + \underset{-2\ +1\ -2\ +1}{C_2H_5OH}_{(aq)} \rightarrow \underset{+3}{Cr^{3+}}_{(aq)} + \underset{+1\ -2}{H_2O}_{(l)} + \underset{0\ +1\ 0\ -2\ -2\ +1}{CH_3COOH}_{(aq)}}}$$

Chromium atoms in $Cr_2O_7{}^{2-}$ are reduced (+6 to +3). The dichromate ion is the oxidizing agent. Carbon atoms in C_2H_5OH are oxidized (–2 to 0). Ethanol is the reducing agent.

The Oxidation Number Method of Balancing Redox Equations

Simple oxidation-reduction equations can be balanced by inspection or by a trial-and-error method. More complex redox equations can be written and balanced if the half-reaction equations are known (see the example on page 626) or if sufficient information is available to construct half-reaction equations (page 630). Complex redox equations can also be balanced using oxidation numbers. This method is an alternative to the half-reaction method. Both the half-reaction and the oxidation number methods are based on the assumption that the number of electrons gained equals the number of electrons lost. In terms of oxidation numbers, this means that the total increase in the oxidation numbers must equal the total decrease in the oxidation numbers.

Step 1

To use the oxidation number method, you must know the reactants and products so that you can write an unbalanced equation. The first step is to identify the oxidation numbers that change. For example, consider the reaction of aluminum with the black tarnish that forms on silver, $Ag_2S_{(s)}$ (Figure 15.17).

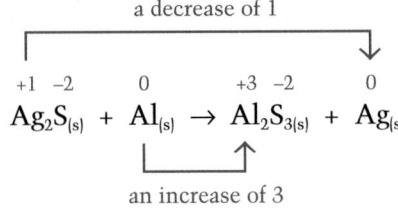

a decrease of 1

$$\overset{+1\ -2}{Ag_2S_{(s)}} + \overset{0}{Al_{(s)}} \rightarrow \overset{+3\ -2}{Al_2S_{3(s)}} + \overset{0}{Ag_{(s)}}$$

an increase of 3

Figure 15.17
The dark tarnish that forms on silver objects is silver sulfide. If the tarnished silver is placed in a hot solution of baking soda in an aluminum dish, or on aluminum foil in a nonmetallic dish, a redox reaction converts the tarnish back to silver metal. This method of cleaning silver is better than polishing, as polishes remove the silver compound.

Step 2

The oxidation number of silver has decreased by 1 and the oxidation number of aluminum has increased by 3. That is, each silver ion has gained 1 electron and each aluminum atom has lost 3 electrons. The oxidation number of sulfur has not changed.

Step 3

Use the formula to determine the number of electrons transferred per formula unit. In one formula unit of Ag_2S there are two ions of silver, so each formula unit of Ag_2S gains two electrons. Each aluminum atom loses three electrons.

■ BLEACHING WOOD PULP ■

The production of pulp and paper is one of Canada's major industries. Thanks to modern chemistry, cellulose fibres from wood pulp can be bleached, dyed, coated, and treated to manufacture paper, paperboard, and countless products such as rayon, photographic films, cellophane, and explosives.

In the production of white papers, a bleaching process is used in which a strong oxidizing agent oxidizes colored organic compounds. The oxidizing agent that has been commonly used for this purpose is chlorine, which not only bleaches the pulp but also breaks down and removes lignin, an organic polymer that binds the wood fibres together. In this process, over 300 reaction by-products result. For example, when chlorine and lignin react in the bleaching process, many different chlorinated carbon compounds (organochlorines) are formed, including chloroform, carbon tetrachloride, chlorophenols, and dioxins.

Many of these by-products are potentially harmful. Research indicates that, although only one of 75 dioxin isomers is extremely toxic, some dioxins can cause immune system suppression and severe reproductive disorders, including birth defects and sterility. Also, certain dioxins are potent carcinogens. Although most dioxins enter the ecosystem from forest fires, these dioxins are not considered toxic. Currently, most toxic dioxins enter the ecosystem as part of the effluent from pulp mills. Traces have also been found in bleached paper products such as diapers, sanitary napkins, paper plates, toilet paper, coffee filters, food packaging, and writing paper. Once in the ecosystem, dioxins are resistant to breakdown and accumulate in animal tissues.

To reduce the emission of organochlorines of all types, several measures are feasible, for example:

- Oxygen pre-bleaching: The use of oxygen gas in the first stage of the bleaching process can reduce organochlorine formation by about 50%. Since oxygen is a less expensive bleaching agent than chlorine, this process can potentially save money for pulp and paper companies.
- Prolonged cooking: The cooking process separates the cellulose fibres from the lignin. The more lignin that is removed in the cooking process, the less that remains to react with chlorine in the bleaching process.
- More thorough washing of pulp: Better washing before bleaching reduces the number and quantity of organic compounds which could form organochlorines.

- Partial replacement of chlorine by chlorine dioxide: Eliminating as much chlorine gas as possible from the bleaching process dramatically reduces dioxin formation.
- Bleaching to a lesser degree: With the realization that many paper products are unnecessarily bleached, several countries have taken steps to encourage consumers to buy unbleached or partially bleached goods.

With exports amounting to more than 11 billion dollars annually, the pulp and paper industry contributes greatly to Canada's economic well-being. The measures taken by governments, industries, and consumers to reduce the emission of organochlorines provide both economic and ecological advantages to society.

- *Why is bleaching used for paper products? Is it necessary?*
- *What are two chemical substances that are used as alternatives to chlorine?*

The number of electrons gained must equal the number of electrons lost, so these two reactants cannot be reacting in a 1:1 ratio. By inspection, you can determine the simplest coefficients to balance the number of electrons transferred.

$$Ag_2S_{(s)} \quad + \quad Al_{(s)} \quad \rightarrow \quad Al_2S_{3(s)} \quad + \quad Ag_{(s)}$$

$1\,e^-/Ag \qquad\qquad 3\,e^-/Al$

$2\,e^-/Ag_2S$

$(\times 3) \qquad\qquad\quad (\times 2)$

In the balanced equation, the reactants will have the coefficients as shown in color.

$$3\,Ag_2S_{(s)} \; + \; 2\,Al_{(s)} \rightarrow Al_2S_{3(s)} \; + \; Ag_{(s)}$$

Step 5

The products can now be balanced by inspection.

$$3\,Ag_2S_{(s)} \; + \; 2\,Al_{(s)} \rightarrow Al_2S_{3(s)} \; + \; 6\,Ag_{(s)}$$

In many reactions, water, hydrogen, or hydroxide ions may be present. The same procedure used on page 630 is followed for aqueous or acidic solutions. Balance the oxygen using water molecules and the hydrogen using hydrogen ions. If the solution is basic, the hydrogen ions are converted to water molecules by adding, to both sides, a number of hydroxide ions equal to the number of hydrogen ions present. Be sure to cancel anything that is the same on both sides and to check the final balancing of atoms and charge. This is illustrated in the following example, where the reactants and products have been determined by diagnostic tests.

EXAMPLE

Balance the chemical equation for the oxidation of ethanol by dichromate ions in a breathalyzer (page 646) to form chromium(III) ions and acetic acid in an acidic solution.

$$16\,H^+_{(aq)} + 2\,Cr_2O_7{}^{2-}_{(aq)} + 3\,C_2H_5OH_{(aq)} \rightarrow 4\,Cr^{3+}_{(aq)} + 3\,CH_3COOH_{(aq)} + 11\,H_2O_{(l)}$$

with oxidation numbers: over $Cr_2O_7{}^{2-}$: +6, −2; over C_2H_5OH: −2 +1 −2 +1; over Cr^{3+}: +3; over CH_3COOH: 0 +1 0 −2 −2 +1; below H^+: +6; below C_2H_5OH: −2; below Cr^{3+}: +3; below CH_3COOH: 0 0

$3e^-/Cr \qquad 2e^-/C$

$6e^-/Cr_2O_7{}^{2-} \qquad 4e^-/C_2H_5OH$

This chemical equation is a good example of a redox equation that cannot be balanced by inspection.

Once the coefficients for the dichromate ions and ethanol are obtained, the oxygen (left 17, right 6) was balanced by adding 11 $H_2O_{(l)}$ to the right side. Then the hydrogen (left 18, right 34) was balanced by adding 16 $H^+_{(aq)}$ to the left side.

SUMMARY: BALANCING REDOX EQUATIONS USING OXIDATION NUMBERS

Step 1: Determine all oxidation numbers and identify the atoms or ions whose oxidation numbers change and write the oxidation numbers below each one.

Step 2: Determine the number of electrons transferred per atom or ion from the change in its oxidation number.

Step 3: Record the number of electrons transferred per mole of oxidizing or reducing agent. (Use the formula subscripts to determine this.)

Steps 5 and 6 parallel the procedure used to write complex half-reaction equations (page 630). Using the oxidation number method for basic solutions, it is possible to use the following steps as an alternative to Steps 5 and 6.

Step 5:   Balance the charge using $OH^-_{(aq)}$.

Step 6:   Balance hydrogen and oxygen by adding $H_2O_{(l)}$.

Step 4: Calculate the simplest whole number coefficients for the oxidizing and reducing agents that will balance the number of electrons transferred.

Step 5: Balance the oxygen using $H_2O_{(l)}$ and balance the hydrogen using $H^+_{(aq)}$.

Step 6: If the solution is basic, add $OH^-_{(aq)}$ to both sides equal in number to the $H^+_{(aq)}$ present. Combine $H^+_{(aq)}$ and $OH^-_{(aq)}$ to form $H_2O_{(l)}$ and cancel anything that is the same on both sides.

Exercise

44. Use the oxidation number method to balance the following oxidation-reduction reaction equations.

(a) $H_{2(g)} + Fe_2O_{3(aq)} \rightarrow FeO_{(s)} + H_2O_{(g)}$

(b) $HBr_{(aq)} + H_2SO_{4(aq)} \rightarrow SO_{2(g)} + Br_{2(aq)}$

(c) $MnO_4^-{}_{(aq)} + CH_3OH_{(l)} \rightarrow Mn^{2+}_{(aq)} + CH_2O_{(aq)}$ (acidic)

(d) $MnO_4^-{}_{(aq)} + SO_3^{2-}{}_{(aq)} \rightarrow SO_4^{2-}{}_{(aq)} + MnO_{2(s)}$ (basic)

(e) $IO_3^-{}_{(aq)} + HSO_3^-{}_{(aq)} \rightarrow SO_4^{2-}{}_{(aq)} + I_{2(aq)}$ (acidic)

(f) $I_{2(aq)} + HSO_3^-{}_{(aq)} \rightarrow I^-_{(aq)} + SO_4^{2-}{}_{(aq)}$ (acidic)

(g) $NH_{3(g)} + O_{2(g)} \rightarrow NO_{2(g)} + H_2O_{(g)}$

(h) $C_2H_5OH_{(l)} + NO_3^-{}_{(aq)} \rightarrow CH_3COOH_{(aq)} + NO_{2(g)}$ (acidic)

(i) $ClO^-_{(aq)} \rightarrow ClO_3^-{}_{(aq)} + Cl^-_{(aq)}$

(j) (enrichment)
$HXeO_4^-{}_{(aq)} \rightarrow XeO_6^{4-}{}_{(aq)} + Xe_{(g)} + O_{2(g)}$ (basic)

45. What three methods are available for balancing oxidation-reduction reaction equations other than the method of inspection? Which method do you prefer, and why?

15.4 REDOX STOICHIOMETRY

The stoichiometric method can be used to predict or analyze the quantity of a chemical involved in a chemical reaction. Many applications of stoichiometry have been illustrated in Chapter 7, involving masses, volumes, and concentrations of reactants and products. For the stoichiometry calculations in Chapter 7, it was necessary to assume that all the reactions are spontaneous, fast, stoichiometric, and quantitative. These same assumptions apply to redox stoichiometry.

There are many industrial and laboratory applications of redox stoichiometry as well. For example, a mining engineer must know the concentration of iron in a sample of iron ore in order to decide whether or not a mine would be profitable. Chemical technicians in industry, monitoring the quality of their companies' products, must determine the concentration of substances such as sodium hypochlorite (NaClO) in bleach, or hydrogen peroxide (H_2O_2) in disinfectants. Hospital laboratory technicians and environmental chemists detect tiny traces of chemicals by a variety of methods. Although much analytical chemistry involves sophisticated equipment, the basic technological process of titration still has an important role (Appendix C, C.3). If a titration is to be performed, the reaction must be not only spontaneous, stoichiometric, and quantitative, but also fast.

In a titration, one reagent (the *titrant*) is slowly added to another (the *sample*) until an abrupt change in a solution property (the *endpoint*) occurs (Figure 15.18). In a redox titration, the endpoint is often a color change. Two oxidizing agents commonly used in redox titrations are permanganate ions and dichromate ions; in acidic solution, they are both strong oxidizing agents and undergo a color change. The permanganate ion, which has an intense purple color in solution, changes to the essentially colorless manganese(II) ion (Figure 15.14, page 626).

$$MnO_4^-{}_{(aq)} + 8\,H^+{}_{(aq)} + 5\,e^- \rightarrow Mn^{2+}{}_{(aq)} + 4\,H_2O_{(l)}$$

Once the sample has completely reacted, the next drop of permanganate added remains unreacted and causes a pink color in the mixture. The color change of the sample (colorless to pink) is the endpoint and corresponds to a slight excess of unreacted permanganate ion. The volume of permanganate solution added when the endpoint is reached is a measurement of the theoretical *equivalence point* — the point at which stoichiometric quantities of reactants have been combined (see page 276).

The dichromate ion is also commonly used in redox titrations; however, its color change is not very distinct — the orange dichromate solution changes gradually to a green chromium(III) solution. A redox indicator is usually added to produce a sharp visible endpoint.

$$Cr_2O_7^{2-}{}_{(aq)} + 14\,H^+{}_{(aq)} + 6\,e^- \rightarrow 2\,Cr^{3+}{}_{(aq)} + 7\,H_2O_{(l)}$$

In any titration, the concentration of the titrant used in an analysis must be accurately known. If the titrant is not a standard solution, the titrant is standardized by calculating its concentration using evidence from an analysis with a primary standard. A *primary standard* is a chemical that can be used directly to prepare a standard solution — a solution of precisely known concentration (page 276).

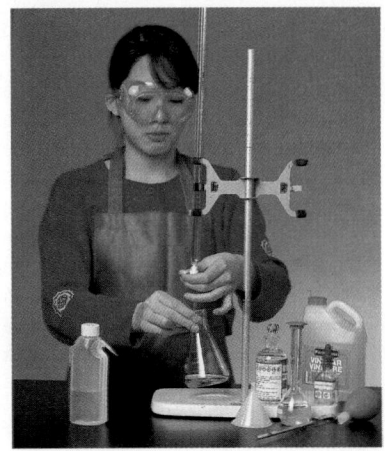

Figure 15.18
Titration is a common experimental design for quantitative chemical analysis.

Problem 15E Standardizing Potassium Permanganate

A solution of potassium permanganate cannot be directly prepared with a precisely known concentration because the permanganate ion reacts with organic and inorganic impurities in the water and with the water itself. Thus, potassium permanganate is not used as a primary standard. The scientific purpose of this problem is to evaluate the necessity for standardizing a potassium permanganate solution and to illustrate the method of a redox titration. Complete the Evaluation of the investigation report. (Decide now on the criteria for judging the accuracy of the prediction. What percent difference is acceptable?)

Problem

What is the concentration of the potassium permanganate solution?

Prediction

According to the accepted method of preparing standard solutions, as well as the laboratory technician who prepared the solution, the molar concentration of the potassium permanganate solution is 0.0125 mol/L or 12.5 mmol/L.

Experimental Design

In order to standardize a freshly prepared solution of potassium permanganate, it is titrated against samples of acidic tin(II) chloride solution. The tin(II) chloride solution is the primary standard.

Evidence

TITRATION OF TIN(II) SOLUTION
(volume of $KMnO_{4(aq)}$ required to react with 10.00 mL of acidic 0.0500 mol/L tin(II) chloride)

Trial	1	2	3	4
Final buret reading (mL)	18.4	35.3	17.3	34.1
Initial buret reading (mL)	1.0	18.4	0.6	17.3
Volume of $KMnO_{4(aq)}$ (mL)	17.4	16.9	16.7	16.8
Endpoint color	dark pink	light pink	light pink	light pink

Analysis

The first endpoint was overshot and was not used in the average for the analysis. At the endpoint, an average of 16.8 mL of permanganate solution was used.

$$\text{OA} \qquad \text{SOA} \quad \text{OA} \qquad \text{OA} \qquad\qquad \text{OA}$$
$$K^+_{(aq)} \quad MnO_4^-{}_{(aq)} \quad H^+_{(aq)} \quad Sn^{2+}_{(aq)} \quad Cl^-_{(aq)} \quad H_2O_{(1)}$$
$$\text{SRA} \qquad \text{RA} \ \ \text{RA} \ \ \text{RA}$$

$$2\,[MnO_4^-{}_{(aq)} + 8\,H^+_{(aq)} + 5\,e^- \rightarrow Mn^{2+}_{(aq)} + 4\,H_2O_{(1)}]$$
$$5\,[Sn^{2+}_{(aq)} \rightarrow Sn^{4+}_{(aq)} + 2\,e^-]$$
$$\overline{2\,MnO_4^-{}_{(aq)} + 16\,H^+_{(aq)} + 5\,Sn^{2+}_{(aq)} \rightarrow 2\,Mn^{2+}_{(aq)} + 8\,H_2O_{(1)} + 5\,Sn^{4+}_{(aq)}}$$

$$\begin{array}{ll} 16.8 \text{ mL} & 10.00 \text{ mL} \\ C & 0.0500 \text{ mol/L} \end{array}$$

$$n_{Sn^{2+}} = 10.00 \text{ mL} \times \frac{0.0500 \text{ mol}}{1 \text{ L}} = 0.500 \text{ mmol}$$

$$n_{MnO_4^-} = 0.500 \text{ mmol} \times \frac{2}{5} = 0.200 \text{ mmol}$$

$$C_{MnO_4^-} = \frac{0.200 \text{ mmol}}{16.8 \text{ mL}} = 0.0119 \text{ mol/L or } 11.9 \text{ mmol/L}$$

$$\text{or } C_{MnO_4^-} = 10.00 \text{ mL Sn}^{2+} \times \frac{0.0500 \text{ mol Sn}^{2+}}{1 \text{ L Sn}^{2+}} \times \frac{2 \text{ mol MnO}_4^-}{5 \text{ mol Sn}^{2+}} \times \frac{1}{16.8 \text{ mL}}$$

$$= 0.0119 \text{ mol/L}$$

According to the evidence gathered and the stoichiometric analysis, the molar concentration of the potassium permanganate solution is 0.0119 mol/L or 11.9 mmol/L.

Problem 15F Analyzing for Tin

Fluoride treatments of children's teeth have been found to significantly reduce tooth decay. When this was first discovered, toothpastes were produced containing tin(II) fluoride. Complete the Analysis of the investigation report.

Problem

What is the concentration of tin(II) ions in a solution prepared for research on toothpaste?

Experimental Design

An acidified tin(II) solution is titrated with a standardized potassium permanganate solution.

Evidence

TITRATION OF TIN(II) SOLUTION			
(volume of 0.0832 mol/L $KMnO_{4(aq)}$ required to react with 10.00 mL of tin(II) solution)			
Trial	1	2	3
Final buret reading (mL)	15.8	28.1	40.6
Initial buret reading (mL)	3.4	15.8	28.1
Volume of $KMnO_{4(aq)}$ (mL)	12.4	12.3	12.5

Problem 15G Analysis of Chromium in Steel

Stainless steel is a corrosion-resistant, esthetically pleasing alloy, normally composed of nickel, chromium, and iron. Complete the Analysis of the investigation report.

Problem

What is the concentration of chromium(II) ions in a solution obtained in the analysis of a stainless steel alloy?

Experimental Design

A standard potassium dichromate solution is used as an oxidizing agent to oxidize chromium(II) ions to chromium(III) ions in an acidic solution (Figure 15.19).

Evidence

TITRATION OF CHROMIUM(II) SOLUTION			
(volume of 0.125 mol/L $K_2Cr_2O_{7(aq)}$ required to react with 10.00 mL of chromium(II) solution)			
Trial	1	2	3
Final buret reading (mL)	17.5	34.9	18.9
Initial buret reading (mL)	0.1	17.5	1.5
Volume of $K_2Cr_2O_{7(aq)}$ (mL)	17.4	17.4	17.4

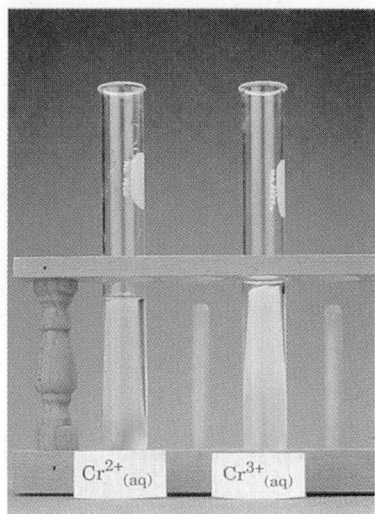

Figure 15.19
The blue $Cr^{2+}_{(aq)}$ solution is oxidized to a green $Cr^{3+}_{(aq)}$ solution.

Problem 15H Analyzing for Iron

Complete the Analysis and Evaluation (of the prediction and thus of the metallurgical process) of the investigation report. (The equation for the reduction of cerium(IV) ion to cerium(III) ion is not shown in the table of redox half-reactions in Appendix F.)

Problem

What is the concentration of iron(II) ions in a solution obtained in an iron ore analysis?

Prediction

According to the required standards for the metallurgical process, the concentration of the iron(II) ions should be 80.0 mmol/L.

Experimental Design

The iron(II) solution is titrated to iron(III) with a standard cerium(IV) ion solution. The indicator shows, as the endpoint, a sharp color change from red to pale blue.

Evidence

TITRATION OF IRON(II) SOLUTION (volume of 0.125 mol/L Ce^{4+} solution required to react with 25.0 mL of $Fe^{2+}_{(aq)}$)				
Trial	**1**	**2**	**3**	**4**
Final buret reading (mL)	15.7	30.7	45.6	40.2
Initial buret reading (mL)	0.6	15.7	30.7	25.3
Volume of Ce^{4+} solution (mL)	15.1	15.0	14.9	14.9
Final indicator color	blue	blue	blue	blue

Exercise

46. Titration is a common experimental design for the quantitative analysis of chemical substances that react under the conditions required by the method of stoichiometry. What are the four requirements or assumptions for designing and analyzing titration experiments?

47. Titration is one of several experimental designs that can be used to determine the quantity of a chemical in a sample. What are some alternative designs available for this purpose?

48. Silver metal can be recycled by reacting nickel metal with waste silver ion solutions. What volume of 0.10 mol/L silver ion solution will react completely with 25.0 g of nickel metal?

49. In a chemical analysis of a chromium alloy, all of the chromium is first converted to chromate ions. A 50.0 mL sample of the chromate ion solution is then reduced in a basic solution to chromium (III) hydroxide by reaction with 22.6 mL of 1.08 M sodium sulfite. In this reaction, the sulfite ions are oxidized to sulfate ions. What is the molar concentration of the chromate ion solution?

50. Pure iron metal may be used as a primary standard for permanganate solutions. A 1.08 g sample of pure iron wire was dissolved in acid, converted to iron(II) ions, and diluted to 250.0 mL. In the titration, an average volume of 13.6 mL of permanganate solution was required to react with 10.0 mL of the acidic iron(II) solution. Calculate the concentration of the permanganate solution.

51. (Enrichment) Potassium dichromate is a common reagent used in the analysis of the iron content of iron ore samples. If each analysis begins with the same mass of the ore, a redox titration can be designed such that the volume of dichromate required corresponds to the percent iron in the ore. This design eliminates any calculations, so rapid, efficient analyses can be carried out by technicians. Starting with a 1.00 g sample of iron ore, the sample is treated to convert all the iron into iron(II) ions and then acidified. Predict the concentration of potassium dichromate required in the analysis so that the volume (in millilitres) equals the percentage of iron in the original sample.

Problem 15I Analyzing for Tin(II) Chloride

Complete the two steps of the Analysis of the investigation report.

Problem

What is the concentration of a tin(II) chloride solution prepared from a sample of tin ore?

Experimental Design

The potassium dichromate solution is first standardized by titration with 10.00 mL of an acidified 0.0500 mol/L solution of the primary standard, iron(II) ammonium sulfate-6-water. The standardized dichromate solution is then titrated against 10.00 mL of the acidified tin(II) chloride solution.

Evidence

TITRATION OF IRON(II) SOLUTION (volume of $K_2Cr_2O_{7(aq)}$ required to react with 10.00 mL of 0.0500 mol/L $Fe^{2+}_{(aq)}$)				
Trial	1	2	3	4
Final buret reading (mL)	13.8	24.4	35.2	45.9
Initial buret reading (mL)	2.3	13.8	24.4	35.2

TITRATION OF TIN(II) SOLUTION (volume of $K_2Cr_2O_{7(aq)}$ required to react with 10.00 mL of $Sn^{2+}_{(aq)}$)				
Trial	1	2	3	4
Final buret reading (mL)	11.8	22.9	33.9	45.0
Initial buret reading (mL)	0.3	11.8	22.9	33.9

15.4 Analysis of a Hydrogen Peroxide Solution

Titration is an efficient, reliable, and precise experimental design used by laboratory technicians for testing the concentration of oxidizing and reducing agents. In this investigation you assume the role of a laboratory technician working in a consumer advocacy laboratory, testing the concentration of a hydrogen peroxide solution (Figure 15.20). The technological purpose of this investigation is to test and evaluate the concentration of the consumer solution of hydrogen peroxide.

Problem

What is the percent concentration of hydrogen peroxide in a consumer product?

Experimental Design

An acidic solution of the primary standard, iron(II) ammonium sulfate-6-water, is prepared and the potassium permanganate solution is standardized by a titration with this primary standard. A 25.0 mL sample of a consumer solution of hydrogen peroxide is diluted to 1.00 L with water (that is, it is diluted by a factor of 40). The standardized potassium permanganate solution is used to titrate the diluted and acidified hydrogen peroxide. The molar concentration of the original hydrogen peroxide is obtained by analysis of the titration evidence, and by using a graph of the information in Table 15.4.

Materials

lab apron
safety glasses
$FeSO_4 \cdot (NH_4)_2SO_4 \cdot 6\,H_2O_{(s)}$
2 mol/L $H_2SO_{4(aq)}$
diluted $H_2O_{2(aq)}$
$KMnO_{4(aq)}$
wash bottle
50 mL buret and clamp
10 mL graduated cylinder
(2) 100 mL beakers
(2) 250 mL beakers
(2) 250 mL Erlenmeyer flasks
100 mL volumetric flask and stopper
10 mL volumetric pipet and bulb
medicine dropper
stirring rod
centigram balance
small funnel
laboratory stand
laboratory scoop

Problem
✔ Prediction
Design
Materials
Procedure
✔ Evidence
✔ Analysis
✔ Evaluation
Synthesis

Figure 15.20
In drugstores, hydrogen peroxide is usually sold as a 3% solution. Hairdressers use a 6% solution. In higher concentrations, peroxides can be explosive.

Table 15.4

MOLAR AND PERCENT CONCENTRATION OF $H_2O_{2(aq)}$	
Molar Concentration (mol/L)	Percent Concentration (%)
0.73	2.5
0.76	2.6
0.79	2.7
0.82	2.8
0.85	2.9
0.88	3.0
0.91	3.1
0.94	3.2
0.97	3.3
1.0	3.4

Procedure

1. (Pre-lab) Calculate the mass of $FeSO_4 \cdot (NH_4)_2SO_4 \cdot 6\,H_2O_{(s)}$ required to prepare 100.0 mL of a 0.0500 mol/L solution.

2. Dissolve the iron(II) compound in about 40 mL of $H_2SO_{4(aq)}$ before preparing the standard solution in the 100 mL volumetric flask.

3. Transfer 10.00 mL of the standard iron(II) solution by pipet into a clean 250 mL Erlenmeyer flask.

4. Titrate the acidic iron(II) sample with $KMnO_{4(aq)}$.

5. Repeat steps 3 and 4 until three consistent volumes (within 0.1 mL) are obtained.

6. Transfer 10.00 mL of the diluted hydrogen peroxide solution by pipet into a clean 250 mL Erlenmeyer flask.

7. Using a 10 mL graduated cylinder, add 5 mL of $H_2SO_{4(aq)}$ to the hydrogen peroxide solution.

8. Titrate the acidic hydrogen peroxide solution with $KMnO_{4(aq)}$.

9. Repeat steps 6 to 8 until three consistent volumes (within 0.1 mL) are obtained.

> **CAUTION**
>
> Sulfuric acid is corrosive and hydrogen peroxide is an irritant to the skin. Avoid skin contact. Excess acidified permanganate is a strong oxidizing agent that should be handled with care and disposed of as your teacher directs.

Problem 15J Analyzing Blood for Alcohol Content

In the chemical analysis of blood samples for alcohol content, the sample is first mixed with excess acidic potassium dichromate solution. The mixture is heated in an oven and left standing, often overnight, to ensure complete oxidation of any ethanol present. To determine the amount of unreacted potassium dichromate left, the solution is titrated with a standard iron(II) ammonium sulfate solution, which acts as a reducing agent. Because the color change of the reactant is difficult to detect at the concentrations used, a few drops of the redox indicator, *o*-phenanthroline, which changes from blue to pink as the endpoint, is added.

In a police laboratory, the result of this *back titration* (titration of the unreacted portion of an excess reagent) is directly converted to a blood alcohol content in g/100 mL. In this problem, you will use two stoichiometric calculations to determine the molar concentration of ethanol in a blood sample. In the first part, the difference between the amount in moles of the potassium dichromate initially added to the blood and the excess (unreacted amount) determined from the titration evidence is determined. This is the amount of potassium dichromate that reacted with the ethanol in the blood sample. In the second part, using the redox equation for the reaction of dichromate with ethanol (page 637), the molar concentration of ethanol is obtained. This molar concentration is converted to a concentration in g/100 mL of blood, using a graph obtained by plotting the data in Table 15.5. The purpose of this problem is to test the accuracy of a breathalyzer test done on a driver. Complete the Analysis and Evaluation of the investigation report.

> Blood samples to be tested for alcohol content are routinely handled in carefully controlled conditions. If there is the possibility that a blood sample has been kept too long or at too high a temperature, it may be argued that fermentation of the blood sugar has occurred. This possible fermentation could produce ethanol in the sample and the validity of the laboratory results would be placed in doubt. When science and technology encounter the legal system, both the accuracy and the precision may be questioned.

Table 15.5

MOLAR CONCENTRATION OF ETHANOL AND BLOOD ALCOHOL CONTENT	
Molar Concentration (mmol/L)	Blood Alcohol Content (g/100 mL)
13.0	0.060
15.2	0.070
17.4	0.080
19.5	0.090
21.7	0.100
23.9	0.110
26.0	0.120
28.2	0.130
30.4	0.140
32.6	0.150

Problem

What is the blood alcohol content in a blood sample?

Prediction

A roadside breathalyzer screening test indicated the driver has a blood alcohol content greater than 0.10 mg/100 mL.

Experimental Design

A 2.70 mL sample of 10.62 mmol/L acidic potassium dichromate is pipetted into a 0.500 mL blood sample, and heated overnight in an oven. The sample is then titrated with a 20.40 mmol/L iron(II) ammonium sulfate solution using o-phenanthroline as an indicator.

Evidence

TITRATION (excess acidic potassium dichromate reacting with iron(II) ammonium sulfate solution)			
Trial	1	2	3
Volume of blood sample (mL)	0.500	0.500	0.500
Final buret reading (mL)	7.24	13.40	19.54
Initial buret reading (mL)	1.12	7.24	13.40
Volume of iron(II) solution (mL)	6.12	6.16	6.14

BLOOD ALCOHOL CONTENT AND THE LAW

Canadian society views drinking and driving as unacceptable behavior. The cost in damaged lives from alcohol-related motor vehicle accidents has prompted police to adopt sophisticated methods for obtaining the evidence they require to prove a driver is impaired. It is a criminal offence if the driver of a vehicle has an alcohol content greater than 0.08 g/100 mL of blood. To prosecute such a driver, however, police must be able to present evidence in court that establishes what the driver's blood alcohol content was at the time of the test.

For many years, in all jurisdictions in Canada, the Breathalyzer has been the standard instrument the police have used to obtain that evidence. This device measures the alcohol content of exhaled breath, which is proportional to the blood alcohol content. Inside the instrument, the breath sample is mixed with a known,

excess quantity of acidic potassium dichromate solution, a process that oxidizes the ethanol. The quantity of unreacted potassium dichromate is then measured using a colorimeter. That quantity is inversely proportional to the alcohol concentration of the breath sample.

In addition to the Breathalyzer, police are now starting to use a new breath analysis instrument in their roadside checks. The Intoxilyzer (see photo) uses infrared absorption spectroscopy to pass infrared light through the breath sample. It then measures how much absorption is caused by the presence of alcohol. The accuracy of the test is better than ±0.005% of the actual blood alcohol content. Over the next decade, it is expected that the Intoxilyzer will gradually replace the Breathalyzer in police efforts to eliminate drinking and driving.

Exercise

52. Examine the list of Canadian scientific and technological achievements in Table 15.6.
    (a) (Discussion) For each achievement, consider whether it is primarily "scientific" or "technological."
    (b) Choose one of the individuals, organizations, or achievements listed. Consult references, and write a report including historical background and a description of the scientific or technological breakthrough. Include an assessment of how the society of the day influenced or was influenced by the achievement. (Or, with your teacher's approval, do the same for another achievement not listed here.)

Table 15.6

CANADIAN SCIENTIFIC AND TECHNOLOGICAL ACHIEVEMENTS	
Maud Menten	develops an equation describing enzyme activity
Frederick Banting and colleagues	discover insulin
Richard Taylor	gathers evidence for the existence of quarks
Atomic Energy of Canada	develops cobalt bomb for radiation therapy
Henry Taube	makes discoveries about electrochemical reactions
Armand Bombardier	invents the snowmobile
James Hillier and Albert Prebus	produce the first commercial electron microscope
Spar Aerospace Limited	designs and builds the Canadarm
Raymond Lemieux	synthesizes sucrose
Brenda Milner	discovers specific function-related areas of the brain
A. E. Gibbs	patents an electrolytic cell used to produce chlorine
Tuzo Wilson	develops the modern theory of plate tectonics
Gerhard Herzberg	studies electronic properties of molecules
Neil Bartlett	synthesizes compounds of a noble gas
James Guillet	develops biodegradable plastic
Cluny McPherson	designs a gas mask used in World War I
Ursula Franklin	pioneers the development of archeometry
Alexander Graham Bell	invents the telephone
Margaret Newton	produces several strains of rust-resistant wheat
Helen Battle	pioneers the laboratory study of marine biology
Charles Fenerty	produces newsprint from wood pulp
John Polanyi	describes molecular motion in chemical reactions
Abraham Gesner	makes kerosene oil
Archibald Huntsman	develops commercial process for freezing fish

See the feature on Canadian Nobel Prize winners on page 533. There is a remarkable number of Canadian Nobel laureates for a country with such a small population.

Electrochemistry

Summary

- Oxidation and reduction were initially associated with the technological processes of corrosion, burning, and metallurgy. In modern theoretical terms, oxidation and reduction occur simultaneously as electrons are transferred from the reducing agent to the oxidizing agent in a redox reaction.

- Oxidation is a process in which electrons are lost by an entity as it reacts with an oxidizing agent that gains the electrons. Reduction is a process in which electrons are gained by an entity as it reacts with a reducing agent that loses the electrons.

- Predictions of redox reactions are made after determining which are the strongest possible oxidizing and reducing agents present in the initial mixture.

- A redox reaction is spontaneous if the oxidizing agent is listed above the reducing agent in a table of relative strengths of oxidizing and reducing agents.

- Predictions of the products of redox reactions and the balanced redox equation can be made using half-reaction equations of the strongest oxidizing and reducing agents chosen from a half-reaction table.

- If the main reactants and products are known, redox reaction equations can also be written by constructing half-reaction equations for aqueous, acidic, or basic solutions.

- Oxidation is defined as an increase in oxidation number and reduction as a decrease in oxidation number. These numbers are used to recognize redox reactions and balance redox equations by balancing the changes in oxidation numbers.

- Redox stoichiometry can be used in analytical procedures to quantitatively determine oxidizing or reducing agents in many industrial, medical, and consumer products or processes.

Key Words

disproportionation
electrochemistry
oxidation
oxidation number
oxidation state
oxidizing agent (OA)
redox reaction
redox spontaneity rule
reducing agent (RA)
reduction
spontaneous

Review

1. In a few words, describe the historical origin of the terms "oxidation" and "reduction."

2. Write a theoretical description of a redox reaction.

3. Explain, in terms of electrons, the role of each of the following in a redox reaction.
   (a) oxidizing agent
   (b) reducing agent

4. If a spontaneous redox reaction occurs, what kinds of evidence might be observed?

5. Using a table of relative strengths of oxidizing and reducing agents such as the one in Appendix F, how can you predict whether or not a combination of substances will react spontaneously?

6. Which of the following combinations would produce a spontaneous redox reaction?
   (a) chromium metal and aqueous cobalt(II) chloride
   (b) nitric acid and iron(III) chloride solution
   (c) oxygen gas bubbled into a sodium bromide solution
   (d) tin(II) nitrate and copper(II) sulfate solutions

7. If a solution to be used as a titrant cannot be prepared with a precisely known concentration, it must be standardized. What does "standardized" mean?

8. What is the oxidation number of
   (a) I in $I_{2(s)}$?
   (b) I in $CaI_{2(s)}$?
   (c) I in $HIO_{(aq)}$?
   (d) H in NH_3?
   (e) H in AlH_3?
   (f) O in CH_3OH?

9. Define each of the following in terms of both electrons and oxidation numbers.
   (a) oxidation
   (b) reduction
   (c) redox reaction

10. Make a list of everything that must be balanced in a net ionic equation representing a redox reaction.

Applications

11. Write and label balanced half-reaction equations for each of the following redox reactions.
    (a) $2 Fe^{3+}_{(aq)} + Ni_{(s)} \rightarrow 2 Fe^{2+}_{(aq)} + Ni^{2+}_{(aq)}$
    (b) $Br_{2(aq)} + 2 I^-_{(aq)} \rightarrow 2 Br^-_{(aq)} + I_{2(s)}$
    (c) $Pd^{2+}_{(aq)} + Sn^{2+}_{(aq)} \rightarrow Pd_{(s)} + Sn^{4+}_{(aq)}$
    (d) Label each reactant in (a), (b), and (c) as an oxidizing or a reducing agent.

12. For each of the following, complete the half-reaction equation and classify as an oxidation or reduction.
    (a) $HClO_{2(aq)} \rightarrow HClO_{(aq)}$ (acidic)
    (b) $Al(OH)_4^-{}_{(aq)} \rightarrow Al_{(s)}$ (basic)
    (c) $Br^-_{(aq)} \rightarrow BrO_4^-{}_{(aq)}$ (acidic)
    (d) $ClO^-_{(aq)} \rightarrow Cl_{2(g)}$ (basic)

13. For the following solutions, list the entities believed to be present and classify them as oxidizing or reducing agents.
    (a) aqueous chlorine solution
    (b) tin(II) nitrate solution
    (c) acidic potassium iodate solution

14. The following Group 13 metals were each placed in solutions of ions of the other three metals. Mixtures in which there was evidence of a chemical reaction are indicated by checkmarks in the following table. Use this evidence to construct a table of half-reaction equations, in order of strength as oxidizing and reducing agents.

	$Al^{3+}_{(aq)}$	$Ga^{3+}_{(aq)}$	$In^{3+}_{(aq)}$	$Tl^+_{(aq)}$
$Al_{(s)}$	—	✔	✔	✔
$Ga_{(s)}$	—	—	✔	✔
$In_{(s)}$	—	—	—	✔
$Tl_{(s)}$	—	—	—	—

15. Use the evidence from the following chemical reactions and the redox spontaneity rule to develop a table of oxidizing and reducing agents in order of strength.

    $2 Ga_{(s)} + 3 Cd^{2+}_{(aq)} \rightarrow 2 Ga^{3+}_{(aq)} + 3 Cd_{(s)}$

    $Ga_{(s)} + Mn^{2+}_{(aq)} \rightarrow$ no evidence of reaction

    $3 Mn^{2+}_{(aq)} + 2 Ce_{(s)} \rightarrow 3 Mn_{(s)} + 2 Ce^{3+}_{(aq)}$

16. Solid copper reacts spontaneously with a solution containing silver ions. Explain this observation by including ideas from the kinetic molecular theory, collision-reaction theory, and redox theory.

17. For each of the following mixtures, list and classify the entities present, predict the half-reaction and net ionic reaction equations, and predict whether or not a spontaneous reaction will be observed.
    (a) Chlorine gas is bubbled into an iron(II) sulfate solution.
    (b) Nickel(II) nitrate solution is mixed with a tin(II) sulfate solution.
    (c) A zinc coating on a drain pipe is exposed to air and water.
    (d) An acidic solution of sodium sulfate is spilled on a steel laboratory stand. (Consider only the iron in the steel.)
    (e) For use in a titration, a sodium hydroxide solution is added to a potassium sulfite solution to make it basic.

18. Predict the redox reaction for each of the following (unbalanced) equations by constructing and labelling oxidation and reduction half-reaction equations.
    (a) $O_{3(g)} + I^-_{(aq)} \rightarrow IO_3^-{}_{(aq)} + O_{2(g)}$ (acidic)
    (b) $Pt_{(s)} + NO_3^-{}_{(aq)} + Cl^-_{(aq)} \rightarrow$
                         $PtCl_6^{2-}{}_{(aq)} + NO_{2(g)}$ (acidic)
    (c) $CN^-_{(aq)} + ClO_2^-{}_{(aq)} \rightarrow CNO^-_{(aq)} + Cl^-_{(aq)}$ (basic)
    (d) $PH_{3(g)} + CrO_4^{2-}{}_{(aq)} \rightarrow Cr(OH)_4^-{}_{(aq)} + P_{4(s)}$ (basic)
    (e) $MnO_4^{2-}{}_{(aq)} \rightarrow Mn^{2+}_{(aq)} + MnO_4^-{}_{(aq)}$ (acidic)
    (f) $ClO^-_{(aq)} \rightarrow ClO_2^-{}_{(aq)} + Cl_{2(g)}$ (basic)

19. Magnesium metal reacts rapidly in hot water. Predict the mass of precipitate that will form if a 2.0 g strip of magnesium reacts completely with water.

20. In a standardization experiment, 25.0 mL of an acidic 0.100 mol/L tin(II) chloride solution required an average volume of 12.7 mL of potassium dichromate solution for complete reaction. Calculate the concentration of the potassium dichromate solution.

21. A student uses a redox titration to determine the concentration of iron(II) ions in an acidic solution. The following evidence shows the volume of 7.50 mmol/L $KMnO_{4(aq)}$ that reacted with 10.0 mL of $Fe^{2+}_{(aq)}$. Calculate the concentration of the iron(II) ions.

Trial	1	2	3
Final buret reading (mL)	16.4	31.4	46.3
Initial buret reading (mL)	1.3	16.4	31.4

22. Potassium metal spontaneously reacts with water.

    (a) Write the half-reaction and net ionic reaction equations for this reaction.

    (b) Describe diagnostic tests (procedure, evidence, analysis) that could be done to test for the predicted products.

23. Three creative chemistry teachers contrived a problem to test students' understanding of redox concepts. The challenge is to identify three unknown solutions (labelled A, B, and C) using only the materials listed below. Assuming all possible spontaneous reactions are rapid and that the nitrate ion is a spectator ion, write a procedure to identify which solution is sodium nitrate, which one is lead(II) nitrate, and which one is calcium nitrate. Describe the expected results. The following materials may be used: 0.25 mol/L solutions of A, B, and C; silver, zinc, and magnesium strips; dropper bottles of 0.25 mol/L aqueous solutions of sodium sulfate, sodium carbonate, and sodium hydroxide; steel wool; test tubes and test tube rack; 50 mL beakers; 400 mL waste beaker.

24. Many natural gas wells, called "sour" gas wells, contain considerable quantities of hydrogen sulfide gas as well as methane. When this mixture burns, hydrogen sulfide is converted to sulfur dioxide. Once in the atmosphere, sulfur dioxide may be converted to sulfur trioxide. Is the sulfur in these two reactions being oxidized or reduced? Defend your answer by referring to the oxidation states of sulfur in the three compounds.

25. Silver(II) oxide, a reagent used in chemical analysis, reacts spontaneously with water according to the following (unbalanced) equation.

$$Ag^{2+}_{(aq)} + H_2O_{(l)} \rightarrow Ag^+_{(aq)} + O_{2(g)}$$

    (a) Identify the oxidation numbers of each atom or ion.

    (b) Classify the reactants as oxidizing or reducing agents.

    (c) Balance the equation.

26. Chromium steel alloys are analyzed using a series of redox reactions. The alloy is initially reacted with perchloric acid which converts the chromium metal into dichromate ions while the perchloric acid is reduced to chlorine gas. The dichromate ions are then reduced to chromium(III) ions by adding an excess of iron(II) solution. The unreacted iron(II) is then titrated with a solution of cerium(IV) ions, which reduces them to cerium(III) ions. Write a balanced redox equation for each step of this procedure.

27. Balance the following chemical equations using the oxidation number method.

    (a) $C_6H_{12}O_{6(s)} + O_{2(g)} \rightarrow CO_{2(g)} + H_2O_{(l)}$

    (b) $Au^{3+}_{(aq)} + SO_{2(aq)} \rightarrow SO_4^{2-}_{(aq)} + Au_{(s)}$ (acidic)

    (c) $BrO_3^-_{(aq)} + C_2H_6O_{(aq)} \rightarrow CO_{2(g)} + Br^-_{(aq)}$

    (d) $Ag_{(s)} + NO_3^-_{(aq)} \rightarrow Ag^+_{(aq)} + NO_{(g)}$ (acidic)

    (e) $HNO_{3(aq)} + SO_{2(g)} \rightarrow H_2SO_{4(aq)} + NO_{(g)}$

    (f) $Zn_{(s)} + BrO_4^-_{(aq)} \rightarrow Zn(OH)_4^{2-}_{(aq)} + Br^-_{(aq)}$ (basic)

28. A commercial kit is available to clean silver by removing the tarnish using a redox reaction. (Assume that silver tarnish is silver sulfide.) A zinc strip is placed in a water softener solution and the tarnished silver is placed so that it is in contact with the zinc strip.

    (a) Write the overall chemical equation and balance it using the simplest possible method.

    (b) Verify, using oxidation numbers, that the chemical equation is balanced.

    (c) Write oxidation and reduction half-reaction equations.

Extensions

29. Vanadium is a very versatile element in terms of its reactivity. Vanadium metal reacts with fluorine to form VF_5, with chlorine to form VCl_4, with bromine to form VBr_3, with iodine to form VI_2, with oxygen to form V_2O_5, and with hydrochloric acid to form VCl_2.

    (a) Identify the oxidation states of vanadium in each of the compounds mentioned.

    (b) What interpretation can be made about the oxidizing power of the chemicals that react with vanadium metal?

    (c) Consult a reference, then describe how the oxidation state of vanadium is related to the colors of the compounds formed.

    (d) Use a reference to write a report on some technological applications of vanadium and its compounds.

30. For the production of pulp from wood, a variety of methods are used, including mechanical and chemical processes. These have advantages and disadvantages that have been widely debated. Collect information about these processes and provide an assessment using technological, economic, and ecological perspectives.

31. Road salt apparently increases the rate at which automobiles rust. Design an experiment to determine how the concentration of a solution or the type of electrolyte in a solution affects the rate of corrosion of iron.

32. Write at least three experimental designs for an analysis to determine the concentration of

silver ions in a waste solution from a photofinishing laboratory.

33. Use your knowledge of redox half-reactions and some research or brainstorming to describe five methods for determining or approximating the position of the beryllium half-reaction in a table of half-reactions.

Table 15.7

CONCENTRATIONS AND FREEZING POINTS OF AQUEOUS SOLUTIONS OF METHANOL

Molar Concentration (mol/L)	Percent by Mass (%)	Freezing Point (°C)
0	0	0
6.035	20.00	−15.0
11.672	40.00	−38.6
16.754	60.00	−74.5

Problem 15K Analyzing Antifreeze

Methanol is used as a windshield washer antifreeze; containers are usually labelled with the freezing point of the solution. A chemical technician can test the validity of the claim using various experimental designs. The experimental design chosen below is the titration of a basic solution of methanol with a standardized solution of potassium permanganate based on the following (unbalanced) chemical equation.

$$CH_3OH_{(aq)} + MnO_4^-{}_{(aq)} \rightarrow$$
$$CO_3^{2-}{}_{(aq)} + MnO_4^{2-}{}_{(aq)}$$

Use the information in Table 15.7 and complete the Analysis of the investigation report.

Problem

What is the freezing point of a sample of windshield washer fluid?

Experimental Design

A potassium permanganate solution is prepared and standardized against an acidic 0.331 mol/L solution of iron(II) ammonium sulfate. The standardized permanganate solution is then titrated against a basic methanol solution, which has been diluted by a factor of 1000.

Evidence

VOLUMES OF POTASSIUM PERMANGANATE USED IN TITRATIONS

	10.00 mL of Acidic $FeSO_4 \cdot (NH_4)_2SO_{4(aq)}$				10.00 mL of Basic $CH_3OH_{(aq)}$		
Trial	1	2	3	4	1	2	3
Final buret reading (mL)	13.3	25.8	38.1	12.9	12.4	24.1	35.8
Initial buret reading (mL)	0.2	13.3	25.8	0.5	0.1	12.4	24.1

Problem 15L Redox Indicators

Redox indicators are one color in oxidizing agent form and a different color in reducing agent form, as listed below. Complete the Analysis of the investigation report.

REDOX INDICATORS

Redox Indicator	Oxidizing Agent Form	Reducing Agent Form
eriogreen	rose	red-yellow
nitroferrion	faint blue	red
methylene blue	colorless	blue
diphenylamine	violet	colorless

Problem

Where do the redox indicators in the above list fit in a table of oxidizing and reducing agents?

Experimental Design

Selected oxidizing agents are allowed to react with the redox indicators.

Evidence

REACTIONS OF REDOX INDICATORS

Oxidizing Agent	Reducing Agent	Color Changes
$IO_3^-{}_{(aq)} + H^+_{(aq)}$	eriogreen	red-yellow to rose
$IO_3^-{}_{(aq)} + H^+_{(aq)}$	nitroferrion	no change
$Ag^+_{(aq)}$	eriogreen	no change
$Ag^+_{(aq)}$	diphenylamine	colorless to violet
$Au^{3+}_{(aq)}$	nitroferrion	red to faint blue
diphenylamine	methylene blue	blue to colorless
$Cu^{2+}_{(aq)}$	methylene blue	no change

16 Voltaic and Electrolytic Cells

What do electric eels and futuristic cars have in common? Both can produce electricity from the energy of redox reactions. Cars of the future may be powered entirely by electricity. Electric eels use redox reactions not only to produce electricity, but also to carry on the processes characteristic of living organisms such as growth, movement, and reproduction.

Electricity was first produced technologically from chemical reactions in a laboratory around 1800, when batteries were invented. Batteries led to many advances in science and technology. The invention of motors, generators, lights, and electric generating stations made electricity a common feature in homes and industries. More recently, much of our society "runs" on batteries — in radios, computers, watches, cellular phones, and many other devices that require portable sources of electricity. Batteries are once again propelling us into the future.

Redox reactions can produce electricity and, conversely, electricity can cause redox reactions. Many materials that we take for granted were virtually unknown until the process of electrolysis made their production possible. Aluminum, chlorine, hydrogen, sodium hydroxide, magnesium, and copper are produced in large quantities by electrolytic processes. In this chapter, you will learn how batteries are made, how electricity can be used to produce chemicals, and how science and technology work together in the development of electrochemical processes.

Before 1800, scientists knew how to produce static electricity by friction between two objects. They discovered ways of storing the charges temporarily, but when the energy was released in the form of an electrical spark, it could not be put to practical use (Figure 16.1). Practical applications of electricity were developed only after 1800, the year in which Alessandro Volta announced his invention of **electric cells**, devices that continuously convert chemical energy into electrical energy.

"We owe almost all our knowledge not to those who have agreed, but to those who have differed." — Charles Colton (1780 – 1832), English clergyman/writer

INVESTIGATION

16.1 Demonstration: A Simple Electric Cell

The purpose of this investigation is to demonstrate an electric cell.

☐	Problem
☐	Prediction
☐	Design
☐	Materials
☐	Procedure
✔	Evidence
✔	Analysis
✔	Evaluation
☐	Synthesis

Problem

What electrical properties are observed when two metals come in contact with a conducting solution?

Prediction

According to the hypothesis of Luigi Galvani (1737 – 1798), electricity will only be produced if metals are in contact with animal tissue. Galvani was the Italian scientist who discovered that an electric current flows when two different metals are in contact with a muscle in a frog's leg.

Experimental Design

Different pairs of metal strips are placed in contact with fruits, vegetables, and inorganic solutions to test Galvani's hypothesis.

Materials

lab apron
safety glasses
paper towel
salt water
strips of metals such as $Zn_{(s)}$, $Cu_{(s)}$, $Pb_{(s)}$, and $Ag_{(s)}$
potato, orange, apple
clothespin
ammeter (sensitive current meter)
voltmeter and connecting wires

Procedure

1. Place a paper towel soaked in salt water between strips of two different metals. Hold this "sandwich" together with a clothespin.

2. Connect the two metals to the terminals of an ammeter and observe the reading.

3. Connect the two metals to the terminals of a voltmeter and observe the reading.

4. Remove the paper towel and insert the two metals into one of the fruits or vegetables (Figure 16.2, page 654) and repeat steps 2 and 3. (The metals should not touch each other.)

5. Repeat steps 1 to 4 using different combinations of metals.

Figure 16.1
A lightning bolt is a large spark — a discharge of electricity — similar to the spark you generate when you touch a metal object after shuffling across a rug on a cold and dry winter day.

Figure 16.2
When two different electrodes are inserted into a grapefruit, an electric current is produced.

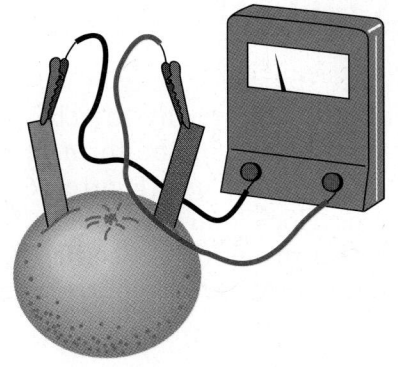

Figure 16.3
A potato clock contains two electric cells; the electric energy is produced by copper and zinc metals in contact with the solution of electrolytes inside the potatoes.

> "I introduced into my ears two metal rods with rounded ends and joined them to the terminals of the apparatus. At the moment the circuit was completed I received a shock in the head — and began to hear a noise — a crackling and boiling. This disagreeable sensation, which I feared might be dangerous, has deterred me so that I have not repeated the experiment."
> — Alessandro Volta (1745 – 1827)

> A single cell of a battery, as we commonly call it, is given the names *voltaic cell*, *galvanic cell*, and *electrochemical cell*. The term used in this book is voltaic cell.

Advances Based on Volta's Battery

The individual cells Volta invented produced very little electricity, so he came up with a better design by joining several cells together. A **battery** is a group of two or more cells connected to each other, in series, like railway cars in a train (Figure 16.3). Volta's first battery consisted of several bowls of brine (aqueous sodium chloride) connected by metals that dipped from one bowl into the next (Figure 16.4). This arrangement of metal strips and electrolytes produced a steady flow of electric current. Volta improved the design of this battery by replacing the strips of metal with flat sheets, and replacing the bowls with paper or leather soaked in brine. As shown in Figure 16.5, Volta stacked cells on top of each other to form a battery, known as a *voltaic pile*. When a loop of wire was attached to the top and bottom of this voltaic pile, a steady electric current flowed. Volta assembled voltaic piles containing more than 100 cells.

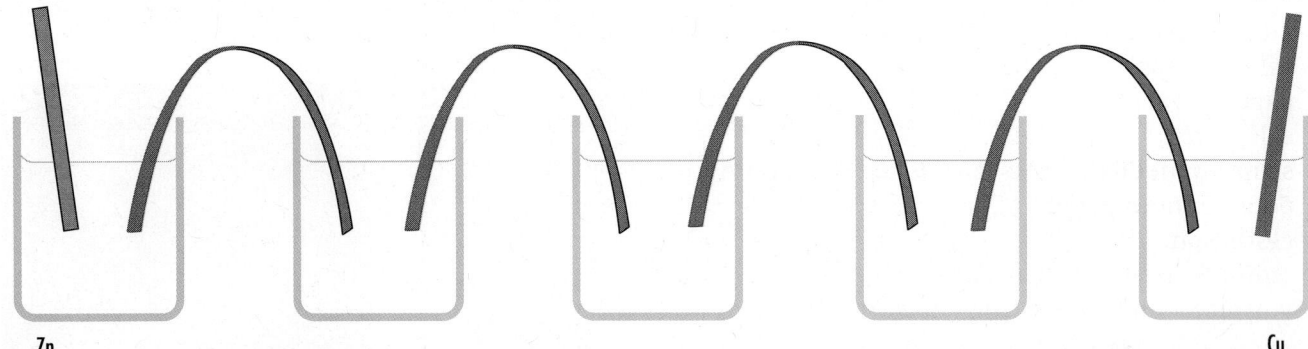

Zn   Cu

Figure 16.4
A version of Volta's first battery. Each bowl contains two different metals, copper and zinc, in an electrolyte, salt water. A series of bowls forms a battery whose total voltage is the sum of the individual voltages of all cells.

Volta's invention was an immediate success because it produced an electric current more simply and more reliably than methods that depended on static charges. It also produced a steady electric current — something no other device could do. In recognition of the work done by Volta, an electric cell that produces electricity from conductors placed in a conducting solution is called a **voltaic cell**. The development of this technology led to many advances in physics (for example, the theory and description of current electricity), in chemistry (for example, the work of Humphry Davy, page 45), and in electrical and chemical engineering.

Chemical Engineering of Cells and Batteries

Each electric cell is composed of two **electrodes**, which are solid conductors, and one *electrolyte*, which is an aqueous conductor (Figure 16.6, page 656). In the cells we buy for home use, the electrolyte is usually a moist paste, containing only enough conducting solution to make the cell function. The electrodes are usually two metals, or graphite and a metal. In some designs, one of the electrodes is the container of the cell. One of the electrodes is marked positive (+) and the other is marked negative (–). By convention, the positive electrode in an electric cell is called the *cathode* and the negative electrode is the *anode*. According to the theory that electricity is the flow of electrons and the evidence from a meter, when a conductor connects the two electrodes, electrons move from the anode of a battery through the conductor and then to the cathode. A battery produces electricity only when there is an external conducting path through which electrons can move.

A voltmeter is a device that measures the difference in electric potential energy, also called the voltage, between any two points in an electric circuit. **Electric potential difference (voltage)**, measured in **volts** (V), is a measure of energy difference per unit electric charge; for example, the electrons transferred via a 1.5 V cell release only one-sixth as much energy as the electrons from a 9 V battery. The voltage of a cell is independent of the size of the cell and depends mainly on the chemical composition of the reactants in the cell. **Electric current**, measured by an ammeter in **amperes** (A), is a measure of the rate of flow of charge past a point in an electrical circuit. The larger the electric cell of a particular kind, the greater the current that can be produced by the cell. The *charge* transferred by a cell or battery is measured in **coulombs** (C) and expresses the total charge transferred by the movement of charged particles. The *power* of a cell or battery is the rate at which it produces electrical energy. Power is measured in *watts* (W), and is calculated as the product of the current and the voltage of the battery. The *energy density*, or specific energy of a battery, is a measure of the quantity of energy stored or supplied per unit mass. Energy density may be measured in joules per kilogram (J/kg). Table 16.1 summarizes electrical quantities and their units of measurement.

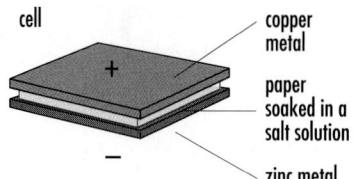

cell — copper metal — paper soaked in a salt solution — zinc metal

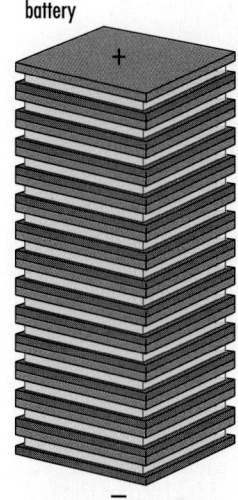

battery

Figure 16.5
Volta's revised cell design, more simple than the first, consisted of a sandwich of two metals separated by paper soaked in salt water (the electrolyte). A cell consisted of a layer of zinc metal separated from a layer of copper metal by the brine-soaked paper. A large pile of cells could be constructed to give more electrical energy.

Table 16.1

ELECTRICAL QUANTITIES AND SI UNITS				
Quantity	**Symbol**	**Meter**	**Unit**	**Unit Symbol**
charge	q	—	coulomb	C
current	I	ammeter	ampere	A (1 A = 1 C/s)
potential difference	V	voltmeter	volt	V (1 V = 1 J/C)
power	P	—	watt	W (1 W = 1 J/s)
energy density		—	joules per kilogram	J/kg

A dam built across a stream or river may stop the flow of water. Each kilogram of water that backs up behind the dam has a certain quantity of potential energy relative to the bottom of the dam. In other words, there is a potential energy difference between a kilogram of water at the top of the dam and a kilogram of water at the bottom of the dam. A voltmeter can be used to measure the height of the "dam" inside a battery; that is, the potential energy difference between a unit number of electrons at the cathode and a unit number of electrons at the anode.

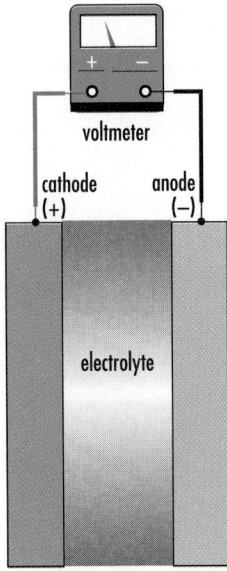

Figure 16.6
A cell always contains two electrodes — an anode and a cathode — and an electrolyte. When testing the voltage of a cell or battery, the red (+) lead of the voltmeter is connected to the positive electrode (cathode), and the black (–) lead is connected to the negative electrode (anode).

Technological Problem Solving

Technological problem solving is similar in some ways to scientific problem solving, but its purpose differs. The purpose of technological problem solving is to find a realistic way around a practical difficulty, to make something work, while the purpose of scientific problem solving is to describe, explain, or predict natural and technological phenomena. Technology and science have a symbiotic relationship. Although scientific knowledge can be used to guide the creation of a technology, the technology created may extend beyond scientific understanding. A systematic trial-and-error process, such as the following one, is often used in technological problem solving (Appendix C, C.1).

- Develop a general design for problem-solving trials; for example, select which variables to manipulate and which to control.
- Follow several prediction-procedure-evidence-analysis cycles, manipulating and systematically studying one variable at a time.
- Complete an evaluation based on criteria such as efficiency, reliability, cost, and simplicity.

Try out this technological problem-solving model in the following investigation.

Problem
✔ Prediction
Design
Materials
✔ Procedure
✔ Evidence
✔ Analysis
✔ Evaluation
Synthesis

INVESTIGATION

16.2 Designing an Electric Cell

The purpose of this investigation is to use everyday materials to simulate the technological development of an electric cell in which an aluminum soft-drink can is one of the electrodes (Figure 16.7). The other electrode is a solid conductor such as graphite from a pencil, an iron nail, or a piece of copper wire or pipe. The electrolyte may be a salt solution or an acidic or basic solution. Although many characteristics of a cell are important for an overall evaluation of performance, only one characteristic, voltage, is investigated here. Check with your teacher if you wish to evaluate other designs and materials. Include safety instructions in your procedure.

Problem

What combination of electrodes and electrolyte gives the largest voltage for an aluminum-can cell?

Experimental Design

(a) Using the same electrolyte and aluminum can as the controlled variables, two or three different materials are employed as the second electrode. The voltage of each cell is measured.

(b) Using the same two electrodes as the controlled variables, two or three possible electrolytes are tested. The voltage of each cell is measured.

(c) Additional combinations are tested, based on the analysis of the initial trials.

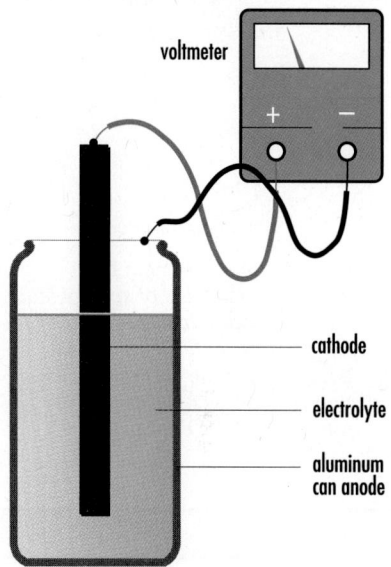

Figure 16.7
An aluminum-can cell is an efficient design since one of the electrodes also serves as the container.

Consumer, Commercial, and Industrial Cells

Since Volta's invention of the electric cell and battery, there have been many advances in electrochemistry and technology. Invented in 1865, the zinc chloride cell is commonly referred to as a *dry cell* because this design was the first to use a sealed container. (The electrolyte is actually a moist paste; if the cell were completely dry it would not work.) These 1.5 V dry cells were used to make the first 9 V battery (Figure 16.8). Both the 1.5 V dry cell and the 9 V battery are simple, reliable, and relatively inexpensive. Other cells, such as the alkaline dry cell and the mercury cell (Table 16.2, page 658), were developed to improve the performance of the original dry cell. One problem with all of these cells is that the chemicals are eventually depleted and irreversible side reactions largely prevent these cells from being recharged. Cells such as these, that cannot in practice be recharged, are called **primary cells**. Two types of cells have been developed that do not have this disadvantage.

Secondary cells can be recharged by using electricity to reverse the chemical reaction that occurs when electricity is produced by the cell. Secondary cells and batteries include the nickel-cadmium (Ni-Cad) cell and the lead-acid battery (Table 16.2 and Figure 16.9). A relatively

Zinc Chloride Dry Cells

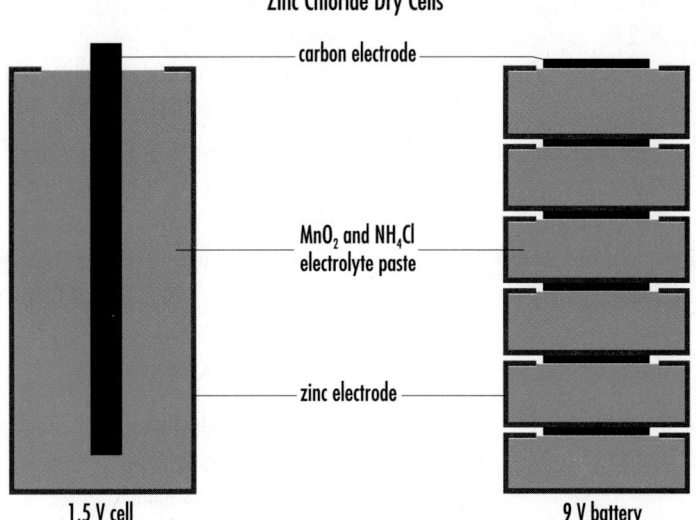

carbon electrode

MnO_2 and NH_4Cl electrolyte paste

zinc electrode

1.5 V cell

9 V battery

Figure 16.8
Like a flashlight D cell, the dry cell on the left has a voltage of 1.5 V. The 9 V battery on the right is made up of six 1.5 V dry cells in series.

recently developed secondary cell with a unique design is the lithium-ion cell, or Molicel (Figure 16.10). All secondary cells are recharged by reversing the chemical reactions that originally produced the electricity. For example, the discharging of a cell in a typical car battery (see lead-acid cell, Table 16.2) produces approximately 2.0 V for the following net equation. This process is analogous to water spontaneously running downhill.

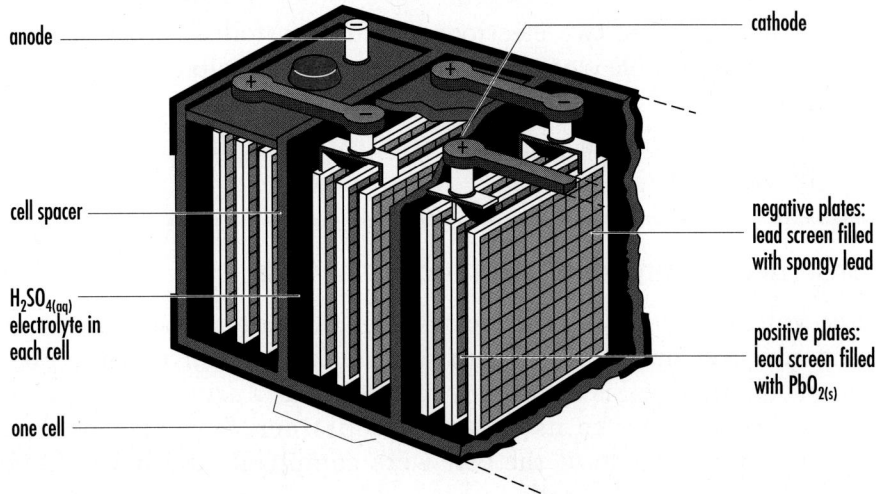

Figure 16.9
The anodes of a lead-acid car battery are composed of spongy lead and the cathodes are composed of lead(IV) oxide on a metal screen. The large electrode surface area is designed to deliver the current required to start a car engine.

Table 16.2

PRIMARY, SECONDARY, AND FUEL CELLS			
Type	**Name of Cell**	**Half-Reactions**	**Characteristics and Uses**
primary cells	dry cell (1.5 V)	$2\,MnO_{2(s)} + 2\,NH_4^+{}_{(aq)} + 2\,e^- \rightarrow Mn_2O_{3(s)} + 2\,NH_{3(aq)} + H_2O_{(l)}$ $Zn_{(s)} \rightarrow Zn^{2+}{}_{(aq)} + 2\,e^-$	• inexpensive, portable, many sizes • flashlights, radios, many other consumer items
	alkaline dry cell (1.5 V)	$2\,MnO_{2(s)} + H_2O_{(l)} + 2\,e^- \rightarrow Mn_2O_{3(s)} + 2\,OH^-{}_{(aq)}$ $Zn_{(s)} + 2\,OH^-{}_{(aq)} \rightarrow ZnO_{(s)} + H_2O_{(l)} + 2\,e^-$	• longer shelf life; higher currents for longer periods compared with dry cell • same uses as dry cell
	mercury cell (1.35 V)	$HgO_{(s)} + H_2O_{(l)} + 2\,e^- \rightarrow Hg_{(l)} + 2\,OH^-$ $Zn_{(s)} + 2\,OH^-{}_{(aq)} \rightarrow ZnO_{(s)} + H_2O_{(l)} + 2\,e^-$	• small cell; constant voltage during its active life • hearing aids, watches
secondary cells	Ni-Cad cell (1.25 V)	$2\,NiO(OH)_{(s)} + 2\,H_2O_{(l)} + 2\,e^- \rightarrow 2\,Ni(OH)_{2(s)} + 2\,OH^-$ $Cd_{(s)} + 2\,OH^-{}_{(aq)} \rightarrow Cd(OH)_{2(s)} + 2\,e^-$	• can be completely sealed; lightweight but expensive • all normal dry cell uses, as well as power tools, shavers, portable computers
	lead-acid cell (2.0 V)	$PbO_{2(s)} + 4\,H^+{}_{(aq)} + SO_4^{2-}{}_{(aq)} + 2\,e^- \rightarrow PbSO_{4(s)} + 2\,H_2O_{(l)}$ $Pb_{(s)} + SO_4^{2-}{}_{(aq)} \rightarrow PbSO_{4(s)} + 2\,e^-$	• very large currents; reliable for many recharges • all vehicles
fuel cells	aluminum-air cell (2 V)	$3\,O_{2(g)} + 6\,H_2O_{(l)} + 12\,e^- \rightarrow 12\,OH^-{}_{(aq)}$ $4\,Al_{(s)} \rightarrow 4\,Al^{3+}{}_{(aq)} + 12\,e^-$	• very high energy density; made from readily available aluminum alloys • designed for electric cars
	hydrogen-oxygen cell (1.2 V)	$O_{2(g)} + 2\,H_2O_{(l)} + 4\,e^- \rightarrow 4\,OH^-{}_{(aq)}$ $2\,H_{2(g)} + 4\,OH^-{}_{(aq)} \rightarrow 4\,H_2O_{(l)} + 4\,e^-$	• lightweight; high efficiency; can be adapted to use hydrogen-rich fuels • vehicles and space shuttle

$$Pb_{(s)} + PbO_{2(s)} + 2\ H_2SO_{4(aq)} \quad \xrightarrow{\text{discharging}} \quad 2\ PbSO_{4(s)} + 2\ H_2O_{(l)}$$

To charge (recharge) this cell requires the input (from the car's generator) of at least 2.0 V to force the products to change back to the reactants. The half-reactions for the lead-acid cell listed in Table 16.2 both need to be reversed to obtain the following net equation. This process is analogous to water at the bottom of a hill being pumped back to the top.

$$2\ PbSO_{4(s)} + 2\ H_2O_{(l)} \quad \xrightarrow{\text{charging}} \quad Pb_{(s)} + PbO_{2(s)} + 2\ H_2SO_{4(aq)}$$

A battery can be recharged if the products are stable with no further side reactions occurring and if the products have the mobility to travel through the electrolyte toward the appropriate electrode.

A **fuel cell**, another solution to the problem of the limited life of a cell, produces electricity by the reaction of a fuel that is continually supplied to keep the cell operating. Considerable technological research and development (R & D) has gone into fuel cell design. An aluminum-air cell (Table 16.2) is actually an aluminum-oxygen cell and has been developed for possible use in electric cars. Air is pumped into the cell and oxygen reacts at the cathode while a replaceable mass of aluminum reacts at the anode. Estimates from prototypes suggest that the aluminum anode will need replacement every 2500 km in an electric car.

The first fuel cell was invented by William Grove in 1839 using hydrogen and oxygen as fuels and sulfuric acid as the electrolyte (Figure 16.11). The major application of a hydrogen-oxygen fuel cell is by NASA in the space shuttle program. NASA's fuel cell is an alkaline cell using potassium hydroxide as the electrolyte (Table 16.2). It operates at 70% efficiency in 16 kW batteries. A commercial variation of a hydrogen-oxygen fuel cell is one using hydrogen-rich fuels such as methane or methanol together with oxygen from the air. (See Ballard Cell feature, page 661.)

Commercially viable fuel cells today are usually acid electrolyte cells such as the phosphoric acid fuel cell, which has reached a size of 400 MW, sufficient for the electrical energy needs of a small city. Fuel cells have several advantages over fossil fuel combustion methods of generating electricity. The efficiency of fuel cells is generally higher (Table 16.3) and the efficiency does not depend on the size of the cell. Environmental problems, such as acid rain, are largely eliminated by fuel cells. However, the cost of fuel cells still remains relatively high.

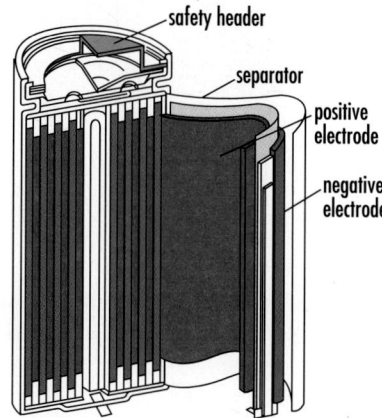

Figure 16.10
Invented and manufactured in British Columbia, the Molicel is a high-energy, rechargeable cell in a unique, jelly-roll design. (See Molicel feature, page 660.)

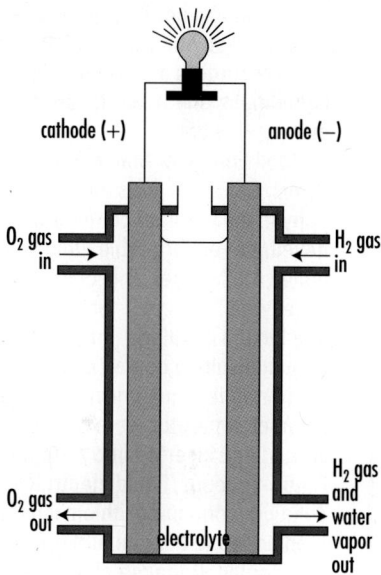

Figure 16.11
Hydrogen and oxygen gases are continuously pumped into the cell and each reacts at a different electrode. The unused gases are removed, filtered, and then recycled.

Table 16.3

EFFICIENCIES OF DIFFERENT TECHNOLOGIES	
Fuel cells	40 – 70 %
Electric power plants	30 – 40 %
Automobile engines	17 – 23 %
Gasoline lawnmower	12 %

Exercise

6. What is the relationship between scientific knowledge and technological problem solving?

7. What steps are involved in technological problem solving?

8. Suppose you decided to develop and market an aluminum-can cell. How and why would you alter the electrolyte?

THE MOLICEL

Portable electronic devices can be found everywhere these days. Laptop computers, cellular telephones, mobile radios, cordless phones, portable disc and tape players, and video recorders all require a small, lightweight, and powerful source of energy — a cell.

Moli Energy of Maple Ridge, British Columbia, was the first company in the world to commercially develop a rechargeable lithium cell, called the Molicel. This cell featured a "jelly roll" design in which sheets of a lithium metal foil (anode), molybdenum disulfide coated on aluminum foil (cathode), and a microporous polymer separator were spirally wound and inserted into a steel can. A non-aqueous, liquid electrolyte solution (containing lithium ions) was then added to the can and the can was sealed. Because of its small atomic size, lithium is very mobile, permitting high current electrochemical cells. Larger rechargeable lithium metal cells, however, have not exhibited the high level of safety required for consumer applications and are not available commercially.

As a result of the need for a powerful, rechargeable, and safe cell system, significant efforts have been put into improving the original design of the lithium cell. Removal of the lithium metal anode was key in this effort. Carbon anodes have been used as a replacement. Moli Energy was the first North American company to commercialize this new cell technology, known as lithium-ion. In this system, the lithium metal anode is replaced by a safer graphitic carbon anode, and the molybdenum disulfide is replaced with a higher voltage lithium cobaltite cathode. The internal cell construction is similar, except that a specially designed safety header is used to seal these cells (see Figure 16.10). The header provides protection from overcharge, short circuit, and thermal abuse.

The Molicel lithium-ion cell transfers its energy using a reversible process called intercalation. Intercalation is the insertion of guest atoms into host materials with reversible and usually minimal physical alteration of the host structure. The lithium-ion cell uses a dual intercalation process as the lithium ions (guests) migrate back and forth, inserting into the lithium cobaltite host on discharge and the graphitic carbon host while charging. The discharging process is shown from left to right in the following set of half-reaction equations and the charging process is shown from right to left. The resulting average cell voltage is 3.6 to 3.7 V, depending on load.

Cathode half-reaction
$$Li_{1-x}CoO_2 + x\,Li^+ + x\,e^- \rightleftharpoons Li_1CoO_2 \ (x<1)$$

Anode half-reaction
$$Li_xC_6 \rightleftharpoons C_6 + x\,Li^+ + x\,e^-$$

The charge and discharge cycles are shown in the diagram.

How does a lithium-ion cell of this design compare with other rechargeable cells? The Molicel lithium-ion system provides three times the voltage and nearly twice the energy density by mass of either a conventional nickel metal hydride or nickel cadmium cell. It also retains its charge five times longer — it has five times the shelf life. As well, unlike other cells, the Molicel does not have to be fully discharged to enhance its cycle life, nor does it contain any environmentally hazardous materials such as mercury, lead, or cadmium.

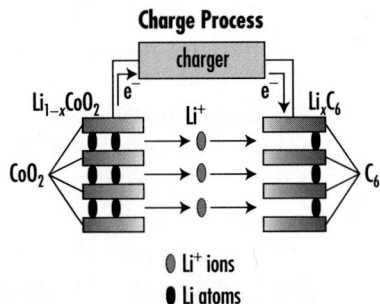

Charge Process

charger

$Li_{1-x}CoO_2$ e^- Li^+ e^- Li_xC_6

CoO_2 C_6

● Li^+ ions
● Li atoms

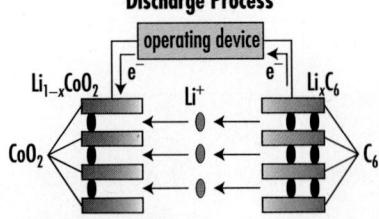

Discharge Process

operating device

$Li_{1-x}CoO_2$ e^- Li^+ e^- Li_xC_6

CoO_2 C_6

- *What is creative about the design of the Molicel?*
- *Describe and explain the advantages of the Molicel.*

Problem 16A Evaluating Batteries

Criteria used to evaluate a battery include its reliability, cost, simplicity of use, safety (leakage), size (volume), shelf life, active life, energy density, power capacity, maintenance, disposal, environmental impact, and ability to be recharged.

Gather some information and complete the Evidence and Analysis of the investigation report.

Problem

Taking all of the preceding criteria into account, what is the best cell or battery for a portable radio, cassette player, or CD player?

▌ THE BALLARD FUEL CELL ▌

Ballard Power Systems of North Vancouver has become a world leader in the development of proton exchange membrane (PEM) fuel cells. The zero-emission engines convert hydrogen, natural gas, and even methanol into electricity, producing water and heat as the main by-products. Research and development of the Ballard fuel cell began in 1983, and since 1990 several prototype systems have been successfully demonstrated.

The Ballard fuel cell consists of an anode and a cathode separated by a polymer membrane electrolyte. Hydrogen fuel admitted through a porous anode is then converted into hydrogen ions (protons) and free electrons in the presence of a catalyst at the anode. An external circuit conducts the free electrons and produces the desired electrical current. Water and heat are produced when the protons, after migrating through the polymer membrane to the cathode, react both with oxygen molecules from the air and with the free electrons from the external circuit. Fuel cells can be connected in series (stacked) to increase the voltage and power output (see photo on the left).

Ballard Power Systems has focused its fuel cell technology on urban transit buses (see photo on the right) and small stationary power generation systems — two applications where clean and efficient power systems are in demand. The world's first zero-emission vehicle powered by a PEM fuel cell was demonstrated by Ballard in 1993. This system was further developed in 1995 to allow a Vancouver regional transit diesel bus to be converted to a Ballard fuel cell power system capable of 205 kW (250 hp).

- *Make a list, in order of importance, of the advantages of a Ballard hydrogen fuel cell for urban buses compared with a diesel engine for buses. Make a list of possible disadvantages.*

- *Another Ballard-type fuel cell being developed uses methanol as a fuel. What is the advantage of methanol over hydrogen or natural gas?*

Voltaic cells developed to serve practical purposes were not explained scientifically until about 100 years after their invention in 1800. However, their use contributed to scientific understanding of redox reactions and, later, this knowledge helped explain reactions inside the cell itself.

From a scientific perspective, the design of a cell "plays a trick" on oxidizing and reducing agents, resulting in electrons passing through an external circuit rather than directly from one substance to another. In

ELECTRIC CARS

Since their invention in 1888, vehicles powered by electricity have waxed and waned in popularity. Many experts predict that electric vehicles will make a breakthrough during the next decade, because in California a combination of political, economic, and environmental factors makes them a viable alternative to gasoline-powered vehicles. The main advantage of electric cars over gasoline-fueled cars is efficiency. Cars powered by gasoline engines are about 15% efficient, but many electric cars are 90% efficient. (Of course, overall efficiency depends on how the electricity and gasoline are produced in the first place.) Other attractive features of electric vehicles are near-silence and minimal maintenance.

A disadvantage of battery powered cars is that early test models could travel only a limited distance before recharging was necessary and this may take several hours. Therefore, many automotive experts argue that electric cars would be feasible in urban areas only. Also, prolonged testing of some electric vehicles has shown that the batteries must be replaced after about 80 000 km, or once every four to five years, which adds to the operating cost.

The most serious obstacle to the widespread use of electric cars is the lack of a powerful, lightweight, inexpensive battery. Scientists are researching alternatives to lead-acid batteries in order to increase the range and utility of electric vehicles. Batteries have been modified to withstand thousands of cycles of deep discharge and recharge. The most promising types — nickel-iron and sodium-sulfur — are steadily undergoing improvement.

Another potential power source for electric cars is the aluminum-air fuel cell, which consists of aluminum plates, an air cathode, and an electrolyte. Electricity is produced as the aluminum oxidizes, and the cell is kept operating by replacing the aluminum plates and adding more electrolyte. A prototype mini-van fitted with an aluminum-air fuel cell has a range of 300 km, compared with 75 km for an electric van powered by lead-acid batteries alone. Another possibility is the solid polymer hydrogen (or methanol) fuel cell. Its discovery led to a four-fold improvement in power, so that liquid and gas fuel cells are potentially feasible batteries for electric cars.

- *Why are electric cars a promising alternative to regular cars for urban commuting?*
- *What are two types of commercial cells that have been tested for use in electric cars?*

Investigations 16.1 and 16.2 you saw that the individual components of a cell — electrodes and electrolytes — determine electrical characteristics such as voltage and current. Why is this so? What happens in different parts of a cell? To answer these questions, chemists use a cell with a different design, with the parts of the cell separated so they can be studied more easily. Each electrode is in contact with an electrolyte, but the electrolytes surrounding each electrode are separated. This is accomplished by a **porous boundary**, a barrier that separates electrolytes while still permitting ions to move through tiny openings between the two solutions. Two common examples of porous boundaries are the *salt bridge* and the *porous cup*, shown in Figure 16.12.

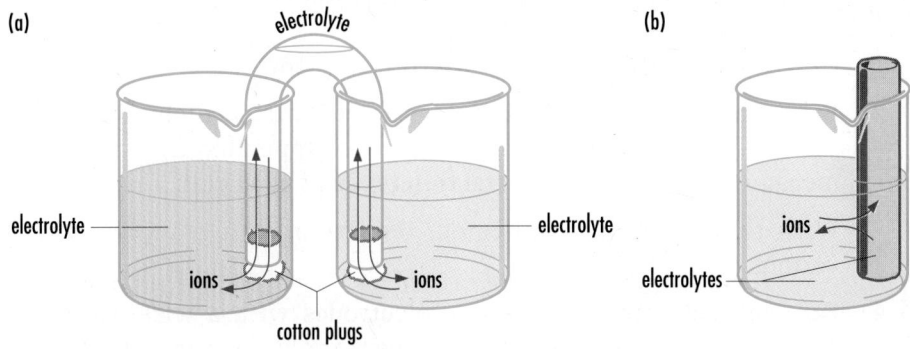

(a)   electrolyte   **(b)**

electrolyte   electrolyte   ions

ions   ions   electrolytes

cotton plugs

Figure 16.12
(a) A salt bridge is a U-shaped tube containing an unreactive aqueous electrolyte such as sodium sulfate. The cotton plug allows ions to move into or out of the ends of the tube when the ends are immersed in electrolytes.
(b) An unglazed porcelain (porous) cup containing one electrolyte sits in a container of a second electrolyte. The two solutions are separated but ions can move in and out of the cup through the pores in the porcelain.

With this design modification, a cell can be split into two parts connected by a porous boundary. Each part, called a **half-cell**, consists of one electrode and one electrolyte. For example, the copper-zinc cell shown in Figure 16.13 has two half-cells, copper metal in a solution of copper ions, and zinc metal in a solution of zinc ions. It can be represented as follows.

$$Cu_{(s)} \mid Cu(NO_3)_{2(aq)} \parallel Zn(NO_3)_{2(aq)} \mid Zn_{(s)}$$

In this notation, a single line (|) indicates a phase boundary such as the interface of an electrode and an electrolyte in a half-cell. A double line (||) represents a physical boundary such as a porous boundary between half-cells. A voltaic cell is an arrangement of two half-cells that can produce electricity spontaneously. Cells such as the one in Figure 16.13 are especially suitable for scientific study.

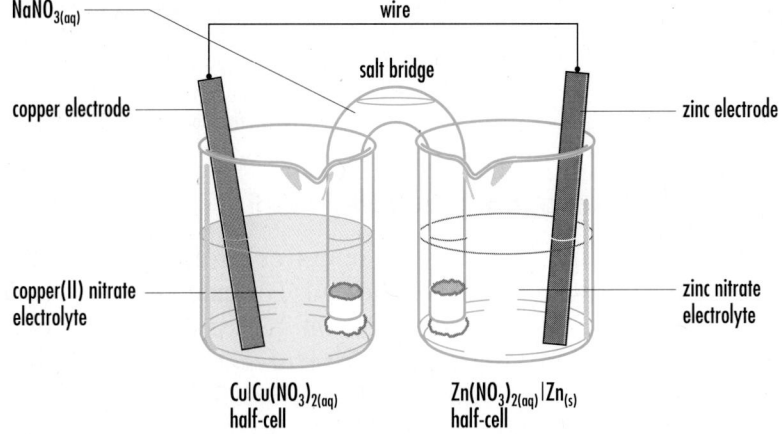

$NaNO_{3(aq)}$ wire

salt bridge

copper electrode   zinc electrode

copper(II) nitrate electrolyte   zinc nitrate electrolyte

$Cu \mid Cu(NO_3)_{2(aq)}$ half-cell $Zn(NO_3)_{2(aq)} \mid Zn_{(s)}$ half-cell

Figure 16.13
The essential parts of a cell are two electrodes and an electrolyte. In this design each electrode is in its own electrolyte, forming a half-cell. The two half-cells are connected by a salt bridge and by an external conductor to make a complete circuit.

- Problem
- Prediction
- Design
- Materials
- Procedure
- ✔ Evidence
- ✔ Analysis
- Evaluation
- Synthesis

CAUTION

Solutions used are toxic and irritant. Avoid contact with skin and eyes.

16.3 Demonstration: A Voltaic Cell

The purpose of this investigation is to demonstrate the design and operation of a voltaic cell used in scientific research.

Problem

What is the design and operation of a voltaic cell?

Experimental Design

An electric cell with only one electrolyte is compared with similar voltaic cells containing the same electrodes but two electrolytes.

Procedure

1. Construct the three cells shown in Figure 16.14.

2. For each design, use a voltmeter to determine which electrode is positive and which is negative (see Appendix C, C.2), and measure the electric potential difference of each cell.

3. With the voltmeter connected, remove and then replace the various parts of the cell.

4. For each cell, connect the two electrodes with a wire. Record any evidence of a reaction after several minutes, and after one or two days. Measure the electric potential difference after several days.

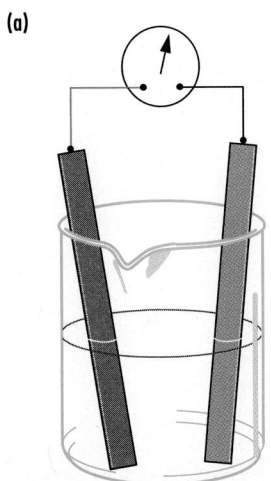

(a)

No porous boundary: $Ag_{(s)}$ | $NaNO_{3(aq)}$ | $Cu_{(s)}$

Figure 16.14
Investigation 16.3 compares three different cell designs.

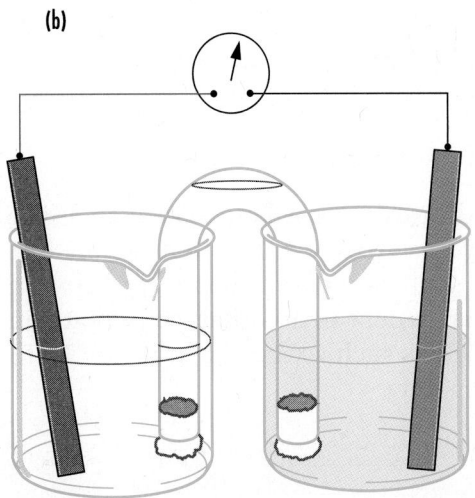

(b)

Salt bridge: $Ag_{(s)}$ | $AgNO_{3(aq)}$ || $Cu(NO_3)_{2(aq)}$ | $Cu_{(s)}$

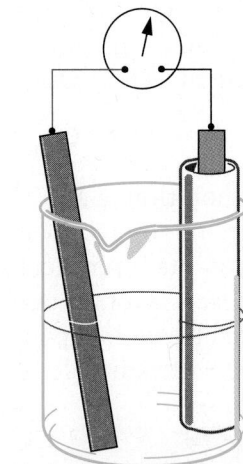

(c)

Porous cup: $Ag_{(s)}$ | $AgNO_{3(aq)}$ || $Cu(NO_3)_{2(aq)}$ | $Cu_{(s)}$

A Theoretical Description of a Voltaic Cell

Observation of a voltaic cell as it operates provides evidence that explains what is happening inside the cell. For example, the study of a silver-copper voltaic cell in Investigation 16.3 provides the evidence listed in Table 16.4 A theoretical interpretation of each point is included in the table and is shown in Figure 16.15.

Table 16.4

EVIDENCE AND INTERPRETATION OF THE SILVER-COPPER CELL	
Evidence	**Interpretation**
The copper electrode decreases in mass and the intensity of the blue color of the electrolyte increases.	Oxidation is occurring. $$Cu_{(s)} \rightarrow Cu^{2+}_{(aq)} + 2\,e^-$$ <center>blue</center>
The silver electrode increases in mass as long silver-colored crystals grow.	Reduction is occurring. $$Ag^+_{(aq)} + e^- \rightarrow Ag_{(s)}$$
A blue color slowly moves up the U-tube from the copper half-cell to the silver half-cell and the solution remains electrically neutral.	Copper(II) ions move toward the cathode. Negative ions (anions) move toward the anode.
A voltmeter indicates that the silver electrode is the cathode (positive) and the copper electrode is the anode (negative).	Electrons move from the copper electrode to the silver electrode.
An ammeter shows that the electric current flows between the copper electrode and the silver electrode.	Electrons leave the copper half-cell and enter the silver half-cell.

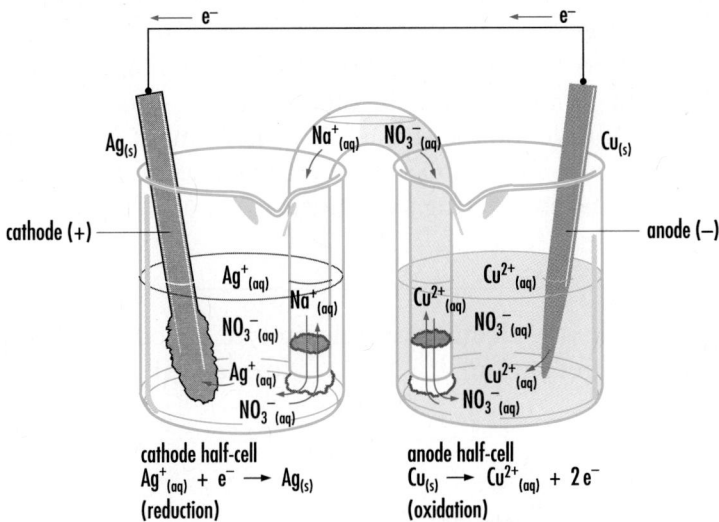

cathode half-cell
$$Ag^+_{(aq)} + e^- \rightarrow Ag_{(s)}$$
(reduction)

anode half-cell
$$Cu_{(s)} \rightarrow Cu^{2+}_{(aq)} + 2\,e^-$$
(oxidation)

Figure 16.15
A theoretical interpretation of the silver-copper cell.

According to the electron transfer theory and the concept of relative strengths of oxidizing and reducing agents, silver ions are the strongest oxidizing agents in the cell; they undergo a reduction half-reaction at the cathode. *The strongest oxidizing agent always undergoes a reduction at the cathode.* Copper atoms, which are the strongest reducing agents in the cell, give up electrons in an oxidation half-reaction and enter the solution at the anode. *The strongest reducing agent always undergoes an oxidation at the anode.* According to theory, the **cathode** is the electrode where reduction occurs and the **anode** is the electrode where oxidation occurs. Electrons released by the oxidation of copper atoms at the anode travel through the connecting wire to the silver cathode. The direction of electron flow

Memory Devices
People often use acronyms or similar devices to help them remember important information. LEO and GER (page 613) are examples of acronyms. One way to help you remember important details of a cell is the expression SOAC/GERC, loosely read as "soak a jerk." Translated, this means the **S**trongest **O**xidizing **A**gent at the **C**athode **G**ains **E**lectrons and is **R**educed at the **C**athode. Another example is "An ox ate a red cat," which helps to recall *An*ode *ox*idation . . . *red*uction *cat*hode. Sometimes the best memory devices are the ones you invent yourself.

can be explained in terms of competition for electrons. According to the table of relative strengths of oxidizing and reducing agents in Appendix F, silver ions are stronger oxidizing agents than copper(II) ions. Silver ions win the tug-of-war for the electrons available from the conducting wire. To write the net equation for the silver-copper voltaic cell, identify the strongest oxidizing and reducing agents. Then follow the same procedure as for reactions in which the two materials are in contact with each other (page 626).

$$\overset{\text{SOA}}{}\qquad\overset{\text{OA}}{}$$

$$\underset{\textbf{RA}}{Ag_{(s)}} \mid \underset{}{Ag^+_{(aq)}} \mid\mid \underset{}{Cu^{2+}_{(aq)}} \mid \underset{\textbf{SRA}}{Cu_{(s)}}$$

reduction at the cathode $2\,[\,Ag^+_{(aq)} + e^- \rightarrow Ag_{(s)}\,]$

oxidation at the anode $Cu_{(s)} \rightarrow Cu^{2+}_{(aq)} + 2\,e^-$

net $Cu_{(s)} + 2\,Ag^+_{(aq)} \rightarrow Cu^{2+}_{(aq)} + 2\,Ag_{(s)}$

The electrical neutrality in the half-cells and the U-tube solution can be explained in terms of the half-reactions and the movement of ions. If cations did not move to the cathode, the removal of silver ions from the solution near the cathode would create a net negative charge around the cathode and the buildup of negative charge would prevent electrons from being transferred. However, chemists explain that cations migrate toward the cathode solution and electrical neutrality is maintained. Likewise, the formation of copper(II) ions at the anode would create a net positive charge but this is balanced by the movement of negative ions to the anode compartment through the salt bridge or porous cup. This explains the evidence of electrical neutrality and the need for the connecting salt bridge.

The cathode and the anode can be represented in cell notations and in cell diagrams using the words *cathode* and *anode* and/or the signs + and –.

electrons

cathode (+) | electrolyte || electrolyte | anode (–)
(reduction)   (oxidation)

anions →
← cations

Voltaic Cells with Inert Electrodes

For cells containing metals and metal ions, the electrodes are usually the metals, and half-reactions take place on the surface of the metals. What happens if an oxidizing or a reducing agent other than these is used? For example, an acidic dichromate solution is a strong oxidizing agent that reacts spontaneously with copper metal. To construct this cell, you can use a copper half-cell as in Figure 16.15 (page 665) but an electrode is required for the dichromate half-cell. You cannot use solid sodium dichromate as an electrode because solid ionic compounds do not conduct electricity and solid sodium dichromate would also dissolve in the solution. You need a solid conductor that will not react in the cell or interfere with the desired cell reaction. In other words, you need an unreactive or inert electrode. Inert electrodes provide a location to connect a wire and a surface on which a half-reaction can occur. A carbon (graphite) rod (Figure 16.16) or platinum metal foil are two commonly used inert electrodes.

Solutions in salt bridges are always inert electrolytes and are usually Group 1 sulfates or nitrates such as sodium sulfate or potassium nitrate.

EXAMPLE

(a) Write the cathode, anode, and net cell reaction equations for the following cell description.
(b) Draw a diagram of the cell, labelling electrodes, electrolytes, electron flow, and ion movement.

$$\overset{\text{SOA}\quad\text{OA}}{\text{C}_{(s)}\ |\ \text{Cr}_2\text{O}_7{}^{2-}{}_{(aq)}, \text{H}^+{}_{(aq)}\ ||\ \overset{\text{OA}}{\text{Cu}^{2+}{}_{(aq)}}\ |\ \text{Cu}_{(s)}}$$
$$\qquad\qquad\qquad\qquad\qquad\qquad\qquad\qquad\text{SRA}$$

cathode $\quad \text{Cr}_2\text{O}_7{}^{2-}{}_{(aq)} + 14\,\text{H}^+{}_{(aq)} + 6\,\text{e}^- \rightarrow 2\,\text{Cr}^{3+}{}_{(aq)} + 7\,\text{H}_2\text{O}_{(l)}$

anode $\qquad\qquad\qquad 3\,[\,\text{Cu}_{(s)} \rightarrow \text{Cu}^{2+}{}_{(aq)} + 2\,\text{e}^-\,]$
_____

net $\quad \text{Cr}_2\text{O}_7{}^{2-}{}_{(aq)} + 14\,\text{H}^+{}_{(aq)} + 3\,\text{Cu}_{(s)} \rightarrow 3\,\text{Cu}^{2+}{}_{(aq)} + 2\,\text{Cr}^{3+}{}_{(aq)} + 7\,\text{H}_2\text{O}_{(l)}$

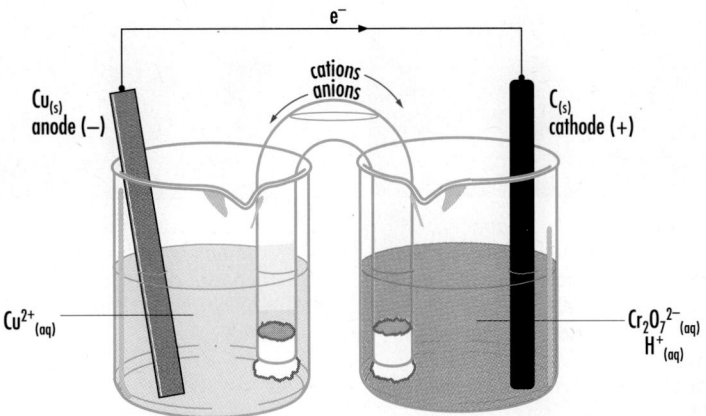

Figure 16.16
A copper-dichromate cell with an inert carbon electrode.

Electric Potential Difference and Concentration

Evidence from Investigation 16.3 shows that the electric potential difference or voltage of a cell decreases slowly as the cell operates. Simultaneously, color changes and precipitate formation occurs. If the cell is left for a long time, the voltage would eventually become zero and no further changes would be observed in the cell. When people refer to a dead cell or battery this is often what is meant.

The electric potential difference or voltage of a cell indicates the tendency for electrons to flow and is a measure of the energy difference per unit electric charge (page 655). The value that is measured by a voltmeter represents a potential or stored energy just like the water behind a lock in a canal has potential energy (Figure 16.17(a)). Connecting the electrodes of a cell in a circuit allows the electrons to flow from the anode to the cathode. This is analogous to opening the valve or sluice and allowing the water to flow from behind the gate to a lower point in front of the gates (Figure 16.17(b)). In both cases stored potential energy is converted to kinetic energy of electrons or water. If a finite quantity of water is available behind a lock and this water is allowed to flow out, then eventually no more water will flow as the level (potential energy) of the water on the two sides of the gate is equalized. An equilibrium is reached with no potential energy difference (Figure 16.17(c)). A similar situation occurs with an operating cell. If electrons are allowed to flow, eventually an equilibrium will be reached when the flow ceases. The rate of the forward reaction, which predominates initially, decreases as the rate of the reverse reaction increases until the two rates become equal. This is the equilibrium condition and no net flow of electrons will occur. At this time the electric potential difference as measured by a voltmeter becomes zero.

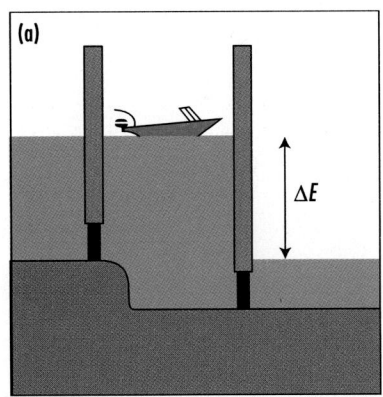

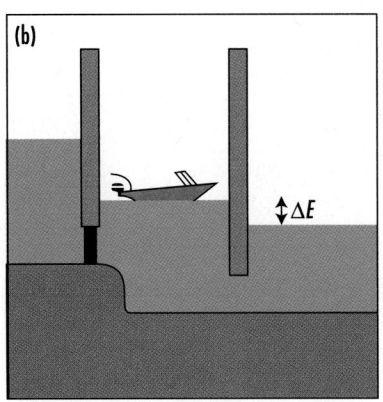

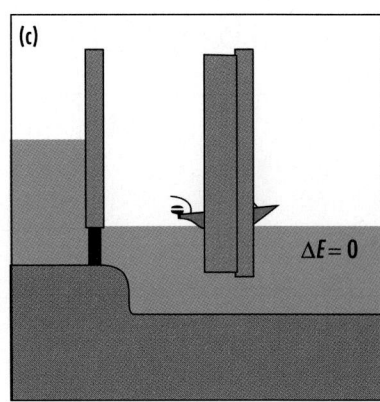

Figure 16.17
(a) The water behind the gates in a lock has a certain potential energy, ΔE, relative to the bottom of the closed outlet.

(b) When the outlet is opened, water spontaneously flows to the lower level on the other side of the gates. Potential energy, ΔE, is converted to kinetic energy of the flowing water. The water flowing through the outlet is analogous to electron flow.

(c) The flow of water ceases when the levels on both sides of the gates become equal. The gates open, and the ship can then exit to the next lock.

Nernst Equation

Standard cell potentials can be readily determined using standard reduction potentials. If the conditions are not standard, the cell potential can be predicted using a relationship discovered by Walter Nernst. The simplified version of the Nernst equation for 25°C is shown below.

$$\Delta E = \Delta E^\circ - \frac{0.0592}{n} \log Q$$

where

ΔE is the cell potential at 25°C and non-standard concentrations

ΔE° is the cell potential at 25°C and standard concentrations

n is the amount in moles of electrons transferred according to the cell reaction

Q is the reaction quotient (see page 516)

SUMMARY: VOLTAIC CELLS

- A voltaic or electrochemical cell consists of two half-cells separated by a porous boundary with solid electrodes connected by an external circuit.

- The cathode is the positive electrode and the location of the reaction of the strongest oxidizing agent present.

- The anode is the negative electrode and the location of the reaction of the strongest reducing agent present.

- Electrons travel in the external circuit from the anode to the cathode.

- Internally, anions move toward the anode and cations move toward the cathode as the cell operates.

- When a cell reaches equilibrium, the electric potential difference (voltage) becomes zero.

Exercise

14. Write an empirical description of each of the following terms: voltaic cell, half-cell, porous boundary, and inert electrode.

15. Write a theoretical definition of a cathode and an anode.

16. Indicate whether the following processes occur at the cathode or at the anode of a voltaic cell.
    (a) reduction half-reaction
    (b) oxidation half-reaction
    (c) reaction of the strongest reducing agent
    (d) reaction of the strongest oxidizing agent

17. When is an inert electrode used?

18. What are the characteristics of the solution in a salt bridge? Provide an example.

19. For each of the following cells, use the given cell notation to identify the strongest oxidizing and reducing agents. Write chemical equations to represent the cathode, anode, and net cell reactions. Draw a diagram of each cell, labelling the electrodes, electrolytes, electron flow, and ion movement.

    (a) $Ag_{(s)}$ | $Ag^+_{(aq)}$ || $Zn^{2+}_{(aq)}$ | $Zn_{(s)}$
    (b) $Cu_{(s)}$ | $Cu^{2+}_{(aq)}$ || $Zn^{2+}_{(aq)}$ | $Zn_{(s)}$
    (c) $Sn_{(s)}$ | $Sn^{2+}_{(aq)}$ || $Cr_2O_7^{2-}_{(aq)}$, $H^+_{(aq)}$ | $C_{(s)}$
    (d) $Al_{(s)}$ | $Al^{3+}_{(aq)}$ || $Na^+_{(aq)}$, $Cl^-_{(aq)}$, $O_{2(g)}$, $H_2O_{(l)}$ | $Pt_{(s)}$

20. Ions move through a porous boundary between the two half-cells of a voltaic cell.
    (a) Why do the ions move? (Answer a series of "why" questions.)
    (b) In what direction do the cations and anions move?

21. Draw and label a diagram for a voltaic cell constructed from some (not all) of the following materials.

strip of cadmium metal	voltmeter
strip of nickel metal	connecting wires
solid cadmium sulfate	glass U-tube
solid nickel(II) sulfate	cotton
solid potassium sulfate	various beakers
distilled water	porous porcelain cup

22. Redesign the voltaic cell in question 21 by changing at least one electrode and one electrolyte. The net reaction should remain the same for the redesigned cell.

Standard Cells and Cell Potentials

This chapter's investigations have shown that the design of a cell affects its operation. To facilitate comparison and scientific study, chemists specify a cell's composition and the conditions under which the cell is constructed and measured. A **standard cell** is a voltaic cell in which each half-cell contains all entities shown in the half-reaction equation at SATP conditions, with a concentration of 1.0 mol/L for the aqueous entities. If a metal is not part of a half-cell, then an inert electrode is used to construct the standard cell.

$$C_{(s)} \mid Cr_2O_7{}^{2-}{}_{(aq)}, H^+{}_{(aq)}, Cr^{3+}{}_{(aq)} \mid\mid Zn^{2+}{}_{(aq)} \mid Zn_{(s)} \qquad \text{at SATP}$$
$$\text{1.0 mol/L} \qquad\qquad\qquad \text{1.0 mol/L}$$

The **standard cell potential** $\Delta E°$ is the maximum electric potential difference of a standard cell; $\Delta E°$ represents the energy difference (per unit of charge) between the cathode and the anode. The degree sign (°) indicates standard 1.0 mol/L and SATP conditions. Based on the idea of competition for electrons, a **standard reduction potential** $E_r°$ represents the ability of a standard half-cell to attract electrons, thus undergoing a reduction. The half-cell with the greater attraction for electrons — that is, the one with the more positive reduction potential — gains electrons from the half-cell with the lower reduction potential. The standard cell potential is the difference between the reduction potentials of the two half-cells.

$$\Delta E° = \underset{\text{cathode}}{E_r°} - \underset{\text{anode}}{E_r°}$$

It is impossible to empirically determine the reduction potential of a single half-cell. A voltmeter can only measure a potential difference, $\Delta E°$. In order to assign values for standard reduction potentials, the standard hydrogen half-cell is internationally regarded as the reference half-cell from which all other reduction potentials are derived. A half-cell such as this, that is chosen as a reference and arbitrarily assigned an electrode potential of exactly zero volts, is called a **reference half-cell**.

Standard Hydrogen Half-Cell

The standard hydrogen half-cell (Figure 16.18) consists of an inert platinum electrode immersed in a 1.00 mol/L solution of hydrogen ions, with hydrogen gas at a pressure of 100 kPa bubbling over the electrode. The pressure and temperature of the cell are kept at SATP conditions. Standard reduction potentials for all other half-cells are measured relative to that of the standard hydrogen half-cell, defined as zero volts.

$$2\,H^+{}_{(aq)} + 2\,e^- \rightleftharpoons H_{2(g)} \qquad E_r° = 0.00\ V$$

A positive reduction potential for a half-cell connected to the hydrogen half-cell means that the oxidizing agent in the half-cell is a stronger oxidizing agent than hydrogen ions and attracts electrons more strongly than hydrogen ions do. A negative reduction potential means that the oxidizing agent in the half-cell connected to the hydrogen half-cell attracts electrons less strongly than hydrogen ions do. The choice of the standard hydrogen half-cell as a reference is an accepted convention. If a different half-cell had been chosen, individual reduction potentials would be different, but their *relative* values would remain the same. In the example below, the reference half-cell changes from hydrogen to copper to zinc, but the electric potential differences remain the same.

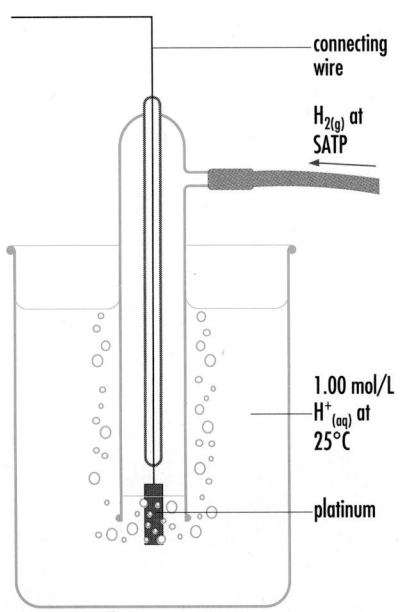

connecting wire

$H_{2(g)}$ at SATP

1.00 mol/L $H^+{}_{(aq)}$ at 25°C

platinum

$Pt_{(s)} \mid H_{2(g)}, H^+{}_{(aq)} \qquad E_r° = 0.00\ V$

Figure 16.18
The standard hydrogen half-cell is used internationally as the reference half-cell in electrochemical research.

The reduction potential of the standard copper half-cell is 0.34 V and the reduction potential of the standard zinc half-cell is –0.76 V.

(a) Determine a revised set of half-cell reduction potentials for the half-reactions if the reference half-cell is changed and one of the half-cells were arbitrarily assigned a value of 0.00 V.

(b) What is the difference between the reduction potentials of the copper and zinc half-cells in each case?

$$\Delta E^\circ = E^\circ_{r\,(cathode)} - E^\circ_{r\,(anode)}$$

(a)

Reduction Half-Reaction	REFERENCE HALF-CELL		
	Hydrogen $E^\circ_r(V)$	Copper $E^\circ_r(V)$	Zinc $E^\circ_r(V)$
$Cu^{2+}_{(aq)} + 2\,e^- \rightarrow Cu_{(s)}$	+0.34	0.00	+1.10
$2\,H^+_{(aq)} + 2\,e^- \rightarrow H_{2(g)}$	0.00	–0.34	+0.76
$Zn^{2+}_{(aq)} + 2\,e^- \rightarrow Zn_{(s)}$	–0.76	–1.10	0.00

(b) In all cases, $\Delta E^\circ = 1.10$ V for a copper-zinc cell.

Measuring Standard Reduction Potentials

The standard reduction potential of a half-cell can be measured by constructing a standard cell using a hydrogen reference half-cell and the half-cell whose reduction potential you want to measure. The cell potential is measured with a voltmeter. The cell shown in Figure 16.19 can be represented as follows.

$$Pt_{(s)} \mid H_{2(g)}, H^+_{(aq)} \parallel Cu^{2+}_{(aq)} \mid Cu_{(s)} \qquad \Delta E^\circ = 0.34 \text{ V}$$

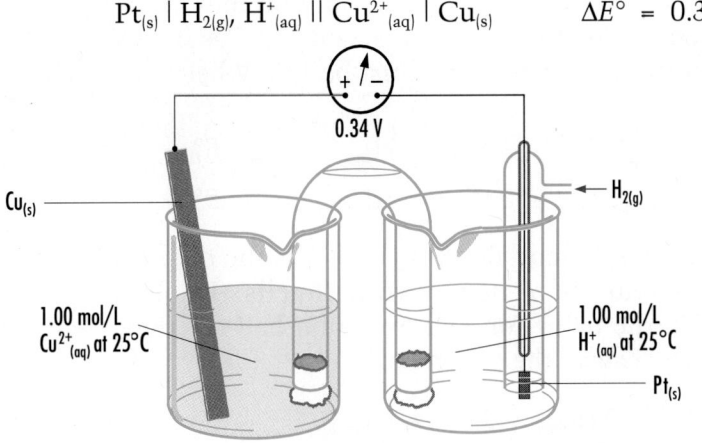

Figure 16.19
A copper-hydrogen standard cell.

The voltmeter shows that the copper electrode is the cathode and is 0.34 V higher in potential than the platinum anode. If the voltmeter is replaced by a connecting wire so that current is allowed to flow, the blue color of the copper(II) ion disappears and the pH of the hydrogen half-cell decreases as the solution becomes more acidic. Based on this evidence, copper(II) ions are being reduced to copper metal and hydrogen molecules are being oxidized to hydrogen ions. Since this redox reaction is spontaneous, copper(II) ions must be stronger oxidizing agents than hydrogen ions.

$$Cu^{2+}_{(aq)} + 2\,e^- \rightleftharpoons Cu_{(s)} \qquad E^\circ_r = 0.34\ V$$

$$2\,H^+_{(aq)} + 2\,e^- \rightleftharpoons H_{2(g)} \qquad E^\circ_r = 0.00\ V$$

The standard cell potential, $\Delta E^\circ = 0.34\ V$, is the difference between the reduction potentials of these two half-cells; $\Delta E^\circ = 0.34\ V - 0.00\ V$.

cathode	$Cu^{2+}_{(aq)} + 2\,e^- \rightarrow Cu_{(s}$
anode	$H_{2(g)} \rightarrow 2\,H^+_{(aq)} + 2\,e^-$

net	$Cu^{2+}_{(aq)} + H_{2(g)} \rightarrow Cu_{(s)} + 2\,H^+_{(aq)}$ $\qquad \Delta E^\circ = 0.34\ V$

Suppose a standard aluminum half-cell is set up with a standard hydrogen half-cell (Figure 16.20).

$$Al_{(s)} \mid Al^{3+}_{(aq)} \parallel H^+_{(aq)}, H_{2(g)} \mid Pt_{(s)} \qquad \Delta E^\circ = 1.66\ V$$

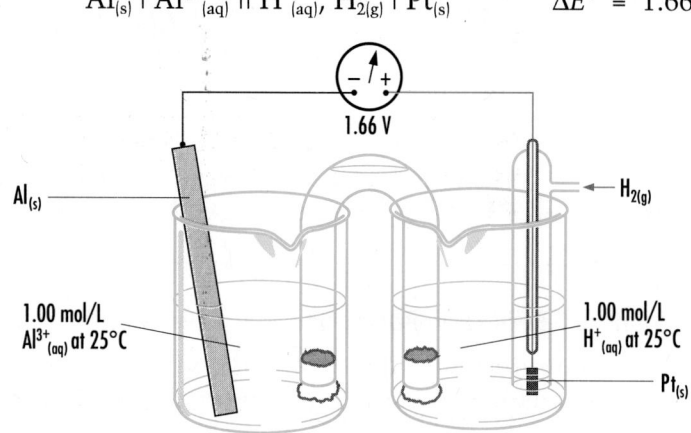

Figure 16.20
An aluminum-hydrogen standard cell.

E° (V)

$+0.34$ $\quad Cu^{2+}_{(aq)} + 2\,e^- \rightleftharpoons Cu_{(s)}$

0.00 $\quad 2\,H^+_{(aq)} + 2\,e^- \rightleftharpoons H_{2(g)}$ $\quad \Big\} \ 0.34\ V$

$\Big\} \ 1.66\ V$

-1.66 $\quad Al^{3+}_{(aq)} + 3\,e^- \rightleftharpoons Al_{(s)}$

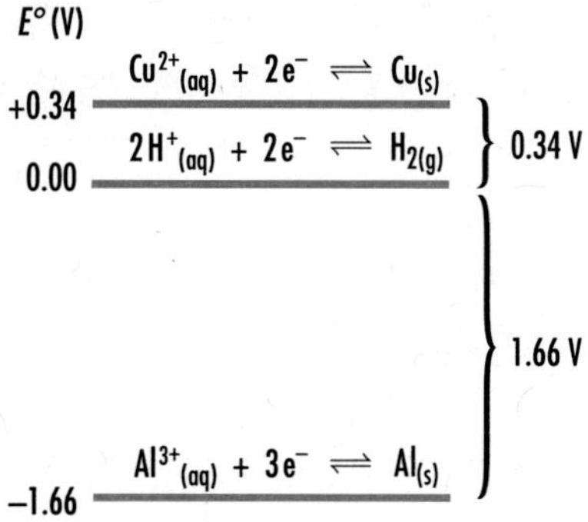

Figure 16.21
Measurements of standard cell potentials show that the reduction potential of $Cu^{2+}_{(aq)}$ is 0.34 V greater than that of $H^+_{(aq)}$, which is 1.66 V greater than that of $Al^{3+}_{(aq)}$. If a standard copper-aluminum cell were measured, we would expect copper to be the cathode, with a reduction potential 2.00 V above that of the aluminum anode.

According to the voltmeter, the platinum electrode is the cathode and the aluminum electrode is the anode. This indicates that hydrogen ions are stronger oxidizing agents than aluminum ions, by 1.66 V. Since the reduction potential of hydrogen ions is defined as 0.00 V, the reduction potential of the aluminum ions must be 1.66 V below that of hydrogen, or –1.66 V.

$$2\,H^+_{(aq)} + 2\,e^- \rightleftharpoons H_{2(g)} \qquad E^\circ_r = 0.00\ V$$

$$Al^{3+}_{(aq)} + 3\,e^- \rightleftharpoons Al_{(s)} \qquad E^\circ_r = -1.66\ V$$

The standard cell potential, $\Delta E^\circ = 1.66\ V$, is the difference between the reduction potentials of these two half-cells. To obtain the net cell reaction, add the reduction and oxidation half-reactions, remembering to balance and cancel the electrons.

cathode	$3\,[\,2\,H^+_{(aq)} + 2\,e^- \rightarrow H_{2(g)}\,]$
anode	$2\,[\,Al_{(s)} \rightarrow Al^{3+}_{(aq)} + 3\,e^-\,]$

net	$6\,H^+_{(aq)} + 2\,Al_{(s)} \rightarrow 3\,H_{2(g)} + 2\,Al^{3+}_{(aq)}$

$$\Delta E^\circ = 0.00\ V - (-1.66\ V) = 1.66\ V$$

Notice that the half-reaction equations were multiplied by factors to balance the electrons, but *the reduction potentials are not altered by the factors used to balance the electrons*. Electric potential represents energy per coulomb of charge (1 V = 1 J/C), or the energy per electron, and does not depend on the total charge transferred in the half-reaction.

In both of these examples, the strongest oxidizing agent reacts at the cathode and the strongest reducing agent reacts at the anode. The measured cell potential is the difference between the reduction potentials at the cathode and at the anode. A positive difference ($\Delta E > 0$) indicates that the net reaction is spontaneous — a requirement for all voltaic cells. In Figure 16.21 the results from the copper-hydrogen and aluminum-hydrogen standard cells are combined. A more extensive list of reduction potentials is found in the table of relative strengths of oxidizing and reducing agents in Appendix F.

Using the table in Appendix F, you can predict standard cell reactions by identifying the strongest oxidizing agent, which reacts at the cathode, and the strongest reducing agent, which reacts at the anode. The standard cell potential is predicted as follows.

$$\Delta E^\circ = \underset{\text{cathode}}{E^\circ_r} - \underset{\text{anode}}{E^\circ_r}$$

This order of subtraction is necessary to confirm the spontaneity from the sign of ΔE. If ΔE is positive, the reaction is spontaneous. (To ensure a correct interpretation, always write the cathode half-reaction first.)

EXAMPLE

A standard dichromate-lead cell is constructed. Write the cell notation, label the electrodes, and determine the cell potential.

$$\underset{\text{cathode}}{C_{(s)} \mid Cr_2O_7^{2-}{}_{(aq)}, H^+{}_{(aq)}, Cr^{3+}{}_{(aq)}} \mid\mid \underset{\text{anode}}{Pb^{2+}{}_{(aq)} \mid Pb_{(s)}}$$

$$\Delta E^\circ = 1.23 \text{ V} - (-0.13 \text{ V}) = 1.36 \text{ V}$$

Oxidation Potentials

According to redox theory, a competition for electrons occurs when reactants combine directly or when reactants are connected in separate half-cells. Different substances have different attractions for electrons, as measured by their reduction potentials. In this competition for electrons, the substance with the stronger attraction (the more positive reduction potential) succeeds in removing electrons from the oxidized form of the weaker substance. In other words, the strongest oxidizing agent removes electrons from the strongest reducing agent. The ease with which a reducing agent gives up its electrons is called its **oxidation potential**, defined as the negative (additive inverse) of the reduction potential. In the dichromate-lead cell discussed above, lead is forced to act as a reducing agent because the dichromate ion is a much stronger oxidizing agent than the lead(II) ion.

$$Pb_{(s)} \rightarrow Pb^{2+}{}_{(aq)} + 2\,e^- \qquad E^\circ_o = -(-0.13 \text{ V}) = +0.13 \text{ V}$$

The standard oxidation potential E°_o for the oxidation of lead is +0.13 V. That is, if the reduction half-reaction is reversed to give an oxidation half-reaction, the reduction potential is reversed to give the oxidation potential.

Exercise

23. For each of the following cells, write the cathode, anode, and net cell reaction equations and calculate the cell potential. Assume standard conditions.
    (a) $Cr_{(s)} | Cr^{2+}_{(aq)} || Sn^{2+}_{(aq)} | Sn_{(s)}$
    (b) $C_{(s)} | SO_4^{2-}_{(aq)}, H^+_{(aq)}, H_2SO_{3(aq)} || Co^{2+}_{(aq)} | Co_{(s)}$
    (c) $Pt_{(s)} | H_{2(g)}, OH^-_{(aq)} || OH^-_{(aq)}, O_{2(g)} | Pt_{(s)}$

24. For each of the following standard cells, refer to the table of relative strengths of oxidizing and reducing agents in Appendix F, to write the standard cell notation. Label electrodes and determine the standard cell potential without writing half-reaction equations.
    (a) lead-copper standard cell
    (b) nickel-zinc standard cell
    (c) iron(III)-hydrogen standard cell

25. One experimental design for determining the position of a half-cell reaction that is not included in a table of oxidizing and reducing agents is shown below. Use the following standard cell, refer to the standard reduction potential of gold in Appendix F, and determine the reduction potential for the indium(III) ion.

    $$Au_{(s)} | Au^{3+}_{(aq)} || In^{3+}_{(aq)} | In_{(s)}$$

    cathode   anode

    $$\Delta E° = 1.84 \text{ V}$$

26. You can determine the identity of an unknown half-cell from the cell potential involving a known half-cell. Use the following evidence and the table of reduction potentials in Appendix F, to determine the reduction potential and the identity of the unknown $X^{2+}_{(aq)} | X_{(s)}$ redox pair.

    $$2\,Ag^+_{(aq)} + X_{(s)} \rightarrow 2\,Ag_{(s)} + X^{2+}_{(aq)}$$

    $$\Delta E° = +1.08 \text{ V}$$

27. Any standard half-cell could have been chosen as the reference half-cell — the zero point of the reduction potential scale. What would be the standard reduction potentials for copper and zinc half-cells, assuming that the standard lithium cell were chosen as the reference half-cell, with its reduction potential defined as 0.00 V?

28. A zinc-iron cell is constructed and allowed to operate until the measured potential difference becomes zero. What interpretation can be made about the chemical system at this point?

Problem 16B Creating a Table of Reduction Potentials

The purpose of this problem is to develop a table of oxidizing agents and reduction potentials from experimental evidence. Complete the Analysis of the investigation report.

Problem

What is the relative strength, in decreasing order, of four oxidizing agents?

Experimental Design

Several cells are investigated; each cell has at least one half-cell in common with one of the other cells. The cell potentials are measured and the positive and negative electrodes of each cell are identified.

Evidence

Positive electrode	Negative electrode	
$C_{(s)} \mid Cr_2O_7{}^{2-}{}_{(aq)}, H^+{}_{(aq)}$	$\parallel Pd^{2+}{}_{(aq)} \mid Pd_{(s)}$	$\Delta E° = +0.28$ V
$Tl_{(s)} \mid Tl^+{}_{(aq)}$	$\parallel Ti^{2+}{}_{(aq)} \mid Ti_{(s)}$	$\Delta E° = +1.29$ V
$Pd_{(s)} \mid Pd^{2+}{}_{(aq)}$	$\parallel Tl^+{}_{(aq)} \mid Tl_{(s)}$	$\Delta E° = +1.29$ V

Problem 16C Series Cells (Enrichment)

Complete the Prediction of the investigation report. Include your reasoning.

Problem

What is the electric potential difference between two cells connected in series?

Experimental Design

Copper-silver and copper-zinc standard cells are connected as shown in Figure 16.22. The electric potential difference between the two cells is measured with a voltmeter.

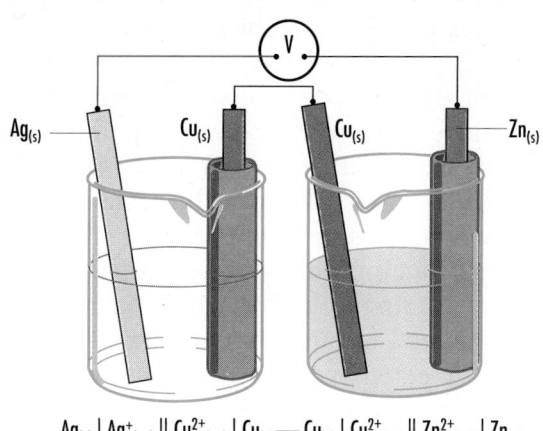

$Ag_{(s)} \mid Ag^+{}_{(aq)} \parallel Cu^{2+}{}_{(aq)} \mid Cu_{(s)} - Cu_{(s)} \mid Cu^{2+}{}_{(aq)} \parallel Zn^{2+}{}_{(aq)} \mid Zn_{(s)}$

Figure 16.22
Two standard cells in series.

■ Problem
✔ Prediction
✔ Design
■ Materials
✔ Procedure
✔ Evidence
✔ Analysis
✔ Evaluation
■ Synthesis

CAUTION

The materials used are toxic and irritant. Avoid skin and eye contact.

INVESTIGATION

16.4 Testing Voltaic Cells

The purpose of this investigation is to test the predictions of cell potentials and the charge on the electrodes of various cells.

Problem

In cells constructed from various combinations of copper, lead, silver, and zinc half-cells, what are the standard cell potentials, and which is the anode and cathode in each case?

Materials

lab apron
safety glasses
voltmeter and connecting wires
U-tube with cotton plugs, porous cups, or filter paper
(4) 100 mL beakers or well plate
distilled water
steel wool
$Cu_{(s)}$, $Pb_{(s)}$, $Ag_{(s)}$, and $Zn_{(s)}$ strips
1.0 mol/L $CuSO_{4(aq)}$, $Pb(NO_3)_{2(aq)}$, $AgNO_{3(aq)}$, $NaNO_{3(aq)}$, and $ZnSO_{4(aq)}$

16.3 CORROSION

Historical periods in the evolution of civilizations are often characterized by their technologies of refining and using metals. The Copper, Bronze, and Steel Ages are examples of this classification. The process of refining a metal is electrochemical in nature and requires energy to recover the pure metal from its naturally occurring compounds (ores). Corrosion is also an electrochemical process and represents the return of a metal to an ore-like state. Because we live in an oxidizing (oxygen) environment, spontaneous oxidation (corrosion) of a metal occurs. In fact we need to continuously produce metals such as iron to replace the iron that is continually lost because of corrosion. Preventing corrosion and dealing with the effects of corrosion is a major economic and technological problem in our society (Figure 15.4, page 608).

As a metal is oxidized, metal atoms lose electrons to form positive ions. A table of relative strengths of oxidizing and reducing agents provides the evidence that metals vary greatly in their ability to be oxidized. Some metals, such as gold and silver, are noble because they are relatively weak reducing agents. On the other hand, Group 1 and 2 metals are very strong reducing agents and therefore easily oxidized. In general, any metal appearing below the oxygen half-reactions on a redox table will be oxidized in our environment. Both iron (including steel) and aluminum are extensively used as structural materials. Why is the corrosion or rusting of iron such a major problem compared with aluminum, which is a much stronger reducing agent? The answer lies primarily in the nature of the oxide that forms on the surface of the metal. A freshly cleaned surface of aluminum rapidly oxidizes in air to form aluminum oxide.

$$4 \, Al_{(s)} + 3 \, O_{2(g)} \rightarrow 2 \, Al_2O_{3(s)}$$

The aluminum oxide that forms adheres tightly to the surface of the metal (Figure 16.23). This prevents further corrosion by effectively sealing any exposed surfaces.

Unfortunately, the iron compounds that form on the surface of exposed iron do not adhere very well and eventually flake off, exposing new iron to be corroded. In addition, the corrosion of iron, as you will see in Investigation 16.5, is a complex process that is significantly affected by the presence of other substances.

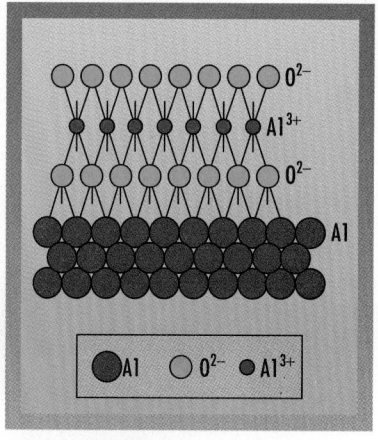

Figure 16.23
Aluminum oxide forms a thin, almost transparent layer that adheres strongly to the aluminum metal underneath.

INVESTIGATION

Investigation 16.5 The Corrosion of Iron

The purpose of this investigation is to test your predictions of factors affecting the rate of corrosion of iron. The knowledge gained from this experience is used to help explain corrosion and to develop methods of corrosion prevention.

Problem

What factors, chemical and electrical, affect the rate of corrosion of iron?

Experimental Design

Several iron nails or pieces of iron wire are thoroughly cleaned with steel wool and acetone. In part 1, the iron is exposed to different conditions in separate test tubes. A clean piece of iron in a dry empty test tube is the control. All test tubes are observed immediately and after one day. In part 2, two iron-carbon cells are connected to 9 V batteries, attaching the electrodes oppositely for the two cells.

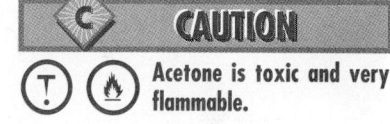

Acetone is toxic and very flammable.

Rusting of Iron

Studies of the corrosion of iron have shown that the presence of both oxygen and water is required and the iron is converted into iron hydroxides and oxides. The first step of the mechanism is thought to be the oxidation of iron at a wet exposed surface (Figure 16.24).

$$Fe_{(s)} \rightarrow Fe^{2+}_{(aq)} + 2 \, e^-$$

Iron(II) ions diffuse through the water on the iron surface while the electrons easily travel through the iron metal which is an electrical conductor. Away from the original iron oxidation, the electrons are picked up by oxygen molecules dissolved in some water on the surface (Figure 16.24).

$$\tfrac{1}{2} O_{2(g)} + H_2O_{(l)} + 2 \, e^- \rightarrow 2 \, OH^-_{(aq)}$$

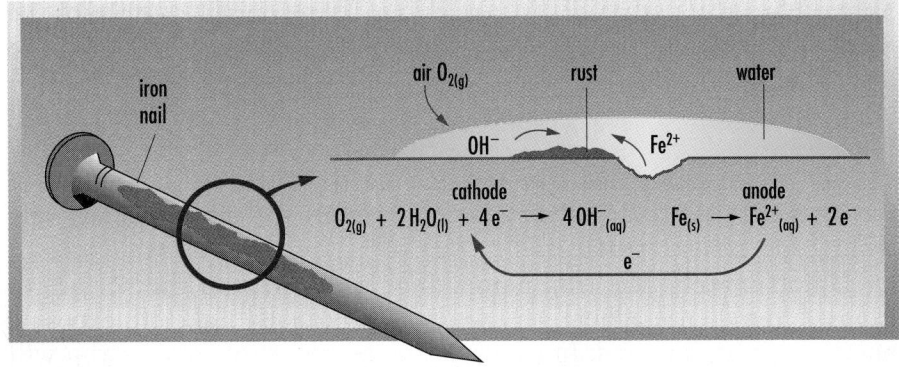

Figure 16.24
The corrosion of iron is a small electrochemical cell with iron oxidation at one location (the anode) and oxygen reduction at another location (the cathode).

Iron(III) hydroxide can be converted to iron(III) oxide-3-water as shown below.

$$2 Fe(OH)_{3(s)} \longrightarrow Fe_2O_3 \cdot 3H_2O_{(s)}$$

In fact, it is difficult to determine how much of the iron(III) exists in rust as the hydroxide or hydrated oxide. Warming this mixture can drive off some of the waters of hydration.

Figure 16.25
Rusting of exposed iron is almost negligible when the relative humidity is less than 50%. This iron pillar in Delhi, India, has existed for about fifteen hundred years because of the very dry and unpolluted environment.

The combination of iron(II) ions and hydroxide ions forms a low solubility precipitate of iron(II) hydroxide, which is further oxidized by oxygen and water to form iron(III) hydroxide, a yellow-brown solid. The familiar red-brown rust is formed by the dehydration of iron(III) hydroxide to form a mixture of iron(III) hydroxide and hydrated iron(III) oxide. The composition of rust varies between that of the hydroxide and the oxide. Therefore, rust is referred to as a hydrated oxide of indeterminate formula, $Fe_2O_3 \cdot xH_2O_{(s)}$.

This simplified mechanism for the rusting of iron can be used to explain why certain conditions promote rusting. If the iron is kept in a dry environment (low humidity) or if water has been deaerated, little or no corrosion occurs (Figure 16.25). Eliminating either water or the oxygen in the water makes the reduction of aqueous oxygen impossible. Iron cannot be oxidized unless a suitable oxidizing agent is present. If oxidizing agents other than oxygen are present, such as certain metal ions, nonmetals, or hydrogen ions, the iron can still be corroded through spontaneous redox reactions. This helps to explain the corrosion of iron in acidic environments, for example, why acid rain corrodes more than natural rain.

In general, electrolytes accelerate rusting. Ships rust more rapidly in seawater than in fresh water and cars rust more rapidly in places where salt is used on roads. Chloride ions from salt are known to inhibit the adherence of protective oxide coatings on many metals, thus exposing more metal to be corroded. Electrolytes, such as sodium chloride, conduct electricity and improve the charge transfer and therefore accelerate the rusting process. It is well known to plumbers that you cannot use steel straps or nails to hold copper pipes in place because the corrosion of iron will be accelerated. Any moisture that is present sets up an electric cell similar in principle to Volta's original discovery of electricity from dissimilar metals (pages 653–655). As the cell operates, the iron corrodes to form rust. The rusting of iron requires the presence of oxygen and water and is accelerated by the presence of acidic solutions, electrolytes, mechanical stresses, and contact with less active metals.

Corrosion Prevention

The prevention or minimizing of the corrosion of iron can be divided into two categories: protective coatings and the method of cathodic protection. In some critical situations, such as a large fuel tank, both methods may be used.

Paint and other similar coatings are a simple method of corrosion prevention. This method works well as long as the complete surface is covered and the coating remains intact. Unfortunately, a scratch or chip in the surface can easily expose a small surface of iron and corrosion begins. Both tin and zinc are used as metallic coatings. Tin coatings, such as the familiar tin cans, adhere well to the iron and provide a strong, shiny coating. The outer surface of the tin coating has a thin, strongly adhering tin oxide that protects the tin. If a crack or break occurs in the tin layer, moisture can collect in the crack and an electric cell with tin and iron electrodes is established. Since iron is more easily oxidized than tin, iron becomes the anode in this cell. The electrons released by the oxidation of iron flow to the tin and corrosion is accelerated. Evidence of this is the typical iron rust on tin cans which have been crushed and left outside. A spontaneous electric cell also arises when a zinc coating on an iron object is broken. However, in this case the zinc is more easily oxidized than the iron. Therefore, zinc is preferentially oxidized, thus preventing the corrosion of the iron. Zinc-plated (galvanized) steel or iron provides a double protection — a protective layer and a preferential corrosion of the zinc.

Cathodic Protection

According to the redox theory of a cell, oxidation is the loss of electrons and occurs at the anode of a cell. Therefore, an effective method of preventing corrosion of iron is **cathodic protection**, forcing the iron to become the cathode by supplying the iron with electrons. This can be done using either an impressed current or a sacrificial anode.

For a battery or DC generator connected in a circuit, electrons flow out of the negative terminal and into the positive terminal of the battery. If the negative terminal is connected to the iron object and the positive terminal to an inert carbon electrode, an electric current is forced to flow to the iron, through an electrolyte such as ground water, to the carbon electrode. The iron is thereby forced to become the cathode and is prevented from corroding. The shiny appearance of the iron nail left connected to the negative terminal of the battery in Investigation 16.5 is evidence that corrosion is prevented. An **impressed current** is an electric current forced to flow toward an iron object by an external potential difference. This method of corrosion prevention requires a constant electric power supply (typically 8 mV) and is used as cathodic protection for pipelines and culverts.

A less common but simpler method of cathodic protection is the use of a **sacrificial anode**. A sacrificial anode is a metal more easily oxidized than iron and connected to the iron object to be protected. The practice of zinc plating (galvanizing) iron objects is a common example of this method. Sacrificial zinc anodes are also connected to the exposed underwater metal surfaces of ships and boats to prevent the corrosion of the iron in the steel. Blocks of magnesium can also be used as sacrificial anodes (Figure 16.26). In all cases, the more active

▌Fighting Corrosion

Concrete structures are reinforced with steel bars (sometimes referred to as re-bar). These bars are made from scrap steel that is melted, reshaped, and then air-cooled. This cooling step allows the carbon in the steel to precipitate, forming microscopic carbide "fingers" that strengthen the steel. Unfortunately, they also act as little batteries. Electrons tend to flow from the iron toward the carbide fingers. As the iron loses electrons, it corrodes. If a concrete structure is in an area prone to corrosion, such as at the waterline or even near ocean spray, this electron loss and corrosion occur even faster, resulting in a general weakening of the steel, and therefore the concrete. The concrete structure may look just as solid, but it is no longer as strong as when it was first built. One solution being researched is to cool the steel with water instead of air. The steel cools more quickly, which seems to reduce the growth of the carbide fingers. The result is a less "electrically-active" steel that may last longer.

metal (appearing below iron on a half-reaction table) is slowly consumed or sacrificed at the anode, forcing the iron object to be the cathode of the cell.

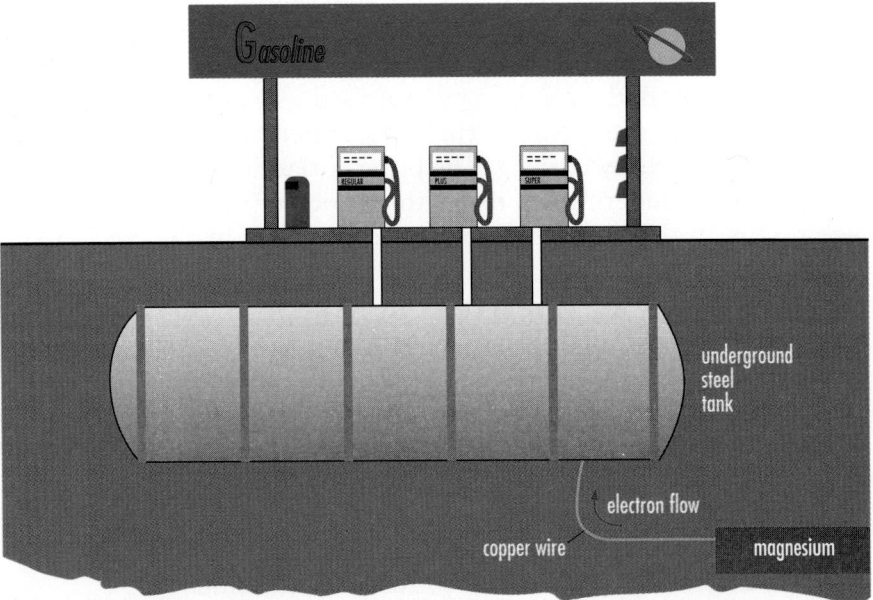

Figure 16.26
Corrosion of iron involves the oxidation of iron at the anode of a cell. If the iron is connected electrically to a metal that is more easily oxidized (a sacrificial anode), then a spontaneous cell develops in which iron is the cathode. The electrolyte of the cell is the moisture in the ground.

Exercise

29. What are the minimum requirements for the corrosion of iron?

30. List some factors that accelerate or promote the corrosion of iron.

31. Write the balanced net ionic equation for the corrosion of iron to iron(II) ions in the presence of oxygen and water.

32. Although the corrosion of iron is the most serious problem, other metals are also corroded in air or other environments. For each of the following situations, use your knowledge of writing and balancing redox equations from Chapter 15 to write and label the half-reaction and net ionic equations.
    (a) Zinc is quite an active metal that oxidizes quickly when exposed to air and water.
    (b) A lead pipe corrodes if used to transport acidic solutions containing insignificant dissolved oxygen.
    (c) In dry air, minute quantities of hydrogen sulfide gas will slowly react with silver objects to produce hydrogen gas and silver sulfide, recognized by the dark tarnish on the surface of the silver.

33. You may have noticed that when rusting appears on a car body, the rust appears around the break or chip in the paint but the damage may extend under the painted surface for some distance.
    (a) What is the evidence for damage extending well beyond the break in the paint?
    (b) Suggest an explanation why the damage may extend far from the location of the rust.

34. Why is a zinc coating on iron better than a tin coating?

35. What are the two methods of cathodic protection and how are they similar?

36. A zinc wire was connected to and buried with the Alaskan pipeline when it was built. Why was this done? Include a brief description of the principles involved.

37. (Internet) Search the internet using the key words "iron corrosion." How many different titles did the search find? Find a general site and list the different classes of iron corrosion.

16.4 ELECTROLYSIS

A standard cell always reacts spontaneously, since two oxidizing-reducing agent pairs are present in every cell. (Non-standard voltaic cells also involve spontaneous reactions.) In other cells, a redox reaction takes place only if electrical energy from a battery or power supply is added continuously to the cell. The decomposition of water, shown in Figure 8.19 on page 313, is an example. The process of supplying electrical energy to force a non-spontaneous redox reaction to occur is called **electrolysis**. An **electrolytic cell** consists of a combination of two electrodes, an electrolyte, and an external power source (see Figure 16.27, page 683). The external power supply acts as an "electron pump"; its electric energy causes an electron transfer inside the electrolytic cell. In an electrolytic cell the chemical reaction is the reverse of that of a voltaic cell; however, most of the scientific principles developed for cells already studied also apply to electrolytic cells (Table 16.5).

$$\text{reactants} \quad \underset{\text{electrolytic cell}}{\overset{\text{voltaic cell}}{\rightleftharpoons}} \quad \text{products} + \text{electrical energy}$$

Table 16.5

COMPARISON OF VOLTAIC AND ELECTROLYTIC CELLS		
	Voltaic Cell	**Electrolytic Cell**
Spontaneity	spontaneous reaction	non-spontaneous reaction
Standard cell potential, $\Delta E°$	positive	negative
Cathode	• positive electrode • strongest oxidizing agent undergoes a reduction	• negative electrode • strongest oxidizing agent undergoes a reduction
Anode	• negative electrode • strongest reducing agent undergoes an oxidation	• positive electrode • strongest reducing agent undergoes an oxidation
Electron movement	anode $\rightarrow$ cathode	anode $\rightarrow$ cathode
Ion movement	anions $\rightarrow$ anode cations $\rightarrow$ cathode	anions $\rightarrow$ anode cations $\rightarrow$ cathode

Positive and Negative
"Positive" and "negative" are labels that are used in many situations, mostly decided by general agreement (convention). For example, "positive" and "negative" can be used to label or describe attitudes, directions, axes on a graph, charges, and electrodes. By convention, a voltaic cell has a cathode labelled positive and an anode labelled negative. In an electrolytic cell, it is customary to reverse these labels. It is best to think of positive and negative for electrodes as labels not as charges.

A secondary cell such as a Ni-Cad cell (Table 16.2, page 658) can be used to illustrate the difference between a voltaic cell and an electrolytic cell. As a secondary cell discharges, electrical energy is spontaneously produced and the cell functions as a voltaic cell. When the cell is recharged, the electrical energy supplied to the cell forces the products to react and re-form the original reactants. During recharging, the secondary cell is functioning as an electrolytic cell.

- ▨ Problem
- ▨ Prediction
- ▨ Design
- ▨ Materials
- ▨ Procedure
- ✔ Evidence
- ✔ Analysis
- ▨ Evaluation
- ▨ Synthesis

INVESTIGATION

16.6 A Potassium Iodide Electrolytic Cell

The purpose of this investigation is to observe the operation of an electrolytic cell and to determine its reaction products.

Problem

What are the products of the reaction during the operation of an aqueous potassium iodide electrolytic cell?

Experimental Design

Inert electrodes are placed in a 0.50 mol/L solution of potassium iodide and a battery or power supply provides a direct current of electricity to the cell. The litmus and halogen diagnostic tests (Appendix C, C.3) are conducted to test the solution near each electrode before and after the reaction.

Materials

lab apron
safety glasses
petri dish
two carbon electrodes
two connecting wires
3 V to 9 V battery or power supply
red and blue litmus paper
ring stand and two utility clamps
small test tube with stopper
dropper bottle of trichlorotrifluoroethane
0.50 mol/L $KI_{(aq)}$

Procedure

1. Set up the $KI_{(aq)}$ cell as shown in Figure 16.27 (or as shown but with one ring stand) but without connecting the power supply.
2. Observe the cell and test the solution with litmus paper and trichlorotrifluoroethane.
3. Use a wire to join the two electrodes and observe the cell.
4. Connect and turn on the power supply.
5. Record all observations at each electrode.
6. Perform both diagnostic tests at each electrode.

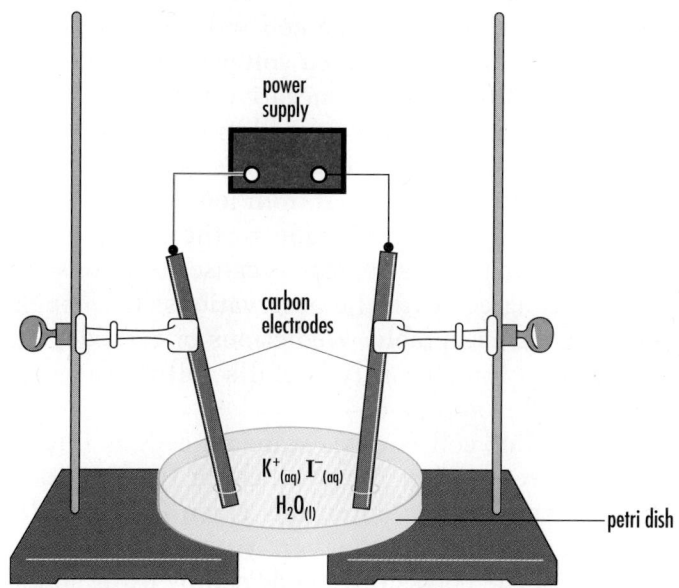

Figure 16.27
A petri dish is a convenient container for the aqueous potassium iodide solution of this electrolytic cell. Carbon rods serve as inert electrodes.

The Potassium Iodide Electrolytic Cell: A Synthesis

In the $KI_{(aq)}$ electrolytic cell, litmus paper did not change color in the initial solution and turned blue only near the electrode from which gas bubbled. At the other electrode, a yellow-brown color and a dark precipitate formed. The yellow-brown substance produced a violet color in a chlorinated hydrocarbon layer. This chemical evidence agrees with the interpretation supplied by the following half-reaction equations. According to the table of relative strengths of oxidizing and reducing agents, water is the stronger oxidizing agent and iodide ions are the stronger reducing agents present in a potassium iodide solution.

OA   **SOA**

$K^+_{(aq)}$, $I^-_{(aq)}$, $H_2O_{(l)}$

SRA   **RA**

cathode $\quad 2\,H_2O_{(l)} + 2\,e^- \rightarrow H_{2(g)} + 2\,OH^-_{(aq)}$

gas bubbles   blue litmus

anode $\quad\quad\quad\quad 2\,I^-_{(aq)} \rightarrow I_{2(s)} + 2\,e^-$

yellow-brown

(purple in chlorinated hydrocarbons)

---

net $\quad\quad 2\,H_2O_{(l)} + 2\,I^-_{(aq)} \rightarrow H_{2(g)} + 2\,OH^-_{(aq)} + I_{2(s)}$

Evidence from the study of this and many other aqueous electrolytic cells suggests that the generalizations that apply to voltaic cells also apply to electrolytic cells. From a theoretical perspective, the strongest oxidizing agent present has the greatest attraction for electrons and gains electrons at the cathode. The strongest reducing agent has the least attraction for electrons and loses electrons at the anode. *The theoretical definitions of cathode and anode are the same in both voltaic and electrolytic cells.*

Observation of a potassium iodide cell indicates that the transfer of electrons is not spontaneous. When a voltage is supplied to the cell, electrons that are supplied from the negative terminal of the battery enter the cathode of the electrolytic cell and are gained by water molecules, which have the more positive reduction potential. Simultaneously, electrons are removed from iodide ions on the surface of the anode by their apparent attraction to the positive terminal of the battery. This explanation is logical because it is consistent with redox concepts and it agrees with the observations. The explanation is judged, therefore, to be acceptable. Predictions of cathode, anode, and overall cell reactions for electrolytic cells follow the same steps outlined for voltaic cells on page 664.

The potassium iodide cell is not a standard cell, as initially water and iodide ions are present, but the products of the reactions are not present. Therefore, the electric potentials given in the table of half-reactions do not apply here. However, you can approximate the potential difference of the potassium iodide cell by using standard half-reaction reduction potentials. Note that the water reduction potential used is for neutral water, $[OH^-_{(aq)}] = 10^{-7}$ M.

cathode $2\,H_2O_{(l)} + 2\,e^- \rightarrow H_{2(g)} + 2\,OH^-_{(aq)}$
anode $2\,I^-_{(aq)} \rightarrow I_{2(s)} + 2\,e^-$

net $2\,H_2O_{(l)} + 2\,I^-_{(aq)} \rightarrow H_{2(g)} + 2\,OH^-_{(aq)} + I_{2(s)}$

$$\Delta E^\circ = E^\circ_{r\,\text{cathode}} - E^\circ_{r\,\text{anode}}$$
$$= -0.41\text{ V} - (+0.54\text{ V})$$
$$= -0.95\text{ V}$$

A negative sign for a cell potential indicates that the chemical process is non-spontaneous. The more negative the cell potential, the more energy is required. To force the cell reactions, electrons must be supplied with a minimum of 0.95 V from an external battery or other power supply.

Cold Fusion

Fusion is the type of nuclear reaction that takes place in the sun at extremely high temperatures. "Cold fusion" is the combining of the nuclei of hydrogen isotopes to yield larger nuclei at room temperature rather than at solar temperatures. A claim was made in 1989 that cold fusion had been achieved during the electrolysis of heavy water, $^2_1H_2O$, with palladium electrodes. Some experimental results indicated a possible net gain in energy. The hypothesis proposed was that hydrogen-2 molecules that had dissolved in the palladium crystal lattice had undergone fusion to produce helium-3, a neutron, and some energy. There has been a lack of supporting evidence from other laboratories and the theoretical hypothesis of cold fusion has not received support from theoreticians. Cold fusion is a story in science with an important lesson — wait for replication of an experimental result before making exorbitant public claims.

Exercise

38. Predict the cathode, anode, and net cell reactions and minimum potential difference for each of the following electrolytic cells.

    (a) $C_{(s)} \mid Ni^{2+}_{(aq)}, I^-_{(aq)} \mid C_{(s)}$

    (b) $Pt_{(s)} \mid Na^+_{(aq)}, OH^-_{(aq)} \mid Pt_{(s)}$

39. What is the minimum electric potential difference of an external power supply that produces chemical changes in the following electrolytic cells?

    (a) $C_{(s)} \mid Cr^{3+}_{(aq)}, Br^-_{(aq)} \mid C_{(s)}$

    (b) $Cu_{(s)} \mid Cu^{2+}_{(aq)}, SO_4^{2-}_{(aq)} \mid Cu_{(s)}$

16.7 Demonstration: Electrolysis

The purpose of this demonstration is to test the method of predicting the products of electrolytic cells.

Problems

What are the products of electrolytic cells containing
- aqueous copper(II) sulfate?
- aqueous sodium sulfate?
- aqueous sodium chloride?

Experimental Design

The electrolysis of the aqueous copper(II) sulfate is carried out in a U-tube, and the electrolysis of aqueous sodium sulfate and sodium chloride is carried out in a Hoffman apparatus (Figure 8.19, page 313) so that any gases produced can be collected. Diagnostic tests with necessary control tests are conducted to determine the presence of the predicted products.

- [ ] Problem
- [x] Prediction
- [ ] Design
- [ ] Materials
- [ ] Procedure
- [x] Evidence
- [x] Analysis
- [x] Evaluation
- [ ] Synthesis

CAUTION

Copper(II) sulfate is toxic and irritant. Avoid skin and eye contact. If you spill copper(II) sulfate solution on your skin, wash the affected area with lots of cool water. During electrolysis, corrosive substances are produced; avoid skin and eye contact. Remember to wash your hands before leaving the laboratory.

Science and Technology of Electrolysis

The major technological application of voltaic cells is the production of electrical energy. Applications of electrolytic cells include the production of elements, the refining of metals, and the plating of metals onto the surface of an object. The study of electrolysis in industry reveals the strong relationship between science and technology.

Production of Elements

Most elements occur naturally combined with other elements in compounds. For example, ionic compounds of sodium, potassium, lithium, magnesium, calcium, and aluminum are abundant, but the corresponding metals do not occur naturally. The explanation for this involves reduction potentials. Even water has a more positive reduction potential than any of these metal ions, so if the metals did exist naturally, a spontaneous reaction would convert them into their ions.

Metals can often be produced by electrolysis of solutions of their ionic compounds, but two difficulties may arise. First, many ionic compounds have low solubility in water and second, water is a stronger oxidizing agent than active metal cations. To overcome these difficulties, a technological design in which water is eliminated may be used. Molten ionic compounds are good electrical conductors and can function as the electrolyte in a cell. In the electrolysis of molten binary ionic compounds, only one oxidizing agent and one reducing agent are present. *Cations are reduced to metals at the cathode and anions are oxidized to nonmetals at the anode.* The production of active metals (strong reducing agents) from their minerals typically involves the electrolysis of molten compounds of the metal, a technology first used in the scientific work of Humphry Davy.

Overpotential

The evidence from Investigation 16.7 shows an unexpected production of chlorine where oxygen is predicted from the oxidation of water. This phenomenon cannot be understood or predicted without further information about electrolytic cells. There are many variables that affect an electrolytic cell, such as concentration gradients within the electrolyte, internal resistance, temperature, nature of the electrodes, and current density. Like other chemical processes, a half-cell reaction at an electrode has an activation energy that varies for different half-reactions and conditions. Therefore, the actual potential required for a particular half-reaction and the reported half-reaction potential may be quite different. The difference is known as the half-cell overpotential and is generally much greater for the production of oxygen than chlorine.

THE CANADIAN DOLLAR

The loonie, Canada's 11-sided, gold-colored dollar coin, is a technological solution to a societal problem. As inflation eroded the value of the dollar during the 1980s, paper dollar bills changed hands so often that they wore out in just a few months. Urban transit systems and vending machine companies began to lobby the government to produce a dollar coin, which would be more convenient than paper money for

their customers.

At the request of the federal government, the Royal Canadian Mint began to design a replacement for the paper dollar bill. The mint wanted to produce a dollar coin with a richer sheen than the shiny metals used in coins of lower value. A competition was held to manufacture a handsome, gold-colored coin, with a lifetime of twenty years, that would resist erosion and corrosion.

Research technologists at the Sherritt Gordon plant in Fort Saskatchewan, Alberta, experimented to produce a bronze-plated nickel blank that would meet the required standards in the continuous process required for mass production.

The nickel blanks for loonies are punched in the same way as the blanks for other coins, from long strips of metal that are more than 99.9% nickel. The electroplating process takes place in large barrels, in an electrolyte consisting of potassium cyanide, potassium stannate, and copper(I) cyanide. The anode in the cell is made of copper and the

electric wires are arranged to make the coins act as the cathode. As the barrel is rotated, the tumbling action gives each coin a uniform exposure to ensure homogeneous plating. A barrel contains approximately 19 000 coin blanks, and has a current of one thousand amperes supplied to it for seven hours. After the electroplating process, each coin has a mass of seven grams. The bronze plate itself is 12.5% tin and 87.5% copper, with a tolerance of 1%. A bronze coating with this particular composition is called *aureate* because of its golden color.

One of the most difficult problems for the researchers was controlling the deposition of the tin, the proportions of which varied from 3% to 33% in early trials. By systematically manipulating variables such as current, surface area of electrodes, spacing between electrodes, and composition of the electrolyte, they achieved optimum conditions.

After the coin blanks have been bronze-plated, they are heated in a hydrogen atmosphere to soften the bronze and bind it to the nickel. They are then polished in a tumbler and sprayed with an anti-stain agent. The coin blanks are shipped to the mint in Winnipeg, Manitoba, where they are stamped with an image on each side. Finally, they are given an 11-sided edge, which enables

visually impaired people to identify the coins by touch. Since the dollar coin lasts much longer than the paper dollar, experts anticipate that in its first twenty years of existence the loonie will save Canadian taxpayers more than $175 million. The original plan was to stamp the coin with the image of a voyageur, but the stamping dies were lost in transit to the mint in Winnipeg and the image of a loon was used instead.

- *What are some advantages and disadvantages of a dollar coin compared with a dollar bill?*

- *If the original stamping die had not been lost, what do you suppose we would call the dollar coin today?*

EXAMPLE

Lithium is the least dense of all metals and it has a very high oxidation potential; both qualities make it an excellent anode for batteries (Figure 16.10, page 659). Lithium can be produced by the electrolysis of molten lithium chloride. Write the cathode, anode, and net cell reactions for this electrolysis. (Note that no electric potentials are listed. The table of oxidizing and reducing agents in Appendix F lists only electric potentials for half-reactions in 1.0 mol/L aqueous solutions at SATP.)

SOA

$Li^{+}_{(l)},\quad Cl^{-}_{(l)}\qquad t > 605°C$

SRA

cathode	$2\,Li^{+}_{(l)}\ +\ 2\,e^{-} \rightarrow 2\,Li_{(l)}$
anode	$2\,Cl^{-}_{(l)} \rightarrow Cl_{2(g)}\ +\ 2\,e^{-}$
net	$2\,Li^{+}_{(l)}\ +\ 2\,Cl^{-}_{(l)} \rightarrow 2\,Li_{(l)}\ +\ Cl_{2(g)}$

Electrolysis of molten ionic compounds is not a simple technology. The high temperatures needed to melt ionic compounds cause problems for cell components and add to the cost of the elements produced. For example, initial efforts to produce aluminum by electrolysis were unproductive because its common ore, $Al_2O_{3(s)}$, melts above 2072°C. No material could be found to hold the molten compound. Then in 1886 two scientists, working independently and knowing nothing of each other's work, made the same discovery. Charles Martin Hall in the United States and Paul Louis Toussaint Héroult in France discovered that $Al_2O_{3(s)}$ dissolves in a molten mineral called cryolite, Na_3AlF_6. In this design the cryolite acts as an inert solvent for the electrolysis of aluminum oxide and forms a molten conducting mixture with a melting point around 1000°C. Aluminum can be produced electrolytically from this molten mixture (Figure 16.28).

Aluminum oxide is obtained from bauxite, an aluminum ore. Once the ore is purified, the aluminum oxide is added to the molten cryolite

> The production of aluminum is important to Canada's economy, although Canada does not have large deposits of aluminum ore. An abundant supply of inexpensive hydroelectric power is used to produce aluminum metal from concentrated imported bauxite. Recycling aluminum from soft drink and beer cans requires only 5% of the energy originally needed to manufacture the aluminum by electrolysis.

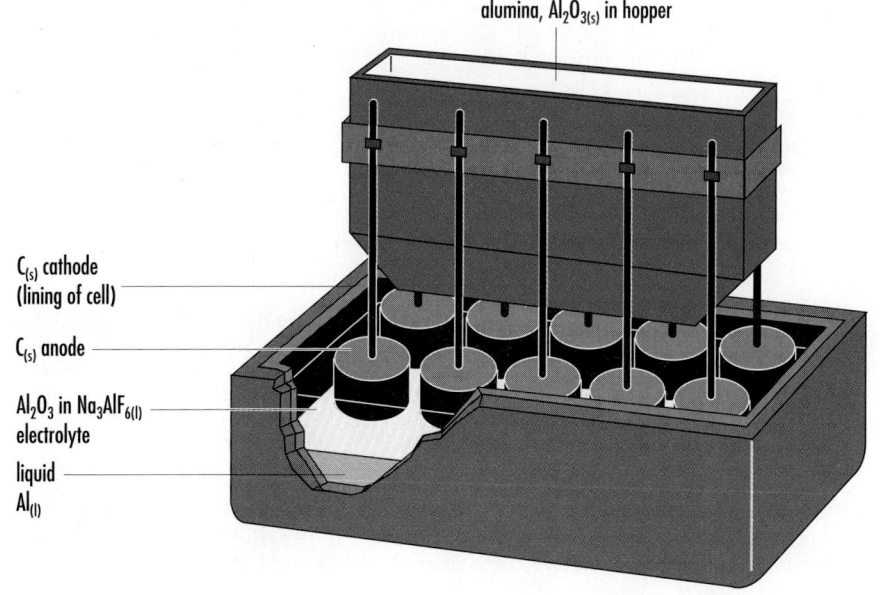

alumina, $Al_2O_{3(s)}$ in hopper

$C_{(s)}$ cathode (lining of cell)

$C_{(s)}$ anode

Al_2O_3 in $Na_3AlF_{6(l)}$ electrolyte

liquid $Al_{(l)}$

Figure 16.28
The Hall-Héroult cell for the production of aluminum. The cathode is the carbon lining of the steel cell. At the cathode, the aluminum ions are reduced to produce liquid aluminum, which collects at the bottom of the cell and is periodically drained away. At the carbon anodes, oxide ions are oxidized to produce oxygen gas. The oxygen produced at the anode reacts with the carbon electrodes, producing carbon dioxide, so these electrodes must be replaced frequently.

in which it dissolves and dissociates. The reactions occurring at the electrodes in a Hall-Héroult cell are summarized below.

$$\text{SOA}$$
$$Al^{3+}_{(cryolite)}, O^{2-}_{(cryolite)}$$
$$\text{SRA}$$

cathode $4\,[\,Al^{3+}_{(cryolite)} + 3\,e^- \rightarrow Al_{(l)}\,]$
anode $3\,[\,2\,O^{2-}_{(cryolite)} \rightarrow O_{2(g)} + 4\,e^-\,]$

net $4\,Al^{3+}_{(cryolite)} + 6\,O^{2-}_{(cryolite)} \rightarrow 4\,Al_{(l)} + 3\,O_{2(g)}$

The overall effect is a decomposition reaction.

$$2\,Al_2O_{3(s)} \rightarrow 4\,Al_{(s)} + 3\,O_{2(g)}$$

Instead of eliminating or replacing water as a solvent in electrolytic production of elements, a third design overcomes the difficulty by simply "overpowering" the reduction of the water. A high voltage encourages the reduction of metal ions over the reduction of water. An example of this design is the electrolysis of aqueous sodium chloride to produce chlorine, hydrogen, and sodium hydroxide (Figure 16.29). This process, called the chlor-alkali process, uses high voltages to force the reduction of aqueous sodium ions to sodium metal, and depends on the much faster rate of this half-reaction compared with the reduction of water. The technology requires large quantities of relatively inexpensive electrical energy, such as that available to the chlor-alkali plant at Fort Saskatchewan, Alberta. However, the use of highly toxic mercury as the cathode is a potential threat to the safety of workers and the environment and requires special precautions.

Chlor-Alkali Cells

In the chlor-alkali cells at Dow Chemical near Fort Saskatchewan, Alberta, a relatively low voltage (3.1 V) and high current (55 kA) are applied to a saturated sodium chloride solution. Suitable raw materials are nearby — large salt beds beneath the surface in the Fort Saskatchewan area, and water from the North Saskatchewan River. Also, energy from fossil fuels is abundant and inexpensive. The sodium and chlorine produced are used in the manufacture of hydrochloric acid, bleaches, plastics, and solvents, as well as chemicals for the pulp and paper industry. This illustrates three requirements of a successful chemical industry — a supply of raw materials, energy and water resources, and a ready market.

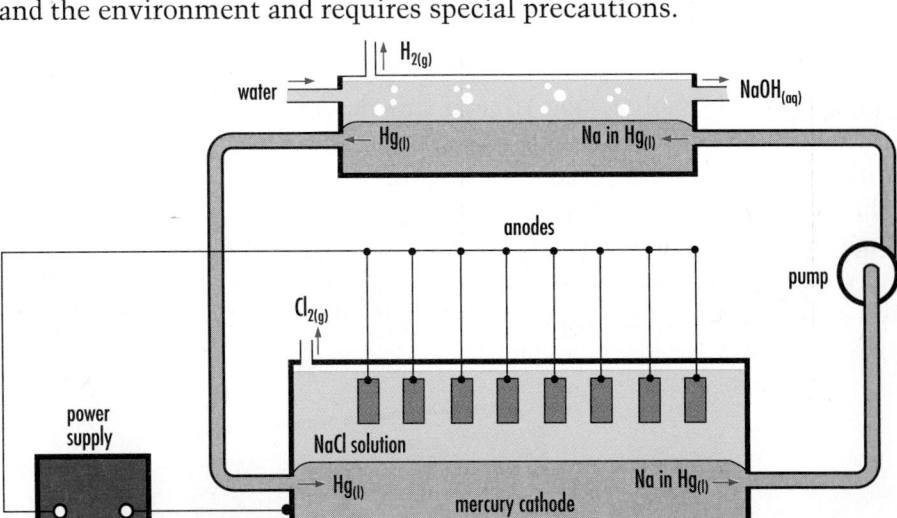

Figure 16.29
Design of a chlor-alkali plant. The sodium metal forms rapidly at the cathode and is dissolved and carried away by a liquid mercury cathode as soon as it forms. Water is later added to the sodium-mercury solution to form hydrogen gas and a sodium hydroxide solution. Chlorine gas is formed and collected at the anodes.

Exercise

40. (a) What are two difficulties associated with the electrolysis of aqueous ionic compounds in the production of active metals?
    (b) What three designs can be used to offset these difficulties?

41. Scandium is a metal with a low density and a melting point that is higher than that of aluminum. These properties are of interest to engineers who design space vehicles. Scandium metal is produced by the electrolysis of molten scandium chloride. List

all entities present and then write the cathode, anode, and net cell reaction equations for this electrolysis.

42. Why should we recycle metals such as aluminum? State several arguments that you might use in a debate.

43. The following statements summarize the steps in the chemical technology of obtaining magnesium from sea water. Write a balanced equation to represent each reaction.
    (a) Slaked lime (solid calcium hydroxide) is added to sea water (ignore all solutes except $MgCl_{2(aq)}$) in a double replacement reaction to precipitate and separate magnesium hydroxide.
    (b) Hydrochloric acid is added to the magnesium hydroxide precipitate.
    (c) After the magnesium chloride product is crystallized, it is melted in preparation for electrolysis. List entities present and write cathode, anode, and net cell reaction equations to describe the electrolysis of molten magnesium chloride.
    (d) An alternative process produces magnesium from dolomite, a mineral containing $CaCO_3$ and $MgCO_3$. Suggest some technological advantages and disadvantages of the dolomite process compared with the sea water process.

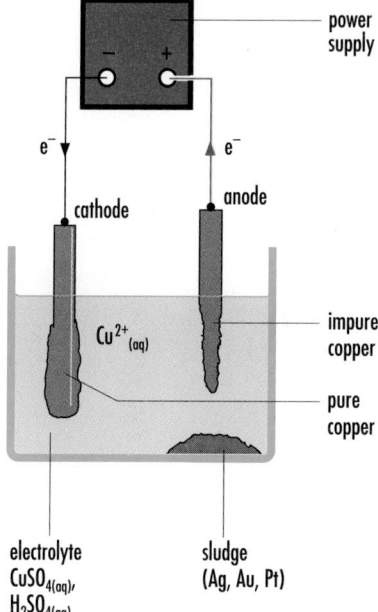

Figure 16.30
When the electrolytic cell is operated at a carefully controlled voltage, only copper and metals more easily oxidized than copper, such as iron and zinc, dissolve at the anode. Only copper is reduced at the cathode. Other impurities in the anode, such as silver, gold, and platinum, do not react; these fall to the bottom of the cell as a sludge called anode mud. Removed from the cell periodically, the anode mud undergoes further processing to extract valuable metals.

Refining of Metals

In the production of metals, the initial product obtained is usually an impure metal. To purify or refine a metal, a variety of methods are used including burning off impurities to obtain iron and distilling to obtain mercury. However, a common purification method, known as **electrorefining**, uses an electrolytic cell to obtain high grade metals at the cathode from an impure metal at the anode. A good example of this process is the electrorefining of copper. The initial smelting process produces copper that is about 99% pure, containing some silver, gold, platinum, iron, and zinc. The presence of impurities lowers the electrical conductivity of the copper, so it must be further purified in an electrolytic process. As shown in Figure 16.30, a slab of impure copper is the anode of an electrolytic cell that contains copper(II) sulfate dissolved in sulfuric acid. The cathode is made of a thin sheet of very pure copper. As the cell operates, copper and some of the other metals in the anode are oxidized, but only copper is reduced at the cathode. A theoretical understanding of oxidation and reduction and electric potentials allows precise control over what is oxidized and what is reduced, so that after the electrolysis process the copper is about 99.95% pure (Figure 16.31). The half-reactions are:

cathode	reduction of copper	$Cu^{2+}_{(aq)} + 2e^- \rightarrow Cu_{(s)}$
anode	oxidation of copper	$Cu_{(s)} \rightarrow Cu^{2+}_{(aq)} + 2e^-$
	oxidation of zinc	$Zn_{(s)} \rightarrow Zn^{2+}_{(aq)} + 2e^-$
	oxidation of iron	$Fe_{(s)} \rightarrow Fe^{2+}_{(aq)} + 2e^-$

Another related method of purifying metals is to reduce metal cations from a molten or aqueous electrolyte at the cathode of an electrolytic cell. This method is known as **electrowinning**. For some active metals, such as Group 1, electrowinning using a molten salt is the only way to obtain the metal (Example, page 687). Many other

Figure 16.31
Each copper cathode shown in the photograph is about 2 m high and requires 28 days to form from the impure copper anodes.

metals such as zinc can be produced by electrowinning an aqueous solution. For example, Cominco's operation at Trail, British Columbia, uses the electrolysis of an acidic zinc sulfate solution with a specially treated lead anode to deposit very pure zinc metal at the cathode.

cathode	$2[Zn^{2+}_{(aq)} + 2\ e^- \rightarrow Zn_{(s)}]$
anode	$2\ H_2O_{(l)} \rightarrow O_{2(g)} + 4\ H^+_{(aq)} + 4\ e^-$

$$net \quad 2\ Zn^{2+}_{(aq)} + 2\ H_2O_{(l)} \rightarrow 2\ Zn_{(s)} + O_{2(g)} + 4\ H^+_{(aq)}$$

Electroplating

Many technological inventions, such as electric cells, preceded scientific understanding of the processes involved. There are modern examples of the reverse — advances in science have produced new technologies, such as transistors and lasers. Some of these processes, for example, chromium plating (Figure 16.32), work well but are not fully understood. Another example is that there is no satisfactory explanation as to why silver deposited in an electrolysis of a silver

■ MICHAEL FARADAY (1791 – 1867) ■

One of ten children, Michael Faraday was born in a village near London, England. Because his family could not afford to provide him with more than an elementary education (his father was a blacksmith), he was apprenticed at age 14 to a bookseller and bookbinder. He spent much of his time reading the books that he handled, including Lavoisier's text on chemistry and the electricity section of the *Encyclopædia Britannica*. Fortunately, Faraday's employer appreciated his thirst for knowledge and allowed him time for reading and for attending scientific lectures.

After hearing Humphry Davy's lecture at the Royal Institution, Faraday applied to Davy for a job as his assistant, expressing an eager desire to "enter into the service of science." To prove his interest, he sent Davy 386 pages of careful notes that he had taken on Davy's lectures, illustrated with colored diagrams and bound in leather. Davy was impressed with the young man's obvious ability and in 1813 he hired Faraday as his laboratory assistant.

Faraday proved himself more than worthy as his diligent laboratory work led to significant accomplishments. By 1825 he had devised methods for liquefying gases under pressure, he had developed a way to produce very cold temperatures in the laboratory,

and he had discovered benzene.

Faraday also continued Davy's work in electrochemistry, coining the terms *electrolysis, electrolyte, electrode, anode, cathode,* and *ion*. His quantitative study of electrolysis

identified the factors that determine the mass of an element produced at the electrode. Faraday's investigations of the magnetic effects of electricity led to the discovery of electromagnetic induction, as well as the invention of the electric transformer and the electric generator. In 1833 he was appointed professor of chemistry at

the Royal Institution.

Besides his talents as a researcher, Faraday was an excellent lecturer who enlivened his public talks with demonstrations. His skill at communicating science to non-experts made his lectures enormously popular and helped pull the Royal Institution out of financial difficulties. Christmas lectures for children were one of Faraday's specialties. *The Chemical History of a Candle* is the published version of six of these lectures.

A deeply religious man, Faraday had strong convictions about the appropriate use of science and technology. He refused to produce a poison gas to help the British fight Russia in the Crimean War.

Faraday accepted numerous awards and medals but did not let his international fame distract him from his love of science. His meticulous notes covering forty-two years of scientific research were published in seven volumes, an indication of the magnitude of his contribution to science and technology.

• *How did Faraday obtain his early science education?*

• *It has often been said that if Nobel Prizes existed at the time, Faraday would likely have earned at least two of these awards. What do you think are Faraday's two most significant accomplishments?*

nitrate solution does not adhere well to any surface, whereas silver plated from silver cyanide solutions does.

Several metals, such as silver, gold, zinc, and chromium, are valuable because of their beauty or their resistance to corrosion. However, products made from these metals in their pure form are either too expensive or they lack suitable mechanical properties, such as strength and hardness. To achieve the best compromise among mechanical properties, appearance, and corrosion resistance, utensils or jewelry may be made of a relatively inexpensive yet strong alloy such as steel, and then coated with another metal or alloy to enhance appearance or corrosion resistance. Plating of a metal at the cathode of an electrolytic cell is a common technology that provides a surface layer covering an object. The design of a process for plating metals is obtained by systematic trial and error, involving the careful manipulation of one possible variable at a time. In this situation, a scientific perspective helps identify variables but cannot usually provide successful predictions.

Figure 16.32
Chromium is best plated from a solution of chromic acid. A thin layer of chromium metal is very shiny and, like aluminum, protects itself from corrosion by forming a tough oxide layer.

INVESTIGATION

16.8 Copper Plating

The purpose of this investigation is to determine the best procedures for plating copper onto various objects. Evaluate both the process and the product.

Problem

Which procedure causes a smooth layer of copper metal to adhere to a conducting object?

Experimental Design

A small metal object, such as a spoon, a key, or a piece of metal, is carefully cleaned. A 0.50 mol/L copper(II) sulfate electrolytic cell that uses an inert electrode as the anode is constructed. The object to be plated is used as the cathode (Figure 16.33). Potentially relevant variables are identified and systematically manipulated. The success of the plating is evaluated by the appearance of the object and by polishing the plated object with steel wool to test adherence of the copper coating.

■ Problem
✔ **Prediction**
■ Design
✔ **Materials**
✔ **Procedure**
✔ **Evidence**
✔ **Analysis**
✔ **Evaluation**
■ Synthesis

CAUTION

Copper(II) sulfate is toxic and irritant. Avoid skin and eye contact. If you spill copper sulfate solution on your skin, wash the affected area with lots of cool water. Remember to wash your hands before leaving the laboratory.

16.5 STOICHIOMETRY OF CELL REACTIONS

In the production of elements, the refining of metals, and electroplating, the quantity of electricity (charge in coulombs) that passes through a cell determines the mass of substances that react or are produced at the electrodes. In SI (Table 16.1, page 655), charge in coulombs is determined from the electric current in amperes (coulombs per second) and the time in seconds, according to the following definition.

$$q = It$$

The relationship between electricity and electrochemical changes was first investigated by Michael Faraday in the 1830s. Based on his work, a constant has been determined that relates the charge transferred to the amount of electricity. The molar charge of electrons (also known

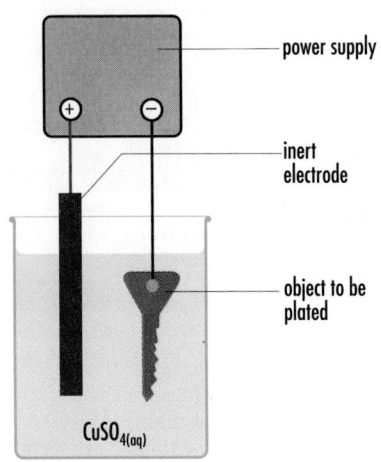

Figure 16.33
An electrolytic cell for copper-plating small objects.

power supply

inert electrode

object to be plated

$CuSO_{4(aq)}$

as the **Faraday constant**, F) was found to be 9.65×10^4 C/mol of electrons. This constant can be used as a conversion factor in converting electric charge to an amount in moles — in the same way that molar mass is used to convert mass to an amount in moles — as the following equation shows.

$$n_{e^-} = \frac{It}{F}$$

EXAMPLE

Convert a current of 1.74 A for 30.0 min into an amount in moles of electrons. Recall that 1 A = 1 C/s (Table 16.1, page 655).

$$n_{e^-} = \frac{1.74 \text{ C}}{1 \text{ s}} \times 30.0 \text{ min} \times \frac{60 \text{ s}}{1 \text{ min}} \times \frac{1 \text{ mol}}{9.65 \times 10^4 \text{ C}} = 0.0325 \text{ mol}$$

or

$$n_{e^-} = \frac{It}{F} = \frac{1.74 \text{ C/s} \times 30.0 \text{ min} \times 60 \text{ s/min}}{9.65 \times 10^4 \text{ C/mol}} = 0.0325 \text{ mol}$$

Half-Cell Calculations

Since the mass of an element produced at an electrode depends on the amount in moles of transferred electrons, a half-reaction equation is necessary in order to do stoichiometric calculations. Either voltaic or electrolytic half-cell reactions may be involved in this process. Separate calculations are carried out for each electrode, although the same charge and therefore the same amount in moles of electrons passes through each electrode in a cell or a group of cells in series. As the following example shows, concepts of stoichiometry used in other calculations also apply to half-cell calculations.

EXAMPLE

What is the mass of copper deposited at the cathode of a copper-refining electrolytic cell operated at 12.0 A for 40.0 min (Figure 16.30, page 689)?

$$\begin{array}{ccccc} Cu^{2+}_{(aq)} & + & 2 \text{ e}^- & \rightarrow & Cu_{(s)} \\ & & 40.0 \text{ min} & & m \\ & & 12.0 \text{ A} & & 63.55 \text{ g/mol} \\ & & 9.65 \times 10^4 \text{ C/mol} & & \end{array}$$

$$n_{e^-} = \frac{12.0 \text{ C}}{1 \text{ s}} \times 40.0 \text{ min} \times \frac{60 \text{ s}}{1 \text{ min}} \times \frac{1 \text{ mol}}{9.65 \times 10^4 \text{ C}} = 0.298 \text{ mol}$$

$$n_{Cu} = 0.298 \text{ mol} \times \frac{1}{2} = 0.149 \text{ mol}$$

$$m_{Cu} = 0.149 \text{ mol} \times \frac{63.55 \text{ g}}{1 \text{ mol}} = 9.48 \text{ g}$$

or

$$m_{Cu} = 40.0 \text{ min} \times \frac{60 \text{ s}}{1 \text{ min}} \times \frac{12.0 \text{ C}}{1 \text{ s}} \times \frac{1 \text{ mol e}^-}{9.65 \times 10^4 \text{ C}} \times \frac{1 \text{ mol Cu}}{2 \text{ mol e}^-} \times \frac{63.55 \text{ g Cu}}{1 \text{ mol Cu}}$$

$$= 9.48 \text{ g}$$

Exercise

44. A student reconstructs Volta's electric battery using sheets of copper and zinc, and a current of 0.500 A is produced for 10.0 min. Predict the mass of zinc oxidized to aqueous zinc ions.

45. Electroplating is a common technological process for coating objects with a metal to enhance the objects' attractiveness or resistance to corrosion.
    (a) A car bumper is plated with chromium using chromium(III) ions in solution. If a current of 54 A flows in the cell for 45 min 30 s, predict the mass of chromium deposited on the bumper.
    (b) For corrosion resistance, a steel bolt is plated with nickel from a solution of nickel(II) sulfate. If 0.25 g of nickel produces a plating of the required thickness and a current of 0.540 A is used, predict how long the process will take. (Hint: Transpose the formula, $n = It/F$.)
    (c) A family wishes to plate an antique teapot with 10.00 g of silver. If the process takes 84 min, predict the average current used.

46. A rapidly developing technology is the production of less expensive, more durable, and more energy-dense voltaic cells; that is, cells with a high energy-to-mass ratio.
    (a) A car battery has a rating of 120 A•h (ampere-hours). This battery can produce a 1.00 A current for 120 h. What mass of lead is oxidized as this battery discharges?
    (b) If an aluminum-oxygen fuel cell has the same rating as the car battery in (a), what mass of aluminum metal would be oxidized?

47. During the electrolysis of molten aluminum chloride in an electrolytic cell, 5.40 g of aluminum are produced at the cathode. Predict the mass of chlorine produced at the anode.

Problem 16D Quantitative Electrolysis

The purpose of this problem is to test the method of stoichiometry in cells. Complete the Prediction, Analysis, and Evaluation of the investigation report.

Problem

What is the mass of tin plated at the cathode of a tin-plating cell by a current of 3.46 A for 6.0 min?

Experimental Design

A steel can is placed in an electroplating cell as the cathode. An electric current of 3.46 A flows through the cell, which contains a 3.25 mol/L solution of tin(II) chloride, for 6.0 min.

Evidence

initial mass of can = 117.34 g
final mass of can = 118.05 g

OVERVIEW

Voltaic and Electrolytic Cells

Summary

- Consumer, commercial, and industrial cells usually consist of two electrodes and one electrolyte. Voltaic cells used in research usually consist of two half-cells, each with an electrode and an electrolyte, separated by a porous boundary.

- In a standard cell, each half-cell contains all the entities shown in the half-reaction equations with 1.0 mol/L of aqueous entities, gases at 100 kPa, and a temperature of 25°C.

- Cell potential indicates the tendency for electrons to flow and is calculated as the difference between the reduction potentials of the strongest oxidizing agent at the cathode and the strongest reducing agent at the anode. Reduction potentials are measured relative to the standard hydrogen half-cell. At equilibrium, the cell potential becomes zero.

- A voltaic cell produces electrical energy from a spontaneous chemical reaction ($\Delta E > 0$), but an electrolytic cell consumes electrical energy in a non-spontaneous reaction ($\Delta E < 0$). In both kinds of cells, reduction of the strongest oxidizing agent occurs at the cathode, oxidation of the strongest reducing agent occurs at the anode, cations move to the cathode, anions move to the anode, and electrons move from the anode to the cathode.

- Corrosion of iron is a major economic and technological problem and requires oxygen and water to convert the iron into iron(III) oxide and hydroxide. Corrosion is accelerated by the presence of various electrolytes and by contact with a less active metal. Corrosion is prevented by using protective coatings and/or cathodic protection.

- Electrolysis is an important commercial and industrial technology. Applications of electrolytic cells include molten salt cells with only one oxidizing and reducing agent, aqueous salt cells with inert electrodes, and aqueous salt cells with reactive electrodes.

- The Faraday constant is used in stoichiometric calculations involving cell half-reactions to convert electrical measurements into amount in moles of electrons.

Key Words

ampere
anode
battery
cathode
cathodic protection
coulomb
electric cell
electric current
electric potential difference
electrode
electrolysis
electrolytic cell
electrorefinishing
electrowinning
Faraday constant
fuel cell
half-cell
impressed current
oxidation potential
porous boundary
primary cell
reference half-cell
sacrificial anode
secondary cell
standard cell
(standard) cell potential
(standard) reduction potential
volt
voltage
voltaic cell

Review

1. What are the three essential parts of an electric cell?

2. What is a key difference between technological and scientific problem solving?

3. What are two technological solutions to the problem of batteries "going dead"?

4. What is the voltage and chemical state of a "dead" cell?

5. List two examples of inert electrodes.

6. What are two common examples of porous boundaries?

7. Identify and describe the components of the standard half-cell used as a reference for reduction potentials.

8. List several specific methods of preventing the corrosion of iron.

9. List three technological applications of electrolytic cells.

10. State two problems that might be encountered in attempting to produce an active metal by electrolysis of aqueous solutions. Suggest some ways to solve these problems.

11. In what important way do voltaic cells and electrolytic cells differ?

12. (a) In a summary table, identify the location of each of the following in a voltaic cell and in an electrolytic cell: reduction and oxidation half-reactions; reaction of the strongest oxidizing and reducing agents at the electrodes; movement of cations, anions, and electrons.
    (b) Describe the spontaneity of the reaction, including position of agents in the table of oxidizing and reducing agents and the value of $\Delta E°$.

Applications

13. What is the predicted cell potential of each of the following standard cells? Include half-cell reaction equations.
    (a) hydrogen-tin cell
    (b) silver-permanganate cell
    (c) tin-zinc cell

14. What is the minimum potential difference that must be applied to the following electrolytic cells to cause a chemical reaction? (Do not write half-cell reaction equations.)
    (a) nickel(II) sulfate electrolyte with inert electrodes
    (b) hydrochloric acid electrolyte with silver electrodes
    (c) tin(II) chloride electrolyte with tin electrodes

15. Why will a standard cell always allow a spontaneous reaction?

16. Suppose that the scientific community decided to use the standard iodine-iodide half-cell as the reference point ($E_r° = 0.00$ V) for measuring cell potentials.
    (a) Calculate the reduction potential of a standard silver-silver ion half-cell.
    (b) Calculate the oxidation potential of a standard zinc-zinc ion half-cell.
    (c) Calculate the electric potential of a standard silver-zinc cell.

17. A cobalt-lead standard cell is constructed and tested.
    (a) Predict which electrode will be the cathode and which one will be the anode.
    (b) List all entities present, write the half-cell and net cell reaction equations, and calculate the cell potential.
    (c) As part of the prediction, sketch and label a cell diagram for a standard lead-cobalt cell. Specify all substances, label important cell components, and show the direction of electron and ion movement.

18. The mercury cell (below) is a special cell for products such as watches and hearing aids.

Using only the following half-reactions from a table of reduction potentials, write the net reaction equation and determine the potential for this cell.

$$ZnO_{(s)} + H_2O_{(l)} + 2\,e^- \rightarrow Zn_{(s)} + 2\,OH^-_{(aq)}$$
$$E^\circ_r = -1.25 \text{ V}$$

$$HgO_{(s)} + H_2O_{(l)} + 2\,e^- \rightarrow Hg_{(l)} + 2\,OH^-_{(aq)}$$
$$E^\circ_r = +0.10 \text{ V}$$

19. Compare chrome-plated steel, tin-plated steel, and galvanized steel in terms of appearance, oxidation of the plated metal, and the protection of the steel from corrosion.

20. In a methane fuel cell, the chemical energy of this compound is converted into electrical energy instead of the heat that would flow during the combustion of methane. Using only the following half-reactions and reduction potentials, write a net reaction equation and determine the approximate potential for the methane fuel cell.

$$CO_3^{2-}{}_{(l)} + 7\,H_2O_{(g)} + 8\,e^- \rightarrow$$
$$CH_{4(g)} + 10\,OH^-_{(l)} \quad E^\circ_r = +0.17 \text{ V}$$

$$O_{2(g)} + 2\,H_2O_{(g)} + 4\,e^- \rightarrow 4\,OH^-_{(l)}$$
$$E^\circ_r = +0.40 \text{ V}$$

21. From the information in this chapter, list two or three examples of situations in which technology preceded scientific explanations.

22. Use the following information and the table of reduction potentials in Appendix F to determine the reduction potential for the gold(I) ion.

$$2[Au^+_{(aq)} + e^- \rightarrow Au_{(s)}]$$
$$\underline{Zn_{(s}} \rightarrow Zn^{2+}_{(aq)} + 2\,e^-}$$
$$2\,Au^+_{(aq)} + Zn_{(s)} \rightarrow 2\,Au_{(s)} + Zn^{2+}_{(aq)}$$
$$\Delta E^\circ = +2.44 \text{ V}$$

23. An experiment is designed to determine the identity of a half-cell by using a known half-cell and measuring the potential difference.
    (a) Use the evidence gathered to determine the reduction potential and the identity of the unknown $X^{2+}_{(aq)} \mid X_{(s)}$ redox pair.

$$Cu^{2+}_{(aq)} + X_{(s)} \rightarrow Cu_{(s)} + X^{2+}_{(aq)}$$
$$\Delta E^\circ = +0.48 \text{ V}$$

    (b) What is the significance of negative value for the reduction potential obtained?

24. In which of the following mixtures must an external voltage be applied to inert electrodes to observe evidence of a redox reaction?
    (a) a solution of cadmium nitrate
    (b) a solution of iron(III) iodide
    (c) solutions of iron(III) bromide and tin(II) sulfate in connected half-cells
    (d) solutions of potassium iodide and zinc nitrate in connected half-cells

25. One technological process for refining zinc metal involves the electrolysis of a zinc sulfate solution.
    (a) Predict the cathode, anode, and net cell reactions for the electrolysis of a zinc sulfate solution.
    (b) What minimum applied voltage is required to operate the cell?

26. Predict the cathode, anode, and net cell reactions for the electrolysis of molten strontium oxide.

27. In a school chemistry experiment, an electrolytic cell is constructed to electroplate nickel metal using 500 mL of a 0.125 mol/L nickel(II) sulfate solution. How much time is needed for the solution to react completely if a 3.50 A current is used?

28. Predict the current required to produce 15 kg of aluminum per hour in an aluminum refinery.

Extensions

29. Metal is a finite resource on our planet and its production affects the environment. Write a short report on the efforts made toward reducing metal waste and re-using and recycling metal in your community.

30. Describe how to connect car batteries to give someone a "boost." Why should the final connection always be made to ground at a distance from both batteries?

31. What is the molar charge of calcium ions?

32. Suggest three or more experimental designs that might determine the identity of an unknown solution believed to contain a transition element cation.

33. Suggest three or more experimental designs that might determine the concentration of an aqueous copper(II) ion solution.

34. A standard copper-dichromate cell is constructed using a porous cup design. According to the table of relative strengths of oxidizing and reducing agents, a cell potential of 0.89 V is expected. According to the evidence obtained, the cell potential, when it became constant, was 1.01 V and the color of the solution in the copper half-cell changed from blue to green as the cell operated. Provide an explanation for these observations and evaluate your explanation using information from references.

35. Using information from this chapter and other references, outline the importance of batteries in our society and speculate on the future of batteries.

36. Find out about the Nernst equation: where and how is it used? Provide an example.

37. Most chemical reactions are explained as being either electron transfer reactions or proton transfer reactions.

    (a) What are the similarities and differences between electron and proton transfer reactions?

    (b) State some evidence for energy changes in both electron and proton transfer reactions.

    (c) Identify a combination of chemicals that might produce either an electron or a proton transfer reaction, and describe some diagnostic tests that could be used to determine which reaction predominates.

38. Classify each of the following reactions as an acid-base reaction, a redox reaction, or one of the types of reactions that are not acid base or redox.

    (a) $Mg_{(s)} + 2 H_2O_{(l)} \rightarrow Mg(OH)_{2(s)} + H_{2(g)}$

    (b) $Al(H_2O)_6^{3+}{}_{(aq)} + H_2O_{(l)} \rightarrow$
    $\qquad H_3O^+{}_{(aq)} + Al(H_2O)_5OH^{2+}{}_{(aq)}$

    (c) $Ag^+{}_{(aq)} + OH^-{}_{(aq)} \rightarrow AgOH_{(s)}$

    (d) $NaHSO_{4(aq)} + NaHCO_{3(aq)} \rightarrow$
    $\qquad Na_2SO_{4(aq)} + CO_{2(g)} + H_2O_{(l)}$

    (e) $N_2H_{4(g)} + O_{2(g)} \rightarrow N_{2(g)} + 2 H_2O_{(l)}$

    (f) $CH_3NH_{2(g)} + HCl_{(g)} \rightarrow CH_3NH_3Cl_{(s)}$

Problem 16E Testing a Voltaic Cell

The purpose of this problem is to test concepts of electrochemistry. Complete the Prediction and Materials of the investigation report and add diagnostic tests to the Experimental Design. Include a complete labelled diagram of the cell to be constructed, and suggest a test for all of the qualitative and quantitative predictions.

Problem

What are the components, reactions, and electric potential difference of a standard lead-dichromate cell?

Experimental Design

A standard cell is constructed from the materials provided. Tests of each prediction are conducted.

Problem 16F Cell Competition

In a Science and Technology Olympics event, a team of students designs and assembles a voltaic cell using only the materials provided. Complete the Prediction of the investigation report, including a labelled diagram and cell reactions.

Problem

Which of the voltaic cells that can be constructed using only the materials provided has the highest possible electric potential difference?

Materials

aquarium
unglazed ceramic vase
carbon rod or pencil
aluminum foil
box of table salt
container of solid copper(II) sulfate
several connecting wires
large container of distilled water

Answers to Overview Questions

CHAPTER 1

Overview

1. Useful attitudes include open-mindedness, a respect for evidence, and a tolerance of reasonable uncertainty.

2. If the subject of the statement is observable or based on observable objects or changes, then it is empirical. If the subject is non-observable, then it is a theoretical statement.

3. Step 1: Turn the air and gas adjustments to the off position.
   Step 2: Connect the burner hose to the gas outlet on the bench.
   Step 3: Turn the bench gas valve to the fully on position.
   Step 4: While holding a lit match above and to one side of the barrel, open the burner gas valve until a small yellow flame results.
   Step 5: Adjust the air flow to obtain a pale blue flame with an inner, darker blue cone.

4. Quickly use the eyewash or nearest running water to rinse the eye for at least 15 min. Get medical attention.

5. Smother the flames with a blanket or piece of clothing.

6. Is the chemical flammable, corrosive, toxic, or hazardous in any way? What special disposal procedures and containers are required? Where are those containers kept?

7. (Sample answers)
   (a) oxygen (element), sodium chloride (compound)
   (b) tap water (solution)
   (c) sand and water

8. (a) An element is a substance that cannot be broken down chemically into simpler substances by heat or electricity. According to theory, an element is composed of only one kind of atom.
   (b) A compound is a substance that can be decomposed chemically by heat or electricity. According to theory, a compound is composed of two or more kinds of atoms.

9. The invention was the battery; it produces an electric current that is passed through a sample in an attempt to cause it to decompose.

10. Science and technology work together. Sometimes science leads technology, and sometimes technology leads science.

11. From the photograph you cannot observe any odors that may be produced, how much heat is generated, or how fast the reaction occurs.

12. (a) observation, qualitative, empirical
    (b) observation, quantitative, empirical
    (c) interpretation, qualitative, theoretical
    (d) observation, qualitative, empirical
    (e) interpretation, qualitative, theoretical
    (f) observation, quantitative, empirical

13. (Discussion)

14. (a) economic
    (b) scientific
    (c) political
    (d) technological, ecological
    (e) scientific
    (f) technological
    (g) ecological, technological

15. A variety of plastic blocks are slowly heated with a weight on top of the block to see if the block is flattened. The manipulated variable is the plastic; the responding variable is the thickness of the block; controlled variables are rate of heating and mass of the weight.

16. Samples of different substances will be heated and an electric current passed through each one. If a sample decomposes, then it is known to be a compound.

17. mechanical: Egg whites and yolks are separated manually.
    filtration: A paper coffee filter is used to separate the ground coffee beans from the coffee.
    extraction: Herbs and spices are soaked in oil, water, or vinegar to extract the flavoring.
    distillation: A water purifier distills tap water to produce purer water.
    (Many other possibilities exist.)

18. Prepare a concentrated aqueous solution using the drink crystals. Use paper chromatography and a variety of solvents to separate as many components as possible.

CHAPTER 2

Overview

1. (a) iron   (e) chlorine
   (b) iodine   (f) mercury
   (c) phosphorus   (g) copper
   (d) sodium   (h) sulfur

2. (a) Fe, Na, Hg, Cu
   (b) metals: Fe, Na, Hg, Cu
       nonmetals: I, P, Cl, S

3. No. According to IUPAC rules, only the first letter can be an uppercase letter.

4. A scientific law describes and predicts in a simpler fashion than competing laws.

5. Metals conduct electricity; nonmetals do not.

6. oxygen, carbon, and hydrogen

7. (a) Dmitri Mendeleyev
   (b) The periodic law (table) was used to successfully predict new elements.

8. (a) Bromine and mercury are liquids at SATP. Helium, nitrogen, oxygen, fluorine, neon, chlorine, argon, krypton, xenon, and radon are gases at SATP.
   (b) The purpose of the staircase line is to separate the metals from the nonmetals.
   (c) magnesium, Mg; lead, Pb; fluorine, F
   (d) 1, 8, 13, 14, 17, 29

9. (See Figure 2.13, page 69.)

10. The most reactive metal is francium and the most reactive nonmetal is fluorine.

11. Many properties, such as melting and boiling points, do not vary smoothly within a family or regularly across a period. Hydrogen does not fit very well in any family. (Other examples are possible; see page 69, Limitations of the Periodic Table.)

12. Hydrogen can behave like a member of Group 1 or Group 17, or in its own unique way (acids).

13. Empirical knowledge is observable; theoretical knowledge is not.

14. theoretical descriptions, hypotheses and definitions, theories, analogies, models

15. Acceptable theories must describe and explain observations, predict future experimental results, and be as simple as possible.

16. (a) (Solid sphere) billiard ball
    (b) (See Figure 2.17 (a), page 75.) raisin bun
    (c) (See Figure 2.17 (b), page 75.) Saturn
    (d) (Figure 2.19, page 76, without red lines) beehive with bees flying out
    (e) (Figure 2.25, page 80, without red lines and arrow) solar system

17. (See Table 2.9, page 76.)

18. Bohr suggested that the properties of the elements can be explained by the arrangement of electrons in specific orbits with certain maximum numbers of electrons (2, 8, 8).

19. Unacceptable theories may be restricted, revised, or replaced.

20. (a) the atomic number
    (b) equal to the number of protons (atomic number)
    (c) equal to the last digit of the group number
    (d) equal to the period number

21. (a) Use the theoretical rule that atoms of the representative elements lose or gain electrons to achieve the same electron arrangement as the nearest noble gas atom.
    (b) 1+, 2+, 3+, 3–, 2–, 1–

22. Mendeleyev was able to *describe* all elements using groups with similar properties, such as Group 1. He was able to *predict* new elements, such as germanium. Finally, the arrangement of rows (periods) and columns (groups) was a *simple* arrangement.

23. According to the Rutherford model, most of the atom is empty space with a tiny, massive, positively charged nucleus. Only a few alpha particles would pass close enough to the nucleus to be deflected at large angles.

24. Theories are tested by their ability to explain and predict. As new experimental evidence was collected that conflicted with an existing theory, it was revised to account for this new information.

25. Ununbiium

26. Chemical and physical properties are similar to those of the noble gases.

27. (a) bleach   (d) computer chips
    (b) window frames   (e) fuel
    (c) respiration   (f) advertising signs

28. (a) $12 p^+$, $12 e^-$, $2 e^-$
    (b) $13 p^+$, $13 e^-$, $3 e^-$
    (c) $53 p^+$, $53 e^-$, $7 e^-$

29. (a) 20 protons, 22 neutrons, 20 electrons
    (b) 38 protons, 52 neutrons, 38 electrons

30. 53 protons, 70 neutrons, 54 electrons

31. Diagrams like Figure 2.31 (page 85) should show the number of protons and the arrangement of electrons as follows.
    (a) K $19 p^+$, $2 e^-$, $8 e^-$, $8 e^-$, $1 e^-$
        K^+ $19 p^+$, $2 e^-$, $8 e^-$, $8 e^-$
    (b) O $8 p^+$, $2 e^-$, $6 e^-$
        O^{2-} $8 p^+$, $2 e^-$, $8 e^-$
    (c) Cl $17 p^+$, $2 e^-$, $8 e^-$, $7 e^-$
        Cl^- $17 p^+$, $2 e^-$, $8 e^-$, $8 e^-$

32. Noble gases are very unreactive and are thought to have full outer energy levels.

33. (a) Representative elements
    (b) Transition elements, boron, carbon, silicon, hydrogen

34. (a) sodium ion, Na^+
    (b) phosphide ion, P^{3-}
    (c) sulfide ion, S^{2-}

35. Diagrams like Figure 2.31 (page 85) should show the number of protons and the arrangement of electrons as follows.
    Mg $12 p^+$, $2 e^-$, $8 e^-$, $2 e^-$
    O $8 p^+$, $2 e^-$, $6 e^-$
    Mg^{2+} $12 p^+$, $2 e^-$, $8 e^-$
    O^{2-} $8 p^+$, $2 e^-$, $8 e^-$

36. (a) C   (i) K   (q) Pb
    (b) Al   (j) O   (r) Fe
    (c) Cu   (k) He   (s) Au
    (d) S   (l) P   (t) Ca
    (e) Cl   (m) Hg   (u) I
    (f) H   (n) Ne   (v) Ni
    (g) N   (o) U
    (h) W   (p) Ag

Problem 2C

Prediction

The formula is predicted to be AlF_3. Each aluminum atom loses three electrons to three fluorine atoms. The aluminum atom forms an Al^{3+} ion and the fluorine atoms form F^- ions with the same number of electrons as the nearest noble gas Ne.

Evaluation

The prediction is judged to be verified because the experimentally determined formula is the same as the predicted formula. The authority used to make the prediction, the restricted quantum mechanics theory of atoms and ions, is judged to be acceptable because the prediction was verified.

CHAPTER 3

Overview

1. (a)

Property	Ionic Compounds	Molecular Compounds
*SATP state	(s) only	(s), (l), or (g)
*conductivity†	high	none

(b)

Property	Acids	Bases
*litmus† color	turns red	turns blue

   (c) * indicates defining properties
       † indicates diagnostic tests

2. There are two kinds of ions (positive and negative); the sum of the charges of all the ions is zero.

3. (a) Effective scientific communication is international, logical, precise, and simple.
   (b) International Union of Pure and Applied Chemistry (IUPAC)
   (c) A chemical formula is international, whereas chemical names are not.

4. The kinds of atoms/ions, the number or ratio of the atoms/ions, and the state of matter should be communicated by a chemical formula.

5. Both terms apply to chemical formulas. Chemical formulas can be determined empirically or theoretically.

6. (a) $NaHSO_{4(s)}$
   (b) $NaOH_{(s)}$
   (c) $CO_{2(g)}$
   (d) $CH_3COOH_{(aq)}$
   (e) $Na_2S_2O_3 \cdot 5H_2O_{(s)}$
   (f) $NaClO_{(s)}$
   (g) $S_{8(s)}$
   (h) $KNO_{3(s)}$
   (i) $H_3PO_{4(aq)}$
   (j) $I_{2(s)}$
   (k) $Al_2O_{3(s)}$
   (l) $KOH_{(s)}$
   (m) $O_{3(g)}$
   (n) $CH_3OH_{(l)}$
   (o) $H_2CO_{3(aq)}$
   (p) $C_3H_{8(g)}$

7. (a) calcium carbonate
   (b) diphosphorus pentaoxide
   (c) magnesium sulfate-7-water
   (d) dinitrogen oxide
   (e) sodium silicate
   (f) calcium hydrogen carbonate
   (g) hydrochloric acid
   (h) copper(II) sulfate-5-water
   (i) sulfuric acid
   (j) calcium hydroxide
   (k) sulfur trioxide
   (l) sodium fluoride

8. (a) Potassium hydroxide and carbonic acid react to form water and potassium carbonate.
   (b) Lead(II) nitrate and ammonium sulfate react to produce lead(II) sulfate and ammonium nitrate.
   (c) Aluminum and iron(II) sulfate react to produce iron and aluminum sulfate.
   (d) Nitrogen dioxide and water react to form nitric acid and nitrogen monoxide.

9. (a) $N_{2(g)} + O_{2(g)} \rightarrow NO_{2(g)}$
   (b) $Fe(CH_3COO)_{3(aq)} + Na_2OOCCOO_{(aq)} \rightarrow$
       $Fe_2(OOCCOO)_{3(s)} + NaCH_3COO_{(aq)}$
   (c) $S_{8(s)} + Cl_{2(g)} \rightarrow S_2Cl_{2(l)}$
   (d) $Cu_{(s)} + AgNO_{3(aq)} \rightarrow Ag_{(s)} + Cu(NO_3)_{2(aq)}$

10. (a) $KBr_{(s)}$
    (b) $AgI_{(s)}$
    (c) $PbO_{(s)}$
    (d) $ZnS_{(s)}$
    (e) $CuO_{(s)}$
    (f) $Li_3N_{(s)}$

11. (a) $CaO_{(s)}$
    (b) $SiO_{2(s)}$
    (c) $Al_2O_{3(s)}$
    (d) magnesium oxide
    (e) iron(III) oxide
    (f) sulfur trioxide

12. (a) $CaO_{(s)} + H_2O_{(l)} \rightarrow Ca(OH)_{2(s)}$
    (b) $CaO_{(s)} + SiO_{2(s)} \rightarrow CaSiO_{3(s)}$
    (c) $CaO_{(s)} + SO_{3(g)} \rightarrow CaSO_{4(s)}$
    (d) $CaO_{(s)} + CO_{2(g)} \rightarrow CaCO_{3(s)}$

Problem 3B

Analysis

According to the evidence gathered in this experiment, the solutions labelled 1, 2, 3, and 4 are $KCl_{(aq)}$, $HCl_{(aq)}$, $C_2H_5OH_{(aq)}$, and $Ba(OH)_{2(aq)}$, respectively. The reasoning is that the evidence indicates an ionic compound, an acid, a molecular compound, and a base, respectively.

Problem 3C

Analysis

According to the evidence gathered in this experiment, the empirical definitions of ionic and molecular compounds are:

ionic: all solids at SATP, electrically conductive in the aqueous and molten states.

molecular: solids, liquids, or gases at SATP, electrically non-conductive in the aqueous and molten states.

Evaluation

The prediction is falsified. Because the current empirical definitions do not agree with the evidence gathered, they are unacceptable and need to be revised to include the evidence that ionic and molecular compounds differ in their conductivity in the molten state.

CHAPTER 4

Overview

1. The central idea of the kinetic molecular theory is that the smallest entities of a substance are in constant motion.

2. Solids have mainly vibrational motion. Liquids have some vibrational, rotational, and translational motion. Gases have mainly translational motion.

3. Reactant particles must collide with a certain minimum energy and orientation before any rearrangement of atoms or ions occurs.

4. 6.02×10^{23}

5. (a) If a burning splint is inserted into a test tube containing an unknown gas, and a squeal or pop sound is heard, then the gas is likely to be hydrogen.
   (b) If an unknown gas is bubbled through limewater, and the mixture turns cloudy, then the gas is likely to be carbon dioxide.

6. The evidence of conservation of mass supports the idea that atoms are conserved in chemical reactions.

7. Coefficients represent the mole ratio of reactants and products in a chemical reaction. Formula subscripts represent a ratio of ions in an ionic compound or the number of atoms per molecule in a molecular compound.

8. step 1: Write all reactant and product chemical formulas including the states of matter.
   step 2: Begin by balancing the atom or ion present in the greatest number.
   step 3: Repeat step 2 to balance each of the remaining atoms or ions.
   step 4: Check the final reaction equation to ensure that all entities are balanced.

9. To be widely accepted, communication systems must be international, precise, logical, and simple.

10. One million — M; one thousand — k; one thousandth — m; one millionth — µ.

11. (a) 47.624 kg   (e) 4.73 mmol
    (b) 28 mL   (f) 97 ms
    (c) 2.000 kmol   (g) 2 g
    (d) 5.026 L   (h) 13.429 mol

12. (a) m
    (b) n
    (c) M

13. (a) g   (d) µmol
    (b) mmol   (e) mL
    (c) g/mol   (f) kg

14. (a) 15 g/mol
    (b) 25.2 s
    (c) 50 km/h

15. (a) 7.342×10^3 mol
    (b) 3.68×10^{-2} L
    (c) 8.1504×10^3 g

16. (a) 3.04%

17. (a) 100.09 g   (c) 119 g
    (b) 8.9 mol   (d) 5.0 mol

18. One amu (atomic mass unit) is defined as one-twelfth of the mass of a carbon-12 atom.

19. (a) 32.00 g/mol   (d) 310.18 g/mol
    (b) 100.09 g/mol   (e) 46.08 g/mol
    (c) 92.02 g/mol   (f) 286.19 g/mol

20. (a) $NaCl_{(s)}$ and $NaCl_{(aq)}$
    (b) $C_{12}H_{22}O_{11(s)}$ and $C_{12}H_{22}O_{11(aq)}$
    (c) $NaHCO_{3(s)}$ and $NaHCO_{3(aq)}$
    (d) $NH_{3(g)}$ and $NH_{3(aq)}$
    (e) $CH_3OH_{(l)}$ and $CH_3OH_{(aq)}$
    (f) $C_3H_{8(g)}$ and $C_3H_{8(g)}$
    (g) $C_6H_{12}O_{6(s)}$ and $C_6H_{12}O_{6(aq)}$
    (h) $C_{(s)}$ and $C_{(s)}$

21. (a) $Cu_{(s)}$ (e) $PbSO_{4(s)}$
    (b) $K_2CO_{3(aq)}$ (f) $Ba(NO_3)_{2(aq)}$
    (c) $CH_{4(g)}$ (g) $HCl_{(aq)}$
    (d) $AgCl_{(s)}$ (h) $(NH_4)_2SO_{3(aq)}$

22. formation: elements → compound
    simple decomposition: compound → elements
    complete combustion: substance + oxygen → most common oxides
    single replacement: element + compound → element + compound

double replacement: compound + compound → compound + compound

23. The type of element (metal or nonmetal) produced in a single replacement reaction is the same as the type of element that reacts.

24. (a) $CO_{2(g)}$ (c) $SO_{2(g)}$
    (b) $H_2O_{(l)}$ (d) $Fe_2O_{3(s)}$

25. In general, elements have a low solubility in water.

26. (a) 17.11 mol   (c) 22.72 mol
    (b) 2.921 mol   (d) 55.49 mol

27. (a) 48.0 g   (c) 301 g
    (b) 44.1 g   (d) 240 g

28. (a) 2.5×10^{23} (c) 3.91×10^{23}
    (b) 4.6×10^{20} (d) 1.77×10^{24}

29. (a) Two moles of solid nickel(II) sulfide and three moles of oxygen gas react to form two moles of solid nickel(II) oxide and two moles of sulfur dioxide gas.
    2:3:2:2
    (b) Two moles of solid aluminum and three moles of aqueous copper(II) chloride react to produce two moles of aqueous aluminum chloride and three moles of solid copper.
    2:3:2:3
    (c) Two moles of liquid hydrogen peroxide react to form two moles of liquid water and one mole of oxygen gas.
    2:2:1

30. (a) simple decomposition
    $2 NaCl_{(s)} \rightarrow 2 Na_{(s)} + Cl_{2(g)}$
    (b) formation
    $4 Na_{(s)} + O_2 \rightarrow 2 Na_2O_{(s)}$
    (c) single replacement
    $2 Na_{(s)} + 2 H_2O_{(l)} \rightarrow H_{2(g)} + 2 NaOH_{(aq)}$
    (d) double replacement
    $AlCl_{3(aq)} + 3 NaOH_{(aq)} \rightarrow Al(OH)_{3(s)} + 3 NaCl_{(aq)}$
    (e) single replacement
    $2 Al_{(s)} + 3 H_2SO_{4(aq)} \rightarrow 3 H_{2(g)} + Al_2(SO_4)_{3(aq)}$
    (f) complete combustion
    $2 C_8H_{18(l)} + 25 O_{2(g)} \rightarrow 16 CO_{2(g)} + 18 H_2O_{(g)}$

31. (a) formation
    $8 Ni_{(s)} + S_{8(s)} \rightarrow 8 NiS_{(s)}$
    8:1:8
    (b) complete combustion
    $2 C_6H_{6(l)} + 15 O_{2(g)} \rightarrow 12 CO_{2(g)} + 6 H_2O_{(g)}$
    2:15:12:6
    (c) single replacement
    $2 K_{(s)} + 2 H_2O_{(l)} \rightarrow H_{2(g)} + 2 KOH_{(aq)}$
    2:2:1:2

32. $Cl_{2(g)} + 2 KI_{(aq)} \rightarrow I_{2(s)} + 2 KCl_{(aq)}$
    If a few millilitres of a chlorinated hydrocarbon solvent are added to a test tube containing the reaction mixture, and the solvent layer appears purple, then iodine has likely been formed.

33. (a) $2 C_2H_{2(g)} + 5 O_{2(g)} \rightarrow 4 CO_{2(g)} + 2 H_2O_{(g)}$
    technological
    (b) $MgCl_{2(l)} \rightarrow Mg_{(s)} + Cl_{2(g)}$
    scientific
    (c) $2 Fe_{(s)} + 3 CuSO_{4(aq)} \rightarrow Fe_2(SO_4)_{3(aq)} + 3 Cu_{(s)}$
    economic
    (d) $2 ZnS_{(s)} + 3 O_{2(g)} \rightarrow 2 ZnO_{(s)} + 2 SO_{2(g)}$
    political
    (e) $2 Pb(C_2H_5)_{4(l)} + 27 O_{2(g)} \rightarrow 2 PbO_{(s)} + 16 CO_{2(g)} + 20 H_2O_{(g)}$
    ecological

Problem 4C

Prediction

According to the double replacement reaction generalization, the products of the reaction of aqueous copper(II) chloride and sodium hydroxide are solid copper(II) hydroxide and aqueous sodium chloride.

$$CuCl_{2(aq)} + 2\,NaOH_{(aq)} \rightarrow Cu(OH)_{2(s)} + 2\,NaCl_{(aq)}$$

Experimental Design

Aqueous solutions of copper(II) chloride and sodium hydroxide are mixed.

Diagnostic Tests

1. If a precipitate forms in the mixture, then solid copper(II) hydroxide likely formed.
2. If the colors of the solutions are observed, and the blue color of the copper(II) chloride solution decreases, then copper(II) chloride likely reacted.
3. If the solution is tested before and after with litmus, and the initial red to blue does not become a final red to blue, then the basic sodium hydroxide likely reacted.
4. If the conductivities of the initial and final mixtures are tested and the conductivity decreases, then the reaction likely occurred as predicted.

CHAPTER 5

Overview

1. conductivity test, litmus test
2. (a) The solute is calcium chloride; the solvent is water.
   (b) The solute is ammonia; the solvent is water.
3. (a) acids, bases, and ionic compounds
   (b) molecular substances
4. Arrhenius studied depression of freezing points and conductivities of solutions.
5. Soluble ionic compounds and bases dissociate and acids ionize.
6. The solubility of ionic compounds can be predicted using a solubility table. The solubility of molecular compounds is either memorized (see Table 4.10, page 171) or known from experience.
7. Hydrogen ions are responsible for acidic properties and hydroxide ions are responsible for basic properties.
8. Solutions make it easy to handle chemicals, to allow chemicals to react, and to control the reactions.
9. Vinegar is used in making pickles. Ammonia solution is used for cleaning windows. Soft drinks are used for refreshment. Gasoline is used as a fuel for cars. (Many other examples are possible.)
10. A net ionic equation shows only those entities (usually ions) that change in a chemical reaction. A non-ionic equation shows all entities as part of complete chemical formulas.
11. Net ionic equations are simplified, often shorter descriptions that more accurately show what is believed to happen in a chemical reaction.
12. The two main types are qualitative (identification of substances present) and quantitative (measurement of the quantity of a substance present).

13. Three examples are litmus test, precipitation test, and color test (solution, flame, or gas discharge).
14. *Dilute* means a relatively small quantity of solute per unit volume, and *concentrated* means a relatively large quantity of solute per unit volume.
15. The concentrated salt solution in a water softener is necessary for effective operation. A high concentration of hydrogen peroxide for use as a disinfectant would be dangerous.
16. All concentration units represent a ratio of quantity of solute to quantity of solution.
17. (a) molar mass
    (b) molar concentration
18. (a) 0.250 mol
    (b) 66 g
    (c) 24.3 g/mol
19. A standard solution is one with a precisely known concentration. Standard solutions are required for most quantitative chemical analyses and for precise control of chemical reactions.
20. The two methods are preparation from a solid and by dilution.
21. (a) volumetric flask
    (b) 10 mL graduated pipet
    (c) 10 mL volumetric pipet
22. According to the solubility rules, the solubility of most solids decreases and the solubility of gases increases as the temperature of the solution drops.
23. Immiscible means that the two liquids do not mix; they form separate layers.
24. Ion concentration depends on the concentration of the compound and the relative number of ions making up the compound.
25. (a) 15
    (b) 4
    (c) –1
26. (a) If the solution is tested for conductivity, and it conducts electricity, then the solution contains an ionic compound. If it does not conduct, then it contains a molecular compound.
    (b) If the solution is tested with both red and blue litmus paper, and the blue litmus turns red, then the solution contains an acid. If the red litmus turns blue, then a base is present.
    (c) If the freezing point of the solution is measured and the freezing point is less than 0°C, then the solution contains a compound.
27. (a) $Sr(OH)_{2(s)} \rightarrow Sr^{2+}_{(aq)} + 2\,OH^-_{(aq)}$
    (b) $K_3PO_{4(s)} \rightarrow 3\,K^+_{(aq)} + PO_4^{3-}_{(aq)}$
    (c) $HBr_{(g)} \rightarrow H^+_{(aq)} + Br^-_{(aq)}$
    (d) $Mg(CH_3COO)_{2(s)} \rightarrow Mg^{2+}_{(aq)} + 2\,CH_3COO^-_{(aq)}$
28. (a) $Ca^{2+}_{(aq)}, Cl^-_{(aq)}, H_2O_{(l)}$
    (b) $C_2H_5OH_{(aq)}, H_2O_{(l)}$
    (c) $NH_4^+_{(aq)}, CO_3^{2-}_{(aq)}, H_2O_{(l)}$
    (d) $Cu_{(s)}, H_2O_{(l)}$
    (e) $Pb(OH)_{2(s)}, H_2O_{(l)}$
    (f) $H^+_{(aq)}, SO_4^{2-}_{(aq)}, H_2O_{(l)}$
    (g) $Al^{3+}_{(aq)}, SO_4^{2-}_{(aq)}, H_2O_{(l)}$
    (h) $S_{8(s)}, H_2O_{(l)}$

29. (a) $Co^{2+}_{(aq)} + 2\,OH^-_{(aq)} \rightarrow Co(OH)_{2(s)}$
    (b) $Ag^+_{(aq)} + I^-_{(aq)} \rightarrow AgI_{(s)}$
    (c) $2\,Ag^+_{(aq)} + Zn_{(s)} \rightarrow 2\,Ag_{(s)} + Zn^{2+}_{(aq)}$
    (d) $H^+_{(aq)} + OH^-_{(aq)} \rightarrow HOH_{(l)}$

30.

Test Solution	$Na^+_{(aq)}$	$Li^+_{(aq)}$	$Ca^{2+}_{(aq)}$	$Ni^{2+}_{(aq)}$	$Cu^{2+}_{(aq)}$	$Fe^{3+}_{(aq)}$
Color	none	none	none	green	blue	yellow-brown
Flame Color	yellow	bright red	yellow-red	—	blue or green	—

31. Add an excess of zinc nitrate solution to the unknown solutions to precipitate any sulfide ions present. Filter and test the filtrate for chloride ions by adding silver nitrate solution.

32. $Sr(OH)_{2(s)}$ or $Ba(OH)_{2(s)}$

33. 15.7 g/100 mL

34. (a) 0.23 mol/L
    (b) 20 mmol
    (c) 0.103 L
    (d) 0.155 mol/L
    (e) 1.21 g

35. 660 mg

36. 12.6 g

37. (a) 2.82 g
    (b) 1. Obtain the 2.82 g of potassium hydrogen tartrate in a clean, dry 100 mL beaker.
       2. Dissolve the solid using about 50 mL of pure water.
       3. Transfer the solution into a clean 100 mL volumetric flask.
       4. Add pure water to the calibration line.
       5. Stopper and mix the solution.

38. 42.8 mL

39. (a) 25.0 mL
    (b) 1. Add approximately 50 mL of pure water to a 100 mL volumetric flask.
       2. Measure 25.00 mL of potassium dichromate solution using a pipet.
       3. Transfer the solution slowly, with mixing, into the volumetric flask.
       4. Add pure water to the calibration line.
       5. Stopper and mix the solution.

40. 53.9 mL

41. 16.9 mg/L

42. 8.64 g/100 mL

43. 1.1 g

44. According to the solubility rules, some solid sodium carbonate will precipitate from the solution. The reasoning is that the solubility decreases as the temperature decreases. Therefore, the excess sodium carbonate will precipitate until the concentration is the same as the solubility at that temperature.

45. (a) $Na_2S_{(s)} \rightarrow 2\,Na^+_{(aq)} + S^{2-}_{(aq)}$
       4.48 mol/L   2.24 mol/L
    (b) $Fe(NO_3)_{2(s)} \rightarrow Fe^{2+}_{(aq)} + 2\,NO_3^-_{(aq)}$
       0.44 mol/L   0.88 mol/L
    (c) $K_3PO_{4(s)} \rightarrow 3\,K^+_{(aq)} + PO_4^{3-}_{(aq)}$
       0.525 mol/L   0.175 mol/L

(d) $Co_2(SO_4)_{3(s)} \rightarrow 2\,Co^{3+}_{(aq)} + 3\,SO_4^{2-}_{(aq)}$
    0.0431 mol/L   0.0862 mol/L   0.129 mol/L

46. (a) 1.66 g
    (b) 43 g

47. (a) $[Cu^{2+}_{(aq)}] = 0.46$ mol/L, $[SO_4^{2-}_{(aq)}] = 0.30$ mol/L, $[NO_3^-_{(aq)}] = 0.32$ mol/L
    (b) $[Al^{3+}_{(aq)}] = 0.143$ mol/L, $[Ca^{2+}_{(aq)}] = 0.0931$ mol/L, $[Cl^-_{(aq)}] = 0.615$ mol/L

48. 45.0 g

49. 0.3 GL

Problem 5E

Analysis

According to the evidence gathered by this experimental design, the four cations present in the solution are $H^+_{(aq)}$, $Ag^+_{(aq)}$, $Cu^+_{(aq)}$, and $Na^+_{(aq)}$. Provide the reasoning in the form of a flow chart.

Evaluation

The experiment could be modified to identify a fifth cation by testing the white, sulfate precipitate with a flame test to determine whether the cation is barium or strontium.

CHAPTER 6

Overview

1. (a) The volume of a gas sample varies inversely with the pressure.
   (b) The volume of a gas sample varies directly with the temperature.
   (c) The volume of a gas sample varies directly with the temperature and inversely with the pressure.

2. The chemical properties vary considerably from the unreactive noble gases to the very reactive halogens. All gases have very similar physical properties.

3. The product of the pressure and volume of a gas is equal to the product of the amount of gas, universal gas constant, and absolute temperature.

4. The behavior of gases becomes similar to an ideal gas as temperature increases and pressure decreases.

5. A law is empirical, for example, the volume-temperature relationship is a law. A theory is based on non-observable ideas such as the random, colliding motion of molecules.

6. Avogadro's idea is theoretical, since it is based on a non-observable concept. Molecules in a gas cannot be seen or counted directly.

7. (a) 273 K
   (b) 294 K
   (c) 310 K

8. (a) 44.02 g/mol   (b) 44.11 g/mol

9. (a) 0.69 mol   (b) 1.70 mmol   (c) 0.465 kmol

10. (a) 4.2 kg   (b) 699 µg   (c) 0.22 Mg

11. (a) 0.21 mol   (b) 0.924 mmol   (c) 3.6 kmol

12. (a) 12.4 kL   (b) 1.4 ML

13. 2.6 kL

14. 74°C

15. A low pressure system refers to an air mass with a pressure lower than normal; a high pressure system refers to higher than normal atmospheric air pressure.

16. (a) 150 L
    (b) 330 L
    (c) 26.1 L/mol
    (d) A measured volume of a cold soft drink will be gently warmed to drive off the carbon dioxide gas that is collected by the downward displacement of water. (Alternative design: Limewater will be added to the soft drink until no further precipitation occurs. The mixture will be filtered and the mass of precipitate measured.)

17. 8.23 L

18. (a) Since Banff is at a higher altitude than Vancouver, there is less atmosphere above Banff than above Vancouver.
    (b) 4.4 L
    (c) The main assumption made is that the temperature is the same when the volumes are measured.

19. 317°C

20. 302 kPa

21. 27%

22. (a) 50 mL
    (b) Assuming sufficient air (oxygen) is available, the first reaction produces greater leavening because 12 volumes of gas are produced, compared to 4 volumes in the second reaction.

23. 0.56 L

24. 2.1 L

25. 170 kL

26. 0.33 mol

27. (a) 44.2 g/mol
    (b) The gas may be $CO_{2(g)}$ (44.01 g/mol) but this is not very certain since other possibilities such as $N_2O_{(g)}$ exist. A diagnostic test would increase the certainty.

28. 1.19 L/mol

29. 14.2 g/L

30. You should be near the floor because methane is less dense (0.648 g/L) than either nitrogen (1.13 g/L) or oxygen (1.29 g/L) at SATP.

31. The hot air has a density of approximately 0.93 g/L, whereas the surrounding air has a density of approximately 1.16 g/L.

Problem 6A

Analysis

According to the evidence collected in this investigation, the mass of sulfur dioxide present in the 20.00 L sample of air is 0.67 g. (Provide reasoning based on the ideal gas law.)

CHAPTER 7

Overview

1. (a) The limiting reagent is the sample.
   (b) Excess reagents are used to be more certain that all of the sample has reacted.

2. (a) The result of a quantitative analysis (e.g., mass of a precipitate) is found along one axis of the graph, extended to the line on the graph, and the unknown value is read from the other axis.
   (b) Using a graph is simpler and more efficient and it may be more reliable than the alternative of doing a stoichiometry calculation.

3. The relationship among reactants and products in a balanced chemical equation is given by the coefficients in units of moles. Since there is no instrument that measures in moles directly, other measurements such as mass and volume are made and then converted to units of moles.

4. (a) gravimetric, gas, and solution stoichiometry
   (b) mass, volume, volume
   (c) molar mass, molar volume, molar concentration

5. Stoichiometric calculations may be found in the Prediction and Analysis sections.

6. The balanced equation provides the mole ratio of the two chemicals being considered.

7. (a) A percent yield is the ratio, expressed as a percentage, of the actual or experimental quantity of product obtained to the maximum possible quantity of product.
   (b) Some possible reasons are the purity of the chemicals used, experimental uncertainties associated with measurements (skills and equipment), inadequate experimental design or procedure, and a non-quantitative reaction.

8. (a) A diagnostic test of the final mixture or filtrate is used to determine the completeness of the reaction.
   (b) The solution changes color.

9. crystallization, filtration, gas collection, titration

10. reproducibility, efficiency, economics, safety, and simplicity

11. Chemical science is international in scope and technology is more localized. The approach in science is more theoretical, whereas in technology the approach is more empirical.

12. Technology is evaluated on the basis of simplicity, reliability, efficiency, and cost.

13. forensic, industrial, and environmental chemists and lab technicians

14. The limiting reagent is the calcium ions and the excess reagent is the sodium oxalate.

15. (a) 1.2 mol of each product
    (b) 0.631 kg
    (c) The decomposition of baking soda produces carbon dioxide, which covers the region around the fire, preventing an adequate supply of oxygen to sustain the fire.

16. (a) $CuSO_4$ — limiting reagent; 0.20 mol of excess Zn
    (b) NaI — limiting reagent; 5.0 mmol of excess Cl_2
    (c) $AlCl_3$ — limiting reagent; 1 g of excess NaOH

17. 3.77 kg

18. 2.13 kg

19. (a) About 11.5 mL would be reasonable.
    (b) A test with litmus will show an excess of sodium hydroxide if the solution tests basic.
    (c) 0.624 g
    (d) 98.0%. The purity of the solution is relatively high.

20. 0.35 L

21. 17.8 mol/L

22. 23.9 mmol/L

23. 46 mL

24. 2.5 ML

25. 104 kL

26. 244 mL

27. 1.57 mol/L

28. (1) A measured volume of oxalic acid is titrated with a standardized sodium hydroxide solution and the equivalence point measured.

    (2) A measured volume of oxalic acid is reacted with excess zinc. The hydrogen gas is collected and its volume, temperature, and pressure measured. (Alternatively, the mass of zinc that reacts is measured.)

    (3) A measured volume of oxalic acid is reacted with excess calcium chloride. The mass of calcium oxalate precipitate is measured.

29. (Design 1) A measured volume of sodium hydroxide will be reacted with an excess of magnesium chloride. The mixture will be filtered to determine the mass of the precipitate.

    (Design 2) A measured volume of sodium hydroxide will be titrated with a standard solution of hydrochloric acid using bromothymol blue indicator to determine the equivalence point.

30. (a) The experimental design is inadequate since litmus is an acid-base indicator and the reactants are neither acids nor bases. The problem cannot be answered.

    (b) The industrial design appears adequate to answer the question since the product, silver sulfate, has low solubility.

    (c) The industrial design is inadequate since the use of the lead(II) compound is unwarranted. Many other substances could be used that are less toxic or harmful if spilled.

    (d) The experimental design is inadequate since the concentration of the hydrochloric acid is not precisely known and concentrated hydrochloric acid is not a primary standard. The question could be answered, but with a low level of certainty.

31. The list could include nomenclature rules, chemical formula theories and rules, states of matter and solubility generalizations, conservation of atoms/ions idea, concept of molar mass and molar volume, mass-amount conversions, molar concentration and volume conversions, ideal gas law, mole ratio concept, certainty and precision rules, international rules for symbols of elements, quantities and numbers. Concept maps will vary.

Problem 7J

Analysis

According to the evidence gathered in this experiment, the concentration of the oxalic acid in the rust-removing solution is 9.71% W/V. (Provide reasoning based on the method of stoichiometry.)

Evaluation

Titration is a good choice as the experimental design for this analysis because it provides an experimental answer to the question. Crystallization is a more efficient design, especially

if a certainty of only two significant digits is required. Because the percent difference is less than 5%, the prediction is considered to be verified. (Provide reasoning showing calculations of 3% difference.) The labelling is considered to be acceptable because the prediction is verified.

CHAPTER 8

Overview

1. (a) Chemical reactivity increases with increasing atomic size in Groups 1 and 2.
   (b) Chemical reactivity decreases with increasing atomic size in Groups 16 and 17.
   (c) Within period 3, chemical reactivity decreases from sodium to silicon and then increases from phosphorus to chlorine. Argon is very unreactive.
   (d) All of the elements in Group 18 have very low reactivity.

2. (a) Atomic radius decreases as one moves from left to right in a given row of the periodic table and increases from top to bottom in a given group of the periodic table.
   (b) Ionization energy increases as one moves from left to right in a given row of the periodic table and decreases from top to bottom in a given group of the periodic table.
   (c) Electronegativity increases as one moves from left to right in a given row of the periodic table and decreases from top to bottom in a given group of the periodic table.

3. Metals react with nonmetals to form ionic compounds. Nonmetals react with other nonmetals to form molecular compounds.

4. Eight

5. The electronegativities of the representative metals are lower than the electronegativities of the representative nonmetals.

6.

Atom	Bonding Electrons	Lone Pairs
(a) ·Ca·	2	0
(b) ·Ȧl·	3	0
(c) ·Ċe·	4	0
(d) ·N̈·	3	1
(e) :S̈·	2	2
(f) :B̈r·	1	3
(g) :N̈e:	0	4

7. (a) :N:::N: + :Ï:Ï: → :Ï:N:Ï:
           :Ï:

$$N \equiv N + I - I \rightarrow I - \underset{|}{\overset{}{N}} - I$$
$$\qquad\qquad\qquad\qquad |$$
$$\qquad\qquad\qquad\qquad I$$

(b) H:Ö:Ö:H → H:Ö:H + :Ö::Ö:

$$H - O - O - H \rightarrow H - O - H + O = O$$

8. The idea of double and triple covalent bonds explains empirical molecular formulas such as $O_{2(g)}$ and $N_{2(g)}$ without changing the assumptions of covalent bonding.

9. The rapid reaction of some substances with bromine is evidence that a double or triple carbon-carbon bond is present in the substance.

10. be able to explain and predict, and is simple

11. Photosynthesis in green plants and the decomposition of water are examples of endothermic chemical changes. The combustion of gasoline in a car engine and the metabolism of fats and carbohydrates in the human body are examples of exothermic chemical changes.

12. The boiling point of a substance is an indication of the strength of its intermolecular forces. The higher the boiling point, the stronger the intermolecular forces.

13. London forces act between all molecules. An example of a substance in which London forces are the only type of intermolecular force is iodine, $I_{2(s)}$. Dipole-dipole forces are present in liquid hydrogen chloride, $HCl_{(l)}$. Hydrogen bonds are present in water, $H_2O_{(l)}$.

14. An ionic compound is a pure substance formed from metals and nonmetals; is a hard, crystalline solid at SATP with a high melting and boiling point; and conducts electricity in molten and aqueous states.

15. Ionic compounds are neutral, three-dimensional structures of oppositely charged ions, held together by the simultaneous attraction of positive and negative ions.

16. (a) $Mg + \ddot{S}: \rightarrow Mg^{2+}[:\ddot{S}:]^{2-}$

(b) $:\ddot{C}l\cdot + \cdot Al\cdot + \cdot\ddot{C}l: \rightarrow Al^{3+}[:\ddot{C}l:]^-_3$

$:\ddot{C}l:$

17. The numbers in a molecular formula indicate the actual number of atoms of each element in the molecule. The numbers in an ionic formula indicate the ratio of the ions in the ionic crystal.

18. (a) London forces
(b) metallic
(c) ionic
(d) London forces, dipole-dipole
(e) covalent network
(f) ionic
(g) London forces, dipole-dipole, hydrogen bonding
(h) London forces, dipole-dipole, hydrogen bonding

19. (a) linear
(b) trigonal planar
(c) tetrahedral
(d) tetrahedral
(e) linear
(f) V-shaped
(g) pyramidal
(h) V-shaped

20. (a) London forces

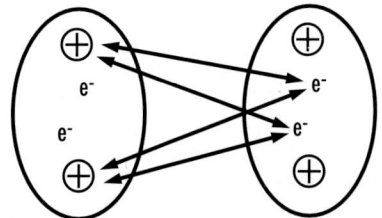

b) dipole-dipole forces

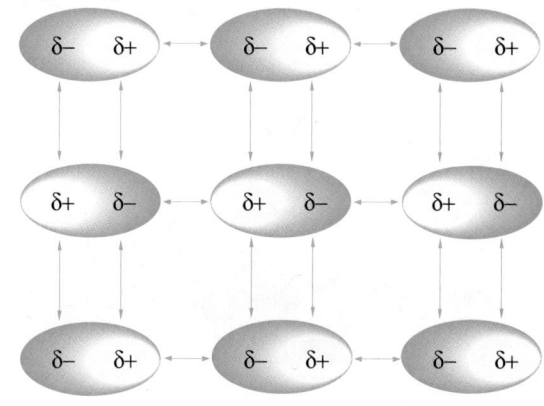

(c) hydrogen bonding

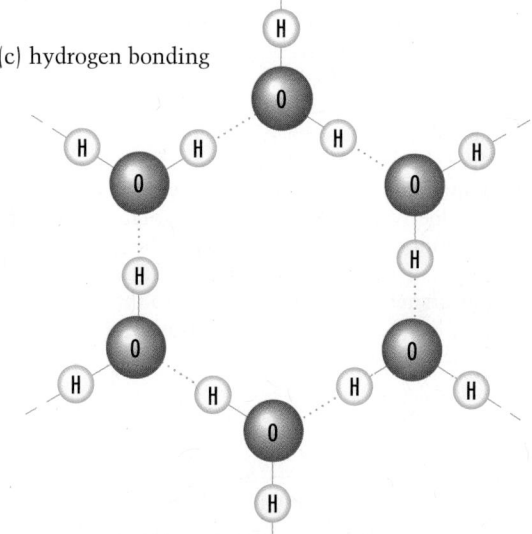

(d) metallic bonding

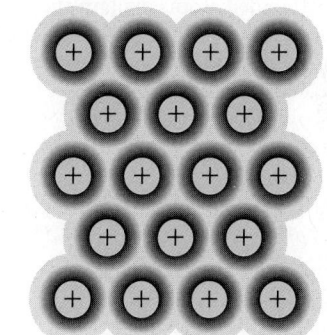

(e) ionic bonding

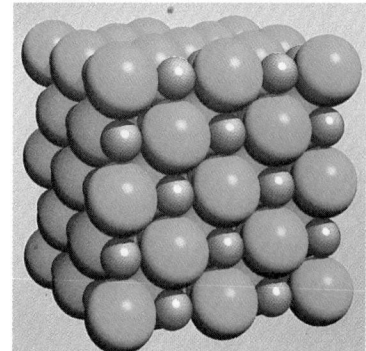

ANSWERS TO OVERVIEW QUESTIONS   **707**

(f) bonding in metalloids

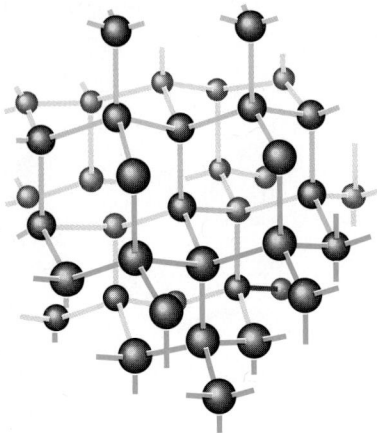

21. (a) Mg: $1s^2\,2s^2\,2p^6\,3s^2$
    K: $1s^2\,2s^2\,2p^6\,3s^2\,3p^6\,4s^1$
    Cr: $1s^2\,2s^2\,2p^6\,3s^2\,3p^6\,4s^1 3d^5$
    Ag: $1s^2\,2s^2\,2p^6\,3s^2\,3p^6\,4s^2\,3d^{10}\,4p^6\,5s^1\,4d^{10}$
    Hg: $1s^2\,2s^2\,2p^6\,3s^2\,3p^6\,4s^2\,3d^{10}\,4p^6\,5s^2\,4d^{10}\,5p^6\,6s^2\,4f^{14}5d^{10}$
    (b) Empty valence orbitals explain electron mobility
    (conductivity). Malleability is explained by the non-
    directional nature of the metallic bonds, which allow
    planes of atoms to slide over each other. The shiny
    appearance is thought to be due to the ability of the
    mobile valence electrons to absorb and re-emit light.

22. (a) B: $1s^2\,2s^2\,2p^1$
    Si: $1s^2\,2s^2\,2p^6\,3s^2\,3p^2$
    Ge: $1s^2\,2s^2\,2p^6\,3s^2\,3p^6\,4s^2\,3d^{10}\,4p^2$
    As: $1s^2\,2s^2\,2p^6\,3s^2\,3p^6\,4s^2\,3d^{10}\,4p^3$
    Sb: $1s^2\,2s^2\,2p^6\,3s^2\,3p^6\,4s^2\,3d^{10}\,4p^6\,5s^2\,4d^{10}\,5p^3$
    (b) Metalloids have many bonding electrons that enable
    them to form strong, directional, covalent bonds with
    neighboring atoms. The three-dimensional arrays of
    covalently bonded atoms explain their hardness and
    high melting points. Their poor electrical
    conductivity is due to the covalent bonding of all the
    valence electrons.

23. (a) N: $1s^2\,2s^2\,2p^3$
    S: $1s^2\,2s^2\,2p^6\,3s^2\,3p^4$
    Br: $1s^2\,2s^2\,2p^6\,3s^2\,3p^6\,4s^2\,3d^{10}\,4p^5$
    Xe: $1s^2\,2s^2\,2p^6\,3s^2\,3p^6\,4s^2\,3d^{10}\,4p^6\,5s^2\,4d^{10}\,5p^6$
    I: $1s^2\,2s^2\,2p^6\,3s^2\,3p^6\,4s^2\,3d^{10}\,4p^6\,5s^2\,4d^{10}\,5p^5$
    (b) N₂: :N⋮⋮⋮N:

    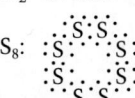

    Br₂: :Br̈:Br̈:

    Xe: :Xë:

    I₂: :Ï:Ï:

    (c) In nonmetal molecules, all the valence orbitals are
    filled so electrons cannot easily move from molecule
    to molecule. Atoms of the same element share the
    bonding electrons equally so there are no dipole-
    dipole forces, and the London forces are relatively
    weak.

24. (a) $Al_{(s)} \rightarrow Al^{3+}_{(s)} + 3\,e^-$
    (b) The aluminum ion has a much smaller radius than
    the aluminum atom (50 pm compared to 143 pm).
    When an atom forms a cation, its radius decreases;
    since the nuclear charge remains the same, removing
    an electron reduces electron-electron repulsion and
    causes the electron cloud to shrink.

25. (a) $Cl_{2(g)} + 2\,e^- \rightarrow 2Cl^-$
    (b) A chloride ion is larger than a chlorine atom (181 pm
    compared to 99 pm). When an atom forms an anion
    its radius increases; since the nuclear charge remains
    the same, the repulsion resulting from the additional
    electrons enlarges the electron cloud.

26. (a) protons and electrons in the same molecule
    (b) protons and electrons in different molecules
    (c) oppositely charged ends of polar molecules
    (d) proton of a hydrogen atom bonded to a nitrogen,
    oxygen, or fluorine atom and a lone pair of electrons
    in another molecule
    (e) positive and negative ions

27. (a) fluorine, chlorine, bromine, iodine
    (b) reduction
    (c) The most reactive, fluorine, has the greatest tendency
    to gain electrons in a reduction half-reaction. In
    fluorine there are fewer inner electrons that shield the
    electrons from the attraction of the nucleus.
    (d) The order of the activity series in Group 17 is
    completely consistent with the electronegativity
    values. As the reactivity decreases, the
    electronegativity decreases.

28. (a) Na_2O, MgO, Al_2O_3 are classified as ionic. SiO_2,
    P_2O_5, SO_2, Cl_2O are classified as molecular.
    (b) Na_2O (2.6), MgO (2.3), Al_2O_3 (2.0), SiO_2 (1.7),
    P_2O_5 (1.4), SO_2 (1.0), Cl_2O (0.5)
    (c) The larger electronegativity differences correspond to
    the ionic compounds and the smaller electronegativity
    differences correspond to the molecular compounds.

29. The molecular formula for nicotine is $C_{10}H_{14}N_2$.

30. (a) H—P—H PH_3 phosphorus trihydride
           |
           H

    (b) Cl—Si—Cl $SiCl_4$ silicon tetrachloride
            |
            Cl

    (c) C=O   CO carbon monoxide
                     (violates octet rule)
    (d) F—B—F BF_3 boron trifluoride
           | (violates octet rule)
           F

31. Both intermolecular forces and covalent bonds are
    explained as a simultaneous attraction of opposite
    charges. Covalent bonds involve shared electrons in
    overlapping orbitals but intermolecular forces do not.

32. (a) $2\,K_{(s)} \rightarrow 2\,K^+_{(s)} + 2\,e^-$
        $Br_{2(l)} + 2\,e^- \rightarrow 2\,Br^-_{(s)}$
        $2\,K_{(s)} + Br_{2(l)} \rightarrow 2\,KBr_{(s)}$
    (b) $2\,Sr_{(s)} \rightarrow 2\,Sr^{2+}_{(s)} + 4\,e^-$
        $O_{2(g)} + 4\,e^- \rightarrow 2\,O^{2-}_{(s)}$
        $2\,Sr_{(s)} + O_{2(g)} \rightarrow 2\,SrO_{(s)}$

33. The high melting and boiling points of ionic compounds are due to the strong simultaneous forces of attraction between the positive and negative ions.

Problem 8G

Prediction

According to molecular theory, the simplest compound is PF_3, as shown by the following electron dot diagram.

$$:\overset{..}{\underset{..}{F}}:\overset{..}{\underset{..}{P}}:\overset{..}{\underset{..}{F}}:$$
$$:\overset{..}{\underset{..}{F}}:$$

Analysis

According to the evidence, the molecular formula is PF_5.

Evaluation

The prediction is judged to be falsified because the predicted answer does not agree with the experimental answer. Experimental uncertainties are not likely to account for the difference. It is quite certain that the molecular theory being tested is unacceptable in this case because of the falsified prediction. The theory needs to be restricted or revised to improve its predictive power.

CHAPTER 9

Overview

1. fuels, petrochemical feedstock
2. Carbon atoms can form combinations of single, double, or triple covalent bonds with up to four other atoms, and they can bond to other carbon atoms to form very large structures.
3. Organic compounds are the basis of all known life forms.
4. A *saturated* organic compound is composed of relatively stable molecules containing no double or triple covalent bonds between carbon atoms.

   An *unsaturated* organic compound is composed of reactive molecules containing double or triple covalent bonds between carbon atoms.
   (a) unsaturated
   (b) saturated
   (c) unsaturated
   (d) saturated
   (e) unsaturated

5. (a) alcohols

   $CH_3-CH_2-CH_2-OH$ 1–propanol

   $CH_3-CH-CH_3$ 2–propanol
   with OH below the CH

   (b) ethers

   $CH_3-CH_2-O-CH_3$ methoxyethane

   (c) aldehydes

   $CH_3-CH_2-C=O$ with H below   propanal

(d) ketones

$CH_3-C=O$ with CH_3 below   propanone

(e) carboxylic acids

$CH_3-CH_2-C=O$ with OH below   propanoic acid

(f) esters

$CH_3-C=O$ with $O-CH_3$ below   methylethanoate

$HC=O$ with $O-CH_2-CH_3$ below   ethylmethanoate

(g) amines

$CH_3-CH_2-CH_2-NH_2$ propylamine

CH_3-CH_2-NH with CH_3 below   methylethylamine

CH_3-N-CH_3 with CH_3 below   trimethylamine

(h) amides

$CH_3-CH_2-C=O$ with NH_2 below   propanamide

6. (Answers will vary.)
   (a) C_nH_{2n+2}, CH_4, methane, fuel for heating
   (b) C_nH_{2n}, C_2H_4, ethene, petrochemical feedstock
   (c) C_nH_{2n-2}, C_2H_2, ethyne, fuel for welding
   (d) C_6H_5R, $C_6H_5CH_3$, methylbenzene, solvent in lacquers
   (e) RX, CCl_2F_2, freon (CFC-12), refrigerant
   (f) ROH, C_2H_5OH, ethanol, gasoline additive
   (g) R_1-O-R_2, $C_2H_5-O-C_2H_5$, ethoxyethane, anesthetic
   (h) $R(H)CHO$, CH_3CHO, ethanal, believed to cause alcohol "hangover"
   (i) R_1COR_2, CH_3COCH_3, propanone, solvent in plastic cements
   (j) $R(H)COOH$, CH_3COOH, ethanoic acid, vinegar
   (k) $R_1(H)COOR_2$, CH_3COOCH_3, methyl ethanoate, manufacture of artificial leather

7. (a) alkane, $CH_3-CH-CH_2-CH_3$ with CH_3 below the CH

   (b) aromatic, CH_3-CH_2- (benzene ring)

   (c) alkyne, $CH_3-CH_2-C\equiv C-CH_2-CH_3$

   (d) alkene, $CH_3-C=C-CH_2-CH_3$ with CH_3 above the first C and CH_3 below the second C

(e) alcohol, $CH_3-CH-CH_3$
   $\quad\quad\quad\quad\quad\quad\;\;|$
   $\quad\quad\quad\quad\quad\quad\;OH$

(f) amine, CH_3-NH_2

(g) ester, $CH_3-\overset{\displaystyle O}{\overset{\|}{C}}-O-CH_3$

8. (a) 1,2-dichloroethane, organic halide
   (b) methylpropene, alkene
   (c) butane, alkane
   (d) ethanal, aldehyde
   (e) butanone, ketone
   (f) ethanamide, amide

9. Polymer molecules are made up of many similar small molecules linked together. Cellulose and starch occur in living systems. Polyethylene is a manufactured polymer.

10.
$$\begin{array}{c} H \quad\quad Cl \\ \backslash\quad\quad/ \\ C=C \\ /\quad\quad\backslash \\ Cl \quad\quad H \end{array}$$
trans–1,2–dichloroethene
non-polar (bond dipoles cancel)

$$\begin{array}{c} H \quad\quad H \\ \backslash\quad\quad/ \\ C=C \\ /\quad\quad\backslash \\ Cl \quad\quad Cl \end{array}$$
cis–1,2–dichloroethene
polar (bond dipoles in same direction)

$$\begin{array}{c} H \quad\quad Cl \\ \backslash\quad\quad/ \\ C=C \\ /\quad\quad\backslash \\ H \quad\quad Cl \end{array}$$
1,1–dichloroethene
polar (bond dipoles in same direction)

11. (a) methoxymethane, ethanol, methoxyethane, 1–propanol, ethoxyethane, 1–butanol

   (b) Within each pair of isomers the alcohol consistently has the higher boiling point. The higher boiling point of the alcohols is likely due to the hydrogen bonding among –OH groups.

12. (a) reforming
   $CH_3-CH_3 \;+\; CH_3-CH=CH-CH_3 \;\rightarrow$
   $$CH_3-CH_2-\overset{\displaystyle CH_3}{\overset{|}{CH}}-CH_2-CH_3$$

   (b) cracking
   $$CH_3-\overset{\displaystyle CH_3}{\overset{|}{CH}}-CH_2-\overset{\displaystyle CH_3}{\overset{|}{CH}}-CH_2-CH_3 \;+\; H-H \;\rightarrow$$
   $$CH_3-CH_2-CH_2-CH_3 \;+\; CH_3-\overset{\displaystyle CH_3}{\overset{|}{CH}}-CH_3$$

   (c) combustion
   $O=C=O \;+\; H-O-H$
   carbon dioxide $\quad\quad$ water

(d) addition
1,2-dichlorocyclohexane

(e) reforming
   propane + pentane → 2-methylheptane + hydrogen

(f) substitution
   cyclopentane + bromine →
   $\quad\quad$ bromocyclopentane + hydrogen bromide
   Br $\quad\quad\quad\quad$ H — Br

(g) esterification
   butanoic acid + 1-propanol →
   $\quad\quad\quad\quad$ propyl butanoate + water
   $$CH_3-CH_2-CH_2-\overset{\displaystyle O}{\overset{\|}{C}}-O-CH_2-CH_2-CH_3$$
   $$H-O-H$$

(h) elimination
   2-chloropropane + hydroxide ion →
   $\quad\quad\quad\quad$ propene + water + chloride ion
   $\quad\quad CH_2=CH-CH_3 \quad\quad H-O-H \quad\quad Cl^-$

13. (a) substitution
   + Br — Br →
   + + H — Br
   1,2,3-tribromobenzene $\quad$ 1,2,4-tribromobenzene $\quad$ hydrogen bromide

(b) addition
   $CH_2=CH-CH_2-CH_3 \;+\; H-O-H \;\rightarrow$
   $HO-CH_2-CH_2-CH_2-CH_3 \;+$
   1-butanol
   $$CH_3-\overset{\displaystyle OH}{\overset{|}{CH}}-CH_2-CH_3$$
   2-butanol

(c) addition
   2-pentyne + hydrogen iodide →
   $$CH_3-\overset{\displaystyle I}{\underset{\displaystyle I}{\overset{|}{\underset{|}{C}}}}-CH_2-CH_2-CH_3 \;+\; CH_3-\overset{\displaystyle I}{\overset{|}{CH}}-\overset{\displaystyle}{\underset{\displaystyle I}{\underset{|}{CH}}}-CH_2-CH_3$$
   2,2-diiodopentane $\quad\quad\quad\quad$ 2,3-diiodopentane

   $$+\; CH_3-CH_2-\overset{\displaystyle I}{\underset{\displaystyle I}{\overset{|}{\underset{|}{C}}}}-CH_2-CH_3$$
   3,3-diiodopentane

(d) elimination
2-chlorohexane + hydroxide ion →

$$CH_2\!=\!CH\!-\!(CH_2)_3\!-\!CH_3 \ +$$
1-hexene

$$CH_3\!-\!CH\!=\!CH\!-\!(CH_2)_2\!-\!CH_3 \ + \ Cl^- \ + \ H\!-\!O\!-\!H$$
2-hexene   chloride ion   water

14. (a) $CH_3\!-\!\overset{\displaystyle O}{\overset{\|}{C}}\!-\!OH \ + \ CH_3\!-\!CH_2\!-\!OH \ \rightarrow$

$$CH_3\!-\!\overset{\displaystyle O}{\overset{\|}{C}}\!-\!O\!-\!CH_2\!-\!CH_3 \ + \ H\!-\!O\!-\!H$$

(b) $CH_2\!=\!CH_2 \ + \ H\!-\!O\!-\!H \ \rightarrow \ CH_3\!-\!CH_2\!-\!OH$

(c) $CH_3\!-\!\underset{\underset{\displaystyle Cl}{|}}{CH}\!-\!CH_3 \ + \ OH^- \ \rightarrow$

$$CH_2\!=\!CH\!-\!CH_3 \ + \ H\!-\!O\!-\!H \ + \ Cl^-$$

(d) $n\,CF_2\!=\!CF_2 \ \rightarrow \ (\!-\!CF_2\!-\!CF_2\!-\!)_n$

(e) $Cl\!-\!CH_2\!-\!CH_2\!-\!Cl \ + \ OH^- \ \rightarrow$

$$CH_2\!=\!CH\!-\!Cl \ + \ H\!-\!O\!-\!H \ + \ Cl^-$$

$$CH_2\!=\!CH\!-\!Cl \ + \ OH^- \ \rightarrow$$

$$CH\!\equiv\!CH \ + \ H\!-\!O\!-\!H \ + \ Cl^-$$

(f) $CH_4 \ + \ Cl\!-\!Cl \ \rightarrow \ CH_3\!-\!Cl \ + \ H\!-\!Cl$

$$CH_3\!-\!Cl \ + \ Cl\!-\!Cl \ \rightarrow \ Cl\!-\!CH_2\!-\!Cl \ + \ H\!-\!Cl$$

$$Cl\!-\!CH_2\!-\!Cl \ + \ F\!-\!F \ \rightarrow \ Cl\!-\!\underset{\underset{\displaystyle F}{|}}{\overset{\overset{\displaystyle F}{|}}{C}H}\!-\!Cl \ + \ H\!-\!F$$

$$Cl\!-\!\underset{\underset{\displaystyle F}{|}}{\overset{\overset{\displaystyle F}{|}}{C}H}\!-\!Cl \ + \ F\!-\!F \ \rightarrow \ Cl\!-\!\underset{\underset{\displaystyle F}{|}}{\overset{\overset{\displaystyle F}{|}}{C}}\!-\!Cl \ + \ H\!-\!F$$

Problem 9C

Analysis

According to the evidence, C_3H_4O is propenal, whose structure is

$$H\!-\!\underset{\underset{\displaystyle H}{|}}{\overset{\overset{\displaystyle H}{|}}{C}}\!=\!C\!-\!\overset{\displaystyle O}{\overset{\|}{C}}\!-\!H$$

Problem 9D

Analysis

According to the evidence, the yield of chloromethane is 78.2% by mass.

CHAPTER 10

Overview

1. combustion of natural gas to heat buildings, cook meals, heat water, etc.
   combustion of gasoline and diesel fuel to power cars and trucks

reaction of carbohydrates in the body to maintain body temperature

2. geothermal energy (hot springs), solar energy (water cycle), nuclear energy (uranium)

3. When heat is transferred between substances, a change in temperature occurs.
   $q = mc\Delta t$, $q = vc\Delta t$, and $q = C\Delta t$

4. 991 kJ

5. phase changes (different state is formed), chemical changes (new substances are formed), and nuclear changes (new elements or subatomic particles are formed)

6. endothermic: fusion (melting) < vaporization < sublimation $(s \rightarrow g)$
   exothermic: solidification < condensation < sublimation $(g \rightarrow s)$

7. Phase, chemical, and nuclear changes all involve changes in potential energy. None of these are believed to involve a kinetic energy change.

8. The molar enthalpy of a chemical change (10^2 kJ/mol to 10^4 kJ/mol) is approximately ten to one hundred times the molar enthalpy of a phase change (10^0 to 10^2 kJ/mol).

9. Nuclear reactions produce the largest quantities of energy because they involve the strongest bonds.

10. 3.19 kJ

11. A positive sign is used to report an endothermic molar enthalpy and a negative sign is used to report an exothermic molar enthalpy. During endothermic reactions, the potential energy of the chemical system increases as energy is gained from the surroundings. During exothermic reactions, the potential energy of the chemical system decreases as energy is lost to the surroundings.

12. (a) insulated container, thermometer, known quantity of water
    (b) law of conservation of energy
    (c) The calorimeter is isolated from the surroundings — no heat is transferred between the calorimeter and the outside environment. A dilute aqueous solution has the same density and specific heat capacity as pure water. Any heat absorbed or released by the calorimeter is negligible.

13. specific heat capacity: c, J/(g·°C)
    heat capacity: C, J/°C
    temperature change: Δt, °C
    heat: q, J
    change in potential energy: ΔE_p, J
    amount of a substance: n, mol
    molar enthalpy: H, J/mol
    enthalpy change: ΔH, J

14.
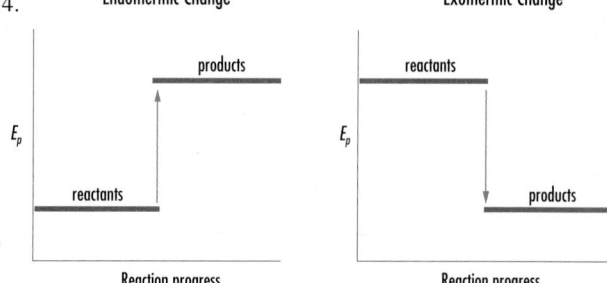

1.1 J/(g·°C)

16. 81 MJ
17. (a) 14 MJ
    (b) Caulk any cracks such as those around windows and doors and make sure fireplace and furnace dampers are completely closed.
18. Steam at 100°C will release the energy of condensation to the surroundings (the person) in addition to the heat transferred when the water at 100°C cools to 35°C.

    $$\Delta E_{total} = 5.66 \text{ kJ} + 0.68 \text{ kJ} = 6.34 \text{ kJ}$$

    Water at 100°C will transfer only 0.68 kJ of heat as it cools from 100°C to 35°C.
19. (a) The heating curve for ice at –25°C to steam at 115°C is shown below.

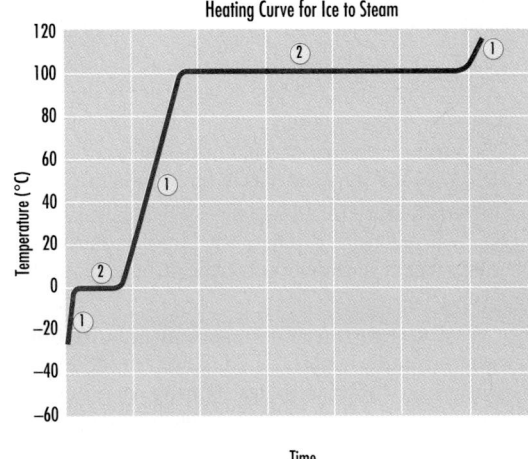

Heating Curve for Ice to Steam

(b) Regions labelled 1 correspond to temperature changes and those labelled 2 correspond to phase changes.
20. A sketch of the cooling curve for chlorine from 25°C to –150°C is shown below. (boiling/condensation point = –35°C, freezing point = –101°C)

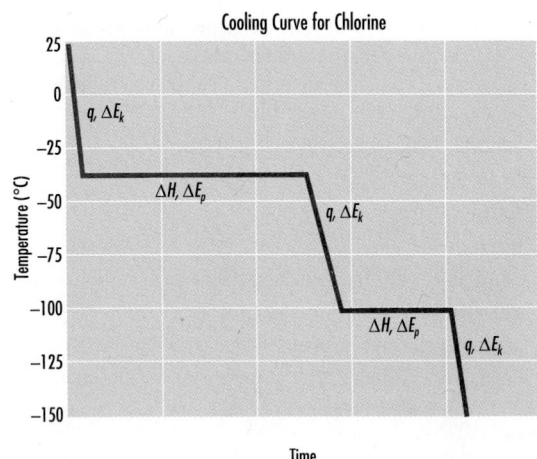

Cooling Curve for Chlorine

21. 1.839 MJ
22. 2.63 GJ
23. (a) The heating curve for pressurized water heated from 85°C to 120°C as water and to 150°C as steam is shown in the next column.

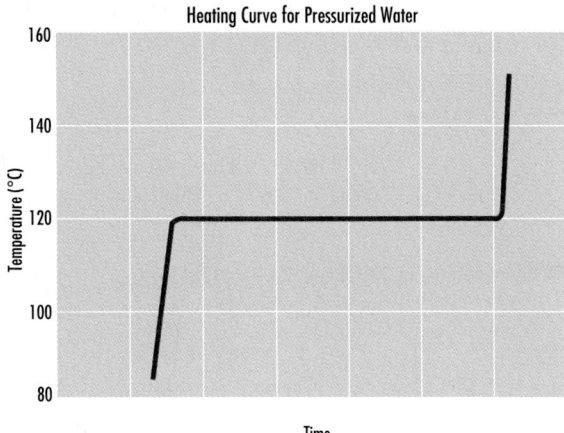

Heating Curve for Pressurized Water

(b) 297 MJ
24. 80.4 kJ
25. (a) 52.0 MJ
    (b) high pressure and low temperature
    (c) 63.9 mol
    (d) 43 m³
26. 25.9 g
27. 0.94 kg
28. 0.37 J/(g·°C)
29. –37.7 kJ/g

Problem 10E

Analysis
According to the evidence, the molar enthalpy of solidification of wax is –59 kJ/mol.

Evaluation
The experimental design is judged to be inadequate even though the Problem was answered. This design is flawed since it is not possible to determine the moment when all of the wax has solidified. It is possible that some liquid wax exists inside some of the solid, or that some of the solid wax cools significantly below its freezing point. In an improved design, the amount of time would be extended, to allow the contents to reach a maximum temperature before a temperature reading is taken.

CHAPTER 11

Overview

1. molar enthalpy of a substance in a specific reaction; enthalpy change, ΔH, of a reaction; enthalpy change as a term in an equation; potential energy diagram
2. If ΔH is negative, then the energy term is on the product side. If ΔH is positive, then the energy term is on the reactant side.
3. (a) When one mole of butane is burned to produce carbon dioxide and water vapor, 2657 kJ of energy is released when the initial and final conditions are at SATP.

   (b) $C_4H_{10(g)} + \frac{13}{2} O_{2(g)} \rightarrow 4 CO_{2(g)} + 5 H_2O_{(g)}$
   $$\Delta H° = -2657 \text{ kJ}$$

   (c) $C_4H_{10(g)} + \frac{13}{2} O_{2(g)} \rightarrow 4 CO_{2(g)} + 5 H_2O_{(g)} + 2657 \text{ kJ}$

   (d) According to the table of molar enthalpies in Appendix F, the value of $H_f°$ for butane is –125.6 kJ/mol. This is not the molar enthalpy of combustion of butane; it represents the quantity of heat released per mole of butane formed from its elements, carbon and hydrogen, as illustrated in the equation below.

$$4\,C_{(s)} + 5\,H_{2(g)} \rightarrow C_4H_{10(g)} \qquad \Delta H_f^\circ = -125.6\ kJ$$

4. The reference zero point of potential energy for chemical reactions is the potential energy of elements in their most stable form at SATP. The standard molar enthalpy of formation of any element in its most stable form at SATP is therefore assigned as zero.

5. During an endothermic reaction, potential energy increases. During an exothermic reaction, potential energy decreases.

6. List any two of the following: Hess's law, enthalpies of formation method, bond energies.

7. Enthalpy changes are independent of the way the system changes from its initial state to its final state.

8. If a chemical equation is reversed, then its ΔH changes its sign. If the coefficients of a chemical equation are altered by multiplying or dividing by a constant factor, then the ΔH is altered in the same way.

9. (a) $\Delta H_r^\circ = \Sigma n H_{fp}^\circ - \Sigma n H_{fr}^\circ$
   (b) The enthalpy change for a chemical reaction equals (the sum of the amount times the molar enthalpy of formation for each product) minus (the sum of the amount times the molar enthalpy of formation for each reactant).
   (c) The summation term for the products represents the total potential energy of the products, while the summation term for the reactants represents the total potential energy of the reactants.

10. (a) Nuclear reactions typically have much larger energy changes than chemical reactions.
    (b) Any endothermic reaction has a positive ΔH and any exothermic reaction has a negative ΔH.

11. (a) Heating: Choose among solar energy, geothermal energy, biomass gas, electricity from nuclear reactions, or others.
    Transportation: Choose among alcohol/gasohol and hydrogen fuels, batteries and fuel cells, or others.
    Industry: Choose among solar energy, nuclear energy, hydroelectricity, or others.
    (b) Heating: improved insulation
    Transportation: car pools and mass transit
    Industry: recovery of waste heat

12. $C_3H_{8(g)} + 5\,O_{2(g)} \rightarrow 3\,CO_{2(g)} + 4\,H_2O_{(g)} + 2.25\ MJ$
    or
    $C_3H_{8(g)} + 5\,O_{2(g)} \rightarrow 3\,CO_{2(g)} + 4\,H_2O_{(g)} \quad \Delta H_c^\circ = -2.25\ MJ$

13. (a) 129 kJ/mol $CO_{2(g)}$
    (b) $2\,NaHCO_{3(s)} \rightarrow Na_2CO_{3(s)} + H_2O_{(g)} + CO_{2(g)}$
    $$\Delta H_r^\circ = 129\ kJ$$

(c)

Decomposition of Baking Soda

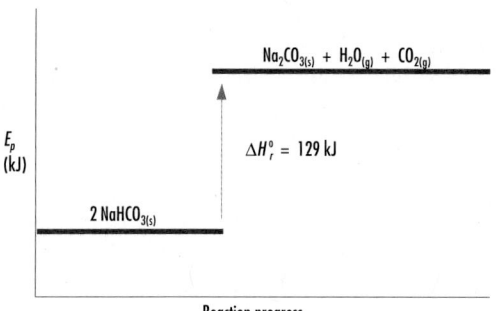

14. (a) $2\,CO_{(g)} + 2\,NO_{(g)} \rightarrow N_{2(g)} + 2\,CO_{2(g)} \quad \Delta H_r^\circ = -746\ kJ$
    (b) –373 kJ/mol $NO_{(g)}$
    (c) 6.21 MJ

15. (a) –400 kJ/mol $N_2H_{4(l)}$
    (b) –133 kJ/mol $O_{2(g)}$
    (c) 33.3 kJ

16. (a) –401.0 kJ

(b)

Roasting of Copper(II) Sulfide

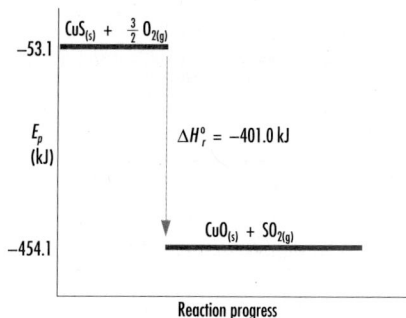

17. (a) –2802.7 kJ
    (b) The equations are the same, although in respiration, liquid water is produced while in combustion, water vapor is produced. Also, respiration occurs more slowly than combustion; therefore, it releases its energy more slowly.

18. (a) $\Delta H_r^\circ = 765.6\ kJ$
    (for $N_{2(g)} + 3\,H_2O_{(l)} \rightarrow 2\,NH_{3(g)} + \frac{3}{2}\,O_{2(g)}$)
    (b) 22.5 MJ
    (c) 6.24 m²
    (d) The major assumption is that all of the collected energy is transferred to the reaction.

19. (a) –40.8 kJ
    (b) –286 kJ (liquid water)
    (c) -1.7×10^9 kJ
    (d) 0.408 cm, 2.86 cm, 1.7×10^7 cm; 6.1×10^5 pages

20. (a) 136.3 kJ

(b)

Cracking of Ethane

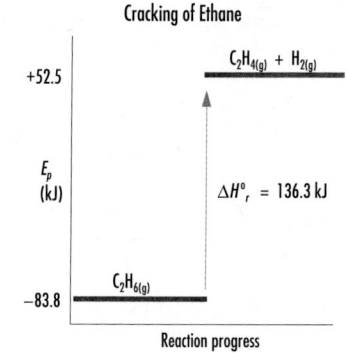

21. –1257.0 kJ/mol
22. acetylene, ethylene, ethane
23. –110.7 kJ
24. (a) $C_2H_{4(g)} + \frac{1}{2}\,O_{2(g)} + H_2O_{(l)} \rightarrow C_2H_4(OH)_{2(l)}$
    (b) –221.5 kJ

(c)

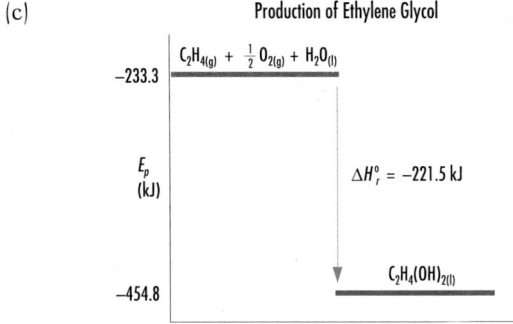

Production of Ethylene Glycol

$C_2H_{4(g)} + \frac{1}{2}O_{2(g)} + H_2O_{(l)}$

−233.3

E_p (kJ)

$\Delta H_r^\circ = -221.5$ kJ

$C_2H_4(OH)_{2(l)}$

−454.8

Reaction progress

(d) 3.57 GJ of energy released

25. (a) −64.6 kJ/mol

   (b) $C_2H_{4(g)} + HCl_{(g)} \rightarrow C_2H_5Cl_{(l)} + 64.6$ kJ

   (c) −104.4 kJ/mol

26. 16 g

27. Three possible designs, in order of increasing certainty, are the determination of the boiling point; the determination of the molar enthalpy of combustion; and the determination of the molecular formula from combustion and mass spectrometer analyses.

Problem 11G

Prediction

According to the table of standard molar enthalpies of formation, the enthalpy change for the reaction is

$$Ca_{(s)} + \frac{1}{2}O_{2(g)} \rightarrow CaO_{(s)} \qquad \Delta H_f^\circ = -634.9 \text{ kJ}$$

Analysis

According to the evidence,

$$Ca_{(s)} + \frac{1}{2}O_{2(g)} \rightarrow CaO_{(s)} \qquad \Delta H_f^\circ = -0.60 \text{ MJ}$$

Problem 11H

Prediction

According to the molar enthalpy of formation method, the standard molar enthalpy of combustion of methylpropane is −2868.8 kJ/mol. Assume $H_2O_{(l)}$ is produced in the bomb calorimeter. [A similar prediction can be reached in three ways: using the molar enthalpy of formation method with data from a table such as that listing standard molar enthalpies in Appendix F; using balanced equations with ΔH_f° values in the Hess's law method; and using reference data about specific bond energies.]

Analysis

According to the evidence gathered by calorimetry and the analysis using the law of conservation of energy, the standard molar enthalpy of combustion of methylpropane is −2.88 MJ/mol. (Provide the reasoning.)

Evaluation

The Hess's law method and the standard molar enthalpy of formation method are equivalent, as they both use the same data, namely, the standard molar enthalpies of formation. Each method predicts the same molar enthalpy of combustion of methylpropane, which leads to a difference of 0.2%. This is an unusually close agreement between the prediction and the empirically determined value. These two predictions are clearly verified and the two methods are considered acceptable. The bond energy method is the most theoretical and provides only an approximate value for the molar energy of combustion. The prediction based on bond energies is inconclusive. The method of bond energies should be restricted to situations where the standard molar enthalpies of formation are not known.

CHAPTER 12

Overview

1. The factors are the nature of the reactants, reactant concentration, temperature, and the presence of a catalyst.

2. Generalization: An increase in surface area increases the rate of reaction.

   Explanation: Larger surface area provides a greater number of particles exposed to the possibility of collision. More collisions per unit of time means more effective collisions and a faster rate of reaction.

3. (a) Rate increases.

   (b) Rate decreases.

   (c) Rate increases.

   (d) Rate unchanged.

4. (a) More concentrated hydrochloric acid has more particles per unit volume so more collisions per second will occur at the surface of the zinc per unit of area. More collisions per unit of time means more effective collisions and a faster rate of reaction.

   (b) A drop in temperature means particles move more slowly, so fewer collisions per second will occur. Fewer collisions per unit of time means fewer effective collisions and a slower rate of reaction.

   (c) A rise in temperature means particles move more rapidly, so more collisions per second will occur. More collisions per unit of time means more effective collisions and a faster rate of reaction.

   (d) The fraction of collisions that are effective at room temperature is negligible. The rise in temperature caused by the igniter means particles move more rapidly, so more collisions per second will occur. More collisions per unit of time means more effective collisions and a faster rate of reaction. In this exothermic reaction, the energy released will be absorbed by surrounding reactant molecules, raising their temperature enough to keep the reaction going continuously.

   (e) The fraction of collisions that are effective at room temperature is negligible. If a spark begins the reaction, however, the energy released will be absorbed by surrounding reactant molecules, raising their temperature enough to keep the reaction going continuously. The reaction rate is explosively fast because the very fine grain dust particles have extremely high surface area.

   (f) The fraction of collisions that are effective at room temperature is negligible. If a catalyst is present, however, the activation energy for the reaction is lowered because the reaction mechanism is changed. The number of collisions per second will not change, but the fraction of collisions that are effective increases greatly, so the rate of reaction increases proportionally.

(g) The fraction of collisions that are effective at room temperature is significant, but the number of collisions when reactants are in solid form is negligible, since collisions can only occur at the points where solid particles are in physical contact. Dissolving the reactants increases the number of particles that are available for collision to a maximum, so the rate of reaction increases proportionally.

5.

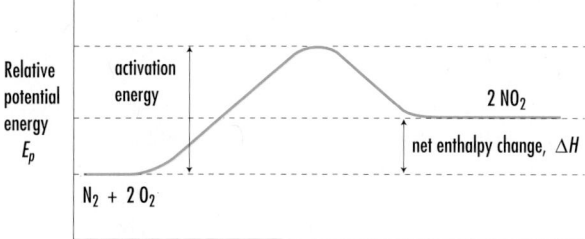

Potential Energy vs Reaction Progress

6. The activation energy must be very low if the reaction is rapid at SATP.

7. Although a higher temperature shifts the equilibrium position left, this is not important in this case, since the equilibrium is shifted right continuously by changing reactant and product concentrations. The higher temperature increases the rate of reaction, so that product can be formed (and removed) faster.

8. If this equation were the reaction mechanism, six particles would have to collide simultaneously in a single event. Collisions of more than three particles are so unlikely that they are essentially impossible.

9. The average rate of reaction is 44.2 mL/30.0 s = 1.47 mL/s. At 30°C the rate would be higher and the time to produce the same volume would be less. At 25°C without the catalyst the rate would be lower and the time to produce the same volume would be greater.

10. A *reactant* is a substance present before, but not after, reaction.

A *product* is a substance present after, but not before, reaction.

A *catalyst* is a substance present both before and after reaction.

An *intermediate product* is a substance present during reaction that is not present either before or after reaction.

11. (a) A
    (b) B
    (c) C

12. (a)

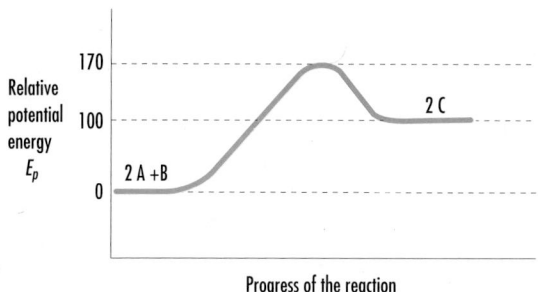

Potential Energy vs Reaction Progress

(b) E_{act} = 85 kJ
(c) ΔH = –50 kJ

Problem 12A

Analysis

Theoretically, the rate law expression will be
$r = k \ [W]^a \ [X]^b \ [Y]^c$.

Comparing the first two trials shows that the effect of doubling the concentration of "W" is no change in the overall reaction rate. This means that exponent "a" in the rate law expression has a value of zero.

Comparing the first and third trials shows that doubling the concentration of "X" doubles the reaction rate. This means that exponent "b" in the rate law expression has a value of one.

Comparing the first and fourth trials shows that doubling the concentration of "Y" quadruples the reaction rate. This means that the exponent "c" in the rate law expression has a value of two.

Empirically, the rate law expression is $r = k \ [X] \ [Y]^2$. Therefore, when values from any trial (in this case trial 1) are substituted, ignoring units, the rate constant may be found. $12.0 = k \ (0.100) \ (0.100)^2$. Therefore $k = 12.0/0.00100 = 1.2 \times 10^4$, so the complete rate law expression is $r = 1.2 \times 10^4 \ [X] \ [Y]^2$.

CHAPTER 13

Overview

1. Chemical equilibrium is a state of a closed system in which all macroscopic properties are constant.

2. Chemical equilibrium is explained by the idea that the rates of forward and reverse changes are equal.

3. These can be described by a percent reaction or by an equilibrium constant.

4. (a) When a soft drink bottle has just been opened it is in a non-equilibrium state. Carbon dioxide gas escapes from the solution as the rate of decomposition of carbonic acid into carbon dioxide and water exceeds the rate at which carbon dioxide gas and water combine to produce carbonic acid.
   (b) When the bottle is sealed and at a constant temperature, it is in an equilibrium state. Carbon dioxide gas and water are in equilibrium with carbonic acid.

5. When a chemical system at equilibrium is disturbed by a change in some property of the system, the system adjusts in a way that opposes the change.

6. Temperature, volume, and concentration are commonly manipulated when an equilibrium system is shifted.

7. Decreasing the volume increases the pressure, and increasing the volume decreases the pressure.

8. (a) If a solution is neutral, the hydrogen ion concentration equals the hydroxide ion concentration.
   (b) If a solution is acidic, the hydrogen ion concentration

is greater than the hydroxide ion concentration.

(c) If a solution is basic, the hydrogen ion concentration is less than the hydroxide ion concentration.

9. Both increasing the concentration of the reactant (by adding more) and decreasing the concentration of the product (by removing some) will increase the yield of the product.

10. A catalyst does not affect the state of equilibrium. It decreases the time required to reach equilibrium.

11. Conductivity and pH tests can distinguish a weak acid from a strong acid, if the temperature and the initial solute concentrations are the same.

12. According to Arrhenius's theory, all bases increase the hydroxide ion concentration of solutions.

13. $K = \dfrac{[C]^c[D]^d}{[A]^a[B]^b}$

$K_W = [H^+_{(aq)}][OH^-_{(aq)}]$

$[H^+_{(aq)}] = \dfrac{p}{100} \times [HA_{(aq)}]$

$pH = -\log[H^+_{(aq)}]$

$pOH = -\log[OH^-_{(aq)}]$

$[H^+_{(aq)}] = 10^{-pH}$

$[OH^-_{(aq)}] = 10^{-pOH}$

$pH + pOH = 14.00$

$K_a = \dfrac{[H^+_{(aq)}][A^-_{(aq)}]}{[HA_{(aq)}]}$

14. (a) $N_{2(g)} + 3 H_{2(g)} \overset{<10\%}{\rightleftharpoons} 2 NH_{3(g)}$

$K = \dfrac{[NH_{3(g)}]^2}{[N_{2(g)}][H_{2(g)}]^3}$

(b) $2 H_{2(g)} + O_{2(g)} \rightarrow 2 H_2O_{(g)}$

$K = \dfrac{[H_2O_{(g)}]^2}{[H_{2(g)}]^2[O_{2(g)}]}$

(c) $CO_{(g)} + H_2O_{(g)} \overset{67\%}{\rightleftharpoons} CO_{2(g)} + H_{2(g)}$

$K = \dfrac{[CO_{2(g)}][H_{2(g)}]}{[CO_{(g)}][H_2O_{(g)}]}$

15. (a) $1.15\ L/mol = \dfrac{[N_2O_{4(g)}]}{[NO_{2(g)}]^2}$ $t = 55°C$

(b) $[N_2O_{4(g)}] = 2.9\ mmol/L$

(c) According to Le Châtelier's principle, if the concentration of nitrogen dioxide is increased, then the equilibrium would shift to the right. The system shifts in such a way as to reduce the concentration of nitrogen dioxide (it reacts to produce dinitrogen tetraoxide).

16. At equilibrium the concentration of the product is much greater than the concentration of the reactants. The large equilibrium constant suggests a quantitative reaction equilibrium.

17. (a) left   (c) right
(b) left   (d) no effect

18. (a) high concentration of $C_2H_{6(g)}$, low concentration of $C_2H_{4(g)}$ and $H_{2(g)}$, high temperature, low pressure

(b) high concentration of $CO_{(g)}$ and $H_{2(g)}$, low concentration of $CH_3OH_{(g)}$, low temperature, high pressure

19. (a) shift left

(b) shift left

(c) shift right

(d) no shift

(e) shift right

20. (a) shift right; forward rate is increased as $[Cu^{2+}_{(aq)}]$ increases

(b) shift left; forward and reverse rates are both decreased with forward rate decreasing more

(c) no shift; forward and reverse rates are both increased by the same amount as solid surface area increases

(d) shift left; reverse rate is increased as $[CO_{2(g)}]$ increases

(e) shift right; reverse rate is decreased as $[Cl^-_{(aq)}]$ decreases

(f) no shift; forward and reverse rates are both decreased by the same amount — all concentrations decrease

(g) shift right; forward rate is increased as $[Fe^{3+}_{(aq)}]$ increases

21. $CaF_{2(s)} \rightleftharpoons Ca^{2+}_{(aq)} + 2 F^-_{(aq)}$

$K_{sp} = [Ca^{2+}_{(aq)}][F^-_{(aq)}]^2$

$[Ca^{2+}_{(aq)}] = [CaF_{2(aq)}] = 0.02676\ g/1.00\ L \times 1\ mol/78.08\ g$
$= 3.43 \times 10^{-4}\ mol/L$

$[F^-_{(aq)}] = 2 \times [Ca^{2+}_{(aq)}] = 2 \times 3.43 \times 10^{-4}\ mol/L$
$= 6.85 \times 10^{-4}\ mol/L$

$K_{sp} = [3.43 \times 10^{-4}][6.85 \times 10^{-4}]^2$
$= 1.61 \times 10^{-10}$ (units ignored)

22. $CaC_2O_{2(s)} \rightleftharpoons Ca^{2+}_{(aq)} + C_2O_4^{2-}_{(aq)}$

$K_{sp} = [Ca^{2+}_{(aq)}][C_2O_4^{2-}_{(aq)}]$

Since, from the balanced equation, $[C_2O_4^{2-}_{(aq)}] = [Ca^{2+}_{(aq)}]$, the K_{sp} expression may be rewritten as

$K_{sp} = [Ca^{2+}_{(aq)}]^2 = 2.3 \times 10^{-9}$

Taking the square root of both sides gives

$[Ca^{2+}_{(aq)}] = 4.8 \times 10^{-5}\ mol/L$, and since
$[Ca^{2+}_{(aq)}] = [CaC_2O_{4(aq)}]$

the molar solubility of calcium oxalate is $4.8 \times 10^{-5}\ mol/L$.

23. (a), (c), and (e)

24. Calculate the value of $Q = \dfrac{[C_2H_{4(g)}]}{[C_2H_{2(g)}][H_{2(g)}]} = \dfrac{(3.2 \times 10^{-4})}{(0.40)(0.020)}$

$Q = 0.040$, which is lower than the value of K, so the reaction will be shifting right at this time, producing ethylene faster than it decomposes.

25. (a) Let c stand for the change in concentration of the hydrogen.

Reaction Progress	$[H_2]$	$[Br_2]$	$[HBr]$
Initial	4.00	4.00	0
Change	$-c$	$-c$	$+2c$
Equilibrium	$4.00 - c$	$4.00 - c$	$2c$

$K = \dfrac{[HBr]^2}{[H_2][Br_2]} = \dfrac{(2c)^2}{(4.00-c)^2} = 12.0$

c can be found by taking the square root of both sides:

$\dfrac{2c}{(4.00-c)} = 3.4641$

$c = 2.54$ mol/L

Reaction Progress	$[H_2]$	$[Br_2]$	$[HBr]$
Initial	4.00	4.00	0
Change	–2.54	–2.54	+5.07
Equilibrium	1.46	1.46	5.07

(b) Let c stand for the change in concentration of the hydrogen.

Reaction Progress	$[H_2]$	$[Br_2]$	$[HBr]$
Initial	6.00	6.00	0
Change	$-c$	$-c$	$+2c$
Equilibrium	$6.00-c$	$6.00-c$	$2c$

$K = \dfrac{[HBr]^2}{[H_2][Br_2]} = \dfrac{(2c)^2}{(6.00-c)^2} = 12.0$

c can be found by taking the square root of both sides:

$\dfrac{2c}{(6.00-c)} = 3.4641$

$c = 3.80$ mol/L

Reaction Progress	$[H_2]$	$[Br_2]$	$[HBr]$
Initial	6.00	6.00	0
Change	–3.80	–3.80	+7.60
Equilibrium	2.20	2.20	7.60

26. (a) $HCN_{(aq)} \overset{0.0078\%}{\rightleftharpoons} H^+_{(aq)} + CN^-_{(aq)}$

27. 7.7×10^{-12} mol/L

28. $[H^+] = 4.0 \times 10^{-8}$ mol/L
    pH = 7.40

29. 3×10^{-6} mol/L

30. by a factor of 1000

31. 0.372

32. 0.25 g

33. pH = 4.27

34. –0.016, 14.02

It is necessary to assume that the percent reaction of hydrogen chloride is the same as a 0.10 mol/L solution and that the volume of the solution is equal to the volume of water used.

35. 2.42, 11.58 (You should use K_a, not percent ionization, as the concentration of the acid is not 0.10 mol/L.)

36. 18 mg

37.
Diagnostic Test	Strong Acid	Weak Acid	Neutral Molecular	Neutral Ionic
Conductivity	high	low	none	high
Litmus	turns red	turns red	no change	no change

38.

$$pH = -\log[H^+_{(aq)}]$$

$$pH \rightleftharpoons [H^+_{(aq)}] = 10^{-pH}$$

$$[H^+_{(aq)}] \qquad [OH^-_{(aq)}] = K_w/[H^+_{(aq)}]$$
$$\rightleftharpoons [OH^-_{(aq)}]$$
$$[H^+_{(aq)}] = K_w/[OH^-_{(aq)}]$$

$$[H^+_{(aq)}] = \dfrac{p}{100} \times [HA_{(aq)}] \qquad [HA_{(aq)}] = [H^+_{(aq)}] \times \dfrac{100}{p}$$

$$[HA_{(aq)}]$$

39. 2.62

As the temperature changes to 37°C, more acetylsalicylic acid might dissolve, increasing the concentration of hydrogen ions and decreasing the pH.

40. 4.77 (from K_a)

41. 1×10^{-3} mol/L

42. 1.3×10^{-10} mol/L

43. 0.4 mmol/L

CHAPTER 14

Overview

1. According to the oxygen concept, all acids contain oxygen. This definition is too restricted and has too many exceptions, notably $HCl_{(aq)}$. According to the hydrogen concept, all acids are compounds of hydrogen. This definition is limited because it does not explain why only certain hydrogen compounds are acids. According to Arrhenius's concept, acids are substances that ionize in aqueous solutions to produce hydrogen ions. This definition is limited to aqueous solutions and cannot explain or predict the properties of many common substances. According to the Brønsted-Lowry concept, acids are substances that donate protons to bases in a chemical reaction. The main limitation of the Brønsted-Lowry concept is the restriction to protons and the inability to explain and predict the acid nature of ions of multi-valent metals.

2. The theory is restricted, revised, or replaced.

3. Test the explanations and predictions made using the theory.

4. According to the evidence, the hydrogen ion exists as a hydrated proton whose simplest representation is the hydronium ion, $H_3O^+_{(aq)}$.

5. According to Arrhenius's concept, a base is a substance that dissociates in aqueous solution to produce hydroxide ions. According to the revised Arrhenius's concept, a base is a substance that reacts with water to produce hydroxide ions. According to the Brønsted-Lowry concept, a base is a proton acceptor that removes protons from an acid.

6. (a) The pH of the nitric acid solution is lower than the pH of the nitrous acid solution.

(b) Water, acting as a base, can quantitatively remove the proton from the HNO_3 molecule. The proton in HNO_2 is more strongly bonded and not as easily given up to the water.

$$H_2O_{(l)} + HNO_{2(aq)} \overset{8.1\%}{\rightleftharpoons} H_3O^+_{(aq)} + NO_2^-_{(aq)}$$

7. If the pH of the two solutions is taken and the pH is different, then the lower pH is the hydrochloric acid.

   If the conductivity of the two solutions is taken and the conductivities are different, the lower conductivity is the acetic acid solution.

   If the odor of the two solutions is checked and the odor of vinegar is detected in one solution, then that solution is the acetic acid solution.

   If an active metal, such as zinc or magnesium, is added to the two solutions, and there is a noticeable difference in the rate of reaction, then the solution with the slower reaction is the acetic acid solution.

   If the freezing point of the two solutions is determined and one solution has a noticeably lower freezing point, then that solution is the hydrochloric acid solution.

8. The position of equilibrium is determined by the result of the competition for protons. Of the forward and reverse reactions, the reaction involving the stronger acid and the stronger base is favored.

9. If the acid is listed above the base in the table of acids and bases, then the products will be favored. If the acid is listed below the base, then the reactants will be favored.

10. $H_2SO_{3(aq)}/HSO_3^-{}_{(aq)}$ and $HSO_3^-{}_{(aq)}/SO_3^{2-}{}_{(aq)}$

11. (a) yellow
    (b) red
    (c) green
    (d) colorless
    (e) yellow

12. (a) Buffering action means a relatively constant pH when small amounts of a strong acid or base are added.

    (b) Buffering action is most noticeable at a volume of titrant that is one-half the first equivalence point or half-way between successive equivalence points for polyprotic acids.

    (c) Quantitative reactions are represented by nearly vertical portions of a pH curve.

    (d) The pH endpoint is the mid-point of the sharp change in pH in an acid-base titration. The equivalence point is the quantity of titrant at the endpoint of the titration.

    (e) The mid-point of the pH range of a suitable indicator should equal the pH endpoint and the indicator should complete its color change while the pH is changing abruptly.

    (f) Non-quantitative reactions do not have a distinct endpoint because the pH changes gradually in the region where the equivalence point is expected.

13. See Figure 14.27, page 574.

14. See Figure 14.28, page 578.

15. (a) F
    (b) T
    (c) T
    (d) T
    (e) T
    (f) F
    (g) T
    (h) T

16. Buffers are used in making cheese, yogurt, and sour cream, in preserving food, and in the production of antibiotics.

17. pH of natural acid rain $\geq$ 5.6. Natural carbon dioxide (forest fires), sulfur dioxide (volcanoes), and nitrogen dioxide (lightning) can react with rain to produce natural acid rain.

18. (a) $HBr_{(aq)} + H_2O_{(l)} \rightarrow Br^-{}_{(aq)} + H_3O^+{}_{(aq)}$ acidic

    (b) $NO_2^-{}_{(aq)} + H_2O_{(l)} \rightleftharpoons HNO_{2(aq)} + OH^-{}_{(aq)}$ basic

    (c) $NH_{3(aq)} + H_2O_{(l)} \rightleftharpoons NH_4^+{}_{(aq)} + OH^-{}_{(aq)}$ basic

    (d) Since the K_a for $HSO_4^-{}_{(aq)}$ is much larger than the K_b for $HSO_4^-{}_{(aq)}$, the reaction with water to produce an acidic solution should predominate.

    $HSO_4^-{}_{(aq)} + H_2O_{(aq)} \rightleftharpoons SO_4^{2-}{}_{(aq)} + H_3O^+{}_{(aq)}$

    $HSO_4^-{}_{(aq)} + H_2O_{(aq)} \rightleftharpoons H_2SO_{4(aq)} + OH^-{}_{(aq)}$

    (e) $CO_{2(g)} + 2\,H_2O_{(l)} \rightleftharpoons H_3O^+{}_{(aq)} + HCO_3^-{}_{(aq)}$

    (f) $CaO_{(s)} + H_2O_{(l)} \rightarrow Ca^{2+}{}_{(aq)} + 2\,OH^-{}_{(aq)}$

    (g) $CH_3COOH_{(aq)} + H_2O_{(l)} \rightleftharpoons H_3O^+{}_{(aq)} + CH_3COO^-{}_{(aq)}$

19. Each substance is tested with litmus paper. The substance tested is the manipulated variable and the color change of the litmus is the responding variable. Temperature and concentration are controlled variables.

20. Solutions of equal concentration of several bases are prepared and the pH is measured for each solution. The manipulated variable is the base, and the responding variable is the pH. The controlled variables are temperature and concentration.

    (second design) Solutions of equal concentration of several bases are prepared and the redox reaction in a $Ag_{(s)} | Ag^+{}_{(aq)} || base | Pt_{(s)}$ cell is measured. The manipulated variable is the base, the responding variable is the rate of oxygen gas production at the anode, and the controlled variables are temperature and initial concentration of the base solution.

    (second or third design) The electrical conductivities of aqueous solutions of equal concentration are measured. The manipulated variable is the base, the responding variable is the electrical conductivity, and the controlled variables are temperature and concentration.

21. (a) $CH_3COOH_{(aq)} + NaOH_{(aq)} \rightarrow H_2O_{(l)} + NaCH_3COO_{(aq)}$
    $CH_3COOH_{(aq)} + Na^+{}_{(aq)} + OH^-{}_{(aq)} \rightarrow$
    $\qquad\qquad\qquad H_2O_{(l)} + Na^+{}_{(aq)} + CH_3COO^-{}_{(aq)}$
    $CH_3COOH_{(aq)} + OH^-{}_{(aq)} \rightarrow H_2O_{(l)} + CH_3COO^-{}_{(aq)}$

    (b) $NaHCO_{3(aq)} + HCl_{(aq)} \rightarrow H_2CO_{3(aq)} + NaCl_{(aq)}$
    $Na^+{}_{(aq)} + HCO_3^-{}_{(aq)} + H^+{}_{(aq)} + Cl^-{}_{(aq)} \rightarrow$
    $\qquad\qquad\qquad H_2CO_{3(aq)} + Na^+{}_{(aq)} + Cl^-{}_{(aq)}$
    $HCO_3^-{}_{(aq)} + H^+{}_{(aq)} \rightarrow H_2CO_{3(aq)}$ (or $CO_{2(g)} + H_2O_{(l)}$)

    (c) $NH_{3(aq)} + H_2CO_{3(aq)} \rightleftharpoons NH_4HCO_{3(aq)}$
    $NH_{3(aq)} + H_2CO_3 \rightleftharpoons NH_4^+{}_{(aq)} + HCO_3^-{}_{(aq)}$
    $NH_{3(aq)} + H_2CO_3 \rightleftharpoons NH_4^+{}_{(aq)} + HCO_3^-{}_{(aq)}$

22.
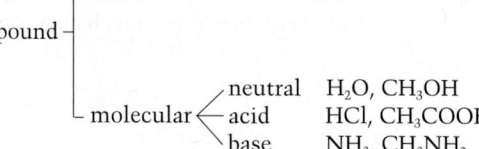

23. (a)
$$\overset{A}{HCOOH_{(aq)}} + \overset{B}{CN^-_{(aq)}} \overset{>50\%}{\rightleftharpoons} \overset{B}{HCOO^-_{(aq)}} + \overset{A}{HCN_{(aq)}}$$
$HCOOH_{(aq)}/HCOO^-_{(aq)}$ and $HCN_{(aq)}/CN^-_{(aq)}$

(b)
$$\overset{B}{HPO_4^{2-}_{(aq)}} + \overset{A}{HCO_3^-_{(aq)}} \overset{<50\%}{\rightleftharpoons} \overset{A}{H_2PO_4^-_{(aq)}} + \overset{B}{CO_3^{2-}_{(aq)}}$$
$HCO_3^-_{(aq)}/CO_3^{2-}_{(aq)}$ and $H_2PO_4^-_{(aq)}/HPO_4^{2-}_{(aq)}$

24. A possible pH is 5.1 (5.0–5.2). The hydronium ion concentration for this pH is 8×10^{-6} mol/L.

25. (a) $\overset{}{HF_{(aq)}} + \overset{}{SO_4^{2-}_{(aq)}} \overset{<50\%}{\rightleftharpoons} \overset{}{F^-_{(aq)}} + \overset{}{HSO_4^-_{(aq)}}$

If the pH of the sulfate solution is measured before and after addition of $HF_{(aq)}$ and the pH decreases, then the $HF_{(aq)}$ probably reacted with the $SO_4^{2-}_{(aq)}$.

(b) $HSO_4^-_{(aq)} + HS^-_{(aq)} \overset{>50\%}{\rightleftharpoons} SO_4^{2-}_{(aq)} + H_2S_{(aq)}$

If the mixture is carefully smelled, and a "rotten egg" odor is noticed, then $H_2S_{(aq)}$ is likely to be present.

(c) $HCOOH_{(aq)} + OH^-_{(aq)} \rightarrow HCOO^-_{(aq)} + H_2O_{(l)}$
If the pH is measured during the titration, and a sharp change in pH is observed, then methanoic acid has reacted quantitatively with the sodium hydroxide.

(d) $H_3O^+_{(aq)} + PO_4^{3-}_{(aq)} \overset{>50\%}{\rightleftharpoons} H_2O_{(l)} + HPO_4^{2-}_{(aq)}$

If the pH of the solution is measured before and after the addition of a strong acid, and the pH remains relatively constant, then the strong acid has reacted.

(e) $HPh_{(aq)} + OH^-_{(aq)} \overset{>50\%}{\rightleftharpoons} Ph^-_{(aq)} + H_2O_{(l)}$

If the color is observed and it changes immediately to red, then the indicator has reacted with the strong base.

(f) (Note: Recognize that the initial reaction is sodium bisulfate and sodium hydroxide reacting to produce water and sodium sulfate. With stoichiometrically equivalent amounts at the equivalence point, the final solution is aqueous sodium sulfate.)

$$\overset{}{Na^+_{(aq)}}, \overset{A}{SO_4^-_{(aq)}}, \overset{}{H_2O_{(l)}}$$
$$\underset{SB}{} \underset{B}{}$$

$SO_4^{2-}_{(aq)} + H_2O_{(l)} \rightleftharpoons OH^-_{(aq)} + HSO_4^-_{(aq)}$
If you obtain a pH curve for the reaction and the pH at the equivalence point is greater than 7, then hydroxide ions were likely produced.

(g) (Note: According to the indicator table, bromothymol blue is blue in the ion ($Bb^-_{(aq)}$) form.)

$$\overset{SA}{Bb^-_{(aq)}}, \overset{A}{CH_3COOH_{(aq)}}, \overset{}{H_2O_{(l)}}$$
$$\underset{SB}{} \underset{B}{}$$

$Bb^-_{(aq)} + CH_3COOH_{(aq)} \rightleftharpoons HBb_{(aq)} + CH_3COO^-_{(aq)}$
If you observe the color of the bromothymol blue and if it changes from blue to yellow, then the bromothymol blue ion has reacted to produce the bromothymol blue molecule.

(h) (Note: The initial reaction of sulfur dioxide gas with water produces sulfurous acid.)

$$\overset{SA}{H_2SO_{3(aq)}}, \overset{A}{H_2O_{(l)}}$$
$$\underset{SB}{}$$
$H_2SO_{3(aq)} + H_2O_{(l)} \rightleftharpoons H_3O^+_{(aq)} + HSO_3^-_{(aq)}$

If the solution that is formed is tested for pH and the pH decreases, then hydronium ions were likely produced.

(i) (Note: Recall that limestone has low solubility and that sulfuric acid is a strong acid.)

$$\overset{SA}{CaCO_{3(s)}}, \overset{A}{H_3O^+_{(aq)}}, \overset{A}{HSO_4^-_{(aq)}}, \overset{}{H_2O_{(l)}}$$
$$\underset{SB}{} \underset{B}{} \underset{B}{}$$

$CaCO_{3(s)} + H_3O^+_{(aq)} \rightleftharpoons H_2O_{(l)} + Ca^{2+}_{(aq)} + HCO_3^-_{(aq)}$
If the pH of the acid is tested before and after adding limestone and the pH increases, then hydronium ions are consumed.

(j) (Note: The initial reaction is of aluminum ions with water to produce the aquo complex ion, $Al(H_2O)_6^{3+}_{(aq)}$.)

$$\overset{SA}{Al(H_2O)_6^{3+}_{(aq)}}, \overset{A}{Cl^-_{(aq)}}, \overset{}{H_2O_{(l)}}$$
$$\underset{B}{} \underset{SB}{}$$

$Al(H_2O)_6^{3+}_{(aq)} + H_2O_{(l)} \rightleftharpoons H_3O^+_{(aq)} + Al(H_2O)_5(OH)^{2+}_{(aq)}$
If the pH is tested before and after adding aluminum chloride and the pH decreases, then hydronium ions are likely produced.

(k) (Note: Reference common chemical names in Appendix G.)

$$\overset{}{Na^+_{(aq)}}, \overset{SA}{CO_3^{2-}_{(aq)}}, \overset{}{H_2O_{(l)}}$$
$$\underset{SB}{} \underset{B}{}$$

$CO_3^{2-}_{(aq)} + H_2O_{(l)} \rightleftharpoons OH^-_{(aq)} + HCO_3^-_{(aq)}$
If the pH is tested before and after and the pH increases, then hydroxide ions are likely produced by the reaction.

(l) (Note: Reference common chemical names in Appendix G.)

$$\overset{A}{Na^+_{(aq)}}, \overset{SA}{HCO_3^-_{(aq)}}, \overset{}{H_2O_{(l)}}$$
$$\underset{SB}{} \underset{B}{}$$

$HCO_3^-_{(aq)} + H_2O_{(l)} \rightleftharpoons OH^-_{(aq)} + H_2CO_{3(aq)}$
If the pH is tested before and after and the pH increases, then hydroxide ions are likely produced by the reaction.

26. According to Le Châtelier's principle, the equilibrium will shift to the
(a) left (and the ammonia odor will increase)
(b) right (and the solution will become (more) blue)
(c) right (and the vinegar odor will decrease)
(d) left (but the pH increases anyway)

27. The solution will be basic.
$CH_3O^-_{(aq)} + H_2O_{(l)} \rightleftharpoons CH_3OH_{(aq)} + OH^-_{(aq)}$

28. $OH^-_{(aq)} + HCO_3^-_{(aq)} \overset{>50\%}{\rightleftharpoons} H_2O_{(l)} + CO_3^{2-}_{(aq)}$

$H_3O^+_{(aq)} + HCO_3^-_{(aq)} \overset{>50\%}{\rightleftharpoons} H_2O_{(l)} + H_2CO_{3(aq)}$

29. (Note that there are many correct solutions to this problem — be creative.) If the solutions are tested with a pH meter, and the pH values are ordered from smallest to largest, then the solutions are sulfuric acid,

hydrochloric acid, acetic acid, ethanediol, ammonia, sodium hydroxide, and barium hydroxide, respectively.

Diagnostic Tests on the Unlabelled Solutions

Litmus	Conductivity	Acid/Base Titration	Analysis
red	low	one volume	$CH_3COOH_{(aq)}$
blue	very high	two volumes	$Ba(OH)_{2(aq)}$
blue	low	one volume	$NH_{3(aq)}$
no change	none	not applicable	$C_2H_4(OH)_{2(aq)}$
red	higher	two volumes	$H_2SO_{4(aq)}$
red	high	one volume	$HCl_{(aq)}$
blue	high	one volume	$NaOH_{(aq)}$

30. (a) one
   (b) $H_3O^+_{(aq)} + SO_3^{2-}_{(aq)} \rightarrow H_2O_{(l)} + HSO_3^-_{(aq)}$
   (c) The endpoint occurs at a pH of 4.0, when 23 mL of hydrochloric acid has been added.
   (d) congo red or methyl orange
   (e) A buffering region on the graph occurs where about 12 mL of $HCl_{(aq)}$ have been added. The entities present are $Na^+_{(aq)}$, $SO_3^{2-}_{(aq)}$, $HSO_3^-_{(aq)}$, $Cl^-_{(aq)}$, and $H_2O_{(l)}$.

31. (a) 10 mL of $HCl_{(aq)}$
   (b) 11.11
   (c) 9.24
   (d) $NH_4^+_{(aq)}$, $Cl^-_{(aq)}$, $H_2O_{(l)}$
   (e) 5.27
   (f) 1.70
   (g) methyl orange through bromothymol blue

32. 0.882 mol/L

33. 0.537 mol/L

34. No. pH 1.69

35. (a)

Acid	Base	Conjugate Acid/Base Pair
C_4H_4NH	$(C_6H_5)_3C^-$	$C_4H_4NH/C_4H_4N^-$, $(C_6H_5)_3CH/(C_6H_5)_3C^-$
CH_3COOH	HS^-	CH_3COOH/CH_3COO^-, H_2S/HS^-
$(C_6H_5)_3CH$	O^{2-}	$(C_6H_5)_3CH/(C_6H_5)_3C^-$, OH^-/O^{2-}
H_2S	$C_4H_4N^-$	H_2S/HS^-, $C_4H_4NH/C_4H_4N^-$

   (b)

Acid	Conjugate Base
CH_3COOH	CH_3COO^-
H_2S	HS^-
C_4H_4NH	$C_4H_4N^-$
$(C_6H_5)_3CH$	$(C_6H_5)_3C^-$
OH^-	O^{2-}

36. (a) There is no third pH endpoint and therefore no third equivalence point in this titration (Figure 14.24, page 572).
   (b) Pure acetic acid is a liquid at SATP and will be driven off by the boiling.
   (c) Removing $OH^-_{(aq)}$ ions by precipitation will cause the equilibrium to shift to the right, producing more $OH^-_{(aq)}$ ions.
   $$NH_{3(aq)} + H_2O_{(l)} \rightleftharpoons NH_4^+_{(aq)} + OH^-_{(aq)}$$
   (d) Hydrochloric acid is not a primary standard.
   (e) Both reactants and products form basic solutions and litmus cannot be used to distinguish among basic solutions.
   (f) Cobalt chloride paper is used in a diagnostic test for the production of water in a strong acid–strong base reaction.

(g) The strength of an acid is determined by titration.

37. Design 1: Prepare a primary standard of $Na_2CO_{3(aq)}$ and use it to titrate the $HCl_{(aq)}$. Calculate the concentration of the $HCl_{(aq)}$ from the reaction equation.
   Design 2: Use a pH meter to measure the pH of the $HCl_{(aq)}$. Calculate the concentration of the $HCl_{(aq)}$ from the pH.
   Design 3: Use indicators to estimate the pH of the $HCl_{(aq)}$. Calculate the concentration of the $HCl_{(aq)}$ from the pH.
   Design 4: Place an excess of $Zn_{(s)}$ in a measured volume of the $HCl_{(aq)}$ and collect the gas produced at a measured temperature and pressure. Calculate the concentration of the $HCl_{(aq)}$ from the amount of gas using the ideal gas law and the reaction equation.
   Design 5: Place a measured mass of $CaCO_{3(s)}$ in a measured volume of the $HCl_{(aq)}$. After the reaction stops, dry the $CaCO_{3(s)}$ and measure its mass. Calculate the concentration of the $HCl_{(aq)}$ from the reaction equation.
   Design 6: Measure the density of the $HCl_{(aq)}$ and determine the concentration from a graph of density and concentration.

38. The electrical conductivity of equal concentration acids is tested.
   The pH of equal concentration acids is determined using pH paper or a pH meter.
   Indicators are used to determine the pH of equal concentration acid solutions.
   The rate of reaction of equal concentration acid solutions with an active metal is determined.
   The freezing point depression of equal concentration acids is determined.

39. $H_2SO_{4(aq)} + Ba(OH)_{2(aq)} \rightarrow 2\,H_2O_{(l)} + BaSO_{4(s)}$
   Place an accurately measured volume of $Ba(OH)_{2(aq)}$ into a beaker and immerse the leads from a conductivity meter in the solution. Measure and record the initial conductivity reading. Titrate with standardized $H_2SO_{4(aq)}$ and record the conductivity readings. A sudden drop in the conductivity serves as the endpoint of the titration. This drop in conductivity corresponds to the removal of most of the ions through the production of water and insoluble barium sulfate. When the $H_2SO_{4(aq)}$ is added in excess, the conductivity increases due to the ions present in the acid.

40. Scientists usually use the simplest theory that will explain the observation.

Problem 14J

Prediction

According to the single replacement reaction generalization and redox concepts, the products of the reaction between solid aluminum and aqueous copper(II) chloride are solid copper and aqueous aluminum chloride.

$$2\,Al_{(s)} + 3\,CuCl_{2(aq)} \rightarrow 3\,Cu_{(s)} + 2\,AlCl_{3(aq)}$$

Analysis

According to the evidence gathered, the products of the reaction between solid aluminum and aqueous copper(II) chloride are solid copper, gaseous hydrogen, and other substances not detected by diagnostic tests.

Evaluation

Although the experimental design was revised in progress to include diagnostic tests on the gas that was unexpectedly produced, the design should have included tests to determine the identity of the solution components and the orange-brown solid. The prediction was falsified. Although copper appeared to be produced as predicted, the gas produced was not predicted. The redox concepts used to make the prediction are judged to be unacceptable because the prediction was falsified.

Synthesis

An explanation that would be consistent with the evidence is that redox and acid-base reactions both occur. The aluminum and copper(II) chloride single replacement redox reaction may be accompanied by another spontaneous redox reaction between aluminum and hydronium ions. The hydronium ions may be produced by the hydrolysis of aqueous copper(II) ions (i.e., by the acid-base reaction of water with copper(II) ions).

CHAPTER 15

Overview

1. Historically, oxidation referred to reactions involving oxygen, while reduction referred to the reduction in mass of a metal ore when a metal is produced.

2. A redox reaction is the transfer of electrons from a reducing agent to an oxidizing agent.

3. (a) An oxidizing agent accepts electrons from another substance, causing that substance to be oxidized.
   (b) A reducing agent donates electrons to another substance, causing that substance to be reduced.

4. Possible evidence of a spontaneous redox reaction includes the formation of a precipitate or gas, a color or odor change, or an energy change.

5. In a table of redox half-reactions, if the oxidizing agent is listed above the reducing agent, the reaction is predicted to be spontaneous. If the oxidizing agent is listed below the reducing agent, the reaction is predicted to be non-spontaneous.

6. (a) spontaneous   (c) non-spontaneous
   (b) non-spontaneous   (d) spontaneous

7. The term *standardized* means the molar concentration has been determined empirically using a primary standard.

8. (a) 0; (b) –1; (c) +1; (d) +1; (e) –1; (f) –2

9. (a) Oxidation is defined as a loss of electrons involving an increase in oxidation number.
   (b) Reduction is defined as a gain of electrons involving a decrease in oxidation number.
   (c) In a redox reaction, an oxidizing agent gains electrons from a reducing agent. The total decrease in the oxidation number of the atoms/ions in the oxidizing agent is balanced by the total increase in the oxidation number of the atoms/ions in the reducing agent.

10. A net ionic equation is balanced in terms of the numbers of different kinds of atoms or ions and total charge.

11. (a) $2\,Fe^{3+}_{(aq)} + 2\,e^- \rightarrow 2\,Fe^{2+}_{(aq)}$ (reduction)
    $Ni_{(s)} \rightarrow Ni^{2+}_{(aq)} + 2\,e^-$ (oxidation)
    OA is $Fe^{3+}_{(aq)}$; **RA** is $Ni_{(s)}$

    (b) $Br_{2(aq)} + 2\,e^- \rightarrow 2\,Br^-_{(aq)}$ (reduction)
    $2\,I^-_{(aq)} \rightarrow I_{2(s)} + 2\,e^-$ (oxidation)
    OA is $Br_{2(aq)}$; **RA** is $I^-_{(aq)}$

    (c) $Pd^{2+}_{(aq)} + 2\,e^- \rightarrow Pd_{(s)}$ (reduction)
    $Sn^{2+}_{(aq)} \rightarrow Sn^{4+}_{(aq)} + 2\,e^-$ (oxidation)
    OA is $Pd^{2+}_{(aq)}$; **RA** is $Sn^{2+}_{(aq)}$

12. (a) $HClO_{2(aq)} + 2\,H^+_{(aq)} + 2\,e^- \rightarrow HClO_{(aq)} + H_2O_{(l)}$
    reduction
    (b) $Al(OH)_4^-{}_{(aq)} + 3\,e^- \rightarrow Al_{(s)} + 4\,OH^-_{(aq)}$
    reduction
    (c) $Br^-_{(aq)} + 4\,H_2O_{(l)} \rightarrow BrO_4^-{}_{(aq)} + 8\,H^+_{(aq)} + 8\,e^-$
    oxidation
    (d) $2\,ClO^-_{(aq)} + 2\,H_2O_{(l)} + 2\,e^- \rightarrow Cl_{2(aq)} + 4\,OH^-_{(aq)}$
    reduction

13. (a) **OA**   **OA**
    $Cl_{2(aq)}$, $H_2O_{(l)}$
            RA

    (b) **OA**   **OA**
    $Sn^{2+}_{(aq)}$, $NO_3^-{}_{(aq)}$, $H_2O_{(l)}$
    RA   **RA**

    (c) **OA**   **OA**   **OA**   **OA**
    $K^+_{(aq)}$ $H^+_{(aq)}$ $IO_3^-{}_{(aq)}$, $H_2O_{(l)}$
                   RA

14. $Tl^+_{(aq)} + e^- \rightleftharpoons Tl_{(s)}$
    $In^{3+}_{(aq)} + 3\,e^- \rightleftharpoons In_{(s)}$
    $Ga^{3+}_{(aq)} + 3\,e^- \rightleftharpoons Ga_{(s)}$
    $Al^{3+}_{(aq)} + 3\,e^- \rightleftharpoons Al_{(s)}$

15. $Cd^{2+}_{(aq)} + 2\,e^- \rightleftharpoons Cd_{(s)}$
    $Ga^{3+}_{(aq)} + 3\,e^- \rightleftharpoons Ga_{(s)}$
    $Mn^{2+}_{(aq)} + 2\,e^- \rightleftharpoons Mn_{(s)}$
    $Ce^{3+}_{(aq)} + 3\,e^- \rightleftharpoons Ce_{(s)}$

16. According to the kinetic molecular theory, aqueous silver ions are in constant motion. According to collision theory, some of these silver ions collide with atoms of copper. According to redox theory, a competition for electrons results in silver ions removing electrons from the copper. This results in the silver ions being reduced to solid silver and the solid copper being oxidized to aqueous copper(II) ions.

17. (a) **SOA**   **OA**   **OA**   **OA**
    $Cl_{2(g)}$, $Fe^{2+}_{(aq)}$, $SO_4^{2-}_{(aq)}$, $H_2O_{(l)}$
          SRA   **RA**

    $Cl_{2(g)} + 2\,e^- \rightarrow 2\,Cl^-_{(aq)}$
    $2\,[Fe^{2+}_{(aq)} \rightarrow Fe^{3+}_{(aq)} + e^-]$
    ———————————————————
    $Cl_{2(g)} + 2\,Fe^{2+}_{(aq)} \rightarrow 2\,Cl^-_{(aq)} + 2\,Fe^{3+}_{(aq)}$
    spontaneous

(b) OA   SOA   OA   OA

$$Ni^{2+}_{(aq)}, \quad NO_3^-_{(aq)}, \quad Sn^{2+}_{(aq)}, \quad SO_4^{2-}_{(aq)}, \quad H_2O_{(l)}$$

SRA   RA

$$Sn^{2+}_{(aq)} + 2\,e^- \rightarrow Sn_{(s)}$$
$$Sn^{2+}_{(aq)} \rightarrow Sn^{4+}_{(aq)} + 2\,e^-$$
$$\overline{2\,Sn^{2+}_{(aq)} \rightarrow Sn_{(s)} + Sn^{4+}_{(aq)}}$$

non-spontaneous

(c)   OA   SOA

$$Zn_{(s)}, \quad H_2O_{(l)}, \quad O_{2(g)}$$

SRA   RA

$$O_{2(g)} + 2\,H_2O_{(l)} + 4\,e^- \rightarrow 4\,OH^-_{(aq)}$$
$$2\,[Zn_{(s)} \rightarrow Zn^{2+}_{(aq)} + 2\,e^-]$$
$$\overline{O_{2(g)} + 2\,H_2O_{(l)} + 2\,Zn_{(s)} \rightarrow 2\,Zn(OH)_{2(s)}}$$

spontaneous

(d) OA SOA   OA   OA   OA

$$H^+_{(aq)}, \quad SO_4^{2-}_{(aq)}, \quad H_2O_{(l)} \quad Fe_{(s)}, \quad Na^+_{(aq)}$$

RA   SRA

$$SO_4^{2-}_{(aq)} + 4\,H^+_{(aq)} + 2\,e^- \rightarrow H_2SO_{3(aq)} + H_2O_{(l)}$$
$$Fe_{(s)} \rightarrow Fe^{2+}_{(aq)} + 2\,e^-$$
$$\overline{Fe_{(s)} + SO_4^{2-}_{(aq)} + 4\,H^+_{(aq)} \rightarrow}$$
$$Fe^{2+}_{(aq)} + H_2SO_{3(aq)} + H_2O_{(l)}$$

spontaneous

(e) OA   SOA   OA

$$Na^+_{(aq)}, \quad H_2O_{(l)} \quad K^+_{(aq)}, \quad SO_3^{2-}_{(aq)}, \quad OH^-_{(aq)}$$

RA   SRA   RA

$$2\,H_2O_{(l)} + 2\,e^- \rightarrow H_{2(g)} + 2\,OH^-_{(aq)}$$
$$SO_3^{2-}_{(aq)} + 2\,OH^-_{(aq)} \rightarrow SO_4^{2-}_{(aq)} + H_2O_{(l)} + 2\,e^-$$
$$\overline{SO_3^{2-}_{(aq)} + H_2O_{(l)} \rightarrow H_{2(g)} + SO_4^{2-}_{(aq)}}$$

spontaneous

18. (a) $3\,[O_{3(g)} + 2\,H^+_{(aq)} + 2\,e^- \rightarrow O_{2(g)} + H_2O_{(l)}]$
      reduction

$$I^-_{(aq)} + 3\,H_2O_{(l)} \rightarrow IO_3^-_{(aq)} + 6\,H^+_{(aq)} + 6\,e^-$$
oxidation
$$\overline{3\,O_{3(g)} + I^-_{(aq)} \rightarrow 3\,O_{2(g)} + IO_3^-_{(aq)}}$$

(b) $Pt_{(s)} + 6\,Cl^-_{(aq)} \rightarrow PtCl_6^{2-}_{(aq)} + 4\,e^-$
        oxidation

$$4\,[NO_3^-_{(aq)} + 2\,H^+_{(aq)} + e^- \rightarrow NO_{2(g)} + H_2O_{(l)}]$$
reduction
$$\overline{Pt_{(s)} + 6\,Cl^-_{(aq)} + 4\,NO_3^-_{(aq)} + 8\,H^+_{(aq)} \rightarrow}$$
$$PtCl_6^{2-}_{(aq)} + 4\,NO_{2(g)} + 4\,H_2O_{(l)}$$

(c) $2\,[CN^-_{(aq)} + H_2O_{(l)} \rightarrow CNO^-_{(aq)} + 2\,H^+_{(aq)} + 2\,e^-]$
     oxidation

$$ClO_2^-_{(aq)} + 4\,H^+_{(aq)} + 4\,e^- \rightarrow Cl^-_{(aq)} + 2\,H_2O_{(l)}$$
reduction
$$\overline{2\,CN^-_{(aq)} + ClO_2^-_{(aq)} \rightarrow 2\,CNO^-_{(aq)} + Cl^-_{(aq)}}$$

(d) $PH_{3(g)} \rightarrow \frac{1}{4}\,P_{4(s)} + 3\,H^+_{(aq)} + 3\,e^-$
        oxidation

$$CrO_4^{2-}_{(aq)} + 4\,H^+_{(aq)} + 3\,e^- \rightarrow Cr(OH)_4^-_{(aq)}$$
reduction
$$\overline{PH_{3(g)} + CrO_4^{2-}_{(aq)} + H^+_{(aq)} \rightarrow \frac{1}{4}\,P_{4(s)} + Cr(OH)_4^-_{(aq)}}$$
$$PH_{3(g)} + CrO_4^{2-}_{(aq)} + H_2O_{(l)} \rightarrow \frac{1}{4}\,P_{4(s)} + Cr(OH)_4^-_{(aq)} + OH^-_{(aq)}$$

(e) $MnO_4^{2-}_{(aq)} + 8\,H^+_{(aq)} + 4\,e^- \rightarrow Mn^{2+}_{(aq)} + 4\,H_2O_{(l)}$
     reduction

$$4\,[MnO_4^{2-}_{(aq)} \rightarrow MnO_4^-_{(aq)} + e^-]$$
oxidation
$$\overline{5\,MnO_4^{2-}_{(aq)} + 8\,H^+_{(aq)} \rightarrow Mn^{2+}_{(aq)} + 4\,MnO_4^-_{(aq)} + 4\,H_2O_{(l)}}$$

(f) $ClO^-_{(aq)} + H_2O_{(l)} \rightarrow ClO_2^-_{(aq)} + 2\,H^+_{(aq)} + 2\,e^-$
       oxidation

$$2\,ClO^-_{(aq)} + 4\,H^+_{(aq)} + 2\,e^- \rightarrow Cl_{2(g)} + 2\,H_2O_{(l)}$$
reduction
$$\overline{3\,ClO^-_{(aq)} + 2\,H^+_{(aq)} \rightarrow ClO_2^-_{(aq)} + Cl_{2(g)} + H_2O_{(l)}}$$
$$3\,ClO^-_{(aq)} + H_2O_{(l)} \rightarrow ClO_2^-_{(aq)} + Cl_{2(g)} + 2\,OH^-_{(aq)}$$

19. 4.8 g

20. 65.6 mmol/L

21. 56.3 mmol/L

22. (a) $2\,H_2O_{(l)} + 2\,e^- \rightarrow H_{2(g)} + 2\,OH^-_{(aq)}$
       $2\,[K_{(s)} \rightarrow K^+_{(aq)} + e^-]$
$$\overline{2\,K_{(s)} + 2\,H_2O_{(l)} \rightarrow H_{2(g)} + 2\,OH^-_{(aq)} + 2\,K^+_{(aq)}}$$

(b) If a gas is collected and exposed to a flame, and a popping sound is heard, then hydrogen gas was likely produced. If a piece of red litmus paper is placed into the reaction mixture and the litmus paper turns blue, then hydroxide ions were likely produced. If a sample of the final solution is placed into a burner flame and a pale violet color is produced, then potassium ions were likely produced.

23. *Procedure*

(a) Clean three small strips of magnesium with steel wool.

(b) Add a few millilitres of each unknown solution into separate clean test tubes.

(c) Place a strip of magnesium metal into each solution and record evidence of reaction.

(d) For each solution that was unreactive with magnesium, add a few millilitres of each solution to separate clean test tubes.

(e) To each of these test tubes, add a few drops of sodium carbonate solution and record evidence of reaction.

(f) Dispose of all solutions into the waste beaker.

Expected Evidence

• Two solutions showed no change with magnesium and a dark precipitate formed on the metal in the third solution.

• One solution showed no change when aqueous sodium carbonate was added and one solution produced a white precipitate.

24. Sulfur is oxidized in each case. In the first reaction, sulfur is oxidized from –2 in $H_2S_{(g)}$ to +4 in $SO_{2(g)}$. In the second reaction, sulfur is further oxidized to +6 in $SO_{3(g)}$.

25. (a) $\overset{+2}{Ag^{2+}}_{(aq)} + \overset{+1\ -2}{H_2O}_{(l)} \to \overset{+1}{Ag^+}_{(aq)} + \overset{0}{O_2}_{(g)}$

(b) **OA** is $Ag^{2+}_{(aq)}$; **RA** is $H_2O_{(l)}$

(c) $2\,Ag^{2+}_{(aq)} + H_2O_{(l)} \to 2\,Ag^+_{(aq)} + \frac{1}{2}\,O_{2(g)} + 2\,H^+_{(aq)}$

or

$4\,Ag^{2+}_{(aq)} + 2\,H_2O_{(l)} \to 4\,Ag^+_{(aq)} + O_{2(g)} + 4\,H^+_{(aq)}$

26. $14\,Cr_{(s)} + H_2O_{(l)} + 12\,HClO_{4(aq)} \to$
$\qquad\qquad 7\,Cr_2O_7{}^{2-}_{(aq)} + 6\,Cl_{2(g)} + 14\,H^+_{(aq)}$

$Cr_2O_7{}^{2-}_{(aq)} + 14\,H^+_{(aq)} + 6\,Fe^{2+}_{(aq)} \to$
$\qquad\qquad 2\,Cr^{3+}_{(aq)} + 6\,Fe^{3+}_{(aq)} + 7\,H_2O_{(l)}$

$Fe^{2+}_{(aq)} + Ce^{4+}_{(aq)} \to Ce^{3+}_{(aq)} + Fe^{3+}_{(aq)}$

27. (a) $C_6H_{12}O_{6(s)} + 6\,O_{2(g)} \to 6\,CO_{2(g)} + 6\,H_2O_{(l)}$

(b) $2\,Au^{3+}_{(aq)} + 3\,SO_{2(g)} + 6\,H_2O_{(l)} \to$
$\qquad\qquad 3\,SO_4{}^{2-}_{(aq)} + 2\,Au_{(s)} + 12\,H^+_{(aq)}$

(c) $2\,BrO_3{}^-_{(aq)} + C_2H_6O_{(aq)} \to 2\,CO_{2(g)} + 2\,Br^-_{(aq)} + 3\,H_2O_{(l)}$

(d) $3\,Ag_{(s)} + NO_3{}^-_{(aq)} + 4\,H^+_{(aq)} \to$
$\qquad\qquad 3\,Ag^+_{(aq)} + NO_{(g)} + 2\,H_2O_{(l)}$

(e) $2\,HNO_{3(aq)} + 3\,SO_{2(g)} + 2\,H_2O_{(l)} \to$
$\qquad\qquad 3\,H_2SO_{4(aq)} + 2\,NO_{(g)}$

(f) $4\,Zn_{(s)} + BrO_4{}^-_{(aq)} + 4\,H_2O_{(l)} + 8\,OH^-_{(aq)} \to$
$\qquad\qquad 4\,Zn(OH)_4{}^{2-}_{(aq)} + Br^-_{(aq)}$

28. (a) $Zn_{(s)} + Ag_2S_{(s)} \to ZnS_{(s)} + 2\,Ag_{(s)}$

(b) $2\,e^-/Zn \qquad 1\,e^-/Ag$
$\quad\ 2\,e^-/Zn \qquad 2\,e^-/Ag_2S$

(c) $Ag^+_{(s)} + e^- \to Ag_{(s)}$
$\quad\ Zn_{(s)} \to Zn^{2+}_{(s)} + 2\,e^-$

Problem 15K

Analysis

–33°C (This answer is obtained using both the certainty rule and the precision rule. If only the certainty rule is used, the answer obtained is –33.3°C. The first trial of the second titration was omitted, as the difference in volume from the average of the other two trials was greater than 0.2 mL.)

Problem 15L

Analysis

Based on the evidence gathered and on the redox reaction spontaneity generalization, the position of the redox indicators in a table of oxidizing and reducing agents is, in order of decreasing strength of oxidizing agent, $Au^{3+}_{(aq)}$, nitroferrion, $IO_3{}^-_{(aq)} + H^+_{(aq)}$, eriogreen, $Ag^+_{(aq)}$, diphenylamine, methylene blue, $Cu^{2+}_{(aq)}$.

CHAPTER 16

Overview

1. The three essential parts of an electric cell are two electrodes and an electrolyte.

2. Technological problem solving involves a systematic trial-and-error approach to develop a product or process. Scientific problem solving usually involves answering questions to test a scientific concept.

3. Batteries could be made rechargeable (that is, they could be secondary cells), or they could be made so that the fuel can be continuously added (that is, they could be fuel cells).

4. A "dead" cell has a voltage of zero and has reached an equilibrium state.

5. Carbon and platinum are two commonly used inert electrodes.

6. Porous boundaries are provided by a porcelain cup and by a salt bridge containing an inert electrolyte.

7. The components of the hydrogen reference half-cell are a 1.00 mol/L hydrogen ion solution and hydrogen gas at 100 kPa bubbling over a platinum electrode, with all components at 25°C.

8. The corrosion of iron can be prevented or minimized by protective coatings and by cathodic protection, which includes impressed currents and sacrificial anodes.

9. Three technological applications of electrolytic cells are the production of elements, the refining of metals, and the plating of metals onto other objects.

10. Problems might arise because some ionic compounds have a low solubility in water or because water is a stronger oxidizing agent than the active metal cations. If the compound has a low solubility in water, it could be dissolved in an ionic compound that has a low melting point. If the metal cation is less reactive than water, electrolysis could be carried out using the molten compound.

11. Voltaic cells convert chemical energy into electrical energy, while electrolytic cells convert electrical energy into chemical energy.

12. (a) *Voltaic/Electrolytic Cells*

	Anode	Cathode
Half-reaction	oxidation	reduction
Agent reacted	reducing agent	oxidizing agent
Anions	move toward	move away
Cations	move away	move toward
Electrons	move away	move toward

(b)

	Voltaic Cell	*Electrolytic Cell*
Agents in redox table	**SOA** above **SRA**	**SOA** below **SRA**
Cell potential	positive	negative

13. (a) cathode $\quad 2\,H^+_{(aq)} + 2\,e^- \to H_{2(g)}$
anode $\qquad Sn_{(s)} \to Sn^{2+}_{(aq)} + 2\,e^-$
$\Delta E° = 0.00\ V - (-0.14\ V) = +0.14\ V$

(b) cathode $\quad MnO_4{}^-_{(aq)} + 8\,H^+_{(aq)} + 5\,e^- \to$
$\qquad\qquad\qquad\qquad\qquad Mn^{2+}_{(aq)} + 4\,H_2O_{(l)}$
anode $\qquad Ag_{(s)} \to Ag^+_{(aq)} + e^-$
$\Delta E° = 1.51\ V - (+0.80\ V) = +0.71\ V$

(c) cathode $\quad Sn^{2+}_{(aq)} + 2\,e^- \to Sn_{(s)}$
anode $\qquad Zn_{(s)} \to Zn^{2+}_{(aq)} + 2\,e^-$
$\Delta E° = -0.14\ V - (-0.76\ V) = +0.62\ V$

14. (a) +1.49 V
(b) +0.80 V
(c) 0.00 V

15. A standard cell contains two pairs of oxidizing and reducing agents. Therefore, an oxidizing agent will always be listed above a reducing agent in a redox table.

16. (a) +0.26 V
(b) +1.30 V
(c) +1.56 V

17. (a) Lead will be the cathode. Cobalt will be the anode.

(b) entities present $Co^{2+}_{(aq)}$, $Co_{(s)}$, $Pb^{2+}_{(aq)}$, $Pb_{(s)}$, $H_2O_{(l)}$

       OA   SOA   OA

       SRA   RA   RA

cathode $Pb^{2+}_{(aq)} + 2e^- \rightarrow Pb_{(s)}$ $E^°_r = -0.13$ V

anode $Co_{(s)} \rightarrow Co^{2+}_{(aq)} + 2e^-$ $E^°_r = -0.28$ V

net $Pb^{2+}_{(aq)} + Co_{(s)} \rightarrow Pb_{(s)} + Co^{2+}_{(aq)}$

                                    $\Delta E^° = +0.15$ V

(c)

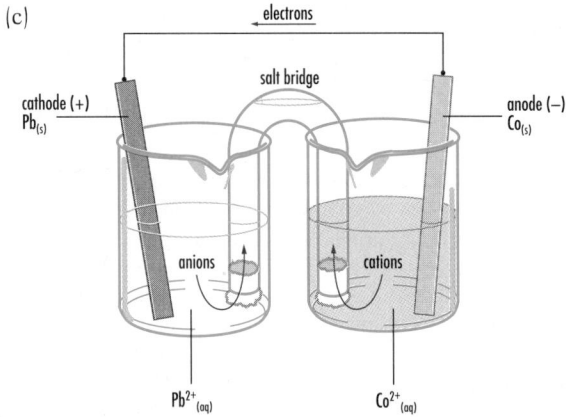

18. $HgO_{(s)} + Zn_{(s)} \rightarrow ZnO_{(s)} + Hg_{(l)}$ $\Delta E^° = +1.35$ V

19. • In terms of appearance from shiniest to dullest, the order is chrome-plated, tin-plated, and galvanized steels.

    • All metal coatings listed are readily oxidized in air to form tough oxide layers.

    • Chromium and zinc are both more easily oxidized than iron and provide the best protection of the steel from corrosion. Tin is less readily oxidized than iron and will accelerate the corrosion of the iron in the steel if a break occurs.

20. $CH_{4(g)} + 2O_{2(g)} + 2OH^-_{(l)} \rightarrow$
                 $CO_3^{2-}_{(l)} + 3H_2O_{(g)}$ $\Delta E^° = +0.23$ V

21. Volta's invention of the electric cell and Humphry Davy's invention of molten salt electrolysis preceded most of modern atomic theory, including redox theory.

22. +1.68 V

23. (a) –0.14 V; the redox pair is probably $Sn^{2+}_{(aq)}/Sn_{(s)}$

    (b) The negative value indicates that tin(II) ions attract electrons less strongly than hydrogen ions do; in other words, tin(II) ions are less easily reduced.

24. (a) and (d)

25. (a) cathode $2[Zn^{2+}_{(aq)} + 2e^- \rightarrow Zn_{(s)}]$

     anode $2H_2O_{(l)} \rightarrow O_{2(g)} + 4H^+_{(aq)} + 4e^-$

     net $2Zn^{2+}_{(aq)} + 2H_2O_{(l)} \rightarrow$
                               $2Zn_{(s)} + O_{2(g)} + 4H^+_{(aq)}$

    (b) 1.99 V

26. cathode $2[Sr^{2+}_{(l)} + 2e^- \rightarrow Sr_{(l)}]$

    anode $2O^{2-}_{(l)} \rightarrow O_{2(g)} + 4e^-$

    net $2Sr^{2+}_{(l)} + 2O^{2-}_{(l)} \rightarrow 2Sr_{(l)} + O_{2(g)}$

27. 57.4 min

28. 45 kA

Problem 16E

Prediction

According to redox concepts and the table of redox half-reactions, a standard lead-dichromate cell has a cell potential of 1.36 V at SATP and the following reactions and components.

cathode $Cr_2O_7^{2-}_{(aq)} + 14H^+_{(aq)} + 6e^- \rightarrow 2Cr^{3+}_{(aq)} + 7H_2O_{(l)}$

anode $3[Pb_{(s)} \rightarrow Pb^{2+}_{(aq)} + 2e^-]$

net $Cr_2O_7^{2-}_{(aq)} + 14H^+_{(aq)} + 3Pb_{(s)} \rightarrow$
           $2Cr^{3+}_{(aq)} + 7H_2O_{(l)} + 3Pb^{2+}_{(aq)}$ $\Delta E^° = +1.36$ V

Materials

lab apron, safety glasses, lead strip, graphite rod, 1.0 mol/L lead(II) nitrate solution, 1.0 mol/L potassium dichromate solution, two beakers, salt bridge (or porous cup), voltmeter, connecting wires

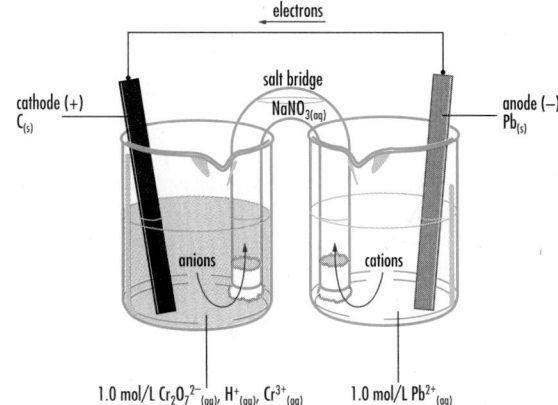

Experimental Design (Diagnostic Tests)

• If a voltmeter is connected to the electrodes (red – $C_{(s)}$, black – $Pb_{(s)}$), and a positive potential of 1.36 V is measured, then carbon is the cathode, lead is the anode, and the half-reactions listed are probably correct.

• If the electrodes of the cell are connected with a wire and an ammeter (red – $C_{(s)}$, black – $Pb_{(s)}$), and a positive current is measured, then the electron flow is from the lead to the carbon electrode.

• If the pH of the dichromate half-cell is measured while the cell is connected with a wire, and the pH increases, then the hydrogen ion concentration is decreasing according to the dichromate half-reaction.

• If the concentration of the lead(II) ions in the lead half-cell is analyzed by precipitation with sodium sulfate, and the concentration is higher than 1.0 mol/L, then lead is undergoing oxidation in this half-cell.

• If a sample of the solution from the dichromate half-cell is analyzed for lead(II) ions by precipitation with sodium sulfate, and a precipitate is observed, then lead(II) cations have moved toward the cathode.

APPENDIX A

Problem 16F

Prediction

According to redox concepts and the table of redox half-reactions, the voltaic cell with the highest cell potential is

$$C_{(s)} \mid Cu^{2+}_{(aq)}, SO_4^{2-}_{(aq)} \parallel Na^+_{(aq)}, Cl^-_{(aq)} \mid Al_{(s)} \qquad \Delta E° = 2.00 \text{ V}$$

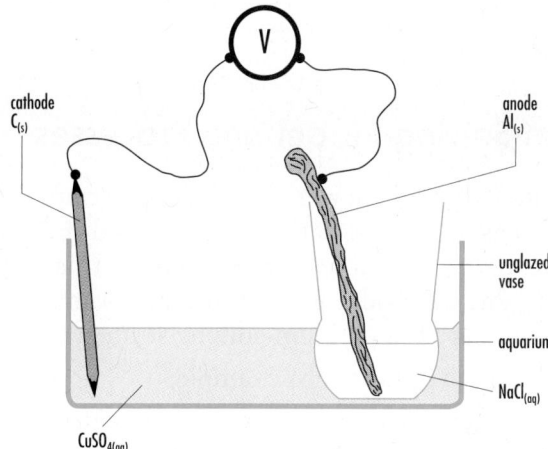

cathode
$C_{(s)}$

anode
$Al_{(s)}$

unglazed
vase

aquarium

$NaCl_{(aq)}$

$CuSO_{4(aq)}$

APPENDIX B

Scientific Problem Solving

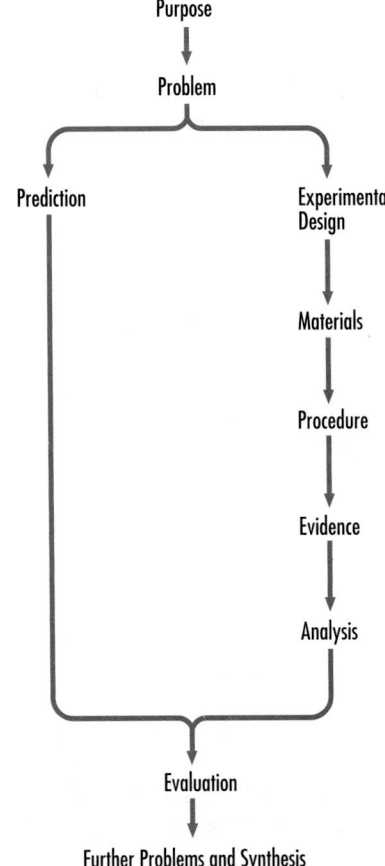

Purpose

↓

Problem

↓

Prediction ← → Experimental Design

↓

Materials

↓

Procedure

↓

Evidence

↓

Analysis

↓

Evaluation

↓

Further Problems and Synthesis

Figure B1
A scientific problem-solving model helps to guide your laboratory work, but does not illustrate the complexity of the work.

✔ Problem
✔ Prediction
✔ Design
✔ Materials
✔ Procedure
✔ Evidence
✔ Analysis
✔ Evaluation
✔ Synthesis

B.1 Scientific Problem-Solving Model and Processes

Scientists ask questions and seek the answers to these questions by applying consistent, logical reasoning to describe, explain, and predict observations, and by doing experiments to test predictions. In this way science progresses using a general model for solving problems and employing specific processes as part of a problem-solving strategy.

Every investigation in science has a *purpose*; for example,

- to develop a scientific concept (a theory, law, generalization, or definition);
- to test a scientific concept;
- to perform a chemical analysis;
- to determine a scientific constant; or
- to test an experimental design, a procedure, or a skill.

Once you know the purpose, you need a problem and a general design. For example, if the purpose is to perform a chemical analysis to determine the quantity of a substance, then possible designs include distillation and precipitation. Once you choose a design, there are many specific questions that you might ask, many possible reactants you might choose, and many other variables you might need to consider.

B.2 Investigation Report

An investigation report is the final result of your problem solving. Your report should follow the model outlined in Figure B1. As a further guide, use the information and instructions for the specific processes listed below. The parts of the investigation report that you are to provide are indicated in the text in a checklist like the one shown below, in the margin.

Purpose

Although this is usually provided, you will be expected to identify the purpose of an investigation before, during, and after your laboratory work.

Problem

The Problem is a specific question to be answered in the investigation. If appropriate, you should state the question in terms of manipulated and responding variables. In most cases the problem is chosen for you.

Prediction

The Prediction is the expected answer to the Problem according to a scientific concept (for example, a theory, law, or generalization) or another authority (for example, a reference source or a label on a bottle). Write your Prediction using the format, "According to [an authority], [answer to the Problem]." Include your qualitative and quantitative reasoning with the Prediction.

Experimental Design

The Experimental Design is a specific plan to answer the Problem, including reacting chemicals and, if applicable, brief descriptions of diagnostic tests, variables, and controls. Write your Experimental Design as a paragraph of one to three sentences.

Materials

This section consists of a complete list of all equipment and chemicals, including sizes and quantities.

Appendix C, page 733, shows laboratory equipment, including common sizes.

Procedure

The Procedure is a detailed set of instructions designed to obtain the evidence needed to answer the Problem. Write a list of numbered steps in the correct sequence, including any waste disposal and safety instructions. You should always wear safety glasses and a lab apron and wash your hands before leaving the laboratory. Sometimes disposable plastic gloves are recommended. In procedures that are written in the textbook, these safety measures are indicated by

 (put on safety glasses)

 (wear a lab apron)

 (wash hands thoroughly)

 (wear disposable plastic gloves)

Evidence

The Evidence includes all qualitative and quantitative observations relevant to answering the Problem. Organize your evidence in tables whenever possible (page 758). Be as precise as possible in your measurements and include any unexpected observations that may affect your answer and its certainty.

It is part of scientific honesty to report all evidence collected and not just the evidence you think is good or "normal."

Analysis

The Analysis includes calculations and interpretations based on the evidence. You may need to differentiate between relevant and irrelevant data. Communicate your work clearly and logically. Conclude the Analysis with a statement of your experimental answer to the Problem, including a phrase such as, "According to the evidence gathered in this experiment, [answer]."

Evaluation

The Evaluation includes your judgment of the processes used to plan and perform the investigation in the laboratory, and of the prediction and the authority used to make the prediction. Write your Evaluation in paragraph form, using the topic sentences suggested below or an adaptation of them. Some of the more important criteria for a judgment are listed as questions; use selected questions to guide your judgments.

This evaluation outline assumes that a prediction is being tested in an experiment. If it is the experimental design that is being tested, then the two sections of the evaluation would be reversed and the percent difference is used to judge the success of the design.

Show as much independent, critical, and creative thought as possible in support of your judgment.

1. Evaluation of the Experiment

See page 282 for more detailed examples of criteria used to evaluate experimental designs.

- "The experimental design [name or describe in a few words] is judged to be adequate/ inadequate because …"

    Were you able to answer the Problem using the chosen experimental design? Are there any obvious flaws in the design? What alternative designs (better or worse) are available? As far as you know, is this design the best available in terms of controls, efficiency, and cost? How great is your confidence in the chosen design?

- "The procedure is judged to be adequate/inadequate because …"

    Were the steps that you used in the laboratory correctly sequenced, and adequate to gather sufficient evidence? What improvements could be made to the procedure? What steps, if not done correctly, would have significantly affected the results?

Common sources of experimental uncertainty are:
- any measurement process
- manipulative skill
- impure reactants or products
- incomplete reaction
- incomplete drying of a product
- loss of solid in a filtration (stuck to glass or passed through filter)
- conditions (e.g., SATP) not controlled
- judgment of color (e.g., indicator)

- "The technological skills are judged to be adequate/inadequate because …"

    Which specialized skills, if any, might have the greatest effect on the experimental results? Was the evidence from repeated trials reasonably similar? Can the measurements be made more precise?

- "Based upon my evaluation of the experiment, I am not/moderately/very certain of my experimental results. The major sources of uncertainty or error are …"

    Do you have sufficient confidence in your experimental results to proceed with your evaluation of the authority being tested? What would be a reasonably acceptable percent difference for this experiment (1%, 5%, or 10%)?

2. Evaluation of the Prediction and Authority Being Tested

In most experiments, a percent difference is a measure of accuracy. In some experiments in which a product is collected and measured, a percent yield is used instead of a percent difference.

$$\% \text{ yield} = \frac{\text{actual quantity obtained}}{\text{predicted (maximum) quantity}} \times 100$$

- "The percent difference between the experimental result and the predicted value is…"

    How does this difference compare with your estimated total uncertainty?

$$\% \text{ difference} = \frac{|\text{ experimental value} - \text{ predicted value }|}{|\text{ predicted value }|} \times 100$$

- "The prediction is judged to be verified/inconclusive/falsified because …"

    Does the predicted answer clearly agree with the experimental answer in your analysis? Can the percent difference be accounted for by the sources of uncertainty listed above?

An authority may be judged unacceptable in one experiment. This does not mean the authority is immediately discarded. Replication by independent workers is always required to refute any accepted theory.

- "The authority being tested [name the authority] is judged to be acceptable/unacceptable because …"

    Was the prediction verified, inconclusive, or falsified? How confident do you feel about your judgment? Is there a need to restrict, reverse, or replace the authority being tested?

Synthesis

Synthesis is the process of creating or integrating ideas. The synthesis process may be completed as a discussion in the classroom and is usually not included in the written report. Often a series of problems are investigated before a synthesis is attempted. Synthesis could include answers to any of the following questions: What restrictions, revisions, or replacements would you recommend in the experimental design, procedure, and skills before doing this experiment again? How would you restrict, revise, or replace the authority used to make the prediction? Does the concept have descriptive and explanatory power as well as predictive power? What scientific concept can be created to describe the results from the experiment? How would you combine the results of this experiment with other knowledge to produce a more unified concept or experimental design?

B.3 Sample Investigation Report: The Reaction of Hydrochloric Acid with Zinc

The purpose of this investigation is to test one of the ideas of the collision-reaction theory.

> ▢ Problem
> ✔ Prediction
> ✔ Design
> ✔ Materials
> ✔ Procedure
> ✔ Evidence
> ✔ Analysis
> ✔ Evaluation
> ▢ Synthesis

Problem

How does changing the concentration of hydrochloric acid affect the time required for the reaction of hydrochloric acid with a fixed quantity of zinc?

Prediction

According to the collision-reaction theory, if the concentration of hydrochloric acid is increased, then the time required for the reaction with zinc will decrease. The reasoning that supports the prediction is that a higher concentration produces more collisions per second between the hydrochloric acid particles and the zinc atoms. More collisions per second would produce more reactions per second and therefore a shorter time required to consume the zinc.

Experimental Design

Different known concentrations of excess hydrochloric acid react with zinc metal. The time for the zinc to completely react is measured for each concentration of acid solution. Variables are:

- manipulated: concentration of hydrochloric acid
- responding: time for the zinc to be consumed
- controlled: temperature of solution, quantity of zinc, surface area of zinc in contact with acid, volume of acid

Materials

lab apron
safety glasses
(4) 10 mL graduated cylinders
(4) 18 × 150 mm test tubes and test-tube rack
clock or watch (precise to the nearest second)
four pieces of a zinc metal strip (5 mm × 5 mm)
stock solutions of $HCl_{(aq)}$: 2.0 mol/L, 1.5 mol/L, 1.0 mol/L, 0.5 mol/L
a solution of a weak base

Procedure

1. Transfer 10 mL of 2.0 mol/L $HCl_{(aq)}$ into an 18×150 mm test tube.
2. Carefully place a piece of $Zn_{(s)}$ into the hydrochloric acid solution and note the starting time of the reaction.
3. Measure and record the time required for all of the zinc to react.
4. Repeat steps 1 to 3 using 1.5 mol/L, 1.0 mol/L, and 0.5 mol/L $HCl_{(aq)}$.
5. Neutralize the acid with a solution of a weak base such as baking soda, then pour it down the sink with large amounts of water.

Evidence

Gas bubbles formed immediately on the surface of the zinc strip when it was placed into the hydrochloric acid solution. The bubbles appeared to form more rapidly when the concentration of the acid was higher.

THE EFFECT OF CONCENTRATION ON REACTION TIME	
Concentration of $HCl_{(aq)}$ (mol/L)	Time for Reaction (s)
2.0	70
1.5	80
1.0	144
0.5	258

Analysis

Figure B2
This graph is part of the Analysis of the sample investigation report.

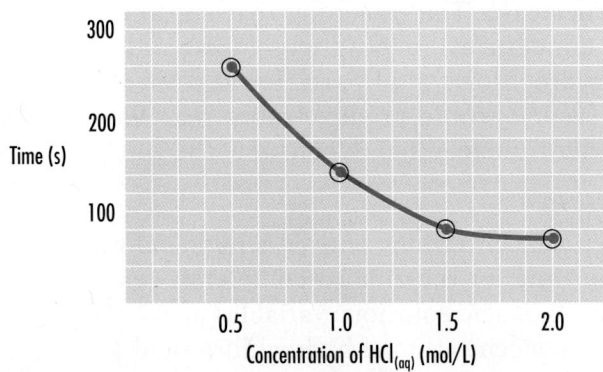

According to the evidence obtained, increasing the concentration of hydrochloric acid decreases the time required for the complete reaction of a fixed quantity of zinc.

Evaluation

The experimental design, reacting zinc with excess hydrochloric acid, is judged to be adequate because this experiment produced the type of evidence needed to answer the problem with a high degree of certainty. An experimental design involving reacting gases would be more difficult to set up and would require more sophisticated and expensive equipment. In my judgment this is a good design — the design is efficient and inexpensive, and all necessary variables are controlled. I am very confident in the results obtained using this experimental design.

The procedure is also judged to be adequate since the steps used in the laboratory are simple and straightforward. Possible improvements to the procedure would include extending the range of concentrations and performing more than one trial for each concentration. Perhaps the reaction should have been done in a container that allowed mixing. With the method used, picking up one of the test tubes or agitating the solution could have affected the results.

The technological skills of the experimenter are judged adequate because the evidence provided a graph where the points formed a distinct pattern with little deviation. The graph of the evidence was virtually identical with the graphs drawn by other groups. This consistency indicates that the technological skills are reproducible and therefore, it is assumed, adequate. Measuring the volume of solution and operating the timer were the procedures requiring the most skill, and these could probably not be improved significantly.

Based upon my evaluation of the experiment, however, I am very certain about the experimental results. Sources of uncertainty in this investigation include the purity of the zinc metal strip, the concentration of the acid, and a little uncertainty in estimating when the last bit of zinc had reacted. The small deviations are consistent with the use of a clock rather than a stopwatch, a graduated cylinder rather than a pipet, and solution concentrations measured to a certainty of only two significant digits. Due to the nature of the experiment, I cannot calculate a percent difference as an estimate of the accuracy of the prediction.

The prediction based on the collision-reaction theory is verified because the qualitative observations and the graph clearly indicate that the reaction time decreases as the concentration increases. There is little deviation from a smooth curve in the graphed results.

The collision-reaction theory is judged to be acceptable because the prediction was verified and because other groups in the class obtained similar results. Although the design called for only one reaction — the reaction of a solution with metal in a single replacement reaction — I feel confident in the ability of the collision-reaction theory to predict the rate of a chemical reaction.

(This is the end of the report requested, as indicated by the checkboxes on page 729.)

Further Problems

(Occasionally you may be asked to suggest Further Problems in an investigation, so an example has been given here.)

Additional investigations studying the effect of concentration on reaction rate using different reactants and reaction types should be conducted. To extend the testing of the collision-reaction theory, problems involving the effect of temperature and surface area on the reaction rate should be investigated.

Synthesis

(In this investigation report, no Synthesis is required.)

APPENDIX C

Technological Problem Solving

The goal of technological problem solving is to develop or revise a product or a process. The product or the process must fulfill its function, but it is not essential to understand why or how it works. Products are evaluated based on criteria such as simplicity, reliability, and cost. Technological processes are evaluated by their efficiency. Ecological and political perspectives are also essential in the assessment of technological products. For example, chlorofluorocarbons may be simple and inexpensive to make, and they may be useful for a particular function, but their effect on the ozone layer in the upper atmosphere must also be considered. Processes such as the chlorine bleaching of wood pulp may be efficient, but they may adversely affect an ecosystem.

Chemistry has always been closely associated with technology. Part of technology is the laboratory equipment, processes, and procedures used in both chemical and technological research and development. In modern chemistry, simple equipment and processes, such as beakers and filtration, are still used but chemistry also depends on sophisticated technology, such as computers, to store and manipulate the evidence collected.

C.1 Model of Technological Problem Solving

A characteristic of technological problem solving is a systematic, trial-and-error manipulation of variables (Figure C1 and page 656). Variables are predicted and tested and the results are evaluated. When the cycle is repeated many times the most effective set of conditions can be determined. Compare this model with the scientific problem-solving model in Figure B1, page 726.

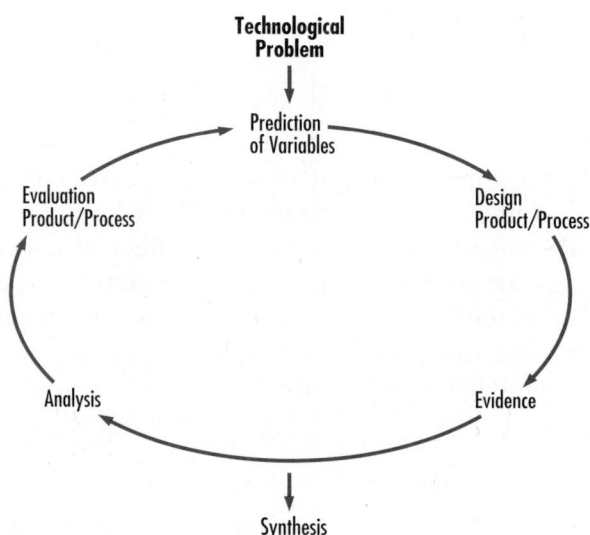

Figure C1
A technological problem-solving model.

C.2 Laboratory Equipment

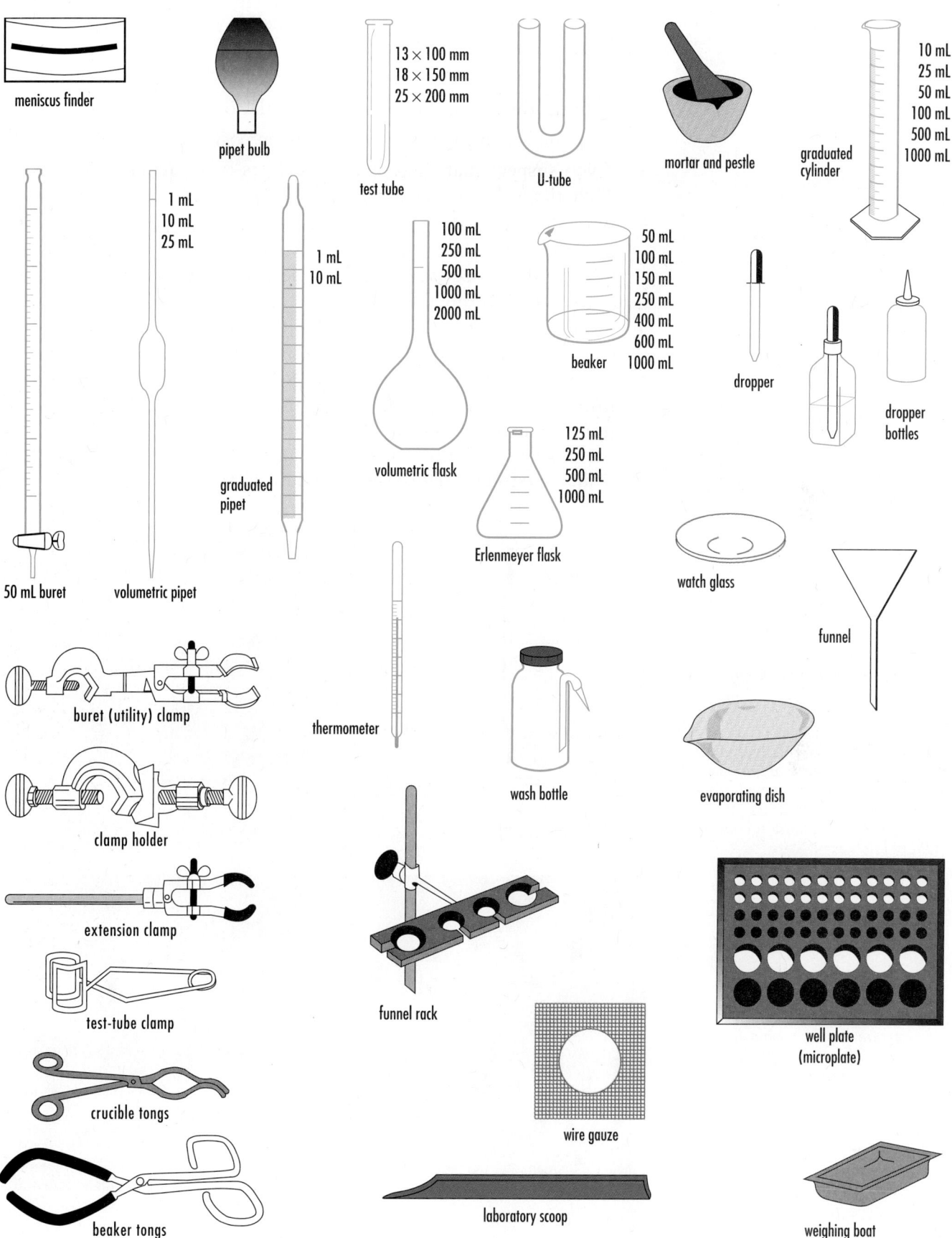

meniscus finder

pipet bulb

test tube

13 × 100 mm
18 × 150 mm
25 × 200 mm

U-tube

mortar and pestle

graduated cylinder

10 mL
25 mL
50 mL
100 mL
500 mL
1000 mL

1 mL
10 mL
25 mL

graduated pipet

1 mL
10 mL

volumetric flask

100 mL
250 mL
500 mL
1000 mL
2000 mL

beaker

50 mL
100 mL
150 mL
250 mL
400 mL
600 mL
1000 mL

dropper

dropper bottles

50 mL buret

volumetric pipet

Erlenmeyer flask

125 mL
250 mL
500 mL
1000 mL

watch glass

funnel

buret (utility) clamp

clamp holder

thermometer

wash bottle

evaporating dish

extension clamp

test-tube clamp

funnel rack

well plate (microplate)

crucible tongs

wire gauze

beaker tongs

laboratory scoop

weighing boat

APPENDIX C

Using a Laboratory Burner

The procedure outlined below should be practiced and memorized. Note the safety caution. You are responsible for your safety and the safety of others near you.

1. Turn the air and gas adjustments to the off position (Figure C2).
2. Connect the burner hose to the gas outlet on the bench.
3. Turn the bench gas valve to the fully on position.
4. If you suspect that there may be any gas leaks, replace the burner. (Give the leaky burner to your teacher.)
5. While holding a lit match above and to one side of the barrel, open the burner gas valve until a small yellow flame results (Figure C3). If a striker is used instead of matches, generate sparks over the top of the barrel (Figure C4).
6. Adjust the air flow and obtain a pale blue flame with a dual cone (Figure C5). In most common types of laboratory burners, rotating the barrel adjusts the air intake. Rotate the barrel slowly. If too much air is added, the flame may go out. If this happens, immediately turn the gas flow off and relight the burner following the procedure outlined above. If your burners have a different kind of air adjustment, revise the procedure accordingly.

APPENDIX C

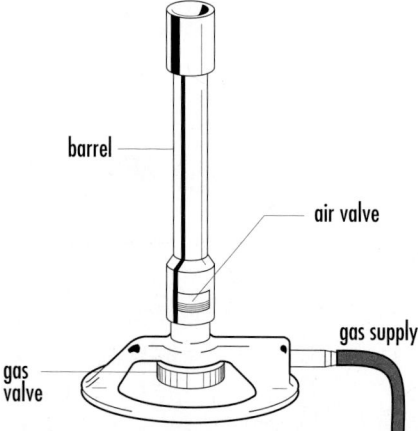

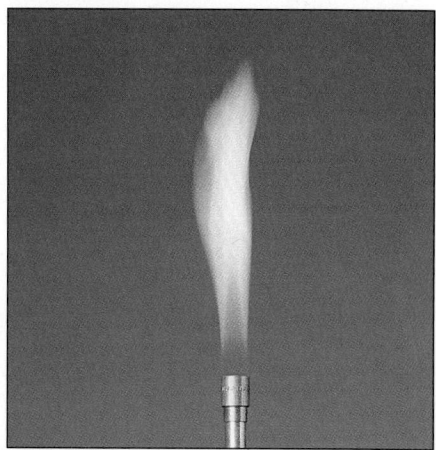

Figure C2
The parts of a common laboratory burner.

Figure C3
A yellow flame is a relatively cool flame and is easier to obtain than a blue flame when lighting a burner. A yellow flame is not used for heating objects because it contains a lot of black soot.

Figure C4
To generate a spark with a striker, pull up and across on the side of the handle containing the flint.

Figure C5
A pale, almost invisible flame is much hotter than a yellow flame. The hottest point is at the tip of the inner blue cone.

7. Adjust the gas valve on the burner to increase or decrease the height of the blue flame. The hottest part of the flame is the tip of the inner blue cone. Usually a 5 to 10 cm flame, which just about touches the object heated, is used.

8. Laboratory burners, when lit, should not be left unattended. If the burner is on but not being used, adjust the air and gas intakes to obtain a small yellow flame. This flame is more visible and therefore less likely to cause problems.

Using a Laboratory Balance

A balance is a sensitive instrument used to measure the mass of an object. There are two types of balances — electronic (Figure C6) and mechanical (Figure C7). All balances must be handled carefully and kept clean. Always place chemicals into a container such as a beaker or plastic boat to avoid contamination and corrosion of the balance pan. To avoid error due to convection currents in the air, allow hot or cold samples to return to room temperature before placing them on the balance. Always record masses showing the correct precision. On a centigram balance, mass is measured to the nearest hundredth of a gram (0.01 g). When it is necessary to move a balance, hold the instrument by the base and steady the beam. Never lift a balance by the beams or pans.

To avoid contaminating a whole bottle of reagent, a scoop should not be placed in the original container of a chemical. A quantity of the chemical should be poured out of the original reagent bottle into a clean, dry beaker or bottle, from which samples can be taken. Another acceptable technique for dispensing a small quantity of chemical is to rotate or tap the chemical bottle.

Using an Electronic Balance

Electronic balances are sensitive instruments requiring care in their use. Be gentle when placing objects on the pan, and remove the pan when cleaning it. Electronic balances are sensitive to small movements and changes in level; do not lean on the counter when using the balance.

1. Place a container or weighing paper on the balance.

2. Reset (tare) the balance so the mass of the container registers as zero.

3. Add chemical until the desired mass of chemical is displayed. The last digit may not be constant, indicating uncertainty due to air currents or the high sensitivity of the balance.

4. Remove the container and sample.

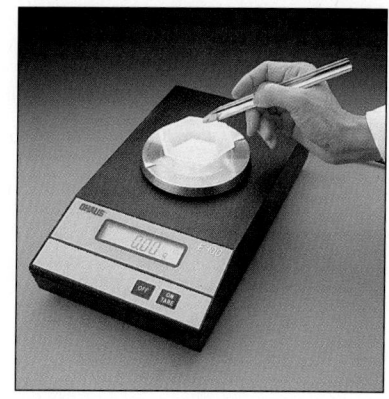

Figure C6
An electronic balance.

Using a Mechanical Balance

Different kinds of mechanical balances are shown in Figures C7 (a) and (b). Some general procedures apply to most of them.

1. Clean and zero the balance. (Turn the zero adjustment screw so that the beam is balanced when the instrument is set to read 0 g and no load is on the pan.)

2. Place the container on the pan.

3. Move the largest beam mass one notch at a time until the beam drops, then move the mass back one notch.

4. Repeat this process with the next smaller mass and continue until all masses have been moved and the beam is balanced. If you are using a dial type balance, the final step will be to turn the dial until the beam balances, as shown in Figure C7 (c).

5. Record the mass of the container.

6. Set the masses on the beams to correspond to the total mass of the container plus the desired sample.

7. Add the chemical until the beam is once again balanced.

8. Remove the sample from the pan and return all beam masses to the zero position. (For a dial type balance, return the dial to the zero position.)

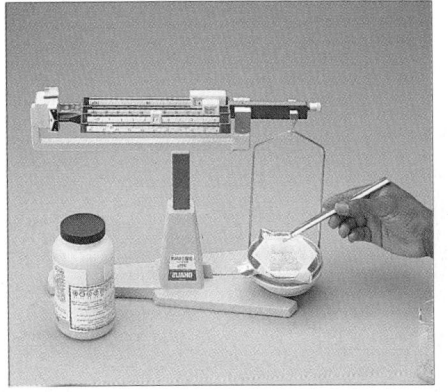
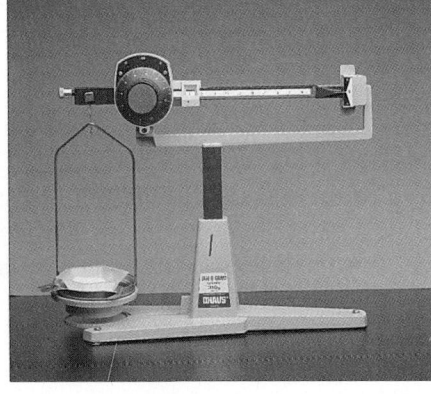
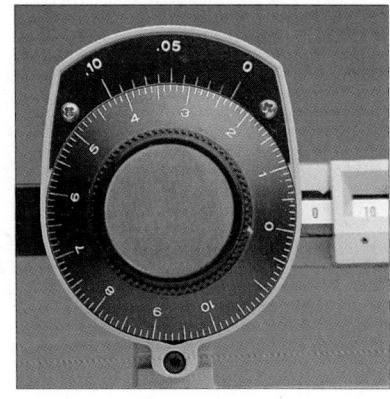

Figure C7
(a) On this type of mechanical balance the sample is balanced by moving masses on several beams.

(b) Another type of mechanical balance has beams for the larger masses and a dial for the final adjustment.

(c) The dial reading on this balance with a vernier scale is 2.34 g. To read the hundredth of a gram, look below the zero on the vernier and then look for the line on the vernier that lines up best with a line on the dial.

Using a Multimeter

A multimeter (Figure C8) is a device that measures a variety of electrical quantities, such as resistance, voltage, and current.

Conductivity Measurements of Solutions

1. Set the dial on the meter to one of the higher values on the ohm (Ω) scale, for example, $R \times 100$ or $R \times 1K$.

2. Touch the two metal probes together to check the battery. If the needle does not deflect significantly (more than one-half scale), have your teacher adjust the meter or replace the battery.

3. Test a sample of pure water as a control and note the movement of the needle.

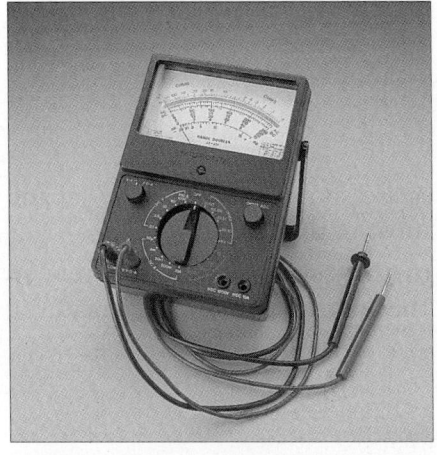
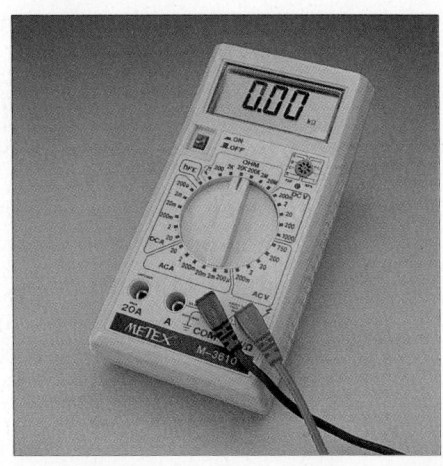

Figure C8
(a) An analog meter has a needle that moves in front of a labelled scale.

(b) A digital meter gives a direct reading with appropriate units.

4. Test your aqueous sample and record the deflection of the needle according to your teacher's instructions.

5. Rinse the probes with pure water before testing another sample.

6. Shut off the meter using either the on/off switch or turn the dial to any setting other than the resistance (ohm) scales.

Voltage Measurements of Cells and Batteries

1. Set the dial to the appropriate value on the direct current volts (DCV) scale; for example, 3 V.

2. The black lead (labelled negative or COM) is normally connected to the anode and the red lead (positive) is connected to the cathode of a voltaic cell.

3. Make a firm contact between each metal probe and an electrode of the cell. (Press firmly with the pointed probe or use leads with an alligator clip.)

4. On analog meters (those with a needle), be sure to read the scale corresponding to the meter value you set in step 1.

5. If the needle attempts to move to the left off the scale or a digital meter registers a negative number, then switch the connections to the cell.

Using a Pipet

A pipet is a specially designed glass tube used to measure precise volumes of liquids. There are two types of pipets and a variety of sizes for each type. A *volumetric pipet* (Figure C9) transfers a fixed volume, such as 10.00 mL or 25.00 mL, accurate to within 0.04 mL. A *graduated pipet* (Figure C10) measures a range of volumes within the limit of the scale, just as a graduated cylinder does. A 10 mL graduated pipet delivers volumes accurate to within 0.1 mL.

Figure C9
A volumetric pipet delivers the volume printed on the label if the temperature is near room temperature.

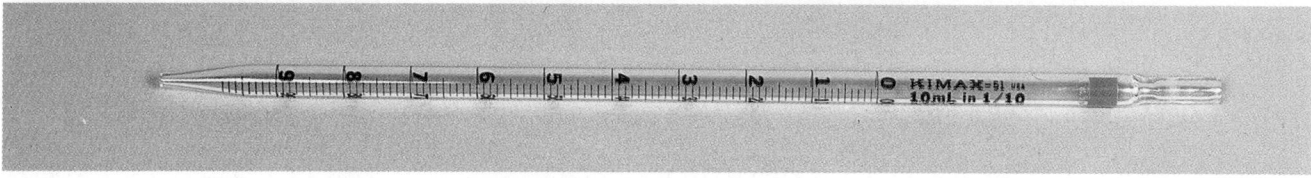

1. Rinse the pipet with small volumes of distilled water using a wash bottle, then with the sample solution. A clean pipet has no visible residue or liquid drops clinging to the inside wall. Rinsing with aqueous ammonia and scrubbing with a pipe cleaner might be necessary to clean the pipet.

2. Hold the pipet with your thumb and fingers near the top. Leave your index finger free.

Figure C10
To use a graduated pipet, you must be able to start and stop the flow of the liquid.

Figure C11
Release the bulb slowly. Pressing down with your thumb placed across the top of the bulb maintains a good seal. Setting the pipet tip on the bottom slows the rise or fall of the liquid.

Figure C12
To allow the liquid to drop slowly to the calibration line, it is necessary for your finger and the pipet top to be dry. Also keep the tip on the bottom to slow down the flow.

Figure C13
A special pipet bulb that can be used to dispense solution.

Figure C14
A vertical volumetric pipet is drained by gravity and then the tip is placed against the inside wall of the container. A small volume is expected to remain in the tip.

3. Place the pipet in the sample solution, resting the tip on the bottom of the container if possible. Be careful that the tip does not hit the sides of the container.

4. Squeeze the bulb into the palm of your hand and place the bulb firmly and squarely on the end of the pipet (Figure C11) with your thumb across the top of the bulb.

5. Release your grip on the bulb until the liquid has risen above the calibration line. (This may require bringing the level up in stages: remove the bulb, put your finger on the pipet, squeeze the air out of the bulb, re-place the bulb, and continue the procedure.)

6. Remove the bulb, placing your index finger over the top. A dispensing bulb remains attached to the pipet (Figure C13).

7. Wipe all solution from the outside of the pipet using a paper towel.

8. While touching the tip of the pipet to the inside of a waste beaker, gently roll your index finger (or rotate the pipet between your thumb and fingers) or squeeze the valve of the dispensing bulb to allow the liquid level to drop until the bottom of the meniscus reaches the calibration line (Figure C12). To avoid parallax errors, set the meniscus at eye level. Stop the flow when the bottom of the meniscus is on the calibration line. Use the bulb to raise the level of the liquid again if necessary.

9. While holding the pipet vertically, touch the pipet tip to the inside wall of a clean receiving container. Remove your finger or adjust the valve and allow the liquid to drain freely until the solution stops flowing.

10. Finish by touching the pipet tip to the inside of the container held at about a 45° angle (Figure C14). Do not shake the pipet. The delivery pipet is calibrated to leave a small volume in the tip.

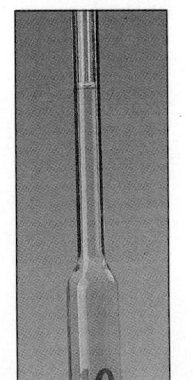

C.3 Laboratory Processes

The processes or experimental procedures listed below are part of common designs used in scientific or technological laboratories.

Crystallization

Crystallization is used to separate a solid from a solution by evaporating the solvent or lowering the temperature. Evaporating the solvent is useful for quantitative analysis of a binary solution; lowering the temperature is commonly used to purify and separate a solid whose solubility is temperature-sensitive. Chemicals that have a low boiling point or decompose on heating cannot be separated by crystallization using a heat source. Fractional distillation is an alternative design for the separation of a mixture of liquids.

1. Measure the mass of a clean beaker or evaporating dish.

2. Place a precisely measured volume of the solution in the container.

3. Set the container aside to evaporate the solution slowly, or warm the container gently on a hot plate or with a laboratory burner.

4. When the contents appear dry, measure the mass of the container and solid (Figure C15).

5. Heat the solid with a hot plate or burner, cool it, and measure the mass again.

6. Repeat step 5 until the final mass remains constant. (Constant mass indicates that all of the solvent has evaporated.)

Figure C15
When the substance has crystallized it may appear dry but small quantities of water may still be present. To be certain the solid is dry, it must be heated until the mass becomes constant.

Filtration

In filtration, solid is separated from a mixture using a porous filter paper. The more porous papers are called qualitative filter papers. Quantitative filter papers allow only invisibly small particles through the pores of the paper.

1. Set up a filtration apparatus (Figure C16): stand, funnel holder, filter funnel, waste beaker, wash bottle, and a stirring rod with a flat plastic or rubber end for scraping.

2. Fold the filter paper along its diameter and then fold it again to form a cone. A better seal of the filter paper on the funnel is obtained if a small piece of the outside corner of the filter paper is torn off (Figure C17).

3. Measure and record the mass of the filter paper after removing the corner.

Figure C16
The tip of the funnel should touch the inside wall of the collecting beaker.

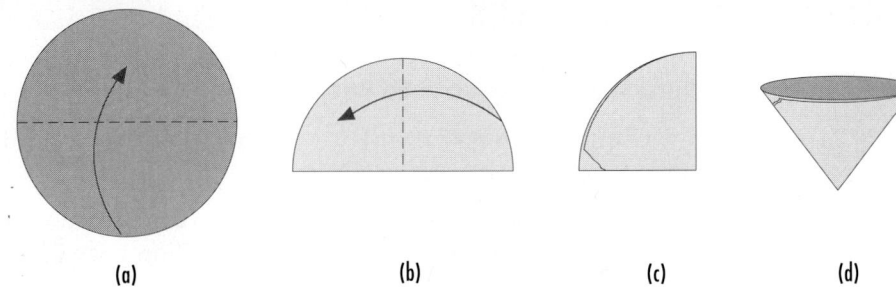

| (a) | (b) | (c) | (d) |

Figure C17
To prepare a filter paper, fold it in half twice and then remove the outside corner as shown.

4. While holding the open filter paper in the funnel, wet the entire paper and seal the top edge firmly against the funnel with the tip of the cone centered in the bottom of the funnel.

5. With the stirring rod touching the spout of the beaker, decant most of the solution into the funnel (Figure C18). Transferring the solid too soon clogs the pores of the filter paper. Keep the level of liquid about two-thirds up the height of the filter paper. The stirring rod should be rinsed each time it is removed.

6. When most of the solution has been filtered, pour the remaining solid and solution into the funnel. Use the wash bottle and the flat end of the stirring rod to clean any remaining solid from the beaker.

7. Use the wash bottle to rinse the stirring rod and the beaker.

8. Wash the solid two or three times to ensure that no solution is left in the filter paper. Direct a gentle stream of water around the top of the filter paper.

9. When the filtrate has stopped dripping from the funnel, remove the filter paper. Press your thumb against the thick (three-fold) side of the filter paper and slide the paper up the inside of the funnel.

10. Transfer the filter paper from the funnel onto a labelled watch glass and unfold the paper to let the precipitate dry.

11. Determine the mass of the filter paper and dry precipitate.

Preparation of Standard Solutions

Laboratory procedures often call for the use of a solution of specific, precise concentration. The apparatus used to prepare such a solution is a volumetric flask. A meniscus finder is useful in setting the bottom of the meniscus on the calibration line (Figure C19).

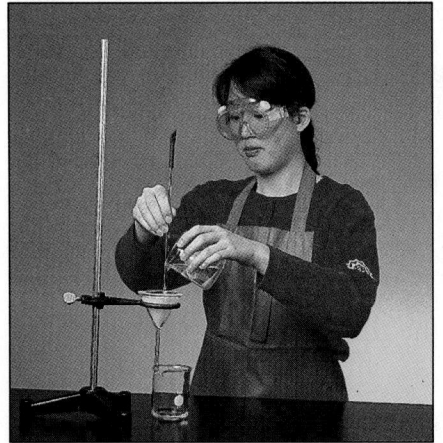

Preparing a Standard Solution from a Solid Reagent

1. Calculate the required mass of solute from the volume and concentration of the solution.

2. Obtain the required mass of solute in a clean, dry beaker or weighing boat. (Refer to "Using a Laboratory Balance" on page 735.)

3. Dissolve the solid in pure water using less than one-half of the final solution volume.

4. Transfer the solution and all water used to rinse the equipment into a clean volumetric flask. (The beaker and any other equipment should be rinsed two or three times with pure water.)

5. Add pure water, using a medicine dropper for the final few millilitres while using a meniscus finder to set the bottom of the meniscus on the calibration line.

6. Stopper the flask and mix the solution by slowly inverting the flask several times.

Figure C18
The separation technique of pouring off clear liquid is called decanting. Pouring along the stirring rod prevents drops of liquid from going down the outside of the beaker when you stop pouring.

Figure C19
Raise the meniscus finder along the back of the neck of the volumetric flask until the meniscus is outlined as a sharp, black line against a white background.

CAUTION

If water is added directly to some solids, there may be boiling or splattering. Always add a solid solute to water.

Preparing a Standard Solution by Dilution

1. Calculate the volume of concentrated reagent required.
2. Add approximately one-half of the final volume of pure water to the volumetric flask.
3. Measure the required volume of stock solution using a pipet. (Refer to "Using a Pipet" on page 737).
4. Transfer the stock solution slowly into the volumetric flask while mixing.
5. Add pure water and then use a medicine dropper and a meniscus finder to set the bottom of the meniscus on the calibration line.
6. Stopper and mix the solution by slowly inverting the flask several times.

Titration

Titration is used in the volumetric analysis of an unknown concentration of a solution. Titration involves adding a solution (the titrant) from a buret to another solution (the sample) in an Erlenmeyer flask until a recognizable endpoint, such as a color change, occurs.

1. Rinse the buret with small volumes of distilled water using a wash bottle. Using a buret funnel, rinse with small volumes of the titrant (Figure C20). (If liquid droplets remain on the sides of the buret after rinsing, scrub the buret with a buret brush. If the tip of the buret is chipped or broken, replace the tip or the whole buret.)
2. Using a small buret funnel, pour the solution into the buret until the level is near the top. Open the stopcock for maximum flow to clear any air bubbles from the tip and to bring the liquid level down to the scale.
3. Record the initial buret reading to the nearest 0.1 mL. Avoid parallax errors by reading volumes at eye level with the aid of a meniscus finder.
4. Pipet a sample of the solution of unknown concentration into a clean Erlenmeyer flask. Place a white piece of paper beneath the Erlenmeyer flask to make it easier to detect color changes.
5. Add an indicator if one is required. Add the smallest quantity necessary (usually 1 to 2 drops) to produce a noticeable color change in your sample.
6. Add the solution from the buret quickly at first, and then slowly, drop-by-drop, near the endpoint (Figure C21). Stop as soon as a drop of the titrant produces a permanent color change in the sample solution. A permanent color change is considered to be a noticeable change that lasts for 10 s after swirling.
7. Record the final buret reading to the nearest 0.1 mL.
8. The final buret reading for one trial becomes the initial buret reading for the next trial. Three trials with results within 0.2 mL are normally required for a reliable analysis of an unknown solution.
9. Drain and rinse the buret with pure water. Store the buret upside down with the stopcock open.

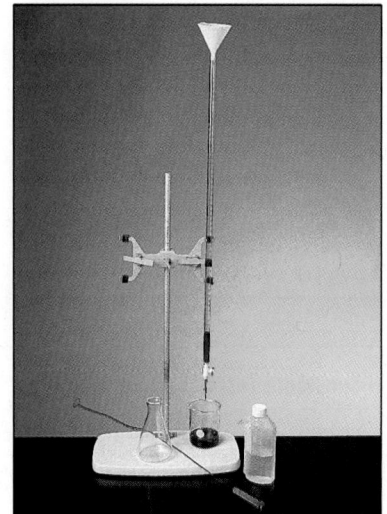

Figure C20
A buret should be rinsed with water and then the titrant before use.

Figure C21
Near the endpoint, continuous gentle swirling of the solution is particularly important.

APPENDIX C

Communication of
Diagnostic Tests
The procedure, evidence, and analysis
information for a diagnostic test can be
communicated in three different formats:
• "If ... and ... then ..." statement
• table
• flowchart

Diagnostic Tests

The tests described in Table C1 are commonly used to detect the presence of a specific substance. All diagnostic tests include a brief procedure, some expected evidence, and an interpretation of the evidence obtained. This is conveniently communicated using the format — "If [procedure] and [evidence], then [analysis]." Diagnostic tests can be constructed using any characteristic empirical property of a substance. For example, diagnostic tests for acids, bases, and neutral substances can be specified in terms of the pH of the solutions. For specific chemical reactions, properties of the products that the reactants do not have, such as the insolubility of a precipitate, the production of a gas, or the color of ions in aqueous solutions, can be used to construct diagnostic tests.

If possible, you should use a control to illustrate that the test does not give the same results with other substances. For example, in the test for oxygen, inserting a glowing splint into a test tube that contains only air is used to compare the effect of air on the splint with a test tube in which you expect oxygen has been collected.

Table C1

SOME STANDARD DIAGNOSTIC TESTS	
Substance Tested	**Diagnostic Test**
water	If cobalt(II) chloride paper is exposed to a liquid or vapor, and the paper turns from blue to pink, then water is likely present.
oxygen	If a glowing splint is inserted into the test tube, and the splint glows brighter or relights, then oxygen gas is likely present.
hydrogen	If a flame is inserted into the test tube, and a squeal or pop is heard, then hydrogen is likely present.
carbon dioxide	If the unknown gas is bubbled into a limewater solution, and the limewater turns cloudy, then carbon dioxide is likely present.
halogens	If a few millilitres of a chlorinated hydrocarbon solvent is added, with shaking, to a solution in a test tube, and the color of the solvent appears to be • light yellow-green, then chlorine is likely present. • orange, then bromine is likely present. • purple, then iodine is likely present.
acid	If strips of blue and red litmus paper are dipped into the solution, and the blue litmus turns red, then an acid is present.
base	If strips of blue and red litmus paper are dipped into the solution, and the red litmus turns blue, then a base is present.
neutral solution	If strips of blue and red litmus paper are dipped into the solution, and neither litmus changes color, then only neutral substances are likely present.
neutral ionic solution	If a neutral solution is tested for conductivity with a multimeter, and the solution conducts a current, then a neutral ionic substance is likely present.
neutral molecular solution	If a neutral solution is tested for conductivity with a multimeter, and the solution does not conduct a current, then a neutral molecular substance is likely present.
	There are thousands of diagnostic tests. You can create some of these, using data from the periodic table (inside front cover of this book); and from the data tables in Appendix F, pages 761 to 766, and on the inside back cover.

C.4 Computers and the Internet

Computer technology — both the equipment (hardware) and the programs (software) to run the equipment — is the most pervasive modern technology today. There is little doubt that basic computer skills are becoming essential for people as consumers, students, employees, and employers. As with all new technologies, a language has developed that is the first barrier people encounter when trying to become computer-literate. Some of the more important and practical terms are explained below. There are differences in terms depending on whether you are using a Macintosh or IBM (PC) computer.

Table C2

IMPORTANT COMPUTER TERMS	
Term	**Definition**
desktop	the start-up screen upon which icons, boxes, and windows appear
window	a rectangular framed area on the desktop within which documents or applications are displayed
menu	a pull-down list of available commands (dimmed ones are not active)
icon	a graphic representation of an object such as a file, program, drive, or actions to be carried out
file or document	an electronic equivalent of a document or set of instructions
open	to retrieve a file from a disk and place it into the window containing the application that you are using
save	to store a file on a disk using the previous name; use Save As if you want to change the name
floppy drive	a disk drive that accepts removable magnetic storage "floppy disks," usually 3.5 inch square
hard drive	a disk drive containing a non-removable large-capacity magnetic storage drive
toolbar	a line of small, square boxes (buttons) containing icons used to perform specific tasks
help	usually a menu containing relevant information to explain terms and procedures for a particular program
cursor	the blinking character appearing on your screen; may take different forms
Function (Fn) key	the group of keys on some keyboards labelled F1, F2, etc., that carry out specific actions
Control (Ctrl) key (Macs also have a Command key)	a special key that is always used simultaneously (like a shift key) with another key to carry out specific actions
Alt key (PC only) Option key (Macs only)	another special key that is always used simultaneously (like a shift key) with another key to carry out specific actions
Enter key	usually the largest key near the right-hand side; used to signal the computer that you have finished typing (also called "return" key)
Tab key	a key used to insert tabs into text or to move from one box to another
Escape (Esc) key	a special key often used to abort an instruction, close a box, or other specific operation
mouse	a small device that usually moves on a flat pad to control the position of the mouse pointer or the cursor; may have 1, 2, or 3 buttons to press that initiate certain actions
click	to position the mouse pointer or cursor over a button or element and then press and release the left button (PC) or the single button (Mac) on the mouse
double-click	same as click except you click twice in rapid succession
select	to position the cursor or pointer at the beginning of one object and then pressing and holding down the mouse button as you move the mouse to highlight several objects
drag	to select an object, hold down mouse button and move the mouse; often used to move files, icons, or borders
scroll	to move the information in a window up/down or left/right using the small square or button in the bars at the right or bottom of the window

Computer Network

A computer network is basically a group of computers that are all connected together. A network often has one or more computers (called *servers*) that contain shared programs or files. Typically in a school, the server controls who uses any computer, what they are allowed to do, and what programs are available. To use any computer that is part of a network you will normally need an *ID* (identification word, usually based on your name) and a *password* (a special code assigned only to you). Both of these must be provided to you by the person in charge of the computer network or your teacher.

Login: This means you are going to "sign in" and are requesting access to a server. Typically, you need to first type your ID (and press the Enter or Tab key if necessary) and then type your password. Note that everything in the password must be exactly correct, including upper or lower case.

Menu: When your login is successful, you will usually be presented with a list of programs or activities from which you choose. If this is a numbered or lettered list, you can simply type the number or letter (and press the Enter key); you then select the desired item by clicking it with your mouse pointer (and pressing the Enter key) or by double-clicking with your mouse pointer.

Logout: This means you are finished with the computer and you are going to sign out. Close or exit any current activities. Usually you just have to type "logout" or "exit." It is not recommended that you turn off the power on the computer you are using in the middle of a program unless it *hangs* (won't respond anymore).

Basic Operation of Windows and Macintosh Systems

1. The main screen contains a group of icons labelled with a group or program name. When you see the group or program that you want, move the mouse pointer onto the icon and double-click the left mouse button (or the single button on Macs). This will either start the program or open a box containing other choices; double-click another choice if necessary.

2. The desired program should now be starting. An hourglass cursor (or clock face on Macs) means that the computer is working and you will have to be patient until the hourglass symbol disappears.

3. Once the program is finished loading, you will usually see a screen with *menu* titles across the top. Each of these, when clicked with the left (or single) mouse button, will show a list of choices that appear to "drop down" from the main title. Choose the job you want or remove the menu box — with a PC, by clicking anywhere outside of the box, and with a Mac by releasing the mouse button. Here are some common menu titles: *File* (opening, saving, closing, printing document files), *Edit* (cutting, copying, pasting selected

text or pictures), *View* (choosing different appearances or formats for what you see on the screen), *Window* (switching from one screen of information to another), and *Help* (the place to go when you want to find out how to do something).

4. Modern programs often have a toolbar across the top of the screen below the menu titles. The *toolbar* is made up of small squares with a word or icon inside; for example, a drawing of a printer. These squares or buttons are meant to replace certain common menu items to speed up your work.

5. One of the main advantages of modern programs is that you can have several boxes or windows open, each doing a different job. You can change the size of the active (highlighted) window by using the small arrow buttons in the top right corner or by moving the cursor to the edge of the window until it changes to a double arrow. Dragging the edge or corner will automatically resize the current window.

Internet Basics

The Internet is like a worldwide computer network. Once you are connected you can obtain information from millions of other computers currently on the network. For example, almost all universities, many government departments, many companies, special interest groups, and millions of individuals are connected. There are many things you can do when connected, such as send and receive e-mail (electronic mail), chat to people, search, look at or browse information on other computers, and send and receive computer files (Figure C22).

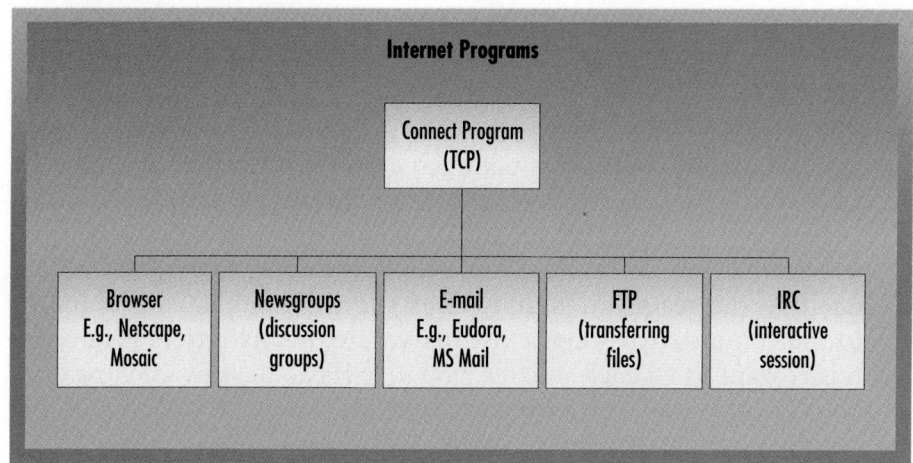

Figure C22
Once you connect using a TCP (transfer control protocol) program you have a choice of many different programs and activities. You may start, stop, and switch to any of these or have several of the programs operating at the same time in different windows.

Once you become familiar with the procedure for using the Internet, you will be able to find information on a variety of topics. By far the most useful program for researching information on the Internet is the browser. The following procedure is aimed specifically at starting and using such a program:

1. To start, you need access to a computer, such as one on your school computer network, that can be connected to the Internet. There will be an icon that you need to choose to open a box containing programs to connect and perform activities once you are connected. Look for and double-click an icon such as Startup, Connect, TCP Manager, etc.

2. The computer will dial a preprogrammed phone number and once connected will ask you for your identification and password. This will likely be different from your network ID and password.

3. Once you are accepted, you can then choose the Internet-related program you want to run. There are two widely used types of programs; e-mail and WWW (World Wide Web) browser.

4. Double-click on the WWW browser icon, such as Netscape or Mosaic. With these programs, you can connect to programs called *search engines* that allow you to search the Internet computers for *web pages* that have information you want. You can also transfer computer files to your own computer, print documents and pictures that you find, and even send e-mail. All of this is possible because every computer connected to the Internet has a unique address and sub-addresses to files stored in its memory. You usually do not need to type these addresses because web pages contain highlighted words or icons that, when you click on them, allow you to jump to a new location (these highlighted words are called hotlinks).

5. Find the menu item or toolbar button that says Internet Search or equivalent. You may need to choose which search engine you want to use; for example, Lycos, Webcrawler, AltaVista. Select one of these and wait for the connection to be completed. When you see an empty box, click in the box and type one or more key words that you want information about; for example, type "iron corrosion." Usually there is a Send Request or Submit Request button to indicate that you have finished typing your request.

6. The search engine will now search its database, which is based on abstracts from computers connected to the Internet, and present you with a list of titles of articles containing your key words. Depending on the search engine used, you may also see some of the text of the articles found. Each title will be highlighted. If you see one that interests you, simply click the title and you will connect to the computer that contains this article. If this article is one you might want to go back to later, you save its address by clicking onto Bookmark as a button or menu item. This must be done while the article is on your screen.

7. Most browsers have Forward (→) and Back (←) buttons that you can click to move between web pages you have already visited.

8. When finished, choose Exit under the *File* menu, choose Disconnect or Bye and log out of the computer. It is important to log out to prevent someone else from using your account and perhaps doing unauthorized things.

9. You should know the accepted-use policy about etiquette, ethics, and legality for the use of computers on your network and the Internet.

APPENDIX D

Safety and Society

Science is a human endeavor, technology has a social purpose, and both have always been part of society. Science, together with technology, affects society in a myriad of ways. Society also affects science and technology, by placing controls on them and expecting solutions to societal problems. Within our society, safety for people and the environment is of paramount importance whether in a chemistry laboratory, chemical industry, or home.

D.1 Laboratory Safety Rules

Safety is always important in a laboratory or in other settings that feature chemicals or technological devices. It is your responsibility to be aware of possible hazards, to know the rules — including ones specific to your classroom — and to behave appropriately. Always alert the teacher in case of any accident.

Glass Safety and Cuts

- Never use glassware that is cracked or chipped. Give such glassware to your teacher or dispose of it as directed. Do not put the item back into circulation.
- Never pick up broken glassware with your fingers. Use a broom and dustpan.
- Do not put broken glassware into garbage containers. Dispose of glass fragments in special containers marked "broken glass."
- If you cut yourself, inform your teacher immediately. Imbedded glass or continued bleeding requires medical attention.

Burns

- In a laboratory where burners or hot plates are being used, never pick up a glass object without first checking the temperature by lightly and quickly touching the item. Glass items that have been heated stay hot for a long time but do not appear to be hot. Metal items such as ring stands and hot plates can also cause burns; take care when touching them.
- Do not use a laboratory burner near wooden shelves, flammable liquids, or any other item that is combustible.
- Before using a laboratory burner, make sure that long hair is always tied back. Do not wear loose clothing (wide long sleeves should be tied back or rolled up).

- Never look down the barrel of a laboratory burner.
- Always pick up a burner by the base, never by the barrel.
- Never leave a lighted bunsen burner unattended.
- If you burn yourself, *immediately* run cold water over the burned area and inform your teacher.

Eye Safety

If you wear contact lenses in the laboratory, there is a danger that a chemical might get behind the lens where it cannot be rinsed out with water. Tell your teacher if you are wearing contact lenses in the laboratory.

- Always wear approved eye protection in a laboratory, no matter how simple or safe the task appears to be. Keep the safety glasses over your eyes, not on top of your head. For certain experiments, full face protection may be necessary.
- Never look directly into the opening of flasks or test tubes.
- If, in spite of all precautions, you get a solution in your eye, quickly use the eyewash or nearest running water. Continue to rinse the eye with water for at least 15 min. This is a very long time — have someone time you. Unless you have a plumbed eyewash system, you will also need assistance in refilling the eyewash container. Have another student inform your teacher of the accident. The injured eye should be examined by a doctor.
- If you must wear contact lenses in the chemistry laboratory, be extra careful; whether or not you wear contact lenses, do not touch your eyes without first washing your hands. It is recommended that you do not wear contact lenses in the laboratory.
- If a piece of glass or other foreign object enters an eye, immediate medical attention is required.

Fire Safety

Immediately inform your teacher of any fires. Very small fires in a container may be extinguished by covering the container with a wet paper towel or a ceramic square which would cut off the supply of air. If anyone's clothes or hair catch fire, the fire can be extinguished by smothering the flames with a blanket or a piece of clothing. Larger fires require a fire extinguisher. (Know how to use the fire extinguisher that is in your laboratory.) If the fire is too large to approach safely with an extinguisher, vacate the location and sound the fire alarm. (School staff will inform the fire department.)

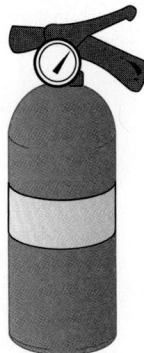

If you use a fire extinguisher, direct the extinguisher at the base of the fire and use a sweeping motion, moving the extinguisher nozzle back and forth across the front of the fire's base. You must use the correct extinguisher for the kind of fire you are trying to control. Each extinguisher is marked with the class of fire for which it is effective. The fire classes are outlined below. Most fire extinguishers in schools are of the ABC type.

- Class A fires involve ordinary combustible materials that leave coals or ashes, such as wood, paper, or cloth. Use water or dry chemical extinguishers on Class A fires. (Carbon dioxide extinguishers are not satisfactory as carbon dioxide dissipates quickly and the hot coals can re-ignite.)

APPENDIX D

- Class B fires involve flammable liquids such as gasoline or solvents. Carbon dioxide or dry chemical extinguishers are effective on Class B fires. (Water is not effective on a Class B fire since the water splashes the burning liquid and spreads the fire.)

- Class C fires involve live electrical equipment, such as appliances, photocopiers, computers, or laboratory electrical apparatus. Carbon dioxide or dry chemical extinguishers are recommended for Class C fires. Carbon dioxide extinguishers are much cleaner than the dry chemical variety. (Using water on live electrical devices can result in severe electrical shock.)

- Class D fires involve burning metals, such as sodium, potassium, magnesium, or aluminum. Sand or salt are usually used to put out Class D fires. (Using water on a metal fire can cause a violent reaction.)

- Class E fires involve a radioactive substance. These involve special considerations at each site.

Electrical Safety

Water or wet hands should never be used near electrical equipment. When unplugging equipment, remove the plug gently from the socket (do not pull on the cord).

Safety Rules

Safety in the laboratory is an attitude and a habit more than it is a set of rules. It is easier to prevent accidents than to deal with the consequences of an accident. Most of the following rules are common sense.

- Always wear eye protection and lab aprons or coats.

- Wear closed shoes (not sandals) when working in the laboratory.

- Place your books and bags away from the work area.

- Do not chew gum, eat, or drink in the laboratory.

- Know potential hazards in the laboratory, including the location of MSDS information and all safety equipment.

- Avoid sudden or rapid motion in the laboratory that may interfere with someone carrying or working with chemicals.

- Ask for assistance when you are not sure how to do a procedural step.

- Do not taste any substance in a laboratory.

- Use accepted techniques for checking odors. Do not inhale the vapors directly from the container. Fan the vapors toward your nose, keeping the container at a distance. Gradually move the container closer until you can detect the odor.

- Never handle any reagent with your hands. Use a laboratory scoop or spoon for handling solids.

- Never use the contents of a bottle that has no label or has an illegible label. Give any containers with illegible labels to your teacher. Always double check the label to ensure that you are using the chemical you need. (Always pour from the side opposite the label on a reagent bottle; your hands and the label are protected as previous drips are always on the side of the bottle opposite the label.)

- When leaving chemicals in containers, ensure that the containers are labelled.
- Know the MSDS information for hazardous chemicals in use.
- Always wash your hands with soap and water before you leave the laboratory.
- Always use a pipet bulb, and never pipet by mouth.
- When heating a test tube over a laboratory burner, use a test-tube holder. Holding the test tube at an angle, facing away from you and others, gently move the test tube backwards and forwards through the flame.
- Never attempt any unauthorized experiments.
- Never work in a crowded area or alone in the laboratory.
- Clean up all spills, even spills of water, immediately.
- Do not forget safety procedures when you leave the laboratory. Accidents can also occur at home or at work.

D.2 Safety Symbols and Information

Educational, Commercial, and Industrial Information

Although MSDS must be supplied with every product sold, current MSDS can also be obtained at several Internet sites. These sites, listed below, are also useful for researching information about chemicals.

- http://www.fisher1.com
- gopher://atlas.chem.utah.edu: 70/11/MSDS

The following site contains an index of MSDS sites as well as a MSDS training page.

- http://www.denison.edu/ sec-safe/safety/msdsres.html

The Workplace Hazardous Materials Information System (WHMIS) provides workers and students with complete and accurate information regarding hazardous products. All chemical products supplied to schools, businesses, and industry must contain standardized labels and be accompanied by Material Safety Data Sheets (MSDS) providing detailed information about the product. Clear and standardized labelling is an important component of WHMIS (Figure D1). These labels must be present on the product's original container or be added to other containers if the product is transferred.

 **Class A:** Compressed gas

 **Class B:** Flammable and combustible material

 **Class C:** Oxidizing material

Class D: Poisonous and Infectious Materials

 Division 1 — Materials causing immediate and serious toxic effect

 Division 2 — Materials causing other toxic effects

 Division 3 — Biohazardous infectious material

 **Class E:** Corrosive material

 **Class F:** Dangerously reactive material

Figure D1
WHMIS symbols.

APPENDIX D

Consumer Information

The Canadian Hazardous Products Act requires manufacturers of consumer products containing chemicals to include a symbol specifying both the nature of the primary hazard and the degree of this hazard. In addition, any secondary hazards, first aid treatment, storage and disposal must be noted. The symbols that are used show the hazard by an illustration and the degree of the hazard by the type of border surrounding the illustration (Figure D2).

D.3 Waste Disposal

Disposal of chemical wastes at home, at school, or at work is a societal issue. To protect the environment, both federal and provincial governments have regulations to control chemical wastes. For example, the WHMIS program (page 27) applies to controlled products that are being handled. (When being transported, they are regulated under the Transport of Dangerous Goods Act, and for disposal they are subject to federal, provincial, and municipal regulations.) Most laboratory waste can be washed down the drain, or, if it is in solid form, placed in ordinary garbage containers. However, some waste must be treated more carefully. Throughout this textbook, special waste disposal problems are noted, but it is your responsibility to dispose of waste in the safest possible manner.

Flammable Substances

Flammable liquids should not be washed down the drain. Special fire-resistant containers are used to store flammable liquid waste. Waste solids that pose a fire hazard should be stored in fireproof containers. Care must be taken not to allow flammable waste to come into contact with any sparks, flames, other ignition sources, or oxidizing materials. The particular method of disposal depends on the nature of the substance.

Corrosive Solutions

Solutions that are corrosive but not toxic, such as acids, bases, or oxidizing agents, can usually be washed down the drain, but care should be taken to ensure that they are properly diluted. Use large quantities of water and continue to pour water down the drain for a few minutes after all the substance has been washed away.

Heavy Metal Solutions

Heavy metal compounds (for example, lead, mercury, or cadmium compounds) should not be flushed down the drain. These substances are cumulative poisons and should be kept out of the environment. A special container is kept in the laboratory for heavy metal solutions. Pour any heavy metal waste into this container. Remember that paper towels used to wipe up solutions of heavy metals, as well as filter papers with heavy metal compounds imbedded in them, should be treated as solid toxic waste.

Disposal of heavy metal solutions is usually accomplished by precipitating the metal ion (for example, as lead(II) silicate) and disposing of the solid. Disposal may be by elaborate means such as deep well burial, or by simpler but accepted means such as delivering the substance to a landfill. Heavy metal compounds should not be placed in school garbage containers. Usually, waste disposal companies

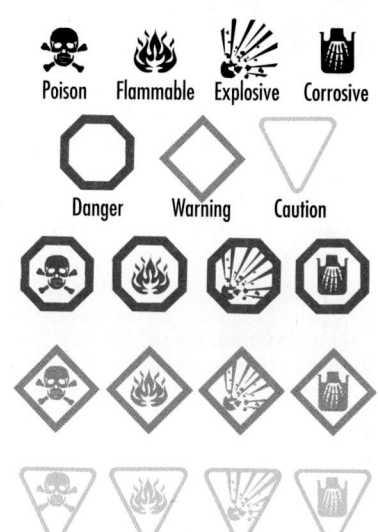

Figure D2
Household Hazardous Product Symbols.

WHMIS symbol for flammable and combustible materials.

WHMIS symbol for corrosive materials.

Acids and bases should always be diluted or neutralized before disposal. To neutralize diluted waste acids, use diluted waste bases, and vice versa. Or, use sodium bicarbonate for neutralizing the acid and use dilute hydrochloric acid for neutralizing the base. Oxidizing agents, such as potassium permanganate, should also be reduced in strength with a 10% aqueous solution of sodium thiosulfate (reducing agent) before washing into the drain.

WHMIS symbol for materials causing an immediate and serious toxic effect.

WHMIS symbol for substances causing other toxic effects that are not immediately dangerous to health.

▌**Perspectives on STS Issues**
Statements of STS issues can be classified for purposes of organizing your knowledge. The following classification system may be helpful.
- scientific
- technological
- ecological
- economic
- political
- legal
- ethical
- social
- militaristic
- aesthetic
- mystical
- emotional

collect materials that require special disposal and dispose of them as required by law.

Toxic Substances

Solutions of toxic substances, such as oxalic acid, should not be poured down the drain, but should be disposed of in the same manner as heavy metal solutions. Solid toxic substances are handled similarly to precipitates of heavy metal.

D.4 STS Decision-Making Model

When controversial issues related to science and technology arise in our society, there is often a heated debate among various special interest groups. Little progress is often made because different parties in the debate often recognize only a single perspective on the issue. Many people now realize that an informed multi-perspective view is more defensible. The following model represents one possible procedure for making an informed decision on a social issue related to science and technology.

1. Identify an STS (science-technology-society) issue. Newspapers, magazines, and news broadcasts are sources of current STS issues. However, some issues like acid rain have been current for some time and only occasionally appear in the news. When identifying an issue for discussion or debate, it is convenient to state the issue as a resolution. For example, "Be it resolved that the use of fossil fuels for heating homes should be eliminated."

2. Design a plan to address the STS issue. Possible designs include individual research, a debate, a town-hall meeting (or role-playing), or participation in an actual hearing or on a committee.

3. Identify and obtain relevant information on as many perspectives as possible. An STS issue will always have scientific and technological perspectives. Other perspectives include ecological, economic, political, legal, ethical, social, militaristic, esthetic, mystical, and emotional. (See the glossary on page 768 for definitions of these perspectives.) Information can be obtained from references and through group discussions. There are many sides to every issue. There can be positive and negative viewpoints about the resolution from every perspective.

4. Generate a number of alternative solutions to the STS problem. Some obvious solutions will arise from the resolution. Other creative solutions often arise from a brainstorming session within a group.

5. Evaluate each solution and decide which is best. One method is to rank on a scale the value of a particular solution from each perspective. For example, a solution might have little economic advantage and be ranked as 1 on a scale of 1 to 5; the solution might have a significant ecological benefit and be ranked as 5, for a total of 6. A different solution might be judged as 3 from the economic perspective and 1 from the ecological perspective, for a total of 4. The solution with the highest total is likely to be approved. Although simplistic, this method facilitates evaluation and illustrates the trade-offs that occur in any real issue.

APPENDIX E

Communication Skills

Communication is essential in science. The international scope of science requires that quantities, chemical symbols, and mathematical tools such as numbers, operations, tables, and graphs, be understood by scientists in different countries with different languages. The way in which scientific knowledge is expressed also reflects the nature of scientific knowledge, in particular the certainty of the knowledge.

E.1 Scientific Language

Science deals with two types of knowledge — empirical (observable "facts") and theoretical (non-observable ideas). Directly observable knowledge is generally considered to be more certain than interpretations or concepts. For example, a candle does not burn unless air is present. In a closed container, a candle flame is extinguished after a short period of time. These are simple and relatively certain facts that can be directly stated. At one time, scientists believed that burning releases a substance called phlogiston, which was absorbed by the air until it could hold no more phlogiston; this is what stopped the burning. This theory, which was firmly believed by many chemists until the 1800s, was eventually replaced by the oxygen theory of combustion. Theories are subject to change and are therefore less certain than the observations upon which they are based.

When observations are interpreted or explained, the language used should reflect some uncertainty or tentativeness. Use phrases such as:

- The evidence suggests that...
- According to the theory of...
- It appears likely that...
- Scientists generally believe that...
- One could hypothesize that...

Avoid the use of the word "prove." Scientific ideas cannot be proven. The evidence may be extensive and reliable, but a theory to explain the evidence will never be 100% certain. In general, the language that you use should reflect the certainty of the information (observations are more certain than scientific concepts) and it should refer to the evidence available to you.

Did the doctor say thermometer or barometer?

"There is one thing certain, namely, that we can have nothing certain..." — Samuel Butler (1835 – 1902)

E.2 SI Symbols and Conventions

The International System of Units, known as SI from the French name, *Système international d'unités*, is the measurement and communication system used internationally by scientists; it is also the legal measurement system in Canada and most countries in the world. Physical quantities are ultimately expressed in terms of seven fundamental SI units, called base units, which cannot be expressed as combinations of simpler units (Table E1).

Table E1

QUANTITIES AND BASE UNITS			
Quantity	**Symbol**	**Unit**	**Symbol**
length	l	metre	m
time	t	second	s
mass	m	kilogram	kg
amount of substance	n	mole	mol
temperature	T	kelvin	K
electric current	I	ampere	A
luminous intensity	I_v	candela	cd

Although the base unit for mass is the kilogram (kg), it is more common in a chemistry laboratory to use the gram (g). Similarly, although the base unit for temperature (T) is kelvin (K), the common temperature (t) unit is degree Celsius (°C).

All other quantities can be expressed in terms of these seven fundamental quantities. For convenience, a unit derived from a combination of base units may be assigned a symbol of its own. Table E2 lists a few of the physical quantities and derived units most commonly encountered in chemistry.

Table E2

COMMON SI QUANTITIES AND UNITS			
Quantity	**Symbol**	**Unit**	**Symbol**
molar mass	M	grams per mole	g/mol
volume	v	litre	L
molar concentration	C	moles per litre	mol/L
pressure	p	pascal	Pa
energy	E	joule	J
power	P	watt	W
heat capacity	C	joules per degree Celsius	J/°C
specific heat capacity	c	joules per gram per degree Celsius	J/(g•°C)
volumetric heat capacity	c	megajoules per cubic metre per degree Celsius	MJ/(m³•°C)
molar enthalpy	H	kilojoules per mole	kJ/mol
enthalpy change	ΔH	kilojoules	kJ
electric charge	q	coulomb	C
electric potential difference (voltage)	E	volt (joules per coulomb)	V

SI Prefixes

Table E3

SI PREFIXES		
Prefix	**Symbol**	**Factor**
giga	G	10^9
mega	M	10^6
kilo	k	10^3
milli	m	10^{-3}
micro	μ	10^{-6}
nano	n	10^{-9}

Next to universality, the most important feature of any system of units is convenience. Units that are inconvenient in actual use tend to cause frustration and fall into disuse. SI has been designed to maximize convenience in a number of ways. A given quantity is always measured in the same base unit regardless of the context in which it is measured. For example, all forms of energy, including energy in food, are measured in joules. When a unit is too large or too small for convenient measurement, the unit is adjusted in size with a prefix. (See Table E3.) Prefixes allow units to be changed in size by multiples of ten. However, except for the use of "centi" in centimetre, only prefixes that change the unit in multiples of a thousand are commonly used.

The Rule of a Thousand

People tend to be most comfortable working with numbers greater than 0.1 and less than 1000. SI measurements that give numerical values outside this range are adjusted by changing the prefix of the unit. Prefixes that change the value of the unit by a factor of 10^3 are most common. Thus, 0.0032 g is reported as 3.2 mg, and 40 102 g is reported as 40.102 kg. The rule of a thousand is commonly used in commercial labelling to avoid numbers of awkward size.

Scientific Notation

Scientific notation is a convenient method for expressing either a very large value or a very small value as a number between 1 and 10 multiplied by a power of 10. For example, the following numbers are expressed in regular notation and scientific notation.

regular notation	scientific notation
1200 L	1.200×10^3 L
0.000 000 998 mol/L	9.98×10^{-7} mol/L

On some calculators, the F $\rightleftharpoons$ E key or the FSE key changes the number in the display into or from scientific notation. To enter a value in scientific notation in your calculator, the EXP or EE key is used to enter the power of ten. Note that the base 10 is not keyed into the calculator. For example, to enter

1.200×10^3 press $\boxed{1}$ $\boxed{\bullet}$ $\boxed{2}$ $\boxed{\text{EXP}}$ $\boxed{3}$

9.98×10^{-7} press $\boxed{9}$ $\boxed{\bullet}$ $\boxed{9}$ $\boxed{8}$ $\boxed{\text{EXP}}$ $\boxed{7}$ $\boxed{+/-}$

All mathematical operations and functions (such as +, −, ×, ÷, log) can be carried out with numbers in scientific notation.

Scientific notation is useful in calculations because it simplifies the cancellation of units and the totalling of powers of ten. However, scientific notation is sometimes overused. SI recommends that, wherever possible, prefixes be used to report measured values. Scientific notation should be reserved for situations where no prefix exists, or where it is essential to use the same unit (for example, comparing a wide range of energy values in kilojoules per gram). A reported value should use a prefix or scientific notation, but not both, unless you are comparing values. Scientific notation should usually use the base unit.

E.3 Quantitative Precision and Certainty

Quantities that have *exact values* are either *defined* quantities (for example, 1 t is defined as exactly 1000 kg, and the SI prefix *kilo*, k, is exactly 1000) or quantities obtained by *counting* (for example, 32 people in a class or any coefficient in a balanced chemical equation).

Here are some common examples of the use of the rule of a thousand.

candy bar	49 g
refined sugar	4 kg
soft drinks	300 mL
gasoline	48.3 L
pain relief tablets	325 mg
vitamin capsules	200 mg
bulk fertilizer	25 t
concrete	7.5 m³
carpet	12.4 m²

For the number of particles in one mole, writing 6.02×10^{23} is acceptable. No prefix is large enough to report this number as a value between 0.1 and 1000.

You can be almost certain about such quantities; there will be a small degree of uncertainty when counting very large numbers.

On the other hand, most quantities are measured by a person using some measuring instrument (for example, measuring the mass of a chemical using a balance). Since every instrument has its limitations and no one can perfectly measure a quantity, there is always some uncertainty about the number obtained. This uncertainty depends on the size of the sample measured, the particular instrument used, and the technological skill of the person doing the measurement.

Accuracy

Accuracy is an expression of how close an experimental value is to the accepted value. The comparison of the two values is often expressed as a percent difference. For example, the accuracy of a prediction based on some authority can be expressed as the absolute value of the difference divided by a predicted value and converted to a percent.

$$\% \text{ difference } = \frac{|\text{experimental value} - \text{predicted value}|}{|\text{predicted value}|} \times 100$$

This expression of accuracy is often used in the Evaluation section of investigation reports.

Precision

Precision is the place value of the last measurable digit and is determined by the instrument. A mass of 17.13 g is more precise than 17.1 g. The precision is determined by the particular system or instrument used; for example, a centigram balance versus a decigram balance.

Accuracy is an expression of how close a value is to the accepted, expected, or predicted value, whereas precision is a measure of the reproducibility or consistency of a result (Figure E1). Accuracy is generally attributed to an error in the system (a *systematic error*); precision is associated with a *random error* of measurement. For example, if you used a balance without zeroing it, you might obtain measurements that have high precision (reproducibility) but low accuracy. The same is true for calibrating a pH meter at pH 7.00 and then making a measurement of a very high or low pH. The systematic error might be high (low accuracy), even though the random error of the measurement is low (high precision).

You may not know how uncertain the last measured digit is. On a centigram balance, the error of measurement in the last digit is usually considered to be ±0.01 g. Measurements such as 12.39 g, 12.40 g, and 12.41 g all have the same precision (hundredths), and may all be equally correct masses for the same object. The precision with which you read a thermometer might be ±0.2°C (for example, 21.0°C, 21.2°C or 21.4°C) and a ruler might be read to ±0.5 mm; you must decide, for example, whether to record 11.0 mm, 11.5 mm or 12.0 mm.

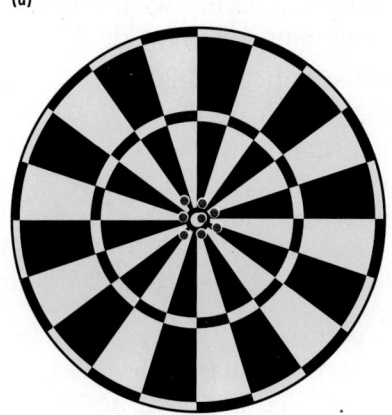

(a)

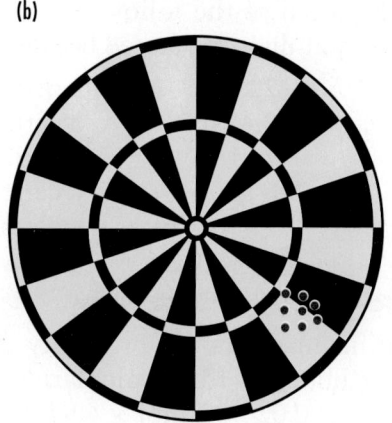

(b)

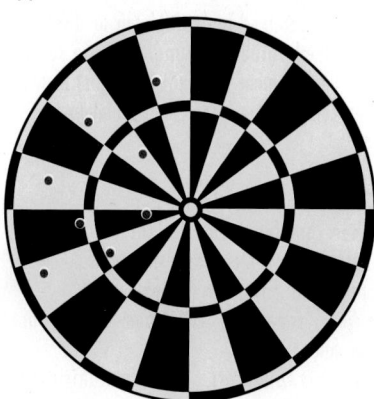

(c)

Rounding

Calculations are usually based on measurements (for example, in the Analysis section of a report). To report a calculated result correctly, you must know the place value at which the result becomes uncertain and a method for rounding the answer. *Rounding* means checking the first digit following the digit that will be rounded. If this digit is less than 5, it and all following digits are discarded. If this digit is 5 or greater, it and all following digits are discarded, and the preceding digit is increased by one.

Precision Rule for Calculations

A result obtained by adding or subtracting measurements is rounded to the same precision as the least precise value. For example, 12.6 g + 2.07 g + 0.142 g totals to 14.812 g on your calculator. This value is rounded to one-tenth of a gram and reported as 14.8 g because the first measurement limits the precision of the final result to tenths of a gram and the rounding rule suggests leaving the 8 as is. The final result is reported to the least number of decimal places in the values added or subtracted.

Certainty

How certain you are about a measurement depends on two factors — the precision of the instrument and the value of the measured quantity. More precise instruments give more certain values; for example, 15°C as opposed to 15.215°C. Consider two measurements with the same precision, 0.4 g and 12.8 g. If the balance used is precise to ±0.2 g, the value 0.4 g could vary by as much as 50%. However, 0.2 g of 12.8 g is a variation of less than 2%. For both factors — precision of instrument and value of the measured quantity — the more digits in a measurement, the more certain you are about the measurement. The **certainty** of any measurement is communicated by the number of significant digits in the measurement. In a measured or calculated value, **significant digits** are all those digits that are certain plus one estimated (uncertain) digit. Significant digits include all digits correctly reported from a measurement, except leading zeros. Leading zeros are the zeros at the beginning of a decimal fraction and are written only to locate the decimal point. For example, 6.20 mL (3 significant digits) has the same number of significant digits as 0.00620 L.

Figure E1
The positions of the darts in each of these figures are analogous to measured or calculated results in a laboratory setting. The results in (a) are precise and accurate, in (b) they are precise but not accurate, and in (c) they are neither precise nor accurate.

Rounding Rule
Check the first digit following the digit that will be rounded. If this first digit is less than 5, it and all following digits are discarded. If this first digit is 5 or greater, it and all following digits are discarded and the preceding digit is increased by one.

Precision Rule
A result obtained by adding or subtracting measured values is rounded to the same precision (number of decimal places) as the least precise value used in the calculation.

Certainty

A result obtained by multiplying or dividing measured values is rounded to the same certainty (number of significant digits) as the least certain value used in the calculation.

Chained Calculations

When completing calculations that involve more than one step, there are two rules that are used for answers in this textbook.
• *Never* round off partial answers in your calculator.
• *Always* round off when communicating partial answers on paper.
    When chained calculations involve both multiplication/division and addition/subtraction, you may be required to store the partial answers in your calculator memory or to use the bracket function on your calculator.

For each of the following measurements, the certainty (number of significant digits) is stated beside the measured or calculated value.

22.07 g	a certainty of 4 significant digits
0.41 mL	a certainty of 2 significant digits
700 mol	a certainty of 3 significant digits
0.020 50 km	a certainty of 4 significant digits
2×10^{40} m	a certainty of 1 significant digit

Certainty Rule for Calculations

Significant digits are primarily used to determine the certainty of a result obtained from calculations using several measured values. For example, 0.024 89 mol × 6.94 g/mol is displayed as 0.1727366 g on a calculator. This is correctly reported as 0.173 g or as 173 mg because the second measured value used (6.94) limits the final result to a certainty of three significant digits.

E.4 Tables and Graphs

Both tables and graphs are used to summarize information and to illustrate patterns or relationships. Preparing tables and graphs requires some knowledge of accepted practice and some skill in designing the table or graph to best describe the information.

Tables

1. Write a descriptive title that communicates the contents or the relationship among the entries in the table.
2. The row or column with the manipulated variable usually precedes the row or column with the responding variable.
3. Label all rows and columns with a heading, including units in parentheses where necessary. Units are not usually written in the main body of the table.

Table E4

THE EFFECT OF CONCENTRATION ON REACTION TIME	
Concentration of $HCl_{(aq)}$ (mol/L)	Time for Reaction (s)
2.0	70
1.5	80
1.0	144
0.5	258

Graphs

1. Write a title on the graph and label the axes.
   (a) The title should be at the top of the graph. A statement of the two variables is often used as a title; for example, "Solubility versus Temperature for Sodium Chloride."

(b) Label the horizontal (x) axis with the name of the manipulated variable and the vertical (y) axis with the name of the responding variable.

(c) Include the unit in parentheses on each axis label, for example, "Time (s)."

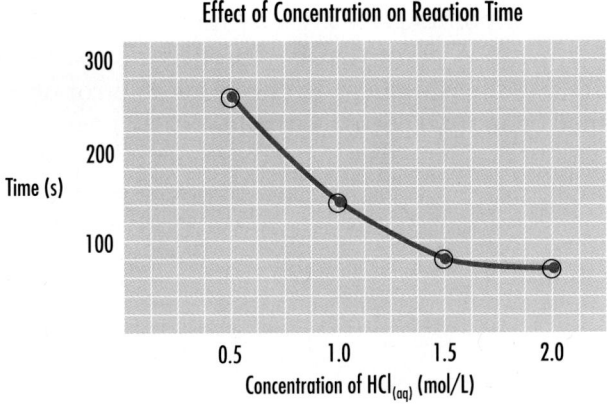

Figure E2
A sample graph.

2. Assign numbers to the scale on each axis.

   (a) As a general rule, the data points should be spread out so that at least one-half of the graph paper is used.

   (b) Choose a scale that is easy to read and has equal divisions. Each division (or square) must represent a small simple number of units of the variable; for example, 0.1, 0.2, 0.5, or 1.0.

   (c) It is not necessary to have the same scale on each axis or to start a scale at zero.

   (d) Do not label every division line on the axis. Scales on graphs are labelled in a way similar to the way scales on rulers are labelled.

3. Plot the data points.

   (a) Locate each data point by making a small dot in pencil. When all points are drawn and checked, draw an X over each point or circle each point in ink. The size of the circle can be used to indicate the precision of the measurement.

   (b) Be suspicious of a data point that is obviously not part of the pattern. Double check the location of such points, but do not eliminate the point from the graph if it does not align with the rest.

4. Draw the best fitting curve.

   (a) Using a sharp pencil, draw a line that best represents the trend shown by the collection of points. Do not force the line to go through each point. Uncertainty of experimental measurements may cause some of the points to be misaligned.

   (b) If the collection of points appears to fall in a straight line, use a ruler to draw the line. Otherwise draw a smooth curve that best represents the pattern of the points.

   (c) Since the data points are in ink and the line is in pencil, it is easy to change the position of the line if your first curve does not fit the points to your satisfaction.

Although a graph is constructed using a limited number of measured values, the pattern may be used to extend the empirical information.

- *Interpolation* is used to find values between measured points on the graph.

- *Extrapolation* is used to find values beyond the measured points on a graph. A dotted line on a graph indicates an extrapolation.

- The scattering of points gives a visual indication of the uncertainty in the experiment. A point that is obviously not part of the pattern may require a re-measurement to check for an error or may indicate the influence of an unexpected variable.

APPENDIX F

Data Tables

			THE ELEMENTS					
Element	Symbol	Atomic Number	Ionization Energy (kJ/mol)	Electro-negativity	Electron Affinity (kJ/mol)	Atomic Radius (pm)	Ionic Radius (pm)	Common Ion Charge
actinium	Ac	89	499	1.1		188	111	3+
aluminum	Al	13	578	1.5	−44	143	50	3+
americium	Am	95	578	1.3		184	99	3+
antimony	Sb	51	834	2.05	−101	141	90	3+
argon	Ar	18	1521		>0*	174		
arsenic	As	33	947	2.00	−77	121	222	3−
astatine	At	85		2.2	−270		227	1−
barium	Ba	56	503	0.9	>0	217	135	2+
berkelium	Bk	97	601	1.3			98	3+
beryllium	Be	4	899	1.5	>0	111	31	2+
bismuth	Bi	83	703	1.9	−106	155	96	3+
boron	B	5	801	2.0	−27	88	20	3+
bromine	Br	35	1140	2.85	−325	114	196	1−
cadmium	Cd	48	868	1.7	>0	149	97	2+
calcium	Ca	20	590	1.0	>0	197	99	2+
californium	Cf	98	608	1.3			98	3+
carbon	C	6	1086	2.60	−122	77	15	4+
cerium	Ce	58	528	1.1	(−48)**	183	103	3+
cesium	Cs	55	376	0.7	−45	265	169	1+
chlorine	Cl	17	1251	3.15	−349	99	181	1−
chromium	Cr	24	653	1.6	−64	125	64	3+
cobalt	Co	27	758	1.8	−68	125	74	2+
copper	Cu	29	745	1.9	−118	128	72	2+
curium	Cm	96	581	1.3			99	3+
dysprosium	Dy	66	572	1.2	(−48)	177	91	3+
einsteinium	Es	99	619	1.3			98	3+
erbium	Er	68	589	1.2	(−48)	176	88	3+
europium	Eu	63	547	1.2	(−48)	204	95	3+
fermium	Fm	100	627	1.3			97	3+
fluorine	F	9	1681	4.00	−328	64	136	1−
francium	Fr	87		0.65	−44	270	176	1+
gadolinium	Gd	64	592	1.1	(−48)	180	94	3+
gallium	Ga	31	579	1.6	−29	122	62	3+
germanium	Ge	32	762	1.90	−116	123	53	4+
gold	Au	79	890	2.4	−223	144	91	3+
hafnium	Hf	72	642	1.3		156	78	4+
helium	He	2	2372		>0	122		
holmium	Ho	67	581	1.2	(−48)	177	89	3+
hydrogen	H	1	1312	2.20	−73	37	10^{-3}/208	1+/1−
indium	In	49	558	1.7	−29	163	81	3+
iodine	I	53	1008	2.65	−295	133	216	1−
iridium	Ir	77	878	2.2	−154	136	64	4+
iron	Fe	26	759	1.8	−24	124	64	3+
krypton	Kr	36	1351		>0	189		
lanthanum	La	57	538	1.1	−48	188	106	3+
lawrencium	Lr	103				188	94	3+

lead	Pb	82	716	1.8	−106	175	120	2+
lithium	Li	3	520	1.0	−60	152	68	1+
lutetium	Lu	71	524	1.2	(−48)	173	85	3+
magnesium	Mg	12	738	1.2	>0	160	65	2+
manganese	Mn	25	717	1.5	>0	124	80	2+
mendelevium	Md	101	635	1.3			114	2+
mercury	Hg	80	1007	1.9	>0	160	110	2+
molybdenum	Mo	42	685	1.8	−96	136	62	6+
neodymium	Nd	60	530	1.2	(−48)	182	100	3+
neon	Ne	10	2081		>0	160		
neptunium	Np	93	597	1.3		152	88	5+
nickel	Ni	28	737	1.8	−111	125	72	2+
niobium	Nb	41	664	1.6	−96	143	70	5+
nitrogen	N	7	1402	3.05	+7	70	171	3−
nobelium	No	102	642	1.3			113	2+
osmium	Os	76	839	2.2	−106	134	65	4+
oxygen	O	8	1314	3.50	−141	66	140	2−
palladium	Pd	46	805	2.2	−58	138	86	2+
phosphorus	P	15	1012	2.15	−72	110	212	3−
platinum	Pt	78	868	2.2	−205	138	70	4+
plutonium	Pu	94	585	1.3		151	90	4+
polonium	Po	84	812	2.0	−183	167	65	4+
potassium	K	19	419	0.8	−48	227	133	1+
praseodymium	Pr	59	523	1.1	(−48)	183	101	3+
promethium	Pm	61	535	1.2	(−48)	181	98	3+
protactinium	Pa	91	568	1.5		161	90	5+
radium	Ra	88	509	0.9		220	140	2+
radon	Rn	86	1037		>0	140	214	
rhenium	Re	75	760	1.9	−14	137	60	7+
rhodium	Rh	45	720	2.2	−116	135	75	3+
rubidium	Rb	37	403	0.8	−47	248	148	1+
ruthenium	Ru	44	711	2.2	−106	133	77	3+
samarium	Sm	62	543	1.2	(−48)	180	96	3+
scandium	Sc	21	631	1.3	>0	161	81	3+
selenium	Se	34	941	2.45	−195	117	198	2−
silicon	Si	14	786	1.90	−134	117		
silver	Ag	47	731	1.9	−126	144	126	1+
sodium	Na	11	496	0.9	−53	186	95	1+
strontium	Sr	38	549	1.0	>0	215	113	2+
sulfur	S	16	1000	2.60	−200	104	184	2−
tantalum	Ta	73	761	1.3	−58	143	68	5+
technetium	Tc	43	702	1.9	−68	136	58	7+
tellurium	Te	52	869	2.30	−190	137	221	2−
terbium	Tb	65	564	1.2	(−48)	178	92	3+
thallium	Tl	81	589	1.8	−29	170	144	1+
thorium	Th	90	587	1.3		180	99	4+
thulium	Tm	69	597	1.2	(−48)	175	87	3+
tin	Sn	50	709	1.8	−121	141	71	4+
titanium	Ti	22	658	1.5	−19	145	68	4+
uranium	U	92	584	1.7		139	83	6+
vanadium	V	23	650	1.6	−48	132	59	5+
wolfram	W	74	770	1.7	−58	137	65	6+
xenon	Xe	54	1170		>0	130		
ytterbium	Yb	70	603	1.1	(−48)	194	86	3+
yttrium	Y	39	616	1.3	0	181	93	3+
zinc	Zn	30	906	1.6	0	133	74	2+
zirconium	Zr	40	660	1.6	−48	160	79	4+

* >0 indicates an endothermic process forming an unstable negative ion.

** bracketed values are estimated.

All values in this table are taken from *Lange's Handbook of Chemistry*, 13th Edition.

Chemical Name	Formula	H_f° (kJ/mol)	Chemical Name	Formula	H_f° (kJ/mol)	Chemical Name	Formula	H_f° (kJ/mol)
acetone	$(CH_3)_2CO_{(l)}$	−248.1	hydrogen chloride	$HCl_{(g)}$	−92.3	phosphorus trichloride (vapor)	$PCl_{3(g)}$	−287.0
aluminum oxide	$Al_2O_{3(s)}$	−1675.7	hydrogen fluoride	$HF_{(g)}$	−273.3	potassium chlorate	$KClO_{3(s)}$	−397.7
ammonia	$NH_{3(g)}$	−45.9	hydrogen iodide	$HI_{(g)}$	+26.5	potassium chloride	$KCl_{(s)}$	−436.7
ammonium chloride	$NH_4Cl_{(s)}$	−314.4	hydrogen peroxide	$H_2O_{2(l)}$	−187.8	potassium hydroxide	$KOH_{(s)}$	−424.8
ammonium nitrate	$NH_4NO_{3(s)}$	−365.6	hydrogen sulfide	$H_2S_{(g)}$	−20.6	propane	$C_3H_{8(g)}$	−104.7
barium carbonate	$BaCO_{3(s)}$	−1216.3	iodine (vapor)	$I_{2(g)}$	+62.4	silicon dioxide	$SiO_{2(s)}$	−910.7
barium hydroxide	$Ba(OH)_{2(s)}$	−944.7	iron(III) oxide	$Fe_2O_{3(s)}$	−824.2	silver bromide	$AgBr_{(s)}$	−100.4
barium oxide	$BaO_{(s)}$	−553.5	iron(II, III) oxide	$Fe_3O_{4(s)}$	−1118.4	silver chloride	$AgCl_{(s)}$	−127.0
barium sulfate	$BaSO_{4(s)}$	−1473.2	lead(II) oxide	$PbO_{(s)}$	−219.0	silver iodide	$AgI_{(s)}$	−61.8
benzene	$C_6H_{6(l)}$	+49.0	lead(IV) oxide	$PbO_{2(s)}$	−277.4	sodium bromide	$NaBr_{(s)}$	−361.1
bromine (vapor)	$Br_{2(g)}$	+30.9	magnesium carbonate	$MgCO_{3(s)}$	−1095.8	sodium chloride	$NaCl_{(s)}$	−411.2
butane	$C_4H_{10(g)}$	−125.6	magnesium chloride	$MgCl_{2(s)}$	−641.3	sodium hydroxide	$NaOH_{(s)}$	−425.6
calcium carbonate	$CaCO_{3(s)}$	−1206.9	magnesium hydroxide	$Mg(OH)_{2(s)}$	−924.5	sodium iodide	$NaI_{(s)}$	−287.8
calcium hydroxide	$Ca(OH)_{2(s)}$	−986.1	magnesium oxide	$MgO_{(s)}$	−601.6	sucrose	$C_{12}H_{22}O_{11(s)}$	−2225.5
calcium oxide	$CaO_{(s)}$	−634.9	manganese(II) oxide	$MnO_{(s)}$	−385.2	sulfur dioxide	$SO_{2(g)}$	−296.8
carbon dioxide	$CO_{2(g)}$	−393.5	manganese(IV) oxide	$MnO_{2(s)}$	−520.0	sulfur trioxide (liquid)	$SO_{3(l)}$	−441.0
carbon disulfide	$CS_{2(l)}$	+89.0	mercury(II) oxide	$HgO_{(s)}$	−90.8	sulfur trioxide (vapor)	$SO_{3(g)}$	−395.7
carbon monoxide	$CO_{(g)}$	−110.5	mercury(II) sulfide	$HgS_{(s)}$	−58.2	sulfuric acid	$H_2SO_{4(l)}$	−814.0
chloroethene	$C_2H_3Cl_{(g)}$	+37.3	methanal (formaldehyde)	$CH_2O_{(g)}$	−108.6	tin(II) oxide	$SnO_{(s)}$	−280.7
chromium(III) oxide	$Cr_2O_{3(s)}$	−1139.7	methane	$CH_{4(g)}$	−74.4	tin(IV) oxide	$SnO_{2(s)}$	−577.6
copper(I) oxide	$Cu_2O_{(s)}$	−168.6	methanoic (formic) acid	$HCOOH_{(l)}$	−425.1	2,2,4-trimethylpentane	$C_8H_{18(l)}$	−259.2
copper(II) oxide	$CuO_{(s)}$	−157.3	methanol	$CH_3OH_{(l)}$	−239.1	urea	$CO(NH_2)_{2(s)}$	−333.5
copper(I) sulfide	$Cu_2S_{(s)}$	−79.5	methylpropane	$C_4H_{10(g)}$	−134.2	water (liquid)	$H_2O_{(l)}$	−285.8
copper(II) sulfide	$CuS_{(s)}$	−53.1	nickel(II) oxide	$NiO_{(s)}$	−239.7	water (vapor)	$H_2O_{(g)}$	−241.8
1,2-dichloroethane	$C_2H_4Cl_{2(l)}$	−126.9	nitric acid	$HNO_{3(l)}$	−174.1	zinc oxide	$ZnO_{(s)}$	−350.5
ethane	$C_2H_{6(g)}$	−83.8	nitrogen dioxide	$NO_{2(g)}$	+33.2	zinc sulfide	$ZnS_{(s)}$	−206.0
1,2-ethanediol	$C_2H_4(OH)_{2(l)}$	−454.8	nitrogen monoxide	$NO_{(g)}$	+90.2			
ethanoic (acetic) acid	$CH_3COOH_{(l)}$	−432.8	nitromethane	$CH_3NO_{2(l)}$	−113.1			
ethanol	$C_2H_5OH_{(l)}$	−235.2	octane	$C_8H_{18(l)}$	−250.1			
ethene (ethylene)	$C_2H_{4(g)}$	+52.5	ozone	$O_{3(g)}$	+142.7			
ethyne (acetylene)	$C_2H_{2(g)}$	+228.2	pentane	$C_5H_{12(l)}$	−173.5			
glucose	$C_6H_{12}O_{6(s)}$	−1273.1	phenylethene (styrene)	$C_6H_5CHCH_{2(l)}$	+103.8			
hexane	$C_6H_{14(l)}$	−198.7	phosphorus pentachloride	$PCl_{5(s)}$	−443.5			
hydrogen bromide	$HBr_{(g)}$	−36.3	phosphorus trichloride (liquid)	$PCl_{3(l)}$	−319.7			

- Standard molar enthalpies (heats) of formation are measured at SATP (25°C and 100 kPa). The values were obtained from *The CRC Handbook of Chemistry and Physics*, 71st Edition.
- The standard molar enthalpies of elements in their standard states are defined as zero.

APPENDIX F

	Oxidizing Agents		Reducing Agents	E°_r (V)

SOA
Strongest
Oxidizing
Agents

Oxidizing Agents	Reducing Agents	E°_r (V)
$F_{2(g)} + 2e^- \rightleftharpoons$	$2F^-_{(aq)}$	+2.87
$PbO_{2(s)} + SO_4^{2-}{}_{(aq)} + 4H^+{}_{(aq)} + 2e^- \rightleftharpoons$	$PbSO_{4(s)} + 2H_2O_{(l)}$	+1.69
$MnO_4^-{}_{(aq)} + 8H^+{}_{(aq)} + 5e^- \rightleftharpoons$	$Mn^{2+}{}_{(aq)} + 4H_2O_{(l)}$	+1.51
$Au^{3+}{}_{(aq)} + 3e^- \rightleftharpoons$	$Au_{(s)}$	+1.50
$ClO_4^-{}_{(aq)} + 8H^+{}_{(aq)} + 8e^- \rightleftharpoons$	$Cl^-{}_{(aq)} + 4H_2O_{(l)}$	+1.39
$Cl_{2(g)} + 2e^- \rightleftharpoons$	$2Cl^-{}_{(aq)}$	+1.36
$2HNO_{2(aq)} + 4H^+{}_{(aq)} + 4e^- \rightleftharpoons$	$N_2O_{(g)} + 3H_2O_{(l)}$	+1.30
$Cr_2O_7^{2-}{}_{(aq)} + 14H^+{}_{(aq)} + 6e^- \rightleftharpoons$	$2Cr^{3+}{}_{(aq)} + 7H_2O_{(l)}$	+1.23
$O_{2(g)} + 4H^+{}_{(aq)} + 4e^- \rightleftharpoons$	$2H_2O_{(l)}$	+1.23
$MnO_{2(s)} + 4H^+{}_{(aq)} + 2e^- \rightleftharpoons$	$Mn^{2+}{}_{(aq)} + 2H_2O_{(l)}$	+1.22
$2IO_3^-{}_{(aq)} + 12H^+{}_{(aq)} + 10e^- \rightleftharpoons$	$I_{2(s)} + 6H_2O_{(l)}$	+1.20
$Br_{2(l)} + 2e^- \rightleftharpoons$	$2Br^-{}_{(aq)}$	+1.07
$Hg^{2+}{}_{(aq)} + 2e^- \rightleftharpoons$	$Hg_{(l)}$	+0.85
$ClO^-{}_{(aq)} + H_2O_{(l)} + 2e^- \rightleftharpoons$	$Cl^-{}_{(aq)} + 2OH^-{}_{(aq)}$	+0.84
$Ag^+{}_{(aq)} + e^- \rightleftharpoons$	$Ag_{(s)}$	+0.80
$NO_3^-{}_{(aq)} + 2H^+{}_{(aq)} + e^- \rightleftharpoons$	$NO_{2(g)} + H_2O_{(l)}$	+0.80
$Fe^{3+}{}_{(aq)} + e^- \rightleftharpoons$	$Fe^{2+}{}_{(aq)}$	+0.77
$O_{2(g)} + 2H^+{}_{(aq)} + 2e^- \rightleftharpoons$	$H_2O_{2(l)}$	+0.70
$MnO_4^-{}_{(aq)} + 2H_2O_{(l)} + 3e^- \rightleftharpoons$	$MnO_{2(s)} + 4OH^-{}_{(aq)}$	+0.60
$I_{2(s)} + 2e^- \rightleftharpoons$	$2I^-{}_{(aq)}$	+0.54
$Cu^+{}_{(aq)} + e^- \rightleftharpoons$	$Cu_{(s)}$	+0.52
$O_{2(g)} + 2H_2O_{(l)} + 4e^- \rightleftharpoons$	$4OH^-{}_{(aq)}$	+0.40
$Cu^{2+}{}_{(aq)} + 2e^- \rightleftharpoons$	$Cu_{(s)}$	+0.34
$SO_4^{2-}{}_{(aq)} + 4H^+{}_{(aq)} + 2e^- \rightleftharpoons$	$H_2SO_{3(aq)} + H_2O_{(l)}$	+0.17
$Sn^{4+}{}_{(aq)} + 2e^- \rightleftharpoons$	$Sn^{2+}{}_{(aq)}$	+0.15
$Cu^{2+}{}_{(aq)} + e^- \rightleftharpoons$	$Cu^+{}_{(aq)}$	+0.15
$S_{(s)} + 2H^+{}_{(aq)} + 2e^- \rightleftharpoons$	$H_2S_{(aq)}$	+0.14
$AgBr_{(s)} + e^- \rightleftharpoons$	$Ag_{(s)} + Br^-{}_{(aq)}$	+0.07
$2H^+{}_{(aq)} + 2e^- \rightleftharpoons$	$H_{2(g)}$	0.00
$Pb^{2+}{}_{(aq)} + 2e^- \rightleftharpoons$	$Pb_{(s)}$	−0.13
$Sn^{2+}{}_{(aq)} + 2e^- \rightleftharpoons$	$Sn_{(s)}$	−0.14
$AgI_{(s)} + e^- \rightleftharpoons$	$Ag_{(s)} + I^-{}_{(aq)}$	−0.15
$Ni^{2+}{}_{(aq)} + 2e^- \rightleftharpoons$	$Ni_{(s)}$	−0.26
$Co^{2+}{}_{(aq)} + 2e^- \rightleftharpoons$	$Co_{(s)}$	−0.28
$H_3PO_{4(aq)} + 2H^+{}_{(l)} + 2e^- \rightleftharpoons$	$H_3PO_{3(aq)} + H_2O_{(l)}$	−0.28
$PbSO_{4(s)} + 2e^- \rightleftharpoons$	$Pb_{(s)} + SO_4^{2-}{}_{(aq)}$	−0.36
$Se_{(s)} + 2H^+{}_{(aq)} + 2e^- \rightleftharpoons$	$H_2Se_{(aq)}$	−0.40
$Cd^{2+}{}_{(aq)} + 2e^- \rightleftharpoons$	$Cd_{(s)}$	−0.40
$Cr^{3+}{}_{(aq)} + e^- \rightleftharpoons$	$Cr^{2+}{}_{(aq)}$	−0.41
$Fe^{2+}{}_{(aq)} + 2e^- \rightleftharpoons$	$Fe_{(s)}$	−0.45
$Ag_2S_{(s)} + 2e^- \rightleftharpoons$	$2Ag_{(s)} + S^{2-}{}_{(aq)}$	−0.69
$Zn^{2+}{}_{(aq)} + 2e^- \rightleftharpoons$	$Zn_{(s)}$	−0.76
$Te_{(s)} + 2H^+{}_{(aq)} + 2e^- \rightleftharpoons$	$H_2Te_{(aq)}$	−0.79
$2H_2O_{(l)} + 2e^- \rightleftharpoons$	$H_{2(g)} + 2OH^-{}_{(aq)}$	−0.83
$Cr^{2+}{}_{(aq)} + 2e^- \rightleftharpoons$	$Cr_{(s)}$	−0.91
$SO_4^{2-}{}_{(aq)} + H_2O_{(l)} + 2e^- \rightleftharpoons$	$SO_3^{2-}{}_{(aq)} + 2OH^-{}_{(aq)}$	−0.93
$Al^{3+}{}_{(aq)} + 3e^- \rightleftharpoons$	$Al_{(s)}$	−1.66
$Mg^{2+}{}_{(aq)} + 2e^- \rightleftharpoons$	$Mg_{(s)}$	−2.37
$Na^+{}_{(aq)} + e^- \rightleftharpoons$	$Na_{(s)}$	−2.71
$Ca^{2+}{}_{(aq)} + 2e^- \rightleftharpoons$	$Ca_{(s)}$	−2.87
$Ba^{2+}{}_{(aq)} + 2e^- \rightleftharpoons$	$Ba_{(s)}$	−2.91
$K^+{}_{(aq)} + e^- \rightleftharpoons$	$K_{(s)}$	−2.93
$Li^+{}_{(aq)} + e^- \rightleftharpoons$	$Li_{(s)}$	−3.04

DECREASING STRENGTH OF OXIDIZING AGENTS

DECREASING STRENGTH OF REDUCING AGENTS

SRA
Strongest
Reducing
Agents

- All E° values are reduction potentials measured relative to the standard hydrogen electrode. E° values are measured using standard half-cells with both the oxidizing and reducing agents present at SATP using 1.0 mol/L solutions.
- Values in this table are taken from *The CRC Handbook of Chemistry and Physics*, 71st Edition.

APPENDIX F

Percent Reaction (%)	Equilibrium Constant, K_a (mol/L)	Acid Name	Acid Formula	Conjugate Base Formula	Conjugate Base Name
100	very large	perchloric acid	$HClO_{4(aq)}$	$ClO_4^-{}_{(aq)}$	perchlorate ion
100	very large	hydroiodic acid	$HI_{(aq)}$	$I^-{}_{(aq)}$	iodide ion
100	very large	hydrobromic acid	$HBr_{(aq)}$	$Br^-{}_{(aq)}$	bromide ion
100	very large	hydrochloric acid	$HCl_{(aq)}$	$Cl^-{}_{(aq)}$	chloride ion
100	very large	sulfuric acid	$H_2SO_{4(aq)}$	$HSO_4^-{}_{(aq)}$	hydrogen sulfate ion
100	very large	nitric acid	$HNO_{3(aq)}$	$NO_3^-{}_{(aq)}$	nitrate ion
—	1.0	hydronium ion	$H_3O^+{}_{(aq)}$	$H_2O_{(l)}$	water
51	5.4×10^{-2}	oxalic acid	$HOOCCOOH_{(aq)}$	$HOOCCOO^-{}_{(aq)}$	hydrogen oxalate ion
30	1.3×10^{-2}	sulfurous acid ($SO_2 + H_2O$)	$H_2SO_{3(aq)}$	$HSO_3^-{}_{(aq)}$	hydrogen sulfite ion
27	1.0×10^{-2}	hydrogen sulfate ion	$HSO_4^-{}_{(aq)}$	$SO_4^{2-}{}_{(aq)}$	sulfate ion
23	7.1×10^{-3}	phosphoric acid	$H_3PO_{4(aq)}$	$H_2PO_4^-{}_{(aq)}$	dihydrogen phosphate ion
12	1.5×10^{-3}	iron(II) ion	$Fe(H_2O)_6^{3+}{}_{(aq)}$	$Fe(H_2O)_5(OH)^{2+}{}_{(aq)}$	—
8.2	7.4×10^{-4}	citric acid	$H_3C_6H_5O_{7(aq)}$	$H_2C_6H_5O_7^-{}_{(aq)}$	dihydrogen citrate ion
8.1	7.2×10^{-4}	nitrous acid	$HNO_{2(aq)}$	$NO_2^-{}_{(aq)}$	nitrite ion
7.8	6.6×10^{-4}	hydrofluoric acid	$HF_{(aq)}$	$F^-{}_{(aq)}$	fluoride ion
4.2	1.8×10^{-4}	methanoic acid	$HCOOH_{(aq)}$	$HCOO^-{}_{(aq)}$	methanoate ion
3.1	1.0×10^{-4}	chromium(III) ion	$Cr(H_2O)_6^{3+}{}_{(aq)}$	$Cr(H_2O)_5(OH)^{2+}{}_{(aq)}$	—
—	$\sim 10^{-4}$	methyl orange	$HMo_{(aq)}$	$Mo^-{}_{(aq)}$	methyl orange ion
—	6.3×10^{-5}	benzoic acid	$C_6H_5COOH_{(aq)}$	$C_6H_5COO^-{}_{(aq)}$	benzoate ion
2.3	5.4×10^{-5}	hydrogen oxalate ion	$HOOCCOO^-{}_{(aq)}$	$OOCCOO^{2-}{}_{(aq)}$	oxalate ion
1.3	1.8×10^{-5}	ethanoic (acetic) acid	$CH_3COOH_{(aq)}$	$CH_3COO^-{}_{(aq)}$	ethanoate (acetate) ion
0.99	9.8×10^{-6}	aluminum ion	$Al(H_2O)_6^{3+}{}_{(aq)}$	$Al(H_2O)_5(OH)^{2+}{}_{(aq)}$	—
—	4.4×10^{-7}	carbonic acid ($CO_2 + H_2O$)	$H_2CO_{3(aq)}$	$HCO_3^-{}_{(aq)}$	hydrogen carbonate ion
—	$\sim 10^{-7}$	bromothymol blue	$HBb_{(aq)}$	$Bb^-{}_{(aq)}$	bromothymol blue ion
0.10	1.1×10^{-7}	hydrosulfuric acid	$H_2S_{(aq)}$	$HS^-{}_{(aq)}$	hydrogen sulfide ion
0.079	6.3×10^{-8}	dihydrogen phosphate ion	$H_2PO_4^-{}_{(aq)}$	$HPO_4^{2-}{}_{(aq)}$	hydrogen phosphate ion
0.079	6.2×10^{-8}	hydrogen sulfite ion	$HSO_3^-{}_{(aq)}$	$SO_3^{2-}{}_{(aq)}$	sulfite ion
0.054	2.9×10^{-8}	hypochlorous acid	$HClO_{(aq)}$	$ClO^-{}_{(aq)}$	hypochlorite ion
—	$\sim 10^{-10}$	phenolphthalein	$HPh_{(aq)}$	$Ph^-{}_{(aq)}$	phenolphthalein ion
0.0078	6.2×10^{-10}	hydrocyanic acid	$HCN_{(aq)}$	$CN^-{}_{(aq)}$	cyanide ion
0.0076	5.8×10^{-10}	ammonium ion	$NH_4^+{}_{(aq)}$	$NH_{3(aq)}$	ammonia
0.0076	5.8×10^{-10}	boric acid	$H_3BO_{3(aq)}$	$H_2BO_3^-{}_{(aq)}$	dihydrogen borate ion
0.0032	1.0×10^{-10}	phenol	$C_6H_5OH_{(aq)}$	$C_6H_5O^-{}_{(aq)}$	phenoxide ion
0.0022	4.7×10^{-11}	hydrogen carbonate ion	$HCO_3^-{}_{(aq)}$	$CO_3^{2-}{}_{(aq)}$	carbonate ion
0.00047	2.2×10^{-12}	hydrogen peroxide	$H_2O_{2(aq)}$	$HO_2^-{}_{(aq)}$	hydrogen peroxide ion
0.00020	4.2×10^{-13}	hydrogen phosphate ion	$HPO_4^{2-}{}_{(aq)}$	$PO_4^{3-}{}_{(aq)}$	phosphate ion
0.00011	1.3×10^{-13}	hydrogen sulfide ion	$HS^-{}_{(aq)}$	$S^{2-}{}_{(aq)}$	sulfide ion
—	1.0×10^{-14}	water	$H_2O_{(l)}$	$OH^-{}_{(aq)}$	hydroxide ion
—	very small	hydroxide ion	$OH^-{}_{(aq)}$	$O^{2-}{}_{(aq)}$	oxide ion
—	very small	ammonia ion	$NH_{3(aq)}$	$NH_2^-{}_{(aq)}$	amide ion

SA Strongest Acids

DECREASING STRENGTH OF ACIDS

DECREASING STRENGTH OF BASES

SB Strongest Bases

APPENDIX F

- The percent reaction of acids with water is for 0.10 mol/L solutions and is only valid for concentrations close to 0.10 mol/L. All measurements of acid strengths were made at SATP. No percent reaction is given for benzoic acid or carbonic acid because these acids have molar solubilities less than 0.10 mol/L at SATP. No percent reaction is given for indicators because indicators are generally used at concentrations lower than 0.10 mol/L.

- Values in this table are taken from *Lange's Handbook of Chemistry*, 13th Edition for 25°C.

Name	Formula	K_{sp}
barium carbonate	$BaCO_{3(s)}$	2.6×10^{-9}
barium chromate	$BaCrO_{4(s)}$	1.2×10^{-10}
barium sulfate	$BaSO_{4(s)}$	1.1×10^{-10}
calcium carbonate	$CaCO_{3(s)}$	5.0×10^{-9}
calcium oxalate	$CaC_2O_{4(s)}$	2.3×10^{-9}
calcium phosphate	$Ca_3(PO_4)_{2(s)}$	2.1×10^{-33}
calcium sulfate	$CaSO_{4(s)}$	7.1×10^{-5}
copper(I) chloride	$CuCl_{(s)}$	1.7×10^{-7}
copper(I) iodide	$CuI_{(s)}$	1.3×10^{-12}
copper(II) iodate	$Cu(IO_3)_{2(s)}$	6.9×10^{-8}
copper(II) sulfide	$CuS_{(s)}$	6.0×10^{-37}
iron(II) hydroxide	$Fe(OH)_{2(s)}$	4.9×10^{-17}
iron(II) sulfide	$FeS_{(s)}$	6.0×10^{-19}
iron(III) hydroxide	$Fe(OH)_{3(s)}$	2.6×10^{-39}
lead(II) bromide	$PbBr_{2(s)}$	6.6×10^{-6}
lead(II) chloride	$PbCl_{2(s)}$	1.2×10^{-5}
lead(II) iodate	$Pb(IO_3)_{2(s)}$	3.7×10^{-13}
lead(II) iodide	$PbI_{2(s)}$	8.5×10^{-9}
lead(II) sulfate	$PbSO_{4(s)}$	1.8×10^{-8}
magnesium carbonate	$MgCO_{3(s)}$	6.8×10^{-6}
magnesium hydroxide	$Mg(OH)_{2(s)}$	5.6×10^{-12}
mercury(I) chloride	$Hg_2Cl_{2(s)}$	1.5×10^{-18}
silver bromate	$AgBrO_{3(s)}$	5.3×10^{-5}
silver bromide	$AgBr_{(s)}$	5.4×10^{-13}
silver carbonate	$Ag_2CO_{3(s)}$	8.5×10^{-12}
silver chloride	$AgCl_{(s)}$	1.8×10^{-10}
silver chromate	$Ag_2CrO_{4(s)}$	1.1×10^{-12}
silver iodate	$AgIO_{3(s)}$	3.2×10^{-8}
silver iodide	$AgI_{(s)}$	8.5×10^{-17}
strontium carbonate	$SrCO_{3(s)}$	5.6×10^{-10}
strontium fluoride	$SrF_{2(s)}$	4.3×10^{-9}
strontium sulfate	$SrSO_{4(s)}$	3.4×10^{-7}
zinc hydroxide	$Zn(OH)_{2(s)}$	7.7×10^{-17}
zinc sulfide	$ZnS_{(s)}$	2.0×10^{-25}

Values in this table are taken from *The CRC Handbook of Chemistry and Physics*, 76th Edition.

APPENDIX G

COMMON CHEMICALS

You live in a chemical world. As one bumper sticker asks, "What in the world isn't chemistry?" Every natural and technologically produced substance around you is composed of chemicals. Many of these chemicals are used to make your life easier or safer, and some of them have life-saving properties. Following is a list of selected common chemicals. The Chemicals marked with an asterisk are to be memorized.

Common Name	Recommended Name	Formula	Common Use/Source
acetic acid*	ethanoic acid	$CH_3COOH_{(aq)}$	vinegar
acetone*	propanone	$(CH_3)_2CO_{(1)}$	nail polish remover
acetylene*	ethyne	$C_2H_{2(g)}$	cutting/welding torch
ASA (Aspirin®)	acetylsalicylic acid	$C_6H_4COOCH_3COOH_{(s)}$	for pain relief medication
baking soda*	sodium hydrogen carbonate	$NaHCO_{3(s)}$	leavening agent
battery acid*	sulfuric acid	$H_2SO_{4(aq)}$	car batteries
bleach	sodium hypochlorite	$NaClO_{(s)}$	bleach for clothing
bluestone	copper(II) sulfate-5-water	$CuSO_4 \cdot 5\,H_2O_{(s)}$	algicide/fungicide
brine*	aqueous sodium chloride	$NaCl_{(aq)}$	water-softening agent
citric acid	2-hydroxy-1,2,3-propanetricarboxylic acid	$C_3H_4OH(COOH)_3$	in fruit and beverages
CFC	chlorofluorocarbon	$C_xCl_yF_{z(1)}$; e.g., $C_2Cl_2F_{4(1)}$	refrigerant
charcoal/graphite*	carbon	$C_{(s)}$	fuel/lead pencils
dry ice*	carbon dioxide	$CO_{2(g)}$	"fizz" in carbonated beverages
ethylene*	ethene	$C_2H_{4(g)}$	for polymerization
ethylene glycol*	1,2-ethanediol	$C_2H_4(OH)_{2(1)}$	radiator antifreeze
freon-12	dichlorodifluoromethane	$CCl_2F_{2(1)}$	refrigerant
Glauber's salt	sodium sulfate-10-water	$Na_2SO_4 \cdot 10\,H_2O_{(s)}$	solar heat storage
glucose*	D-glucose; dextrose	$C_6H_{12}O_{6(s)}$	in plants and blood
grain alcohol*	ethanol (ethyl alcohol)	$C_2H_5OH_{(1)}$	beverage alcohol
gypsum	calcium sulfate-2-water	$CaSO_4 \cdot 2\,H_2O_{(s)}$	wallboard
lime (quicklime)*	calcium oxide	$CaO_{(s)}$	masonry
limestone*	calcium carbonate	$CaCO_{3(s)}$	chalk and building materials
lye (caustic soda)*	sodium hydroxide	$NaOH_{(s)}$	oven/drain cleaner
malachite	copper(II) hydroxide carbonate	$Cu(OH)_2 \cdot CuCO_{3(s)}$	copper mineral
methyl hydrate*	methanol (methyl alcohol)	$CH_3OH_{(1)}$	gas-line antifreeze
milk of magnesia	magnesium hydroxide	$Mg(OH)_{2(s)}$	antacid (for indigestion)
MSG	monosodium glutamate	$NaC_5H_8NO_{4(s)}$	flavor enhancer
muriatic acid*	hydrochloric acid	$HCl_{(aq)}$	in concrete etching
natural gas*	methane	$CH_{4(g)}$	fuel
PCBs	polychlorinated biphenyls	$(C_6H_xCl_y)_2$; e.g., $(C_6H_4Cl_2)_{2(1)}$	in transformers
potash*	potassium chloride	$KCl_{(s)}$	fertilizer
road salt*	calcium chloride or sodium chloride	$CaCl_{2(s)}$ or $NaCl_{2(s)}$	melts ice
rotten-egg gas*	hydrogen sulfide	$H_2S_{(g)}$	in natural gas
rubbing alcohol	2-propanol	$CH_3CHOHCH_{3(1)}$	for massage
sand (silica)	silicon dioxide	$SiO_{2(s)}$	in glass making
slaked lime*	calcium hydroxide	$Ca(OH)_{2(s)}$	limewater
soda ash*	sodium carbonate	$Na_2CO_{3(s)}$	in laundry detergents
sugar*	sucrose	$C_{12}H_{22}O_{11(s)}$	sweetener
table salt*	sodium chloride	$NaCl_{(s)}$	seasoning
washing soda*	sodium carbonate-10-water	$Na_2CO_3 \cdot 10\,H_2O_{(s)}$	water softener
vitamin C	ascorbic acid	$H_2C_6H_6O_{6(s)}$	vitamin supplement

APPENDIX G

GLOSSARY

A

absolute zero the lowest possible temperature; 0 K or −273.15°C

accuracy the closeness of an experimental value to an accepted or predicted value; usually expressed as a percent difference

acid a substance that forms a conducting, aqueous solution that turns blue litmus paper red, neutralizes bases, and reacts with active metals to form hydrogen gas (see also *Arrhenius's acid-base theory* and *Brønsted-Lowry definitions*)

acid-base indicator see *indicator*

acid ionization constant (K_a) equilibrium constant for the ionization of an acid; also known as the acid dissociation constant

acid rain rain with a pH less than 5.6

acidic a solution that turns blue litmus red and has a pH less than 7; $[H^+_{(aq)}] > [OH^-_{(aq)}]$

actinides elements with atomic numbers from 90 to 103

activated complex the structural arrangement of particles representing the highest potential energy (changeover) point in a chemical reaction step

activation energy the minimum energy with which particles must collide in order for the collision to be effective, resulting in rearrangement of bonds

active metal a metal that spontaneously reacts with water or an acid to produce hydrogen gas; a metal that is a stronger reducing agent than hydrogen

activity series a list of substances (usually metals) in order of their reactivity with another, controlled chemical (usually an acid)

addition polymerization a reaction in which unsaturated monomers combine with each other to form a polymer

addition reaction a type of organic reaction of alkenes and alkynes in which a small molecule is added to a double bond or triple bond

alcohols a family of organic compounds characterized by the presence of a hydroxyl functional group; R–OH

aldehydes a family of organic compounds characterized by a terminal carbonyl functional group, RHO (see also Table 9.1, page 342)

aliphatic compound organic compound with chain of single, double, or triple bonds

aliphatic hydrocarbon a member of the alkane, alkene, or alkyne family including "cyclo-" compounds but not aromatics

alkali metals the family of elements corresponding to Group 1 of the periodic table of the elements

alkaline see *basic*

alkaline-earth metals a family of elements in Group 2 of the periodic table of the elements

alkanes a hydrocarbon family of molecules that contain only carbon-carbon single bonds; C_nH_{2n+2}

alkenes a hydrocarbon family of molecules that contain at least one carbon-carbon double bond; usually C_nH_{2n}

alkyl group (or alkyl branch) an alkane with one hydrogen atom removed, acting as a branch of a larger molecule: C_nH_{2n+1}

alkynes a hydrocarbon family of molecules that contain at least one carbon-carbon triple bond; usually C_nH_{2n-2}

alloy a chemical or physical combination of two or more elements that has metallic properties

alpha particle the nucleus of a helium atom (usually two protons and two neutrons)

ambient conditions surrounding or room conditions

amides a family of organic compounds characterized by the presence of a carbonyl functional group bonded to a nitrogen atom (see also Table 9.1, page 342)

amines a family of organic compounds characterized by the presence of single-bonded nitrogen atoms as the functional group (see also Table 9.1, page 342)

amino acids a family of organic compounds that are structurally bi-functional, containing both an amine group (–NH₂) and a carboxyl group (–COOH)

amount of matter (n) SI quantity for the number of particles in a substance in units of moles (mol)

ampere (A) SI base unit for electric current; one coulomb per second (1 C/s)

amphiprotic a substance capable of acting as an acid or a base in different chemical reactions

amphoteric ions that could, theoretically, form an acidic or basic solution (see *amphiprotic*)

analogy a method of communicating an idea by comparison to a more familiar situation

analysis section of a report of scientific work;

768   GLOSSARY

manipulations, calculations, and interpretations of evidence in order to answer the question stated in the Problem of an investigation

analytical chemistry a branch of chemistry concerned with analyzing samples for the type and quantity of chemicals present

anhydrous the form of a substance without any water of hydration

anion a historical name for a negatively charged ion

anode the electrode in a cell where the oxidation half-reaction occurs

anomaly a departure from a regular rule; irregularity

aqueous (aq) a solution that has water as the solvent

aromatics a family of organic compounds including benzene and all other carbon compounds that have benzene-like structures and properties

Arrhenius's acid-base theory acids ionize in aqueous solutions to produce hydrogen ions; bases (ionic hydroxides) dissociate to produce hydroxide ions

assumption untested statement(s) presumed to be correct without proof or demonstration in order to develop or apply a theory, law, or generalization

atmosphere (atm) a non-SI unit of pressure; 1 atm = 101.325 kPa

atom the smallest part of an element that is representative of the element; a neutral particle composed of a nucleus containing protons and neutrons, and with the number of electrons equal to the number of protons

atomic mass historically defined as the mass of an element that combines with one gram of hydrogen

atomic mass unit (amu) a unit of mass defined as one-twelfth of the mass of a carbon-12 atom

atomic number a characteristic number for an element; believed to represent the number of protons in the nucleus of an atom of that element

atomic spectrum the characteristic series of colored lines produced when light emitted by an element energized by heat or electricity passes through a spectroscope

Avogadro's constant (N_A) 6.02×10^{23} entities/mol; the number of entities in one mole

Avogadro's number the number of entities in one mole; 6.02×10^{23}

Avogadro's theory equal volumes of gases at the same temperature and pressure contain the same number of molecules

B

baking soda sodium hydrogen carbonate

balanced chemical equation one in which the total number of each kind of atom or ion in the reactants is equal to the total number of the same kind of atom or ion in the products

barometer a device for measuring atmospheric pressure

base compound forming a conducting, aqueous solution that turns red litmus paper blue and neutralizes acids (see also *Arrhenius's acid-base theory* and *Brønsted-Lowry definitions*)

base ionization constant (K_b) equilibrium constant for the ionization of a base

base unit a fundamental SI unit of measurement that cannot be expressed in terms of simpler units

basic a solution that turns red litmus blue and has a pH greater than 7; $[OH^-_{(aq)}] > [H^+_{(aq)}]$

battery a set of two or more voltaic cells joined to produce an electric current

bauxite aluminum ore; aluminum oxide-2-water

binary acid a substance containing hydrogen and one other nonmetallic element

binary ionic compound a compound that contains only two kinds of monatomic ions

binary molecular compound a compound that contains only two kinds of nonmetal atoms

Bohr model of atoms describes an atom as a nucleus surrounded by orbiting electrons with specific energy levels

bomb calorimeter an apparatus for measuring the quantity of heat absorbed or released by reactants placed in an inner compartment surrounded by water

bond energy the energy required to break a bond or released when a bond is formed (in kJ/mol)

bonding capacity the maximum number of single covalent bonds formed by an atom; determined by the number of bonding electrons available

bonding electron a single, unpaired electron in a valence orbital

Boyle's law the volume of a gas varies inversely with the pressure if the amount of gas and temperature are kept constant

branch any group of carbon atoms that are not part of the main structure of an organic molecule

brass an alloy of copper and zinc

brine a common term for aqueous sodium chloride

Brønsted-Lowry definitions an acid is defined to be a proton donor and a base is defined to be a proton acceptor; a neutralization reaction is competition for protons that results in a proton transfer

bronze an alloy of copper and tin

Brownian motion random, erratic motions of microscopic particles caused by molecular collisions

buffer a mixture of a conjugate acid-base pair that maintains a nearly constant pH when diluted or when a strong acid or base is added

buret a long, graduated tube equipped with a stopcock to measure solution volumes in a titration

burning see *combustion*

by-product an additional product (other than the primary product) that may also have a useful purpose

C

calorimeter an insulated container with a measured quantity of water; an isolated system used to determine quantities of heat transferred

calorimetry the technological process of determining quantities of heat transferred by using a calorimeter

carbohydrates organic compounds composed of carbon, hydrogen, and oxygen whose most common members have the formula $C_x(H_2O)_y$

carbon cycle the movement of carbon throughout the environment by photosynthesis, respiration, and the burning of fossil fuels

carbonic acid acid formed from dissolved carbon dioxide gas (e.g., carbonated drinks); $H_2CO_{3(aq)}$

carbonyl group a functional group containing a carbon atom joined with a double bond to an oxygen atom; $C=O$

carboxyl group a functional group containing a carbonyl group and a hydroxyl group; COOH

carboxylic acid a family of organic compounds characterized by the presence of a carboxyl group; RCOOH (see also Table 9.1, page 342)

catalyst a chemical substance that increases the rate of a chemical reaction without being altered or consumed

cathode the electrode in a cell where the reduction half-reaction occurs

cathodic protection a method of corrosion protection in which the iron is forced to become the cathode by supplying the iron with electrons (using an impressed current or sacrificial anode)

cation a historical name for a positively charged ion

cell a device containing one or more electrolytes and two electrodes; used to convert chemical energy into electrical energy or vice versa

cell potential (ΔE) difference between the reduction potentials of the cathode and anode when no current is flowing

cell reaction the net chemical reaction in an electrochemical cell

cellulose a polymer of beta-glucose molecules that forms the framework of plants

certainty an expression of level of confidence; communicated by a number of significant digits for quantitative values

certainty rule a result obtained by multiplying or dividing measured or calculated values is rounded to the same certainty (number of significant digits) as the least certain value used in the calculation

charge a property of an atom or group of atoms representing an excess or deficiency of electrons and measured in positive or negative units; quantity of electricity (q) measured in coulombs (C)

Charles' law the volume of a gas varies directly with its temperature in kelvins if the amount of gas and pressure are constant

chemical bond the electrical attraction that holds atoms or ions together in a compound

chemical change see *chemical reaction*

chemical decomposition see *simple decomposition*

chemical energy energy transferred in a chemical reaction; potential energy contained in chemical bonds

chemical equation international method of communicating the type, relative number, and state of matter of reactants and products in a chemical reaction

chemical formula a group of symbols representing the number and type of atoms or ions in a chemical substance

chemical kinetics the study of particle interactions and energies at a molecular level during chemical reactions

chemical reaction a change in which new substances with different properties are formed, as evidenced by changes in color, energy, odor, or state

chemical reaction equilibrium a dynamic equilibrium of forward and reverse rates of reaction in a closed chemical reaction system

chemical system a group of chemicals being studied, separated from the surroundings by a boundary

chemical technology the study and application of skills, processes, and equipment for the production and use of chemicals

chemistry the study of the composition, properties, and changes in matter

chlorophyll the green coloring matter in a plant that acts as a catalyst in photosynthesis

classical system an old system of suffixes (e.g., *-ic* and *-ous*) used to name chemicals (not recommended by IUPAC)

closed system one in which no substance can enter or leave

coal fossil fuel made up chiefly of carbon

coefficient the number of molecules or formula units of a chemical involved in a chemical reaction

collision-reaction theory the idea that chemical reactions involve collisions and rearrangements of particles

combined gas law the product of the pressure and

volume of a gas sample is proportional to its absolute temperature

combustion the rapid reaction of a chemical with oxygen to produce oxides and heat; complete combustion produces the most common oxides

commercial pertaining to the production and marketing of goods; smaller in scale than industrial

common ion effect the shift in an equilibrium of ions in solution caused by adding a solute compound containing one of these ions

complete combustion the reaction of an element or compound with oxygen to produce the most common oxides

compound a pure substance that can be separated into elements by heat or electricity; a substance containing atoms of more than one element in a definite fixed proportion

concentrated solution a homogeneous mixture with a relatively high ratio of solute to solution; e.g., a saturated solution

concentration the ratio of the quantity of solute to the quantity of solution or solvent

concept in science, a theory, law, definition, or generalization created to describe a natural phenomenon

condensation the change of state from a gas to a liquid; an organic reaction in which water is formed

condensation polymerization a reaction in which two different monomers join to form a polymer and release a small molecule such as water or hydrogen chloride

condensation reaction an organic reaction in which two smaller molecules combine to form a larger molecule, with water as the other product

conductivity a measure of the ability of a pure substance or mixture to conduct an electric current

conjugate acid an acid formed by adding a proton (H^+) to a base

conjugate acid-base pair two substances whose formulas differ only by one H^+ unit

conjugate base a base formed by removing a proton (H^+) from an acid

control a substance or procedure that does not change or that is used as a comparison in an experiment

controlled variable (fixed variable) any factor that could vary but is held constant so as not to affect the outcome of an experiment

cooling curve a temperature-time graph showing temperature changes of a system as heat is transferred from the system

coordinate covalent bond a covalent bond in which one of the atoms donates both electrons

corrosion the adverse reaction of man-made items with chemicals in the environment (usually metals or alloys reacting to form oxides, carbonates, or sulfides)

coulomb (C) SI unit for the quantity of electric charge transferred in one second by one ampere of current

covalent bond the simultaneous attraction of two nuclei for a shared pair of electrons

covalent bond radius one-half the distance between the nuclei of the two atoms in a molecule

cracking a type of organic reaction in which hydrocarbons are broken down into smaller molecules by means of heat (thermal cracking) or catalysts (catalytic cracking)

cryolite sodium hexafluoroaluminum mineral, $Na_3AlF_{6(s)}$, used as a molten solvent in the Hall-Héroult process for making aluminum metal

crystal lattice a continuous, three-dimensional, repeating pattern of ions, atoms, or molecules in a solid

crystallization the process of obtaining a solid by evaporating the solvent or cooling a concentrated solution

current (I) the rate of transfer of electric charge measured in amperes (A)

cyclic hydrocarbon a hydrocarbon whose molecules have a closed ring structure

cycloalkanes a family of compounds that contain a ring of singly-bonded carbon atoms

D

Dalton's model of atoms describes atoms as tiny, featureless, neutral spheres (analogy: billiard balls)

dependent variable see *responding variable*

diagnostic test a short and specific laboratory procedure with expected evidence and analysis used as an empirical test to detect the presence of a chemical

diatomic composed of two atoms

diffusion the spontaneous mixing of one substance with another

dilute solution a homogeneous mixture that has relatively little solute per unit volume of solution

dilution the process of decreasing the concentration of a solution, usually by adding more solvent

dipole a partial separation of positive and negative charges within a molecule, due to electronegativity differences

dipole-dipole force a type of intermolecular bond caused by the attractions of oppositely charged ends of polar molecules

diprotic the ability to donate or accept two protons (H^+)

disaccharide a sugar that contains two simple sugar molecules bonded together (e.g., sucrose)

discharging the spontaneous conversion of chemical energy into electrical energy in a cell

dispersion the distribution of particles of a substance in a medium (solvent)

disproportionation the reaction of a single substance with itself to produce two different substances

dissociation the separation of an ionic compound into individual ions in a solvent

distillation the process of vaporizing and then condensing a liquid

double bond an attraction between atoms in a molecule due to the sharing of two pairs of electrons in a covalent bond

double replacement the reaction of two ionic compounds in which cations and anions rearrange, producing two new compounds

double salt an ionic compound containing two kinds of cations or anions

dry cell originally used to describe the zinc-chloride cell but now used to describe any sealed electric cell with semi-solid contents

ductile able to be pulled or formed into a wire or tube

dynamic equilibrium a balance between forward and reverse processes occurring at the same rate (see also *equilibrium*)

E

$\Delta E°$ see *standard cell potential*

$E_r°$ see *standard reduction potential*

ecological pertaining to the relationships between living organisms and the environment

economic perspective focusing on the production, distribution, and consumption of wealth

electric cell a device for converting chemical energy continuously into electrical energy

electric current (I) the rate of flow of charge past a point in an electrical circuit

electric potential difference (V) the difference in potential energy per coulomb of charge (J/C) between the anode and cathode of a cell and measured in volts (V); also known as voltage

electrochemical cell a cell that either converts chemical energy into electrical energy or electrical energy into chemical energy by a redox reaction

electrochemistry the branch of chemistry that studies electron transfers in chemical reactions

electrode a solid electrical conductor (usually a metal or carbon rod) in a cell where the electrical connections are made; the site of oxidation (electron loss) or reduction (electron gain) half-reactions

electrolysis the process of using electrical energy to produce non-spontaneous chemical reactions in a cell

electrolyte a solute that forms a solution that conducts an electric current; a substance that ionizes in water to form individual ions

electrolytic cell a cell in which an electrolysis occurs

electron (e^-) a small, negatively charged subatomic particle; has a specific energy within an atom

electron affinity the energy change that occurs when an electron is accepted by a neutral atom in the gaseous state

electron dot diagram a representation of the Lewis model for an atom or molecule that uses the chemical symbol to represent the nucleus and inner electrons and dots to represent the valence electrons

electronegativity a number that describes the relative ability of an atom to attract a pair of bonding electrons in its valence level

electronic structure the arrangement of electrons in the energy levels of an atom

electroplating the process of depositing a metal at the cathode of a cell

electrorefining a process for producing pure metals using an electrolytic cell to deposit the pure metal at the cathode and using the impure metal at the anode

electrowinning a process for producing pure metals using an electrolytic cell that has an inert anode and a cathode at which metal cations are reduced from a molten or aqueous electrolyte

element a pure substance that cannot be further decomposed chemically; composed of only one kind of atom

elimination reaction a type of organic reaction in which a saturated compound is converted into an unsaturated compound by the removal of a hydrogen atom or a hydroxyl group

emotional perspective focusing on feelings as opposed to logic or reason; e.g., fear, love, joy, hate

empirical relating to past experience or experiments

empirical definition a statement that defines an object or process in terms of observable properties

empirical formula a chemical formula determined by experiment

endothermic change a change in which energy (usually in the form of heat) is absorbed by a system from the surroundings, resulting in an increase in the potential energy of the system

endpoint a point of a titration at which a sharp change in a measurable and characteristic property occurs; e.g, a color change

energy a property of a substance or system that relates to its ability to do work

energy density the quantity of energy stored or supplied per unit mass of a battery; also called specific energy

energy level a specific energy an electron can have in an atom or ion

energy resource a natural substance that provides energy (e.g., fossil fuels)

enthalpy the total kinetic and potential energy of a system under constant pressure

enthalpy change (ΔH) the change in total internal energy of a system when the pressure or volume of the system is held constant

entity any particle or particle-like object, e.g., atoms, ions, or molecules; single thing

entropy the quality of randomness, or freedom of movement, in a particle system, or, the quantity of disorder in a system; entropy tends to increase to a maximum in all systems

enzyme complex molecules (proteins) that act to catalyze organic chemical reactions, particularly in living things

equilibrium a state of a closed system in which all measurable properties are constant (see *dynamic equilibrium*)

equilibrium constant (K) the value obtained from the mathematical combination of equilibrium concentrations using the equilibrium law

equilibrium law an expression equal to the product of the equilibrium concentrations of chemical products divided by the product of the equilibrium concentrations of the chemical reactants, where each concentration is raised to the power of the coefficient of that entity in the balanced chemical equation

equivalence point the measured quantity of titrant recorded when the endpoint occurs; the point at which chemically equivalent amounts have reacted

equilibrium shift a change in a system that is initially at equilibrium so that the quantity of either reactants or products appears to be increasing

Erlenmeyer flask a conical flask with a large, flat bottom used to mix a sample during a titration

ester a family of organic compounds characterized by the presence of a carbonyl group bonded to an oxygen atom (see also Table 9.1, page 342)

esterification condensation reaction of a carboxylic acid and an alcohol to produce an ester and water

esthetic perspective focusing on beauty

ether a family of organic compounds characterized by the presence of an oxy functional group R_1–O–R_2

ethical perspective focusing on standards of right and wrong

ethylene glycol a common name for 1,2-ethanediol, used as a radiator antifreeze; $C_2H_4(OH)_{2(l)}$

evaluation section of a report of scientific work; judgment of the processes used to plan and perform an investigation, and of the prediction and the authority used to make the prediction

evidence section of a report of scientific work; qualitative or quantitative observations relevant to answering the problem in an investigation

excess reagent the reactant that is present in more than the required amount for complete reaction

exothermic change a change in which energy (usually in the form of heat) is released from a system into the surroundings, resulting in a decrease in potential energy of the system

experimental design section of a report of scientific work; a specific plan used to obtain the answer to the problem in an experiment, including reacting chemicals, and, if applicable, diagnostic tests, variables, and controls

exponential notation see *scientific notation*

extent of equilibrium the empirical measure of the state of an equilibrium describing relative quantities of reactants and products; e.g., percent reaction, equilibrium constant

extrapolation estimation of values beyond the range of measured points on a graph obtained by extending the line or curve joining the measured points

F

family a group of substances with similar properties; e.g., a family of elements or an organic family

Faraday constant (F) the quantity of charge transferred by one mole of electrons; also known as the molar charge of electrons; 9.65×10^4 C/mol

filtrate the solution that flows through a filter paper

filtration the process of separating a low solubility solid from a liquid using a filter

fission the splitting of atomic nuclei into smaller nuclei and neutrons releasing large quantities of energy

flame test a diagnostic test based on the characteristic colors of ions in a flame

formation the reaction of two or more elements to produce a compound

formula subscript number of ions or atoms present in one molecule or formula unit of a substance

formula unit the smallest amount of a substance that has the composition given by the chemical formula

forward reaction the chemical reaction proceeding from left to right as written in a balanced equation, where reactants are changing to products

fossil fuels an energy resource believed to be the accumulated remains of plants and animals from past geological periods (coal, tar sands, oil, and natural gas)

fractionation a process involving the separation by distillation of different components (fractions) of a liquid mixture

fuel cell chemical cell that produces electricity directly by the reaction of a fuel that is continuously added to the cell

functional group a characteristic arrangement of atoms within a molecule that determines the most important chemical and physical properties of the compound

fusion (1) a physical change commonly known as melting; (2) a nuclear change in which small nuclei combine to form a larger nucleus accompanied by the release of very large quantities of energy

G

galvanic cell a cell that operates spontaneously to produce electricity; also known as a voltaic cell

gas a substance that fills and assumes the shape of its container, diffuses rapidly, mixes readily with other gases, and is highly compressible

gasohol a mixture of gasoline and alcohol used as an automobile fuel

generalization a statement that summarizes a relatively small number of empirical results

geothermal energy a natural energy source using heat from inside the Earth

Glauber's salt mineral obtained from salt lakes; sodium sulfate-10-water

glycerol a common name for 1,2,3-propanetriol; a colorless, odorless, viscous liquid used in foods, cosmetics, and explosives; $C_3H_5(OH)_{3(l)}$

graduated pipet a precise device consisting of a narrow glass tube with regular markings; used to measure liquid volumes

gravimetric pertaining to mass measurements

ground state the most stable state of an atom with all electrons in the lowest allowed energy levels

group a column of elements in the modern periodic table

H

ΔH see *enthalpy change*

half-cell an electrode-electrolyte combination forming one-half of a complete cell

half-reaction a balanced chemical equation representing either a loss or gain of electrons by a substance

halogens family of elements corresponding to Group 17 of the periodic table of the elements

heat energy transferred between two systems

heat capacity (C) the quantity of energy required to change the temperature of a substance by exactly one degree Celsius (see *specific heat capacity*)

heat of reaction the quantity of heat released or absorbed when stoichiometric amounts react; the difference in energy content or potential energy of products and reactants; also known as enthalpy change

heating curve a temperature-time graph showing temperature changes of a system as heat is transferred into the system

heavy metal a transition metal or a toxic metal that accumulates in living systems; usually referring to mercury, lead, or cadmium

Hess's law the algebraic addition of chemical equations yields a net equation whose enthalpy of reaction is the algebraic sum of the individual enthalpies of reaction; $\Delta H_{net} = \Sigma \Delta H$

heterogeneous mixture a non-uniform mixture consisting of more than one phase

homogeneous mixture a uniform mixture consisting of only one phase

Hund's rule no electron pairing takes place in *p*, *d*, or *f* orbitals until each electron of the given set contains one electron

hydrate a compound that decomposes at a relatively low temperature to produce water and another substance; a compound containing loosely bonded water molecules

hydrocarbon a molecular compound containing only carbon and hydrogen atoms

hydrocarbon derivative a molecular compound of carbon and at least one other element that is not hydrogen

hydrogen bond a special, relatively strong dipole-dipole force between molecules containing F–H, O–H, or N–H bonds

hydrogenation an addition reaction involving the addition of hydrogen to convert carbon-carbon double bonds or triple bonds in unsaturated compounds into single carbon-carbon bonds of saturated compounds

hydrolysis the reaction of an entity with water to produce acidic or basic solutions (hydronium or hydroxide ions)

hydronium ion a hydrated proton, conventionally represented as $H_3O^+_{(aq)}$

hydroxyl group an –OH functional group characteristic of alcohols

hypothesis a preliminary concept that requires further testing before being accepted

I

ideal gas a hypothetical gas that obeys all gas laws under all conditions; it is assumed that no forces act among the gas particles except during collisions

ideal gas law $pv = nRT$; the product of the pressure and volume of a gas is directly proportional to the amount and the absolute temperature of the gas

immiscible two liquids that form separate layers instead of dissolving

impressed current an electric current that is forced to flow toward an iron object by applying an external potential difference

independent variable see *manipulated variable*

indicator a chemical substance that changes color when another substance, such as an acid or base, is added

industrial involving large-scale production of substances, usually from natural raw materials

inert chemically non-reactive

inorganic pertaining to compounds other than those based on molecular compounds of carbon

insoluble having negligible solubility

intermediate product substances formed in a step in the progress of a chemical reaction which are reacted in a later step and are not present as final products

inter-metallic compound a pure substance composed of different metals

intermolecular forces weak forces or bonds acting among molecules

interpolation estimating values between measured points on a graph

interpretation any knowledge obtained indirectly to describe or explain a substance or process

ion single particle or group of particles having a net positive or negative charge

ionic bond the simultaneous attraction among positive and negative ions

ionic compound a pure substance formed from a metal and a nonmetal; crystalline solid at SATP; has relatively high melting point; and conductor of electricity in molten or aqueous states

ionic crystals neutral, three-dimensional structures of positive and negative ions arranged in a repeating pattern

ionic formula a group of chemical symbols representing the simplest whole-number ratio of ions in the compound

ionic radius the radius of an anion or a cation in an ionic crystal

ionization the process of converting an atom or molecule to an ion

ionization constant for water (K_w) equilibrium constant for the dissociation of water; also known as the ion product constant; 1.0×10^{-14} mol^2/L^2

isoelectronic having the same number of electrons

isolated system one in which neither matter nor energy can transfer in or out

isomers compounds with the same molecular formula but with different structures

isotope a variety of atoms of an element; atoms of this variety have the same number of protons as atoms of other varieties of this element, but a different number of neutrons

IUPAC International Union of Pure and Applied Chemistry; the organization that establishes the conventions used by chemists

J

joule (J) SI unit of energy

K

K_a see *acid ionization constant*

K_b see *base ionization constant*

K_w see *ionization constant for water*

Kelvin temperature scale (K) a temperature scale with zero kelvins (0 K) at absolute zero and the same size divisions as the Celsius temperature scale

ketones a family of organic compounds characterized by the presence of a carbonyl group bonded to two carbon atoms (see also Table 9.1, page 342)

kinetic energy a form of energy related to the motion of a particle

kinetic molecular theory a current theory of matter based on the idea that all matter is made up of particles in continuous random motion; the average kinetic energy of the particles depends on the temperature

L

lanthanides (or rare earth elements) the elements with atomic numbers 58 to 71

law (scientific law) a statement of a major concept that summarizes a large body of empirical knowledge

law of combining volumes the volumes of gaseous reactants and products are in simple, whole-number ratios when the gases are measured at the same temperature and pressure

law of conservation of energy the total energy remains constant during a physical or chemical change in an isolated system

law of conservation of mass the total initial mass equals the total final mass for any physical or chemical change

law of constant heat summation see *Hess's law*

law of definite composition elements combine to form a specific compound in definite proportions by mass

law of multiple proportions when two different elements combine with each other to form more than one chemical compound, the different masses of one element that combine with the same mass of the other element are in a ratio of small whole numbers

Le Châtelier's principle when a chemical system at equilibrium is disturbed by a change in a property, the system adjusts in a way that opposes this change

lead storage battery a series of cells with lead as the anode and lead(IV) oxide as the cathode in a sulfuric acid electrolyte; commonly used to produce electricity in automobiles

legal perspective focusing on the laws of a country

Lewis acid-base theory a modern acid-base theory in which an acid is considered to be an electron-pair acceptor and a base is considered to be an electron-pair donor

Lewis model a model of the distribution of electrons in valence orbitals

lime common name for calcium oxide

limestone mineral name for calcium carbonate

limiting reagent the reactant that is completely consumed in a chemical reaction

litmus a plant dye commonly used as an acid-base indicator

London force a type of intermolecular bond due to the attraction of electrons in one molecule by positive nuclei of atoms in nearby molecules; also known as dispersion force

lone pair a pair of electrons in a filled valence orbital of an atom

lye common name for sodium hydroxide

M

malachite copper mineral; geological name for copper(II) hydroxide carbonate

malleable the ability to be formed or stretched by hammering or rolling

manipulated variable the variable that is systematically changed in an experiment to see what effect the change will have on another variable

mass (m) SI quantity of matter in a substance in units of grams

mass number the sum of the number of protons and neutrons in an atom of an element

materials section of a report of scientific work; a list of all equipment and chemicals, including sizes and quantities

matter anything that has mass and occupies space

meniscus the curved surface of a liquid

metal element that is shiny, silvery, and a flexible solid at SATP; most metals are good conductors of heat and electricity and they tend to form positive ions

metallic bond the simultaneous attraction between positive nuclei and valence electrons in a metal

metalloid element near the staircase line in the periodic table; solids with very high melting points that are either non-conductors or semiconductors of electricity

metallurgy the science and technology of extracting metals from their naturally occurring compounds and adapting these metals for useful purposes

metaphor an implied comparison between two different things

militaristic perspective focusing on warfare or on military force or power

minerals naturally occurring chemical compounds

miscible liquids that mix in all proportions and have no maximum concentration

mixture two or more pure substances (elements or compounds) mixed together

model a mental or physical diagram or apparatus used to simplify the description of an abstract idea

molar (M) a non-SI unit for molar concentration; 1 M = 1 mol/L

molar bond energy see *bond energy*

molar charge (Q) the quantity of charge on one mole of charged entities; in the case of electrons, this quantity is called the faraday

molar concentration (C) the amount of solute (in moles) in one litre of solution

molar enthalpy (H) the quantity of energy absorbed or released per mole of a substance undergoing a change; also known as molar heat

molar mass (M) the mass of one mole of a substance

molar solubility (C) the molar concentration of a saturated solution

molar volume (V) the volume of one mole of a gas at STP (22.4 L/mol) or at SATP (24.8 L/mol)

molarity a non-SI quantity equivalent to molar concentration

mole (mol) SI base unit for the amount of a substance; one mole is the number of entities corresponding to Avogadro's number (6.02×10^{23}); the number of carbon atoms in exactly 12 g of a carbon-12 sample; the unit of stoichiometry

mole ratio the ratio of the amount in moles of reactants and/or products in a chemical reaction; the central step in stoichiometry

molecular compound a pure substance formed from nonmetals; solid, liquid, or gas at SATP, relatively low melting point, and non-conducting in any state

molecular formula a group of chemical symbols representing the number and kind of atoms covalently bonded to form a single molecule

molecular prefixes a list of prefixes recommended by IUPAC to specify the number of atoms in a molecule or water molecules in a hydrate

molecule a particle containing a fixed number of covalently bonded nonmetal atoms

molten liquid state of a pure substance; usually applied to substances that melt above room temperature

monatomic ion a positively or negatively charged particle formed from a single atom; also known as a simple ion

monomer the smallest repeating unit of a polymer

monoprotic the ability to donate or accept one proton (H^+)

monosaccharides sugars that contain a single simple sugar unit in each molecule (e.g., glucose and fructose)

moral perspective focusing on questions of right and wrong; ethical

MSDS Material Safety Data Sheets; describe in detail the properties, safe handling procedures, and disposal techniques for chemicals

multi-valent the ability of an atom to form a variety of ions

mystical perspective focusing on spiritual or supernatural meanings

N

Nagaoka's model of atoms pictures an atom as a positively charged sphere with an encircling ring of electrons (analogy: Saturn and its rings)

net ionic equation a chemical equation that represents only those ions or neutral substances specifically involved in an overall chemical reaction

neutral may refer to being either electrically neutral or neutral in an acidic/basic sense; an atom, molecule, or ionic formula unit is considered to be (electrically) neutral when its net charge is zero; a compound is considered to be neutral when it forms a solution that is neither acidic nor basic

neutral compound a compound that forms a solution that is not acidic or basic

neutral solution an aqueous solution that is not acidic or basic, i.e., does not affect the color of litmus paper; a solution that is always electrically neutral

neutralization a double replacement reaction of an acid with a base to produce water and a salt of the acid

neutralize to produce a neutral solution by reacting an acid with a base

neutrons (n) uncharged, subatomic particles present in the nuclei of most atoms

Ni-Cad a commercial, rechargeable cell or battery consisting of nickel(III) oxide hydroxide and cadmium metal in a potassium hydroxide electrolyte

noble gases the family of elements corresponding to Group 18 of the periodic table of the elements

nomenclature a system of naming chemical substances

non-electrolyte a solute in a solution that does not conduct an electric current; a substance that does not produce ions in solution

nonmetal element that is not flexible, and does not conduct electricity; nonmetals tend to form negative ions

non-polar substance a substance that is not affected by nearby charged objects

non-spontaneous referring to a chemical change that does not occur unless an external energy source is used

nuclear change a nuclear reaction in which elements are changed into different elements (see *fission* and *fusion*)

nucleon any particle in the nucleus of an atom

nucleus the central region of an atom that contains most of the mass and all of the positive charge of the atom

O

observation direct information obtained using one of your five senses

octet (of electrons) four completely filled valence orbitals, which represents a very stable arrangement

octet rule a maximum of eight electrons can occupy orbitals in the valence level of an atom

open system one in which both energy and matter can transfer in or out

orbital according to the theory of quantum mechanics, a region of space where there is a high probability of finding electrons of a particular energy; the orbital may contain a maximum of two paired electrons

ore a naturally occurring compound from which a useful element or compound can be extracted

organic pertaining to molecular compounds of carbon; also commonly used to refer to living substances

organic chemistry the study of the molecular compounds of carbon, excluding the oxides of carbon

organic halide a molecular compound of carbon containing one or more halogen atoms bonded to a carbon atom

oxidation a chemical process involving a loss of electrons; an increase in oxidation number; historically used to describe any reaction involving oxygen

oxidation half-reaction the part of a redox reaction in which atoms, ions, or molecules lose electrons

oxidation number a positive or negative number corresponding to the oxidation state of an atom

oxidation potential the electric potential of an oxidation half-reaction

oxidation state the net charge that an atom would have if the electron pairs in a chemical bond belonged entirely to the more electronegative atom

oxidizing agent a substance that causes the oxidation of another substance; a substance that removes electrons from another substance

oxyacid acid containing hydrogen, oxygen, and a third element

P

Pascal (Pa) SI unit of pressure (1 N/m^2)

percent difference a measurement of accuracy found by taking the difference between an experimental and a predicted value and multiplying by 100

percent reaction the actual yield of product in an equilibrium compared to the maximum possible yield, expressed as a percent

percent yield the ratio, expressed as a percent, of the actual or experimental quantity of product obtained (actual yield) to the maximum possible quantity of product (predicted or theoretical yield) derived from a stoichiometry calculation

period a horizontal row of elements in the periodic table whose properties change from metallic to nonmetallic from left to right

periodic law chemical and physical properties of elements repeat themselves at regular intervals when the elements are arranged in order of increasing atomic number

periodic table an empirical model of the periodic law usually drawn with vertical groups or families and horizontal periods

perspectives points of view about an issue

petrochemicals chemicals made from petroleum (oil and natural gas)

petroleum a complex liquid mineral composed primarily of hydrocarbons

pH a quantity representing the acidity of a solution; the additive inverse (negative) of the logarithm of the concentration of hydrogen (or hydronium) ions in a solution

pH curve a graph of pH of a reaction solution versus the volume of titrant added

pH meter a device used to measure pH; based on the cell potential of a silver-silver chloride glass electrode and a saturated calomel (dimercury(I) chloride) electrode

phase change a physical change, such as melting or boiling, involving no change in chemical composition

phase equilibrium a dynamic equilibrium of forward and reverse rates of a change of phase of a single substance in a closed system

phenyl group the name for a benzene ring acting as a branch; C_6H_5—

photosynthesis the formation of carbohydrates and oxygen from carbon dioxide, water, and sunlight, catalyzed by chlorophyll in the green parts of a plant

physical change any change in which the chemical composition does not change; no new chemicals are formed

pipet a glass tube used to measure precise volumes of a liquid

pK a quantity representing the strength of an acid by expressing the dissociation in aqueous solution; the additive inverse (negative) of the logarithm of the acid ionization constant, K_a

pOH a quantity representing the basicity of a solution; the additive inverse (negative) of the logarithm of the concentration of hydroxide ions in a solution

polar covalent bond a bond resulting from the unequal sharing of a pair of electrons

polar molecule a molecule that has a slightly uneven charge distribution, with oppositely charged ends

political perspective focusing on government and legislation

polyatomic ion a group of atoms with a net positive or negative charge on the whole group

polymer a long chain molecule made up of many small identical units

polymerization a type of chemical reaction involving the formation of very large molecules from many small molecules (monomers)

polyprotic Brønsted-Lowry acids or bases that have the ability to donate or accept more than one proton

porous boundary a barrier or physical boundary that separates two electrolytes in a cell while still permitting ions to move through tiny openings between the two solutions

position of an equilibrium any method that describes whether, at equilibrium, the quantity of products is greater than the quantity of reactants (known as products favored) or whether the quantity of reactants is greater than the quantity of products (known as reactants favored)

potash mineral containing potassium compounds (especially potassium chloride)

potential energy a stored form of energy; the chemical potential energy of a substance depends on the relative position of the particles of the substance

potential energy diagram a diagram of the potential energy of the reactants and products in a chemical reaction; used to determine enthalpy changes during a chemical reaction

power (P) the rate at which energy is transferred

precipitate a low solubility solid formed from a solution

precipitation the formation of a low solubility solid from a mixture; a common type of double replacement reaction

precision the place value of the last digit obtained in a measurement; the number of decimal places in a measurement

precision rule a result obtained by adding or subtracting measured or calculated values is rounded to the same precision (number of decimal places) as the least precise value used in the calculation

prediction section of a report of scientific work; the part of a scientific problem-solving model in which the answer to the problem is obtained based on a scientific concept (theory, law, generalization) or some other authority (e.g., a reference)

pressure (*p*) force per unit area

primary cell a cell that cannot be recharged, usually due to irreversible side reactions

primary standard a chemical available in a pure and stable form that is used to determine precisely the concentration of a reagent

problem section of a report of scientific work; a specific question to be answered in an investigation

procedure section of a report of scientific work; a step-by-step set of directions designed to obtain the evidence needed to answer the Problem of an investigation

products the substances produced by a chemical reaction; substances whose chemical formulas appear to the right of the arrow in a chemical equation

protein natural polymers of amino acids forming the basic material of living things

protons (p⁺) positively charged, subatomic particles found in the nuclei of atoms

proton acceptor see *Brønsted-Lowry definitions*

proton donor see *Brønsted-Lowry definitions*

pseudo-scientific falsely represented as scientific knowledge or process

pure substance homogeneous matter that has a definite set of physical and chemical properties, and that cannot be separated by physical changes; elements and compounds

purpose section of a report of scientific work; the aim or goal of an investigation

Q

qualitative describes a quality or change in matter that has no numerical value expressed

qualitative chemical analysis the identification of substances present in a sample

quantitative describes a quantity of matter or degree of change in matter

quantitative chemical analysis the determination of the quantity of a substance in a sample

quantitative reaction a reaction in which more than 99% of the limiting reagent is consumed

quantum a specific, indivisible quantity; e.g., quantum of energy

quantum mechanics model of atoms a mathematical model of atoms in which electrons are described in terms of their energies and probability patterns

quicklime a common name for calcium oxide

R

radiation energy or subatomic particles emitted by a substance

radioactive atom an atom that spontaneously emits radiation

radioisotope a radioactive isotope of an element

random error an uncertainty that is non-systematic and related to measuring or sampling errors

rare earth elements see *lanthanides*

rate constant a constant in the rate law expression that is specific to the given reaction and system conditions

rate-determining step the slowest step in a reaction mechanism, one that controls the rate of the overall reaction

rate law the generalization that the rate of a chemical reaction is exponentially proportional to the concentrations of the reactants

reactants the substances being combined in a reaction; substances whose chemical formulas appear on the left side of the arrow in a chemical equation

reaction coordinate the x-axis on a potential energy diagram

reaction mechanism individual equations representing single events in a chemical reaction, describing the step-by-step progress of the reaction

reaction quotient (*Q*) the value obtained from the equilibrium law applied to a system that is not at equilibrium, allowing prediction of reaction direction

reaction rate rate of reaction is usually expressed as the change in concentration of a reactant or product per unit of time

reagent a chemical, usually relatively pure, used in a reaction

recharging the non-spontaneous conversion of electrical energy to chemical energy in a cell

redox reaction a contraction of "reduction-oxidation"; a chemical reaction involving a transfer of electrons

redox spontaneity rule a spontaneous redox reaction occurs only if the oxidizing agent is above the reducing agent in a table of relative strengths of oxidizing and reducing agents; a spontaneous redox reaction occurs if the net cell potential is positive

reducing agent a substance that causes the reduction of another substance; a substance that loses or donates electrons to another substance

reduction a chemical process involving a gain of electrons; a decrease in the oxidation number; historically used to describe a reaction producing a metal from its naturally occurring compound

reduction half-reaction the part of a redox reaction in which atoms, ions, or molecules gain electrons

reduction potential a measure of the tendency of a given half-reaction to occur as a gain of electrons

reference energy state a convention that defines elements in their most stable form at SATP to have a potential energy of zero

reference half-cell a hydrogen electrode at SATP; assigned a half-cell potential of zero volts

refining industrial processes of separating, purifying, and altering raw materials

reforming a type of chemical reaction used in petroleum refining in which larger or branched molecules are built up from smaller molecules using heat (thermal reforming) or catalysts (catalytic reforming)

relative atomic mass the mass of each element that would react with a fixed mass (for example, 16 g) of oxygen

representative elements (main group) the elements that best follow the periodic law; Groups 1, 2, and 13 to 18 in the periodic table of the elements

responding variable the property that is measured as a result of the systematic manipulation of another variable

restricted quantum mechanics theory of atoms a simplified theory of electron structure restricted to the representative elements

restricted quantum mechanics theory of ions atoms of the representative elements lose or gain electrons to achieve the same electron structure of the nearest noble gas atom

restriction a stated limitation of a theory, law, or generalization

reverse reaction the chemical reaction proceeding from right to left as written in a balanced equation, where products are changing to reactants

roasting reactions of metal sulfides with oxygen to form metal oxides and sulfur dioxide

rotational motion (of a particle) a type of motion that involves spinning or turning

rule of a thousand states that a final answer in a question is expressed as a value between 0.1 and 1000

Rutherford's model of atoms (or nuclear model) pictures an atom with a tiny, positively charged nucleus around which the electrons move in various orbits

S

sacrificial anode a metal that is more easily oxidized than iron and that is connected to the iron object to be protected

salt an ionic compound whose cation is not H^+ and whose anion is not OH^-; also the common name for sodium chloride

salt bridge a tube or connection containing an electrolyte that connects two half-cells

SATP standard ambient temperature and pressure; 25°C and 100 kPa (see also *STP*)

saturated organic compound a relatively stable organic molecule having no double or triple covalent bonds between carbon atoms

saturated solution a solution that is in equilibrium with undissolved solute and contains the maximum amount of dissolved solute at specified conditions

science the study of the natural world in an attempt to describe, predict, and explain changes and substances

scientific pertaining to the research and explanation of natural phenomena; must be testable empirically

scientific law see *law*

scientific notation a system of reporting numbers using a number from 1 to 10 and a power of ten to indicate magnitude

scientific skills the thinking processes necessary to solve problems in science

secondary cell a rechargeable cell

serendipity the quality or faculty of making accidental, fortunate discoveries

significant digits all digits in a measured or calculated value that are certain plus one uncertain (estimated) digit

simple decomposition the breakdown (reaction) of a compound into its elements

single bond an attraction between atoms in a molecule due to the sharing of a single pair of electrons in a covalent bond

single replacement the reaction of an element with a compound to produce a new element and a new compound

slaked lime common name for calcium hydroxide

smelting extracting metals from minerals using heat

social perspective focusing on society and human relations

soda ash common name for sodium carbonate

solubility concentration of a saturated solution of a solute in a solvent

solubility equilibrium a dynamic equilibrium of equal rates of dissolving and crystallization in a saturated solution in a closed system

solubility product constant (K_{sp}) the value obtained from the equilibrium law applied to a saturated solution

soluble having high solubility

solute a substance that is dissolved in a solvent

solution a homogeneous mixture of dissolved substances containing at least one solute and one solvent

solvent medium in which a solute is dissolved; usually the liquid component of the solution

specific energy see *energy density*

specific heat capacity (*c*) quantity of energy required to change the temperature of a unit mass of a substance by exactly one degree Celsius

spectator an entity, such as an ion, that does not change or take part in a chemical reaction

spontaneous referring to a chemical change that occurs naturally without any external energy source

stability, thermal see *thermal stability*

standard cell an electrochemical cell constructed using two half-cells, each containing 1 mol/L of oxidized and reduced entities at SATP

standard cell potential ($\Delta E°$) the maximum electric potential of a standard cell

standard hydrogen electrode a standard hydrogen half-cell at SATP with an inert platinum electrode immersed in 1 mol/L hydrogen ions and with hydrogen gas bubbling over the electrode

standard molar enthalpy the quantity of heat energy transferred per mole of a substance in a reaction with the initial and final states at SATP

standard reduction potential ($E_r°$) the reduction potential of a half-cell in which all ion concentrations are 1 mol/L at SATP

standard solution a solution with a precisely known concentration

standard state a set of conditions established by convention, such as SATP

starch food stored by plants during photosynthesis; a polymer of glucose

state of a system set of characteristic empirical properties of a system

state of matter the physical form of a substance, such as solid (s), liquid (l), gas (g), or aqueous solution (aq)

stock solution an initial, usually concentrated, solution from which samples are taken for a dilution

stoichiometric a condition in which a reaction can be represented by a fixed mole ratio from a balanced chemical equation

stoichiometry the method used to calculate the quantities of substances in a chemical reaction

STP standard temperature and pressure; 0°C and 101.3 kPa (see also *SATP*)

strong acid an acid with an ionization of more than 99%; a Brønsted-Lowry acid that has a very weak attraction for its proton

strong base an ionic hydroxide according to Arrhenius; a Brønsted-Lowry base that has a strong attraction for a proton

strong electrolyte a substance that exists completely as ions in a solution

structural diagram a model showing the covalent bonds between atoms in a molecule

STS science, technology, and society; a concept that describes the complex interaction among science, technology, and society

subatomic within an atom

sublimation a change in state directly from solid to gas or gas to solid

subscript see *formula subscript* and *state of matter*

substitution a type of organic reaction in which a hydrogen atom is replaced by another atom or group of atoms; reactions of alkanes and aromatics with halogens to produce organic halides and hydrogen halides

sugar a common name for sucrose; any of several crystalline compounds of carbon, hydrogen, and oxygen that have a sweet taste and are soluble in water

supersaturated an unstable solution with a concentration higher than its normal solubility at the specified conditions

surroundings the environment around a chemical system

suspension a heterogeneous mixture containing finely divided particles

synthesis a scientific skill that involves combining various empirical and theoretical knowledge to produce a new or better description or explanation

system see *chemical system*

systematic error an uncertainty that is inherently part of a measuring system or design

T

tar sands a type of fossil fuel made up of very fine sand particles coated with heavy oil

technological pertaining to the development and use of machines, instruments, and processes that have a social purpose

technology the skills, processes, and equipment required to make useful products or to perform useful tasks

temperature the quantity measured with a thermometer; an indirect measure of the average kinetic energy of molecules

ternary describes an ionic or molecular compound whose chemical formula includes three kinds of element symbols

theoretical relating to explanations; non-observable ideas

theoretical definition a statement that defines an object or process in terms of unobservable properties

theory a comprehensive set of ideas based on general principles that explain a large number of observations

thermal energy the energy of motion of molecules

thermal stability the resistance of a compound to decompose when heated

thermodynamics the quantitative study of the energy changes in physical and chemical systems

Thomson's model of atoms pictures an atom as a positively charged sphere in which tiny negatively charged electrons are embedded (analogy: raisin bun)

titrant the solution in a buret during a titration

titration the precise addition of a solution in a buret into a measured volume of a sample solution

tonne 1000 kg or 1 Mg

transition elements the elements in Groups 3 to 12 of the periodic table of the elements

transition point the pH at which an indicator changes color

translational motion (of a particle) motion in a straight line

transuranic elements the elements with atomic numbers beyond uranium (93 or greater)

triglycerides esters of long-chain carboxylic acids and glycerol; principal component of animal fats and plant oils

triple bond an attraction between atoms in a molecule due to the sharing of three pairs of electrons in a covalent bond

U

universal gas constant (R) the proportionality constant in the ideal gas law; 8.314 L•kPa/(mol•K)

unsaturated organic compound reactive organic molecules containing double or triple covalent bonds between carbon atoms

V

valence electrons the electrons in the outermost (highest) energy levels of an atom

van der Waals forces weak intermolecular attractions, including London forces and dipole forces

van der Waals radius one-half the distance of closest approach between the nuclei of two non-bonded atoms

vaporization the conversion of a liquid into a gas

vibrational motion (of a particle) a back and forth or oscillating motion in a confined space

vinegar a common name for an approximately 1 mol/L solution of acetic acid ($CH_3COOH_{(aq)}$)

volt (V) SI unit of electric potential difference (1 J/C)

voltage a common name for electric potential difference

voltaic cell a cell that spontaneously produces electricity by redox reactions; also known as a galvanic cell

voltmeter a device used to measure electric potential difference in units of volts

volumetric pertaining to volume

volumetric flask a flask with a long, narrow neck used to prepare a precise volume of a solution

volumetric heat capacity (c) quantity of energy required to change the temperature of a unit volume of a substance by exactly one degree Celsius

volumetric pipet a glass tube used to measure precise volumes of a liquid; also known as a delivery pipet

VSEPR theory a theory for predicting and explaining molecular shapes on the basis of the number of valence electron pairs that surround the central atom

W

washing soda a common name for sodium carbonate decahydrate

water gas an industrial fuel composed of a mixture of hydrogen and carbon monoxide

weak acid an acid that partially ionizes in solution but exists primarily in the form of molecules

weak base a base that has a weak attraction for protons

weak electrolyte a substance whose aqueous solution is a poor conductor of electricity; e.g., weak acids

weight the force of gravity on an object

WHMIS Workplace Hazardous Materials Information System; used to communicate chemical hazards to, for example, custodians, lab technicians, teachers, students, and firefighters

wood alcohol a common name for methanol; $CH_3OH_{(l)}$

Y

yield the ratio, expressed as a percent, of the quantity of a substance obtained in an experiment compared to the quantity predicted from stoichiometric calculations

INDEX

Multimeter, use of, 736–37
Multiple proportions, law of, 73
Multi-step energy calculations, 448–50
Multi-valent metals, 110–111
   ions of, 105

N

Nagaoka, H., 74, 75
Natural gas, 436
Negative label, 681
Nernst equation, 668
Net ionic equations, 194–96, 575
Neutral substances, 102, 188
Neutralization, 172, 216
Neutrons, 76
Newlands, John Alexander, 63
Nicholson, John, 79
Nickel-cadmium (Ni-Cad) cell, 657, 658, 682
Nitrogen, 314
Nitrogen dioxide, 134
Nitrogen monoxide, 134
Nitrogen oxides, 119
Nobel prize winners, Canadian, 533
Noble gases, 69, 81, 260
   and atomic theories, 291, 300
Noddack, Ida, 69
Nomenclature, 107
Non-electrolytes, 187–88
Nonmetal oxides:
   hydrolysis of, 547–48
Nonmetals, 60
Nonspontaneous reaction, 621
Northern lights, 198
Nuclear change, 135–36
Nuclear energy, 395
Nuclear fission, 444
Nuclear fusion, 443
Nuclear power, 447
Nuclear radiation, 77
Nuclear reactions, 405, 443–44
Nucleus of atom, 75–77, 82

O

Observation, defined, 33
Octet rule, 302
Odor change, evidence of chemical reaction, 138
Odors, 372
Ohmmeter, 188
Oil spills, 359
Open system, 401

Optical analysis, 281
Orbital diagrams, 297–99
Orbitals, 297–99
Ores, 607
Organic chemistry, defined, 337
Organic compounds. See Carbon compounds
Organic halides, 362–64
Organochlorines, 636
Ostwald process, 441
Overpotential, 685
Oxalic acid, 371, 550, 590
Oxidation, 329, 607–608, 631, 634
   in half-cell, 665–66, 675
   theoretical definition of, 611–13
Oxidation half-reaction, 329
Oxidation numbers, 632–33
   balancing redox equations, 635–38
   common oxidation numbers (table), 632
   determining, 633
   and redox reactions, 634
Oxidation potentials, 673
Oxidation-reduction reactions. See Redox reactions
Oxidation states, 631–38
   defined, 632
Oxidizing agent, 608, 612, 618
Oxygen, 265
   molecule, representation, 307
   transportation throughout the body, 504
Ozone, 134, 362

P

Parent chain, 347, 348
Pasteur, Louis, 23, 238
Pauling, Linus, 303
PCBs (polychlorinated biphenyls), 362
Pentane, 339, 347, 437
Percent reaction, 491–92, 497
Percent yield, 261, 364, 484–85
Percentage composition, 309
Period, in periodic table, 67
Periodic law, 63, 79–81
Periodic table, 63–70, inside front cover
   and atomic structure, 82–83
   Mendeleyev's, 65–66
   modern, 67–70
Periodicity, 293
Perspectives, 49
Petrochemicals, 360, 361
Petroleum refining, 344–46

pH, 216
   changes during acid-base reactions, 567–83
   curves, 567, 568, 569–70, 572, 574–76
   test strips, 559
   using indicators to determine pH, 560
pH meter, 524–25
pH scale, 522–23, 525
Pharmacist (career), 108
Phase changes, 401, 403–404, 405–406, 408, 409–10
Phase equilibrium, 487–88
Phenolphthalein indicator, 558
Phenyl group, 357
Phosphoric acid, 270
Photosynthesis, 319, 431, 466
Physical changes, 30, 135–36
Pipet, 207, 737–38
Planck, Max, 79
Plasma, 405
Plastics technologist (career), 382
Pliny the Elder, 72
Plunkett, Roy, 238
pOH:
   definition, 523
   relationship with pH, 524
Polanyi, John, 474, 533
Polar covalent bond, 321
Polar molecules, 321–23
Political perspective, 49
   on acid rain, 179
Polyalcohols, 366
Polyamides, 381
Polyatomic ions, 104–105, 112
Polyesters, 380–81
Polyethylene, 380
Polymerization, 379, 383
Polymers, 379
   addition polymers, 380
   condensation polymers, 380
   natural polymers, 383
Polypropylene, 381
Polyprotic substances, 570–72
   definition, 571
   polyprotic acids, 571
   polyprotic bases, 571
Polyunsaturated fats, 351
Pool water acidity, 559
Popper, Sir Karl, 545
Porous boundary, 663
Porous cup, 663, 664
Portable technology, 661
Position of equilibrium, 563

CREDITS

Triaminic Cough/Cold Products; Spic & Span, appear courtesy of Procter & Gamble Inc.; Stanley Pharmaceuticals Ltd.; Sun-Rype Products Ltd.; **SIX:** **p. 224** FORD; **6.3** Burndy Library; **6.4** Mary Evans Picture Library; **p. 231** Dr. George Gornacz/Science Photo Library; **p. 235** David Guyon, The BOC Group PLC/Science Photo Library/Masterfile; **6.10** Martin Bond/Science Photo Library; **p. 238 (left)** Science Photo Library; **(right)** VELCRO Canada Inc.; **6.12** Anne Rippy/The Image Bank; **6.13** Canapress Photo; **6.14** Imperial Oil Ltd.; **p. 242 (top)** Air Resources Branch/BC Environment; **(bottom)** Courtesy of Ebco-Hamilton Partners; **p. 243** Courtesy Science Kit & Boreal Laboratories; **SEVEN:** **p. 250, 7.6** Dr. E. R. Degginger; **7.1** Will & Deni McIntyre/Photo Researchers; **7.2, 7.3, 7.7, 7.10, 7.15** Richard Seimens; **7.5** devries mikkelsen/First Light;**p. 258** Alcan Smelters and Chemicals Ltd., Kitimat, B.C.; **p. 260** RCMP Forensic Laboratory, Edmonton, AB.; **7.8** Dave Starrett; **7.9** Hudson Bay Diecasting; **7.11** Canadian Tire Corp.; **7.13** Potash & Phosphate Institute; **7.14** General Chemical; **p. 280** Canapress/Reprinted with the permission of Prof. Owen Beattie, University of Alberta; **p. 283** Visuals Unlimited/SIU; **p. 285** Henry Birks & Sons; **EIGHT:** **p. 290, p. 310 (left)** Tourism B.C.; **8.9** Materials Research Society; **8.10** Royal Ontario Museum; **p. 305** Ferranti Electronics/A. Sternberg/Science Photo Library; **8.13** Sinclair Stammers/Science Photo Library/Masterfile; **p. 308 (left)** Mary Evans Picture Library; **(right)** Sidney Moulds/Science Photo Library; **p. 310 (right), 8.21, 8.32** Richard Seimens; **8.23** Masterfile; **8.24** Bill Brooks/Masterfile; **p. 326** Courtesy Dr. Ronald Gillespie/McMaster University; **8.31** John Mead/Science Photo Library; **8.33** Carolina Biological Supply Company; **p. 331** Courtesy Michael Smith/University of British Columbia; **NINE:** **p. 336** Astrid & Hanns-Freider Michler/Science Photo Library; **9.1, 9.29, 9.31a** Dr. E. R. Degginger; **9.2, 9.5, 9.13, 9.20, 9.24, 9.27** Richard Seimens; **9.7** Douglas E. Walker/Masterfile; **p. 343** Courtesy Prof. R. Lemieux/University of Alberta; **9.10** Tom Carroll/Masterfile; **9.16** Michael Coyne/The Image Bank; **9.20** Canadian Tire Corporation, Limited; **9.21** John Hyde/State of Alaska, Dept. of Environmental Conservation; **p. 362** NASA; **9.22** Chiquita Banana; **9.25** L. Skoogfors/Canapress Photo; **9.27** Crisco, appear courtesy of Procter & Gamble Inc.; Mazola Corn Oil; Best Foods Inc.; **p. 376 (top)** Ronald Royer/Science Photo Library; **(bottom)** Physics Dept., Imperial College/Science Photo Library; **9.30a** Drug Trading Co. Ltd.; **9.30b** Imperial Oil Ltd.; **9.31b** St. Bartholomew's Hospital/Science Photo Library; **p. 382** Courtesy of Tracy DesLaurier; **9.33** Sinclair Stammers/Science Photo Library; **p. 385** University of Toronto; **TEN:** **p. 294** Mike Dobel/Masterfile; **10.1** Masterfile; **10.2, 10.4, 10.13, 10.19** Dr. E.R. Degginger; **10.3** Courtesy of B.C. Hydro; **10.9, 10.11, 10.15, 10.16, 10.17, 10.26** Richard Seimens; **10.10** Dave Starrett; **10.14** Herzburg Institute of Astrophysics, NRCC; **10.15** Corning Incorporated, Corning, New York; **p. 408** Ron Watts/First Light; **10.21** Ontario Ministry of Natural Resources; **p. 412** Courtesy of Mark Archer; **p. 415** Science Photo Library; **10.29** APEGGA; **ELEVEN:** **p. 426** Ulli Seer/Image Bank; **11.2** FORD; **11.3** General Chemical; **11.4** Dale Sanders/Masterfile; **11.9** Rich Carlson/ICG Auto Propane; **11.13** Photo Researchers; **11.14** Ontario Ministry of the Environment; **11.18** Imperial Oil Ltd.; **11.19** Dr. E.R. Degginger; **11.20** NASA; **11.22** EMR-CANADA; **11.23** Mercedes-Benz Canada; **11.24** Trans-Canada PipeLines;**p. 445 (left)** Novosti/Science Photo Library; **(right)** Burndy Library; **p. 447** Ontario Hydro; **TWELVE:** **p. 458** Visuals Unlimited/Bill Beatty; **12.3, 12.4, 12.6, 12.7, 12.8, 12.12, 12.25** Dave Starrett; **12.5** Courtesy of Science Kit & Boreal Laboratories; **12.9** David Young-Wolff/Photo Edit; **12.10** Alastair Shay/Oxford Scientific Films; **12.11** Ontario Ministry of Natural Resources; **12.12** Courtesy of Shopper's Drug Mart; **12.13** Courtesy of Delphi Automotive Systems; **12.16 (left)** NASA; **(bottom)** UPI/Bettman Newsphotos; **p. 474** Brian Willer/Macleans; **p. 477** Canadian International Grains Institute; **THIRTEEN:** **p. 484** Ken Davies/Masterfile; **13.1, 13.3, 13.7, 13.8, 13.9, 13.15, 13.17, 13.18, 13.22, 13.23, p. 538** Richard Seimens; **13.2, 13.16 (top), (middle), (bottom), 13.20** Dave Starrett; **p. 492** Courtesy LifeScan Canada Ltd., A Johnson & Johnson Company; **p. 502** The Edgar Fahs Smith Collection; **p. 505** Potash & Phosphate Institute; **13.19** Beckman Instruments; **13.21** Yoav Levy/Phototake, NYC; **p. 533** Canapress Photos; **FOURTEEN:** **p. 540, 14.32** Ontario Ministry of the Environment; **14.3** University of Laval; **p. 545 (left)** London School of Economics; **(right)** Buston/Canapress Photo; **p. 553** The Royal Society of Chemistry; **14.6, 14.7, 14.8, p. 559, 14.19** Dave Starrett; **14.10, 14.12, 14.13, 14.15, 14.17, 14.18, 14.30** Richard Seimens; **14.7** Arm & Hammer ® Church & Dwight Ltd./Ltée; **p. 581** Courtesy of Kathleen Kaminsky; **14.31** Simon Fraser/Science Photo Library; **p. 585 (bottom)** Canapress Photo; **p. 586** Imperial Oil Ltd.; **FIFTEEN:** **p. 606** Steve Dunwell/Image Bank; **15.2** Al Harvey/The Slide Farm; **15.3** Shashinka Photo Inc.; **15.4** Edward M. Gifford/Masterfile; **15.5** From Chemistry: A Human Venture by Stan Percival and Ross Wilson Toronto, ON: Irwin Publishing, 1988. Photo by Cary Smith. Used with permission of the publisher; **15.6** Richard Seimens; **p. 611** Goltepe/Kestal Excavations; **15.8** CP RAIL; **p. 620** Bettmann Archives; **p. 623** University of Toronto; **15.13, 15.14, 15.15, 15.17, 15.18, 15.19, 15.20** Richard Seimens; **p. 627** Alberta Women's Secretariat; **p. 636** Abitibi-Price Inc.; **p. 646** CMI, Inc.; **SIXTEEN:** **p. 652** J. Carmichael Jr./Image Bank; **16.1** Keith Kent/Science Photo Library; **16.3** Monogram Models; **p. 661 (top), (bottom)** Ballard Power Systems Inc.; **p. 662** General Motors; **16.25** India Tourist Office, New York; **p. 686** Sherritt Gordon Ltd.; **16.31** Kennecott; **p. 690** Bettmann Archives; **16.32** Jerry Kobalenko/First Light; **p. 695** Eveready Canada Inc.; **APPENDIX C:** **C2, C3, C4, C5, C6, C7, C8, C11, C12, C14, C16, C18, C19, C20, C21** Richard Seimens; **C9, C10, C15** Dave Starrett; **APPENDIX F:** **pp. 763, 764, and 766** reprinted with permission from CRC Handbook of Chemistry and Physics, 71st Edition. Copyright CRC Press, Inc. Boca Raton, FL; **pp. 761, 762, and 765** adapted from Lange's

The publishers wish to express their thanks to Harry Ainlay Composite High School, Port Credit Secondary School, and Mary Ward Catholic Secondary School for allowing photographs to be taken in their schools.

ION COLORS

Ion	Solution Color	Ion	Flame Color
Groups 1, 2, 17	colorless	Li^+	bright red
Cr^{2+}	blue	Na^+	yellow
Cr^{3+}	green	K^+	violet
Co^{2+}	pink		
Cu^+	green	Ca^{2+}	yellow-red
Cu^{2+}	blue	Sr^{2+}	bright red
Fe^{2+}	pale green	Ba^{2+}	yellow-green
Fe^{3+}	yellow-brown		
Mn^{2+}	pale pink	Cu^{2+}	blue (halides)
Ni^{2+}	green		green (others)
CrO_4^{2-}	yellow		
$Cr_2O_7^{2-}$	orange	Pb^{2+}	light blue-grey
MnO_4^-	purple	Zn^{2+}	whitish green

SPECIFIC HEAT CAPACITIES OF PURE SUBSTANCES

Substance	Specific Heat Capacity* $(J/(g \cdot °C))$	Substance	Specific Heat Capacity* $(J/(g \cdot °C))$
aluminum	0.900	nickel	0.444
calcium	0.653	potassium	0.753
copper	0.385	silver	0.237
gold	0.129	sodium	1.226
hydrogen	14.267	sulfur	0.732
iron	0.444	tin	0.213
lead	0.159	zinc	0.388
lithium	3.556	ice, $H_2O_{(s)}$	2.01
magnesium	1.017	water, $H_2O_{(l)}$	4.19
mercury	0.138	steam, $H_2O_{(g)}$	2.01

*Elements at SATP state.

CONCENTRATED (SATURATED) REAGENTS

Reagent (• strong acids)	Formula	Concentration (mol/L)	Concentration (mass %)
acetic acid	$CH_3COOH_{(aq)}$	17.4	99.5
ammonia	$NH_{3(aq)}$	14.8	28
carbonic acid	$H_2CO_{3(aq)}$	0.039	0.17
• hydrochloric acid	$HCl_{(aq)}$	11.6	36
• nitric acid	$HNO_{3(aq)}$	15.4	69
phosphoric acid	$H_3PO_{4(aq)}$	14.6	85
sodium hydroxide	$NaOH_{(aq)}$	19.1	50
sulfurous acid	$H_2SO_{3(aq)}$	0.73	6
• sulfuric acid	$H_2SO_{4(aq)}$	17.8	95

SI PREFIXES

Prefix	Symbol	Factor
giga	G	10^9
mega	M	10^6
kilo	k	10^3
milli	m	10^{-3}
micro	μ	10^{-6}
nano	n	10^{-9}

DEFINED (EXACT) QUANTITIES

$$1\ t = 1000\ kg = 1\ Mg$$
$$STP = 0°C\ and\ 101.325\ kPa$$
$$(use\ 0°C\ and\ 101\ kPa)$$
$$SATP = 25°C\ and\ 100\ kPa$$
$$0°C = 273.15\ K\ (use\ 273\ K)$$
$$1\ atm = 101.325\ kPa\ (use\ 101\ kPa)$$

MEASURED (UNCERTAIN) QUANTITIES

$$N_A = 6.02 \times 10^{23}/mol$$
$$R = 8.31\ kPa \cdot L/(mol \cdot K)$$
$$F = 9.65 \times 10^4\ C/mol$$
$$K_W = 1.0 \times 10^{-14}\ (mol/L)^2$$
$$H_{fusion\ H_2O} = +6.03\ kJ/mol$$
$$H_{vap\ H_2O} = +40.8\ kJ/mol$$
$$c = 3.00 \times 10^8\ m/s$$
$$V_{STP} = 22.4\ L/mol$$
$$V_{SATP} = 24.8\ L/mol$$
$$d_{H_2O} = 1.00\ g/mL$$

VOLUMETRIC HEAT CAPACITIES

Substance	Volumetric Heat Capacity $(MJ/(m^3 \cdot °C))$
air	0.0012
brick/rock	1.9
concrete	2.1
ethylene glycol (50%)	3.7
water	4.19